MULTILINGUAL ILLUSTRATED DICTIONARY OF AQUATIC ANIMALS AND PLANTS

ANIMALES Y PLANTAS ACUÁTICOS
VANDDYR OG -PLANTER
WASSERTIERE UND -PFLANZEN
ΥΔΡΟΒΙΑ ΖΩΑ ΚΑΙ ΦΥΤΑ
AQUATIC ANIMALS AND PLANTS
ANIMAUX ET PLANTES AQUATIQUES
ANIMALI E PIANTE ACQUATICI
WATERDIEREN EN -PLANTEN
ANIMAIS E PLANTAS AQUÁTICOS
VESIELÄIMET JA -KASVIT
VATTENDJUR OCH -VÄXTER

Second edition

European Commission

MULTILINGUAL
ILLUSTRATED
DICTIONARY
OF AQUATIC ANIMALS
AND PLANTS

ANIMALES Y PLANTAS ACUÁTICOS
VANDDYR OG -PLANTER
WASSERTIERE UND -PFLANZEN
ΥΔΡΟΒΙΑ ΖΩΑ ΚΑΙ ΦΥΤΑ
AQUATIC ANIMALS AND PLANTS
ANIMAUX ET PLANTES AQUATIQUES
ANIMALI E PIANTE ACQUATICI
WATERDIEREN EN -PLANTEN
ANIMAIS E PLANTAS AQUÁTICOS
VESIELÄIMET JA -KASVIT
VATTENDJUR OCH -VÄXTER

Second edition

Fishing News Books

Office for Official Publications
of the European Communities

Published by Fishing News Books
A division of Blackwell Science Ltd
Osney Mead, Oxford OX2 0EL, United Kingdom

and

Office for Official Publications of the European Communities
2, rue Mercier, L-2985 Luxembourg

Printed in Belgium

British Library
Cataloguing in publication data

A catalogue record for this book
is available from the British Library

ISBN 0-85238-240-5

Office for Official Publications of the European Communities
2, rue Mercier, L-2985 Luxembourg

ISBN 92-828-1886-1

1998 – LI + 548 pp. – 17.6 × 25 cm

First edition published jointly by Fishing News Books
and the Office for Official Publications of the European Communities, 1993
Second edition published 1998

Preface

To understand one another, we need to speak each other's language. While the rich diversity of languages is one of the European Union's glories, when it comes to a field as complex and varied as fish and other aquatic organisms it is easy to lose one's bearings. This glossary should therefore be of inestimable value, not only for translators and scientists but for everyone working in the fishing and aquaculture industry.

In these sectors, perhaps more than in others, problems need to be solved by mutual understanding between the people working in them in the different Union countries: the fishermen themselves, the handlers and processors, the wholesalers and retailers, and of course the consumers. At stake is the survival of our fishing communities. Only together will we meet and overcome the difficult times ahead.

Overcapacity among our fishing fleets is cutting into the profitability of boats and encouraging cheating, the taking of immature fish is stopping stocks from renewing themselves, the globalisation of our markets is putting pressure on traditional business strategies – not to mention the challenges to our natural environment. All of these have to be tackled head on by the common fisheries policy (CFP).

The CFP's job is to set the 'rules of the game' in such a way as to elicit responsible behaviour, so that we can go on exploiting the wealth of the sea without wasting its resources or exhausting them altogether. The CFP needs to ensure that the industry has a future, that it will become more integrated and up to date, so that consumers can go on buying the good quality, healthy products they want at reasonable prices.

To achieve this, the CFP promotes conservation, surveillance and monitoring and research, it supports processing, marketing and promotional campaigns, it sets limits on fleet capacity, it arranges cooperation agreements with non-Community countries, and so on. In all of this, the various partners involved have to understand one another. Every effort to promote mutual comprehension is therefore to be heartily welcomed and this glossary is no exception.

The second edition covers more than 1 500 groups and individual species of fish, crustaceans, shellfish, molluscs, mammals and aquatic plants, giving their most commonly used names in all 11 languages of the Community plus their scientific name in Latin. Wherever possible, an illustration has been included as well.

This impressive volume would not have been feasible without the help of research institutes in the various member countries. It represents a convergence of interests between the world of research and the commercial fishing industry which, although far apart in their day-to-day work, share the same desire for a better understanding and use of the resources of the sea. The glossary is also the fruit of collaboration between research institutes and the Commission's departments dealing with terminology, statistics and fisheries policy.

A big thank you, then, to all those who have contributed to it. Your painstaking work has produced more than a beautiful-looking book – it has helped in a highly original way to advance the European enterprise.

Emma Bonino
Member of the European Commission

Prefacio

Para comprenderse, es imprescindible tener un lenguaje común. Si bien la pluralidad de lenguas constituye uno de los grandes patrimonios de la Unión Europea, en un universo de tanta diversidad como el de los recursos acuáticos resulta bastante fácil perderse en una maraña de términos.

La presente obra representa por tanto un valiosísimo instrumento, no sólo para lingüistas o científicos, sino también para todos los que trabajan en los sectores de la pesca y de la acuicultura.

En estos sectores, quizá más que en otros, las soluciones a los problemas dependen de la comprensión mutua y de la cooperación entre profesionales de los distintos países de la Unión Europea, de pescadores a distribuidores y consumidores, pasando por mariscadores y transformadores.

De ello depende incluso el futuro de la actividad pesquera en nuestras regiones, pues las dificultades que se nos plantean sólo se pueden afrontar y resolver de forma común.

El exceso de capacidad de las flotas pesqueras, que merma la rentabilidad de los buques e incita al fraude, la captura de peces juveniles e inmaduros, que compromete la renovación de las poblaciones, la globalización de los mercados, que desbarata las estrategias tradicionales de las empresas, y las cuestiones medioambientales son los retos a los que debe hacer frente la política pesquera común (PPC).

Ahora bien, la PPC debe fijar unas reglas de juego capaces de aportar soluciones «responsables» que, a la par que garantizan una explotación que evite el despilfarro y el agotamiento de los recursos marinos, aseguren el futuro y el desarrollo de una industria moderna e integrada y permitan satisfacer las necesidades de los consumidores, quienes exigen productos de calidad, sanos y a precios razonables.

Para ello, la PPC incluye medidas de conservación, control, investigación, ayuda a las actividades de transformación, comercialización y promoción, limitación de las capacidades de la flota y cooperación con terceros países, y muchas otras intervenciones para las que es necesario el entendimiento entre todas las partes interesadas.

Por este motivo, toda iniciativa que, como el presente glosario, contribuya a una mejor comprensión mutua debe ser aplaudida.

En esta segunda edición, se han catalogado más de mil quinientas especies o grupos de especies de peces, crustáceos, moluscos, mamíferos y plantas según sus denominaciones más corrientes en las once lenguas comunitarias, a las que se ha añadido el nombre científico (en latín), acompañadas de ilustraciones siempre que ha sido posible.

Este impresionante volumen no habría salido a la luz sin la aportación de los institutos de investigación de los distintos países de la Unión. Constituye un perfecto ejemplo de convergencia de intereses entre el mundo científico y el de los profesionales de la pesca que, aunque muy diferentes en cuanto a la naturaleza de su trabajo, comparten el afán de conocer y explotar mejor los recursos marinos. También es producto de la colaboración entre esos institutos de investigación y los servicios de la Comisión Europea encargados de la terminología, las estadísticas y la política pesquera.

Quiero expresar nuestro agradecimiento a todos los que han participado en esta obra, pues mediante su minuciosa investigación y su trabajo han contribuido a mucho más que a la publicación de un hermoso libro: han colaborado de una forma original en el progreso de la construcción europea.

Emma Bonino
Miembro de la Comisión Europea

Forord

For at kunne forstå hinanden må vi tale samme sprog. Ganske vist er sprogenes mangfoldighed et af EU's store aktiver, men man kommer let til kort, når man har at gøre med et så rigt og varieret fagområde som »vanddyr og -planter«.

Dette værk er derfor et vigtigt arbejdsredskab, ikke blot for sprogfolk og videnskabsmænd, men også for alle dem, som arbejder i sektoren for fiskeri og akvakultur.

Det gælder måske endnu mere i denne sektor end i andre sektorer, at hvis problemerne skal løses, er alle sektorens aktører i de forskellige EU-lande nødt til at samarbejde og forstå hinanden – både fiskere, distributører, engroshandlere, forarbejdningsvirksomheder og forbrugere.

Det drejer sig om selve fiskeriets fremtid i vore regioner, fordi de nuværende problemer kun kan imødegås og løses i fællesskab.

Den fælles fiskeripolitik står over for en række udfordringer: overkapaciteten i fiskerflåden, der nedsætter fartøjernes rentabilitet og forleder til svig; fiskeriet på umodne ungfisk, der bringer fornyelsen af bestandene i fare; globaliseringen af markederne, der truer virksomhedernes traditionelle strategier samt de miljømæssige problemer.

Den fælles fiskeripolitiks formål er at fastsætte nogle spilleregler, som kan udmøntes i »ansvarlige« løsninger, der sikrer et fiskeri uden spild af og rovdrift på havets ressourcer. På den måde kan fiskeripolitikken sikre udviklingen af et moderne og integreret erhverv og dets fortsatte eksistens, et erhverv, som kan tilfredsstille efterspørgslen fra forbrugere, der kræver gode og sunde produkter til rimelige priser.

I dette øjemed er der fastsat foranstaltninger til bevarelse, kontrol, forskning, støtte til forarbejdning, afsætning, fremme af forbruget, begrænsning af fiskerflådens kapacitet, samarbejde med tredjelande mv., som kræver, at alle involverede parter forstår hinanden.

Derfor er alle initiativer, der går i retning af en bedre gensidig forståelse, som dette glossar, yderst velkomne.

I den foreliggende udgave er der medtaget over 1 500 arter og grupper af arter af fisk, krebsdyr, skaldyr, bløddyr, pattedyr og planter under deres mest almindelige betegnelser på de elleve EU-sprog sammen med det videnskabelige navn (på latin). De er så vidt muligt illustreret.

Dette imponerende værk havde ikke kunnet udarbejdes uden hjælp fra de forskellige EU-landes forskningsinstitutter. Det er et eksempel på interessesammenfald mellem de videnskabelige kredse og fiskerierhvervet, som, selv om de har meget forskellige arbejdsfelter, er fælles om ønsket om bedre at kende og udnytte havets ressourcer. Værket er også resultatet af et samarbejde mellem disse forskningsinstitutter og Kommissionens afdelinger for terminologi, statistik og fiskeripolitik.

Tak til alle dem, der har bidraget til dette værk. Gennem deres omhyggelige forskning og arbejde har de deltaget i mere end udarbejdelsen af en meget vellykket bog: De har også på deres egen måde været med til at fremme udviklingen i EU.

Emma Bonino
Medlem af Europa-Kommissionen

Vorwort

Verständigung erfordert eine gemeinsame Sprache. Zwar ist die Sprachenvielfalt einer der größten Schätze der Europäischen Union, aber wenn wir auf die vielfältigen Schätze unserer Gewässer zu sprechen kommen, sind wir schnell mit unserem Latein am Ende. Dieses Werk bietet eine große Verständigungshilfe, nicht nur für Linguisten und Forscher, sondern für alle in der Fischerei und Aquakultur Tätigen. In diesen Wirtschaftszweigen vielleicht mehr noch als in anderen hängt die Lösung von Problemen von der Verständigung und der Zusammenarbeit der Wirtschaftsteilnehmer der Mitgliedstaaten der Europäischen Union ab, von den Fischern über die Seefischgroßhändler und Verarbeiter bis zum Einzelhandel.

Denn die Probleme der Zukunft der Fischerei in unseren Regionen können wir nur gemeinsam anpacken und bewältigen.

Die Überkapazitäten der Fischereiflotte, die die Rentabilität der Schiffe belastet und Betrug Tür und Tor öffnet, die Befischung von Jungfischen, die die Erneuerung der Bestände gefährdet, die Globalisierung der Märkte, die die überkommenen Strategien der Unternehmen in Frage stellt, aber auch die Umweltinteressen sind allesamt Herausforderungen an die gemeinsame Fischereipolitik (GFP).

Die gemeinsame Fischereipolitik muß vielmehr die Spielregeln festsetzen, mit denen verantwortungsvolle Lösungen ins Werk gesetzt werden können. Sie muß für eine Bewirtschaftung Sorge tragen, die den Raubbau und die Überfischung der Meeresschätze verhindert, die das künftige Gedeihen eines modernen, integrierten Industriezweigs ermöglicht und den Erwartungen des Verbrauchers gerecht wird, der hochwertige, gesunde und preisgünstige Erzeugnisse verlangt.

Dazu sieht sie Erhaltungs-, Kontroll-, Forschungs- und Stützungsmaßnahmen für die Verarbeitung, die Vermarktung und das Marketing vor, Maßnahmen zur Beschränkung der Flottenkapazität, zur Zusammenarbeit mit Drittländern und vieles mehr. Voraussetzung dafür ist aber, daß sich die Wirtschaftsteilnehmer miteinander verständigen können.

Somit ist jedwede Initiative für eine bessere Verständigung, wie das vorliegende Glossar, höchst willkommen.

Diese zweite Auflage umfaßt ein Verzeichnis der geläufigsten Namen von über 5 000 Arten oder Artengruppen von Fischen, Krustentieren, Schalentieren, Weichtieren, Säugetieren und Pflanzen in den elf Amtssprachen der Europäischen Union mit ihrem wissenschaftlichen Namen (Latein). Wann immer möglich, ist eine Abbildung enthalten.

Dieses eindrucksvolle Werk wäre ohne die Mithilfe von Forschungseinrichtungen der verschiedenen Mitgliedstaaten der Europäischen Union undenkbar gewesen. Sie ist beispielhaft für die Konvergenz der wissenschaftlichen und wirtschaftlichen Interessen im Fischereisektor, die zwar arbeitsbedingt unterschiedlich, jedoch gleichermaßen darauf gerichtet sind, die Fischereiressourcen besser zu verstehen und zu bewirtschaften.

Dieses Glossar ist ferner das Ergebnis der Zusammenarbeit dieser Forschungseinrichtungen und der für Terminologie, Statistik und Fischereipolitik zuständigen Dienststellen der Kommission.

Allen Mitwirkenden gebührt unser Dank, denn mit ihren sorgfältigen Recherchen und ihrem Fleiß haben sie weit mehr zustande gebracht als ein sehr schönes Lexikon, haben sie doch in originärer Weise einen Baustein zum europäischen Aufbauwerk beigetragen.

Emma Bonino
Mitglied der Europäischen Kommission

Πρόλογος

Για την ύπαρξη επικοινωνίας απαραίτητη προϋπόθεση αποτελεί η χρήση μιας κοινής γλώσσας. Επομένως, εάν η πολλαπλότητα των γλωσσών αποτελεί πλούτο για την Ευρωπαϊκή Ένωση, όταν πρόκειται να εξετασθεί ένας τομέας τόσο περίπλοκα πλούσιος όπως είναι ο τομέας των υδάτινων πόρων, είναι αρκετά εύκολο να χαθεί κανείς γλωσσικά.

Το παρόν έργο είναι επομένως ένα πολύτιμο εργαλείο. Όχι μόνον για τους γλωσσολόγους ή τους επιστήμονες αλλά και για όλους αυτούς που εργάζονται στους τομείς της αλιείας και της υδατοκαλλιέργειας.

Στους τομείς αυτούς, πράγματι, περισσότερο ίσως απ' ό,τι σε άλλους, οι λύσεις των προβλημάτων εξαρτώνται από την αμοιβαία κατανόηση και τη συνεργασία μεταξύ των επαγγελματικών φορέων των διαφόρων χωρών της Ευρωπαϊκής Ένωσης, από τους αλιείς μέχρι τους διανομείς και τους καταναλωτές, διαμεσολαβούντων των ιχθυοπωλών και των μεταποιητών.

Είναι το ίδιο το μέλλον της αλιευτικής δραστηριότητας που διακυβεύεται στις περιοχές μας, διότι τα προβλήματα που συναντούμε δεν μπορούν να αντιμετωπισθούν και να επιλυθούν, παρά μόνο από κοινού.

Η υπερικανότητα των αλιευτικών στόλων που μειώνει την απόδοση των σκαφών και ωθεί προς την απάτη, η αλιεία νεαρών ανώριμων ψαριών που θέτει σε κίνδυνο την ανανέωση των αποθεμάτων, η σφαιρική αντιμετώπιση των αγορών που αμφισβητεί τις παραδοσιακές στρατηγικές των επιχειρήσεων, αλλά και τους περιβαλλοντικούς στόχους, συγκαταλέγονται επίσης στις προκλήσεις που πρέπει να αντιμετωπίσει η κοινή αλιευτική πολιτική (ΚΑΠ).

Αλλά η κοινή αλιευτική πολιτική πρέπει να καθορίσει και τους κανόνες του παιχνιδιού που θα βοηθήσουν στην εξεύρεση υπευθύνων λύσεων, εξασφαλίζοντας την εκμετάλλευση που αποτρέπει τη σπατάλη, και μάλιστα την εξάντληση του θαλάσσιου πλούτου, διασφαλίζοντας το μέλλον και την ανάπτυξη μιας σύγχρονης και ολοκληρωμένης βιομηχανίας και επιτρέποντας την ικανοποίηση της ζήτησης των καταναλωτών οι οποίοι απαιτούν καλά προϊόντα, υγιεινά και σε λογικές τιμές.

Για να το επιτύχει αυτό, προβλέπει μέτρα διαχείρισης, ελέγχου, έρευνας και στήριξης των δραστηριοτήτων μεταποίησης, εμπορίας, προώθησης, μέτρα περιορισμού της ικανότητας του στόλου, συνεργασίας με τις τρίτες χώρες και, για να μη μακρηγορώ, μέτρα για τα οποία είναι αναγκαίο να υπάρξει ομοφωνία με όλους τους ενδιαφερόμενους εταίρους.

Γι' αυτόν το λόγο πρέπει να επιδοκιμάζεται κάθε πρωτοβουλία που αποβλέπει στην καλύτερη αμοιβαία κατανόηση, όπως είναι το παρόν γλωσσάριο.

Στη δεύτερη αυτή έκδοση, συγκεντρώνονται περισσότερα από 1500 είδη ή ομάδες ειδών ψαριών, καρκινοειδών, οστρακόδερμων, μαλακίων, θηλαστικών και φυτών με τις τρέχουσες ονομασίες τους, σε ένδεκα κοινοτικές γλώσσες στις οποίες προστίθεται η επιστημονική ονομασία στα λατινικά. Κάθε φορά που είναι δυνατόν, υπάρχει και εικονογράφηση.

Το εντυπωσιακό αυτό έργο δεν θα μπορούσε να πραγματοποιηθεί χωρίς τη συνεισφορά των ερευνητικών ιδρυμάτων των διαφόρων χωρών της Ένωσης. Αποτελεί παράδειγμα σύγκλισης των ενδιαφερόντων του επιστημονικού κόσμου και των επαγγελματικών φορέων του τομέα της αλιείας που, αν και διαφορετικοί, λόγω φύσης και αντικειμένου εργασίας, έχουν την ίδια μέριμνα να γνωρίσουν καλύτερα και να εκμεταλλευτούν με τον καλύτερο δυνατό τρόπο τους αλιευτικούς πόρους. Το γλωσσάριο αυτό αποτελεί επίσης αποτέλεσμα συνεργασίας

μεταξύ των ερευνητικών ιδρυμάτων και των υπηρεσιών της Ευρωπαϊκής Επιτροπής που είναι επιφορτισμένοι με την ορολογία, την κατάρτιση στατιστικών στοιχείων και την πολιτική της αλιείας.

Ευχαριστώ όλους όσοι συνέβαλαν στην εργασία αυτή, γιατί μέσω της σχολαστικής έρευνάς τους και της εργασίας τους θα έχουν συμβάλει σε ένα έργο που σημαίνει κάτι περισσότερο από την υλοποίηση ενός ωραίου βιβλίου, θα έχουν συμβάλει με τρόπο πρωτότυπο στην προώθηση της ευρωπαϊκής οικοδόμησης.

Emma Bonino
Μέλος της Ευρωπαϊκής Επιτροπής

Préface

Pour pouvoir se comprendre, il est indispensable d'utiliser un langage commun. Or, si la diversité des langues est une des principales richesses de l'Union européenne, lorsqu'il s'agit de traiter d'un domaine tout aussi riche en diversité que celui des ressources aquatiques, il est assez facile d'y perdre son latin.

Le présent ouvrage est donc un outil précieux non seulement pour les linguistes ou les scientifiques, mais également pour tous ceux qui travaillent dans les secteurs de la pêche et de l'aquaculture.

Dans ces secteurs, en effet, plus peut-être que dans d'autres, les solutions aux problèmes dépendent de la compréhension mutuelle et de la coopération entre les professionnels des différents pays de l'Union européenne, des pêcheurs aux distributeurs, et jusqu'aux consommateurs, en passant par les mareyeurs et les transformateurs.

Il en va de l'avenir même de l'activité de pêche dans nos régions, car les problèmes auxquels nous sommes confrontés ne peuvent être affrontés et résolus qu'en commun.

La surcapacité des flottes de pêche, qui réduit la rentabilité des navires et pousse à la fraude, la capture de jeunes poissons immatures, qui met en péril le renouvellement des stocks, la globalisation des marchés, qui remet en cause les stratégies traditionnelles des entreprises, mais également les enjeux environnementaux, sont autant de défis que la politique commune de la pêche (PCP) doit relever.

La PCP doit fixer les règles du jeu susceptibles d'apporter des solutions «responsables», assurant une exploitation qui évite le gaspillage voire l'épuisement des richesses de la mer, garantissant l'avenir et le développement d'une industrie moderne et intégrée et permettant la satisfaction des demandes des consommateurs qui exigent des produits bons, sains et à des prix raisonnables.

Pour ce faire, elle prévoit des mesures de conservation, de contrôle, de recherche, de soutien aux activités de transformation, de commercialisation, de promotion, des mesures de limitation des capacités de la flotte, de coopération avec les pays tiers, et j'en passe, pour lesquelles il est nécessaire que l'on s'entende entre tous les partenaires concernés.

C'est pourquoi toute initiative qui va dans le sens d'une meilleure compréhension mutuelle, comme le présent glossaire, doit être saluée.

Dans cette deuxième édition, plus de mille cinq cents espèces ou groupes d'espèces de poissons, de crustacés, de coquillages, de mollusques, de mammifères et de plantes sont recensées suivant leurs appellations les plus courantes dans les onze langues communautaires, auxquelles est ajouté le nom scientifique (en latin). Une illustration est fournie chaque fois que possible.

Cet ouvrage impressionnant n'aurait pu être réalisé sans l'apport d'instituts de recherche de différents pays de l'Union. Il constitue un exemple de convergence d'intérêts entre le monde scientifique et celui des professionnels de la pêche, qui, bien que différents par la nature de leur travail, se retrouvent dans le souci de mieux connaître et de mieux exploiter les ressources de la mer.

Il est également le résultat d'une collaboration entre ces instituts de recherche et les services de la Commission européenne chargés de la terminologie, des statistiques et de la politique de la pêche.

Que tous ceux qui y ont contribué en soient remerciés, car, à travers leur recherche méticuleuse et leur travail, ils auront participé à bien plus qu'à la réalisation d'un très beau livre, ils auront contribué de manière originale à faire avancer la construction européenne.

Emma Bonino
Membre de la Commission européenne

Prefazione

Elemento indispensabile per la comprensione reciproca è l'uso di un linguaggio comune. Ebbene, se è vero che la diversità delle lingue costituisce una delle principali ricchezze dell'Unione europea, quando si affronta un campo tanto ricco di diversità quanto quello delle risorse acquatiche, si corre il rischio di non venirne a capo. Questo libro rappresenta dunque uno strumento prezioso non solo per i linguisti o gli scienziati, ma anche per tutti coloro che lavorano nei settori della pesca e dell'acquacoltura.

È in questi settori infatti, forse più che in altri, che le soluzioni ai diversi problemi dipendono dalla comprensione reciproca e dalla cooperazione tra gli operatori dei differenti paesi dell'Unione europea: dai pescatori ai distributori e ai consumatori, passando per i grossisti e i trasformatori.

La posta in gioco è l'avvenire stesso dell'attività della pesca nelle nostre regioni, poiché i problemi attuali possono essere affrontati e risolti solo cooperando.

La sovraccapacità delle flotte da pesca che riduce la redditività dei pescherecci e incita alla frode, la cattura di pesci ancora giovani e immaturi che compromette il rinnovamento degli stock, la globalizzazione dei mercati che rimette in discussione le strategie tradizionali delle imprese così come i problemi ambientali sono alcune tra le sfide che la politica comune della pesca deve raccogliere.

Tale politica deve fissare le regole del gioco che consentano di trovare soluzioni «responsabili», che assicurino uno sfruttamento razionale evitando lo spreco, o addirittura, l'esaurimento delle risorse marine, che garantiscano la nascita e lo sviluppo di un'industria moderna e integrata e che permettano di soddisfare la richiesta, da parte dei consumatori, di prodotti di buona qualità, sani e a prezzi ragionevoli.

A tal fine la politica della pesca prevede misure di conservazione, di controllo, di ricerca e di sostegno alle attività di trasformazione, commercializzazione e promozione nonché misure di limitazione della capacità delle flotte, di cooperazione con i paesi terzi, ecc., sulle quali è necessario che tutte le parti interessate concordino.

Per tal motivo è benvenuta qualsiasi iniziativa che, come il presente glossario, possa contribuire ad una comprensione reciproca.

In questa seconda edizione figurano più di millecinquecento specie o gruppi di specie di pesci, crostacei, molluschi, mammiferi e piante con le loro denominazioni più diffuse nelle undici lingue comunitarie, alle quali è aggiunto il nome scientifico (in latino). Ove è possibile, vi è inserita la relativa illustrazione.

Per la realizzazione di quest'opera imponente è stato fondamentale il contributo di istituti di ricerca di diversi paesi dell'Unione. Essa costituisce un esempio di convergenza di interessi tra gli ambienti scientifici e coloro che operano nel settore della pesca; nonostante la diversità del lavoro svolto, essi sono accomunati dall'esigenza di conoscere più a fondo le risorse del mare e sfruttarle al meglio. È frutto, inoltre, della collaborazione tra i suddetti istituti di ricerca e i servizi della Commissione europea che si occupano della terminologia, delle statistiche e della politica della pesca.

Ringrazio tutti coloro che vi hanno collaborato: il loro impegno e la loro meticolosa ricerca non solo hanno permesso la realizzazione di un libro bellissimo, ma hanno anche contribuito in maniera originale al progetto europeo.

Emma Bonino
Membro della Commissione europea

Voorwoord

Onderlinge verstaanbaarheid vergt een gemeenschappelijk begrippenapparaat. De veelheid van talen, die een van de belangrijkste rijkdommen is van de Europese Unie, is er anderzijds de oorzaak van dat men, bij het werk met betrekking tot de natuurlijke hulpbronnen van de zee, met een zo grote verscheidenheid aan soorten, gemakkelijk uit de koers raakt.

Dit glossarium is dan ook een kostbaar instrument, niet alleen voor vertalers of wetenschapsmensen, maar voor iedereen die te maken heeft met de visserij en de aquacultuur. In deze sectoren zijn de oplossingen voor de problemen namelijk, meer wellicht dan in andere, afhankelijk van wederzijds begrip en samenwerking tussen het bedrijfsleven in de verschillende landen van de Europese Unie, van de vissers tot de viswinkel via de groothandel en de verwerkende bedrijven, en zelfs tot de consument.

Hierbij staat de toekomst van de visserij in onze regio's op het spel: de problemen waarmee wij worden geconfronteerd kunnen wij alleen samen oplossen.

De overcapaciteit van de visserijvloten, die de rentabiliteit van de schepen drukt en fraude bevordert, de vangst van ondermaatse vis, die een gevaar is voor de vernieuwing van de visbestanden, de mondialisering van de markten, die een bedreiging vormt voor de traditionele strategieën van de ondernemingen, maar ook de situatie van het milieu, dat zijn allemaal uitdagingen waaraan het gemeenschappelijk visserijbeleid (GVB) het hoofd moet bieden.

Het GVB moet spelregels geven die verantwoorde oplossingen bieden, die zorgen voor een bevissing waarbij verspilling en zelfs volledige uitputting van de rijkdommen van de zee wordt voorkomen, die waarborgen geven voor de toekomst en voor de ontwikkeling van een moderne, geïntegreerde bedrijfsketen, en waardoor tot tevredenheid kan worden voldaan aan de vraag van de consument, die tegen redelijke prijzen producten van goede kwaliteit wil.

Om al die doelstellingen te bereiken, worden in het kader van het GVB maatregelen genomen voor behoud van de visstand, controle, onderzoek, verwerking, afzet en afzetbevordering, beperking van de vlootcapaciteit, samenwerking met derde landen, enzovoort, waarover tussen alle betrokken partijen overeenstemming dient te bestaan.

Daarom verdient ieder initiatief dat, zoals dit glossarium, een beter onderling begrip bevordert, de grootste waardering.

In deze tweede editie is, in de elf talen van de Europese Unie, de meest gangbare benaming opgenomen van meer dan 1 500 soorten of groepen van vissen, schaal- en schelpdieren, weekdieren, zoogdieren en planten, terwijl daarvoor ook de wetenschappelijke naam in het Latijn is vermeld. Waar mogelijk is tevens een afbeelding gegeven.

Dit indrukwekkende glossarium had niet tot stand kunnen worden gebracht zonder de medewerking van onderzoekinstituten uit de verschillende landen van de Unie. Het is een voorbeeld van een gemeenschappelijk belang van de wetenschap enerzijds en de visserijsector anderzijds die, hoewel de aard van hun werk verschilt, elkaar vinden bij het streven naar meer kennis over en een betere exploitatie van de rijkdommen van de zee.

Het is ook het resultaat van samenwerking tussen deze onderzoekinstituten en de diensten van de Europese Commissie voor terminologie, statistiek en visserijbeleid.

Allen die aan dit glossarium hebben meegewerkt, verdienen een woord van dank. Via hun nauwgezet onderzoek hebben zij niet alleen bijgedragen aan de totstandkoming van een heel verzorgd boek, maar ook een heel eigen bijdrage geleverd aan de opbouw van Europa.

Emma Bonino
lid van de Europese Commissie

Prefácio

Para nos compreendermos, é indispensável que utilizemos uma linguagem comum. Ora, se é verdade que a diversidade das línguas constitui uma das principais riquezas da União Europeia, quando se trata de um domínio tão rico pela sua diversidade como o dos recursos aquáticos, é fácil perder o rumo.

O presente glossário constitui, pois, um instrumento precioso, não só para os linguistas e cientistas, como para todos os que trabalham nos sectores das pescas e da aquicultura.

Com efeito, nestes sectores, talvez mais do que noutros, as soluções para os problemas dependem da compreensão mútua e da cooperação entre os profissionais dos vários países da União Europeia, dos pescadores aos distribuidores, passando pelos grossistas e transformadores, sem esquecer os consumidores.

É o próprio futuro da actividade da pesca nas nossas regiões que está em causa: os problemas que se nos deparam só podem ser enfrentados e resolvidos em comum.

A política comum da pesca (PCP) deve fazer face a desafios tão diversos como a sobrecapacidade das frotas de pesca, que reduz a rendibilidade dos navios e incita à fraude, a captura de peixes imaturos, que ameaça a renovação das unidades populacionais, a globalização dos mercados, que põe em causa as estratégias tradicionais das empresas, e ainda as questões ambientais.

A PCP deve fixar as regras do jogo susceptíveis de proporcionar soluções «responsáveis», assegurando uma exploração que evite o desperdício ou mesmo o esgotamento das riquezas marinhas, garantindo o futuro e o desenvolvimento de uma indústria moderna e integrada e permitindo satisfazer a procura dos consumidores, que exigem produtos sãos e de boa qualidade, a preços razoáveis.

Para o efeito, a política comum da pesca prevê medidas de conservação, de controlo, de investigação e de apoio às actividades de transformação, comercialização e promoção, bem como medidas de limitação das capacidades da frota e de cooperação com os países terceiros – para só citar algumas – que exigem uma concertação entre todos os parceiros interessados.

Qualquer iniciativa que, como o presente glossário, contribua para uma melhor compreensão mútua deve, pois, ser saudada.

Nesta segunda edição, são recenseadas mais de mil e quinhentas espécies ou grupos de espécies de peixes, crustáceos, moluscos, mamíferos e plantas, sob as suas designações mais correntes nas 11 línguas comunitárias, com indicação do nome científico (em latim). Sempre que possível, é igualmente apresentada uma ilustração.

Esta publicação impressionante não poderia ter sido realizada sem a contribuição de institutos de investigação de vários países da União, constituindo um exemplo de convergência de interesses entre o mundo científico e o dos profissionais da pesca que, embora diferentes pela natureza do trabalho envolvido, coincidem na preocupação de melhor conhecer e explorar os recursos marinhos. Constitui, igualmente, o resultado da colaboração entre os institutos de investigação e os serviços da Comissão Europeia incumbidos da terminologia, estatísticas e política da pesca.

A todos os que colaboraram e que, através da sua meticulosa investigação e do seu trabalho, não só participaram na realização de uma excelente obra como contribuíram de forma original para fazer progredir a construção europeia, um muito obrigado!

Emma Bonino
Membro da Comissão Europeia

Esipuhe

Voidaksemme ymmärtää toisiamme meidän on löydettävä yhteinen kieli. Vaikka kielten moninaisuus onkin osaltaan Euroopan unionin kallisarvoisinta pääomaa, toiminta kalavarojen kaltaisella äärimmäisen monimuotoisella alalla aiheuttaa helposti hämmennystä.

Siksi tämä teos onkin arvokas väline sekä kielitieteilijöille ja tiedemiehille että kaikille kalastuksen ja vesiviljelyn alalla työskenteleville.

Ratkaisujen löytäminen nimenomaan tämän alan ongelmiin edellyttää keskinäistä ymmärrystä ja yhteistyötä Euroopan unionin kaikkien jäsenvaltioiden ammatillisten piirien välillä – kalastajista tukkukauppiaisiin, jalostajiin ja jakeluportaaseen, unohtamatta myöskään kuluttajia.

Kyseessä on jopa kalastustoiminnan tulevaisuus unionissa, sillä vastassamme on ongelmia, jotka voidaan selvittää ainoastaan yhdessä.

Kalastuslaivastojen ylikapasiteetti heikentää alusten kannattavuutta ja johtaa helposti petoksiin; alamittaisten nuorten kalojen pyynti vaarantaa kalakantojen uudistumisen; markkinoiden globalisoituminen uhkaa viedä pohjan yritysten perinteisiltä strategioilta... Kaikki nämä ovat haasteita, joihin yhteisen kalastuspolitiikan on kyettävä vastaamaan. Lisäksi on otettava huomioon ympäristökysymykset.

Yhteisen kalastuspolitiikan tehtävänä on vahvistaa pelisäännöt, joiden avulla voidaan löytää vastuullisia ratkaisuja ja varmistaa meren rikkauksien kohtuullinen hyödyntäminen siten, että vältetään niiden ehtyminen. On turvattava nykyaikaisen yhtenäisen elinkeinon tulevaisuus ja kehittyminen, jotta kuluttajille voidaan toimittaa näiden edellyttämiä hyvälaatuisia, terveellisiä ja kohtuuhintaisia tuotteita.

Yhteiseen kalastuspolitiikkaan sisältyykin säilyttämis-, valvonta-, tutkimus- ja tukitoimenpiteitä, joita sovelletaan jalostuksen, markkinoinnin ja menekinedistämisen aloilla. Siinä myös rajoitetaan kalastuslaivastokapasiteettia ja säännellään yhteistyötä kolmansien maiden kanssa. Näiden toimenpiteiden toteuttaminen edellyttää kaikkien osapuolten välistä yhteisymmärrystä.

Tästä syystä kaikki sellaiset aloitteet, kuten tämä sanasto, joilla voidaan lisätä keskinäistä ymmärrystä, ovat tervetulleita.

Tähän toiseen painokseen on koottu yli 1 500 kala-, äyriäis-, simpukka-, nisäkäs- ja kasvilajin tai lajiryhmän yleisimmin käytetyt nimitykset yhteisön yhdellätoista virallisella kielellä sekä niiden tieteelliset (latinankieliset) nimet. Aina kun suinkin mahdollista, mukana on myös kuva.

Tämä vaikuttava saavutus ei olisi ollut mahdollinen ilman unionin eri jäsenvaltioiden tutkimuslaitosten panosta. Teos on esimerkki tieteen ja ammattikalastuksen etujen lähentymisestä. Vaikka näiden alojen toiminta on luonteeltaan varsin erilaista, niitä yhdistää pyrkimys meren rikkauksien perusteellisempaan tuntemukseen ja tasapainoisempaan hyödyntämiseen.Teos on myös jäsenvaltioiden tutkimuslaitosten ja termityöstä, tilastoista ja kalastuspolitiikasta vastaavien komission yksiköiden välisen yhteistyön tulos.

Kiitän lämpimästi kaikkia työhön osallistuneita. Perinpohjaisella tutkimustyöllään ja ponnistuksillaan he eivät ole olleet yksinomaan mukana luomassa erittäin hienoa teosta, vaan he ovat myös edistäneet omaperäisellä tavalla Euroopan rakentamista.

Emma Bonino
Euroopan komission jäsen

Förord

För att förstå varandra måste man använda ett gemensamt språk. Språkmångfalden är en av Europeiska unionens viktigaste rikedomar, men när det handlar om ett område som har så rik mångfald som fiskeriresurserna är det lätt att det uppstår språkliga oklarheter.

Detta uppslagsverk är därför ett ovärderligt redskap. Inte bara för översättare, tolkar och vetenskapsmän, utan också för alla dem som arbetar inom fiskeri- och vattenbrukssektorn.

Kanske är det så att lösningen på problemen inom denna sektor mer än inom andra sektorer är beroende av ömsesidig förståelse och samarbete mellan de yrkesverksamma i Europeiska unionens olika länder, från fiskare och distributörer till konsumenter, via fiskgrossister och beredningsföretag.

Detta är också viktigt för fiskets framtid i våra regioner, eftersom de problem som vi ställs inför endast kan bemötas och lösas gemensamt.

Fiskeflottornas överkapacitet som minskar fartygens inkomster och främjar bedrägeri, fångsten av småfisk som äventyrar fiskbeståndens återhämtning, globaliseringen av marknaderna som får till följd att företagens traditionella strategier ifrågasätts, men också miljöfrågorna är utmaningar som väger lika tungt inom den gemensamma fiskeripolitiken.

Den gemensamma fiskeripolitiken skall fastställa de spelregler som kan leda till "ansvarsfyllda" lösningar genom att säkerställa en hantering där man undviker att slösa med och uttömma havets rikedomar och där man garanterar framtid och utveckling för en modern och integrerad industri som kan möta kraven från konsumenterna som vill ha bra och hälsosamma produkter till rimliga priser.

För att uppnå detta mål föreskriver den gemensamma fiskeripolitiken åtgärder för bevarande, kontroll, forskning och stöd till verksamheten inom bearbetning, saluföring, marknadsföring samt åtgärder för begränsning av flottans kapacitet och samarbete med tredje land. Uppräkningen av de områden där alla berörda parter måste komma överens kunde göras längre.

Det är därför som alla initiativ i riktning mot bättre ömsesidig förståelse, som till exempel det här uppslagsverket, bör välkomnas.

Denna andra utgåva tar upp mer än ettusenfemhundra arter eller artgrupper av fiskar, skaldjur, snäckor, mollusker, däggdjur och växter som anges med sin vanligaste benämning på de elva gemenskapsspråken, följd av den vetenskapliga benämningen (på latin). Så ofta som möjligt ges en illustration.

Detta imponerande verk hade aldrig kunnat förverkligas utan medverkan från forskningsinstituten i unionens olika länder. Den är ett exempel på att den vetenskapliga världen och yrkesvärlden inom fisket har sammanfallande intressen och att dessa två världar vars arbete skiljer sig väsentligt åt båda är angelägna om att lära känna och utnyttja havets resurser på ett bättre sätt. Uppslagsverket är i lika hög grad resultatet av ett samarbete mellan dessa forskningsinstitut och de enheter vid kommissionen som ansvarar för terminologi, statistik och fiskeripolitik.

Jag vill rikta ett tack till alla dem som har varit delaktiga, ty genom deras ytterst noggranna forskning och deras arbetsinsats har de inte bara förverkligat en mycket vacker bok utan också på ett originellt sätt bidragit till att främja Europabygget.

Emma Bonino
Ledamot av kommissionen

Contents

v **Preface**
vi Prefacio
vii Forord
viii Vorwort
ix Πρόλογος
xi Préface
xii Prefazione
xiii Voorwoord
xiv Prefácio
xv Esipuhe
xvi Förord

xvii **Contents**

xviii **Preliminary remarks**
xx Observaciones preliminares
xxii Indledende bemærkninger
xxiv Einleitende Bemerkungen
xxvi Εισαγωγικές παρατηρήσεις
xxviii Remarques préliminaires
xxx Osservazioni preliminari
xxxii Toelichting
xxxiv Observações prévias
xxxvi Yleistä
xxxviii Inledande anmärkningar

xl **Members of the expert working group**

xlii **Acknowledgements**

xliii **Classification of aquatic animals and plants**
xliii Clasificación de animales y plantas acuáticos
xliv Klassifikation af vanddyr og -planter
xliv Klassifikation der Wassertiere und -pflanzen
xlv Ταξινόμηση υδρόβιων ζώων και φυτών
xlv Classification des animaux et des plantes aquatiques
xlvi Classificazione degli animali e piante acquatici
xlvi Classificatie van waterdieren en waterplanten
xlvii Classificação das plantas e animais aquáticos
xlvii Vesieläinten ja -kasvien luokitus
xlviii Klassificering av vattendjur och -växter

1 **Corpus**

385 **Index of scientific names**

397 **Index of family names**

399 **Index**
417 Índice ES
427 Indeks DA
443 Index DE
463 Ελληνικό ευρετήριο
471 Index FR
483 Indice IT
491 Index NL
505 Índice PT
515 Suomenkielinen hakemisto
527 Svenskt register

541 **Index of interagency 3-alpha identifiers**

545 **Picture credits**

Preliminary remarks

Coverage

The dictionary contains information on 1 532 species of fish, crustaceans, molluscs, aquatic mammals, seaweeds and other fishery products.

The dictionary has been compiled by a working group of national fishery experts convened by the European Commission and is based on contents of the Commission's multilingual Eurodicautom database and on a listing supplied by the Food and Agriculture Organisation (FAO) of all the species appearing in the commercial catch statistics. The members of this expert working group are listed on page XL. The working group reviewed this listing on the principle that each item should be precisely identifiable, at least to the level of the taxonomic family. More general items were excluded. The list of species was extended to include a number of species which, although not of commercial interest, were considered by the expert group to be of interest to resource managers and conservationists.

Presentation

The presentation of each item in the dictionary is as follows:

Reference number	Family name	Three-alpha identifier

SC	Scientific name(s)
ES	Spanish name(s)
DA	Danish name(s)
DE	German name(s)
EL	Greek name(s)
EN	English name(s)
FR	French name(s)
IT	Italian name(s)
NL	Dutch name(s)
PT	Portuguese name(s)
FI	Finnish name(s)
SV	Swedish name(s)

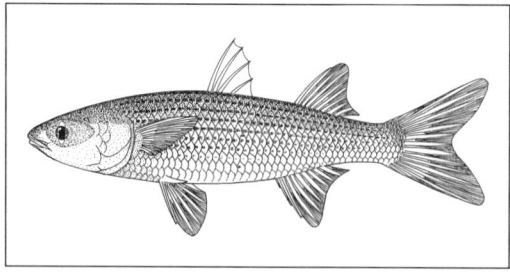

Notes on the entries

Reference number: the number accorded to the species to facilitate the use of the indexes. It has no significance outside this publication.

Family: the family to which the species has been assigned, based on the taxonomic classification used in the 'taxonomic authority list' ([1]). Those parts of the classification used in this publication are shown on page XLVIII. The items in this dictionary are in the order in which the family to which they belong appears in the classification. An index of the family names is included in this publication (page 397). The ordering within the family groups is for items covering the whole family to come first (e.g. *Merluccidae*), followed by identified species within each genus (e.g. *Merluccius hubbsi*), followed by non-specified items within a genus (e.g. *Merluccius* spp.).

Scientific (Latin) name: the name indicated is that by which the species is known in scientific literature. Where more than one is given, the first is the common name and the others are those by which the species has been known but which are considered as superseded.

Three-alpha identifier: this 'inter-agency three-alpha identifier' was assigned to the species by the coordinating working party on fishery statistics for use in international statistical publications. However, this identifier is now widely used outside of the statistical sector in documentation related to fisheries (for example, in fishing logbooks). The depository agency for these identifiers is the FAO (to whom application should be made for the allocation of additional items). Identifiers are generally only available for those species individually identified in fishery statistics: in many cases, commercial species which are aggregated for statistical purposes have not been allocated individual identifiers. The important concept in the allocation of this identifier is that, while scientific and 'common' names are prone to frequent changes, the identifier, once allocated, is permanent.

Names in the 11 official EU languages: the names selected by the expert working group are those considered most appropriate to the species in each language. In many cases alternative names have been given but it is stressed that these alternatives are not an exhaustive list of the regional variations or variations due to the stage in the life cycle, the season of the year, the fishery in which it was taken, the degree of treatment after capture, etc. As a general rule, the name refers to the species as it was taken out of the water. If several alternative names appear in a language, it is always the first one which the experts consider to be the most suitable.

It has been impossible to find existing names for all species. The species are not caught and traded universally and thus, for example, it is not unusual to find no Danish name for a species which is only caught and traded in the south-east Pacific Ocean. However, where no names were readily available, the expert working group attempted to complete the list by making literal translations of the names in one of the other languages. Even this has not been possible in all cases.

Drawings: efforts have been made to include an illustration for each of the species. These illustrations are not intended to enable the precise identification of a species but are included to provide a useful indication of the general characteristics of the species. Where an item refers to a family or a group of species an illustration has been selected which shows the major characteristics of that grouping.

Indexes: Indexes have been compiled for each of the languages, the scientific names and the three-alpha identifiers. These indexes are 'intelligent', for example, 'Atlantic herring' will be found under 'Atlantic herring' and 'herring, Atlantic'.

Some difficulties were experienced in developing the indexes for certain languages due to the formulation of compound names. For example, the English name of the family *Sparidae* may be found in literature written as 'sea breams', 'sea-breams' and 'seabreams'. In order that the various forms of compound names of related species are in adjacent positions in the indexes, a harmonised format has been selected for use in the indexes. This may differ from the form in the body of the dictionary where the form given is that most frequently encountered in literature. However, it must be stressed that, for many names, there appear to be no hard and fast rules as to the correct form of the names.

([1]) Taxonomic authority list (1988), *Aquatic sciences and fisheries information system reference series*, No 8, FAO, Rome, 465 pp.

Observaciones preliminares

Selección de términos

Este glosario abarca 1 532 especies de peces, crustáceos, moluscos, mamíferos acuáticos, plantas marinas y otros productos de la pesca.

Teniendo como punto de partida Eurodicautom, la base de datos plurilingüe de la Comisión, y una lista, facilitada por la FAO, de todas las especies que figuran en las estadísticas de pesca comercial, este glosario ha sido elaborado por un grupo de trabajo, reunido a instancias de la Comisión Europea y compuesto por expertos nacionales en el campo de la pesca, cuyos miembros figuran relacionados en la página XL. El grupo de trabajo ha revisado las listas con el siguiente criterio: cada especie debe poder ser identificada con precisión, por lo menos a nivel de la familia taxonómica. Los términos más genéricos han sido excluidos. La lista de las especies se ha ampliado con aquellas que, aun no representando un interés comercial, han sido consideradas útiles para las personas encargadas de la protección de los recursos haliéuticos.

Presentación de los términos

Para cada especie figuran las siguientes indicaciones:

Número de referencia	Nombre de familia	Identificador alfa-3

SC	*Nombre(s) científico(s)*
ES	Nombre(s) español(es)
DA	Nombre(s) danés(es)
DE	Nombre(s) alemán(es)
EL	Nombre(s) griego(s)
EN	Nombre(s) inglés(es)
FR	Nombre(s) francés(es)
IT	Nombre(s) italiano(s)
NL	Nombre(s) neerlandés(es)
PT	Nombre(s) portugués(es)
FI	Nombre(s) finés(es)
SV	Nombre(s) sueco(s)

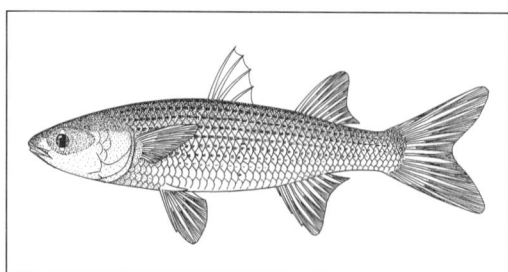

He aquí algunas observaciones a propósito de estos datos

Número de referencia: Número asignado a las especies para facilitar la utilización de los índices. No tiene ningún otro significado fuera del contexto de la presente publicación.

Familia: Familia a la que pertenece la especie según la clasificación taxonómica utilizada en la «Taxonomic Authority List»[1] (lista taxonómica de referencia, publicada por la FAO). Los elementos de esta clasificación utilizados en la presente publicación figuran en la página XLVIII. En este glosario, los términos aparecen clasificados según el orden de la familia a la que pertenece la especie. El índice de los nombres de familia figura en la página 397. En la clasificación correspondiente a cada grupo de familia, aparece en primer lugar el nombre genérico de la familia (por ejemplo, *Merluccidae*), seguido de los nombres de las especies identificadas dentro de cada género (por ejemplo, *Merluccius hubbsi*) y, por último, los nombres de los elementos no específicos que forman parte de un mismo género (por ejemplo, *Merluccius* spp.).

Nombre científico (latino): El nombre que se indica en el glosario es aquel por el que se conoce la especie en la literatura científica. Cuando en esta sección figuran varios nombres, el primer nombre que aparece es el utilizado más corrientemente; los otros corresponden a denominaciones actualmente consideradas como obsoletas, pero que todavía se utilizan.

Identificador alfa-3: Este «identificador alfa-3», utilizado por distintos organismos, ha sido asignado a las especies por el Grupo de trabajo de coordinación de las estadísticas de la pesca en el Atlántico, debido a su uso en las publicaciones estadisticas internacionales. Actualmente se utiliza muy frecuentemente este identificador, fuera del campo de la estadística, en documentos referentes a la pesca (por ejemplo, en los diarios de pesca). El organismo depositario de estos identificadores es la FAO (a quien hay que dirigirse para posibles inclusiones en la lista). Generalmente, sólo se puede disponer de identificadores para las especies identificadas individualmente en las estadísticas de pesca: en numerosos casos, las especies comerciales reagrupadas bajo el mismo epígrafe por necesidades estadísticas no han recibido identificador individual. Insistimos, no obstante, sobre el hecho de que, si bien los nombres científicos y «comunes» son susceptibles de frecuentes cambios, el identificador no varía nunca una vez que ha sido asignado.

Nombres en las once lenguas oficiales de la UE: Los nombres seleccionados por el grupo de trabajo han sido considerados como los más apropiados para cada especie en cada una de las lenguas. No se han considerado de forma sistemática las variantes regionales o las diferentes denominaciones que corresponden a un período cíclico de la vida, a una época del año, al lugar de pesca, al tipo de tratamiento después de la captura, etc. De manera general, el nombre indicado corresponde a la especie tal y como es capturada. Si en una lengua aparecen varias denominaciones; siempre es la primera la considerada por los expertos como la más apropiada.

No ha sido siempre posible encontrar un nombre que ya existiera para cada especie. En efecto, las especies no son siempre capturadas y comercializadas en todo el mundo, y no es extraño, por ejemplo, no hallar ningún nombre danés para una especie capturada y comercializada únicamente en el sudeste del Océano Pacífico. Sin embargo, cuando no se ha podido encontrar un nombre, el grupo de trabajo ha intentado completar la lista haciendo traducciones literales a partir del nombre en una u otra lengua. A pesar de todo, esto no ha sido siempre posible.

Ilustraciones: En la medida de lo posible, cada especie aparece ilustrada. Las ilustraciones no tienen por objeto permitir la identificación precisa, sino facilitar una indicación de las características generales de la especie. Cuando se trata de una familia o de un grupo de especies, la ilustración muestra los rasgos principales de dicho grupo.

Índice: Se han establecido índices para cada lengua, para los nombres científicos y para los identificadores alfa-3. Son de tipo analítico: por ejemplo, «Atlantic herring» figura a la vez bajo «Atlantic herring» y bajo «Herring, Atlantic».

La elaboración de los índices ha presentado algunas dificultades en ciertas lenguas a causa de las palabras compuestas. Por ejemplo, el nombre inglés de la familia *Sparidae* puede revestir las formas «sea breams», «sea-breams» y «seabreams». Para que las diferentes formas de los nombres compuestos de las especies de la misma familia figuren en los índices de forma agrupada, se ha elegido un formato armonizado. Éste puede ser diferente de la forma adoptada para el corpus del glosario, que es la que se encuentra más frecuentemente en la literatura. Sin embargo, hay que señalar que, para muchos nombres, no parecen existir reglas precisas respecto a la ortografía correcta.

---

[1] Taxonomic authority list (1988), *Aquatic sciences and fisheries information system reference series*, n° 8, FAO, Roma, 465 pp.

Indledende bemærkninger

Indhold

Glossaret indeholder oplysninger om 1 532 fiskearter, krebsdyr, bløddyr, vandpattedyr, alger og andre fiskeriprodukter.

Basismaterialet er sammensat af termer fra Europa-Kommissionens flersprogede terminologidatabank, Eurodicautom, og fra FAO's lister over de arter, der forekommer i kommercielle fangststatistikker. Det er bearbejdet af en arbejdsgruppe bestående af sagkyndige fra de enkelte medlemslande samt medarbejdere ved Kommissionen. På side XL findes en liste over medlemmerne af denne arbejdsgruppe. Ved udarbejdelsen af glossaret har arbejdsgruppen arbejdet ud fra det princip, at hver enkelt term skal kunne identificeres ned til det taksonomiske familieniveau. Mere generelle termer er udeladt. Listen blev udvidet til at omfatte en række arter, som godt nok ikke er af handelsmæssig interesse, men som gruppen anså for relevante i forbindelse med bevaring af bestanden af truede arter.

Præsentation

De enkelte termposter præsenteres som vist nedenfor:

Reference-nummer	Familienavn	3-alfa-kode

SC	*videnskabelige (latinske) navne*
ES	spanske navne
DA	danske navne
DE	tyske navne
EL	græske navne
EN	engelske navne
FR	franske navne
IT	italienske navne
NL	nederlandske navne
PT	portugisiske navne
FI	finske navne
SV	svenske navne

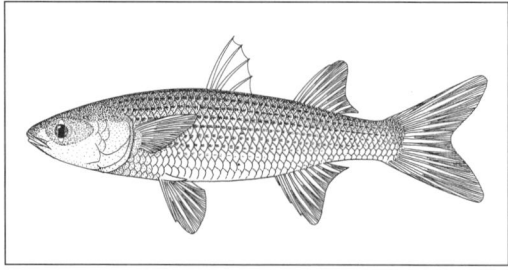

Til disse termposter er der følgende bemærkninger

Referencenummer: Det nummer, som arten har fået tildelt for at lette brugen af indekserne. Det har ingen betydning uden for denne publikation.

Familie: Den familie, som arten tilhører, baseret på den taksonomiske klassifikation, der anvendes i FAO's »Taxonomic Authority List« (¹). De dele af klassifikationen, der er benyttet i denne publikation, er vist på side XLVIII. Arterne er opført i den rækkefølge, som den familie, de tilhører, har i klassifikationen. Der findes et indeks over familienavne på side 397. Inden for familiegrupperne anføres først termposter, der dækker hele familien (f. eks. *Merluccidae*), derefter følger identificerede arter inden for hver slægt (f. eks. *Merluccius hubbsi*), og til sidst termposter for ikke-specificerede arter inden for en slægt (f. eks. *Merluccius spp.*).

Videnskabeligt (latinsk) navn: Det navn, der er anført, er det, som arten kendes under i faglitteraturen. Hvis der er anført mere end ét, er det første navn det gængse, og de andre navne er dem, der anses for forældede, selv om de stadig bruges.

3-alfa-kode: Denne »Inter-Agency 3-Alpha Identifier« (3-alfa-kode) er en identifikator, som er tildelt arterne af Den Koordinerende Arbejdsgruppe vedrørende Statistikker over Fiskeriet i Atlanterhavet til brug i internationale statistiske publikationer. Nu anvendes denne kode dog også i vid udstrækning uden for den statistiske sektor i dokumenter vedrørende fiskeri (f.eks. i fiskernes logbøger). Det er FAO, der administrerer disse koder, og det er til denne organisation, der skal rettes anmodninger om tildeling af yderligere koder. Normalt foreligger der kun koder for de arter, der identificeres individuelt i fiskeristatistikkerne. En del af de arter, der indgår i handelen, samles i grupper af statistiske hensyn, og har derfor ikke fået tildelt en kode. Et vigtigt princip i forbindelse med tildelingen af denne kode er, at mens de videnskabelige og »gængse« navne eventuelt ændres, så forbliver koden den samme, når den først er tildelt.

Navne på de 11 officielle EU-sprog: De navne, som arbejdsgruppen har anført, er dem, der anses for at være de gængse for arterne på hvert enkelt sprog. I mange tilfælde er der givet alternative navne, men det skal understreges, at disse alternativer ikke er en udtømmende liste over regionale varianter eller varianter, der refererer til stadium i livscyklus, årstid, fangststed, bearbejdningsgrad efter fangst osv. Almindeligvis refererer navnet til arten, således som den findes i vandet. Hvis der på et sprog er anført flere navne, står det foretrukne navn altid først.

Det har ikke været muligt at finde eksisterende navne på alle EU-sprog for alle arter. Det er ikke alle arter, der fanges og indgår i handelen overalt, og det er derfor ikke usædvanligt, hvis der f.eks. ikke findes et dansk navn for en art, som kun fanges og indgår i handelen i den sydøstlige del af Stillehavet. Når der ikke umiddelbart forelå et navn, har arbejdsgruppen forsøgt at fuldstændiggøre glossaret ved at lave en direkte oversættelse af navnet på et af de andre sprog. Dette har dog heller ikke kunnet lade sig gøre i alle tilfælde.

Tegninger: Så vidt det har været muligt, findes der en illustration af hver enkelt art. Hensigten med disse illustrationer er ikke at give en præcis identifikation af arten, men at give læseren en idé om artens generelle karakteristika. Til termposter, der refererer til en familie eller artsgruppe, er der valgt en illustration, som viser pågældende gruppes væsentligste karakteristika.

Indekser: Der er udarbejdet indeks for hvert sprog, for de videnskabelige (latinske) navne og for 3-alfa-koderne. Disse indekser er »intelligente«: Dvs. at f.eks »Atlantisk Menhaden« vil kunne findes under »Atlantisk Menhaden« og under »Menhaden, Atlantisk«.

Der har været visse vanskeligheder med at udarbejde indekserne for nogle af sprogene på grund af sammensatte navne. For eksempel findes det engelske navn for familien *Sparidae* skrevet i litteraturen som »sea breams«, »sea-breams« og »seabreams«. For at de forskellige former af sammensatte navne på beslægtede arter kan komme til at stå efter hinanden i indekserne, er der valgt et harmoniseret format til brug i indekserne. Dette kan være forskelligt fra glossarets korpus, hvor der er valgt den form, der hyppigst mødes i litteraturen. Det skal dog understreges, at der ikke synes at være faste regler for, hvad der er navnenes korrekte form.

(¹) Taxonomic authority list (1988), *Aquatic sciences and fisheries information system reference series*, No 8, FAO, Rome, 465 s.

Einleitende Bemerkungen

Umfang

Das Glossar enthält Angaben zu 1 532 Arten von Fischen, Krebstieren, Weichtieren, Wassersäugetieren, Algen und anderen Fischereierzeugnissen.

Ausgangspunkt für dieses Glossar waren die in Eurodicautom, der terminologischen Datenbank der Kommission, gespeicherten sowie die in den Fangstatistiken der FAO angeführten Arten. Eine aus nationalen Experten und Mitarbeitern der zuständigen Dienststellen der Europäischen Kommission zusammengesetzte Arbeitsgruppe, deren Mitglieder auf Seite XL genannt sind, erarbeitete das vorliegende Glossar nach dem Prinzip, daß jedes Stichwort zumindest auf der Ebene der taxonomischen Familie genau identifizierbar sein sollte. Weniger spezifische Stichwörter wurden nicht aufgenommen. Diese Liste wurde erweitert um eine Anzahl weniger kommerziell als vielmehr für die Erhaltung der Bestände relevanter Arten.

Darstellung

Die Einträge sind folgendermaßen dargestellt:

Bezugs- nummer	**Familie**	Drei-Alpha- Kode

SC	*Wissenschaftliche Bezeichnung(en)*
ES	Spanische Bezeichnung(en)
DA	Dänische Bezeichnung(en)
DE	Deutsche Bezeichnung(en)
EL	Griechische Bezeichnung(en)
EN	Englische Bezeichnung(en)
FR	Französische Bezeichnung(en)
IT	Italienische Bezeichnung(en)
NL	Niederländische Bezeichnung(en)
PT	Portugiesische Bezeichnung(en)
FI	Finnische Bezeichnung(en)
SV	Schwedische Bezeichnung(en)

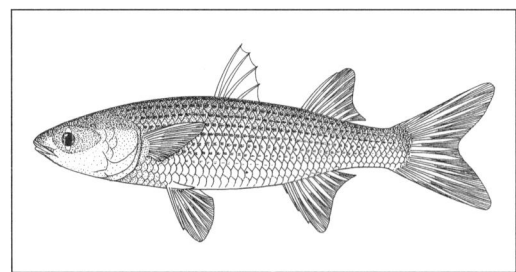

Bei diesen Einträgen ist folgendes zu beachten:

Bezugsnummer: Dabei handelt es sich um die Nummer, die der Art im Index zugeordnet ist. Außerhalb dieser Veröffentlichung ist die Nummer ohne Bedeutung.

Familie: Die Zuordnung der einzelnen Arten entspricht der „Taxonomic Authority List" ([1]) (taxonomische Referenzliste, veröffentlicht von der FAO). Die Teile der in dieser Veröffentlichung benutzten taxonomischen Klassifikation sind auf Seite XLVIII wiedergegeben. Die Arten sind in der gleichen Reihenfolge angeordnet, in der die Familien, zu denen sie gehören, in der Klassifikation erscheinen. Ein Index der Familien befindet sich auf Seite 397. Innerhalb der Familiengruppen erscheinen zuerst die Einträge, die die ganze Familie umfassen (z. B. Merluccidae), danach folgen die innerhalb einer Gattung identifizierten Arten (z. B. Merluccius hubbsi) und zuletzt die nicht identifizierten Arten innerhalb einer Gattung (z. B. Merluccius spp.).

Wissenschaftliche bzw. lateinische Bezeichnung: In wissenschaftlichen Veröffentlichungen ist die Art unter dem hier angegebenen Namen zu finden. Wird mehr als eine Bezeichnung angeführt, so ist die erste die geläufige, während die anderen allgemein als überholt gelten, jedoch nach wie vor verwendet werden.

Drei-Alpha-Kode: Dieser „Inter-Agency 3-Alpha Identifier" wurde den einzelnen Arten von der Arbeitsgruppe für die Koordinierung der Statistiken über Fischerei im Atlantik zur Verwendung in internationalen statistischen Veröffentlichungen zugeteilt. Dieser Kode wird jedoch inzwischen auch oft in fischereibezogenen Dokumenten (z. B. in Logbüchern von Fischereifahrzeugen) außerhalb des statistischen Bereichs benutzt. Die FAO registriert die vergebenen Kodes (Anträge auf Kodierung weiterer Arten sind an die FAO zu richten). Im allgemeinen stehen nur für die Arten Kodes zur Verfügung, die in Fischereistatistiken einzeln aufgeführt werden; in vielen Fällen sind kommerzielle Arten, die zu statistischen Zwecken zusammengefaßt worden sind, nicht einzeln kodiert. Die Vergabe eines Kodes ist deshalb sinnvoll, weil der einmal zugeteilte Kode nicht mehr geändert wird, im Gegensatz zu den wissenschaftlichen und gebräuchlichen Bezeichnungen.

Bezeichnungen in den elf Amtssprachen der Europäischen Union: Die Arbeitsgruppe hat die in den einzelnen Sprachen für die einzelnen Arten als adäquat geltenden Bezeichnungen ausgewählt. Es wurde kein Versuch unternommen, einige der zahlreichen regionalen Varianten, die für ein Entwicklungsstadium, die Jahreszeit, das Fanggebiet, die Art der Verarbeitung nach dem Fang usw. stehen, ebenfalls aufzunehmen. Als allgemeine Regel gilt, daß der Name sich auf den Zustand der Art zum Zeitpunkt des Fangs bezieht. Wird in einer der Sprachen mehr als eine Bezeichnung angegeben, so wird immer die an erster Stelle genannte von den Sachverständigen als adäquat angesehen.

Es erwies sich als unmöglich, Bezeichnungen für alle Arten in allen Amtssprachen zu finden, weil die Arten nicht überall gefangen und in den Handel gebracht werden. Es ist deshalb nicht verwunderlich, daß es keine dänische Bezeichnung für eine Art gibt, die nur im Südostpazifik gefangen und verkauft wird. In den Fällen, in denen keine Bezeichnung ausfindig zu machen war, versuchte die Arbeitsgruppe jedoch, das Verzeichnis durch wörtliche Übersetzung der in einer der anderen Sprachen vorhandenen Bezeichnung zu ergänzen. Doch auch das war nicht in allen Fällen möglich.

Abbildungen: Nach Möglichkeit sind die Einträge mit einer Illustration versehen. Diese Abbildungen sind nicht zur exakten Identifizierung jeder einzelnen Art gedacht, sondern sollen lediglich eine Vorstellung der allgemeinen Charakteristika der verschiedenen Arten vermitteln. Begriffe für eine Familie oder Gruppe sind mit einer Abbildung versehen, die die für diese Gruppe typischen Merkmale aufweist.

Indizes: Für jede Sprache sowie für die wissenschaftlichen Bezeichnungen und den Drei-Alpha-Kode gibt es jeweils einen Index. Die Indizes für die einzelnen Sprachen sind „intelligent", d. h., „Atlantischer Hering" z. B. ist im deutschen Index sowohl unter „Atlantisch" als auch unter „Hering" zu finden.

Bei einigen Sprachen ergaben sich Schwierigkeiten bei der Erstellung der Indizes aufgrund der zusammengesetzten Namen. So findet man z. B. die englische Bezeichnung für den lateinischen Namen der Familie *Sparidae* in folgenden Schreibweisen: „sea breams", „sea-breams" und „seabreams". Damit die verschiedenen Schreibweisen der Namen zusammengehöriger Arten auch in der richtigen Reihenfolge erscheinen, hat man sich für eine einheitliche Darstellung in den Indizes entschieden, die nicht immer mit der im Korpus erscheinenden identisch ist, da dort die in der Literatur am häufigsten verwendete Schreibweise erscheint. Es muß jedoch darauf verwiesen werden, daß es anscheinend keine fest vorgeschriebene Schreibweise für alle diese Namen gibt.

---

[1] Taxonomic Authority List (1988), *Aquatic Sciences and Fisheries Information System Reference Series*, Nr. 8, FAO, Rom, 465 S.

Εισαγωγικές παρατηρήσεις

Περιεχόμενο

Το γλωσσάριο αυτό περιέχει πληροφορίες για 1532 είδη ψαριών, οστρακοειδών, μαλακίων, υδρόβιων ζώων, φυκιών και άλλων προϊόντων αλιείας.

Αφετηρία του γλωσσαρίου αυτού αποτέλεσε το Eurodicautom, η ηλεκτρονική τράπεζα ορολογίας της Επιτροπής, καθώς και τα είδη τα αναφερόμενα στις στατιστικές αλιευμάτων της FAO. Μια ομάδα εργασίας, αποτελούμενη από εθνικούς εμπειρογνώμονες σε θέματα αλιείας, που συγκάλεσε η Επιτροπή των ΕΚ, τα μέλη της οποίας αναφέρονται στη σελίδα XL, κατάρτισε τον κατάλογο των ειδών με γενικό κριτήριο την αρχή ότι κάθε είδος πρέπει να είναι δυνατό να προσδιορίζεται με ακρίβεια τουλάχιστον στο επίπεδο της ταξινομικής του οικογένειας. Άλλες γενικότερες κατηγορίες αποκλείσθηκαν. Ο κατάλογος αυτός επεκτάθηκε με έναν αριθμό ειδών μικρότερου εμπορικού ενδιαφέροντος, επειδή η ομάδα εργασίας θεώρησε ότι παρουσιάζουν ενδιαφέρον για τους αρμόδιους διαχείρισης των αλιευτικών πόρων και τους οικολόγους.

Παρουσίαση

Για το κάθε λήμμα αναφέρονται τα εξής στοιχεία:

Αριθμός αναφοράς	**Ονομασία οικογένειας**	Αναγνωριστικός κωδικός τριών γραμμάτων
SC	Επιστημονική (λατινική) ονομασία	
ES	Ισπανική ονομασία	
DA	Δανική ονομασία	
DE	Γερμανική ονομασία	
EL	Ελληνική ονομασία	
EN	Αγγλική ονομασία	
FR	Γαλλική ονομασία	
IT	Ιταλική ονομασία	
NL	Ολλανδική ονομασία	
PT	Πορτογαλική ονομασία	
FI	Φινλανδική ονομασία	
SV	Σουηδική ονομασία	

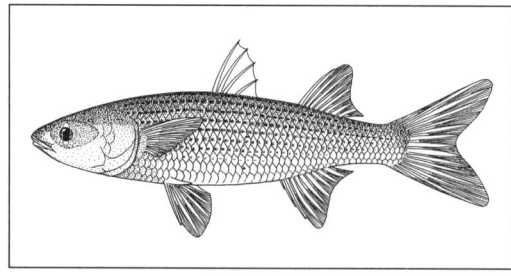

Σχετικά με τα στοιχεία αυτά σημειώνονται τα εξής

Αριθμός αναφοράς: ο αριθμός που δίδεται στο κάθε είδος για να διευκολύνεται η χρήση των ευρετηρίων. Έξω από το πλαίσιο αυτής της έκδοσης ο αριθμός αυτός δεν έχει απολύτως καμία σημασία.

Ονομασία οικογένειας: η οικογένεια που αποδίδεται σε κάθε είδος αντιστοιχεί στην «Taxonomic authority list»(¹) (ταξινομικός κατάλογος αναφοράς που δημοσιεύει η FAO). Τα μέρη της

(¹) Taxonomic authority list (1988), *Aquatic sciences and fisheries information system reference series*, No 8, FAO, Rome, 465 pp.

ταξινόμησης που χρησιμοποιούνται στη δημοσίευση αυτή αναφέρονται στη σελίδα XLVIII. Τα είδη του γλωσσαρίου αναφέρονται με τη σειρά ταξινόμησης της οικογένειας στην οποία ανήκουν. Το ευρετήριο των οικογενειών περιλαμβάνεται στη σελίδα 397. Μέσα στις ομάδες των οικογενειών τα είδη που καλύπτουν το σύνολο της οικογένειας αναφέρονται πρώτα (π.χ. Merluccidae), ενώ στη συνέχεια ακολουθούν τα είδη που εντοπίζονται μέσα σε κάθε γένος (π.χ. Merluccius hubbsi), ακολουθούμενα από τα μη καθοριζόμενα είδη μέσα σ' ένα γένος (π.χ. Merluccius spp).

Επιστημονική (λατινική) ονομασία: αναφέρεται η ονομασία με την οποία ένα είδος είναι γνωστό στην επιστημονική βιβλιογραφία. Όταν δίνονται περισσότερες από μία ονομασίες, η τρέχουσα είναι η πρώτη, ενώ οι επόμενες είναι παλαιότερες ονομασίες του είδους αυτού, που θεωρούνται γενικά απαρχαιωμένες.

Αναγνωριστικός κωδικός τριών γραμμάτων: ο αναγνωριστικός αυτός κωδικός τριών γραμμάτων δόθηκε στο κάθε είδος από την ομάδα εργασίας συντονισμού των στατιστικών αλιείας Ατλαντικού για να χρησιμεύσει στις διεθνείς στατιστικές δημοσιεύσεις. Ωστόσο, ο αναγνωριστικός αυτός κωδικός χρησιμοποιείται σήμερα ευρύτατα και εκτός του στατιστικού τομέα, στα έγγραφα που έχουν σχέση με την αλιεία (για παράδειγμα, στα ημερολόγια αλιείας). Το εξουσιοδοτημένο όργανο στο οποίο κατατίθενται τα σύμβολα αυτά είναι η οργάνωση FAO (στην οποία πρέπει να γίνεται αίτηση για τη χορήγηση άλλων κωδικών). Σε γενικές γραμμές, αναγνωριστικοί κωδικοί υπάρχουν μόνον για τα είδη που προσδιορίζονται στις στατιστικές αλιείας μεμονωμένα· σε πολλές περιπτώσεις, σε ορισμένα εμπορικά είδη τα οποία ομαδοποιούνται για στατιστικούς λόγους, δεν δίνεται αναγνωριστικός κωδικός. Η βασική αρχή η οποία διέπει την εισαγωγή αναγνωριστικού κωδικού είναι ότι, ενώ οι επιστημονικές και οι «κοινές» ονομασίες υπόκεινται συχνά σε μεταβολές, ο αναγνωριστικός κωδικός, από τη στιγμή που θα χορηγηθεί, παραμένει αμετάβλητος.

Ονομασίες στις ένδεκα επίσημες γλώσσες της ΕΚ: η ομάδα εργασίας επέλεξε τις ονομασίες που έκρινε καταλληλότερες για το κάθε είδος στην κάθε γλώσσα. Σε πολλές περιπτώσεις αναφέρονται εναλλακτικές ονομασίες, αλλά δεν κατονομάζονται όλες οι παραλλαγές των περιφερειακών ονομασιών ή οι παραλλαγές που οφείλονται στο στάδιο του κύκλου της ζωής, στην εποχή του έτους, στο ιχθυοτροφείο στο οποίο έγινε η αλίευση, στο βαθμό επεξεργασίας μετά την αλίευση, κλπ. Γενικός κανόνας είναι ότι η ονομασία αφορά ένα είδος όπως αλιεύθηκε.

Η εξεύρεση γνωστών ονομασιών σε όλες τις κοινοτικές γλώσσες για όλα τα είδη αποδείχθηκε αδύνατη. Επειδή η αλιεία και η εμπορία των διαφόρων αυτών ειδών δεν γίνεται σε παγκόσμια κλίμακα, αντιμετωπίστηκε συχνά η εξής δυσκολία, όπως, για παράδειγμα: δεν υπήρχαν δανικές ονομασίες για τα υδρόβια είδη των οποίων η αλιεία και η εμπορία περιορίζεται στην περιφέρεια του νοτιοανατολικού Ειρηνικού. Στις περιπτώσεις που σε ορισμένες γλώσσες δεν υπήρχαν αμέσως διαθέσιμες ονομασίες για ορισμένα είδη, η ομάδα εργασίας προσπάθησε να συμπληρώσει τον κατάλογο με την κατά λέξη μετάφραση των επιστημονικών ονομασιών σε κάθε μία από τις υπόλοιπες γλώσσες. Αλλά ακόμη κι αυτό δεν ήταν δυνατό σε όλες τις περιπτώσεις.

Σχεδιαγράμματα: καταβλήθηκαν προσπάθειες να περιληφθούν εικόνες για κάθε είδος. Οι εικόνες αυτές δεν έχουν σκοπό την αναγνώριση του είδους με ακρίβεια αλλά περιλαμβάνονται για να παρέχουν χρήσιμες πληροφορίες σχετικά με τα γενικά χαρακτηριστικά των ειδών. Όπου ένα λήμμα αναφέρεται σε μια οικογένεια ή μια ομάδα ειδών, η εικόνα που έχει επιλεγεί δείχνει τα γενικά χαρακτηριστικά της ομάδας.

Ευρετήρια: καταρτίστηκαν ευρετήρια με βάση την κάθε γλώσσα, τις επιστημονικές ονομασίες και τα αναγνωριστικά σύμβολα τριών γραμμάτων. Τα ευρετήρια αυτά είναι «έξυπνα», δηλαδή, για παράδειγμα: ο όρος «Atlantic herring» μπορεί να αναζητηθεί στο αγγλικό ευρετήριο είτε στο λήμμα «Atlantic» είτε στο λήμμα «herring».

Σε ορισμένες γλώσσες αντιμετωπίσθηκαν δυσκολίες κατά την κατάρτιση των ευρετηρίων λόγω της διατύπωσης των σύνθετων λέξεων. Για παράδειγμα η αγγλική ονομασία της οικογένειας *Sparidae* αναφέρεται στη βιβλιογραφία ως «sea bream», «sea-bream» και «seabream». Για να ευρίσκονται σε γειτονικές θέσεις στο ευρετήριο οι διάφορες μορφές των σύνθετων ονομασιών των αντίστοιχων ειδών έχει επιλεγεί μια εναρμονισμένη μορφή η οποία και χρησιμοποιείται στα ευρετήρια. Η μορφή αυτή ενδέχεται να διαφέρει από τη μορφή του κύριου σώματος του γλωσσαρίου, όπου έχει δοθεί η πιο συνήθης που συναντάται στη βιβλιογραφία. Εντούτοις, πρέπει να τονισθεί ότι, για πολλές ονομασίες, φαίνεται ότι δεν υπάρχουν καθορισμένοι κανόνες για την ορθή μορφή των ονομασιών.

Remarques préliminaires

Sélection des termes

Ce glossaire couvre 1 532 espèces de poissons, de crustacés, de mollusques, de mammifères aquatiques, de plantes marines et d'autres produits de la pêche.

Partant de Eurodicautom, la base de données multilingues de la Commission, et d'une liste, fournie par la FAO, de toutes les espèces figurant dans les statistiques de pêche commerciale, ce glossaire a été élaboré par un groupe de travail réuni à l'initiative de la Commission européenne et composé d'experts nationaux dans le domaine de la pêche. La liste des membres de ce groupe de travail figure à la page XL. Le groupe de travail a revu les listes sur la base du critère suivant: chaque espèce doit pouvoir être identifiée avec précision au moins au niveau de la famille taxinomique. Les termes plus génériques ont été exclus. La liste des espèces a été étendue à celles qui, bien que ne présentant aucun intérêt commercial, ont été considérées utiles aux personnes chargées de la protection des ressources halieutiques.

Présentation des termes

Pour chaque espèce figurent les indications suivantes:

Numéro de référence	Nom de famille	Identificateur alpha-3
SC	*nom(s) scientifique(s)*	
ES	nom(s) espagnol(s)	
DA	nom(s) danois	
DE	nom(s) allemand(s)	
EL	nom(s) grec(s)	
EN	nom(s) anglais	
FR	nom(s) français	
IT	nom(s) italien(s)	
NL	nom(s) néerlandais	
PT	nom(s) portugais	
FI	nom(s) finnois	
SV	nom(s) suédois	

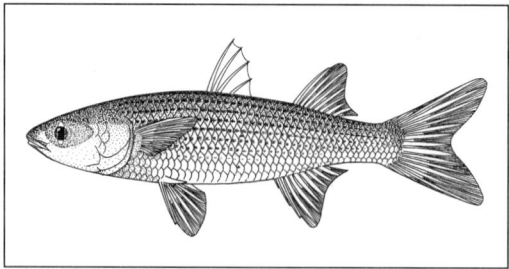

Quelques remarques au sujet de ces données sont mentionnées ci-après.

Numéro de référence: numéro attribué aux espèces pour faciliter l'utilisation des index. Il n'a aucune signification en dehors de la présente publication.

Famille: famille à laquelle appartient l'espèce selon la classification taxinomique utilisée dans la *Taxonomic authority list* ([1]) (liste taxinomique de référence, publiée par la FAO). Les éléments de cette classification utilisés dans la présente publication figurent à la page XLVIII. Dans ce glossaire, les termes apparaissent dans l'ordre selon lequel leur famille d'appartenance figure dans la classification. Un index des noms de famille est inclus dans cet ouvrage (p. 397). À l'intérieur des groupes de familles, les espèces **couvrant l'ensemble** de la famille **sont mentionnées** en premier lieu (par exemple *Merluccidae*), viennent ensuite les espèces identifiées à l'intérieur de chaque genre (par exemple *Merluccius hubbsi*), suivies des espèces non spécifiées à l'intérieur d'un genre (par exemple *Merliccius ssp.*).

Nom scientifique (latin): le nom indiqué est celui sous lequel l'espèce est connue dans la littérature scientifique. Lorsque plusieurs noms figurent dans cette rubrique, le premier est le nom couramment utilisé; les autres correspondent à des dénominations actuellement considérées comme obsolètes, mais qui sont encore utilisées.

Identificateur alpha-3: cet «identificateur interinstitutions alpha-3» a été attribué aux espèces par le groupe de travail de coordination des statistiques de la pêche dans l'Atlantique pour les besoins des publications statistiques internationales. Cet identificateur est actuellement largement utilisé en dehors du domaine statistique dans les documents concernant la pêche (par exemple dans les journaux de pêche). L'organisme dépositaire de ces identificateurs est la FAO (à qui il convient de s'adresser pour obtenir des adjonctions à la liste). Des identificateurs ne sont généralement disponibles que pour les espèces identifiées individuellement dans les statistiques de la pêche: dans de nombreux cas, les espèces commerciales regroupées sous la même rubrique pour les besoins de la statistique n'ont pas reçu d'identificateur individuel. Insistons toutefois sur le fait que, si les noms scientifiques et «communs» sont sujets à des changements fréquents, l'identificateur ne varie plus dès lors qu'il a été attribué.

Noms dans les onze langues officielles de l'Union européenne: les noms retenus par le groupe de travail sont considérés comme les plus appropriés pour chaque espèce dans chacune des langues. Il n'a pas été tenu compte systématiquement des variantes régionales ou des différentes dénominations correspondant à un stade du cycle de vie, à une époque de l'année, au lieu de pêche, au type de traitement après capture, etc. D'une manière générale, le nom indiqué correspond à l'espèce telle qu'elle a été retirée de l'eau. Si, dans une langue, plusieurs dénominations sont mentionnées, c'est toujours la première qui est considérée par les experts comme la plus appropriée.

Il n'a pas toujours été possible de trouver un nom existant pour chaque espèce. Les espèces ne sont en effet pas toujours capturées et commercialisées dans le monde entier et il n'est pas rare, par exemple, de ne trouver aucun nom danois pour une espèce capturée et mise sur le marché uniquement dans le sud-est de l'océan Pacifique. Cependant, lorsqu'aucun nom n'a pu être repéré, le groupe de travail a tenté de compléter la liste en faisant des traductions littérales à partir du nom de l'une ou de l'autre langue. Malgré tout, cela n'a pas toujours été possible.

Illustrations: dans toute la mesure du possible, chaque espèce est illustrée. Les illustrations n'ont pas pour but de permettre l'identification précise, mais de fournir une indication des craractéristiques générales de l'espèce. Lorsqu'il s'agit d'une famille ou d'un groupe d'espèces, l'illustration montre les traits principaux de ce groupe.

Index: des index ont été établis pour chaque langue, pour les noms scientifiques et pour les identificateurs alpha-3. Ils sont de type analytique: par exemple, «Atlantic herring» figure à la fois sous «Atlantic herring» et «Herring, Atlantic».

L'élaboration des index a présenté quelques difficultés pour certaines langues à cause des mots composés. Par exemple, le nom anglais de la famille *Sparidae* peut revêtir les formes «sea breams», «seabreams» et «seabreams». Afin que les différentes formes des noms composés des espèces de la même famille figurent dans les index de façon groupée, un format harmonisé a été choisi. Celui-ci peut être différent de la forme retenue pour le corpus du glossaire, qui est celle que l'on rencontre le plus fréquemment dans la littérature. Cependant, il faut souligner que, pour de nombreux noms, il ne semble pas exister de règles précises concernant l'orthographe correcte.

([1]) Taxonomic authority list (1988), *Aquatic sciences and fisheries information system reference series*, n° 8, FAO, Rome, 465 p.

Osservazioni preliminari

Copertura

Il glossario contiene informazioni su 1 532 specie di pesci, crostacei, molluschi, mammiferi acquatici, alghe e altri prodotti della pesca.

Il glossario è stato elaborato da un gruppo di lavoro di esperti nazionali e di funzionari dei servizi competenti della Commissione europea che si sono basati su quanto contenuto nella base di dati multilingue Eurodicautom della Commissione e su un elenco FAO di tutte le specie figuranti nelle statistiche delle catture commerciali. I membri del gruppo di lavoro di esperti sono elencati a pagina XL. Il gruppo di lavoro ha elaborato il glossario seguendo il criterio generale per cui ogni voce deve poter essere identificata con precisione se non altro a livello della famiglia tassonomica. Voci di carattere più generale sono state escluse. Il gruppo di lavoro ha ritenuto di dover allargare l'elenco inserendovi alcune specie che, pur non rivestendo interesse commerciale, sono importanti ai fini della protezione e conservazione delle risorse ittiche.

Presentazione

Ogni voce del glossario è così presentata:

Numero di riferimento	Nome della famiglia	Codice a tre lettere

SC	*Nome scientifico*
ES	Nome spagnolo
DA	Nome danese
DE	Nome tedesco
EL	Nome greco
EN	Nome inglese
FR	Nome francese
IT	Nome italiano
NL	Nome neerlandese
PT	Nome portoghese
FI	Nome finnico
SV	Nome svedese

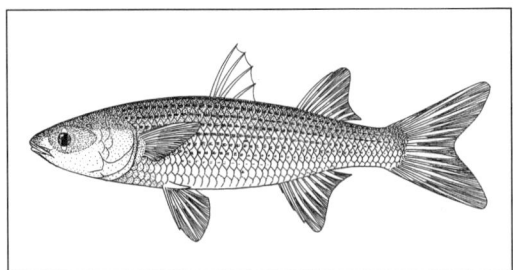

Osservazioni utili

Numero di riferimento: il numero attribuito alle specie serve a facilitare la consultazione degli indici. Non ha riscontro alcuno al di fuori della presente pubblicazione.

Famiglia: famiglia alla quale la specie è assegnata in base alla classificazione tassonomica figurante nell'elenco ufficiale pubblicato dalla FAO[1]. Le parti di tale elenco usate nella presente pubblicazione sono riprodotte a pagina XLVIII. Le singole voci figurano nel glossario nell'ordine in cui le famiglie di appartenenza sono ordinate nella classificazione di cui sopra. L'indice dei nomi di famiglia figura a pagina 397. I gruppi di famiglie sono così ordinati: innanzitutto figurano le voci che coprono un'intera famiglia (ad es., Merluccidae), seguite dalle specie individuate nell'ambito di un singolo genere (ad es., Merluccius hubbsi), per finire con le voci non meglio specificate nell'ambito di un genere (ad es., Merluccius spp.).

Nome scientifico (latino): si riporta il nome con il quale la specie è nota nella letteratura scientifica. In caso di più nomi, il primo è il nome abituale e gli altri sono i nomi con i quali le specie erano note in passato, ma che ora sono da considerarsi desueti.

Codice a tre lettere: il codice internazionale a tre lettere è stato attribuito alle specie dal gruppo di lavoro per il coordinamento delle statistiche della pesca nell'Atlantico ad uso delle pubblicazioni internazionali di statistica. Il codice trova tuttavia attualmente ampio impiego anche al di fuori del campo statistico in varie documentazioni relative alla pesca (ad esempio, nei registri di bordo della pesca). L'organizzazione depositaria di tali codici è la FAO (alla quale è opportuno rivolgersi per ogni richiesta di aggiunta di voci). I codici sono generalmente disponibili soltanto per le specie singolarmente indicate nelle statistiche della pesca: in molti casi non è stato attribuito un proprio codice alle specie che sono state aggregate per fini statistici. Elemento essenziale nell'attribuzione del codice è che, mentre i nomi scientifico e «volgare» sono soggetti a frequenti variazioni, il codice, una volta fissato, resta immutato.

Nomi nelle undici lingue ufficiali dell'UE: i nomi selezionati dal gruppo di lavoro di esperti sono quelli considerati più consoni alle specie nelle varie lingue. In alcuni casi sono riportati nomi alternativi sottolineando però che non si tratta di un elenco esaustivo delle varianti regionali o delle varianti attribuibili alla fase del ciclo vitale, alla stagione dell'anno, alla zona in cui è avvenuta la pesca, al grado di trattamento dopo la cattura ecc. Di norma, il nome si riferisce alla specie così come è stata catturata.

In caso di più denominazioni, la prima riportata è quella giudicata più consona dagli esperti.

È stato impossibile trovare un nome in tutte le lingue per ogni singola specie. Non tutte le specie sono catturate e commercializzate in tutto il mondo: può quindi succedere che non esista un nome danese per una specie catturata e commercializzata nel solo Pacifico sud-orientale. In caso di mancanza di ogni denominazione in una lingua, il gruppo di lavoro di esperti ha tuttavia cercato di completare l'elenco traducendo letteralmente il nome da un'altra lingua, ma nemmeno questo è stato sempre possibile.

Illustrazioni: per ogni specie si è cercato di accludere un'illustrazione. Loro scopo non è quello di rendere possibile l'esatta identificazione di una specie, ma di fornire alcune utili indicazioni sulle caratteristiche generali della specie in questione. Quando la voce si riferisce ad una famiglia o ad un gruppo di specie, è stata scelta un'illustrazione che mostri le principali caratteristiche del raggruppamento.

Indici: sono stati compilati gli indici nazionali, dei nomi scientifici e del codice a tre lettere. Si tratta di indici «intelligenti»: l'inglese «Atlantic herring» potrà ad esempio essere ricercato sotto «Atlantic herring» o sotto «herring, Atlantic».

Si è incontrata qualche difficoltà nell'elaborare gli indici di alcune lingue in seguito all'uso di nomi composti. Il nome inglese della famiglia *Sparidae* appare ad esempio nella letteratura sotto tre diverse forme: «sea breams», «sea-breams» e «seabreams». Per gli indici è stata quindi adottata una forma armonizzata in modo che le varie forme dei nomi composti delle specie correlate non perdano ogni rapporto di contiguità. Tale forma armonizzata può differire dalla forma riportata nel corpo del glossario dove figurano le forme di uso più comune. Va tuttavia sottolineato che, per molti nomi, non sembrano esistere precise regole ortografiche.

---

[1] Taxonomic Authority List (1988) – *Aquatic Sciences and Fisheries Information System Reference Series*, n. 8, FAO, Roma, 465 pagg.

Toelichting

Inhoud

Dit glossarium bevat gegevens over 1 532 soorten: vissen, zeezoogdieren, schelpdieren, weekdieren, zeewieren, algen en andere visserijproducten.

Als basis voor het corpus van deze woordenlijst dienden de soortbenamingen zoals opgenomen in Eurodicautom, de terminologische gegevensbank van de Commissie, en de benamingen ontleend aan de visserijstatistieken van de FAO. Het glossarium werd samengesteld door een werkgroep bestaande uit deskundigen uit de lidstaten en ambtenaren van de Commissie. Een overzicht van de werkgroep is te vinden op blz. XL. Bij de samenstelling van het glossarium werd uitgegaan van het principe dat iedere soort ten minste tot op het niveau van de taxonomische familie geïdentificeerd moet zijn. Minder gebruikelijke soortbenamingen werden niet opgenomen. De lijst is aangevuld met een aantal soorten die eerder op grond van hun behoud dan om hun commerciële betekenis van belang zijn.

Presentatie

De soorten zijn op volgende wijze weergegeven:

Referentie- nummer	Familie- naam	Drie-Alpha- Code

SC	*Wetenschappelijke benaming(en)*
ES	Spaanse benaming
DA	Deense benaming
DE	Duitse benaming
EL	Griekse benaming
EN	Engelse benaming
FR	Franse benaming
IT	Italiaanse benaming
NL	Nederlandse benaming
PT	Portugese benaming
FI	Finse benaming
SV	Zweedse benaming

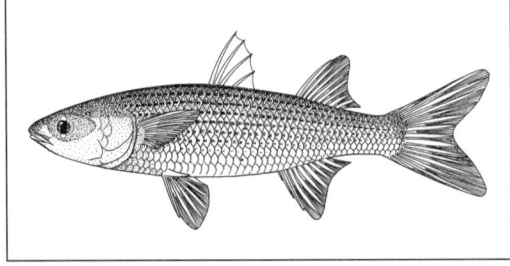

Opmerking

Referentienummer: het referentienummer wordt in het register aan de soortnaam toegevoegd. Dit nummer is uitsluitend binnen deze publicatie van belang.

Familie: hieronder wordt verstaan de familie waartoe een soort behoort op grond van de taxonomische classificatie van de „Taxonomic Authority List"[1] (taxonomische referentielijst van de FAO). De in dit glossarium gebruikte delen van de taxonomische classificatie zijn vermeld op blz. XLVIII. De soorten zijn opgenomen in de volgorde waarin de families waartoe zij behoren in de classificatie zijn gerangschikt. Het register op familienaam bevindt zich op blz. 397. Binnen de familiegroepen staan eerst die termen die de familie als geheel aanduiden (bijv. Merluccidae); daarna komen de volledige soortnamen binnen elk geslacht (bijv. Merluccius hubbsi), en ten slotte de niet nader gedefinieerde soorten binnen een geslacht (bijv. Merluccius ssp.).

Wetenschappelijke (Latijnse) benaming: dit is de benaming waaronder de vissoort in de wetenschappelijke literatuur bekend staat. Indien meer dan één benaming vermeld wordt, is de eerst vermelde benaming de gangbare en de volgende de achterhaalde benaming, die overigens nog wel gehanteerd wordt.

De Drie-Alpha-Code: dit is de „Inter-Agency 3-Alpha Identifier", een drielettercode die door de „Coordinating Working Party of Atlantic Fishery Statistics" aan de afzonderlijke vissoorten wordt toegekend voor gebruik in internationale statistische publicaties. Deze code wordt echter inmiddels ook op brede schaal gebruikt buiten de statistische sector in documentatie betreffende de visserij (bijvoorbeeld in visserslogboeken). De eenmaal toegekende codes worden door de FAO geregistreerd (aanvraag tot codering van „nieuwe" soorten moet bij de FAO ingediend worden). Codes worden doorgaans uitsluitend toegekend aan die soorten die in de statistieken afzonderlijk vermeld worden; in veel gevallen zijn handelssoorten die voor de statistiek zijn samengevoegd, niet van een code voorzien. Toekenning van codes is zinvol omdat zij, in tegenstelling tot de wetenschappelijke en de gangbare benamingen, niet meer gewijzigd worden.

Benamingen in de elf officiële EG-talen: voor ieder van de officiële talen heeft de werkgroep de voor elke soort meest adequate benaming gekozen. Er werd afgezien van het opnemen van enkele van de talrijke regionale varianten die het ontwikkelingsstadium, het seizoen, het vanggebied, de verwerkingswijze, enzovoort, van de vangst weerspiegelen. Algemeen geldt dat de benaming gebaseerd is op de toestand van de soort op het moment van de vangst. Wanneer in één van de talen meer dan één benaming wordt gegeven, geldt de eerst vermelde benaming als de meest adequate.

Het is niet mogelijk gebleken voor alle soorten de benamingen in alle talen van de Gemeenschap te vinden, omdat niet alle soorten overal gevangen en verhandeld worden. Het mag daarom geenszins verbazing wekken dat er bijvoorbeeld geen Deense benaming bestaat voor een soort die alleen in het zuidoostelijke gebied van de Stille Oceaan wordt gevangen en verhandeld. In die gevallen waarin geen overeenkomstige benaming werd gevonden, heeft de werkgroep haar toevlucht genomen tot een letterlijke vertaling van een authentieke benaming uit één van de andere talen. Ook dit bleek niet in alle gevallen haalbaar.

Illustraties: waar mogelijk werden aan de soorten illustraties toegevoegd. Deze zijn niet alleen nuttig voor het nauwkeurig identificeren van de soort, maar geven ook een indruk van de typische kenmerken van de soort. Begrippen met betrekking tot een bepaalde familie of groep werden eveneens van illustraties voorzien om de typische kenmerken duidelijk te maken.

Registers: registers werden gemaakt op de volgende onderdelen: de talen, de wetenschappelijke benamingen, en de „Drie-Alpha-Code". De namen in de registers zijn verwisselbaar, dat wil zeggen dat, bijvoorbeeld, „Atlantische haring" zowel onder „Atlantische" als onder „haring" kan worden gevonden.

In sommige talen is het opstellen van de registers niet zonder problemen. Zo bestaan er bijvoorbeeld voor de Engelse aanduiding voor de Latijnse benaming van de familie der *Sparidae* de volgende schrijfwijzen: „sea breams", „sea-breams", en „seabreams". Om de verschillende schrijfwijzen van bij elkaar behorende soorten in de juiste volgorde te krijgen werd gekozen voor een uniforme presentatie die overigens niet altijd overeenkomt met die in het corpus, omdat daar immers de meest frequente schrijfwijze aangehouden wordt. Er dient op gewezen te worden dat er niet altijd één voorgeschreven schrijfwijze bestaat.

---

[1] Taxonomic Authority List (1988), *Aquatic sciences and fisheries information system reference series*, No 8, FAO, Rome, 465 pp.

Observações prévias

Cobertura

O glossário contém informação sobre 1 532 espécies de peixes, crustáceos, moluscos, algas marinhas e outros produtos da pesca.

O ponto de partida para este glossário foi a base de dados terminológicos da Comissão (Eurodicautom) e uma lista de espécies fornecida pela FAO, que um grupo de trabalho criado pela Comissão das CE, constituído por peritos de pesca nacionais, desenvolveu (os membros deste grupo encontram-se indicados na página XL). O grupo de trabalho estabeleceu esta lista utilizando um critério geral segundo o qual cada item deveria ser identificado de forma precisa, pelo menos ao nível da família taxonómica. Os itens de carácter mais geral foram, assim, excluídos. A presente lista poderá ser aumentada no futuro, tendo em conta questões menos comerciais, destacando-se como principal preocupação a conservação das espécies.

Apresentação

A apresentação de cada entrada no glossário é a seguinte:

Número	**Nome**	Identificador
de referência	**da família**	alfabético
		com três letras

SC	*Nome científico*
ES	Nome espanhol
DA	Nome dinamarquês
DE	Nome alemão
EL	Nome grego
EN	Nome inglês
FR	Nome francês
IT	Nome italiano
NL	Nome neerlandês
PT	Nome português
FI	Nome finlandês
SV	Nome sueco

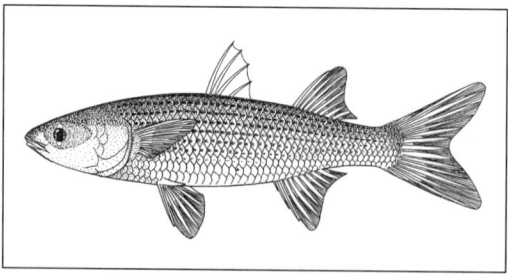

São de notar os seguintes aspectos sobre estas entradas

Número de referência: é o número atribuído à espécie nos índices. Não possui qualquer significado fora desta publicação.

Família: a família em que se inclui a espécie em questão, com base na classificação utilizada pela «Taxonomic Authority List» ([1]) (lista de referência da FAO). Os elementos da classificação utilizados encontram-se na página XLVIII. As espécies neste glossário são apresentadas na ordem em que a sua família aparece na classificação. Na página 397 é apresentado um índice dos nomes das famílias. No interior dos grupos de famílias, primeiro são apresentados os itens que abrangem toda a família (por exemplo, *Merlucidae*), seguidos das espécies identificadas em cada género (por exemplo, *Merluccius hubbsi*), por último, dos itens não especificados num género (por exemplo, *Mercluccius ssp.*).

Nome científico (latim): o nome indicado é aquele pelo qual a espécie é conhecida na literatura científica. Sempre que se apresenta mais do que um nome, o primeiro corresponde ao nome actual e os restantes são os nomes pelos quais a espécie tem sido conhecida, mas que se consideram, geralmente, ultrapassados.

Identificador alfabético com três letras: este identificador foi atribuído às espécies pelo Grupo de Trabalho Coordenador das Estatísticas Pesqueiras do Atlântico, para ser utilizado por publicações estatísticas internacionais. Todavia, é, hoje em dia, amplamente utilizado fora do sector estatístico, no âmbito da documentação das pescas (por exemplo, nos livros de pesca das embarcações pesqueiras). A organização depositária destes identificadores é a FAO, que tem de autorizar a introdução de novos itens. Normalmente, os identificadores apenas se encontram disponíveis para as espécies individualmente identificadas nas estatísticas da pesca: verifica-se com frequência que não são atribuídos identificadores às espécies comerciais agrupadas para fins estatísticos. Importa destacar, ao atribuir este identificador, que enquanto os nomes científicos e «comuns» são susceptíveis de ser alterados frequentemente, o identificador é atribuído com carácter permanente.

Nomes nas 11 línguas oficiais da UE: os nomes seleccionados pelo grupo de trabalho são considerados os mais apropriados para a espécie em cada língua. Não se incluíram as variações regionais ou as que se devem às diferentes fases do ciclo de vida, às estações do ano, às campanhas em que a espécie foi capturada, ao grau de tratamento, etc. Em regra geral, o nome refere-se à espécie tal qual é retirada da água. Sempre que se considerar mais de que um nome na mesma língua, o primeiro é o mais adequado, segundo os peritos.

Foi impossível fazer corresponder a todas as espécies nomes existentes em todas as línguas comunitárias. As espécies não são capturadas e comercializadas a nível mundial e assim, por exemplo, pode não existir nome dinamarquês para uma espécie só capturada e comercializada na região sudeste do oceano Pacífico. Todavia, sempre que não se encontraram nomes disponíveis, o grupo de trabalho procurou completar a lista, procedendo a traduções literais dos nomes existentes numa outra língua, o que, mesmo assim, não foi possível em todos os casos.

Ilustrações: sempre que possível foram incluídas ilustrações para cada espécie. Estas são apresentadas com o intuito de dar uma indicação das características gerais de cada espécie, sem pretender fazer uma identificação precisa da mesma. Sempre que um item se refere a uma família ou a um grupo de espécies, foi seleccionada uma ilustração que apresenta as principais características desse grupo.

Índices: foram compilados índices para cada uma das línguas, para os nomes científicos e para os identificadores alfabéticos de três caracteres. Estes índices são «inteligentes», ou seja: por exemplo, «Atlantic herring» pode ser encontrado no índice inglês em «Atlantic» ou em «herring».

Deparou-se com algumas dificuldades ao criar os índices para certas línguas devido à formulação dos nomes compostos. Por exemplo, a família dos *Sparidae* pode ser encontrada na literatura especializada como «sea breams», «sea-breams» e «seabreams». Para conseguir que as várias formas de nomes compostos das espécies relacionadas se encontrem em posições adjacentes, foi escolhido um formato harmonizado para os índices. Este facto pode levar a que a forma consagrada no glossário seja diferente daquela que frequentemente se encontra na literatura especializada. No entanto, deve salientar-se que para a maioria dos nomes parecem não existir normas rígidas.

---

[1] Taxonomic Authority List (1988), *Aquatic sciences and fisheries information system reference series*, n.º 8, FAO, Roma, 465 p.

Yleistä

Sisältö

Sanasto kattaa yhteensä 1 532 kala-, äyriäis-, nilviäis-, vesinisäkäs- ja vesikasvilajia sekä kalastustuotetta.

Euroopan komission aloitteesta kokoonkutsuttu, kalastusalan kansallisista asiantuntijoista koostuva työryhmä on laatinut tämän sanaston. Sanasto perustuu komission monikielisen tietokannan Eurodicautomin sisältöön sekä FAO:n laatimaan luetteloon lajeista, jotka esiintyvät kaupallisen kalastuksen tilastoissa. Luettelo työryhmän jäsenistä on sivulla XL. Työryhmän lähtökohtana luetteloa tarkistettaessa oli, että jokainen laji tulee voida tunnistaa täsmällisesti vähintään systemaattisen luokittelun mukaan. Yleisnimikkeet on rajattu tämän sanaston ulkopuolelle. Lajiluettelo on laajennettu kattamaan myös ne lajit, joilla ei ole kaupallista merkitystä mutta jotka asiantuntijaryhmä katsoi kalakannan suojelusta vastaaville henkilöille hyödyllisiksi.

Termien esitystapa

Lajit esitellään seuraavin merkinnöin:

Viitenumero **Heimo** Kolmikirjaiminen tunnus

SC	*tieteellinen nimi (nimet)*
ES	espanjankielinen nimi
DA	tanskankielinen nimi
DE	saksankielinen nimi
EL	kreikankielinen nimi
EN	englanninkielinen nimi
FR	ranskankielinen nimi
IT	italiankielinen nimi
NL	hollanninkielinen nimi
PT	portugalinkielinen nimi
FI	suomenkielinen nimi
SV	ruotsinkielinen nimi

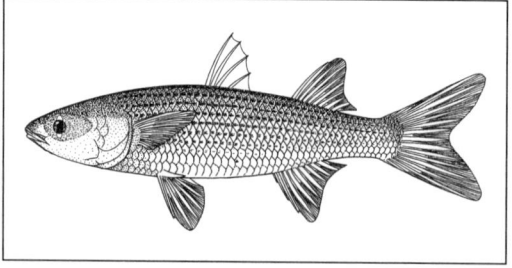

Huomautuksia

Viitenumero: Numero, joka on annettu kullekin lajille hakemiston käytön helpottamiseksi. Viitenumerolla ei ole mitään merkitystä tämän julkaisun ulkopuolella.

Heimo: Heimo, johon laji kuuluu FAO:n julkaisussa "Taxonomic authority list" (¹) esiintyvän taksonomisen luokituksen mukaan. Tässä julkaisussa käytetyt kyseisen luokituksen osat esitetään sivulla XLVIII. Sanastoon otettujen lajien järjestys on sama kuin niiden heimojen järjestys luokituksessa. Teokseen sisältyy myös hakemisto heimoista (s. 397). Heimoryhmien sisällä järjestys määräytyy siten, että koko heimoa tarkoittava nimi tulee ensimmäisenä (esim. *Merluccidae*), sen jälkeen tulevat kunkin suvun eritellyt lajit (esim. *Merluccius hubbsi*) ja lopulta sukuun sisältyvät erittelemättömät lajinimikkeet (esim. *Merluccius* spp.).

Tieteellinen (latinankielinen) nimi: Annettu nimi on sama, jolla lajiin viitataan tieteellisessä kirjallisuudessa. Mikäli tieteellisen nimen kohdalla esiintyy useampi kuin yksi nimi, ensimmäinen on yleinen nimi ja muut nimiä, joilla lajit tunnetaan mutta jotka ovat jo vanhentuneita.

Kolmikirjaiminen tunnus: Kyseessä on se tunnus (Inter-Agency 3-alpha identifier), jonka Atlantin valtameressä tapahtuvan kalastuksen tilastoinnin yhteistyöstä vastaava työryhmä on antanut lajeille kansainvälisiä tilastojulkaisuja varten. Kyseistä tunnusta käytetään nykyään laajasti tilastotieteen ulkopuolella kalastukseen liittyvissä asiakirjoissa (esimerkiksi kalastusalan lehdissä). Tunnukset rekisteröi FAO (jonne tulee myös osoittaa hakemukset mahdollisista lisäyksistä). Tunnuksia käytetään yleensä vain kalastusalan tilastoissa yksittäisesti tunnistettavien lajien kohdalla: useissa tapauksissa kaupalliset lajit, jotka esiintyvät tilastoissa saman otsikon alla, eivät ole saaneet omaa tunnusta. On tärkeää huomata, että annettu tunnus ei muutu tieteellisten ja yleisnimien vaihtuessa.

Nimet Euroopan unionin yhdellätoista virallisella kielellä: Työryhmä on valinnut nimet, joita pidetään kussakin kielessä lajeille sopivimpina. Monessa tapauksessa lajeille on annettu myös vaihtoehtoisia nimiä. Tässä luettelossa ei ole kuitenkaan otettu järjestelmällisesti huomioon alueellisia nimimuunnelmia tai lajin elämänkierron vaihetta, vuodenaikaa, kalastuspaikkaa tai saaliin käsittelytapaa vastaavia nimityksiä. Yleisesti ottaen voidaan todeta, että annettu nimi viittaa lajiin siinä muodossa, jossa tämä nostetaan vedestä.

Mikäli jossakin kielessä yhden lajin kohdalla esiintyy useita nimiä, ensimmäinen niistä on asiantuntijoiden mukaan suositeltavin.

Aina ei kuitenkaan ole ollut mahdollista löytää käytössä olevaa nimeä jokaiselle lajille. Koska kaikkia lajeja ei välttämättä pyydystetä eikä niillä käydä kansainvälistä kauppaa, ei ole lainkaan harvinaista, ettei esimerkiksi kaakkoisella Tyynellä valtamerellä pyydystetyille ja siellä markkinoiduille lajeille ole lainkaan tanskankielistä nimeä. Tapauksissa, joissa lajille ei ole olemassa yhtään nimeä, asiantuntijatyöryhmä on kuitenkin yrittänyt täydentää luetteloa kääntämällä sanatarkasti vieraskielisen nimen. Tämä ei kuitenkaan ole aina ollut mahdollista.

Kuvitukset: Kaikki lajit on pyritty kuvittamaan. Kuvan perusteella ei välttämättä voida aina tunnistaa lajeja, vaan niiden avulla pyritään antamaan hyödyllistä tietoa lajien yleisistä tunnuspiirteistä. Tietyn heimon tai suvun kohdalla kuva esittää kyseisen heimon tai suvun pääasialliset tunnuspiirteet.

Hakemistot: Kaikilla kielillä, tieteellisillä nimillä ja kolmikirjaimisilla tunnuksilla on omat hakemistonsa. Hakemistoista on pyritty tekemään mahdollisimman kattavat: esimerkiksi 'kummeliturska' löytyy sekä hakusanalla 'kummeliturska' että 'turska, kummeli-'.

Erityisesti germaanisten kielten ja suomen kielen kohdalla yhdyssanat vaikeuttavat hakemistojen luomista. Esimerkiksi englanninkielinen nimi *Sparidae*-heimolle esiintyy kirjallisuudessa sekä muodossa 'sea-bream' että 'seabream'. Hakemistoja varten on valittu yhdenmukaistettu muoto, jotta kaikki saman heimon eri yhdyssanamuodot esiintyisivät hakemistoissa oikeilla paikoillaan. Muoto voi erota itse sanastosta löytyvästä muodosta, joka puolestaan on usein kirjallisuudessa esiintyvä muoto. On kuitenkin huomattava, että useille nimille ei ole olemassa täsmällisiä oikeinkirjoitussääntöjä.

(¹) Taxonomic authority list (1988), *Aquatic sciences and fisheries information system reference series*, No 8, FAO, Rome, 465 s.

Inledande anmärkningar

Innehåll

Ordboken innehåller uppgifter om 1 512 arter av fiskar, kräftdjur, blötdjur, marina däggdjur, alger och andra fiskeriprodukter.

Grundmaterialet har sammanställts av termer från kommissionens flerspråkiga termdatabas Eurodicautom och från FN:s livsmedels- och jordbruksorganisations (FAO) listor över arter som ingår i kommersiell fångststatistik. Ordboken har bearbetats av en arbetsgrupp bestående av fiskeexperter från de olika medlemsstaterna samt av anställda vid kommissionen. På sidan XL finns en lista över medlemmarna i denna arbetsgrupp. När ordboken utarbetades, utgick arbetsgruppen från principen att varje enskild term åtminstone skulle kunna identifieras taxonomiskt ned till familjenivå. Mer allmänna termer har utelämnats. Listan utvidgades till att omfatta ett antal arter som, trots att de inte är av kommersiellt intresse, ansågs relevanta för bevarandet av fiskeresurserna samt för naturvården.

Uppställning

Varje post är uppställd på följande sätt:

Referens- nummer	Familje- namn	3-bokstavs- kod

SC	*vetenskapligt namn*
ES	spanskt namn
DA	danskt namn
DE	tyskt namn
EL	grekiskt namn
EN	engelskt namn
FR	franskt namn
IT	italienskt namn
NL	nederländskt namn
PT	portugisiskt namn
FI	finskt namn
SV	svenskt namn

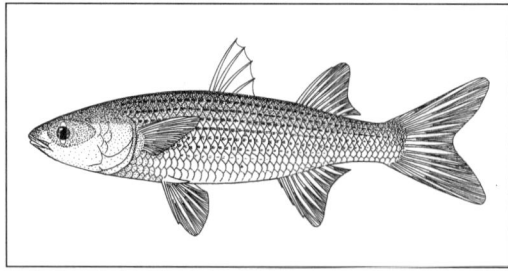

Anmärkningar rörande dessa poster

Referensnummer: Det nummer som arten tilldelats för att göra det lättare att använda registren. Numret är endast av betydelse för denna publikation.

Familj: Den familj som arten tillhör, som grundar sig på klassificeringen i FAO:s *Taxonomic Authority List*([1]). De delar av klassificeringen som används i denna publikation anges på sidan XLVIII. Arterna har ställts upp enligt den familjeordning som finns i klassificeringen. Ett register över familjenamn finns på sidan 397. Ordningsföljden inom varje familj är följande: Först anges namn på familjegruppsnivå (t.ex. *Merluccidae*), därefter namn specificerade till artnivå (t.ex. *Merluccius hubbsi*), och slutligen inom varje släkte namn endast specificerade till släktesnivå (t.ex. *Merluccius spp.*).

Vetenskapliga (latinska) namn: Det namn som anges är det som arten är känd under i facklitteraturen. Om fler än ett namn anges, är det första namnet det vanligaste, och de andra namnen de som anses föråldrade, trots att de ibland fortfarande används.

3-bokstavskod: Denna *Inter-Agency 3-Alpha Identifier* (3-bokstavskod), som används för att identifiera arterna, har tilldelats dem av Samordnande arbetsgruppen för statistik över fiske i Atlanten för användning i internationella publikationer om statistik. Nu används denna kod även i stor utsträckning utanför det statistiska området i dokumentation om fiske (till exempel i loggböcker). Det är FAO som administrerar dessa koder och det är till denna organisation som ansökningar om tilldelning av ytterligare koder skall riktas. Normalt finns endast koder för de arter som anges separat i fiskestatistiken. En del av de arter som är föremål för handel samlas av statistiska hänsyn i grupper, och har därför inte fått en individuell kod. Skälet till att dessa koder har skapats är att de vetenskapliga namnen och trivialnamnen ofta kan ändras men koden, när den en gång har tilldelats, alltid förblir densamma.

Namn på de elva officiella EU-språken: De namn som arbetsgruppen har angett är de som anses vara de lämpligaste för arterna på varje enskilt språk. I många fall har ett alternativt namn angetts, med det bör betonas att dessa alternativ inte utgör en uttömmande lista över regionala namn eller speciella namn beroende på ett stadium i livscykel, årstid, fiskevatten, bearbetningsgrad efter fångst osv. Generellt sett syftar namnet på det tillstånd arten har i vattnet. Om flera olika namn har angetts på ett språk, är det alltid det som anges först som experterna anser vara det lämpligaste.

Det har inte varit möjligt att hitta befintliga namn på alla gemenskapens språk för alla arter. Det är inte alla arter som fångas eller ingår i handeln överallt, och det är därför inte ovanligt att det t.ex. inte finns ett danskt namn för en art som endast fångas och ingår i handeln i den sydöstra delen av Stilla havet. När inga namn omedelbart fanns att tillgå, försökte arbetsgruppen komplettera listan genom att göra en direkt översättning av namnet på ett av de andra språken. Detta har dock inte heller kunnat göras i alla enskilda fall.

Teckningar: Så långt det har varit möjligt har en illustration tagits med för varje art. Avsikten med dessa illustrationer är inte att exakt identifiera arten, utan att ge läsaren en uppfattning om artens allmänna kännetecken. Till poster som hänvisar till en familj eller en grupp av arter har en illustration valts som visar de viktigaste karaktärsdragen hos denna grupp.

Register: Ett register har utarbetats för varje språk, för de vetenskapliga namnen och för 3-bokstavskoderna. Dessa register är *intelligenta*, dvs. *brasiliansk rovansjovis* kan hittas under *brasiliansk rovansjovis, rovansjovis, brasiliansk o*ch under *ansjovis, brasiliansk rov-*.

En del svårigheter uppstod när registren skulle utarbetas för några av språken på grund av sammansatta namn. Till exempel anges engelska namn för familjen *Sparidae* i litteraturen som *sea bream*, *sea-bream* och *seabream*. Så att de olika formerna av sammansatta namn på besläktade arter skall kunna stå efter varandra i registren, har ett harmoniserat format valts för användning i registren. Detta kan skilja sig från ordbokens textmassa, där den form som oftast stöts på i litteraturen har valts. Det bör dock betonas att det inte verkar finnas fasta regler för vad som är namnens korrekta form.

([1]) Taxonomic authority list (1988), *Aquatic sciences and fisheries information system reference series*, nr 8, FAO, Rom, 465 s.

Members of the expert working group

European Commission

David CROSS, Statistical Office

Irmgard FIAMOZZI, Terminology

Ekaterini ANTONOPOULOU, Administrative trainee

Liisa PEURA, Administrative trainee

National experts

España

Miguel IBÁÑEZ, Instituto Bidebieta, Donostia-San Sebastián

Danmark

Jørgen NIELSEN, Zoologisk Museum, København

Deutschland

Martina SCHNEIDER, Frankfurt/Main (formerly FAO)
Walther W. KÜHNHOLD, Bundesforschungsanstalt für Fischerei, Hamburg
Michael TURKAY, Forschungsinstitut Senckenberg, Frankfurt/Main

Elláda

Costas PAPACONSTANTINOU, National Centre for Marine Research, Athens

United Kingdom

Alwyne WHEELER, formerly of the British Museum (Natural History), London

France

Jacques MAIGRET, Muséum national d'histoire naturelle, Paris

Italia

Carlo FROGLIA, CNR, Istituto di ricerche sulla pesca marittima, Ancona

België

Guy HOUVENAGHEL, Laboratoire d'océanographie biologique et d'aquaculture, ULB, Bruxelles

Portugal

José GONÇALVES SANCHES, Instituto Nacional de Investigação das Pescas, Lisboa

Nederland

Dr. S.J. De Groote, RIVO, IJmuiden

Finland

Hannu Lehtonen, Finnish Game and Fisheries Research Institute, Helsinki

Sverige

Laura Píriz, Swedmar, Göteborg
Sven Kullander, Naturhistoriska riksmuséet, Stockholm

Acknowledgements

We would like to thank the FAO, Unesco, Fishing News Books (Oxford), and all the others who have authorised the reproduction of illustrations from their publications.

Classification of aquatic animals and plants

The classification used is based on that of the 'taxonomic authority list'([1]). However, only those families represented in the body of this glossary have been listed below.

The code indicated alongside each entry is the FAO taxonomic code which comprises eight digits to the level of the genus. The classification listed below only descends to the level of the family (five digits of the FAO code) which is given for each entry in the body of the glossary. The first digit is used to separate the main groups of aquatic organisms, the next two to provide a separation at about the level of the order. The last two digits (in brackets) refer to the family.

Clasificación de animales y plantas acuáticos

La clasificación utilizada es la de la «Taxonomic Authority List»[1] (lista taxonómica de referencia, publicada por la FAO). Sin embargo, en la lista facilitada a continuación sólo aparecen las familias representadas en el corpus de este glosario.

El código indicado junto a cada entrada es el código taxonómico FAO, que consta de ocho dígitos hasta el nivel de género. La clasificación que figura a continuación sólo desciende hasta el nivel de la familia (cinco dígitos del código FAO), que aparece con cada entrada del corpus del glosario. El primer dígito se utiliza para separar los grupos principales y los dos siguientes como separación aproximadamente a nivel del orden. Los dos últimos dígitos (entre paréntesis) se refieren a la familia.

---

[1] Taxonomic authority list (1988), *Aquatic sciences and fisheries information system reference series*, No 8, FAO, Rome, 465 pp.

Klassifikation af vanddyr og -planter

Den benyttede klassifikation er »Taxonomic Authority List« ([1]). Nedenstående liste indeholder dog kun de familier, der forekommer i dette glossar.

Den kode, der findes sammen med hver termpost, er FAO's taksonomiske kode, som består af 8 cifre på slægtsniveau. Nedenstående klassifikation går kun ned til familieniveau (5 cifre af FAO's kode), som er det niveau, der er anført ved hver termpost i glossaret. Det første ciffer giver en opdeling i hovedgrupper, de næste to giver en underopdeling, som omtrent svarer til ordensniveau. De sidste to cifre (i parentes) henviser til familien.

Klassifikation der Wassertiere und -pflanzen

Die hier zugrunde gelegte Klassifikation ist die „Taxonomic Authority List" ([1]) (von der FAO veröffentlichte Referenzliste). Aufgeführt werden jedoch nur die in diesem Band erscheinenden Familien.

Jeder Eintrag ist auf der Ebene der Gattung durch den achtstelligen taxonomischen FAO-Kode gekennzeichnet. Die nachstehend aufgeführte Klassifikation reicht für alle im Korpus enthaltenen Arten nur bis zur Familie (5 Ziffern des FAO-Kodes). Die erste Ziffer dient zur Unterscheidung der Hauptgruppen, die folgenden zwei unterscheiden auf der Ebene der Ordnung (Abweichungen sind dabei möglich), während die letzten beiden (in Klammern) sich auf die Familie beziehen.

---

[1] Taxonomic authority list (1988), *Aquatic sciences and fisheries information system reference series*, No 8, FAO, Rome, 465 pp.

Ταξινόμηση υδρόβιων ζώων και φυτών

Χρησιμοποιήθηκε η ταξινόμηση της «Taxonomic Authority List»([^1]). Εντούτοις, μόνο οι οικογένειες που περιλαμβάνονται στο κύριο σώμα του γλωσσαρίου αυτού αναφέρονται παρακάτω.

Ο κωδικός που αναγράφεται δίπλα από κάθε λήμμα είναι ο ταξινομικός κωδικός της FAO ο οποίος περιλαμβάνει οκτώ ψηφία στο επίπεδο του γένους. Η παρακάτω ταξινόμηση κατέρχεται μόνο ώς το επίπεδο της οικογένειας (πέντε ψηφία του κωδικού FAO) η οποία για κάθε λήμμα αναγράφεται στο κύριο σώμα του γλωσσαρίου. Το πρώτο ψηφίο χρησιμοποιείται για να διαχωρίσει τις κύριες ομάδες υδρόβιων οργανισμών, ενώ τα δύο επόμενα ως διαχωρισμός κατά προσέγγιση σε επίπεδο συνομοταξίας. Τα τελευταία δύο ψηφία (μέσα σε παρενθέσεις) αναφέρονται στην οικογένεια.

Classification des animaux et des plantes aquatiques

La classification utilisée est celle de la *Taxonomic authority list*([^1]) (liste taxinomique de référence, publiée par la FAO). Cependant, seules les familles figurant dans le corpus de ce glossaire apparaissent dans la liste ci-dessous.

Le code indiqué pour chaque entrée est le code taxinomique de la FAO, qui comprend huit positions jusqu'au niveau du genre. La classification ci-dessous s'arrête au niveau de la famille (cinq positions du code FAO), qui est mentionnée pour chaque entrée dans le corpus de ce glossaire. La première position est utilisée pour identifier les groupes principaux, les deux suivantes marquent la séparation approximativement au niveau de l'ordre. Les deux dernières positions (entre parenthèses) se réfèrent à la famille.

[^1]: Taxonomic authority list (1988), *Aquatic sciences and fisheries information system reference series*, No 8, FAO, Rome, 465 pp.

Classificazione degli animali e piante acquatici

La classificazione usata è quella della «Taxonomic Authority List» [1]. Sono qui elencate soltanto le famiglie riportate nel corpo del glossario.

Il codice che accompagna ogni voce è il codice tassonomico FAO che comprende 8 cifre a livello del genere. Per tutte le voci del glossario la nostra classificazione si arresta al livello della famiglia (5 cifre del codice FAO). La prima cifra opera una distinzione fra i principali gruppi degli organismi acquatici, le due successive si collocano al livello dell'ordine, le ultime due (fra parentesi) si riferiscono alla famiglia.

Classificatie van waterdieren en waterplanten

De gebruikte classificatie is die van de „Taxonomic Authority List" [1] (Taxonomische Referentielijst). In onderstaande lijst zijn echter uitsluitend die families opgenomen die ook in het corpus van het glossarium vermeld worden.

De achter elk trefwoord aangegeven code is de taxonomische code van de FAO. Deze heeft acht posities en gaat tot op het niveau van het genus. De onderstaande classificatie gaat niet dieper dan het niveau van de familie (vijf posities in de FAO-code), deze wordt in het corpus van het glossarium bij elk trefwoord vermeld. De eerste positie wordt gebruikt om de hoofdgroep aan te duiden, de volgende twee posities onderscheiden tot op het niveau van de orde (hierin zijn afwijkingen mogelijk). De laatste twee posities (tussen haakjes) geven de familie aan.

---

[1] Taxonomic authority list (1988), *Aquatic sciences and fisheries information system reference series*, No 8, FAO, Rome, 465 pp.

Classificação das plantas e animais aquáticos

A classificação usada é a da «Taxonomic Authority List» [1]. No entanto, só são apresentadas em seguida as famílias referidas neste glossário.

O código indicado em cada entrada é o código taxonómico da FAO, que inclui oito dígitos até ao nível do género. A classificação abaixo referida apenas desce até ao nível da família (cinco dígitos do código FAO) que é dado a cada entrada neste glossário. O primeiro dígito é usado para separar os principais grupos de organismos aquáticos, os seguintes para distinguir a nível da ordem, ou aproximadamente. Os dois últimos dígitos (entre parênteses) referem-se à família.

Vesieläinten ja -kasvien luokitus

Julkaisussa käytetty luokitus perustuu FAO:n luetteloon "Taxonomic authority list" [1]. Seuraavassa on kuitenkin lueteltu vain tässä sanastossa esiintyvät heimot.

Sanastossa lajin rinnalla esitetään FAO:n taksonominen, kahdeksannumeroinen lajikoodi. Alla oleva lista ulottuu vain heimotasolle vastaten FAO-koodin viittä ensimmäistä numeroa. Jokaisella sanastoon otetulla lajilla on oma koodinsa. Ensimmäinen numero viittaa yleensä pääjaksoon tai luokkaan; kaksi seuraavaa numeroa karkeasti sijoittumiseen lahkon tasolla. Kaksi viimeistä numeroa (sulkeissa) viittavat heimoon.

---

[1] Taxonomic authority list (1988), *Aquatic sciences and fisheries information system reference series*, No 8, FAO, Rome, 465 pp.

Klassificering av vattendjur och -växter

Den klassificering som har använts är FAO:s *Taxonomic Authority List* ([1]). Listan nedan omfattar emellertid bara de familjer som är representerade i denna ordbok.

Den kod som anges vid varje post är FAO:s taxonomiska kod som består av 8 siffror på släktesnivå. Klassificeringen nedan sträcker sig endast ned till familjenivå (5 siffror av FAO:s kod), som är den nivå som anges vid varje post i ordboken. Den första siffran används för att göra en uppdelning i huvudgrupper, de följande två för en uppdelning som ungefär motsvarar ordningsnivå. De två sista siffrorna (inom parentes) hänvisar till familjen.

Pisces

1.02	**Petromyzoniformes**	
	1.02(01)	*Petromyzonidae*
1.05	**Hexanchiformes**	
	1.05(01)	*Chlamydoselachidae*
	1.05(02)	*Hexanchidae*
1.06	**Lamniformes**	
	1.06(01)	*Cetorhinidae*
	1.06(02)	*Odontaspididae*
	1.06(06)	*Alopiidae*
	1.06(08)	*Lamnidae*
1.07	**Orectolobiformes**	
	1.07(01)	*Stegostomatidae*
	1.07(02)	*Orectolobidae*
	1.07(03)	*Ginglymostomatidae*
	1.07(05)	*Rhincodontidae*
1.08	**Scyliorhinoidei**	
	1.08(01)	*Scyliorhinidae*
	1.08(02)	*Carcharhinidae*
	1.08(03)	*Sphyrnidae*
	1.08(04)	*Triakidae*
1.09	**Squaliformes**	
	1.09(01)	*Squalidae*
	1.09(03)	*Squatinidae*
	1.09(05)	*Oxynotidae*
	1.09(06)	*Echinorhinidae*
1.10	**Rajiformes**	
	1.10(01)	*Rhinobatidae*
	1.10(02)	*Pristidae*
	1.10(04)	*Rajidae*
	1.10(05)	*Dasyatidae*
	1.10(07)	*Myliobatidae*
	1.10(08)	*Mobulidae*
	1.10(10)	*Gymnuridae*
1.11	**Torpediniformes**	
	1.11(01)	*Torpedinidae*
1.12	**Chimaeriformes**	
	1.12(01)	*Chimaeridae*
	1.12(03)	*Callorhynchidae*
1.13	**Ceratodiformes**	
	1.13(01)	*Ceratodontidae*
1.14	**Lepidosireniformes**	
	1.14(02)	*Protopteridae*
1.16	**Polypteriformes**	
	1.16(01)	*Polypteridae*
1.17	**Acipenseriformes**	
	1.17(01)	*Acipenseridae*
	1.17(02)	*Polyodontidae*
1.18	**Amiiformes**	
	1.18(01)	*Amiidae*
1.19	**Lepisosteiformes**	
	1.19(01)	*Lepisosteidae*
1.21	**Clupeoidei**	
	1.21(01)	*Elopidae*
	1.21(02)	*Megalopidae*
	1.21(03)	*Albulidae*
	1.21(05)	*Clupeidae*
	1.21(06)	*Engraulidae*
	1.21(07)	*Alepocephalidae*
1.22		
	1.22(01)	*Chirocentridae*
	1.22(02)	*Chanidae*
1.23	**Salmonoidei**	
	1.23(01)	*Salmonidae*
	1.23(02)	*Thymallidae*
	1.23(03)	*Plecoglossidae*
	1.23(04)	*Osmeridae*
	1.23(05)	*Argentinidae*

([1]) Taxonomic authority list (1988), *Aquatic sciences and fisheries information system reference series*, No 8, FAO, Rome, 465 pp.

1.24		**Esocoidei**
	1.24(02)	*Umbridae*
	1.24(03)	*Esocidae*
1.27		**Notopteroidei**
	1.27(02)	*Notopteridae*
1.32		**Scopeliformes**
	1.32(01)	*Mochokidae*
	1.32(02)	*Aulopidae*
	1.32(08)	*Myctophidae*
	1.32(11)	*Harpadontidae*
	1.32(12)	*Chlorophthalmidae*
1.38		**Characoidei**
	1.38(01)	*Characidae*
1.40		**Cyprinoidei**
	1.40(01)	*Catostomidae*
	1.40(02)	*Cyprinidae*
	1.40(05)	*Cobitidae*
1.41		**Siluroidei**
	1.41(01)	*Ariidae*
	1.41(06)	*Plotosidae*
	1.41(07)	*Siluridae*
	1.41(10)	*Ictaluridae*
	1.41(18)	*Clariidae*
	1.41(20)	*Synodontidae*
	1.41(21)	*Malapteruridae*
1.43		**Anguilloidei**
	1.43(02)	*Anguillidae*
	1.43(06)	*Muraenidae*
	1.43(09)	*Muraenesocidae*
	1.43(13)	*Congridae*
	1.43(15)	*Ophichthidae*
1.47		**Beloniformes**
	1.47(01)	*Belonidae*
	1.47(02)	*Scomberesocidae*
	1.47(04)	*Exocoetidae*
1.48		**Gadiformes**
	1.48(02)	*Moridae*
	1.48(03)	*Bregmacerotidae*
	1.48(04)	*Gadidae*
	1.48(05)	*Merlucciidae*
	1.48(06)	*Macrouridae*
1.50		**Gasterosteiformes**
	1.50(01)	*Gasterosteidae*
1.51		**Syngnathiformes**
	1.51(03)	*Macrorhamphosidae*
	1.51(06)	*Syngnathidae*
1.53		**Lampridoidei**
	1.53(01)	*Lampridae*
1.55		**Trachypteroidei**
	1.55(01)	*Regalecidae*
	1.55(02)	*Trachipteridae*
1.57		**Cyprinodontiformes**
	1.57(02)	*Cyprinodontidae*
1.61		**Beryciformes**
	1.61(02)	*Berycidae*
	1.61(05)	*Trachichthyidae*
	1.61(11)	*Holocentridae*
1.62		**Zeiformes**
	1.62(01)	*Zeidae*
	1.62(03)	*Caproidae*
	1.62(04)	*Oreosomatidae*

1.64		**Sphyraenoidei**
	1.64(01)	*Sphyraenidae*
1.65		**Mugiloidei**
	1.65(01)	*Mugilidae*
	1.65(02)	*Atherinidae*
1.66		**Polynemiformes**
	1.66(01)	*Polynemidae*
1.67		**Ophiocephaliformes**
	1.67(01)	*Channidae*
1.69		**Perciformes**
1.70		**Percoidei**
	1.70(00)	*Caesionidae*
	1.70(01)	*Centropomidae*
	1.70(02)	*Serranidae*
	1.70(04)	*Theraponidae*
	1.70(05)	*Rhiniodontidae*
	1.70(06)	*Moronidae*
	1.70(08)	*Percichthyidae*
	1.70(10)	*Centrarchidae*
	1.70(11)	*Priancanthidae*
	1.70(12)	*Apogonidae*
	1.70(14)	*Percidae*
	1.70(15)	*Sillaginidae*
	1.70(17)	*Malacanthidae*
	1.70(19)	*Lactariidae*
	1.70(20)	*Pomatomidae*
	1.70(22)	*Rachycentridae*
	1.70(23)	*Carangidae*
	1.70(25)	*Formionidae*
	1.70(26)	*Menidae*
	1.70(27)	*Bramidae*
	1.70(28)	*Coryphaenidae*
	1.70(29)	*Arripidae*
	1.70(30)	*Emmelichthyidae*
	1.70(32)	*Lutjanidae*
	1.70(33)	*Nemipteridae*
	1.70(35)	*Leiognathidae*
	1.70(36)	*Haemulidae*
	1.70(37)	*Sciaenidae*
	1.70(38)	*Lethrinidae*
	1.70(39)	*Sparidae*
	1.70(40)	*Centracanthidae*
	1.70(41)	*Mullidae*
	1.70(43)	*Coracinidae*
	1.70(49)	*Ephippidae*
	1.70(50)	*Drepanidae*
	1.70(51)	*Scatophagidae*
	1.70(52)	*Chaetodontidae*
	1.70(59)	*Cichlidae*
	1.70(60)	*Cepolidae*
	1.70(62)	*Pomacentridae*
	1.70(63)	*Labridae*
	1.70(65)	*Scaridae*
	1.70(70)	*Cheilodactylidae*
	1.70(71)	*Latridae*
	1.70(72)	*Trichodontidae*
	1.70(76)	*Mugiloididae*
	1.70(78)	*Trachinidae*
	1.70(79)	*Percophidae*
	1.70(88)	*Uranoscopidae*
	1.70(92)	*Nototheniidae*
	1.70(94)	*Channichthyidae*
	1.70(95)	*Pentapodidae*
	1.70(99)	*Gerreidae*

1.71		**Blennioidei**
	1.71(01)	*Blenniidae*
	1.71(02)	*Anarhichadidae*
	1.71(08)	*Clinidae*
	1.71(11)	*Pholidae*
	1.71(15)	*Zoarcidae*
1.72		
	1.72(02)	*Ophidiidae*
	1.72(03)	*Fierasferidae*
	1.72(04)	*Ammodytidae*
1.73		
	1.73(01)	*Callionymidae*
	1.73(03)	*Siganidae*
1.74		
	1.74(02)	*Acanthuridae*
	1.74(03)	*Gempylidae*
	1.74(04)	*Trichiuridae*
1.75		**Scombroidei**
	1.75(01)	*Scombridae*
	1.75(03)	*Istiophoridae*
	1.75(04)	*Xiphiidae*
1.76		
	1.76(03)	*Stromateidae*
	1.76(04)	*Nomeidae*
	1.76(05)	*Anabantidae*
	1.76(08)	*Centrolophidae*
	1.76(09)	*Osphronemidae*
	1.76(10)	*Belontiidae*
	1.76(11)	*Helostomatidae*
1.77		**Gobioidei**
	1.77(01)	*Eleotridae*
	1.77(02)	*Gobiidae*
1.78		**Cottoidei**
	1.78(01)	*Scorpaenidae*
	1.78(02)	*Triglidae*
	1.78(07)	*Hexagrammidae*
	1.78(08)	*Anoplopomatidae*
	1.78(09)	*Platycephalidae*
	1.78(13)	*Cottidae*
	1.78(19)	*Agonidae*
	1.78(20)	*Cyclopteridae*
	1.78(22)	*Peristediidae*
	1.78(23)	*Liparidae*
1.79		**Dactylopteriformes**
	1.79(01)	*Dactylopteridae*
1.81		**Pleuronectiformes**
1.82		**Psettodoidei**
	1.82(01)	*Psettodidae*
1.83		**Pleuronectoidei**
	1.83(01)	*Bothidae*
	1.83(02)	*Pleuronectidae*
	1.83(03)	*Soleidae*
	1.83(04)	*Cynoglossidae*
	1.83(05)	*Scophthalmidae*
	1.83(06)	*Citharidae*
1.87		**Echeneiformes**
	1.87(01)	*Echeneididae*
1.89		**Balistoidei**
	1.89(03)	*Balistidae*
	1.89(04)	*Monacanthidae*

1.90		
	1.90(02)	*Tetraodontidae*
1.91		**Moloidei**
	1.91(01)	*Molidae*
1.92		**Gobiesociformes**
	1.92(01)	*Gobiesocidae*
1.93		**Batrachoidiformes**
	1.93(01)	*Batrachoididae*
1.95		
	1.95(01)	*Lophiidae*

2.00   Crustacea

2.13		**Thoracica**
	2.13(01)	*Scalpellidae*
	2.13(02)	*Lepadidae*
	2.13(03)	*Balanidae*
2.25		**Stomatopoda**
	2.25(01)	*Squillidae*
2.26		**Euphausiacea**
	2.26(01)	*Euphausiidae*
2.28		**Natantia**
	2.28(01)	*Penaeidae*
	2.28(02)	*Aristeidae*
	2.28(04)	*Pandalidae*
	2.28(07)	*Sergestidae*
	2.28(09)	*Atyidae*
	2.28(12)	*Palaemonidae*
	2.28(23)	*Crangonidae*
	2.28(28)	*Sicyoniidae*
	2.28(29)	*Solenoceridae*
2.29		**Reptantia**
	2.29(01)	*Palinuridae*
	2.29(03)	*Astacidae*
	2.29(05)	*Paguridae*
	2.29(09)	*Cancridae*
	2.29(10)	*Xanthidae*
	2.29(11)	*Portunidae*
	2.29(13)	*Grapsidae*
	2.29(15)	*Scyllaridae*
	2.29(19)	*Galatheidae*
	2.29(20)	*Lithodidae*
	2.29(21)	*Majidae*
	2.29(27)	*Parastacidae*
	2.29(42)	*Nephropidae*
	2.29(43)	*Geryonidae*

3.00   Mollusca

3.05		**Gasteropoda**
3.07		**Scutibranchiata**
	3.07(03)	*Haliotidae*
	3.07(04)	*Trochidae*
	3.07(05)	*Turbinidae*
	3.07(08)	*Patellidae*
3.09		
	3.09(05)	*Muricidae*
	3.09(08)	*Buccinidae*
	3.09(09)	*Melongenidae*

L

3.10		**Taenioglossa**	
	3.10(03)	*Littorinidae*	
	3.10(16)	*Strombidae*	

3.16   **Prionodesmacea**
- 3.16(04)   *Arcidae*
- 3.16(05)   *Unionidae*
- 3.16(06)   *Pteriidae*
- 3.16(07)   *Ostreidae*
- 3.16(08)   *Pectinidae*
- 3.16(10)   *Mytilidae*
- 3.16(14)   *Margaritanidae*
- 3.16(15)   *Glycimeridae*
- 3.16(16)   *Pinnidae*

3.17   **Teleodesmacea**
- 3.17(04)   *Arcticidae*
- 3.17(09)   *Cardiidae*
- 3.17(10)   *Mactridae*
- 3.17(11)   *Veneridae*
- 3.17(15)   *Donacidae*
- 3.17(16)   *Solenidae*
- 3.17(17)   *Myidae*
- 3.17(19)   *Pholadidae*

3.21   **Dibranchia**
- 3.21(02)   *Sepiidae*
- 3.21(03)   *Sepiolidae*
- 3.21(04)   *Loliginidae*
- 3.21(05)   *Ommastrephidae*
- 3.21(09)   *Octopodidae*

4.00   Mammalia

4.06   **Pinnipedia**
- 4.06(01)   *Otariidae*
- 4.06(02)   *Odobenidae*
- 4.06(03)   *Phocidae*

4.14   **Sirenia**
- 4.14(01)   *Dugongidae*
- 4.14(02)   *Trichechidae*

4.22   **Odontoceti**
- 4.22(02)   *Ziphiidae*
- 4.22(03)   *Physeteridae*
- 4.22(04)   *Delphinidae*
- 4.22(06)   *Monodontidae*

4.23   **Mystacoceti**
- 4.23(02)   *Balaenopteridae*
- 4.23(03)   *Balaenidae*
- 4.23(04)   *Eschrichtiidae*

5.00   Amphibia

5.12   **Diplasiocoela**
- 5.12(01)   *Ranidae*

5.30   Reptilia

5.31   **Chelonia**
- 5.31(01)   *Dermochelyidae*
- 5.31(07)   *Chelonidae*
- 5.31(12)   *Emydidae*

5.36   **Loricata**
- 5.36(01)   *Crocodylidae*

5.94   Tunicata

5.96   **Ascidiacea**
- 5.96(09)   *Pyuridae*

6.13   Porifera

6.15   **Noncalcarea**
- 6.15(01)   *Spongidae*

6.16   Coelenterata

6.17   **Hydrozoa**
- 6.17(41)   *Rhizostomidae*

6.19   **Anthozoa**
- 6.19(01)   *Coralliidae*

6.54   Arthropoda

6.56   **Xiphosura**
- 6.56(01)   *Limulidae*

6.89   Echinodermata

6.91   **Asteroidea**
- 6.91(05)   *Asteriidae*

6.93   **Echinoidea**
- 6.93(02)   *Strongylocentrotidae*
- 6.93(04)   *Echinidae*

6.94   **Holothurioidea**
- 6.94(03)   *Stichopodidae*

7.00   Algae

7.71   **Phaeophyceae**

7.77   **Laminariales**
- 7.77(02)   *Laminariaceae*
- 7.80   Fucales

7.81   **Rhodophyceae**

7.85   **Gelidiales**
- 7.85(01)   *Gelidiaceae*

7.86   **Cryptonemiales**
- 7.86(06)   *Cruoriaceae*
- 7.86(09)   *Corallinaceae*

7.87   **Gigartinales**
- 7.87(16)   *Gigartinaceae*

Corpus

1 PETROMYZONTIDAE

SC *Petromyzontidae*
ES lampreas
DA lampret-familien; lampretter; niøjer; negenøjer
DE Neunaugen; Lampreten
EL λάμπρενες
EN lampreys
FR lamproies
IT lamprede
NL prikken
PT lampreias
FI tonnikalat
SV nejonögon

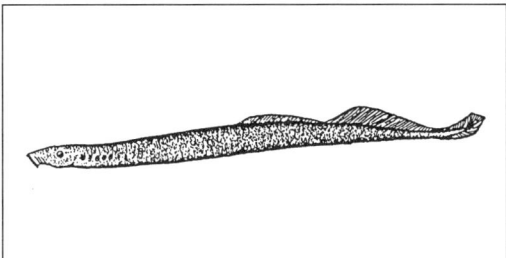

3 PETROMYZONTIDAE

SC *Eudontomyzon mariae* (Berg, 1931)
ES lamprea ucraniana
DA ukrainsk lampret
DE Ukrainische Lamprete; Ukrainisches Bachneunauge
EL λάμπρενα της Ουκρανίας
EN Ukranian lamprey
FR lamproie d'Ukraine
IT lampreda dell'Ucraina
NL Oekraïense prik
PT lampreia da Ucrânia
FI ukrainannahkiainen
SV ukrainskt nejonöga

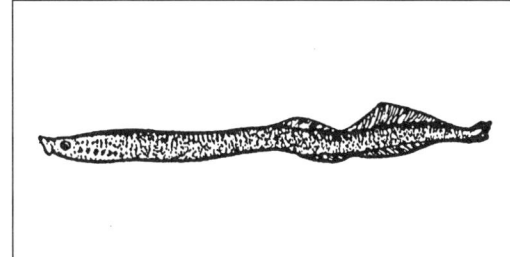

2 PETROMYZONTIDAE

SC *Eudontomyzon danfordi* (Regan, 1911)
ES lamprea del Danubio
DA Donau-lampret
DE Donau-Lamprete; Donau-Neunauge
EL λάμπρενα του Δούναβη
EN Carpathian lamprey
FR lamproie du Danube
IT lampreda del Danubio
NL Donauprik
PT lampreia do Danúbio
FI tonavannahkiainen
SV karpatiskt nejonöga; donaunejonöga

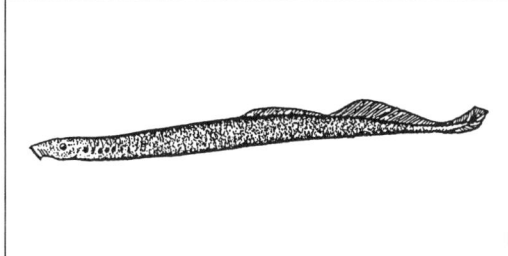

4 PETROMYZONTIDAE

SC *Eudontomyzon vladykovi* (Olive and Zanandrea, 1959)
ES lamprea de Vladikov
DA Vladykovs lampret
DE Vladikov-Lamprete; Vladikov-Bachneunauge
EL λάμπρενα του Βλαντύκωφ
EN Vladikov's lamprey
FR lamproie de Vladykov
IT lampreda di Vladikov
NL Vladikovs prik
PT lampreia de Vladykov
FI vladikovinnahkiainen
SV donaunejonöga

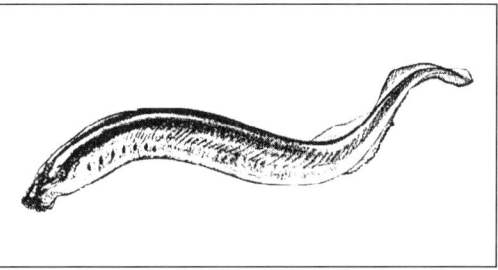

5 PETROMYZONTIDAE LAR

SC	*Lampetra fluviatilis* (Linnaeus, 1758); *Petromyzon fluviatilis*
ES	lamprea de río
DA	flodlampret; flodniøje; flodnegenøje
DE	Flußneunauge; Flußpricke; Uhl; Uhlen; Pricke; Perel
EL	λάμπρενα· λάμπρινα
EN	river lamprey; freshwater lamprey; lampern; mud lamprey; lamprey-eel
FR	lamproie de rivière
IT	lampreda di fiume
NL	rivierprik
PT	lampreia do rio
FI	nahkiainen; jokinahkiainen
SV	flodnejonöga

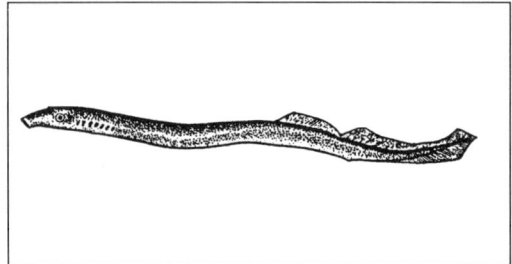

7 PETROMYZONTIDAE

SC	*Petromyzon marinus* (Linnaeus, 1758)
ES	lamprea marina; lamprea de mar; pegatimón
DA	havlampret; havnegenøje
DE	Meerneunauge; Meerpricke; Seelamprete; Lamprete; Neunaugenkönig
EL	λάμπρενα· λάμπρινα
EN	sea lamprey
FR	lamproie marine; grande lamproie marine
IT	lampreda di mare; lampreda marina
NL	zeeprik
PT	lampreia do mar
FI	merinahkiainen
SV	havsnejonöga

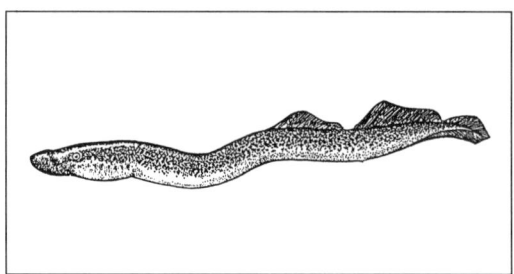

6 PETROMYZONTIDAE

SC	*Lampetra planeri* (Bloch, 1784)
ES	lampreilla
DA	bæklampret; bæknegenøje
DE	Gewöhnliches Bachneunauge; Bachneunauge
EL	μικρολάμπρενα
EN	brook lamprey
FR	petite lamproie; lamproie de Planer; chatouille
IT	lampreda di ruscello
NL	beekprik
PT	lampreia de esteiro
FI	pikkunahkiainen
SV	bäcknejonöga

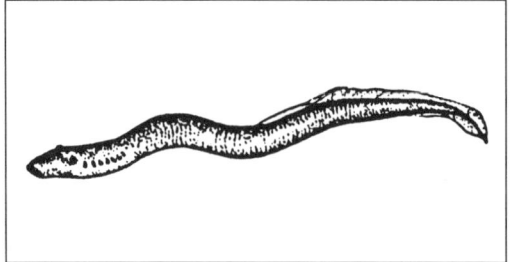

8 CHLAMYDOSELACHIDAE

SC	*Chlamydoselachus anguineus* (Garman, 1884)
ES	clámide; tiburón lagarto
DA	kravehaj
DE	Kragenhai; Krausenhai
EL	ερπετοκαρχαρίας
EN	frilled shark
FR	requin lézard
IT	squalo serpente
NL	franjehaai
PT	tubarão-cobra
FI	kaulushai
SV	kråshaj

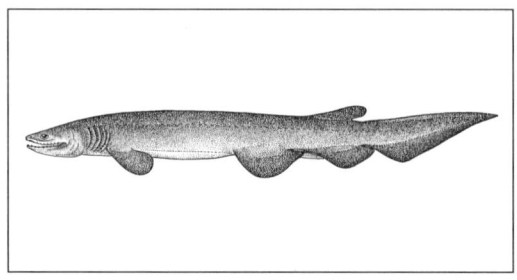

9 HEXANCHIDAE

SC	*Hexanchidae*
ES	cañabotas
DA	seksgællede hajer
DE	Grauhaie
EL	καρχαρίας· αλέτρι
EN	cowsharks
FR	hexanchidés
IT	squali
NL	zeskieuwige haaien
PT	tubarões-albafar
FI	kidushait; kidushait-heimo
SV	kamtandhajar

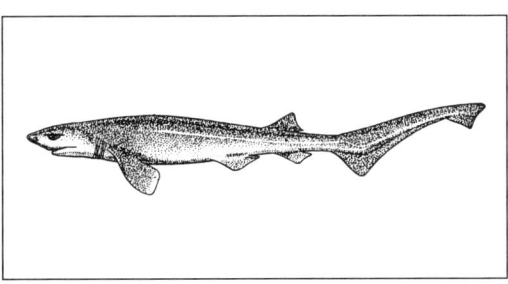

10 HEXANCHIDAE

SC	*Heptranchias perlo* (Bonnaterre, 1788)
ES	boquidulce
DA	syvgællet haj
DE	Perlon; Spitzkopfsiebenkiemer
EL	λύτρι· καρχαρίας· σκυλόψαρο
EN	seven-gilled shark; sharpnose seven-gill shark
FR	perlon cendre; requin perlon
IT	squalo manzo
NL	scherpneuskoehaai
PT	boca doce
FI	kapeapäähai
SV	pärlhaj

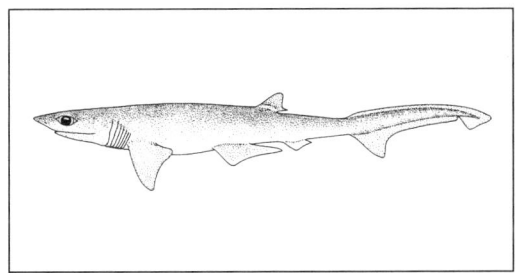

11 HEXANCHIDAE   SBL

SC	*Hexanchus griseus* (Bonnaterre, 1788); *Notidanus griseus*
ES	cañabota; cañabota gris; tiburón de peinetas
DA	seksgællet haj
DE	Grauhai; Sechskiemer
EL	αλέτρι· καρχαρίας
EN	bluntnose six-gill shark; six-gill shark; six-gilled shark
FR	requin griset; requin gris
IT	squalo capopiatto
NL	zeskieuwige koehaai
PT	tubarão-albafar; albafar
FI	kidushai
SV	sexbågig kamtandhaj

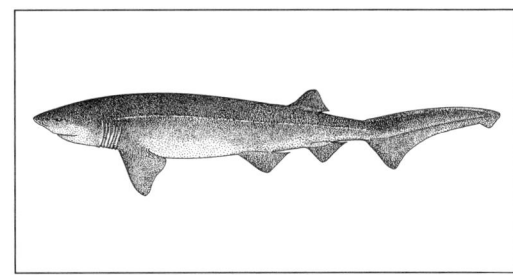

12 CETORHINIDAE   BSK

SC	*Cetorhinus maximus* (Gunncrus, 1765)
ES	peregrino; tiburón ballena; marrajo gigante
DA	brugde
DE	Riesenhai
EL	προσκυνητής· παπάς· καρχαρίας· σκυλόψαρο
EN	basking shark
FR	requin pèlerin; pèlerin; flaneur
IT	squalo elefante
NL	reuzenhaai
PT	tubarão-frade; peixe-frade
FI	jättiläishai
SV	brugd

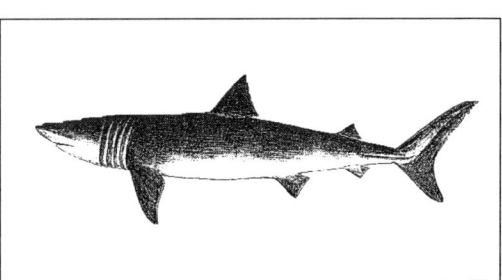

13   ODONTASPIDIDAE

SC	*Eugomphodus taurus* (Rafinesque, 1810);
	Carcharias taurus; *Odontaspis taurus*
ES	tiburón toro
DA	sandtigerhaj
DE	Sandhai; Sandtigerhai
EL	καρχαρίας· ταυροκαρχαρίας
EN	sand shark
FR	requin-taureau; odontaspide-taureau
IT	squalo toro
NL	gespikkelde scheurtandhaai
PT	tubarão-toiro
FI	hietahai
SV	oxhaj

15   ALOPIIDAE   PTH

SC	*Alopias pelagicus* (Nakamura, 1935)
ES	zorro pelágico
DA	Stillehavs-rævehaj
DE	Indopazifischer Fuchshai
EL	αλεπόσκυλος του πελάγους
EN	pelagic thresher
FR	requin-renard pélagique; renard de mer
IT	squalo volpe pelagico
NL	kleintand-voshaai
PT	tubarão-raposo do Índico
FI	ulappakettuhai
SV	stillahavsrävhaj

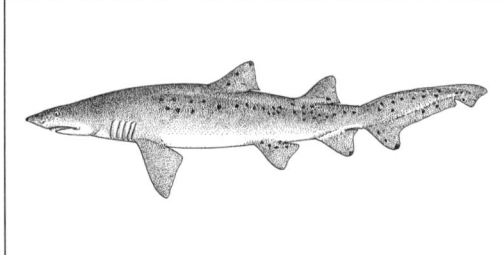

14   ODONTASPIDIDAE

SC	*Odontaspis ferox* (Risso, 1810); *Carcharias ferox*
ES	sobrajo
DA	skjoldtandhaj
DE	Schildzahnhai
EL	καρχαρίας· αγριοκαρχαρίας· σκυλόψαρο
EN	ragged-tooth shark
FR	requin féroce
IT	cagnaccio
NL	knopstaart-scheurtandhaai
PT	tubarão-areia
FI	hietahai-laji
SV	rovhaj

16   ALOPIIDAE   BTH

SC	*Alopias superciliosus* (Lowe, 1839)
ES	zorro ojón
DA	storøjet rævehaj
DE	Großäugiger Fuchshai
EL	αλεπόσκυλος μεγαλομάτης
EN	big-eye thresher
FR	renard à gros yeux
IT	squalo volpe occhione
NL	grootoog-voshaai
PT	tubarão-raposo olhudo; raposo olhudo; zorro de olhos grandes
FI	karibiankettuhai
SV	storögd rävhaj

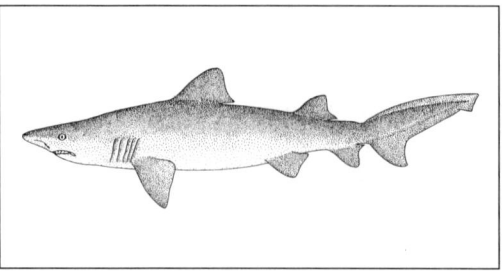

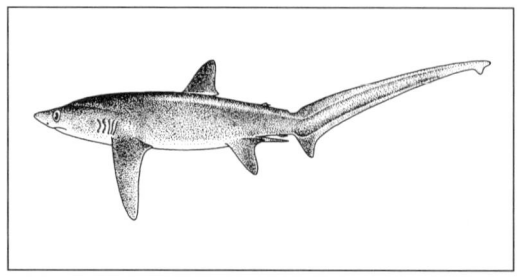

17 ALOPIIDAE ALV

SC	*Alopias vulpinus* (Bonnaterre, 1788)
ES	tiburón zorro
DA	almindelig rævehaj
DE	Drescherhai; Fuchshai; Drescher
EL	αλεπόσκυλος· σκυλόψαρο
EN	thresher shark
FR	requin-renard
IT	pesce volpe; squalo volpe; pesce bandiera
NL	voshaai
PT	tubarão-raposo
FI	kettuhai
SV	rävhaj

19 LAMNIDAE WSH

SC	*Carcharodon carcharias* (Linnaeus, 1758); *Carcharodon rondeletti*
ES	jaquetón blanco; jaquetón; tiburón blanco
DA	store hvide haj; menneskehaj
DE	Weißhai; Menschenhai
EL	καρχαρίας· σμπρίλλιας· σκυλόψαρο· σμπρίλλιος
EN	great white shark; white shark; maneater; white pointer; maneater shark
FR	grand requin blanc; requin blanc; requin; requin mangeur d'homme; carcharodonte
IT	pescecane
NL	witte haai
PT	tubarão de São Tomé
FI	valkohai
SV	vithaj

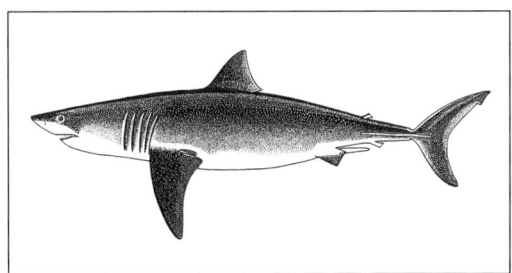

18 LAMNIDAE MSK

SC	*Lamnidae*
ES	cailones; jaquetones
DA	sildehaj-familien
DE	Heringshaie
EL	καρχαρίες
EN	mackerel sharks; porbeagles
FR	requins-taupes
IT	squali
NL	makreelhaaien; haringhaaien
PT	tubarões-sardo
FI	sillihait; sillihait-heimo
SV	jättehajar

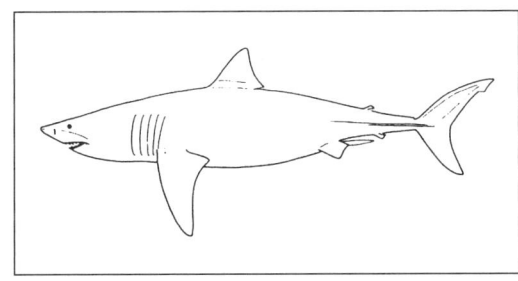

20 LAMNIDAE SMA

SC	*Isurus oxyrhinchus* (Rafinesque, 1810); *Isurus glaucus*; *Oxyrhina spallanzani*
ES	marrajo
DA	makrelhaj; mako-haj
DE	Makrelenhai; Mako
EL	καρχαρίας· ρυγχοκαρχαρίας
EN	mako shark; short-finned mako
FR	lamie à nez pointu; requin-taupe bleu
IT	squalo mako; smeriglio mako
NL	makreelhaai
PT	tubarão-anequim; anequim
FI	makrillihai
SV	makrillhaj; mako

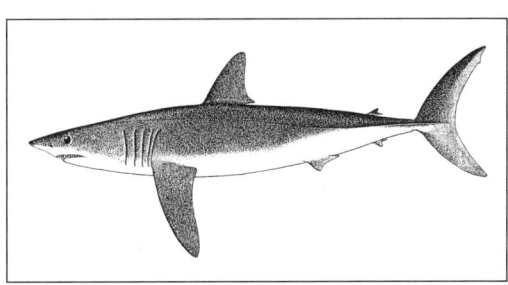

21	**LAMNIDAE**	LMA

SC	*Isurus paucus* Guitart (Manday, 1966)
ES	marrajo carite
DA	langfinnet mako; makrelhaj
DE	Langflossen-Makohai
EL	μακρυπτερυγο-ρυγχοκαρχαρίας· καρχαρίας· σκυλόψαρο
EN	long-fin mako
FR	petit requin-taupe
IT	smeriglio mako
NL	langvin-makreelhaai
PT	tubarão-anequim de gadanha
FI	pitkäevämakrillihai
SV	långfenad mako

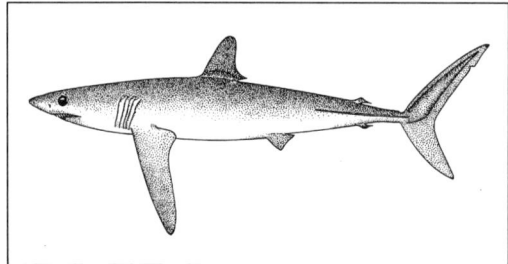

23	**LAMNIDAE**	POR

SC	*Lamna nasus* (Bonnaterre, 1788); *Lamna cornubica*
ES	marrajo sardinero; cailón marrajo
DA	almindelig sildehaj
DE	Heringshai
EL	λάμια· καρχαρίας
EN	porbeagle; porbeagle shark; mackerel shark
FR	taupe commun; requin-taupe commun; taupe; touille; lamnie; maraîche; muzeraille
IT	smeriglio; talpa
NL	haringhaai
PT	tubarão sardo; sardo
FI	sillihai
SV	håbrand; sillhaj

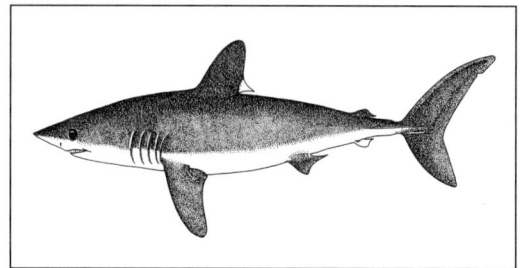

22	**LAMNIDAE**	LMD

SC	*Lamna ditropis* (Hubbs and Follett, 1947)
ES	tiburón salmón
DA	Stillehavs-sildehaj
DE	Pazifischer Heringshai
EL	λάμια του Ειρηνικού· καρχαρίας
EN	salmon shark
FR	requin-taupe saumon
IT	smeriglio del Pacifico
NL	Pacifische haringhaai
PT	tubarão-sardo do Japão
FI	tyynenmerensillihai
SV	laxhaj; stillahavslaxhaj

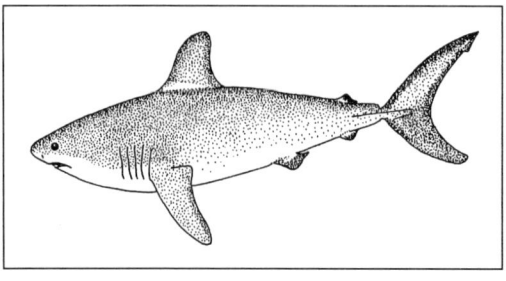

24	**STEGOSTOMATIDAE**

SC	*Stegostoma fasciatum* (Hermann, 1783); *Stegostoma tigrinum*
ES	tiburón cebra
DA	Stillehavs-zebrahaj
DE	Zebrahai
EL	ζεμπρασκυλόψαρο· ζεμπρακαρχαρίας
EN	zebra shark
FR	requin zébré
IT	squalo striato
NL	zebrahaai
PT	tubarão-zebra
FI	seeprapartahai
SV	sebrahaj

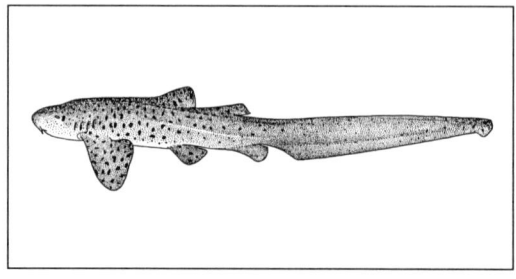

25 ORECTOLOBIDAE

SC	*Orectolobus maculatus* (Bonnaterre, 1788)
ES	tiburón alfombra
DA	australsk skæghaj; wobbegong
DE	Australischer Ammenhai
EL	κηλιδοκαρχαρίας
EN	wobbegong; spotted wobbegong
FR	requin tapis; requin tapis tacheté
IT	squalo nutrice
NL	tapijthaai
PT	tubarão-tapete
FI	rengaspartahai
SV	fläckig wobbegong

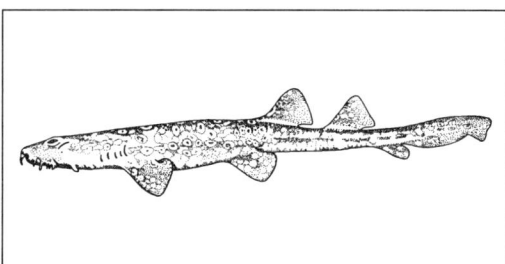

27 RHINCODONTIDAE

SC	*Rhincodon typus* (Smith, 1828); *Rhiniodon typus*
ES	tiburón ballena
DA	hvalhaj
DE	Walhai
EL	φαλαινοκαρχαρίας
EN	whale-shark
FR	requin-baleine
IT	squalo balena
NL	walvishaai
PT	tubarão-baleia
FI	valashai
SV	valhaj

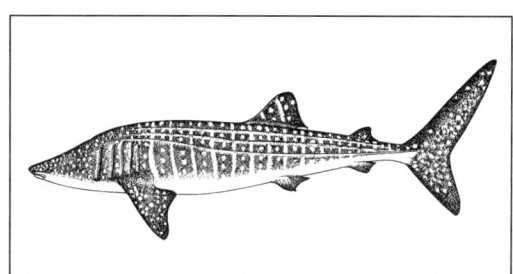

26 GINGLYMOSTOMATIDAE   GNC

SC	*Ginglymostoma cirratum* (Bonnaterre, 1788)
ES	tiburón nodriza
DA	nursehaj
DE	Ammenhai
EL	καρχαρίας
EN	nurse shark
FR	requin nourrice; dormeur
IT	squalo nutrice
NL	nursehaai
PT	tubarão dormedor
FI	atlantinpartahai
SV	amhaj

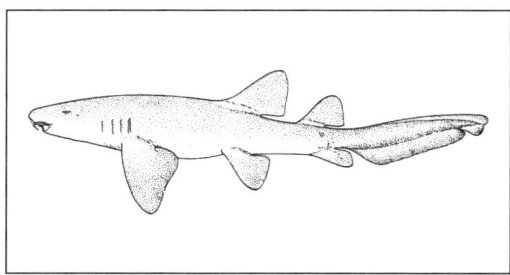

28 SCYLIORHINIDAE   SYX

SC	*Scyliorhinidae*
ES	pintarrojas
DA	rødhaj-familien
DE	Katzenhaie
EL	σκυλάκια
EN	dogfishes; catsharks
FR	roussettes
IT	gattucci
NL	hondshaaien
PT	pata-roxas e leitões
FI	punahait; punahait-heimo
SV	rödhajar

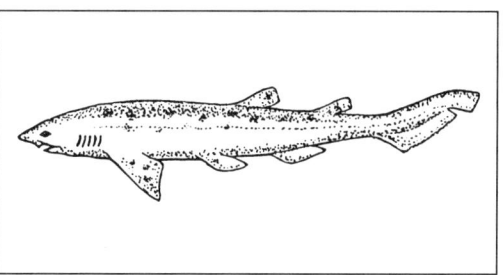

29 SCYLIORHINIDAE

SC *Apristurus brunneus* (Gilbert, 1892)
ES tiburón gato marrón
DA brun kattehaj
DE Brauner Katzenhai
EL καφέ σκυλάκι
EN brown cat shark
FR requin-chat brun; holbiche brune
IT gattuccio bruno
NL bruine hondshaai
PT pata-roxa castanha
FI
SV hågäl

31 SCYLIORHINIDAE

SC *Scyliorhinus canicula* (Linnaeus, 1758)
ES pintarroja
DA småplettet rødhaj
DE Kleingefleckter Katzenhai; Kleiner Katzenhai
EL γάτος· σκυλοψαράκι· σκυλάκι· γατοψαράκι
EN dogfish
FR petite roussette; petit chat de mer
IT gattuccio
NL hondshaai
PT pata-roxa
FI pistepunahai
SV småfläckig rödhaj

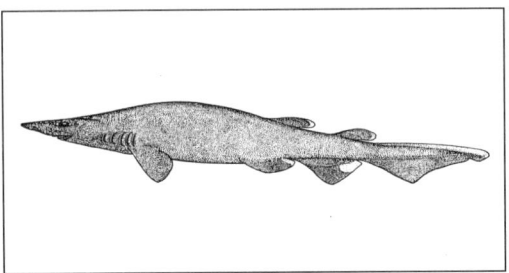

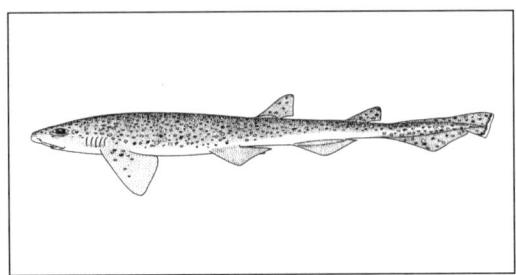

30 SCYLIORHINIDAE   SHO

SC *Galeus melastomus* (Rafinesque, 1810); *Pristiurus melanostomus*
ES pintarroja bocanegra; colayo; bocanegra; olayo; colaina; gazapa;
DA ringhaj
DE Fleckhai; Sägeschwanz
EL σκυλάκι· γάτος· μελανόστομος
EN black-mouthed dogfish; black-mouth catshark
FR chien espagnol
IT boccanegra; boccanera
NL zwartmond-hondshaai
PT leitão
FI rengashai
SV hågäl

32 SCYLIORHINIDAE   SYT

SC *Scyliorhinus stellaris* (Linnaeus, 1758)
ES alitán
DA storplettet rødhaj
DE Großgefleckter Katzenhai; Großer Katzenhai
EL γάτος· σκυλοψαράκι· σκυλάκι· γατοψαράκι
EN nursehound
FR grande roussette; grand chat de mer
IT gattopardo
NL kathaai
PT pata-roxa-gata
FI täpläpunahai
SV storfläckig rödhaj

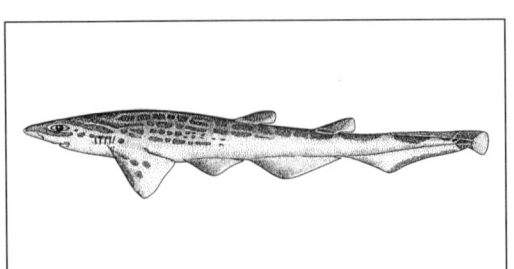

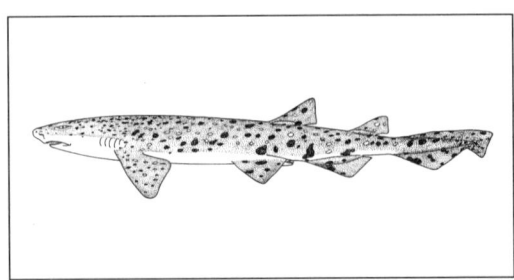

| 33 | **SCYLIORHINIDAE** | SCL | 35 | **CARCHARHINIDAE** | ALS |

33 SCYLIORHINIDAE SCL

SC *Scyliorhinus* spp.
ES pintarrojas; alitanes
DA rødhaj-slægt
DE Katzenhaie
EL σκυλάκια· σκυλόψαρα
EN catsharks; nursehounds
FR roussettes
IT gattucci
NL hondshaaien
PT pata-roxas
FI punahai-suku
SV rödhajar

35 CARCHARHINIDAE ALS

SC *Carcharhinus albimarginatus* (Rüppell, 1837)
ES tiburón de aletas plateadas
DA sølvtippet haj
DE Silberspitzen-Hai
EL καρχαρίας ασπροπτερύγιος
EN silvertip shark
FR requin à pointes blanches des récifs
IT squalo pinne bianche
NL zilvertiphaai
PT tubarão de pontas prateadas
FI hopeatippahai
SV silverspetshaj

 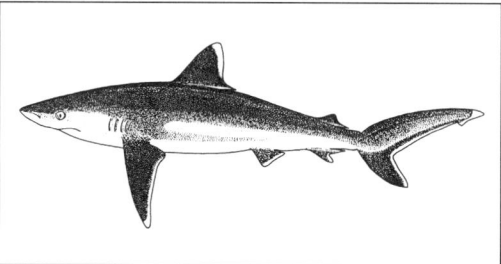

34 CARCHARHINIDAE RSK

SC *Carcharhinidae*
ES tiburones
DA blinkhindehaj-familien
DE Blauhaie
EL καρχαρίες
EN requiem sharks; sand sharks
FR requins; requins-tigres
IT carcarinidi
NL blauwe haaien
PT carcarinídeos; tubarões-marracho
FI ihmishait; ihmishait-heimo
SV gråhajar

36 CARCHARHINIDAE BRO

SC *Carcharhinus brachyurus* (Günther, 1870)
ES tiburón cobrizo
DA kobberhaj
DE Kupferhai
EL χαλκοχροοκαρχαρίας
EN copper shark
FR requin cuivré
IT squalo bronzeo
NL koperhaai
PT tubarão-cobre
FI kuparihai
SV kopparhaj

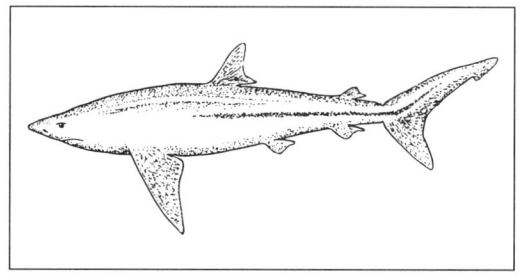 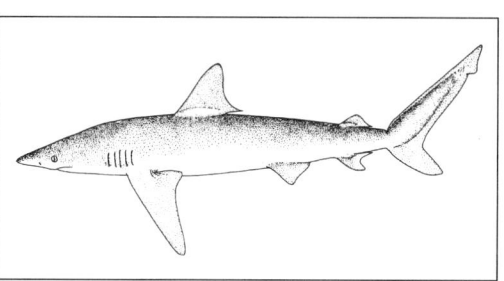

37	**CARCHARHINIDAE**	OCS

SC Carcharhinus longimanus (Poey, 1861)
ES tiburón oceánico
DA hvidtippet haj
DE Langflossen-Hai; Weißspitzen-Menschenhai
EL μακροπτερυγοκαρχαρίας
EN oceanic white-tip shark; white-tip shark
FR requin océanique; rameur; aileron blanc du large; requin à longues nageoires
IT squalo alalunga
NL witpunthaai
PT tubarão de pontas brancas
FI valkopilkkahai
SV årfenshaj

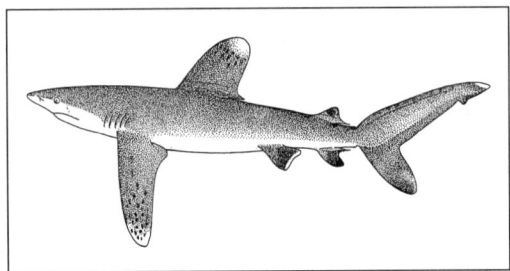

38	**CARCHARHINIDAE**	BLR

SC Carcharhinus melanopterus (Quoy and Gaimard, 1824)
ES tiburón de aletas negras
DA sorttippet haj
DE Schwarzspitzen-Riffhai
EL μαυροπτερυγοκαρχαρίας
EN black-tip reef shark
FR requin à pointes noires
IT squalo pinne nere
NL zwartvin-rifhaai
PT tubarão negro
FI mustaevähai
SV svartspetshaj

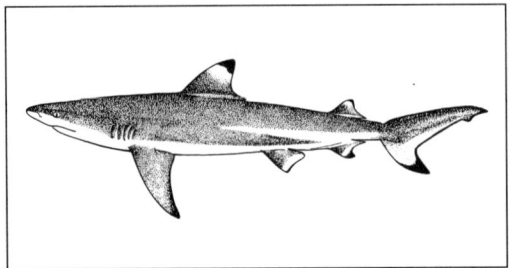

39	**CARCHARHINIDAE**	DUS

SC Carcharhinus obscurus (Lesueur, 1818)
ES melgacho; tiburón lamia
DA mørkhaj
DE Sandbankhai
EL σταχτοκαρχαρίας
EN dusky shark
FR requin de sable; requin obscur
IT squalo grigio
NL donkere haai
PT tubarão-faqueta
FI sumuhai
SV mörkhaj

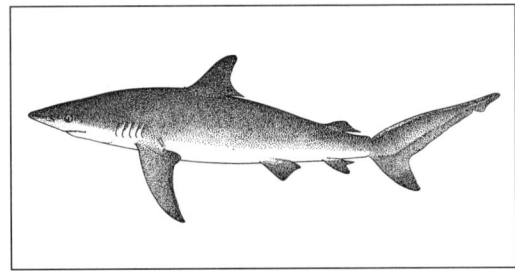

40	**CARCHARHINIDAE**	CCP

SC Carcharhinus plumbeus (Nardo, 1827)
ES tiburón gris
DA brunhaj
DE Atlantischer Braunhai
EL σταχτοκαρχαρίας· καρχαρίας· σκυλόψαρο
EN sandbar shark
FR requin gris
IT squalo grigio
NL grijze haai
PT tubarão corre-costa
FI ihmishai-laji
SV högfenad haj

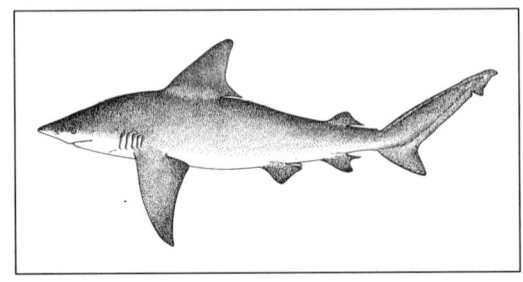

41   CARCHARHINIDAE   NGB

SC	*Negaprion brevirostris* (Poey, 1868)
ES	tiburón limón
DA	citronhaj
DE	Zitronenhai
EL	κίτρινος καρχαρίας
EN	lemon shark
FR	requin citron
IT	squalo limone
NL	citroenhaai
PT	tubarão-limão
FI	sitruunahai
SV	citronhaj

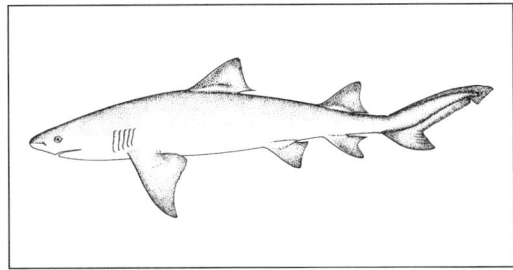

42   CARCHARHINIDAE   BSH

SC	*Prionace glauca* (Linnaeus, 1758); *Carcharhinus glaucus*; *Carcharias glaucus*; *Squalus glaucus*
ES	tintorera; tiburón azul; tintorera-marrajo; tintoleta
DA	blåhaj
DE	Großer Blauhai; Blauhai
EL	γλαυκοκαρχαρίας· καρχαρίας· σκυλόψαρο
EN	blue shark; great blue shark; blue whaler
FR	peau bleue; requin bleu
IT	verdesca
NL	blauwe haai
PT	tintureira; guelha
FI	sinihai
SV	blåhaj

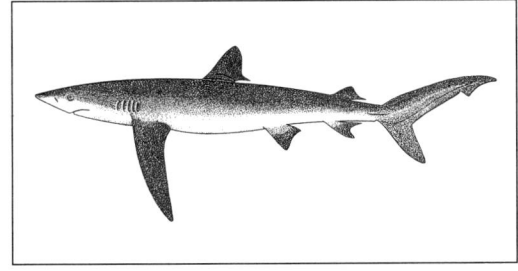

43   CARCHARHINIDAE

SC	*Rhizoprionodon longurio* (Jordan and Gilbert, 1882)
ES	tiburón narigudo del Pacífico
DA	pacifisk spidssnudet haj
DE	Pazifischer Spitzmaulhai
EL	οξυρυγχοκαρχαρίας του Ειρηνικού
EN	Pacific sharp-nose shark
FR	requin à nez pointu du Pacifique; requin bironche
IT	squalo musoguzzo
NL	Pacifische melkhaai
PT	tubarão bicudo do Pacífico
FI	tyynenmerennokkahai
SV	stillahavsspetsnoshaj

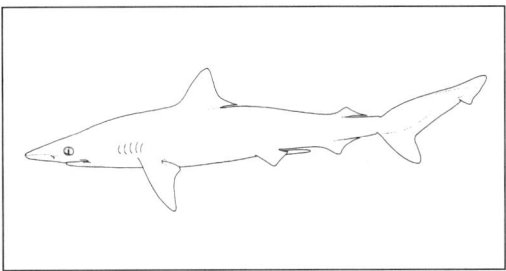

44   CARCHARHINIDAE   RHT

SC	*Rhizoprionodon terraenovae* (Richardson, 1836)
ES	tiburón narigudo atlántico
DA	atlantisk spidssnudet haj
DE	Atlantischer Spitzmaulhai
EL	οξυρυγχοκαρχαρίας του Ατλαντικού
EN	Atlantic sharp-nose shark
FR	requin à nez pointu de l'Atlantique; requin aiguille
IT	squalo musoguzzo
NL	Atlantische melkhaai
PT	tubarão bicudo
FI	nokkahai
SV	vitprickig spetsnoshaj

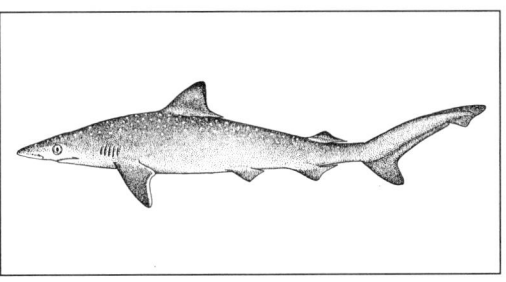

45 SPHYRNIDAE   SPY

SC	*Sphyrnidae*
ES	tiburones martillo; peces martillo
DA	hammerhaj-familien
DE	Hammerhaie
EL	ζύγαινες
EN	hammerhead sharks; bonnetheads
FR	requins-marteaux
IT	squali martello; pesci martello
NL	hamerhaaien
PT	tubarões-martelo; cornudas
FI	vasarahait; vasarahait-heimo
SV	hammarhajar

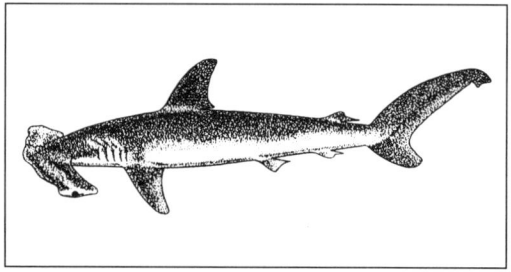

47 SPHYRNIDAE   SPZ

SC	*Sphyrna zygaena* (Linnaeus, 1758); *Zygaena malleus*
ES	pez martillo; cornuda; cornudilla; guardia civil; tiburón martillo
DA	almindelig hammerhaj
DE	Hammerhai; Glatter Hammerhai
EL	ζύγαινα· σφύρνα· πατερίτσα
EN	smooth hammerhead; hammerhead shark; hammerhead; common hammerhead
FR	requin-marteau commun; requin-marteau; marteau; maillet; requin-marteau lisse
IT	pesce martello
NL	hamerhaai
PT	tubarão-martelo; martelo
FI	vasarahai
SV	hammarhaj

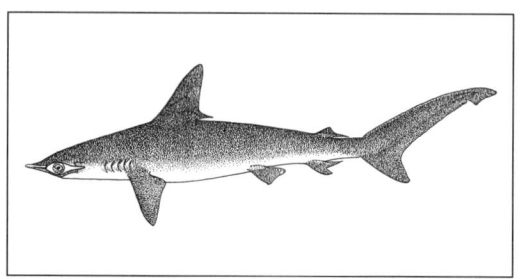

46 SPHYRNIDAE   SPL

SC	*Sphyrna lewini* (Griffith and Smith, 1834)
ES	cornuda común
DA	indo-pacifisk hammerhaj
DE	Indopazifischer Hammerhai
EL	ζύγαινα· πατερίτσα
EN	scalloped hammerhead
FR	requin-marteau halicorne
IT	pesce martello
NL	geschulpte hamerhaai
PT	tubarão-martelo recortado
FI	kampavasarahai
SV	flerhornig hammarhaj

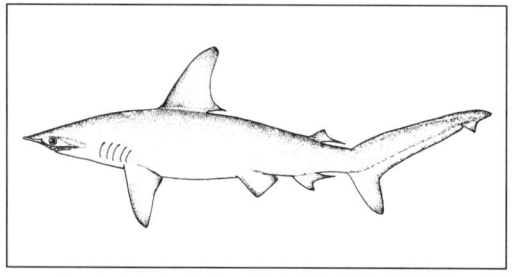

48 SPHYRNIDAE   SPN

SC	*Sphyrna* spp.
ES	tiburones martillo
DA	hammerhaj-slægten
DE	Hammerhaie
EL	ζύγαινες· πατερίτσες
EN	hammerhead sharks
FR	requins-marteaux
IT	pesci martello
NL	hamerhaaien
PT	tubarões-martelo; cornudas; peixes-martelo
FI	vasarahai-suku
SV	hammarhajar

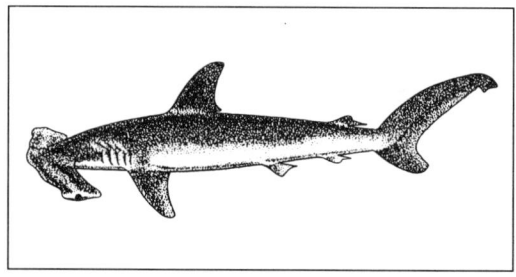

49 TRIAKIDAE   TRK

SC	*Triakidae*
ES	musolas; cazones; tollos
DA	glathaj-familien
DE	Glatthaie; Hundshaie; Marderhaie
EL	γαλέοι
EN	houndsharks; smoothhounds; topes
FR	émissoles
IT	palombi; triachidi
NL	ruwe en gladde haaien
PT	cações e pernas de moça
FI	silohait; silohait-heimo
SV	glatthajar

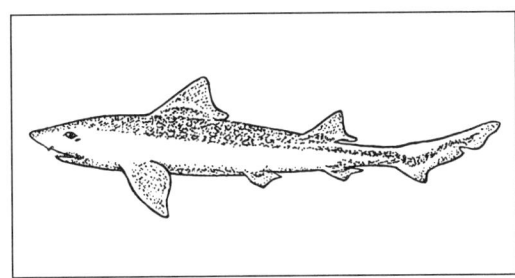

51 TRIAKIDAE

SC	*Mustelus antarcticus* (Günther, 1870)
ES	musola antártica
DA	antarktisk glathaj
DE	Australischer Glatthai
EL	ανταρκτογαλέος
EN	gummy shark
FR	chien de mer antarctique; mustellus gommé
IT	palombo antartico
NL	Antarctische gladde haai
PT	cação antárctico
FI	silohai-laji
SV	gummihaj; sydlig hundhaj

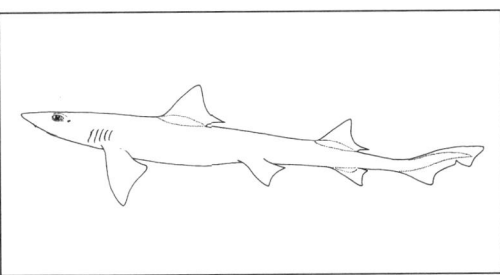

50 TRIAKIDAE   GAG

SC	*Galeorhinus galeus* (Linnaeus, 1758); *Eugaleus galeus*; *Galeus vulgaris*
ES	cazón; tollo; pez peine; tolle; lija; gato; galeo
DA	gråhaj
DE	Hundshai; Biethai
EL	γαλέος· δροσίτης· σκυλογαλέος· σκύλος
EN	tope shark; tope; flake; soupfin shark
FR	requin hâ; milandre; hâ; chien de mer
IT	canesca
NL	ruwe haai
PT	perna de moça; tubarão-perna de moça
FI	harmaahai
SV	gråhaj

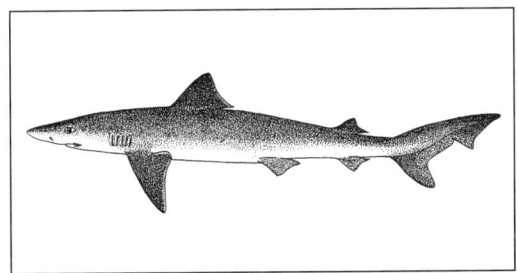

52 TRIAKIDAE   SDS

SC	*Mustelus asterias* (Cloquet, 1821); *Mustelus vulgaris*; *Mustelus stellatus*
ES	musola dentuda; caella; tolla
DA	plettet glathaj
DE	Nördlicher Glatthai; Gefleckter Glatthai
EL	αστρογαλέος· δροσίτης· γαλέος· σκυλογαλέος
EN	starry smooth-hound; stellate smooth-hound
FR	émissole tachetée
IT	palombo stellato; palombo
NL	gevlekte gladde haai
PT	cação pintado
FI	silohai-laji
SV	nordlig hundhaj

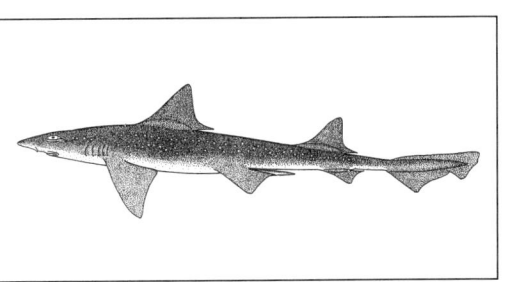

| **53** | **TRIAKIDAE** | SMD | **55** | **TRIAKIDAE** | SDV |

SC	*Mustelus mustelus* (Linnaeus, 1758); *Mustelus laevis*
ES	musola; boca blanca; cazón; musola pintada; musola vera; jaqueta; pique; tollo
DA	almindelig glathaj
DE	Südlicher Glatthai; Mittelmeer-Glatthai
EL	γκριζογαλέος· δροσίτης· γαλέος· σκυλογαλέος
EN	smoothhound
FR	émissole lisse; émissole; émissole commune; moutelle commune
IT	palombo
NL	gladde haai
PT	cação liso; caneja
FI	kärppähai
SV	sydlig hundhaj

SC	*Mustelus* spp.
ES	musolas
DA	glathaj-slægt
DE	Glatthaie
EL	γαλέοι
EN	smoothhounds
FR	émissoles; chiens de mer
IT	palombi
NL	gladde haaien
PT	cações; canejas
FI	silohai-suku
SV	hundhajar

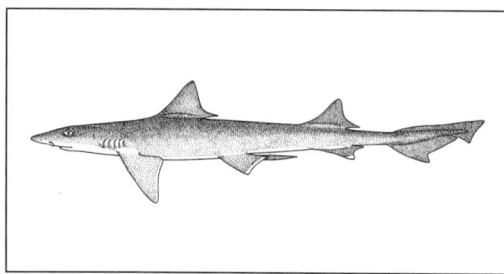

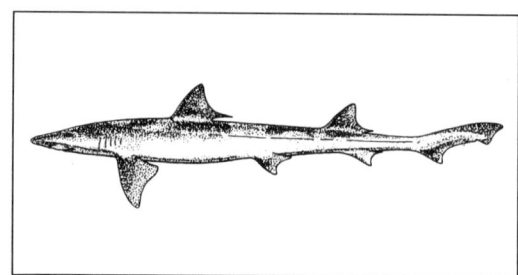

| **54** | **TRIAKIDAE** | SDP | **56** | **TRIAKIDAE** | LES |

SC	*Mustelus schmitti* (Springer, 1940)
ES	gatuso
DA	patagonisk glathaj
DE	Patagonischer Glatthai
EL	γαλέος της Παταγωνίας
EN	Patagonian smoothhound
FR	émissole de Patagonie; émissole gatuso
IT	palombo atlantico
NL	Patagonische gladde haai
PT	cação da Patagónia
FI	patagoniansilohai
SV	argentinsk hundhaj

SC	*Triakis semifasciata* (Girard, 1854)
ES	tiburón leopardo
DA	leopardhaj
DE	Kalifornischer Leopardhai
EL	καρχαρίας λεοπάρδαλη
EN	leopard shark
FR	requin-léopard
IT	palombo maculato
NL	luipaardhaai
PT	tubarão-leopardo
FI	leopardihai
SV	leopardhaj

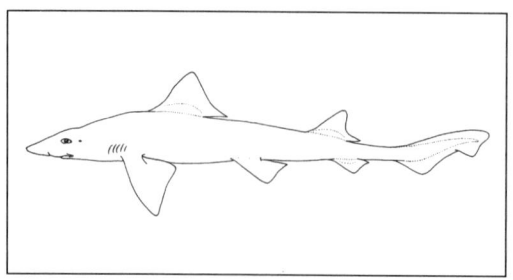

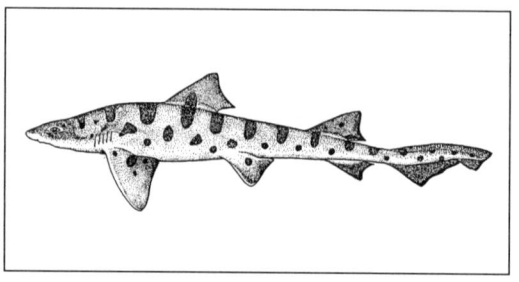

57	**SQUALIDAE**	DGX	59	**SQUALIDAE**	GUQ

57 SQUALIDAE DGX

SC *Squalidae*
ES mielgas; galludos
DA pighaj-familien
DE Dornhaie
EL καρχαρίες· σκυλόψαρα
EN spurdogs
FR squales; requins épineux; aiguillats
IT spinaroli; squalidi
NL haaien en doornhaaien
PT esqualídeos
FI piikkihait; piikkihait-heimo
SV pigghajar

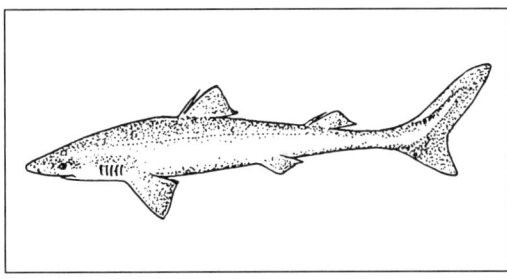

59 SQUALIDAE GUQ

SC *Centrophorus squamosus* (Bonnaterre, 1788)
ES quelvacho negro
DA mørk pighaj
DE Düsterer Dornhai
EL αγκαθίτης του Ατλαντικού
EN leaf-scale gulper shark
FR squale-chagrin de l'Atlantique
IT sagrì
NL donkere doornhaai
PT lixa; lixa de escama; xara branca
FI piikkihai-laji
SV brun pigghaj

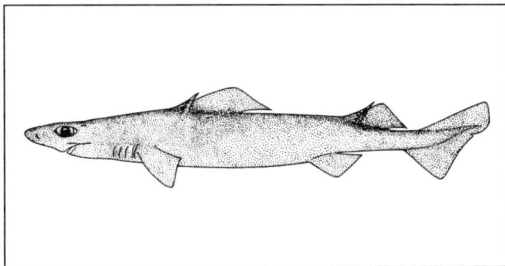

58 SQUALIDAE GUP

SC *Centrophorus granulosus* (Bloch and Schneider, 1801)
ES quelvacho; quelve
DA ru pighaj
DE Rauher Dornhai
EL αγκαθίτης· κεντρόνι· κοκκοαγκαθίτης
EN gulper shark; rough shark
FR squale-chagrin commun; requin chagrin; centrophore; centrophore granuleux
IT sagrì
NL ruwe doornhaai
PT barroso; lixa de lei
FI piikkihai-laji
SV sorghaj

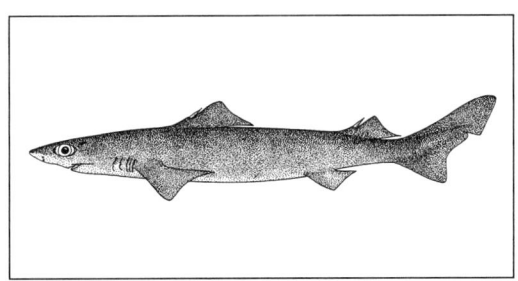

60 SQUALIDAE SCK

SC *Dalatias licha* (Bonnaterre, 1788); *Scymnorhinus licha*
ES carocho; negra; negrita; pailona
DA chokolade-haj
DE Schokoladenhai
EL σκυλόψαρο· μαύρο σκυλόψαρο· σκυμνοσκυλόψαρο
EN Darkie Charlie; kitefin shark
FR squale liche; gatte
IT zigrino
NL zwarte haai
PT gata; gata-lixa; lixa de pau
FI piikkihai-laji
SV chokladhaj

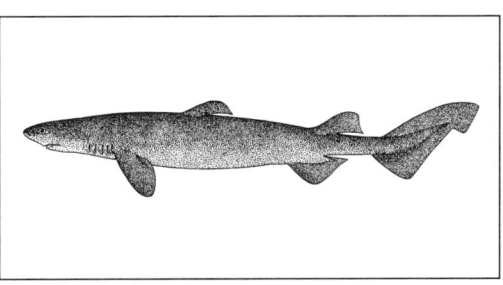

61 SQUALIDAE

SC *Etmopterus spinax* (Linnaeus, 1758); *Spinax niger*
ES negrito; negrita; cochino; negra
DA sorthaj
DE Schwarzer Dornhai; Schwarzer Hundfisch
EL μακροαγκαθίτης
EN velvet belly; lantern shark; black centrina
FR sagre commun; épineux de fond; aiguillat-sagre; aiguillat noir
IT sagrì nero; pesce diavolo
NL zwarte doornhaai
PT lixinha da fundura; lixinha
FI pikkuhai
SV blåkäxa

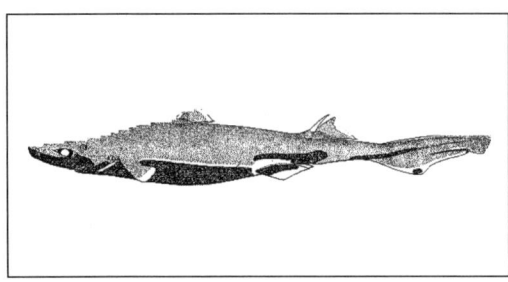

63 SQUALIDAE   GSK

SC *Somniosus microcephalus* (Bloch and Schneider, 1801)
ES tiburón boreal; tollo de Groenlandia
DA havkal; Grønlandshaj; ishaj
DE Eishai; Grundhai; Grönlandhai
EL μαυροσκυλόψαρο της Γροιλανδίας
EN Greenland shark; ground shark; sleeper shark
FR laimargue du Groenland; laimargue; apocalle; requin du Groenland
IT squalo di Groenlandia; lemargo
NL Groenlandse haai
PT tubarão da Gronelândia; lobo
FI holkeri
SV håkäring

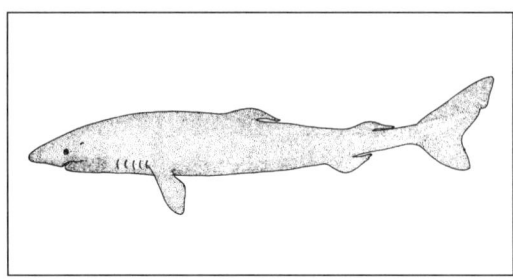

62 SQUALIDAE   SHL

SC *Etmopterus* spp.
ES negritos
DA pighaj-slægt
DE Schwarze Dornhaie
EL αγκαθίτες
EN lantern sharks
FR sagres
IT pesci diavolo
NL zwarte doornhaaien
PT lixinhas da fundura
FI piikkihai-suku
SV blåkäxor

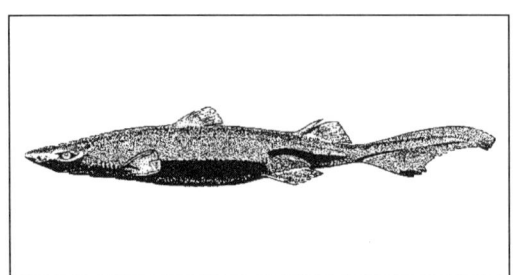

64 SQUALIDAE   DGS

SC *Squalus acanthias* (Linnaeus, 1758); *Acanthias vulgaris*; *Spinax acanthias*
ES mielga; galludo; pinchorro; galludo de pintilla; ferrón
DA almindelig pighaj
DE Dornhai; Grundhai
EL κεντρόνι· σκύλος· σκυλόψαρο· στικτοκεντρόνι
EN spurdog; piked dogfish; spiny; spring dogfish; common spiny; grayfish; spiny dogfish; dogfish
FR aiguillat commun; aiguillat; chien de mer; chien piquet; chien à dard; chien; squales méditerranéens; chien-piquet; spinax; requin épineux
IT spinarolo
NL doornhaai
PT galhudo malhado; melga; cação galhudo;
FI piikkihai
SV pigghaj

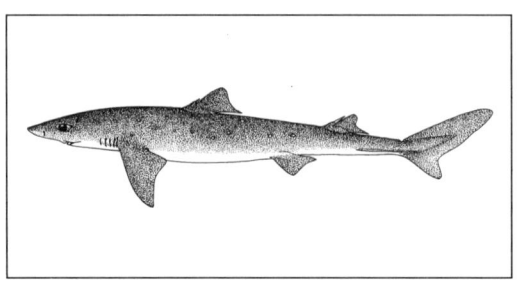

65 SQUALIDAE

SC	*Squalus blainvillei* (Risso, 1826); *Squalus fernandinus*
ES	galludo
DA	Blainvilles pighaj
DE	Blainvilles Dornhai
EL	κεντρόνι· σκύλος· σκυλόψαρο· στικτοκεντρόνι
EN	Blainville's spurdog; long-nose spurdog
FR	aiguillat galludo; aiguillat-coq
IT	spinarolo bruno; spinarolo
NL	Blainville's doornhaai
PT	galhudo
FI	piikkihai-laji
SV	galludohaj; medelhavspigghaj

67 SQUATINIDAE ASK

SC	*Squatinidae*
ES	angelotes
DA	havengel-familien
DE	Meerengel
EL	ἀγγελοι· ῥίνες
EN	angel sharks; sand devils
FR	anges de mer
IT	squadri; squatinidi
NL	zee-engelen
PT	anjos
FI	merienkelit; merienkelit-heimo
SV	havsänglar

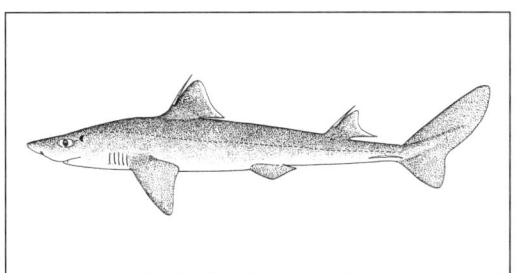

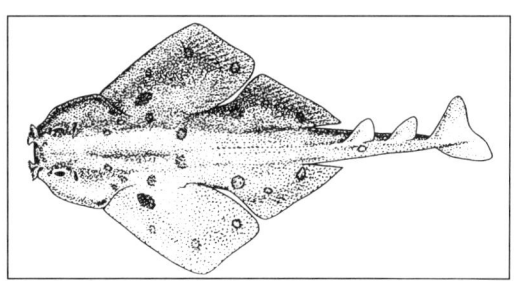

66 SQUALIDAE DOP

SC	*Squalus megalops* (Macleay, 1881)
ES	galludo chato
DA	storøjet pighaj
DE	Großaugen-Dornhai
EL	κοντόρυγχος αγκαθίτης
EN	short-nose spurdog
FR	aiguillat à nez court
IT	spinarolo muso corto
NL	kortsnuit-doornhaai
PT	galhudo de focinho curto
FI	piikkihai-laji
SV	kortnosad pigghaj

68 SQUATINIDAE AGN

SC	*Squatina squatina* (Linnaeus, 1758); *Squatina angelus*
ES	angelote; pez ángel; villano
DA	havengel; munkefisk
DE	Engelhai; Gemeiner Meerengel
EL	ἀγγελος· ῥίνα· βιολί
EN	angel shark; angelfish; monkfish
FR	ange de mer commun; ange de mer; requin-raie; angelot; angel
IT	squadro; pesce angelo
NL	zeeëngel
PT	anjo
FI	merienkeli
SV	havsängel

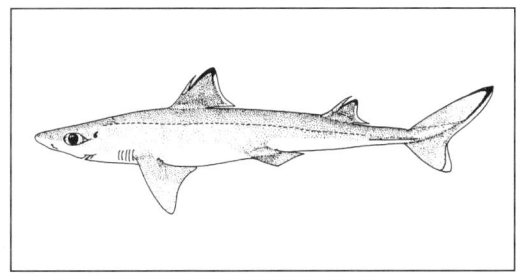

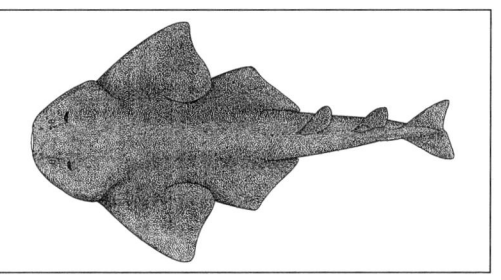

69 OXYNOTIDAE

SC	*Oxynotidae*
ES	cerdos marinos
DA	trekanthaj-familien
DE	Meersäue
EL	γουρουνόψαρα
EN	rough sharks
FR	centrines
IT	pesci porco
NL	zeevarkenhaaien
PT	peixes-porco
FI	hai-heimo
SV	trekantshajar

71 ECHINORHINIDAE SHB

SC	*Echinorhinus brucus* (Bonnaterre, 1788); *Echinorhinus spinosus*
ES	pez clavo; pez tachuela
DA	sømhaj
DE	Stachelhai; Alligatorhai; Nagelhai
EL	καρχαρίας· αχινοσκυλόψαρο
EN	bramble shark; spiny shark; spinous shark
FR	squale bouclé; chenille; requin bouclé
IT	ronco; ronco spinoso
NL	braamhaai
PT	tubarão-prego; peixe-prego
FI	okahai
SV	tagghaj

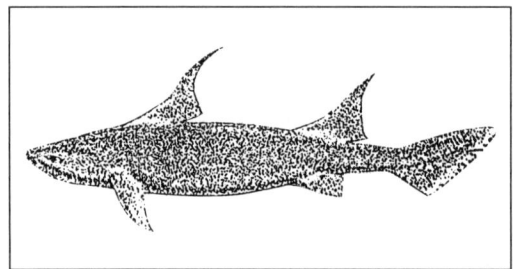

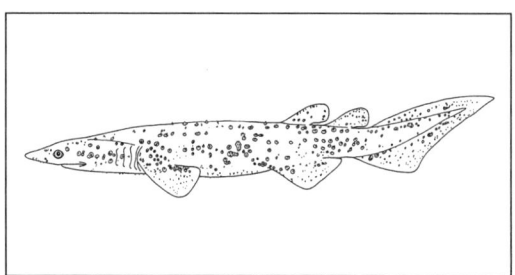

70 OXYNOTIDAE OXY

SC	*Oxynotus centrina* (Linnaeus, 1758); *Centrina salviani*
ES	cerdo marino; cerdito marino
DA	trekanthaj
DE	Meersau; Schweinhai
EL	γουρουνόψαρο· κεντρόνι
EN	humantin; angular rough shark; centrina shark; prickly dogfish
FR	centrine commune; centrine
IT	pesce porco
NL	zeevarkenhaai
PT	peixe-porco; porco marinho; porco
FI	hai-laji
SV	trekantshaj

72 RHINOBATIDAE GTF

SC	*Rhinobatidae*
ES	guitarras
DA	hajrokker; guitarfisk
DE	Geigenrochen
EL	ρινόβατοι· κιθάρες
EN	guitarfishes
FR	guitares de mer
IT	pesci violino; pesci chitarra
NL	gitaarroggen
PT	violas
FI	kitararauskut; kitararauskut-heimo
SV	hajrockor

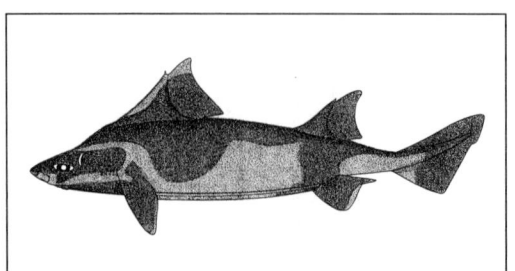

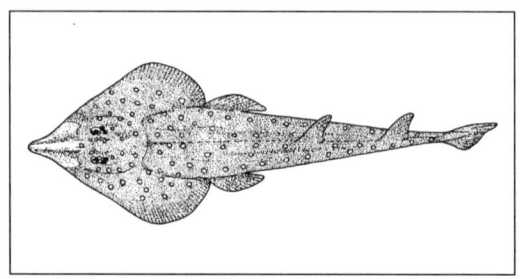

73   RHINOBATIDAE   GUB

SC   *Rhinobatos albomaculatus* (Norman, 1930)
ES   guitarra pecosa
DA   hvidplettet guitarfisk
DE   Weißflecken-Geigenrochen
EL   κιθάρα· σελάχι ρινόβατος
EN   white-spotted guitarfish
FR   raie-guitare tachetée
IT   pesce chitarra
NL   gevlekte gitaarrog
PT   viola malhada
FI   valkopilkkukitararausku
SV   hajrocka

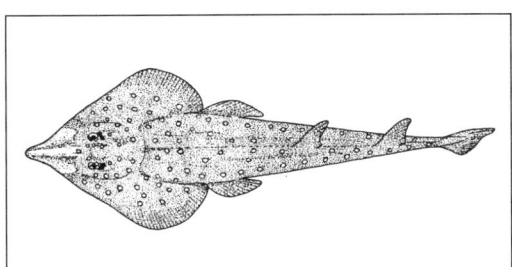

74   RHINOBATIDAE   GUD

SC   *Rhinobatos percellens* (Walbaum, 1792)
ES   guitarra chola
DA   sydlig guitarfisk
DE   Südlicher Geigenrochen
EL   σελάχι ρινόβατος· κιθαρόψαρο
EN   chola guitarfish
FR   raie-guitare chola
IT   pesce chitarra
NL   chola-gitaarrog
PT   viola do Golfo
FI   kitararausku-laji
SV   sydlig hajrocka

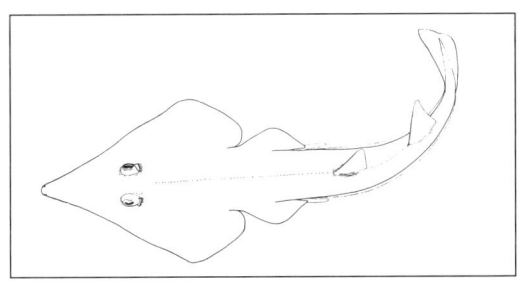

75   RHINOBATIDAE   GUF

SC   *Rhinobatos planiceps* (Garman, 1880)
ES   guitarra
DA   peruansk guitarfisk
DE   Peruanischer Geigenrochen
EL   σελάχι ρινόβατος· κιθαρόψαρο
EN   Peruvian guitarfish
FR   raie-guitare du Pérou
IT   pesce chitarra
NL   Peruaanse gitaarrog
PT   viola do Peru
FI   perunkitararausku
SV   peruansk hajrocka

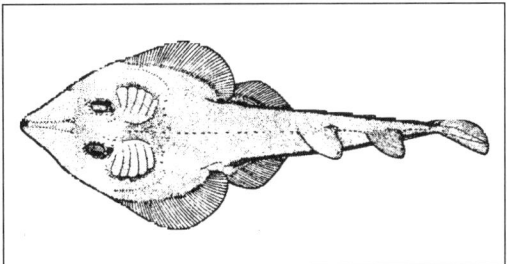

76   RHINOBATIDAE

SC   *Rhinobatos rhinobatos* (Linnaeus, 1758);
      Rhinobatos columnae
ES   guitarra; rebeca; viola
DA   almindelig guitarfisk
DE   Geigenrochen; Gemeiner Geigenrochen
EL   σελάχι ρινόβατος· κιθαρόψαρο
EN   guitarfish; common guitarfish
FR   guitare; violon de mer; guitare de mer commune
IT   pesce violino
NL   gitaarrog
PT   viola
FI   kitararausku
SV   hajrocka

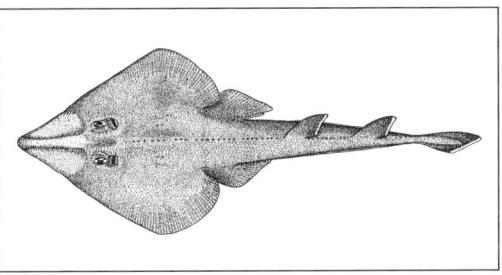

77	**RHINOBATIDAE**	GUZ	**79**	**PRISTIDAE**	

SC	*Rhinobatos* spp.		SC	*Pristis pectinata* (Latham, 1794)
ES	guitarras		ES	pez sierra
DA	guitarfisk-slægt		DA	småtandet savrokke
DE	Geigenrochen		DE	Sägerochen; Sägefisch; Westlicher Sägefisch
EL	σελάχι· ρινόβατος· κιθαρόψαρο		EL	πριονόψαρο
EN	guitarfishes		EN	sawfish; common sawfish; smalltooth sawfish
FR	raies-guitares		FR	poisson-scie; requin-scie; poisson-scie trident
IT	pesci chitarra		IT	pesce sega
NL	gitaarroggen		NL	zaagvis
PT	violas; guitarras; rebecas		PT	espadarte-serra; peixe-serra
FI	kitararausku-suku		FI	saharausku
SV	hajrockor		SV	småtandad sågfisk

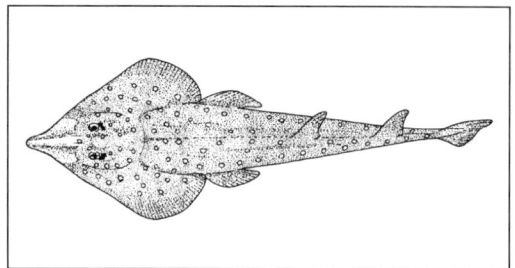 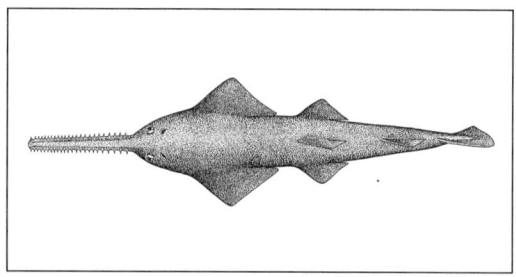

78	**PRISTIDAE**	SAW	**80**	**RAJIFORMES**	SRX

SC	*Pristidae*		SC	*Rajiformes*
ES	peces sierra		ES	rayas
DA	savfisk-familien; savrokke-familien		DA	rokker
DE	Sägefische; Sägerochen		DE	Rochen
EL	πριονόψαρα		EL	σελάχια
EN	sawfishes		EN	skates and rays
FR	poissons-scies		FR	raies
IT	pesci sega		IT	raiformi
NL	zaagvissen		NL	rogachtigen
PT	espadartes-serra; peixes-serra		PT	rajiformes; raias e afins
FI	saharauskut; saharauskut-heimo		FI	rauskut
SV	sågfiskar		SV	rockor

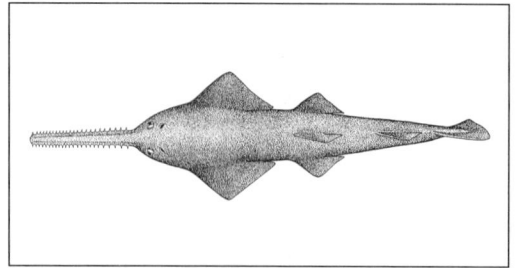 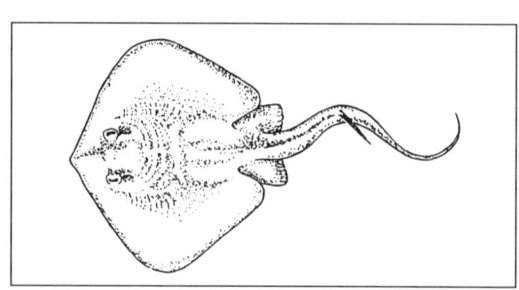

81 RAJIDAE RAJ

SC	*Rajidae*
ES	rayas
DA	rokke-familien
DE	Echte Rochen
EL	σελάχια· βάτοι· ράσες
EN	skates and rays
FR	raies
IT	razze; rajidi
NL	roggen
PT	raias
FI	rauskut; rauskut-heimo
SV	rockor

83 RAJIDAE

SC	*Raja asterias* (Delaroche, 1809); *Raja punctata*
ES	raya radiada; raya estrellada
DA	stjernerokke
DE	Mittelmeer-Sternrochen; Sternrochen
EL	αστρόβατος· σελάχι· βάτος· ράσα
EN	Mediterranean starry ray
FR	raie étoilée
IT	razza stellata; razza
NL	Middellandse-Zeesterrog
PT	raia pintada; raia estrelada
FI	tähtirausku
SV	stjärnrocka

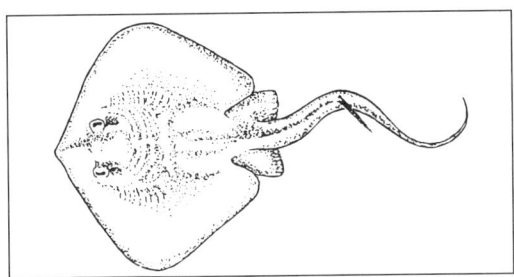

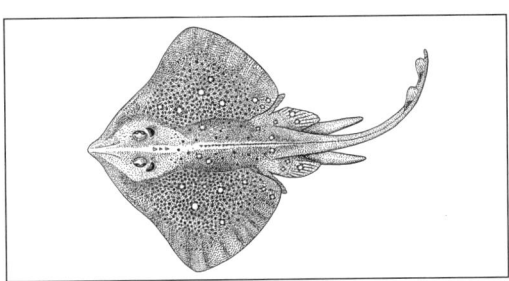

82 RAJIDAE

SC	*Raja alba* (Lacepède, 1803); *Raja marginata*; *Rostroraja alba*
ES	raya blanca; raya bramante; raya alba
DA	spidsrokke
DE	Spitzrochen; Bandrochen
EL	λευκόβατος· σελάχι· βάτος· ράσα
EN	white skate; bottlenose skate; bordered skate; white-bellied skate
FR	raie blanche; raie bordée
IT	razza bianca
NL	spitsneusrog
PT	raia taigora
FI	pullonnokkarausku
SV	grårocka

84 RAJIDAE RJB

SC	*Raja batis* (Linnaeus, 1758); *Raja macrorhynchus*; *Dipturus batis*
ES	noriega
DA	skade
DE	Glattrochen; Spiegelrochen
EL	γκριζόβατος· σελάχι· βάτος· ράσα
EN	skate; flapper skate; blue skate; common European skate
FR	pocheteau gris; raie grise; pocheteau blanc; raie cendrée
IT	razza bavosa
NL	vleet
PT	raia oirega
FI	silorausku
SV	slätrocka

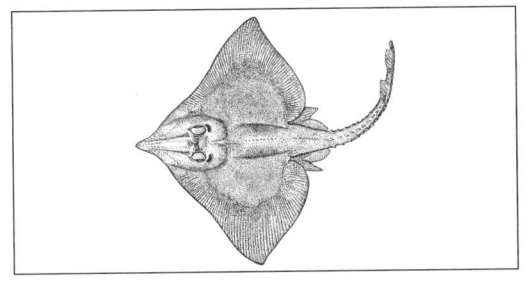

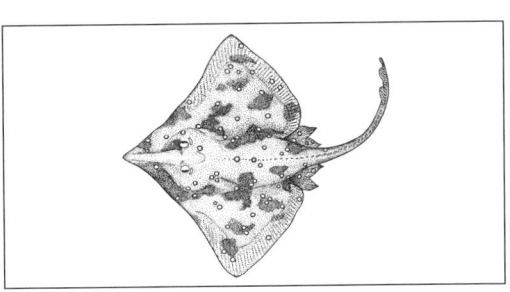

85 RAJIDAE

SC	*Raja brachyura* (Lafont, 1873)
ES	raya boca de rosa; escrita
DA	blond rokke
DE	Blonde
EL	ξανθόβατος· σελάχι· βάτος· ράσα
EN	blond ray
FR	raie lisse
IT	razza a coda corta
NL	blonde rog
PT	raia pontuada
FI	pilkkurausku
SV	ljusrocka

87 RAJIDAE   RJC

SC	*Raja clavata* (Linnaeus, 1758)
ES	raya común; raya de clavos; pez de Mahoma común
DA	sømrokke
DE	Nagelrochen; Keulenrochen; Steinrochen
EL	καλκανόβατος· σελάχι· βάτος· ράσα
EN	thornback ray; roker
FR	raie bouclée
IT	razza chiodata; razza
NL	stekelrog
PT	raia lenga; raia-pinta
FI	okarausku
SV	knaggrocka

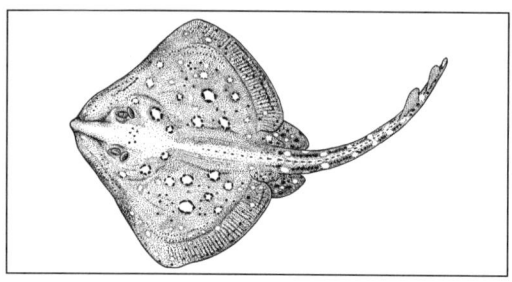

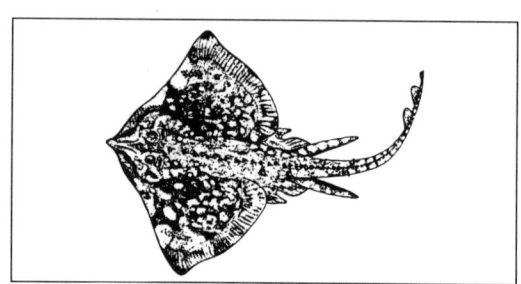

86 RAJIDAE

SC	*Raja circularis* (Couch, 1838); *Leucoraja circularis*
ES	raya falsa-vela
DA	sandrokke
DE	Sandrochen
EL	στρογγυλόβατος· σελάχι· βάτος· ράσα
EN	sandy ray
FR	raie circulaire; raie ronde
IT	razza rotonda
NL	zandrog
PT	raia de São Pedro
FI	hietarausku
SV	sandrocka

88 RAJIDAE

SC	*Raja erinacea* (Mitchill, 1825)
ES	raya de Canadá
DA	pindsvinerokke
DE	Igelrochen
EL	μικρόβατος· σελάχι· βάτος· ράσα
EN	little skate
FR	raie-hérisson
IT	razza
NL	Canadese rog
PT	raia de verão
FI	rausku-laji
SV	igelkottsrocka

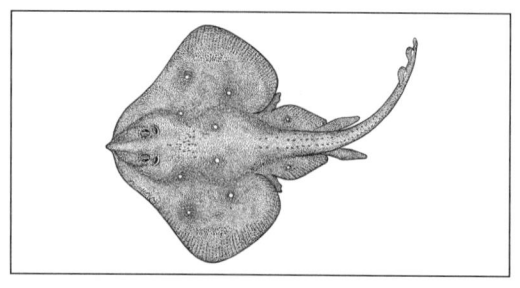

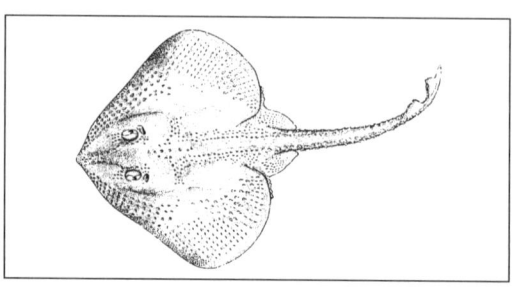

89 RAJIDAE RJF

SC	*Raja fullonica* (Linnaeus, 1758); *Raja chagrine*; *Leucoraja fullonica*
ES	raya cardadora
DA	gøgerokke
DE	Chagrinrochen; Schnabelrochen; Walkerrochen
EL	ακανθόβατος· σελάχι· βάτος· ράσα
EN	shagreen ray; Fuller's ray
FR	raie-chardon
IT	razza spinosa
NL	kaardrog
PT	raia pregada
FI	käkirausku
SV	näbbrocka; parfläckig rocka

91 RAJIDAE

SC	*Raja microocellata* (Montagu, 1818)
ES	raya cimbreira
DA	småøjet rokke; lysspættet rokke
DE	Kleinäugiger Rochen; Hellfleckiger Rochen
EL	μικροστιγματόβατος· σελάχι· βάτος· ράσα
EN	painted ray; small-eyed ray
FR	raie bâtarde; raie mêlée
IT	razza
NL	kleinoogrog
PT	raia zimbreira
FI	palettirausku
SV	småögd rocka

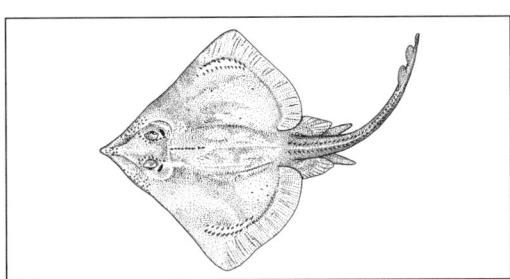

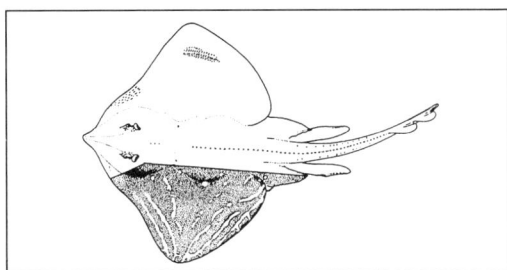

90 RAJIDAE

SC	*Raja lintea* (Fries, 1839)
ES	raya vela
DA	hvidrokke
DE	Weißrochen
EL	ρυγχόβατος· σελάχι· βάτος· ράσα
EN	sharpnose skate; sailray
FR	raie-voile
IT	razza bianca atlantica; razza
NL	witte rog
PT	raia nevoeira
FI	valkorausku
SV	vitrocka

92 RAJIDAE

SC	*Raja miraletus* (Linnacus, 1758)
ES	raya de espejos; levirraya
DA	spejlrokke
DE	Vieräugiger Spiegelrochen
EL	ματόβατος· σελάχι· βάτος· ράσα
EN	brown ray
FR	raie-miroir
IT	razza quattrocchi; razza
NL	spiegelrog
PT	raia de quatro olhos
FI	peilirausku
SV	spegelrocka

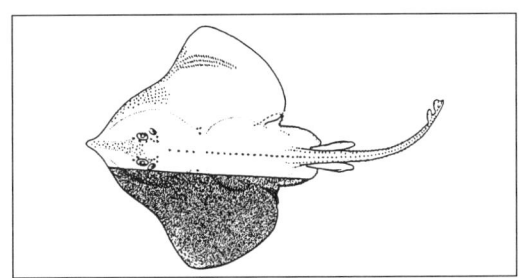

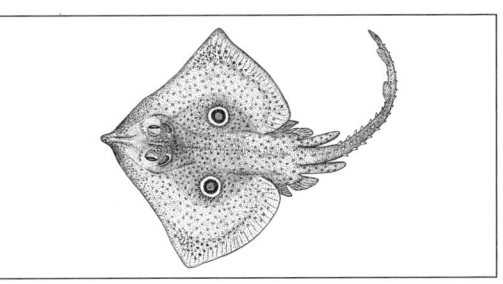

93 RAJIDAE RJM

SC	*Raja montagui* (Fowler, 1910)
ES	raya pintada; raya fina
DA	storplettet rokke
DE	Fleckrochen; Gefleckter Rochen
EL	σελάχι· βάτος· ράσα
EN	spotted ray; homelyn ray
FR	raie douce; raie de montague
IT	razza maculata
NL	gevlekte rog
PT	raia manchada
FI	pisterausku
SV	fläckrocka

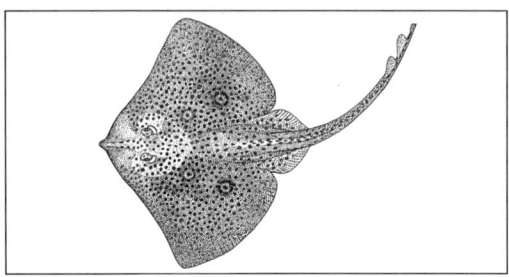

94 RAJIDAE RJN

SC	*Raja naevus* (Müller and Henle, 1841); *Leucoraja naevus*
ES	raya santiguesa; raya de San Pedro; raya basta
DA	pletrokke
DE	Kuckucksrochen
EL	σελάχι· βάτος· κούκος· ράσα
EN	cuckoo ray; butterfly skate
FR	raie fleurie
IT	razza fiorita; razza cuculo
NL	grootoogrog
PT	raia de dois olhos
FI	marmorirausku
SV	blomrocka; gökrocka

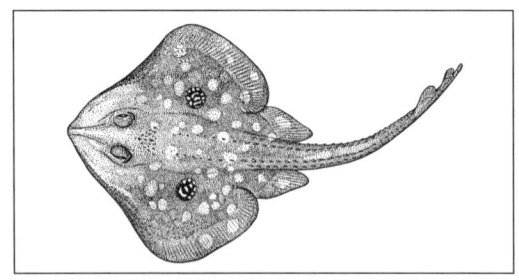

95 RAJIDAE

SC	*Raja ocellata* (Mitchill, 1815)
ES	raya manchada americana
DA	vinterrokke
DE	Winterrochen
EL	χειμωνιάτικος βάτος· χειμωνιάτικο σελάχι· χειμωνιάτικη ράσα
EN	winter skate; big skate; eyed skate
FR	raie ocellée
IT	razza occhiata; razza
NL	winterrog
PT	raia inverneira
FI	täplärausku
SV	vinterrocka

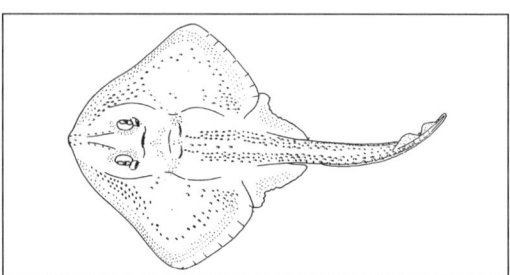

96 RAJIDAE RJO

SC	*Raja oxyrinchus* (Linnaeus, 1758); *Dipturus oxyrinchus*
ES	picón; raya picuda; raya aguda; rayón
DA	plovjernsrokke
DE	Spitzschnauzenrochen; Spitzrochen
EL	νονά· αετός· σελάχι· βάτος· ράσα
EN	long-nosed skate; long-nose skate
FR	pocheteau noir; raie-capucin
IT	razza monaca
NL	scherpsnuitrog
PT	raia bicuda
FI	vannasrausku
SV	plogjärnsrocka

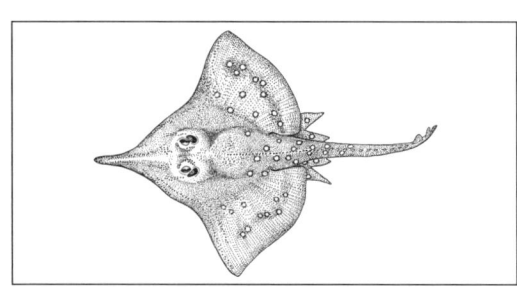

97 RAJIDAE

SC *Raja radiata* (Donovan, 1808); *Amblyraja radiata*
ES raya radiante; raya estrellada; raya radiada
DA tærbe
DE Atlantischer Sternrochen
EL ακτινόβατος· σελάχι· βάτος· ράσα
EN tarry skate; thorny skate; Atlantic prickly skate
FR raie radiée; raie épineuse
IT razza stellata
NL sterrog
PT raia repregada
FI kynsirausku
SV klorocka

99 RAJIDAE

SC *Bathyraja spinicauda* (Jensen, 1914)
ES raya de cola espinosa
DA tornhalet rokke
DE Grönlandrochen
EL σελάχι της Γροιλανδίας ·ράσα· σελάχι· βάτος
EN spiny-tail skate
FR raie à queue épineuse
IT razza
NL Groenlandse rog
PT raia da Gronelândia
FI rausku-laji
SV taggsvansrocka

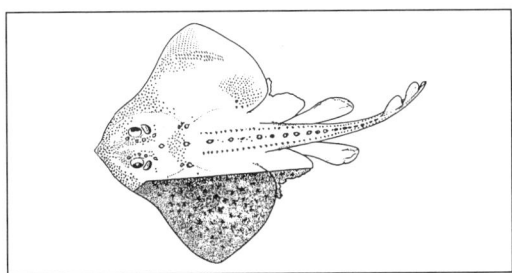

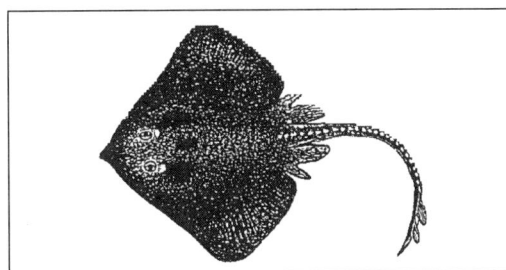

98 RAJIDAE

SC *Raja radula* (Delaroche, 1809)
ES raya áspera
DA ru rokke
DE Rauher Rochen
EL τραχιβάτος· σελάχι· βάτος· ράσα
EN rough ray
FR raie-rape
IT razza scuffina; razza spinosa
NL ruwe rog
PT raia áspera
FI rausku-laji
SV strävrocka

100 RAJIDAE

SC *Raja stellulata* (Jordan and Gilbert, 1880)
ES raya del Pacífico
DA Stillehavs-rokke
DE Pazifischer Rochen
EL βάτος του Ειρηνικού· σελάχι· βάτος· ράσα
EN starry skate
FR raie du Pacifique
IT razza
NL Pacifische rog
PT raia do Pacífico
FI rausku-laji
SV stillahavsrocka

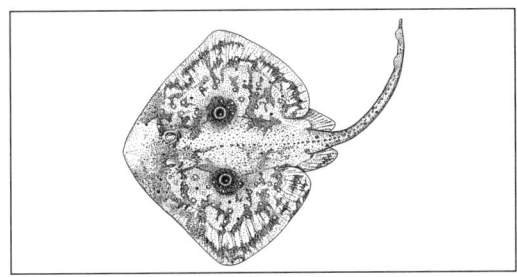

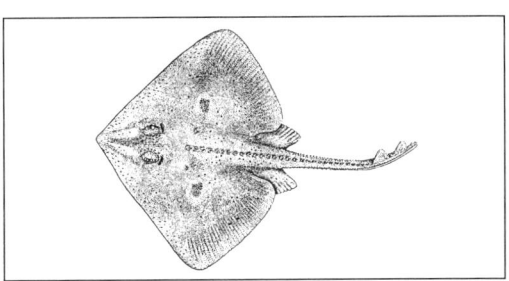

101 RAJIDAE

SC	*Raja undulata* (Lacépède, 1802)
ES	raya mosaica
DA	broget rokke
DE	Bänderrochen; Stekenrochen; Marmorrochen; Scheckenrochen
EL	κυματόβατος· σελάχι· βάτος· ράσα
EN	undulate ray; painted ray; marbled ray
FR	raie ondulée; raie brunette; raie mosaïque
IT	razza ondulata
NL	golfrog
PT	raia curva
FI	aaltorausku
SV	brokrocka

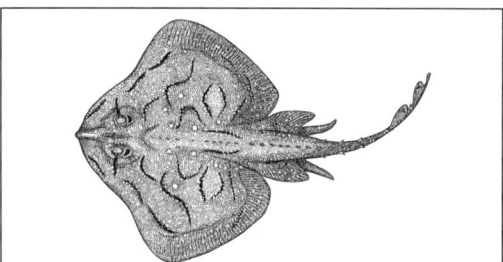

103 DASYATIDAE   STT

SC	*Dasyatidae*
ES	pastinacas; bastangas
DA	pigrokke-familien; pilrokke-familien
DE	Stechrochen; Stachelrochen
EL	τρυγόνες· σελάχια
EN	stingrays; butterfly rays
FR	pastenagues
IT	trigoni
NL	pijlstaartroggen
PT	uges
FI	keihäsrauskut; keihäsrauskut-heimo
SV	spjutrockor

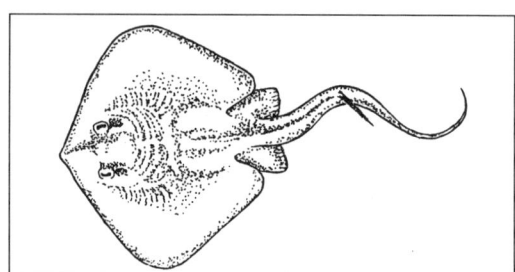

102 RAJIDAE   SKA

SC	*Raja* spp.
ES	rayas
DA	rokke-slægt
DE	Rochen im engeren Sinne
EL	σελάχια· βάτοι· ράσες
EN	skates
FR	raies; pocheteaux
IT	razze
NL	roggen
PT	raias
FI	rausku-suku
SV	rockor

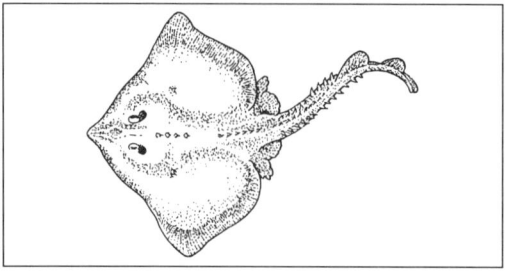

104 DASYATIDAE   WST

SC	*Dasyatis akajei* (Muller and Henle, 1841)
ES	raya látigo del Pacífico; pastinaca
DA	Stillehavs-pilrokke
DE	Pazifischer Stechrochen; Pazifischer Peitschenrochen
EL	τρυγόνα· σελάχι
EN	whip stingray; whip ray
FR	pastenague du Pacifique
IT	pastinaca; trigone; trigone del Pacifico
NL	Pacifische pijlstaartrog
PT	uge do Pacífico
FI	punakeihäsrausku
SV	stillahavsspjutrocka

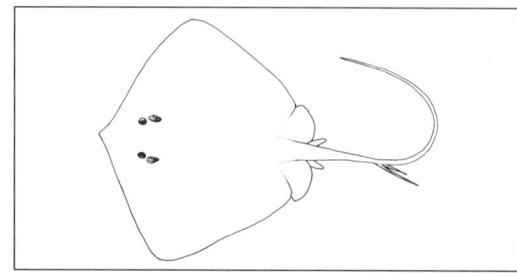

105 DASYATIDAE

SC	*Dasyatis centroura* (Mitchill, 1815)
ES	pastinaca espinosa
DA	ruhalet pilrokke
DE	Stachelschwanz-Stechrochen; Brucko
EL	τρυγόνα· σελάχι
EN	rough-tail stingray
FR	pastenague épineuse
IT	trigone spinoso
NL	stekelpijlstaartrog
PT	uge de cardas
FI	pohjankeihäsrausku
SV	taggspjutsrocka

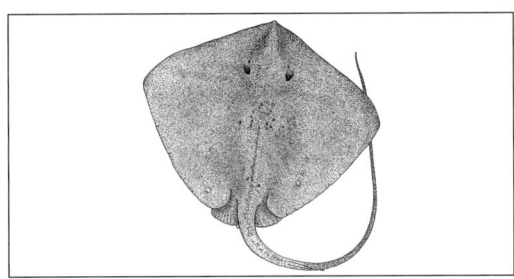

107 DASYATIDAE

SC	*Dasyatis violacea* (Bonaparte, 1832)
ES	chucho pelágico
DA	violet pilrokke
DE	Violetter Stechrochen
EL	μπλε τρυγόνα· σελάχι
EN	pelagic stingray
FR	pastenague violette
IT	trigone viola
NL	pelagische pijlstaartrog
PT	uge violeta
FI	sinikeihäsrausku
SV	violett spjutrocka

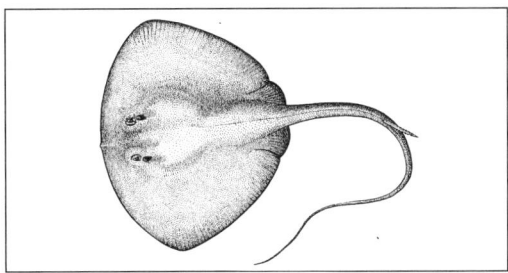

106 DASYATIDAE

SC	*Dasyatis pastinaca* (Linnaeus, 1758)
ES	pastinaca
DA	pilrokke; pigrokke; europæisk pilrokke
DE	Gewöhnlicher Stechrochen; Stechrochen
EL	παπλωματάς· σελάχι· τρυγόνα
EN	stingray
FR	pastenague commune; terre
IT	pastinaca
NL	pijlstaartrog
PT	uge
FI	keihäsrausku
SV	spjutrocka

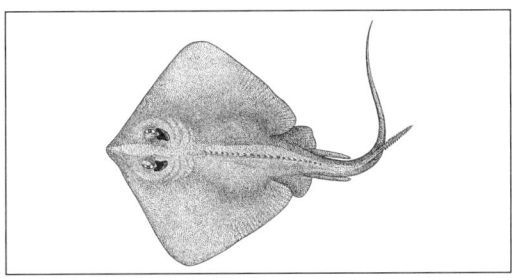

108 MYLIOBATIDAE   EAG

SC	*Myliobatidae*
ES	águilas de mar
DA	ørnerokke-familien
DE	Adlerrochen
EL	αετοί· αετόψαρα
EN	eagle rays
FR	aigles de mer
IT	aquile di mare
NL	vleermuisroggen
PT	ratões
FI	kotkarauskut; kotkarauskut-heimo
SV	örnrockor

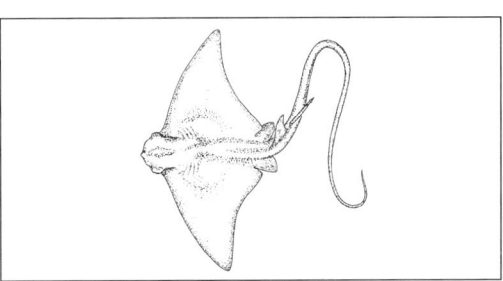

109 MYLIOBATIDAE

SC *Myliobatis aquila* (Linnaeus, 1758)
ES águila marina
DA almindelig ørnerokke; ørnerokke
DE Adlerrochen; Gewöhnlicher Adlerrochen
EL αετόψαρο· αετός· μαύρο αετόψαρο
EN eagle ray
FR aigle de mer; mourine
IT aquila di mare
NL arendsrog
PT ratão-águia
FI kotkarausku
SV örnrocka

111 MOBULIDAE   MAN

SC *Mobulidae*
ES mantas
DA djævlerokke-familien
DE Teufelsrochen; Mantas
EL διαβολόψαρα
EN mantas; devilfish; devil rays
FR mantes; diables de mer
IT mante; diavoli di mare
NL manta's
PT jamantas; diabos-do-mar
FI paholaisrauskut; paholaisrauskut-heimo
SV djävulsrockor

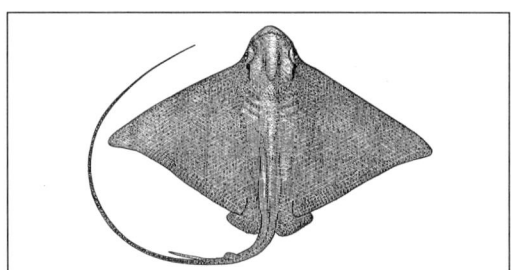

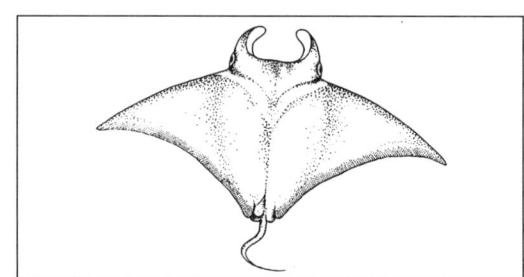

110 MYLIOBATIDAE

SC *Pteromylaeus bovinus* (Geoffroy St-Hilaire, 1817);
    Myliobatis bovina
ES pez obispo
DA afrikansk ørnerokke
DE Afrikanischer Adlerrochen
EL χελιδόνι
EN bull ray
FR mourine-vachette; aigle-vachette
IT vaccarella
NL bulrog
PT ratão-bispo
FI juovakotkarausku
SV korocka

112 MOBULIDAE

SC *Manta birostris* (Donndorff, 1798)
ES manta
DA atlantisk djævlerokke
DE Manta; Großer Teufelsrochen
EL διαβολόψαρο του Ατλαντικού
EN Atlantic manta
FR mante atlantique; raie manta
IT manta; diavolo di mare
NL manta
PT manta
FI paholaisrausku
SV manta

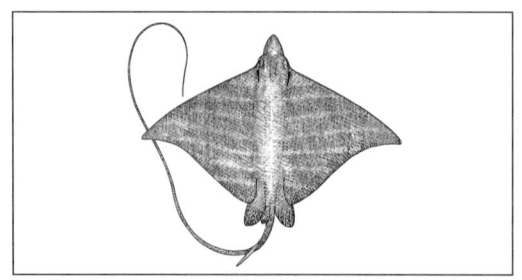

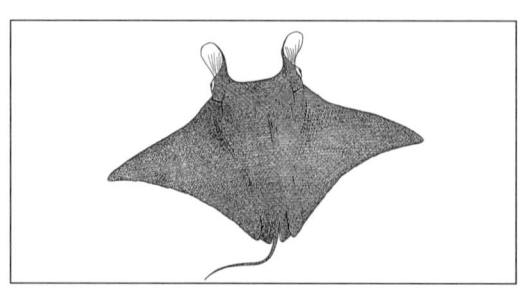

113 MOBULIDAE

SC	*Manta hamiltoni* (Newman, 1849)
ES	manta del Pacífico
DA	Stillehavs-djævlerokke
DE	Pazifischer Manta
EL	διάβολος του Ειρηνικού
EN	Pacific manta
FR	mante du Pacifique
IT	manta
NL	Pacifische manta
PT	manta do Pacífico
FI	tyynenmerenpaholaisrausku
SV	stillahavsmanta

115 GYMNURIDAE

SC	*Gymnura altavela* (Linnaeus, 1758); *Pteroplatea altavela*
ES	mantenilla
DA	sommerfuglerokke
DE	Schmetterlingsrochen
EL	πλατυσέλαχο· σελάχι· πεταλούδα
EN	spiny butterfly ray
FR	mourine bâtarde; pastenague ailée
IT	altavela
NL	vlinderrog
PT	uje-manta
FI	isoperhosrausku
SV	marmorerad fjärilsrocka

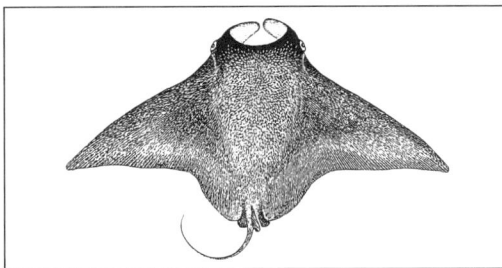

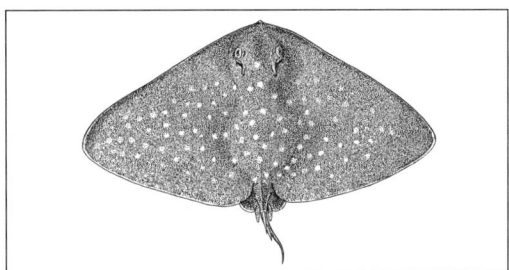

114 MOBULIDAE   DIA

SC	*Mobula mobular* (Bonnaterre, 1788), *Mobula olfersi*; *Mobula hypostoma*
ES	manta; vaca marina; diablo de mar
DA	lille djævlerokke
DE	Kleiner Teufelsrochen
EL	διαβολόψαρο· σελάχι· σελαχοκεφαλόψαρο
EN	devilfish
FR	diable de mer; raie cornue; mante méditerranéenne
IT	diavolo di mare
NL	kleine duivelsrog
PT	jamanta; diabo-do-mar
FI	sarvirausku
SV	mindre djävulsrocka

116 GYMNURIDAE   RBY

SC	*Gymnura* spp.
ES	rayas mariposa
DA	sommerfuglerokke-slægt
DE	Schmetterlingsrochen
EL	πλατυσέλαχο· πεταλούδα· σελάχι
EN	butterfly rays
FR	pastenagues ailées
IT	altavele
NL	vlinderroggen
PT	ujes-manta
FI	perhosrausku-suku
SV	fjärilsrockor

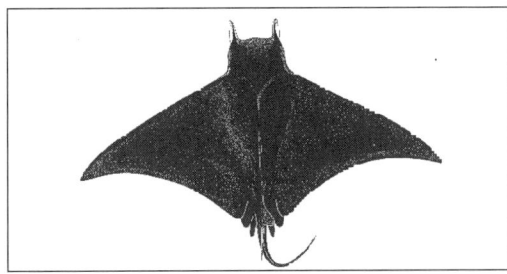

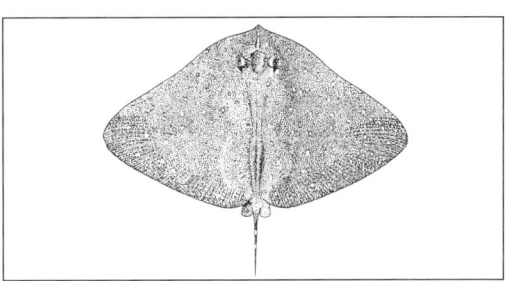

117 TORPEDINIDAE

SC	*Torpedinidae*
ES	torpedos; tembladeras; tremolinas; tremielgas
DA	elektriske rokker; el-rokker; el-rokke-familien
DE	Zitterrochen; Echte Zitterrochen
EL	μουδιάστρες
EN	electric rays; torpedo rays
FR	torpilles
IT	torpedini; tremoli
NL	sidderroggen
PT	tremelgas
FI	sähkörauskut; sähkörauskut-heimo
SV	darrockor

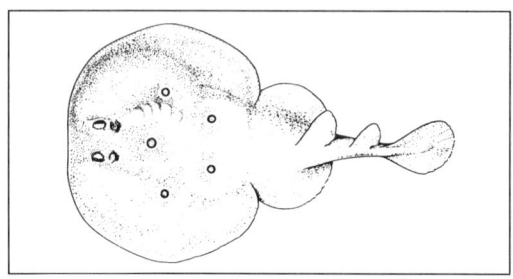

119 TORPEDINIDAE

SC	*Torpedo nobiliana* (Bonaparte, 1835)
ES	tremielga negra; torpedo; tembladera
DA	sort elektrisk rokke
DE	Schwarzer Zitterrochen
EL	μαυρομουδιάστρα· μουδιάστρα
EN	dark electric ray; Atlantic torpedo; dark torpedo
FR	torpille noire; raie électrique noire
IT	torpedine nera
NL	Atlantische sidderrog
PT	tremelga negra; tremedeira
FI	sysisähkörausku
SV	darrocka

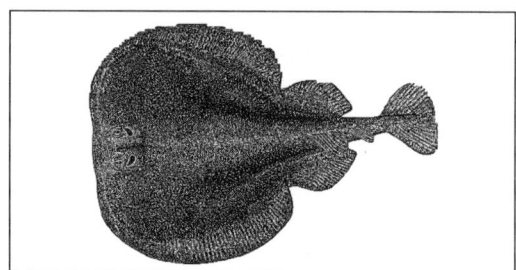

118 TORPEDINIDAE

SC	*Torpedo marmorata* (Risso, 1810)
ES	tremielga; torpedo; tembladera; temblón
DA	marmoreret elektrisk rokke
DE	Marmel-Zitterrochen; Marmor-Zitterrochen
EL	μαρμαρομουδιάστρα· μουδιάστρα
EN	marbled electric ray
FR	torpille marbrée; raie électrique marbrée; tremble
IT	torpedine marezzata
NL	gemarmerde sidderrog
PT	tremelga marmoreada; dormideira
FI	marmorisähkörausku
SV	marmorerad darrocka

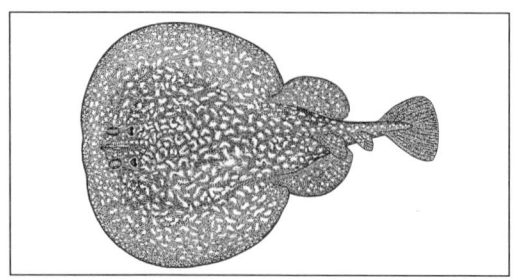

120 TORPEDINIDAE

SC	*Torpedo torpedo* (Linnaeus, 1758)
ES	tremielga
DA	almindelig elektrisk rokke
DE	Augenzitterrochen; Gefleckter Zitterrochen
EL	μαυρομουδιάστρα· μουδιάστρα
EN	common torpedo; eyed electric ray
FR	torpille tachetée; torpille ocellée
IT	torpedine occhiuta
NL	gevlekte sidderrog
PT	tremelga de olhos
FI	silmäsähkörausku
SV	ögondarrocka

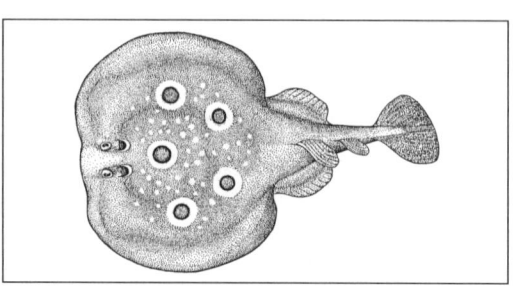

121 TORPEDINIDAE TOE

SC *Torpedo* spp.
ES torpedos; tembladeras; tremielgas
DA elektrisk rokke-slægt
DE Echte Zitterrochen
EL μουδιάστρα
EN torpedo rays
FR torpilles; raies électriques
IT torpedini
NL sidderroggen
PT tremelgas
FI sähkörausku-suku
SV darrockor

123 CHIMAERIDAE

SC *Chimaera monstrosa* (Linnaeus, 1758)
ES borrico; quimera
DA havmus
DE Seeratte; Spöke; Seekatze
EL χίμαιρα· γάτος
EN ratfish; rabbitfish
FR rat de mer; chimère commune
IT chimera
NL draakvis
PT ratazana
FI sillikuningas
SV havsmus

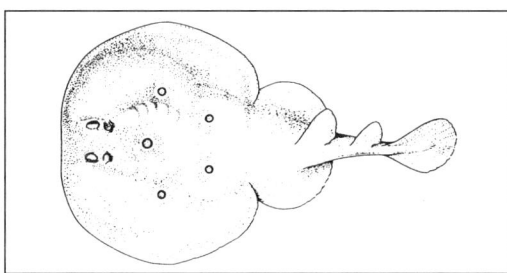

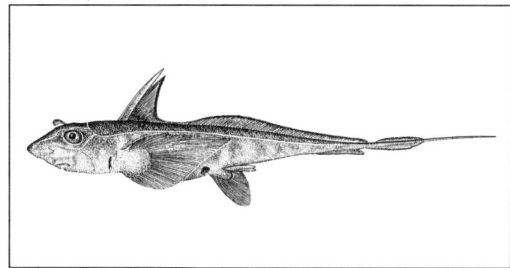

122 CHIMAERIDAE

SC *Chimaeridae*
ES quimeras; borricos; peces rata
DA kimærer; havmus-familien
DE Seeratten; Seekatzen; Chimären
EL χίμαιρες· γάτοι
EN ratfishes; rabbitfishes
FR chimères
IT chimere
NL draakvissen
PT ratazanas; peixes-rato
FI sillikuninkaat; sillikuninkaat-heimo
SV havsmusfiskar

124 CHIMAERIDAE RAT

SC *Hydrolagus colliei* (Lay and Bennett, 1839)
ES quimera americana
DA havrotte
DE Amerikanische Spöke; Chimäre; Amerikanische Chimäre
EL χίμαιρα
EN ratfish; spotted ratfish
FR chimère d'Amérique
IT chimera elefante
NL Amerikaanse draakvis
PT quimera americana; ratazana americana
FI amerikansillikuningas
SV vitfläckig havsmus

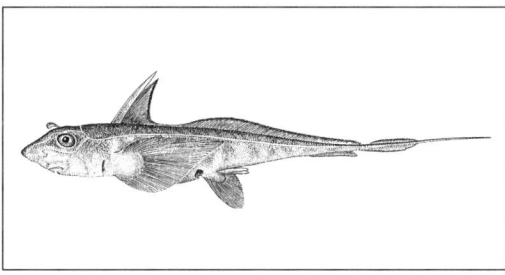

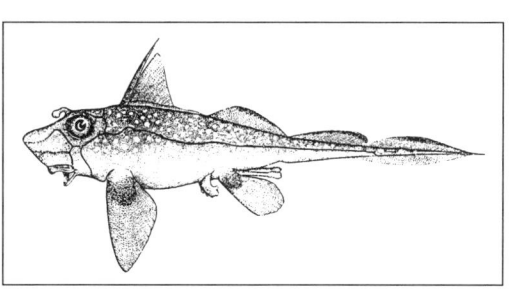

125 CALLORHYNCHIDAE   CHM

SC	*Callorhynchus capensis* (Dumaril, 1865)
ES	quimera del Cabo
DA	Kap-elefantfisk
DE	Totenkopfchimäre
EL	χίμαιρα του Ακρωτηρίου
EN	Cape elephantfish
FR	chimère du Cap
IT	chimera tapiro
NL	josef; Kaapse olifantzeerat
PT	peixe-elefante
FI	kärsäkuningas-laji
SV	kap plognos

127 CERATODONTIDAE

SC	*Neoceratodus forsteri* (Krefft, 1870)
ES	pulmonado australiano
DA	australsk lungefisk
DE	Australischer Lungenfisch
EL	δίπνοος της Αυστραλίας· δίπνοος
EN	Australian lungfish
FR	barramunda; poisson pulmoné
IT	neceratodo di Forster; dipnoo australiano
NL	Australische longvis
PT	peixe-casulo australiano
FI	australiankeuhkokala
SV	australisk lungfisk

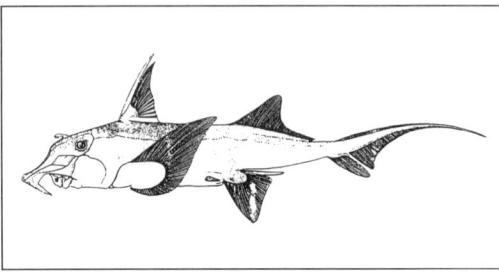

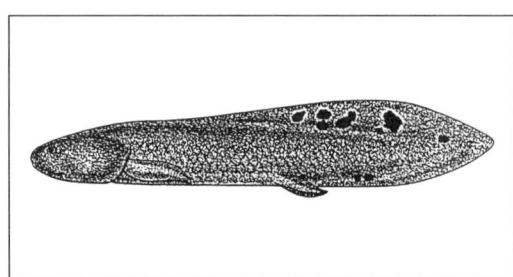

126 CALLORHYNCHIDAE   ELF

SC	*Callorhynchus* spp.
ES	quimeras
DA	elefantfisk-slægten
DE	Pflugnasenchimären
EL	χίμαιρα
EN	plownose chimaeras; elephantfishes
FR	chimères-éléphants; poissons-éléphants
IT	chimere tapiro
NL	olifantzeeratten
PT	peixes-elefante
FI	kärsäkuningas-suku
SV	plognosade havsmusfiskar

128 POLYPTERIDAE

SC	*Polypterus bichir* (Geoffroy St-Hilaire, 1802)
ES	bichir
DA	bikir
DE	Flösselhecht; Nilflösselhecht
EL	πολύπτερος· μπιχίρ
EN	bichir
FR	bichir
IT	bichir
NL	bichir
PT	peixe-manel
FI	niilinhauki
SV	bichir; nilfengädda

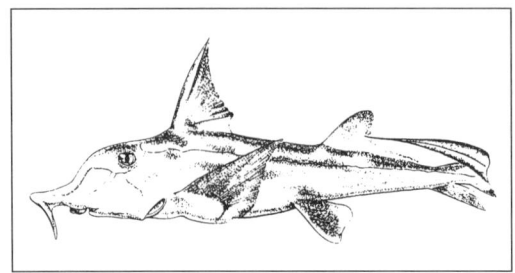

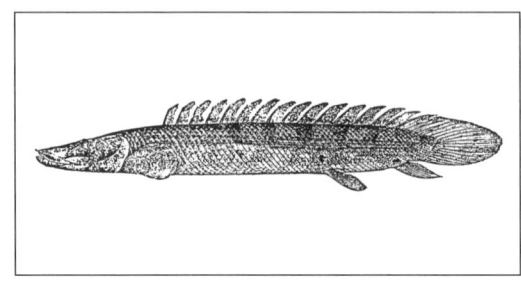

129 PROTOPTERIDAE   FLU

SC *Protopterus* spp.
ES protópteros
DA afrikansk lungefisk-slægt
DE Afrikanische Lungenfische
EL πρωτόπτεροι
EN African lungfishes
FR protoptères
IT prototteri
NL Afrikaanse longvissen
PT peixes-casulo africanos
FI afrikankeuhkokala-suku
SV afrikanska lungfiskar

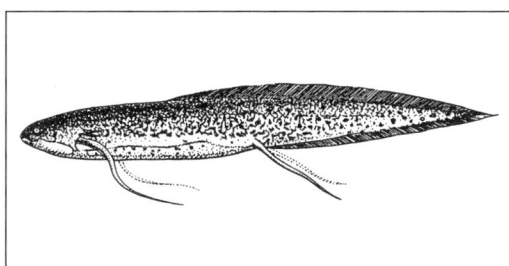

131 ACIPENSERIDAE   APG

SC *Acipenser gueldenstaedti* (Brandt, 1833)
ES esturión del Danubio
DA waxdick
DE Waxdick
EL στουριόνι του Δούναβη· μουρούνα του
    Δούναβη
EN osetr; Danube sturgeon
FR esturgeon russe; osetr
IT storione danubiano
NL Donausteur
PT esturjão do Danúbio
FI venäjänsampi
SV rysk stör

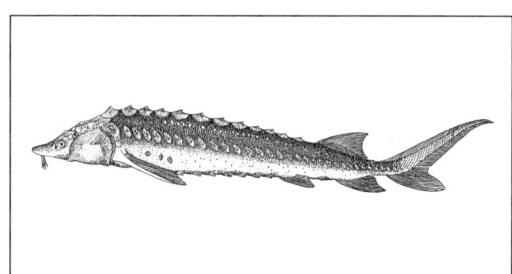

130 ACIPENSERIDAE   STU

SC *Acipenseridae*
ES esturiones
DA stør-familien
DE Störe
EL στουριόνι
EN sturgeons
FR esturgeons
IT acipenseridi, storioni
NL steuren
PT esturjões
FI sammet-heimo
SV störfiskar

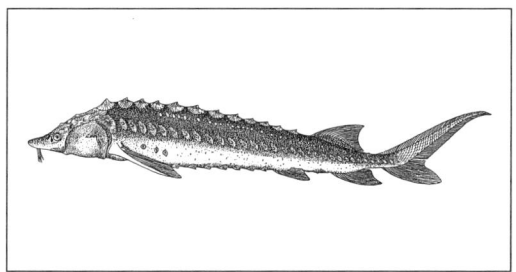

132 ACIPENSERIDAE

SC *Acipenser medirostris* (Ayres, 1854)
ES esturión verde
DA grøn stør
DE Grüner Stör
EL πράσινο στουριόνι· μουρούνα
EN green sturgeon
FR esturgeon vert
IT storione verde; storione del Pacifico
NL groene steur
PT esturjão verde
FI vihersampi
SV grön stör

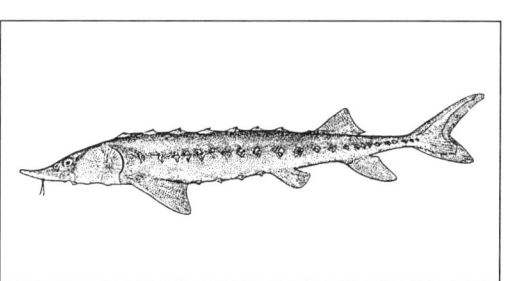

133 ACIPENSERIDAE

SC	*Acipenser naccarii* (Bonaparte, 1830)
ES	esturión adriático
DA	adriatisk stør
DE	Adriastör; Mittelmeerstör
EL	ξυρήχι· στουριόνι· μουρούνα
EN	Adriatic sturgeon
FR	esturgeon de Naccare; esturgeon de l'Adriatique
IT	cobice
NL	Adriatische steur
PT	esturjão adriático
FI	adriansampi
SV	adriatisk stör

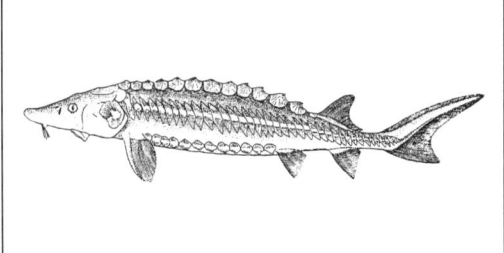

135 ACIPENSERIDAE APR

SC	*Acipenser ruthenus* (Linnaeus, 1758)
ES	esterlete
DA	sterlet
DE	Sterlet
EL	στουριόνι· στέρλετ
EN	sterlet; Siberian sterlet
FR	sterlet; esturgeon du Danube
IT	sterleto
NL	sterlet
PT	esturjão do Volga
FI	sterletti
SV	sterlett

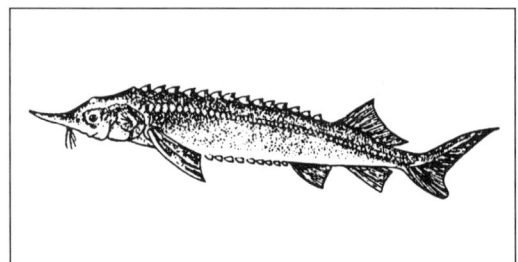

134 ACIPENSERIDAE

SC	*Acipenser nudiventris* (Lovetzky, 1828)
ES	esturión de vientre desnudo; esturión barba de flecos
DA	glatdick
DE	Glattdick; Dick
EL	στουριόνι· μουρούνα· γλαντίκ
EN	ship; fringe-barbel sturgeon
FR	esturgeon à ventre nu; glatdick; esturgeon à barbillons frangés
IT	glatdick
NL	glatdicksteur
PT	esturjão-ventre nu
FI	sampi-laji
SV	glattdick

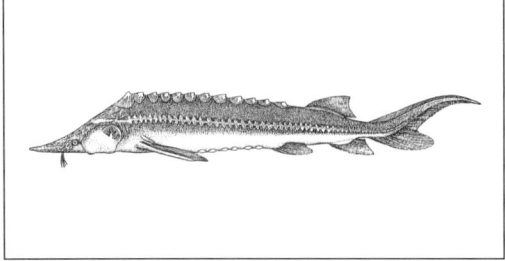

136 ACIPENSERIDAE

SC	*Acipenser stellatus* (Pallas, 1771)
ES	esturión estrellado
DA	stjernehus
DE	Sternhausen
EL	στουριόνι· αστροστούριονο
EN	sevruga; starry sturgeon
FR	sevruga; esturgeon étoilé
IT	storione stellato
NL	stersteur
PT	esturjão estrelado
FI	tähtisampi
SV	stjärnstör

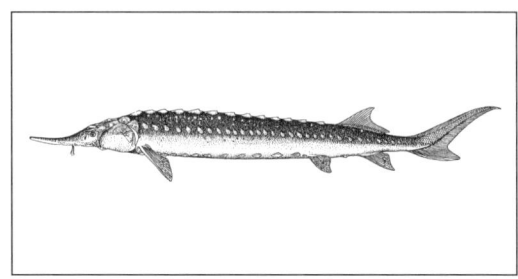

137 ACIPENSERIDAE

SC *Acipenser sturio* (Linnaeus, 1758)
ES esturión
DA almindelig stør
DE Baltischer Stör; Stör
EL ακιπίσιος· στουριόνι· ξυρήχι
EN sturgeon
FR esturgeon commun; esturgeon d'Europe
    occidentale
IT storione; storione comune
NL steur
PT esturjão
FI sampi
SV stör

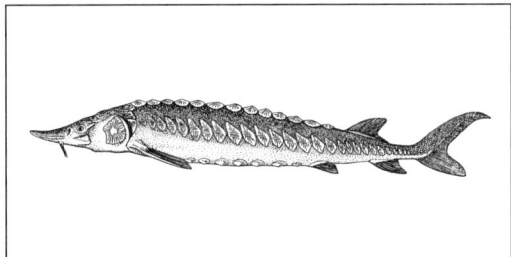

138 ACIPENSERIDAE

SC *Acipenser transmontanus* (Richardson, 1836)
ES esturión blanco
DA hvid stør
DE Weißer Stör
EL ασπροστούριονο
EN white sturgeon
FR esturgeon blanc
IT storione bianco
NL witte steur
PT esturjão branco
FI valkosampi
SV vit stör

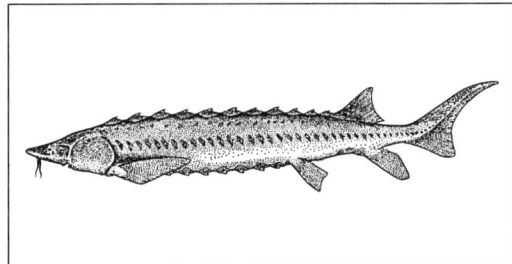

139 ACIPENSERIDAE

SC *Huso huso* (Linnaeus, 1758)
ES esturión beluga
DA hus
DE Europäischer Hausen; Hausen
EL μουρούνα· στουριόνι· μπελούγκα
EN beluga
FR grand esturgeon; béluga; huiron
IT storione ladano
NL huso; beluga
PT esturjão-beluga
FI kitasampi
SV hus; belugastör

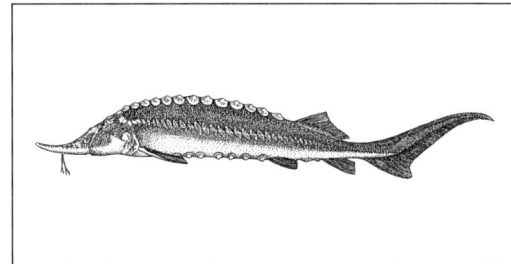

140 POLYODONTIDAE

SC *Polyodontidae*
ES peces espátula
DA spadestør-familien
DE Löffelstöre; Vielzähner; Schaufelrüssler
EL πολύδοντοι
EN spoonbills; paddlefishes
FR spatules
IT pesci spatola
NL lepelsteuren
PT peixes-espátula
FI lapasammet; lapasammet-heimo
SV skedstör

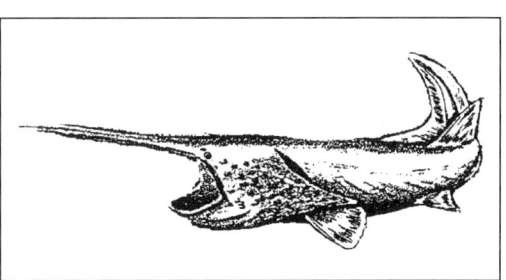

141 POLYODONTIDAE

SC	*Psephurus gladius* (Martens, 1861)
ES	pez espátula chino
DA	kinesisk spadestør
DE	Chinesischer Schwertstör; Schwertstör
EL	πολύδοντος της Κίνας
EN	Chinese paddlefish
FR	polyodon de Chine
IT	pesce spatola cinese
NL	Chinese lepelsteur
PT	peixe-espátula chinês
FI	miekkasampi
SV	svärdstör

143 LEPISOSTEIDAE

SC	*Lepisosteus osseus* (Linnaeus, 1758)
ES	lepisósteo óseo; gasper picudo
DA	langnæset pansergedde
DE	Schlanker Knochenhecht; Knochenhecht; Alligatorfisch
EL	λεπισόστεος
EN	long-nose gar
FR	brochet-lance à long nez; lépisostée osseux
IT	luccio alligatore
NL	langneus-beensnoek
PT	bicudo de focinho comprido
FI	luuhauki
SV	långnosad bengädda

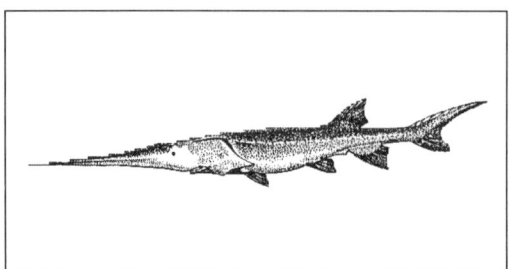

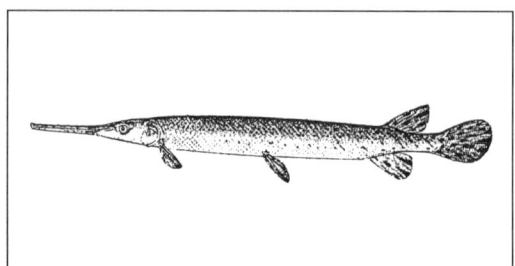

142 AMIIDAE

SC	*Amia calva* (Linnaeus, 1766)
ES	amia; alcaraz
DA	amia; dyndfisk
DE	Kahlhecht; Amerikanischer Schlammfisch
EL	άμια
EN	bowfin
FR	amie chauve
IT	amia
NL	moddersnoek
PT	peixe-castor
FI	amia
SV	bågfena

144 LEPISOSTEIDAE

SC	*Lepisosteus platystomus* (Rafinesque, 1820)
ES	lepisósteo chato; gaspar chato
DA	kortnæset pansergedde
DE	Kurznasen-Knochenhecht
EL	λεπισόστεος
EN	short-nose gar
FR	brochet-lance à nez plat; lépisostée à nez plat
IT	luccio alligatore
NL	kortneus-beensnoek
PT	bicudo de focinho curto
FI	pikkuluuhauki
SV	kortnosad bengädda

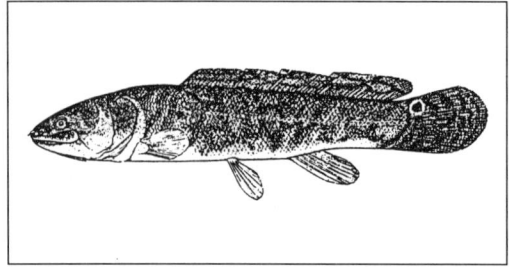

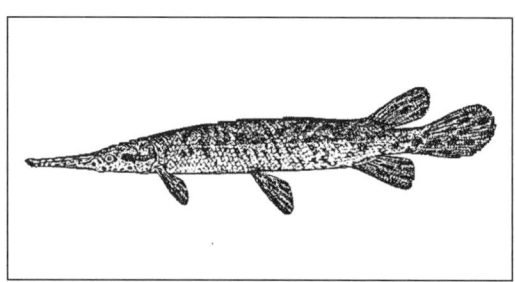

145 ELOPIDAE

SC	*Elopidae*
ES	tarpones; peces lagarto
DA	tipunder-familien
DE	Frauenfische; Tarpone
EL	ταρπόνια
EN	ten-pounders; tarpons
FR	grandes écailles; guinées
IT	tarponi
NL	tienponders
PT	elopídeos
FI	hopeakalat; hopeakalat-heimo
SV	tiopundare

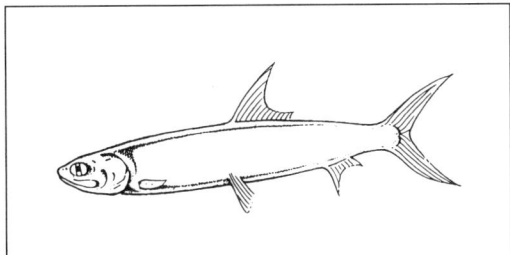

147 ELOPIDAE

SC	*Elops* spp.
ES	malachos
DA	tipunder-slægten
DE	Tarpone
EL	ταρπόνι
EN	tarpons
FR	guinées; bananes
IT	tarponi
NL	tienponders
PT	fateixas
FI	hopeakala-suku
SV	tiopundare

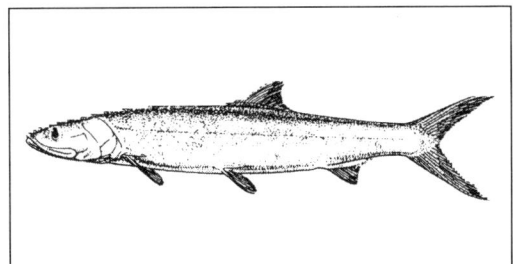

146 ELOPIDAE

SC	*Elops saurus* (Linnnaeus, 1766)
ES	malacho; banano; chiro
DA	tipunder
DE	Tarpon
EL	ταρπόνι· μπανανόψαρο
EN	ladyfish
FR	guinée machète; tarpon banane
IT	tarpone
NL	tienponder
PT	fateixa-torpedo; fateixa
FI	hopeakala
SV	ladyfisk

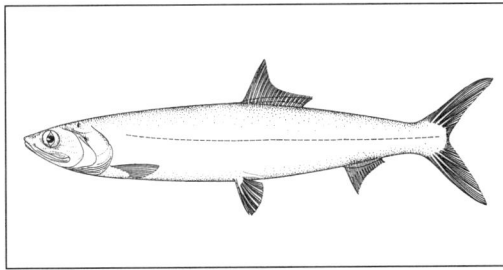

LAD 148 MEGALOPIDAE

SC	*Megalopidae*
ES	tarpones
DA	tarpon-familien
DE	Tarpune
EL	ταρπόνια
EN	tarpons
FR	tarpons
IT	tarponi
NL	tarpoenen
PT	tarpões
FI	tarponit; tarponit-heimo
SV	tarponfiskar

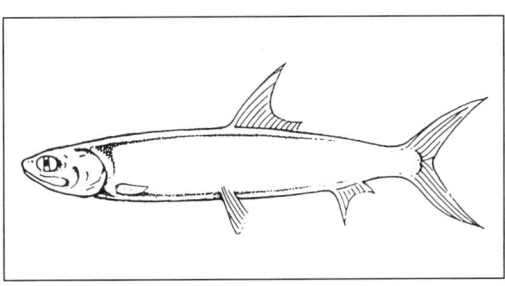

149 MEGALOPIDAE

TAI

SC	*Megalops cyprinoides* (Broussonet, 1782)
ES	tarpón Indo-Pacífico
DA	Stillehavs-tarpon
DE	Ochsenauge
EL	ινδικό ταρπόνι
EN	Indo-Pacific tarpon
FR	tarpon du Pacifique
IT	tarpone indiano
NL	ossenoog-tarpoen
PT	tarpão do Indo-Pacífico
FI	pikkutarponi
SV	stillahavstarpon

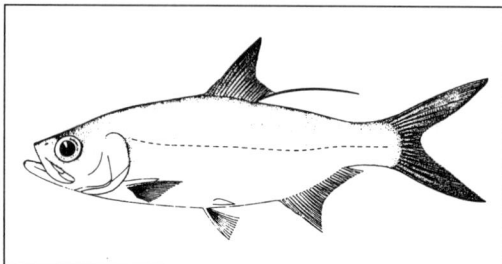

150 MEGALOPIDAE

TAR

SC	*Tarpon atlanticus* (Valenciennes, 1847); *Megalops atlanticus*
ES	tarpón; tarpón atlántico
DA	atlantisk tarpon
DE	Atlantischer Tarpun; Tarpun
EL	ταρπόνι
EN	tarpon
FR	tarpon de l'Atlantique; tarpon argenté; grande écaille
IT	tarpone
NL	tarpoen
PT	tarpão do Atlântico; tarpão; peixe-prata do Atlântico
FI	tarponi
SV	tarpon

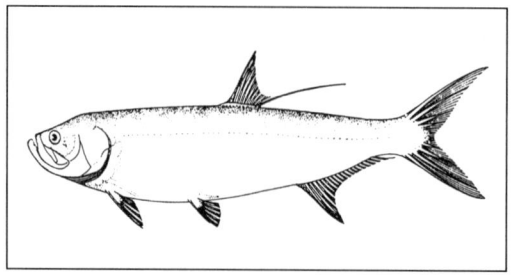

151 ALBULIDAE

ALU

SC	*Albulidae*
ES	peces banano
DA	albulid-familien
DE	Damenfische; Grätenfische
EL	αλβουλίδες
EN	bonefishes
FR	albules; bananes de mer
IT	tarponi
NL	gratenvissen
PT	flechas
FI	naiskalat; naiskalat-heimo
SV	beningar

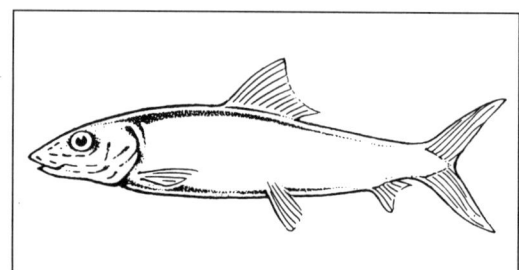

152 ALBULIDAE

BOF

SC	*Albula vulpes* (Linnaeus, 1758)
ES	macabijo; alburno
DA	damefisk
DE	Damenfisch; Grätenfisch
EL	μπανάνα της θάλασσας· ταρπόνι
EN	bonefish; ladyfish
FR	banane de mer; albule commun
IT	tarpone
NL	gratenvis
PT	flecha; peixe-banana
FI	naiskala
SV	bening

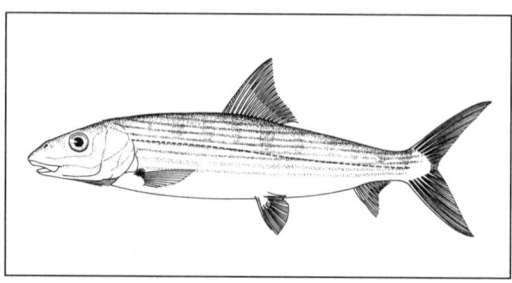

153	ALBULIDAE	BNF	155	CLUPEIDAE	BBH

153 ALBULIDAE BNF

SC	*Pterothrissus belloci* (Cadenat, 1937)
ES	macab badejo
DA	langfinnet damefisk
DE	Großflossen-Grätenfisch
EL	γκίσου
EN	long-fin bonefish
FR	banane gisu; banane à longues nageoires
IT	gisu africano
NL	grootvin-gratenvis
PT	falso badejo
FI	naiskala-laji
SV	långfensbening

155 CLUPEIDAE BBH

SC	*Alosa aestivalis* (Mitchill, 1814)
ES	sábalo del Canadá; alosa
DA	canadisk stamsild
DE	Kanadische Alse
EL	φρίσσα του καλοκαιριού
EN	blue-back herring; shad herring; blue-back shad
FR	alose d'été du Canada; alose d'été
IT	alosa canadese
NL	Canadese elft
PT	alosa azul
FI	sinisilli
SV	blå staksill

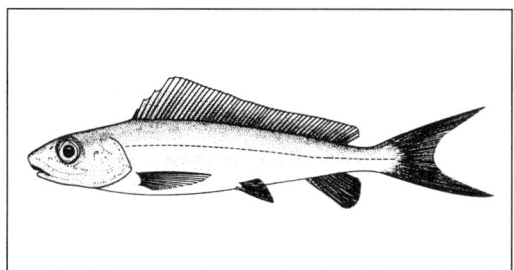

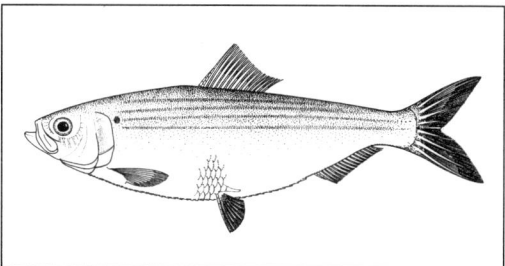

154 CLUPEIDAE CLP

SC	*Clupeidae*
ES	clupeidos
DA	silde-familien
DE	Heringe; Clupeiden
EL	κλουπεοειδή
EN	herrings; sardines
FR	clupéidés; clupes
IT	clupeidi; pesci azzurri
NL	haringen
PT	clupeídeos
FI	sillit; sillit-heimo
SV	sillfiskar

156 CLUPEIDAE ASD

SC	*Alosa alosa* (Linnaeus, 1758); *Alosa vulgaris*; *Clupea alosa*
ES	sábalo; sábalo común; trisa; arencón; alosa
DA	majsild
DE	Gewöhnliche Alse; Gewöhnlicher Maifisch; Alse; Maifisch; Alose
EL	φρίσσα· σαρδελομάνα
EN	allis shad; allice shad; alewife; rock herring
FR	alose vraie; alose; alose vulgaire; grande alose; poisson de mai; altacke
IT	alaccia; alosa
NL	elft
PT	sável
FI	pilkkusilli
SV	majfisk; majsill

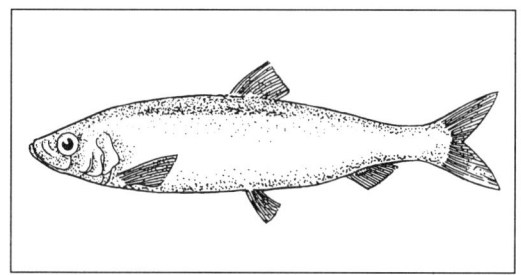

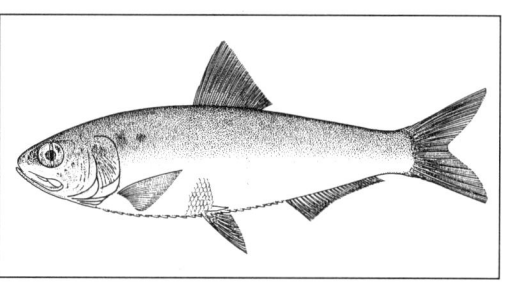

157	**CLUPEIDAE**	TSD

SC	*Alosa fallax* (Lacepède, 1803); *Alosa fallax fallax*; *Alosa finta*; *Clupea finta*; *Paralosa fallax*
ES	alosa; saboga
DA	stavsild
DE	Finte; Maifisch; Pergel; Elf; Staffhering; Elben
EL	φρίσσα· σαρδελομάνα
EN	twaite shad
FR	alose feinte; poisson de mai
IT	alosa; agone
NL	fint
PT	savelha; saboga
FI	täpläsilli
SV	staksill

159	**CLUPEIDAE**	SHH

SC	*Alosa mediocris* (Mitchill, 1814)
ES	alosa mediocre
DA	vestatlantisk stamsild
DE	Westatlantische Alse
EL	φρίσσα
EN	hickory shad
FR	alose médiocre; matowacca
IT	alosa
NL	West-Atlantische fint
PT	sável de salto
FI	hikkorisilli
SV	hickorysill

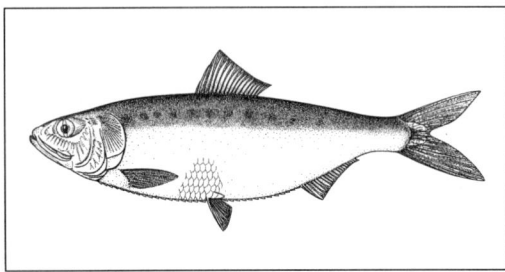

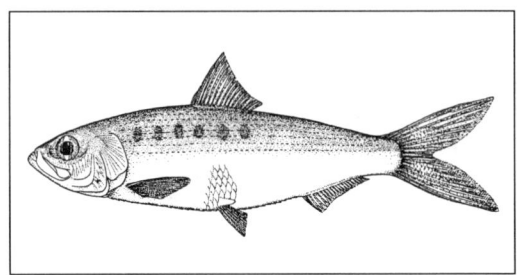

158	**CLUPEIDAE**	

SC	*Alosa fallax nilotica* (Lacépède, 1803)
ES	alosa mediterránea
DA	Nil-stavsild
DE	Mittelmeer-Finte
EL	φρίσσα· σαρδελομάνα
EN	Nile twaite shad
FR	alose feinte nilotique; alose feinte du Nil
IT	cheppia
NL	Middellandse-Zeefint
PT	savelha do Mediterrâneo
FI	niilintäpläsilli
SV	medelhavsstaksill

160	**CLUPEIDAE**	SHC

SC	*Alosa pontica* (Eichwald, 1838); *Caspialosa pontica*
ES	sábalo del Mar Negro
DA	kaspisk stamsild
DE	Donauhering
EL	φρίσσα του Πόντου
EN	Black Sea shad; Pontic shad
FR	alose de la mer noire
IT	alosa del Mar Nero
NL	Zwarte-Zee-elft
PT	sável do mar Negro
FI	mustanmerensilli
SV	svartahavsstaksill

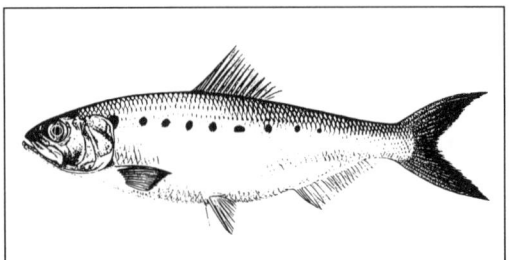

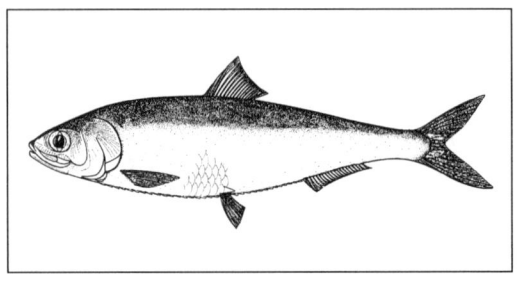

161 CLUPEIDAE ALE

SC *Alosa pseudoharengus* (Wilson, 1811); *Pomolobus pseudoharengus*
ES pinchagua
DA flodsild
DE Nordamerikanischer Flußhering
EL ποταμόρεγγα
EN alewife; river herring
FR gasparot; gaspareau
IT falsa aringa atlantica
NL rivierharing
PT alosa cinzenta
FI harmaasilli
SV gumsill

163 CLUPEIDAE SHZ

SC *Alosa* spp.
ES sábalos comunes; sabogas
DA stamsild-slægten
DE Maifische
EL φρίσσα· σαρδελομάνα
EN allis shads; twaite shads; shads
FR aloses feintes; aloses vraies
IT alose
NL elften; finten
PT sáveis e savelhas
FI kantasilli-suku
SV staksillar

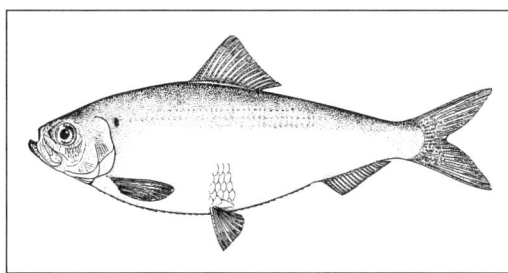

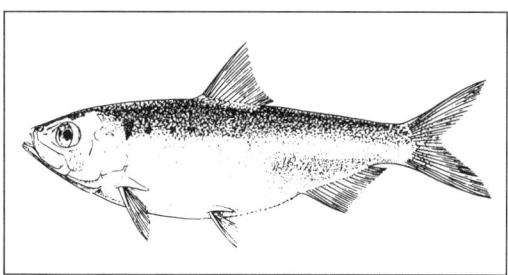

162 CLUPEIDAE SHA

SC *Alosa sapidissima* (Wilson, 1811)
ES sábalo americano
DA amerikansk stamsild
DE Amerikanischer Maifisch; Amerikanische Alse
EL τριχιός της Αμερικής
EN American shad
FR alose savoureuse; alose canadienne
IT alaccia americana
NL Amerikaanse elft
PT sável americano
FI amerikankantasilli
SV vit staksill

164 CLUPEIDAE CHG

SC *Anodontostoma chacunda* (Hamilton-Buchanan, 1822)
ES sábalo chacunda
DA chacunda-kråsesild
DE Chacunda
EL χακούντα
EN chacunda gizzard shad
FR alose chacunda
IT ciacunda
NL chacunda-elft
PT sável chacunda
FI kupusilli-laji
SV chacunda

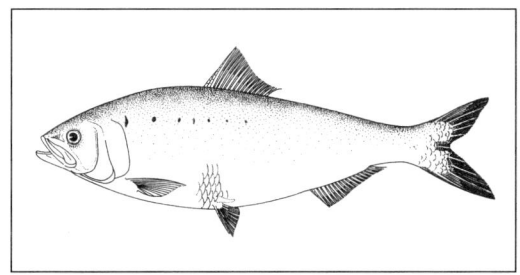

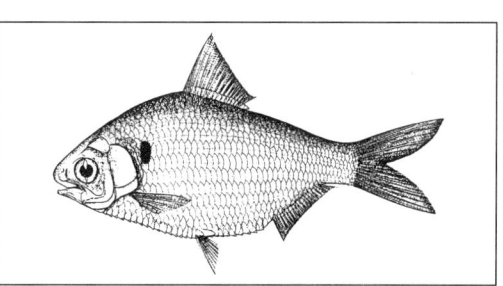

165 CLUPEIDAE

MHG

SC	*Brevoortia patronus* (Goode, 1878)
ES	lacha escamuda
DA	Gulf-menhaden
DE	Golf-Menhaden
EL	λεπιδοφόρα φρίσσα
EN	Gulf menhaden; large scale menhaden; bunker; pogy; mossbunker; shad
FR	menhaden écailleux
IT	menhaden messicana
NL	Mexicaanse menhaden
PT	menhadem escamudo
FI	lahtimenhaden
SV	gulfmenhaden

167 CLUPEIDAE

SC	*Brevoortia* spp.
ES	lachas
DA	menhaden-slægten
DE	Menhaden
EL	μενάχεν
EN	menhadens
FR	menhaden
IT	menhaden
NL	menhaden
PT	menhadens
FI	menhaden-suku
SV	menhaden

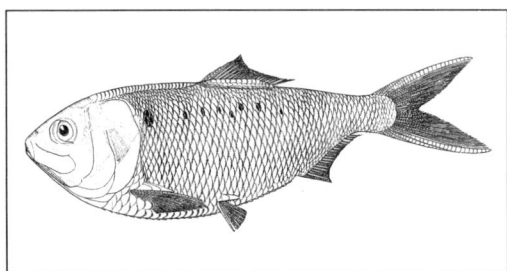

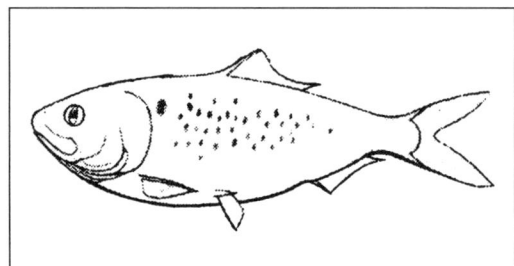

166 CLUPEIDAE

MHA

SC	*Brevoortia tyrannus* (Latrobe, 1802)
ES	lacha tirana
DA	atlantisk menhaden
DE	Nordwestatlantischer Menhaden; Menhaden; Bunker
EL	φρίσσα του Ατλαντικού
EN	Atlantic menhaden; menhaden; shad; bunker; pogy; mossbunker
FR	menhaden tyran; menhaden
IT	alaccia americana; menhaden
NL	menhaden
PT	menhadem
FI	menhaden
SV	atlantisk menhaden

168 CLUPEIDAE

DAS

SC	*Clupanodon thrissa* (Linnaeus, 1758)
ES	alosa chata
DA	kinesisk kråsesild
DE	Chinesische Alse
EL	κοντόρυγχη φρίσσα
EN	Chinese gizzard shad
FR	alose à museau court
IT	alosa cinese
NL	Chinese elft
PT	sável chato
FI	kiinankupusilli
SV	vimpelsill

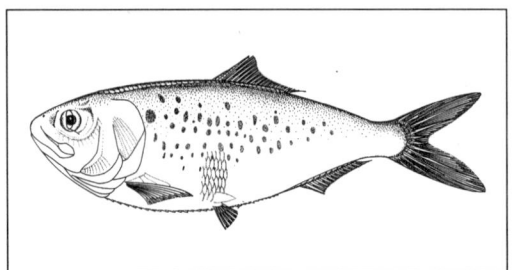

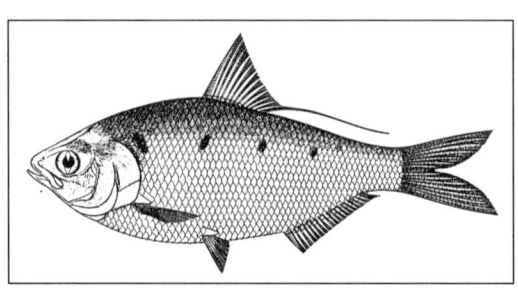

169 CLUPEIDAE HER

SC   *Clupea harengus* (Linnaeus, 1758)
ES   arenque
DA   sild; atlantisk sild
DE   Atlantischer Hering; Hering
EL   ρέγγα
EN   Atlantic herring; herring; digby; mattie; sild; yawling; sea herring
FR   hareng de l'Atlantique; hareng commun
IT   aringa
NL   haring
PT   arenque
FI   silli
SV   sill; strömming

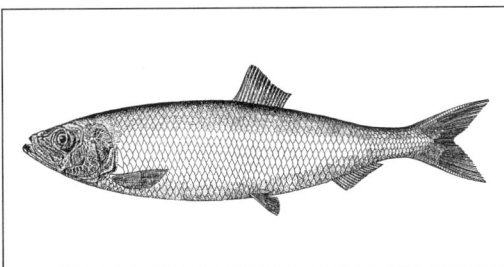

170 CLUPEIDAE HEP

SC   *Clupea pallasii* (Valenciennes, 1847); *Clupea harengus pallasi*
ES   arenque del Pacífico
DA   Stillehavs-sild
DE   Pazifischer Hering
EL   ρέγγα του Ειρηνικού
EN   Pacific herring
FR   hareng du Pacifique
IT   aringa; aringa del Pacifico
NL   Pacifische haring
PT   arenque do Pacífico
FI   tyynenmerensilli
SV   stillahavssill

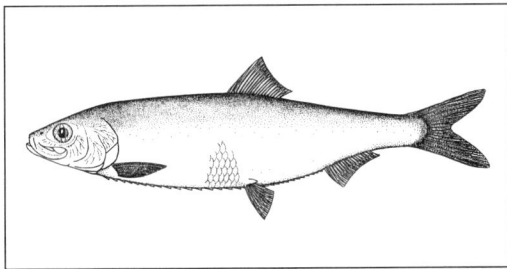

171 CLUPEIDAE CLA

SC   *Clupeonella cultriventris* (Nordmann, 1840); *Clupeonella delicatula*
ES   clupeonella; espadilla del Caspio
DA   kaspisk sild
DE   Kilka; Tyulka-Sardelle
EL   παπαλίνα της Μαύρης Θάλασσας· κίλκα
EN   clupeonella; kilka; Black Sea sprat; Tyulka sprat; Azovtyulka
FR   clupéonelle; kilka
IT   papalina del Caspio
NL   kilka
PT   espadilha do mar Negro
FI   silli-laji
SV   kaspisk skarpsill

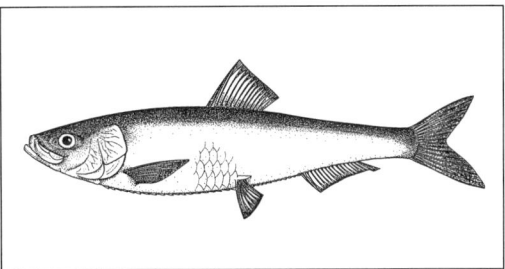

172 CLUPEIDAE SHG

SC   *Dorosoma cepedianum* (Lesueur, 1818)
ES   sábalo molleja
DA   amerikansk kråsesild
DE   Fadenflossige Alse
EL   φρίσσα της Αμερικής
EN   gizzard shad; American gizzard shad
FR   alose-noyer; alose à gésier; alose américaine
IT   alosa americana
NL   draadvinnige elft
PT   sável de papo
FI   kupusilli
SV   vimpelsill

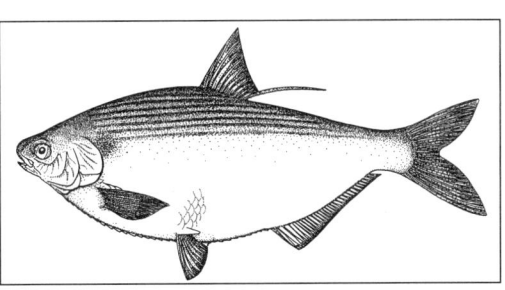

SC   *Dussumieria acuta* (Valenciennes, 1847);
      Dussumieria productissima
ES   sardina arco iris
DA   regnbuesardin
DE   Regenbogen-Rundhering
EL   σαρδέλα της Ανατολής
EN   round herring; rainbow sardine; slender rainbow
      sardine
FR   sardine arc-en-ciel
IT   dussumeria
NL   regenboog-rondeharing
PT   oió
FI   kirjosardiini
SV   regnbågssardin

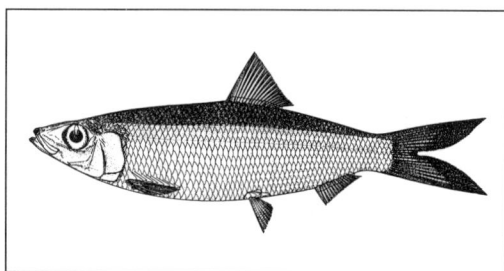

SC   *Ethmalosa fimbriata* (Bowdich, 1825)
ES   sábalo africano
DA   bonga-stamsild
DE   Bonga-Hering
EL   γοπόρεγγα
EN   bonga shad
FR   ethmalose d'Afrique; sardinelle jaune
IT   alaccia larga
NL   bonga-elft
PT   galucha; quilucha
FI   kantasilli-laji
SV   bongasill

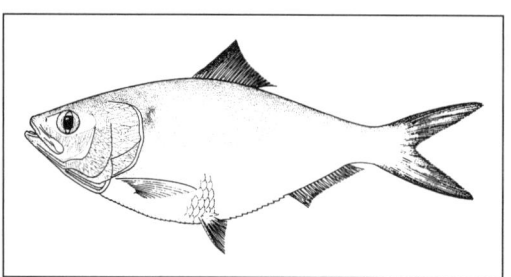

SC   *Ethmidium maculatum* (Valenciennes, 1847)
ES   machete
DA   Stillehavs-menhaden
DE   Pazifischer Menhaden
EL   φρίσσα της Χιλής
EN   Pacific menhaden
FR   menhaden du Pacifique Sud-Est
IT   alaccia cilena
NL   Pacifische menhaden
PT   menhadem do Pacífico Sudeste
FI   tyynenmerenmenhaden
SV   stillahavsmenhaden

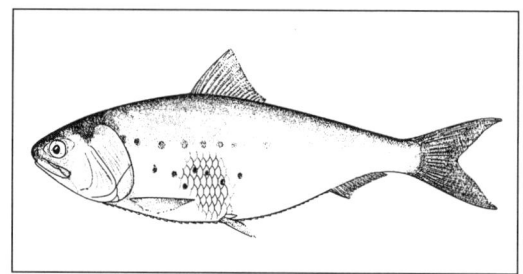

SC   *Etrumeus acuminatus* (Gilbert, 1891)
ES   arenque redondo de California
DA   californisk rundsild
DE   Kalifornischer Rundhering
EL   στρογγυλόρεγγα της Καλιφόρνιας
EN   Californian round herring
FR   shadine de Californie
IT   aringa tonda di California
NL   Californische ronde haring
PT   arenque redondo da Califórnia
FI   kalifornianpyörösilli
SV   kalifornisk rundsill

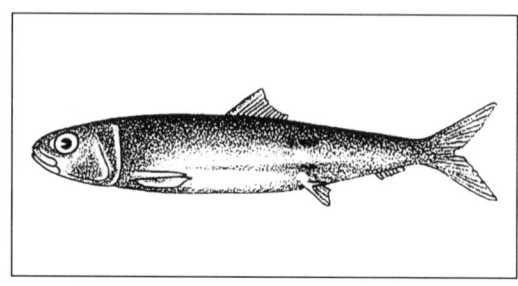

177 CLUPEIDAE

SC	*Etrumeus sadina* (Mitchill, 1814)
ES	arenque redondo atlántico
DA	atlantisk rundsild
DE	Atlantischer Rundhering
EL	στρογγυλόρεγγα του Ατλαντικού
EN	Atlantic round herring
FR	shadine de l'Atlantique
IT	aringa tonda dell'Atlantico
NL	Atlantische ronde haring
PT	arenque redondo do Atlântico
FI	atlantinpyörösilli
SV	atlantisk rundsill

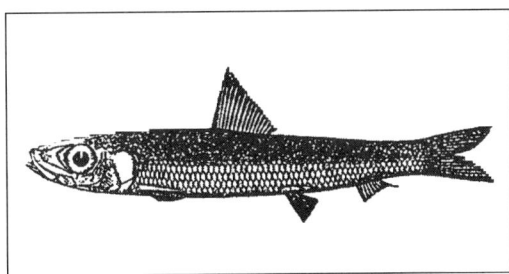

179 CLUPEIDAE   EMX

SC	*Etrumeus* spp.
ES	arenques redondos
DA	rundsild-slægten
DE	Rundheringe
EL	στρογγυλόρεγγες
EN	round herrings; big-eye herrings
FR	shadines
IT	aringhe tonde
NL	ronde haringen
PT	arenques redondos
FI	pyörösilli-suku
SV	rundsillar

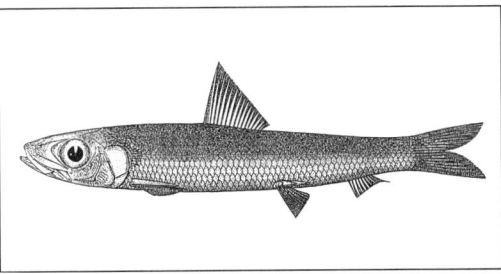

178 CLUPEIDAE   RRH

SC	*Etrumeus teres* (De Kay, 1842)
ES	sardineta canalera
DA	rødøjet rundsild
DE	Gemeiner Rundhering
EL	στρογγυλόρεγγα
EN	round herring; red-eye round herring
FR	shadine à yeux rouges
IT	aringa tonda
NL	ronde haring
PT	arenque redondo
FI	punasilmäsilli
SV	rundsill

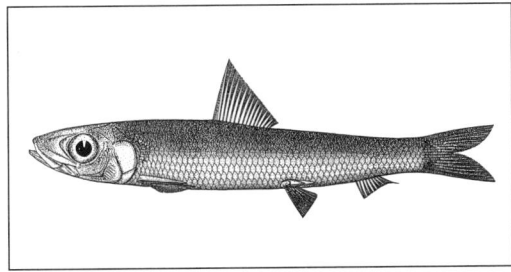

180 CLUPEIDAE   SAS

SC	*Harengula* spp.
ES	sardinetas
DA	silde-slægt
DE	Kleinheringe
EL	σαρδέλες με λέπια
EN	scaled sardines
FR	harengules
IT	alacce centramericane
NL	kleinharingen
PT	sardinetas
FI	valesardiini-suku
SV	sillingar

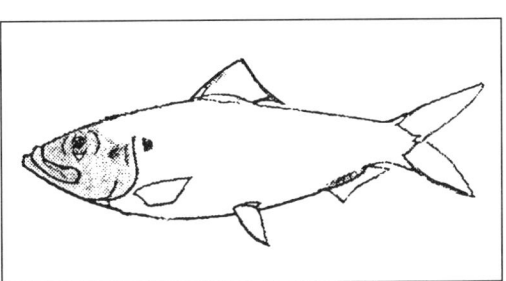

181 CLUPEIDAE

SC	*Herklotsichthys punctatus* (Rüppell, 1837)
ES	arenque punteado
DA	plettet sild
DE	Rotmeer-Hering
EL	στικτόρεγγα
EN	spotted herring
FR	hareng tacheté
IT	sardelletta macchiata
NL	gevlekte haring
PT	arenque pontuado
FI	silli-laji
SV	rödahavssill

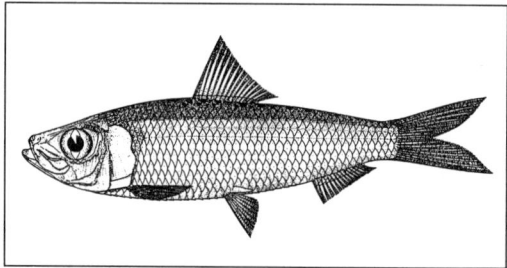

183 CLUPEIDAE   ILI

SC	*Ilisha africana* (Bloch, 1795)
ES	sardineta africana
DA	vestafrikansk stamsild
DE	Westafrikanische Ilisha
EL	σαρδέλα της Δυτικής Αφρικής
EN	West African ilisha
FR	alose-rasoir
IT	ilissa africana
NL	West-Afrikaanse ilisha
PT	capasseca
FI	afrikanhilsa
SV	afrikansk buksill

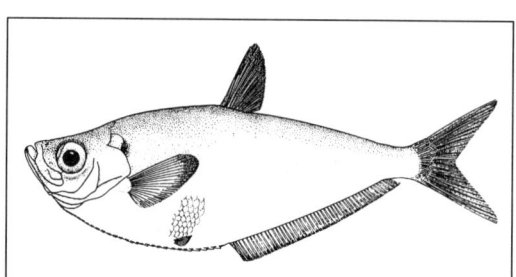

182 CLUPEIDAE   HIX

SC	*Hilsa kelee* (Cuvier, 1829)
ES	sábalo kelee
DA	kelee-stamsild
DE	Kelee-Alse
EL	φρίσσα-κέλε
EN	kelee shad
FR	alose kelee
IT	alosa kelee
NL	keleeharing
PT	pala chata
FI	hilsa
SV	kelee staksill

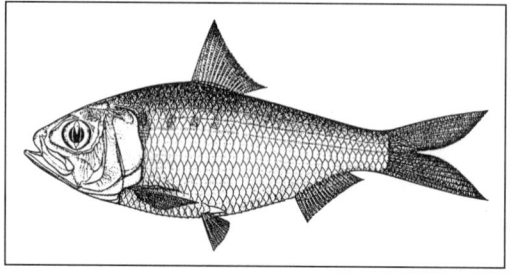

184 CLUPEIDAE   EIL

SC	*Ilisha elongata* (Bennett, 1830)
ES	ilisa
DA	slank stamsild
DE	Schlank-Ilisha
EL	μακρόσωμη σαρδέλα
EN	elongate ilisha
FR	alose gracile
IT	ilissa maggiore
NL	slanke ilisha
PT	choupaque
FI	hilsa-laji
SV	kinesisk buksill

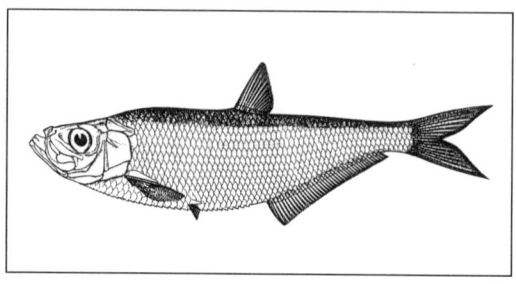

185 CLUPEIDAE THP

SC	*Opisthonema libertate* (Günther, 1867)
ES	machuelo del Pacífico; arenque de hebra
DA	Stillehavs-trådsild
DE	Pazifischer Fadenhering
EL	νηματόρεγγα του Ειρηνικού
EN	Pacific thread herring; deep-body thread herring
FR	chardin du Pacifique; faux hareng du Pacifique
IT	alaccia vessillifera
NL	Pacifische draadvinnige haring
PT	machete do Pacífico; plumuda do Pacífico
FI	tyynenmerensiimasilli
SV	stillahavstrådsill

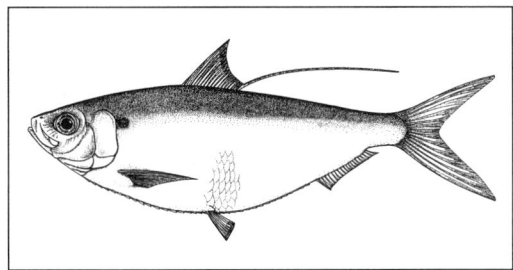

187 CLUPEIDAE THX

SC	*Opisthonema* spp.
ES	falsos arenques
DA	trådsild-slægt
DE	Fadenheringe
EL	ρεγγάκια
EN	thread herrings
FR	faux harengs
IT	alacce vessillifere
NL	draadvinnige haringen
PT	machetes; plumudas
FI	siimasilli-suku
SV	trådsillar

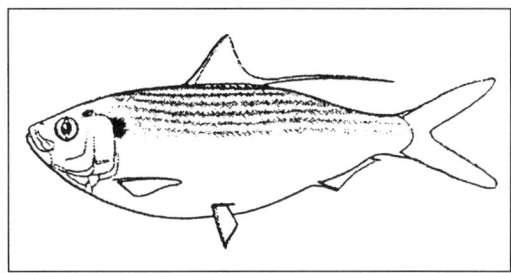

186 CLUPEIDAE THA

SC	*Opisthonema oglinum* (Lesueur, 1818)
ES	machuelo hebra atlántico
DA	atlantisk trådsild
DE	Atlantischer Fadenhering
EL	νηματόρεγγα του Ατλαντικού
EN	Atlantic thread herring
FR	chardin; faux hareng de l'Atlantique
IT	alaccia vessillifera
NL	Atlantische draadvinnige haring
PT	machete do Atlântico; plumuda do Atlântico
FI	siimasilli
SV	atlantisk trådsill

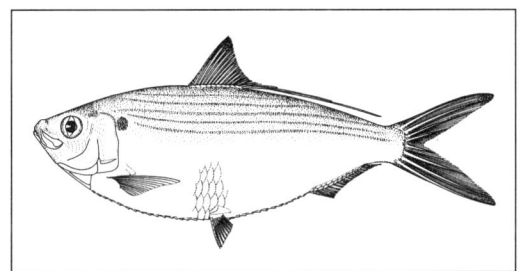

188 CLUPEIDAE PEO

SC	*Pellona ditchela* (Valenciennes, 1847)
ES	sardineta índica
DA	pellona-sild
DE	Indische Pellona
EL	πελόνα της Ινδίας
EN	Indian pellona
FR	alose-caille indienne
IT	pellona indiana
NL	Indische pellona
PT	sardinata índica
FI	silli-laji
SV	indisk buksill

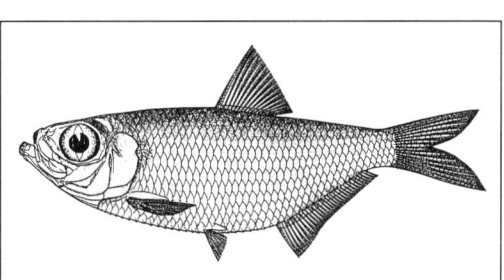

| **189** | **CLUPEIDAE** | PIL | **191** | **CLUPEIDAE** | FRS |

SC	*Sardina pilchardus* (Walbaum, 1792); *Clupea pilchardus*; *Alosa pilchardus*
ES	sardina; sardina europea; parrocha
DA	europæisk sardin; sardin
DE	Pilchard; Sardine
EL	σαρδέλα
EN	European sardine; sardine; pilchard; European pilchard
FR	sardine européenne; sardine; sardine commune
IT	sardina; sardella
NL	sardien; pelser
PT	sardinha; sardinha europeia
FI	sardiini
SV	sardin

SC	*Sardinella fimbriata* (Valenciennes, 1847)
ES	sardinela
DA	asiatisk sardinel
DE	Asiatische Kleinsardine
EL	τριχιός· φρίσσα
EN	fringe-scale sardinella
FR	ethmalose; sardinelle jaune
IT	alaccia
NL	Aziatische sardinella
PT	sardinela franjada
FI	sardiini-laji
SV	bengalisk sardinell

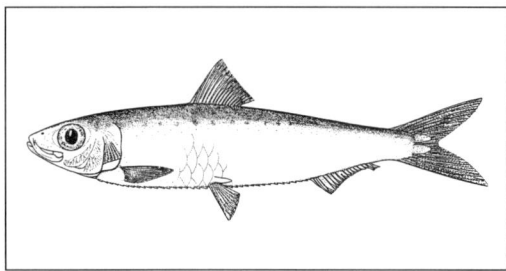

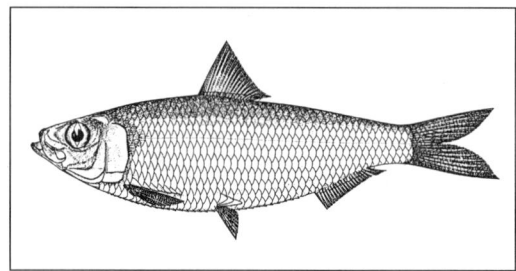

| **190** | **CLUPEIDAE** | SAA | **192** | **CLUPEIDAE** | IOS |

SC	*Sardinella aurita* (Valenciennes, 1847)
ES	alacha; sardina de ley
DA	rund sardinel
DE	Ohrensardine; Sardinelle
EL	τριχιός· φρίσσα
EN	round sardinella; gilt sardine; sardinella; alache
FR	sardinelle ronde; allache
IT	alaccia; sardella d'Africa
NL	gouden sardinella
PT	sardinela lombuda
FI	pyörösardiini
SV	rund sardinell

SC	*Sardinella longiceps* (Valenciennes, 1847)
ES	sardinela de la India
DA	storhovedet sardinel
DE	Großkopfsardine
EL	τριχιός της Ινδίας
EN	Indian oil sardine; oil sardine
FR	sardinelle des Indes
IT	alaccia
NL	Indische sardinella
PT	sardinela da Índia
FI	sardiini-laji
SV	indisk sardinell

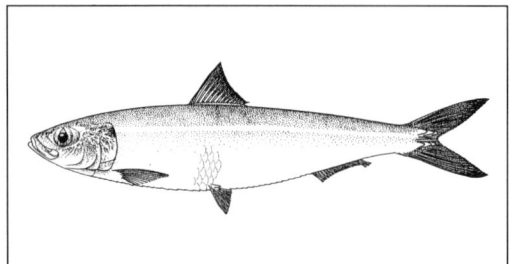

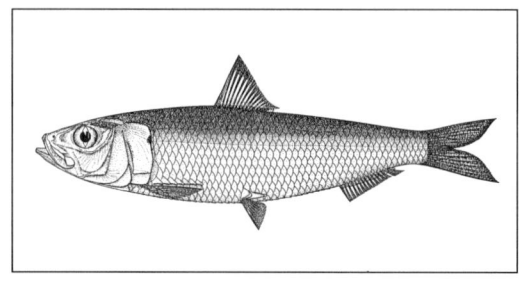

193	**CLUPEIDAE**	SAE

SC	*Sardinella maderensis* (Lowe, 1839); *Sardinella granigera*; *Sardinella eba*
ES	machuelo
DA	Madeira-sardinel
DE	Madeira-Sardinelle
EL	τριχιός της Αφρικής
EN	short-body sardinella; short-bodied sardine; Madeiran sardinella
FR	sardinelle plate; grande allache
IT	alaccia
NL	Madeira-sardinella
PT	sardinela da Madeira
FI	pilkkapyrstösardiini
SV	afrikansk sardinell

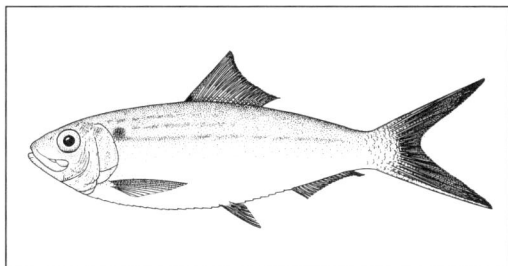

195	**CLUPEIDAE**	SIX

SC	*Sardinella* spp.
ES	alachas; machuelos
DA	sardineller
DE	Sardinellen
EL	τριχιοί· φρίσσα
EN	sardinellas
FR	sardinelles; allaches
IT	alacce
NL	sardinella's
PT	sardinelas
FI	sardiini-suku
SV	sardineller

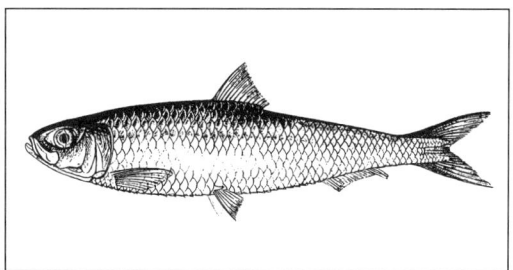

194	**CLUPEIDAE**	JSS

SC	*Sardinella zunasi* (Bleeker, 1854); *Harengula zunasi*
ES	sardineta del Japón
DA	japansk sardinel
DE	Pazifischer Kleinhering
EL	τριχιός της Ιαπωνίας
EN	Japanese sardinella
FR	harengule du Japon
IT	alaccia zunasi
NL	Pacifische sardinella
PT	sardineta do Japão
FI	sardiini-laji
SV	japansk sardinell

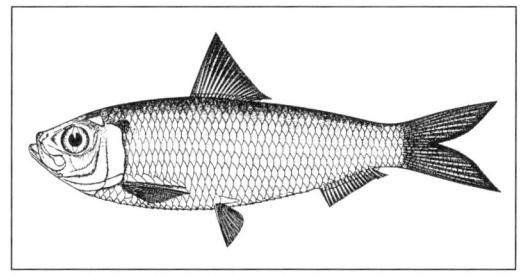

196	**CLUPEIDAE**	CPI

SC	*Sardinops caeruleus* (Girard, 1854)
ES	sardina de California
DA	californisk sardin
DE	Kalifornische Sardine; Pazifische Sardine
EL	σαρδέλα της Καλιφόρνιας
EN	Californian pilchard; Californian sardine
FR	pilchard de Californie; sardine du Pacifique
IT	sardina di California
NL	Pacifische sardine
PT	sardinopa da Califórnia
FI	kaliforniansardiini
SV	kalifornisk sardin

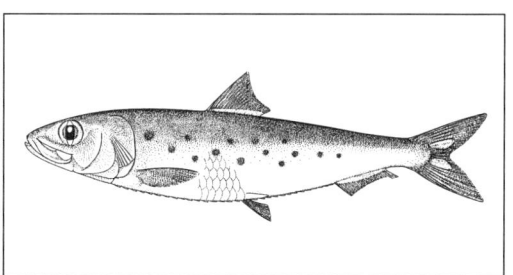

197 CLUPEIDAE JAP

SC   *Sardinops melanostictus* (Schlegel, 1846)
ES   sardina japonesa
DA   japansk sardin
DE   Japanische Sardine
EL   σαρδέλα της Ιαπωνίας
EN   Japanese pilchard; Japanese sardine
FR   pilchard du Japon
IT   sardina giapponese
NL   Japanse sardine
PT   sardinopa japonesa; sardinopa do Japão
FI   japaninsardiini
SV   japansk sardin

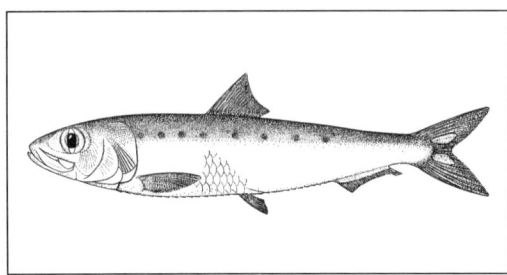

199 CLUPEIDAE PIA

SC   *Sardinops ocellatus* (Pappe, 1854)
ES   sardina sudafricana
DA   sydafrikansk sardin
DE   Südafrikanische Sardine
EL   σαρδέλα της Νότιας Αφρικής
EN   Southern African pilchard
FR   pilchard d'Afrique du Sud; sardine d'Afrique du Sud
IT   sardina del Sudafrica
NL   Zuid-Afrikaanse pelser; Zuid-Afrikaanse sardine
PT   sardinopa da África do Sul; sardinha da África do Sul; sardinha do Cabo
FI   silmäsardiini
SV   sydafrikansk sardin

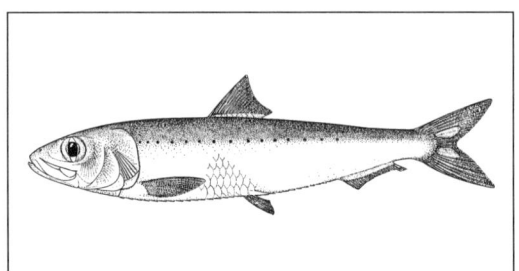

198 CLUPEIDAE

SC   *Sardinops neopilchardus* (Steindachner, 1879)
ES   sardina australiana
DA   australsk sardin
DE   Australische Sardine
EL   σαρδέλα της Αυστραλίας
EN   picton herring; herring
FR   pilchard d'Australie; sardine australienne
IT   sardina australiana
NL   Australische pelser; Australische sardine
PT   sardinopa da Austrália
FI   australiansardiini
SV   australisk sardin

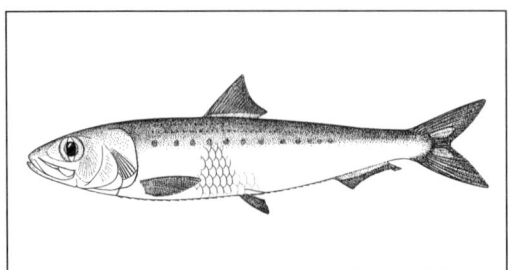

200 CLUPEIDAE CHP

SC   *Sardinops sagax* (Jenyns, 1842)
ES   sardina chilena
DA   chilensk sardin
DE   Südamerikanische Sardine
EL   σαρδέλα του Περού
EN   Chilean pilchard; Peruvian sardine; South American pilchard
FR   sardine chilienne; pilchard du Chili
IT   sardina del Cile
NL   Chileense sardine
PT   sardinopa chilena; sardinha do Chile
FI   perunsardiini
SV   sydamerikansk sardin

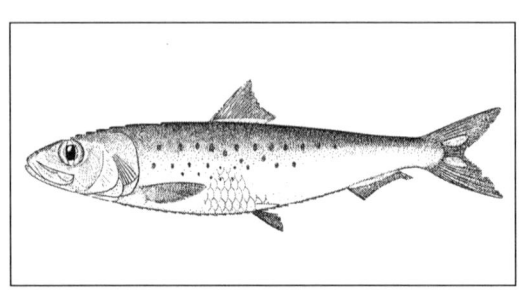

201	**CLUPEIDAE**	SRH

SC *Spratelloides gracilis* (Temminck and Schlegel, 1846)
ES espadín grácil
DA sølvstribet rundsild
DE Silberstreifen-Rundhering
EL παπαλινάκι
EN silver-stripe round herring
FR sprat gracile
IT spratello
NL zilverstreep-rondeharing
PT arenque redondo prateado
FI silli-laji
SV silverrandig rundsill

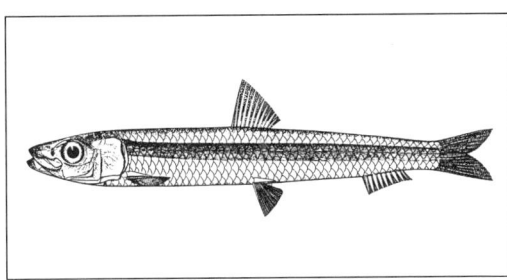

202	**CLUPEIDAE**	SPR

SC *Sprattus sprattus* (Linnaeus, 1758); *Clupea sprattus*
ES espadín; trancho; sardineta
DA brisling; europæisk brisling
DE Sprotte; Sprott; Brisling; Breitling
EL παπαλίνα
EN sprat; brisling; European sprat
FR sprat; esprot
IT papalina; spratto
NL sprot
PT espadilha; lavadilha
FI kilohaili
SV skarpsill; vassbuk

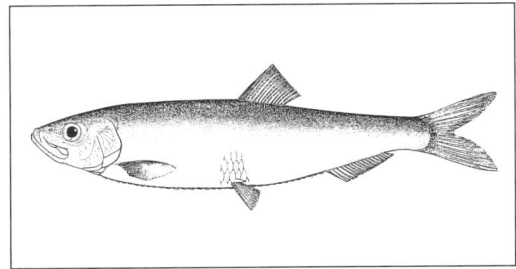

203	**CLUPEIDAE**	DAG

SC *Stolothrissa* spp.; *Limnothrissa* spp.
ES dagas
DA sildefisk-slægter
DE Dagas
EL ντόγκες
EN dagaas
FR dagaas
IT daga
NL dagaas
PT dagas
FI silli
SV daga-sillar; kapenta

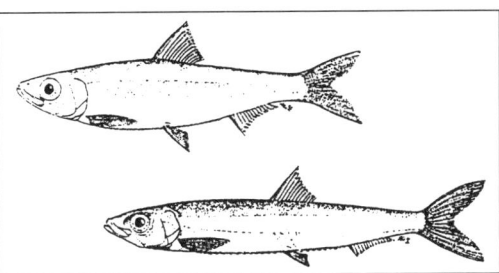

204	**CLUPEIDAE**	CKI

SC *Strangomera bentincki* (Norman, 1936); *Clupea bentincki*
ES sardina del sur
DA chilensk sild
DE Chilenischer Hering
EL ρέγγα της Χιλής
EN Chilean herring; Araucanian herring
FR hareng du Chili
IT aringa cilena
NL Chileense haring
PT espadilha chilena
FI chilensilli
SV chilensk sill

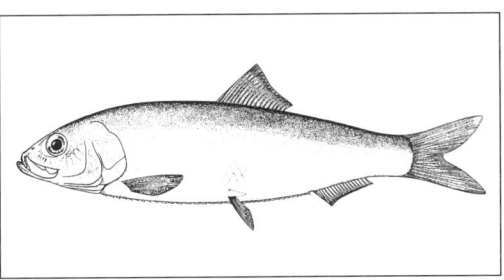

205 CLUPEIDAE HIL

SC	*Tenualosa ilisha* (Hamilton-Buchanan, 1822); *Hilsa ilisha*
ES	sábalo hilsa
DA	hilsa
DE	Ilisha-Alse
EL	φρίσσα των Ινδιών
EN	hilsa shad
FR	alose hilsa
IT	alosa indiana
NL	hilsa-elft
PT	pala
FI	hilsa
SV	indisk staksill

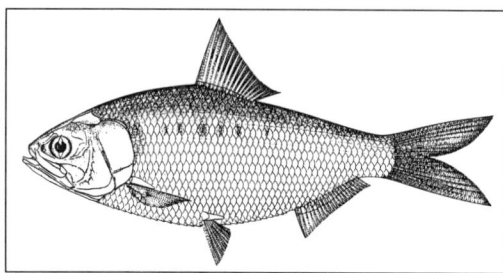

206 CLUPEIDAE REE

SC	*Tenualosa reevesi* (Richardson, 1846); *Hilsa reevesi*
ES	alosa de Reeves
DA	Reeves' stamsild
DE	Reeves Alse
EL	φρίσσα των Ινδιών
EN	Reeves' shad
FR	alose de Reeves
IT	alosa indiana
NL	Reeves' elft
PT	pala de estação
FI	hilsa-laji
SV	kinesisk staksill

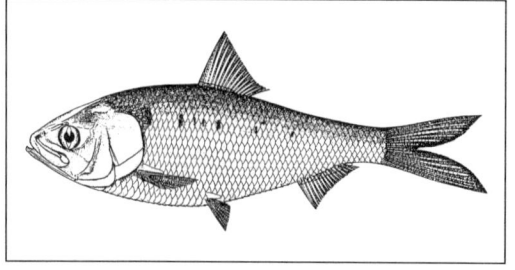

207 CLUPEIDAE TOL

SC	*Tenualosa toli* (Valenciennes, 1847); *Hilsa toli*
ES	sábalo toli
DA	toli-stamsild
DE	Toli-Alse
EL	φρίσσα των Ινδιών
EN	toli shad
FR	alose toli
IT	alosa indiana
NL	toli-elft
PT	pala-toli
FI	hilsa-laji
SV	toli-staksill

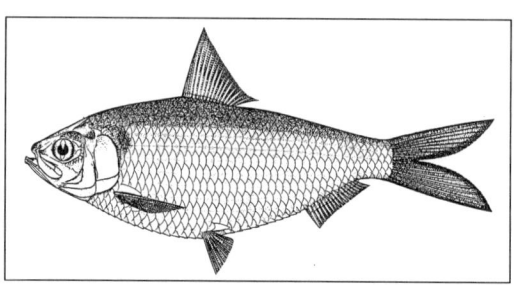

208 ENGRAULIDAE DIA

SC	*Engraulidae*
ES	anchoas; boquerones
DA	ansjos-familien
DE	Sardellen; Anchovis
EL	γαύροι
EN	anchovies
FR	engraulidés; anchois
IT	acciughe; alici; engraulidi
NL	ansjovissen
PT	biqueirões; enchovas
FI	sardellit; sardellit-heimo
SV	ansjovisfiskar

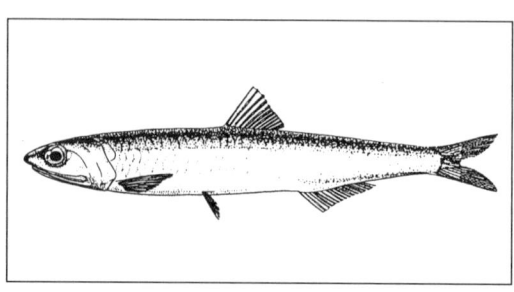

209 ENGRAULIDAE   ENP

SC *Anchoa hepsetus* (Linnaeus, 1758); *Engraulis hepsetus*
ES anchoa listrada; anchoa tropical
DA stribet ansjos
DE Breitstreifensardelle
EL γραμμωτός γαύρος
EN striped anchovy; broad-striped anchovy
FR anchois tropical; anchois de Guinée;
    anchois de l'Atlantique Sud; piquitinga
IT sardoncino americano
NL Zuid-Atlantische ansjovis
PT biqueirão listrado
FI raitasardelli
SV bredrandig ansjovis

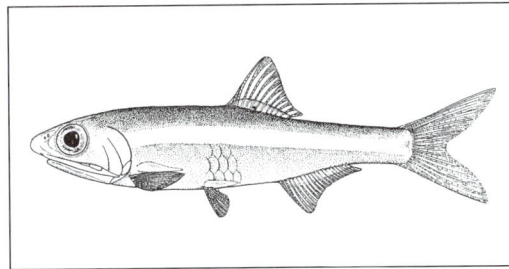

210 ENGRAULIDAE   ANB

SC *Anchoa mitchilli* (Valenciennes, 1848)
ES anchoa de caleta
DA nordvestatlantisk ansjos
DE Nordwestatlantische Sardelle
EL γαύρος της Αμερικής
EN bay anchovy
FR anchois baie; anchois américain
IT sardoncino americano
NL Amerikaanse ansjovis
PT biqueirão de baía
FI länsiatlantinsardelli
SV amerikansk ansjovis

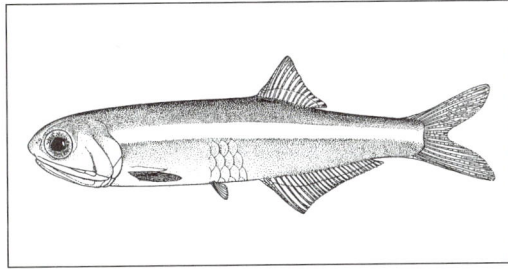

211 ENGRAULIDAE   AVA

SC *Cetengraulis edentulus* (Cuvier, 1829)
ES anchoveta rabo amarillo
DA atlantisk ansjos
DE Atlantische Anchoveta
EL ψευτόγαυρος του Ατλαντικού
EN Atlantic anchoveta
FR anchois à queue jaune
IT pseudacciuga atlantica
NL Atlantische ansjovis
PT biqueirão-rabo amarelo; enchova-rabo amarelo
FI sardelli-laji
SV atlantisk grovansjovis

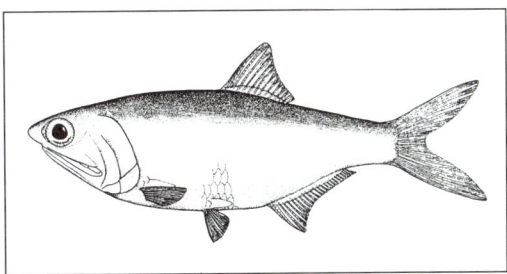

212 ENGRAULIDAE   VEP

SC *Cetengraulis mysticetus* (Günther, 1867)
ES anchoveta
DA Stillehavs-ansjos
DE Pazifische Anchoveta
EL ψευτόγαυρος του Ειρηνικού
EN Pacific anchoveta
FR anchois du Pérou; anchovette
IT pseudacciuga del Pacifico
NL Pacifische ansjovis
PT biqueirão do Pacífico Central; enchova do Pacífico Central
FI sardelli-laji
SV stillahavsgrovansjovis

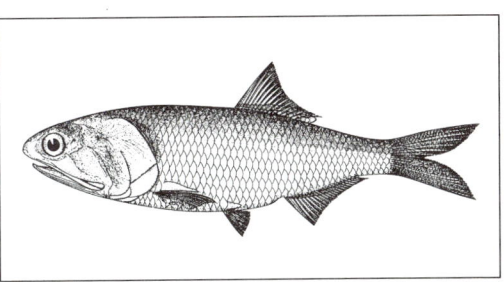

53

213 ENGRAULIDAE RAA

SC	*Coilia mystus* (Linnaeus, 1758)
ES	anchoa cola de rata
DA	langhalet ansjos
DE	Grenadieranchovis
EL	ποντικόγαυρος
EN	rat-tail anchovy; Osbeck's grenadier anchovy
FR	anchois grenadier
IT	pesce topo pelagico
NL	grenadier-ansjovis
PT	biqueirão-cauda de rato; enchova-cauda de rato
FI	sardelli-laji
SV	råttsvans

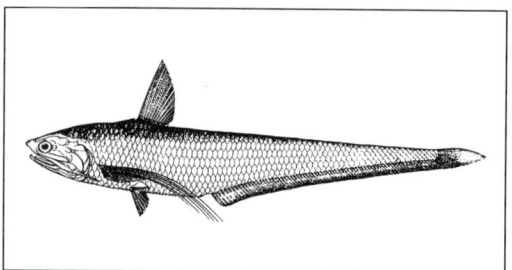

215 ENGRAULIDAE

SC	*Engraulis australis* (Shaw, 1790)
ES	anchoa australiana
DA	australsk ansjos
DE	Australische Sardelle
EL	γαύρος της Αυστραλίας
EN	Australian anchovy
FR	anchois d'Australie
IT	acciuga d'Australia
NL	Australische ansjovis
PT	biqueirão da Austrália; enchova da Austrália
FI	australiansardelli
SV	australisk ansjovis

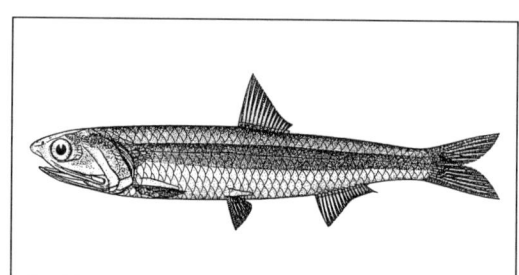

214 ENGRAULIDAE ANA

SC	*Engraulis anchoita* Hubbs and Marini, 1935
ES	anchoita
DA	argentinsk ansjos
DE	Argentinische Sardelle
EL	γαύρος της Αργεντινής
EN	anchoita; Argentine anchovy
FR	anchois d'Argentine
IT	acciuga d'Argentina; acciuga
NL	Argentijnse ansjovis
PT	biqueirão argentino; anchoveta argentina; enchova argentina
FI	argentiinansardelli
SV	anchoita

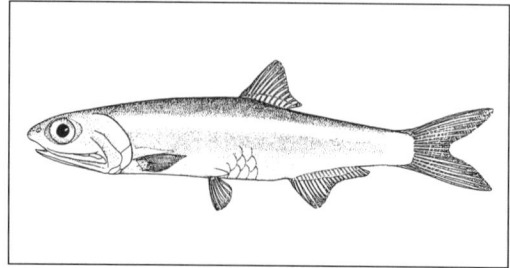

216 ENGRAULIDAE ANC

SC	*Engraulis capensis* Gilchrist, 1913
ES	anchoa del Cabo
DA	Kap-ansjos
DE	Südafrikanische Sardelle
EL	γαύρος του Ακρωτηρίου
EN	stet anchovy; Southern African anchovy
FR	anchois du Cap
IT	acciuga del Sudafrica; acciuga
NL	Kaapse ansjovis
PT	biqueirão do Cabo; biqueirão do Indo-Pacífico
FI	kapinsardelli
SV	kapansjovis

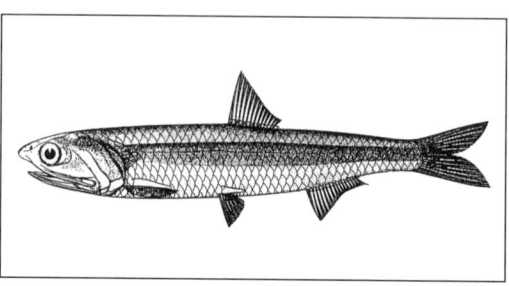

217 ENGRAULIDAE   ANE

SC	*Engraulis encrasicolus* (Linnaeus, 1758)
ES	anchoa europea; boquerón; bocarte
DA	europæisk ansjos
DE	Europäische Sardelle; Sardelle; Anchovis
EL	γαύρος
EN	European anchovy; anchovy
FR	anchois; anchois européen; anchois commun
IT	acciuga; alice; sardone
NL	ansjovis
PT	biqueirão; enchova; anchova
FI	sardelli
SV	ansjovis; sardell

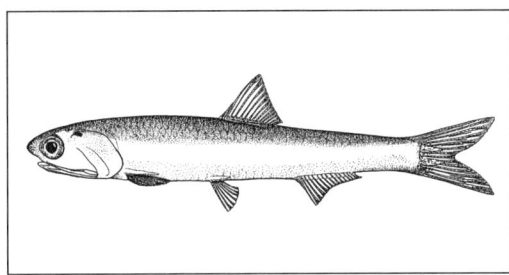

219 ENGRAULIDAE   NPA

SC	*Engraulis mordax* (Girard, 1856)
ES	anchoa del Pacífico norte; anchoa del Pacífico
DA	nordpacifisk ansjos
DE	Amerikanische Sardelle
EL	γαύρος του Ειρηνικού
EN	North Pacific anchovy; Northern anchovy; Californian anchoveta
FR	anchois du Pacifique Nord; anchois du Nord; anchois du Pacifique
IT	acciuga del Nord Pacifico
NL	Noord-Pacifische ansjovis
PT	biqueirão do Pacífico Norte
FI	kaliforniansardelli
SV	stillahavsansjovis

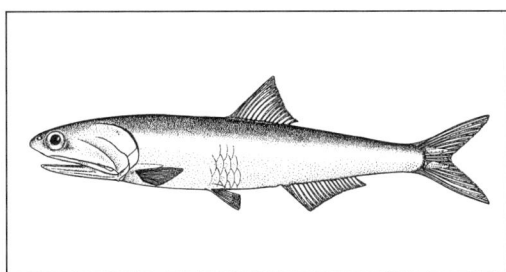

218 ENGRAULIDAE   JAN

SC	*Engraulis japonicus* (Temminck and Schlegel, 1846); *Engraulis japonica*
ES	anchoa japonesa
DA	japansk ansjos
DE	Japanische Sardelle
EL	γαύρος της Ιαπωνίας
EN	Japanese anchovy
FR	anchois du Japon; anchois japonais
IT	acciuga; acciuga del Giappone
NL	Japanse ansjovis; Kaapse ansjovis
PT	biqueirão japonês
FI	japaninsardelli
SV	japansk ansjovis

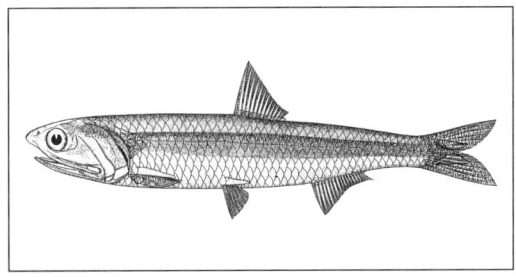

220 ENGRAULIDAE   VET

SC	*Engraulis ringens* (Jenyns, 1842)
ES	anchoveta del Perú
DA	peruansk ansjos
DE	Peru-Sardelle; Anchoveta
EL	γαύρος του Περού
EN	anchoveta; Peruvian anchovy
FR	anchois du Pérou; anchovetta; anchois péruvien
IT	acciuga del Cile
NL	Peruaanse ansjovis
PT	biqueirão do Peru; anchoveta do Peru
FI	perunsardelli
SV	anchoveta

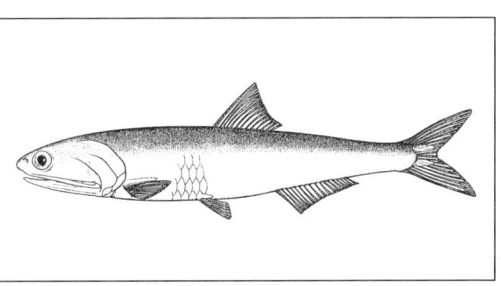

221 ENGRAULIDAE ANR

SC	*Lycengraulis grossidens* (Agassiz, 1829); *Lycengraulis olidus*
ES	anchoíta de río
DA	flodansjos
DE	Atlantische Säbelzahn-Sardelle
EL	ποταμόγαυρος
EN	river anchoita
FR	anchois d'eau douce
IT	acciuga dentata
NL	rivier-ansjovis
PT	biqueirão do rio; enchova do rio
FI	sardelli-laji
SV	brasiliansk rovansjovis

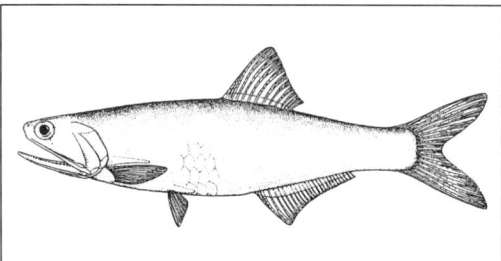

223 ALEPOCEPHALIDAE ALC

SC	*Alepocephalus bairdii* (Goode and Bean, 1879)
ES	alepocéfalo
DA	Bairds glathovedfisk
DE	Glattkopf
EL	αλεποκέφαλος
EN	Baird's smooth-head
FR	alépocéphale
IT	alepocefalo
NL	glijkop
PT	celindra; triste-linda
FI	
SV	Bairds släthuvudfisk

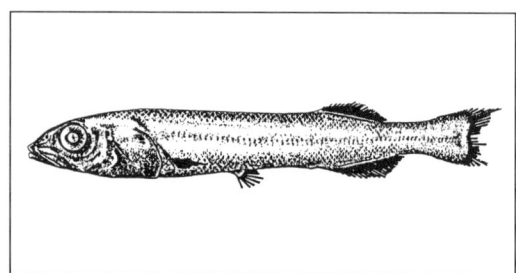

222 ENGRAULIDAE STO

SC	*Stolephorus* spp.
ES	anchoas «stolephorus»
DA	ansjos-slægt
DE	Anchovis
EL	γαύροι
EN	stolephorus anchovies
FR	anchois indopacifique
IT	stolefori
NL	stoleforus-ansjovissen
PT	biqueirões «Stolephorus»
FI	sardelli-suku
SV	ansjoveller

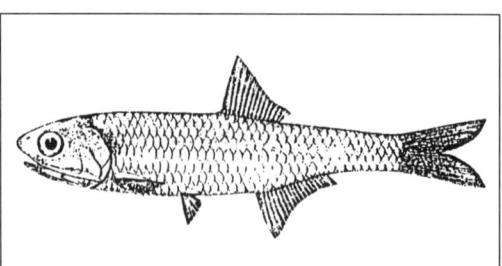

224 CHIROCENTRIDAE

SC	*Chirocentridae*
ES	arenques lobo
DA	rovsild-familien
DE	Wolfsheringe
EL	λυκόρεγγες
EN	wolf-herrings
FR	harengs-loups
IT	dorab
NL	wolfharingen
PT	espadelas
FI	susisillit; susisillit-heimo
SV	vargsillar

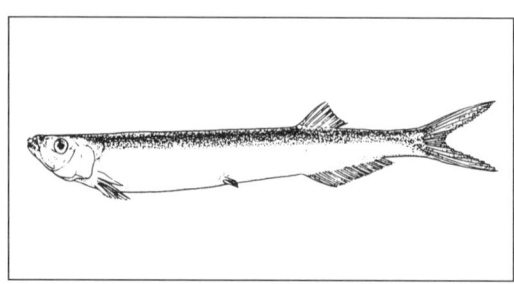

225 CHIROCENTRIDAE　　DOB

SC	*Chirocentrus dorab* (Forsskål, 1775)
ES	arenque lobo de la India
DA	indisk rovsild
DE	Großer Wolfshering; Indischer Hering
EL	λυκόρεγγες
EN	wolf-herring; dorab wolf-herring
FR	chirocentre dorab; hareng-loup des Indes
IT	dorab
NL	Indische wolfharing
PT	espadela
FI	susisilli
SV	vargsill

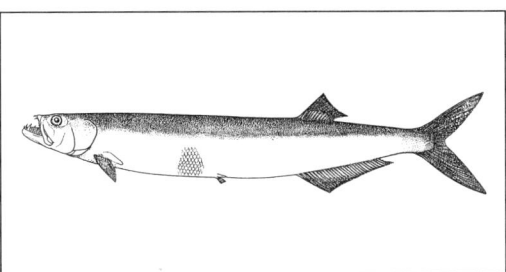

227 CHANIDAE

SC	*Chanidae*
ES	sabalotes
DA	mælkefisk-familien
DE	Milchfische
EL	χανίδες
EN	milkfishes
FR	chanidés
IT	cefaloni
NL	melkvissen
PT	peixes-leite
FI	maitokalat; maitokalat-heimo
SV	mjölkfiskar

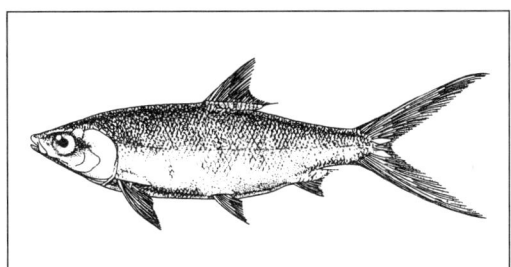

226 CHIROCENTRIDAE　　DOS

SC	*Chirocentrus* spp.
ES	arencones
DA	rovsild-slægten
DE	Wolfsheringe
EL	λυκόρεγγες
EN	wolf-herrings
FR	chirocentres; harengs-loups
IT	dorab
NL	wolfharingen
PT	espadelas
FI	susisilli-suku
SV	vargsillar

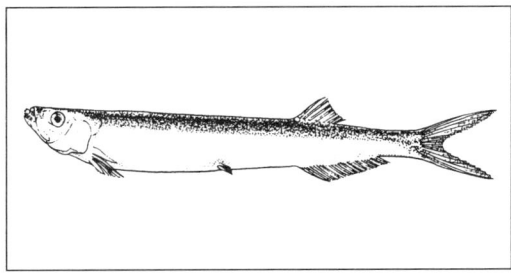

228 CHANIDAE　　MIL

SC	*Chanos chanos* (Forskal, 1775)
ES	chanos; sabalote
DA	mælkefisk
DE	Milchfisch
EL	γαλατόψαρο
EN	milkfish; bandeng; bandang
FR	chanos; chanidé
IT	cefalone
NL	melkvis
PT	peixe-leite
FI	maitokala
SV	mjölkfisk

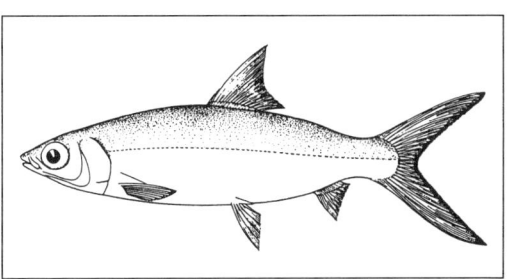

SC	*Coregonus albula* (Linnaeus, 1758); *Coregonus vandesius*
ES	coregono blanco
DA	heltling
DE	Kleine Maräne; Zwergmaräne; Silbermaräne; Marenke; Zollfisch
EL	λευκοκορέγονος
EN	vendace; whitefish; European whitefish
FR	corégone blanc; petite marène
IT	coregone bianco
NL	kleine marene
PT	coregono branco
FI	muikku
SV	siklöja

SC	*Coregonus artedii* (Lesueur, 1818); *Leucichthys artedii*
ES	coregono de Artedi; arenque de lago
DA	amerikansk helt
DE	Amerikanische Kleine Maräne; Felchen; Maräne
EL	λιμνόρεγγα
EN	lake cisco; lake herring; cisco; tullibee; chub; lakefish
FR	cisco de lac; cisco de l'Est; hareng de lac
IT	coregone americano
NL	ciscomarene
PT	coregono de artedi; arenque de lago;
FI	amerikanmuikku
SV	amerikansk siklöja

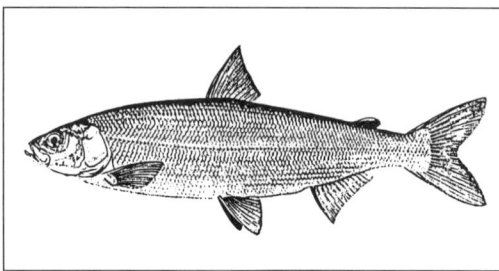

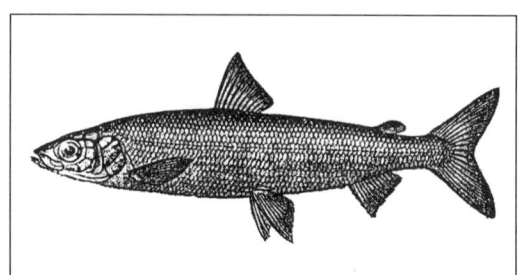

230 SALMONIDAE

SC	*Coregonus altior*; *Coregonus elegans*
ES	coregono elegante
DA	pollan-helt
DE	Pollan-Maräne; Felchen; Maräne
EL	κορέγονος
EN	pollan; freshwater herring
FR	corégone élégante
IT	coregone
NL	pollanmarene
PT	coregono das ilhas
FI	pollansiika
SV	pollan

SC	*Coregonus clupeaformis* (Mitchill, 1818)
ES	coregono de lago
DA	sø-helt
DE	Nordamerikanisches Felchen
EL	λιμνοκορέγονος
EN	lake whitefish; common whitefish
FR	corégone de lac
IT	coregone dei Grandi Laghi
NL	Amerikaanse marene
PT	coregono de lago
FI	sillisiika
SV	kanadasik

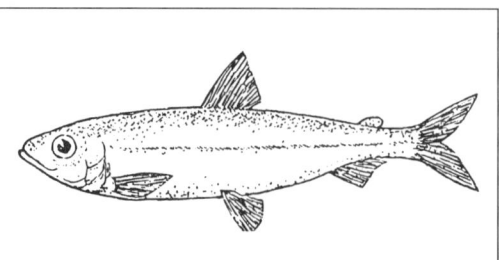

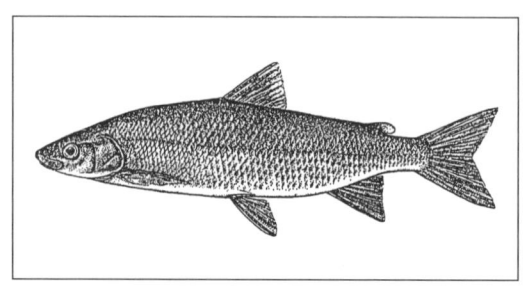

233 SALMONIDAE

SC	*Coregonus hiemalis*
ES	coregono del Leman
DA	Léman-helt
DE	Leman-Felchen
EL	κορέγονος
EN	Lake Geneva whitefish
FR	petite féra; corégone du Léman
IT	coregone
NL	Zwitserse marene
PT	coregono do lago Leman
FI	Geneve-järven siika
SV	genevesik

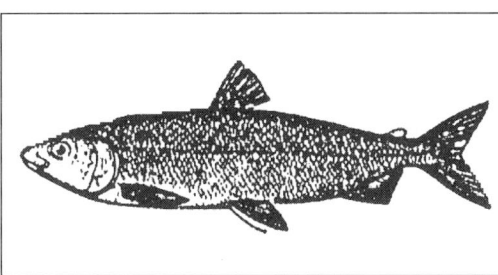

235 SALMONIDAE

SC	*Coregonus macrophthalmus* (Nusslin, 1882)
ES	coregono de ojos grandes
DA	storøjet helt
DE	Gangfisch
EL	κορέγονος
EN	Lake Neuchâtel whitefish
FR	bondelle
IT	bondella
NL	grootoog-marene
PT	coregono de olhos grandes
FI	siika-laji
SV	bondelle-sik

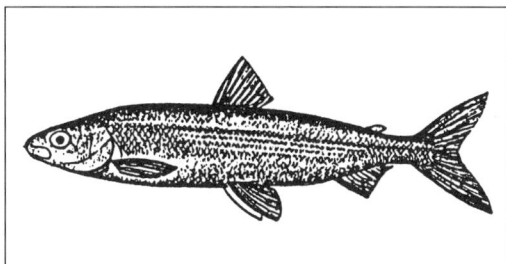

234 SALMONIDAE   PLN

SC	*Coregonus lavaretus* (Linnaeus, 1758)
ES	lavareto
DA	almindelig helt
DE	Große Maräne; Lavaret; Bodenrenke; Wandermaräne; Ostseeschnepel; Große Schwebrenke; Madümaräne; Blaufelchen; Schleischnäpel; Coregone; Edelmaräne
EL	γαλάζιος κορέγονος
EN	powan; Gwyniad; pollan; whitefish
FR	corégone lavaret; lavaret; lavaret du Bourget; féra; bezole
IT	coregone lavarello
NL	grote marene
PT	coregono lavareda
FI	siika
SV	lavaretsik

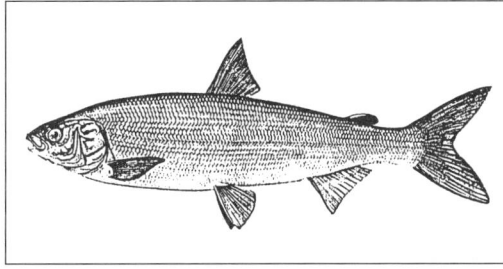

236 SALMONIDAE

SC	*Coregonus nasus* (Pallas, 1776); *Coregonus fera*
ES	coregono narigudo
DA	stor bundhelt
DE	Sandfelchen; Große Bodenrenke; Große Maräne; Weißfelchen
EL	κορέγονος
EN	large-bottom pollan
FR	grand pollan; grande féra; gravanche
IT	coregone
NL	grote houting
PT	coregono da vasa
FI	kuonosiika
SV	bodensik

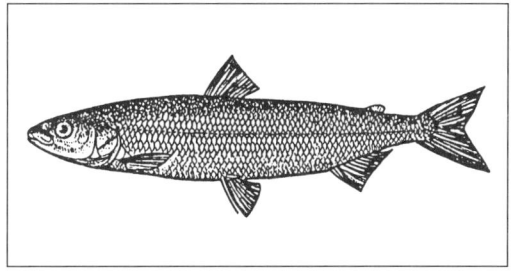

237 **SALMONIDAE**

SC	*Coregonus oxyrinchus* (Linnaeus, 1758); *Coregonus lavaretus oxyrinchus*
ES	coregono
DA	snæbel
DE	Schnepel; Schnäpel; Kleine Schwebrenke; Gangfisch; Nordseeschnäpel; Edelmaräne; Peipusmaräne; Elbel; Silberfelchen; Renke
EL	ρυγχοκορέγονος
EN	houting
FR	corégone oxyringue; hauting; bondelle; corégone
IT	coregone musino; coregone
NL	houting
PT	coregono bicudo
FI	järvisiika
SV	nordsjösik

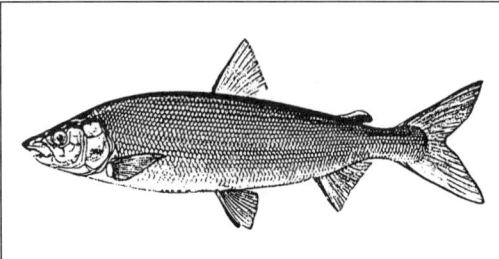

239 **SALMONIDAE**

SC	*Coregonus pidchianoides* (Pravdin, 1931); *Coregonus lavaretus pidchian*
ES	coregono de Pidchian; coregono de Pidchian
DA	bundhelt
DE	Kilch; Kleine Bodenrenke; Kropffelchen; Kröpfling
EL	κορέγονος του βυθού
EN	bottom pollan
FR	pollan; kilch
IT	coregone
NL	Arctische houting
PT	coregono do fundo
FI	pohjasiika
SV	bodensik

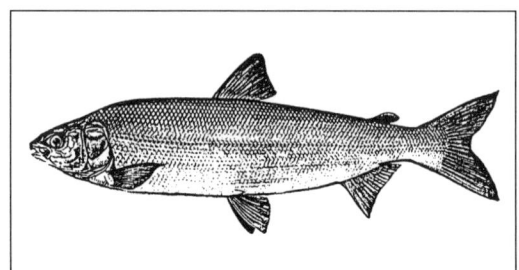

238 **SALMONIDAE**

SC	*Coregonus peled* (Gmelin, 1789)
ES	coregono peled
DA	stor stævhelt
DE	Peledmaräne
EL	κορέγονος
EN	big powan
FR	grand powan
IT	coregone
NL	peled
PT	coregono da Sibéria
FI	peledsiika
SV	peledsik

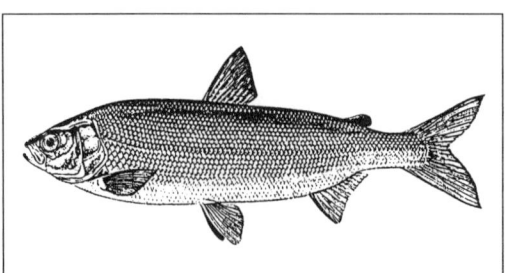

240 **SALMONIDAE**

SC	*Coregonus wartmanni*
ES	coregono azul
DA	blåhelt
DE	Blaufelchen; Große Schwebrenke; Bläuling; Ahlbock; Edelfisch; Stubben; Seelen; Hägling; Heuerling
EL	γαλάζιος κορέγονος
EN	blue whitefish
FR	corégone bleue
IT	coregone azzurro
NL	Bodenmeerhouting
PT	coregono azul
FI	tuppisiika; murokas; riika; reeska
SV	azursik

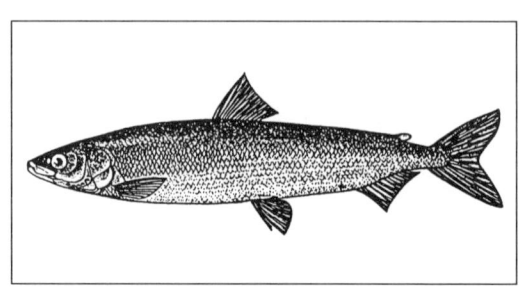

241 SALMONIDAE WHF

SC *Coregonus* spp.
ES coregonos
DA helt-slægt
DE Felchen; Moränen; Renken; Schnäpel
EL κορέγονοι
EN whitefishes
FR corégones
IT coregoni
NL houtingen
PT coregonos
FI siika-suku
SV sikar

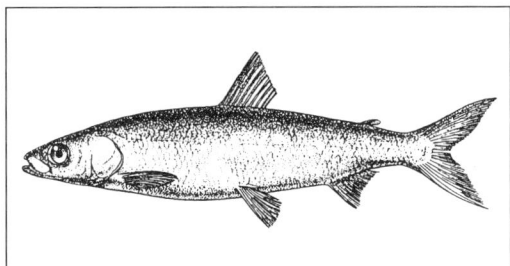

243 SALMONIDAE

SC *Oncorhynchus aguabonita* (Jordan, 1892); *Salmo aguabonita* Jordan,
ES trucha dorada
DA guldørred
DE Goldforelle
EL χρυσός σολομός
EN golden trout
FR truite dorée
IT trota
NL goudforel
PT truta dourada
FI kultataimen
SV guldöring

242 SALMONIDAE

SC *Hucho hucho* (Linnaeus, 1758); *Salmo hucho*
ES salmón del Danubio
DA Donau-laks
DE Huchen; Sibirischer Huchen; Donaulachs; Rotfisch
EL σολομός του Δούναβη
EN Danube salmon; huchen
FR huchon; saumon du Danube; huchon du Danube
IT salmone del Danubio
NL Donauzalm
PT salmão do Danúbio
FI jokilohi
SV huchen

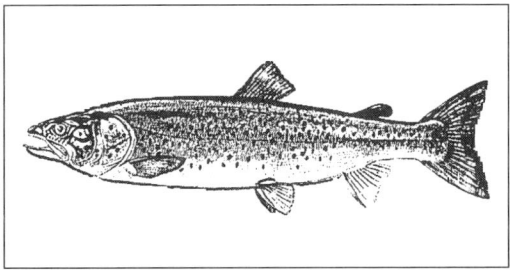

244 SALMONIDAE

SC *Oncorhynchus apache* (Miller, 1972); *Salmo apache*
ES trucha de Arizona
DA apacheørred
DE Arizona-Forelle
EL πέστροφα της Αριζόνας
EN Arizona trout
FR truite apache
IT trota apache
NL Arizonaforel
PT truta do Arizona
FI arizonanlohi
SV apachelax; apacheöring

245 SALMONIDAE

SC	*Oncorhynchus chrysogaster* (Needham Gard, 1964); *Salmo crysogaster*
ES	trucha dorada Mejicana
DA	mexikansk ørred
DE	Mexikanische Goldforelle
EL	πέστροφα του Μεξικού
EN	Mexican golden trout
FR	truite dorée mexicaine
IT	trota dorata messicana
NL	Mexicaanse gouden forel
PT	truta dourada mexicana
FI	meksikonlohi
SV	mexikansk guldöring

DRAWING NOT AVAILABLE

247 SALMONIDAE   PIN

SC	*Oncorhynchus gorbuscha* (Walbaum, 1792)
ES	salmón rosado; salmón jorobado
DA	pukkellaks
DE	Buckellachs; Rosa Lachs; Roter Lachs; Humback-Lachs
EL	ϱός σολομός
EN	pink salmon; humpback salmon; gorbuscha
FR	saumon rose; saumon à bosse
IT	salmone rosa; salmone del Pacifico
NL	roze zalm; pink zalm
PT	salmão rosa
FI	kyttyrälohi
SV	puckellax; pinklax

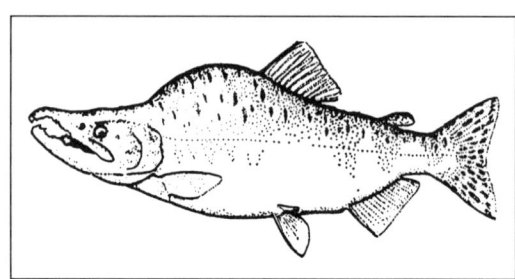

246 SALMONIDAE

SC	*Onchorynchus gilae* (Miller, 1950); *Salmo gilae*
ES	salmón gila
DA	gila-ørred
DE	Gila-Forelle
EL	σολομός
EN	gila trout
FR	saumon bossu
IT	trota
NL	gilaforel
PT	truta-gila
FI	gilataimen
SV	gila-öring

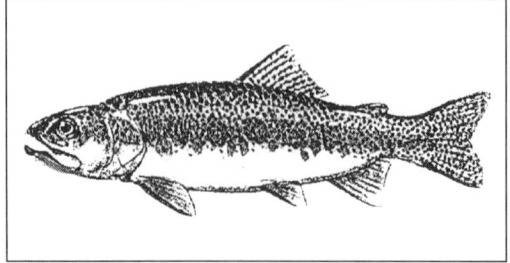

248 SALMONIDAE   CHU

SC	*Oncorhynchus keta* (Walbaum, 1792)
ES	keta; salmón chum; salmón keta
DA	ketalaks
DE	Keta-Lachs; Hundelachs; Hundslachs
EL	σολομός κέτα
EN	chum salmon; dog salmon; keta salmon; qualla; calico salmon; hum; fall salmon
FR	saumon keta; saumon chum; saumon-chien
IT	salmone keta; salmone del Pacifico
NL	ketazalm; chumzalm
PT	salmão-cão
FI	koiralohi
SV	hundlax; ketalax

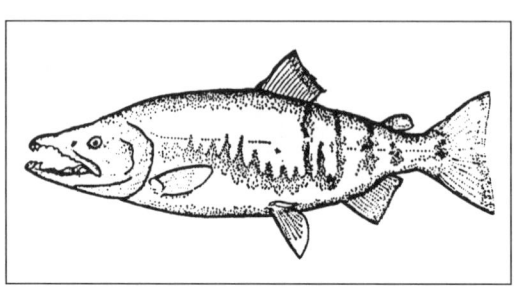

249 SALMONIDAE COH

SC	*Oncorhynchus kisutch* (Walbaum, 1792)
ES	salmón plateado; salmón cóho
DA	koho; sølvlaks
DE	Silberlachs; Kisutch-Lachs
EL	ασημένιος σολομός· κόχο
EN	coho; silver salmon; blueback; medium red salmon; jack salmon; silverside; coho salmon
FR	saumon coho; saumon argenté
IT	salmone argentato; salmone del Pacifico
NL	cohozalm
PT	salmão prateado
FI	hopealohi
SV	coho; coholax; silverlax

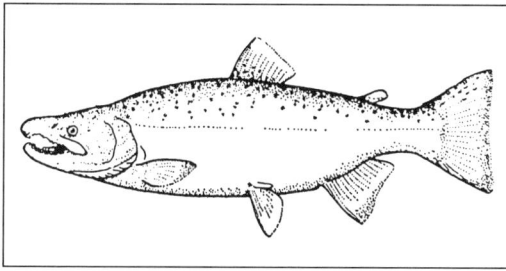

251 SALMONIDAE TRR

SC	*Oncorhynchus mykiss* (Walbaum, 1792); *Salmo gairdneri*; *Salmo irideus*
ES	trucha arco iris
DA	regnbueørred
DE	Regenbogenforelle; Amerikanische Forelle; Stahlkopfforelle; Purpurforelle; Teichforelle
EL	ιριδίζουσα πέστροφα· πέστροφα
EN	rainbow trout; steelhead trout; steelhead
FR	truite arc-en-ciel; truite américaine
IT	trota iridea; trota arcobaleno; trota americana
NL	regenboogforel
PT	truta arco-íris
FI	kirjolohi
SV	regnbåge

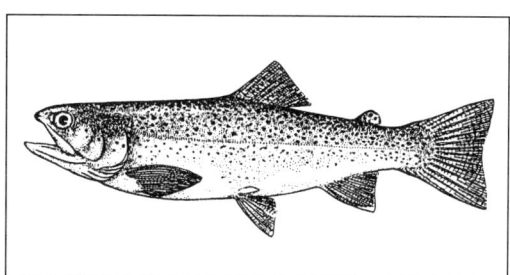

250 SALMONIDAE CHE

SC	*Oncorhynchus masu* (Brevoort, 1856); *Oncorhynchus masou*
ES	salmón japonés
DA	Japan-laks; masu-laks
DE	Japan-Lachs; Masu-Lachs; Sako; Japanischer Lachs
EL	σολομός της Ιαπωνίας
EN	Japanese salmon; Japanese cherry salmon; cherry salmon; masu; masu salmon
FR	saumon japonais; saumon du Japon; saumon masou
IT	salmone giapponese; salmone del Pacifico
NL	Japanse zalm; masouzalm
PT	salmão japonês
FI	masulohi
SV	Masulax

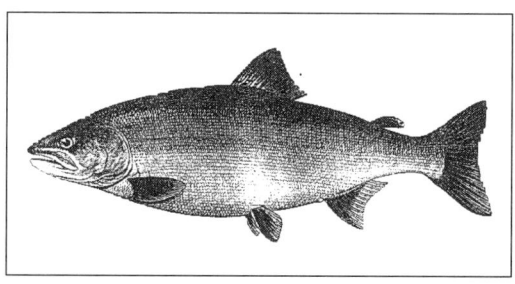

252 SALMONIDAE SOC

SC	*Oncorhynchus nerka* (Walbaum, 1792)
ES	salmón rojo; salmón
DA	kokanee-laks
DE	Rotlachs; Blaurücken; Blaurückenlachs; Roter Lachs
EL	κόκκινος σολομός
EN	sockeye salmon; red salmon; blueback; quinalt
FR	saumon sockeye; saumon rouge
IT	salmone rosso; salmone del Pacifico
NL	sockeye zalm; rode zalm
PT	salmão vermelho
FI	punalohi; intiaanilohi
SV	indianlax; sockeyelax

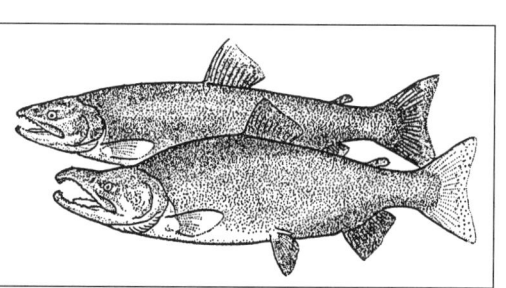

63

253 SALMONIDAE

SC	*Oncorhynchus rhodurus* (Jordan and McGregor, 1925); *Salmo rhodurus*
ES	salmón de Biwa
DA	Biwa-ørred
DE	Biwa-Forelle
EL	μπίβα
EN	Biwa
FR	truite biwamasou
IT	trota giapponese
NL	Amagozalm
PT	salmão de Biwa
FI	Biwa-lohi
SV	Biwa-lax

DRAWING NOT AVAILABLE

254 SALMONIDAE CHI

SC	*Oncorhynchus tshawytscha* (Walbaum, 1792)
ES	salmón real; salmón chinook
DA	kongelaks
DE	Königslachs; Quinnat
EL	σολομός του Ειρηνικού· μαύρος σολομός
EN	chinook; spring salmon; king salmon; Pacific salmon; quinnat; black salmon; chub salmon; tyee; quinnat salmon; chinook salmon
FR	saumon du Pacifique; chinook; quinnat; saumon royal
IT	salmone reale; salmone del Pacifico
NL	chinook zalm
PT	salmão real
FI	kuningaslohi
SV	kungslax

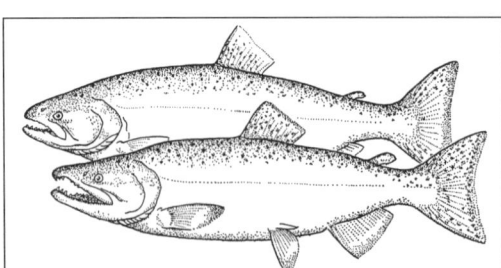

255 SALMONIDAE

SC	*Salmo clarki* (Richardson, 1836)
ES	trucha clarki
DA	cutthroat-ørred
DE	Cutthroat-Forelle
EL	κηλιδόστικτος σολομός
EN	cutthroat trout
FR	truite fardée; cou coupé
IT	trota
NL	cutthroatforel
PT	truta feroz
FI	punakurkkulohi
SV	strupsnittsöring

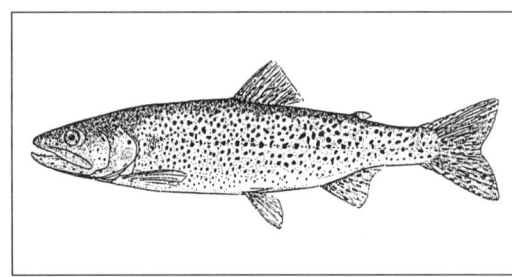

256 SALMONIDAE SAL

SC	*Salmo salar* (Linnaeus, 1758)
ES	salmón del Atlántico; salmón;
DA	atlantisk laks
DE	Lachs; Echter Lachs; Salm; Atlantischer Lachs
EL	σολομός του Ατλαντικού
EN	Atlantic salmon; salmon
FR	saumon de l'Atlantique; saumon commun; saumon atlantique
IT	salmone atlantico; salmone
NL	zalm
PT	salmão do Atlântico
FI	lohi
SV	lax

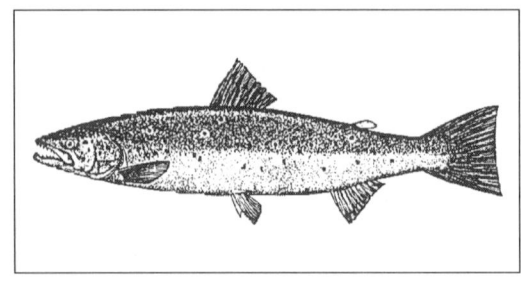

257 SALMONIDAE

TRS

SC	*Salmo trutta* (Linnaeus, 1758); *Salmo trutta trutta*; *Trutta marina*
ES	trucha marina; trucha; reo; trucha marisca; trucha de mar
DA	havørred
DE	Meerforelle; Lachsforelle; Weißforelle; Strandlachs; Silberlachs; Schwarzlachs
EL	πέστροφα
EN	sea trout; brown trout
FR	truite de mer; truite brune; truite de Dieppe
IT	trota di mare; salmotrota
NL	zeeforel
PT	truta marisca; truta marinha; truta sapeira
FI	taimen; meritaimen
SV	öring

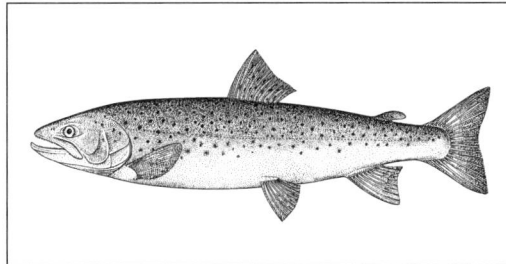

258 SALMONIDAE

SC	*Salmo trutta fario* (Linnaeus, 1758); *Trutta fluviatilis*
ES	trucha común; trucha de río; trucha de arroyo
DA	bækørred
DE	Bachforelle; Flußforelle; Wildforelle; Bergforelle; Alpenforelle; Steinforelle; Schwarzforelle; Weißforelle
EL	πέστροφα
EN	river trout; brook trout; brown trout
FR	truite commune; truite de rivière; truite fario
IT	trota; trota fario; trota di fiume
NL	beekforel
PT	truta comum
FI	purotaimen
SV	öring

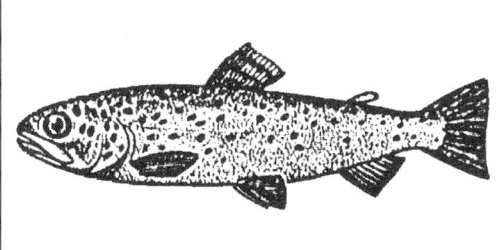

259 SALMONIDAE

SC	*Salmo trutta lacustris* (Linnaeus, 1758)
ES	trucha lacustre
DA	søørred
DE	Seeforelle; Lachsforelle; Schwebforelle; Grundforelle; Maiforelle; Blauforelle; Herbstforelle; Silberforelle; Ferchen; Illanke; Rheinanke
EL	λιμνοπέστροφα
EN	lake trout
FR	truite de lac
IT	trota di lago
NL	meerforel
PT	truta do lago
FI	järvitaimen
SV	öring

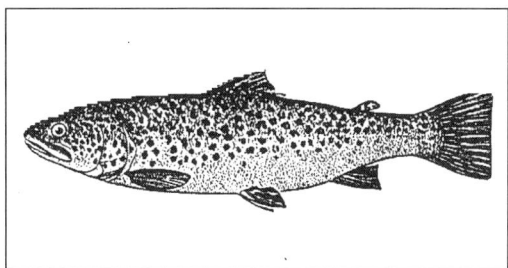

260 SALMONIDAE

TRO

SC	*Salmo* spp.
ES	truchas
DA	lakse-slægt
DE	Forellen
EL	πέστροφες
EN	trouts
FR	truites
IT	trote
NL	zalmachtigen
PT	trotas
FI	lohi-suku
SV	laxar och öringar

DRAWING NOT AVAILABLE

261	**SALMONIDAE**

SC *Coregoninae*
ES coregonos
DA helt-familien
DE Felchen; Maränen; Renken; Schnäpel
EL κορέγονοι
EN whitefishes
FR corégones
IT coregoni
NL houtingen en marenen
PT coregonos
FI siiat
SV sikfiskar

263	**SALMONIDAE**	ACH

SC *Salvelinus alpinus* (Linnaeus, 1758); *Salmo salvelinus*
ES trucha alpina; salvelino
DA fjeldørred
DE Wandersaibling; Seesaibling; Saibling; Rotforelle; Rotfisch; Salmerin; Ritter; Schwarzreuter
EL αρκτοσαλβελίνος
EN Arctic charr; mountain trout; salmon trout; ilkalupik; char
FR omble chevalier; omble
IT salmerino alpino
NL riddervis
PT salvelino árctico
FI nieriä
SV fjällröding

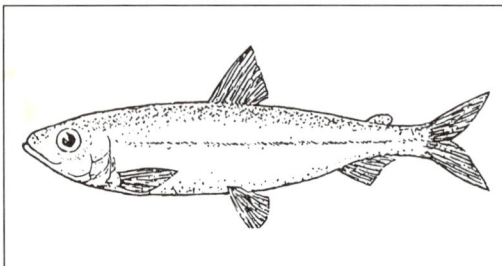

262	**SALMONIDAE**

SC *Salmonidae*
ES salmones y truchas
DA lakse-familien
DE Lachse, Forellen und Saiblinge
EL σολομοί· πέστροφες
EN salmons, trouts and charrs
FR saumons, truites et ombles
IT salmonidi; salmoni, trote e salmerini
NL zalmen en forellen
PT salmonídeos
FI lohet; lohet-heimo
SV laxfiskar

264	**SALMONIDAE**	SVF

SC *Salvelinus fontinalis* (Mitchill, 1815)
ES salvelino; trucha de fontana
DA kildeørred
DE Bachsaibling
EL σαλβελίνος· λιμνοπέστροφα
EN brook trout; brook charr; speckled trout; red trout; salmontrout; squaretail
FR omble de fontaine; saumon de fontaine; truite de lac; omble moucheté; truite mouchetée; truite rouge; truite de ruisseau
IT salmerino di fontana
NL bronforel
PT truta das fontes
FI puronieriä
SV bäckröding

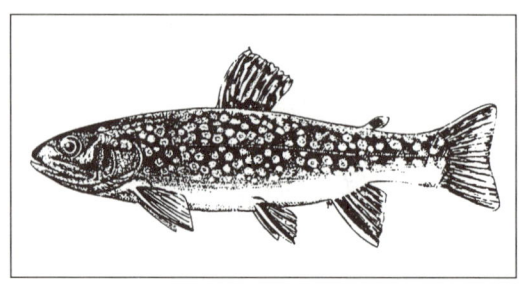

265 SALMONIDAE VAR

SC	*Salvelinus malma* (Walbaum, 1792)
ES	salvelino
DA	Malma-ørred
DE	Malma-Saibling
EL	σαλβελίνος του Ειρηνικού· πεστροφοσολομός
EN	Dolly Varden; Dolly Varden trout; salmon trout; bull trout
FR	dolly varden; omble du Pacifique
IT	salmerino
NL	Malmaforel
PT	salvelino do Pacífico
FI	härkänieriä
SV	dolly varden

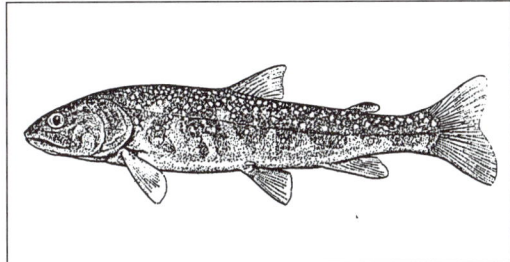

267 SALMONIDAE CHR

SC	*Salvelinus* spp.
ES	salvelinos
DA	røddinger
DE	Saiblinge
EL	σαλβελίνοι· λιμνοπέστροφες
EN	charrs; chars
FR	ombles
IT	salmerini
NL	riddervissen
PT	salvelinos
FI	nieriä-suku
SV	rödingar

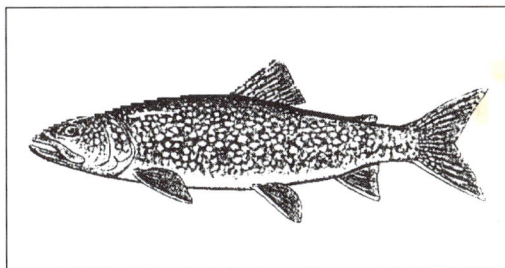

266 SALMONIDAE LAT

SC	*Salvelinus namaycush* (Walbaum, 1792); *Christivomer namaycush*
ES	trucha lacustre; trucha lacustre americana
DA	Canada-rødding; amerikansk søørred
DE	Amerikanischer Seesaibling; Amerikanische Seeforelle
EL	λιμνοπέστροφα της Αμερικής
EN	lake trout; grey trout; touladi; char; togue; namay-cush; Great Lake trout; American lake trout; American lake char
FR	omble; truite de lac; touladi; togue; christivomer; truite grise; truite de lac d'Amérique; omble du Canada
IT	salmerino di lago
NL	Amerikaanse meerforel
PT	salvelino lacustre; truta do lago
FI	harmaanieriä
SV	kanadaröding

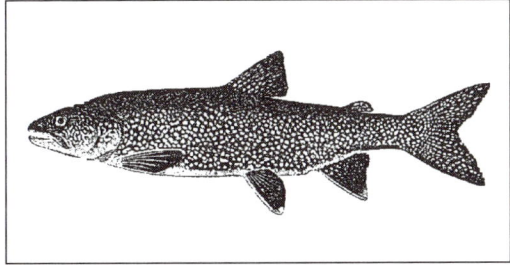

268 THYMALLIDAE TLV

SC	*Thymallus thymallus* (Linnaeus, 1758); *Thymallus vexillifer*
ES	tímalo
DA	stalling
DE	Europäische Äsche; Äsche; Asche; Mailing; Springer; Strommaräne; Perpel; Auch; Harr; Zeitasch
EL	θύμαλος
EN	grayling
FR	ombre commun; oumbre; umbra
IT	temolo
NL	vlagzalm
PT	peixe-sombra
FI	harjus
SV	harr

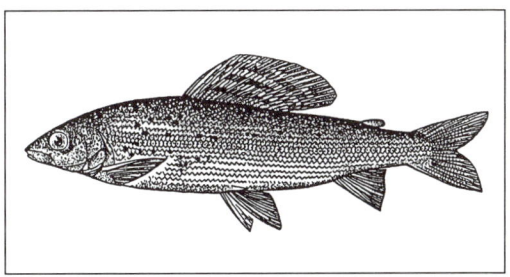

| **269** | **PLECOGLOSSIDAE** | PCA | **271** | **OSMERIDAE** | PSM |

269 PLECOGLOSSIDAE   PCA

SC *Plecoglossus altivelis* (Temminck and Schlegel, 1846)
ES ayu
DA ayu
DE Ayu
EL αγιού
EN ayu; ayu sweetfish
FR ayu
IT ayu
NL ayu
PT peixe doce
FI aju
SV ayu

271 OSMERIDAE   PSM

SC *Hypomesus olidus* (Pallas, 1811)
ES eperlano de estanque
DA ferskvandssmelt
DE Süßwasser-Stint
EL μικρόστομος επερλάνος
EN pond smelt
FR éperlan à petite bouche
IT sperlano
NL vijverspiering
PT eperlano de tanque
FI lampikuore
SV sötnors

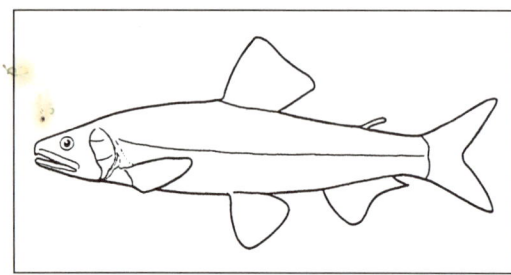

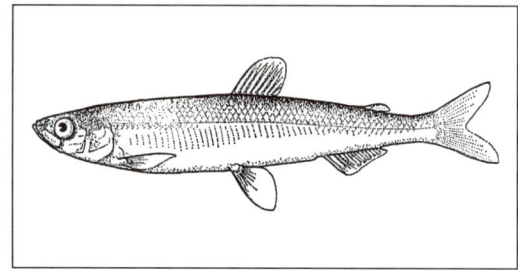

270 OSMERIDAE

SC *Osmeridae*
ES eperlanos
DA smelt-familien
DE Stinte
EL επερλάνοι· οσμερίδες
EN smelts
FR éperlans
IT sperlani
NL spieringen
PT eperlanos
FI kuoreet; kuoreet-heimo
SV norsfiskar

272 OSMERIDAE   SUS

SC *Hypomesus pretiosus* (Girard, 1854)
ES eperlano del Pacífico
DA Stillehavs-smelt
DE Kleinmäuliger Kalifornischer Seestint
EL επερλάνος του Ειρηνικού
EN surf smelt
FR éperlan du Pacifique
IT sperlano del Pacifico
NL Californische spiering
PT eperlano do Pacífico
FI tyrskykuore
SV surfnors

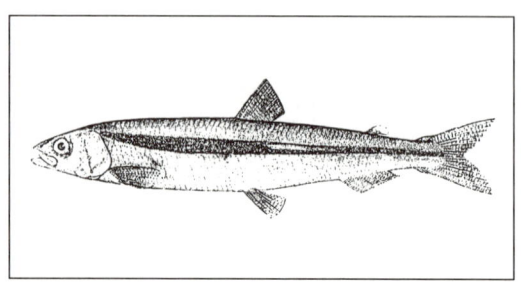

273	**OSMERIDAE**	CAP	275	**OSMERIDAE**	SMR

273   **OSMERIDAE**   CAP

SC   *Mallotus villosus* (Müller, 1776)
ES   capelán
DA   lodde
DE   Lodde; Kapelan
EL   καπελάνος
EN   capelin; caplin
FR   capelan; capelan atlantique
IT   capelin
NL   lodde
PT   capelim
FI   villakuore
SV   lodda

275   **OSMERIDAE**   SMR

SC   *Osmerus mordax* (Mitchill, 1814)
ES   eperlano arco iris
DA   amerikansk smelt
DE   Regenbogen-Stint
EL   ιριδίζων επερλάνος· επερλάνος της Αμερικής
EN   rainbow smelt; American smelt
FR   éperlan arc-en-ciel; éperlan d'Amérique
IT   sperlano; eperlano
NL   Amerikaanse spiering
PT   eperlano arco-íris
FI   amerikankuore
SV   regnbågsnors

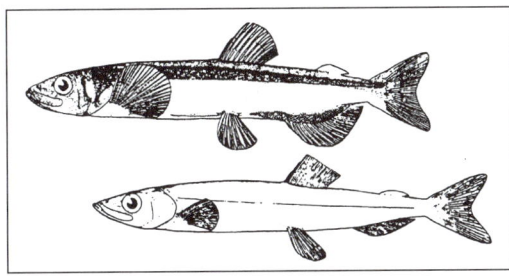

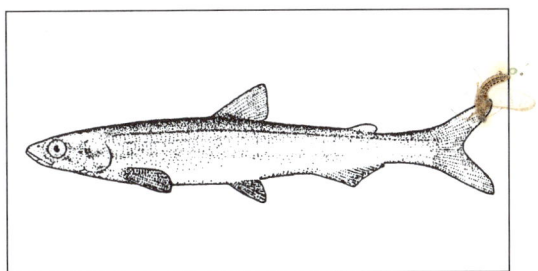

274   **OSMERIDAE**   SME

SC   *Osmerus eperlanus* (Linnaeus, 1758)
ES   eperlano; eperlán
DA   smelt; europæisk smelt
DE   Stint; Spierling; Wanderstint; Seestint; Heilstint; Binnenstint
EL   επερλάνος
EN   smelt; sparling; European smelt
FR   éperlan; éperlan d'Europe
IT   eperlano; sperlano
NL   spiering
PT   eperlano europeu
FI   kuore
SV   nors

276   **OSMERIDAE**

SC   *Osmerus mordax dentex* (Steindachner, 1870); *Osmerus dentex*
ES   eperlano ártico
DA   arktisk smelt
DE   Asiatischer Stint
EL   επερλάνος της Αρκτικής
EN   Arctic smelt; Asiatic smelt; rainbow smelt; boreal smelt
FR   éperlan arctique; éperlan; éperlan de l'Arctique
IT   sperlano; eperlano
NL   Arctische spiering
PT   eperlano árctico
FI   kuore-laji
SV   arktisk regnbågsnors

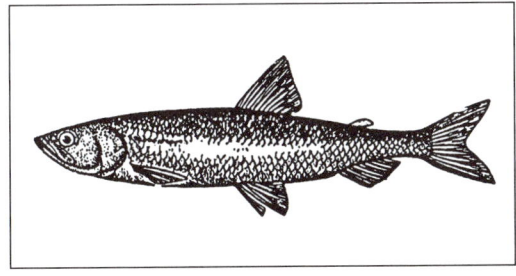

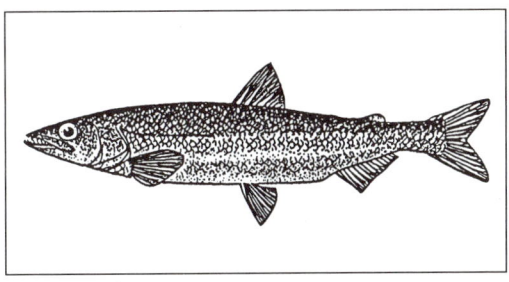

277 OSMERIDAE SMX

SC	*Osmerus* spp.; *Hypomesus* spp.
ES	eperlanos
DA	smelt-slægter
DE	Osmerus-Stinte
EL	επερλάνοι
EN	smelts
FR	éperlans
IT	sperlani
NL	spieringen
PT	eperlanos
FI	kuore
SV	norsar

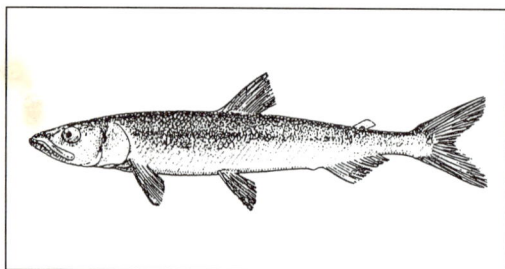

279 ARGENTINIDAE ARG

SC	*Argentinidae*
ES	peces plata; argentinas
DA	guldlaks-familien
DE	Glasaugen; Goldlachse
EL	γουρλομάτηδες
EN	argentines
FR	argentines
IT	argentine; argentinidi
NL	zilversmelten
PT	argentinídeos
FI	hopeakuoreet; hopeakuoreet-heimo
SV	guldlaxfiskar

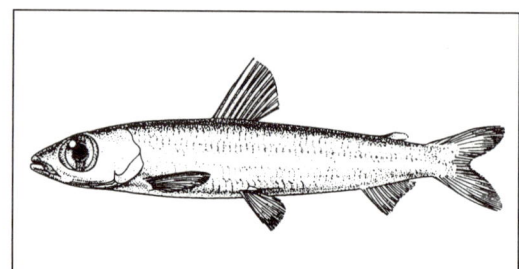

278 OSMERIDAE EUL

SC	*Thaleichthys pacificus* (Richardson, 1836)
ES	eulacón; eperlón del Pacífico
DA	kærtefisk
DE	Kerzenfisch
EL	επερλάνος
EN	eulachon; candlefish
FR	eulachon; eulakane
IT	sperlano striato
NL	kaarsvis
PT	eulacom
FI	kynttiläkuore
SV	ljusfisk

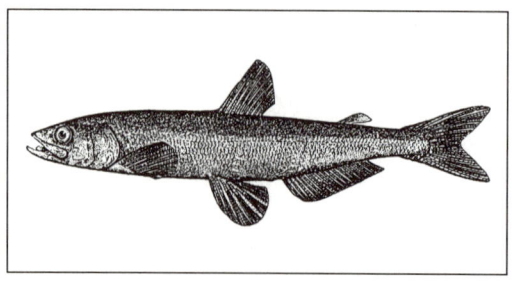

280 ARGENTINIDAE

SC	*Argentina sialis* (Gilbert, 1890)
ES	argentina; pejerrey; pez de plata
DA	Stillehavs-guldlaks
DE	Pazifisches Glasauge; Glasauge; Goldlachs
EL	γουρλομάτης
EN	Pacific argentine
FR	argentine du Pacifique
IT	argentina
NL	Pacifische zilversmelt
PT	argentina do Pacífico
FI	kuore-laji
SV	stillahavsguldlax

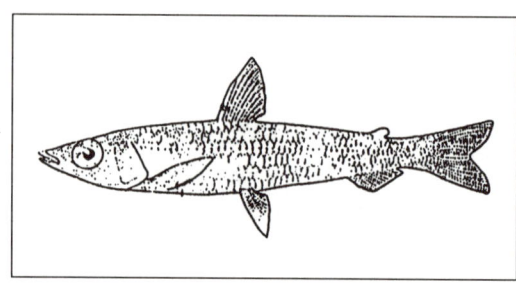

281 ARGENTINIDAE

SC	*Argentina silus* (Ascanius, 1775)
ES	pejerrey; argentina; pez de plata grande
DA	guldlaks
DE	Goldlachs; Glasauge; Großauge
EL	γουρλομάτης του Ατλαντικού
EN	great silver smelt; smelt; herring smelt; Atlantic argentine
FR	grande argentine
IT	argentina
NL	grote zilversmelt
PT	argentina dourada
FI	kultakuore
SV	guldlax; strömsill

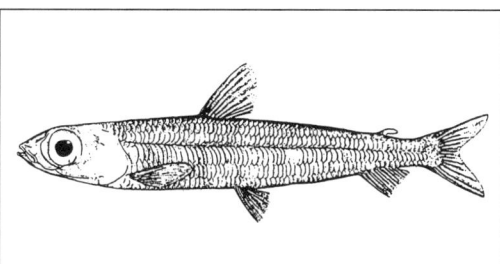

283 ARGENTINIDAE   DES

SC	*Glossanodon semifasciatus* (Kishinouye, 1904); *Argentina semifasciata*
ES	argentina del Pacífico
DA	japansk guldlaks
DE	Japanisches Glasauge; Goldlachs
EL	γουρλομάτης του Ειρηνικού
EN	Pacific argentine; deepsea smelt
FR	argentine du Japon
IT	argentina
NL	Japanse zilversmelt
PT	argentina do Japão
FI	kuore-laji
SV	japansk guldlax

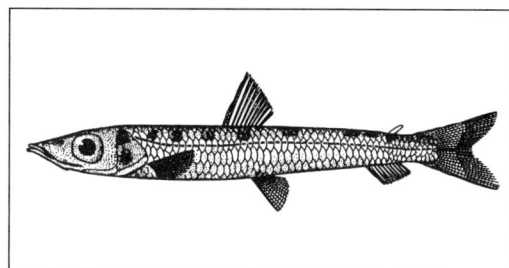

282 ARGENTINIDAE

SC	*Argentina sphyraena* (Linnaeus, 1758)
ES	pez plata; pez plata pequeño
DA	strømsild
DE	Glasauge
EL	γουρλομάτης
EN	lesser silver smelt
FR	argentine; petite argentine
IT	argentina
NL	kleine zilversmelt
PT	argentina branca; biqueirão branco
FI	hopeakuore
SV	silverfisk

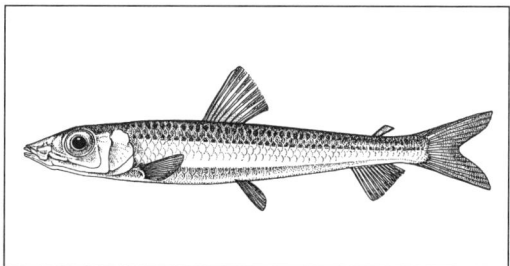

284 UMBRIDAE

SC	*Umbra krameri* (Walbaum, 1792)
ES	umbra común
DA	europæisk hundefisk
DE	Europäischer Hundsfisch; Hundsfisch
EL	λασποκυπρίνος της Ευρώπης
EN	European mud-minnow
FR	poisson-chien d'Europe
IT	umbra
NL	Europese hondsvis
PT	peixinho da lama europeu
FI	koirakala
SV	hundfisk

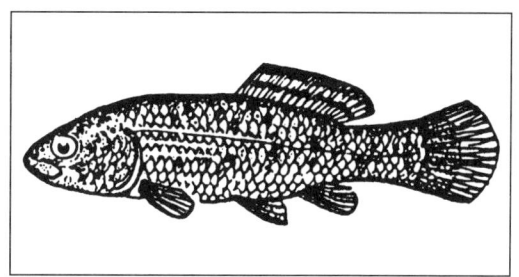

285 UMBRIDAE

SC	*Umbra pygmaea* (De Kay, 1842)
ES	umbra pigmea
DA	lille hundefisk; amerikansk hundefisk
DE	Amerikanischer Hundsfisch; Zwerghundsfisch
EL	λασποκυπρίνος της Αμερικής
EN	American mud-minnow; Eastern mud-minnow
FR	petit poisson-chien d'Amérique
IT	umbra americana
NL	Amerikaanse hondsvis
PT	peixinho da lama americano
FI	pikkukoirakala
SV	dvärghundfisk

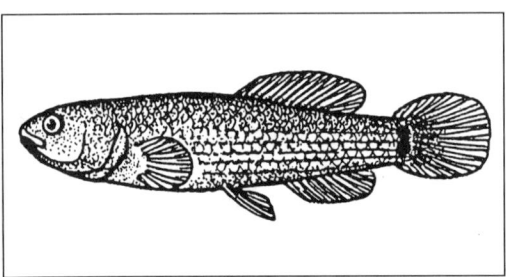

286 ESOCIDAE

SC	*Esocidae*
ES	lucios
DA	gedde-familien
DE	Hechte
EL	τούρνες
EN	pikes; pickerels
FR	brochets
IT	lucci
NL	snoeken
PT	lúcios
FI	hauet; hauet-heimo
SV	gäddfiskar

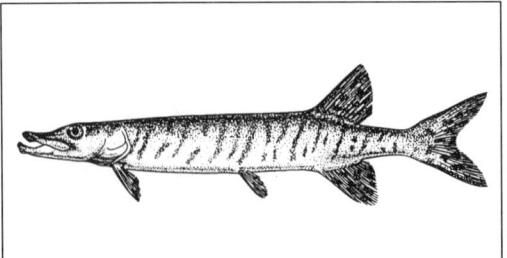

287 ESOCIDAE

SC	*Esox americanus vermiculatus* (LeSueur, 1846); *Esox vermiculatus*
ES	lucio listado
DA	græsgedde
DE	Maß-Hecht
EL	τούρνα της Αμερικής
EN	grass pickerel
FR	brochet vermiculé
IT	luccio nordamericano
NL	Amerikaanse grassnoek
PT	lúcio listado
FI	amerikanhauki
SV	gräspickerell

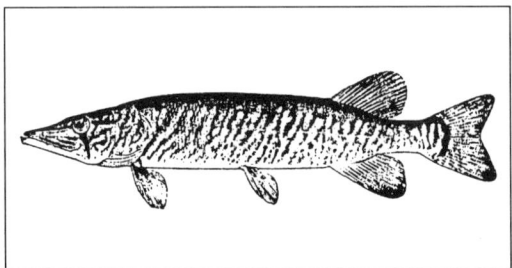

288 ESOCIDAE   FPI

SC	*Esox lucius* (Linnaeus, 1758)
ES	lucio
DA	gedde; almindelig gedde
DE	Hecht; Flußhecht
EL	τούρνα
EN	pike; jack; Northern pike
FR	grand brochet; brochet du Nord; brochet européen; bec; brochet commun; becquet; buché
IT	luccio
NL	snoek
PT	lúcio
FI	hauki
SV	gädda

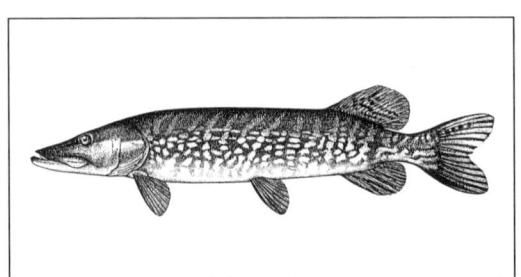

289 ESOCIDAE

SC	*Esox masquinongy* (Mitchill, 1824)
ES	lucio masquinongy
DA	muskellunge
DE	Muskellunge
EL	τούρνα
EN	muskellunge; maskinonge
FR	maskinongé
IT	luccio nordamericano
NL	maskulonge
PT	lúcio mosqueado
FI	jättihauki
SV	maskalung

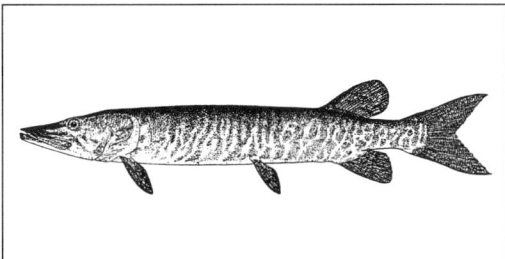

291 ESOCIDAE   AMU

SC	*Esox reicherti* (Dybowski, 1869)
ES	lucio del Pacífico
DA	Amur-gedde
DE	Amur-Hecht
EL	τούρνα του Ειρηνικού
EN	Amur pike
FR	brochet du Pacifique; brochet à points noirs
IT	luccio del Pacifico
NL	Amoer-snoek
PT	lúcio do Pacífico
FI	täplähauki
SV	amurgädda

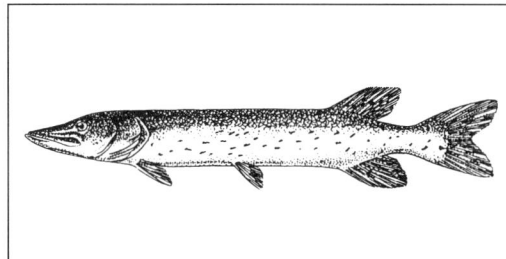

290 ESOCIDAE

SC	*Esox niger* (Lesueur, 1818)
ES	lucio negro
DA	østamerikansk gedde
DE	Kettenhecht
EL	μαύρη τούρνα
EN	chain pickerel
FR	brochet maillé; brochet noir américain
IT	luccio nordamericano
NL	zwarte snoek
PT	lúcio malhado
FI	mustahauki
SV	kedjepickerell

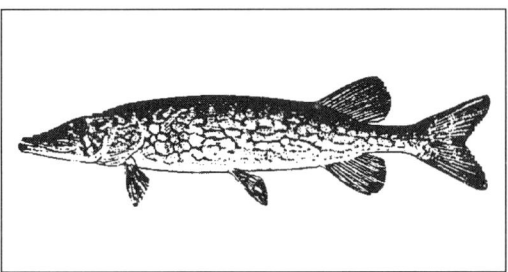

292 NOTOPTERIDAE   FKN

SC	*Notopterus* spp.
ES	peces cuchillo
DA	knivbladfisk-slægt
DE	Fähnchen-Messerfische
EL	μαχαιρόψαρα
EN	knifefishes
FR	poissons-couteaux
IT	pesci coltello
NL	mesvissen
PT	peixes-faca
FI	teräkala-suku
SV	knivfiskar

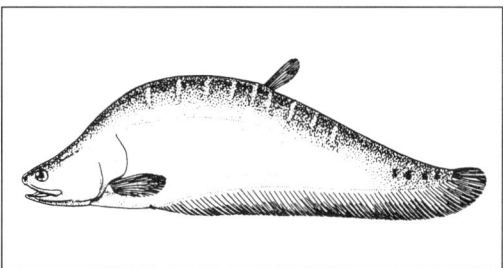

293	**MOCHOKIDAE**	CSY	**295**	**MYCTOPHIDAE**	LAN

SC	*Synodontis* spp.
ES	sinodontis
DA	malle-slægt
DE	Fiederbartwelse
EL	γατόψαρα
EN	upsidedown catfishes
FR	poissons-chats nettoyeurs
IT	pesci pulitori
NL	rugzwemmende meerval
PT	gatos cabeçudos
FI	ripsimonni-suku
SV	fjädertömsmalar

SC	*Lampanyctodes hectoris* (Günther, 1876)
ES	pez linterna
DA	Hektors prikfisk
DE	Hektor-Laternenfisch
EL	λυχναρόψαρο
EN	lanternfish
FR	poisson-lanterne
IT	pesce lanterna
NL	lantaarnvis
PT	peixe-lanterna
FI	valokala-laji
SV	hector-prickfisk

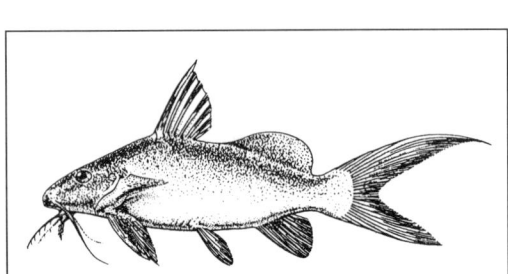

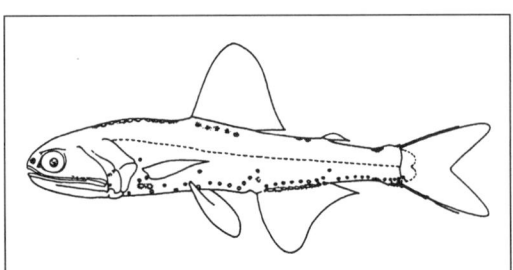

294	**MYCTOPHIDAE**	LXX	**296**	**HARPADONTIDAE**

SC	*Myctophidae*
ES	peces linterna
DA	prikfisk-familien
DE	Laternenfische
EL	λυχναρόψαρα
EN	lanternfishes
FR	poissons-lanternes
IT	mictofidi; pesci lanterna
NL	lantaarnvissen
PT	peixes-lanterna
FI	valokalat; valokalat-heimo
SV	prickfiskar

SC	*Harpadontidae*
ES	bombay
DA	Bombay-and-familien
DE	Bombay-Enten
EL	μπαμαλίδες
EN	Bombay ducks
FR	bombay
IT	bumali
NL	Bombay-eenden
PT	bumblins
FI	bummalot; bummalot-heimo
SV	bombay-ankor

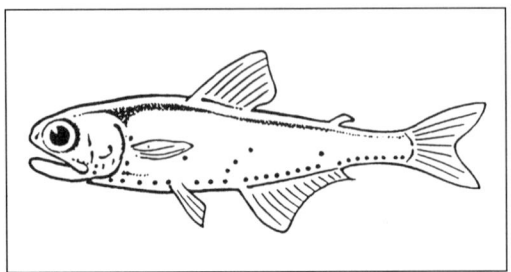

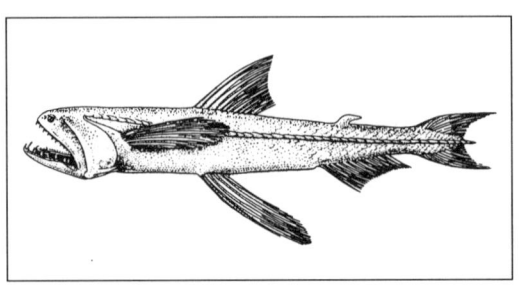

297 HARPADONTIDAE   BUC

SC *Harpadon nehereus* (Hamilton-Buchanan, 1822)
ES bumalo
DA Bombay-and
DE Bombay-Ente
EL μπάμαλο
EN Bombay duck; bumalo; bummalow
FR scopélidé; bombay duck; bumalo; bummalow
IT bumalo
NL Bombay-eend
PT bumblim
FI bummalo
SV bombay-anka

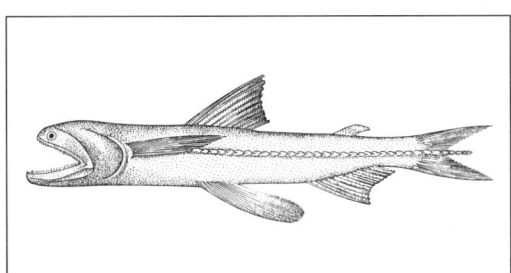

299 CHLOROPHTHALMIDAE

SC *Chlorophthalmus agassizi* (Bonaparte, 1840);
    Chlorophthalmus productus
ES cloroftalmo; ojiverde
DA grønøje
DE Grünauge
EL πρασινομάτης· γουρλομάτης
EN green-eye; short-nose green-eye
FR éperlan du large; yeux-verts à nez court
IT occhione; occhiverdi
NL groenoog
PT olho verde
FI vihersilmä
SV grönögonfisk

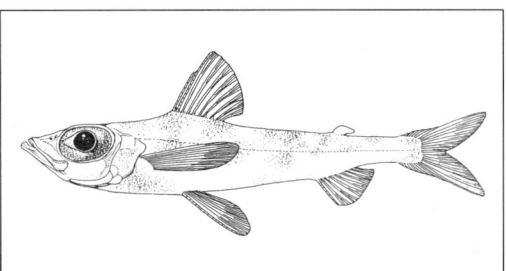

298 CHLOROPHTHALMIDAE   GRE

SC *Chlorophthalmidae*
ES cloroftálmidos
DA grønøje-familien
DE Grünaugen
EL πρασινομάτηδες· γουρλομάτηδες
EN spiderfish; greeneyes
FR yeux-verts
IT occhioni; occhiverdi
NL groenogen
PT olhos verdes
FI vihersilmät; vihersilmät-heimo
SV grönögonfiskar

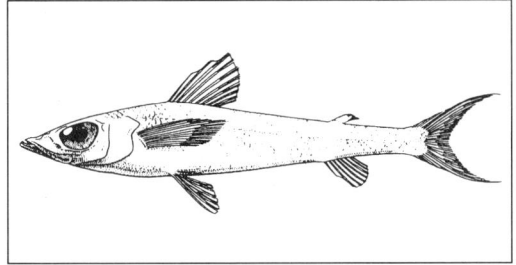

300 CHARACIDAE   CHA

SC *Characidae*
ES carácidos
DA karpelaks-familien
DE Salmler
EL χαρακίδες
EN characins; tetras
FR characidés piranhas et hydrocyons; tétra
IT caracidi
NL karperzalmen
PT caracinídeos
FI tetrat; tetrat-heimo
SV laxkarpar; tetror

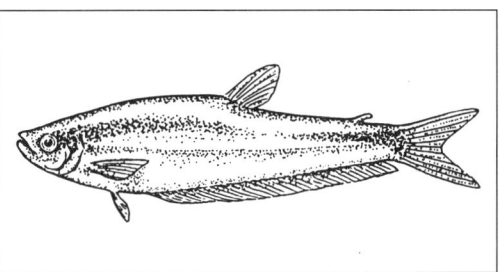

301 CATOSTOMIDAE

SC Catostomidae
ES chupadores
DA sugekarpe-familien
DE Sauger
EL γατοστόματα
EN suckers
FR cyprins-sucets
IT catostomidi
NL zuigkarpers
PT chupadores
FI imukarpit; imukarpit-heimo
SV sugkarpar

303 CYPRINIDAE FCY

SC Cyprinidae
ES ciprínidos; carpas
DA karpe-familien
DE Karpfenfische; Weißfische
EL κυπρινοειδή
EN carps; minnows
FR cyprinidés
IT ciprinidi
NL karpers
PT ciprinídeos
FI särkikalat; särkikalat-heimo
SV karpfiskar

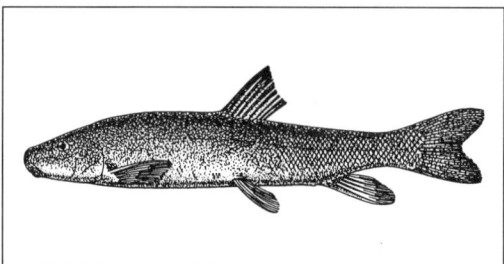 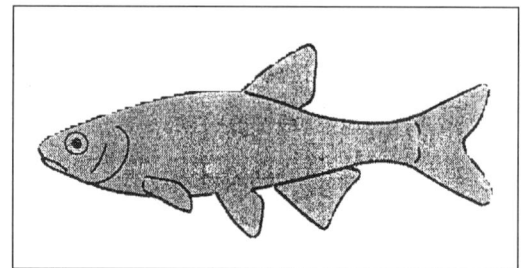

302 CATOSTOMIDAE BUF

SC Ictiobus spp.
ES peces toro
DA sugekarpe-slægt
DE Sauger
EL βοϊδόψαρο
EN buffalofishes
FR poissons-taureaux; cyprins-sucets
IT pesci bisonte
NL zuigkarper
PT peixes-búfalo
FI imukarppi-suku
SV buffelfiskar

304 CYPRINIDAE

SC Abramis ballerus (Linnaeus, 1758)
ES zope
DA brasenflire
DE Zope; Spitzpleinzen; Schwuppe; Pleinzen;
   Schwabe; Spitzer; Sporn; Flixe
EL ζόπα
EN zope
FR zope
IT zope
NL brasemblei
PT brema azul
FI sulkava
SV faren

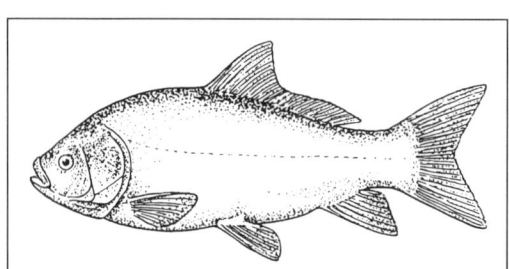 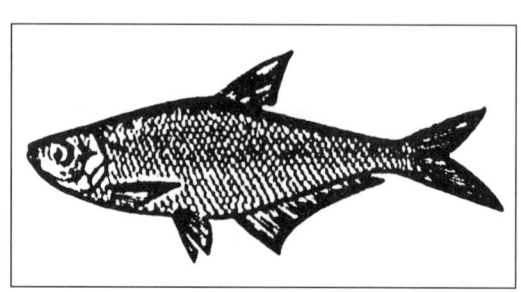

305 CYPRINIDAE   FBM

SC	*Abramis brama* (Linnaeus, 1758)
ES	brema común
DA	brasen
DE	Brachsen; Brassen; Blei; Breitling; Bressen; Bresen; Brassenplieten; Schlaffke; Halbfisch; Scheibpleinzen; Brachmen
EL	λεστιά
EN	freshwater bream; common bream; bream
FR	brème d'eau douce; brème commune
IT	abramide
NL	brasem
PT	brema
FI	lahna
SV	braxen

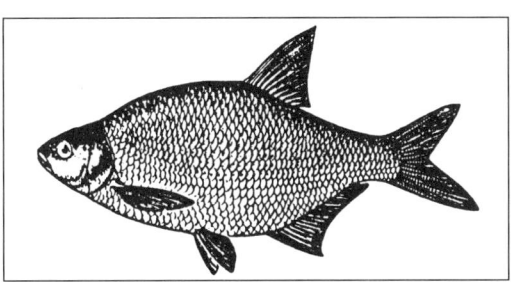

307 CYPRINIDAE   FBR

SC	*Abramis* spp.
ES	bremas
DA	brasen-slægten
DE	Brachsen; Brassen
EL	λεστιές
EN	freshwater breams; breams
FR	brèmes d'eau douce; brèmes
IT	abramidi
NL	brasems
PT	bremas
FI	lahna-suku
SV	braxar

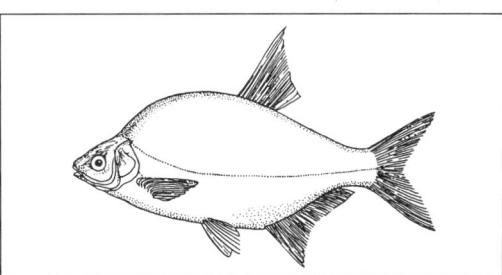

306 CYPRINIDAE

SC	*Abramis sapa* (Pallas, 1811)
ES	brema del Danubio
DA	Donau-brasen
DE	Zobel; Scheibpleinzen; Halbbrachsen; Donaubrachsen; Kanov; Sape; Pleinzen
EL	λεστιά του Δούναβη
EN	Danube bream
FR	brème du Danube
IT	abramide del Danubio
NL	Donaubrasem
PT	brema do Danúbio
FI	sapa
SV	donaubraxen

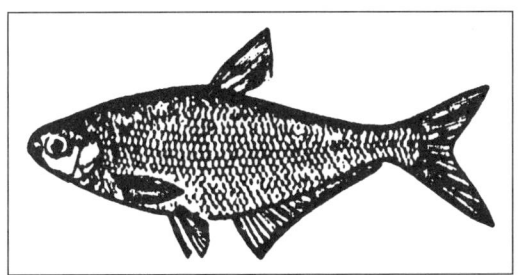

308 CYPRINIDAE

SC	*Alburnoides bipunctatus* (Bloch, 1728); *Spirlinus bipunctatus*
ES	alburno bimaculado
DA	strømløje
DE	Schneider; Alandblecke; Laube; Riemling; Schusslaube; Reitblecke; Streifling; Flecke
EL	πλατίτσα
EN	schneider
FR	ablette de rivière; spirlin; ablette grise; lavette
IT	alburno di fiume
NL	gestippelde alver
PT	alburno pintado
FI	punasalakka
SV	strömlöja

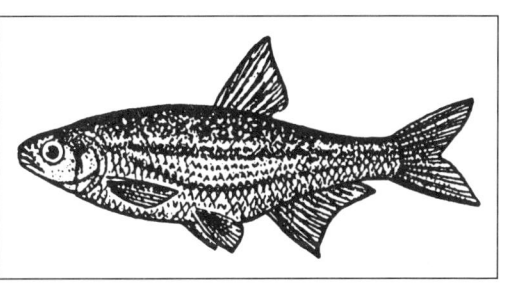

309 CYPRINIDAE

SC *Alburnus albidus* (Costa, 1838)
ES alburno italiano
DA italiensk løje
DE Weißer Ukelei; Laube; Zumpel
EL τσιρώνι· μπελοβίτσα
EN Italian bleak
FR ablette italienne; ablette blanche; ovelle; mirandelle
IT alborella del Vulture
NL witte alver
PT alburno italiano
FI italiansalakka
SV vitlöja

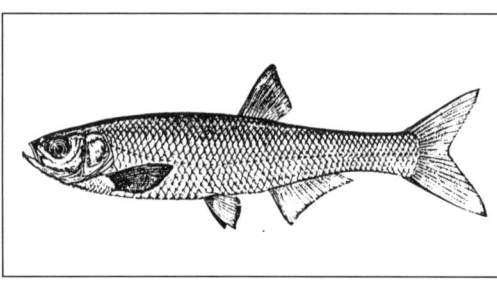

311 CYPRINIDAE

SC *Aspius aspius* (Linnaeus, 1758)
ES aspio
DA asp
DE Rapfen; Schied; Raap; Rappe; Mülpe; Zalat; Rotschiedl
EL ασπρόψαρο· ασπρογρίβαδο
EN asp
FR aspe
IT aspio
NL roofblei
PT áspio
FI toutain
SV asp

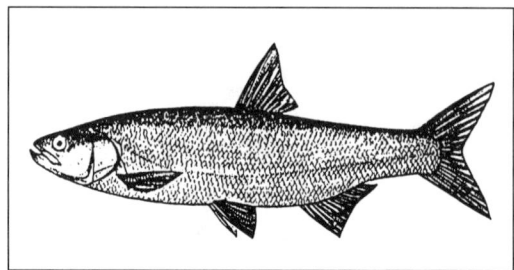

310 CYPRINIDAE

SC *Alburnus alburnus* (Linnaeus, 1758); *Alburnus lucidus*
ES alburno
DA løje; almindelig løje
DE Laube; Ukelei; Albe; Albola; Blicke; Maiblecke; Uckelin; Laugele; Alwe; Lauel; Spitzlaube; Schupper
EL σίρκο· μπελοβίτσα
EN bleak
FR ablette
IT alborella
NL alver
PT alburno
FI salakka
SV löja

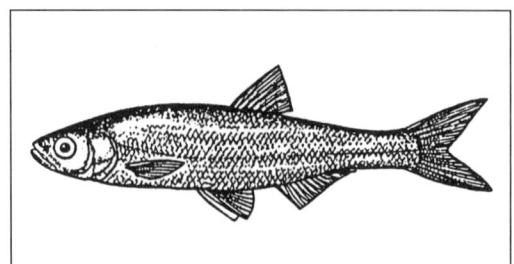

312 CYPRINIDAE

SC *Aulopyge huegelii* (Heckel, 1841)
ES gobio dálmata
DA dalmatisk barbe-grundling
DE Barbengründling
EL μουστακογωβιός της Δαλματίας
EN Dalmatian barbel-gudgeon
FR goujon-barbeau de Dalmatie
IT barbo dalmata
NL Dalmatische barbeelgrondel
PT barbo da Dalmácia
FI dalmatiantörö
SV nakenbarb

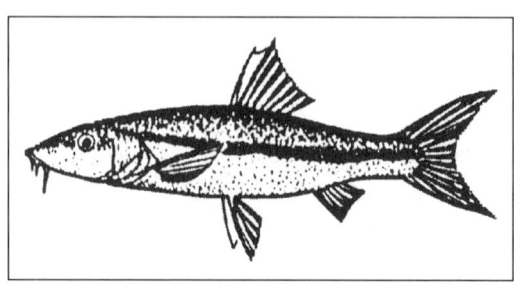

313 CYPRINIDAE

SC	*Barbus albanicus* (Steindachner, 1895)
ES	barbo albanés
DA	albansk barbe
DE	Albanische Barbe
EL	στρωσίδι· μαρίτσα
EN	Albanian barbel
FR	barbeau d'Albanie
IT	barbo albanese
NL	Albanese barbeel
PT	barbo da Albânia
FI	albanianbarbi
SV	albansk barb

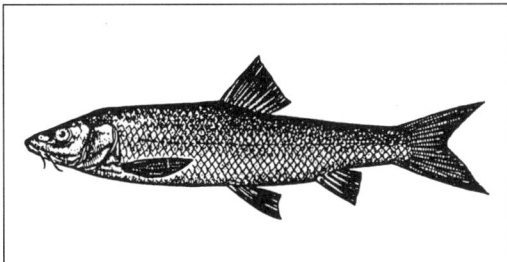

315 CYPRINIDAE

SC	*Barbus comiza* (Steindachner, 1866)
ES	barbo comiza
DA	iberisk barbe
DE	Iberische Barbe
EL	μπριάνα της Ισπανίας
EN	Iberian barbel
FR	barbeau ibérique
IT	barbo iberico
NL	Iberische barbeel
PT	barbo-cumba
FI	espanjanbarbi
SV	iberisk barb

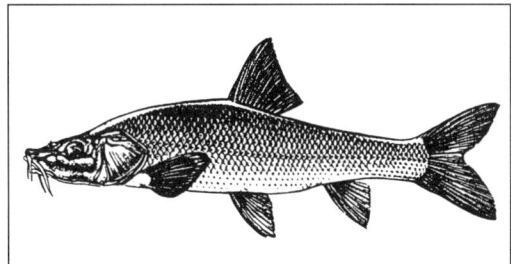

314 CYPRINIDAE   PTB

SC	*Barbus barbus* (Linnaeus, 1758)
ES	barbo común; barbito; becut
DA	flodbarbe
DE	Barbe; Barben; Barbine
EL	μπριάνα· ποταμολαύρακο· μουστακάτο
EN	barbel
FR	barbeau commun; barbeau
IT	barbo
NL	barbeel
PT	barbo
FI	jokibarbi
SV	flodbarb

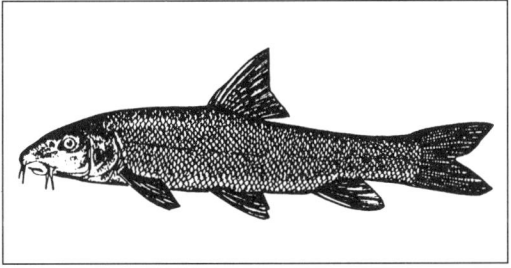

316 CYPRINIDAE

SC	*Barbus meridionalis* (Risso, 1826)
ES	barbo de montaña
DA	Middelhavs-barbe
DE	Hundsbarbe
EL	μπριάνα· χαμουζαίος
EN	Mediterranean barbel
FR	barbeau méridional; canin; barp; durgan
IT	stronazza; barbo canino
NL	mediterrane barbeel
PT	barbo meridional
FI	välimerenbarbi
SV	medelhavsbarb

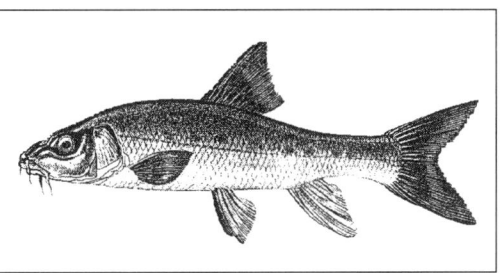

317 CYPRINIDAE

SC	*Barbus meridionalis petenyi* (Heckel, 1847)
ES	barbo del Danubio
DA	semling
DE	Semling
EL	μπρένα
EN	Danubian barbel
FR	barbeau du Danube
IT	barbo del Danubio
NL	Donaubarbeel
PT	barbo do Danúbio
FI	tonavanbarbi
SV	donaubarb

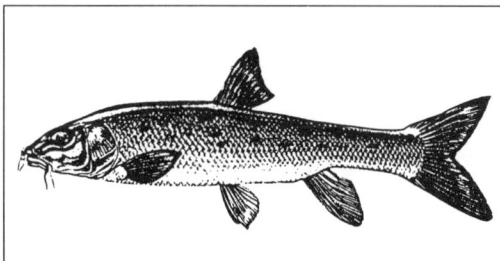

319 CYPRINIDAE

SC	*Blicca bjoerkna* (Linnaeus, 1758); *Abramis bjoerkna*
ES	brema blanca
DA	flire
DE	Blicke; Güster; Pliete; Halbbrachsen
EL	ασημένια λεστιά
EN	silver bream
FR	brème bordelière
IT	blicca
NL	kolblei
PT	brema prateada
FI	pasuri
SV	björkna

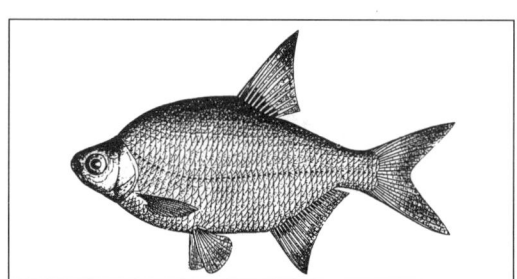

318 CYPRINIDAE

SC	*Barbus plebejus* (Valenciennes, 1829)
ES	barbo italiano
DA	italiensk barbe
DE	Italienische Barbe
EL	μπριάνα· μουστακάτο
EN	Italian barbel
FR	barbeau italien
IT	barbo
NL	Italiaanse barbeel
PT	barbo italiano
FI	italianbarbi
SV	italiensk barb

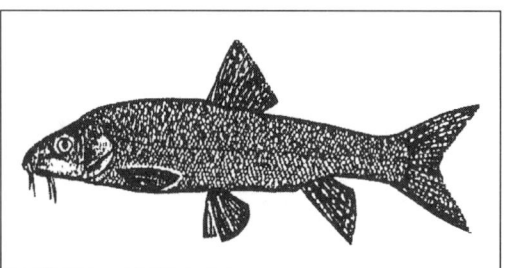

320 CYPRINIDAE CGO

SC	*Carassius auratus* (Linnaeus, 1758)
ES	pez rojo; pez dorado
DA	guldfisk; sølvkarusse
DE	Goldfisch; Silberkarausche
EL	χρυσόψαρο
EN	goldfish
FR	carassin doré; carassin rouge; poisson rouge; cyprin doré; carassin vulgaire
IT	ciprino dorato; pesce rosso; carassio dorato
NL	goudvis; giebel
PT	peixe-encarnado; peixe-dourado
FI	kultakala
SV	guldfisk

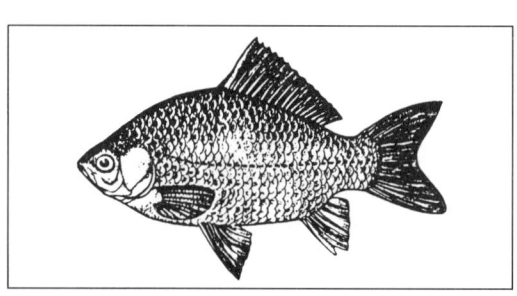

321 CYPRINIDAE FCC

SC *Carassius carassius* (Linnaeus, 1758)
ES carpín; carpa gibel; panzón
DA damkarusse; karusse; søkarusse
DE Karausche; Giebel; Bauernkarpfen; Moorkarpfen;
    Gareisle; Goldkarpfe; Steinkarpfe; Potkarpfe;
    Schneiderkarpfe; Geibel; Mölenke; Kotschebel;
    Karras; Burretschel; Gareisel; Boretsch; Breitling
EL πεταλούδα
EN crucian carp
FR carassin commun; carouche; carreau; cyprin;
    carassin
IT carassio; carassio comune
NL kroeskarper
PT pimpão comum
FI ruutana
SV ruda

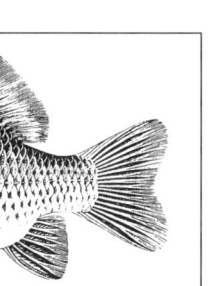

322 CYPRINIDAE

SC *Chalcalburnus chalcoides* (Güldenstädt, 1772)
ES alburno del Danubio
DA Donau-løje
DE Mairenke; Schemaja
EL αλάια του Δούναβη
EN Danube bleak
FR ablette du Danube
IT alburno danubiano
NL Donau-alver
PT alburno do Danúbio
FI tonavansalakka
SV donaulöja

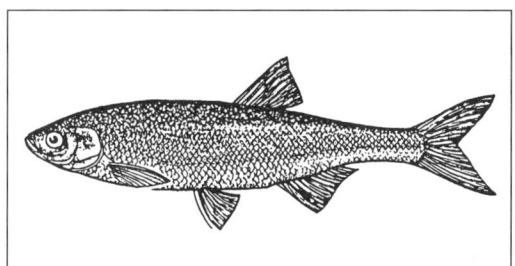

323 CYPRINIDAE

SC *Chondrostoma genei* (Bonaparte, 1839)
ES condrostoma sudeuropeo
DA sydeuropæisk næsling
DE Lau
EL συρτάρι
EN souffie; South European nase
FR nase de l'Europe du Sud
IT lasca
NL Laskasneep
PT boga sul-europeia
FI huulisärki
SV lasca-näsling

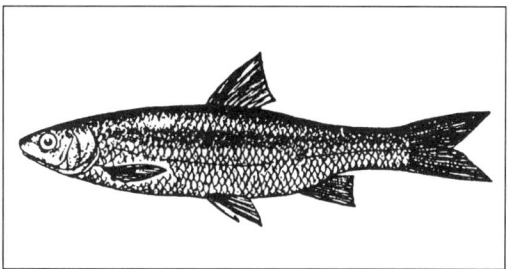

324 CYPRINIDAE

SC *Chondrostoma kneri* (Heckel, 1843)
ES condrostoma dálmata
DA dalmatisk næsling
DE Dalmatinischer Näsling
EL συρτάρι της Δαλματίας
EN Dalmatian nase
FR nase dalmate
IT lasca dalmatina
NL Dalmatische sneep
PT boga da Dalmácia
FI dalmatiannokkasärki
SV dalmatisk näsling

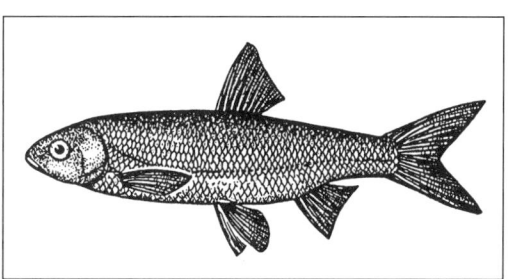

325 CYPRINIDAE

SC	*Chondrostoma nasus* (Linnaeus, 1758)
ES	condrostoma común
DA	næse
DE	Nase
EL	συρτάρι· γουρουνομύτης
EN	nase; common nase
FR	nase commun; hotu
IT	naso
NL	sneep
PT	boga do Danúbio
FI	nokkasärki
SV	näsling

327 CYPRINIDAE

SC	*Chondrostoma polylepis* (Steindachner, 1865)
ES	boga de río
DA	iberisk næsling
DE	Iberischer Näsling
EL	συρτάρι της Ισπανίας
EN	Iberian nase
FR	nase ibérique
IT	lasca iberica
NL	Iberische sneep
PT	boga de boca direita
FI	espanjannokkasärki
SV	boga; boganäsling

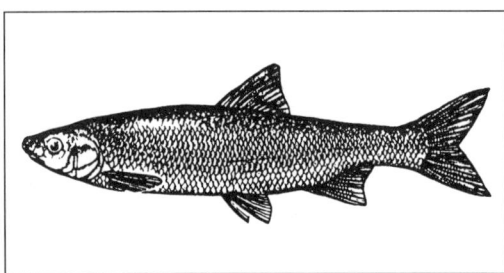

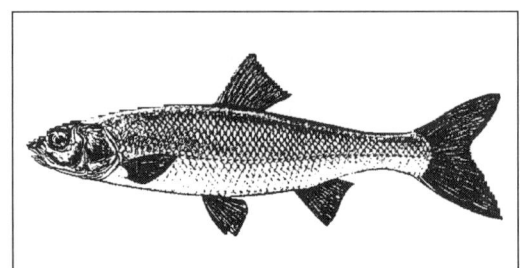

326 CYPRINIDAE

SC	*Chondrostoma phoxinus* (Heckel, 1843)
ES	condrostoma foxino
DA	elritsenæsling
DE	Elritzennäsling
EL	συρτάρι
EN	minnow-nase
FR	nase-vairon
IT	sanguinerola; lasca
NL	elritssneep
PT	boga da Jugoslávia
FI	pikkusuomunokkasärki
SV	bäcknäsling

328 CYPRINIDAE

SC	*Chondrostoma soetta* (Bonaparte, 1840)
ES	condrostoma italiano
DA	italiensk næsling
DE	Italienischer Näsling
EL	συρτάρι της Ιταλίας
EN	Italian nase
FR	nase d'Italie
IT	savetta
NL	Italiaanse sneep
PT	boga italiana
FI	italiannokkasärki
SV	savetta; savetta-näsling

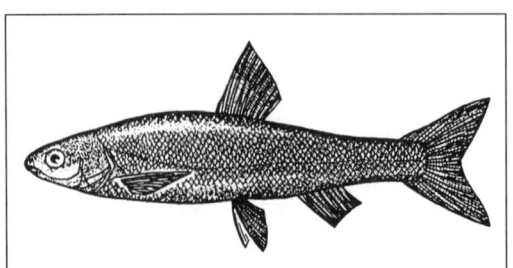

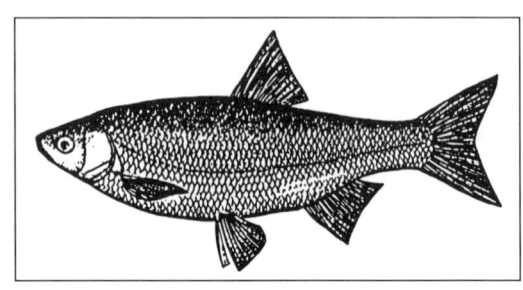

329 CYPRINIDAE

SC	*Chondrostoma toxostoma* (Vallot, 1837)
ES	madrilla
DA	sydvesteuropæisk næsling
DE	Südwesteuropäischer Näsling; Strömer
EL	συρτάρι
EN	Southwest European nase; soufie
FR	nase de l'Europe du Sud-Ouest; soffie; seuffe; suiffe; blageon
IT	lasca
NL	Franse sneep
PT	boga de boca curva
FI	pikkunokkasärki
SV	trubbnäsling

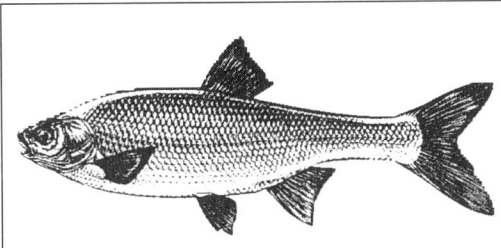

330 CYPRINIDAE   MUC

SC	*Cirrhinus molitorella* (Cuvier and Valenciennes, 1844)
ES	carpa de fango
DA	mudderkarpe
DE	Schlammkarpfen
EL	λασποκυπρίνος
EN	mud carp
FR	carpe de vase
IT	carpa di fango
NL	modderkarper
PT	carpa da lama
FI	särkikala-laji
SV	dykarp

DRAWING NOT AVAILABLE

331 CYPRINIDAE   FCG

SC	*Ctenopharyngodon idella* (Valenciennes, 1844)
ES	carpa china
DA	græskarpe
DE	Graskarpfen
EL	χορτοκυπρίνος
EN	grass carp; white amur
FR	carpe herbivore
IT	carpa erbivora; amur
NL	graskarper
PT	carpa do limo
FI	ruohokarppi
SV	gräskarp

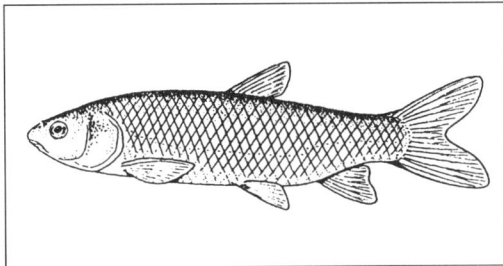

332 CYPRINIDAE   FCP

SC	*Cyprinus carpio* (Linnaeus, 1758)
ES	carpa
DA	karpe; almindelig karpe
DE	Karpfen; Flußkarpfen
EL	κυπρίνος· γριβάδι· σαζάνι
EN	common carp; carp; koi carp; mirror carp
FR	carpe commune; carpe-miroir; carpe
IT	carpa; carpa a specchi
NL	karper; spiegelkarper
PT	carpa; carpa comum; carpa-espelho
FI	karppi
SV	karp

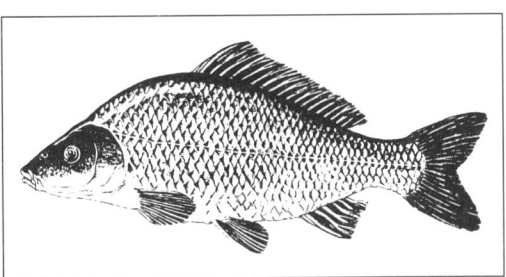

333 CYPRINIDAE

SC	*Gobio gobio* (Linnaeus, 1758); *Gobio fluviatilis*
ES	gobio
DA	grundling; almindelig grundling; sandhest
DE	Gründling; Greßling; Grundel; Gewöhnlicher Gründling; Kreßling; Giffer; Giefen
EL	χρύσκος· σέτσκα
EN	gudgeon
FR	goujon commun; touret; tregan; goiffon
IT	gobione
NL	riviergrondel
PT	góbio
FI	törö
SV	sandkrypare

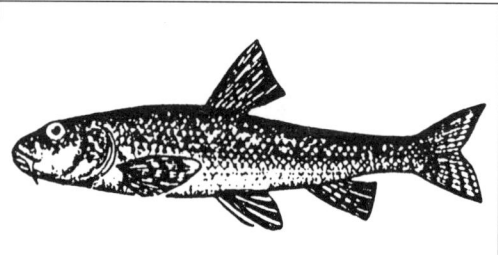

335 CYPRINIDAE   SVC

SC	*Hypophthalmichthys molitrix* (Valenciennes, 1844)
ES	carpa plateada
DA	sølvkarpe
DE	Gewöhnlicher Tolstolob
EL	ασημοκυπρίνος
EN	silver carp
FR	carpe argentée
IT	carpa argentata
NL	zilverkarper
PT	carpa prateada
FI	hopeapaksuotsa
SV	silverkarp

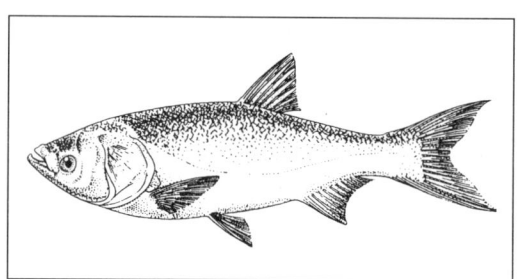

334 CYPRINIDAE

SC	*Gobio uranoscopus* (Agassiz, 1828)
ES	gobio del Danubio
DA	Donau-grundling
DE	Steingreßling; Steingründling
EL	σέτσκα του Δούναβη
EN	Danube gudgeon
FR	goujon du Danube
IT	gobione del Danubio
NL	Donaugrondel
PT	góbio do Danúbio
FI	kivitörö
SV	donausandkrypare

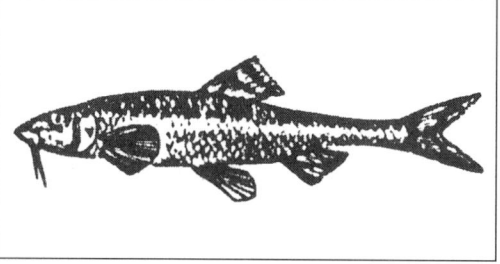

336 CYPRINIDAE   BIC

SC	*Hypophthalmichthys nobilis* (Richardson, 1844); *Aristichys nobilis*
ES	carpa cabezona
DA	marmorkarpe
DE	Edler Tolstolob
EL	κινέζικος κυπρίνος
EN	bighead carp
FR	carpe à grosse tête
IT	carpa testa grossa
NL	grootkopkarper
PT	carpa cabeçuda
FI	marmoripaksuotsa
SV	marmorkarp

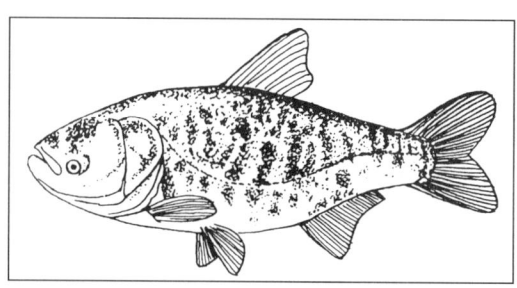

337 CYPRINIDAE

SC *Labeo* spp.
ES labeos
DA hajbarbe-slægten
DE Fransenlipper
EL χειλόψαρα
EN rhinofishes
FR labéos
IT labei
NL franjelipbarbelen
PT beiçudos
FI huulibarbi-suku
SV labeo

339 CYPRINIDAE

SC *Leucaspius delineatus* (Heckel, 1843)
ES alburno rayado
DA regnløje
DE Moderlieschen; Sonnenfischchen; Zwerglaube
EL τσιρωνάκι· τσίμα
EN moderlieschen
FR able de Heckel
IT alburno di Heckel
NL vetje
PT alburno raiado
FI allikkosalakka
SV groplöja

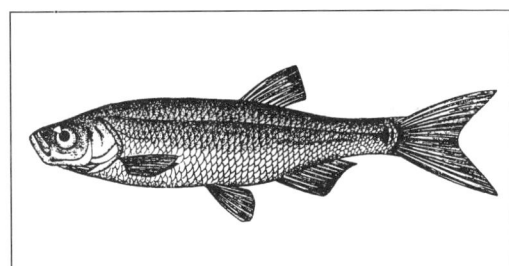

338 CYPRINIDAE

SC *Leptobarbus hoëveni* (Bleeker, 1851)
ES barbo de Hoven
DA Hovens barbe
DE Hovens Karpfen
EL κυπρίνος του Χόβεν
EN Hoven's carp
FR barbus d'Hoven
IT barbo di Hoven
NL Hovens karper
PT barbo de Hoven
FI siaminbarbi
SV slankbarb

340 CYPRINIDAE

SC *Leuciscus borysthenicus* (Kessler, 1859)
ES leucisco de Kessler
DA bobyrez
DE Bobyrez
EL κέφαλος της Μαύρης Θάλασσας
EN bobyrez
FR chevaine de la mer Noire; bobyrez
IT bobirez; cavedano russo
NL Zwarte-Zeekopvoorn
PT escalo do mar Negro
FI bobyretsi
SV bobyrez

DRAWING NOT AVAILABLE

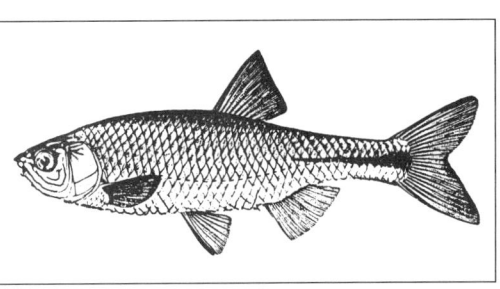

341 CYPRINIDAE

SC	*Leuciscus cephalus* (Linnaeus, 1758); *Squalius cephalus*
ES	leucisco cabezudo; gallego; cacho; molinero
DA	almindelig døbel; døbel
DE	Döbel; Dickkopf; Aitel; Abt; Hartkopf; Deibel; Rohrkarpfen
EL	κέφαλος· κλένι· δροσίνα
EN	chub; skelly; graining
FR	chevesne commun; chevaine commun; meunier; chabot
IT	cavedano
NL	kopvoorn
PT	escalo
FI	turpa
SV	färna

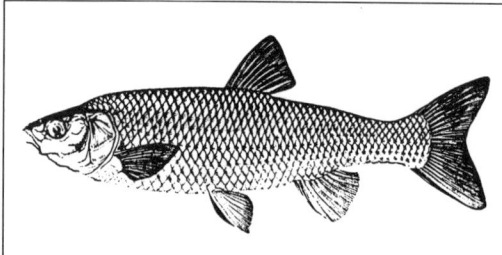

343 CYPRINIDAE FID

SC	*Leuciscus idus* (Linnaeus, 1758); *Idus idus*
ES	cacho; cachuelo
DA	rimte; emde
DE	Aland; Orfe; Nerfling; Schwarznerfling; Goldorfe; Gängling; Kühling; Jessen; Jeese; Geese
EL	χρυσοκέφαλος
EN	ide; orfe
FR	ide mélanote
IT	ido
NL	winde
PT	escalo prateado
FI	säyne
SV	id

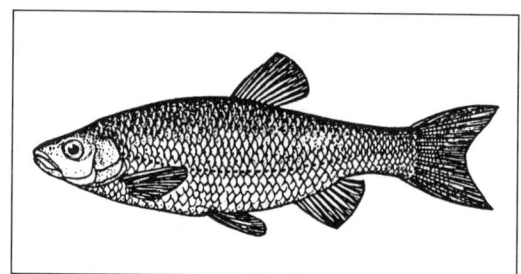

342 CYPRINIDAE

SC	*Leuciscus cephalus cabeda* (Risso, 1826)
ES	cacho; cachuelo
DA	Middelhavs-døbel
DE	Döbel
EL	κέφαλος· κλένι
EN	Mediterranean chub
FR	chevesne méridional; chevaine méridional; meunier
IT	cavedano
NL	Middellandse-Zeekopvoorn
PT	escalo do norte
FI	turpa
SV	färna

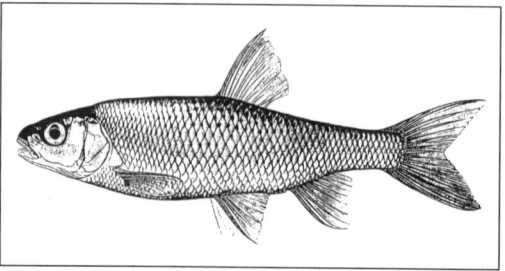

344 CYPRINIDAE FIE

SC	*Leuciscus leuciscus* (Linnaeus, 1758); *Squalius leuciscus*; *Squalius rostratus*; *Leuciscus rostratus*
ES	leucisco común
DA	strømskalle; almindelig strømskalle
DE	Hasel; Häsling; Rüßling; Nesling; Schnutt
EL	κέφαλος
EN	dace
FR	vandoise; dard; meunier argenté; cabotin
IT	leucisco
NL	serpeling
PT	escalo do sul
FI	seipi
SV	stäm

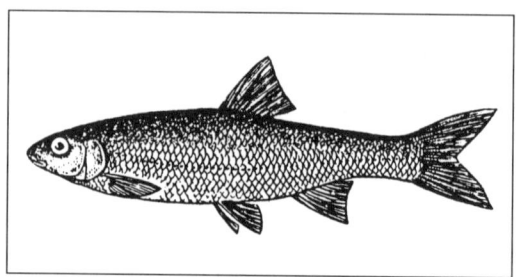

345 CYPRINIDAE

SC   *Leuciscus souffia* (Risso, 1826); *Telestes agassizii*
ES   leucisco italiano
DA   strømling
DE   Strömer; Rießling; Grieslauge; Lauge; Gangfisch
EL   κέφαλος της Ιταλίας
EN   vairone; stroemling
FR   blageon; suiffe; gandoise; aubour
IT   vairone
NL   sufia-voorn
PT   escalo italiano
FI   siniseipi
SV   strömling

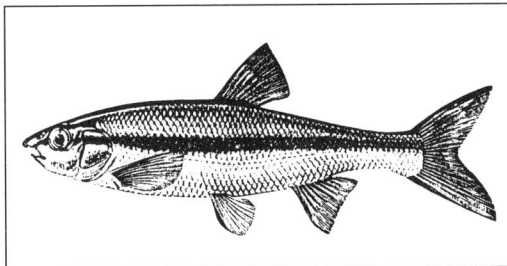

347 CYPRINIDAE   BKC

SC   *Mylopharyngodon piceus* (Richardson, 1845)
ES   carpa negra
DA   sortkarpe
DE   Schwarzer Karpfen
EL   μαυροκυπρίνος
EN   black carp
FR   carpe noire
IT   carpa nera
NL   zwarte karper
PT   carpa negra
FI   mustakarppi
SV   svartkarp

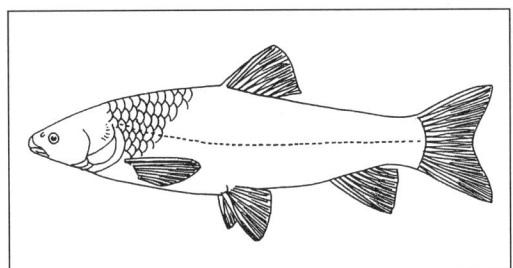

346 CYPRINIDAE   WUB

SC   *Megalobrama amblycephala* (Yih, 1955)
ES   carpa de Wuchang
DA   Wuchang-brasen
DE   Chinesischer Brachsen
EL   λεστιά της Κίνας
EN   Wuchang bream
FR   carpe de Wuchang
IT   carpa Wuchang
NL   Wuchangbrasem
PT   brema chinesa
FI   särkikala-laji
SV   wuchangbraxen

DRAWING NOT AVAILABLE

348 CYPRINIDAE

SC   *Pachychilon pictum* (Heckel and Knev, 1858)
ES   rutilo de Albania
DA   albansk skalle
DE   Albanische Plötze
EL   πλατίτσα της Αλβανίας
EN   moranec; Albanian roach
FR   gardon d'Albanie
IT   triotto albanese
NL   Albanese blankvoorn
PT   bordalinho da Albânia
FI   albaniansärki
SV   marmormört

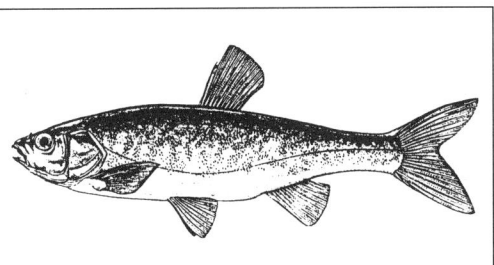

349 CYPRINIDAE WAB

SC	*Parabramis pekinensis* (Basilewski, 1855)
ES	brema de Pekín
DA	Peking-brasen
DE	Peking-Brachsen
EL	λεστιά του Πεκίνου
EN	white amur bream
FR	brème de Pékin
IT	abramide cinese
NL	Pekingbrasem
PT	brema branca
FI	pekinginlahna
SV	pekingbraxen

DRAWING NOT AVAILABLE

350 CYPRINIDAE FSC

SC	*Pelecus cultratus* (Linnaeus, 1758)
ES	peleco
DA	sabelkarpe
DE	Ziege; Sichling; Messerfisch; Säbelfisch; Sichelfisch; Zicke; Messerkarpf
EL	πελεκοκυπρίνος
EN	sabre carp; sichel
FR	rasoir
IT	pesce rasoio
NL	sabelbliek
PT	peixe-sabre
FI	miekkasärki
SV	skärkniv

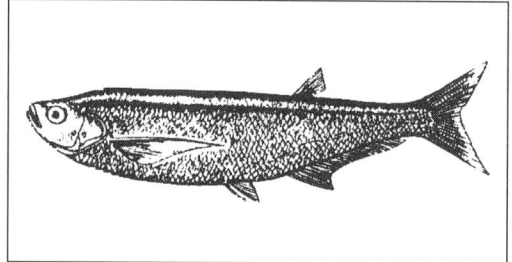

351 CYPRINIDAE

SC	*Phoxinus percnurus* (Pallas, 1811)
ES	foxino de fangal
DA	sump-elritse
DE	Sumpf-Elritze
EL	κοκκινόγαστρος
EN	swamp minnow
FR	vairon des marais
IT	sanguinerola di palude
NL	moeras-elrits
PT	peixinho do pântano
FI	suomutu
SV	sumpelritsa

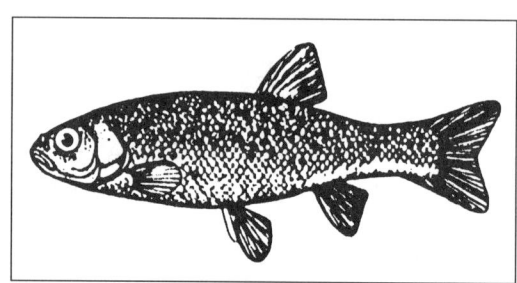

352 CYPRINIDAE

SC	*Phoxinus phoxinus* (Linnaeus, 1758)
ES	foxino común; chipa
DA	elritse
DE	Elritze; Pfrille; Wibling; Erling; Brutt; Elderitz; Grümpel; Rümpahen; Pierling; Wettling
EL	κοκκινόγαστρος
EN	minnow; pink
FR	véron; vairon
IT	sanguinerola
NL	elrits
PT	peixinho de engodo
FI	mutu
SV	elritsa; kvidd

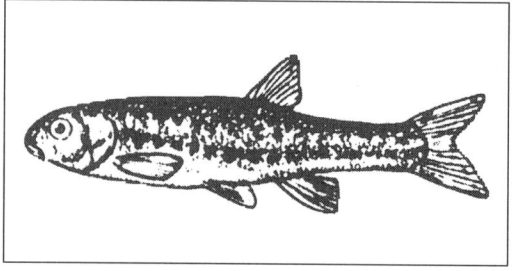

353 CYPRINIDAE FJB

SC	*Puntius javanicus* (Bleeker, 1855); *Barbus javanicus*
ES	barbo de Java
DA	Java-barbe
DE	Java-Barbe
EL	μπριάνα της Ιάβας
EN	Java barb
FR	barbeau de Java
IT	barbo di Giava
NL	Javaanse barbeel
PT	barbo de Java
FI	jaavanbarbi
SV	sunda-barb; java-barb

DRAWING NOT AVAILABLE

354 CYPRINIDAE FAB

SC	*Puntius* spp.; *Barbus* spp.
ES	barbos de Asia
DA	barbe-slægter
DE	Zierbarben
EL	μπριάνες της Ασίας
EN	Asian barbs
FR	barbeaux d'Asie
IT	barbi asiatici
NL	Aziatische barbelen
PT	barbos da Ásia
FI	barbi
SV	barber

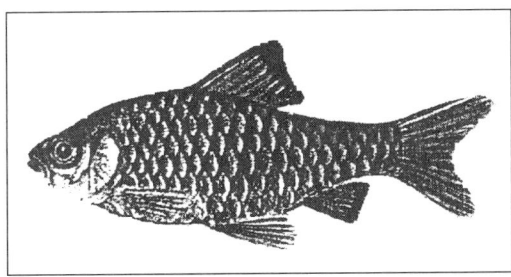

355 CYPRINIDAE ENA

SC	*Rastrineobola argentea* (Pellegrin, 1904)
ES	ciprino plateado
DA	Victoria-karpe
DE	Victoriasee-Karpfen
EL	ασημοκυπρίνος
EN	silver cyprinid
FR	cyprin argenté
IT	argentone
NL	Victoriameer-karper
PT	sipa do lago Vitória
FI	särkikala-laji
SV	mukene

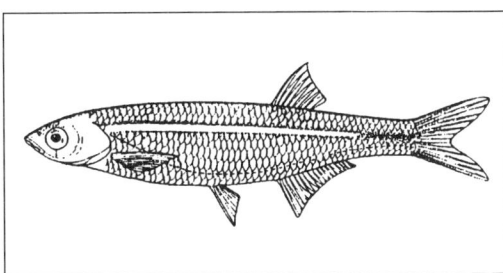

356 CYPRINIDAE

SC	*Rhodeus amarus* (Bloch, 1782); *Rhodeus sericeus amarus*
ES	ródeo
DA	bitterling; blåfisk
DE	Bitterling; Gewöhnlicher Bitterling
EL	βαβούκι· μουρμουρίτσα
EN	bitterling
FR	bouvière; péteuse
IT	rodeo amaro
NL	bittervoorn
PT	peixe-amargo
FI	katkerokala
SV	bitterling

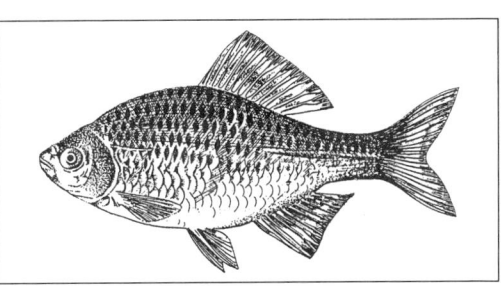

357 CYPRINIDAE

SC	*Rutilus frisii* (Nordmann, 1840); *Luciscus meidingeri*
ES	rutilo del Mar Negro
DA	Sortehavs-skalle
DE	Schwarzmeerplötze
EL	πλατίτσα της Μαύρης Θάλασσας
EN	Black Sea roach
FR	gardon de la mer Noire
IT	triotto russo
NL	parel-blankvoorn
PT	bogardo do mar Negro
FI	helmisärki
SV	svartahavsmört

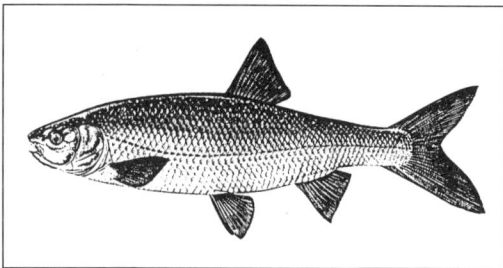

359 CYPRINIDAE

SC	*Rutilus rubilio* (Bonaparte, 1837)
ES	rutilo sudeuropeo
DA	sydeuropæisk skalle
DE	Südeuropäische Plötze
EL	δρομίτσα
EN	South European roach
FR	gardon de l'Europe du Sud
IT	rovella
NL	Adriatische blankvoorn
PT	ruivaca do sul
FI	etelänsärki
SV	adriamört

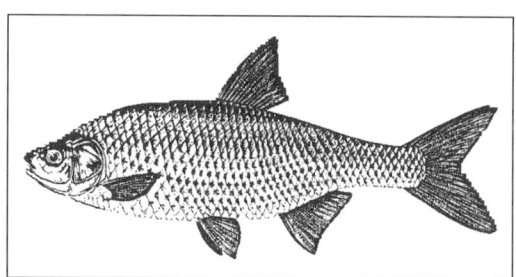

358 CYPRINIDAE

SC	*Rutilus pigus* (Lacepède, 1804); *Leuciscus pigus*
ES	rutilo del Danubio
DA	Donau-skalle
DE	Pigo
EL	πλατίτσα του Δούναβη
EN	Danube roach
FR	gardon galant
IT	pigo
NL	Donauvoorn
PT	bogardo do Danúbio
FI	tonavansärki
SV	pigomört

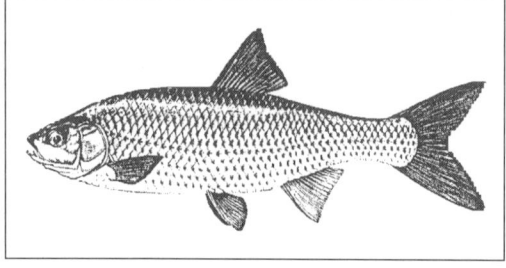

360 CYPRINIDAE   FRO

SC	*Rutilus rutilus* (Linnaeus, 1758); *Gardonus rutilis*; *Leuciscus rutilus*
ES	bermejuela; coloradilla; rutilo; calandino; pardilla
DA	almindelig skalle; skalle
DE	Plötze; Rotauge; Rötel; Rotkarpfen; Weißfisch; Schmal; Bleier; Ridder; Rottez; Schwall; Schral
EL	πλατίτσα· ασπρίτσα
EN	roach
FR	gardon commun; guidon blanc; rousse; gardon
IT	triotto rosso
NL	blankvoorn
PT	pardelha dos Alpes; ruivaca
FI	särki
SV	mört

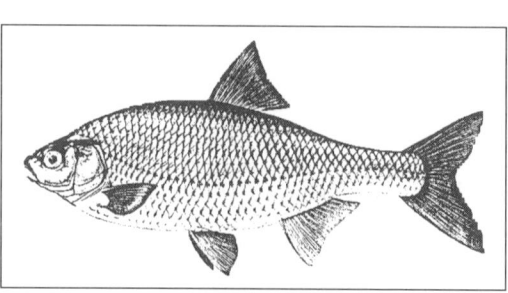

361 CYPRINIDAE FRX

SC   *Rutilus* spp.
ES   rutilos
DA   karpefisk-slægter
DE   Rotaugen
EL   δρομίτσες· πλατίτες
EN   roaches
FR   gardons
IT   triotti
NL   blankvoorns
PT   pardelhas, bogardos e ruivacas
FI   särki-suku
SV   mörtar

363 CYPRINIDAE

SC   *Scardinius graecus* (Stephanidis, 1937)
ES   escardinio griego
DA   græsk rudskalle
DE   Griechische Rotfeder
EL   καλαμίθρα
EN   Greek rudd
FR   rotengle grec
IT   scardola greca
NL   Griekse rietvoorn
PT   escardínio grego
FI   kreikansorva
SV   grekisk sarv

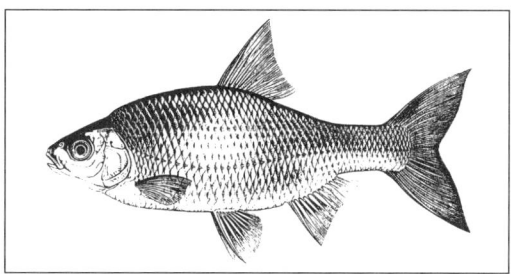

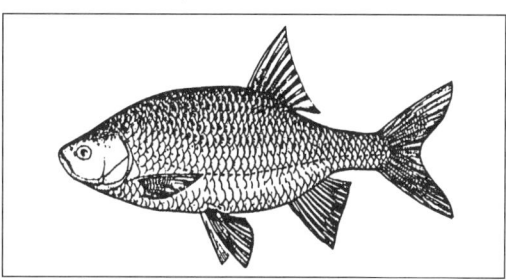

362 CYPRINIDAE

SC   *Scardinius erythrophthalmus* (Linnaeus, 1758)
ES   escardinio
DA   rudskalle; almindelig rudskalle
DE   Rotfeder; Rothasel; Maifisch; Rötel; Rotflosser; Rotkarpfen
EL   κοκκινοφτέρα· αγριοτσιρώνι
EN   rudd; red-eye
FR   rotengle; gardon rouge; charin; platelle
IT   scardola
NL   ruisvoorn; rietvoorn
PT   escardínio olho-vermelho
FI   sorva
SV   sarv

364 CYPRINIDAE FTE

SC   *Tinca tinca* (Linnacus, 1758)
ES   tenca; tinca; aguijón
DA   suder
DE   Schleie; Schlei; Schleihe; Schuster; Schlüpfling; Schlammler; Grünschleie; Schleiforelle
EL   γλήνι
EN   tench
FR   tanche; tanche commune; tenca; tiche
IT   tinca
NL   zeelt
PT   tenca
FI   suutari
SV   sutare

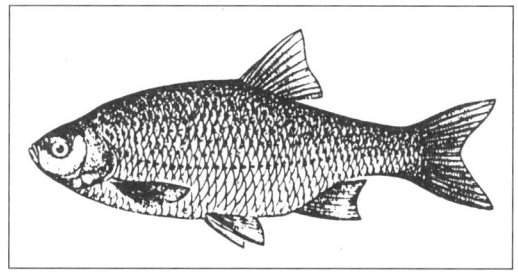

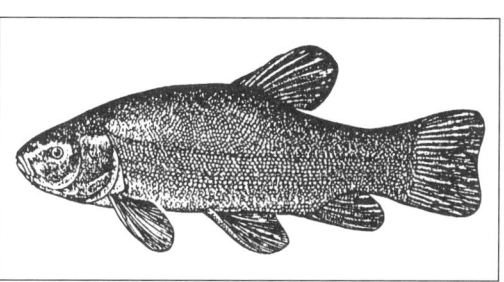

365 CYPRINIDAE

SC *Vimba vimba* (Linnaeus, 1758); *Abramis vimba*
ES vimba
DA vimme
DE Zährte; Rußnase; Seerüßling; Blaunase; Meernase; Näsling; Näse
EL μαυρομάτα
EN vimba bream
FR zaerthe
IT vimba; abramide russo
NL blauwneus
PT brema do rio Weser
FI vimpa
SV vimma

367 COBITIDAE

SC *Cobitis elongata* (Heckel and Kner, 1858)
ES colmilleja larga
DA balkansk pigsmerling
DE Großer Steinbeißer; Balkan-Steinbeißer
EL φιδόψαρο των Βαλκανίων
EN Balkan loach
FR loche des Balkans
IT cobite dei Balcani
NL Balkan-modderkruiper
PT verdemã dos Balcãs
FI balkaninnuoliainen
SV balkannissöga

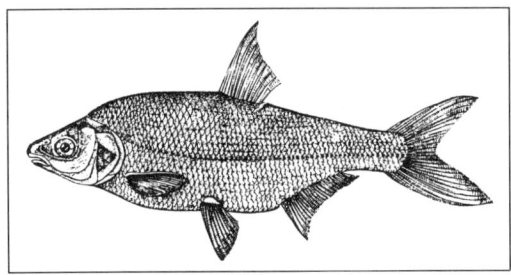

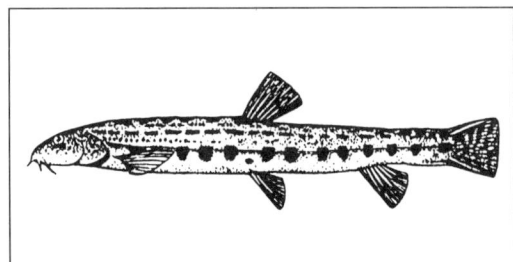

366 COBITIDAE

SC *Cobitis aurata* (De Filippi, 1856)
ES colmilleja dorada
DA gylden pigsmerling
DE Gold-Steinbeißer
EL φιδόψαρο· στέρβα
EN golden loach
FR loche dorée
IT cobite dorato
NL gouden modderkruiper
PT verdemã dourada
FI kultanuoliainen
SV guldnissöga

368 COBITIDAE

SC *Cobitis larvata* (De Filippi, 1859)
ES colmilleja de cejas
DA Bergatino pigsmerling
DE Italienischer Steinbeißer
EL φιδόψαρο του Μπεργαντίνου
EN Bergatino loach
FR loche de Bergatino
IT cobite mascherato
NL Bergatino-modderkruiper
PT verdemã de Bergatino
FI bergatinonnuoliainen
SV bergantino-nissöga

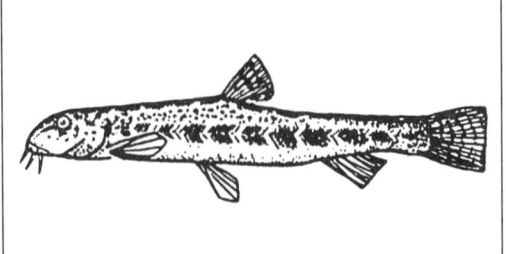

369 COBITIDAE

SC	*Cobitis romanica* (Bacescu, 1943)
ES	colmilleja rumana
DA	rumænsk pigsmerling
DE	Rumänischer Steinbeißer
EL	φιδόψαρο της Ρουμανίας
EN	Romanian loach
FR	loche roumaine
IT	cobite rumeno
NL	Roemeense modderkruiper
PT	verdemã romana
FI	romaniannuoliainen
SV	rumänskt nissöga

371 COBITIDAE

SC	*Misgurnus fossilis* (Linnaeus, 1758); *Cobitis fossilis*
ES	misgurno
DA	dyndsmerling
DE	Schlammpeitzger; Schlammbeißer; Wetterfisch
EL	φιδόψαρο
EN	weatherfish
FR	loche d'étang
IT	cobite di stagno
NL	grote modderkruiper
PT	verdemã de tanque
FI	mutakala
SV	slampiskare

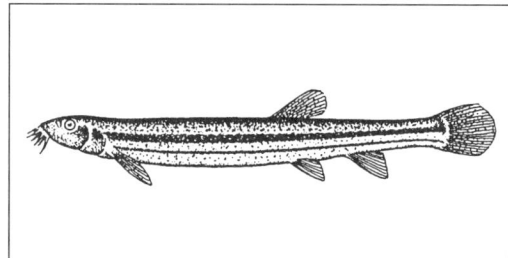

370 COBITIDAE

SC	*Cobitis taenia* (Linnaeus, 1758); *Acanthopsis taenia*
ES	colmilleja; aranya; gatet; locha
DA	pigsmerling; almindelig pigsmerling
DE	Europäischer Steinbeißer; Steinbeißer; Dorngrundel
EL	φιδόψαρο· στέρβα
EN	spined loach; groundling; bibbaud loach
FR	loche de rivière; loche épineuse; perce-pierre; mord-pierre
IT	cobite comune
NL	kleine modderkruiper
PT	verdemã
FI	rantaneula
SV	nissöga

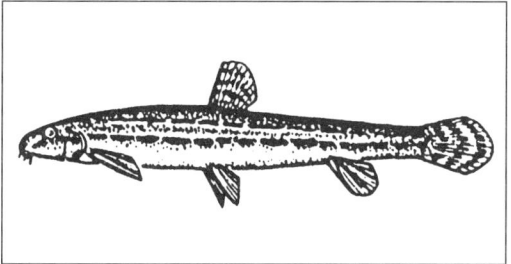

372 COBITIDAE

SC	*Nemacheilus barbatulus* (Linnaeus, 1758); *Cobitis barbatulus*; *Nemachilus barbatula*; *Orthrias barbatulus*
ES	locha de roca; lobo; sarbo; zarbo; sarbito
DA	smerling; almindelig smerling
DE	Bartgrundel; Schmerle; Gewöhnliche Schmerle; Bachschmerle
EL	βραχοφιδόψαρο
EN	stone loach
FR	loche franche; loque; barbette; chatouille
IT	cobite barbatello; barbatello
NL	bermpje
PT	verdemã da pedra
FI	kivennuoliainen
SV	grönling

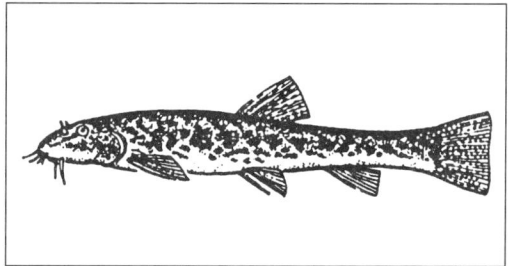

373 COBITIDAE

SC	*Sabanejewia conspersa* (Cantoni, 1882); *Cobitis conspersa*
ES	colmilleja italiana
DA	italiensk pigsmerling
DE	Venezianischer Steinbeißer
EL	φιδόψαρο της Ιταλίας
EN	Italian loach
FR	loche italienne
IT	cobite
NL	Venetiaanse modderkruiper
PT	verdemã italiana
FI	italiannuoliainen
SV	bergantino-nissöga

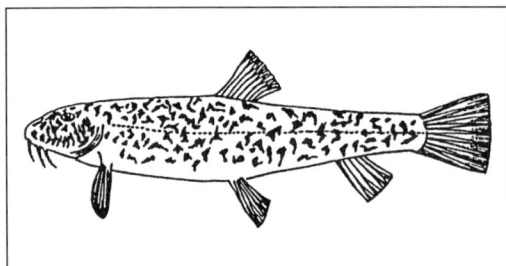

375 ARIIDAE SMC

SC	*Arius heudeloti* (Valenciennes, 1840)
ES	bagre bocalisa
DA	glatmundet havmalle
DE	Glattmaul-Kreuzwels
EL	θαλασσινό γατόψαρο
EN	smooth-mouth sea catfish
FR	machoiron banderille
IT	pesce-gatto marino
NL	gladbek-zeemeerval
PT	bagre-boca lisa
FI	merimonni-laji
SV	västafrikansk havsmal

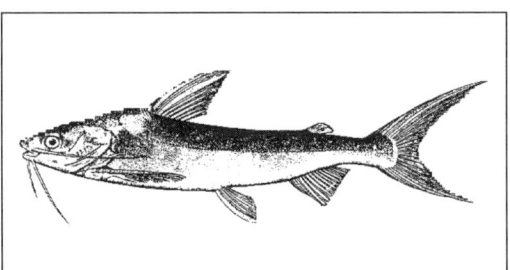

374 ARIIDAE CAX

SC	*Ariidae*
ES	bagres marinos
DA	havmalle-familien
DE	Kreuzwelse; Meerwelse
EL	θαλασσινά γατόψαρα
EN	sea catfishes
FR	machoirons; poissons-chats de mer
IT	pescigatto di mare
NL	zeemeervallen
PT	bagres; bagres marinhos
FI	merimonnit; merimonnit-heimo
SV	havsmalar

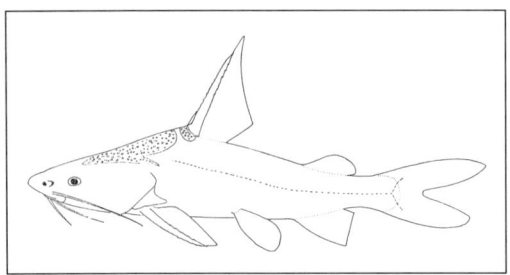

376 ARIIDAE CAO

SC	*Arius maculatus* (Thunberg, 1792)
ES	bagre moteado
DA	plettet havmalle
DE	Gefleckter Kreuzwels
EL	στικτογατόψαρο
EN	spotted catfish
FR	machoiron tacheté
IT	pescegatto marino
NL	gevlekte zeemeerval
PT	bagre malhado
FI	merimonni-laji
SV	fläckig havsmal

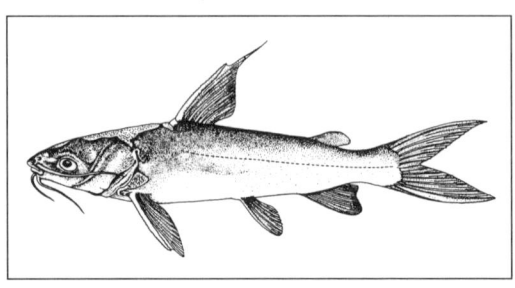

377 ARIIDAE   GAT

SC	*Galeichthys feliceps* (Valenciennes, 1840)
ES	bagre barba blanca
DA	hvid havmalle
DE	Katzen-Kreuzwels
EL	θαλασσινό γατόψαρο
EN	white baggar
FR	barbillon blanc; poisson-chat à barbillon blanc
IT	pesce gatto marino
NL	witte zeemeerval
PT	bagre-barba branca
FI	merimonni-laji
SV	kapmal

379 SILURIDAE   CAG

SC	*Kryptopterus* spp.
ES	peces gato de cristal
DA	glasmalle-slægten
DE	Indische Glaswelse
EL	κρυσταλλογατόψαρα
EN	glass catfishes
FR	poissons-chats de verre; silures de verre
IT	pesci cristallo
NL	glasmeervallen
PT	gatos transparentes
FI	lasimonni-suku
SV	glasmalar

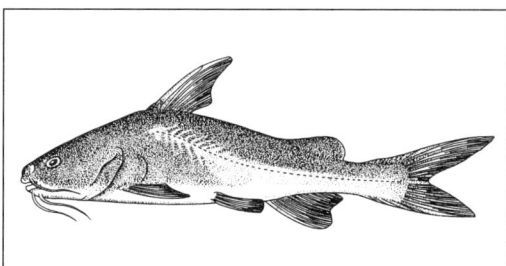

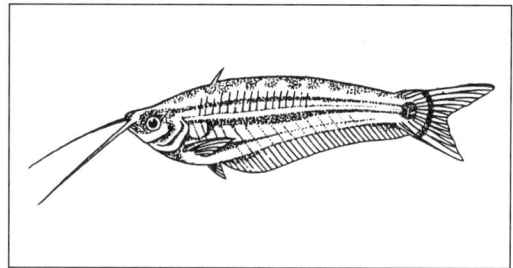

378 PLOTOSIDAE   CAE

SC	*Plotosus* spp.
ES	anguilas peces-gato
DA	koralmaller
DE	Korallenwelse
EL	χελογατόψαρα
EN	catfish eels
FR	poissons-chats anguilles; gonzui
IT	plutosi
NL	koraalmeervallen
PT	gatos marinhos
FI	korallimonni-suku
SV	korallmalar

380 SILURIDAE

SC	*Siluridae*
ES	siluros
DA	malle-familien
DE	Welse; Echte Welse
EL	γατόψαρα
EN	wels catfishes
FR	siluridés; silures
IT	siluri
NL	meervallen
PT	siluros
FI	monnit; monnit-heimo
SV	egentliga malar

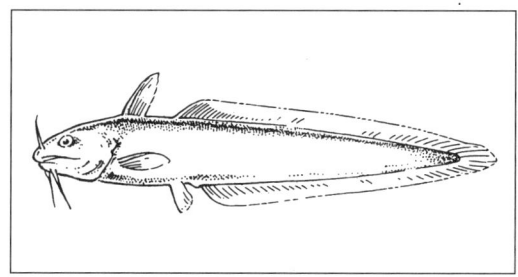

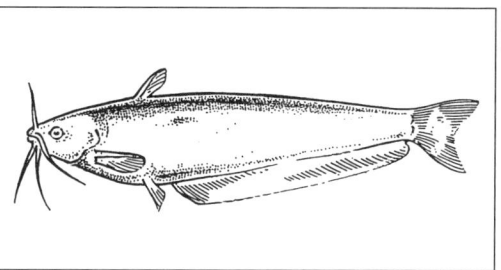

381 SILURIDAE

SC	*Silurus aristotelis* (Agassiz, 1856); *Parasilurus aristotelis*
ES	siluro griego
DA	græsk malle
DE	Aristoteles-Wels
EL	γλανίδι· γλανός
EN	Greek sheatfish
FR	silure grec
IT	siluro greco
NL	Aristoteles-meerval
PT	siluro grego
FI	kreikanmonni
SV	aristotelesmal

383 ICTALURIDAE

SC	*Ictaluridae*
ES	bagres; cotos
DA	dværgmalle-familien
DE	Zwergwelse
EL	γατόψαρα
EN	catfishes
FR	poissons-chats
IT	pesci gatto
NL	dwergmeervallen
PT	peixes-gato
FI	piikkimonnit; piikkimonnit-heimo
SV	dvärgmalar

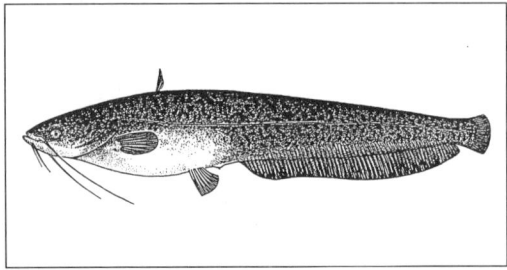

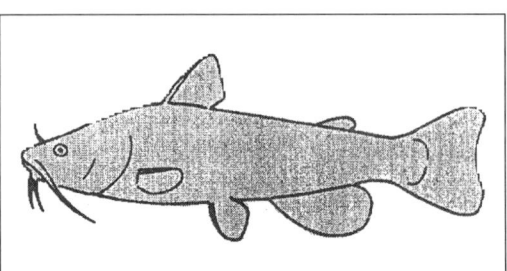

382 SILURIDAE   SOM

SC	*Silurus glanis* (Linnaeus, 1758)
ES	siluro; siluro europeo
DA	malle; europæisk malle
DE	Wels; Waller; Flußwels
EL	γουλιανός
EN	sheatfish; Danubian wels; wels catfish; wels; som catfish
FR	silure glane; salut
IT	siluro
NL	meerval
PT	siluro europeu
FI	monni
SV	mal

384 ICTALURIDAE   ITM

SC	*Ictalurus melas* (Rafinesque, 1820)
ES	coto negro
DA	sort dværgmalle
DE	Schwarzer Zwergwels
EL	γατόψαρο
EN	black bullhead
FR	poisson-chat d'Amérique; silure d'Amérique
IT	pesce gatto
NL	zwarte dwergmeerval
PT	peixe-gato negro
FI	piikkimonni
SV	svart dvärgmal; svart kattfisk

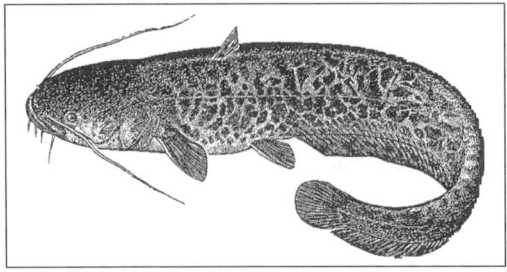

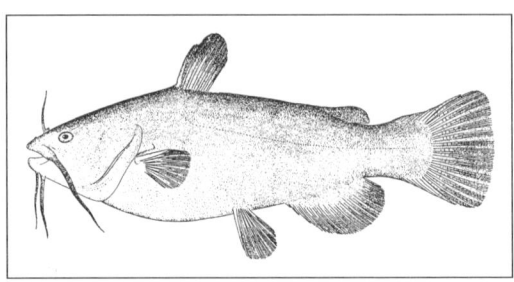

385 ICTALURIDAE   ITE

SC	*Ictalurus nebulosus* (Lesueur, 1819)
ES	coto pardo
DA	brun dværgmalle
DE	Amerikanischer Zwergwels; Katzenwels; Amerikanischer Wels; Katzenfisch; Zwergwels
EL	γατόψαρο
EN	brown bullhead
FR	poisson-chat noir; silure noir
IT	pesce gatto
NL	bruine dwergmeerval
PT	peixe-gato castanho
FI	piikkimonni
SV	brun dvärgmal; brun kattfisk

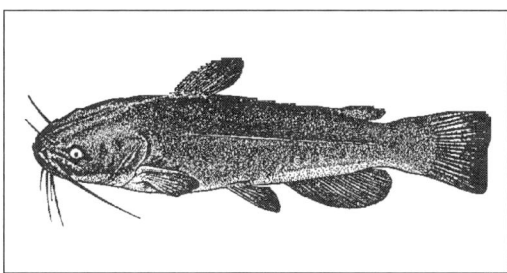

387 ICTALURIDAE   CAF†

SC	*Ictalurus* spp.
ES	bagres; cotos
DA	nordamerikansk malle-slægt
DE	Katzenwelse
EL	γατόψαρα
EN	bullhead catfishes; catfishes
FR	poissons-chats américains
IT	pesci gatto
NL	dwergmeervallen
PT	peixes-gato americanos; gatos da América do Norte
FI	piikkimonni-suku
SV	dvärgmalar

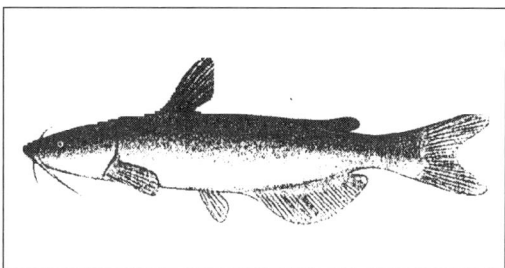

386 ICTALURIDAE   ITP

SC	*Ictalurus punctatus* (Rafinesque, 1818)
ES	coto punteado
DA	plettet dværgmalle
DE	Getüpfelter Gabelwels
EL	γατόψαρο
EN	channel catfish
FR	poisson-chat tacheté; silure tacheté
IT	pesce gatto puntado
NL	channel-dwergmeerval
PT	peixe-gato pontuado
FI	pilkkupiikkimonni
SV	prickig dvärgmal

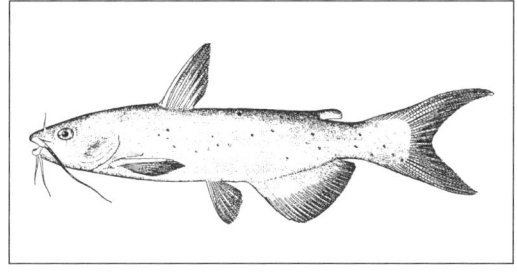

388 CLARIIDAE   CTO

SC	*Clarias* spp.
ES	clarias
DA	ålemalle-slægt
DE	Raubwelse; Luftatmende Welse
EL	κλαρίας
EN	air-breathing catfishes; torpedo-shaped catfishes
FR	clarias; poissons-chats
IT	
NL	Afrikaanse meervallen
PT	gatos de cabeça chata; clárias
FI	ankeriasmonni-suku
SV	ålmalar

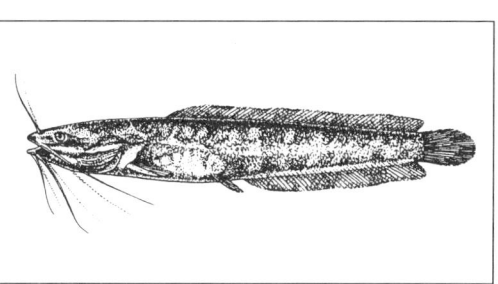

389 AULOPIDAE

SC	*Aulopus filamentosus* (Bloch, 1792)
ES	lagarto real
DA	flagfinnefisk
DE	Fadensegelfisch
EL	γουρλομάτης
EN	Mediterranean flagfin
FR	limbert
IT	lacerto
NL	vlagvinvis
PT	lagarto-do-mar
FI	auloppi-laji
SV	flaggödlefisk

391 SYNODONTIDAE LIG

SC	*Saurida tumbil* (Bloch, 1795)
ES	lagarto
DA	stor øglefisk
DE	Großer Eidechsenfisch
EL	γιγαντοσκαρμός
EN	greater lizardfish
FR	anoli de mer; lézard géant
IT	pesce ramarro indiano
NL	hagedisvis
PT	lagarto-verde
FI	sisiliskokala-laji
SV	stor ödlefisk

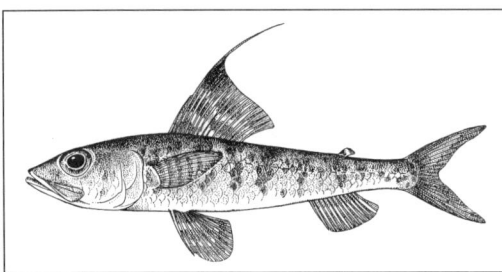

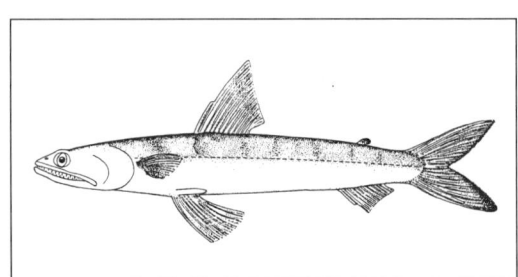

390 SYNODONTIDAE LIX

SC	*Synodontidae*
ES	lagartos
DA	øglefisk-familien
DE	Eidechsenfische
EL	σκαρμοί
EN	lizardfishes
FR	anolis de mer; anolis
IT	pesci ramarro; pesci lucertola
NL	hagedisvissen
PT	lagartos; lagartos-do-mar
FI	sisiliskokalat
SV	ödlefiskar

392 SYNODONTIDAE LIB

SC	*Saurida undosquamis* (Richardson, 1848); *Saurida grandisquamis*
ES	lagarto; lagarto escamoso
DA	storskællet øglefisk
DE	Großschuppen-Eidechsenfisch; Gefleckter Eidechsenfisch
EL	σκαρμός της Ανατολής
EN	brushtooth lizardfish; large-scale lizardfish
FR	anoli de mer; poisson-lézard à grandes écailles; lézard à dents en brosse
IT	pesce ramarro orientale
NL	groteschubbenhagedisvis
PT	lagarto escamudo
FI	sisiliskokala
SV	storfjällad ödlefisk

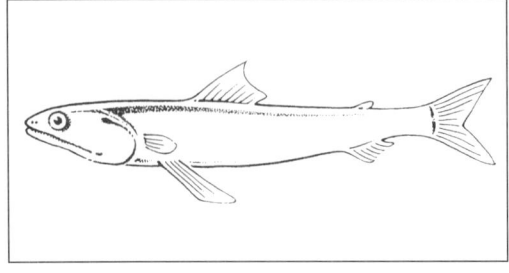

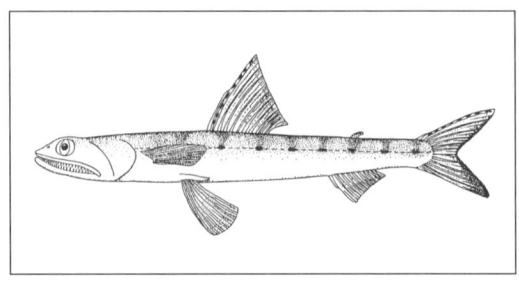

393 SYNODONTIDAE

SC   *Synodus saurus* (Linnaeus, 1758)
ES   lagarto; dragó; salta murades
DA   atlantisk øglefisk
DE   Atlantischer Eidechsenfisch; Eidechsenfisch
EL   σκαρμός
EN   lizardfish; Atlantic lizardfish
FR   lézard; poisson-lézard de l'Atlantique
IT   pesce ramarro; pesce lucertola
NL   Atlantische hagedisvis
PT   lagarto-da-costa
FI   sisiliskokala
SV   ödlefisk

395 ANGUILLIDAE

SC   *Anguillidae*
ES   anguilas
DA   ferskvandsåle-familien
DE   Aale; Flußaale; Echte Aale
EL   χέλια
EN   river eels; eels
FR   anguilles
IT   anguille; anguillidi
NL   palingen; alen
PT   enguias
FI   ankeriaat; ankeriaat-heimo
SV   ålfiskar

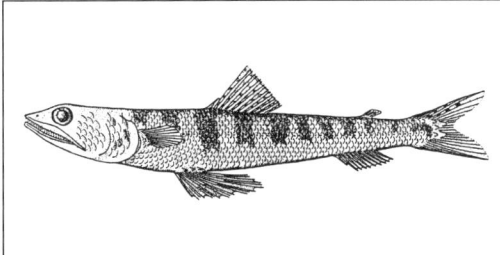

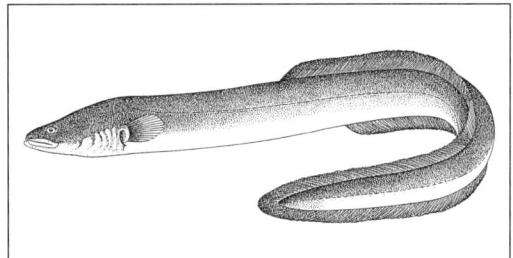

394 MALAPTERURIDAE

SC   *Malapterurus electricus* (Gmelin, 1789)
ES   siluro eléctrico
DA   elektrisk malle; el-malle
DE   Elektrischer Wels
EL   ηλεκτρογατόψαρο
EN   electric catfish
FR   poisson-chat électrique; silure électrique; poisson tonnerre
IT   tordo elettrico
NL   siddermeerval
PT   peixe-choque
FI   sähkömonni
SV   darrmal

396 ANGUILLIDAE   ELE

SC   *Anguilla anguilla* (Linnaeus, 1758); *Anguilla vulgaris*
ES   anguila europea; anguila
DA   europæisk ferskvandsål; ål; almindelig ål
DE   Europäischer Aal; Flußaal; Aal; Europäischer Flußaal
EL   χέλι
EN   European eel; river eel; common eel; eel; freshwater eel
FR   anguille d'Europe; anguille de rivière; anguille; andouille; anguille commune; anguille argentée
IT   anguilla; ragano; anguilla argentina; anguilla gialla; capitone
NL   paling; aal
PT   enguia europeia; enguia; irós; eirós
FI   ankerias
SV   ål

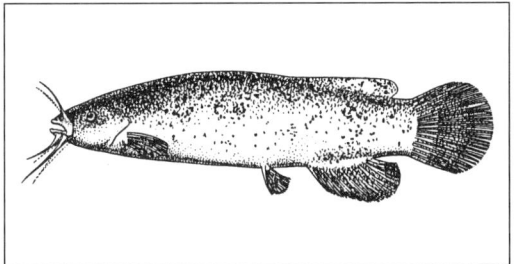

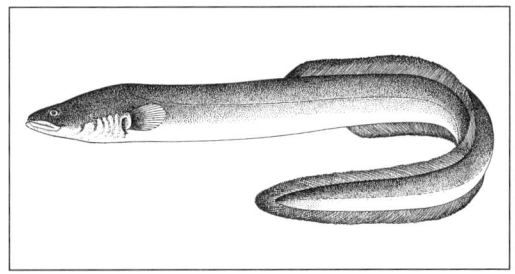

| 397 | ANGUILLIDAE | ELU | 399 | ANGUILLIDAE | ELJ |

397 ANGUILLIDAE ELU

SC	*Anguilla australis* (Richardson, 1841)
ES	anguila australiana
DA	australsk ål
DE	Australischer Aal
EL	χέλι της Αυστραλίας
EN	Australian eel; short-finned eel
FR	anguille d'Australie; anguille australienne
IT	anguilla australiana
NL	Australische paling
PT	enguia australiana
FI	australianankerias
SV	australisk ål

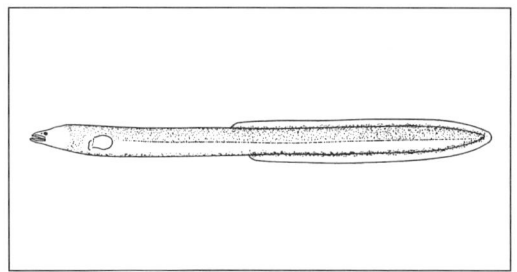

399 ANGUILLIDAE ELJ

SC	*Anguilla japonica* (Temminck and Schlegel, 1846)
ES	anguila japonesa
DA	japansk ål
DE	Japanischer Aal
EL	χέλι της Ιαπωνίας
EN	Japanese eel
FR	anguille du Japon; anguille japonaise
IT	anguilla giapponese
NL	Japanse paling
PT	enguia japonesa
FI	japaninankerias
SV	japansk ål

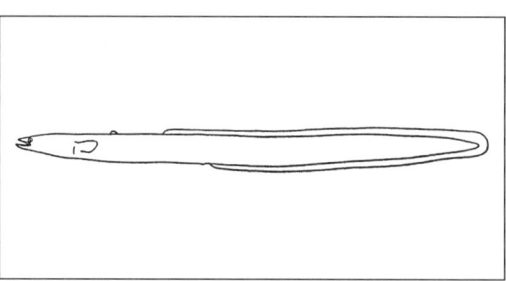

398 ANGUILLIDAE

SC	*Anguilla diffenbachii* (Gray, 1846)
ES	anguila de Nueva Zelanda
DA	newzealandsk ål
DE	Neuseeland-Aal
EL	χέλι της Νέας Ζηλανδίας
EN	New Zealand eel
FR	anguille de Nouvelle-Zélande
IT	anguilla neozelandese
NL	Nieuw-Zeelandse paling
PT	enguia da Nova Zelândia
FI	uudenseelanninankerias
SV	nyzeeländsk ål

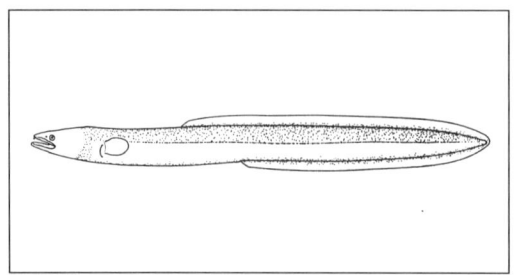

400 ANGUILLIDAE ELA

SC	*Anguilla rostrata* (Lesueur, 1817)
ES	anguila americana
DA	amerikansk ål
DE	Amerikanischer Aal
EL	χέλι της Αμερικής
EN	American eel
FR	anguille américaine
IT	anguilla americana
NL	Amerikaanse paling
PT	enguia americana
FI	amerikanankerias
SV	amerikansk ål

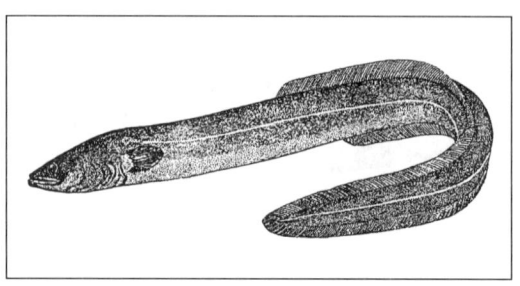

401 ANGUILLIDAE ELX

SC	*Anguilla* spp.
ES	anguilas
DA	ferskvandsåle-slægten
DE	Echte Aale; Flußaale
EL	χέλια
EN	eels
FR	anguilles
IT	anguille
NL	palingen; alen
PT	enguias
FI	ankeriaat-suku
SV	ålar

403 MURAENIDAE

SC	*Gymnothorax funebris* (Ranzani, 1840); *Lycodontis funebris*
ES	morena verde
DA	grøn muræne
DE	Westatlantische Muräne
EL	πράσινη σμέρνα
EN	green moray
FR	murène verte
IT	murena verde
NL	groene murene
PT	moreia verde
FI	vihermureena
SV	grön muräna

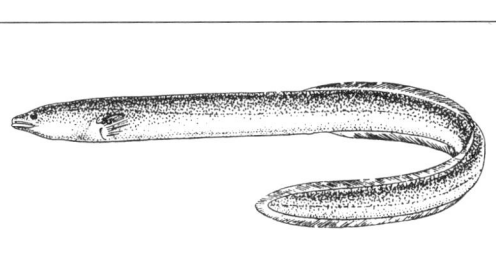

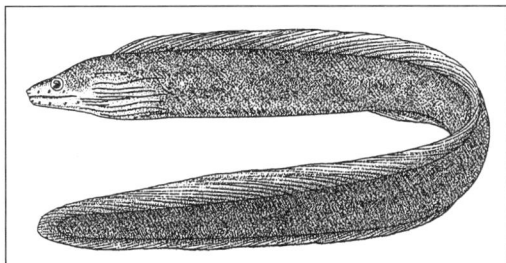

402 MURAENIDAE MUI

SC	*Muraenidae*
ES	morenas
DA	muræne-familien
DE	Muränen
EL	σμέρνες
EN	morays; moray eels
FR	murènes
IT	murene
NL	murenen
PT	moreias
FI	mureenat; mureenat-heimo
SV	muränfiskar

404 MURAENIDAE

SC	*Muraena helena* (Linnaeus, 1758)
ES	morena
DA	Middelhavs-muræne
DE	Mittelmeer-Muräne
EL	σμέρνα
EN	Mediterranean moray; European moray; moray; murry
FR	murène; murène de Méditerranée
IT	murena
NL	murene
PT	moreia
FI	mureena
SV	muräna

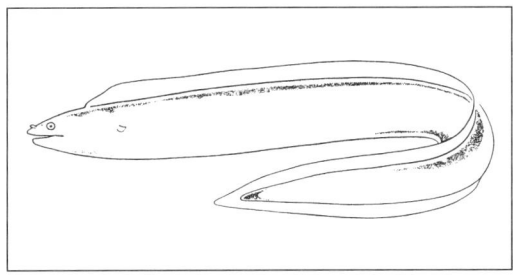

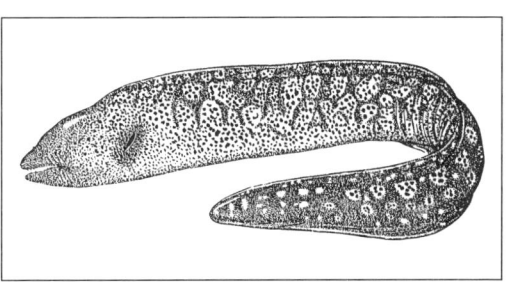

405 MURAENESOCIDAE GPC

SC	*Cynoponticus ferox* (Costa, 1846)
ES	morenocio de Guinea
DA	Guinea-geddemuræne
DE	Guinea-Messerzahnaal
EL	μουγγρί της Γουινέας
EN	Guinean pike conger
FR	congre brochet de Guinée
IT	cinopontico; grongo canino
NL	Guinea-snoekaal
PT	congro branco africano; congro branco
FI	haukiankerias-laji
SV	sorgkantsmuräna

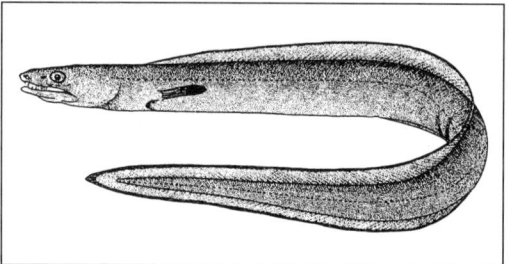

406 MURAENESOCIDAE DPC

SC	*Muraenesox cinereus* (Forsskål, 1775)
ES	morenocio dentón
DA	Batavia-geddemuræne
DE	Batavia-Hechtmuräne; Hechtmuräne; Batavia-Putjekanipa
EL	φίδι της θάλασσας
EN	dagger-tooth pike-conger; sharp-toothed eel
FR	murènesoce du Japon; murène brochet cendrée; murènesoce-dague
IT	murena del Giappone
NL	Batavia-snoekaal
PT	congro bicudo do Japão
FI	hopeahaukiankerias
SV	gäddmuräna

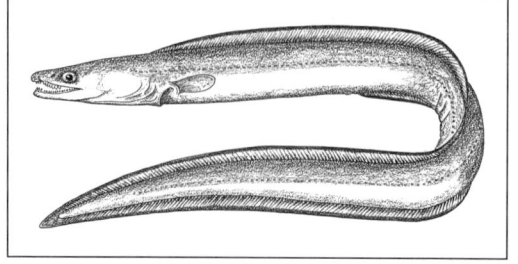

407 MURAENESOCIDAE PCX

SC	*Muraenesox* spp.
ES	morenocios
DA	geddemuræne-slægt
DE	Messerzahnaale
EL	φίδια της θάλασσας
EN	pike-congers
FR	murènes brochets
IT	lucciomurene
NL	snoekalen
PT	congros bicudos
FI	haukiankerias-suku
SV	gäddmuränor

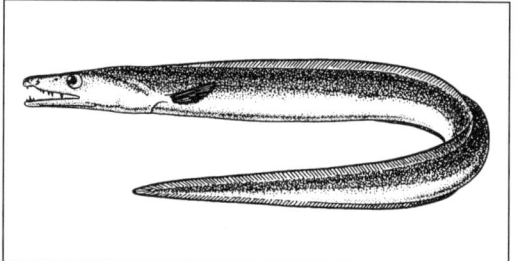

408 CONGRIDAE COX

SC	*Congridae*
ES	congrios
DA	havål-familien
DE	Meeraale
EL	μουγγριά
EN	conger eels; congers
FR	congres
IT	gronghi; congridi
NL	congeralen
PT	congros
FI	meriankeriaat; meriankeriaat-heimo
SV	havsålsfiskar

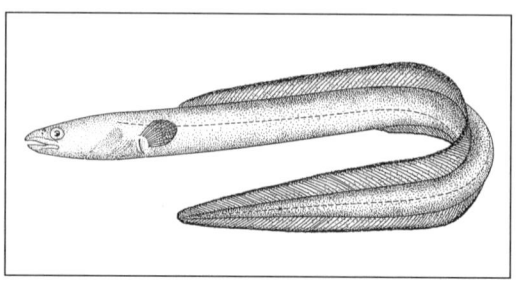

409 CONGRIDAE   ELS

SC	*Astroconger myriaster* (Brevoort, 1856); *Conger myriaster*
ES	congrio del Pacífico
DA	Stillehavs-havål
DE	Weißflecken-Congeraal
EL	μουγγρί της Ιαπωνίας
EN	white-spotted conger
FR	congre du Pacifique nord-ouest; congre à points blancs
IT	grongo giapponese
NL	Pacifische conger
PT	congro malhado do Pacífico
FI	meriankerias-laji
SV	vitfläckshavsål

DRAWING NOT AVAILABLE

411 CONGRIDAE   COA

SC	*Conger oceanicus* (Mitchill, 1818)
ES	congrio americano
DA	amerikansk havål
DE	Amerikanischer Meeraal
EL	μουγγρί της Αμερικής
EN	American conger; conger eel
FR	congre d'Amérique
IT	grongo americano
NL	Amerikaanse congeraal
PT	congro americano
FI	amerikanmeriankerias
SV	amerikansk havsål

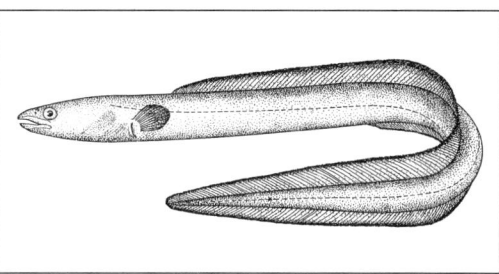

410 CONGRIDAE   COE

SC	*Conger conger* (Linnaeus, 1758); *Conger vulgaris*
ES	congrio
DA	almindelig havål; havål
DE	Meeraal; Congeraal; Conger; Seeaal; Gemeiner Meeraal
EL	μουγγρί· δρόγγος· κόγγρος
EN	European conger; conger; conger eel
FR	congre commun; congre
IT	grongo
NL	congeraal
PT	congro; safio
FI	meriankerias
SV	havsål

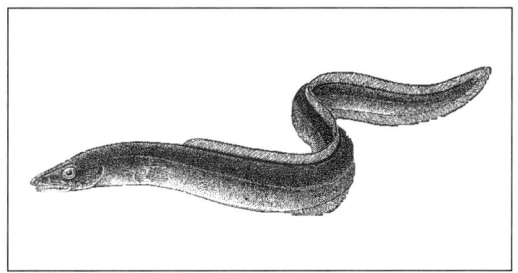

412 CONGRIDAE   COS

SC	*Conger orbignyanus* (Valenciennes, 1839)
ES	congrio argentino
DA	argentinsk havål
DE	Argentinischer Meeraal
EL	μουγγρί της Αργεντινής
EN	Argentine conger
FR	congre argentin
IT	grongo d'Argentina
NL	Argentijnse congeraal
PT	congro argentino
FI	argentiinanmeriankerias
SV	argentinsk havsål

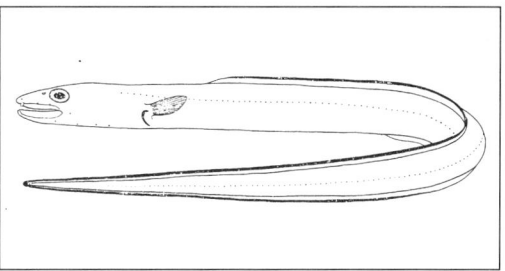

413 OPHICHTHIDAE

SC   *Echelus myrus* (Linnaeus, 1758)
ES   serpiente de mar
DA   ormeål
DE   Kurzschnauziger Schlangenaal
EL   φίδι της θάλασσας
EN   worm-eel
FR   anguille demoiselle
IT   miro
NL   wormaal
PT   cobra-de-orelhas
FI   käärmeankerias-laji
SV   trubbnosig ormål

415 BELONIDAE   BEN

SC   *Belonidae*
ES   agujas; agujones
DA   hornfisk-familien
DE   Hornhechte
EL   ζαργάνες
EN   needlefishes; garfishes
FR   orphies; aiguilles; aiguillettes
IT   aguglie; belonidi
NL   gepen
PT   agulhas e agulhetas
FI   nokkakalat; nokkakalat-heimo
SV   näbbgäddfikar

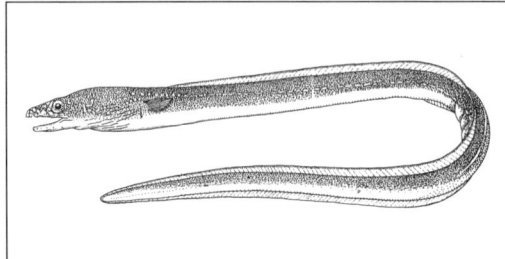

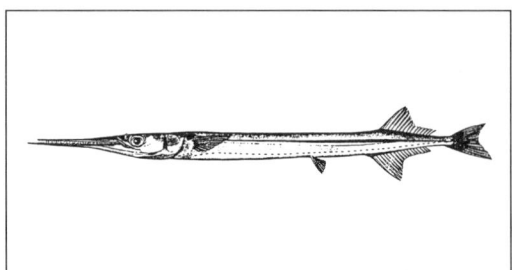

414 OPHICHTHIDAE

SC   *Ophisurus serpens* (Linnaeus, 1758)
ES   serpiente de mar
DA   slangeål
DE   Mittelmeer-Schlangenaal
EL   φίδι της θάλασσας· φιδόχελο
EN   snake-eel; serpent eel
FR   anguille serpent; serpent de mer
IT   biscia di mare; serpente di mare
NL   zeeslang
PT   cobra-do-mar
FI   käärmeankerias-laji
SV   spetsnosig ormål

416 BELONIDAE   GAR

SC   *Belone belone* (Linnaeus, 1761); *Belone vulgaris*; *Belone acus*; *Belone euxini*
ES   aguja; agujeta; saltón
DA   almindelig hornfisk; hornfisk
DE   Hornhecht; Hornfisch; Grünknochen; Europäischer Hornhecht
EL   βελονίδι· σάργομος· ζαργάνα
EN   garfish; garpike; hornpike; billfish; greenbone; mackerel guide; sea needle; gar; sea gar; needle-fish
FR   aiguille de mer; orphie; aiguillette
IT   aguglia; agora
NL   geep
PT   agulha; peixe-agulha
FI   nokkakala
SV   horngädda; näbbgädda

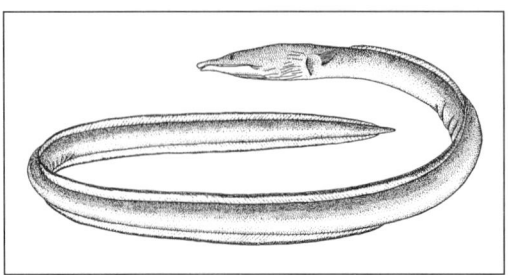

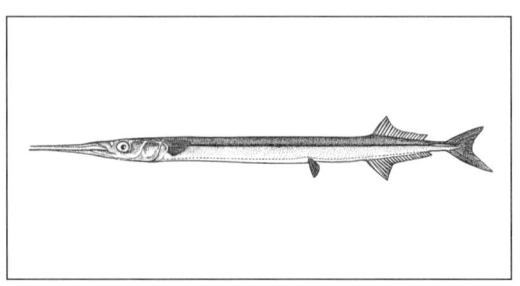

417 BELONIDAE NFA

SC	*Strongylura marina* (Walbaum, 1792)
ES	agujón verde
DA	atlantisk hornfisk
DE	Atlantischer Hornhecht
EL	ζαργάνα του Ατλαντικού
EN	Atlantic needlefish
FR	aiguillette verte
IT	aguglia americana
NL	Atlantische geep
PT	agulheta verde
FI	atlantinnokkakala
SV	atlantisk nålgädda

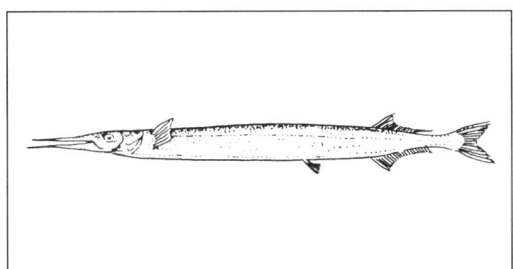

418 BELONIDAE AND

SC	*Tylosurus acus* (Lacepède, 1803); *Tylosaurus acus melanotus*
ES	marao ojón
DA	spids hornfisk
DE	Nadel-Hornhecht
EL	γιγαντοζαργάνα
EN	agujon needlefish
FR	aiguille voyeuse
IT	aguglia maggiore
NL	agujon-geep
PT	agulheta imperial
FI	sininokkakala
SV	nålgädda

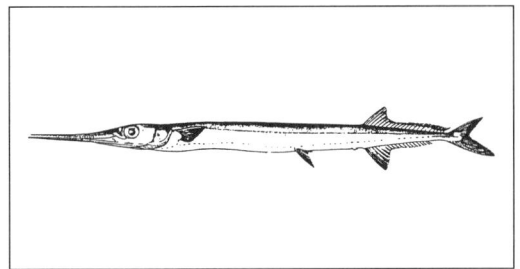

419 BELONIDAE NED

SC	*Tylosurus* spp.
ES	maraos
DA	hornfisk-slægt
DE	Hornhechte
EL	γιγαντοζαργάνες
EN	needlefishes
FR	aiguilles
IT	aguglie maggiori
NL	gepen
PT	agulhetas
FI	nokkakala-suku
SV	nålgäddor

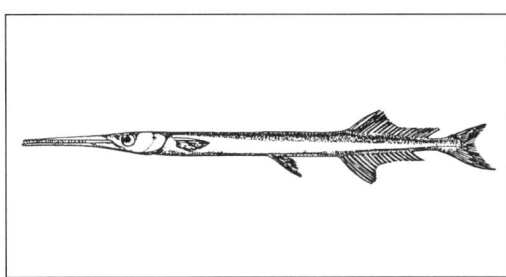

420 SCOMBERESOCIDAE SAX

SC	*Scomberesocidae*
ES	papardas
DA	makrelgedde-familien
DE	Makrelenhechte
EL	λουτσοζαργάνες
EN	sauries; skippers
FR	balaous
IT	costardelle; scomberesocidi
NL	makreelgepen
PT	agulhões
FI	makrillihauet; makrillihauet-heimo
SV	makrillgäddfiskar

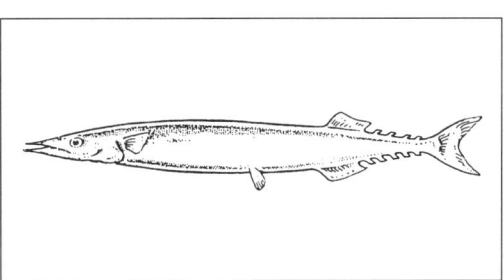

421 SCOMBERESOCIDAE　　SAP

SC	*Cololabis saira* (Brevoort, 1856)
ES	paparda del Pacífico
DA	Stillehavs-makrelgedde
DE	Kurzschnabel-Makrelenhecht
EL	λουτσοζαργάνα του Ειρηνικού
EN	Pacific saury; mackerel-pike; skipper
FR	balaou du Japon; balaou japonais; samma
IT	costardella balau; costardella saira
NL	Japanse makreelgeep
PT	agulhão do Japão
FI	saira
SV	stillahavsmakrillgädda

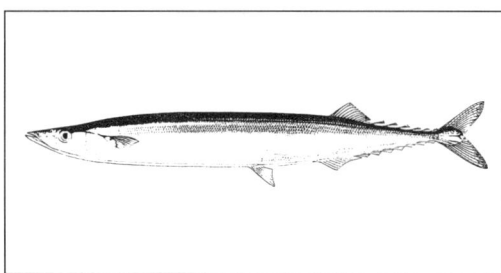

422 SCOMBERESOCIDAE　　SAU

SC	*Scomberesox saurus* (Walbaum, 1792)
ES	paparda; paparda del Atlántico; agujilla; relazón; agujón; saltarín
DA	makrelgedde; almindelig makrelgedde
DE	Makrelenhecht; Atlantischer Makrelenhecht
EL	λουτσοζαργάνα· βελονίδα· ζαργάνα
EN	Atlantic saury; saury pike; skipper; billfish; needle-nose
FR	balaou de l'Atlantique; balaou atlantique; balaou; aiguille de mer
IT	costardella
NL	makreelgeep
PT	agulhão
FI	makrillihauki
SV	makrillgädda

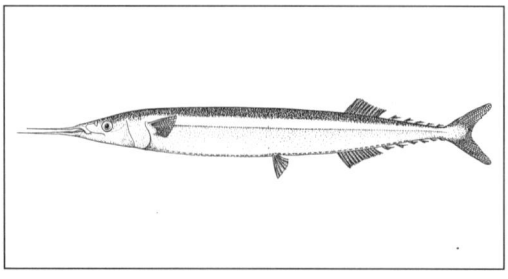

423 EXOCOETIDAE　　FLY

SC	*Exocoetidae*
ES	peces voladores; voladores
DA	flyvefisk-familien
DE	Fliegende Fische
EL	χελιδονόψαρα
EN	flying fishes
FR	poissons volants; exocets
IT	pesci volanti; pesci rondine
NL	vliegende vissen
PT	peixes-voadores
FI	liitokalat; liitokalat-heimo
SV	flygfiskar

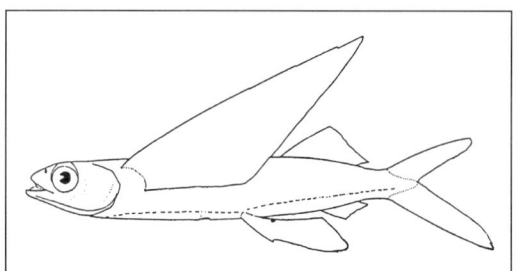

424 EXOCOETIDAE　　JFL

SC	*Cheilopogon agoo* (Temmnick and Schlegel, 1846); *Cypselurus agoo*; *Prognichthys agoo*
ES	volador japonés
DA	japansk flyvefisk
DE	Japanischer Fliegender Fisch
EL	χελιδονόψαρο της Ιαπωνίας
EN	Japanese flying fish
FR	poisson volant du Japon
IT	rondone di mare giapponese
NL	Japanse vliegende vis
PT	peixe-voador do Japão
FI	japaninliitokala
SV	japansk flygfisk

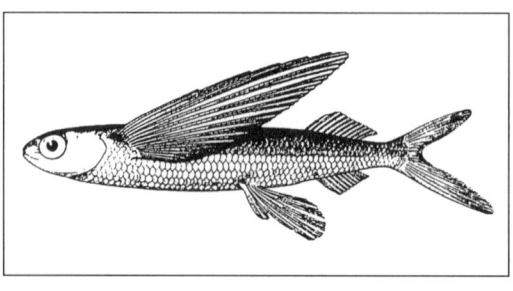

425 EXOCOETIDAE

SC *Cypselurus* spp.
ES peces voladores
DA flyvefisk-slægt
DE Fliegende Fische; Flugfische
EL χελιδονόψαρα
EN flying fishes
FR poissons volants; exocets
IT pesci volanti
NL vliegende vissen
PT peixes-voadores; voadores
FI liitokala-suku
SV flygfiskar

427 EXOCOETIDAE

SC *Hemiramphidae*
ES agujetas
DA halvnæb-familien
DE Halbschnabelhechte
EL ημίραμφοι
EN halfbeaks
FR demi-becs
IT mezzobecchi; emiranfidi
NL halfsnavelbekken
PT meias-agulhas
FI nokkakalat; nokkakalat-heimo
SV halvnäbbfiskar

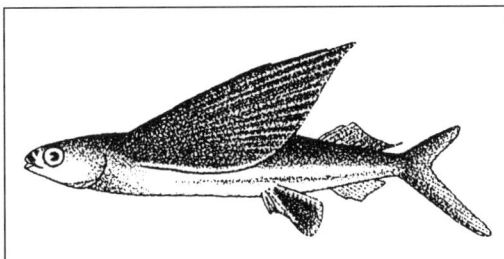

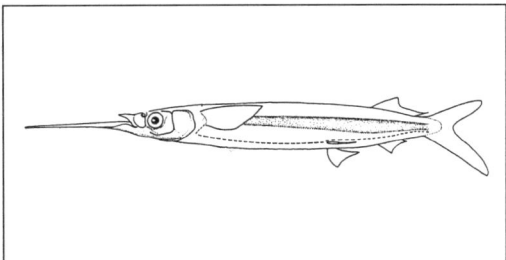

426 EXOCOETIDAE

SC *Exocoetus volitans* (Linnaeus, 1758)
ES golondrina
DA flyvefisk
DE Fliegender Fisch; Flugfisch
EL χελιδονόψαρο
EN flying fish
FR exocet volant; poisson volant
IT pesce volante
NL vliegende vis
PT peixe-voador
FI liitokala
SV flygfisk

428 EXOCOETIDAE

SC *Hemiramphus australensis* (Seale, 1906)
ES agujeta austral
DA australsk halvnæb
DE Australischer Halbschnäbler
EL ημίραμφος της Αυστραλίας
EN sea garfish
FR demi bec d'Australie
IT mezzobecco
NL Australische halfsnavelbek
PT meia-agulha austral
FI australiannokkakala
SV australisk halvnäbb

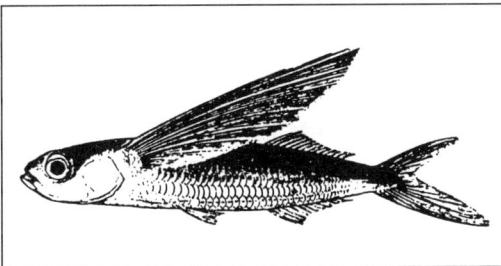

DRAWING NOT AVAILABLE

107

429 EXOCOETIDAE BHA

SC	*Hemiramphus balao* (Lesueur, 1823)
ES	agujeta balaju
DA	balao-halvnæb
DE	Balao-Halbschnäbler
EL	ημίραμφος
EN	balao halfbeak; balao
FR	demi-bec balaou
IT	mezzobecco
NL	balao-halfsnavelbek
PT	meia-agulha preta
FI	balao
SV	balao-halvnäbb

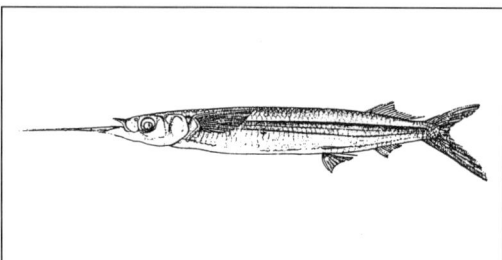

431 EXOCOETIDAE HAJ

SC	*Hemiramphus sajori* (Temminck and Schlegel, 1846); *Hyporhamphus sajori*
ES	agujeta del Japón
DA	japansk halvnæb
DE	Japanischer Halbschnäbler
EL	γιαπωνέζικος ημίραμφος
EN	Japanese halfbeak
FR	demi-bec du Japon
IT	mezzobecco giapponese
NL	Japanse halfsnavelbek
PT	meia-agulha do Japão
FI	japaninnokkakala
SV	japansk halvnäbb

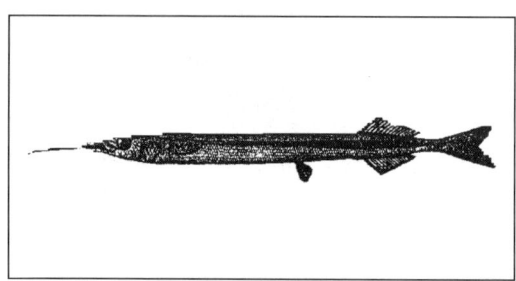

430 EXOCOETIDAE BAL

SC	*Hemiramphus brasiliensis* (Linnaeus, 1758)
ES	agujeta brasileña
DA	brasiliansk halvnæb
DE	Brasilianischer Halbschnäbler
EL	ημίραμφος της Βραζιλίας
EN	ballyhoo halfbeak; ballyhoo
FR	demi-bec brésilien
IT	mezzobecco brasiliano; mezzobecco
NL	Braziliaanse halfsnavelbek
PT	meia-agulha brasileira
FI	brasiliannokkakala
SV	brasiliansk halvnäbb

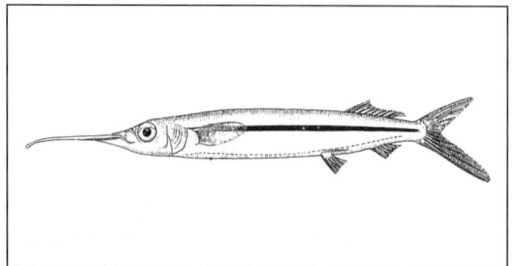

432 EXOCOETIDAE

SC	*Hemiramphus saltator* (Gilbert and Starks, 1904)
ES	aguja; balao saltador
DA	langfinnet halvnæb
DE	Langflossen-Halbschnäbler
EL	μακρόπτερος ημίραμφος
EN	long-fin halfbeak
FR	demi-bec à longues nageoires
IT	mezzobecco
NL	langvin-halfsnavelbek
PT	meia-agulha saltadora
FI	nokkakala-laji
SV	långfenad halvnäbb

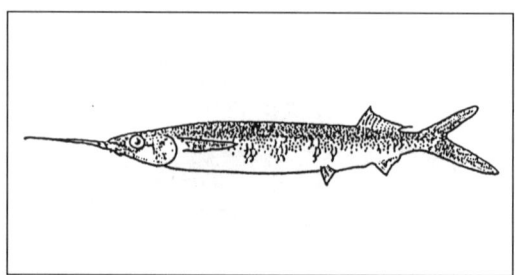

433 EXOCOETIDAE · HAX

SC	*Hemiramphus* spp.
ES	agujetas
DA	halvnæb-slægt
DE	Halbschnäbler
EL	ημίραμφοι
EN	halfbeaks
FR	demi-becs
IT	mezzobecchi
NL	halfsnavelbekken
PT	meias-agulhas; meios-bicos; ponteiros
FI	nokkakala-suku
SV	halvnäbbar

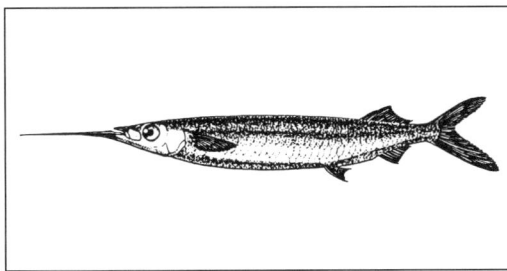

435 GADIFORMES · GAD

SC	*Gadiformes*
ES	gadiformes
DA	torskefisk
DE	Dorschartige
EL	γάδοι
EN	gadiforms
FR	gadiformes
IT	gadiformi
NL	kabeljauwachtigen
PT	gadiformes
FI	turskakalat-lahko
SV	torskartade fiskar

DRAWING NOT AVAILABLE

434 EXOCOETIDAE · FFV

SC	*Hirundichthys affinis* (Günther, 1886)
ES	volador golondrina
DA	firvinget flyvefisk
DE	Indopazifischer Fliegender Fisch
EL	τετραπτέρυγο χελιδονόψαρο
EN	four-wing flying fish
FR	exocet hirondelle
IT	rondine di mare
NL	viervleugelige vliegende vis
PT	peixe-voador-andorinha
FI	liitokala-laji
SV	fyrvingad flygfisk

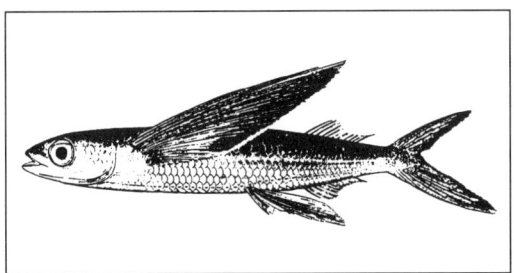

436 MORIDAE · MOR

SC	*Moridae*
ES	moras
DA	dybhavstorske-familien
DE	Tiefseedorsche
EL	μορίδες
EN	morid cods
FR	mores
IT	moridi
NL	diepzeekabeljauwen
PT	moras
FI	siimaturskat; siimaturskat-heimo
SV	moratorskfiskar

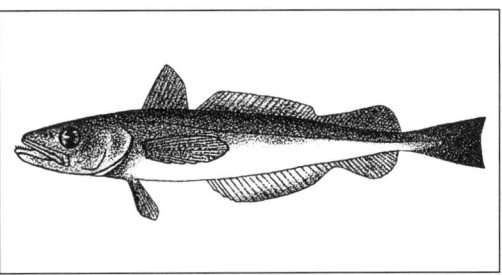

437 MORIDAE   ANT

SC	*Antimora rostrata* (Günther, 1878)
ES	mollera azul
DA	blå antimora
DE	Blauhecht
EL	γαλάζια αντιμόρα
EN	blue antimora
FR	antimore bleue
IT	antimora blu
NL	blauwe diepzeekabeljauw
PT	mora azul
FI	sinisiimaturska
SV	blå antimora

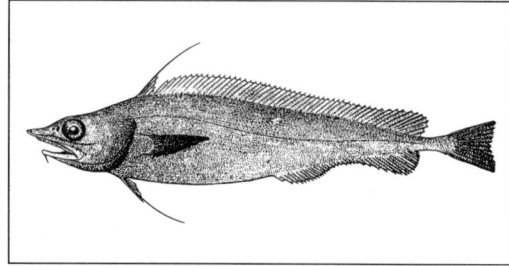

438 MORIDAE   NEC

SC	*Pseudophycis bacchus* (Bloch and Schneider, 1801); *Physiculus bacchus*
ES	brotolilla
DA	newzealandsk rødtorsk
DE	Neuseeland-Eisfisch
EL	κόκκινος μπακαλιάρος της Νέας Ζηλανδίας
EN	New Zealand red cod
FR	morue rouge de Nouvelle-Zélande
IT	busbana neozelandese
NL	Nieuw-Zeelandse diepzeekabeljauw
PT	abrótea da Nova Zelândia
FI	siimaturska-laji
SV	röd moratorsk

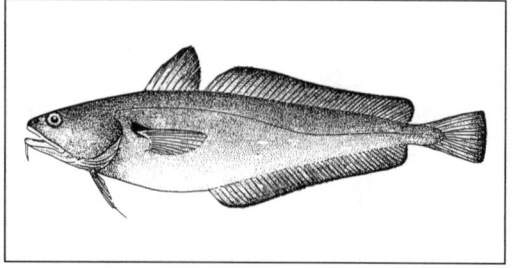

439 BREGMACEROTIDAE   UNC

SC	*Bregmaceros mcclellandi* (Thompson, 1840)
ES	bregmacero
DA	næsehornstorsk
DE	Einhorndorsch
EL	μπρεγμακέρος
EN	unicorn cod
FR	bregmaceros de l'océan Indien
IT	bregmacero
NL	eenhoornkabeljauw
PT	penacheiro índico
FI	täpläantenniturska
SV	indisk djuplånga

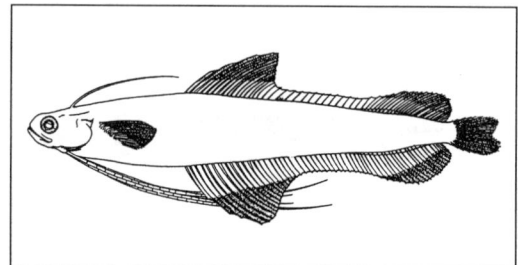

440 GADIDAE

SC	*Gadidae*
ES	gádidos
DA	torskefisk-familien
DE	Dorschfische
EL	μπακαλιάροι
EN	codfishes
FR	gades
IT	gadidi
NL	kabeljauwen
PT	gadídeos
FI	turskat; turskat-heimo
SV	torskfiskar

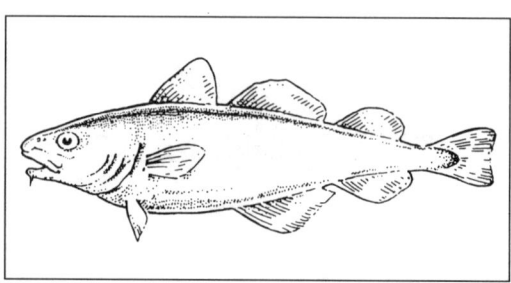

441 GADIDAE   POC

SC	*Boreogadus saida* (Lepechin, 1774)
ES	bacalao polar
DA	polartorsk
DE	Polardorsch
EL	βορεομπακαλιάρος
EN	Polar cod; Arctic cod
FR	saïda; morue polaire
IT	merluzzo artico
NL	poolkabeljauw
PT	bacalhau polar
FI	jäämerenseiti
SV	polartorsk

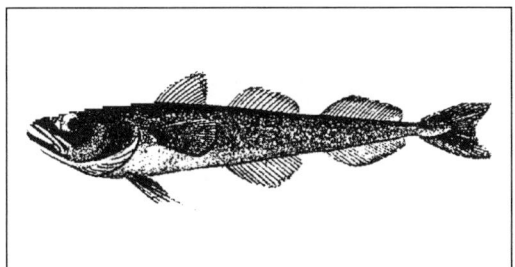

443 GADIDAE

SC	*Ciliata mustela* (Linnaeus, 1758)
ES	mustela
DA	femtrådet havkvabbe
DE	Fünfbärtelige Seequappe
EL	γαϊδουρόψαρο· ποντικός
EN	five-bearded rockling
FR	motelle à cinq barbillons; mustelle à cinq barbillons
IT	motella a cinque baffi
NL	vijfdradige meun
PT	laibeque de cinco barbilhos
FI	viisiviiksimade
SV	femtömmad skärlånga

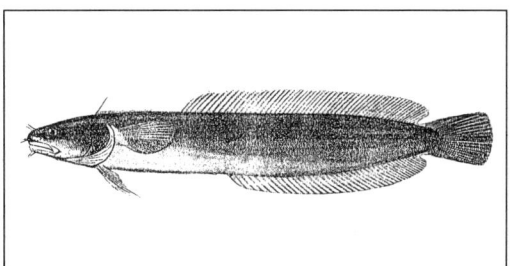

442 GADIDAE   USK

SC	*Brosme brosme* (Ascanius, 1772); *Brosmius brosme*
ES	brosmio
DA	brosme
DE	Brosme; Lumb
EL	μπρόσμιος
EN	tusk; torsk; cusk
FR	brosmes; tusk
IT	brosmio; brosme
NL	torsk
PT	bolota
FI	keila
SV	lubb

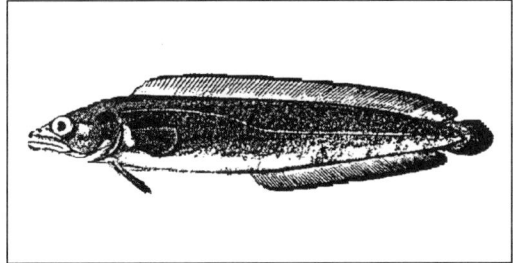

444 GADIDAE   SAF

SC	*Eleginus gracilis* (Tilesius, 1810)
ES	bacalao oriental
DA	østlig navaga
DE	Fernöstliche Navaga; Wachna; Wachnja
EL	μπακαλιάρος της Ανατολής· ναβάγκα
EN	saffron cod; Far Eastern navaga; Northern cod; wachna cod
FR	navaga jaune
IT	navaga
NL	oostelijke navaga
PT	bacalhau do Extremo Oriente
FI	idännavaga
SV	saffranstorsk

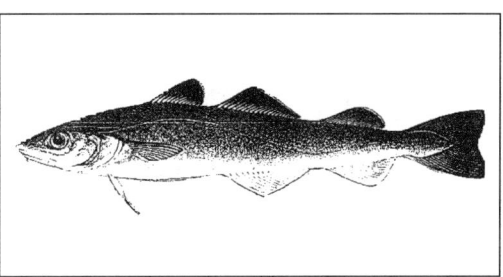

445 GADIDAE COW

SC	*Eleginus navaga* (Pallas, 1814); *Gadus navaga*
ES	bacalao del Ártico
DA	atlantisk navaga
DE	Navaga; Europäische Navaga; Wachna; Wachnja
EL	μπακαλιάρος της Αρκτικής· ναβάγκα
EN	wachna cod; navaga; Arctic cod; Atlantic navaga
FR	morue arctique
IT	navaga
NL	navaga-kabeljauw
PT	bacalhau árctico; bacalhau do Árctico
FI	navaga
SV	navagatorsk

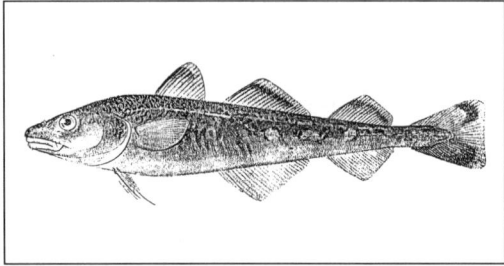

447 GADIDAE COD

SC	*Gadus morhua* (Linnaeus, 1758); *Gadus morrhua*; *Gadus callarias*
ES	bacalao
DA	atlantisk torsk; torsk
DE	Dorsch; Kabeljau
EL	μπακαλιάρος
EN	Atlantic cod; cod; codfish; codling
FR	cabillaud; morue fraîche
IT	merluzzo bianco; merluzzo
NL	kabeljauw
PT	bacalhau do Atlântico; bacalhau
FI	turska
SV	torsk

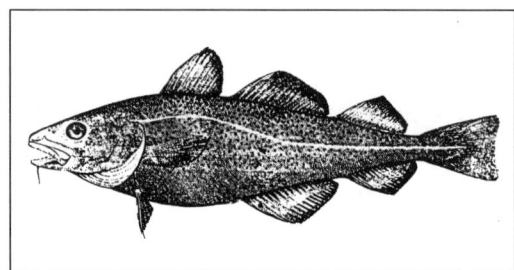

446 GADIDAE PCO

SC	*Gadus macrocephalus* (Tilesius, 1810)
ES	bacalao del Pacífico; bacalao de Alaska
DA	Stillehavs-torsk
DE	Pazifischer Kabeljau
EL	μπακαλιάρος του Ειρηνικού
EN	Pacific cod; gray cod; grayfish
FR	morue grise; morue du Pacifique
IT	merluzzo del Pacifico; merluzzo bianco
NL	Pacifische kabeljauw
PT	bacalhau do Pacífico Norte; bacalhau do Alasca; bacalhau do Pacífico
FI	tyynenmerenturska
SV	stillahavstorsk

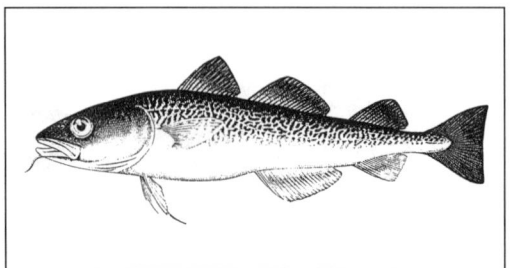

448 GADIDAE GRC

SC	*Gadus ogac* (Richardson, 1836); *Gadus morhua ogac*
ES	bacalao de Groenlandia
DA	uvak; fjordtorsk
DE	Grönland-Dorsch; Grönland-Kabeljau; Fjord-Dorsch; Kabeljau; Ogac
EL	μπακαλιάρος της Γροιλανδίας
EN	Greenland cod
FR	morue ogac; ogac; morue de roche; morue du Groenland
IT	merluzzo bianco
NL	Groenlandse kabeljauw
PT	bacalhau da Gronelândia
FI	grönlanninturska
SV	uvak

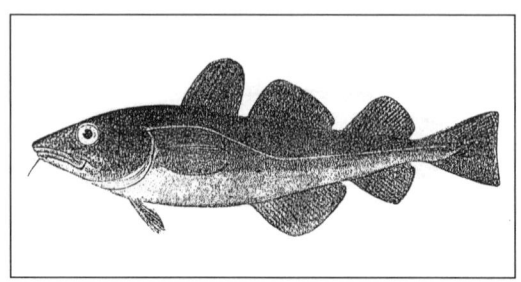

449 GADIDAE

SC	*Gadiculus argenteus* (Guichenot, 1849); *Gadiculus thori*
ES	faneca plateada
DA	sølvtorsk
DE	Silberdorsch; Nördlicher Silberdorsch
EL	μπακαλιαράκι· γουρλομάτης
EN	silver pout
FR	merlan argenté; gadicule
IT	pesce fico
NL	zilverwijting
PT	badejinho; badejinho do norte
FI	hopeaturska
SV	silvertorsk

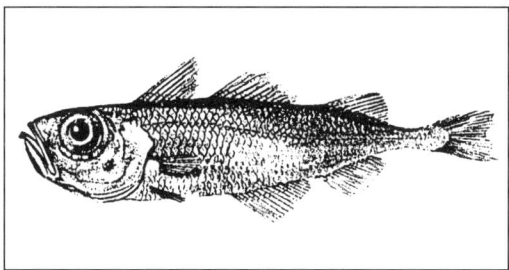

451 GADIDAE   ROL

SC	*Gaidropsarus* spp.
ES	barbadas
DA	havkvabbe-slægt
DE	Dreibärtelige Seequappen
EL	γαϊδουρόψαρα
EN	rocklings
FR	motelles
IT	motelle
NL	meunen
PT	laibeques; larotes
FI	merimade-suku
SV	skärlångor

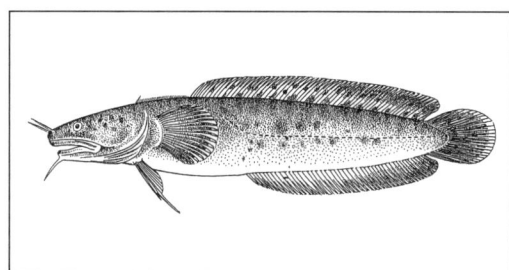

450 GADIDAE

SC	*Gaidropsarus vulgaris* (Cloquet, 1824); *Motella tricirrata*; *Onos tricirrata*
ES	lota; mollareta
DA	tretrådet havkvabbe
DE	Dreibärtelige Seequappe; Seequappe
EL	κηλιδογαϊδουρόψαρο
EN	three-bearded rockling; three-beard rockling
FR	motelle à trois barbillons; mustelle à trois barbillons; motelle commune
IT	motella maculata
NL	driedradige meun
PT	laibeque de três barbilhos
FI	kolmiviiksimade
SV	tretömmad skärlånga

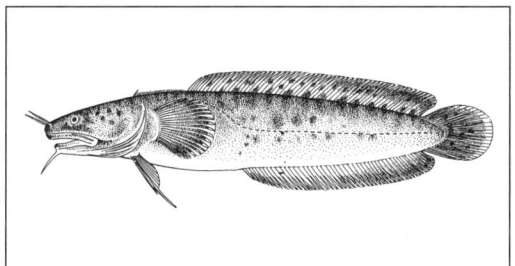

452 GADIDAE   FBU

SC	*Lota lota* (Linnaeus, 1758); *Lota lacustris*; *Lota maculosa*
ES	lota de río
DA	ferskvandskvabbe; knude
DE	Quappe; Rutte; Trüsche; Aalquappe; Aalraupe; Aalruppe; Aalrutte; Rufunkel; Rufolken; Quappau; Treische; Aalrauke
EL	παχύχελο
EN	burbot; eel pout
FR	lotte de rivière
IT	bottatrice
NL	kwabaal
PT	lota do rio
FI	made
SV	lake

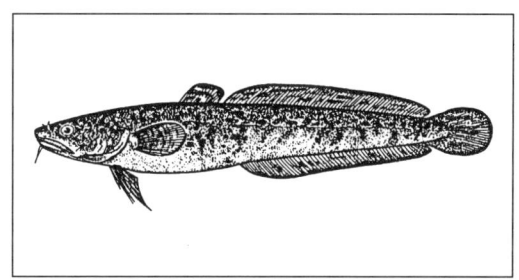

453	**GADIDAE**	HAD

SC *Melanogrammus aeglefinus* (Linnaeus, 1758);
    Gadus aeglefinus
ES eglefino; liba
DA kuller
DE Schellfisch
EL μπακαλιάρος
EN haddock; chat; jumbo
FR églefin; aiglefin; ânon; morue noire; morue
    Saint-Pierre
IT eglefino
NL schelvis
PT arinca
FI kolja
SV kolja

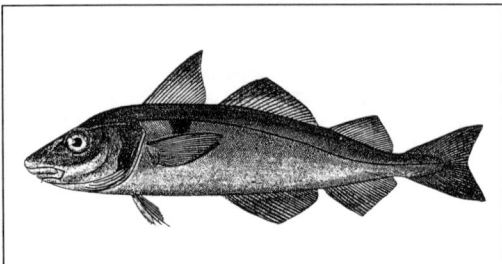

454	**GADIDAE**	WHG

SC *Merlangius merlangus* (Linnaeus, 1758); *Gadus
    merlangus*
ES merlán; plegonegro; liba; sarreta
DA hvilling
DE Wittling; Merlan
EL νταούκι του Ατλαντικού
EN whiting; marling
FR merlan; merlan commun; valet
IT merlano; molo
NL wijting
PT badejo
FI valkoturska
SV vitling

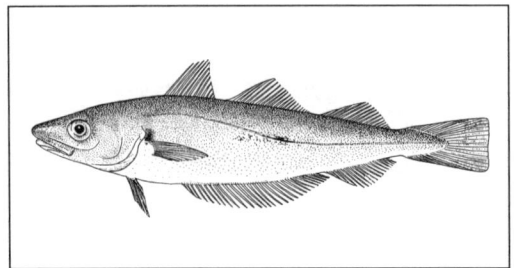

455	**GADIDAE**	COP

SC *Merlangius merlangus euxinus* (Nordmann, 1830)
ES capelán del Mar Negro
DA Sortehavs-hvilling
DE Schwarzmeer-Wittling
EL νταούκι
EN Black Sea whiting
FR merlan de la mer Noire
IT merlano; molo
NL Zwarte-Zeewijting
PT badejo do mar Negro
FI mustanmerenvalkoturska
SV vitling

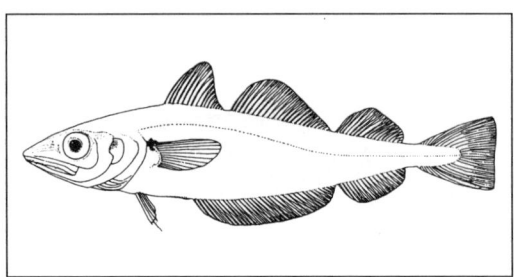

456	**GADIDAE**	TOM

SC *Microgadus tomcod* (Walbaum, 1792)
ES tomcod
DA atlantisk tomcod
DE Atlantischer Tomcod; Frostfisch
EL μπακαλιαράκι του Ατλαντικού
EN Atlantic tomcod
FR poulamon atlantique
IT tomcod
NL Atlantische tomcod
PT tomecode
FI jääturska
SV atlantisk frostfisk

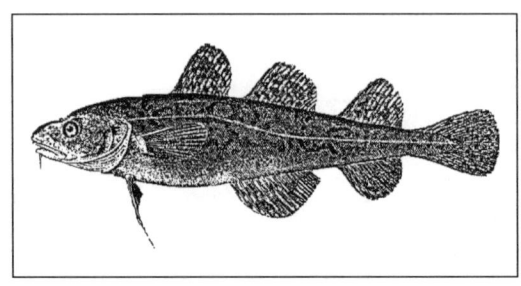

457 GADIDAE

SC	*Microgadus* spp.
ES	tomcod
DA	tomcod
DE	Tomcod
EL	μπακαλιαράκια
EN	tomcod
FR	poulamon
IT	tomcod
NL	tomcod
PT	tomecode
FI	jääturska-suku
SV	frostfiskar

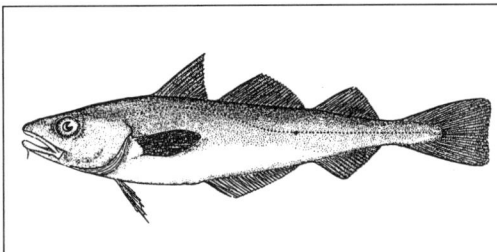

459 GADIDAE WHB

SC	*Micromesistius poutassou* (Risso, 1826); *Gadus poutassou*
ES	bacaladilla; perlita; bacalada; pez lirio; maira; lirio
DA	sortmund; blåhvilling
DE	Blauer Wittling; Blauwittling
EL	προσφυγάκι· τσμπλάκι
EN	blue whiting; poutassou
FR	merlan bleu; poutassou
IT	melù; potassolo
NL	blauwe wijting
PT	verdinho
FI	mustakitaturska
SV	blåvitling; kolmule

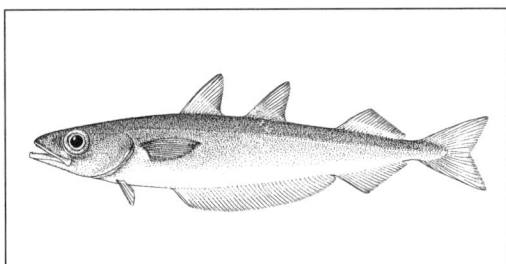

458 GADIDAE POS

SC	*Micromesistius australis* (Norman, 1937)
ES	polaca austral
DA	sydlig sortmund
DE	Südlicher Wittling
EL	προσφυγάκι της Αυστραλίας
EN	Southern blue whiting
FR	merlan bleu austral
IT	melù australe
NL	zuidelijke blauwe wijting
PT	verdinho austral
FI	etelänmustakitaturska
SV	sydlig blåvitling

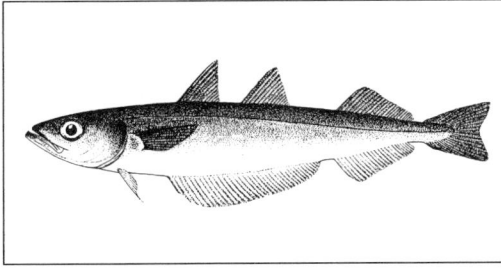

460 GADIDAE BLI

SC	*Molva dipterygia dipterygia* (Pennant, 1784); *Molva dypterygia*; *Molva byrkelange*
ES	maruca azul; arbitán; maruca
DA	byrkelange
DE	Blauleng
EL	ποντικόψαρο· μουρούνα
EN	blue ling
FR	lingue bleue
IT	molva azzurra
NL	blauwe leng
PT	donzela azul; maruca azul; lingue azul
FI	tylppäpyrstömolva
SV	birkelånga

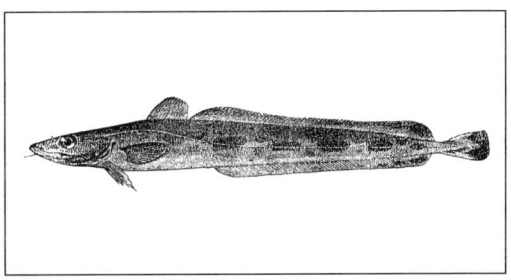

461 GADIDAE SLI

SC *Molva dipterygia macrophthalma* (Rafinesque, 1810); *Molva macrophthalma*; *Molva elongata*
ES arbitán
DA Middelhavs-lange
DE Mittelmeer-Leng
EL ποντικόψαρο· μουρούνα
EN Mediterranean ling; Spanish ling
FR lingue méditerranéenne; lingue espagnole
IT molva occhiona; molva
NL Middellandse-Zeeleng
PT donzela do Mediterrâneo; maruca da pedra; lingue da pedra
FI luikeromolva
SV medelhavslånga

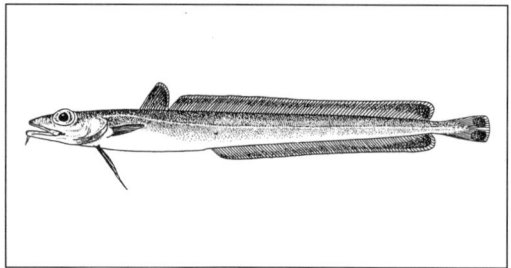

463 GADIDAE GFB

SC *Phycis blennoides* (Brünnich, 1768); *Urophycis blennoides*
ES brótola de fango; locha; escolar; fura; mullosa
DA skælbrosme; almindelig skælbrosme
DE Gabeldorsch; Meertrüsche
EL σαλουβάρδος· ποντικός· λασποσαλούβαρδος
EN greater forkbeard
FR phycis de fond; mostelle de vase; mostelle de fond
IT musdea bianca; mustella
NL gaffelkabeljauw
PT abrótea-do-alto
FI suomuturska
SV fjällbrosme

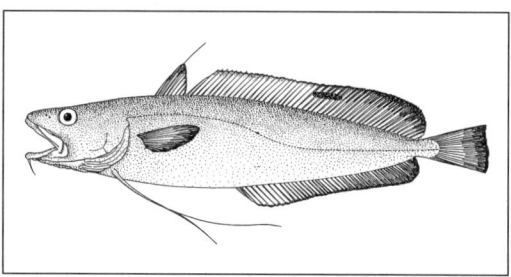

462 GADIDAE LIN

SC *Molva molva* (Linnaeus, 1758); *Lota molva*
ES maruca; barruenda; juliana; tachuela; pez de bacalao; guitarra
DA lange
DE Leng; Lengfisch; Heller Leng
EL ποντίκι
EN ling
FR lingue commune; morue longue; lingue; morue longue commune; grande lingue
IT molva
NL leng
PT donzela; maruca; lingue
FI molva
SV långa

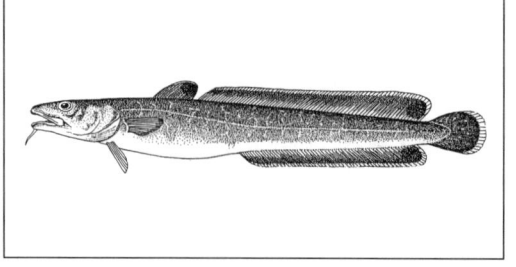

464 GADIDAE FOR

SC *Phycis phycis* (Linnaeus, 1766)
ES brótola de roca; alfaneca; escolano; locha; barbada
DA Middelhavs-skælbrosme
DE Mittelmeer-Gabeldorsch; Südliche Meerschleie; Mittelmeer-Trüsche
EL σαλούβαρδος· ποντικός· πετροσαλούβαρδος· σαραβάνος
EN forkbeard
FR phycis de roche; mostelle de roche
IT musdea; mostella
NL Middellandse-Zeegaffelkabeljauw
PT abrótea-da-costa
FI luikeroturska
SV klippbrosme

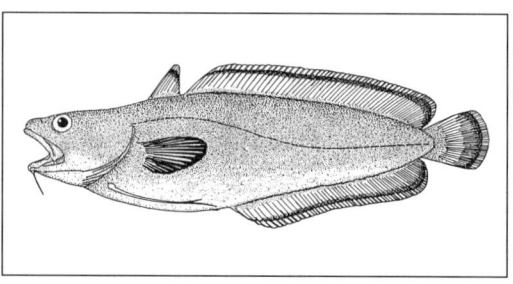

465 GADIDAE FOX

SC   *Phycis* spp.
ES   brótolas
DA   skælbrosme-slægten
DE   Ostatlantische Gabeldorsche
EL   σαλούβαρδοι
EN   forkbeards
FR   mostelles
IT   musdee
NL   gaffelkabeljauwen
PT   abróteas
FI   luikeroturska-suku
SV   fjällbrosmar

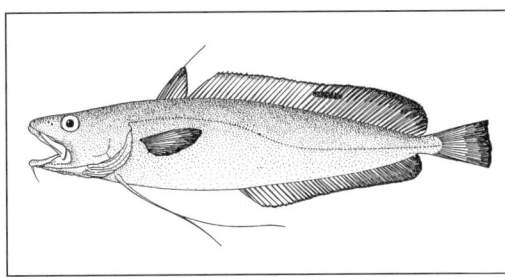

467 GADIDAE POK

SC   *Pollachius virens* (Linnaeus, 1758); *Gadus virens*;
     Merlangus virens
ES   carbonero; fogonero
DA   sej; gråsej
DE   Seelachs; Köhler; Blaufisch
EL   μαύρος μπακαλιάρος
EN   saithe; coalfish; black cod; black pollack; sillock;
     pollack
FR   lieu noir; colin noir
IT   merluzzo carbonaro
NL   koolvis
PT   escamudo; paloco
FI   seiti
SV   gråsej

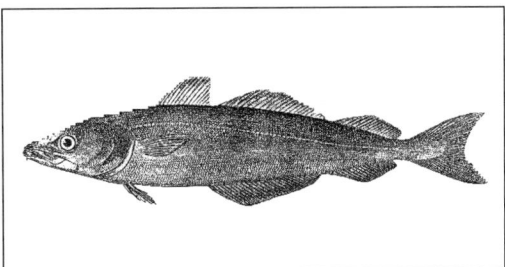

466 GADIDAE POL

SC   *Pollachius pollachius* (Linnaeus, 1758); *Gadus*
     pollachius; *Merlangius pollachius*
ES   abadejo; serreta
DA   lubbe; blåsej
DE   Pollack; Heller Seelachs; Klamottendorsch;
     Steinköhler
EL   κίτρινος μπακαλιάρος
EN   pollack lythe; pollock; green pollack
FR   lieu jaune; colin jaune
IT   merluzzo giallo; pollack
NL   pollak
PT   juliana
FI   lyyraturska
SV   bleka; lyrtorsk

468 GADIDAE

SC   *Raniceps raninus* (Linnaeus, 1758)
ES   pez rana
DA   sortvels
DE   Froschquappe; Froschquabbe; Froschdorsch
EL   βατραχομπακαλιάρος
EN   tadpole fish
FR   trident; grenouille de mer
IT   musdea bianca; mostella; musdea
NL   vorskwab
PT   rainúnculo negro
FI   mustaturska
SV   paddtorsk

DRAWING NOT AVAILABLE

469 GADIDAE

SC	*Rhinonemus cimbrius* (Linnaeus, 1766)
ES	barbuda de cuatro barbillas
DA	firtrådet havkvabbe
DE	Vierbärtelige Seequappe
EL	τετραμούστακο γαϊδουρόψαρο
EN	four-bearded rockling
FR	motelle à quatre barbillons
IT	motella a quattro baffi
NL	vierdradige meun
PT	peixe-chumbo
FI	neliviiksimade
SV	fyrtömmad skärlånga

471 GADIDAE NOP

SC	*Trisopterus esmarki* (Nilsson, 1855); *Gadus esmarki*
ES	faneca noruega
DA	sperling
DE	Stintdorsch
EL	σύκο της Νορβηγίας
EN	Norway pout
FR	tacaud norvégien
IT	busbana norvegese
NL	kever
PT	faneca-noruega
FI	harmaaturska
SV	vitlinglyra

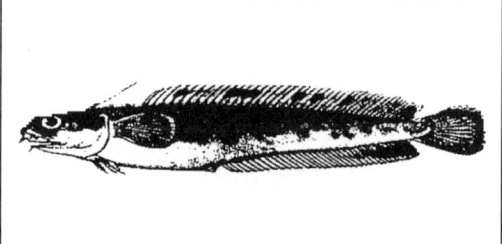

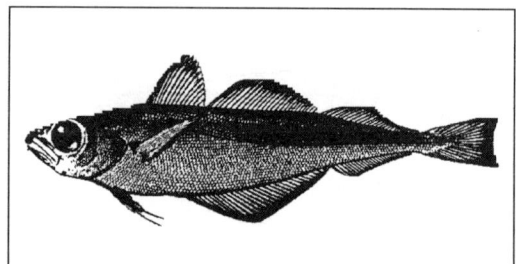

470 GADIDAE ALK

SC	*Theragra chalcogramma* (Pallas, 1811)
ES	colín de Alaska; abadejo de Alaska
DA	Alaska-sej
DE	Pazifischer Pollack; Alaska-Pollack; Pazifischer Polardorsch
EL	μπακαλιάρος της Αλάσκας
EN	Alaska pollack; walleye pollack
FR	morue du Pacifique occidental; lieu de l'Alaska
IT	merluzzo dell'Alasca; pollack d'Alasca
NL	Alaskapollak
PT	escamudo do Alasca
FI	tyynenmerenseiti
SV	alaskapollack

472 GADIDAE BIB

SC	*Trisopterus luscus* (Linnaeus, 1758); *Gadus luscus*
ES	faneca; paneka; palenca
DA	skægtorsk
DE	Französischer Dorsch; Franzosendorsch
EL	σύκο του Ατλαντικού
EN	pout; bib; pouting; whiting-pout
FR	tacaud; plouse; gode; tacaud commun
IT	busbana francese; merluzzetto bruno
NL	steenbolk
PT	faneca
FI	partaturska
SV	skäggtorsk

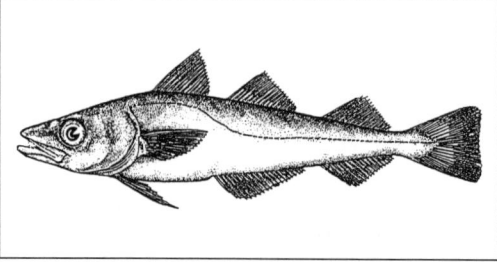

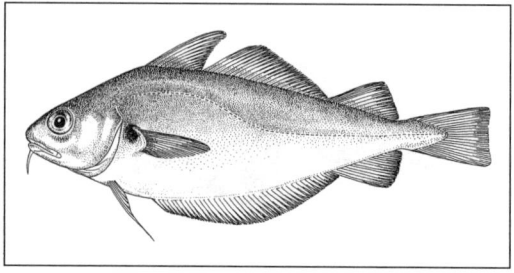

| 473 | **GADIDAE** | POD | 475 | **GADIDAE** | HKR |

473 GADIDAE   POD

SC   *Trisopterus minutus capelanus* (Lacépède, 1800);
     Trisopterus minutus; *Gadus minutus*
ES   capellán; mollera; romero; carajuelo
DA   glyse
DE   Zwergdorsch
EL   σύκο
EN   poor cod
FR   capelan; capelan de Méditerranée; capelan
     de France
IT   merluzzo capellano; merluzzetto; capellano;
     busbana
NL   dwergbolk
PT   fanecão
FI   pikkuturska
SV   glyskolja

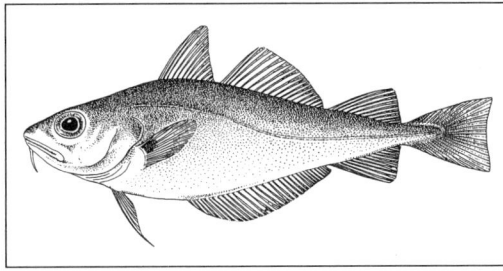

475 GADIDAE   HKR

SC   *Urophycis chuss* (Walbaum, 1792)
ES   locha roja; locha
DA   rød skægbrosme
DE   Roter Gabeldorsch
EL   κόκκινος μπακαλιάρος
EN   red hake; squirrel hake
FR   merluche-écureuil
IT   musdea atlantica
NL   Atlantische gaffelkabeljauw
PT   abrótea vermelha; linguiça
FI   punasuomuturska
SV   skäggbrosme

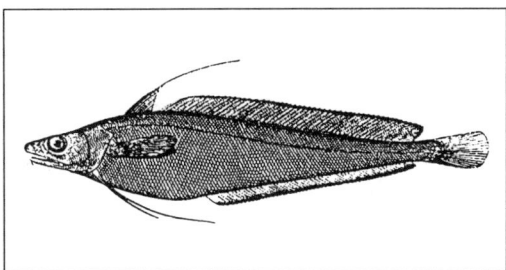

474 GADIDAE   HKU

SC   *Urophycis brasiliensis* (Kaup, 1858)
ES   brótola brasileña
DA   brasiliansk skægbrosme
DE   Brasilianischer Gabeldorsch
EL   μπακαλιάρος της Βραζιλίας
EN   Brazilian codling
FR   mostelle brésiliènne; merluche brésilienne
IT   musdea brasiliana
NL   Braziliaanse gaffelkabeljauw
PT   abrótea brasileira
FI   brasiliansuomuturska
SV   brasiliansk brosme

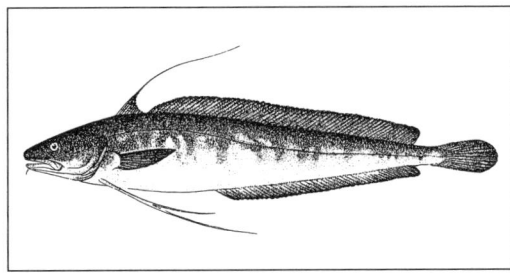

476 GADIDAE   HKW

SC   *Urophycis tenuis* (Mitchill, 1814)
ES   locha blanca; locha
DA   hvid skægbrosme
DE   Weißer Gabeldorsch
EL   λευκός μπακαλιάρος
EN   white hake
FR   merluche blanche
IT   musdea americana
NL   witte heek
PT   abrótea branca
FI   valkosuomuturska
SV   vitbrosme

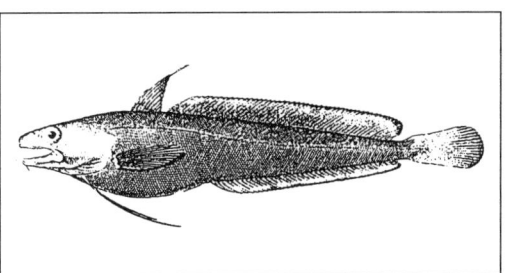

477	**MERLUCCIIDAE**	HKZ

SC　*Merlucciidae*
ES　merluzas
DA　kulmule-familien
DE　Seehechte
EL　μπακαλιάροι
EN　hakes
FR　merlus; colins
IT　merluzzi; naselli; merluccidi
NL　heken
PT　pescadas
FI　kummeliturskat; kummeliturskat-heimo
SV　kummelfiskar

479	**MERLUCCIIDAE**	GRN

SC　*Macruronus novaezealandiae* (Hector, 1871)
ES　cola de rata azul
DA　newzealandsk langhale
DE　Neuseeländischer Grenadier
EL　γρεναδιέρος της Νέας Ζηλανδίας
EN　blue grenadier
FR　grenadier bleu de Nouvelle-Zélande
IT　merluzzo granatiere
NL　blauwe grenadier
PT　granadeiro azul
FI　hoki
SV　hoki

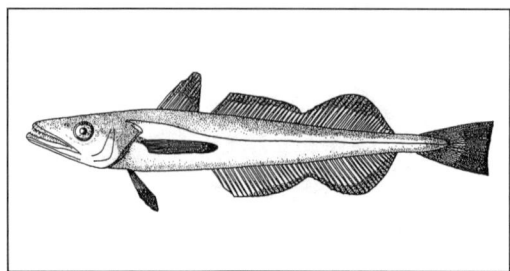

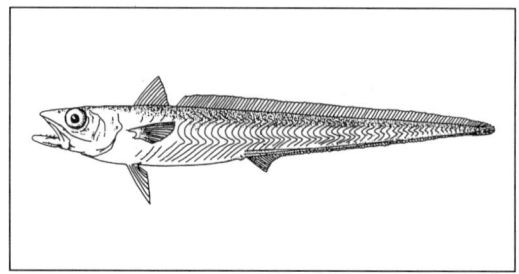

478	**MERLUCCIIDAE**	GRM

SC　*Macruronus magellanicus* (Lönnberg, 1907)
ES　merluza de cola
DA　patagonisk langhale
DE　Patagonischer Grenadier
EL　γρεναδιέρος της Παταγωνίας
EN　Patagonian grenadier
FR　grenadier de Patagonie
IT　merluzzo granatiere
NL　Patagonische grenadier
PT　granadeiro da Patagónia
FI　patagonianhoki
SV　chilensk hoki

480	**MERLUCCIIDAE**	GRS

SC　*Macruronus* spp.
ES　colas de rata
DA　langhale-slægt
DE　Grenadierfische; Grenadiere
EL　μπλε γρεναδιέροι
EN　blue grenadiers
FR　grenadiers bleus
IT　merluzzi granatieri
NL　grenadiers
PT　granadeiros
FI　hoki-suku
SV　hoki

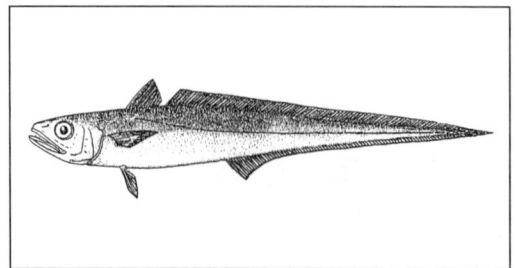

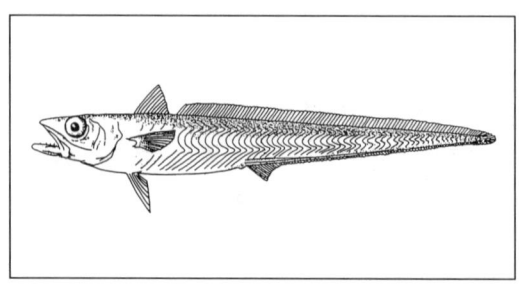

481 MERLUCCIIDAE HOF

SC	*Merluccius albidus* (Mitchill, 1817)
ES	merluza blanca de altura
DA	sølvkulmule
DE	Großäugiger Seehecht
EL	μπακαλιάρος
EN	offshore silver hake
FR	merlu argenté du large
IT	nasello
NL	diepzeezilverheek
PT	pescada-prateada do alto
FI	kummeliturska-laji
SV	vitkummel

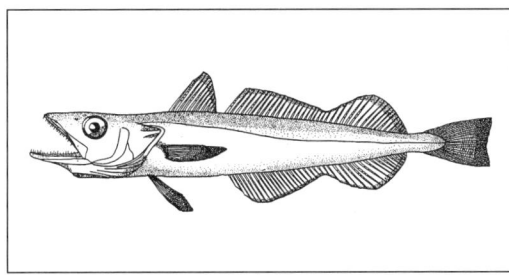

482 MERLUCCIIDAE HKN

SC	*Merluccius australis* (Hutton, 1872)
ES	merluza sureña; merluza austral
DA	sydlig kulmule
DE	Südlicher Seehecht
EL	μπακαλιάρος της Αυστραλίας
EN	Southern hake
FR	merlu austral
IT	nasello; merluzzo
NL	Australische heek
PT	pescada da Nova Zelândia
FI	etelänkummeliturska
SV	sydkummel

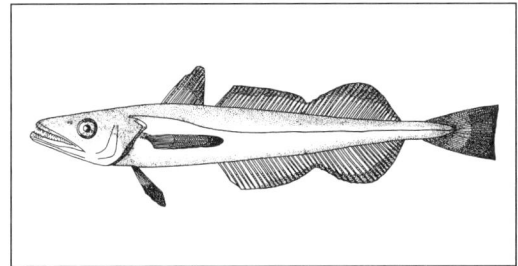

483 MERLUCCIIDAE HKS

SC	*Merluccius bilinearis* (Mitchill, 1814)
ES	merluza de Boston; merluza norteamericana; merluza atlántica
DA	nordvestatlantisk kulmule
DE	Nordamerikanischer Seehecht; Silberhecht
EL	μπακαλιάρος του Ατλαντικού
EN	silver hake; whiting; offshore hake
FR	merlu argenté d'Amérique du Nord; merlu argenté
IT	nasello atlantico; nasello; merluzzo
NL	zilverheek
PT	pescada-prateada
FI	hopeakummeliturska
SV	silverkummel

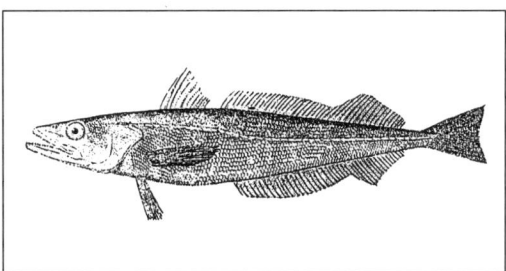

484 MERLUCCIIDAE HKC

SC	*Merluccius capensis* (Castelnau, 1861)
ES	merluza costera; merluza del Cabo
DA	sydafrikansk kulmule
DE	Kaphecht
EL	μπακαλιάρος της Νότιας Αφρικής
EN	Cape hake; shallow-water hake
FR	merlu côtier; merlu du Cap
IT	nasello del Capo; nasello; merluzzo
NL	ondiepwaterheek
PT	pescada da África do Sul
FI	kapinkummeliturska
SV	kapkummel

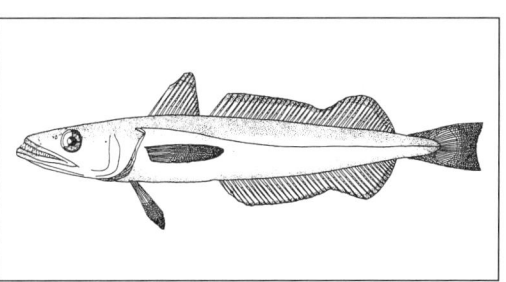

485 MERLUCCIIDAE PHA

SC	*Merluccius gayi* (Guichenot, 1848)
ES	merluza chilena
DA	chilensk kulmule
DE	Chilenischer Seehecht; Pazifischer Seehecht
EL	μπακαλιάρος του Περού
EN	Chilean hake; Peruvian hake; South Pacific hake
FR	merlu du Chili
IT	nasello del Cile; nasello; merluzzo
NL	Chileense heek
PT	pescada do Chile
FI	perunkummelifturska
SV	chilensk kummel

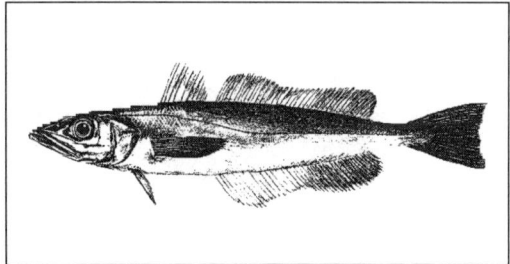

487 MERLUCCIIDAE HKE

SC	*Merluccius merluccius* (Linnaeus, 1758)
ES	merluza europea; merluza; pescadilla; pijota; pijotilla
DA	europæisk kulmule; kulmule
DE	Europäischer Seehecht; Seehecht; Hechtdorsch
EL	μπακαλιάρος
EN	European hake; hake
FR	merlu européen; merlu; merluche; merlu commun; colin; merluchon; colinot
IT	nasello; merluzzo argentato
NL	heek
PT	pescada-branca; pescada; marmota; pescadinha
FI	kummelifturska
SV	kummel

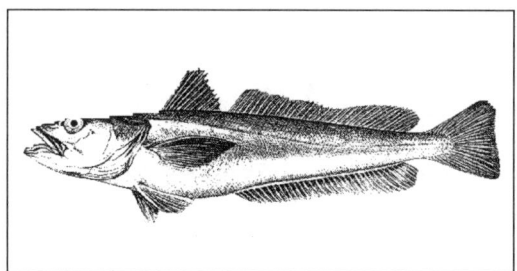

486 MERLUCCIIDAE HKP

SC	*Merluccius hubbsi* (Marini, 1932)
ES	merluza argentina; merluza sudamericana
DA	sydvestatlantisk kulmule
DE	Patagonischer Seehecht; Argentinischer Seehecht
EL	μπακαλιάρος της Αργεντινής
EN	Argentine hake; South-west Atlantic hake
FR	merlu d'Argentine; merlu sud-américain
IT	nasello; merluzzo
NL	Argentijnse heek
PT	pescada argentina
FI	argentiinankummelifturska
SV	argentinsk kummel

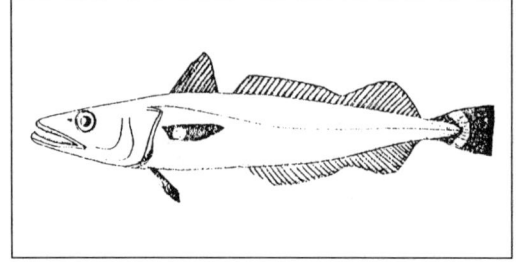

488 MERLUCCIIDAE HKO

SC	*Merluccius paradoxus* (Franca, 1960)
ES	merluza de altura
DA	dybvandskulmule
DE	Tiefenwasser-Kapseehecht
EL	μπακαλιάρος του Ακρωτηρίου
EN	deepwater hake; deepwater Cape hake
FR	merlu profond
IT	nasello; merluzzo
NL	diepwaterheek
PT	pescada do Sudoeste Africano
FI	kummelifturska-laji
SV	djupkapkummel

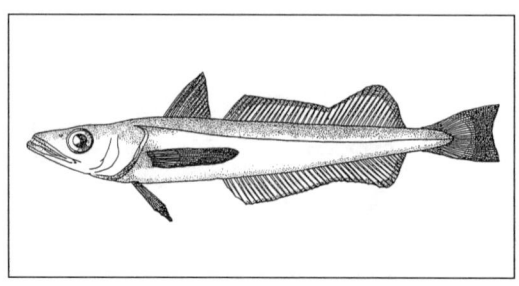

489	**MERLUCCIIDAE**	HKB

SC	*Merluccius polli* (Cadenat, 1950); *Merluccius cadenati*
ES	merluza angolense; merluza de Benguela
DA	Benguela-kulmule
DE	Benguela-Seehecht
EL	μπακαλιάρος της Αγκόλας
EN	Benguela hake
FR	merlu d'Afrique tropicale
IT	nasello; merluzzo
NL	Benguelaheek
PT	pescada de Angola
FI	kummeliturska-laji
SV	svart kummel

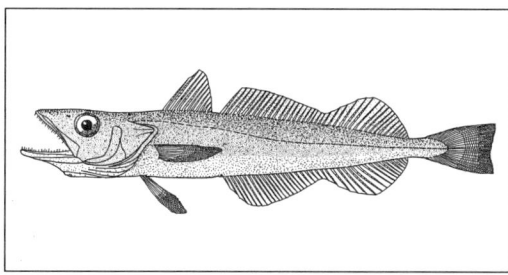

490	**MERLUCCIIDAE**	NHA

SC	*Merluccius productus* (Ayres, 1855)
ES	merluza del Pacífico norte; merluza pacífica norteamericana
DA	Stillehavs-kulmule
DE	Nordpazifischer Seehecht
EL	μπακαλιάρος του Ειρηνικού
EN	North Pacific hake; Pacific hake
FR	merlu du Pacifique nord; merlu du Pacifique
IT	nasello del Pacifico; nasello; merluzzo
NL	Pacifische heek
PT	pescada do Pacífico Norte; pescada da Califórnia; pescada do Alasca
FI	kaliforniankummeliturska
SV	stillahavskummel

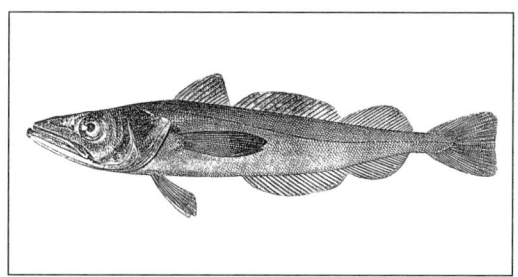

491	**MERLUCCIIDAE**	HKM

SC	*Merluccius senegalensis* (Cadenat, 1950)
ES	merluza senegalesa; merluza del Senegal
DA	senegalsk kulmule
DE	Senegalesischer Seehecht
EL	μπακαλιάρος της Σενεγάλης
EN	Senegalese hake; black hake
FR	merlu du Sénégal; merlu noir
IT	merluzzo senegalese; nasello; merluzzo
NL	Senegalese heek
PT	pescada-negra
FI	senegalinkummeliturska
SV	senegalkummel

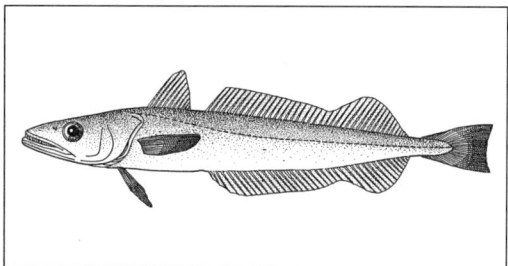

492	**MERLUCCIIDAE**	HKX

SC	*Merluccius* spp.
ES	merluzas
DA	kulmule-slægten
DE	Seehechte
EL	μπακαλιάρος
EN	hake
FR	merlus
IT	naselli; merluzzi
NL	heken
PT	pescadas
FI	kummeliturska-suku
SV	kumlar

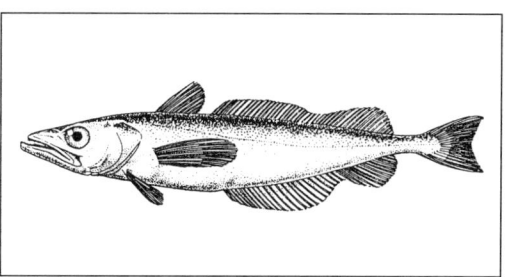

493   MACROURIDAE

SC   *Coelorhynchus coelorhynchus* (Risso, 1810);
     Coelorinchus coelorinchus; *Macrourus orhynchus*;
     Macrourus atlanticus; *Coelorhynchus carminatus*
ES   pez rata; ratón
DA   sortplettet langhale
DE   Schwarzfleck-Grenadierfisch
EL   κηλιδόμαυρος γρεναδιέρος· ποντικουρόψαρο·
     κορδέλα
EN   black-spot grenadier; rat-tail
FR   rat; grenadier
IT   pesce sorcio
NL   gevlekte grenadiervis
PT   lagartixa-do-mar
FI   lestikala-laji
SV   spiritist

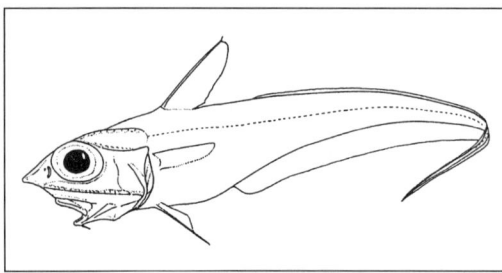

495   MACROURIDAE

SC   *Hymenocephalus italicus* (Giglioli, 1884)
ES   rata italiana
DA   italiensk langhale
DE   Kurzschnauzen-Grenadierfisch
EL   κορδέλα· ποντικουρόψαρο
EN   Italian grenadier
FR   rat; grenadier d'Italie
IT   pesce topino
NL   Italiaanse grenadiervis
PT   lagartixa-prateada
FI   italianlestikala
SV   glashuvudfisk

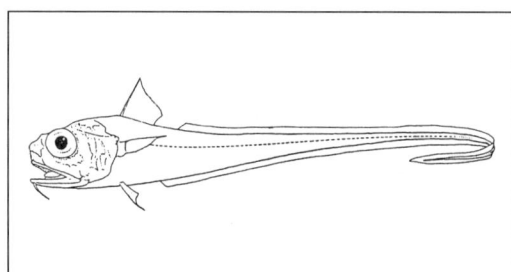

494   MACROURIDAE   RNG

SC   *Coryphaenoides rupestris* (Gunnerus, 1765);
     Macrurus rupestris
ES   granadero
DA   skolæst
DE   Grenadierfisch; Langschwanz
EL   γρεναδιέρος των βράχων· ποντικουρόψαρο·
     κορδέλα
EN   round-nose grenadier; rock grenadier
FR   grenadier de roche
IT   granatiere
NL   grenadiervis
PT   lagartixa-da-rocha
FI   lestikala
SV   skoläst

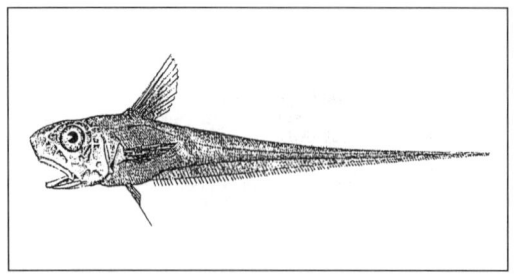

496   MACROURIDAE

SC   *Macrouridae*
ES   granaderos
DA   langhale-familien
DE   Grenadierfische; Tiefsee-Langschwänze;
     Rattenschwänze
EL   γρεναδιέροι· κορδέλες· ποντικουρόψαρα
EN   grenadiers; rat-tails
FR   grenadiers et rats
IT   pesci ratti; macruridi
NL   grenadiervissen
PT   lagartixas e granadeiros
FI   lestikalat; lestikalat-heimo
SV   skolästfiskar

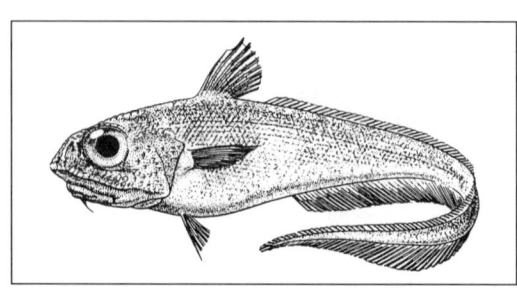

497 MACROURIDAE RHG

SC	*Macrourus berglax* (Lacépède, 1801)
ES	granadero de roca
DA	nordlig skolæst
DE	Nordatlantik-Grenadier
EL	γρεναδιέρος· ποντικουρόψαρο· κορδέλα
EN	rough-head grenadier
FR	grenadier à tête rude
IT	granatiere
NL	noordelijke grenadiervis
PT	lagartixa-cabeça áspera
FI	isolestikala
SV	långstjärt

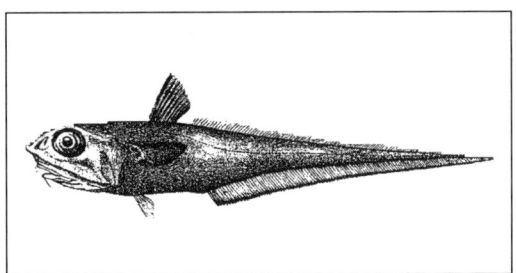

498 MACROURIDAE GRV

SC	*Macrourus* spp.
ES	granaderos
DA	langhale-slægt
DE	Grenadierfische
EL	γρεναδιέροι· ποντικουρόψαρα
EN	grenadiers
FR	grenadiers
IT	granatieri
NL	grenadiervissen
PT	lagartixas
FI	lestikala-suku
SV	långstjärtar

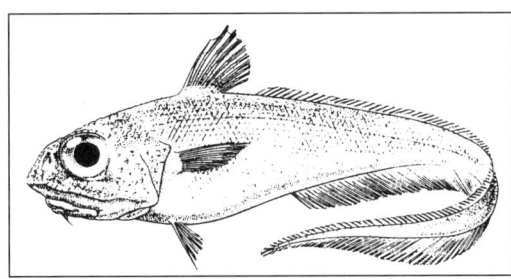

499 GASTEROSTEIDAE

SC	*Gasterosteidae*
ES	espinosos
DA	hundestejle-familien
DE	Stichlinge
EL	αγκαθερά
EN	sticklebacks
FR	épinoches
IT	spinarelli; gasterosteidi
NL	stekelbaarzen
PT	esgana-gatas
FI	piikkikalat; piikkikalat-heimo
SV	spiggfiskar

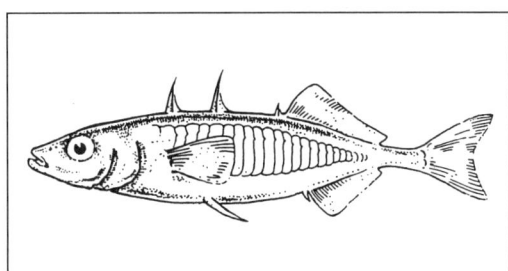

500 GASTEROSTEIDAE

SC	*Gasterosteus aculeatus* (Linnaeus, 1758)
ES	espinoso; venenoso; salpa xurel
DA	trepigget hundestejle; stor hundestejle
DE	Dreistachliger Stichling; Gemeiner Stichling
EL	αγκαθερό
EN	three-spined stickleback; three-spine stickleback
FR	épinoche à trois épines; épinoche aiguillonnée; picot; savetier; arselet
IT	spinarello
NL	driedoornige stekelbaars
PT	esgana-gata; espinelha; peixe-espinho
FI	kolmipiikki
SV	storspigg

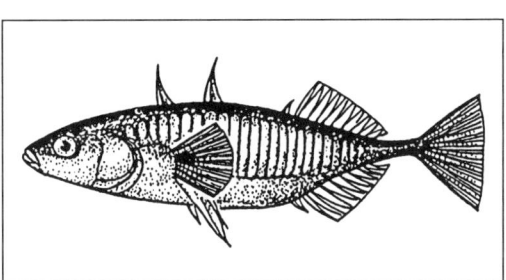

501 GASTEROSTEIDAE

SC	*Pungitius pungitius* (Linnaeus, 1758); *Pygosteus pungitius*
ES	espinosillo
DA	nipigget hundestejle
DE	Neunstachliger Stichling; Zwergstichling
EL	αγκαθερό
EN	nine-spined stickleback
FR	épinochette
IT	spinarello minore
NL	tiendoornige stekelbaars
PT	espinho
FI	kymmenpiikki
SV	småspigg

503 MACRORHAMPHOSIDAE SNI

SC	*Macroramphosidae*
ES	trompeteros
DA	sneppefisk-familien
DE	Schnepfenfische
EL	μπεκατσόψαρα· ψάρια τρομπέτες
EN	snipefishes
FR	bécasses
IT	pesci trombetta
NL	snipvissen
PT	trombeteiros
FI	torvikalat; torvikalat-heimo
SV	snäppfiskar

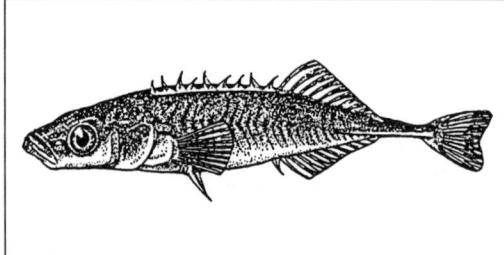

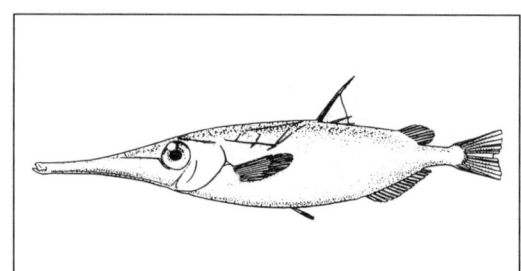

502 GASTEROSTEIDAE

SC	*Spinachia spinachia* (Linnaeus, 1758)
ES	espinoso
DA	tangsnarre
DE	Seestichling; Meerstichling
EL	αγκαθερό της θάλασσας
EN	fifteen-spined stickleback
FR	épinoche; épinoche de mer
IT	spinarello marino
NL	zeestekelbaars
PT	esgana-gata-marinha
FI	vaskikala
SV	tångspigg

504 MACRORHAMPHOSIDAE SNS

SC	*Macroramphosus scolopax* (Linnaeus, 1758)
ES	trompetero
DA	sneppefisk
DE	Schnepfenfisch
EL	μπεκατσόψαρο· τρομπέτα
EN	snipefish; long-spine snipefish; slender snipefish
FR	bécasse de mer
IT	pesce trombetta
NL	snipvis
PT	trombeteiro; apara-lápis
FI	torvikala
SV	snäppfisk

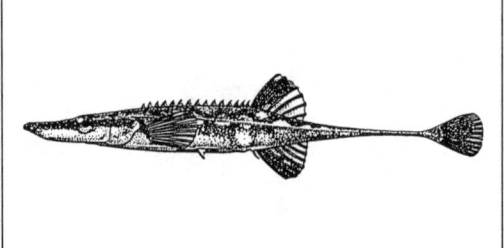

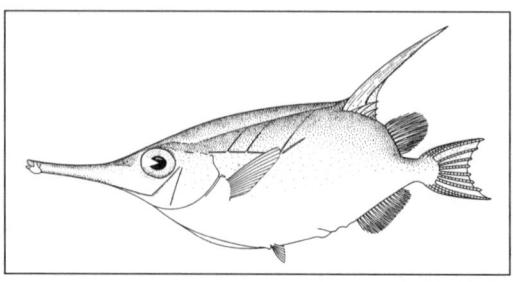

126

505 SYNGNATHIDAE

SC   *Hippocampus* spp.
ES   caballitos de mar
DA   søheste-slægt
DE   Seepferdchen
EL   ιππόκαμποι
EN   seahorses
FR   chevaux de mer; hippocampes; chevaux marins
IT   cavallucci marini
NL   zeepaardjes
PT   cavalos-marinhos
FI   merihevonen-suku
SV   sjöhästar

507 SYNGNATHIDAE

SC   *Syngnathus rostellatus* (Nilsson, 1855)
ES   aguja de mar armada
DA   lille tangnål
DE   Kleine Seenadel
EL   σακοράφα· κατουρλίδα
EN   Nilsson's pipefish
FR   petite aiguille de mer; petit syngnathe; syngnathe de Dumeril
IT   pesce ago minore
NL   kleine zeenaald
PT   marinha-cabeça chata
FI   pikkuneula
SV   mindre kantnål

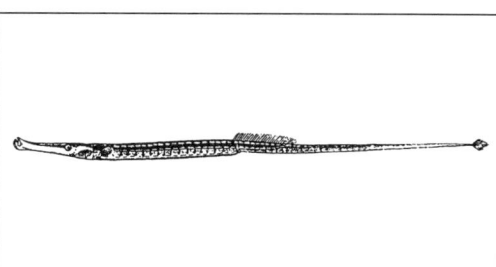

506 SYNGNATHIDAE

SC   *Syngnathus acus* (Linnaeus, 1758)
ES   mula
DA   stor tangnål
DE   Große Seenadel; Mittelmeer-Seenadel
EL   σακοράφα· κατουρλίδα
EN   great pipefish
FR   grande aiguille de mer; syngnathe commun; aiguille de mer commune; syngnathe aiguille; vipère de mer
IT   pesce ago
NL   grote zeenaald
PT   marinha comum
FI   isoneula
SV   större kantnål

508 SYNGNATHIDAE

SC   *Syngnathus typhle* (Linnaeus, 1758); *Siphonostomus typhle*
ES   aguja mula
DA   almindelig tangnål
DE   Grasnadel; Pfeifenfisch; Schmalschnäuzige Seenadel
EL   σακοράφα· κατουρλίδα
EN   deep-snouted pipefish
FR   siphonostome; vipère de mer; aiguille de mer
IT   pesce ago cavallino
NL   trompetterzeenaald
PT   marinha-focinho grosso
FI   särmäneula
SV   tångsnälla

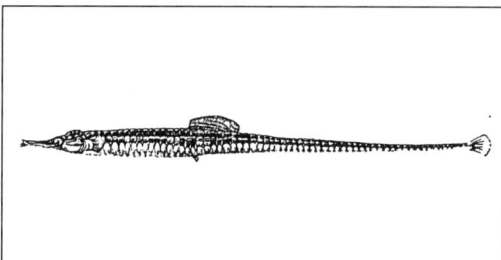

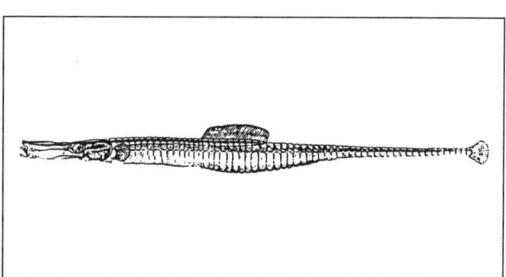

509 LAMPRIDAE

SC   *Lampris guttatus* (Brünnich, 1788); *Lampris regius*; *Lampris luna*
ES   luna real
DA   glansfisk
DE   Gotteslachs
EL   φεγγαρόψαρο
EN   opah
FR   lampris
IT   pesce re
NL   koningsvis
PT   peixe-cravo
FI   kiiltolahna
SV   glansfisk

511 TRACHIPTERIDAE

SC   *Trachipterus trachipterus* (Gmelin, 1789); *Trachipterus taenia*
ES   traquíptero
DA   Middelhavs-vågmær
DE   Spanfisch
EL   χαρτόψαρο
EN   Mediterranean dealfish
FR   poisson ruban; trachyptère; argentin
IT   pesce nastro
NL   bandvis
PT   peixe-tábua
FI   viikatekala-laji
SV   vågmär

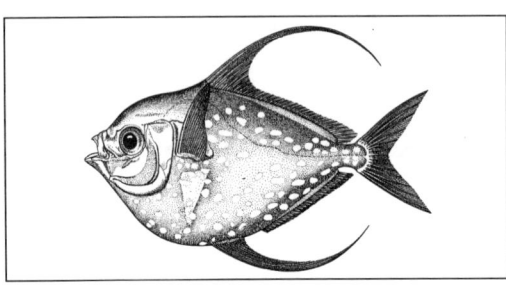

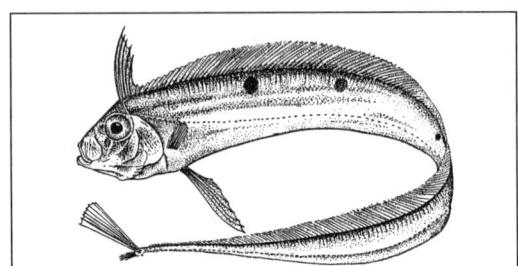

510 REGALECIDAE

SC   *Regalecus glesne* (Ascanius, 1772)
ES   pez remo
DA   sildekonge
DE   Riemenfisch; Bandfisch
EL   βασιλιάς της ρέγγας
EN   oarfish
FR   roi des harengs; régalec
IT   re di aringhe
NL   haringkoning
PT   relangueiro
FI   airokala
SV   sillkung

512 CYPRINODONTIDAE

SC   *Cyprinodontidae*
ES   ciprinodóntidos
DA   æglæggende tandkarper
DE   Zahnkarpfen
EL   κυπρινόδοντοι
EN   killifishes
FR   killi; cyprinodontes; fondules
IT   ciprinodonti
NL   eierleggende tandkarpers
PT   ciprinodontídeos
FI   hammaskarpit; hammaskarpit-heimo
SV   tandkarpar

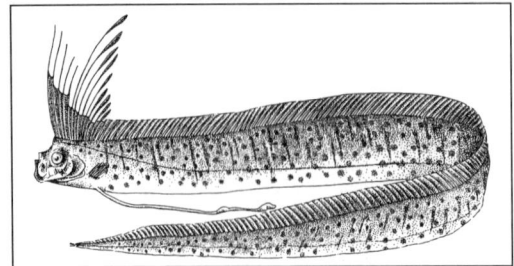

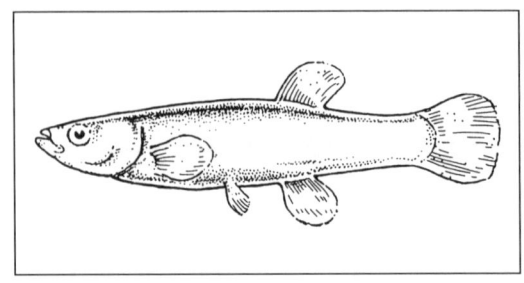

513 CYPRINODONTIDAE

SC *Aphanius fasciatus* (Nardo, 1827)
ES farfet sudeuropeo
DA sydeuropæisk tandkarpe
DE Zebrakärpfling
EL ζαμπαρόλα
EN South European toothcarp
FR cyprinodonte d'Europe du Sud
IT nono
NL Zuid-Europese tandkarper
PT peixinho ligeiro sul-europeu
FI seeprahammaskarppi
SV medelhavskilli

515 CYPRINODONTIDAE

SC *Valencia hispanica* (Valenciennes, 1846);
    Fundulus hispanicus
ES samarugo
DA Valencia-tandkarpe
DE Valencia-Kärpfling
EL ζουρνάς
EN Spanish toothcarp
FR cyprinodonte de Valence; fundule d'Espagne
IT nono andaluso
NL Spaanse tandkarper
PT peixinho ligeiro andaluz
FI espanjanhammaskarppi
SV valencia-tandkarp

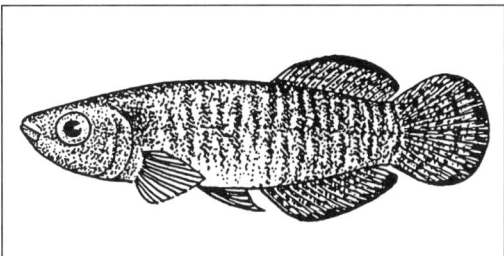

514 CYPRINODONTIDAE

SC *Aphanius iberus* (Valenciennes, 1846); *Cyprinodon iberus*
ES farfet común
DA spansk tandkarpe
DE Spanienkärpfling
EL ζαμπαρόλα της Ισπανίας
EN Spanish toothcarp
FR cyprinodonte d'Espagne
IT nono iberico
NL Spaanse tandkarper
PT peixinho ligeiro espanhol
FI turkoosihammaskarppi
SV spansk killi

516 POECILIIDAE

SC *Gambusia affinis* (Baird and Girard, 1853);
    Gambusia holbrooki
ES gambusia
DA gambusia
DE Koboldkärpfling; Gambuse
EL κουνουπόψαρο
EN mosquitofish
FR gambusie
IT gambusia
NL gambusia
PT gambúsia
FI moskiittokala
SV moskitfisk

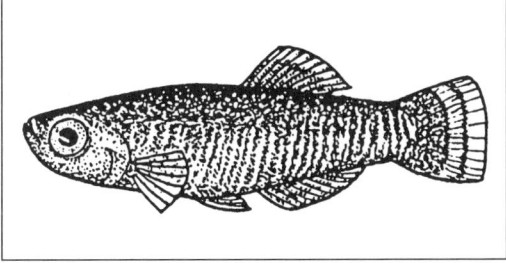

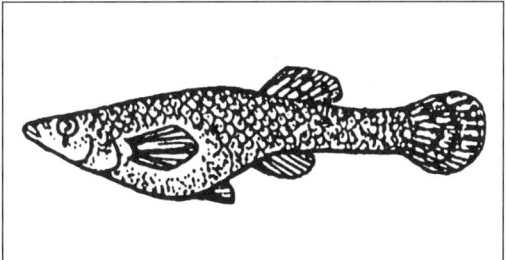

517 BERYCIDAE BRX

SC	*Berycidae*
ES	palometas
DA	berycid-familien
DE	Schleimköpfe
EL	μπερυτσίδες
EN	alfonsinos
FR	bérycidés; béryx
IT	berici
NL	beryciden
PT	imperadores
FI	limapäät; limapäät-heimo
SV	beryxfiskar

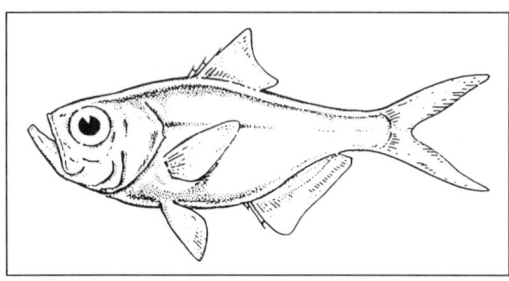

519 BERYCIDAE ALF

SC	*Beryx spp*
ES	alfonsinos
DA	berycid-slægt
DE	Nordischer Schleimkopf; Kaiserbarsch
EL	μπέρυχες
EN	alfonsinos
FR	béryx
IT	berici
NL	beryciden
PT	imperadores; alfonsins
FI	limapää-suku
SV	beryxar

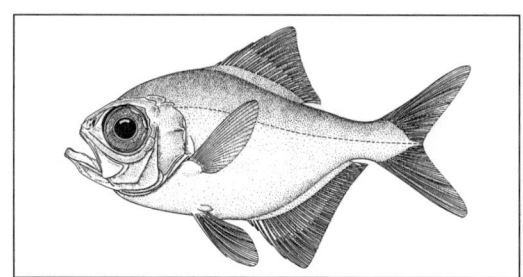

518 BERYCIDAE

SC	*Beryx decadactylus* (Cuvier, 1829)
ES	palometa roja; besugo americano; rey del besugo
DA	nordisk beryx
DE	Nordischer Schleimkopf; Kaiserbarsch
EL	κόκκινος μπέρυξ
EN	red bream
FR	béryx rouge; béryx; béryx commun
IT	berice rosso
NL	beryx
PT	imperador; alfonsim
FI	pohjanlimapää
SV	nordisk beryx

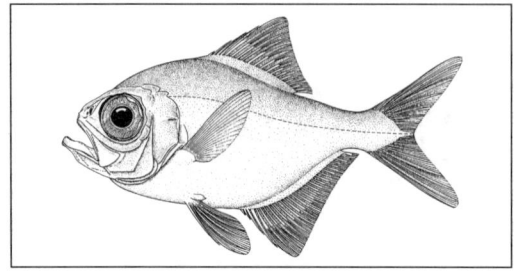

520 TRACHICHTHYIDAE ORY

SC	*Hoplostethus atlanticus* (Collett, 1889)
ES	reloj anaranjado
DA	orange savbug
DE	Atlantischer Sägebauch
EL	καθρεπτόψαρο του Ατλαντικού
EN	orange roughy
FR	hoplostète orange; poisson-montre
IT	pesce specchio atlantico
NL	Atlantische slijmkop
PT	olho-de-vidro laranja
FI	keltaroussi
SV	atlantisk soldatfisk

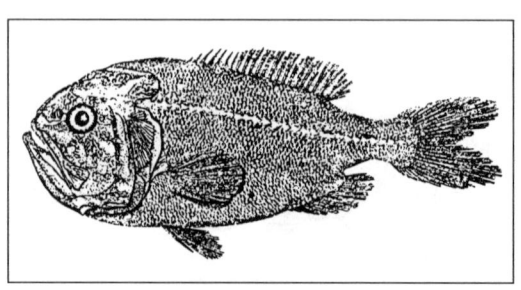

521 TRACHICHTHYIDAE

SC	*Hoplostethus mediterraneus* (Cuvier, 1829)
ES	reloj
DA	Middelhavs-savbug
DE	Mittelmeer-Kaiserbarsch
EL	καθρεπτόψαρο
EN	rosy soldierfish
FR	poisson-montre; hoplostète de Méditerranée
IT	pesce specchio
NL	Middellandse-Zeeslijmkop
PT	olho-de-vidro
FI	välimerenroussi
SV	medelhavssoldatfisk

523 ZEIDAE ZEX

SC	*Zeidae*
ES	peces de San Pedro
DA	sanktpetersfisk-familien
DE	Petersfische
EL	χριστόψαρα
EN	dories; John Dories
FR	zéidés; saint-pierre
IT	pesci San Pietro; zeidi
NL	zonnevissen
PT	galos; peixes-galo
FI	pietarinkalat; pietarinkalat-heimo
SV	sanktpersfiskar

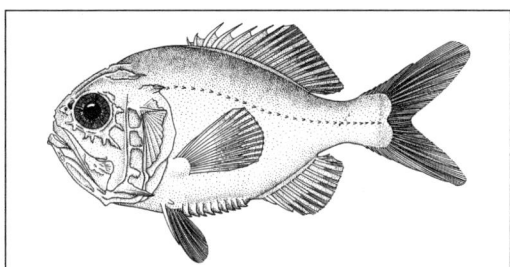

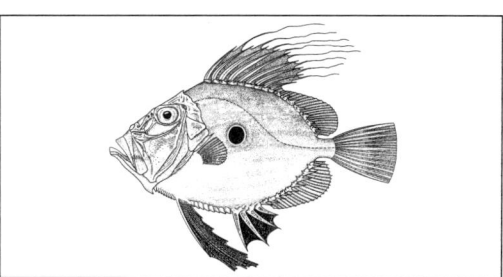

522 HOLOCENTRIDAE

SC	*Sargocentron rubrum* (Forsskål, 1775); *Holocentrus ruber*
ES	soldado rojo
DA	rød egernfisk
DE	Roter Soldatenfisch
EL	κοκκινόψαρο
EN	red soldierfish
FR	poisson-soldat; soldat rouge
IT	sergente rosso
NL	rode soldaatvis
PT	esquilo vermelho
FI	juovaoravakala
SV	röd soldatfisk

524 ZEIDAE JOS

SC	*Zenopsis conchifera* (Lowe, 1852); *Zenopsis ocellata*
ES	pez de San Pedro americano
DA	amerikansk sanktpetersfisk
DE	Amerikanischer Petersfisch
EL	χριστόψαρο της Αμερικής
EN	American John Dory
FR	zéé bouclée d'Amérique; faux saint-pierre
IT	San Pietro d'America
NL	Amerikaanse zonnevis
PT	galo branco; peixe-galo branco
FI	amerikanpietarikala
SV	vit sanktpersfisk

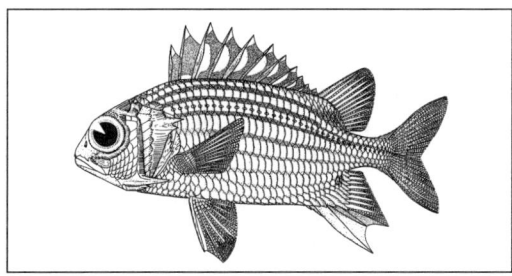

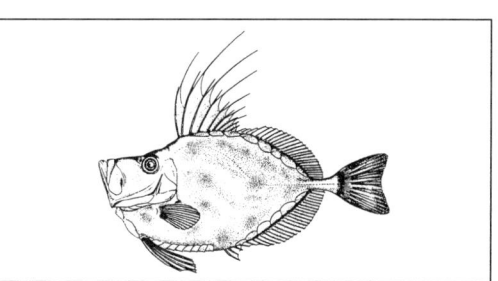

525 ZEIDAE JOD

SC	*Zeus faber* (Linnaeus, 1758); *Zeus pungio*; *Zeus japonicus*
ES	pez de San Pedro; gallo; San Martín; gallo cristo
DA	sanktpetersfisk; atlantisk sanktpetersfisk
DE	Petersfisch; Heringskönig; Peterfisch; Sonnenfisch
EL	χριστόψαρο
EN	Atlantic John Dory; John Dory; dory; Peter fish
FR	saint-pierre; jean doré; poule de mer dorée; saint-christophe; zée
IT	pesce San Pietro
NL	zonnevis
PT	galo negro; peixe-galo
FI	pietarinkala
SV	sanktpersfisk

527 CAPROIDAE

SC	*Capros aper* (Linnaeus, 1758)
ES	ochavo
DA	havgalt
DE	Eberfisch
EL	βασιλάκης· κότα
EN	boarfish
FR	sanglier
IT	pesce tamburo
NL	evervis
PT	pimpim
FI	karjukala
SV	trynfisk

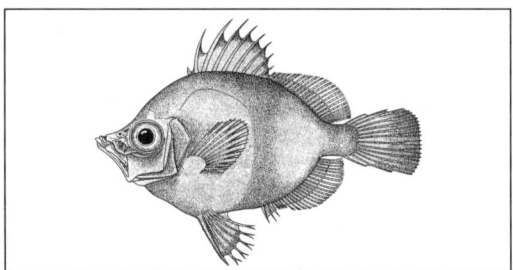

526 CAPROIDAE BOR

SC	*Caproidae*
ES	óreos
DA	havgalt-familien
DE	Eberfische
EL	καπροειδή
EN	boarfishes
FR	sangliers
IT	pesci tamburo; caproidi
NL	evervissen
PT	pimpins; mini-saias; periquitos
FI	karjukalat; karjukalat-heimo
SV	trynfiskar

528 OREOSOMATIDAE ORD

SC	*Oreosomatidae*
ES	ochavos
DA	oreo-familien
DE	Oreos
EL	ψευτόκοτες
EN	oreo dories
FR	oréos
IT	orei
NL	valse evervis
PT	falsos pimpins
FI	oreot; oreot-heimo
SV	guldkroppar

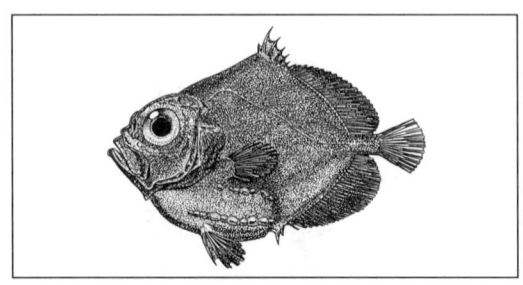

529 SPHYRAENIDAE BAZ

SC	*Sphyraenidae*
ES	picudas; espetones; barracudas
DA	barracuda-familien
DE	Pfeilhechte; Barrakudas
EL	λούτσοι· μπαρακούντα
EN	barracudas; sea pikes
FR	bécunes; brisures; brochets de mer; barracudas
IT	barracuda; lucci marini; sfirenidi
NL	barracuda's
PT	bicudas; barracudas
FI	barrakudat; barrakudat-heimo
SV	barrakudafiskar

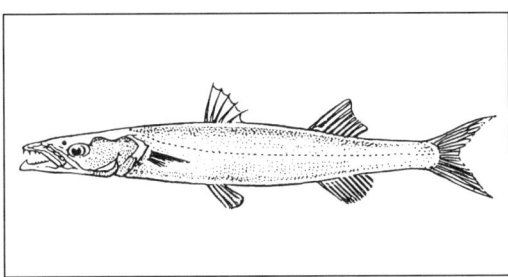

530 SPHYRAENIDAE BAG

SC	*Sphyraena afra* (Peters, 1844)
ES	espetón de Guinea
DA	Guinea-barracuda
DE	Guinea-Barrakuda
EL	μπαρακούντα της Γουινέας
EN	Guinean barracuda
FR	bécune de Guinée; barracuda de Guinée
IT	barracuda di Guinea
NL	Guinese barracuda
PT	bicuda da Guiné; barracuda da Guiné
FI	guineanbarrakuda
SV	afrikansk barrakuda

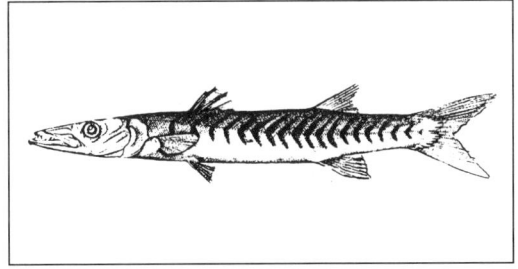

531 SPHYRAENIDAE

SC	*Sphyraena argentea* (Girard, 1854)
ES	barracuda; picuda; sula de playa
DA	Stillehavs-barracuda
DE	Kalifornischer Barrakuda
EL	μπαρακούντα της Καλιφόρνιας
EN	Pacific barracuda; California barracuda
FR	barracuda du Pacifique
IT	barracuda della California
NL	Californische barracuda
PT	bicuda da Califórnia; barracuda da Califórnia
FI	hopeabarrakuda
SV	stillahavsbarrakuda

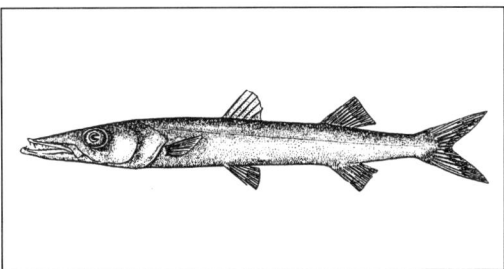

532 SPHYRAENIDAE GBA

SC	*Sphyraena barracuda* (Walbaum, 1792)
ES	picuda barracuda
DA	stor barracuda
DE	Atlantischer Barrakuda; Atlantischer Pfeilhecht
EL	μπαρακούντα
EN	barracuda; great barracuda
FR	barracuda; bécune; barracuda atlantique
IT	barracuda maggiore
NL	barracuda
PT	bicuda gigante; barracuda gigante
FI	barrakuda
SV	jättebarrakuda

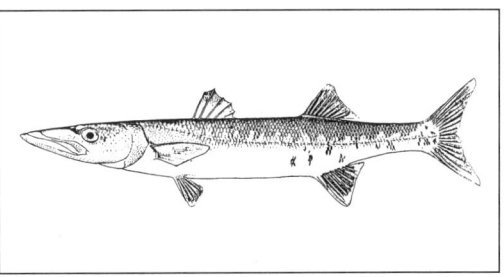

533 SPHYRAENIDAE

SC *Sphyraena chrysotaenia* (Klunzinger, 1884);
    Sphyraena obtusata
ES barracuda oriental
DA gulfinnet barracuda
DE Gelbflossen-Pfeilhecht
EL μπαρακούντα της Ανατολής
EN yellow-finned barracuda; blunt-jaw barracuda
FR bécune orientale
IT barracuda orientale
NL oostelijke barracuda
PT bicuda amarela; barracuda amarela
FI keltaeväbarrakuda
SV gulfenad barrakuda

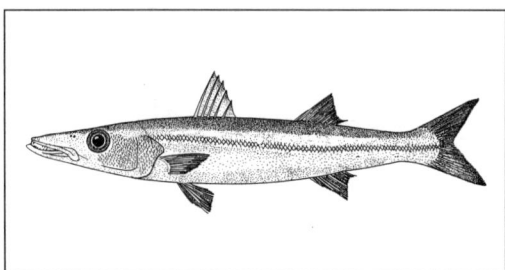

535 SPHYRAENIDAE

SC *Sphyraena sphyraena* (Linnaeus, 1758)
ES espetón; barracuda; picudo; espet; peto
DA europæisk barracuda
DE Mittelmeer-Barrakuda; Mittelmeer-Pfeilhecht
EL λούτσος
EN barracuda
FR brochet de mer; spet; bécune; barracuda européen
IT luccio marino; luccio di mare; barracuda
NL Middellandse-Zeebarracuda
PT bicuda; barracuda
FI pikkubarrakuda
SV barrakuda

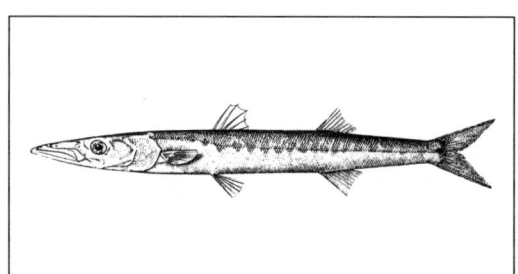

534 SPHYRAENIDAE

SC *Sphyraena jello* (Cuvier, 1829)
ES barracuda india
DA indisk barracuda
DE Indischer Barrakuda; Indischer Pfeilhecht
EL μπαρακούντα της Ινδίας
EN Indian barracuda
FR bécune indienne
IT barracuda indiano
NL Indische barracuda
PT bicuda de barras; barracuda de barras
FI keihäsbarrakuda
SV indisk barrakuda

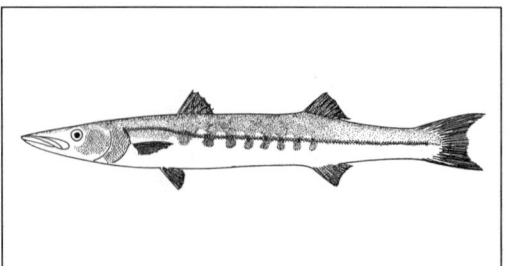

536 SPHYRAENIDAE   BAR

SC *Sphyraena* spp.
ES picudas; espetones
DA barracuda-slægten
DE Pfeilhechte; Barrakudas
EL μπαρακούντες
EN barracudas
FR bécunes; barracudas; brochets de mer
IT barracuda
NL barracuda's
PT bicudas; barracudas
FI barrakuda-suku
SV barrakudor

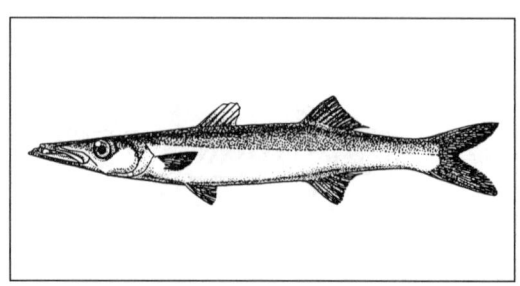

537 MUGILIDAE
MUL

SC	*Mugilidae*
ES	lisas; mugílidos; corcones
DA	multe-familien
DE	Meeräschen
EL	κέφαλοι
EN	mullets; grey mullets
FR	muges; mulets; mugilidés
IT	muggini; cefali; mugilidi
NL	harders
PT	tainhas
FI	keltit; keltit-heimo
SV	multefiskar

539 MUGILIDAE
MUA

SC	*Joturus pichardi* (Poey, 1860)
ES	bobo
DA	bobo-multe
DE	Bobo-Meeräsche
EL	κέφαλος μπόμπο
EN	bobo mullet
FR	mulet bobo
IT	cefalo boccasotto
NL	boboharder
PT	tainha-bobo
FI	keltti-laji
SV	dummulte; kubamulte

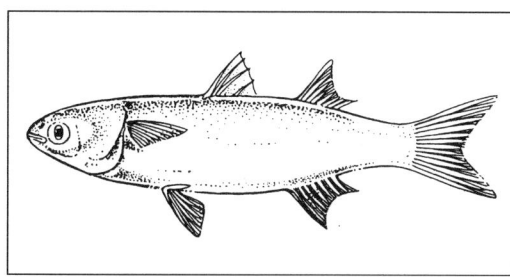

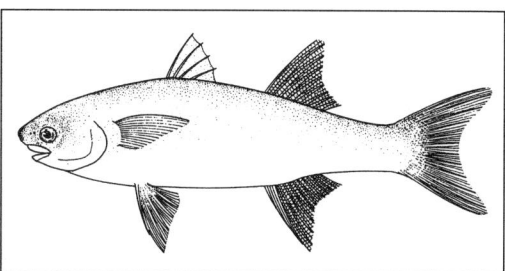

538 MUGILIDAE

SC	*Chelon labrosus* (Risso, 1826); *Mugil provensalis*; *Mugil chelo*; *Crenimugil labrosus*; *Mugil labrosus labrosus*
ES	lisa; corcón; lisa negra; mugle
DA	tyklæbet multe
DE	Dicklippige Meeräsche; Grauäsche
EL	βελανίτσα
EN	thick-lipped grey mullet; thick-lip grey mullet
FR	mulet lippu; muge à grosse lèvre; muge à grosses lèvres
IT	cefalo; bosega; cefalo bosega; cerina
NL	diklipharder
PT	tainha-liça; tainha-negrão
FI	paksuhuulikeltti
SV	tjockläppad multe

540 MUGILIDAE
MGA

SC	*Liza aurata* (Risso, 1810); *Mugil auratus*
ES	galupe; lisa; dabeta; lisa dorada; laban
DA	guldmulte
DE	Meeräsche; Goldäsche; Goldmeeräsche
EL	μυξινάρι
EN	golden grey mullet; long-finned grey mullet; golden mullet; glory mullet
FR	mulet doré; mulet daurin; muge doré; muge daurin
IT	cefalo dorato; muggine dorato; lotregano; cefalo
NL	goudharder
PT	tainha-garrento; tainha-amarela
FI	kultakeltti
SV	guldmulte

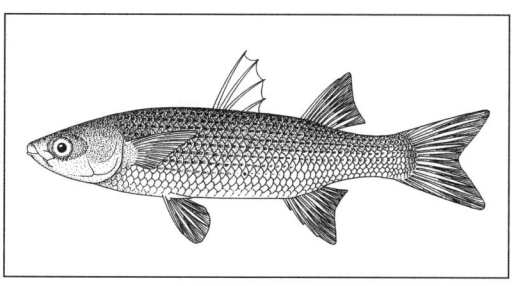

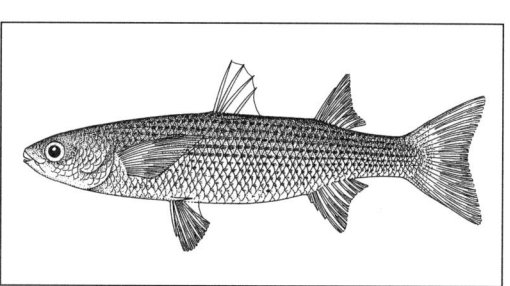

541 MUGILIDAE SOY

SC	*Liza haematochila* (Temminck and Schlegel, 1845)
ES	lisa japonesa
DA	japansk multe
DE	So-iny-Meeräsche
EL	κεφαλόπουλο
EN	so-iny mullet
FR	mulet
IT	cefalo so-iny
NL	so-iny harder
PT	tainha-lábio vermelho
FI	keltti-laji
SV	japansk multe

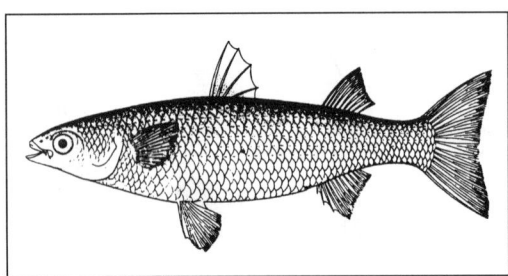

542 MUGILIDAE MGC

SC	*Liza ramada* (Risso, 1826); *Mugil ramada*; *Mugil capito*
ES	morragute; capitón; daplata
DA	tyndlæbet multe
DE	Dünnlippige Meeräsche; Gemeine Meeräsche
EL	μαυράκι
EN	thin-lip grey mullet; thin-lipped grey mullet
FR	mulet ramada; mulet capiton; muge ramada; mulet porc; muge capiton
IT	cefalo calamita; cefalo; botolo; calamita; caustel
NL	dunlipharder
PT	tainha-fataça
FI	ohuthuulikeltti
SV	tunnläppad multe

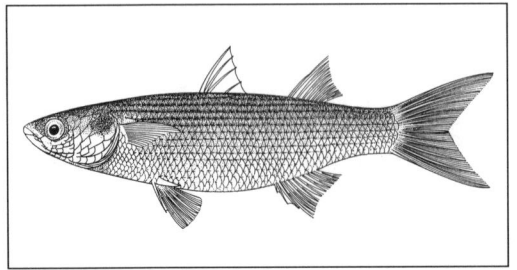

543 MUGILIDAE

SC	*Liza saliens* (Risso, 1810); *Mugil saliens*
ES	galúa; galúa blanca; galúa negra
DA	springmulte
DE	Springmeeräsche; Kleine Meeräsche
EL	γάστρος
EN	leaping grey mullet
FR	mulet sauteur; muge sauteur
IT	cefalo verzelata; musino; cefalo; verzelata
NL	springharder
PT	tainha-de-salto
FI	keltti-laji
SV	hoppmulte

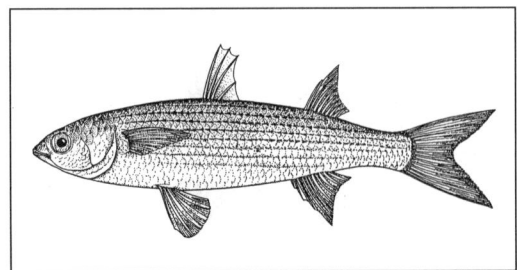

544 MUGILIDAE MUO

SC	*Mugil capurrii* (Perugia, 1892)
ES	galúa africana
DA	afrikansk multe
DE	Afrikanische Meeräsche
EL	κέφαλος της Αφρικής
EN	narrow-head mullet; leaping African mullet
FR	mulet sauteur d'Afrique
IT	cefalo saltatore
NL	Afrikaanse harder
PT	tainha africana
FI	afrikankeltti
SV	afrikansk multe

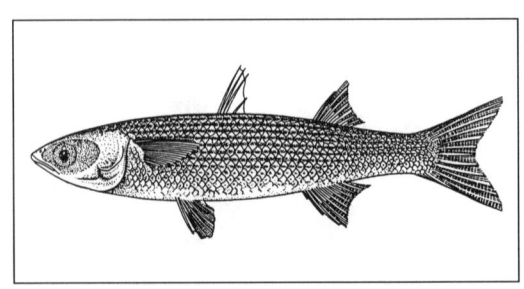

545 MUGILIDAE MUF

SC	*Mugil cephalus* (Linnaeus, 1758)
ES	lisa pardete; mugil común; pardete; albur; mujol
DA	stribet multe
DE	Großkopf-Meeräsche; Gestreifte Meeräsche; Gewöhnliche Meeräsche
EL	μπάφα· κέφαλος· κεφαλόπουλο
EN	common grey mullet; striped mullet; flat-head grey mullet
FR	mulet cabot; mulet commun; mulet céphale; muge cabot; muge commun; muge céphale; mulet à grosse tête; carida; sautereau
IT	cefalo; muggine; volpina
NL	grootkopharder
PT	tainha-olhalvo; ilhalvo; mugem; muginha
FI	juovakeltti
SV	grå multe

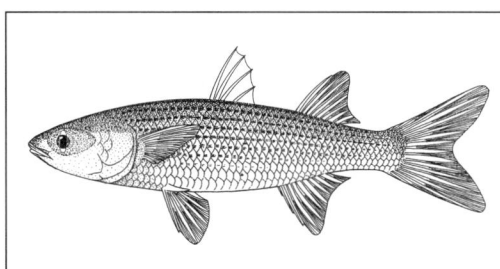

547 MUGILIDAE

SC	*Oedalechilus labeo* (Cuvier, 1829); *Mugil labeo*
ES	caluga; labeo
DA	læbemulte
DE	Graue Meeräsche
EL	γρέντζος
EN	lesser grey mullet; box-lip mullet
FR	mulet labéon; sabornié; muge labéon
IT	cefalo labbrone
NL	grijze harder
PT	tainha-sabão
FI	keltti-laji
SV	läppmulte

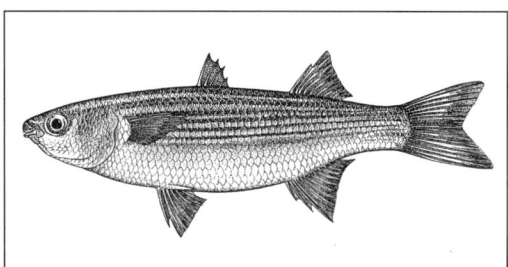

546 MUGILIDAE MUB

SC	*Mugil liza* (Valenciennes, 1836); *Mugil brasiliensis*
ES	lebranche
DA	brasiliansk multe
DE	Brasilianische Meeräsche
EL	κέφαλος της Βραζιλίας
EN	Brazilian mullet
FR	mulet du Brésil
IT	cefalo lebranche
NL	Braziliaanse harder
PT	tainha brasileira
FI	brasiliankeltti
SV	brasiliansk multe

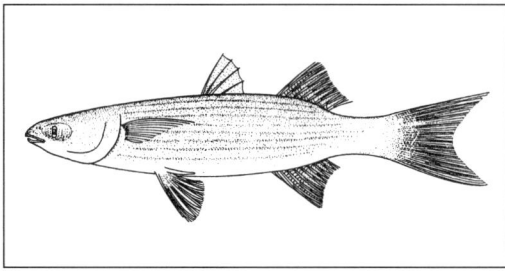

548 ATHERINIDAE SIL

SC	*Atherinidae*
ES	pejerreyes; aterínidos
DA	stribefisk-familien
DE	Ährenfische
EL	αθερίνες
EN	silversides; sandsmelts
FR	athérines; joëls; cabassons; prêtres
IT	aterinidi; latterini; acquadelle
NL	koornaarvissen
PT	peixes-rei
FI	hopeakyljet; hopeakyljet-heimo
SV	silversidefiskar

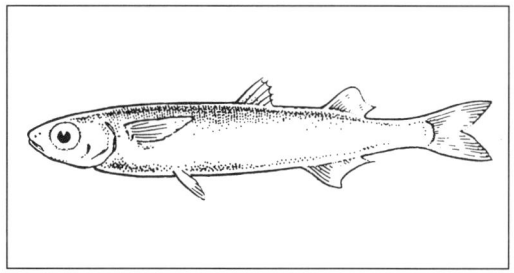

549 ATHERINIDAE

SC	*Atherina boyeri* (Risso, 1810)
ES	pejerrey
DA	Boyers stribefisk; lille stribefisk
DE	Boyers Ährenfisch; Kleiner Ährenfisch
EL	αθερίνα· σουβλίτης
EN	Boyer's sandsmelt; small sandsmelt; Caspian sand-smelt
FR	petite athérine; prêtre; joël
IT	latterino; acquadella
NL	kleine koornaarvis
PT	peixe-rei do Mediterrâneo
FI	boyerinhopeakylki
SV	liten silversida

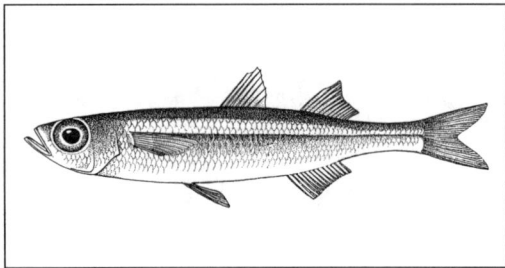

551 ATHERINIDAE

SC	*Atherina presbyter* (Cuvier, 1829); *Hepsetia presbyter*
ES	pejerrey; abichón
DA	stribefisk; almindelig stribefisk
DE	Ährenfisch; Streifenfisch
EL	αθερίνα· σουβλομύτης
EN	sandsmelt
FR	athérine; prêtre; poisson d'argent
IT	latterino
NL	koornaarvis
PT	peixe-rei
FI	hopeakylki
SV	prästfisk

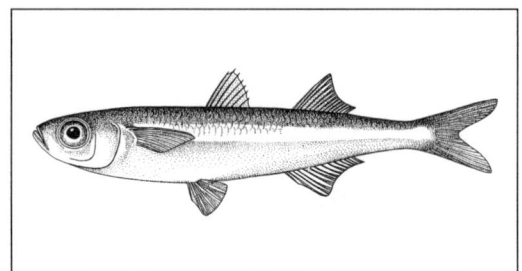

550 ATHERINIDAE

SC	*Atherina hepsetus* (Linnaeus, 1758)
ES	chucleto; xasclet; abichón; pejerrey
DA	stor stribefisk
DE	Großer Ährenfisch
EL	αθερίνα· σουβλίτης
EN	sandsmelt; silverside
FR	siouclet; prêtre; arbusseau; cabasson
IT	latterino; latterino sardaro
NL	grote koornaarvis
PT	peixe-rei do alto; piarda
FI	isohopeakylki
SV	stor silversida

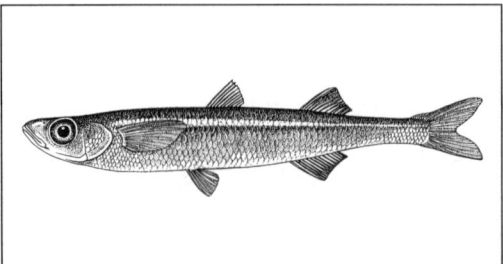

552 ATHERINIDAE   SSA

SC	*Menidia menidia* (Linnaeus, 1766); *Menidia notata*
ES	pejerrey del Atlántico
DA	atlantisk stribefisk
DE	Gezeiten-Ährenfisch
EL	πρασινοαθερίνα
EN	Atlantic silverside
FR	prêtre capucette; capucette
IT	latterino menidia
NL	Atlantische koornaarvis
PT	peixe-rei verde
FI	hopeakylki-laji
SV	menidia

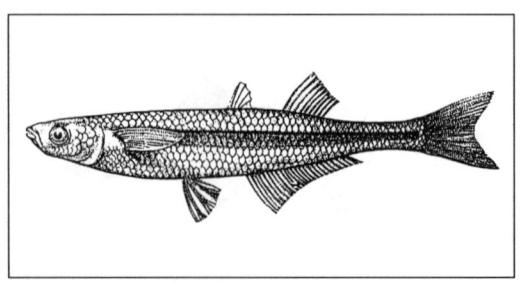

553 POLYNEMIDAE THF

SC	*Polynemidae*
ES	barbudos
DA	trådfinnefisk-familien
DE	Fadenfische
EL	πολυνημίδες
EN	threadfins; tasselfishes
FR	capitaines; barburs
IT	capitani; polinemidi
NL	kapiteinvissen
PT	barbudos; barbinhos; peixes-barba
FI	rihmaevät
SV	trådfensfiskar

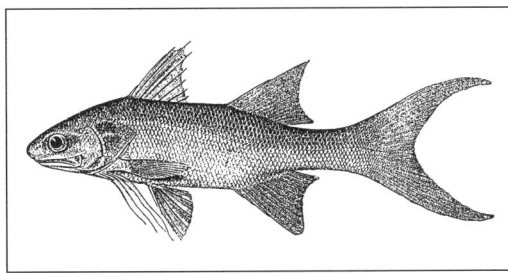

555 POLYNEMIDAE PET

SC	*Pentanemus quinquarius* (Linnaeus, 1758)
ES	barbudo real
DA	konge-trådfinnefisk
DE	Königsfadenfisch
EL	βασιλικό μυστακόψαρο
EN	royal threadfin
FR	capitaine royal
IT	capitano reale
NL	koningskapiteinvis
PT	barbudo real
FI	rihmaevä-laji
SV	kungskäggsfisk

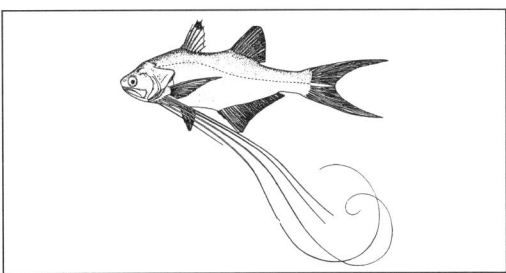

554 POLYNEMIDAE FOT

SC	*Eleutheronema tetradactylum* (Shaw, 1804)
ES	barbudo del Indo-Pacífico
DA	indo-pacifisk trådfinnefisk
DE	Riesenfadenfisch
EL	μυστακόψαρο του Ινδικού
EN	four-finger threadfin
FR	barbur de l'Indo-Pacifique; capitaine de l'Indo-Pacifique
IT	capitano orientale
NL	reuzenkapiteinvis
PT	barbudo do Indo-Pacífico
FI	jättirihmaevä
SV	jättetrådfisk

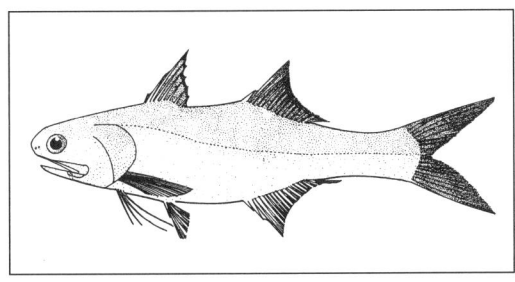

556 POLYNEMIDAE

SC	*Polynemus quadrifilis* (Cuvier and Valenciennes, 1829)
ES	barbudo rayado; barbudo gigante africano
DA	firstrålet trådfinnefisk; stor trådfinnefisk
DE	Fingerfisch; Kapitänsfisch
EL	μυστακόψαρο γίγας
EN	threadfin; five-rayed threadfin; giant African threadfin
FR	capitaine à quatre rayons; gros capitaine
IT	grancapitano
NL	grote kapiteinvis
PT	barbudo gigante
FI	rihmaevä-laji
SV	kaptensfiskar

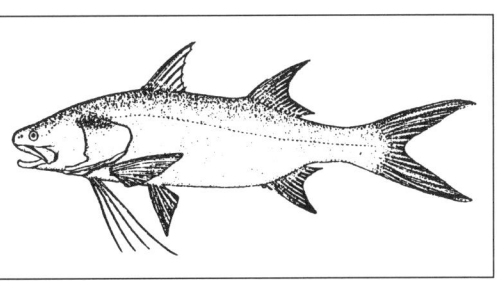

557 CHANNIDAE FIS

SC	*Channa micropeltes*
ES	cabeza de serpiente rojo
DA	indonesisk slangehovedfisk
DE	Indonesischer Schlangenkopffisch
EL	οφιοκέφαλος της Ινδονησίας
EN	Indonesian snakehead
FR	tête de serpent d'Indonésie
IT	pesce testa di serpente
NL	Indonesische slangenkopvis
PT	cabeça-de-cobra da Indonésia
FI	juovakäärmeenpääkala
SV	röd ormhuvudfisk

DRAWING NOT AVAILABLE

559 CHANNIDAE FSS

SC	*Channa striatus* (Bloch, 1797)
ES	cabeza de serpiente cabrío
DA	stribet slangehovedfisk
DE	Gestreifter Schlangenkopffisch
EL	γραμμωτός οφιοκέφαλος
EN	striped snakehead
FR	tête de serpent strié
IT	pesce testa di serpente
NL	gestreepte slangenkopvis
PT	cabeça-de-cobra listado
FI	raitakäärmeenpääkala
SV	randig ormhuvudfisk

DRAWING NOT AVAILABLE

558 CHANNIDAE FSN

SC	*Channa* spp.; *Ophicephalus* spp.
ES	cabezas de serpiente
DA	slangehovedfisk-slægt
DE	Schlangenkopffische
EL	οφιοκέφαλοι
EN	snakeheads; murruls
FR	poissons-tête de serpent
IT	pesci testa di serpente
NL	slangenkopvissen
PT	cabeças-de-cobra
FI	käärmeenpääkala
SV	ormhuvudfiskar

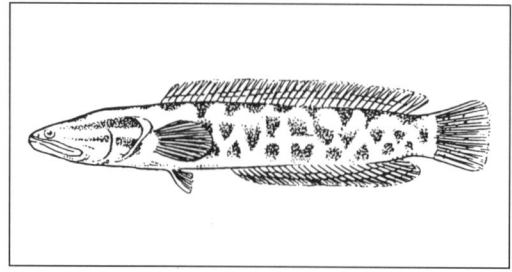

560 CENTROPOMIDAE

SC	*Centropomidae*
ES	centropómidos
DA	centropomider
DE	Snooks; Glasbarsche
EL	κεντροπομίδες
EN	snooks
FR	brochets de mer; perches de verre; perches cristal
IT	centropomidi
NL	zeesnoeken
PT	centropomídeos
FI	lasiahvenet; lasiahvenet-heimo
SV	nilabborrfiskar

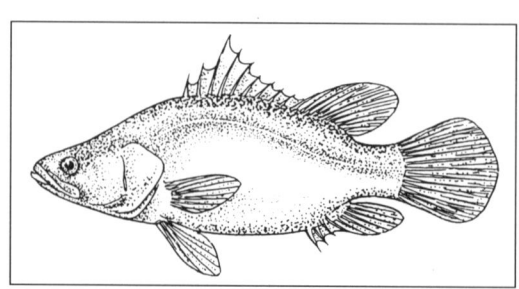

561	**CENTROPOMIDAE**	SNO		**563**	**CENTROPOMIDAE**	GLA

SC	*Centropomus undecimalis* (Bloch, 1792)		SC	*Chanda* spp.
ES	róbalo blanco		ES	peces de cristal
DA	almindelig robalo		DA	glasfisk-slægt
DE	Olivgrüner Snook		DE	Glasbarsche
EL	ψευδολαβράκι της Αμερικής		EL	υαλόψαρα
EN	snook; common snook		EN	glassfishes
FR	loubine		FR	poissons de verre
IT	pseudospigola americana		IT	pesci cristallo
NL	olijfgroene zeesnoek		NL	glasbaarzen
PT	falso robalo branco		PT	peixes-vidro
FI	robalo		FI	lasiahven-suku
SV	snook		SV	glasabborrar

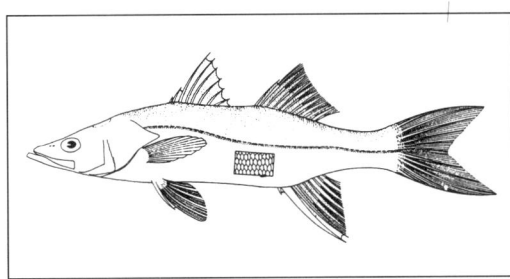

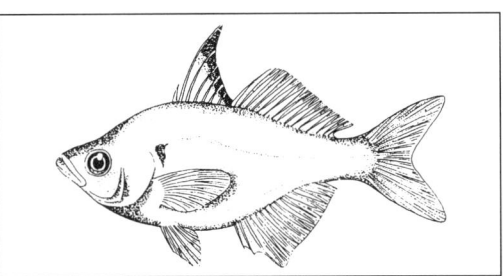

562	**CENTROPOMIDAE**	ROB		**564**	**CENTROPOMIDAE**	GIP

SC	*Centropomus* spp.		SC	*Lates calcarifer* (Bloch, 1790)
ES	róbalos		ES	perca gigante
DA	robalo-slægten		DA	barramundi
DE	Snooks; Robalos; Schaufelkopfbarsche; Glasbarsche		DE	Barramundi
EL	ψευδολαβράκι		EL	γιγαντόπερκα· μπαραμούντι
EN	snooks; robalos		EN	barramundi; giant sea perch
FR	crossies; robalos		FR	perche barramundi
IT	pseudospigole		IT	barramundi
NL	zeesnoeken		NL	barramundibaars
PT	falsos robalos		PT	perca gigante
FI	lasiahven-suku		FI	barramundi
SV	snookar		SV	barramundi

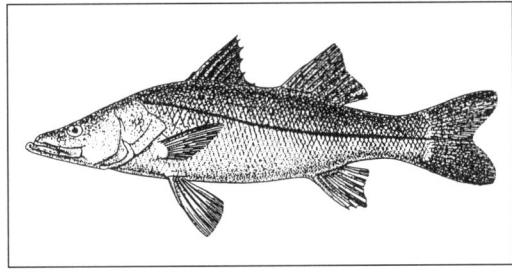

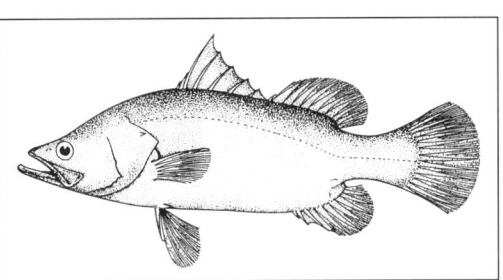

565	**CENTROPOMIDAE**	NIP	**567**	**SERRANIDAE**	BSX

SC	*Lates niloticus* (Linnaeus, 1758)
ES	perca del Nilo
DA	nilaborre
DE	Nilbarsch
EL	πέρκα του Νείλου
EN	Nile perch
FR	perche du Nil
IT	persico del Nilo
NL	Nijlbaars
PT	perca do Nilo
FI	niilinahven
SV	nilabborre

SC	*Serranidae*
ES	serranos
DA	koralbars-familien; havaborre-familien
DE	Sägebarsche
EL	σερανίδες
EN	groupers; seabasses; seaperches
FR	serranidés; bars, serrans et mérous
IT	serranidi
NL	zaagbaarzen
PT	serranídeos; garoupas, meros
FI	meriahvenet; meriahvenet-heimo
SV	havsabborrfiskar

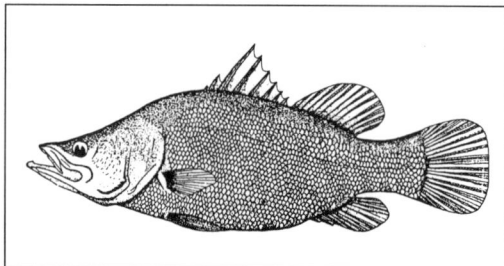 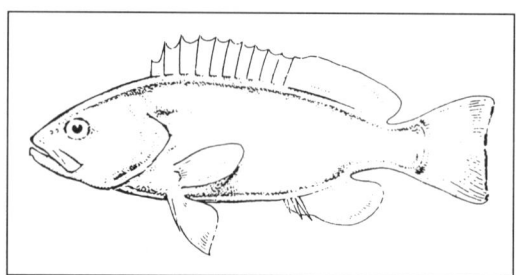

566	**CENTROPOMIDAE**	PEX	**568**	**SERRANIDAE**	BSZ

SC	*Lates* spp.; *Luciolates* spp.
ES	percas
DA	centropomid-slægt
DE	Nilbarsche
EL	πέρκες του Νείλου
EN	Nile perches
FR	perches africaines
IT	persici tropicali
NL	Nijlbaarzen
PT	percas africanas
FI	niilinahven
SV	nilabborrar

SC	*Acanthistius brasilianus* (Cuvier, 1828)
ES	mero sureño
DA	argentinsk havaborre
DE	Argentinischer Zackenbarsch
EL	χάνος της Αργεντινής
EN	Argentine seabass
FR	serran d'Argentine
IT	cernia
NL	Argentijnse zeebaars
PT	serrano argentino
FI	meriahven-laji
SV	argentinsk havsaborre

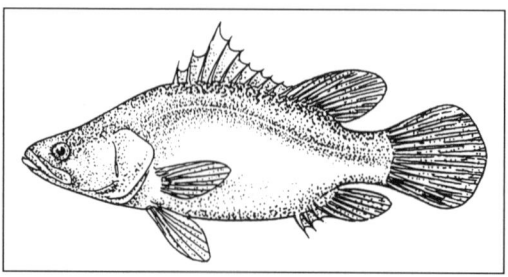

DRAWING NOT AVAILABLE

569 SERRANIDAE

SC	*Anthias anthias* (Linnaeus, 1758)
ES	tres colas
DA	svalehale-bars
DE	Roter Fahnenbarsch; Rötling
EL	κοκκινόχανος
EN	swallow-tail sea perch
FR	barbier rose; anthias
IT	castagnola rossa
NL	rode vlagbaars
PT	canário-do-mar
FI	saha-ahven
SV	svalabborre

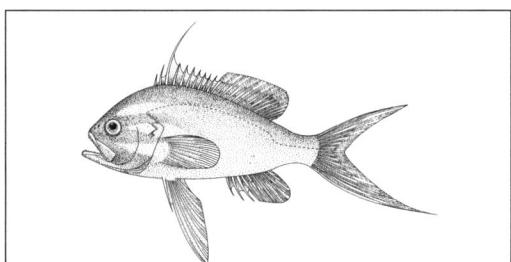

570 SERRANIDAE

SC	*Callanthias ruber* (Ratinesque, 1810)
ES	callantias; papagallo
DA	rød barberfisk
DE	Papageien-Fahnenbarsch
EL	παπαγαλόψαρο
EN	barberfish
FR	barbier rouge
IT	canario rotondo
NL	rode barbiervis
PT	canarinho-do-mar
FI	meriahven-laji
SV	papegojabborre

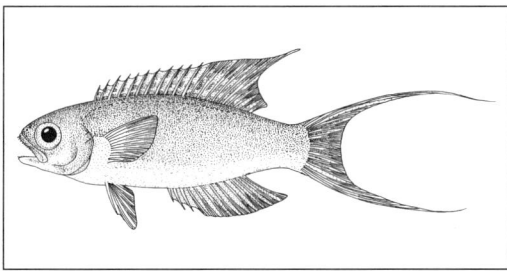

571 SERRANIDAE   BSB

SC	*Centropristis striata* (Linnaeus, 1758)
ES	serrano estriado
DA	sort havaborre
DE	Schwarzer Sägebarsch
EL	μαυρόπερκα
EN	black seabass
FR	fanfre noir
IT	perchia nera; perchia striata
NL	zwarte zeebaars
PT	serrano estriado
FI	kalliomeriahven
SV	svart havsabborre

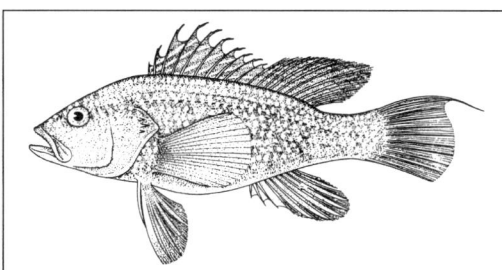

572 SERRANIDAE   PES

SC	*Diplectrum formosum* (Linnaeus, 1766)
ES	serrano arenero
DA	sandhavaborre
DE	Sandbarsch
EL	αμμόπερκα
EN	sand perch
FR	serran de sable
IT	perchia americana
NL	zandbaars
PT	serrano-da-areia
FI	hietameriahven
SV	sandhavsabborre

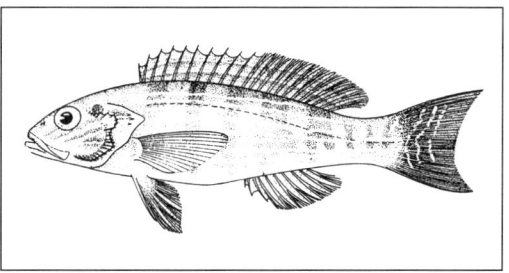

573　SERRANIDAE　　　　　　　GPW

SC　*Epinephelus aeneus* (Geoffroy St.-Hilaire, 1817)
ES　cherna de ley
DA　hvid havaborre
DE　Weißer Zackenbarsch
EL　σφυρίδα
EN　white grouper
FR　mérou blanc; tiof
IT　cernia bianca; cernia
NL　witte zaagbaars
PT　garoupa legítima
FI　valkomeriahven
SV　vit grouper

575　SERRANIDAE　　　　　　　GPS

SC　*Epinephelus analogus* (Gill, 1863)
ES　mero moteado
DA　plettet havaborre
DE　Pazifischer Fleckenbarsch
EL　κηλιδόστικτη σφυρίδα
EN　spotted grouper
FR　mérou cabrilla
IT　cernia
NL　Pacifische gevlekte zaagbaars
PT　garoupa mosqueada
FI　meriahven-laji
SV　fläckig grouper

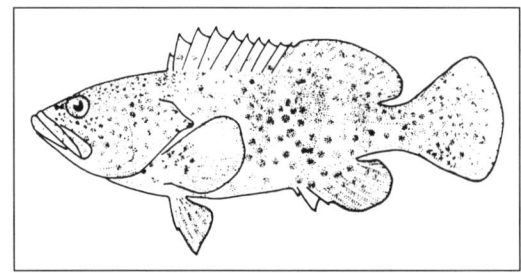

574　SERRANIDAE

SC　*Epinephelus alexandrinus* (Valenciennes, 1828)
ES　mero de Alejandría
DA　gylden havaborre
DE　Goldener Zackenbarsch
EL　στείρα
EN　golden grouper
FR　fausse badèche; mérou d'Alexandrie
IT　cernia dorata; cernia
NL　gouden zaagbaars
PT　mero amarelo
FI　meriahven-laji
SV　guldgrouper

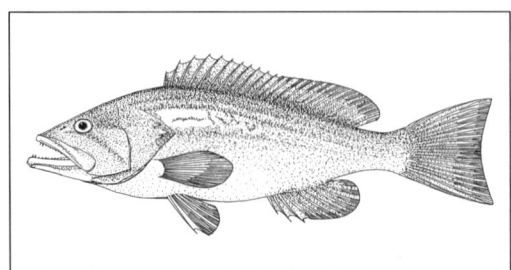

576　SERRANIDAE

SC　*Epinephelus caninus* (Valenciennes, 1843)
ES　mero negro
DA　grå havaborre
DE　Grauer Zackenbarsch
EL　σφυρίδα· βλαχοσφυρίδα
EN　dogtooth grouper
FR　mérou noir
IT　cernia nera; cernia
NL　zwarte zaagbaars
PT　mero gigante
FI　meriahven-laji
SV　grå havsabborre

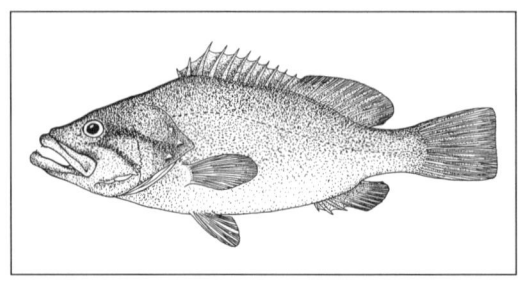

577   SERRANIDAE   GPD

SC	*Epinephelus guaza* (Linnaeus, 1758); *Epinephlus marginatus*; *Epinephelus gigas*; *Serranus gigas*; *Cerna gigas*
ES	mero
DA	kæmpeaborre
DE	Riesen-Zackenbarsch; Brauner Zackenbarsch
EL	ροφός
EN	dusky sea perch; dusky grouper
FR	mérou commun; mérou des provençaux; mérou sombre; mérou de Méditerranée
IT	cernia mediterranea; cernia
NL	grote zaagbaars
PT	mero legítimo; mero
FI	tummameriahven
SV	brun havsabborre

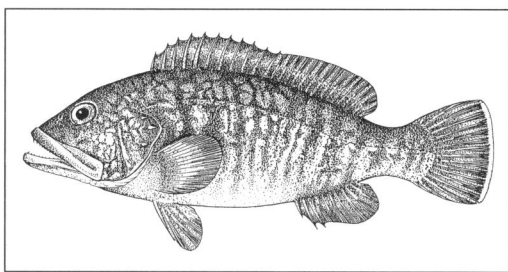

579   SERRANIDAE   MAR

SC	*Epinephelus malabaricus* (Bloch and Schneider, 1801)
ES	mero malabárico
DA	Malabar-havaborre
DE	Malabar-Zackenbarsch
EL	ροφός
EN	malabar grouper
FR	mérou de malabar
IT	cernia
NL	Malabar-zaagbaars
PT	garoupa-malabar
FI	malabarinmeriahven
SV	malabargrouper

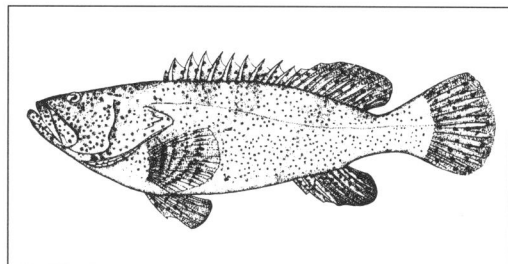

578   SERRANIDAE

SC	*Epinephelus itajara* (Lichtenstein, 1822); *Promicrops itajara*
ES	mero gigante
DA	jødefisk
DE	Judenfisch
EL	γιγαντοροφός
EN	jewfish
FR	tétarde
IT	cernia gigante
NL	reuzenzaagbaars
PT	mero-tigre
FI	raitameriahven
SV	judefisk

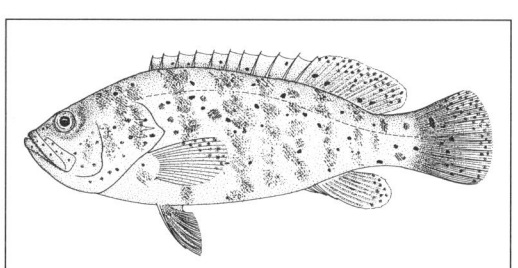

580   SERRANIDAE   GPR

SC	*Epinephelus morio* (Valenciennes, 1828)
ES	mero americano; garoupa
DA	rød havaborre
DE	Roter Zackenbarsch; Roter Grouper
EL	στικτοροφός
EN	red grouper
FR	mérou rouge
IT	cernia
NL	rode zaagbaars
PT	mero americano
FI	punameriahven
SV	röd grouper

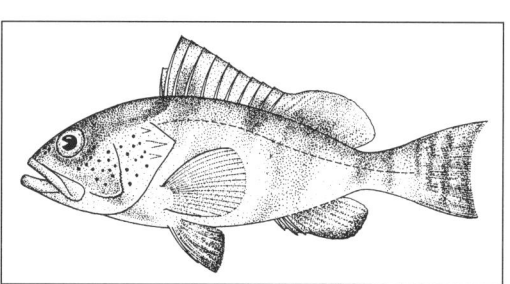

581 SERRANIDAE GPN

SC	*Epinephelus striatus* (Bloch, 1792)
ES	cherna criolla
DA	Nassau-koralbars
DE	Nassau-Barsch
EL	ροφός του Νασσάου
EN	Nassau grouper
FR	mérou rayé
IT	cernia di Nassau
NL	Nassau-zaagbaars
PT	mero crioulo
FI	nassaunmeriahven
SV	nassaugrouper

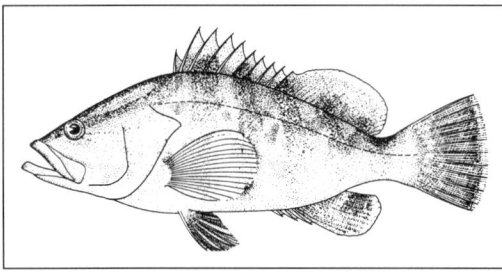

583 SERRANIDAE GPB

SC	*Mycteroperca* spp.
ES	cunas
DA	havaborre-slægt; koralbars-slægt
DE	Zackenbarsche
EL	πίγγες
EN	Brazilian groupers
FR	badèches
IT	cernie
NL	koraalbaarzen
PT	garoupas-badejo
FI	meriahven-suku
SV	groupers

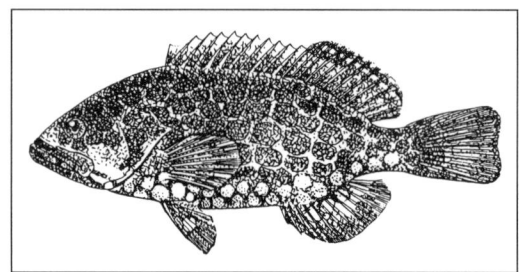

582 SERRANIDAE GPX

SC	*Epinephelus* spp.
ES	meros
DA	havaborre-slægt; koralbars-slægt
DE	Zackenbarsche
EL	σφυρίδες· ροφοί
EN	groupers
FR	mérous
IT	cernie
NL	zaagbaarzen
PT	garoupas e meros
FI	meriahven-suku
SV	groupers

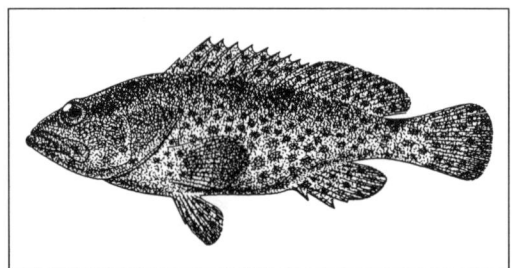

584 SERRANIDAE BAP

SC	*Paralabrax humeralis* (Cuvier and Valenciennes, 1828)
ES	cabrilla loca
DA	peruansk havaborre
DE	Peruanischer Felsbarsch
EL	πέρκα του Περού
EN	Peruvian rock bass
FR	bar du Pérou
IT	perchia del Perù
NL	Peruaanse zeebaars
PT	robalo do Peru
FI	perunmeriahven
SV	peruansk grouper

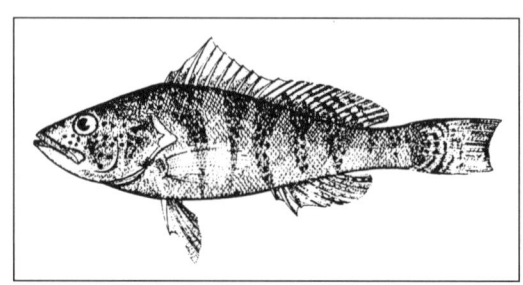

585	**SERRANIDAE**	WRF

SC	*Polyprion americanus* (Bloch and Schneider, 1801); *Polyprion cernium*
ES	cherna
DA	vragfisk; atlantisk vragfisk
DE	Wrackbarsch; Atlantischer Wrackbarsch
EL	βλάχος
EN	stone bass; wreckfish; wreck bass
FR	cernier commun; cernier brun; mérou des Basques; cernier atlantique; fanfre
IT	cernia di fondale; cernia; dotto
NL	Atlantische wrakbaars
PT	cherne
FI	hylkyahven
SV	vrakfisk

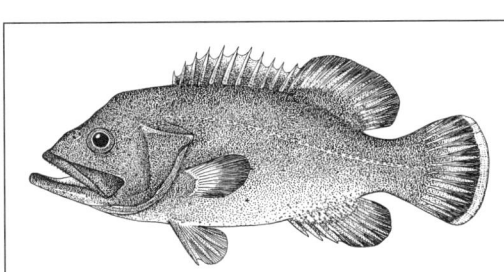

587	**SERRANIDAE**	CBR

SC	*Serranus cabrilla* (Linnaeus, 1758)
ES	cabrilla
DA	gedebars
DE	Ziegenbarsch; Sägebarsch
EL	χάνος
EN	comber
FR	serran chevrette
IT	perchia; sciarrano
NL	geitenbaars
PT	serrano-alecrim
FI	pastelliahven
SV	större medelhavsabborre

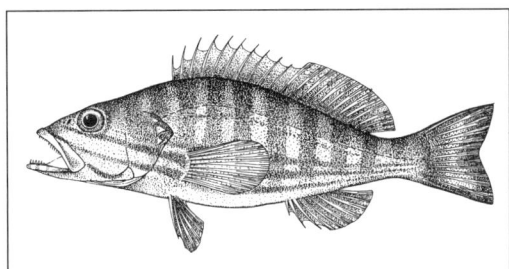

586	**SERRANIDAE**	WHA

SC	*Polyprion oxygeneios* (Bloch and Schneider, 1801); *Hectoria oxygeneios*
ES	cherna hapuku; cherna de Juan Fernández
DA	newzealandsk vragfisk
DE	Hapuku-Wrackbarsch
EL	βλάχος της N. Ζηλανδίας
EN	hapuku wreckfish
FR	cernier de Nouvelle-Zélande
IT	dotto neozelandese
NL	hapuku-wrakbaars
PT	cherne da Nova Zelândia
FI	hylkyahven-laji
SV	nyzeeländsk vrakfisk

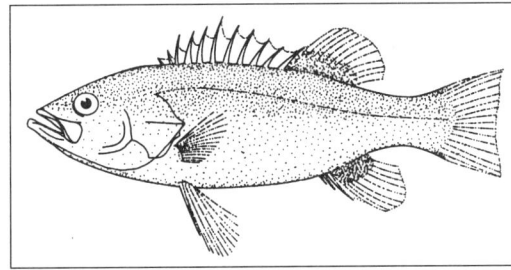

588	**SERRANIDAE**	

SC	*Serranus hepatus* (Linnaeus, 1766)
ES	merillo
DA	dværgbars
DE	Zwergbarsch
EL	χανάκι· πέρκα· καψομούλα
EN	brown comber
FR	serran hépate; tambour
IT	sacchetto; sciarrano
NL	dwergzaagbaars
PT	serrano-ferreiro
FI	ruskomeriahven
SV	brun medelhavsabborre

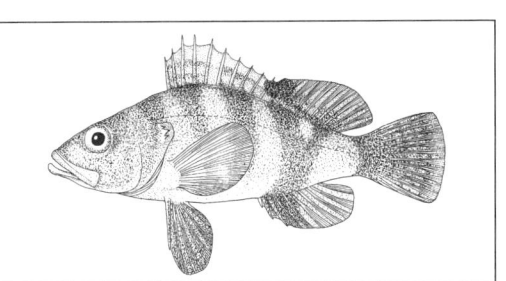

589 SERRANIDAE

SC	*Serranus scriba* (Linnaeus, 1758)
ES	serrano
DA	skriftbars
DE	Schriftbarsch
EL	πέρκα· πέρδικα
EN	painted comber
FR	serran écriture
IT	sciarrano
NL	schriftbaars
PT	serrano riscado
FI	harlekiiniahven
SV	skriftabborre

591 SERRANIDAE

SC	*Stereolepis gigas* (Ayres, 1859)
ES	mero gigante del Pacífico
DA	californisk havaborre
DE	Kalifornischer Judenfisch
EL	γιγαντορροφός του Ειρηνικού
EN	giant sea bass
FR	mérou géant du Pacifique; poisson juif
IT	cernia gigante del Pacifico
NL	Californische grote zaagbaars
PT	mero gigante do Pacífico
FI	tummameriahven
SV	kalifornisk havsabborre

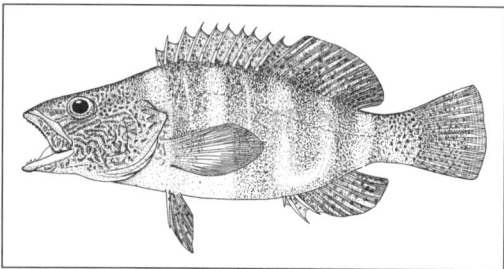

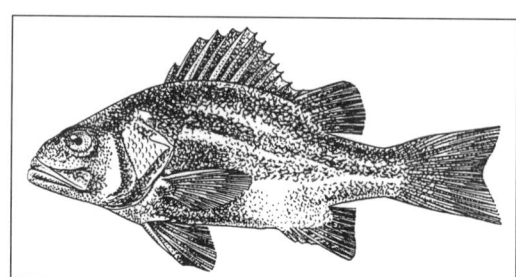

590 SERRANIDAE   BAS

SC	*Serranus* spp.
ES	serranos
DA	havaborre-slægten
DE	Sägebarsche
EL	πέρκες
EN	combers
FR	serrans
IT	sciarrani
NL	zaagbaarzen
PT	serranos; serrões
FI	meriahven-suku
SV	havsabborrar

592 THERAPONIDAE   THE

SC	*Theraponidae*
ES	terapónidos
DA	teraponid-familien
DE	Tigerbarsche
EL	θεραπονίδες
EN	therapon perches
FR	poissons-tigres
IT	teraponidi
NL	doornvissen
PT	teraponídeos
FI	tiikeriahvenet; tiikeriahvenet-heimo
SV	teraponfiskar

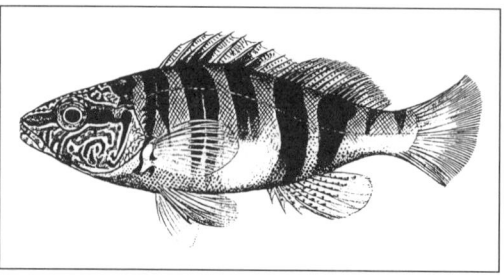

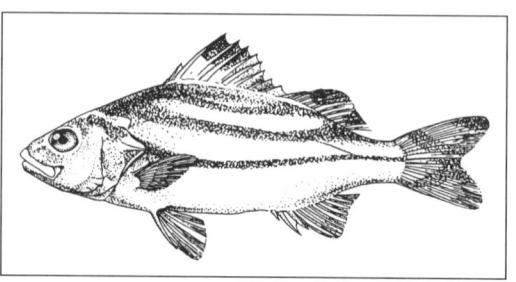

593	**MORONIDAE**	**BSS**

SC   *Dicentrarchus labrax* (Linnaeus, 1758); *Morone labrax*; *Roccus labrax*; *Labrax lupus*
ES   lubina; robaliza; róbalo; magallón
DA   almindelig bars; havbars
DE   Wolfsbarsch; Seebarsch; Meerbarsch
EL   λαβράκι
EN   European seabass; bass; sea perch; white sea perch
FR   bar; loup; loubine; perche de mer; barreau; bar commun
IT   spigola; branzino
NL   zeebaars
PT   robalo legítimo; robalo
FI   meribassi
SV   havsabborre

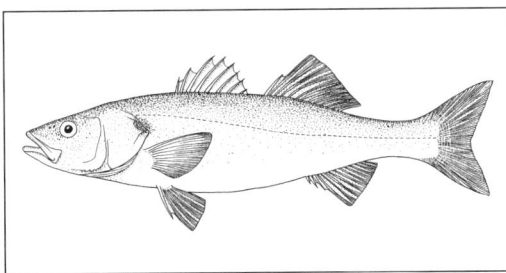

594	**MORONIDAE**	**SPU**

SC   *Dicentrarchus punctatus* (Bloch, 1792); *Labrax punctatus*; *Morone punctata*
ES   baila
DA   plettet bars
DE   Gefleckter Streifenbarsch
EL   λαβράκι
EN   sea spotted bass
FR   bar tacheté
IT   spigola macchiata; spigola puntata
NL   gevlekte zeebaars
PT   robalo-baila
FI   täpläbassi
SV   fläckig havsabborre

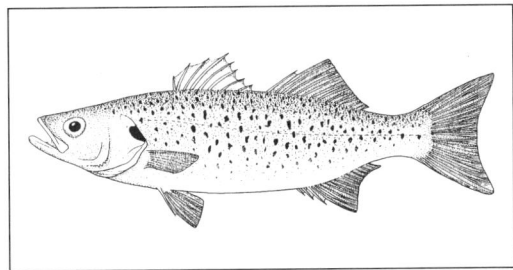

595	**MORONIDAE**	**BSE**

SC   *Dicentrarchus spp*
ES   lubinas
DA   bars-slægt
DE   Seebarsche
EL   λαβράκι
EN   Seabasses
FR   bars
IT   spigole
NL   zeebaarzen
PT   robalos
FI   meriahvenet
SV   havsabborrar

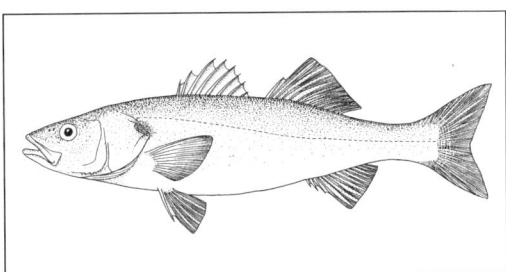

596	**MORONIDAE**	**PEW**

SC   *Morone americana* (Gmelin, 1789); *Roccus americanus*
ES   lubina blanca americana
DA   amerikansk bars
DE   Amerikanischer Streifenbarsch
EL   λαβράκι της Αμερικής
EN   white perch
FR   bar américain; bar blanc d'Amérique; perche blanche
IT   persicospigola americana
NL   Amerikaanse zeebaars
PT   robalo do norte
FI   amerikanbassi
SV   vitabborre

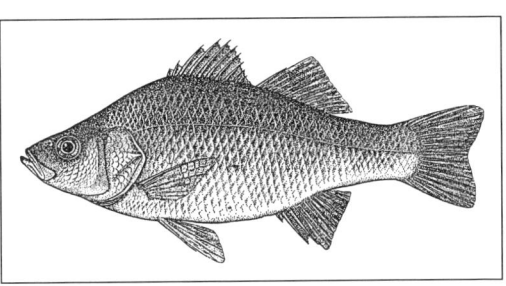

149

597 MORONIDAE

SC *Morone chrysops* (Rafinesque, 1820); *Roccus chrysops*
ES lubina blanca
DA hvid bars
DE Weißer Sägebarsch
EL ασπρολάβρακο
EN white bass
FR bar blanc
IT persicospigola bianca
NL witte zeebaars
PT robalo branco
FI valkobassi
SV vit havsabborre

599 PERCICHTHYIDAE   BAJ

SC *Lateolabrax japonicus* (Cuvier, 1828)
ES serrano japonés
DA japansk havaborre
DE Japanischer Barsch
EL γιαπωνέζικο λαβράκι
EN Japanese seabass
FR bar du Japon
IT spigola giapponese
NL Japanse zeebaars
PT robalo japonês
FI japaninbassi
SV japansk havsabborre

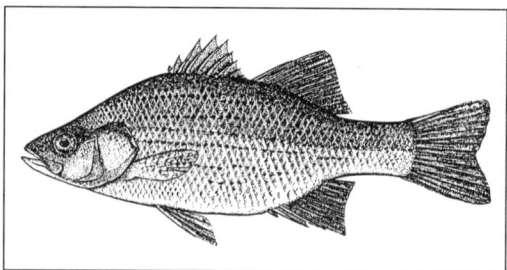

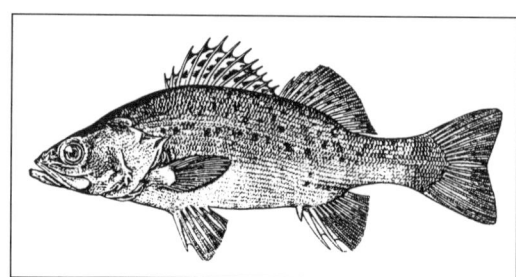

598 MORONIDAE   STB

SC *Morone saxatilis* (Walbaum, 1792); *Roccus saxatilis*
ES lubina americana; lubina estriada
DA stribet bars
DE Felsenbarsch
EL γραμμωτό λαβράκι
EN striped bass
FR bar d'Amérique
IT persicospigola striata
NL gestreepte zeebaars
PT robalo-muge; robalo riscado
FI juovabassi
SV strimmig havsabborre

600 CENTRARCHIDAE

SC *Ambloplites rupestris* (Rafinesque, 1817)
ES perca de roca
DA stenbars
DE Steinbarsch
EL κοκκινομάτα πέρκα
EN rock bass; red-eye
FR perche des roches
IT persico occhi rossi
NL steenbaars
PT perca da rocha
FI kiviahven
SV klippabborre

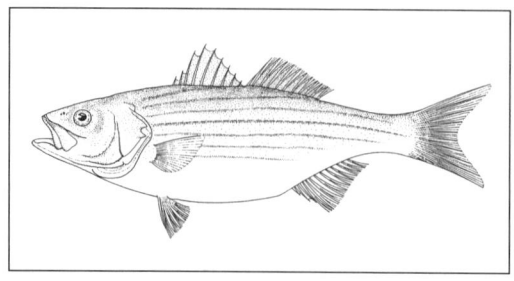

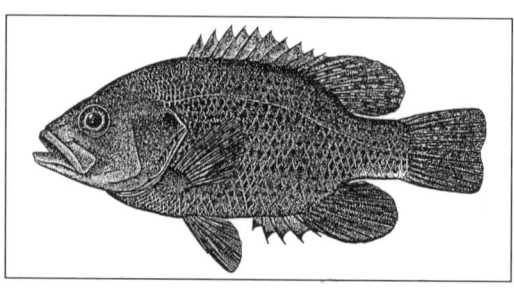

601 CENTRARCHIDAE

SC	*Lepomis gibbosus* (Linnaeus, 1758); *Eupomotis gibbosus*
ES	pez sol
DA	solaborre
DE	Sonnenbarsch
EL	ηλιόπερκα
EN	pumpkin-seed
FR	perche-soleil
IT	persico sole
NL	zonnebaars
PT	perca-sol
FI	aurinkoahven
SV	solabborre

603 CENTRARCHIDAE   MPS

SC	*Micropterus salmoides* (Lacepède, 1801); *Ambloplites salmoides*
ES	perca americana
DA	stormundet ørredaborre
DE	Forellenbarsch
EL	μεγαλόστομη πέρκα
EN	large-mouth bass; large mouthed black bass
FR	*black-bass* à grande bouche; achigan; perche noire d'Amérique
IT	persicotrota
NL	forelbaars
PT	achigã
FI	isobassi
SV	öringabborre

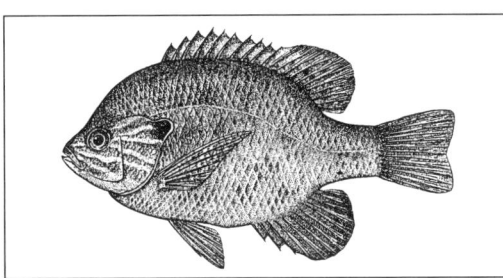

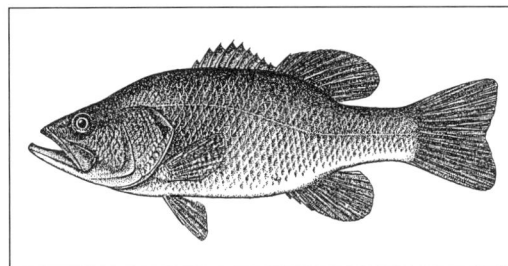

602 CENTRARCHIDAE

SC	*Micropterus dolomieui* (Lacépède, 1801)
ES	perca negra
DA	småmundet ørredaborre
DE	Schwarzbarsch
EL	μικρόστομη πέρκα
EN	small-mouth bass
FR	perche noire
IT	persico di Dolomieu
NL	zwarte baars
PT	achigã-boca pequena
FI	pikkubassi
SV	svartabborre

604 CENTRARCHIDAE

SC	*Pomoxis annularis* (Rafinesque, 1818)
ES	crapet
DA	hvid solaborre
DE	Weißer Crappie
EL	ασπρόπερκα
EN	white crappie
FR	crapet calicot
IT	persico bianco
NL	witte crappie
PT	perca anelada
FI	hopea-ahven
SV	vit solabborre

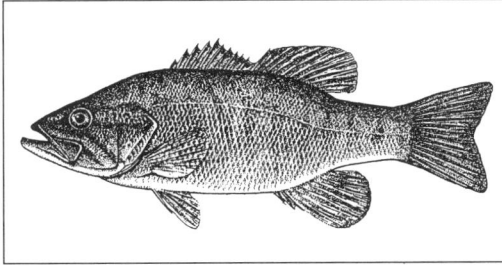

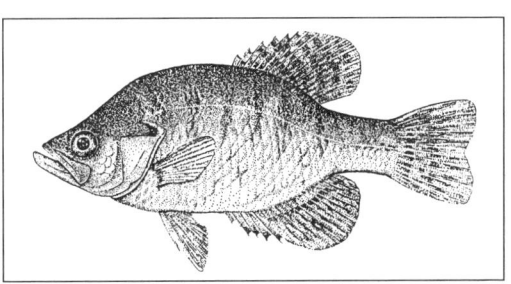

605 CENTRARCHIDAE

SC	*Pomoxis nigromaculatus* (Lesueur, 1829); *Centrarchus hexacanthus*
ES	perca plateada
DA	sort solaborre
DE	Silberbarsch; Schwarzer Crappie
EL	αργυρόπερκα
EN	black crappie
FR	perche argentée; bachelier noir
IT	persico argentato
NL	zilverbaars
PT	perca prateada
FI	mustapilkkuahven
SV	svart solabborre

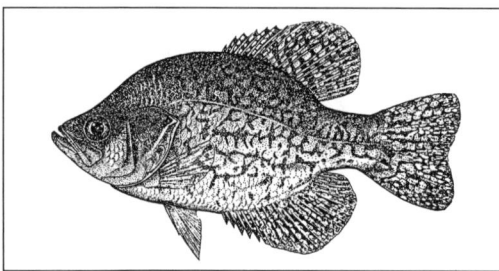

607 PRIACANTHIDAE　　　　BIR

SC	*Priacanthus macracanthus* (Cuvier, 1829)
ES	catalufa del Pacífico
DA	Stillehavs-catalufa
DE	Australischer Großaugenbarsch
EL	μεγαλόφθαλμος του Ειρηνικού
EN	red bigeye
FR	beauclaire du Pacifique
IT	catalufa
NL	Pacifische grootoogbaars
PT	fura-vasos do Pacífico
FI	suurisilmä-laji
SV	röd storögonfisk

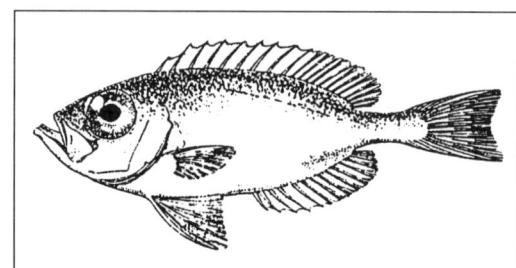

606 PRIACANTHIDAE　　　　PRI

SC	*Priacanthidae*
ES	priacántidos
DA	storøjefisk
DE	Großaugenbarsche
EL	πριακάνθιδοι· μεγαλόφθαλμοι
EN	bigeyes
FR	juifs; beauclaires
IT	priacantidi; catalufe
NL	grootoogvissen
PT	fura-vasos
FI	suurisilmät; suurisilmät-heimo
SV	storögonfiskar

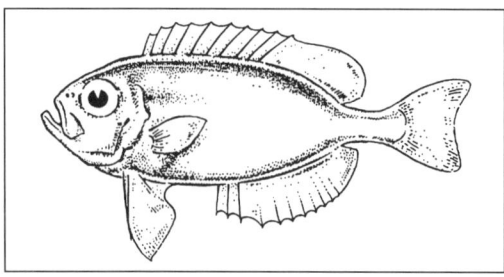

608 PRIACANTHIDAE　　　　BIG

SC	*Priacanthus* spp.
ES	catalufas
DA	catalufa-slægt
DE	Großaugenbarsche
EL	μεγαλόφθαλμοι
EN	bigeyes
FR	beauclaires; catalufas; juifs
IT	catalufe
NL	grootoogbaarzen
PT	fura-vasos
FI	suurisilmä-suku
SV	storögonfiskar

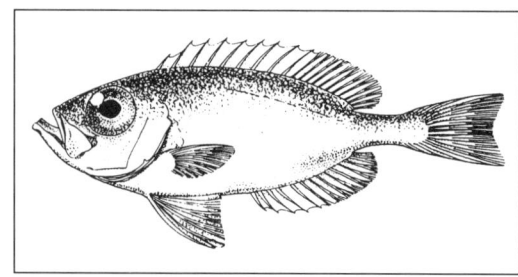

609	**APOGONIDAE**	APO	

SC   *Apogonidae*
ES   apogónidos
DA   kardinalfisk-familien
DE   Kardinalfische
EL   απογονίδες· κρεμμύδια
EN   cardinalfishes
FR   apogons; cardinaux
IT   re di triglie; apogonidi
NL   kardinaalvissen
PT   apogonídeos
FI   kardinaaliahvenet; kardinaaliahvenet-heimo
SV   djuphavsabborrar

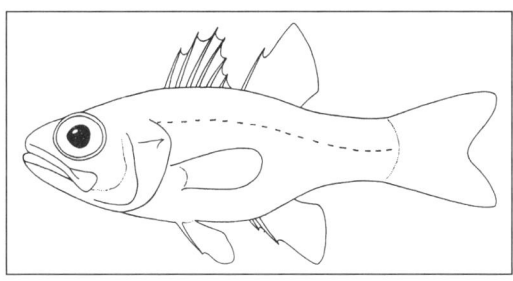

611   APOGONIDAE   EPI

SC   *Epigonus telescopus* (Risso, 1810)
ES   boca negra
DA   teleskop-kardinalfisk
DE   Teleskop-Kardinalfisch
EL   μαύρο κρεμμύδι
EN   black cardinalfish
FR   apogon noir
IT   re di triglie nero
NL   zwarte kardinaalvis
PT   olhudo
FI   kardinaaliahven-laji
SV   teleskopabborre

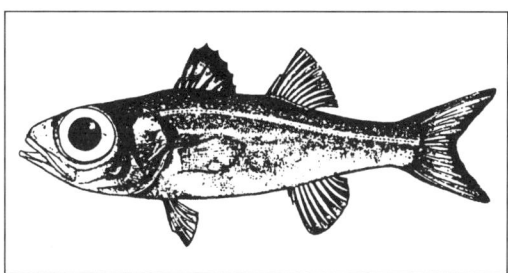

610   APOGONIDAE

SC   *Apogon imberbis* (Linnaeus, 1758)
ES   reyezuelo
DA   kardinalfisk
DE   Meerbarbenkönig
EL   κρεμμύδι· τσιμπούκι
EN   cardinalfish
FR   roi des rougets; apogon; castagnole rouge
IT   re di triglie
NL   kardinaalvis
PT   alcarraz
FI   kardinaaliahven-laji
SV   kardinalfisk

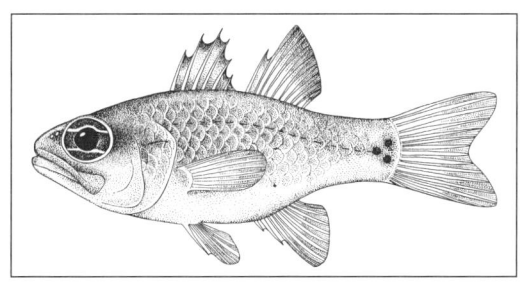

612   APOGONIDAE   SYN

SC   *Synagrops japonicus* (Doderlein, 1884)
ES   cardenal japonés
DA   japansk kardinalfisk
DE   Japanischer Kardinalfisch
EL   γιαπωνέζικο κρεμμύδι
EN   Japanese splitfin
FR   cardinal à bouche noire
IT   re di triglie orientale
NL   Japanse kardinaalvis
PT   dentinho do Cabo
FI   japaninkardinaaliahven
SV   japansk djuphavsabborre

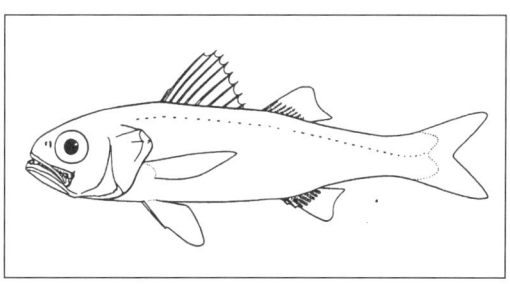

153

613   PERCIDAE

SC	*Percidae*
ES	percas
DA	aborre-familien
DE	Echte Barsche; Barsche
EL	πέρκες
EN	perches
FR	perches
IT	percidi
NL	zoetwaterbaarzen
PT	percas e afins
FI	ahvenet; ahvenet-heimo
SV	abborrfiskar

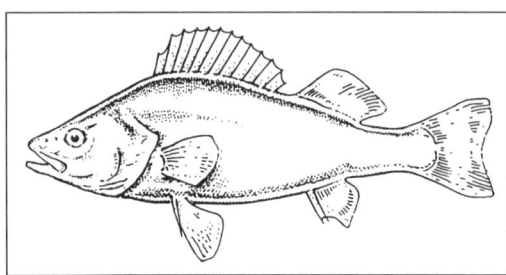

615   PERCIDAE

SC	*Gymnocephalus schraetzer* (Linnaeus, 1758); *Acerina schraetzer*
ES	acerina del Danubio
DA	Donau-hork
DE	Schrätzer; Schratz; Schratzen; Schraitzer; Schillschrätzer
EL	ακερίνα του Δούναβη
EN	Danube ruffe
FR	grémille du Danube
IT	acerina del Danubio
NL	gestreepte pos
PT	perca do Danúbio
FI	tonavankiiski
SV	donaugärs

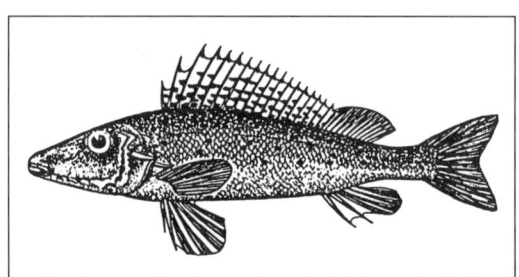

614   PERCIDAE

SC	*Gymnocephalus cernuus* (Linnaeus, 1758); *Acerina cernua*; *Gymnoecephalus cernua*
ES	acerina
DA	hork; almindelig hork
DE	Kaulbarsch; Stur; Pfaffenlaus; Stuhr; Kutt; Rotzbarsch; Schroll; Tork
EL	ακερίνα
EN	ruffe
FR	grémille; perche goujonnière
IT	acerina
NL	pos
PT	perca-caboz
FI	kiiski
SV	gärs

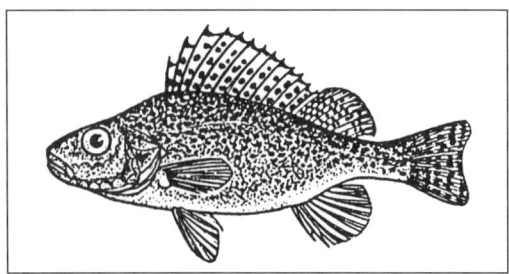

616   PERCIDAE   FPY

SC	*Perca flavescens* (Mitchill, 1814)
ES	perca canadiense
DA	gul aborre
DE	Amerikanischer Flußbarsch; Gelbbarsch
EL	κιτρινόπερκα
EN	American yellow perch
FR	perche canadienne
IT	persico dorato
NL	Amerikaanse baars
PT	perca americana
FI	kelta-ahven
SV	gulabborre

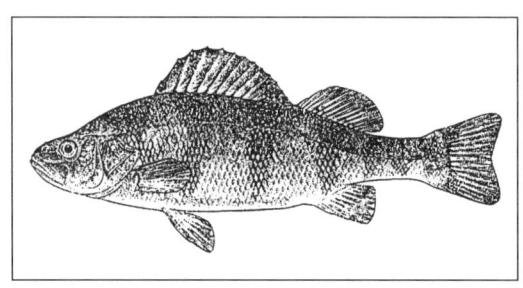

617 PERCIDAE   FPE

SC   *Perca fluviatilis* (Linnaeus, 1758)
ES   perca
DA   almindelig aborre; aborre
DE   Flußbarsch; Barsch; Bars; Perschke; Bersig;
     Bürschling; Egli; Kretzer; Schratz; Stachelferzer;
     Krätzer; Schranzen; Streifbarsch
EL   ποταμόπερκα
EN   perch; European perch
FR   perche commune; perchaude; pichette; hurlin;
     perdrix de rivière; perchat; perchelle
IT   pesce persico; perca
NL   baars
PT   perca europeia
FI   ahven
SV   abborre

619 PERCIDAE

SC   *Stizostedion canadense* (Smith, 1836)
ES   lucioperca canadiense
DA   amerikansk sandart
DE   Kanadischer Zander
EL   ποταμολάβρακο του Καναδά
EN   sauger
FR   sandre canadien
IT   sandra canadese
NL   Canadese snoekbaars
PT   picão canadiano
FI   hietakuha
SV   kanadagös

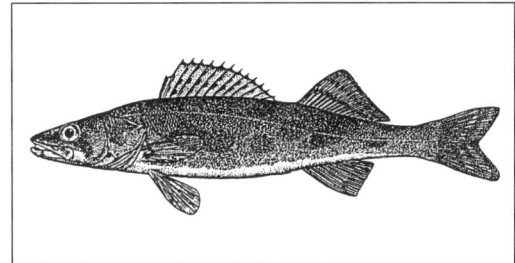

618 PERCIDAE

SC   *Romanichtys valsanicola* (Dumitrescu, Banarescu
     and Stoica, 1957)
ES   perca rumana
DA   rumænsk ulkeaborre
DE   Groppenbarsch
EL   πέρκα της Ρουμανίας
EN   Romanian bull-head perch
FR   perche-chabot roumaine
IT   perca rumena
NL   asprete
PT   perca romena
FI   simppuahven
SV   forsabborre

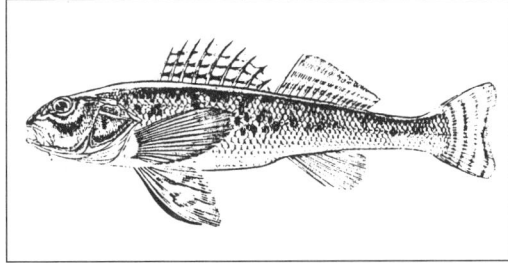

620 PERCIDAE   FPP

SC   *Stizostedion lucioperca* (Linnaeus, 1758);
     Lucioperca lucioperca; *Stizostedion vitreum glau-
     cum*; *Sander lucioperca*
ES   lucioperca
DA   sandart; almindelig sandart
DE   Zander; Hechtbarsch; Schill; Amaul; Sandbarsch;
     Sandart; Ogosch
EL   ποταμολάβρακο
EN   zander; pike-perch
FR   sandre; perche brochet; perche du Rhin; sandart
IT   sandra; luccioperca
NL   snoekbaars
PT   lucioperca
FI   kuha
SV   gös

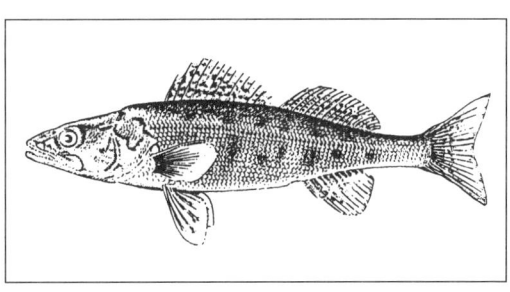

621 PERCIDAE   STV   **623 PERCIDAE**

SC   *Stizostedion vitreum vitreum* (Mitchill, 1818)
ES   lucioperca americana
DA   hvidøjet sandart
DE   Amerikanischer Zander; Glasaugenbarsch
EL   ποταμολάβρακο της Αμερικής
EN   walleye
FR   doré jaune; doré commun; sandre américain
IT   sandra americana
NL   Amerikaanse snoekbaars
PT   picão verde
FI   valkosilmäkuha
SV   amerikansk gös

SC   *Zingel streber* (Siebold, 1863); *Aspro streber*
ES   aspro menor
DA   streber; almindelig streber
DE   Streber; Spindelfisch; Zegel; Pfeiferl; Streukatze
EL   στρέμπερ
EN   streber
FR   streber
IT   streber
NL   streber
PT   peixe-fuso menor
FI   tonavantsingeli
SV   streber

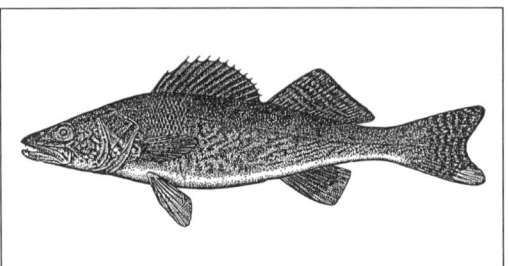

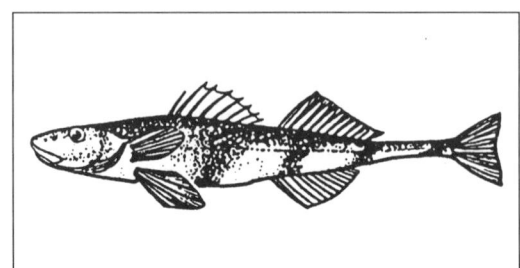

622 PERCIDAE   **624 PERCIDAE**

SC   *Zingel asper* (Linnaeus, 1758); *Aspro asper*
ES   aspro común
DA   Rhône-streber
DE   Rhône-Streber
EL   στρέμπερ του Ροδανού
EN   Rhône streber
FR   apron du Rhône
IT   asprone
NL   asper
PT   peixe-fuso
FI   hyppijätsingeli
SV   asper

SC   *Zingel zingel* (Linnaeus, 1758); *Aspro zingel*
ES   aspro mayor
DA   zingel
DE   Zingel; Zindel; Zink; Zinne
EL   ζίνγκελ
EN   zingel
FR   apron
IT   zinghel; aspro
NL   zingel
PT   peixe-fuso maior
FI   tsingeli
SV   zingel

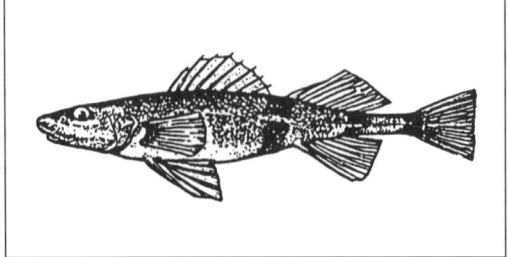

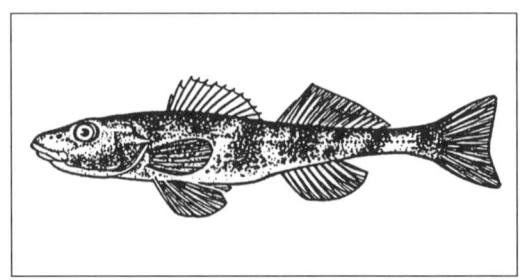

625 SILLAGINIDAE WHS

SC	*Sillaginidae*
ES	sillagínidos
DA	sillago-familien
DE	Weißlinge
EL	σιλάγγοι
EN	sillago-whitings
FR	pêche-madames
IT	sillaginidi
NL	sillaginiden
PT	silaginídeos
FI	kuoreahvenet; kuoreahvenet-heimo
SV	sillaginider

627 MALACANTHIDAE TIL

SC	*Lopholatilus chamaeleonticeps* (Goode and Bean, 1879)
ES	blanquillo camello
DA	teglfisk
DE	Blauer Ziegelbarsch
EL	πλακολεπιδόψαρο
EN	tilefish
FR	tile; tile chameau
IT	tile gibboso
NL	blauwe tegelvis
PT	peixe-paleta camelo
FI	tiilikala-laji
SV	blå tegelabborre

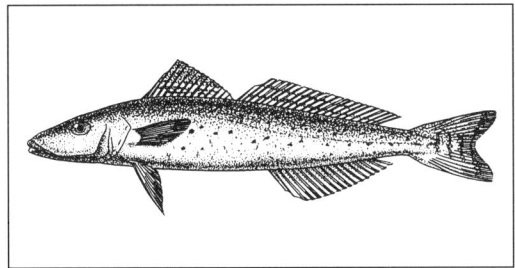

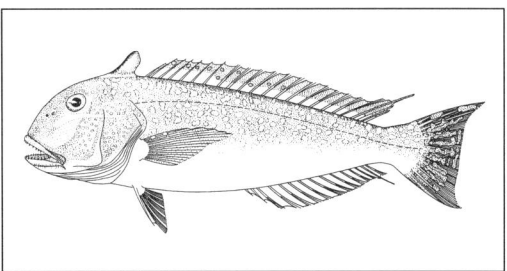

626 MALACANTHIDAE

SC	*Malacanthidae*
ES	blanquillos; paletas
DA	teglfisk-familien
DE	Ziegelbarsche
EL	πλακολεπιδόψαρα
EN	tilefishes
FR	branchostégidés malacanthidés; tiles
IT	tili
NL	tegelvissen
PT	peixes-paleta
FI	tiilikalat; tiilikalat-heimo
SV	tegelabborrfiskar

628 LACTARIIDAE TRF

SC	*Lactarius lactarius* (Bloch and Schneider, 1801)
ES	pez blanco
DA	hvidfisk
DE	Lactarius
EL	ασπρόψαρο
EN	false trevally; white fish
FR	poisson blanc
IT	lattario
NL	lactariusvis
PT	peixe-algodão
FI	valkoahven-laji
SV	silverabborre

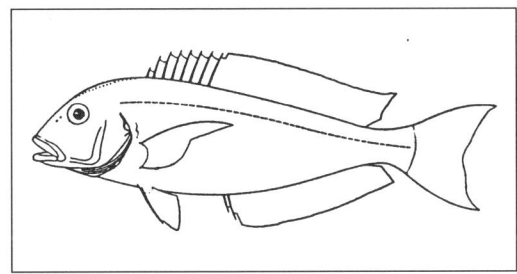

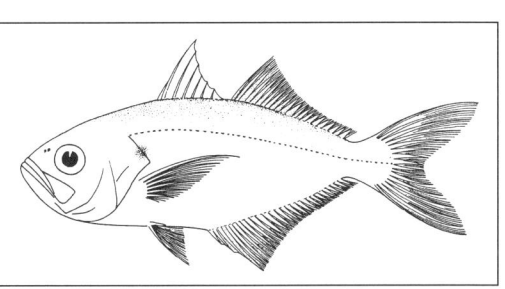

629	**POMATOMIDAE**	POT		631	**RACHYCENTRIDAE**	CBA

SC	*Pomatomidae*
ES	anjovas
DA	pomatomider
DE	Blaufische; Blaubarsche
EL	γοφάρια
EN	bluefishes
FR	pomatomidés; tassergals; coupe fil
IT	pesci serra; pomatomidi
NL	blauwbaarzen
PT	anchovas
FI	sinikalat; sinikalat-heimo
SV	blåfiskar

SC	*Rachycentron canadum* (Linnaeus, 1766)
ES	cobia
DA	sergentfisk
DE	Königsbarsch
EL	κόμπια
EN	cobia
FR	mafou
IT	cobia
NL	cobia
PT	fogueteiro-galego
FI	okakala
SV	cobia

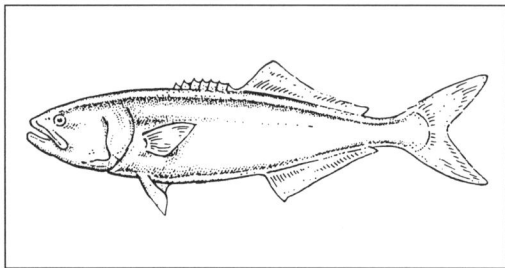

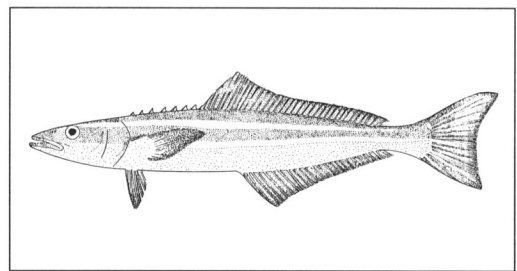

630	**POMATOMIDAE**	BLU		632	**CARANGIDAE**	CGX

SC	*Pomatomus saltator* (Linnaeus, 1766); *Pomatomus saltatrix*; *Temnodon saltator*
ES	anjova; chova; cova; anjora
DA	blåbars
DE	Blaufisch; Blaubarsch
EL	γοφάρι
EN	bluefish
FR	tassergal; coupe fil
IT	pesce serra
NL	blauwbaars
PT	anchova
FI	sinikala
SV	blåfisk

SC	*Carangidae*
ES	carángidos
DA	hestemakrel-familien
DE	Bastardmakrelen; Stachelmakrelen
EL	κοκκάλια
EN	carangids; jacks
FR	carangidés; carangues et chinchards
IT	carangidi
NL	horsmakrelen
PT	carangídeos
FI	piikkimakrillit; piikkimakrillit-heimo
SV	taggmakrillfiskar

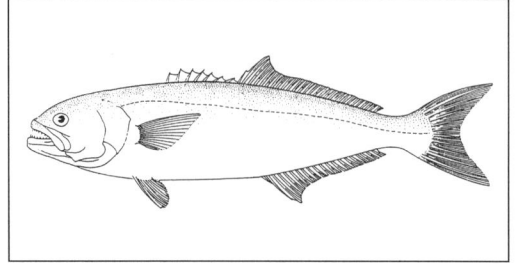

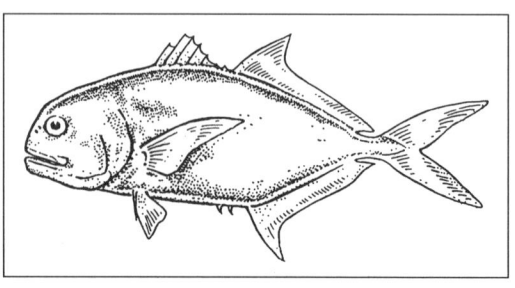

633 CARANGIDAE ALA

SC	*Alectis alexandrinus* (Geoffroy St-Hilaire, 1817)
ES	jurel de Alejandría
DA	Alexandria-trådfisk
DE	Alexandria-Fadenmakrele
EL	κοκκάλι της Αλεξάνδρειας
EN	Alexandria pompano
FR	cordonnier bossu; scyris d'Alexandrie
IT	carango d'Alessandria
NL	Alexandrië-pompano
PT	xaréu-enxada
FI	alexandrianrihmapompano
SV	alexandriatrådmakrill

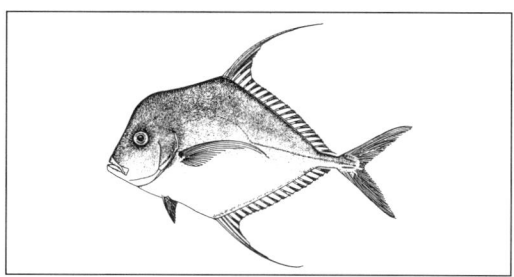

635 CARANGIDAE RUB

SC	*Caranx crysos* (Mitchill, 1815); *Caranx fusus*
ES	jurel; jurel azul; cojiúna negra
DA	blå hestemakrel
DE	Blaue Stachelmakrele
EL	κοκκάλι
EN	blue runner
FR	carangue coubali; carangue commune
IT	carango mediterraneo; carango dorato
NL	blauwe horsmakreel
PT	xaréu azul
FI	sinipiikkimakrilli
SV	blålöpare

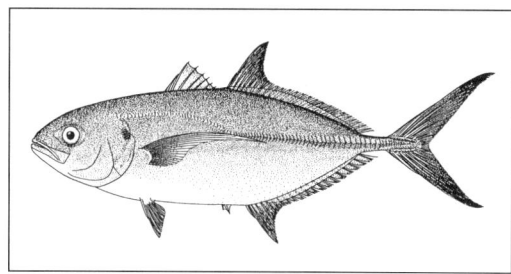

634 CARANGIDAE

SC	*Alepes djedaba* (Forsskål, 1775); *Caranx kalla*; *Caranx djedaba*
ES	seriola dorada
DA	gylden hestemakrel
DE	Goldene Stachelmakrele
EL	χρυσό κοκκάλι
EN	golden scad; shrimp scad
FR	sélar subari
IT	carango calla
NL	gouden horsmakreel
PT	xaréu-rei
FI	rihmapompano-laji
SV	räkmakrill

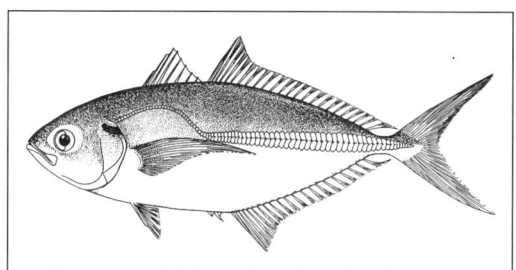

636 CARANGIDAE CVJ

SC	*Caranx hippos* (Linnaeus, 1766)
ES	seriola caballo
DA	crevalle
DE	Pferde-Stachelmakrele
EL	κοκκάλι
EN	crevalle jack; Samson fish; sea kingfish
FR	sériole cheval; carangue cheval; grande carangue
IT	carango cavallo
NL	paardenhorsmakreel
PT	xaréu-macoa
FI	hevospiikkimakrilli
SV	gullöpare

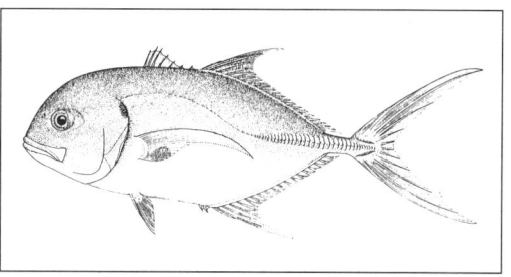

637	**CARANGIDAE**	HMY	**639**	**CARANGIDAE**	TRE

SC	*Caranx rhonchus* (Geoffroy St.-Hilaire, 1817)		SC	*Caranx* spp.	
ES	jurel amarillo		ES	jureles; pámpanos	
DA	gul hestemakrel		DA	hestemakrel-slægt	
DE	Gelbe Stachelmakrele		DE	Stachelmakrelen; Jacks	
EL	κοκκάλι		EL	κοκκάλι	
EN	yellow horse mackerel		EN	jacks; crevalles	
FR	carangue jaune		FR	carangues	
IT	carango ronco		IT	carangi	
NL	gele horsmakreel		NL	horsmakrelen	
PT	charro amarelo; charro francês		PT	xaréus e charros	
FI	keltapiikkimakrilli		FI	piikkimakrilli-suku	
SV	gul taggmakrill		SV	taggmakrillar	

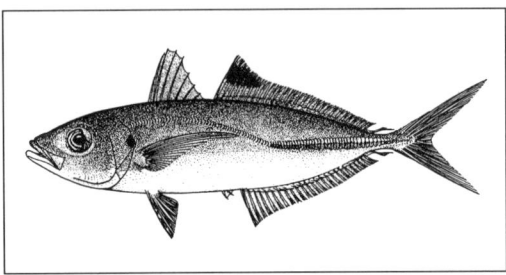 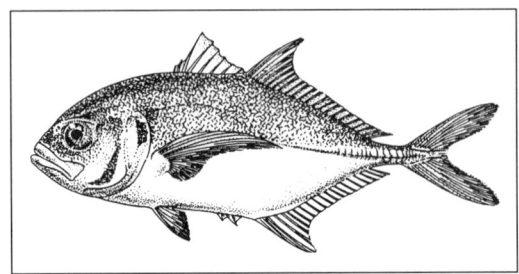

638	**CARANGIDAE**	CXR	**640**	**CARANGIDAE**	BUA

SC	*Caranx ruber* (Bloch, 1793)		SC	*Chloroscombrus chrysurus* (Linnaeus, 1766)	
ES	cojiúna carbonera		ES	casabe	
DA	sortrygget hestemakrel		DA	en art hestemakrel	
DE	Schwarzrücken-Stachelmakrele		DE	Schwanzfleck-Stachelmakrele	
EL	κόκκινο κοκκάλι		EL	αλουμινόψαρο	
EN	bar jack		EN	Atlantic bumper	
FR	carangue rouge		FR	sapater	
IT	carango bandanera		IT		
NL	zwarte horsmakreel		NL	bijlhorsmakreel	
PT	xaréu carvoeiro		PT	prato de alumínio	
FI	juovapiikkimakrilli		FI	puskurikala	
SV	svartryggad taggmakrill		SV	svansfläckstaggmakrill	

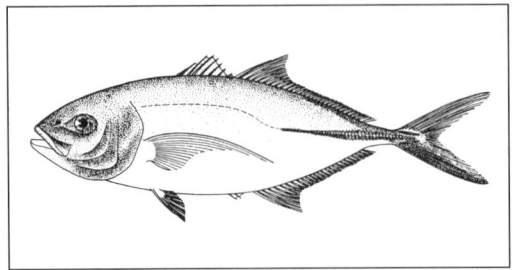 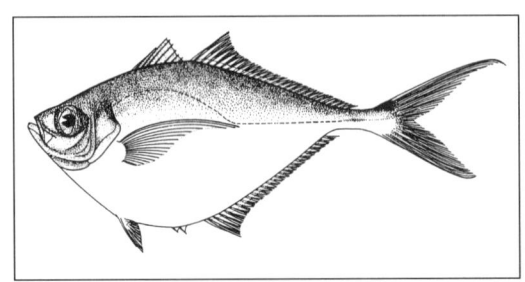

641 CARANGIDAE — MSD

SC	*Decapterus macarellus* (Cuvier, 1833)
ES	macarela caballa
DA	gulhalet hestemakrel
DE	Gelbschwanz-Stachelmakrele
EL	δεκάπτερο σαυρίδι
EN	mackerel scad
FR	comète maquereau; saurel maquereau
IT	sugarotto; suro; sugarello
NL	geelstaart-stekelmakreel
PT	charro-olho largo
FI	punapyrstöpiikkimakrilli
SV	kometmakrill

643 CARANGIDAE — WEC

SC	*Decapterus punctatus* (Cuvier, 1829)
ES	macarela chuparaco; surela
DA	plettet hestemakrel
DE	Gefleckte Stachelmakrele
EL	στρογγυλοσαύριδο
EN	round scad
FR	comète; saurel tacheté
IT	sugarotto; suro; sugarello
NL	gevlekte stekelmakreel
PT	charro moiro
FI	pilkkupiikkimakrilli
SV	fläckig taggmakrill

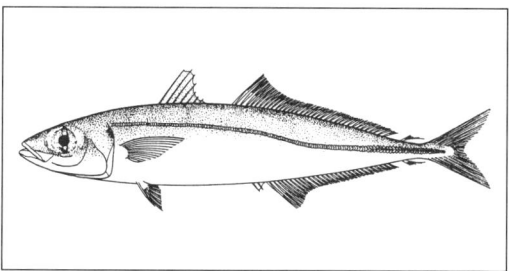

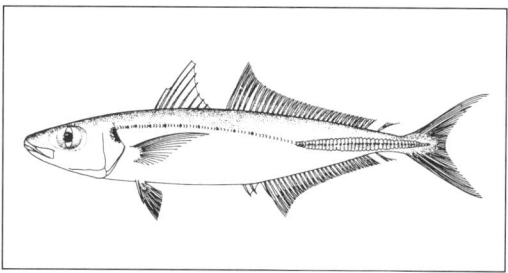

642 CARANGIDAE — RSA

SC	*Decapterus maruadsi* (Temminck and Schlegel, 1842)
ES	macarela japonesa
DA	langfinnet hestemakrel
DE	Langflossen-Stachelmakrele
EL	σαυρίδι της Ιαπωνίας
EN	round scad
FR	comète japonaise; saurel japonais
IT	sugarotto
NL	Japanse stekelmakreel
PT	charro japonês
FI	piikkimakrilli-laji
SV	japansk taggmakrill

644 CARANGIDAE — RUS

SC	*Decapterus russelli* (Rüppell, 1830)
ES	macarela de Russel
DA	indisk hestemakrel
DE	Indische Stachelmakrele
EL	σαυρίδι του Ινδικού
EN	Indian scad
FR	comète de Russel; saurel de Russel
IT	sugarotto
NL	Russels stekelmakreel
PT	charro de Russel
FI	piikkimakrilli-laji
SV	indisk taggmakrill

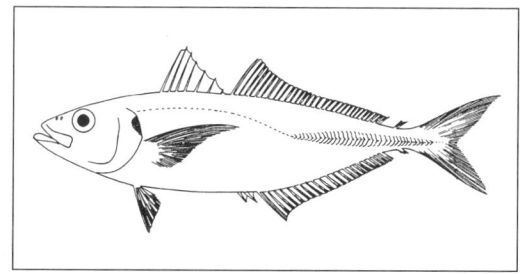

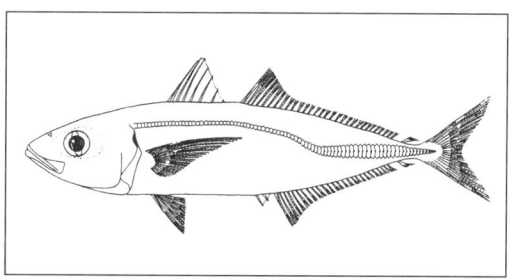

645 CARANGIDAE SDX

SC	*Decapterus* spp.
ES	macarelas
DA	hestemakrel-slægt
DE	Stachelmakrelen
EL	σαυρίδια
EN	scads
FR	comtes; chinchards ronds; saurel
IT	sugarotti
NL	stekelmakrelen
PT	charros
FI	piikkimakrilli-suku
SV	taggmakrillar

647 CARANGIDAE LEE

SC	*Lichia amia* (Linnaeus, 1758)
ES	palometón
DA	stor gaffelmakrel
DE	Große Gabelmakrele
EL	λίτσα
EN	leerfish
FR	liche amie; amia
IT	leccia
NL	grote gaffelmakreel
PT	palombeta
FI	amiapompano
SV	snegelfisk

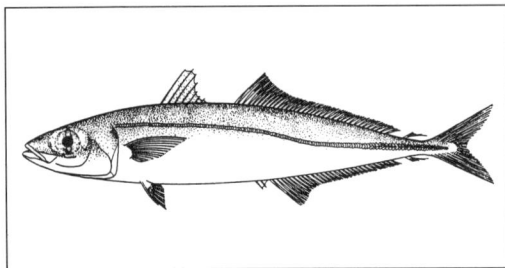

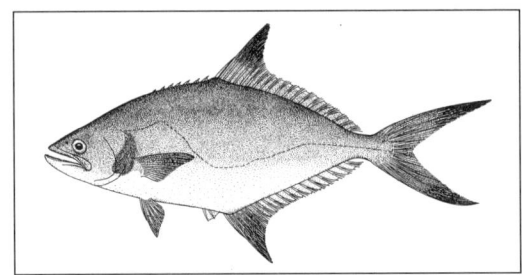

646 CARANGIDAE RRU

SC	*Elagatis bipinnulata* (Quoy and Gaimard, 1824)
ES	macarela salmón
DA	regnbue-hestemakrel
DE	Regenbogen-Stachelmakrele
EL	κομήτης
EN	rainbow runner
FR	arc-en-ciel; comète saumon
IT	cometa
NL	regenboog-stekelmakreel
PT	fogueteiro arco-íris
FI	sateenkaariharppoja
SV	regnbågslöpare

648 CARANGIDAE LEX

SC	*Lichia* spp.
ES	palometones
DA	hestemakrel-slægt
DE	Lichia
EL	λίτσες
EN	leerfishes
FR	liches
IT	lecce
NL	lichia
PT	palombetas
FI	pompano-suku
SV	snegelfiskar

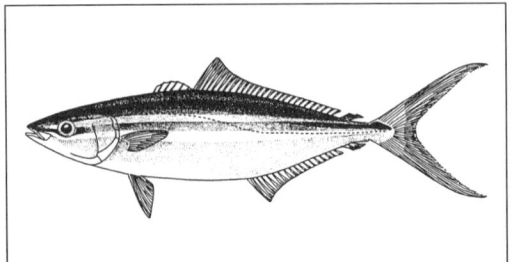

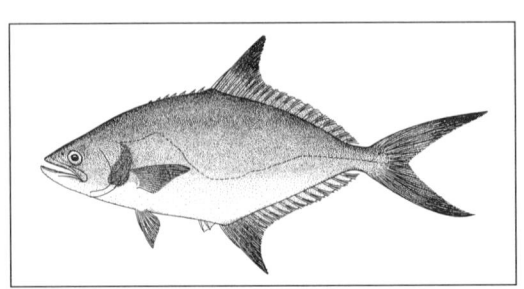

649	**CARANGIDAE**	HAS

SC	*Megalaspis cordyla* (Linnaeus, 1758)
ES	panga; tetenkel
DA	torpedo-hestemakrel
DE	Torpedo-Stachelmakrele
EL	λίτσα με λέπια
EN	torpedo scad
FR	saurel torpille
IT	leccia scagliosa
NL	torpedo-stekelmakreel
PT	torpedo
FI	piikkimakrilli-laji
SV	torpedlöpare

651	**CARANGIDAE**	PAO

SC	*Parona signata* (Jenyns, 1841)
ES	parona
DA	parona
DE	Parona
EL	παρόνα
EN	parona leatherjack
FR	sauteur parone
IT	parona
NL	parona
PT	parona
FI	piikkimakrilli-laji
SV	paronamakrill

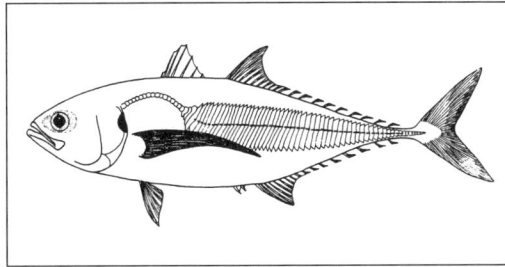

DRAWING NOT AVAILABLE

650	**CARANGIDAE**	

SC	*Naucrates ductor* (Linnaeus, 1758)
ES	pez piloto
DA	lodsfisk
DE	Lotsenfisch; Pilotfisch
EL	κολαούζος
EN	pilotfish
FR	poisson pilote
IT	pesce pilota; fanfano
NL	loodsmannetje
PT	peixe-piloto
FI	luotsikala
SV	lotsfisk

652	**CARANGIDAE**	TRZ

SC	*Pseudocaranx dentex* (Bloch and Schneider, 1801);
	Caranx georgianus
ES	jurel de Nueva Zelanda
DA	newzealandsk hestemakrel
DE	Neuseeland-Stachelmakrele
EL	κοκκάλι της Νέας Ζηλανδίας
EN	New Zealand trevally
FR	carangue de Nouvelle-Zélande
IT	carango di Nuova Zelanda
NL	Nieuw-Zeelandse horsmakreel
PT	xaréu bicudo
FI	uudenseelanninpiikkimakrilli
SV	gulbandad hästmakrill

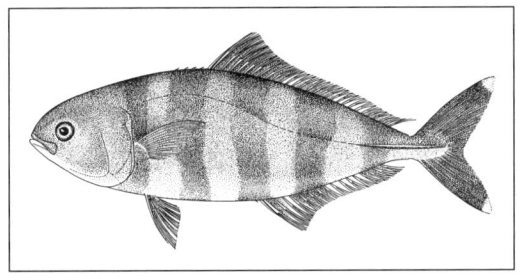

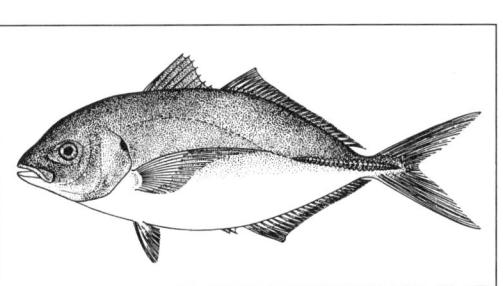

653	**CARANGIDAE**	QUE	**655**	**CARANGIDAE**	TRY

SC	*Scomberoides* spp.; *Chorinemus* spp.		SC	*Selaroides leptolepis* (Cuvier, 1833)
ES	caballas reales		ES	chicharro de rayas amarillas
DA	hestemakrel-slægter		DA	guldstribet selar
DE	Königinnen-Stachelmakrele		DE	Goldband-Selar
EL	βασιλικό σαυρίδι		EL	κιτρινοσταύριδο
EN	queenfishes		EN	yellow-stripe trevally; yellow-stripe scad
FR	chinchards royaux		FR	sélar à rayures jaunes
IT	lecce salterine		IT	suro bandagialla
NL	koninginne-horsmakrelen		NL	gestreepte selar
PT	cavalas reais		PT	charro de riscas amarelas
FI	piikkimakrilli		FI	piikkimakrilli-laji
SV	kungstaggmakrillar		SV	guldbandsselar

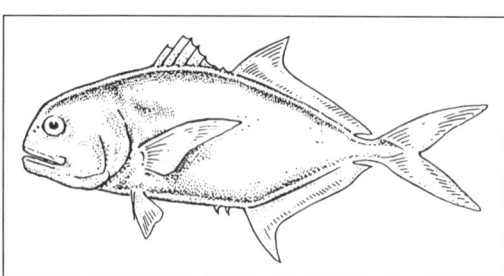 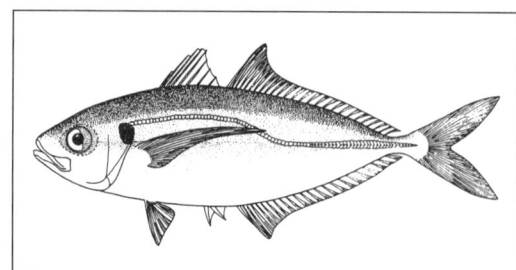

654	**CARANGIDAE**	BIS	**656**	**CARANGIDAE**	LUK

SC	*Selar crumenophthalmus* (Bloch, 1793)		SC	*Selene dorsalis* (Gill, 1862)
ES	chicharro ojón		ES	jorobado africano
DA	storøjet selar		DA	afrikansk hestehoved
DE	Großäugiger Selar		DE	Afrikanischer Pferdekopf
EL	μεγαλομάτης σούρος		EL	φεγγαρόψαρο της Αφρικής
EN	big-eye scad		EN	African lookdown; lookdownfish
FR	sélar coulisson		FR	mussolini africain
IT	suro occhione		IT	carango piatto
NL	grootoogselar		NL	afkijkvis
PT	charro preto		PT	corcovado africano
FI	isosilmäpiikkimakrilli		FI	afrikanmurjottajakala
SV	storögonselar		SV	afrikansk hästhuvudmakrill

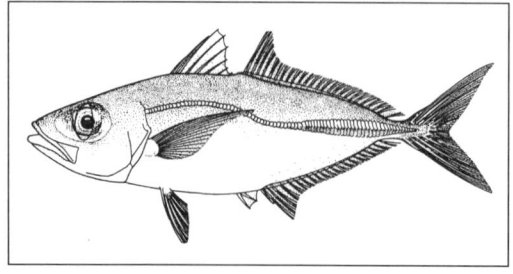 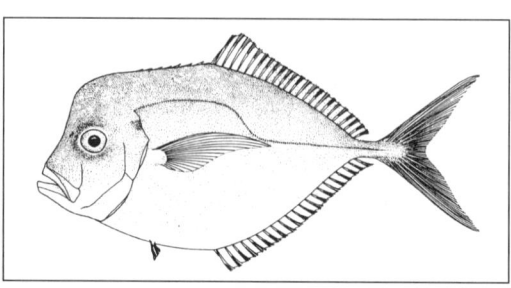

657	**CARANGIDAE**	MOA

SC *Selene setapinnis* (Mitchill, 1815)
ES jorobado lamparosa
DA atlantisk hestehoved
DE Atlantischer Pferdekopf
EL φεγγαρόψαρο του Ατλαντικού
EN Atlantic moonfish
FR mussolini
IT carango piatto
NL Atlantische afkijkvis
PT corcovado do Golfo
FI murjottajakala-laji
SV atlantisk hästhuvudmakrill

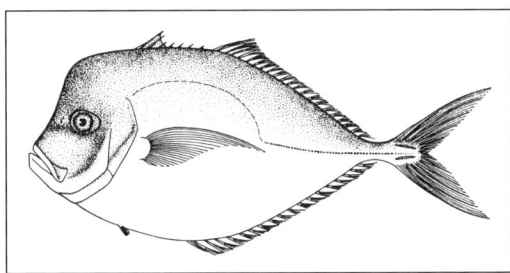

659	**CARANGIDAE**	YTC

SC *Seriola lalandi* (Valenciennes, 1833); *Regificola grandis*
ES seriola australiana
DA australsk ravfisk
DE Australische Gelbschwanzmakrele
EL μαγιάτικο της Αυστραλίας
EN yellow-tail kingfish; yellow-tail amberjack
FR sériole australienne
IT ricciola del Sudafrica; ricciola australe
NL reuzengeelstaart
PT charuteiro-limão
FI australianpiikkimakrilli
SV sydseriola

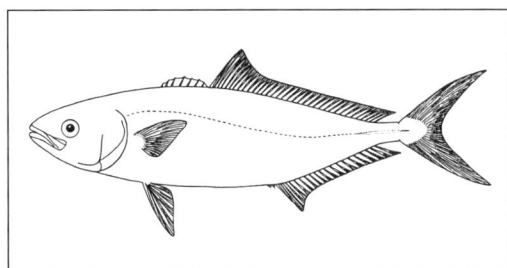

658	**CARANGIDAE**	AMB

SC *Seriola dumerili* (Risso, 1810)
ES medregal coronado
DA stor ravfisk
DE Bernsteinfisch; Gelbschwanz; Grünel; Gelbschwanzmakrele
EL μαγιάτικο· κυνηγός
EN greater amberjack
FR sériole; sériole couronne
IT ricciola
NL grote geelstaart
PT charuteiro-catarino
FI isopiikkimakrilli
SV seriola

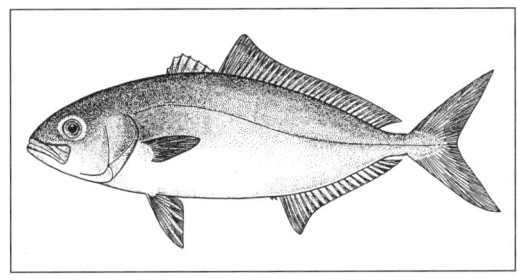

660	**CARANGIDAE**	AMJ

SC *Seriola quinqueradiata* (Temminck and Schlegel, 1845)
ES pez limón del Japón
DA japansk ravfisk
DE Japanische Seriola
EL μαγιάτικο της Ιαπωνίας
EN Japanese amberjack
FR sériole du Japon
IT ricciola giapponese
NL Japanse geelstaart
PT charuteiro do Japão
FI piikkimakrilli-laji
SV japansk seriola

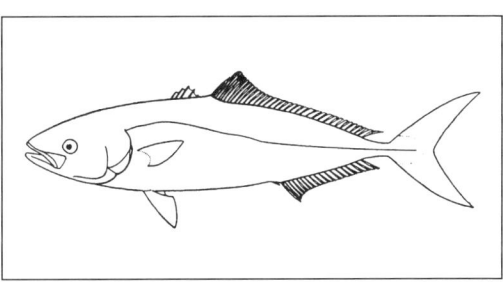

661 CARANGIDAE AMX

SC *Seriola* spp.
ES peces limón
DA ravfisk-slægten
DE Seriola; Stachelmakrelen
EL μαγιάτικο
EN amberjacks
FR sérioles
IT ricciole
NL geelstaarten
PT charuteiros; seríolas
FI piikkimakrilli-suku
SV serioler

663 CARANGIDAE POP

SC *Trachinotus ovatus* (Linnaeus, 1758); *Trachynotus glaucus*
ES palometa
DA gaffelmakrel
DE Gabelmakrele; Bläuel
EL λίτσα
EN derbio; pompano
FR liche glauque; palomine
IT leccia stella
NL gaffelmakreel
PT sereia-camochilo
FI nuijapompano
SV långfenad pompano; blå gaffelmakrill

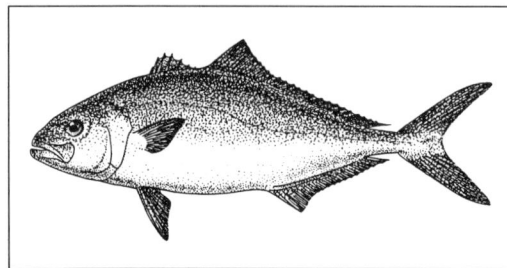

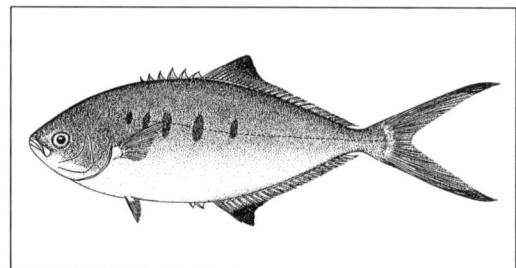

662 CARANGIDAE POM

SC *Trachinotus carolinus* (Linnaeus, 1766)
ES pámpano amarillo
DA atlantisk pampano
DE Gemeiner Pampano
EL λίτσα της Φλόριντα
EN Florida pompano; common pompano
FR pompano sole; pompano de Floride
IT leccia dei Caraibi
NL gele pompano
PT sereia da Florida
FI pompano
SV atlantisk pampano

664 CARANGIDAE POX

SC *Trachinotus* spp.
ES pámpanos; palometas
DA hestemakrel-slægt
DE Pampanos
EL λίτσες
EN pompanos
FR trachynotes; pompanos
IT lecce
NL pompano's
PT sereias
FI pompano-suku
SV gaffelmakrillar

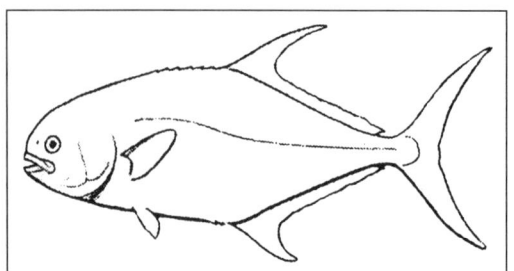

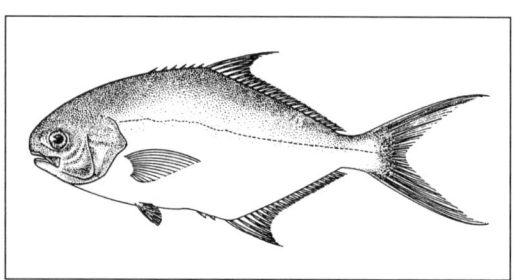

665 CARANGIDAE HMG

SC	*Trachurus declivis* (Jenyns, 1841)
ES	jurel verde
DA	grønrygget hestemakrel
DE	Grünrücken-Bastardmakrele
EL	πράσινο σαυρίδι
EN	green-back horse mackerel
FR	chinchard à dos vert
IT	sugarello australe
NL	groene horsmakreel
PT	carapau verde
FI	piikkimakrilli-laji
SV	grön taggmakrill

DRAWING NOT AVAILABLE

667 CARANGIDAE RSC

SC	*Trachurus lathami* (Nichols, 1920)
ES	chicharro garetón
DA	ru hestemakrel
DE	Rauhe Bastardmakrele
EL	αγριοσαύριδο
EN	rough scad
FR	chinchard frappeur
IT	suro americano
NL	ruwe horsmakreel
PT	carapau rugoso
FI	piikkimakrilli-laji
SV	sträv taggmakrill

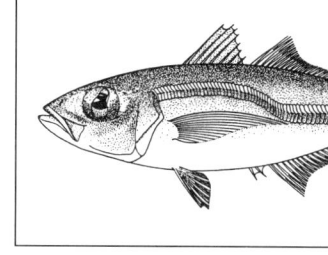

666 CARANGIDAE JJM

SC	*Trachurus japonicus* (Temminck and Schlegel, 1845)
ES	jurel japonés
DA	japansk hestemakrel
DE	Japanische Bastardmakrele
EL	σαυρίδι της Ιαπωνίας
EN	Japanese jack mackerel
FR	chinchard du Japon
IT	sugarello giapponese; suro; sugarello
NL	Japanse horsmakreel
PT	carapau do Japão
FI	japaninpiikkimakrilli
SV	japansk taggmakrill

DRAWING NOT AVAILABLE

668 CARANGIDAE HMM

SC	*Trachurus mediterraneus* (Steindachner, 1868)
ES	jurel mediterráneo
DA	Middelhavs-hestemakrel
DE	Mittelmeerstöcker; Mittelmeer-Bastardmakrele
EL	ασπροσαύριδο· σαυρίδι· σαμπανιός
EN	Mediterranean scad; Mediterranean horse mackerel
FR	saurel; chinchard à queue jaune; chinchard de la Méditerranée
IT	sugarello; suro
NL	Middellandse-Zeehorsmakreel
PT	carapau do Mediterrâneo
FI	välimerenpiikkimakrilli
SV	medelhavstaggmakrill

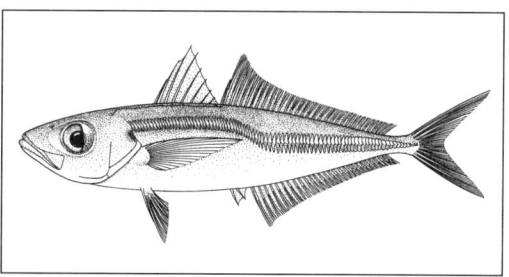

669 CARANGIDAE   CJM

SC   *Trachurus murphyi* (Nichols, 1920)
ES   jurel chileno
DA   chilensk hestemakrel
DE   Chilenische Bastardmakrele
EL   σαυρίδι της Χιλής
EN   Chilean jack mackerel
FR   chinchard du Chili
IT   sugarello cileno
NL   Chileense horsmakreel
PT   carapau chileno
FI   chilenpiikkimakrilli
SV   chilensk taggmakrill

671 CARANGIDAE   PJM

SC   *Trachurus symmetricus* (Ayres, 1855)
ES   jurel del Pacífico
DA   Stillehavs-hestemakrel
DE   Pazifische Bastardmakrele
EL   σαυρίδι του Ειρηνικού
EN   Pacific jack mackerel
FR   chinchard du Pacifique
IT   suro del Pacifico; suro; sugarello
NL   Pacifische horsmakreel
PT   carapau do Pacífico
FI   tyynenmerenpiikkimakrilli
SV   stillahavstaggmakrill

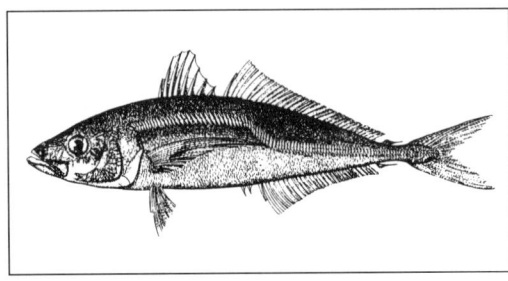
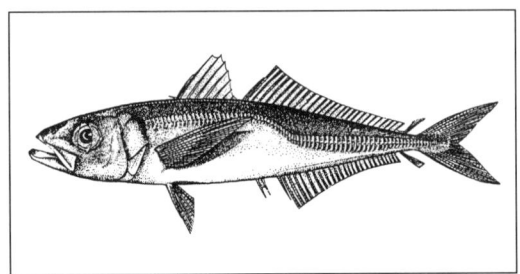

670 CARANGIDAE   JAA

SC   *Trachurus picturatus* (Bowdich, 1825)
ES   chicharro; jurel
DA   blå hestemakrel
DE   Blaue Bastardmakrele; Blauer Stöcker
EL   μαυροσαύριδο της Αυστραλίας
EN   offshore jack mackerel; blue jack mackerel
FR   chinchard du large; chinchard bleu; saurel
IT   suro; sugarello pittato
NL   blauwe horsmakreel
PT   carapau negrão
FI   piikkimakrilli-laji
SV   blå taggmakrill

672 CARANGIDAE   HOM

SC   *Trachurus trachurus* (Linnaeus, 1758)
ES   jurel
DA   hestemakrel; almindelig hestemakrel
DE   Stöcker; Bastardmakrele
EL   σαυρίδι· σαμπανιός
EN   scad; Atlantic horse mackerel
FR   saurel; chinchard; chinchard commun; chinchard
     d'Europe
IT   suro; sugarello
NL   horsmakreel
PT   carapau; chicharro
FI   piikkimakrilli
SV   taggmakrill

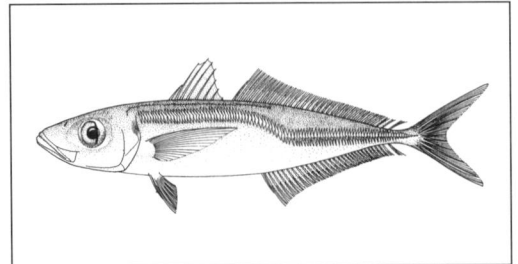
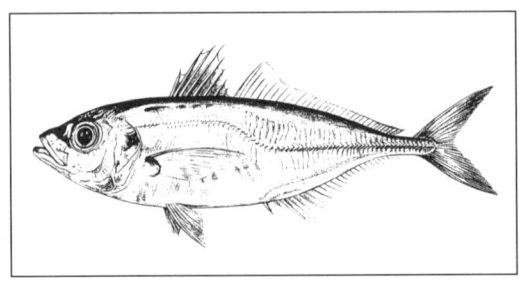

673	**CARANGIDAE**	HMC	

673 CARANGIDAE HMC

SC	*Trachurus trachurus capensis* (Castelnau, 1861); *Trachurus capensis*
ES	jurel del Cabo
DA	sydafrikansk hestemakrel
DE	Kap-Bastardmakrele
EL	σαυρίδι του Ακρωτηρίου
EN	Cape horse mackerel
FR	chinchard du Cap
IT	suro del Sudafrica
NL	Kaapse horsmakreel; maasbanker
PT	carapau do Cabo
FI	kapinpiikkimakrilli
SV	sydafrikansk taggmakrill

675 CARANGIDAE JAX

SC	*Trachurus* spp.
ES	jureles
DA	hestemakrel-slægt
DE	Bastardmakrelen
EL	σαυρίδι· σαμπανιός
EN	scads and horse mackerels; jack and horse macke-rels
FR	chinchards
IT	suri; sugarelli
NL	horsmakrelen
PT	carapaus
FI	piikkimakrilli-suku
SV	taggmakrillar

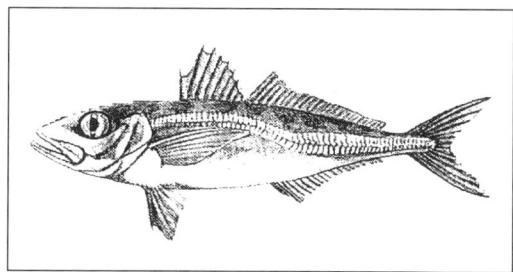

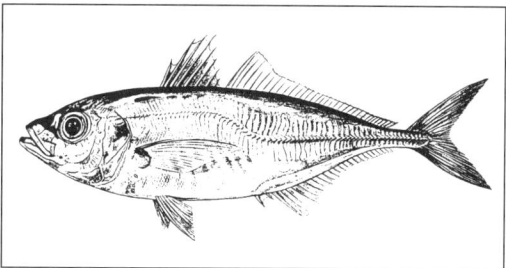

674 CARANGIDAE HMZ

SC	*Trachurus trecae* (Cadenat, 1949)
ES	jurel cunene
DA	Cunene-hestemakrel
DE	Cunene-Bastardmakrele
EL	σαυρίδι κούνενε
EN	Cunene horse mackerel
FR	chinchard noir
IT	suro cunene
NL	Cunene-horsmakreel
PT	carapau do Cunene
FI	piikkimakrilli-laji
SV	cunenetaggmakrill

676 CARANGIDAE

SC	*Vomer* spp.
ES	jorobados
DA	hestemakrel-slægt
DE	Pferdeköpfe
EL	φεγγαρόψαρα
EN	moonfishes
FR	assiettes-mussolini
IT	pesci ascia; mussolini
NL	paardenkoppen
PT	corcovados
FI	piikkimakrilli-suku
SV	månmakrillar

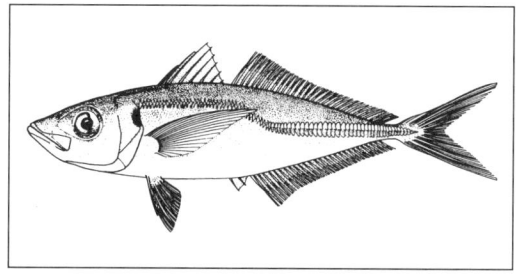

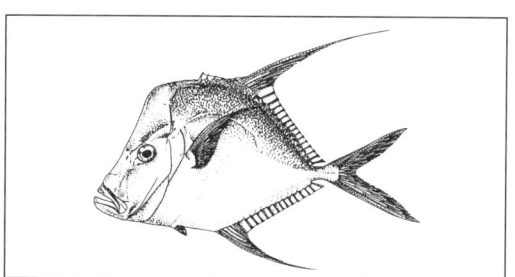

677	**FORMIONIDAE**	POB

SC	*Formio niger* (Bloch, 1795); *Parastromateus niger*
ES	makulu; palometa negra
DA	sort pomfret
DE	Schwarzer Pomfret
EL	
EN	black pomfret
FR	castanoline noire; pomfret noir
IT	fieto nero
NL	zwarte pomfret
PT	falso pampo
FI	okamakrilli
SV	svart pomfret

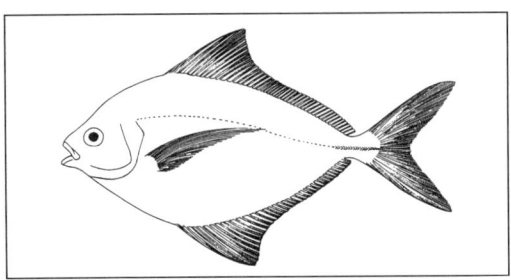

678	**MENIDAE**	MOO

SC	*Mene maculata* (Bloch and Schneider, 1801)
ES	pez luna de la India
DA	plettet månefisk
DE	Fleckenmondfisch
EL	μπλε φεγγαρόψαρο
EN	moonfish
FR	assiette; lure; musso
IT	luna blu
NL	gevlekte maanvis
PT	peixe-roda
FI	kirveskala
SV	indisk bukfisk

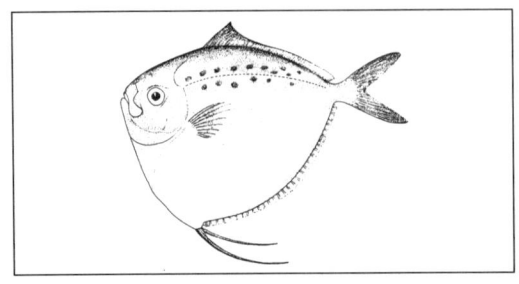

679	**BRAMIDAE**	BRZ

SC	*Bramidae*
ES	palometas; brámidos; castañolas
DA	havbrasen-familien
DE	Brachsenmakrelen
EL	καστανόψαρα
EN	breams
FR	grandes castagnoles; hirondelles de mer
IT	pesci castagna
NL	bramen
PT	xaputas; chaputas
FI	merilahnat; merilahnat-heimo
SV	havsbraxnar

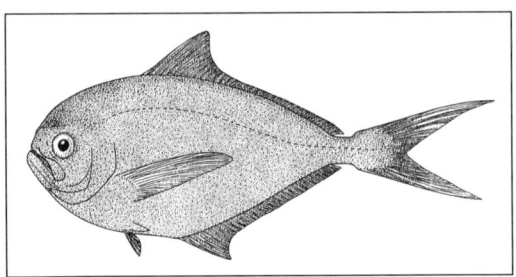

680	**BRAMIDAE**	POA

SC	*Brama brama* (Bonnaterre, 1788); *Brama raji*; *Brama raii*
ES	japuta; palometa negra; castaña; castañeta; zapatero; parpada
DA	havbrasen
DE	Brachsenmakrele
EL	καστανόψαρο
EN	Ray's bream; Atlantic pomfret
FR	brème de mer; hirondelle; grande castagnole
IT	pesce castagna
NL	braam
PT	xaputa; chaputa
FI	merilahna
SV	havsbraxen

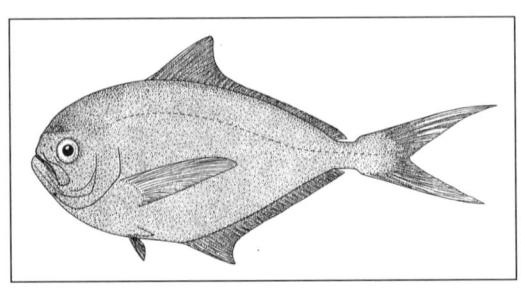

681 CORYPHAENIDAE DOL

SC *Coryphaena hippurus* (Linnaeus, 1758)
ES lampuga
DA guldmakrel
DE Gemeine Goldmakrele; Goldmakrele
EL κυνηγός· λαμπούγγα· σύρτης
EN dolphinfish; common dolphinfish
FR coryphène; grande coryphène
IT lampuga; corifena
NL goudmakreel; dolfijnvis
PT doirado; sapatorra
FI dolfiini
SV guldmakrill

683 ARRIPIDAE ASA

SC *Arripis trutta* (Schneider, 1801)
ES salmón australiano
DA australsk laks
DE Australischer Lachs
EL σολομός της Αυστραλίας
EN Australian salmon; ruff
FR saumon australien
IT salmone australiano
NL Australische zalm
PT peixe-grosa australiano
FI arripi
SV Kahawai; laxabborre

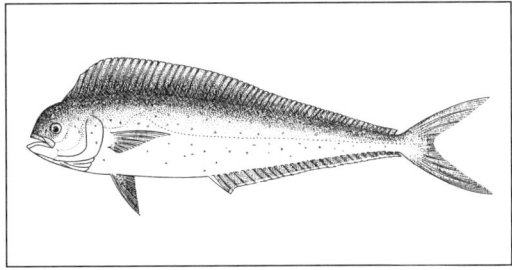

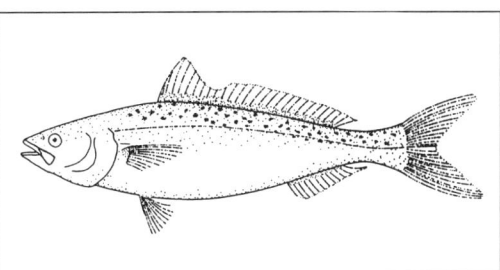

682 ARRIPIDAE RUF

SC *Arripis georgiana* (Valenciennes, 1831)
ES salmón de Georgia
DA Georgia-laks
DE Georgia-Lachs
EL σολομός της Γεωργίας
EN ruff
FR saumon australien de Georgie
IT salmone australiano
NL Georgia-zalm
PT peixe-grosa da Geórgia
FI arripi-laji
SV georgialaxabborre

684 EMMELICHTHYIDAE EMT

SC *Emmelichthyidae*
ES emelíchtidos; andoderros
DA huefisk-familien
DE
EL εμμελιχθίδες
EN rovers; bonnetfishes; ladyfishes
FR mendoles et picarels
IT emelittidi
NL rondlopers
PT peixes-rubi; charros ingleses
FI kupusuut; kupusuut-heimo
SV bahyttfiskar

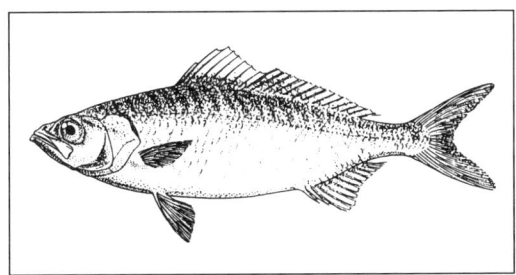

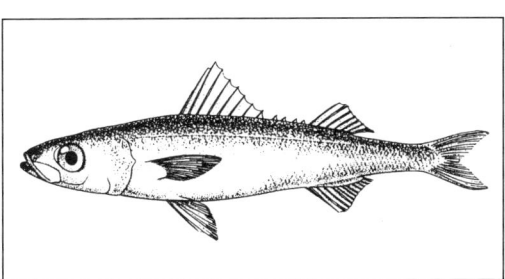

685 EMMELICHTHYIDAE　　　　EMM

SC	*Emmelichthys nitidus* (Richardson, 1845)
ES	andoderro del Cabo
DA	Kap-huefisk
DE	
EL	
EN	bonnetmouth; red-bait; Cape bonnetfish
FR	andorrève du Cap
IT	
NL	zuidelijke rondloper
PT	peixe-rubi do Índico
FI	kupusuu-laji
SV	kapbahytt

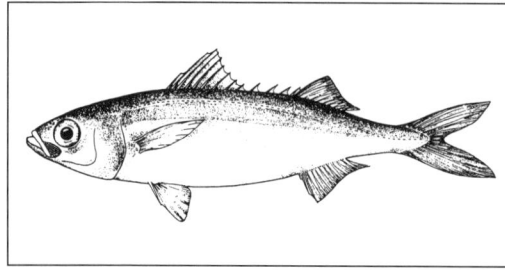

686 CAESIONIDAE　　　　FUS

SC	*Caesio* spp.
ES	fusileros
DA	musketerfisk
DE	Füsilierfische
EL	τουφεκόψαρα
EN	fusiliers
FR	vivaneaux nains; happeurs nains; vivanette
IT	pesci fucilieri
NL	fuseliers
PT	fuzileiros
FI	
SV	fysiljärfiskar

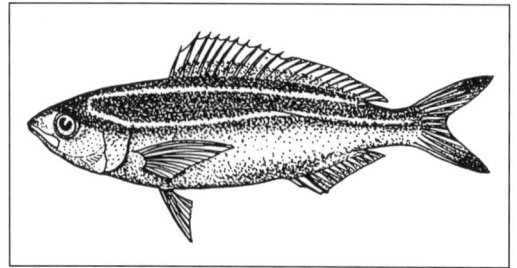

687 LUTJANIDAE　　　　SNX

SC	*Lutjanidae*
ES	lutjánidos
DA	snapper-familien
DE	Schnapper
EL	λουτιάνιδοι
EN	snappers; jobfishes
FR	lutjanidés; vivaneaux
IT	lutianidi; lutiani
NL	snappers
PT	lutjanídeos
FI	napsijat; napsijat-heimo
SV	gruntfiskar; snapperfiskar

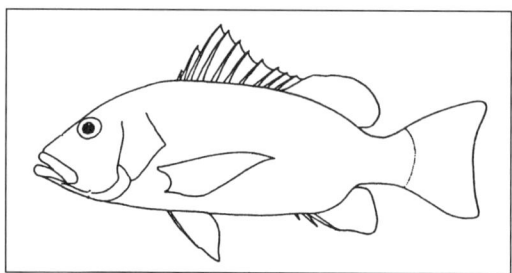

688 LUTJANIDAE　　　　RES

SC	*Lutjanus argentimaculatus* (Forsskål, 1775)
ES	pargo de mangle
DA	mangrovesnapper
DE	Mangroven-Schnapper
EL	λουτιάνος
EN	mangrove red snapper
FR	vivaneau des mangroves
IT	lutiano delle mangrovie
NL	mangrovensnapper
PT	luciano-do-mangal
FI	mangrovenapsija
SV	mangrovesnapper

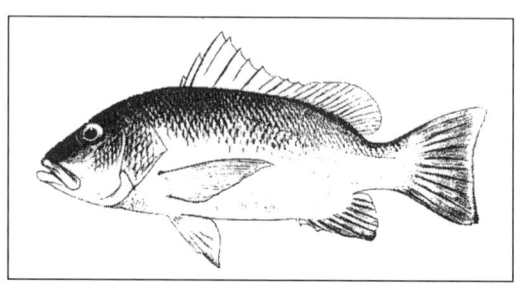

689 LUTJANIDAE   HUS

SC	*Lutjanus argentiventris* (Peters, 1869)
ES	huachinango
DA	gul snapper
DE	Gelber Schnapper
EL	λουτιάνος
EN	amarillo snapper; yellow snapper
FR	vivaneau à ventre d'argent
IT	lutiano
NL	gele snapper
PT	luciano-ventre prateado
FI	napsija-laji
SV	gul snapper

691 LUTJANIDAE   SNC

SC	*Lutjanus purpureus* (Poey, 1867); *Lutjanus aya*
ES	pargo colorado
DA	sydlig snapper
DE	Südlicher Schnapper
EL	κόκκινος λουτιάνος
EN	Southern red snapper
FR	vivaneau rouge
IT	lutiano rosso
NL	zuidelijke rode snapper
PT	luciano vermelho
FI	purppuranapsija
SV	sydlig rödsnapper

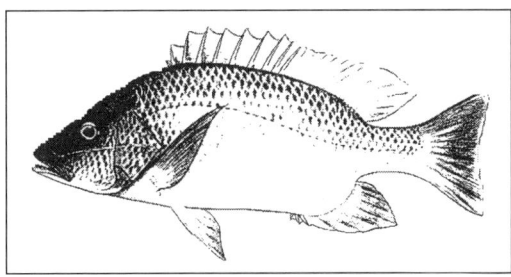

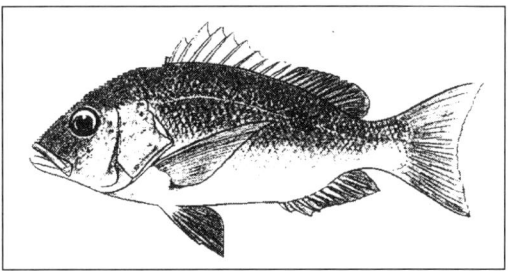

690 LUTJANIDAE   SNR

SC	*Lutjanus campechanus* (Poey, 1860)
ES	pargo del Golfo
DA	nordlig snapper
DE	Nördlicher Schnapper
EL	κόκκινος λουτιάνος
EN	red snapper; Northern red snapper
FR	vivaneau campèche
IT	lutiano rosso
NL	noordelijke rode snapper
PT	luciano do Golfo
FI	napsija-laji
SV	röd snapper

692 LUTJANIDAE   SNL

SC	*Lutjanus synagris* (Linnaeus, 1758)
ES	pargo biajaiba
DA	pletsnapper
DE	Rotschwanzschnapper
EL	ραβδολουτιάνος
EN	lane snapper
FR	vivaneau à queue rouge
IT	lutiano striato
NL	roodstaartsnapper
PT	luciano riscado
FI	juovanapsija
SV	randig snapper

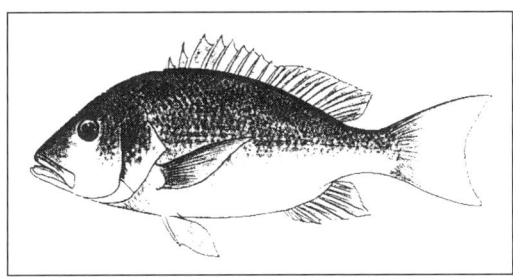

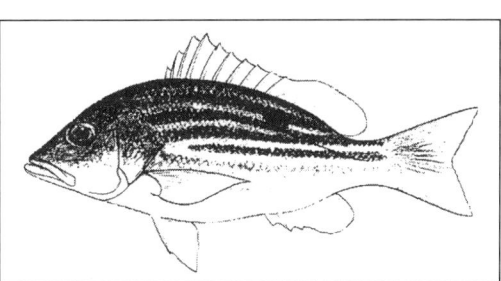

| **693** | **LUTJANIDAE** | SNA | **695** | **NEMIPTERIDAE** | THD |

SC	*Lutjanus* spp.		SC	*Nemipteridae*	
ES	pargos		ES	nemiptéridos	
DA	snapper-slægt		DA	nemipterid-familien	
DE	Schnapper		DE	Scheinschnapper	
EL	λουτιάνοι		EL	νεμιπτερίδες	
EN	snappers		EN	thread-fin breams; monocle breams; dwarf breams	
FR	vivaneaux		FR	cohana	
IT	lutiani		IT	nemipteridi	
NL	snappers		NL	vlinderbrasems	
PT	lucianos; castanholas		PT	nemipterídeos	
FI	napsija-suku		FI	viiriahvenet; viiriahvenet-heimo	
SV	snappers		SV	vimpelsnapperfiskar	

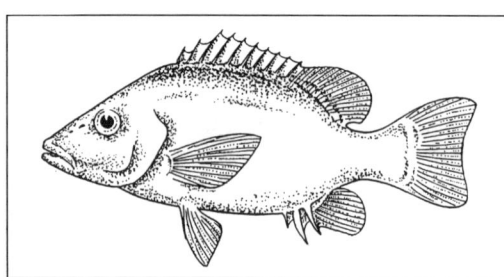 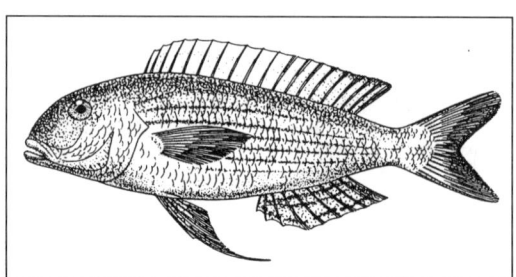

| **694** | **LUTJANIDAE** | SNY | **696** | **NEMIPTERIDAE** | THG |

SC	*Ocyurus chrysurus* (Bloch, 1791)		SC	*Nemipterus virgatus* (Houttuyn, 1782)	
ES	rabirrubia		ES	nemíptero	
DA	gulhalet snapper		DA	gylden nemipterid	
DE	Gelbschwanzschnapper		DE	Goldener Scheinschnapper	
EL	λουτιάνος με κίτρινη ουρά		EL	κιτρινονεμίπτερος	
EN	yellow-tail snapper		EN	golden thread-fin bream	
FR	vivaneau à queue jaune		FR	cohana doré	
IT	lutiano coda gialla		IT	nemiptero	
NL	geelstaartsnapper		NL	gouden vlinderbrasem	
PT	luciano-cauda-amarela		PT	falso besugo doirado	
FI	keltapyrstönapsija		FI	viiriahven-laji	
SV	gulssjärtssnapper		SV	gul vimpelsnapper	

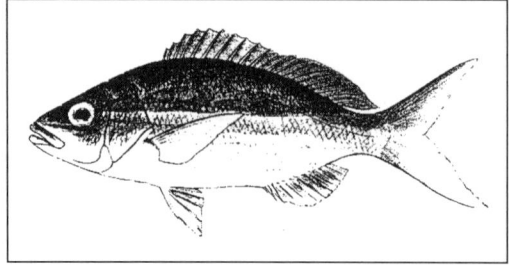 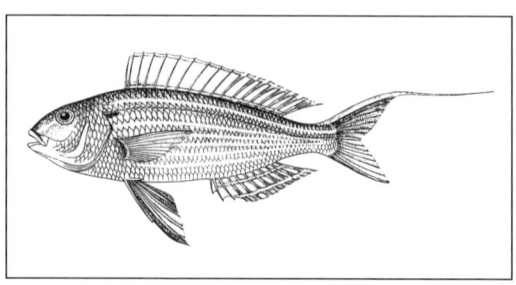

697 NEMIPTERIDAE THB

SC *Nemipterus* spp.
ES nemiptéridos
DA nemipterid-slægt
DE Scheinschnapper
EL νεμίπτεροι
EN thread-fin breams
FR cohana
IT nemipteri
NL vlinderbrasems
PT falsos besugos
FI viiriahven-suku
SV vimpelsnappers

699 LEIOGNATHIDAE PON

SC *Leiognathidae*
ES mojarras
DA ponyfisk-familien
DE Ponyfische
EL λειογναθίδες
EN ponyfishes; slipmouths
FR blanches; sapsap
IT leiognatidi
NL glipvissen
PT peixes-pónei
FI lima-ahvenet; lima-ahvenet-heimo
SV ponnyfiskar

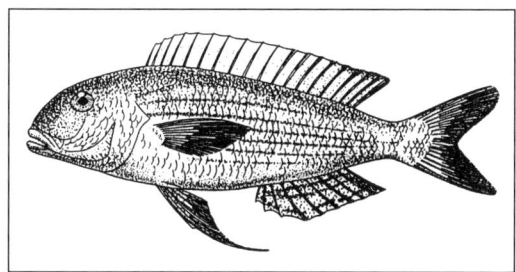

698 NEMIPTERIDAE MOB

SC *Scolopsis* spp.
ES peces luna
DA nemipterid-slægt
DE Monokel-Brassen
EL
EN monocle breams
FR mamilas
IT scoloperi
NL monocle-stekelwang
PT lunetas; peixes-luneta
FI viiriahven-suku
SV monokelsnappers

700 LEIOGNATHIDAE

SC *Leiognathus klunzingeri* (Steindachner, 1898)
ES pez poni
DA Klunzingers ponyfisk
DE Klunzingers Ponyfisch
EL σαπουνόψαρο
EN ponyfish
FR poisson visqueux; sapsap; blanches
IT pesce pony
NL Klunzingers glipvis
PT peixe-pónei
FI lima-ahven-laji
SV skäckponnyfisk

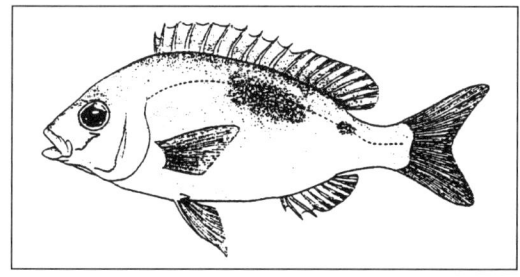

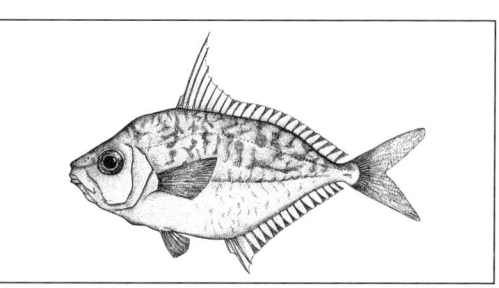

701 LEIOGNATHIDAE   POY

SC	*Leiognathus* spp.
ES	mojarras
DA	ponyfisk-slægt
DE	Ponyfische
EL	σαπουνόψαρα
EN	ponyfishes; slipmouths
FR	blanches; sapsap
IT	pesci pony
NL	glipvissen
PT	peixes-pónei
FI	lima-ahven-suku
SV	ponnyfiskar

703 HAEMULIDAE   GRB

SC	*Brachydeuterus auritus* (Valenciennes, 1831)
ES	burro ojón
DA	storøjet gryntefisk
DE	Großaugen-Angola-Meerbrasse
EL	γουργουρόψαρο ματάς
EN	bigeye grunt
FR	pelon; friture à écaille
IT	otoperca
NL	grootoogknorvis
PT	colo-colo; roncador colo-colo; roncador de olhos grandes
FI	röhkijäkala-laji
SV	glogrymta

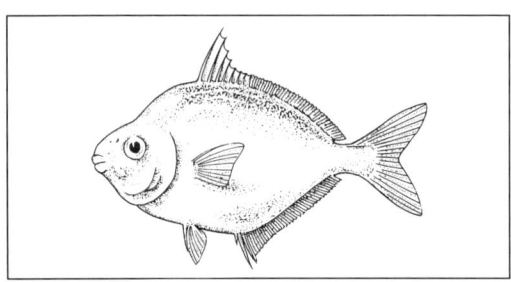

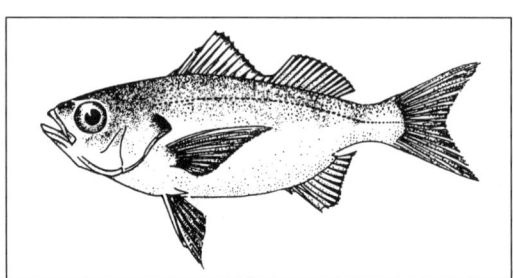

702 HAEMULIDAE

SC	*Anisotremus virginicus* (Linnaeus, 1758)
ES	dorada americana
DA	svinefisk
DE	Schweinsfisch
EL	τσιπούρα της Αμερικής
EN	porkfish
FR	daurade américaine; poisson-cochon
IT	burro della Virginia
NL	Amerikaanse knorvis
PT	roncador listado americano
FI	röhkijäkala
SV	grisfisk

704 HAEMULIDAE   BRG

SC	*Conodon nobilis* (Linnaeus, 1758)
ES	ronco canario
DA	stribet gryntefisk
DE	Gestreifte Süßlippe
EL	γουργουρόψαρο
EN	barred grunt
FR	cagna rayé
IT	grugnolo spinoso
NL	gestreepte knorvis
PT	roncador-canário
FI	röhkijäkala-laji
SV	bandgrymta

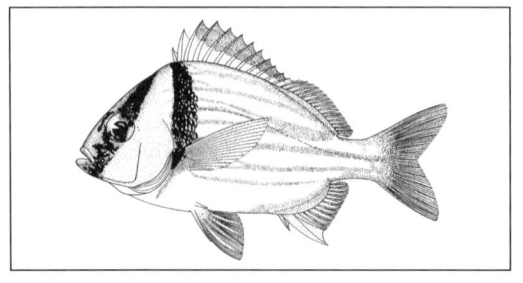

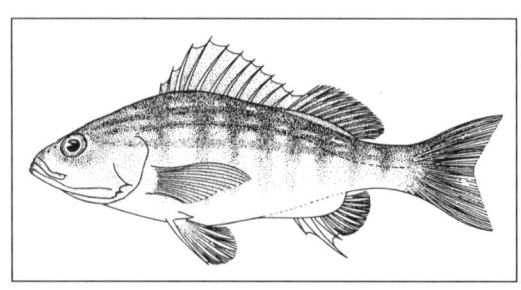

705　HAEMULIDAE　　　　　　　　　GRX

SC	*Haemulidae*
ES	roncadores
DA	gryntefisk-familien
DE	Grunzer; Süßlippen
EL	βουτυρόψαρα
EN	grunts; sweetlips
FR	pomadasydés; tambours; grondeurs; diagrammes
IT	emulidi; burri; pesci burri
NL	knorvissen
PT	roncadores
FI	murisijat; murisijat-heimo
SV	grymtor

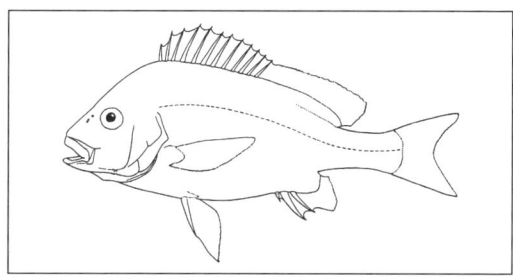

707　HAEMULIDAE　　　　　　　　　PIG

SC	*Orthopristis chrysoptera* (Linnaeus, 1766)
ES	corocoro burro
DA	gulfinnet gryntefisk
DE	Gelbflossen-Süßlippe
EL	βουτυρόψαρο
EN	pigfish
FR	goret-mule
IT	pesce burro maculato
NL	varkenvis
PT	roncador mexicano
FI	korokoro
SV	åsnegrymta

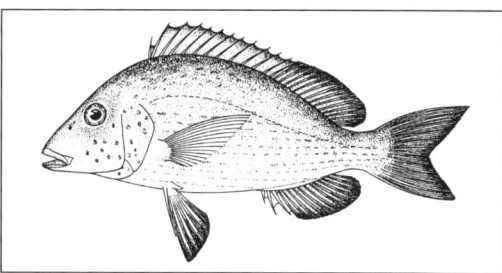

706　HAEMULIDAE　　　　　　　　　GRP

SC	*Isacia conceptionis* (Valenciennes, 1830)
ES	roncador del Pacífico sudoriental
DA	sydøstpacifisk gryntefisk
DE	Südostpazifische Süßlippe
EL	βουτυρόψαρο του Ειρηνικού
EN	Southeast Pacific grunt
FR	grondeur du Pacifique sud-est
IT	cabinza
NL	Zuidoost-Pacifische knorvis
PT	roncador do Pacífico Sudeste
FI	murisija-laji
SV	chilegrymta

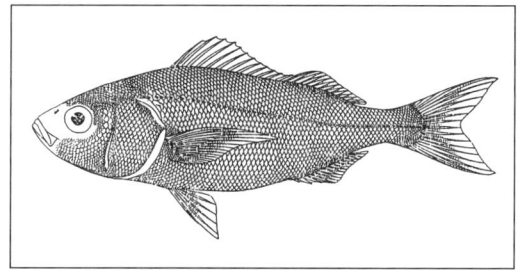

708　HAEMULIDAE　　　　　　　　　GRA

SC	*Parapristipoma octolineatum* (Valenciennes, 1833)
ES	burro listado
DA	stribet gryntefisk
DE	Afrikanische Streifensüßlippe
EL	βουτυρόψαρο της Αφρικής
EN	African striped grunt
FR	grondeur rayé; pristipome à quatre bandes
IT	pesce burro striato
NL	Afrikaanse gestreepte knorvis
PT	riscado; roncador riscado
FI	murisija-laji
SV	strimgrymta

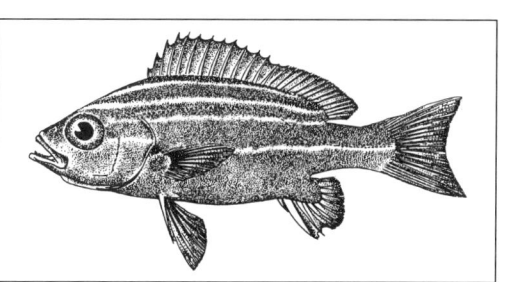

709 HAEMULIDAE   GBL

SC	*Plectorhinchus macrolepis* (Boulenger, 1899)
ES	burro labiogrueso
DA	tyklæbet gryntefisk
DE	Dicklippiger Grunzer
EL	χονδρόχειλο βουτυρόψαρο
EN	biglip grunt
FR	diagramme à grosses lèvres
IT	pesce burro
NL	diklipknorvis
PT	roncador-batata
FI	paksuhuulimurisija
SV	fläskläpp

711 HAEMULIDAE   GRL

SC	*Pomadasys argenteus* (Forsskål, 1775)
ES	roncador plateado
DA	sølv-gryntefisk
DE	Silberne Süßlippe
EL	ασημοβουτυρόψαρο
EN	lined silver grunt; silver grunt
FR	grondeur argenté
IT	grugnolo
NL	zilverknorvis
PT	roncador prateado
FI	hopeamurisija
SV	silvergrymta

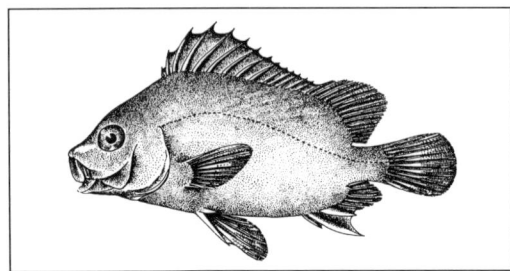

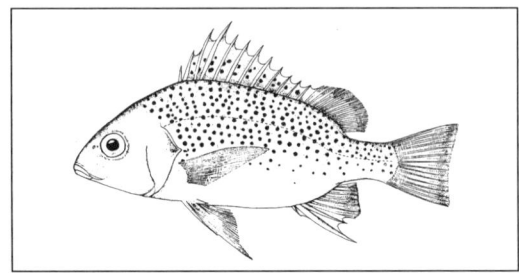

710 HAEMULIDAE   GBR

SC	*Plectorhinchus mediterraneus* (Guichenot, 1850)
ES	burro
DA	Middelhavs-gryntefisk
DE	Westmediterrane Süßlippe
EL	βουτυρόψαρο
EN	rubberlip grunt
FR	diagramme gris
IT	pesce burro
NL	Middellandse-Zeeknorvis
PT	pombo; pargo-mulato
FI	välimerenmurisija
SV	gummiläpp

712 HAEMULIDAE   BGR

SC	*Pomadasys incisus* (Bowdich, 1825)
ES	ronco mestizo
DA	bastardgryntefisk
DE	Bastard-Süßlippe
EL	ψευτοβουτυρόψαρο
EN	bastard grunt
FR	grondeur métis; croco
IT	grugnolo
NL	bastaardknorvis
PT	roncador-bravura
FI	murisija-laji
SV	blandgrymta

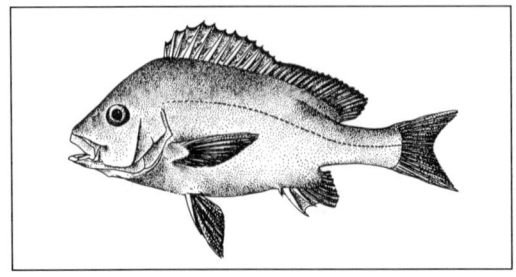

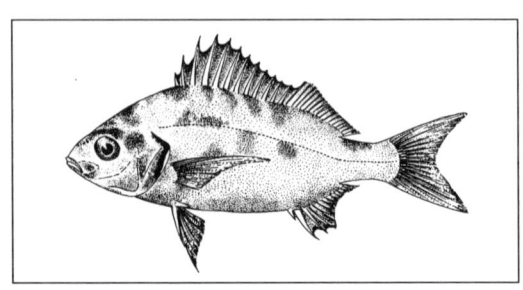

713 HAEMULIDAE BUR

SC	*Pomadasys jubelini* (Cuvier, 1830)
ES	ronco sompat
DA	sompatgryntefisk
DE	Sompat-Süßlippe
EL	βουτυρόψαρο
EN	Atlantic spotted grunt; sompat grunt
FR	sompat; carpe blanche
IT	grugnolo
NL	gespikkelde knorvis
PT	roncador de pintas
FI	murisija-laji
SV	prickgrymta

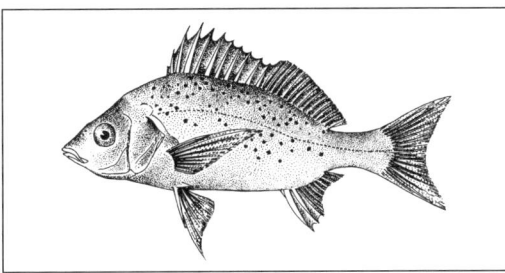

714 HAEMULIDAE

SC	*Pomadasys* spp.
ES	roncadores; burros
DA	gryntefisk-slægt
DE	Süßlippen
EL	βουτυρόψαρα
EN	grunts
FR	grondeurs; tambours
IT	grugnoli
NL	knorvissen
PT	roncadores
FI	murisija-suku
SV	sötläppar

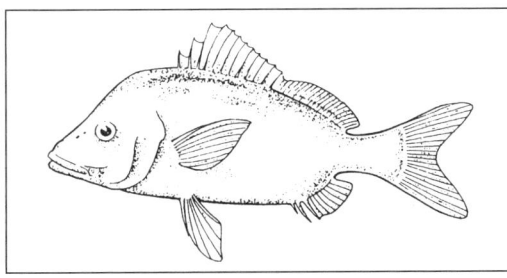

715 SCIAENIDAE CDX

SC	*Sciaenidae*
ES	esciénidos; corvinas; verrugatos
DA	trommefisk-familien; ørnefisk-familien
DE	Umberfische
EL	σιαινίδες
EN	croakers; drums; meagres
FR	sciaenidés; corbs; maigres; ombrines; courbines
IT	scienidi; ombrine; corvine
NL	ombervissen
PT	escienídeos; corvinas e afins
FI	rumpukalat; rumpukalat-heimo
SV	havsgösfiskar; vekor; vekfiskar

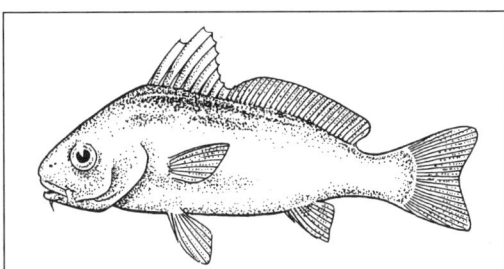

716 SCIAENIDAE

SC	*Aplodinotus grunniens* (Rafinesque, 1819)
ES	umbrina de agua dulce
DA	flodtrommefisk
DE	Süßwasser-Trommelfisch
EL	μυλοκόπι του γλυκού νερού
EN	freshwater drum
FR	malachigan d'eau douce
IT	ombrina d'acqua dolce
NL	zoetwatertrommelvis
PT	corvina de água doce
FI	jokirumpukala
SV	sötveka

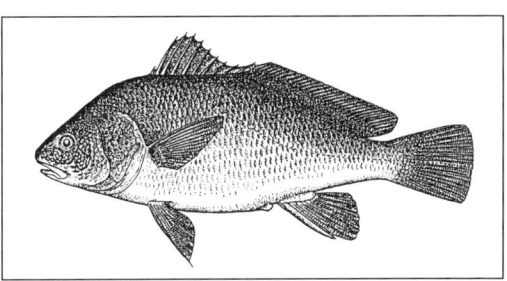

179

717 SCIAENIDAE KOB

SC	*Argyrosomus hololepidotus* (Lacepède, 1801)
ES	corvina africana
DA	afrikansk ørnefisk
DE	Afrikanischer Adlerfisch
EL	μαγιάτικο
EN	kob; Southern meagre
FR	courbine dorée; maigre doré
IT	bocca d'oro
NL	kob
PT	corvina africana
FI	rumpukala-laji
SV	afrikansk havsgös

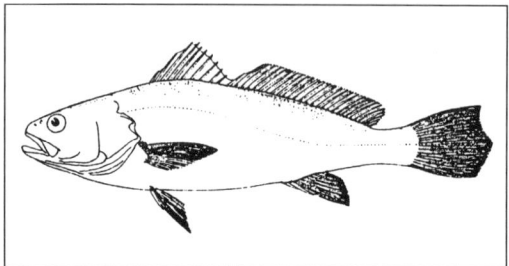

719 SCIAENIDAE AWE

SC	*Atractoscion aequidens* (Cuvier, 1830); *Atractoscion atelodus*
ES	corvinata prieta
DA	afrikansk trommefisk
DE	Afrikanischer Umberfisch
EL	μυλοκόπι της Αφρικής
EN	African weakfish; geelback croaker
FR	téraglin
IT	ombrina bianca; tiraghlin
NL	geelbek
PT	corvina de boca amarela; corvina-teraglim
FI	kapinrumpukala
SV	afrikansk vekfisk

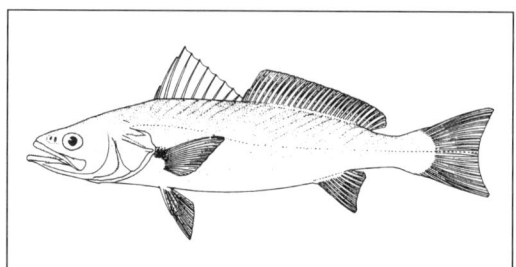

718 SCIAENIDAE MGR

SC	*Argyrosomus regius* (Asso, 1801); *Sciaena aquila*; *Argyrosoma regium*
ES	corvina; corbina; corvinato; pardilleja
DA	almindelig ørnefisk
DE	Adlerfisch; Adlerlachs; Schattenfisch
EL	μαγιάτικο
EN	meagre; maigre; croaker; shadefish
FR	courbine; maigre; haut-bar; aigle; maigre commun
IT	bocca d'oro; ombrina bocca d'oro
NL	ombervis
PT	corvina legítima; corvina
FI	kotkakala
SV	havsgös

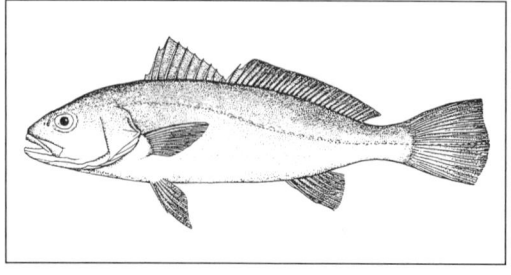

720 SCIAENIDAE

SC	*Atractoscion nobilis* (Ayres, 1860); *Cynoscion nobilis*
ES	corvinata blanca
DA	hvid trommefisk
DE	Prächtiger Umberfisch
EL	λευκομυλοκόπι της Καλιφόρνιας
EN	white weakfish; white seabass
FR	acoupa blanc
IT	ombrina bianca
NL	witte ombervis
PT	corvinata branca
FI	rumpukala-laji
SV	vitveka

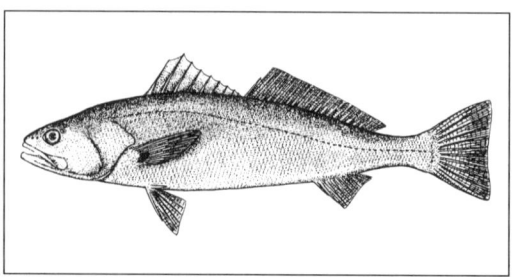

721	**SCIAENIDAE**	CRL

SC	*Atrobucca nibe* (Jordan and Thompson, 1911)
ES	corvina negra
DA	sortmundet trommefisk
DE	Schwarzmaul-Umberfisch
EL	μαυρόστομος σκιός
EN	black-mouth croaker; black croaker
FR	maigre noir
IT	corvina boccanegra
NL	langvinkob
PT	roncadeira negra
FI	rumpukala-laji
SV	kolgapskväkare

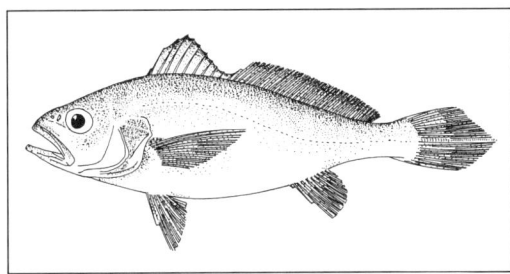

722	**SCIAENIDAE**	LYC

SC	*Collichthys crocea* (Richardson, 1846); *Pseudosciaena crocea*
ES	corvina japonesa
DA	gul trommefisk
DE	Asiatischer Gelber Umberfisch
EL	μεγαλόφθαλμος σκιός
EN	large yellow croaker
FR	tambour à gros yeux
IT	corvina giapponese
NL	gele trommelvis
PT	roncadeira amarela maior
FI	rumpukala-laji
SV	gulveka

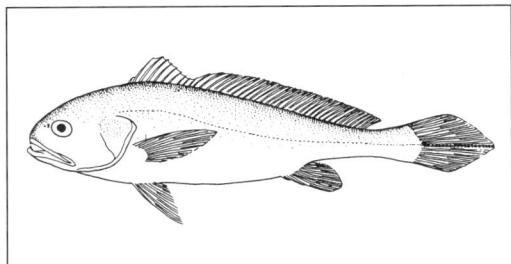

723	**SCIAENIDAE**	WEP

SC	*Cynoscion analis* (Jenyns, 1842)
ES	corvinata ayanque
DA	peruansk trommefisk
DE	Peruanischer Umberfisch
EL	μυλοκόπι του Περού
EN	Peruvian weakfish
FR	acoupa du Pérou
IT	ombrina dentata
NL	Peruaanse ombervis
PT	corvinata do Peru
FI	perunveltto
SV	peruveka

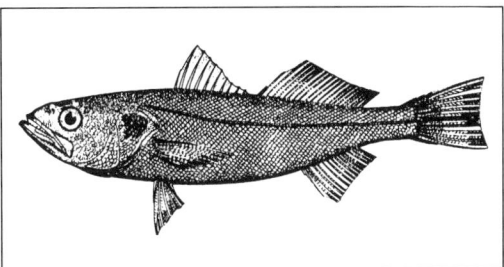

724	**SCIAENIDAE**	WEM

SC	*Cynoscion macdonaldi* (Gilbert, 1890); *Totoaba macdonaldi*
ES	corvinata totoaba
DA	totuava
DE	Macdonalds Umberfische; Riesen-Totoaba
EL	μυλοκόπι Μακ Ντόναλντ
EN	Macdonald weakfish
FR	acoupa de Macdonald
IT	totoaba; acoupa di Macdonald
NL	Macdonalds ombervis
PT	corvinata gigante; corvinata-totuava
FI	isoveltto
SV	kalifornisk vekfisk; jätteveka

DRAWING NOT AVAILABLE

725 SCIAENIDAE SWF

SC	*Cynoscion nebulosus* (Cuvier, 1830)
ES	corvinata pintada
DA	plettet trommefisk
DE	Gefleckter Umberfisch
EL	στικτομυλοκόπι
EN	spotted sea trout; spotted weakfish
FR	acoupa pintade
IT	ombrina dentata
NL	gevlekte ombervis
PT	corvinata pintada
FI	pilkkuveltto
SV	prickveka

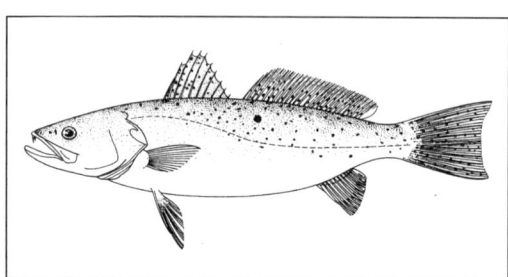

727 SCIAENIDAE WKB

SC	*Cynoscion steindachneri* (Jordan, 1889)
ES	corvinata pescada
DA	småtandet trommefisk
DE	Steindachners Umberfisch
EL	μυλοκόπι μικρόδοντο
EN	small-tooth weakfish
FR	acoupa à petites dents; acoupa de Steindachner; coupa tident
IT	ombrina dentata
NL	Steindachners ombervis
PT	corvinata-pescada
FI	veltto-laji
SV	småtandad veka

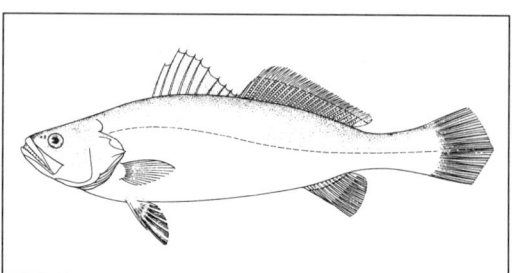

726 SCIAENIDAE STG

SC	*Cynoscion regalis* (Bloch and Schneider, 1801)
ES	corvinata real
DA	kongetrommefisk
DE	Königs-Corvina
EL	βασιλικό μυλοκόπι
EN	weakfish; squeteague; grey weakfish
FR	acoupa royal
IT	ombrina dentata
NL	koningsombervis
PT	corvinata real
FI	veltto
SV	vekfisk; kungsveka

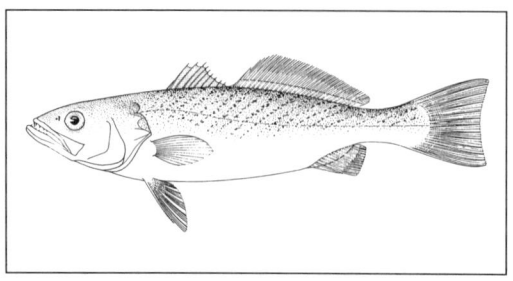

728 SCIAENIDAE WKS

SC	*Cynoscion striatus* (Cuvier, 1830)
ES	corvinata pescadilla
DA	stribet trommefisk
DE	Gestreifter Umberfisch
EL	γραμμωτό μυλοκόπι
EN	striped weakfish
FR	acoupa rayé
IT	ombrina
NL	gestreepte ombervis
PT	corvinata riscada
FI	juovaveltto
SV	strimveka

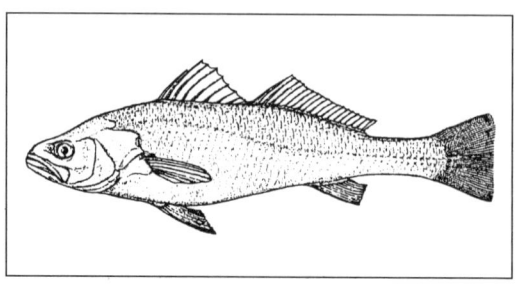

729 SCIAENIDAE WKX

SC	*Cynoscion* spp.
ES	corvinatas
DA	trommefisk-slægt
DE	Umberfische
EL	μυλοκόπια
EN	weakfishes
FR	acoupas; sciènes
IT	ombrine dentate
NL	ombervissen
PT	corvinatas
FI	veltto-suku
SV	vekfiskar

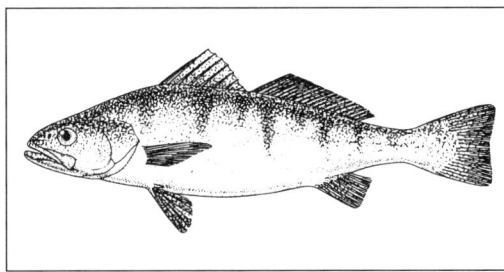

730 SCIAENIDAE KIC

SC	*Genyonemus lineatus* (Ayres, 1855)
ES	umbrina del Pacífico
DA	Stillehavs-trommefisk
DE	Weißer Kalifornia-Umberfisch
EL	άσπρο μυλοκόπι
EN	white croaker
FR	sciène du Pacifique; tambour du Pacifique
IT	ombrina
NL	Californische witte ombervis
PT	rabeta branca
FI	rumpukala-laji
SV	vitkväkare

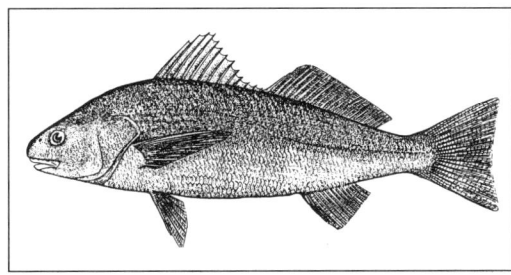

731 SCIAENIDAE SPT

SC	*Leiostomus xanthurus* (Lacépède, 1803)
ES	verrugato croca
DA	punkttrommefisk
DE	Punkt-Umberfisch
EL	γραμμωτός σκιός
EN	spot; spot croaker
FR	tambour croca
IT	corvina striata
NL	puntombervis
PT	roncadeira de pinta
FI	rumpukala-laji
SV	slätkväkare

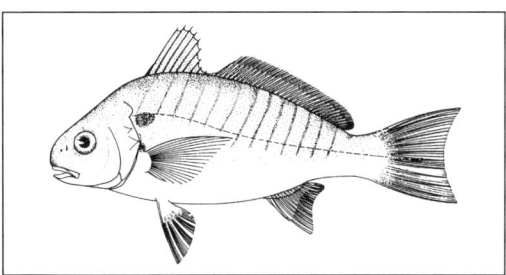

732 SCIAENIDAE WKK

SC	*Macrodon ancylodon* (Bloch and Schneider, 1801)
ES	pescadilla real
DA	sydamerikansk trommefisk
DE	Südamerikanischer Königs-Umberfisch
EL	βασιλικός σκιός
EN	king weakfish
FR	acoupa chasseur
IT	corvina lupo
NL	Zuid-Amerikaanse koningsombervis
PT	rabeta caçadora
FI	rumpukala-laji
SV	kungsveka

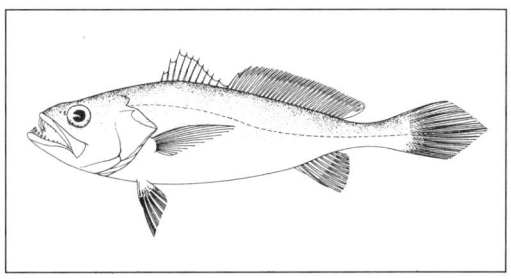

733	**SCIAENIDAE**	KGB

SC	*Menticirrhus americanus* (Linnaeus, 1758)
ES	lambe caletero
DA	amerikansk trommefisk
DE	Amerikanischer Umberfisch
EL	αμερικάνικο μυλοκόπι
EN	Southern kingfish; Southern king-croaker
FR	bourrugue de crique
IT	ombrina americana
NL	Amerikaanse koningsombervis
PT	cangueira-cachorro
FI	rumpukala-laji
SV	slät kungskväkare

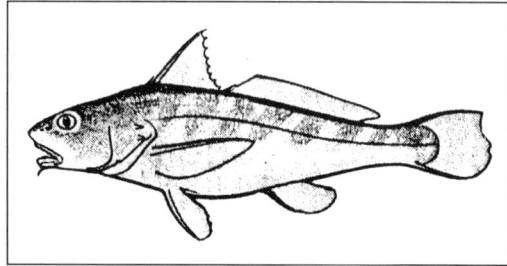

734	**SCIAENIDAE**	KGG

SC	*Menticirrhus littoralis* (Holbrook, 1855)
ES	lambe verrugato
DA	Golf-trommefisk
DE	Golf-Umberfisch
EL	μυλοκόπι του Κόλπου
EN	Gulf kingcroaker
FR	bourrugue du golfe
IT	ombrina americana
NL	Golf-koningsombervis
PT	cangueira do Golfo
FI	rumpukala-laji
SV	bränningskväkare

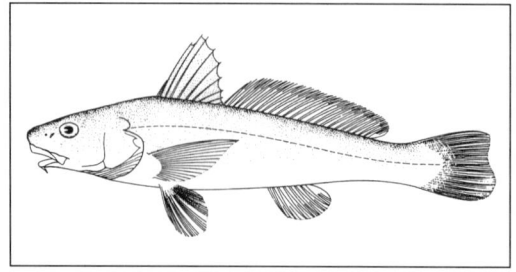

735	**SCIAENIDAE**	KGF

SC	*Menticirrhus saxatilis* (Bloch and Schneider, 1801)
ES	lambe zorro
DA	nordlig trommefisk
DE	Königs-Umberfisch
EL	μυλοκόπι του Βορρά
EN	Northern kingfish
FR	bourrugue-renard; ombrine royale
IT	ombrina americana
NL	koningsombervis
PT	cangueira-zorra
FI	pohjanrumpukala
SV	randig kungskväkare

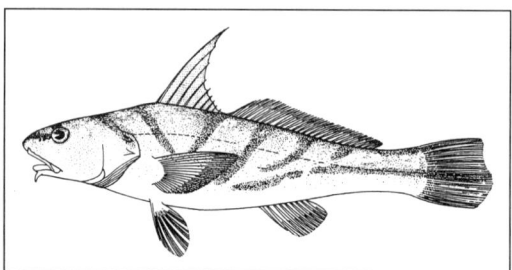

736	**SCIAENIDAE**	KIX

SC	*Menticirrhus* spp.
ES	lambes
DA	trommefisk-slægt
DE	Stumme Umberfische
EL	αμερικάνικο μυλοκόπι
EN	king-croakers
FR	bourrugues
IT	ombrine americane
NL	koningsombervissen
PT	cangueiras
FI	rumpukala-suku
SV	kungskväkare

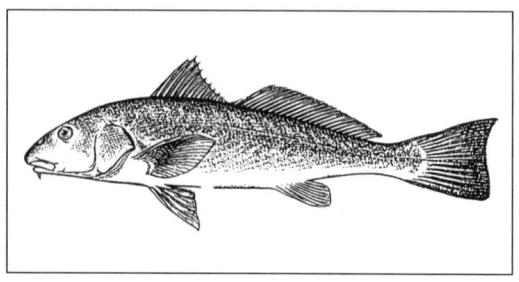

737	**SCIAENIDAE**	CKM		739	**SCIAENIDAE**	CRX

SC	*Micropogonias furnieri* (Desmarest, 1823)
ES	corvinón rayado
DA	hvidmundet trommefisk
DE	Weißmaul-Umberfisch
EL	λευκόστομο μυλοκόπι
EN	white-mouth croaker
FR	tambour rayé
IT	ombrina
NL	witmondombervis
PT	rabeta marisqueira
FI	rumpukala-laji
SV	vitmunnad kväkare

SC	*Micropogonias* spp.
ES	esciénidos
DA	trommefisk-slægt
DE	Umberfische
EL	μυλοκόπι του Ειρηνικού
EN	croakers
FR	tambours
IT	ombrine
NL	ombervissen
PT	rabetas
FI	rumpukala-suku
SV	kväkarfiskar

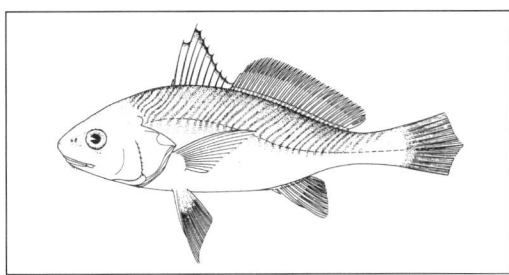
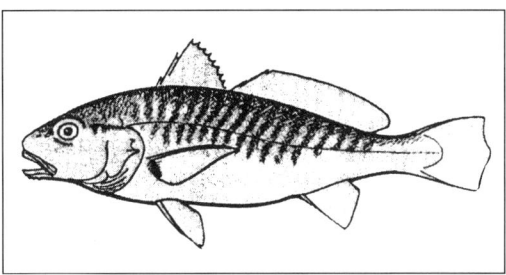

738	**SCIAENIDAE**	CKA		740	**SCIAENIDAE**	YED

SC	*Micropogonias undulatus* (Linnaeus, 1766)
ES	corvinón brasileño
DA	atlantisk trommefisk
DE	Atlantischer Umberfisch
EL	μυλοκόπι της Βραζιλίας
EN	Atlantic croaker
FR	tambour du Brésil
IT	ombrina
NL	Atlantische ombervis
PT	rabeta brasileira
FI	aaltorumpukala
SV	stubbkväkare

SC	*Nibea albiflora* (Richardson, 1846)
ES	corvina amarilla
DA	hvidblomstret trommefisk
DE	Weißblumen-Umberfisch
EL	μυλοκόπι της Ιαπωνίας
EN	white flower croaker; yellow drum
FR	tambour jaune
IT	ombrina giapponese
NL	gele ombervis
PT	roncadeira amarela
FI	rumpukala-laji
SV	blomnibe

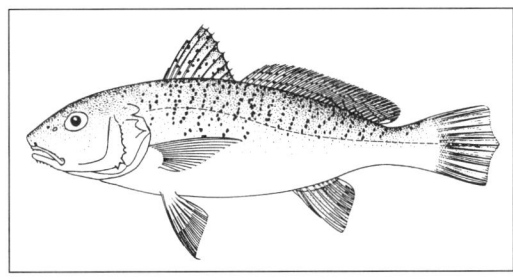
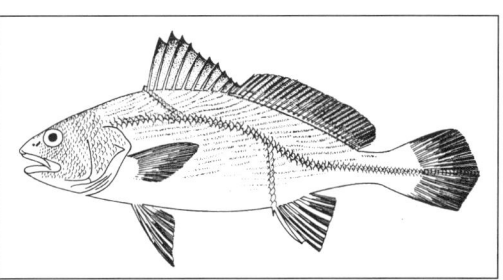

741 SCIAENIDAE   HOC

SC	*Nibea mitsukurii* (Jordan and Snyder, 1901)
ES	corvina honnibe
DA	japansk trommefisk
DE	Mitsukuri-Umberfisch
EL	μυλοκόπι της Ιαπωνίας
EN	honnibe croaker; yellow drum
FR	tambour honnibe; tambour du Japon
IT	ombrina giapponese
NL	Japanse ombervis
PT	roncadeira japonesa
FI	rumpukala-laji
SV	nibe

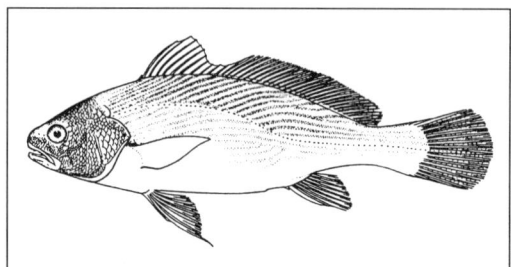

742 SCIAENIDAE   LKR

SC	*Otolithes ruber* (Bloch and Schneider, 1801)
ES	corvina tigre
DA	stortandet trommefisk
DE	Hundszahn-Umberfisch
EL	μυλοκόπι τίγρης
EN	long-tooth croaker; tiger-tooth croaker
FR	otolithe tigre; grande verrue tigrée
IT	ombrina tigre
NL	tijgerombervis
PT	rainha dentuda
FI	rumpukala-laji
SV	tigerhandveka

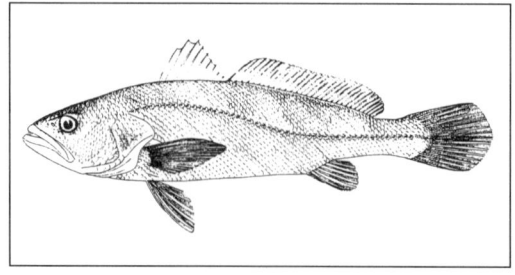

743 SCIAENIDAE   PDR

SC	*Paralonchurus peruanus* (Steindachner, 1875)
ES	coco
DA	peruansk trommefisk
DE	Coco-Umberfisch
EL	μυλοκόπι του Περού
EN	Peruvian drum
FR	tambour du Pérou
IT	ombrina peruviana
NL	Peruaanse trommelvis
PT	roncadeira do Peru
FI	perunrumpukala
SV	peruansk trumfisk

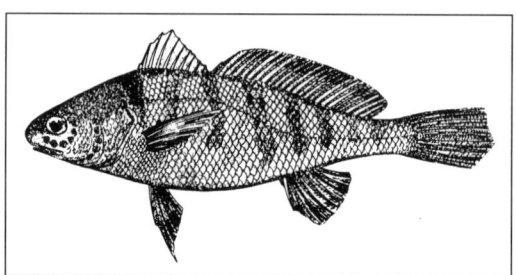

744 SCIAENIDAE   CRV

SC	*Pennahia argentata* (Houttuyn, 1782)
ES	corvina plateada
DA	sølvtrommefisk
DE	Silberne Pennahia
EL	ασημοσκιός
EN	silver croaker
FR	maigre argenté
IT	corvina argentea
NL	zilvertrommelvis
PT	roncadeira prateada
FI	hopearumpukala
SV	silverkväkare

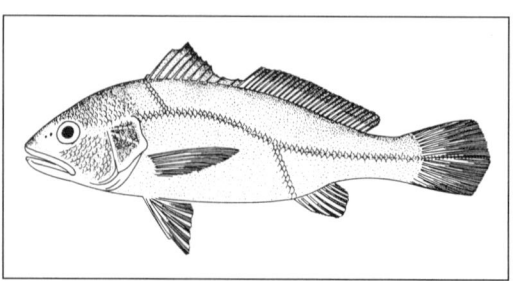

745 SCIAENIDAE BDM

SC	*Pogonias cromis* (Linnaeus, 1766)
ES	corvinón negro
DA	sort trommefisk
DE	Trommelfisch
EL	μαύρο μυλοκόπι
EN	black drum
FR	grand tambour; grondeur noir
IT	ombrina nera
NL	zwarte trommelvis
PT	corvinão negro
FI	mustarumpukala
SV	svart trumfisk

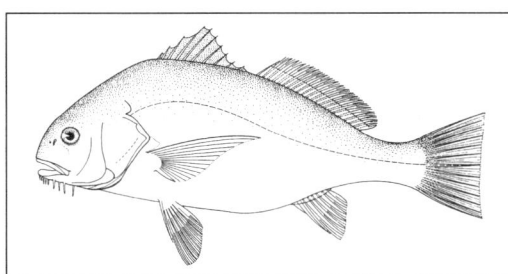

747 SCIAENIDAE LSY

SC	*Pseudosciaena polyactis* (Bleeker, 1877)
ES	verrugato de Japón
DA	lille japansk trommefisk
DE	Kleine Gelbcorvina
EL	γιαπωνέζικος σκιός
EN	lesser yellow croaker
FR	fausse courbine du Japon
IT	corvina giapponese
NL	kleine Japanse trommelvis
PT	roncadeira amarela menor
FI	rumpukala-laji
SV	orange kväkare

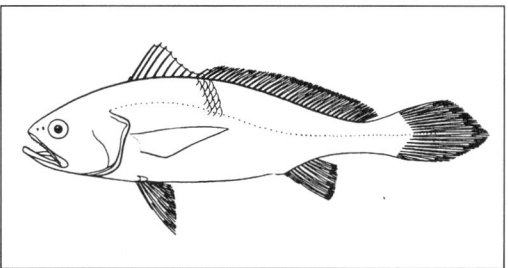

746 SCIAENIDAE CRY

SC	*Pseudosciaena manchurica* (Jordan and Thompson, 1911)
ES	verrugato de Manchuria
DA	manchurisk trommefisk
DE	Mandschurei-Corvina
EL	κίτρινος σκιός
EN	yellow croaker
FR	fausse courbine; courbine de Mandchourie
IT	corvina giapponese
NL	Mandschoeria-trommelvis
PT	roncadeira da Manchúria
FI	mantsurianrumpukala
SV	manchurisk kväkare

DRAWING NOT AVAILABLE

748 SCIAENIDAE CKL

SC	*Pseudotolithus brachygnatus* (Bleeker, 1863)
ES	corvina reina
DA	dronningetrommefisk
DE	Kurzkiefer-Umberfisch
EL	μυλοκόπι γκάπο
EN	law croaker
FR	otolithe gabo
IT	ombrina
NL	kortkaak-ombervis
PT	rainha-de-lei
FI	rumpukala-laji
SV	drottningkväkare

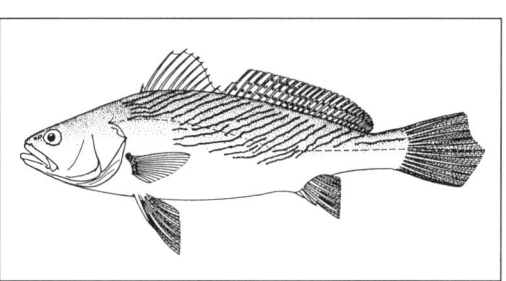

749	**SCIAENIDAE**	PSE

SC *Pseudotolithus elongatus* (Bowdich, 1825)
ES corvina bobo
DA bobo-trommefisk
DE Bobo-Umberfisch
EL μυλοκόπι μπόμπο
EN bobo croaker
FR otolithe bossu
IT ombrina
NL bobo-ombervis
PT rainha-bobo
FI rumpukala-laji
SV bobokväkare

751	**SCIAENIDAE**	PTY

SC *Pseudotolithus typus* (Bleeker, 1863)
ES corvina bosoro
DA langhovedet trommefisk
DE Langkopf-Umberfisch
EL μυλοκόπι
EN longneck croaker
FR otolithe nanka; otolithe commun
IT ombrina
NL langnek-ombervis
PT rainha branca
FI rumpukala-laji
SV långhalskväkare

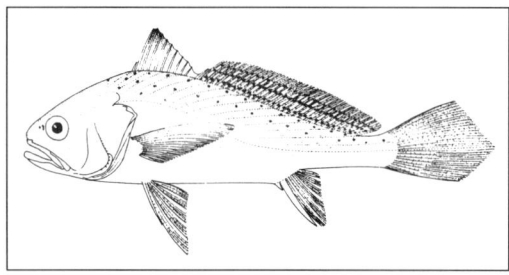

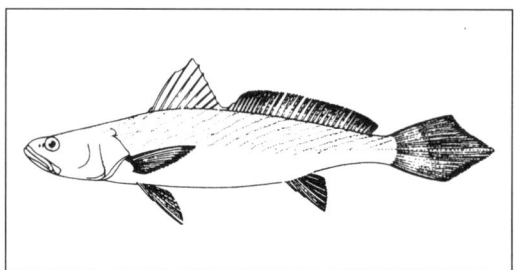

750	**SCIAENIDAE**	PSS

SC *Pseudotolithus senegalensis* (Valenciennes, 1833)
ES corvina casava
DA Senegal-trommefisk
DE Senegal-Umberfisch
EL λευκό μυλοκόπι
EN cassava croaker
FR otolithe du Sénégal
IT ombrina bianca; ombrina
NL Senegalese ombervis
PT rainha-senegal
FI senegalinrumpukala
SV kassavakväkare

752	**SCIAENIDAE**	CKW

SC *Pseudotolithus* spp.
ES corvinas
DA trommefisk-slægt
DE Umberfische
EL μυλοκόπι της Δυτικής Αφρικής
EN West African croakers
FR capitaines de mer; otolithes
IT ombrine
NL ombervissen
PT rainhas; corvinas-rainha
FI rumpukala-suku
SV kväkare

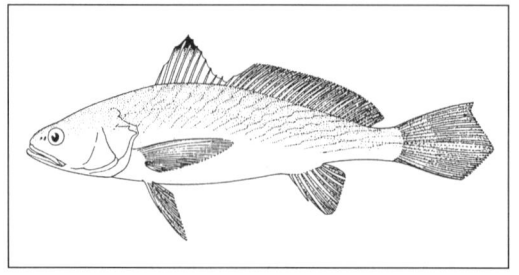

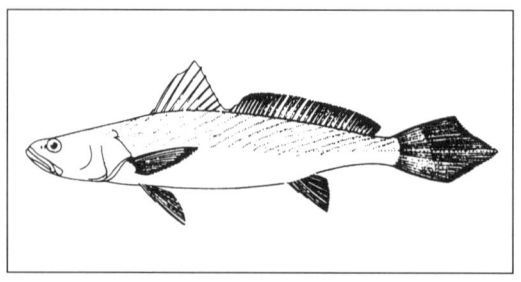

753	**SCIAENIDAE**	DRS

SC	*Pteroscion peli* (Bleeker, 1863)
ES	rabeta
DA	boe-trommefisk
DE	Boe-Umberfisch
EL	γελαδοσκιός
EN	boe drum
FR	friture
IT	corvina tonda
NL	boe-ombervis
PT	rabeta africana
FI	rumpukala-laji
SV	knubbig trumfisk

755	**SCIAENIDAE**	CBM

SC	*Sciaena umbra* (Linnaeus, 1758); *Corvina nigra*
ES	corvallo
DA	brun ørnefisk
DE	Meerrabe; Seerabe
EL	σκιός
EN	brown meagre
FR	corb commun; corb; corbeau; loup des roches
IT	corvo; corvina
NL	zeeraaf
PT	roncadeira preta
FI	rumpukala
SV	brunveka

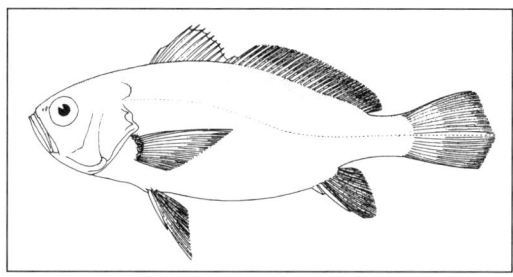

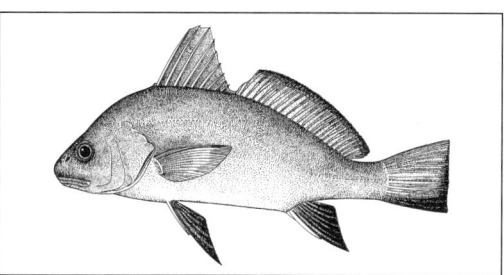

754	**SCIAENIDAE**	MUY

SC	*Sciaena antarctica* (Castelnau, 1872)
ES	verrugato austral
DA	australsk trommefisk
DE	Antarktischer Umberfisch
EL	σκιός της Ανταρκτικής
EN	mulloway
FR	courbine australe
IT	corvina australe
NL	Antarctische ombervis
PT	roncadeira austral
FI	rumpukala-laji
SV	sydveka

756	**SCIAENIDAE**	DRU

SC	*Sciaena* spp.
ES	corvinas
DA	trommefisk-slægt
DE	Umberfische
EL	σκιός
EN	drums
FR	courbines; maigres; ombrines
IT	corvine
NL	ombervissen
PT	roncadeiras
FI	rumpukala-suku
SV	trumfiskar

DRAWING NOT AVAILABLE

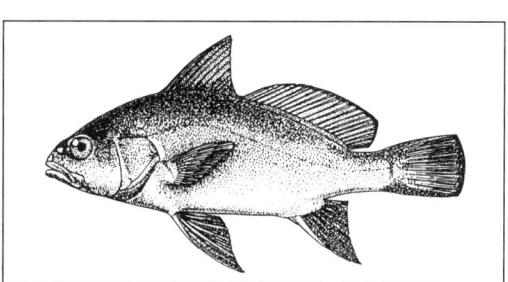

| **757** | **SCIAENIDAE** | RDM | | **759** | **SCIAENIDAE** | COB |

757 SCIAENIDAE   RDM

SC   *Sciaenops ocellatus* (Linnaeus, 1766)
ES   corvinón ocelado
DA   rød trommefisk
DE   Augenfleck-Umberfisch
EL   στικτομυλοκόπι
EN   red drum
FR   tambour rouge
IT   ombrina ocellata
NL   rode ombervis
PT   corvinão-de-pintas
FI   punarumpukala
SV   röd trumfisk

759 SCIAENIDAE   COB

SC   *Umbrina cirrosa* (Linnaeus, 1758)
ES   verrugato; verrugato común; verrugato de piedra;
     corvinato; berruguete
DA   almindelig trommefisk
DE   Umberfisch; Schattenfisch
EL   μυλοκόπι· δόλεμος
EN   corb; croaker; shi drum
FR   ombrine; ombrine commune; ombrine côtière
IT   ombrina
NL   Shi-ombervis
PT   calafate-de-riscas
FI   partarumpukala
SV   skuggfisk

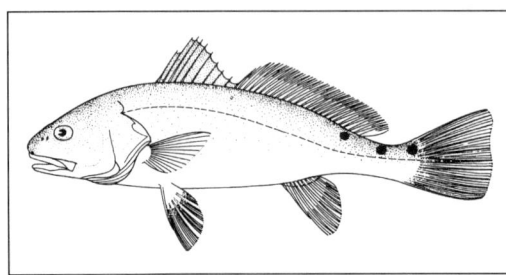

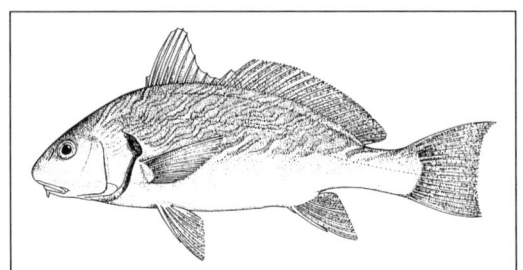

758 SCIAENIDAE   CKY

SC   *Umbrina canosai* (Berg, 1895)
ES   verrugato pargo
DA   argentinsk trommefisk
DE   Argentina-Schattenfisch
EL   μυλοκόπι της Αργεντινής
EN   Argentine croaker
FR   ombrine d'Argentine
IT   ombrina d'Argentina
NL   Argentijnse ombervis
PT   calafate da Argentina
FI   argentiinanrumpukala
SV   argentinsk skuggfisk

760 SCIAENIDAE   UBS

SC   *Umbrina* spp.
ES   verrugatos
DA   trommefisk-slægt
DE   Schattenfische
EL   μυλοκόπι
EN   drums
FR   ombrines
IT   ombrine
NL   ombervissen
PT   calafates
FI   rumpukala-suku
SV   skuggfiskar

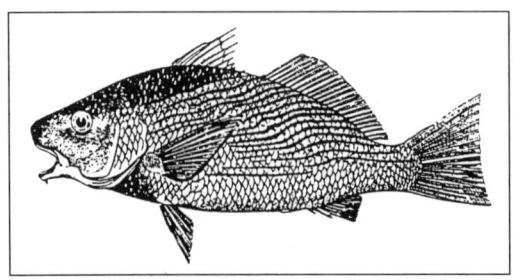

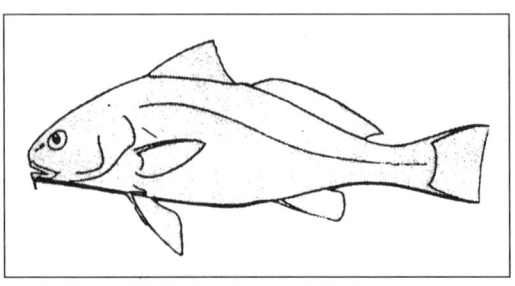

761 LETHRINIDAE LBR

SC	*Gymnocranius* spp.
ES	dentones de ojos grandes
DA	lethrinid-slægt
DE	Imperatorfische; Großaugen-Brassen
EL	ψάρια αυτοκράτορες
EN	large-eye breams
FR	empereurs à gros yeux
IT	pesci imperatore
NL	kaalneuzen
PT	ronquinhas
FI	putsari-suku
SV	revbleckor

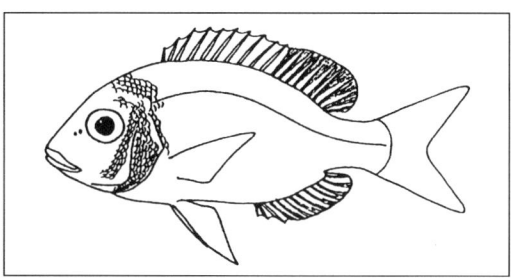

762 LETHRINIDAE EMP

SC	*Lethrinidae*
ES	letrínidos
DA	lethrinid-familien
DE	Imperatorfische
EL	λυθρινίδες
EN	emperors; scavengers
FR	lethrinidés; empereurs
IT	pesci imperatore; letrinidi
NL	keizers
PT	passarinhos e ronquinhas
FI	putsarit; putsarit-heimo
SV	revbleckefiskar

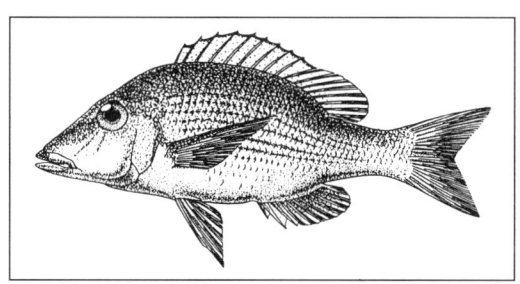

763 SPARIDAE SBX

SC	*Sparidae*
ES	espáridos
DA	havrude-familien; blankesten-familien
DE	Meerbrassen
EL	σπαρίδες
EN	sea breams; porgies
FR	dorades; daurades
IT	sparidi
NL	zeebrasems
PT	esparídeos
FI	hammasahvenet; hammasahvenet-heimo
SV	havsrudefiskar

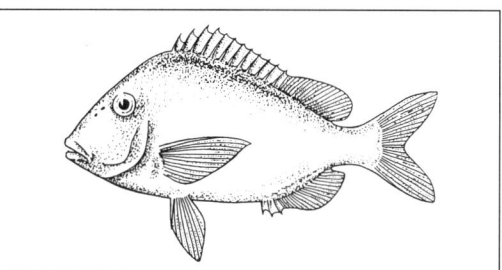

764 SPARIDAE SPH

SC	*Archosargus probatocephalus* (Walbaum, 1792)
ES	sargo chopa
DA	fårehovedhavrude
DE	Schafskopf-Brasse
EL	αμερικάνικος σαργός
EN	sheepshead
FR	rondeau mouton
IT	sarago americano
NL	schaapskop-zeebrasem
PT	sargo-choupa
FI	hammasahven-laji
SV	fårhuvudfisk

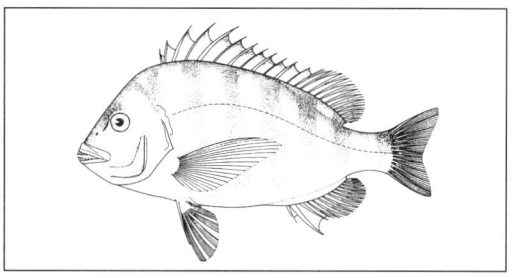

765   SPARIDAE   SLF

SC   *Argyrozona argyrozona* (Valenciennes, 1830)
ES   dentón carpintero
DA   tømrerhavrude
DE   Tischler-Seebrasse
EL   αργυροσυναγρίδα
EN   carpenter; carpenter seabream
FR   denté charpentier
IT   dentice argentato
NL   kapenaar
PT   carpinteiro; dentão-carpinteiro
FI   hammasahven-laji
SV   snickarblecka

767   SPARIDAE   PRG

SC   *Calamus* spp.
ES   plumas
DA   havrude-slægt
DE   Porgys
EL   φαγγριά
EN   porgies
FR   daubenets; béliers
IT   pagri dei Caraibi
NL   porgy's
PT   plumas; pargos azuis
FI   hammasahven-suku
SV   porgier

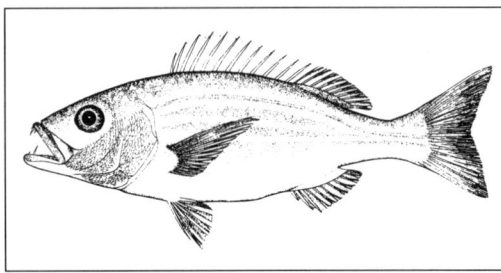

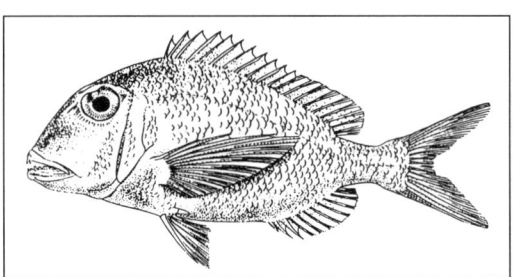

766   SPARIDAE   BOG

SC   *Boops boops* (Linnaeus, 1758); *Box vulgaris*; *Box boops*
ES   boga; boga de mar; boba; catalufa
DA   okseøjefisk
DE   Gelbstriemen; Ochsenauge
EL   γόπα
EN   bogue; ox-eye
FR   bogue
IT   boga; boba; vopa
NL   bokvis
PT   boga do mar; boga
FI   boga
SV   oxögonfisk

768   SPARIDAE   SLD

SC   *Cheimerius nufar* (Valenciennes, 1830)
ES   dentón nufar
DA   nufar-havrude
DE   Nufar-Seebrasse
EL   συναγρίδα-νούφαρ
EN   soldier; santer seabream
FR   denté nufar
IT   dentice nufar
NL   nufar-zeebrasem
PT   guerreiro-de-barras
FI   hammasahven-laji
SV   nufarblecka

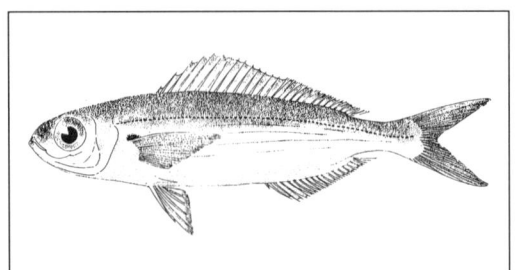

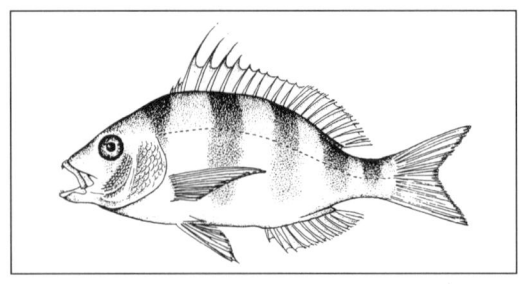

769 SPARIDAE   RSN

SC	*Chrysoblephus gibbiceps* (Valenciennes, 1830)
ES	pargo del Çabo
DA	stumpsnudet havrude
DE	Rote Stumpfnasenbrasse
EL	φαγγρί του Ναπολέοντα
EN	red stumpnose
FR	spare gibbeux
IT	pagro napoleone
NL	rode stompneus
PT	marreco do Cabo; pargo marreco do Cabo
FI	hammasahven-laji
SV	röd knölblecka

771 SPARIDAE   DEA

SC	*Dentex angolensis* (Poll and Maul, 1953)
ES	dentón angolés
DA	angolansk havrude
DE	Angola-Zahnbrasse
EL	συναγρίδα της Αγκόλας
EN	Angola dentex
FR	denté angolais
IT	dentice
NL	Angola-tandbrasem
PT	dentão de Angola
FI	angolanhammasahven
SV	angolansk tandbraxen

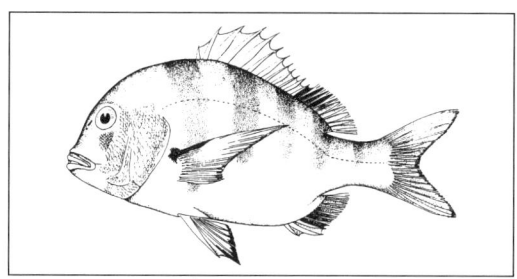

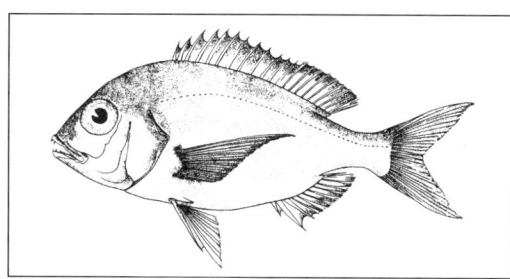

770 SPARIDAE   RSX

SC	*Chrysoblephus* spp.
ES	pargos
DA	havrude-slægt
DE	Stumpfnasenbrassen; Seebrassen
EL	φαγγριά του Ναπολέοντα
EN	red stumpnoses; dageraads
FR	spares
IT	pagri napoleoni
NL	dageraads; stompneuzen
PT	marrecos; pargos marrecos
FI	hammasahven-suku
SV	knölbleckor

772 SPARIDAE   DEN

SC	*Dentex canariensis* (Steindachner, 1881); *Cheimerius canariensis*
ES	chacarona de Canarias; dentón de Canarias
DA	kanarisk havrude
DE	Kanarische Zahnbrasse
EL	συναγρίδα των Καναρίων
EN	Canary dentex
FR	denté à tache rouge; denté canarien
IT	dentice
NL	Kanarische tandbrasem
PT	dentão-quissanga; dentão das Canárias
FI	kanarianhammasahven
SV	kanarietandbraxen

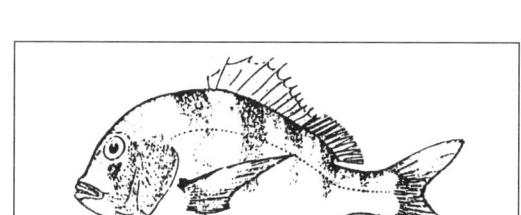

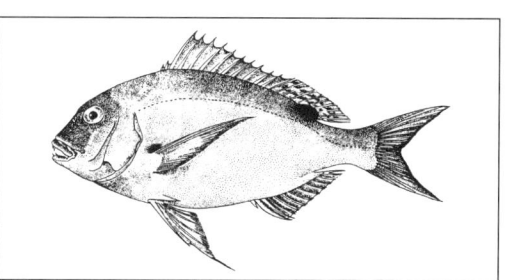

773 SPARIDAE   DNC

SC   *Dentex congoensis* (Poll, 1954)
ES   dentón congolés
DA   kongolesisk havrude
DE   Kongo-Zahnbrasse
EL   συναγρίδα του Κονγκό
EN   Congo dentex
FR   denté congolais
IT   dentice
NL   Kongo-tandbrasem
PT   dentão do Congo
FI   kongonhammasahven
SV   kongotandbraxen

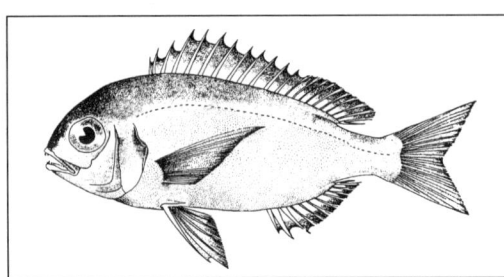

775 SPARIDAE   DEP

SC   *Dentex gibbosus* (Rafinesque, 1810)
ES   sama de pluma
DA   rød havrude
DE   Rosa Zahnbrasse
EL   τσαούσης· φαγγρί κορονάτο
EN   pink dentex
FR   denté bossu
IT   dentice corazziere; dentice
NL   roze tandbrasem
PT   capatão-de-bandeira; dentão-de-bandeira; pargo-de-bandeira
FI   hammasahven-laji
SV   rosa tandbraxen

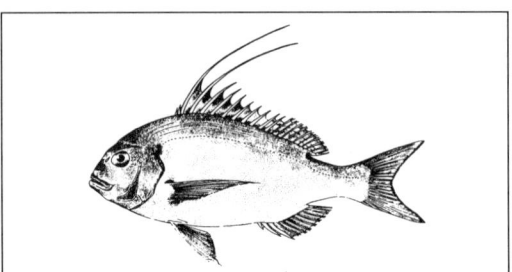

774 SPARIDAE   DEC

SC   *Dentex dentex* (Linnaeus, 1758); *Dentex vulgaris*
ES   dentón; dentón común; sama dorada; pargo testudo; capitón
DA   tandbrasen
DE   Zahnbrasse
EL   συναγρίδα
EN   dentex; sea bream; dog's tooth bream; common dentex
FR   denté commun; denté
IT   dentice; dentice mediterraneo; dentale
NL   tandbrasem
PT   capatão legítimo; dentão
FI   hammasahven
SV   tandbraxen

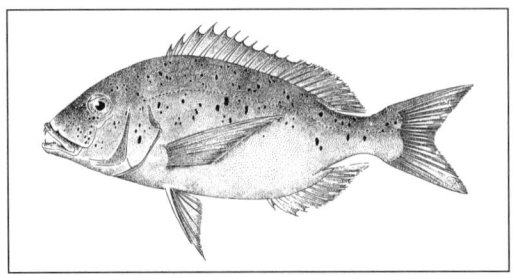

776 SPARIDAE   DEL

SC   *Dentex macrophthalmus* (Bloch, 1791); *Polysteganus macrophthalmus*
ES   cachucho; calé; rubiel
DA   storøjet havrude
DE   Großaugenzahnbrasse
EL   μπαλάς
EN   large-eyed dentex; large-eye dentex
FR   denté aux gros yeux
IT   dentice occhione; dentice
NL   grootoogtandbrasem
PT   cachucho
FI   isosilmähammasahven
SV   storögd tandbraxen

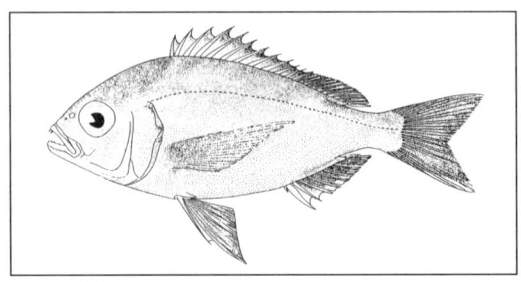

194

777 SPARIDAE DEM

SC	*Dentex maroccanus* (Valenciennes, 1830); *Polysteganus maroccanus*
ES	sama marroquí
DA	marokkansk havrude
DE	Marokko-Zahnbrasse
EL	συναγρίδα του Μαρόκου
EN	Morocco dentex
FR	denté du Maroc
IT	dentice marocchino
NL	Marokkaanse tandbrasem
PT	cachucho-dentão; dentão de Marrocos
FI	marokonhammasahven
SV	röd tandbraxen

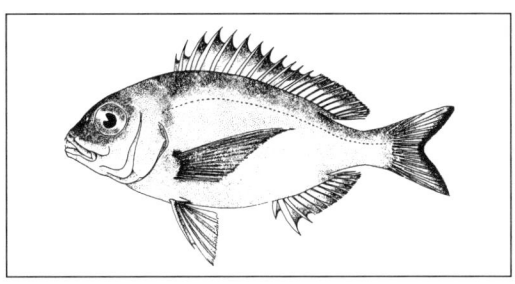

779 SPARIDAE SBZ

SC	*Diplodus cervinus* (Lowe, 1838); *Diplodus trifasciatus*
ES	sargo breado
DA	zebrahavrude
DE	Zebrabrasse
EL	ζεμπρασαργός
EN	zebra seabream
FR	sar tambour; sar à grosses lèvres
IT	sarago faraone; sarago; sargo
NL	zebra-ringbrasem
PT	sargo-veado
FI	seeprasargi
SV	sebrablecka

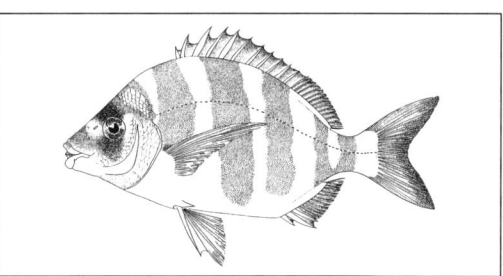

778 SPARIDAE ANN

SC	*Diplodus annularis* (Linnaeus, 1758)
ES	raspallón
DA	sorthalet havrude; ringhavrude
DE	Ringelbrasse
EL	σπάρος
EN	annular seabream
FR	sparaillon
IT	sparaglione; sarago sparaglione; sbaro
NL	ringbrasem
PT	sargo-alcorraz
FI	rengassargi
SV	stjärtbandsblecka

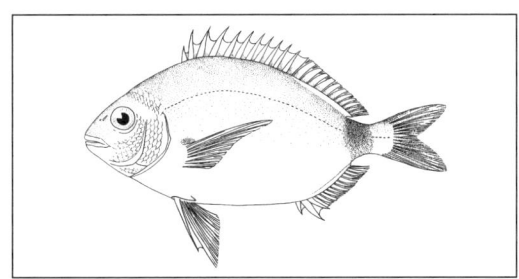

780 SPARIDAE DIH

SC	*Diplodus holbrooki* (Bean, 1878)
ES	sargo americano
DA	Holbrooks havrude
DE	Holbrooks-Brasse
EL	σπάρος της Φλόριντα
EN	spot-tail pinfish; spot-tail seabream
FR	sar cotonnier
IT	sparaglione americano
NL	Holbrooks ringbrasem
PT	sargo da Florida
FI	hammasahven-laji
SV	sjögräsblecka

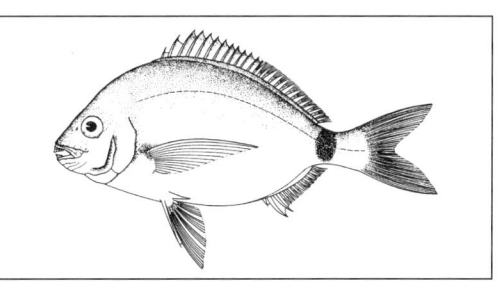

781 SPARIDAE SHR

SC *Diplodus puntazzo* (Cetti, 1777);
   Puntazzo puntazzo
ES sargo picudo
DA spidssnudet havrude
DE Spitzbrasse
EL ούγαινα· μυτάκι
EN sharp-snout seabream
FR sar à museau pointu; charax bec fin
IT sarago pizzuto; sargo pizzuto
NL spitse ringbrasem
PT sargo bicudo
FI hammasahven-laji
SV spetsnosblecka

783 SPARIDAE CTB

SC *Diplodus vulgaris* (Geoffroy St-Hilaire, 1817)
ES sargo mojarra; mojarra, muxarra
DA tvebåndet havrude
DE
EL καραγκιόζης· αυλιάς
EN common two-banded seabream
FR sar à tête noire
IT sarago testa nera; sarago fasciato; sarago
NL zwartkop-ringbrasem
PT sargo-safia
FI kaulussargi
SV tvåbandsblecka

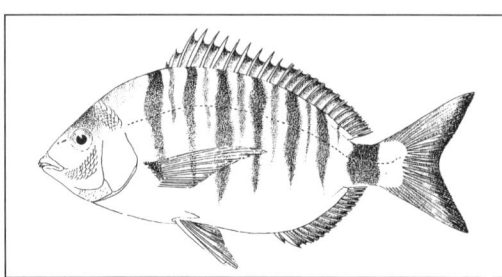

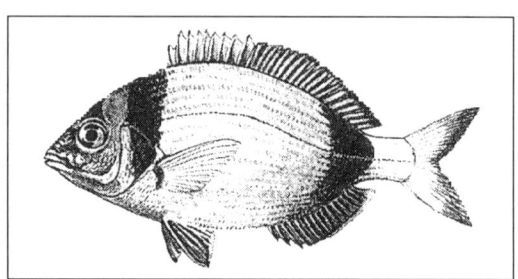

782 SPARIDAE SWA

SC *Diplodus sargus* (Linnaeus, 1758)
ES sargo marroquí
DA sorthale
DE Geißbrasse; Bindenbrasse; Große Geißbrasse
EL σαργός
EN white seabream
FR sar commun; sargue commun;
IT sarago maggiore; sarago; sargo
NL witte ringbrasem
PT sargo legítimo
FI isosargi
SV vitblecka

784 SPARIDAE SRG

SC *Diplodus* spp.
ES sargos; raspallones
DA havrude-slægt
DE Seebrassen
EL σπαροειδή
EN sargo breams
FR sars; sparaillons
IT saraghi
NL ringbrasemachtigen
PT sargos
FI sargit
SV stjärtfläcksbleckor

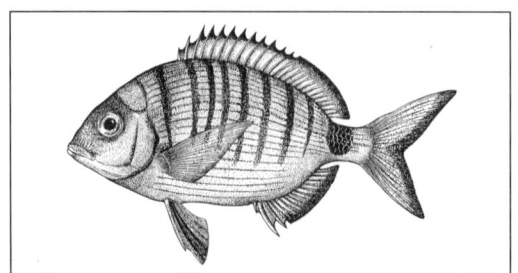

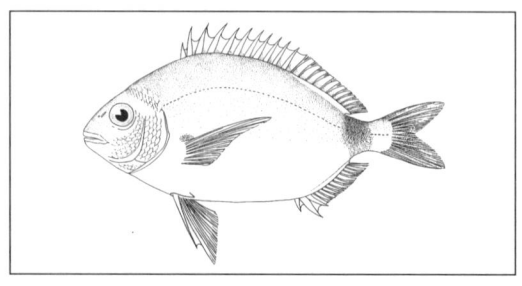

785 SPARIDAE SNW

SC	*Lithognathus lithognatus* (Cuvier, 1830)
ES	erla
DA	hvid stenbrasen
DE	Südafrikanische Marmorbrasse
EL	μουρμούρα της Αφρικής
EN	white steenbras
FR	marbré; mourme; perdrix de mer
IT	mormora africana
NL	witte steenbaars
PT	ferreira-branca
FI	pagelli-laji
SV	vit sandblecka

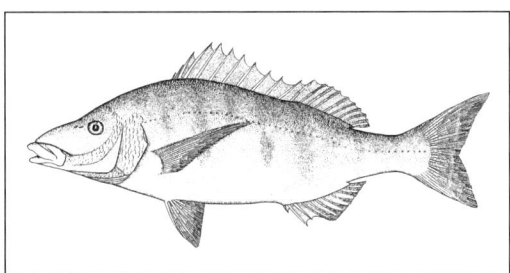

787 SPARIDAE STW

SC	*Lithognathus* spp.
ES	herreras; erlas
DA	havrude-slægt
DE	Streifenbrassen
EL	μουρμούρες
EN	white steenbrasses
FR	marbrés
IT	mormore
NL	steenbaarzen
PT	ferreiras
FI	hammasahven-suku
SV	sandbleckor

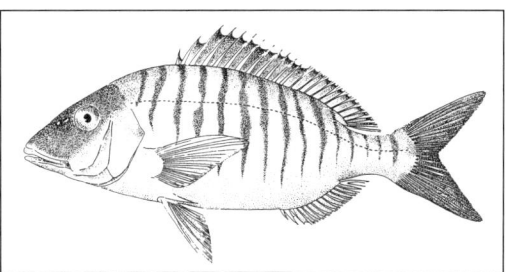

786 SPARIDAE SSB

SC	*Lithognathus mormyrus* (Linnaeus, 1758); *Pagellus mormyrus*
ES	herrera
DA	stribet blankesten
DE	Marmorbrasse
EL	μουρμούρα
EN	striped sea bream; sand steenbras
FR	marbré
IT	marmora; mormora
NL	zandsteenbaars
PT	ferreira
FI	marmoripagelli
SV	randig sandblecka

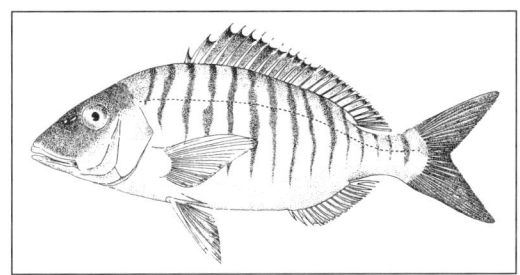

788 SPARIDAE SBS

SC	*Oblada melanura* (Linnaeus, 1758)
ES	oblada
DA	saddel-havrude
DE	Brandbrasse
EL	μελανούρι
EN	saddled bream
FR	oblade
IT	occhiata
NL	zadelbaars
PT	dobradiça
FI	satulasargi
SV	oblada

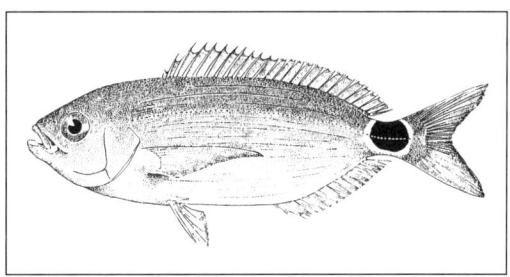

789 SPARIDAE CPP

SC *Pachymetopon* spp.
ES hotentotes
DA havrude-slægt
DE Meerbrassen; Seebrassen
EL χαλκοτσιπούρα
EN copper breams
FR
IT tanute sudafricane
NL hottentot-zebrasems
PT hotentotes
FI hammasahven-suku
SV kopparbleckor

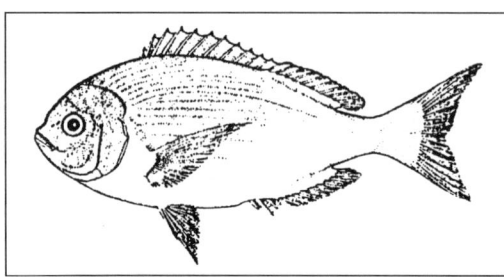

790 SPARIDAE SBA

SC *Pagellus acarne* (Risso, 1826)
ES aligote; besugo blanco; pancho; besugo chato
DA akarnanisk blankesten
DE Achselbrasse; Meerbrasse; Rotbrasse; Nordische Meerbrasse; Spanische Meerbrasse
EL μουσμούλι
EN axillary seabream
FR pageot acarné; pageot blanc; bogaravelle; pageot bâtard
IT pagello mafrone; pagello bastardo; pagello
NL Spaanse zeebrasem
PT besugo
FI pagelli-laji
SV pagell

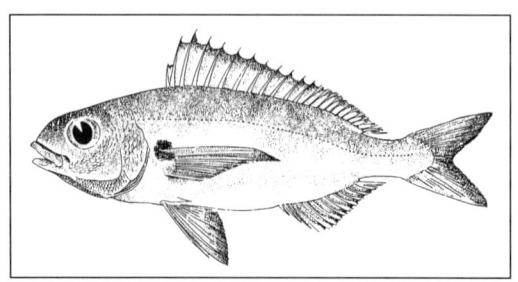

791 SPARIDAE PAR

SC *Pagellus bellottii* (Steindachner, 1882); *Pagellus bellottii bellottii*
ES breca; colorada; pagel; garapello
DA Belottis blankesten
DE Rote Pandora
EL λυθρίνι της Αφρικής
EN red pandora
FR pageot rouge africain
IT pagello rosso
NL rode pandora
PT bica-buço; buço; tico-tico
FI pagelli-laji
SV prickpagell

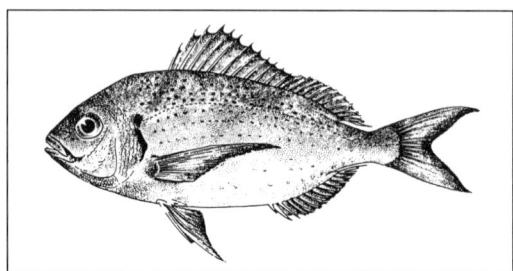

792 SPARIDAE TJO

SC *Pagellus bellotti natalensis* (Steindachner, 1902); *Pagellus natalensis*
ES breca; pagel
DA Natal-blankesten
DE Natal-Meerbrasse
EL λυθρίνι
EN red tjor tjor; Natal pandora
FR pageot du Natal
IT pagello rosso
NL Natal-zebrasem
PT besugo do Cabo; bica do Cabo
FI pagelli-laji
SV natalpagell

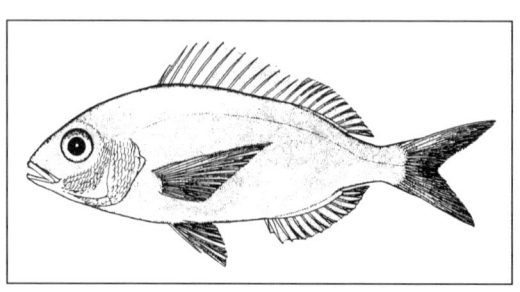

793 SPARIDAE   SBR

SC   *Pagellus bogaraveo* (Brünnich, 1768); *Pagellus centrodontus*
ES   besugo del Cantábrico; besugo de Laredo; goraz; pancho; besugo del Norte; bagaravel; aligote bogaraveo
DA   spidstandet blankesten
DE   Nordischer Meerbrassen; Meerbraum
EL   κεφαλάς
EN   red sea bream; red bream; common sea bream; dorade; chad; blackspot seabream
FR   dorade commune; dorade rose; brème de mer; bogaravelle; dorade; pironneau; rousseau
IT   occhialone; rovello; pagello
NL   zeebrasem
PT   goraz
FI   pilkkupagelli
SV   fläckpagell

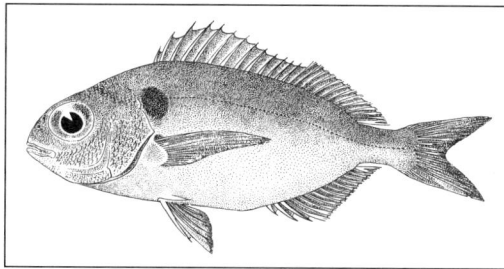

795 SPARIDAE   PAX

SC   *Pagellus* spp.
ES   brecas
DA   blankesten-slægt
DE   Meerbrassen; Seebrassen
EL   λυθρίνια
EN   red seabreams; pandoras
FR   pageots
IT   pagelli
NL   zeebrasems
PT   besugos, bicas e gorazes
FI   pagellit
SV   pageller

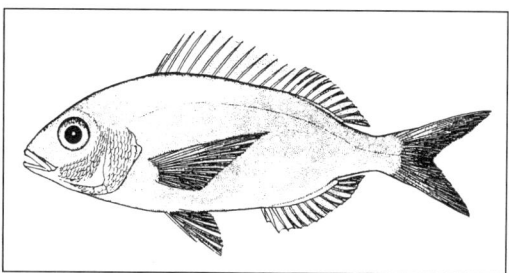

794 SPARIDAE   PAC

SC   *Pagellus erythrinus* (Linnaeus, 1758); *Pagellus canariensis*
ES   breca; pagel; garapello; pajel
DA   rød blankesten
DE   Rotbrassen; Roter Meerbrassen
EL   λυθρίνι
EN   pandora; Spanish sea bream; common pandora
FR   pageot rouge; pageau commun
IT   pagello fragolino; fragolino; pagello; luvaro
NL   rode zeebrasem
PT   bica
FI   punapagelli
SV   rödpagell

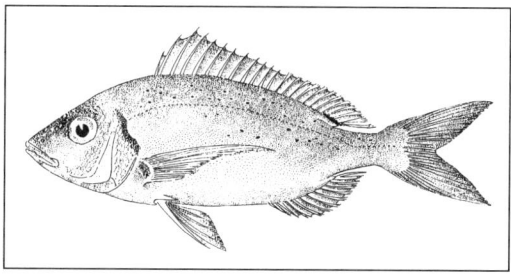

796 SPARIDAE   RER

SC   *Petrus rupestris* (Valenciennes, 1830)
ES   dentón rupestre
DA   rød stenbrasen
DE   Gelbrote Meerbrasse
EL   συναγρίδα του Ακρωτηρίου
EN   red steenbras
FR   denté du Cap
IT   dentice lupo
NL   rode steenbrasem
PT   vermelhão; dentão do Cabo
FI   hammasahven-laji
SV   stenblecka

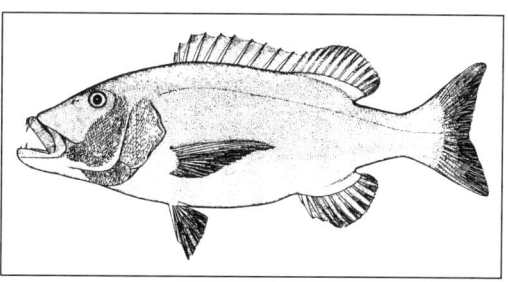

797	**SPARIDAE**	SCM	**799**	**SPARIDAE**	PGA

SC | *Polysteganus praeorbitalis* (Günther, 1859)
ES | dentón listado
DA | Natal-havrude
DE | Natal-Meerbrasse
EL | γραμμωτή συναγρίδα
EN | Scotsman seabream
FR | denté du Natal
IT | dentice striato
NL | Schot-zeebrasem
PT | escocês; dentão do Natal
FI | hammasahven-laji
SV | natalblecka

SC | *Pterogymnus laniarius* (Valenciennes, 1830)
ES | panga
DA | panga-havrude
DE | Panga-Meerbrasse
EL | πάνγκα
EN | panga; seabream
FR | panga; spare panga
IT | panga
NL | panga
PT | panga
FI | hammasahven-laji
SV | pangablecka

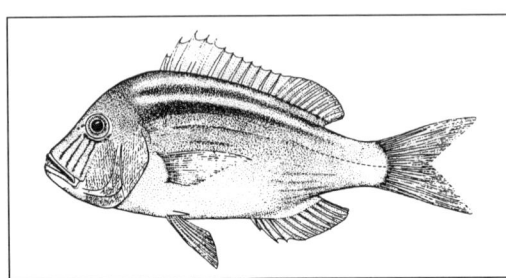 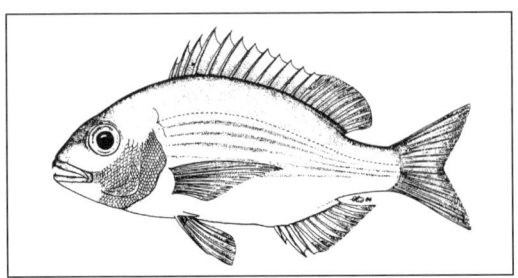

798	**SPARIDAE**	SEV	**800**	**SPARIDAE**	WSN

SC | *Polysteganus undulosus* (Regan, 1908)
ES | dentón ocelado
DA | plettet havrude
DE |
EL | γραμμωτή συναγρίδα
EN | seventy-four seabream
FR | denté maculé
IT | dentice striato
NL | gevlekte zeebrasem
PT | escocês-de-pinta
FI | hammasahven-laji
SV | fläckblecka

SC | *Rhabdosargus globiceps* (Cuvier, 1830)
ES | sargo austral
DA | sydlig havrude
DE | Weiße Stumpfnase
EL | φαγγρί κεφαλάς
EN | white stumpnose
FR | sar austral
IT | pagro testatonda
NL | witte stompneus
PT | sargo austral
FI | hammasahven-laji
SV | sydblecka

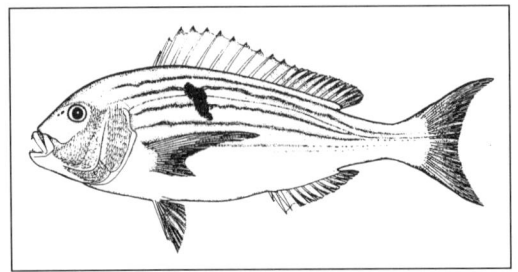 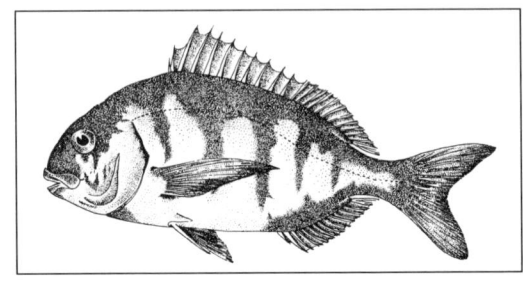

801 SPARIDAE SLM

SC	*Sarpa salpa* (Linnaeus, 1758); *Box salpa*
ES	salpa; salema
DA	guldstribet havrude
DE	Goldstrieme
EL	σάλπα
EN	bogue, saupe; goldline; salema; strepie
FR	saupe
IT	salpa
NL	gestreepte bokvis
PT	salema
FI	juovaboga
SV	salpa

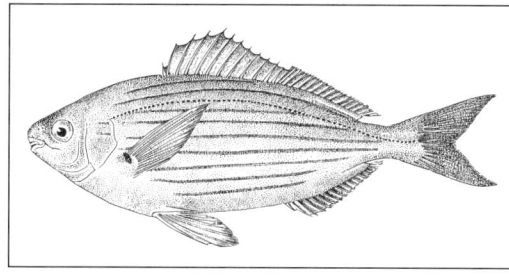

803 SPARIDAE REA

SC	*Pagrus auriga* (Valenciennes, 1843); *Sparusa auriga*
ES	hurta
DA	rødbåndet havrude
DE	Rotgebänderte Meerbrasse
EL	βασιλικό φαγγρί
EN	red-banded seabream
FR	pagre rayé; pagre royal
IT	pagro reale; pagro
NL	roodgestreepte zeebrasem
PT	pargo-sêmola; pargo de riscas; pargo-tereso
FI	hammasahven-laji
SV	rödbandad rödbraxen

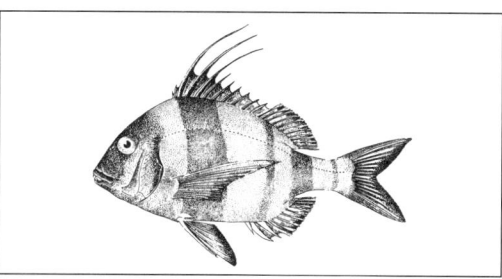

802 SPARIDAE SBG

SC	*Sparus aurata* (Linnaeus, 1758); *Chrysophrys aurata*
ES	pargo dorado; dorada
DA	guldbrasen
DE	Goldbrasse
EL	τσιπούρα
EN	gilt-head seabream
FR	dorade royale; dorade vraie; daurade vraie; dorade dorée
IT	orata
NL	goudbrasem
PT	dourada; doirada
FI	kultaotsa-ahven
SV	guldsparid

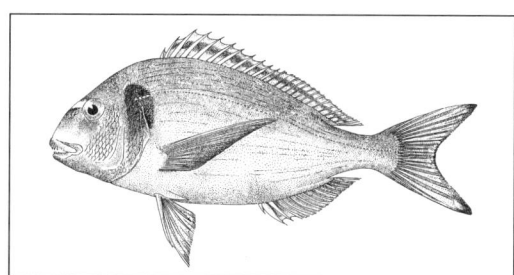

804 SPARIDAE BSC

SC	*Pagrus caeruleostictus* (Valenciennes, 1830); *Sparus caeruleostictus; Sparus ehrenbergii*
ES	pargo zapata
DA	blåplettet havrude
DE	Blaugefleckte Meerbrasse
EL	φαγγρί με μπλε στίγματα
EN	blue-spotted seabream
FR	pagre; pagre à points bleus
IT	pagro; pagro azzurro
NL	blauwgevlekte zeebrasem
PT	pargo ruço; pargo de pintas azuis
FI	hammasahven-laji
SV	blåfläckig rödbraxen

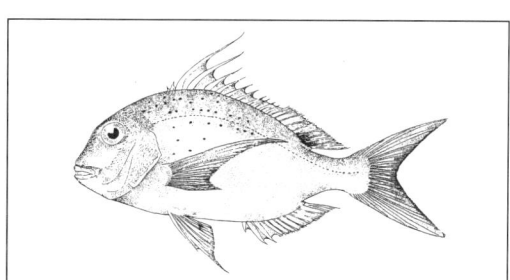

805 SPARIDAE JPG

SC *Pagrus major* (Temmick and Schlegel, 1843);
 Chrysophrys major; *Sparus major*
ES dorada gigante
DA japansk guldbrasen
DE Japanische Goldbrasse; Roter Tai; Akadei
EL φαγγρί της Ιαπωνίας
EN madai; Japanese seabream
FR daurade japonaise; daurade géante
IT orata del Giappone
NL Japanse zeebrasem
PT dourada do Japão
FI punahammasahven
SV japansk rödbraxen

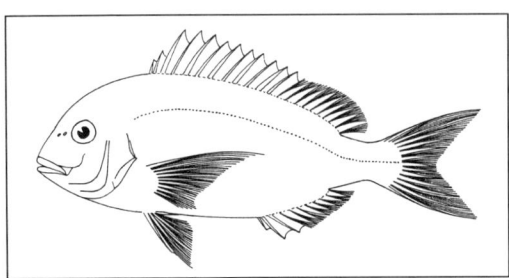

807 SPARIDAE

SC *Pagrus* spp.
ES pargos
DA havrude-slægt
DE Meerbrassen
EL φαγγριά
EN pargo breams
FR pagres
IT pagri
NL zeebrasems
PT pargos
FI hammasahvenet
SV rödbraxnar

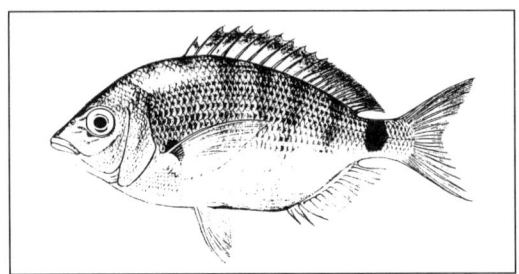

806 SPARIDAE RPG

SC *Pagrus pagrus* (Linnaeus, 1758); *Pagrus vulgaris*;
 Sparus pagrus
ES pargo
DA blankesten; almindelig blankesten
DE Gewöhnliche Sackbrasse
EL φαγγρί· μερτζάνι
EN Couch's sea bream; common seabream; red porgy
FR pagre commun
IT pagro mediterraneo; pagro
NL gewone zeebrasem
PT pargo legítimo; pargo; pargo verdadeiro
FI pargo
SV rödbraxen

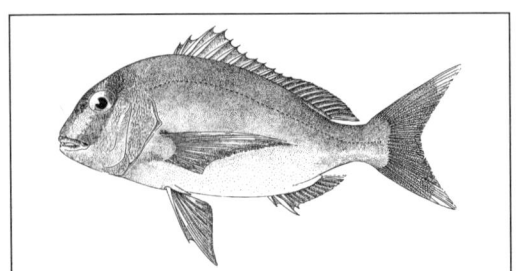

808 SPARIDAE BRB

SC *Spondyliosoma cantharus* (Linnaeus, 1758);
 Cantharus cantharus; *Cantharus griseus*;
 Cantharus lineatus
ES chopa; jargueta; pañoso
DA almindelig havrude
DE Streifenbrasse; Brandbrasse; Seekarpfen
EL σκαθάρι
EN black sea bream; sea bream; old wife
FR dorade grise; griset; canthare gris
IT tanuta; cantaro
NL zeekarper
PT choupa
FI meriruutana
SV havsruda

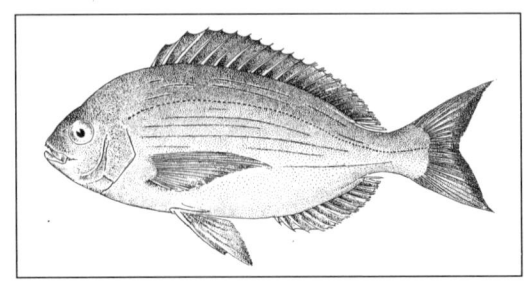

809 SPARIDAE SCP

SC	*Stenotomus chrysops* (Linnaeus, 1766)
ES	sargo de América del Norte
DA	nordlig skælfisk
DE	Nordamerikanische Brasse
EL	σαργός της Αμερικής
EN	scup
FR	spare doré
IT	sarago americano
NL	scup
PT	sargo da América do Norte
FI	amerikanhammasahven
SV	scup

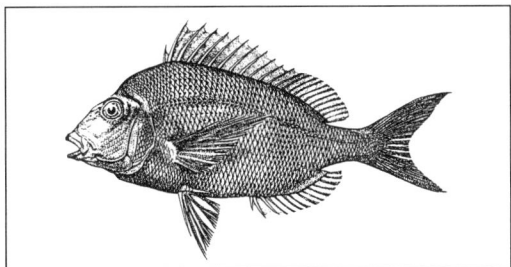

811 CENTRACANTHIDAE

SC	*Pterosmaris melanura* (Valenciennes, 1830)
ES	sucla
DA	haleplettet picarel
DE	Schwanzfleck-Pikarel
EL	αφρικανική μαρίδα
EN	blackspot picarel
FR	picarel de l'Atlantique Sud-Est; picarel à points noirs
IT	menola africana
NL	zwartgevlekte pikarel
PT	trombeiro malha-redonda
FI	mustatäpläpikarelli
SV	svansfläckspikarell

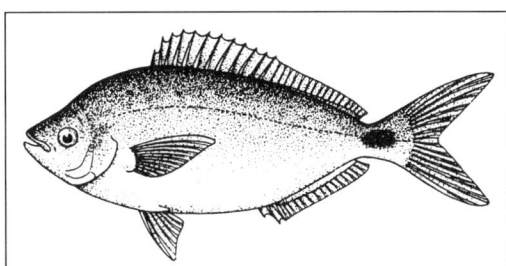

810 CENTRACANTHIDAE CEZ

SC	*Centracanthidae*
ES	chuclas
DA	picarel-familien
DE	Laxierfische; Schnauzenbrassen
EL	κεντρακάνθινοι
EN	picarels
FR	picarels; mendoles
IT	centracantidi; zerri; mennole
NL	pikarellen
PT	centracantídeos
FI	pikarellit; pikarellit-heimo
SV	pikareller

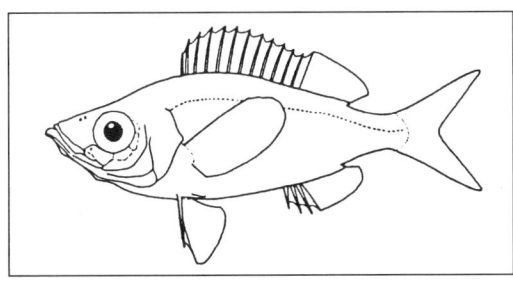

812 CENTRACANTHIDAE BPI

SC	*Spicara maena* (Linnaeus, 1758); *Maena vulgaris*
ES	chucla
DA	sortplettet picarel
DE	Laxierfisch
EL	μένουλα
EN	mendole; blotched picarel
FR	mendole commune; chuscle
IT	mennola; menola
NL	mendole-pikarel
PT	trombeiro-choupa
FI	isopikarelli
SV	menolapikarell

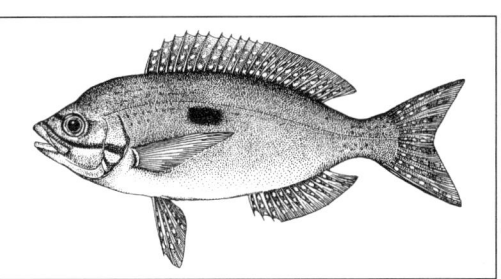

813 CENTRACANTHIDAE

SC	*Smaris smaris* (Linnaeus, 1758); *Smaris maurii;* *Sparus alcedo; Smaris vulgaris*
ES	caramel; picarel
DA	slank picarel
DE	Pikarel
EL	μαρίδα
EN	picarel
FR	gavaron; jarret; picarel
IT	zerro; menola
NL	pikarel
PT	trombeiro-boga
FI	pikarelli-laji
SV	pikarell

815 MULLIDAE   MUM

SC	*Mullidae*
ES	salmonetes
DA	mulle-familien
DE	Meerbarben
EL	μουλίδες
EN	goatfishes; red mullets
FR	rougets barbets
IT	triglie; mullidi
NL	zeebarbelen
PT	salmonetes
FI	mullot; mullot-heimo
SV	mullusfiskar

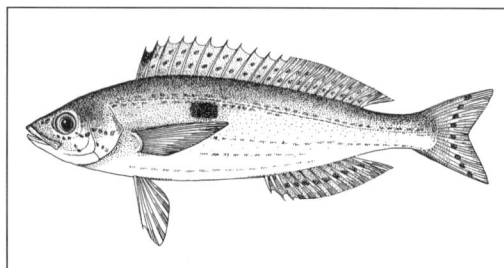

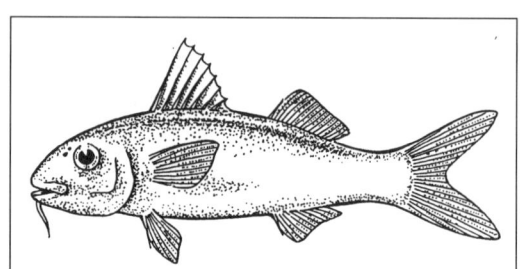

814 CENTRACANTHIDAE   PIC

SC	*Spicara* spp.
ES	chuclas; carameles
DA	picarel-slægt
DE	Schnauzenbrassen; Pikarels
EL	μαρίδες
EN	picarels
FR	mendoles; picarels
IT	mennole; zerri
NL	pikarellen
PT	trombeiros
FI	pikarellit
SV	pikareller

816 MULLIDAE

SC	*Mullus auratus* (Jordan and Gilbert, 1882)
ES	salmonete dorado
DA	guldmulle
DE	Nördlicher Ziegenfisch
EL	χρυσομπάρμπουνο
EN	golden goatfish
FR	rouget doré
IT	triglia dorata
NL	goudmul
PT	salmonete dourado
FI	kultamullo
SV	guldmulle

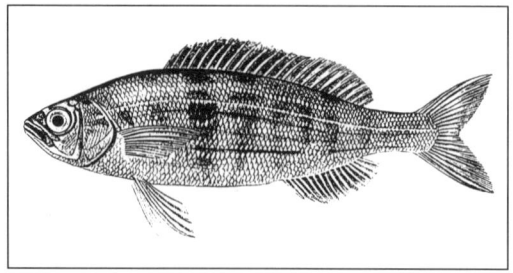

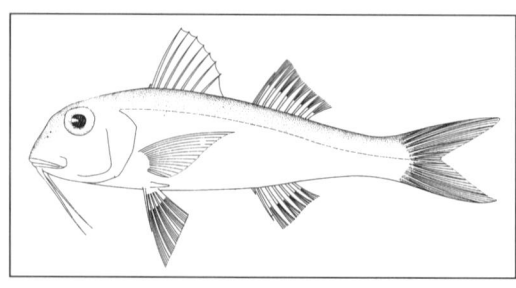

817 MULLIDAE MUT

SC	*Mullus barbatus* (Linnaeus, 1758)
ES	salmonete de fango; salmonete; barbo de mar; salmonete de malacasta; mijarco; igüelo
DA	rød mulle; rødskæg
DE	Gewöhnliche Meerbarbe; Steilstirnige Meerbarbe; Rote Meerbarbe
EL	κουτσομούρα
EN	striped mullet; red mullet
FR	rouget de vase surmulet; rouget barbet
IT	triglia di fango; triglia; triglia bianca; agostinella
NL	gestreepte zeebarbeel
PT	salmonete da vasa
FI	pikkumullo
SV	röd mulle

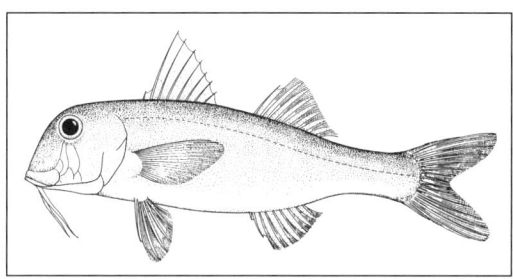

819 MULLIDAE MUX

SC	*Mullus* spp.
ES	salmonetes
DA	mulle-slægt
DE	Meerbarben
EL	μουλίδες
EN	red mullets; surmullets
FR	rougets
IT	triglie
NL	zeebarbelen
PT	salmonetes
FI	mullot
SV	mullar

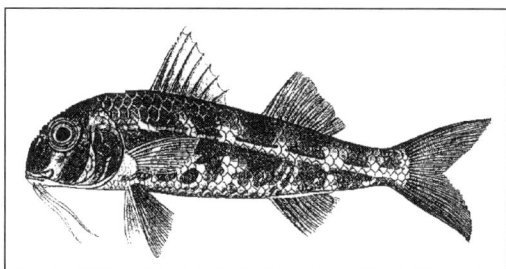

818 MULLIDAE MUR

SC	*Mullus surmuletus* (Linnaeus, 1758)
ES	salmonete de roca; salmonete rayado; salmonete de buena casta
DA	stribet mulle; europæisk mulle
DE	Streifenbarbe; Rotbart; Gestreifte Meerbarbe
EL	μπαρμπούνι
EN	red mullet; surmullet; striped red mullet
FR	rouget de roche; surmulet; rouget barbet de roche
IT	triglia di scoglio
NL	mul
PT	salmonete legítimo; salmonete vermelho
FI	keltajuovamullo
SV	mulle

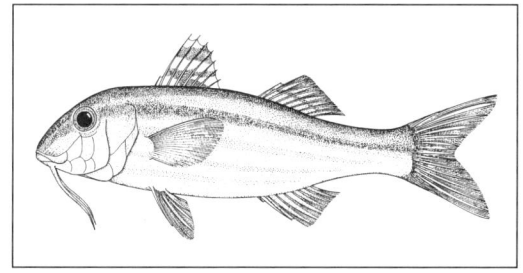

820 MULLIDAE GOA

SC	*Pseudupeneus prayensis* (Cuvier, 1829)
ES	salmonete barbudo
DA	vestafrikansk mulle
DE	Westafrikanische Meerbarbe
EL	μπαρμπούνι της Δυτικής Αφρικής
EN	West African goatfish
FR	rouget du Sénégal
IT	triglia dentata; triglia canina
NL	West-Afrikaanse mul
PT	salmonete barbudo; salmonete branco
FI	länsiafrikanmullo
SV	västafrikansk mulle

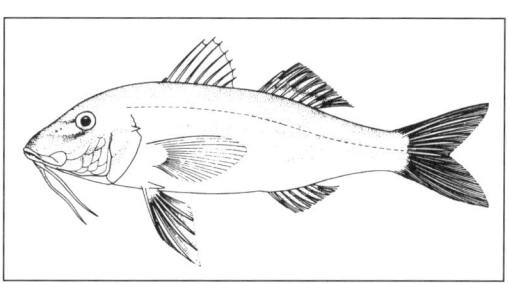

821 MULLIDAE

SC	*Upeneus asymmetricus* (Lachner, 1954)
ES	falso salmonete de roca
DA	sortstribet mulle
DE	Gold-Meerbarbe
EL	κοκκινομπάρμπουνο· μπαρμπούνι της Ανατολής· λοχίας
EN	golden-striped goatfish; black-striped goatfish
FR	faux rouget de roche
IT	triglia rossa
NL	goudbandmul
PT	salmonete da rocha
FI	mullo-laji
SV	bandmulle

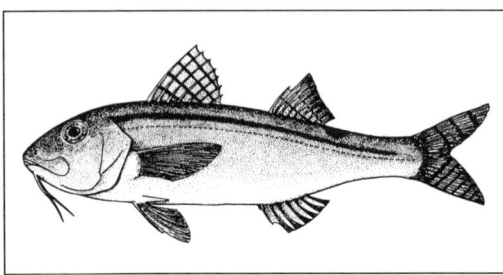

822 MULLIDAE   GOX

SC	*Upeneus* spp.
ES	falsos salmonetes
DA	mulle-slægt
DE	Meerbarben
EL	λοχίες· κοκκινομπαρμπούνια
EN	goatfishes
FR	faux rougets
IT	triglie rosse
NL	goudbandmullen
PT	salmonetes da rocha
FI	mullot
SV	stillahavsmullar

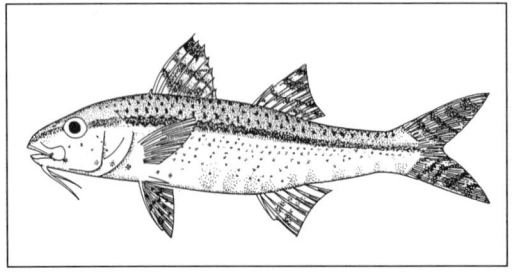

823 CORACINIDAE   COT

SC	*Coracinidae*
ES	dambas
DA	galeonsfisk-familien
DE	Galjoen-Fische
EL	ντάμπα
EN	galjoens
FR	galjoins
IT	coracinidi; damba
NL	galjoenvissen
PT	galeões
FI	kaljuunakalat; kaljuunakalat-heimo
SV	galjonsfiskar

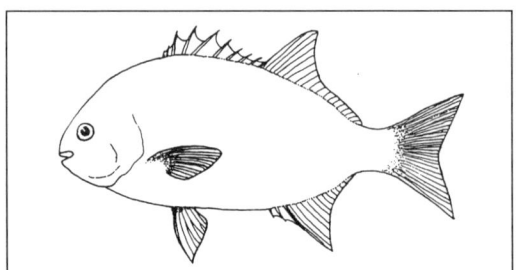

824 CORACINIDAE   GAJ

SC	*Coracinus capensis* (Cuvier, 1831)
ES	damba
DA	Kap-galeonsfisk
DE	Südafrikanischer Galjoen
EL	ντάμπα
EN	galjoen
FR	galjoin franc
IT	damba
NL	galjoen
PT	galeão
FI	kaljuunakala
SV	kapgaljonsfisk

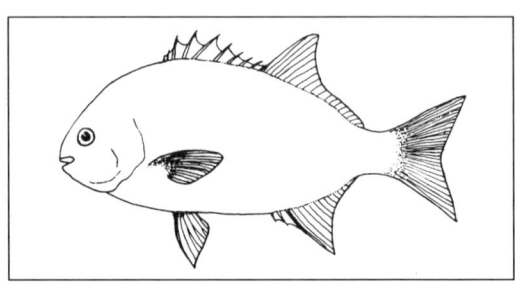

825 EPHIPPIDAE SPA

SC	*Ephippidae*
ES	pagualas
DA	spadefisk-familien
DE	Spatenfische
EL	εφίπιδοι
EN	spadefishes
FR	poissons disques; chèvres de mer
IT	efipidi
NL	spadevissen
PT	efipídeos; enxadas e afins; peixes-enxada e afins
FI	levykalat; levykalat-heimo
SV	spadfiskar

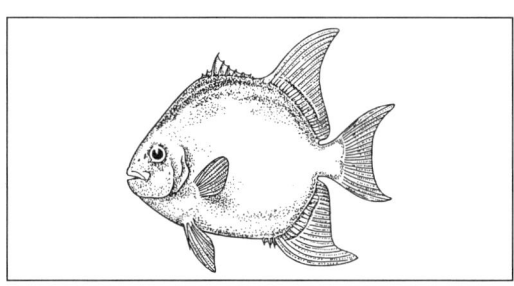

827 DREPANEIDAE SIC

SC	*Drepane africana* (Osorio, 1892)
ES	catemo africano
DA	afrikansk spadefisk
DE	Afrikanischer Sichelflosser
EL	αφρικάνικο δρέπανο
EN	African sicklefish
FR	disque africain; drépane ailé; poisson chameau; forgeron ailé
IT	drepana
NL	Afrikaanse sikkelvis
PT	enxada africana
FI	afrikanlevykala
SV	afrikansk spadfisk

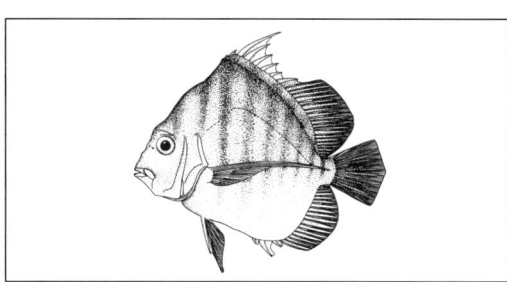

826 EPHIPPIDAE BAT

SC	*Platax* spp.
ES	peces murciélago
DA	flagermusfisk-slægten
DE	Fledermausfische
EL	ψάρια νυχτερίδες
EN	batfishes
FR	poissons battoires; poissons chauve-souris
IT	pesci pipistrello; platax
NL	vleermuisvissen
PT	peixes-morcego
FI	levykalat
SV	fladdermusfiskar

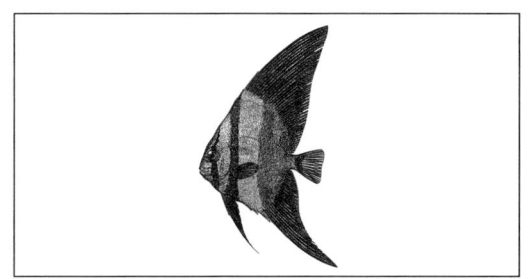

828 DREPANEIDAE SPS

SC	*Drepane punctata* (Linnaeus, 1758)
ES	concertina
DA	plettet spadefisk
DE	Geperlter Sichelflosser
EL	κηλιδόστικτο δρέπανο
EN	spotted sicklefish
FR	disque ponctué
IT	drepana
NL	gevlekte sikkelvis
PT	enxada do Indo-Pacífico
FI	pilkkulevykala
SV	prickig spadfisk

829 SCATOPHAGIDAE SCT

SC *Scatophagus* spp.
ES argós
DA argusfisk-slægt
DE Argusfische
EL σκατοφάγος
EN scats
FR argus; scatophages
IT pesci argo
NL argusvissen
PT remexidos
FI arguskalat
SV argusfiskar

831 CICHLIDAE CIX

SC *Cichlidae*
ES cíclidos
DA cichlide-familien
DE Buntbarsche; Cichliden
EL κικλίδες
EN cichlids
FR ciclidés
IT ciclidi
NL cichliden
PT ciclídeos; tilápias e afins
FI kirjoahvenet; kirjoahvenet-heimo
SV cichlider; ciklider

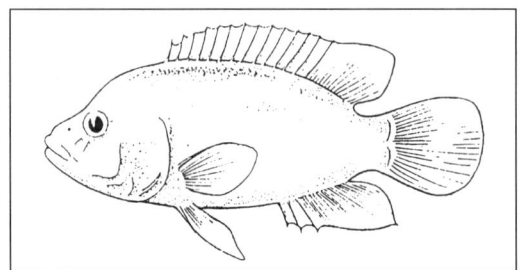

830 CHAETODONTIDAE BUS

SC *Chaetodontidae*
ES peces mariposa
DA fanefisk; skælfinnefisk-familien
DE Schmetterlingsfische; Borstenzähner
EL ψάρια πεταλούδες
EN butterflyfishes
FR poissons papillons
IT pesci angelo
NL vlindervissen
PT borboletas; peixes-borboleta
FI perhokalat; perhokalat-heimo
SV fjärilsfiskar

832 CICHLIDAE TLM

SC *Oreochromis mossambicus* (Peters, 1852)
ES tilapia de Mozambique
DA Mozambique-cichlide
DE Moçambique-Buntbarsch
EL τιλάπια της Μοζαμβίκης
EN Mozambique tilapia
FR tilapia du Mozambique
IT tilapia del Mozambico
NL Mozambique-tilapia
PT tilápia de Moçambique
FI mosambikintilapia
SV moçambiquetilapia

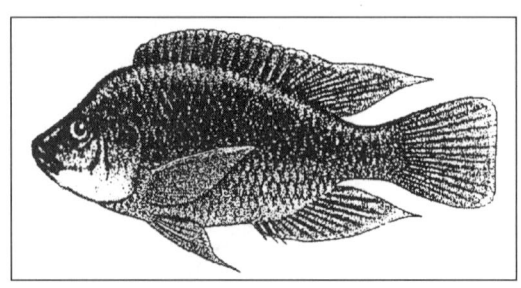

833 CICHLIDAE

SC	*Oreochromis niloticus* (Linnaeus, 1758)
ES	tilapia del Nilo
DA	Nil-cichlide
DE	Nil-Buntbarsch
EL	τιλάπια του Νείλου
EN	Nile tilapia
FR	tilapia du Nil
IT	tilapia del Nilo
NL	nijltilapia
PT	tilápia do Nilo
FI	niilintilapia
SV	niltilapia

TLN

835 POMACENTRIDAE

SC	*Pomacentridae*
ES	pomacéntridos
DA	jomfrufisk-familien
DE	Riffbarsche
EL	πομακεντρίδες
EN	damselfishes
FR	pomacentridés; demoiselles
IT	pomacentridi
NL	pomacentriden
PT	pomacentrídeos; castanhetas e afins
FI	koralliahvenet; koralliahvenet-heimo
SV	frökenfiskar

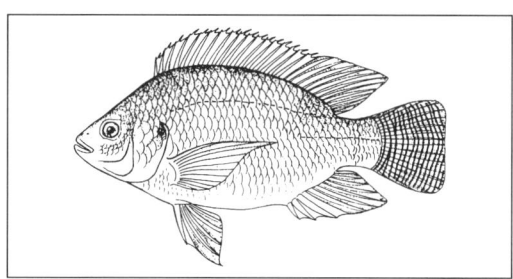

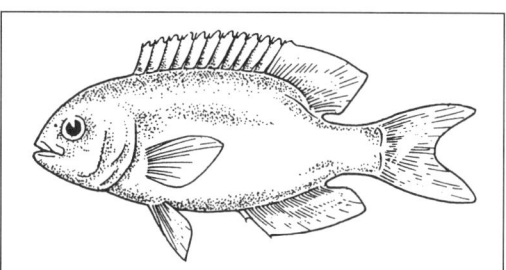

834 CEPOLIDAE

SC	*Cepola rubescens* (Linnaeus, 1766); *Cepola macrophthalma*
ES	cinta
DA	rød båndfisk
DE	Bandfisch
EL	κορδέλα
EN	red bandfish
FR	jarretière; demoiselle; cépole rougeâtre
IT	cepola
NL	rode bandvis
PT	suspensório
FI	liekkikala
SV	bandfisk

836 POMACENTRIDAE

SC	*Chromis chromis* (Linnaeus, 1758)
ES	castañuela
DA	jomfrufisk
DE	Mönchfisch
EL	καλογρίτσα
EN	damselfish
FR	petite castagnole
IT	castagnola
NL	monnikvis
PT	castanheta
FI	ruskoneitokala
SV	frökenfisk

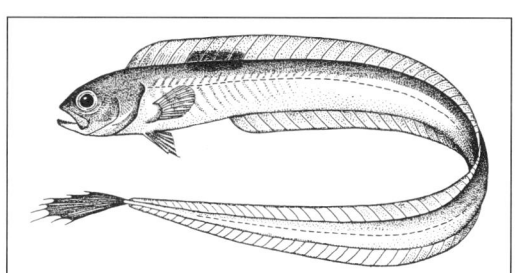

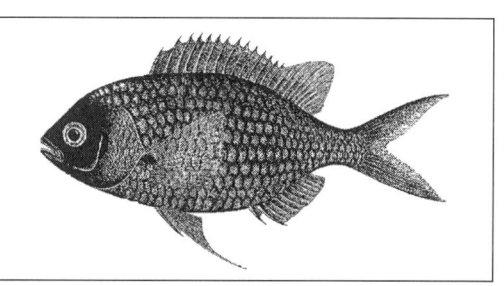

837 LABRIDAE

SC *Labridae*
ES lábridos
DA læbefisk-familien
DE Lippfische
EL χειλούδες· καπίνες
EN wrasses; hogfishes
FR labres; girelles et vieilles
IT labridi
NL lipvissen
PT bodiões
FI huulikalat; huulikalat-heimo
SV läppfiskar

WRA **839 LABRIDAE**

SC *Ctenolabrus rupestris* (Linnaeus, 1758)
ES tabernero
DA havkarusse
DE Klippenbarsch
EL λαπίνα· κατραβάνος
EN goldsinny wrasse
FR labre rupestre; fausse-vieille
IT tordo dorato
NL kliplipvis
PT bodião rupestre
FI kivihuulikala
SV stensnultra

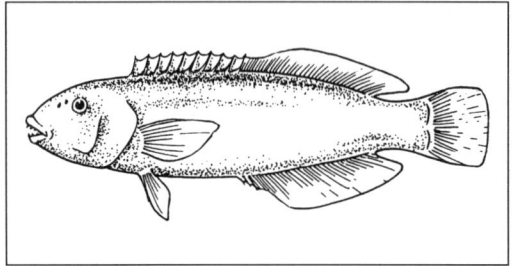

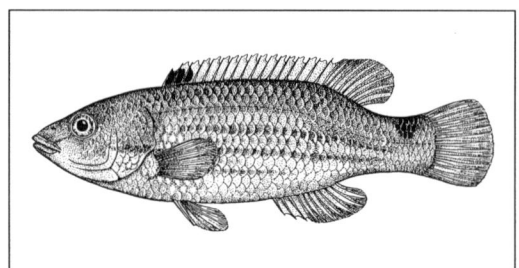

838 LABRIDAE

SC *Coris julis* (Linnaeus, 1758)
ES doncella
DA junkergylte
DE Meerjunker
EL γύλος
EN rainbow wrasse
FR girelle commune
IT donzella
NL regenbooglipvis
PT judia
FI koruhuulikala
SV junkergirella

840 LABRIDAE

SC *Labrus bergylta* (Ascanius, 1767)
ES maragota
DA berggylt
DE Gefleckter Lippfisch
EL χειλού παπαγάλος
EN ballan wrasse
FR grande vieille; vieille commune
IT tordo marvizzo
NL gevlekte lipvis
PT bodião reticulado
FI viherhuulikala
SV berggylta

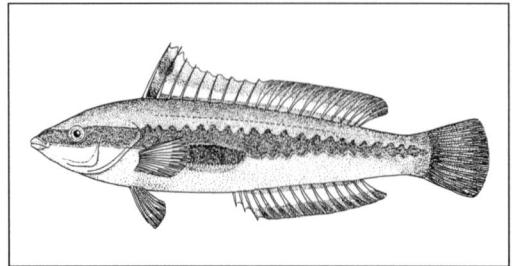

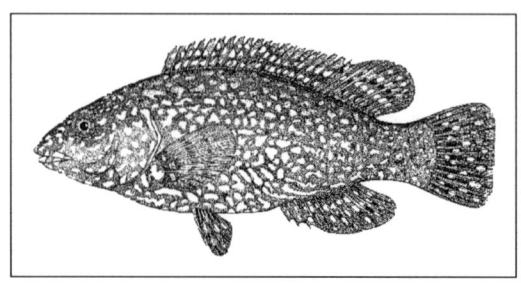

841 LABRIDAE

SC | *Labrus bimaculatus* (Linnaeus, 1758); *Labrus mixtus*
ES | gallano
DA | blåstak (han); rødnæb (hun)
DE | Kuckucks-Lippfisch; Bunter Lippfisch
EL | στικτοχειλού· χειλού· λαπίνα
EN | cuckoo wrasse
FR | petite vieille; coquette; labre melé
IT | tordo fischietto
NL | koekoeklipvis
PT | bodião-canário
FI | sinihuulikala
SV | blågylta

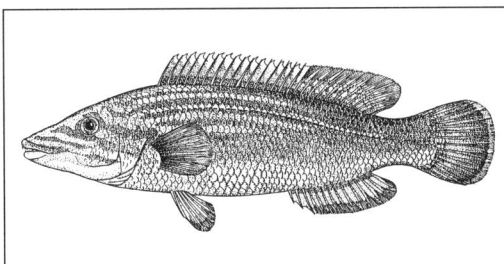

843 LABRIDAE

SC | *Labrus viridis* (Linnaeus, 1758); *Labrus turdus*
ES | bodión verde
DA | grøn læbefisk
DE | Grüner Lippfisch; Amsellippfisch; Meerdrossel
EL | πρασινοχειλού· χειλού· λαπίνα
EN | green wrasse
FR | labre vert; vieille verte; tourdero
IT | tordo
NL | groene lipvis
PT | bodião-tordo
FI | viherhuulikala
SV | gröngylta

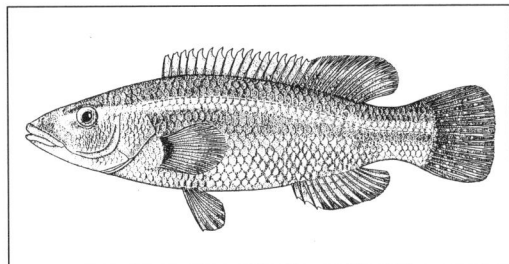

842 LABRIDAE

SC | *Labrus merula* (Linnaeus, 1758)
ES | merlo
DA | brun læbefisk
DE | Brauner Lippfisch
EL | μαυροχειλού· χειλού· λαπίνα
EN | brown wrasse
FR | merle
IT | tordo nero
NL | bruine lipvis
PT | bodião fusco
FI | ruskohuulikala
SV | brungylta

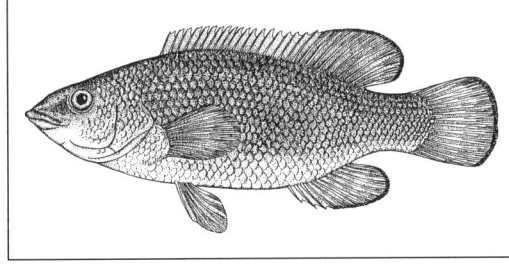

844 LABRIDAE

SC | *Lachnolaimus maximus* (Walbaum, 1792)
ES | pez perro; loro gallo; pargo de pluma
DA | vestatlantisk læbefisk
DE | Westatlantischer Lippfisch
EL | γουρουνόψαρο
EN | hogfish; hogsnapper
FR | capitaine; vieille américaine; vieille de l'Ouest-Atlantique
IT | tordo piumato
NL | bultkopvis
PT | bodião de pluma
FI | porsaskala
SV | grisgylta

845 LABRIDAE

SC *Symphodus cinereus* (Bonnaterre, 1788);
   Crenilabrus cinereus; *Crenilabrus griseus*
ES magnote
DA grå savgylte
DE Grauer Lippfisch
EL λαπίνα· χειλού
EN grey wrasse
FR crénilabre cendré; crénilabre tigré
IT tordo grigio
NL grauwe lipvis
PT bodião cinzento
FI harmaahuulikala
SV gråsnultra

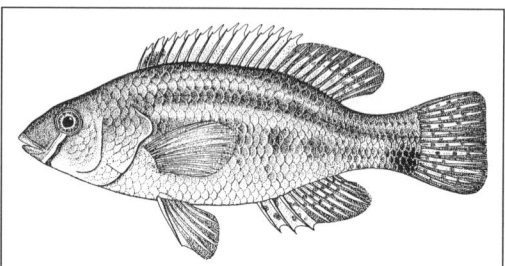

847 LABRIDAE

SC *Symphodus melops* (Linnaeus, 1758); *Crenilabrus
   melops*
ES porredana
DA savgylte; almindelig savgylte
DE Goldmaid
EL λαπίνα· χειλού
EN corkwing wrasse
FR crénilabre commun; roupier; courlazo; crénilabre
   melops
IT tordo occhionero
NL zwartooglipvis
PT bodião vulgar
FI rantahuulikala
SV skärsnultra

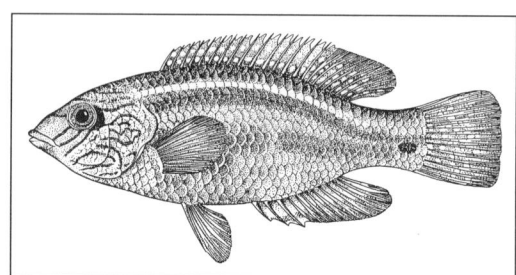

846 LABRIDAE

SC *Symphodus mediterraneus* (Linnaeus, 1758);
   Crenilabrus mediterraneus
ES vaqueta
DA Middelhavs-læbefisk
DE Mittelmeerlippfisch
EL κοκκινολαπίνα· λαπίνα
EN axillary wrasse
FR crénilabre méditerranéen
IT tordo rosso
NL Middellandse-Zeelipvis
PT bodião do Mediterrâneo
FI välimerenhuulikala
SV rödsnultra

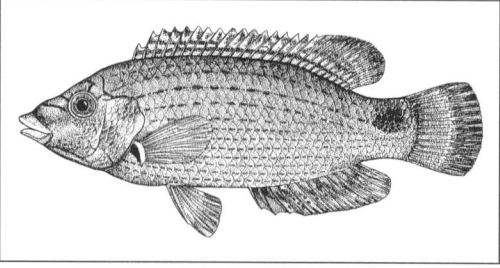

848 LABRIDAE

SC *Symphodus ocellatus* (Forsskål, 1775); *Crenilabrus
   ocellatus*
ES tordo de roca
DA øjeplettet savgylte
DE Augenfleck-Lippfisch; Augenlippfisch
EL μαυρολαπίνα· λαπίνα· χειλού
EN spotted wrasse; ocellated wrasse
FR crénilabre ocellé
IT tordo ocellato
NL ooglipvis
PT bodião de pinta
FI silmähuulikala
SV ögonsnultra

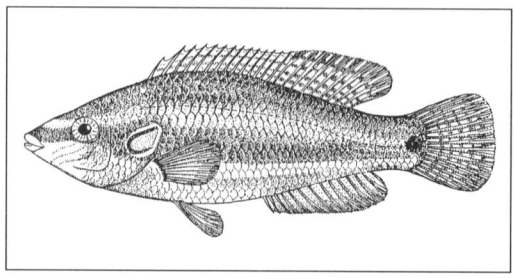

849 LABRIDAE

SC	*Symphodus roissali* (Risso, 1810); *Crenilabrus roissali*; *Symphodus quinquemaculatus*; *Crenilabrus quinquemaculatus*
ES	tordo de cinco manchas
DA	femplettet savgylte
DE	Fünffleckiger Lippfisch
EL	λαπίνα· χειλού
EN	five-spot wrasse
FR	crénilabre à cinq taches; crénilabre de Roissal
IT	tordo verde
NL	vijfvlekkige lipvis
PT	bodião manchado
FI	täplähuulikala
SV	fläcksnultra

851 LABRIDAE

SC	*Symphodus tinca* (Linnaeus, 1758); *Crenilabrus tinca*; *Symphodus pavo*; *Crenilabrus pavo*
ES	tordo
DA	påfuglesavgylte
DE	Pfauenlippfisch; Meerschlei
EL	λαπίνα· χειλού
EN	peacock wrasse
FR	crénilabre paon; crénilabre tanche
IT	tordo pavone; laggion
NL	pauwlipvis
PT	bodião-pavão
FI	mustehuulikala
SV	påfågelssnultra

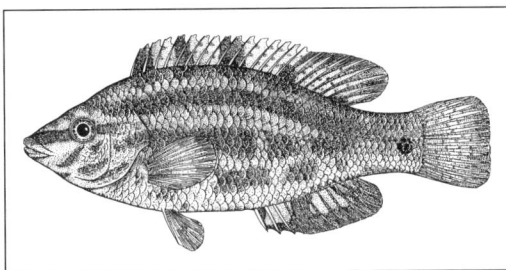

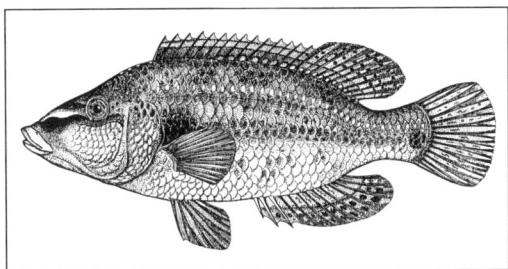

850 LABRIDAE

SC	*Symphodus rostratus* (Bloch, 1797)
ES	tordo picudo
DA	langsnudet savgylte
DE	Schnauzenlippfisch
EL	λαπίνα· χειλού
EN	cannadelle
FR	sublet; cannadelle
IT	tordo musolungo
NL	snavellipvis
PT	bodião das ervas
FI	kuonohuulikala
SV	långsnultra

852 LABRIDAE   TAU

SC	*Tautoga onitis* (Linnaeus, 1758)
ES	tautoga negra
DA	tautog læbefisk
DE	Tautog; Austernfisch
EL	τάουτογκ
EN	tautog
FR	tautogue noir
IT	tautoga
NL	tautog-lipvis
PT	bodião da ostra
FI	huulikala-laji
SV	tautog

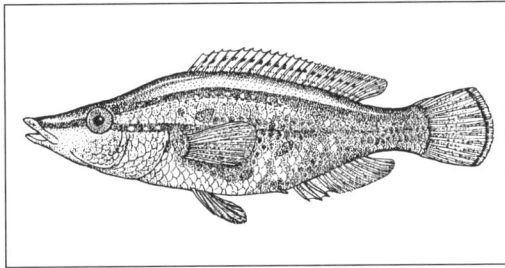

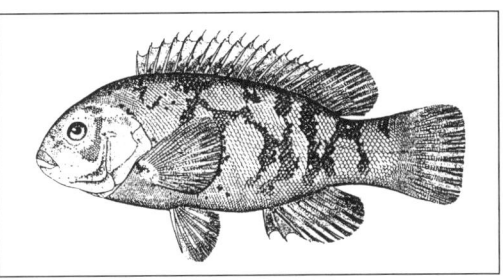

853	**LABRIDAE**	CUN	855	**LABRIDAE**

853 LABRIDAE CUN

SC	*Tautogolabrus adspersus* (Walbaum, 1792)
ES	tautoga americana
DA	amerikansk læbefisk
DE	Amerikanischer Lippfisch
EL	λαπίνα της Αμερικής
EN	cunner
FR	tanche-tautogue
IT	tordo americano
NL	Amerikaanse lipvis
PT	bodião do norte
FI	huulikala-laji
SV	luring

855 LABRIDAE

SC	*Xyrichthys novacula* (Linnaeus, 1758)
ES	raó; rahó; raor
DA	kløvefisk
DE	Schermesserfisch
EL	κατσούλα· κατέργαρος· παπαγάλος
EN	cleaver wrasse
FR	rason
IT	pesce pettine
NL	parelscheermesvis
PT	mordedor
FI	huulikala-laji
SV	rakfisk

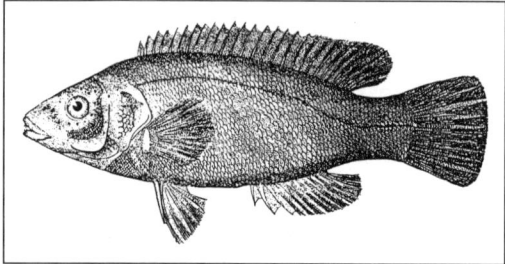

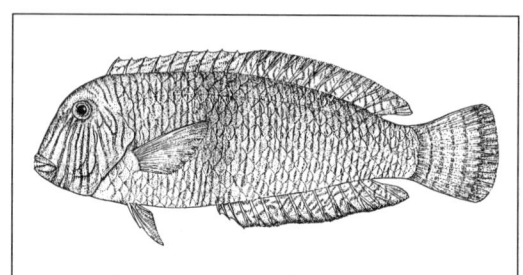

854 LABRIDAE

SC	*Thalassoma pavo* (Linnaeus, 1758); *Labrus pavo*
ES	pez verde
DA	regnbuegylte
DE	Meerpfau
EL	γαϊτανούρι· φιάμουλα
EN	rainbow wrasse
FR	girelle paon
IT	donzella pavonina
NL	pauwlipvis
PT	bodião verde
FI	riikinkukkohuulikala
SV	regnbågsgylta

856 SCARIDAE PRR

SC	*Sparisoma cretense* (Linnaeus, 1758); *Scarus cretensis*
ES	loro viejo; vieja
DA	papegøjefisk
DE	Seepapagei
EL	σκάρος
EN	parrot fish
FR	scare de Grèce; perroquet de Méditerranée; perroquet vieillard
IT	scaro; pesce pappagallo
NL	Griekse papegaaivis
PT	papagaio velho
FI	papukaijakala
SV	papegojfisk

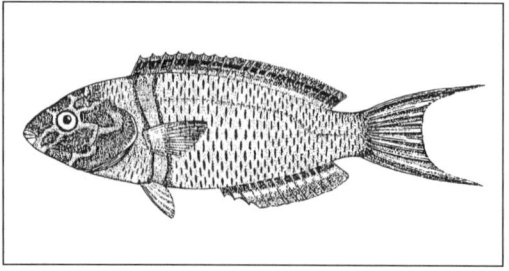

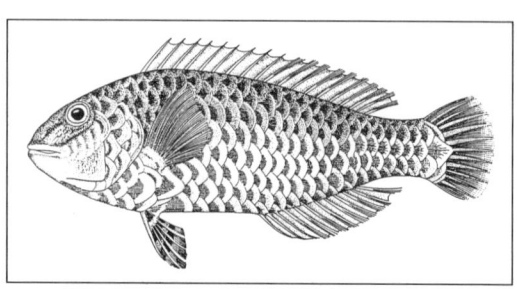

857 CHEILODACTYLIDAE CTA

SC *Cheilodactylus bergi* (Norman, 1937)
ES castañeta
DA Bergs falkefisk
DE Bergs Morwong
EL κόκκινος ψευτοσαργός
EN castaneta
FR castanette pontude
IT pseudosarago rosa
NL Bergs vingervin
PT peixe-bobo bicudo
FI
SV vit fingerfena

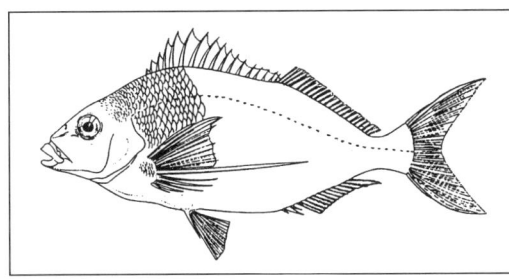

859 CHEILODACTYLIDAE TAK

SC *Nemadactylus macropterus* (Block and Schneider, 1801); *Cheilodactylus macropterus*
ES pintadilla cola larga
DA storfinnet falkefisk
DE Großflossen-Morwong
EL ψευτοσαργός
EN tarakihi
FR castanette «Tarakihi»
IT pseudosarago
NL grootvinnige vingervin
PT peixe-bobo «taraki»
FI
SV tarakihi

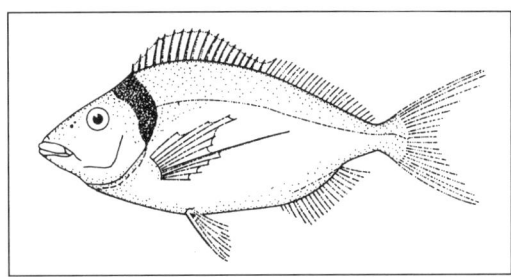

858 CHEILODACTYLIDAE HAW

SC *Cheilodactylus variegatus* (Cuvier and Valenciennes, 1833)
ES pintadilla
DA varieret falkefisk
DE Prächtiger Morwong
EL ψευτοσαργός
EN pintadilla
FR castanette variable
IT pseudosarago
NL veelkleurige vingervin
PT peixe-bobo pintado
FI
SV pintadilla

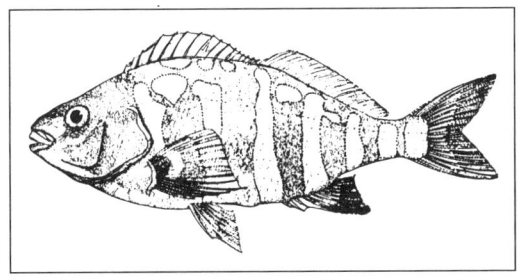

860 CHEILODACTYLIDAE MOW

SC *Nemadactylus* spp.
ES morwongos
DA falkefisk-slægt
DE Morwongs
EL ψευτοσαργοί
EN morwongs
FR vivaneaux royaux d'Australie
IT morwong; pseudosaraghi
NL vingervinnen
PT peixes-bobo
FI
SV fingerfenor

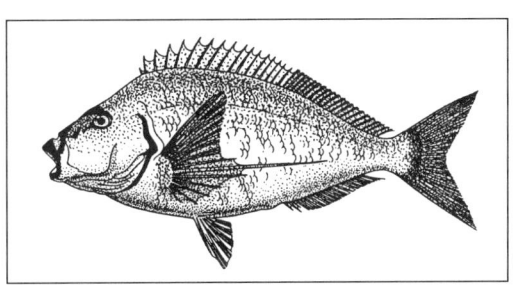

215

861 LATRIDAE TRU

SC *Latridae*
ES trompeteros
DA trompeterfisk-familien
DE Trompeterfische
EL ψάρια τρομπέτες
EN trumpeters
FR mokis
IT moki; latridi
NL Pacifische trompetvissen
PT trompetes
FI
SV mokifiskar; trumpetarfiskar

863 TRICHODONTIDAE JAS

SC *Arctoscopus japonicus* (Steindachner, 1881)
ES pez de arena japonés
DA japansk sandfisk
DE Japanischer Sandfisch
EL αρκτοσκόπος
EN Japanese sandfish; sailfin sandfish
FR poisson de sable japonais
IT
NL Japanse zandvis
PT peixe-areia japonês
FI japaninhietatähystäjä
SV japansk sandfisk

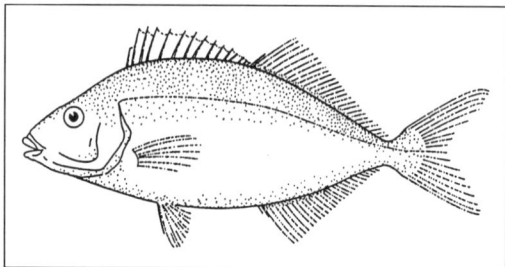

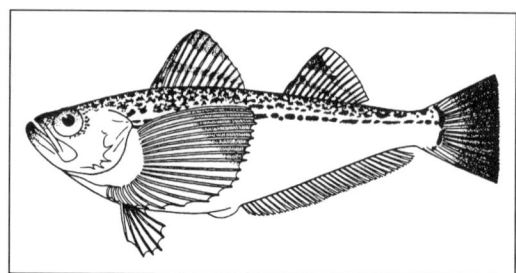

862 LATRIDAE BMO

SC *Latridopsis ciliaris* (Bloch and Schneider, 1801)
ES trompetero
DA blå trompeterfisk
DE Trompeterfisch
EL μπλέ τρομπέτα
EN blue moki
FR moki bleu
IT moki
NL blauwe Nieuw-Zeelandse trompetvis
PT trompete da Nova Zelândia
FI
SV blå moki

864 MUGILOIDIDAE NEB

SC *Parapercis colias* (Bloch and Schneider, 1801)
ES paraperca azul; namorado
DA newzealandsk flodbars
DE Neuseeland-Flußbarsch
EL γαλάζιος μπακαλιάρος της Νέας Ζηλανδίας
EN New Zealand blue cod
FR morue bleue de Nouvelle-Zélande
IT paraperca blu
NL Nieuw-Zeelandse blauwe kabeljauw
PT nedopa da Nova Zelândia
FI hietakala-laji
SV blå sandabborre

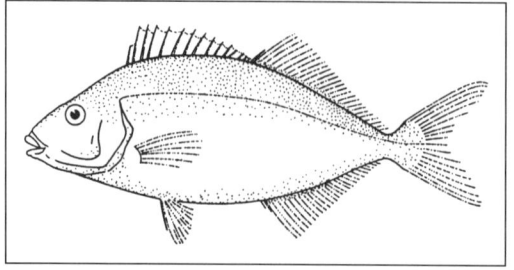

DRAWING NOT AVAILABLE

865 MUGILOIDIDAE

SPB

SC	*Pinguipes* spp.
ES	percas de arena brasileñas
DA	brasiliansk sandaborre-slægt
DE	Brasilianische Sandbarsche
EL	αμμόπερκα της Βραζιλίας
EN	Brazilian sandperches
FR	perches de sable
IT	morati
NL	Braziliaanse zandbaars
PT	nedopas do Brasil
FI	hietakalat
SV	sandabborrar

867 TRACHINIDAE

SC	*Echiichthys vipera* (Cuvier, 1829); *Trachinus vipera*
ES	escorpión
DA	lille fjæsing
DE	Kleines Petermännchen; Viperqueise; Zwergpetermännchen
EL	δράκαινα
EN	lesser weever
FR	petite vive
IT	tracina vipera; tracina ragno
NL	kleine pieterman
PT	peixe-aranha menor
FI	pikkulouhikala
SV	mindre fjärsing

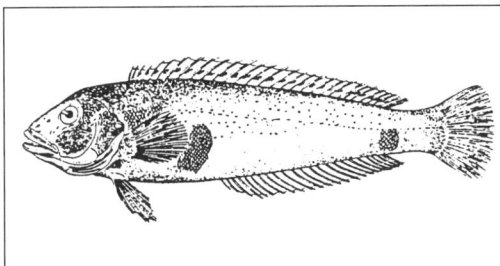

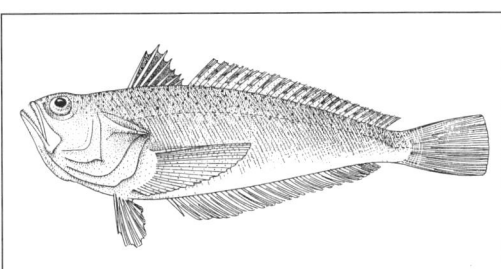

866 TRACHINIDAE

TRA

SC	*Trachinidae*
ES	arañas; traquínidos; escorpiones
DA	fjæsing-familien
DE	Drachenfische; Eigentliche Drachenfische
EL	δράκαινες
EN	weevers; weever fishes
FR	vives
IT	tracine; pesci ragno; trachinidi
NL	pietermannen
PT	peixes-aranha
FI	louhikalat; louhikalat-heimo
SV	fjärsingfiskar

868 TRACHINIDAE

SC	*Trachinus araneus* (Cuvier, 1829)
ES	araña
DA	plettet fjæsing
DE	Petermännchen; Spinnenqueise
EL	δράκαινα
EN	spotted weever
FR	vive araignée
IT	tracina ragno
NL	gevlekte pieterman
PT	peixe-aranha pontuado
FI	täplälouhikala
SV	fläckig fjärsing

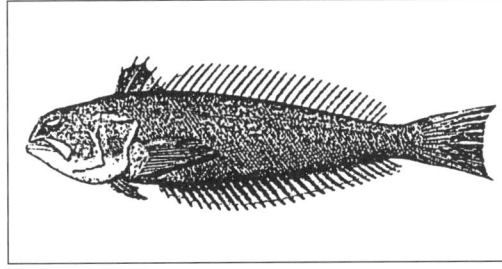

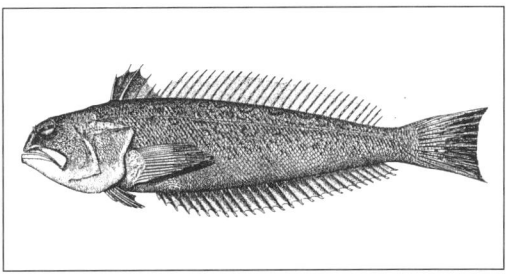

869 **TRACHINIDAE**   WEG

SC	*Trachinus draco* (Linnaeus, 1758)
ES	araña blanca; escorpión; peje araña
DA	stor fjæsing
DE	Petermännchen; Petersmann; Großes Petermännchen; Queise
EL	δράκαινα
EN	greater weever
FR	grande vive; vive commune
IT	tracina drago; tracina; ragno
NL	grote pieterman
PT	peixe-aranha maior; aranha grande
FI	louhikala
SV	fjärsing

871 **TRACHINIDAE**   WEX

SC	*Trachinus* spp.
ES	arañas
DA	fjæsing-slægten
DE	Petermännchen; Viperqueisen
EL	δράκαινες
EN	weevers; weever fishes
FR	vives
IT	tracine; pesci ragno
NL	pietermannen
PT	peixes-aranha
FI	louhikalat
SV	fjärsingar

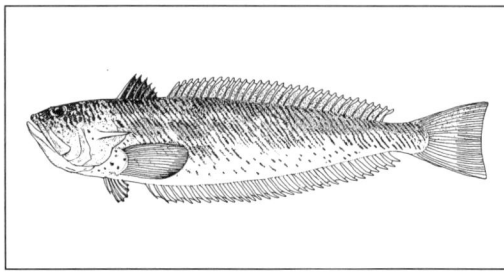

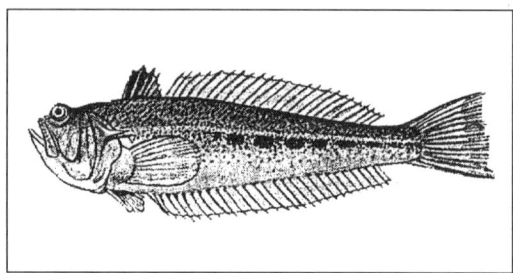

870 **TRACHINIDAE**

SC	*Trachinus radiatus* (Cuvier, 1829)
ES	víbora
DA	ru fjæsing
DE	Strahlenpetermännchen
EL	δράκαινα
EN	streaked weever
FR	vive à tête rayonnée
IT	tracina; ragno; tracina raggiata
NL	gestreepte pieterman
PT	peixe-aranha raiado
FI	sädelouhikala
SV	randig fjärsing

872 **PERCOPHIDAE**

SC	*Percophidae*
ES	picos de pato; peces palo
DA	fladhoved-familien
DE	Krokodilfische
EL	περκοφιδίδες
EN	flatheads
FR	gobies à tête plate; becs de canard
IT	pesci palo
NL	eendenbekken
PT	percofidídeos
FI	lattapääkalat; lattapääkalat-heimo
SV	anknäbbfiskar

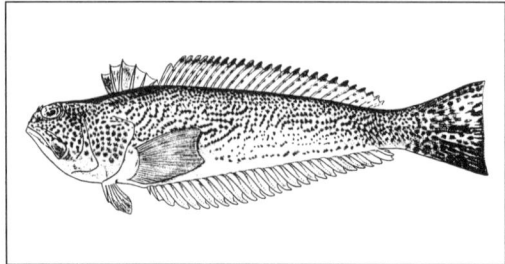

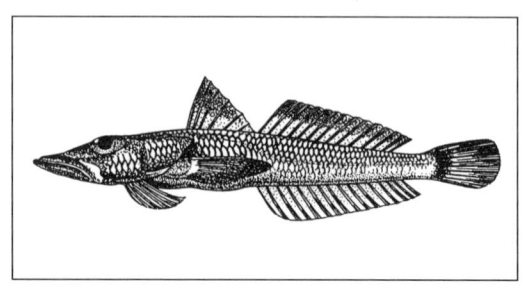

873   PERCOPHIDAE   FLA

SC	*Percophis brasilianus* (Quoy and Gaimard, 1825)
ES	pez palo
DA	brasiliansk fladhoved
DE	Brasilianischer Plattkopf
EL	πλατυκέφαλος της Βραζιλίας
EN	Brazilian flathead
FR	platête du Brésil
IT	pesce palo
NL	Braziliaanse eendenbek
PT	cabeça chata do Brasil
FI	brasilianlattapääkala
SV	brasiliansk anknäbbfisk

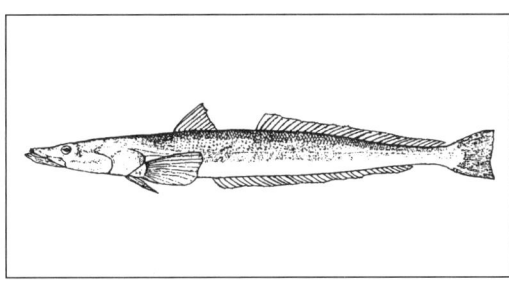

874   URANOSCOPIDAE

SC	*Uranoscopus scaber* (Linnaeus, 1758)
ES	rata
DA	europæisk stjernekigger
DE	Himmelsgucker; Sterngucker; Gemeiner Himmelsgucker
EL	λύχνος
EN	stargazer
FR	rascasse blanche; uranoscope
IT	pesce prete; lucerna mediterranea
NL	sterrenkijker
PT	cabeçudo
FI	taivaantähystäjä
SV	stjärnkikare

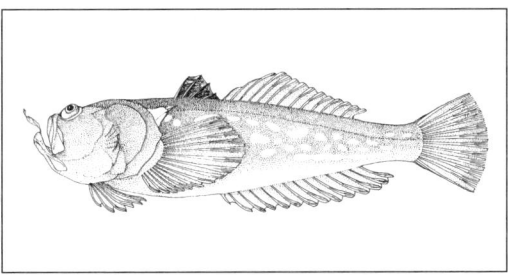

875   URANOSCOPIDAE   URA

SC	*Uranoscopus* spp.
ES	peces rata
DA	stjernekigger-slægt
DE	Himmelsgucker
EL	λύχνοι
EN	stargazers
FR	uranoscopes; rascasses blanches
IT	pesci prete
NL	sterrenkijkers
PT	cabeçudos; peixes cabeçudos; masca-tabaco
FI	taivaantähystäjät
SV	stjärnkikare

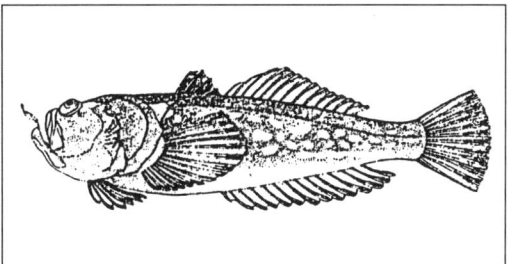

876   NOTOTHENIIDAE   NOX

SC	*Nototheniidae*
ES	nototénidos
DA	isfisk-familie; antarktiske torsk
DE	Antarktis-Eisfische
EL	νοτοθενίδες
EN	notothenids; Antarctic cods; noties; Antarctic rock-cods
FR	notothéniidés; morues antarctiques; bocasses; légines colandres; bocassettes
IT	nototenidi
NL	zuidpoolkabeljauwen
PT	nototenídeos; nototénias e afins
FI	antarktiset ahvenet; antarktiset ahvenet -heimo
SV	notingar

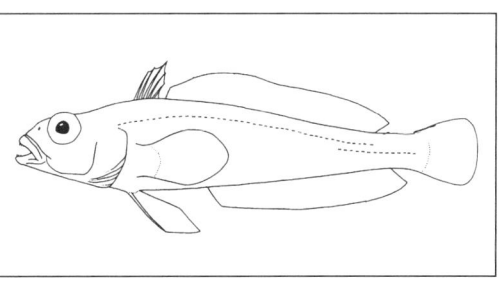

877	**NOTOTHENIIDAE**	TOP

SC	*Dissostichus eleginoides* (Smitt, 1898)
ES	merluza negra
DA	sort patagonisk isfisk
DE	Schwarzer Seehecht; Schwarzer Zahnfisch
EL	μαύρος μπακαλιάρος
EN	Patagonian toothfish
FR	légine australe
IT	austromerluzzo
NL	zwarte Patagonische ijsheek
PT	marlonga negra
FI	hammasjääahven
SV	tandnoting

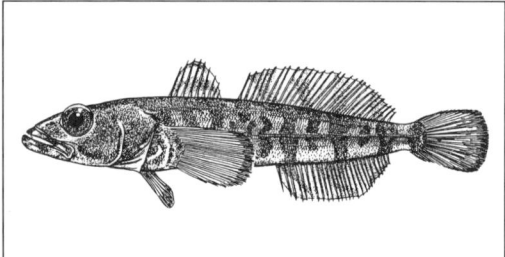

879	**NOTOTHENIIDAE**	TOT

SC	*Dissostichus* spp.
ES	merluzas australes
DA	isfisk-slægt
DE	Zahnfische
EL	μπακαλιάροι της Ανταρκτικής
EN	Antarctic toothfishes
FR	légines
IT	austromerluzzi
NL	Antarctische ijsheken
PT	marlongas
FI	antarktiset ahvenet -suku
SV	tandnotingar

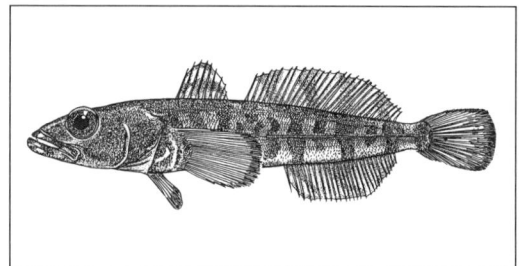

878	**NOTOTHENIIDAE**	TOA

SC	*Dissostichus mawsoni* (Norman, 1937)
ES	merluza austral
DA	antarktisk isfisk
DE	Antarktischer Zahnfisch
EL	μπακαλιάρος της Ανταρκτικής
EN	Antarctic toothfish
FR	légine antarctique
IT	austromerluzzo
NL	Antarctische ijsheek
PT	marlonga do Antárctico
FI	jääahven-laji
SV	antarktisk tandnoting

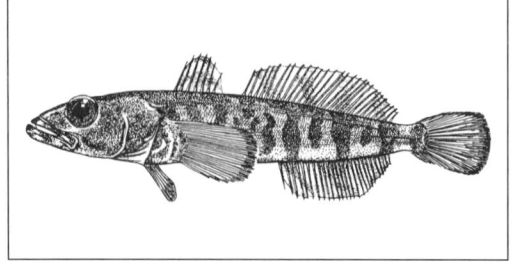

880	**NOTOTHENIIDAE**	BLP

SC	*Eleginops maclovinus* (Cuvier, 1830)
ES	róbalo patagónico
DA	patagonisk isfisk
DE	Patagonischer Zahnfisch
EL	σαλιάρα της Παταγωνίας
EN	Patagonian mullet; Patagonian blennie
FR	blennie de Patagonie
IT	robalo patagonico
NL	Patagonische ijsbaars
PT	babosa da Patagónia
FI	jääahven-laji
SV	patagonisk multenoting

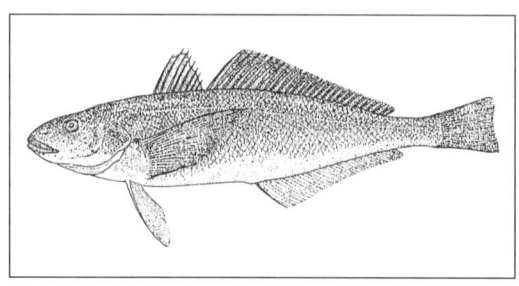

881 NOTOTHENIIDAE NOC

SC	*Notothenia coriiceps* (Richardson, 1844)
ES	nototenia negra
DA	sort isfisk
DE	Schwarze Notothenia
EL	νοτοθένια
EN	black rockcod; broad-headed notothenia
FR	bocasse noire; notothenia noir
IT	nototenia
NL	zwarte zuidpoolkabeljauw
PT	nototénia negra
FI	jääahven-laji
SV	svartnoting

883 NOTOTHENIIDAE NOG

SC	*Notothenia gibberifrons* (Lönnberg, 1905)
ES	nototenia cabezota
DA	pukkel-isfisk
DE	Grüne Notothenia
EL	νοτοθένια
EN	bumphead notothenia; humped rockcod
FR	bocasse bossue
IT	nototenia
NL	groene zuidpoolkabeljauw
PT	nototénia cabeça-chata
FI	jääahven-laji
SV	slanknoting

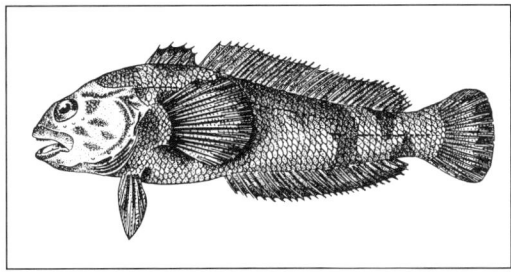

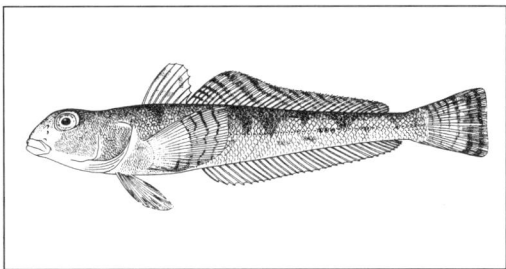

882 NOTOTHENIIDAE NON

SC	*Notothenia coriiceps neglecta* (Nybelin, 1951)
ES	nototenia amarilla
DA	gulbuget isfisk
DE	Gelbbauch-Notothenia
EL	νοτοθένια
EN	smoothhead notothenia; yellow belly rockcod
FR	bocasse jaune
IT	nototenia
NL	gele zuidpoolkabeljauw
PT	nototénia do Antárctico
FI	jääahven-laji
SV	gulbuksnoting

884 NOTOTHENIIDAE NOK

SC	*Notothenia kempi* (Norman, 1937)
ES	nototenia de ojos rayados
DA	øjestribet isfisk
DE	Augenstreifen-Notothenia
EL	νοτοθένια
EN	striped-eyed rockcod
FR	bocasse aux yeux rayés
IT	nototenia
NL	streepoog-zuidpoolkabeljauw
PT	nototénia olhos raiados
FI	jääahven-laji
SV	streckögd noting

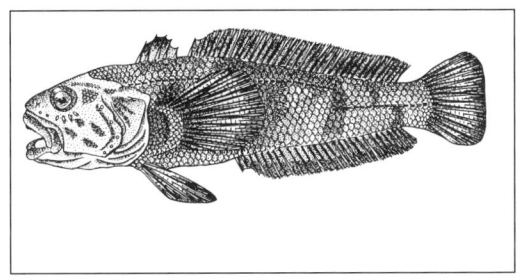

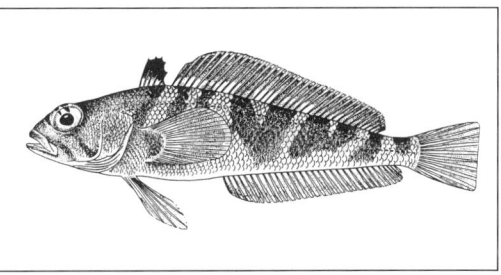

885	**NOTOTHENIIDAE**	NOR

SC *Notothenia rossii* (Richardson, 1844)
ES nototenia jaspeada
DA marmor-isfisk
DE Marmorbarsch
EL νοτοθένια
EN marbled notothenia; marbled rockcod
FR bocasse marbrée
IT nototenia
NL gemarmerde zuidpoolkabeljauw
PT nototénia marmoreada
FI jääahven-laji
SV marmorerad noting

887	**NOTOTHENIIDAE**	NOD

SC *Nototheniops nudifrons* (Lönnberg, 1905);
   Notothenia nudifrons
ES nototenia
DA barhovedet isfisk
DE Gelbflossen-Notothenia
EL νοτοθένια
EN naked-head notothenia; yellow-fin notie
FR bocassette dégarnie
IT nototenia
NL kale zuidpoolkabeljauw
PT nototénia cabeça lisa
FI jääahven-laji
SV gulfenad noting

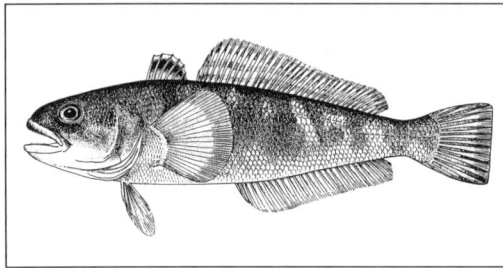

886	**NOTOTHENIIDAE**	NOS

SC *Notothenia squamifrons* (Günther, 1880)
ES nototenia gris
DA grå isfisk
DE Graue Notothenia
EL νοτοθένια
EN scaled notothenia; grey rockcod
FR bocasse grise; colin austral
IT nototenia
NL grijze zuidpoolkabeljauw
PT nototénia escamuda
FI jääahven-laji
SV grånoting

888	**NOTOTHENIIDAE**	TRH

SC *Pagothenia hansoni* (Boulenger, 1902);
   Trematomus hansoni
ES trama rayada
DA stribet isfisk
DE Gestreifte Notothenia
EL νοτοθένια
EN striped rockcod; green rockcod
FR bocasson rayé
IT nototenia
NL gestreepte rotskabeljauw
PT raboto-do-Antárctico
FI jääahven-laji
SV grön noting; blåhuvad noting

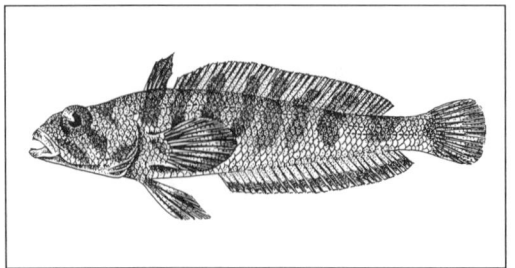

889	**NOTOTHENIIDAE**	NOM

SC   *Paranotothenia magellanica* (Forster, 1801)
ES   nototenia de Magallanes
DA   Magellan-isfisk
DE   Blaue Notothenia
EL   νοτοθένια
EN   Magellanic notothenia; Magellanic rockcod
FR   bocasse de Magellan
IT   nototenia
NL   Magellaan-rotskabeljauw
PT   nototénia da Patagónia
FI   jääahven-laji
SV   magellannoting

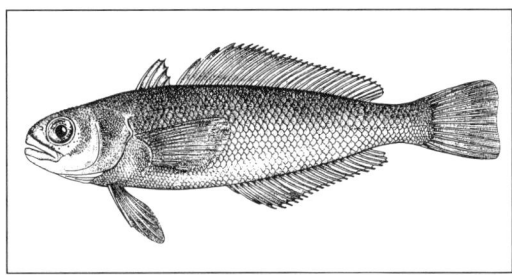

890	**NOTOTHENIIDAE**	NOT

SC   *Patagonotothen brevicauda* (Lönnberg, 1905)
ES   trama patagónica
DA   patagonisk stentorsk; korthalet isfisk
DE   Kurzschwanz-Notothenia
EL   νοτοθένια
EN   Patagonian rockcod
FR   bocasse de Patagonie à queue courte
IT   nototenia
NL   kortestaart Patagonische rotskabeljauw
PT   nototénia rabo-curto
FI   jääahven-laji
SV   kortstjärtsnoting

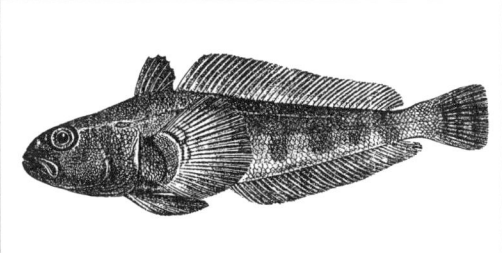

891	**NOTOTHENIIDAE**

SC   *Patagonotothen brevicauda guntheri* (Norman, 1937); *Notothenia guentheri*
ES   trama de Günther
DA   Günthers isfisk
DE   Günthers Notothenia
EL   νοτοθένια
EN   Gunther's notothenia
FR   bocasse de Patagonie
IT   nototenia
NL   Gunthers Patagonische rotskabeljauw
PT   nototénia de Günther
FI   jääahven-laji
SV   Günthers noting

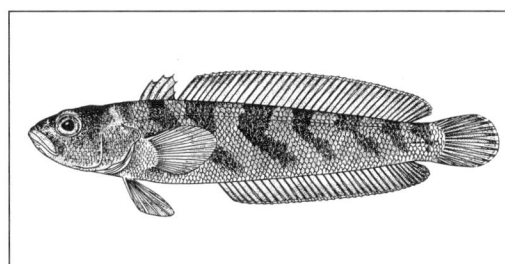

892	**NOTOTHENIIDAE**	PAT

SC   *Patagonotothen longipes ramsay* (Regan, 1913); *Notothenia ramsayi*
ES   nototenia de Ramsay
DA   Ramsays isfisk
DE   Ramsays Notothenia
EL   νοτοθένια
EN   Ramsay's icefish
FR   bocassette de Ramsay
IT   nototenia
NL   Ramsays rotskabeljauw
PT   nototénia de Ramsay
FI   jääahven-laji
SV   Ramsays noting

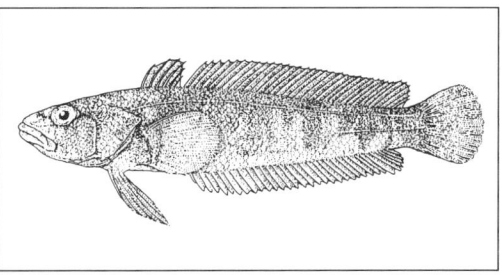

893 NOTOTHENIIDAE ANS

SC *Pleuragramma antarcticum* (Boulenger, 1902)
ES diabelo antártico
DA antarktisk sølvfisk
DE Antarktischer Silberfisch
EL νοτοθένια
EN Antarctic sidestripe; Antarctic silverfish
FR rascasse verte de l'Antarctique; calandre
    antarctique
IT nototenia
NL Antarctische zilvervis
PT peixe-calhandra antárctico
FI jääahven-laji
SV silvernoting

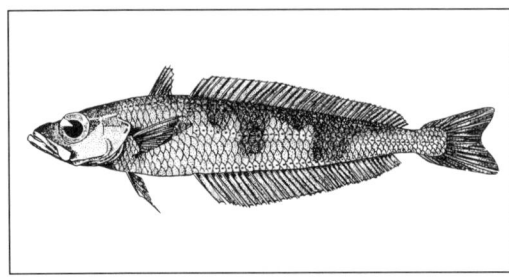

895 CHANNICHTHYIDAE ICX

SC *Channichthyidae*
ES peces hielo
DA isfisk-familie
DE Eisfische
EL παγόψαρα
EN icefishes
FR poissons des glaces
IT cannittidi; pesci del ghiaccio
NL ijsvissen
PT peixes-gelo
FI jääkalat; jääkalat-heimo
SV isfiskar

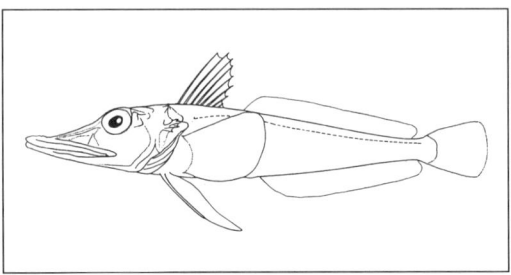

894 NOTOTHENIIDAE TRT

SC *Trematomus* spp.
ES bacalaos antárticos
DA isfisk-slægt
DE Hochantarktische Nototheniiden
EL μπακαλιάροι της Ανταρκτικής
EN Antarctic cods; Antarctic rockcods
FR bocassons; morues antarctiques
IT nototenie
NL Antarctische kabeljauwen
PT rabotos
FI jääahvenet
SV notingar

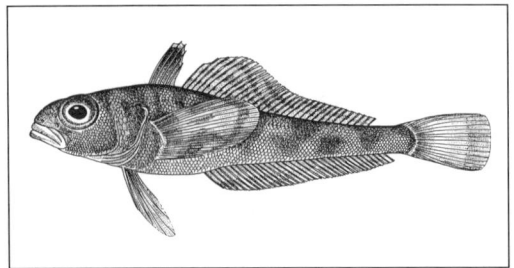

896 CHANNICHTHYIDAE SSI

SC *Chaenocephalus aceratus* (Lönnberg, 1906)
ES pez hielo austral
DA Scotia-isfisk
DE Scotia-See-Eisfisch
EL παγόψαρο της Σκωτίας
EN Scotia Sea icefish; blackfin icefish
FR grande gueule antarctique; poisson des glaces
    à nageoires noires
IT pesce del ghiaccio
NL Scotiazee ijsvis
PT peixe-gelo austral
FI jääkala
SV svartfenad isfisk

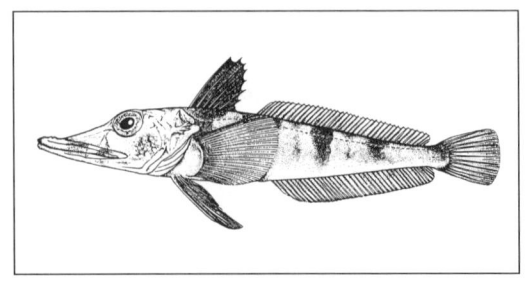

897 CHANNICHTHYIDAE WIC

SC	*Chaenodraco wilsoni* (Regan, 1914)
ES	pez hielo de Wilson
DA	Wilsons isfisk
DE	Wilsons Eisfisch
EL	παγόψαρο
EN	Wilson's icefish; spiny icefish
FR	grande gueule épineuse; poisson des glaces épineux
IT	pesce del ghiaccio
NL	Wilsons ijsvis
PT	peixe-gelo de Wilson
FI	krokotiilikala
SV	fläckig isfisk

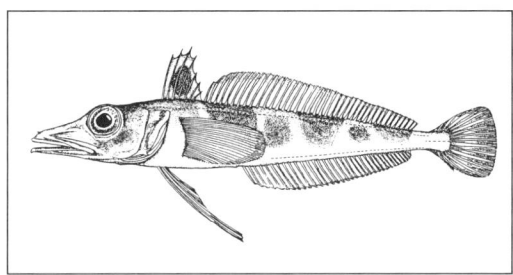

899 CHANNICHTHYIDAE LIC

SC	*Channichthys rhinoceratus* (Richardson, 1844)
ES	pez hielo narigudo
DA	langsnudet isfisk
DE	Langschnauzen-Eisfisch
EL	ρυγχοπαγόψαρο
EN	long-snouted icefish; unicorn icefish
FR	grande gueule; poisson des glaces à long nez
IT	pesce del ghiaccio
NL	langsnuit-ijsvis
PT	peixe-gelo bicudo
FI	jääkala-laji
SV	noshörningsfisk

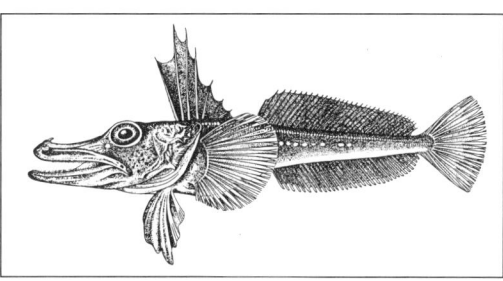

898 CHANNICHTHYIDAE ANI

SC	*Champsocephalus gunnari* (Lönnberg, 1905)
ES	pez hielo común
DA	båndet isfisk
DE	Bändereisfisch
EL	παγόψαρο της Ανταρκτικής
EN	Antarctic icefish; mackerel icefish
FR	poisson des glaces antarctique
IT	pesce del ghiaccio
NL	ijsvis
PT	peixe-gelo do Antárctico
FI	makrillijääkala
SV	Gunnars isfisk

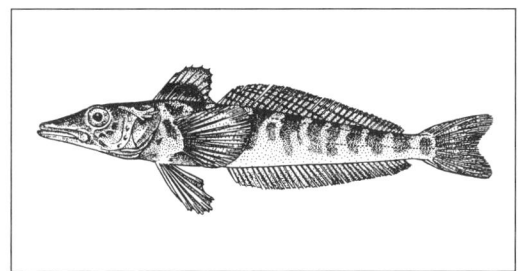

900 CHANNICHTHYIDAE KIF

SC	*Chionodraco rastrospinosus* (De Witt and Hureau, 1979)
ES	pez hielo ocelado
DA	øjeplettet isfisk
DE	Augenfleck-Eisfisch
EL	παγόψαρο του Κάθλιν
EN	Kathleen's icefish; ocellated icefish
FR	grande gueule ocellée; poisson des glaces ocellé
IT	pesce del ghiaccio ocellato
NL	grootoogijsvis
PT	peixe-gelo catarino
FI	jääkala-laji
SV	ögonisfisk

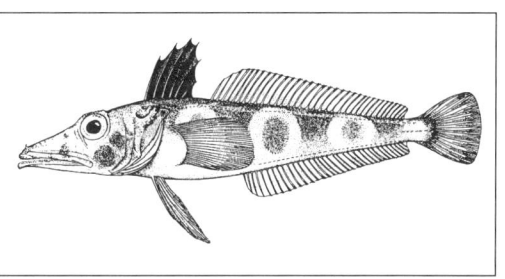

901 CHANNICHTHYIDAE SGI

SC	*Pseudochaenichthys georgianus* (Norman, 1937)
ES	pez hielo de Georgia
DA	Georgia-isfisk
DE	South-Georgia-Eisfisch
EL	παγόψαρο
EN	South Georgia icefish
FR	poisson glace de Georgie
IT	pesce del ghiaccio
NL	Georgia-ijsvis
PT	peixe-gelo da Geórgia do Sul
FI	tummajääkala
SV	krokodilisfisk

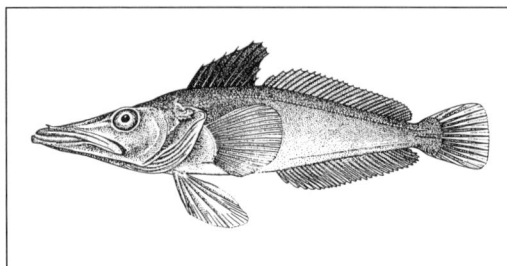

903 GERREIDAE MOJ

SC	*Gerres* spp.
ES	mojarras
DA	gerrid-slægt
DE	Mojarras; Silberlinge
EL	μοϊάρες
EN	mojarras; silver-biddies
FR	blanches
IT	mojarre
NL	plooibekken
PT	beicinhos
FI	mojarat
SV	silveller

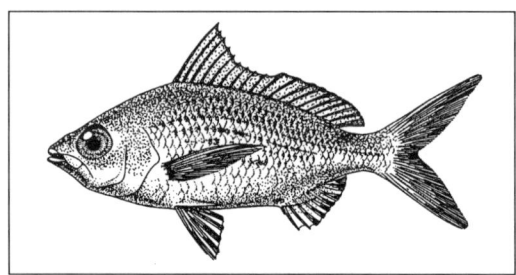

902 GERREIDAE MFF

SC	*Eucinostomus melanopterus* (Bleeker, 1863)
ES	mojarrita de ley
DA	sortfinnet gerrid
DE	Schwarzflossen-Mojarra; Schwarzflossen-Silberling
EL	μοϊάρα
EN	flagfin mojarra
FR	blanche drapeau
IT	mojarra
NL	zeilvinplooibek
PT	beicinho-prata
FI	mojara-laji
SV	vimpelsilvell

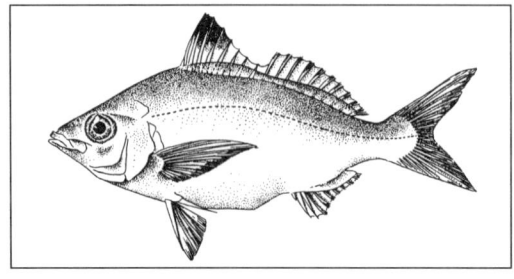

904 BLENNIIDAE BLE

SC	*Blenniidae*
ES	babosas
DA	slimfisk-familien
DE	Schleimfische
EL	σαλιάρες
EN	combtooth blennies
FR	blennies; baveuses
IT	blennidi; bavose
NL	naakte slijmvissen
PT	marachombas; blénios
FI	limakalat; limakalat-heimo
SV	slemfiskar

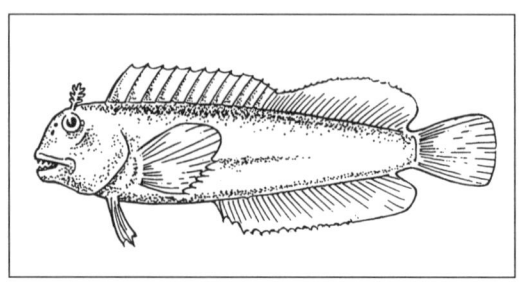

905 BLENNIIDAE

SC	*Blennius ocellaris* (Linnaeus, 1758)
ES	torillo; babosa ocelada
DA	øjeplettet tangkvabbe
DE	Seeschmetterling
EL	σαλιάρα· σαλιάρης· γλινός· λεβέρα
EN	butterfly blenny
FR	baveuse-papillon; blennie-papillon; blennie ocellée
IT	bavosa occhiuta
NL	vlinderslijmvis
PT	marachomba-borboleta
FI	purjelimakala
SV	havsfjäril

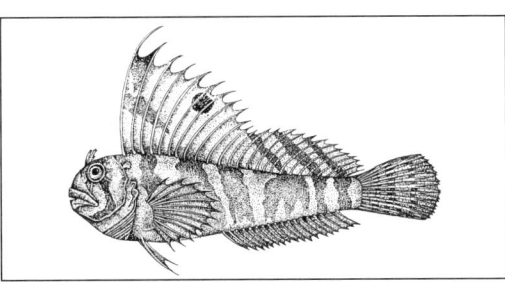

906 BLENNIIDAE

SC	*Blennius* spp.
ES	babosas
DA	tangkvabbe-slægt; slimfisk-slægt
DE	Schleimfische
EL	σαλιάρες· σαλιάρης
EN	blennies
FR	blennies; baveuses
IT	bavose
NL	slijmvissen
PT	marachombas
FI	limakalat
SV	slemfiskar

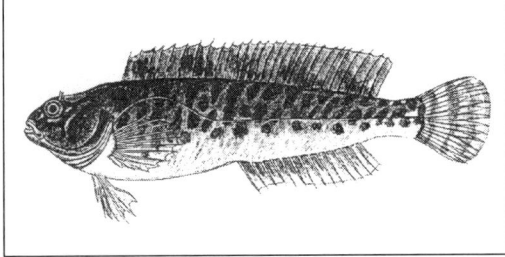

907 BLENNIIDAE

SC	*Lipophrys pavo* (Risso, 1810)
ES	gallerbo
DA	påfugleslimfisk
DE	Pfauenschleimfisch
EL	σαλιάρα· σαλιάρης· γλινός· παπαγάλος
EN	peacock blenny
FR	baveuse; blennie-paon
IT	bavosa; bavosa pavone
NL	pauwslijmvis
PT	marachomba-pavão
FI	riikinkukkolimakala
SV	påfågelsslemfisk

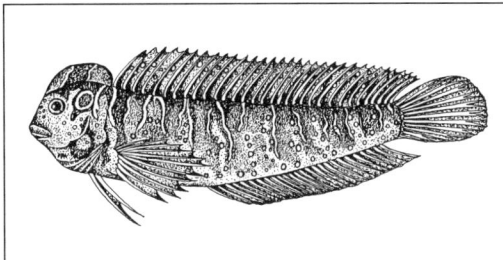

908 BLENNIIDAE

SC	*Lipophrys pholis* (Linnaeus, 1758)
ES	babosa atlántica
DA	almindelig tangkvabbe
DE	Atlantischer Schleimfisch
EL	σαλιάρα· σαλιάρης
EN	shanny
FR	mordocet; blennie palmicorne
IT	bavosa
NL	slijmvis
PT	marachomba-frade
FI	limakala
SV	skyggfisk

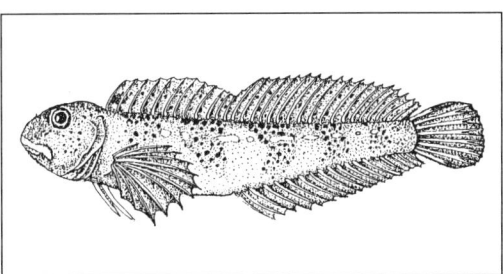

227

909 BLENNIIDAE

SC	*Parablennius gattorugine* (Brünnich, 1768)
ES	cabruza; babosa
DA	stribet slimfisk
DE	Gestreifter Schleimfisch
EL	σαλιάρα· ταινιοσαλιάρα· σαλιάρης· γλινός· λεβέρα
EN	tompot blenny
FR	baveuse; blennie gattorugine
IT	bavosa; bavosa ruggine
NL	gehoornde slijmvis
PT	marachomba babosa
FI	isolimakala
SV	slemfisk

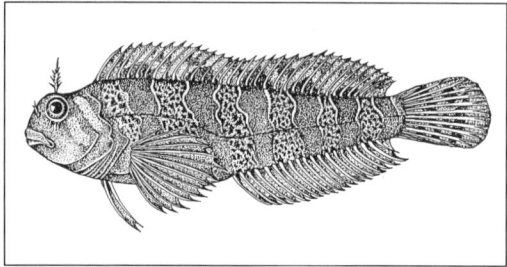

911 BLENNIIDAE

SC	*Salaria fluviatilis* (Asso, 1801); *Blennius fluviatilis*
ES	blenio de río; babosa de río; fraile
DA	ferskvandstangkvabbe
DE	Süßwasser-Schleimfisch; Fluß-Schleimfisch
EL	ποταμοσαλιάρα· σαλιάρα· μαρκόβα· σκυλόψαρο
EN	freshwater blenny
FR	blennie fluviatile; cagnette; baveuse
IT	cagnetto
NL	rivierslijmvis
PT	marachomba de água doce
FI	jokilimakala
SV	blodslemfisk

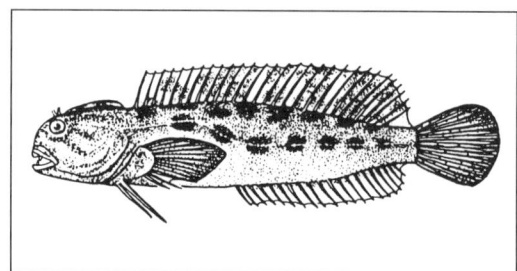

910 BLENNIIDAE

SC	*Parablennius sanguinolentus* (Pallas, 1811)
ES	babosa sanguinolenta
DA	blodstribet slimfisk
DE	Blutstriemen-Schleimfisch
EL	σαλιάρα· σαλιάρης· γλινός· λεβέρα
EN	blood-striped blenny
FR	blennie sanguine
IT	bavosa sanguigna
NL	roodgestreepte slijmvis
PT	marachomba salpicado
FI	harmaalimakala
SV	blodslemfisk

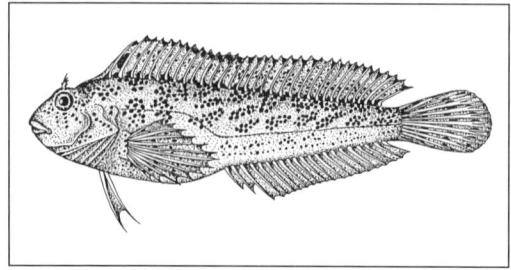

912 BLENNIDAE

SC	*Tripterygion tripteronotum* (Risso, 1810); *Tripterygion nasus*
ES	babosa de tres aletas
DA	trehale-slimfisk
DE	Spitzkopfschleimfisch; Dreiflossenschleimfisch
EL	σαλιάρα
EN	triple-fin blenny
FR	triptérygion à bec
IT	peperoncino; bavosa peparuolo
NL	drievinnige slijmvis
PT	cabrito
FI	limakala-laji
SV	trippelfena

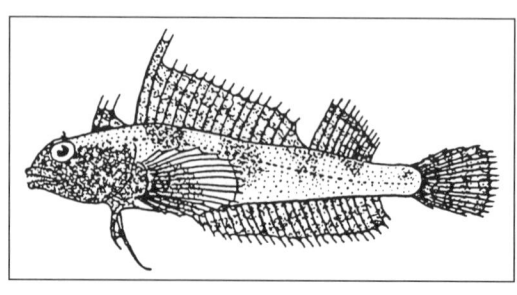

913   ANARHICHADIDAE

SC	*Anarhichadidae*
ES	peces perro
DA	havkat-familien
DE	Seewölfe
EL	λυκόψαρα
EN	wolffishes
FR	poissons-loups
IT	anaricadidi; lupi di mare; gattomare
NL	zeewolven
PT	peixes-lobo
FI	merikissat; merikissat-heimo
SV	havskattfiskar

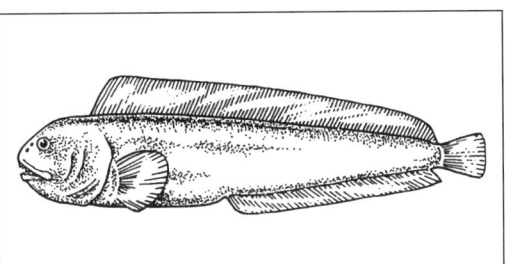

915   ANARHICHADIDAE   CAA

SC	*Anarhichas lupus* (Linnaeus, 1758)
ES	perro del Norte
DA	stribet havkat; almindelig havkat
DE	Gestreifter Katfisch; Steinbeißer; Seewolf; Katfisch; Gestreifter Seewolf; Austernfisch
EL	λυκόψαρο· λυκόψαρο του Ατλαντικού
EN	Atlantic catfish; rockfish; wolffish; Atlantic wolf-fish
FR	loup atlantique; loup anarhique à peau mince; loup de mer à peau mince; poisson-loup; loup marin; loup de mer
IT	lupo di mare; bavosa lupa; gattomare
NL	zeewolf
PT	peixe-lobo riscado
FI	merikissa
SV	havskatt

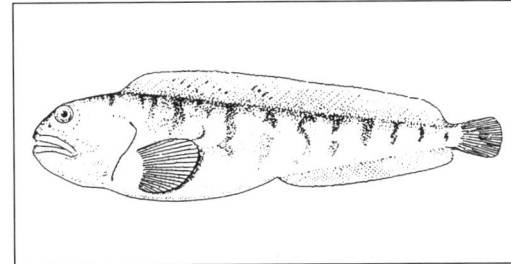

914   ANARHICHADIDAE

SC	*Anarhichas denticulatus* (Krøyer, 1845); *Anarhichas latifrons*
ES	perro azul
DA	blå havkat; bredpandet havkat
DE	Blauer Katfisch; Blauer Seewolf; Wasserkatze
EL	λυκόψαρο
EN	Northern wolffish
FR	loup de mer bleu; loup gélatineux
IT	bavosa lupa
NL	blauwe zeewolf
PT	peixe-lobo azul
FI	sinimerikissa
SV	blå havskatt

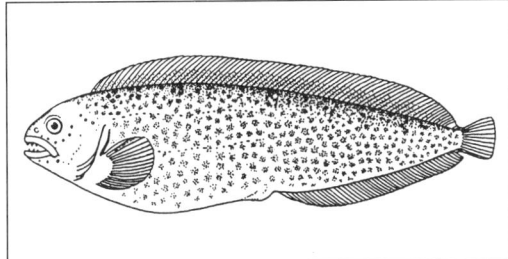

916   ANARHICHADIDAE   CAS

SC	*Anarhichas minor* (Olafsen, 1772)
ES	perro pintado
DA	plettet havkat
DE	Gefleckter Katfisch; Gefleckter Seewolf
EL	μικρό λυκόψαρο
EN	spotted sea cat
FR	petit loup de mer; loup de mer tacheté
IT	bavosa lupa
NL	gevlekte zeewolf
PT	peixe-lobo malhado
FI	kirjomerikissa
SV	fläckig havskatt

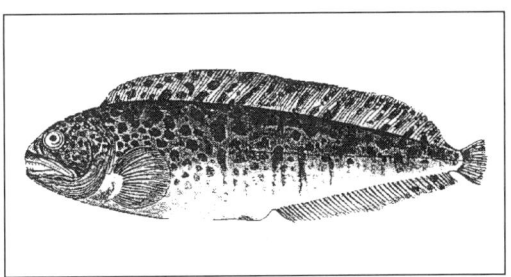

917 ANARHICHADIDAE

SC   *Anarhichas* spp.
ES   perritos del norte
DA   havkat-slægten
DE   Seewölfe; Katfische
EL   λυκόψαρα
EN   wolffishes; catfishes
FR   loups de mer
IT   bavose lupe
NL   zeewolven
PT   peixes-lobo
FI   merikissat
SV   havskatter

919 PHOLIDAE

SC   *Pholis gunnellus* (Linnaeus, 1758); *Centronotus gunnellus*
ES   pez mantequilla
DA   tangspræl
DE   Butterfisch; Messerfisch
EL   σαλιάρα
EN   butterfish
FR   gonelle; papillon de mer
IT   bavosa
NL   botervis
PT   peixe-gonela; gonela da rocha
FI   teisti
SV   tejstefisk

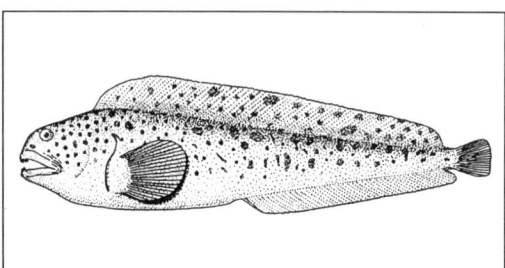

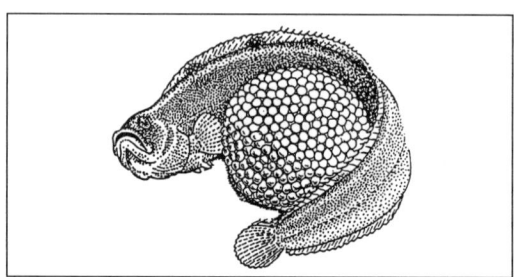

918 CLINIDAE

SC   *Clinitrachus argentatus* (Risso, 1810)
ES   clínido plateado
DA   sølv-slimfisk
DE   Haubenschleimfisch
EL   σαλιάρα των φυκών
EN   Mediterranean clinid
FR   cline argenté
IT   bavosella d'alga
NL   zilverslijmvis
PT   peixe-macaco
FI   luikero-laji
SV   silverslemfisk

920 ZOARCIDAE

SC   *Zoarcidae*
ES   zoárcidos
DA   ålekvabbe-familien
DE   Aalmuttern; Gebärfische
EL   φουσκοχειλόχελα
EN   eelpouts
FR   lycodes; loquettes
IT   zoarcidi
NL   puitalen
PT   peixes-carneiro
FI   kivinilkat; kivinilkat-heimo
SV   tånglakefiskar

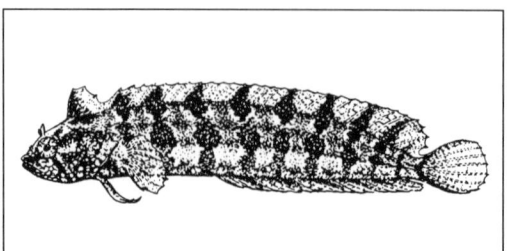

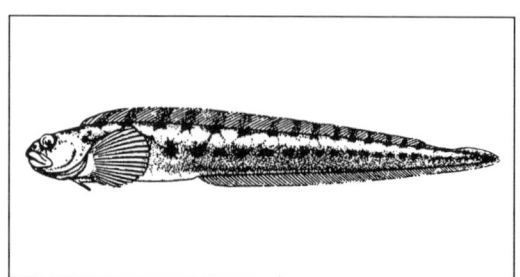

921   ZOARCIDAE   ELZ

SC	*Lycodes* spp.
ES	licodes
DA	ålekvabbe-slægt
DE	Wolfsfische
EL	λυκόδοντας
EN	eelpouts
FR	loquettes
IT	licodi
NL	puitalen
PT	peixes-carneiro do Árctico
FI	kivinilkka-suku
SV	ålbrosmar

923   ZOARCIDAE   ELP

SC	*Zoarces viviparus* (Linnaeus, 1758)
ES	babosa vivípara
DA	ålekvabbe
DE	Aalmutter
EL	φουσκοχειλόχελο
EN	eelpout; viviparous blenny; viviparous eelpout
FR	loquette; blennie vivipare
IT	blennio viviparo
NL	puitaal
PT	peixe-carneiro europeu
FI	kivinilkka
SV	tånglake

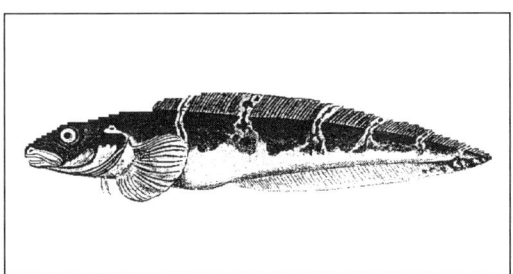

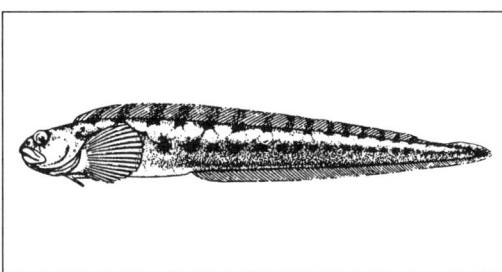

922   ZOARCIDAE   OPT

SC	*Macrozoarces americanus* (Schneider, 1801)
ES	babosa vivípara americana
DA	vestatlantisk ålekvabbe
DE	Nordamerikanische Aalmutter
EL	προβατόψαρο
EN	ocean pout
FR	loquette d'Amérique
IT	blennio viviparo americano
NL	Amerikaanse puitaal
PT	peixe-carneiro americano
FI	amerikankivinilkka
SV	västatlantisk ålbrosme

924   OPHIDIIDAE   BRD

SC	*Brotula barbata* (Bloch and Schneider, 1801)
ES	brótula de barbas
DA	skægget brotula
DE	Bärtige Brotula
EL	μπρότουλα με μουστάκια
EN	bearded brotula
FR	brotule barbe
IT	brotola
NL	brotula
PT	falsa abrótea
FI	partanilkka
SV	skäggbrotula

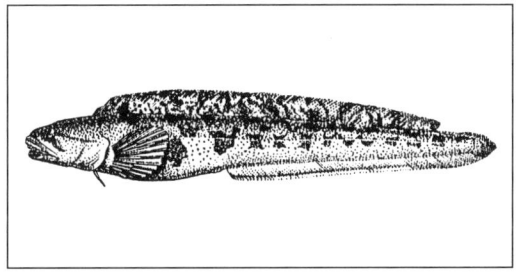

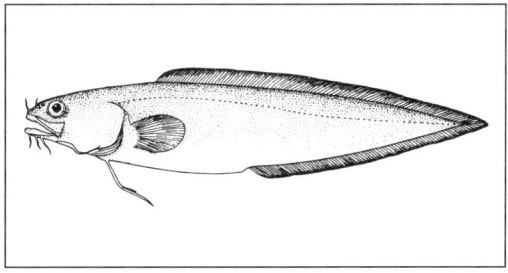

925 OPHIDIIDAE CUS

SC	*Genypterus blacodes* (Schneider, 1801)
ES	rosada
DA	rosa kingklip
DE	Rosa Kingklip
EL	κοκκινοφίδιο
EN	pink cusk-eel
FR	abadèche rose
IT	abadeco
NL	roze koningsklip
PT	maruca da Argentina; abadejo rosado
FI	punarihmanilkka
SV	golden kingklip

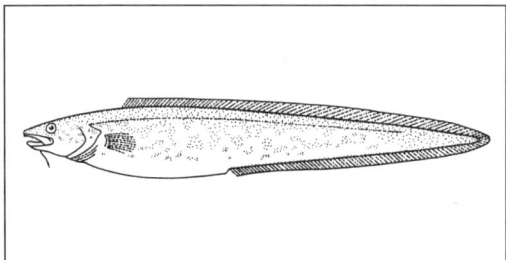

927 OPHIDIIDAE CUC

SC	*Genypterus chilensis* (Guichenot, 1848)
ES	rosada chilena
DA	chilensk kingklip
DE	Chilenischer Kingklip
EL	κοκκινοφίδιο της Χιλής
EN	red cusk-eel
FR	abadèche rouge; abadèche du Chili
IT	abadeco rosso
NL	Chileense koningsklip
PT	maruca do Chile; abadejo vermelho
FI	chilenrihmanilkka
SV	röd kingklip

DRAWING NOT AVAILABLE

926 OPHIDIIDAE KCP

SC	*Genypterus capensis* (Smith, 1847)
ES	rosada del Cabo
DA	sydafrikansk kingklip
DE	Südafrikanischer Kingklip
EL	οφίδιο του Ακρωτηρίου
EN	kingklip
FR	abadèche du Cap
IT	abadeco del Sudafrica
NL	Kaapse koningsklip
PT	maruca da África do Sul; abadejo do Cabo; abrótea sul-africana
FI	kapinrihmanilkka
SV	kapkingklip

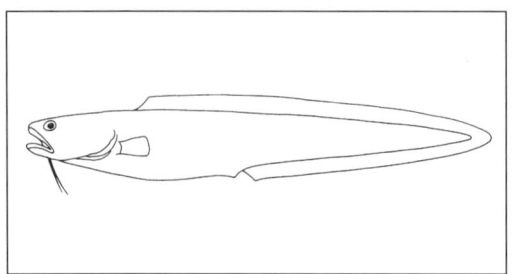

928 OPHIDIIDAE CEX

SC	*Genypterus* spp.
ES	rosadas
DA	kingklip-slægten
DE	Kingklips
EL	οφίδια
EN	cusk-eels
FR	abadèches
IT	abadechi
NL	koningsklippen
PT	marucas; abadejos
FI	partanilkka-suku
SV	kingklip

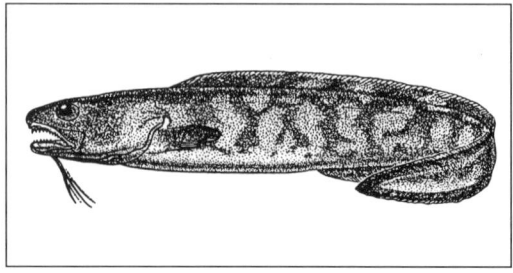

929 OPHIDIIDAE

SC	*Ophidion barbatum* (Linnaeus, 1758)
ES	doncella barbada
DA	slangekvabbe
DE	Bartmännchen
EL	γιλάρι
EN	snake blenny
FR	donzelle
IT	galletto
NL	baardmannetje
PT	peixe-cobrelo
FI	partanilkka
SV	ormfisk

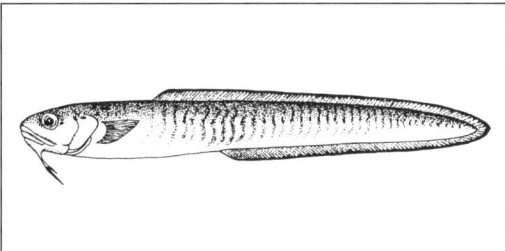

931 AMMODYTIDAE

SC	*Ammodytidae*
ES	lanzones; aguaciosos
DA	tobis-familien
DE	Sandaale; Sandspierlinge
EL	αμμόχελα
EN	sandeels; sandlances
FR	lançons; équilles
IT	cicerelli; ammoditidi
NL	zandspieringen
PT	galeotas; frachões; sandilhos
FI	tuulenkalat; tuulenkalat-heimo
SV	tobisfiskar

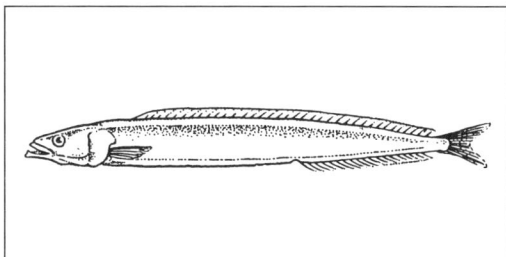

930 FIERASFERIDAE

SC	*Carapus acus* (Brunnich, 1768); *Fierasfer acus*
ES	pez de cristal; pez holoturia
DA	fierasfer
DE	Nadelfisch
EL	κρυσταλλόψαρο
EN	pearlfish
FR	aurin; fierasfer
IT	galiotto
NL	Middellandse-Zeeparelvis
PT	peixe-pérola
FI	neulakala
SV	nålfisk

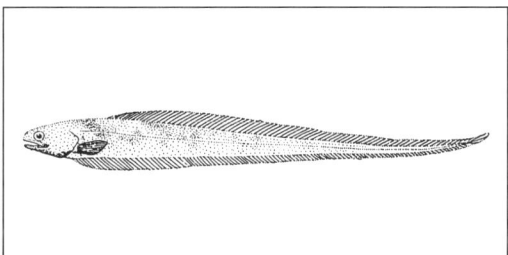

932 AMMODYTIDAE

SC	*Ammodytes americanus* (De Kay, 1842)
ES	lanzón americano
DA	amerikansk tobis
DE	Amerikanischer Sandaal
EL	αμμόχελο της Αμερικής
EN	American sandlance
FR	lançon d'Amérique
IT	cicerello americano
NL	Amerikaanse zandspiering
PT	galeota americana
FI	amerikantuulenkala
SV	amerikansk tobis

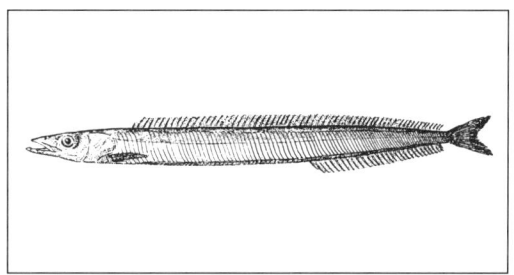

933 AMMODYTIDAE

SC	*Ammodytes dubius* (Reinhardt, 1838)
ES	lanzón del norte; aguacioso
DA	nordlig tobis
DE	Westatlantischer Sandaal
EL	αμμόχελο του Βορρά
EN	Northern sandlance
FR	lançon du Nord
IT	cicerello boreale
NL	IJslandse zandspiering
PT	galeota da Islândia
FI	tuulenkala-laji
SV	nordlig tobis

DRAWING NOT AVAILABLE

935 AMMODYTIDAE

SC	*Ammodytes marinus* (Raitt, 1934)
ES	lanzón del norte
DA	havtobis
DE	Nordischer Sandaal
EL	αμμόχελο
EN	Raitt's sandeel
FR	lançon nordique; équille du Nord
IT	cicerello del largo
NL	Noorse zandspiering
PT	galeota do norte; sandilho do norte
FI	merituulenkala
SV	havstobis

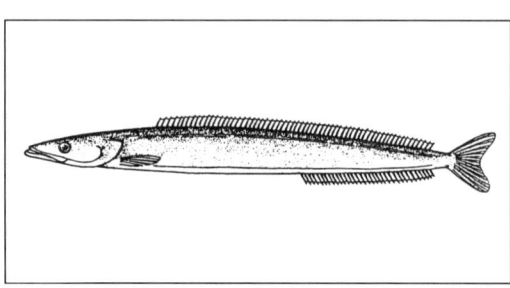

934 AMMODYTIDAE

SC	*Ammodytes hexapterus* (Pallas, 1811)
ES	lanzón del Pacífico
DA	Stillehavs-tobis
DE	Pazifischer Sandaal
EL	αμμόχελο του Ειρηνικού
EN	Pacific sandlance
FR	lançon du Pacifique
IT	cicerello del Pacifico
NL	Pacifische zandspiering
PT	galeota do Pacífico
FI	tyynenmerentuulenkala
SV	stillahavstobis

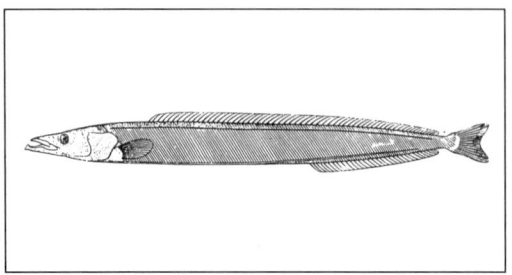

PAS 936 AMMODYTIDAE

SC	*Ammodytes tobianus* (Linnaeus, 1758); *Ammodytes lancea*
ES	pequeño lanzón
DA	kysttobis; sandål
DE	Kleiner Sandaal; Sandspierling; Tobiasfisch; Tobieschen
EL	μικροαμμόχελο
EN	sandeel
FR	petit lançon; équille; ammodyte
IT	ammodite tobiano; cicerello minore
NL	zandspiering
PT	galeota menor
FI	pikkutuulenkala
SV	kusttobis; blåtobis

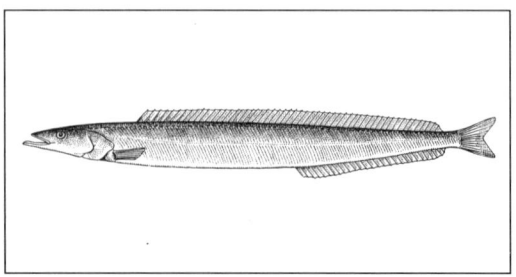

937 AMMODYTIDAE SAN

SC   *Ammodytes* spp.
ES   lanzones
DA   tobis-slægt
DE   Sandaale
EL   αμμόχελο
EN   sandeels; sandlances
FR   lançons; équilles
IT   cicerelli
NL   zandspieringen
PT   galeotas
FI   tuulenkala-suku
SV   tobisar

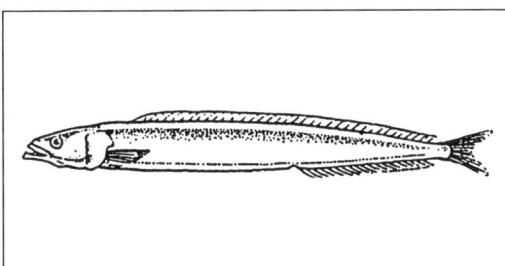

939 AMMODYTIDAE

SC   *Hyperoplus lanceolatus* (Le Sauvage, 1824);
      Ammodytes lanceolatus
ES   lanzón; aguacioso; sula; lansoya
DA   tobiskonge
DE   Großer Sandaal; Großer Sandspierling;
      Sandspierling
EL   αμμόχελο
EN   greater sandeel; sandlance; launce
FR   grand lançon; équille élancée
IT   ammodite lanceolato; cicerello
NL   smelt
PT   galeota maior; frachão; lingueirão
FI   isotuulenkala
SV   tobiskung

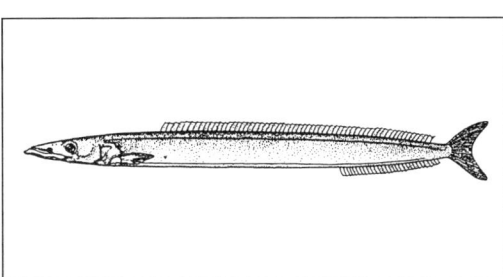

938 AMMODYTIDAE

SC   *Gymnammodytes cicerellus* (Rafinesque, 1810);
      Ammodytes cicerellus
ES   lanzón mediterráneo; barrinaire; sensu; sonso; enfú
DA   Middelhavs-nøgentobis
DE   Nacktsandaal; Mittelmeersandaal
EL   αμμόχελο
EN   Mediterranean sandeel; smooth sandlance
FR   cicerelle; lançon; équille
IT   cicerello
NL   naakte zandspiering
PT   galeota da areia; sandilho da areia
FI   välimerentuulenkala
SV   medelhavstobis

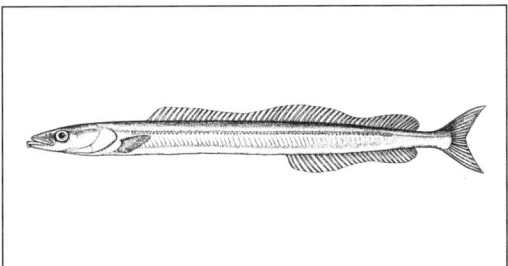

940 AMMODYTIDAE KLA

SC   *Hypoptychus dybowskii* (Steindachner, 1880)
ES   lanzón de Corea
DA   koreansk sandål
DE   Koreanischer Sandaal
EL   αμμόχελο της Κορέας
EN   Korean sandeel
FR   lançon de Corée
IT   cicerello coreano
NL   Koreaanse zandspiering
PT   galeota da Coreia
FI   koreantuulenkala
SV   koreaspigg

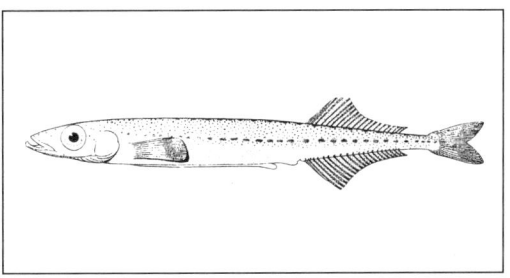

941 CALLIONYMIDAE

SC	*Callionymus lyra* (Linnaeus, 1758)
ES	lagarto
DA	stribet fløjfisk
DE	Gemeiner Leierfisch; Leierfisch; Europäischer Leierfisch
EL	τζιτζίκι· κροκοδείλι
EN	dragonet
FR	dragonet; dragonet lyre; doucet; chiqueur; callionyme lyre
IT	dragoncello
NL	pitvis
PT	peixe-pau-lira
FI	isomerikokki
SV	randig sjökock

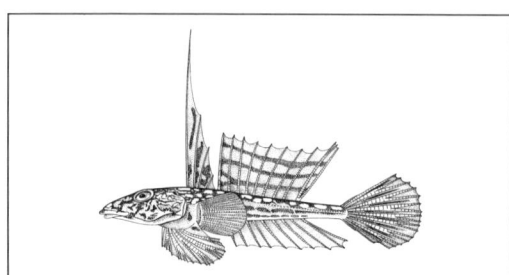

943 SIGANIDAE

SC	*Siganus rivulatus* (Forsskål, 1775); *Teuthis sigan*
ES	siguro
DA	kaninfisk
DE	Kaninchenfisch
EL	άσπρη αγριόσαλπα· γερμανός
EN	marbled spine-foot
FR	germanos; poisson-lapin
IT	sigano
NL	konijnvis
PT	macua
FI	kaniinikala-laji
SV	kaninfisk

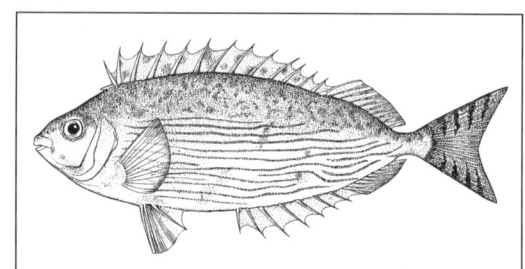

942 CALLIONYMIDAE

SC	*Callionymus maculatus* (Rafinesque, 1810)
ES	lagarto manchado
DA	plettet fløjfisk
DE	Gefleckter Leierfisch
EL	τζιτζίκι· κροκοδείλι
EN	spotted dragonet
FR	dragonet tacheté; callionyme tacheté
IT	dragoncello macchiato
NL	gevlekte pitvis
PT	peixe-pau malhado
FI	pikkumerikokki
SV	fläckig sjökock

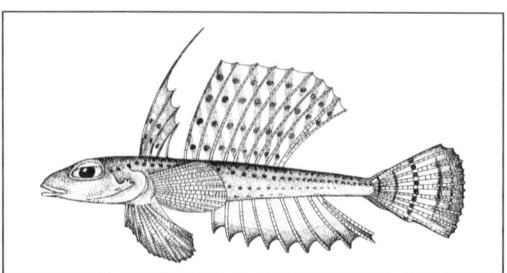

944 SIGANIDAE   SPI

SC	*Siganus* spp.
ES	siganos
DA	kaninfisk-slægt
DE	Kaninchenfische
EL	αγριόσαλπες
EN	spinefeet; rabbitfishes
FR	sigans; poissons-lapins
IT	sigani
NL	konijnvissen
PT	macuas
FI	kaniinikalat
SV	kaninfiskar

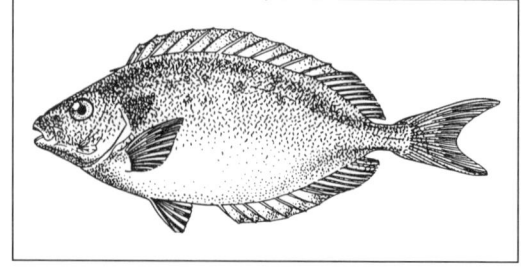

945 ACANTHURIDAE   SUR

SC	*Acanthuridae*
ES	navajones
DA	kirurgfisk-familien
DE	Doktorfische; Chirurgenfische; Seebader
EL	ψάρια χειρουργοί
EN	surgeonfishes; doctorfishes
FR	poissons-chirurgiens; acanthuridés
IT	pesci chirurgo
NL	doktersvissen
PT	unhas; peixes-canivete; peixes-cirurgião
FI	välskärikalat; välskärikalat-heimo
SV	kirurgfiskar

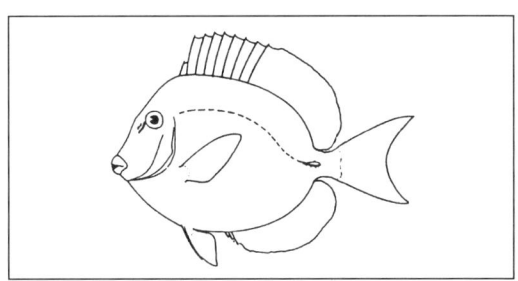

946 ACANTHURIDAE   MDO

SC	*Acanthurus monroviae* (Steindachner, 1876)
ES	navajón coniveto
DA	Monrovia-kirurgfisk
DE	Monrovia-Doktorfisch
EL	ψάρι χειρουργός
EN	Monrovian surgeonfish; Monrovian doctorfish
FR	chirurgien chas-chas
IT	pesce chirurgo
NL	Monrovische doktersvis
PT	unha
FI	monrovianvälskärikala
SV	monroviakirurgfisk

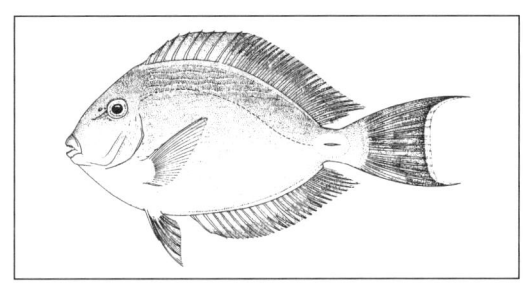

947 GEMPYLIDAE   GEP

SC	*Gempylidae*
ES	escolares
DA	slangemakrel-familien
DE	Schlangenmakrelen
EL	τζεμπιλίδες
EN	snake mackerels; escolars; oilfishes
FR	gempylidés; escoliers
IT	ruvetti
NL	slangmakrelen
PT	escolares e senucas
FI	käärmemakrillit; käärmemakrillit-heimo
SV	havsgäddfiskar

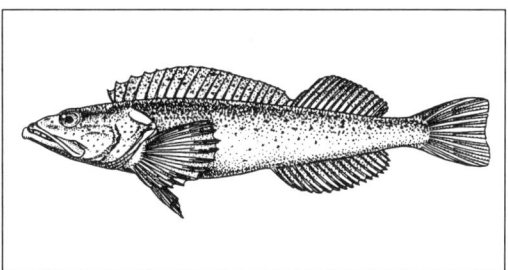

948 GEMPYLIDAE

SC	*Lepidocybium* spp.
ES	escolares
DA	slangemakrel-slægt
DE	Escolar-Schlangenmakrelen
EL	εσκολάρ
EN	escolars
FR	escoliers
IT	ruvetti
NL	scholieren
PT	escolares pretos
FI	käärmemakrillit
SV	havsgäddor

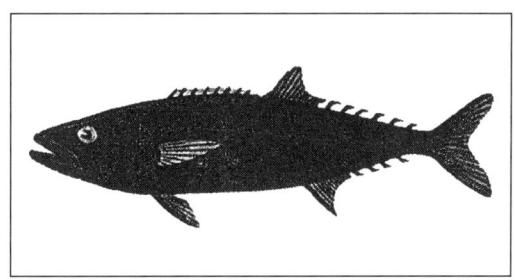

949 GEMPYLIDAE

SC	*Rexea solandri* (Cuvier, 1831)
ES	escolar real
DA	kongeslangemakrel
DE	Königs-Schlangenmakrele
EL	βασιλικός εσκολάρ
EN	gemfish
FR	escolier royal
IT	tirsite
NL	koningsslangmakreel
PT	escolar real
FI	käärmemakrilli-laji
SV	kungshavsgädda

951 GEMPYLIDAE SNK

SC	*Thyrsites atun* (Euphrasen, 1791)
ES	sierra
DA	atun
DE	Atun
EL	
EN	snoek; barracouta
FR	thyrsite; escolier; escolar
IT	tirsite
NL	snoekmakreel
PT	senuca; foguete
FI	kuta
SV	blå havsgädda

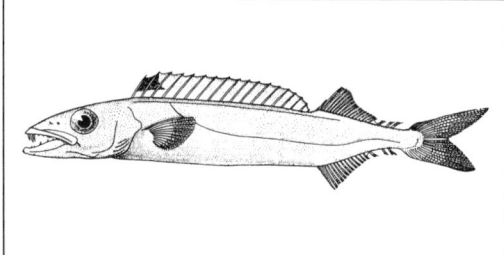

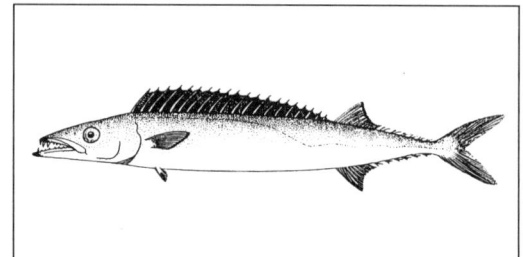

950 GEMPYLIDAE OIL

SC	*Ruvettus pretiosus* (Cocco, 1829)
ES	escolar; buvego
DA	oliefisk
DE	Ölfisch; Rizinusöl-Fisch; Schlangenmakrele
EL	μαυρόψαρο
EN	oilfish; scourer
FR	rouvet; escolier espagnol
IT	ruvetto
NL	olievis
PT	escolar; chocolate
FI	öljykala
SV	oljefisk

952 GEMPYLIDAE

SC	*Thyrsites* spp.
ES	sierras
DA	slangemakrel-slægt
DE	Buttermakrelen; Snoeks
EL	
EN	barracoutas; snoeks
FR	thyrsites
IT	tirsiti
NL	snoekmakrelen
PT	senucas
FI	käärmekalat
SV	havsgäddor

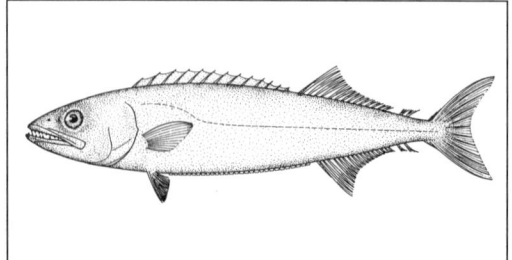

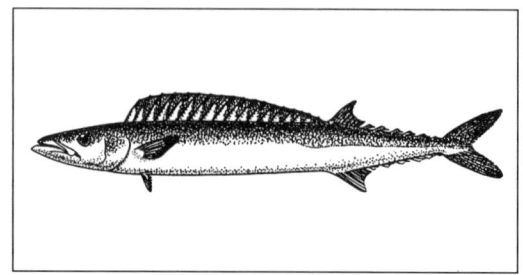

953 GEMPYLIDAE

SC	*Thyrsitoides marleyi* (Fowler, 1929)
ES	escolar de Marley
DA	Marleys slangemakrel
DE	Marleys Schlangenmakrele
EL	
EN	Marley's snake mackerel
FR	escolier gracile
IT	
NL	zwarte snoekmakreel
PT	senuca preta
FI	käärmekala-laji
SV	svart havsgädda

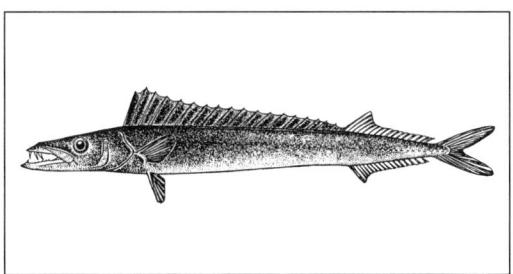

955 TRICHIURIDAE   CUT

SC	*Trichiuridae*
ES	peces sable; sables
DA	hårhale-familien
DE	Haarschwänze; Rinkfische
EL	σπαθόψαρα· ασημόψαρα
EN	scabbardfishes; cutlassfishes; hairtails
FR	sabres; trichiures; ceintures d'argent; poissons-sabres
IT	pesci sciabola; trichiuridi
NL	haarstaarten
PT	peixes-espada e lírios
FI	huotrakalat; huotrakalat-heimo
SV	hårstjärtfiskar

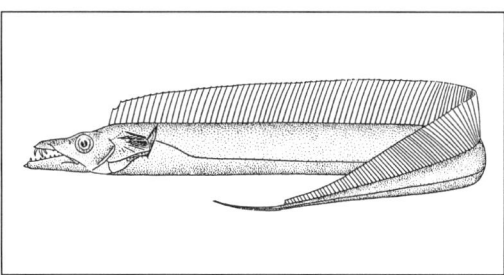

954 GEMPYLIDAE   WSM

SC	*Thyrsitops lepidopoides* (Cuvier, 1832)
ES	escolar sierra
DA	hvid slangemakrel
DE	Weißer Atun
EL	
EN	white snake mackerel; sierra
FR	escolier blanc
IT	tirsite
NL	witte snoekmakreel
PT	escolar-serra
FI	valkokuta
SV	vit havsgädda

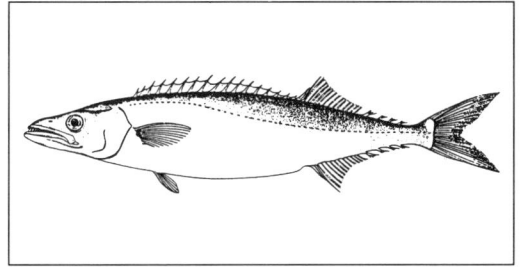

956 TRICHIURIDAE   BSF

SC	*Aphanopus carbo* (Lowe, 1839)
ES	sable negro
DA	sort sabelfisk
DE	Kurzflossen-Haarschwanz
EL	μαύρο σπαθόψαρο
EN	black scabbardfish
FR	sabre noir
IT	pesce sciabola nero
NL	zwarte haarstaart
PT	peixe-espada preto; espada preto
FI	mustahuotrakala
SV	dolkfisk

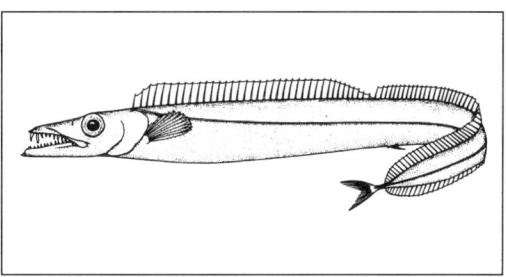

957   TRICHIURIDAE   SFS

SC   *Lepidopus caudatus* (Euphrasen, 1788)
ES   pez cinto; pez sable; sable; espada; aguja paladar
DA   strømpebåndsfisk
DE   Degenfisch; Strumpfbandfisch
EL   σπαθόψαρο· ασημόψαρο
EN   scabbardfish; silver scabbardfish; frostfish
FR   sabre argenté; coutelas; lépidope argentin;
      jarretière; bonnat
IT   pesce sciabola
NL   zilveren haarstaart
PT   peixe-espada
FI   hopeahuotrakala
SV   strumpebandsfisk

959   TRICHIURIDAE   LHT

SC   *Trichiurus lepturus* (Linnaeus, 1758)
ES   pez sable; sable; paire; espada
DA   hårhale
DE   Degenfisch; Haarschwanz
EL   λεπτουροσπαθόψαρο· σπαθόψαρο·
      ασημόψαρο
EN   cutlassfish; large-eyed hairtail; large-head hairtail
FR   poisson-sabre commun; trichiure argenté;
      poisson ceinture argenté
IT   pesce coltello
NL   degenvis
PT   lírio; espada-lírio
FI   huotrakala
SV   hårstjärt

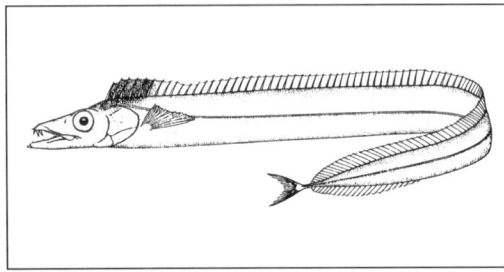

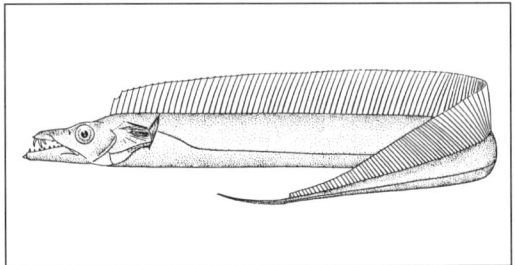

958   TRICHIURIDAE

SC   *Lepidopus xantusi* (Goode and Bean, 1895)
ES   pez sable del Pacífico
DA   strømpebåndsfisk
DE   Degenfisch
EL   σπαθόψαρο· ασημόψαρο
EN   scabbardfish
FR   sabre du Pacifique; jarretière
IT   pesce sciabola del Pacifico
NL   Pacifische haarstaart
PT   peixe-espada do Pacífico
FI   huotrakala-laji
SV   stillahavsstrumpebandsfisk

960   SCOMBROIDEI   TUX

SC   *Scombroidei*
ES   peces parecidos a los atunes
DA   makrelfisk
DE   Thunfischartige
EL   σκομβροειδή
EN   tuna-like fishes
FR   poissons type thon
IT   scomberoidi
NL   tonijnachtigen
PT   atuns e afins
FI   makrillit
SV   makrillika fiskar

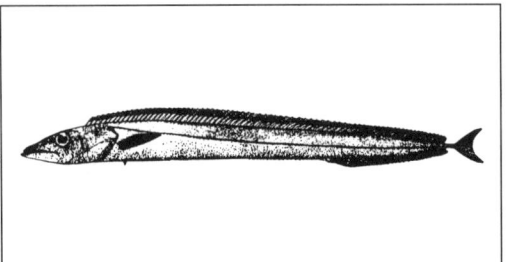

DRAWING NOT AVAILABLE

961 SCOMBRIDAE — MTX

SC	*Scombridae*
ES	escómbridos
DA	makrel- og tunfisk-familien
DE	Makrelen und Thunfische
EL	σκουμπριά· τόνοι· παλαμίδες
EN	tunas and mackerels
FR	thons; maquereaux; bonites et thazards
IT	scombridi, sgombri, tonni, palamite e maccarelli
NL	makrelen en tonijnen
PT	escombrídeos
FI	makrillit; makrillit-heimo
SV	makrillfiskar

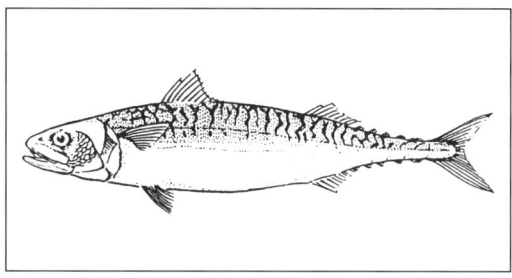

963 SCOMBRIDAE — SLT

SC	*Allothunnus fallai* (Serventy, 1948)
ES	atún lazón
DA	slank tun
DE	Schlankthun
EL	λεπτότονος
EN	slender tuna
FR	thon élégant
IT	tonnina
NL	slanke tonijn
PT	atum-foguete
FI	tonnikala-laji
SV	slank tonfisk

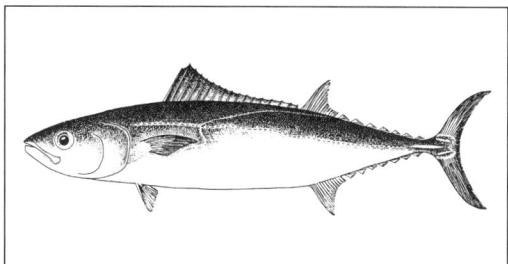

962 SCOMBRIDAE — WAH

SC	*Acanthocybium solandri* (Cuvier, 1832)
ES	peto
DA	wahoo
DE	Wahoo; Peto
EL	γουάχο
EN	wahoo
FR	thazard bâtard; wahoo
IT	waho; maccarello striato
NL	wahoo
PT	serra da Índia
FI	raitamakrilli
SV	wahoo

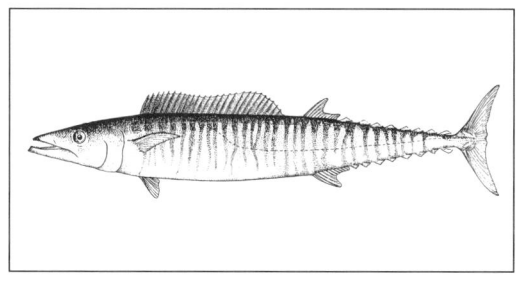

964 SCOMBRIDAE — BLT

SC	*Auxis rochei* (Risso, 1810)
ES	melva; judío; canutero
DA	fregatmakrel
DE	Melvera-Fregattmakrele
EL	κοπάνι· βαρελάκι
EN	frigate mackerel; plain bonito; bullet mackerel; bullet tuna
FR	melva; bonitou; auxide
IT	tombarello; biso
NL	kogeltonijn
PT	judeu
FI	auksidi-laji
SV	auxid

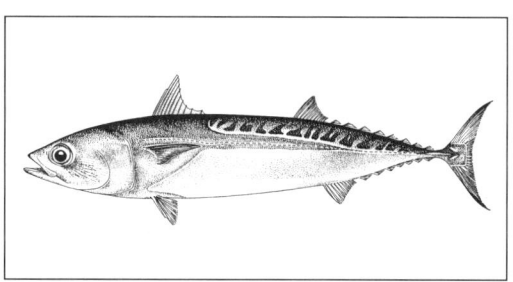

965	**SCOMBRIDAE**	FRI

SC *Auxis thazard* (Lacepède, 1800)
ES melva tazard
DA auxide
DE Fregattmakrele; Unechter Bonito; Makrelthunfisch
EL κοπάνι· βαρελάκι· κοπανάκι
EN frigate tuna
FR thazard; auxide
IT biso; tombarello
NL fregattonijn
PT judeu liso
FI auksidi
SV fregattauxid

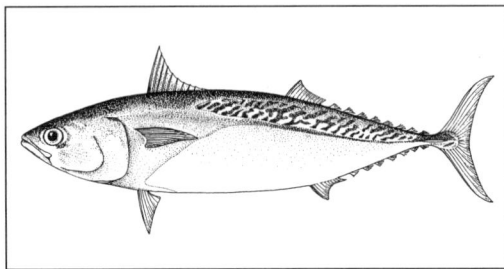

967	**SCOMBRIDAE**	KAW

SC *Euthynnus affinis* (Cantor, 1849)
ES bacoreta oriental
DA lille thunnin
DE Pazifische Thonine
EL τονάκι της Ανατολής
EN kawakawa
FR thonine orientale
IT tonnetto
NL Pacifische tonijn
PT merma oriental
FI boniitti-laji
SV kawakawa

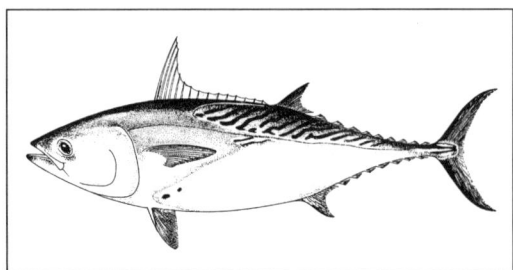

966	**SCOMBRIDAE**	LEB

SC *Cybiosarda elegans* (Whitley, 1935)
ES bonito saltador
DA plettet bonit
DE Gefleckter Bonito
EL στικτοπαλαμίδα
EN leaping bonito
FR bonite à dos tacheté; bonite élégante
IT palamita maculata
NL gevlekte bonito
PT bonito saltador
FI boniitti-laji
SV fläckig pelamid; fläckig bonit

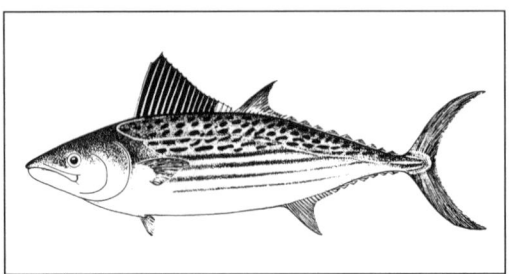

968	**SCOMBRIDAE**	LTA

SC *Euthynnus alletteratus* (Rafinesque, 1810);
    Thynnus thonina; *Euthynnus quadripunctatus*
ES bacoreta; albacora; toñina; sarda
DA thunnin
DE Falscher Bonito; Thonine; Kleiner Thun;
    Gefleckter Thunfisch
EL καρβούνι· τονίνα· λεκατίκι· τάσκα
EN little tuna; little tunny; mackerel tuna; false alba-
    core; Atlantic black skipjack
FR thonine; fausse bonite; ravil
IT tonnetto; alletterato
NL kleine tonijn
PT merma
FI tunniina
SV tunnina

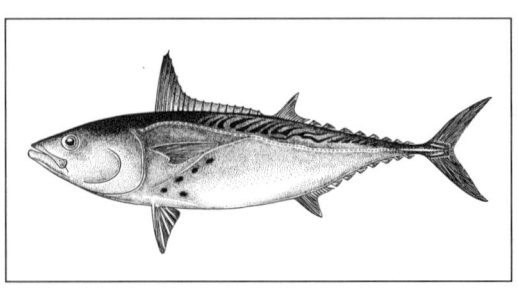

969 SCOMBRIDAE BKJ

SC	*Euthynnus lineatus* (Kishinouye, 1920)
ES	barrilete negro
DA	sort thunnin
DE	Schwarze Thonine
EL	μαύρο καρβούνι
EN	black skipjack
FR	thonine noire
IT	tonnetto
NL	zwarte skipjack
PT	merma negra
FI	mustaboniitti
SV	svart tunnina

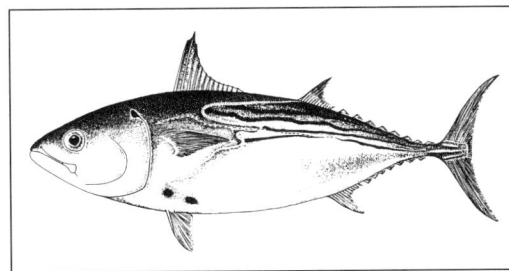

970 SCOMBRIDAE BUK

SC	*Gasterochisma melampus* (Richardson, 1845)
ES	atún chauchera
DA	sommerfugletun
DE	Schmetterlings-Thunfisch
EL	παλαμίδα με λέπια
EN	butterfly kingfish
FR	thon-papillon
IT	palamita squamosa
NL	vlindertonijn
PT	serra-borboleta
FI	tonnikala-laji
SV	fjärilstonfisk

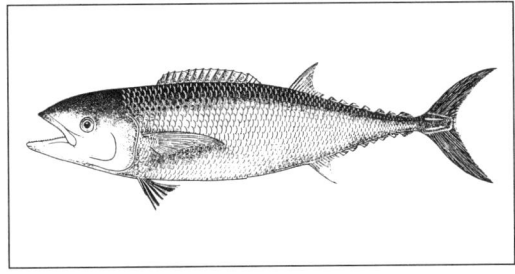

971 SCOMBRIDAE SHM

SC	*Grammatorcynus bicarinatus* (Quoy and Gaimard, 1824)
ES	carite-cazón
DA	hajmakrel
DE	Haimakrele
EL	σκομπροκαρχαρίας
EN	shark mackerel
FR	thazard requin
IT	ammoniosgombro
NL	haaimakreel
PT	serra-cação
FI	haimakrilli
SV	hajmakrill

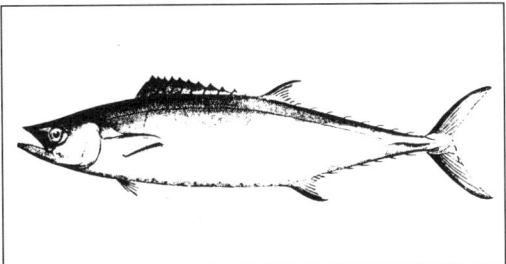

972 SCOMBRIDAE DBM

SC	*Grammatorcynus bilineatus* (Rüppel, 1836)
ES	carite-cazón pintado
DA	dobbeltlinjet makrel
DE	Doppellinien-Makrele
EL	δίγραμμο σκουμπρί
EN	double-lined mackerel
FR	thazard kusara
IT	ammoniosgombro
NL	dubbellijnmakreel
PT	serra-cação pintada
FI	haimakrilli-laji
SV	dubbellinjemakrill

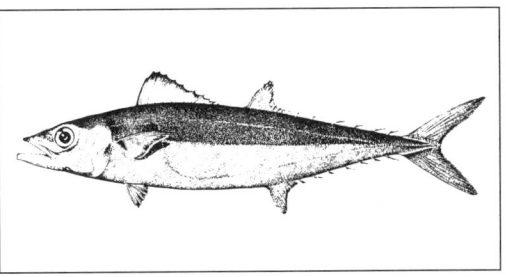

973   SCOMBRIDAE   SKJ

SC	*Katsuwonus pelamis* (Linnaeus, 1758); *Euthynnus pelamis; Thynnus pelamys*
ES	listado; bonito de altura; palomida; conejo; serrucho
DA	bugstribet bonit
DE	Echter Bonito; Gestreifter Thunfisch; Bonito; Bauchstreifiger Bonito
EL	παλαμίδα· λακέρδα
EN	skipjack; bonito; stripe-bellied bonito; striped tuna; oceanic bonito; skipjack tuna
FR	bonite à ventre rayé; bonite vraie
IT	tonnetto striato; tonno
NL	skipjack
PT	gaiado; bonito de barriga listada
FI	boniitti
SV	bonit

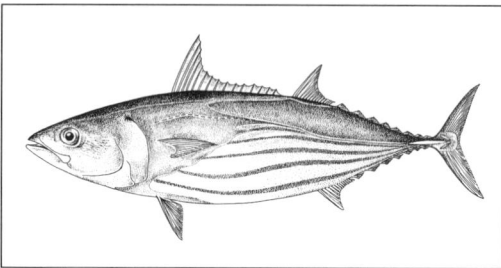

975   SCOMBRIDAE   RAB

SC	*Rastrelliger brachysoma* (Bleeker, 1850)
ES	caballa rechoncha
DA	kort dværgmakrel
DE	Kurze Zwergmakrele
EL	σκουμπρί των Ινδιών
EN	short mackerel
FR	maquereau trapu
IT	sgombro indiano
NL	korte makreel
PT	cavala curta
FI	lyhytkääpiömakrilli
SV	kortmakrill

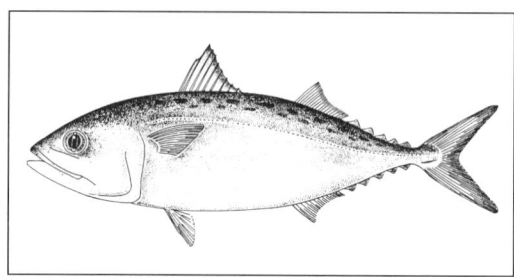

974   SCOMBRIDAE   BOP

SC	*Orcynopsis unicolor* (Geoffroy St.-Hilaire, 1817); *Sarda unicolor; Pelamys bonapartei*
ES	casarte ojón
DA	ustribet pelamide
DE	Ungestreifte Pelamide
EL	κοπάνι
EN	bonito; dog-tooth tuna; plain bonito
FR	palomète; maquereau unicolore; bonite orientale; bonite grosyeux
IT	palamita bianca
NL	ongestreepte bonito
PT	bonito-dente de cão; palmeta
FI	juovaton sarda
SV	blank pelamid

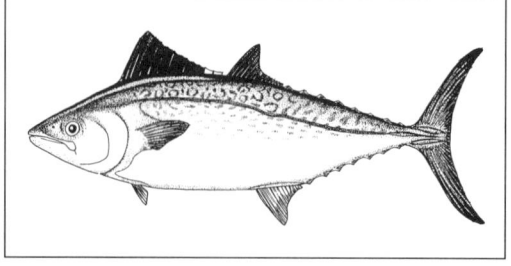

976   SCOMBRIDAE   RAF

SC	*Rastrelliger faughni* (Matsui, 1967)
ES	caballa isleña
DA	ø-dværgmakrel
DE	Insel-Zwergmakrele
EL	σκουμπρί των Ινδιών
EN	island mackerel
FR	maquereau des îles
IT	sgombro indiano
NL	eilandmakreel
PT	cavala das ilhas
FI	kääpiömakrilli-laji
SV	filippinermakrill

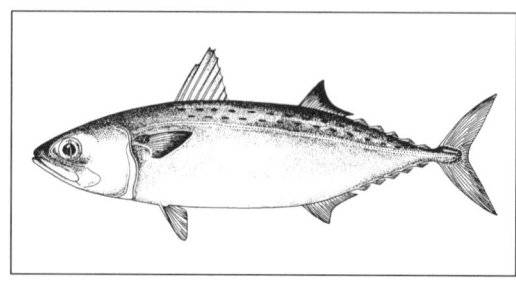

977 SCOMBRIDAE   RAG

SC	*Rastrelliger kanagurta* (Cuvier, 1816)
ES	caballa de la India
DA	indisk dværgmakrel
DE	Indische Zwergmakrele
EL	σκουμπρί των Ινδιών
EN	Indian mackerel
FR	maquereau des Indes
IT	sgombro indiano
NL	Indische makreel
PT	cavala do Índico
FI	intianmakrilli
SV	indisk dvärgmakrill

979 SCOMBRIDAE   BAU

SC	*Sarda australis* (Macleay, 1880)
ES	bonito austral
DA	australsk pelamide
DE	Australischer Bonito
EL	παλαμίδα της Αυστραλίας
EN	Australian bonito
FR	bonite d'Australie
IT	tonnetto orientale; palamita orientale
NL	Australische bonito
PT	bonito austral
FI	etelänsarda
SV	sydlig pelamid

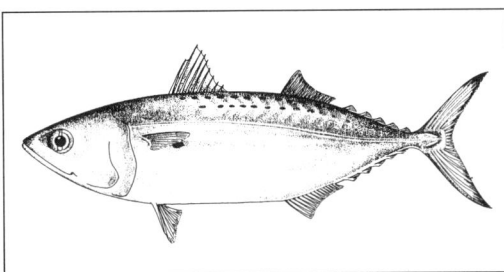

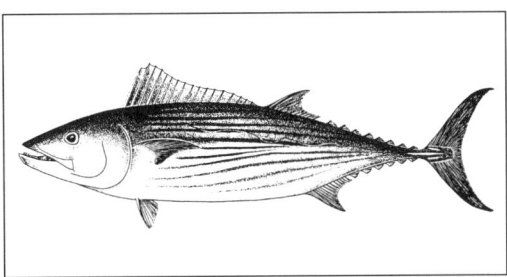

978 SCOMBRIDAE   RAX

SC	*Rastrelliger* spp.
ES	caballas del Indo-Pacífico
DA	dværgmakrel-slægten
DE	Indische Makrelen; Zwergmakrelen; Indopazifische Zwergmakrelen
EL	σκουμπριά των Ινδιών
EN	Indian mackerels
FR	maquereaux du Pacifique
IT	sgombri indiani
NL	dwergmakrelen
PT	cavalas do Índico
FI	kääpiömakrilli-suku
SV	dvärgmakrillar

980 SCOMBRIDAE   BEP

SC	*Sarda chiliensis* (Cuvier, 1831)
ES	bonito del Pacífico oriental
DA	chilensk bonit
DE	Chilenische Pelamide
EL	παλαμίδα του Ειρηνικού
EN	Pacific bonito; Eastern Pacific bonito
FR	bonite du Pacifique
IT	tonnetto cileno; palamita cilena; bonito
NL	Pacifische bonito
PT	bonito do Pacífico
FI	chilensarda
SV	chilensk bonit

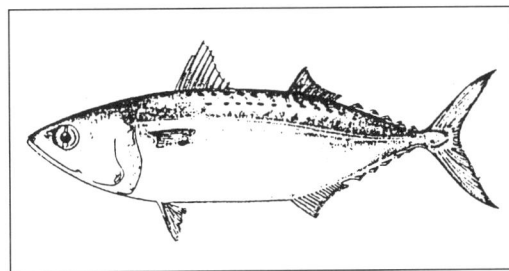

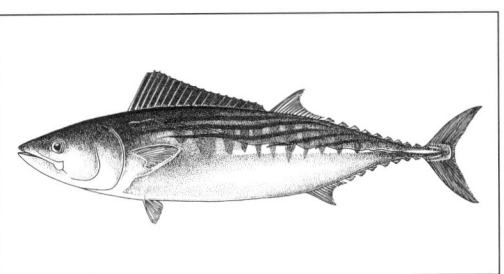

981	**SCOMBRIDAE**	BIP	

SC	*Sarda orientalis* (Temminck and Schlegel, 1844)
ES	bonito mono
DA	Stillehavs-bonit
DE	Westpazifische Pelamide
EL	παλαμίδα της Ανατολής
EN	striped bonito
FR	bonite orientale
IT	palamita orientale; tonnetto
NL	gestreepte bonito
PT	bonito do Indo-Pacífico
FI	juovasarda
SV	strimmig bonit

983 **SCOMBRIDAE** MAA

SC	*Scomber australasicus* (Cuvier, 1831)
ES	caballa pintoja
DA	indo-pacifisk makrel
DE	Indopazifische Makrele
EL	στικτοσκουμπρί
EN	spotted chub mackerel; blue mackerel
FR	maquereau tacheté
IT	sgombro maculato
NL	gevlekte makreel
PT	cavala pintada
FI	makrilli-laji
SV	stillahavsmakrill

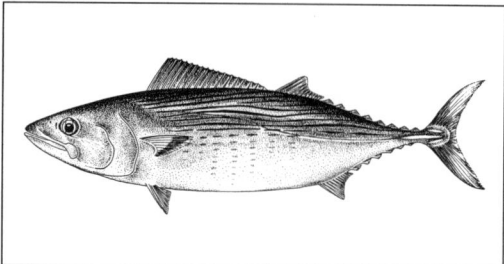

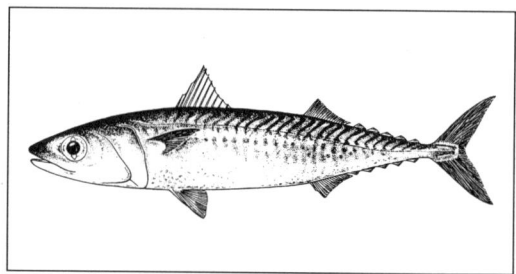

982 **SCOMBRIDAE** BON

SC	*Sarda sarda* (Bloch, 1793); *Pelamys sarda*
ES	bonito; sarda
DA	rygstribet pelamide
DE	Pelamide; Bonito
EL	παλαμίδα· ρίχι
EN	Atlantic bonito; pelamid; belted bonito; short-finned tunny
FR	bonite à dos rayé; bonite commune; pelamide sarde
IT	palamita; tonnetto
NL	bonito
PT	sarrajão; bonito; bonito de dorso listado; serrajão
FI	sarda
SV	ryggstrimmig pelamid

984 **SCOMBRIDAE** MAS

SC	*Scomber japonicus* (Houttuyn, 1782); *Scomber colias*; *Pneumatophorus colias*
ES	estornino
DA	spansk makrel
DE	
EL	κολιός
EN	chub mackerel; Spanish mackerel; thimble-eyed mackerel; Southern mackerel
FR	maquereau espagnol
IT	sgombro; lanzardo; lacerto
NL	Spaanse makreel
PT	cavala
FI	japaninmakrilli
SV	spansk makrill

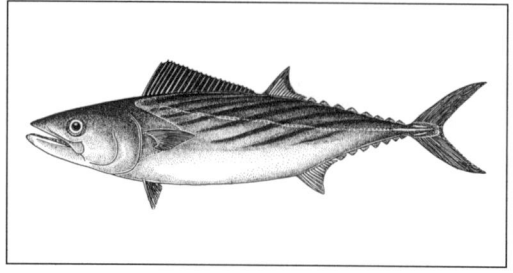

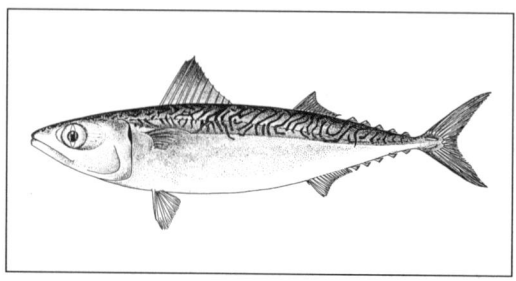

985 SCOMBRIDAE — MAC

SC	*Scomber scombrus* (Linnaeus, 1758); *Scomber scomber*
ES	caballa; caballa del Atlántico; verdel; sarda
DA	almindelig makrel
DE	Europäische Makrele; Gewöhnliche Makrele; Makrele; Gemeine Makrele
EL	σκουμπρί
EN	Atlantic mackerel; common mackerel; mackerel
FR	maquereau; maquereau commun
IT	sgombro; maccarello
NL	makreel
PT	sarda
FI	makrilli
SV	makrill

987 SCOMBRIDAE — BRS

SC	*Scomberomorus brasiliensis* (Collette, Russo and Zavalla-Camin, 1978)
ES	carite brasileño
DA	brasiliansk kongemakrel
DE	Serra-Makrele
EL	σκουμπρί της Βραζιλίας
EN	Serra Spanish mackerel
FR	thazard du Brésil
IT	maccarello reale maculato
NL	serra-koningsmakreel
PT	serra brasileira
FI	kuningasmakrilli-laji
SV	brasiliansk kungsmakrill

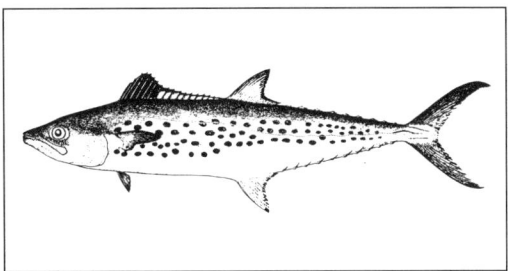

986 SCOMBRIDAE — MAZ

SC	*Scomber* spp.
ES	caballas
DA	makrelfisk-slægt
DE	Makrelen
EL	σκουμπριά
EN	mackerels
FR	maquereaux
IT	sgombri
NL	makrelen
PT	cavalas e sardas
FI	makrilli-suku
SV	makrillfiskar

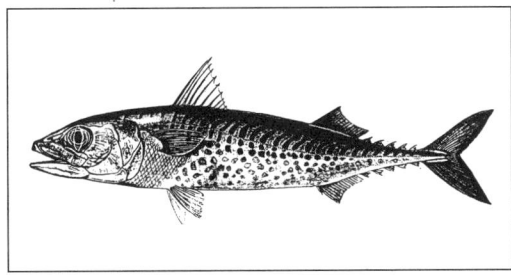

988 SCOMBRIDAE — KGM

SC	*Scomberomorus cavalla* (Cuvier, 1829)
ES	carite lucio
DA	atlantisk kongemakrel
DE	Königsmakrele
EL	βασιλικό σκουμπρί
EN	king mackerel
FR	maquereau royal; thazard
IT	maccarello reale
NL	koningsmakreel
PT	serra real
FI	kuningasmakrilli
SV	kungsmakrill

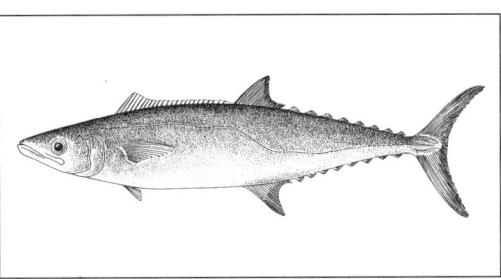

989 SCOMBRIDAE COM

SC	*Scomberomorus commerson* (Lacepède, 1802); *Cybium commersoni*
ES	carite estriado
DA	indisk kongemakrel
DE	Indische Königsmakrele; Spanische Makrele
EL	βασιλικό σκουμπρί
EN	narrow-barred Spanish mackerel; narrow-barred king mackerel
FR	thazard de Commerson
IT	maccarello reale
NL	Indische koningsmakreel
PT	serra-tigre
FI	juovamakrilli
SV	strimmig kungsmakrill

991 SCOMBRIDAE GUT

SC	*Scomberomorus guttatus* (Bloch and Schneider, 1801)
ES	carite del Indo-Pacífico
DA	indo-pacifisk kongemakrel
DE	Indopazifische Königsmakrele
EL	βασιλικό στικτοσκουμπρί
EN	Indo-Pacific king mackerel
FR	thazard ponctué de l'Indo-Pacifique
IT	maccarello reale maculato
NL	Indo-Pacifische koningsmakreel
PT	serra-leopardo
FI	kuningasmakrilli-laji
SV	prickig kungsmakrill

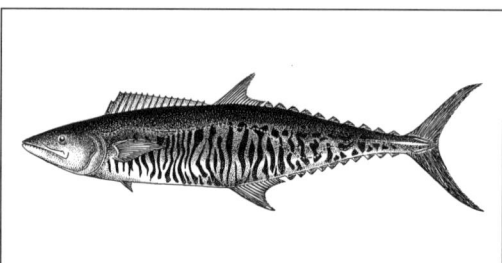 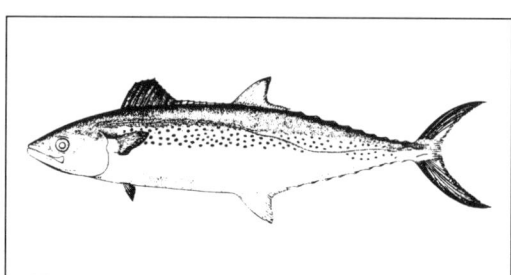

990 SCOMBRIDAE MOS

SC	*Scomberomorus concolor* (Lockington, 1879)
ES	carite de Monterrey
DA	Monterey-kongemakrel
DE	Monterey-Makrele
EL	σκουμπρί της Καλιφόρνιας
EN	Monterey Spanish mackerel
FR	thazard de Monterey
IT	maccarello reale di California
NL	Monterey-koningsmakreel
PT	serra de Monterrey
FI	kuningasmakrilli-laji
SV	monterreymakrill

992 SCOMBRIDAE KOS

SC	*Scomberomorus koreanus* (Kishinouye, 1915)
ES	carite coreano
DA	koreansk kongemakrel
DE	Koreanische Makrele
EL	σκουμπρί της Κορέας
EN	Korean seerfish
FR	thazard coréen
IT	maccarello reale coreano
NL	Koreaanse koningsmakreel
PT	serra coreana
FI	koreanmakrilli
SV	koreansk kungsmakrill

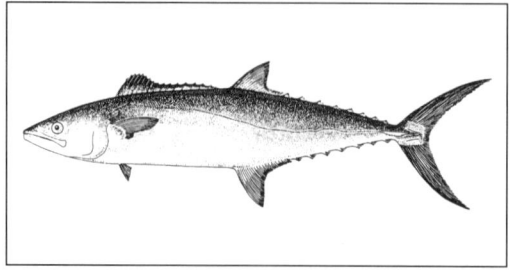 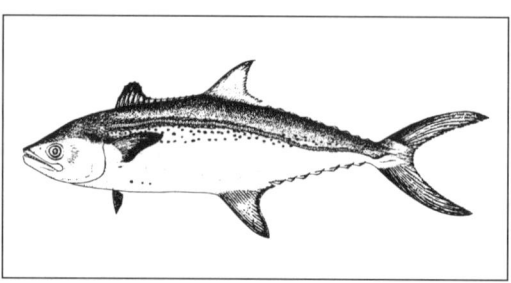

993 SCOMBRIDAE STS

SC	*Scomberomorus lineolatus* (Cuvier, 1831)
ES	carite rayado
DA	stribet kongemakrel
DE	Gestreifte Makrele
EL	γραμμωτό σκουμπρί
EN	streaked seerfish
FR	thazard cirrus
IT	maccarello reale striato
NL	gestreepte koningsmakreel
PT	serra raiada
FI	kuningasmakrilli-laji
SV	streckad kungsmakrill

995 SCOMBRIDAE PAP

SC	*Scomberomorus multiradiatus* (Munro, 1964)
DA	Papua-kongemakrel
DE	Papua-Makrele
EL	σκουμπρί των Παπούα
EN	Papuan seerfish
FR	thazard papou
IT	maccarello reale papuano
NL	Papua-koningsmakreel
PT	serra papuense
FI	papuankuningasmakrilli
SV	papuakungsmakrill

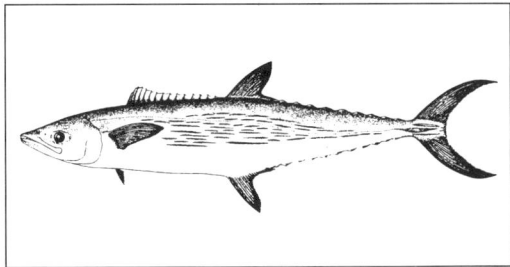

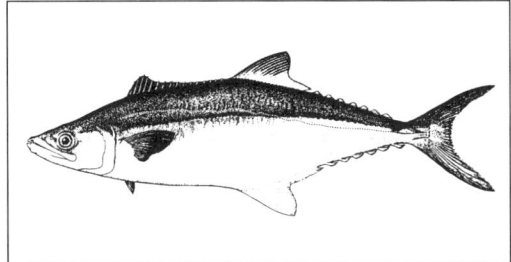

994 SCOMBRIDAE SSM

SC	*Scomberomorus maculatus* (Mitchill, 1815)
ES	carite atlántico
DA	plettet kongemakrel
DE	Gefleckte Königsmakrele
EL	σκουμπρί της Ισπανίας
EN	Atlantic Spanish mackerel
FR	thazard atlantique
IT	maccarello reale maculato
NL	gevlekte koningsmakreel
PT	serra espanhola
FI	pilkkumakrilli
SV	fläckig kungsmakrill

996 SCOMBRIDAE ASM

SC	*Scomberomorus munroi* (Colette and Russo, 1980)
ES	carite australiano
DA	australsk kongemakrel
DE	Australische Fleckenmakrele
EL	σκουμπρί της Αυστραλίας
EN	Australian spotted mackerel
FR	thazard australien
IT	maccarello reale maculato
NL	Australische koningsmakreel
PT	serra australiana
FI	australianmakrilli
SV	australisk kungsmakrill

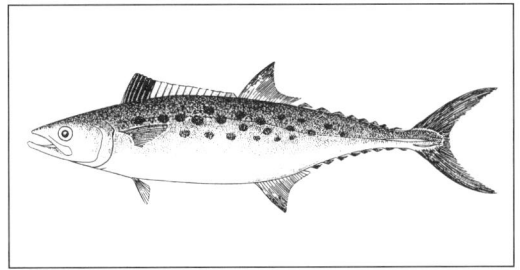

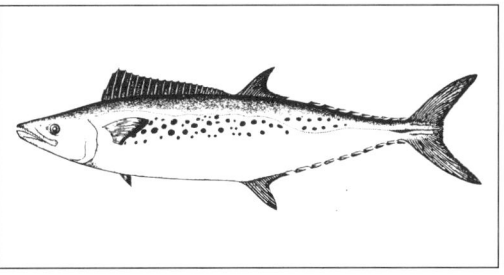

997	**SCOMBRIDAE**	NPH

SC	*Scomberomorus niphonius* (Cuvier, 1831)
ES	carite oriental
DA	japansk kongemakrel
DE	Japanische Makrele
EL	σκουμπρί της Ιαπωνίας
EN	Japanese Spanish mackerel
FR	thazard oriental; thazard du Japon
IT	maccarello reale giapponese
NL	Japanse koningsmakreel
PT	serra oriental
FI	japaninmakrilli
SV	japansk kungsmakrill

999	**SCOMBRIDAE**	QUM

SC	*Scomberomorus queenslandicus* (Munro, 1943)
ES	carite de Queensland
DA	Queensland-kongemakrel
DE	Queensland-Makrele
EL	σκουμπρί του Κουίνσλαντ
EN	Queensland school mackerel
FR	thazard du Queensland
IT	maccarello reale del Queensland
NL	Queensland-koningsmakreel
PT	serra do Indo-Pacífico
FI	queenslandinmakrilli
SV	queenslandmakrill

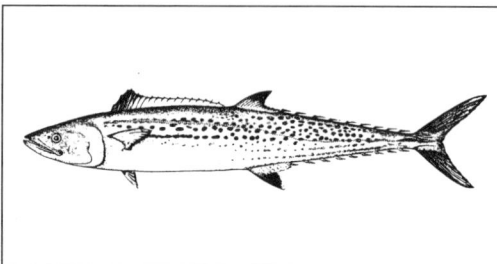

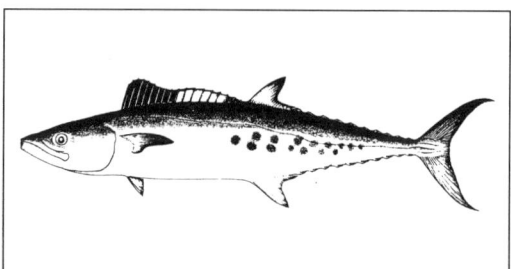

998	**SCOMBRIDAE**	KAK

SC	*Scomberomorus plurilineatus* (Fourmanoir, 1966)
ES	carite kanadi
DA	mangestribet kongemakrel
DE	Kanadi-Makrele
EL	βασιλικό σκουμπρί
EN	Kanadi kingfish
FR	thazard kanadi
IT	maccarello reale Kanadi
NL	Kanadi-koningsmakreel
PT	serra riscada
FI	kuningasmakrilli-laji
SV	kanadikungsmakrill

1000	**SCOMBRIDAE**	CER

SC	*Scomberomorus regalis* (Bloch, 1793)
ES	carite chinigua
DA	prægtig kongemakrel
DE	Falsche Königsmakrele
EL	βασιλικό σκουμπρί
EN	cero
FR	thazard franc
IT	maccarello reale atlantico
NL	valse koningsmakreel
PT	serra malhada
FI	keromakrilli
SV	karibisk kungsmakrill

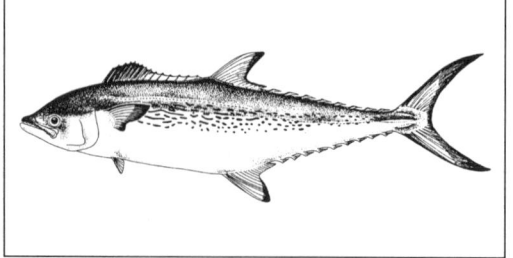

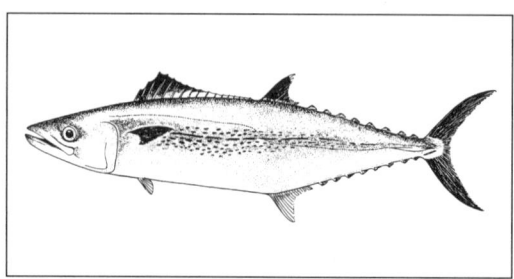

1001 SCOMBRIDAE   BBM

SC   *Scomberomorus semifasciatus* (Macleay, 1884)
ES   carite tigre
DA   bredbåndet kongemakrel
DE   Tigermakrele
EL   σκουμπρί τίγρης
EN   broad-barred king mackerel
FR   thazard tigre
IT   maccarello reale australiano
NL   tijger-koningsmakreel
PT   serra de faixas
FI   tiikerimakrilli
SV   bredbandad kungsmakrill

1003 SCOMBRIDAE   CHY

SC   *Scomberomorus sinensis* (Lacepède, 1800)
ES   carite indochino
DA   kinesisk kongemakrel
DE   Chinesische Königsmakrele
EL   σκουμπρί της Κίνας
EN   Chinese seerfish
FR   thazard nébuleux
IT   maccarello reale cinese
NL   Chinese koningsmakreel
PT   serra chinesa
FI   kiinankuningasmakrilli
SV   kinesisk kungsmakrill

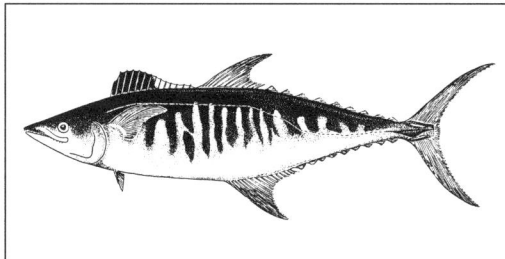

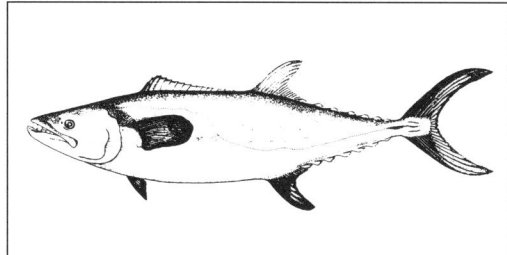

1002 SCOMBRIDAE   SIE

SC   *Scomberomorus sierra* (Jordan and Starks, 1895)
ES   carite sierra
DA   Stillehavs-kongemakrel
DE   Ostpazifische Königsmakrele
EL   σκουμπρί του Ειρηνικού
EN   Pacific sierra
FR   thazard sierra; thazard du Pacifique
IT   maccarello reale maculato
NL   sierra-koningsmakreel
PT   serra do Pacífico
FI   tyynenmerenkuningasmakrilli
SV   sierra

1004 SCOMBRIDAE   KGX

SC   *Scomberomorus* spp.
ES   carites
DA   kongemakrel-slægten
DE   Spanische Makrelen; Königsmakrelen
EL   βασιλικά σκουμπριά
EN   seerfishes
FR   thazards; maquereaux bonites
IT   maccarelli reali
NL   koningsmakrelen
PT   serras; cavalas gigantes
FI   kuningasmakrilli-suku
SV   kungsmakrillar

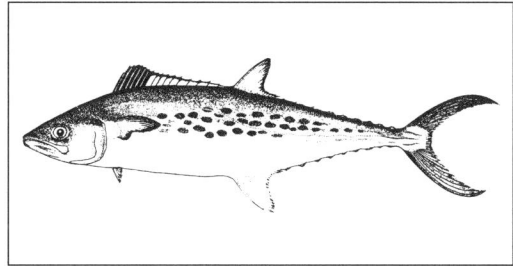

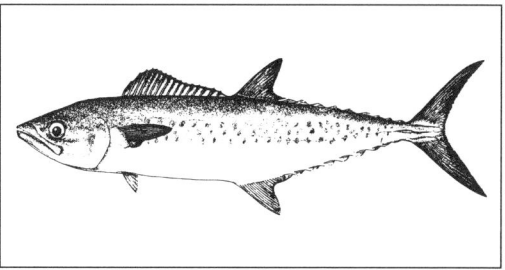

1005 SCOMBRIDAE   MAW

SC	*Scomberomorus tritor* (Cuvier, 1831)
ES	carite pintado
DA	vestafrikansk kongemakrel
DE	Ostatlantische Königsmakrele
EL	σκουμπρί της Γουινέας
EN	West African Spanish mackerel
FR	maquereau bonite; thazard d'Afrique de l'Ouest
IT	maccarello reale di Guinea
NL	Oost-Atlantische koningsmakreel
PT	serra branca
FI	kuningasmakrilli-laji
SV	afrikansk kungsmakrill

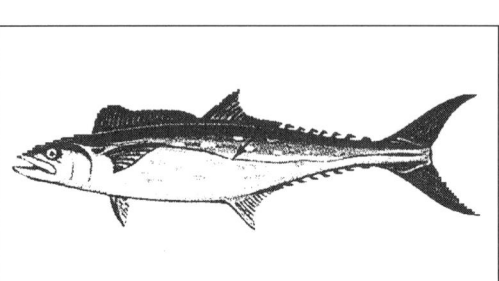

1007 SCOMBRIDAE   ALB

SC	*Thunnus alalunga* (Bonnaterre, 1788); *Germo alalunga*
ES	bonito del norte; atún blanco; atún aleta larga; bonito; albacora
DA	langfinnet tun; hvid tun
DE	Weißer Thun; Germon; Weißer Thunfisch; Langflossenthun;
EL	τόνος μακρύπτερος· τονάκι· όρκυνος
EN	long-finned tuna; long-finned tunny; white tuna; Pacific albacore; long-finned albacore; warman; albacore
FR	germon; thon blanc; germon atlantique
IT	tonno bianco; alalunga; tonno
NL	witte tonijn
PT	atum voador; voador
FI	valkotonnikala
SV	långfenad tonfisk; albacor

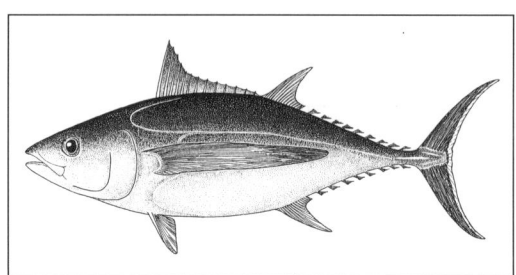

1006 SCOMBRIDAE

SC	*Thunnidae*
ES	túnidos
DA	tunfisk-familien
DE	Thun
EL	τόνοι
EN	tunny; tunas
FR	thons
IT	tonni
NL	tonijnen
PT	tunídeos
FI	tonnikalat; tonnikalat-heimo
SV	tonfiskar

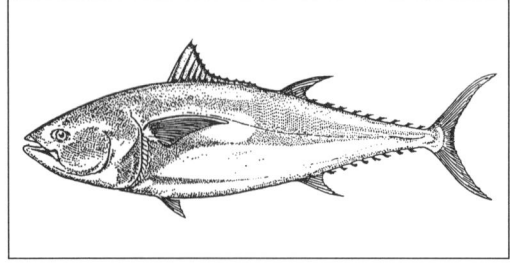

1008 SCOMBRIDAE   YFT

SC	*Thunnus albacares* (Bonnaterre, 1788); *Neothunnus albacora; Neothunnus macopterus; Thunnus albacores*
ES	rabil; atún aleta amarilla; junco
DA	gulfinnet tun
DE	Gelbflossenthun; Albacore
EL	τόνος κιτρινόπτερος
EN	yellowfin tunny; Allison's tuna
FR	thon à nageoires jaunes; thon albacore; albacore
IT	tonno albacora; tonno; tonno pinnagialla
NL	geelvintonijn
PT	atum albacora; albacora
FI	keltaevätonnikala
SV	gulfenad tonfisk; albacora

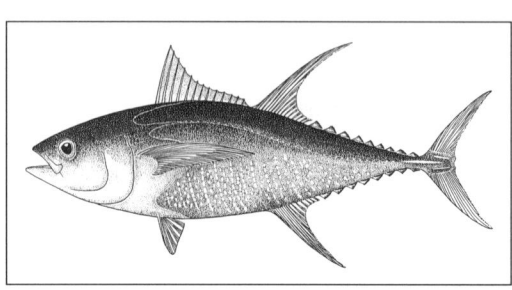

1009 SCOMBRIDAE BLF

SC	*Thunnus atlanticus* (Lesson, 1830) ·
ES	atún de aleta negra
DA	sortfinnet tun
DE	Schwarzflossenthun
EL	τόνος μαυρόπτερος
EN	blackfin tuna
FR	thon à nageoires noires
IT	tonno pinna nera
NL	zwartvintonijn
PT	atum barbatana negra
FI	mustaevätonnikala
SV	karibisk tonfisk; svart tonfisk

1011 SCOMBRIDAE BET

SC	*Thunnus obesus* (Lowe, 1839); *Parathunnus obesus*
ES	patudo; tuna
DA	storøjet tun
DE	Großaugenthun; Dickleibiger Thun; Großäugiger Thun
EL	τόνος μεγαλόφθαλμος
EN	bigeye tuna
FR	patudo; thon obèse; thon à gros œil
IT	tonno obeso; tonno
NL	grootoogtonijn
PT	atum patudo; patudo
FI	isosilmätonnikala
SV	storögd tonfisk

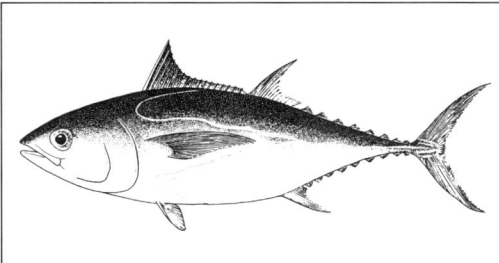

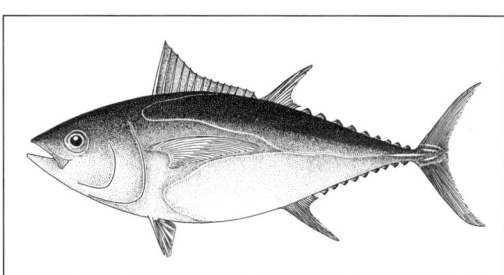

1010 SCOMBRIDAE SBF

SC	*Thunnus maccoyii* (Castelnau, 1872)
ES	atún del sur
DA	sydlig tun
DE	Südlicher Blauflossenthun
EL	τόνος
EN	Southern bluefin tuna
FR	thon rouge du Sud
IT	tonno
NL	zuidelijke blauwvintonijn
PT	atum do Sul
FI	eteläntonnikala
SV	sydlig tonfisk

1012 SCOMBRIDAE BFT

SC	*Thunnus thynnus* (Linnaeus, 1758)
ES	atún rojo; atún; cimarrón; arroaz; cachorreta; albacora; atuarro
DA	atlantisk tun; almindelig tun
DE	Roter Thun; Thunfisch; Gewöhnlicher Thunfisch; Nördlicher Roter Thun; Großer Thun
EL	τόνος
EN	Atlantic bluefin tuna; bluefin tuna; tunny; tuna; Atlantic tuna; common tunny; Northern bluefin tuna
FR	thon rouge; thon rouge commun; thon rouge du Nord
IT	tonno rosso; tonno
NL	tonijn
PT	atum rabilho; atum; rabilho; rabilo
FI	tonnikala
SV	tonfisk

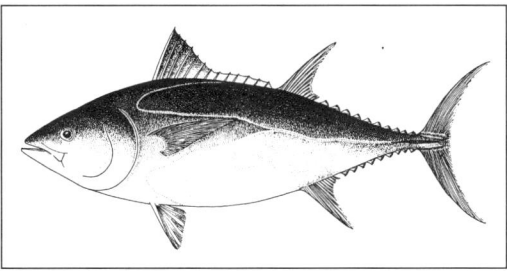

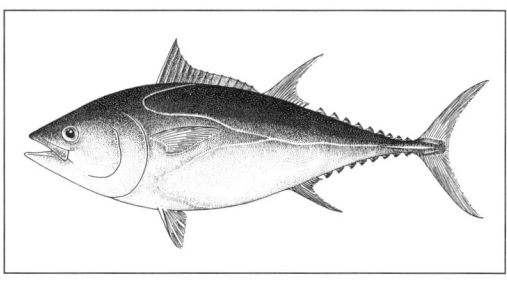

1013 SCOMBRIDAE   LOT

SC	*Thunnus tonggol* (Bleeker, 1851)
ES	atún tongol
DA	tongol tun
DE	Langschwanzthun
EL	μακρύουρος τόνος
EN	longtail tuna
FR	thon mignon
IT	tonno
NL	tongoltonijn
PT	atum tongol
FI	tonnikala-laji
SV	långstjärtad tonfisk

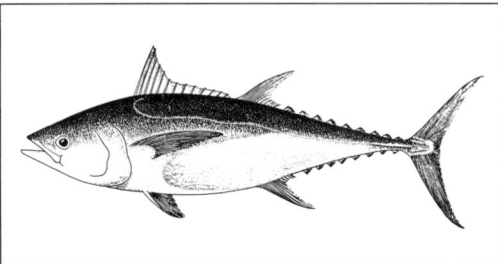

1015 ISTIOPHORIDAE   BIL

SC	*Istiophoridae*
ES	marlines; espadones
DA	sejlfisk-familien; spydfisk; marliner
DE	Fächerfische; Segelfische
EL	ιστιοφόροι
EN	billfishes; sailfishes; marlins; spearfishes; sail bearers
FR	voiliers; makaires; marlins
IT	pesci vela; pesci lancia; marlin; istioforidi
NL	zeilvissen
PT	espadins e veleiros
FI	purjekalat; purjekalat-heimo
SV	segelfiskar; spjutfiskar

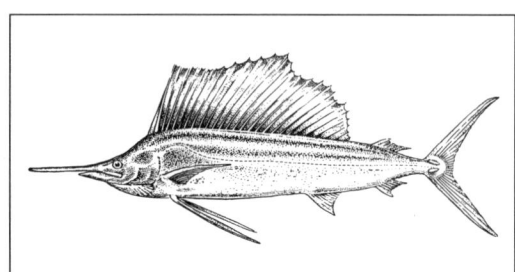

1014 SCOMBRIDAE   TUS

SC	*Thunnus* spp.
ES	atunes
DA	tun-slægt
DE	Thunfische
EL	τόνοι
EN	tunas
FR	thons
IT	tonni
NL	tonijnen
PT	atuns
FI	tonnikala-suku
SV	tonfiskar

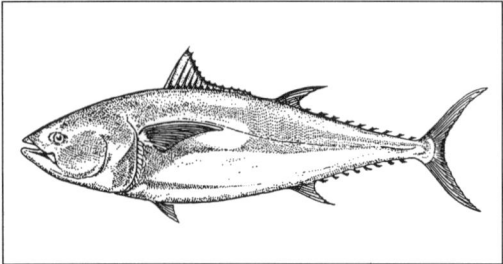

1016 ISTIOPHORIDAE   SAI

SC	*Istiophorus albicans* (Latreille, 1804); *Istiophorus americanus*
ES	pez vela del Atlántico
DA	atlantisk sejlfisk
DE	Atlantischer Segelfisch; Atlantischer Fächerfisch
EL	ιστιοφόρος του Ατλαντικού
EN	Atlantic sailfish
FR	voilier de l'Atlantique
IT	pesce vela atlantico
NL	Atlantische zeilvis
PT	veleiro do Atlântico
FI	atlantinpurjekala
SV	atlantisk segelfisk

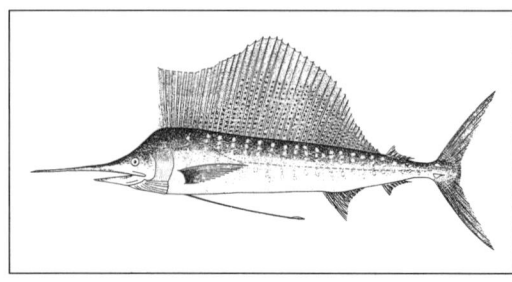

1017 ISTIOPHORIDAE   SFA

SC	*Istiophorus platypterus* (Shaw and Nodder, 1792)
ES	pez vela del Pacífico
DA	Stillehavs-sejlfisk
DE	Indopazifischer Segelfisch; Pazifischer Segelfisch; Pazifischer Fächerfisch
EL	ιστιοφόρος του Ειρηνικού
EN	Pacific sailfish; Indo-Pacific sailfish
FR	voilier du Pacifique
IT	pesce vela del Pacifico
NL	Pacifische zeilvis
PT	veleiro do Pacífico; veleiro do Indo-Pacífico
FI	purjekala
SV	segelfisk

1018 ISTIOPHORIDAE   BLM

SC	*Makaira indica* (Cuvier, 1832); *Tetrapturus indicus*
ES	aguja negra
DA	sort marlin
DE	Schwarzer Marlin
EL	μαύρο μάρλιν
EN	black marlin
FR	makaire noir
IT	marlin nero
NL	zwarte marlijn
PT	espadim negro; marlim negro
FI	mustamarliini
SV	svart marlin

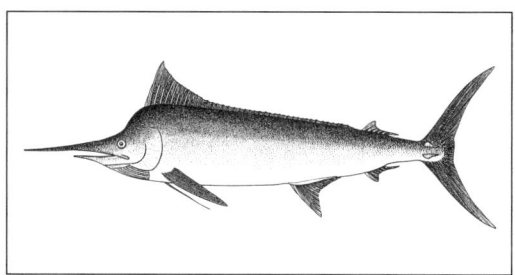

1019 ISTIOPHORIDAE   BUM

SC	*Makaira nigricans* (Lacépède, 1801)
ES	aguja azul
DA	blå marlin
DE	Blauer Marlin
EL	γαλάζιο μάρλιν
EN	blue marlin; Atlantic blue marlin
FR	makaire bleu
IT	marlin azzurro
NL	blauwe marlijn
PT	espadim azul do Atlântico
FI	purjemarliini
SV	blå marlin

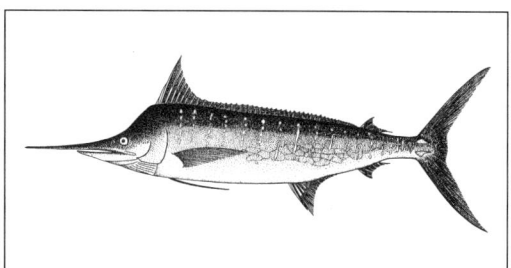

1020 ISTIOPHORIDAE

SC	*Makaira* spp.
ES	marlines
DA	marlin-slægten
DE	Marline
EL	μάρλιν
EN	marlins
FR	makaires; marlins
IT	pesci lancia; marlin
NL	marlijnen
PT	espadins; marlins
FI	marliini-suku
SV	marlins

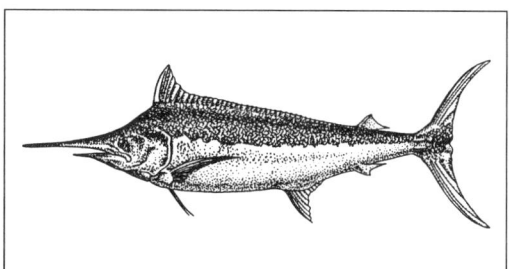

1021 ISTIOPHORIDAE   WHM

SC	*Tetrapturus albidus* (Poey, 1860); *Makaira albida; Makaira marlina*
ES	aguja blanca
DA	hvid marlin
DE	Weißer Marlin
EL	λευκό μάρλιν
EN	white marlin; Atlantic white marlin
FR	makaire blanc
IT	marlin bianco; aguglia imperiale
NL	witte marlijn
PT	espadim branco do Atlântico
FI	valkomarliini
SV	vit marlin

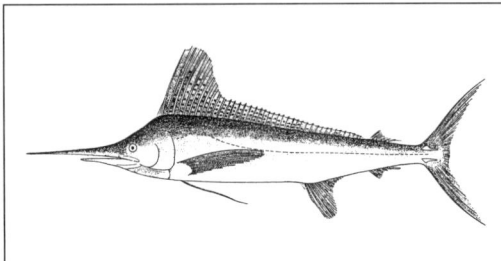

1023 ISTIOPHORIDAE   MLS

SC	*Tetrapturus audax* (Philippi, 1887); *Makaira audax*
ES	marlín rayado
DA	stribet marlin
DE	Gestreifter Marlin
EL	γραμμωτό μάρλιν
EN	striped marlin
FR	makaire strié
IT	pesce lancia striato; marlin striato
NL	gestreepte marlijn
PT	espadim raiado
FI	juovamarliini
SV	randig marlin

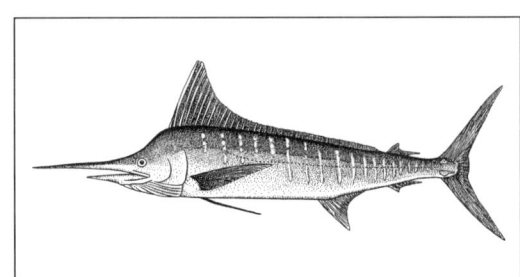

1022 ISTIOPHORIDAE   SSP

SC	*Tetrapturus angustirostris* (Tanaka, 1914)
ES	marlín trompa corta
DA	kortnæbbet spydfisk
DE	Kurzschnauziger Speerfisch
EL	κοντόρυγχο μάρλιν
EN	short-bill spearfish
FR	makaire à rostre court
IT	marlin inerme
NL	kortbekspeervis
PT	espadim de bico curto; marlim de bico curto
FI	lyhytnokkamarliini
SV	kortnosad spjutfisk

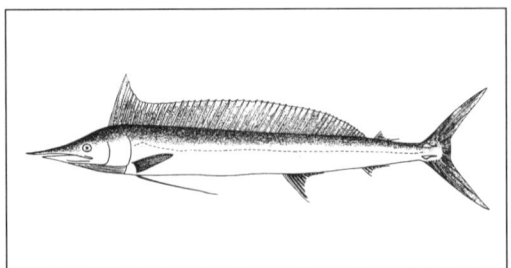

1024 ISTIOPHORIDAE   MSP

SC	*Tetrapturus belone* (Rafinesque, 1810)
ES	marlín; emperador
DA	Middelhavs-spydfisk
DE	Langschwänziger Speerfisch; Speerfisch
EL	μάρλιν της Μεσογείου
EN	Mediterranean spearfish
FR	marlin; poisson-pique; marlin de Méditerranée
IT	aguglia imperiale; aguglia imperiale mediterranea
NL	speervis
PT	espadim do Mediterrâneo
FI	marliini
SV	medelhavsmarlin; medelhavsspjutfisk

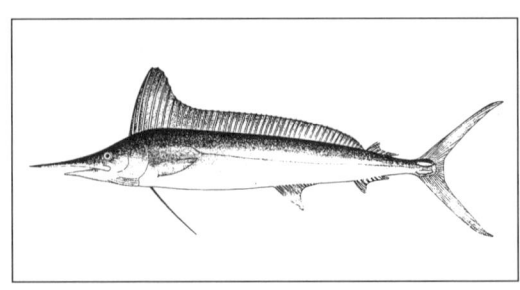

1025 ISTIOPHORIDAE RSP

SC	*Tetrapturus georgei* (Lowe, 1840)
ES	marlín peto
DA	rundskællet spydfisk
DE	Rundschuppen-Speerfisch
EL	βασιλική ζαργάνα
EN	round-scale spearfish
FR	makaire-épée
IT	aguglia imperiale
NL	degenspeervis
PT	espadim-peto; marlim-peto
FI	pyörösuomumarliini
SV	rundfjällig spjutfisk

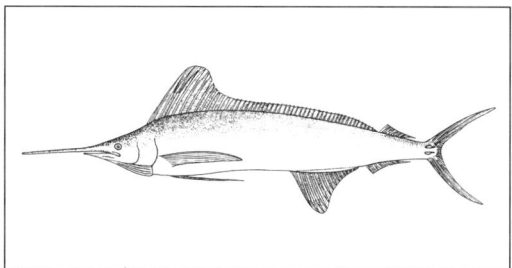

1027 ISTIOPHORIDAE

SC	*Tetrapturus* spp.
ES	marlines
DA	spydfisk-slægten
DE	Speerfische; Marline
EL	μάρλιν
EN	spearfishes
FR	marlins
IT	aguglie imperiali; marlin
NL	speervissen; marlijnen
PT	espadins
FI	marliini-suku
SV	spjutfiskar

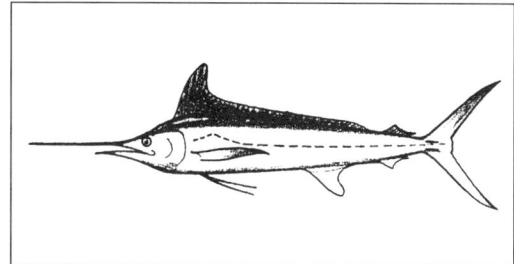

1026 ISTIOPHORIDAE SPF

SC	*Tetrapturus pfluegeri* (Robins and de Sylva, 1963)
ES	aguja picuda
DA	langnæbbet spydfisk
DE	Langschnauziger Speerfisch
EL	βασιλική ζαργάνα
EN	long-bill spearfish
FR	makaire-bécune
IT	aguglia imperiale
NL	langbekspeervis
PT	espadim bicudo; marlim bicudo
FI	pitkäkuonomarliini
SV	långnosad spjutfisk

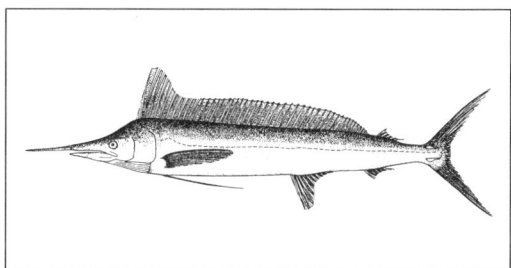

1028 XIPHIIDAE XIP

SC	*Xiphiidae*
ES	peces espada; xífidos
DA	sværdfisk-familien
DE	Schwertfische
EL	ξιφιοί
EN	swordfishes
FR	espadons; xiphiidés
IT	pesci spada
NL	zwaardvissen
PT	espadartes
FI	miekkakalat; miekkakalat-heimo
SV	svärdfiskar

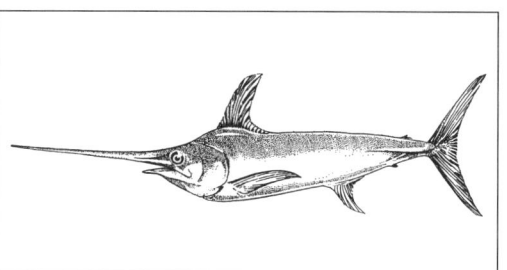

1029 XIPHIIDAE SWO

SC	*Xiphias gladius* (Linnaeus, 1758)
ES	pez espada; emperador; espadarte; aguja palar
DA	sværdfisk
DE	Schwertfisch
EL	ξιφιός· ξιφίας
EN	swordfish; broadbill
FR	espadon; poisson-épée; gladiateur
IT	pesce spada
NL	zwaardvis
PT	espadarte
FI	miekkakala
SV	svärdfisk

1031 STROMATEIDAE SIP

SC	*Pampus argenteus* (Euphrasen, 1788)
ES	palometa plateada
DA	sølv-smørfisk
DE	Silberne Pampel
EL	ψευδολίτσα
EN	silver pomfret
FR	stromaté argenté; aileron argenté
IT	pampo argenteo
NL	zilverpomfret
PT	pampo prateado
FI	hopeavoikala
SV	silversmörfisk

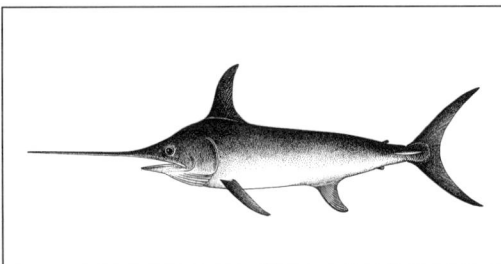

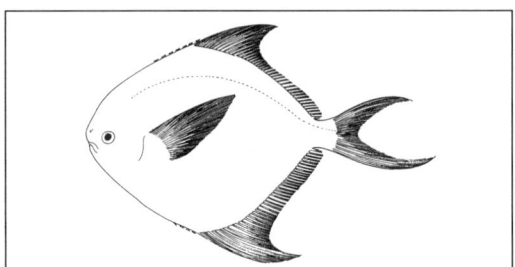

1030 STROMATEIDAE BUX

SC	*Stromateidae*
ES	pámpanos
DA	smørfisk-familien
DE	Erntefische; Butterfische
EL	ψευδολίτσες
EN	butterfishes; harvestfishes; silverfishes
FR	fiatoles; stromatés; stromatéidés
IT	fieti
NL	grootbekken
PT	pampos e pâmpanos
FI	voikalat; voikalat-heimo
SV	smörfiskar

1032 STROMATEIDAE CPO

SC	*Pampus chinensis* (Euphrasen, 1788)
ES	palometa china
DA	kinesisk smørfisk
DE	Chinesische Pampel
EL	ψευδολίτσα της Κίνας
EN	Chinese pomfret; Chinese silver pomfret
FR	stromaté chinois; aileron chinois
IT	pampo cinese
NL	Chinese pomfret
PT	pampo chinês
FI	kiinanvoikala
SV	kinesisk smörfisk

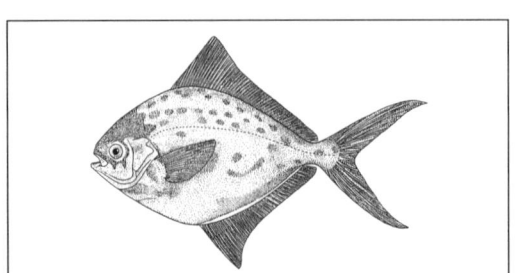

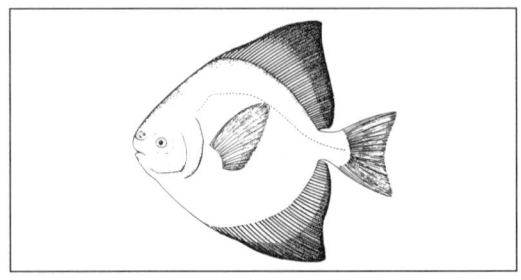

1033 STROMATEIDAE   HVF

SC	*Peprilus paru* (Linnaeus, 1758); *Peprilus alepidotus*
ES	palometa pámpano
DA	amerikansk smørfisk
DE	Amerikanischer Butterfisch
EL	ψευδολίτσα αμερικάνικη
EN	harvestfish; North Atlantic harvestfish
FR	stromaté lune
IT	fieto americano
NL	grootbek
PT	pâmpano-lua
FI	amerikanvoikala
SV	skördefisk

1035 STROMATEIDAE   BUT

SC	*Peprilus triacanthus* (Peek, 1804)
ES	pez mantequilla americano
DA	dollarfisk
DE	Amerikanischer Butterfisch
EL	ψευδολίτσα του Ατλαντικού
EN	butterfish; Atlantic butterfish
FR	stromaté à fossettes
IT	fieto americano
NL	Amerikaanse grootbek
PT	peixe-manteiga americano
FI	amerikanvoikala
SV	fläckig smörfisk

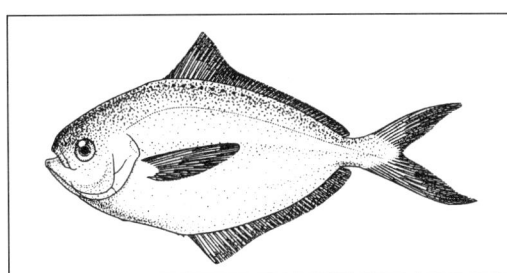

1034 STROMATEIDAE   PPO

SC	*Peprilus simillimus* (Ayres, 1860)
ES	palometa del Pacífico
DA	Stillehavs-smørfisk
DE	Pazifischer Butterfisch
EL	ψευδολίτσα της Καλιφόρνιας
EN	Pacific pompano
FR	stromaté du Pacifique
IT	fieto della California
NL	Pacifische grootbek
PT	pâmpano do Pacífico
FI	tyynenmerenvoikala
SV	stillahavssmörfisk

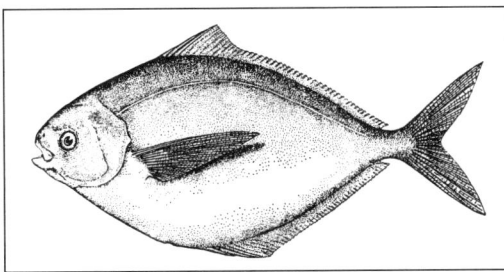

1036 STROMATEIDAE   BTG

SC	*Peprilus* spp.
ES	pámpanos del Golfo
DA	smørfisk-slægt
DE	Butterfische
EL	ψευδολίτσα της Φλόριντας
EN	Gulf butterfishes; harvestfishes
FR	stromatés du Golfe
IT	fieti americani
NL	grootbekken
PT	pâmpanos do golfo
FI	voikalat
SV	fläckiga smörfiskar

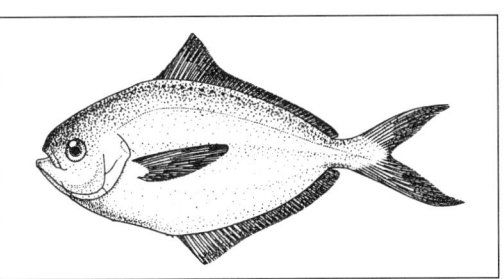

1037 STROMATEIDAE BLB

SC   *Stromateus fiatola* (Linnaeus, 1758); *Fiatola fas-
        ciata; Seserinus microchirus; Stromateus capensis*
ES   pámpano
DA   blå smørfisk
DE   Deckfisch; Pampelfisch; Gemeine Pampel
EL   ψευδολίτσα
EN   butterfish; pomfret; blue butterfish
FR   fiatole
IT   fieto
NL   pomfret
PT   pampo-godinho
FI   voikala-laji
SV   smörfisk

1039 ANABANTIDAE FPC

SC   *Anabas testudineus* (Bloch, 1795)
ES   perca trepadora
DA   klatrefisk
DE   Kletterfisch
EL   ψάρι ορειβάτης
EN   climbing perch
FR   anabas
IT   pesce rampicante
NL   labyrintvis
PT   perca trepadora
FI   kiipijäkala
SV   klätterfisk

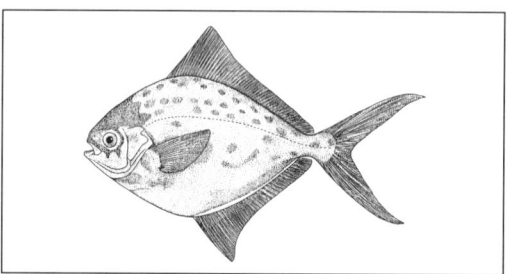 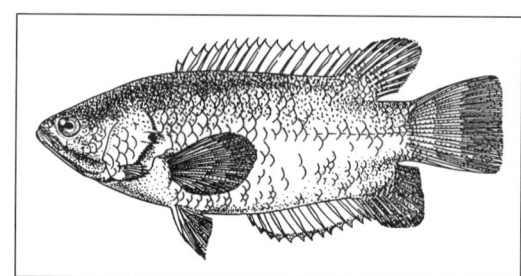

1038 NOMEIDAE DRI

SC   *Ariomma indica* (Day, 1870); *Psenes indicus*
ES   pámpano índico
DA   indisk smørfisk
DE   Indische Schwebmakrele
EL
EN   Indian driftfish
FR   ariome indienne
IT   fieto indiano
NL   Indische drijfvis
PT   peixe-deriva do Índico
FI   intianpolyyppikala
SV   indisk drivfisk

1040 CENTROLOPHIDAE CEN

SC   *Centrolophidae*
ES   centrolófidos
DA   sortfisk-familien
DE   Schwarzfische
EL   μαυρόψαρα· κεντρόλοφοι
EN   ruffs; barrelfishes
FR   centrolophes; roufles
IT   centrolofidi
NL   roeffen
PT   centrolofídeos
FI   meduusakalat; meduusakalat-heimo
SV   svartfiskar

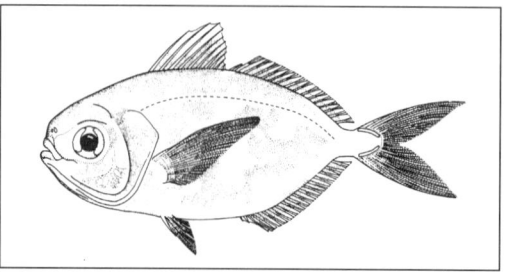 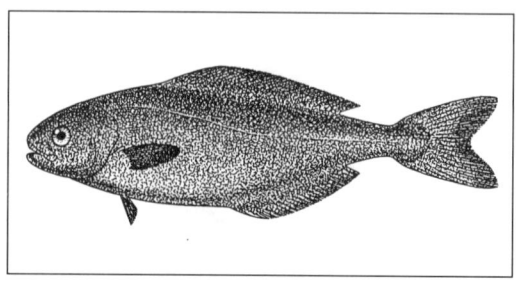

1041 CENTROLOPHIDAE

SC   *Centrolophus niger* (Gmelin, 1788)
ES   romerillo; centrolofo
DA   sortfisk; almindelig sortfisk
DE   Schwarzfisch
EL   μαυρόψαρο
EN   blackfish
FR   centrolophe noir
IT   ricciola di fondale
NL   zwarte vis
PT   liro preto
FI   mustameduusakala
SV   svartfisk

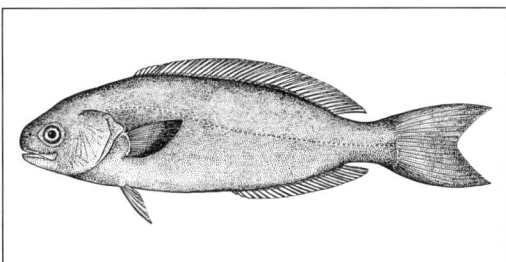

1043 CENTROLOPHIDAE   BWA

SC   *Hyperoglyphe antarctica* (Carmichael, 1818)
ES   pez nariz azul antártico
DA   antarktisk sortfisk
DE   Antarktischer Schwarzfisch
EL
EN   bluenose warehou; Antarctic butterfish
FR   rouffe à nez bleu
IT   ricciola di fondale australe
NL   Antarctische zwarte vis
PT   liro antárctico
FI   etelänmeduusakala
SV   antarktisk svartfisk

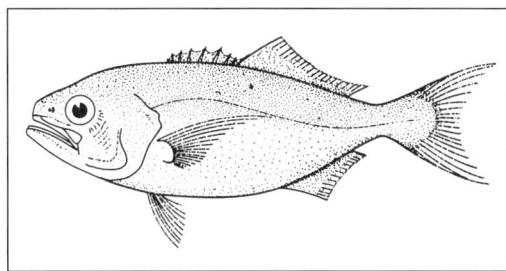

1042 CENTROLOPHIDAE   BSP

SC   *Seriolella* spp.
ES   cojinovas
DA   smørfisk-slægt
DE   Schwarzfische
EL
EN   South Pacific breams
FR   carangues du Pacifique-Sud
IT   seriolelle
NL   Zuid-Pacifische zwarte vissen
PT   seriolelas
FI   meduusakalat
SV   svartfiskar

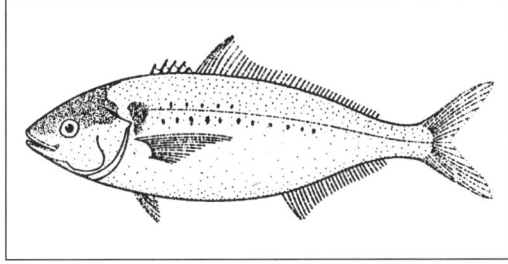

1044 CENTROLOPHIDAE   BUP

SC   *Psenopsis anomala* (Temminck and Schlegel, 1844)
ES   pámpano del Pacífico
DA   japansk sortfisk
DE   Asiatischer Schwarzfisch
EL
EN   Pacific rudderfish
FR   stromaté du Japon
IT   pampo giapponese
NL   Japanse zwarte vis
PT   peixe-leme do Pacífico
FI   japaninmeduusakala
SV   japansk gurami

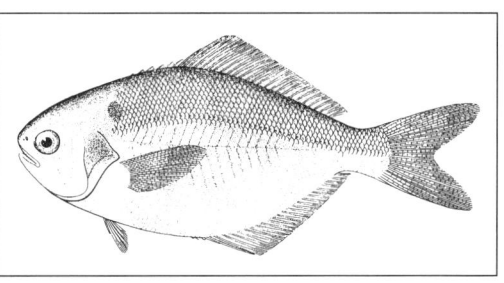

1045 OSPHRONEMIDAE FGG

SC	*Osphronemus goramy* (Lacepède, 1801)
ES	gurami gigante
DA	gurami
DE	Knurrender Gurami; Gurami
EL	γκουράμι
EN	giant gourami
FR	gourami géant; gourami
IT	gourami; gurami
NL	goerami
PT	gurami gigante
FI	gurami
SV	jättegurami

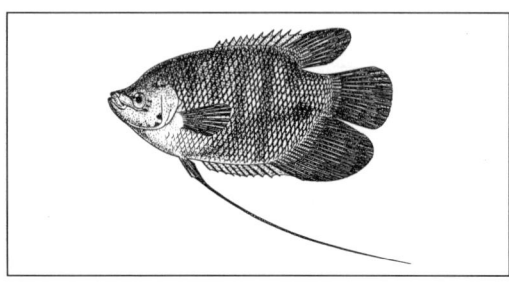

1047 HELOSTOMATIDAE FGO

SC	*Helostoma temminckii* (Cuvier, 1829)
ES	gurami besador
DA	kyssegurami
DE	Küsser; Küssender Gurami
EL	γκουράμι
EN	kissing gourami
FR	kissing gourami vert
IT	baciatore
NL	zoengoerami
PT	gurami beijador
FI	pusukala
SV	kyssgurami

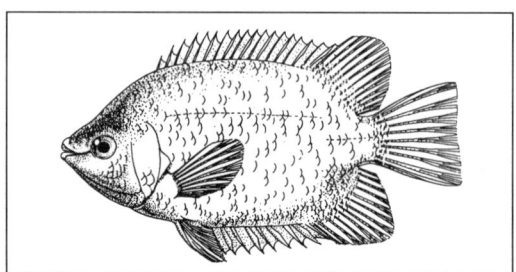

1046 BELONTIIDAE FGS

SC	*Trichogaster pectoralis* (Regan, 1910)
ES	gurami piel de serpiente
DA	slangeskinds-gurami
DE	Schaufelfadenfisch
EL	γκουράμι
EN	snakeskin gourami
FR	gourami-peau de serpent
IT	gurami scaglioso
NL	slangenhuidgoerami
PT	gurami pele-de-cobra
FI	seeprarihmakala
SV	ormskinnsgurami

DRAWING NOT AVAILABLE

1048 ELEOTRIDAE FGB

SC	*Eleotridae*
ES	gobios durmientes
DA	søvnkutling-familien
DE	Schläfergrundeln
EL	ελεοτρίδοι
EN	sleeper gobies
FR	gobies dormeurs
IT	eleotridi
NL	slaapgrondels
PT	dormidores
FI	torkkujatokot; torkkujatokot-heimo
SV	sovfiskar; sömnfiskar

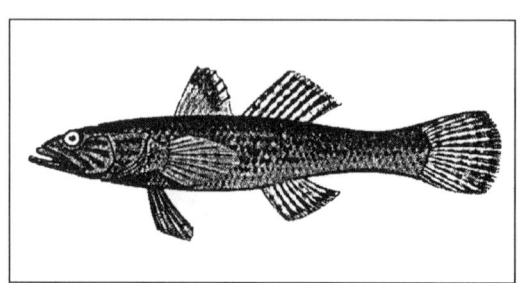

1049 GOBIIDAE

GPA

SC *Gobiidae*
ES góbidos
DA kutling-familien
DE Meergrundeln; Grundeln
EL γωβιοί
EN gobies
FR gobies
IT gobidi; ghiozzi
NL grondels
PT cabozes
FI tokot; tokot-heimo
SV smörbultar

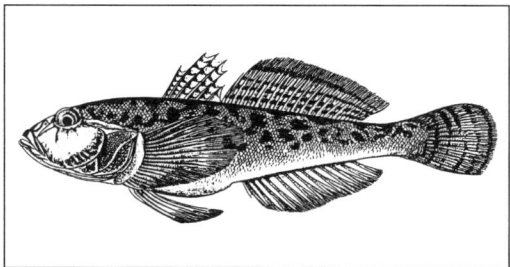

1051 GOBIIDAE

SC *Gobius auratus* (Risso, 1810)
ES gobio dorado
DA guldkutling
DE Goldgrundel
EL χρυσογωβιός
EN golden goby
FR gobie doré
IT ghiozzo dorato
NL goudgrondel
PT caboz dourado
FI kultatokko
SV guldbult

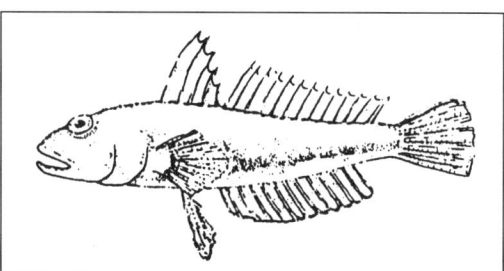

1050 GOBIIDAE

SC *Aphia minuta* (Risso, 1810); *Brachyochirus pellu-cidus; Latrunculus pellucidus; Aphya pellucida*
ES chanquete; jonquillo; yonquillo
DA glaskutling
DE Glasküling; Glasgrundel
EL γωβιουδάκι άφια
EN transparent goby
FR nounat; nonnat
IT rossetto
NL glasgrondel
PT caboz transparente
FI kuultotokko
SV klarbult

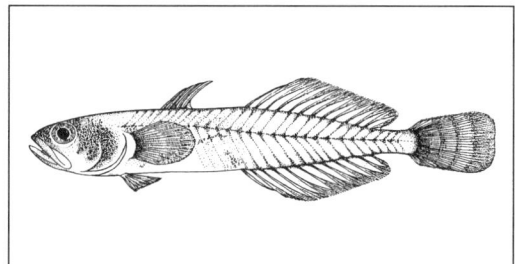

1052 GOBIIDAE

SC *Gobius cobitis* (Pallas, 1811); *Gobius capito*
ES gobio gigante
DA kæmpekutling
DE Große Meergrundel
EL γωβιός
EN giant goby
FR gobie céphalote
IT ghiozzo testone
NL grote grondel
PT caboz cabeçudo
FI isotokko
SV större smörbult

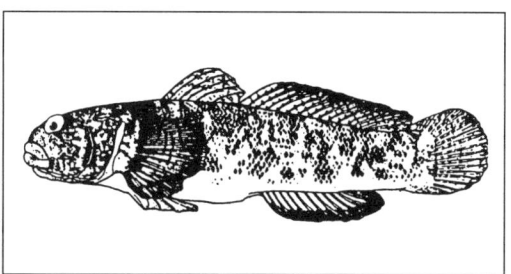

1053 GOBIIDAE GBN

SC	*Gobius niger* (Linnaeus, 1758); *Gobius jozo;* *Gobius niger jozo*
ES	chaparrudo; gobio negro; lorcha
DA	sort kutling
DE	Schwarzküling; Schwarzgrundel; Swaltküling
EL	γωβιός· μαυρογωβιός· ρετρογωβιός
EN	black goby
FR	gobie noir; boulereau
IT	ghiozzo nero
NL	zwarte grondel
PT	caboz negro
FI	mustatokko
SV	svartsmörbult

1055 GOBIIDAE FGX

SC	*Gobius* spp.
ES	góbidos; chaparrudos
DA	kutling-slægt
DE	Grundeln
EL	γωβιοί
EN	freshwater gobies
FR	gobies
IT	ghiozzi
NL	grondels
PT	cabozes
FI	tokot
SV	smörbultar

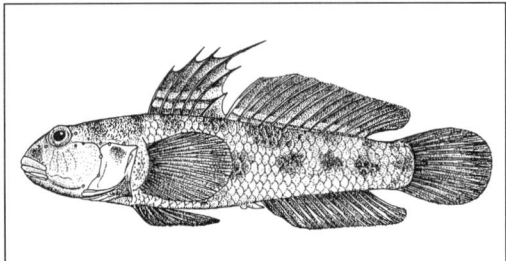 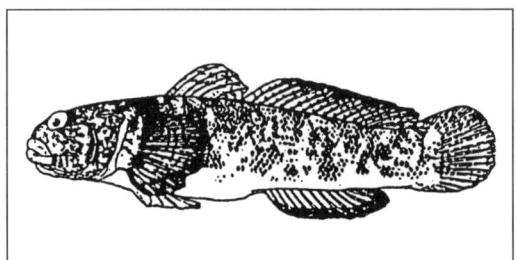

1054 GOBIIDAE

SC	*Gobius paganellus* (Linnaeus, 1758)
ES	paganel
DA	klippekutling
DE	Felsgrundel; Felsengrundel
EL	γωβιός
EN	rock goby
FR	gobie paganel
IT	ghiozzo paganello
NL	rotsgrondel
PT	caboz da rocha
FI	luototokko
SV	klippsmörbult

1056 GOBIIDAE

SC	*Knipowitschia panizzae* (Verga, 1841)
ES	góbido italiano
DA	Panizzas kutling
DE	Ghiozzo; Panizza-Grundel
EL	γωβιός
EN	lagoon goby; Panizza's goby
FR	gobie de Panizza; gobie de lagune
IT	ghiozzetto di laguna
NL	Panizza's grondel
PT	caboz italiano
FI	tokko-laji
SV	gardastubb

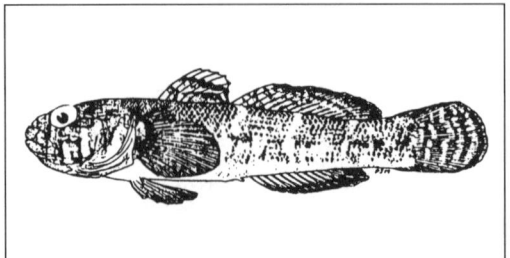 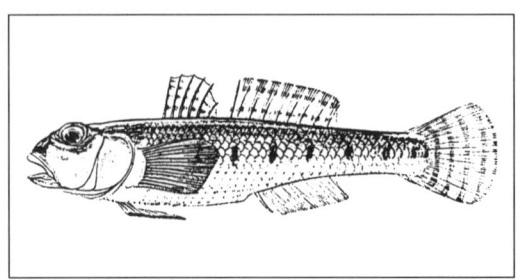

1057 GOBIIDAE

SC	*Neogobius kessleri* (Günther, 1861)
ES	gobio de Kesler
DA	Kesslers kutling
DE	Kessler-Grundel
EL	γωβιός
EN	bighead goby
FR	gobie de Kessler
IT	ghiozzo di Kessler
NL	Kesslers grondel
PT	caboz de Kessler
FI	tokko-laji
SV	storhuvudsbult

1059 GOBIIDAE

SC	*Pomatoschistus minutus* (Pallas, 1770); *Gobius minutus*
ES	cabuchino; gobio de arena
DA	sandkutling
DE	Sandküling; Sandgrundel
EL	γωβιός
EN	sand goby
FR	gobie buhotte; gobie de sable; bourgette
IT	ghiozzetto minuto; marsione
NL	dikkopje
PT	caboz da areia
FI	hietatokko
SV	sandstubb

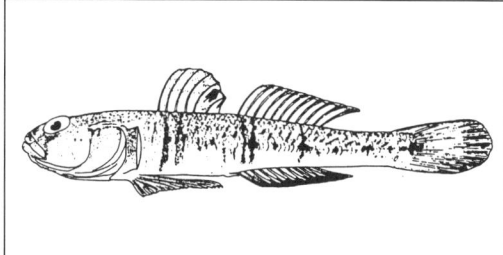

1058 GOBIIDAE

SC	*Pomatoschistus canestrinii* (Ninni, 1883)
ES	gobio de Canestrini
DA	Canestrinis kutling
DE	Canestrini-Grundel
EL	γωβιός του κανεστρίνι
EN	Canestrini's goby
FR	gobie de Canestrini
IT	ghiozzetto cenerino
NL	Canestrinigrondel
PT	caboz do Adriático
FI	tokko-laji
SV	adriatisk stubb; adriastubb

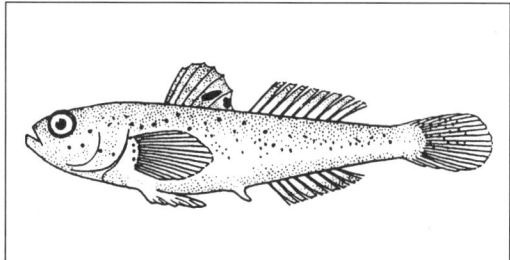

1060 GOBIIDAE

SC	*Proterorhinus marmoratus* (Pallas, 1811)
ES	gobio del Mar Negro
DA	broget Sortehavs-kutling
DE	Marmorgrundel; Marmorierte Meergrundel
EL	γωβιός της Μαύρης Θάλασσας
EN	tubenose goby; mottled Black Sea goby
FR	gobie de la mer Noire
IT	ghiozzo del mar Nero
NL	Zwarte-Zeegrondel
PT	caboz do mar Negro
FI	tokko-laji
SV	marmorsmörbult

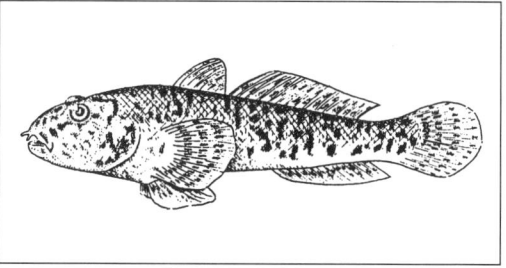

1061 GOBIIDAE

SC	*Zosterisessor ophiocephalus* (Pallas, 1811)
ES	gobio ofiocéfalo
DA	slangehovedkutling
DE	Schlangenkopfgrundel
EL	γωβιός
EN	grass goby
FR	gobie-lotte
IT	ghiozzo gò
NL	slangenkopgrondel
PT	caboz cabeça-de-cobra
FI	levätokko
SV	gräsbult

1063 SCORPAENIDAE　　　BRF

SC	*Helicolenus dactylopterus* (Delaroche, 1809)
ES	gallineta; rascacio rubio
DA	blåkæft
DE	Blaumaul
EL	λειψός· σεβαστός· σκορπιομάνα
EN	blue-mouth; black-belly rosefish; Jacopever
FR	sébaste-chèvre
IT	scorfano di fondale
NL	blauwkeeltje
PT	cantarilho legítimo
FI	sinisuusimppu
SV	blåkäft

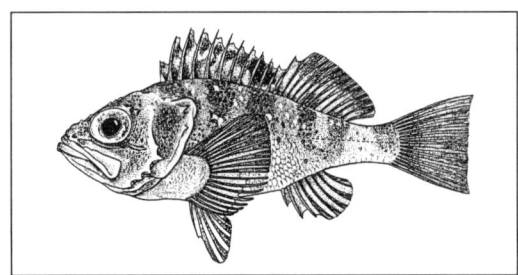

1062 SCORPAENIDAE　　　SCO

SC	*Scorpaenidae*
ES	rascacios; cabrachos; gallinetas
DA	skorpionfisk; dragehovedfisk-familien
DE	Drachenköpfe; Skorpionsfische
EL	σκορπιοί· σκορπίνες
EN	scorpionfishes
FR	rascasses; scorpènes; poissons-scorpions
IT	scorfani; scarpene; scorpenidi
NL	schorpioenvissen
PT	cantarilhos e rascassos; rascassos
FI	skorpionikalat; skorpionikalat-heimo
SV	drakhuvudfiskar

1064 SCORPAENIDAE　　　ROK

SC	*Helicolenus* spp.
ES	rascacios
DA	blåkæft-slægten
DE	Drachenköpfe
EL	κοκκινόψαρα του Ατλαντικού
EN	rosefishes
FR	sébastes
IT	scorfani di fondale
NL	roodbaarzen
PT	cantarilhos
FI	skorpionikala-suku
SV	blåkäftar

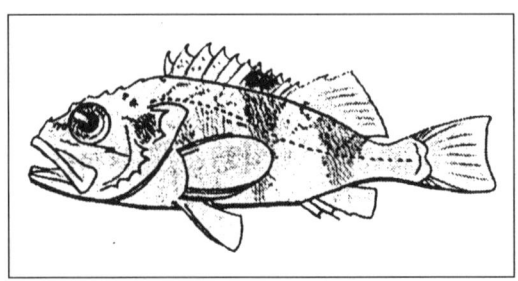

1065 SCORPAENIDAE POI

SC	*Pontinus kuhlii* (Bowdich, 1825)
ES	rascacio de Kuhl
DA	Kuhls dragehovedfisk
DE	Kuhls Drachenkopf
EL	σκορπιός των κοραλλιών
EN	Kuhl's scorpionfish
FR	rascasse de Kuhl
IT	scorfano corallino
NL	Kuhls schorpioenvis
PT	cantarilho-requeime; requeime
FI	skorpionikala-laji
SV	vimpeldrakhuvudfisk

1067 SCORPAENIDAE BBS

SC	*Scorpaena porcus* (Linnaeus, 1758)
ES	rascacio
DA	sort dragehovedfisk
DE	Meersau; Kleiner Drachenkopf
EL	σκορπιός
EN	black scorpionfish
FR	rascasse noire; rascasse brune
IT	scorfano nero
NL	bruine schorpioenvis
PT	rascasso de pintas
FI	tummaskorpionikala
SV	svart skorpionfisk

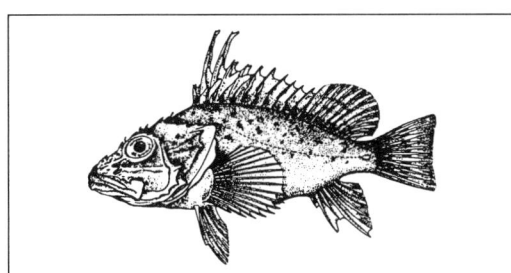

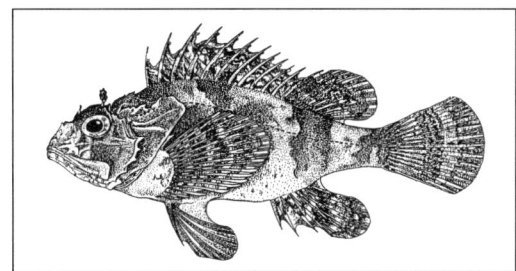

1066 SCORPAENIDAE

SC	*Scorpaena notata* (Rafinesque, 1810); *Scorpaena ustulata*
ES	escórpora
DA	lille dragehovedfisk
DE	Kleiner Roter Drachenkopf
EL	σκορπιός· σκορπίνα
EN	little scorpionfish
FR	rascasse pustuleuse; garde-écueil; guignol
IT	scorfanotto
NL	kleine schorpioenvis
PT	rascasso-escorpião
FI	okaskorpionikala
SV	vårthavssugga

1068 SCORPAENIDAE RSE

SC	*Scorpaena scrofa* (Linnaeus, 1758)
ES	cabracho
DA	orange dragehovedfisk
DE	Roter Drachenkopf; Großer Drachenkopf; Europäische Meersau
EL	σκορπιός· σκορπίνα
EN	red scorpionfish
FR	rascasse rouge; chapon
IT	scorfano rosso; scorfano mediterraneo
NL	rode schorpioenvis
PT	rascasso vermelho
FI	skorpionikala
SV	havssugga

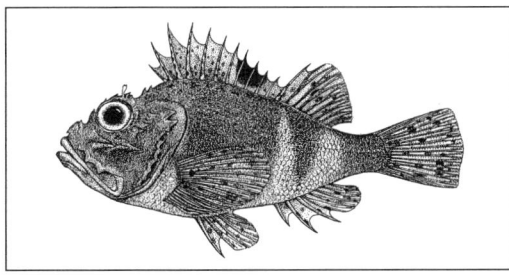

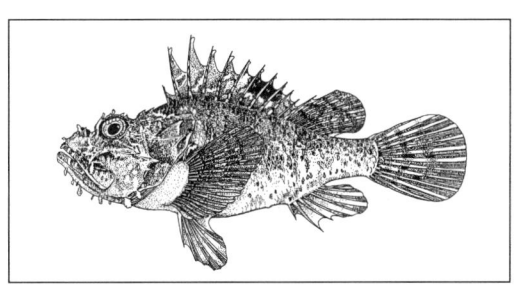

1069 SCORPAENIDAE SCS

SC	*Scorpaena* spp.
ES	rascacios
DA	dragehovedfisk-slægt
DE	Drachenköpfe
EL	σκορπιοί
EN	scorpionfishes
FR	rascasses
IT	scorfani
NL	schorpioenvissen
PT	rascassos
FI	skorpionikalat
SV	skorpionfiskar

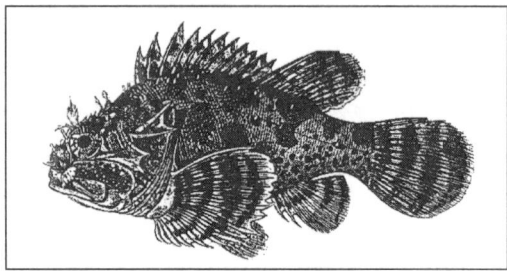

1071 SCORPAENIDAE REC

SC	*Sebastes capensis* (Gmelin, 1788)
ES	gallineta del Cabo
DA	sydafrikansk rødfisk
DE	Kap-Drachenkopf
EL	σκορπάκι του Ακρωτηρίου
EN	Jacopever; Cape redfish
FR	rascasse du Cap; sébaste du Cap
IT	scorfanotto sudafricano
NL	Kaapse roodbaars
PT	cantarilho do Cabo
FI	kapinpuna-ahven
SV	kapkungsfisk

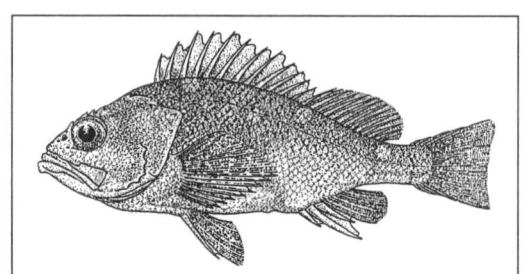

1070 SCORPAENIDAE OPP

SC	*Sebastes alutus* (Gilbert, 1890); *Sebastodes alutus*
ES	gallineta del Pacífico
DA	Stillehavs-rødfisk
DE	Pazifischer Rotbarsch
EL	κοκκινόψαρο
EN	Pacific ocean perch
FR	sébaste du Pacifique
IT	sebaste
NL	Pacifische roodbaars
PT	cantarilho do Pacífico
FI	tyynenmerenpuna-ahven
SV	stillahavskungsfisk

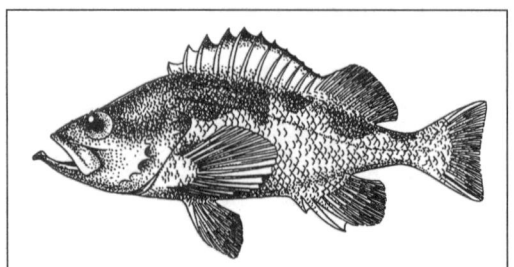

1072 SCORPAENIDAE WRO

SC	*Sebastes entomelas* (Jordan and Gilbert, 1880)
ES	gallineta rocote
DA	enke-rødfisk
DE	Witwen-Drachenkopf
EL	σκούρο κοκκινόψαρο
EN	widow rockfish
FR	sébaste veuf
IT	sebaste bruno
NL	weduweroodbaars
PT	cantarilho viúvo
FI	puna-ahven-laji
SV	änkeuer

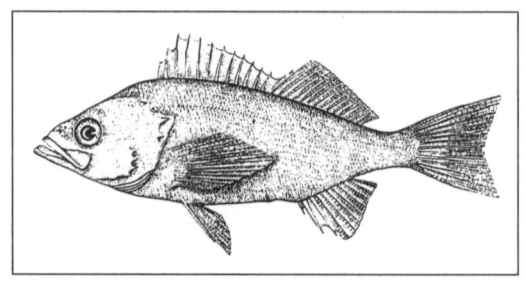

1073 SCORPAENIDAE REN

SC	*Sebastes fasciatus* (Störer, 1854)
ES	pez escorpión americano
DA	amerikansk rødfisk
DE	Akadischer Rotbarsch
EL	κοκκινόψαρο της Αμερικής
EN	Acadian redfish
FR	sébaste rose
IT	sebaste
NL	Amerikaanse roodbaars
PT	cantarilho americano
FI	punasimppu
SV	akadisk kungsfisk

1075 SCORPAENIDAE REG

SC	*Sebastes marinus* (Linnaeus, 1758); *Sebastes norvegicus*
ES	gallineta nórdica
DA	stor rødfisk
DE	Rotbarsch; Goldbarsch; Tiefenbarsch; Großer Rotbarsch
EL	κοκκινόψαρο της Νορβηγίας
EN	redfish; Norway haddock; golden redfish
FR	sébaste; sébaste atlantique; rascasse de Norvège; rascasse du Nord; grande sébaste
IT	scorfano di Norvegia; sebaste di Norvegia
NL	roodbaars
PT	peixe-vermelho; cantarilho dos mares do Norte
FI	punasimppu; puna-ahven
SV	större kungsfisk; uer; rödfisk

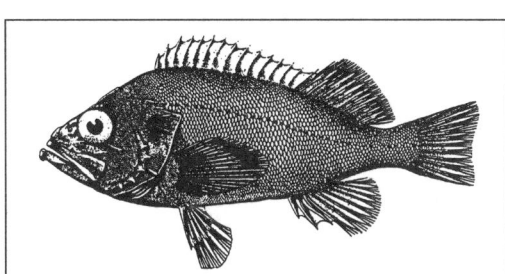

1074 SCORPAENIDAE YRO

SC	*Sebastes flavidus* (Ayres, 1862)
ES	gallineta cola amarilla
DA	gulhalet rødfisk
DE	Gelbschwanz-Drachenkopf
EL	κιτρινοπτέρυγο κοκκινόψαρο
EN	yellowtail rockfish
FR	sébaste à queue jaune
IT	sebaste a pinne gialle
NL	geelstaartroodbaars
PT	cantarilho-rabo amarelo
FI	puna-ahven-laji
SV	gulfenad kungsfisk

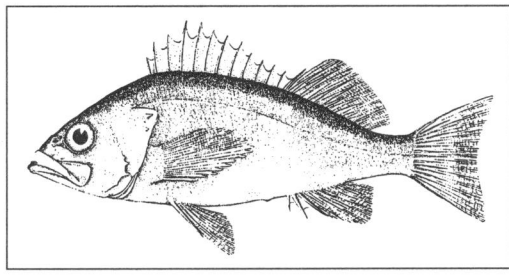

1076 SCORPAENIDAE REB

SC	*Sebastes mentella* (Travin, 1951)
ES	gallineta nórdica
DA	dybhavsrødfisk
DE	Tiefenbarsch; Schnabelbarsch
EL	κοκκινόψαρο του βυθού
EN	deepwater redfish; beaked redfish
FR	sébaste du large; sébaste
IT	sebaste; scorfano atlantico
NL	diepzeeroodbaars
PT	peixe-vermelho da fundura
FI	puna-ahven-laji
SV	djuphavskungsfisk

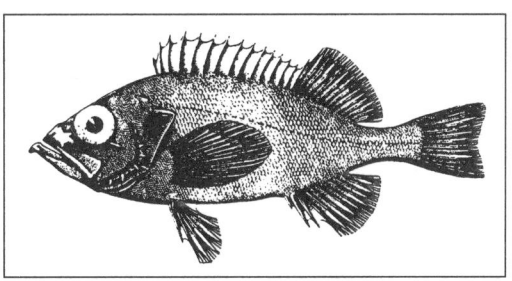

1077 SCORPAENIDAE RED

SC Sebastes spp.
ES gallinetas; chancharros
DA rødfisk-slægten
DE Rotbarsche; Goldbarsche
EL κοκκινόψαρα
EN redfishes; rockfishes
FR sébastes
IT scorfani; sebasti
NL roodbaarzen
PT cantarilhos do Norte; peixes-vermelhos
   do Atlântico
FI punasimput
SV kungsfiskar

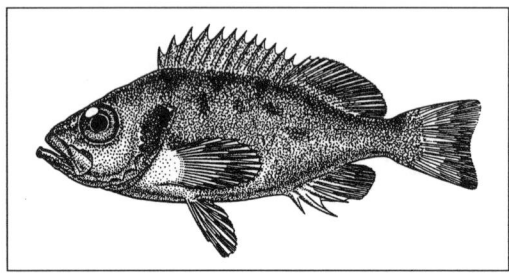

1079 TRIGLIDAE GUR

SC Aspitrigla cuculus (Linnaeus, 1758); Trigla pini;
   Chelidonichthys cuculus; Trigla cuculus
ES arete; cuco; cuchillo; escacho; bobo; peona;
   gallineta
DA tværstribet knurhane
DE Kuckucks-Knurrhahn; Seekuckuck
EL ασπιδοκαπόνι· καπόνι· κούκος
EN red gurnard; cuckoo gurnard; soldier
FR rouget-grondin; grondin rose; grondin rouge;
   grondin-pin
IT capone coccio; gallinella; cappone; capone impe-
   riale
NL Engelse poon
PT cabra vermelha
FI punakurnusimppu
SV rödknot

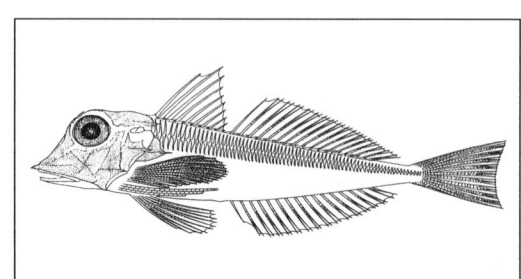

1078 TRIGLIDAE GUX

SC Triglidae
ES rubios; tríglidos
DA knurhane-familien
DE Knurrhähne
EL καπόνια
EN gurnards
FR grondins; trigles; rougets-grondins
IT triglidi; caponi
NL ponen
PT cabras e ruivos; triglídeos
FI kurnusimput; kurnusimput-heimo
SV knotfiskar; knorrhanefiskar

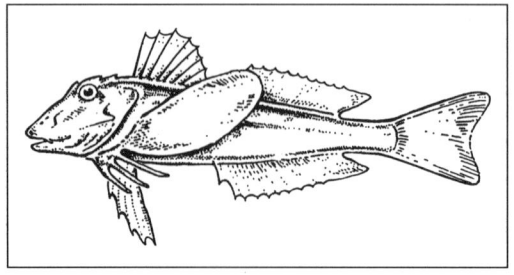

1080 TRIGLIDAE

SC Aspitriglia obscura (Linnaeus, 1764); Trigla
   obscura
ES arete oscuro
DA dunkel knurhane
DE Rauher Knurrhahn
EL γκριζοκαπόνι· καπόνι
EN long-finned gurnard
FR grondin sombre; grondin obscur; grondin morrude
IT capone cavota; gallinella; cappone
NL donkere poon
PT cabra de bandeira
FI töyhtökurnusimppu
SV silverbandad knot

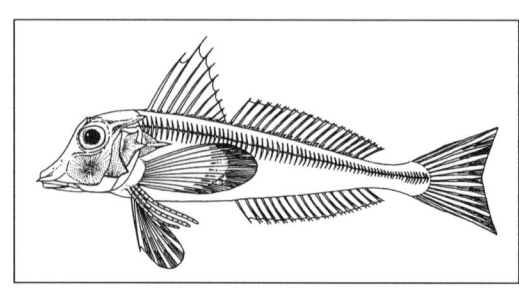

1081 TRIGLIDAE GUC

SC	*Chelidonichthys capensis* (Cuvier, 1829)
ES	rubio del Cabo
DA	Kap-knurhane
DE	Kap-Knurrhahn
EL	καπόνι του Ακρωτηρίου
EN	Cape gurnard
FR	grondin du Cap
IT	capone; gallinella
NL	Kaapse poon
PT	cabra do Cabo
FI	kapinkurnusimppu
SV	kapknorrhane; kapknot

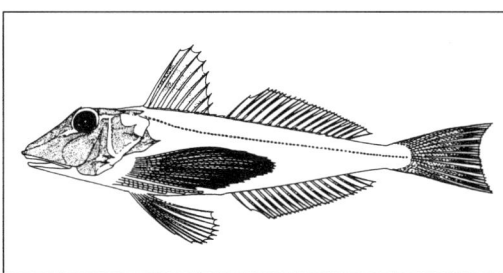

1083 TRIGLIDAE GUI

SC	*Chelidonichthys* spp.
ES	rubios
DA	knurhane-slægt
DE	Knurrhähne
EL	καπόνια του Ινδικού
EN	gurnards
FR	grondins
IT	caponi
NL	Indo-Pacifische ponen
PT	cabras
FI	kurnusimput
SV	knotar; knorrhanar

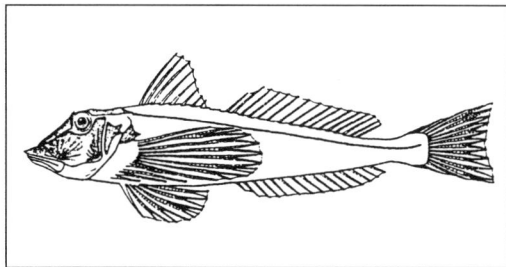

1082 TRIGLIDAE KUG

SC	*Chelidonichthys kumu* (Lesson, 1829)
ES	rubio kumu
DA	blåfinnet knurhane
DE	Blauflossen-Knurrhahn
EL	καπόνι κούμου
EN	blue-fin gurnard
FR	grondin-aile bleue
IT	capone; gallinella
NL	blauwvinpoon
PT	cabra-kumu
FI	sinieväkurnusimppu
SV	blåfenad knorrhane; blåfenad knot

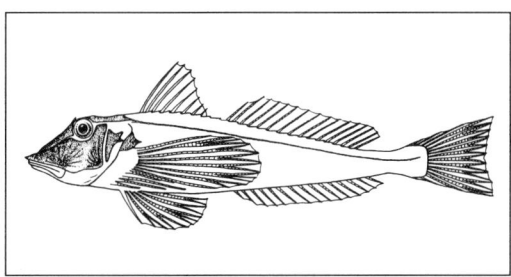

1084 TRIGLIDAE GUG

SC	*Eutrigla gurnardus* (Linnaeus, 1758); *Trigla gurnardus*
ES	borracho; perlón; cuco de altura; cuco americano; crego; clérigo
DA	grå knurhane
DE	Grauer Knurrhahn
EL	καπόνι
EN	grey gurnard; gurnard
FR	grondin gris; trigle gris; gurnard
IT	capone gorno; gallinella; cappone
NL	grauwe poon
PT	cabra morena
FI	kyhmykurnusimppu
SV	knot; knorrhane

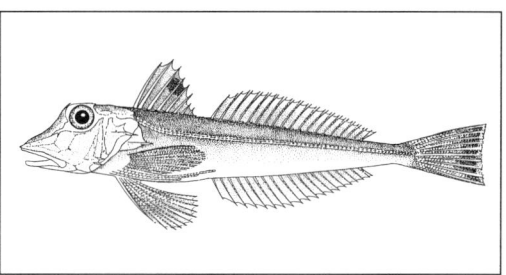

1085 TRIGLIDAE

SC	*Lepidotrigla cavillone* (Lacepède, 1801); *Lepidotrigla asper; Trigla aspera*
ES	cabete
DA	ru knurhane
DE	Stachel-Knurrhahn
EL	λεπιδοκαπόνι· καπόνι
EN	large-scaled gurnard
FR	cavillone commun
IT	caviglione
NL	schubpoon
PT	ruivo
FI	kurnusimppu-laji
SV	fjällknot

1087 TRIGLIDAE   SRA

SC	*Prionotus* spp.
ES	rubios americanos
DA	knurhane-slægt
DE	Nordamerikanische Knurrhähne
EL	αμερικάνικα καπόνια
EN	Atlantic searobins
FR	grondins d'Amérique
IT	caponi americani
NL	Amerikaanse ponen
PT	ruivos americanos
FI	kurnusimput
SV	knorrhanar

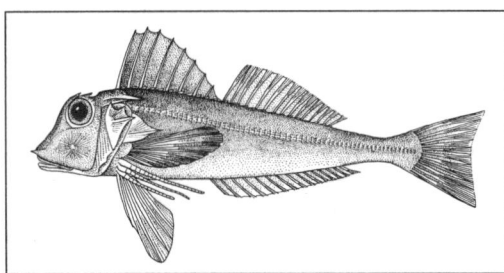

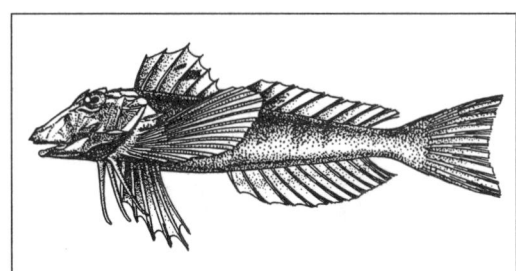

1086 TRIGLIDAE   LEP

SC	*Lepidotrigla dieuzeidei* (Audoin, 1973)
ES	cabete espinudo
DA	pigget knurhane
DE	Stachliger Knurrhahn
EL	λεπιδοκαπόνι· καπόνι
EN	spiny gurnard
FR	grondin de Dieuzeide
IT	caviglione
NL	gedoornde poon
PT	ruivo espinhoso
FI	piikkikurnusimppu
SV	taggknot

1088 TRIGLIDAE   BEG

SC	*Pterygotrigla polyommata* (Richardson, 1838)
ES	rubio picudo
DA	spidssnudet knurhane
DE	Spitzmaul-Knurrhahn
EL	οξύρρυγχο καπόνι
EN	latchet; sharp-beak gurnard
FR	grondin à museau pointu
IT	capone pizzuto
NL	spitssnuitpoon
PT	cabra bicuda
FI	kurnusimppu-laji
SV	spetsknot

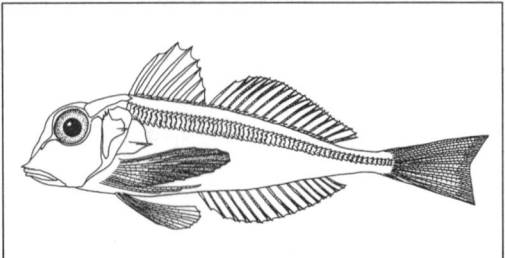

DRAWING NOT AVAILABLE

1089 TRIGLIDAE GUU

SC	*Trigla lucerna* (Linnaeus, 1758); *Trigla corax; Trigla hirundo*
ES	bejel
DA	rød knurhane
DE	Roter Knurrhahn; Seeschwalbenfisch
EL	χελιδονάς· καπόνι
EN	tub gurnard
FR	grondin rouge; galinette; grondin perlon; trigle-hirondelle
IT	capone gallinella; cappone; gallinella
NL	rode poon
PT	cabra-cabaço; cabaço
FI	isokurnusimppu
SV	fenknot

1091 TRIGLIDAE

SC	*Trigloporus lastoviza* (Brünnich, 1768); *Trigla lineata*
ES	rubio
DA	båndet knurhane
DE	Gestreifter Knurrhahn; Gestreifter Seehahn
EL	γραμμοκαπόνι· καπόνι· κούκος
EN	streaked gurnard
FR	grondin strié
IT	capone ubriaco; gallinella; cappone
NL	gestreepte poon
PT	cabra riscada
FI	kurnusimppu-laji
SV	tvärbandad knot

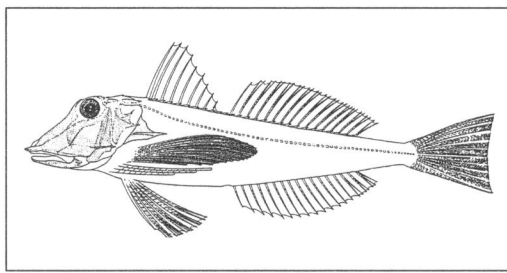

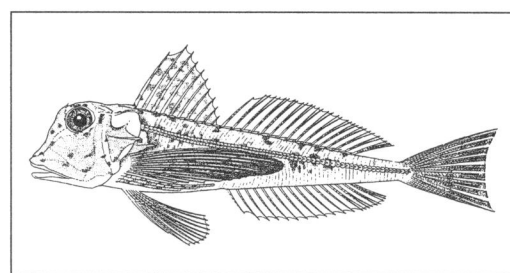

1090 TRIGLIDAE GUN

SC	*Trigla lyra* (Linnaeus, 1758)
ES	garneo
DA	langpigget knurhane
DE	Leierknurrhahn; Pfeifenfisch; Leierhahn
EL	καπόνι
EN	piper; piper gurnard
FR	grondin lyre
IT	cappone lira; cappone; lira; gallinella
NL	lierpoon
PT	cabra-lira
FI	piikkikurnusimpu
SV	lyrknot

1092 HEXAGRAMMIDAE CLI

SC	*Ophiodon elongatus* (Girard, 1854)
ES	bacalao largo; lorcha
DA	lingcod
DE	Langer Grünling; Langer Terpug
EL	γιλάρι
EN	lingcod
FR	rascasse verte; morue-lingue
IT	ofiodonte
NL	lingcod
PT	lorcha
FI	vihersimppu-laji
SV	grönfisk

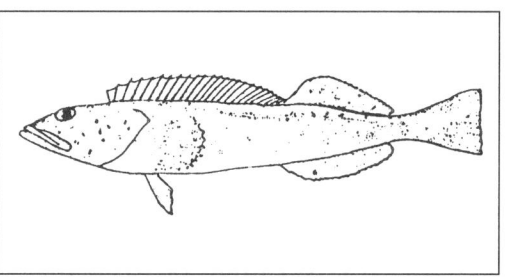

1093 HEXAGRAMMIDAE ATK

SC *Pleurogrammus azonus* (Jordan and Metz, 1913)
ES lorcha de Atka
DA Atka-makrel
DE Terpug
EL
EN Atka mackerel
FR maquereau de Atka; rascasse verte de Atka; terpug
IT terpugo
NL Atka
PT lorcha de Atka; lingue de Atka
FI vihersimppu-laji
SV atkagrönlånga

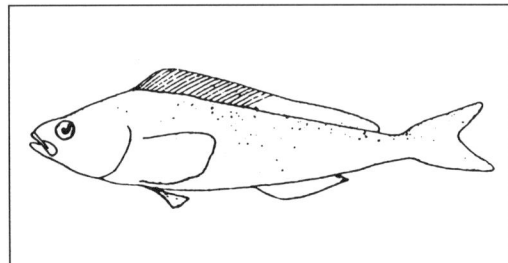

1095 PLATYCEPHALIDAE FLH

SC *Platycephalidae*
ES platicefálidos
DA fladhovedet ulke-familie
DE Flachköpfe
EL πλατυκέφαλοι
EN flat-head gurnards
FR platycéphales; grondins à tête plate
IT platicefalidi
NL platkoppen
PT sapateiros
FI lättäsimput; lättäsimput-heimo
SV platthuvudfiskar

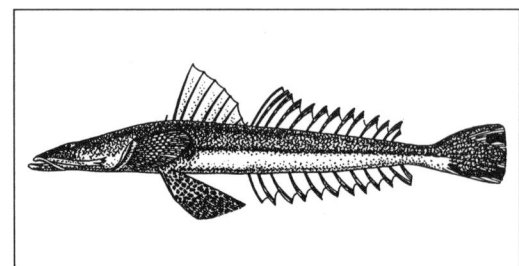

1094 ANOPLOPOMATIDAE SAB

SC *Anoplopoma fimbria* (Pallas, 1811)
ES bacalao negro
DA fakkelfisk
DE Kohlenfisch
EL
EN sablefish
FR rascasse noire; rascasse charbon; morue charbonnière
IT merluzzo dell'Alasca; fimbria
NL zwarte kabeljauw
PT peixe-carvão do Pacífico
FI silosimppu
SV kolfisk

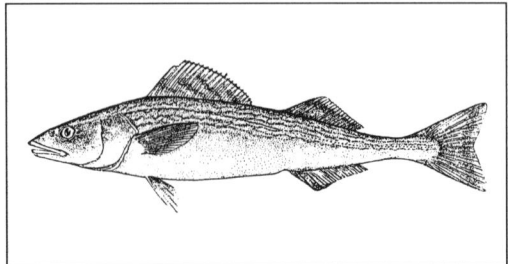

1096 PLATYCEPHALIDAE FLI

SC *Platycephalus indicus* (Linnaeus, 1758)
ES platicéfalo índico
DA indisk fladhoved-ulk
DE Sandflachkopf
EL πλατυκέφαλος
EN Indo-Pacific flathead
FR platycéphale indien
IT testapiatta
NL Indische platkop
PT sapateiro do Indo-Pacífico
FI lättäsimpu-laji
SV indisk platthuvudfisk

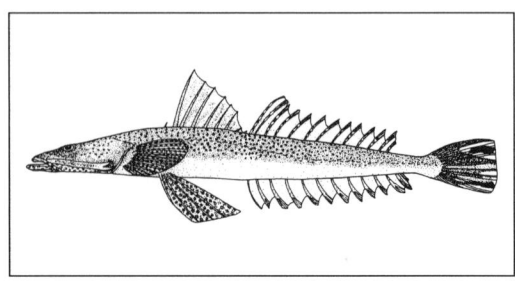

274

1097 COTTIDAE

SC	*Cottus gobio* (Linnaeus, 1758)
ES	coto común; cavilat
DA	hvidfinnet ferskvandsulk; flodulk
DE	Groppe; Kaulkopf; Koppe; Tulzbull
EL	βοϊδοκεφαλόψαρο
EN	bullhead; miller's thumb
FR	chabot de rivière; cabot; bavard; cafard; testard
IT	scazzone
NL	rivierdonderpad
PT	escorpião de água doce
FI	kivisimppu
SV	stensimpa

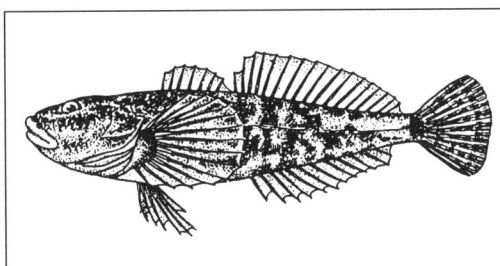

1099 COTTIDAE

SC	*Myoxocephalus quadricornis* (Linnaeus, 1758)
ES	escorpión de cuatro cuernos
DA	hornulk
DE	Vierhörniger Seeskorpion
EL	βοϊδοκεφαλόψαρο του Βορρά
EN	four-horn sculpin
FR	chabot à quatre cornes
IT	scazzone boreale
NL	gehoornde rivierdonderpad
PT	escorpião de quatro cornos
FI	härkäsimppu
SV	hornsimpa

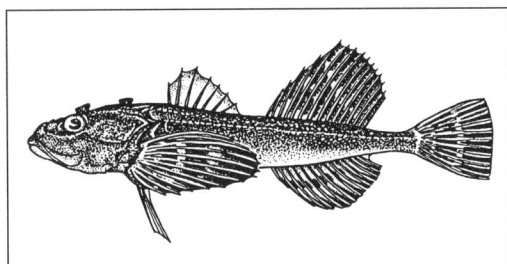

1098 COTTIDAE

SC	*Myoxocephalus octodecemspinosus* (Mitchill, 1815)
ES	escorpión de dieciocho espinas
DA	amerikansk ulk
DE	Amerikanischer Seeskorpion
EL	βοϊδοκεφαλόψαρο της Αμερικής
EN	long-horn sculpin
FR	chabot à dix-huit épines; diable de mer; scorpion de mer
IT	scazzone americano
NL	Amerikaanse zeedonderpad
PT	escorpião americano
FI	amerikanhärkäsimppu
SV	taggrötsimpa

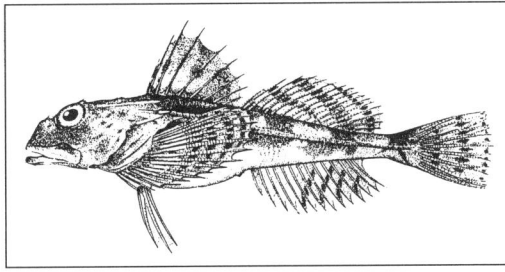

1100 COTTIDAE

SC	*Myoxocephalus scorpius* (Linnaeus, 1758)
ES	coto escorpión; charrasco
DA	almindelig ulk
DE	Seeskorpion
EL	βοϊδοκεφαλόψαρο
EN	short-horn sculpin
FR	chabot-scorpion
IT	scazzone marino
NL	zeedonderpad
PT	escorpião
FI	isosimppu
SV	rötsimpa

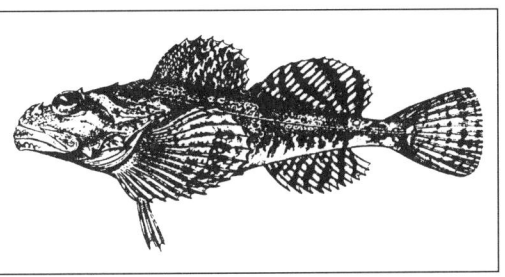

1101 COTTIDAE

SCU

SC	*Myoxocephalus* spp.
ES	escorpiones
DA	ulke-slægt
DE	Seeskorpione
EL	βοϊδοκεφαλόψαρα
EN	sculpins
FR	chabots; diables de mer; scorpions de mer; têtards
IT	scazzoni
NL	zeedonderpadden
PT	escorpiões
FI	simput
SV	rötsimpor

1103 AGONIDAE

SC	*Agonus cataphractus* (Linnaeus, 1758)
ES	agono; ratón de mar
DA	panserulk
DE	Steinpicker
EL	κερατάς
EN	hook-nose
FR	souris de mer; aspidophore armé
IT	aspidoforo corazzato
NL	harnasmannetje
PT	ladrão armado
FI	partasimppu
SV	skäggsimpa

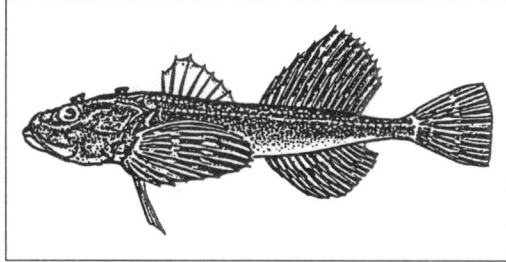

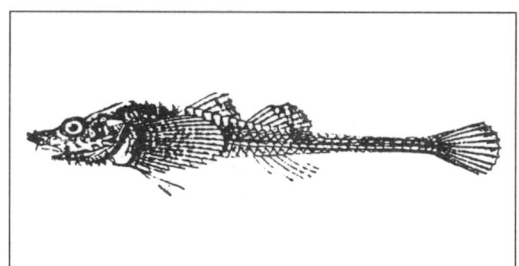

1102 COTTIDAE

SC	*Taurulus bubalis* (Euphrasen, 1786); *Enophris bubalis*
ES	escorpión marino
DA	langtornet ulk
DE	Seeskorpion; Seebull; Groppen
EL	βοϊδοκεφαλόψαρο
EN	sea scorpion
FR	chabot de mer; diable de mer; scorpion de mer; poisson têtard
IT	scazzone marino
NL	groene zeedonderpad
PT	escorpião roco
FI	piikkisimppu
SV	oxsimpa

1104 CYCLOPTERIDAE

SC	*Cyclopteridae*
ES	lompos; ciclópteros
DA	stenbider-familien
DE	Lumpfische; Seehasen
EL	κοτόψαρα
EN	lumpsuckers; lumpfishes
FR	cycloptères; cycloptéridés; lompes
IT	ciclotteri
NL	snotdolven; slakdolven
PT	peixes-lapa
FI	imukalat; imukalat-heimo
SV	sjuryggsfiskar

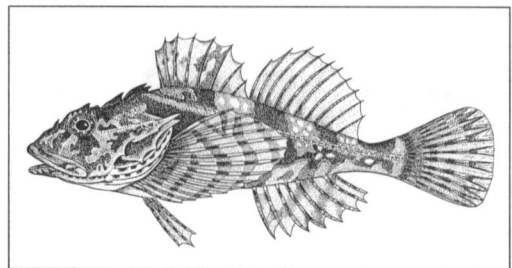

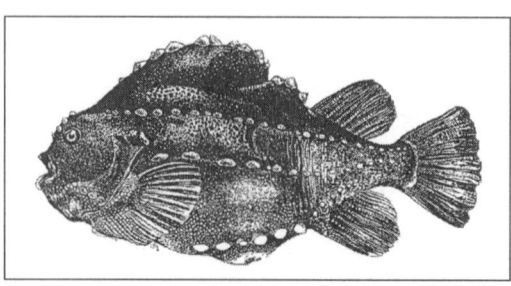

1105 CYCLOPTERIDAE   LUM

SC   *Cyclopterus lumpus* (Linnaeus, 1758)
ES   ciclóptero; lompo
DA   stenbider; kvabso (hun)
DE   Seehase; Lump; Lumpfisch
EL   κοτόψαρο
EN   lumpfish; lumpsucker; henfish; sea hen; paddle-
     cock
FR   lompe; mollet; gros seigneur; lièvre de mer;
     poule de mer;
IT   ciclottero; lompo
NL   snotdolf
PT   peixe-lapa
FI   rasvakala
SV   sjurygg; stenbit

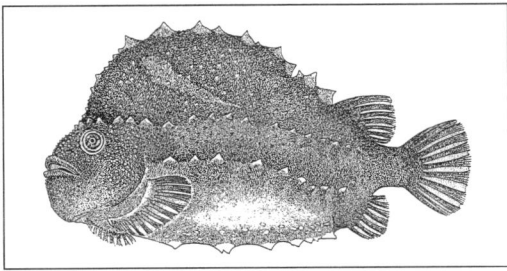

1107 LIPARIDAE

SC   *Liparis atlanticus* (Jordan and Evermann, 1898)
ES   liparis atlántico
DA   atlantisk ringbug
DE   Westatlantischer Scheibenbauch
EL   σαλιγκάρι του Ατλαντικού
EN   Atlantic sea snail
FR   limace de mer atlantique
IT   lipara atlantica
NL   West-Atlantische slakdolf
PT   peixe-caracol do Atlântico
FI   atlantinimukala
SV   atlantisk ringbuk

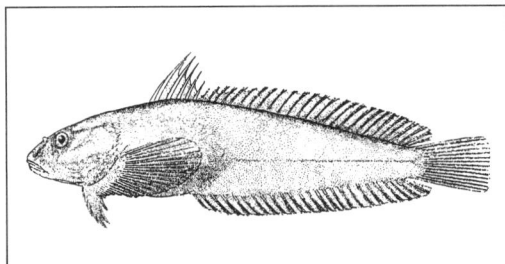

1106 PERISTEDIIDAE

SC   *Peristedion cataphractum* (Linnaeus, 1758)
ES   armado
DA   panserknurhane
DE   Panzerhahn; Panzerfisch
EL   κερατόψαρο· κερατάς
EN   armed gurnard
FR   malarmat; mallarmat; galinette
IT   pesce forca
NL   gepantserde poon
PT   cabra-de-casca
FI   panssarikurnusimppu
SV   pansarhane

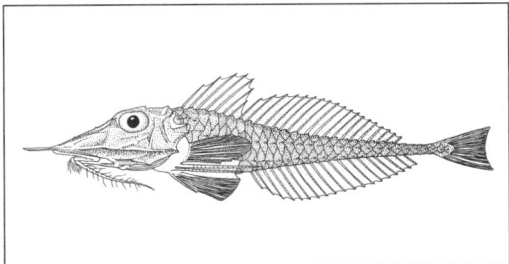

1108 LIPARIDAE

SC   *Liparis liparis* (Linnaeus, 1766); *Liparis vulgaris*
ES   liparis
DA   finnebræmmet ringbug
DE   Großer Scheibenbauch
EL   θαλάσσιο σαλιγκάρι
EN   sea snail
FR   limace barrée; limace de mer; marmotte; sucet
IT   lipara striata
NL   slakdolf
PT   peixe-caracol comum
FI   imukala
SV   ringbuk

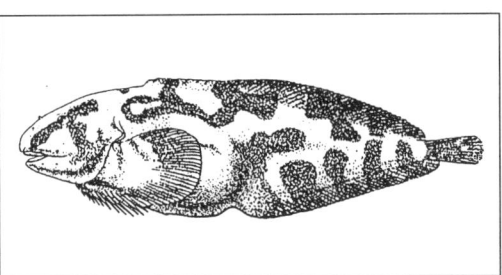

1109 DACTYLOPTERIDAE

SC	*Dactylopterus volitans* (Linnaeus, 1758)
ES	chicharra
DA	flyveknurhane
DE	Flughahn
EL	χελιδονόψαρο
EN	flying gurnard
FR	grondin volant
IT	civetta di mare; pesce civetta
NL	vliegende poon
PT	cabrinha de leque
FI	perhossimppu
SV	flygsimpa; flygknot

1111 PSETTODIDAE   SOT

SC	*Psettodes belcheri* (Bennett, 1831)
ES	lenguado espinudo de altura; perro
DA	plettet pigflynder
DE	Gefleckter Ebarme
EL	ουρόστικτο τάρμποτ
EN	spot-tail spiny turbot
FR	flétan tacheté d'Afrique; turbot tropical tacheté
IT	losanga atlantica
NL	gevlekte doornige heilbot
PT	palma espinhosa
FI	piikkikampela-laji
SV	fläckig tandvar

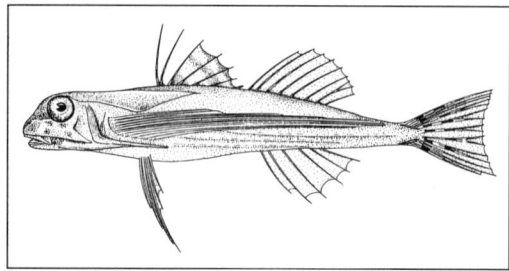

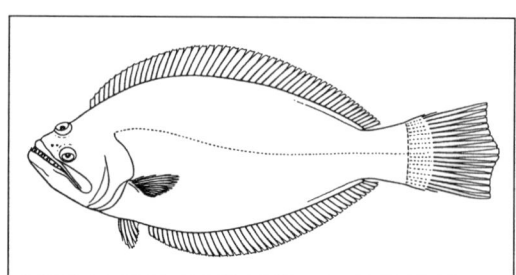

1110 PLEURONECTIFORMES   FLX

SC	*Pleuronectiformes*
ES	peces planos
DA	fladfisk; flynderfisk
DE	Plattfische
EL	πλατύψαρα· καλκάνια
EN	flatfishes
FR	poissons plats
IT	pleuronettiformi, pesci piatti
NL	platvissen
PT	peixes-chatos
FI	kampelakalat-lahko
SV	plattfiskar; flatfiskar

1112 PSETTODIDAE   PSB

SC	*Psettodes bennetti* (Steindachner, 1870)
ES	lenguado espinudo
DA	vestafrikansk pigflynder
DE	Westafrikanischer Ebarme
EL	τάρμποτ του Ατλαντικού
EN	spiny turbot
FR	turbot tropical; flétan d'Afrique
IT	losanga atlantica
NL	doornige heilbot
PT	palma
FI	piikkikampela-laji
SV	taggvar

DRAWING NOT AVAILABLE

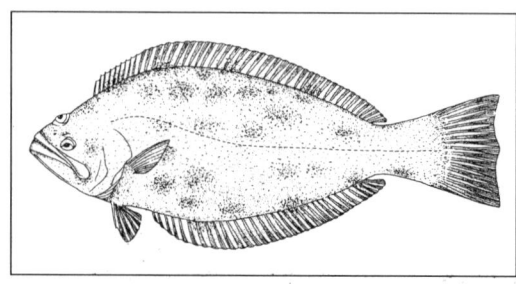

1113 PSETTODIDAE HAI

SC	*Psettodes erumei* (Bloch and Schneider, 1801)
ES	halibut índico
DA	indo-pacifisk pigflynder
DE	Indopazifischer Ebarme
EL	τάρμποτ του Ινδικού
EN	adalah; Indian halibut
FR	flétan tropical indo-pacifique
IT	losanga indiana
NL	Indo-Pacifische heilbot
PT	palma do Índico
FI	alkukampela
SV	tandvar

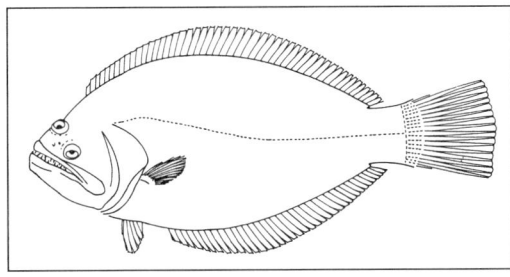

1114 BOTHIDAE LEF

SC	*Bothidae*
ES	peludas
DA	hvarre-familien
DE	Butte
EL	πησί
EN	left-eye flounders
FR	faux turbots; cardeaux
IT	botidi; suacie; pataracce; zanchette
NL	botachtigen
PT	cartas
FI	vasensilmäkampelat; vasensilmäkampelat-heimo
SV	tungevarar

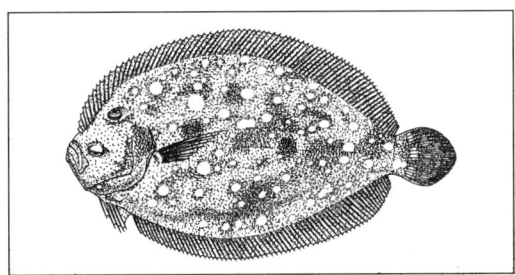

1115 BOTHIDAE MSF

SC	*Arnoglossus laterna* (Walbaum, 1792)
ES	peluda
DA	almindelig tungehvarre
DE	Lammzunge
EL	ζαγκέτα· γλώσσα
EN	scaldfish
FR	arnoglosse; fausse limande
IT	suacia; zanchetta
NL	schurftvis
PT	carta do Mediterrâneo
FI	suomukampela
SV	tungevar

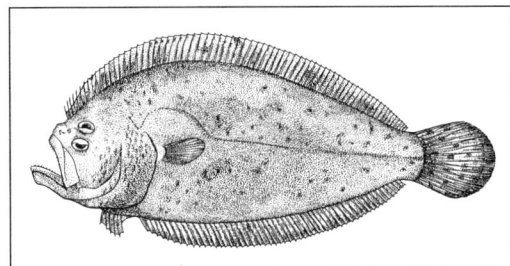

1116 BOTHIDAE

SC	*Arnoglossus thori* (Kyle, 1913); *Arnoglossus mol-toni*
ES	peludilla
DA	Thors tungehvarre
DE	Thors-Lammzunge
EL	ζαγκέτα
EN	Grohman's scaldfish
FR	arnoglosse de Thor; fausse limande
IT	suacia mora
NL	Thors-schurftvis
PT	carta pontuada
FI	töyhtökampela-laji
SV	Thorsvar

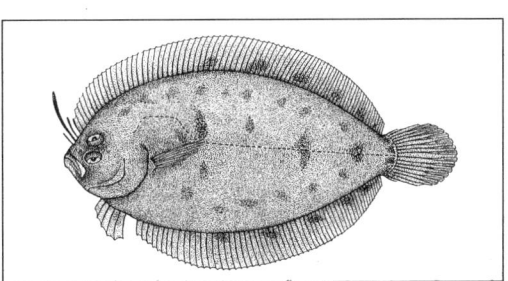

1117 BOTHIDAE

SC   *Bothus podas podas* (Delaroche, 1809)
ES   podas
DA   bredøjet hvarre
DE   Weitäugiger Butt
EL   πησί· καλκάνι
EN   wide-eyed flounder
FR   faux turbot; platophrys
IT   rombo di rena
NL   grootoogbot
PT   carta de olhos grandes
FI   kukkakampela
SV   storögd var

1119 BOTHIDAE

SC   *Paralichthys oblongus* (Mitchill, 1815)
ES   falso halibut de cuatro ocelos
DA   firplettet hvarre
DE   Vierfleckflunder
EL   τετράστικτη χωματίδα
EN   four-spot flounder
FR   cardeau à quatre ocelles
IT   rombo quadripuntato
NL   viervlekkige bastaardheilbot
PT   carta de quatro olhos; carta-alabote de quatro olhos
FI   piikkikampela-laji
SV   fyrfläckig var

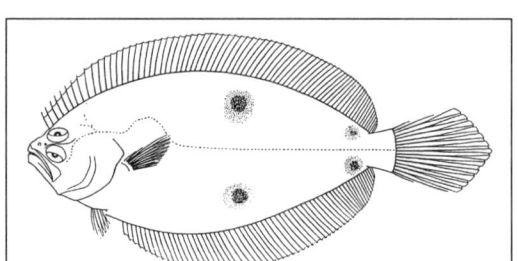

1118 BOTHIDAE   FLS

SC   *Paralichthys dentatus* (Linnaeus, 1766)
ES   falso halibut del Canadá
DA   sommerhvarre
DE   Sommerflunder
EL   χωματίδα του καλοκαιριού
EN   summer flounder
FR   cardeau d'été
IT   rombo dentato
NL   zomerbot
PT   carta de verão; falso alabote de Canadá; carta-alabote de verão
FI   kesäkampela
SV   sommarvar

1120 BOTHIDAE   BAH

SC   *Paralichthys olivaceus* (Temminck and Schlegel, 1846)
ES   falso halibut del Japón
DA   olivengrøn hvarre
DE   Olivgrüne Flunder
EL   χιράμι· πράσινη χωματίδα
EN   olive flounder; bastard halibut
FR   hirame; cardeau olivâtre
IT   hirame
NL   olijfgroene bastaardheilbot
PT   falso alabote japonês; carta-alabote japonesa
FI   töyhtökampela-laji
SV   japansk var

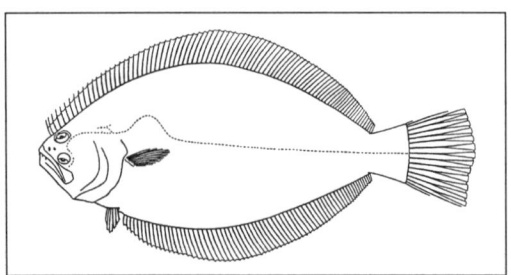

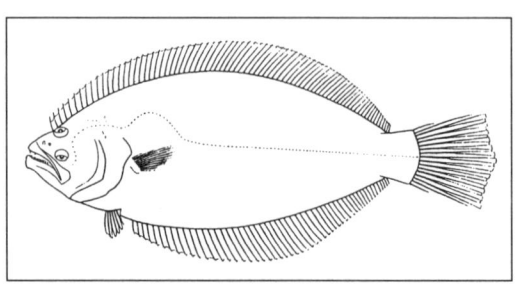

1121 BOTHIDAE   BAX

SC	*Paralichthys* spp.
ES	falsos halibuts
DA	hvarre-slægt
DE	Steinbutte
EL	χωματίδες
EN	flukes; bastard halibuts
FR	cardeaux
IT	hirame
NL	bastaardheilbotten
PT	falsos alabotes; cartas-alabote
FI	vasensilmäkampela-suku
SV	varar

1123 PLEURONECTIDAE   PLZ

SC	*Pleuronectidae*
ES	pleuronéctidos
DA	rødspætte-familien
DE	Schollen
EL	χωματίδες· γλώσσες· καλκάνια
EN	flounders; right-eye flounders
FR	pleuronectidés; limandes, plies et flétans
IT	pleuronettidi
NL	schollen
PT	solhas e alabotes; pleuronectídeos
FI	oikeasilmäkampelat; oikeasilmäkampelat-heimo
SV	flundrefiskar

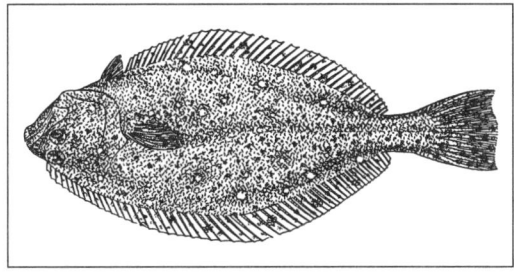

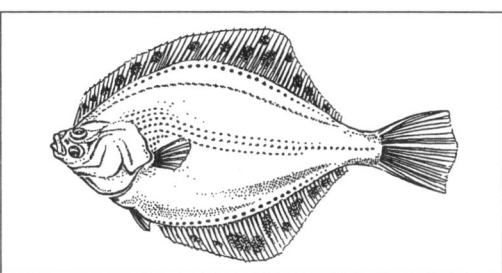

1122 BOTHIDAE   CIF

SC	*Pseudorhombus cinnamoneus* (Temminck and Schleger, 1846)
ES	rémol del Índico
DA	kanelhvarre
DE	Zimtflunder
EL	χωματίδα της Αυστραλίας
EN	cinnamon flounder
FR	rombou d'Australie
IT	suacia orientale
NL	Australische bot
PT	carta japonesa
FI	piikkikampela-laji
SV	kanelvar

1124 PLEURONECTIDAE   KAF

SC	*Atheresthes evermanni* (Jordan and Starks, 1904)
ES	halibut japonés
DA	japansk hellefisk
DE	Pfeilzahn-Heilbutt; Asiatischer Pfeilzahnheilbutt
EL	καλκάνι· χάλμπατ
EN	Kamchatka halibut
FR	flétan du Pacifique; faux flétan du Japon
IT	passera canina
NL	Japanse heilbot
PT	alabote japonês
FI	kampela-laji
SV	japansk flundra

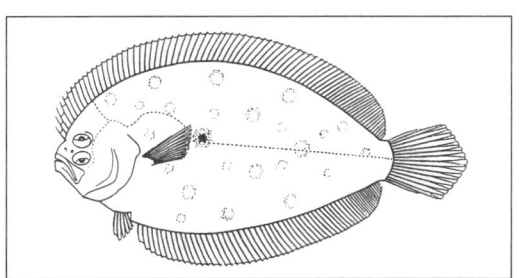

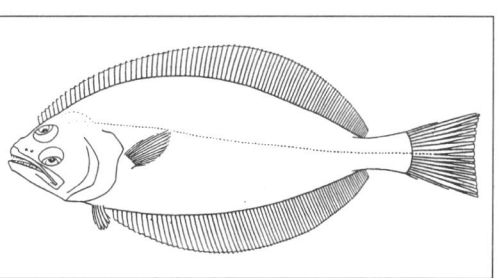

1125 PLEURONECTIDAE ARF

SC	*Atheresthes stomias* (Jordan and Gilbert, 1880)
ES	halibut del Pacífico
DA	Stillehavs-hellefisk
DE	Amerikanischer Pfeilzahnheilbutt
EL	καλκάνι του Ειρηνικού· χάλιμπατ
EN	arrow-tooth flounder
FR	faux flétan du Pacifique
IT	passera canina
NL	Pacifische heilbot
PT	alabote-dente curvo
FI	kampela-laji
SV	piltandsflundra

1127 PLEURONECTIDAE

SC	*Eopsetta jordani* (Lockington, 1879)
ES	rodaballo de California
DA	California-flynder
DE	Kalifornische Scholle
EL	καλκάνι της Καλιφόρνιας
EN	petrale sole
FR	plie de Californie
IT	passera della California
NL	Californische schol
PT	solha da Califórnia
FI	kampela-laji
SV	kalifornisk flundra

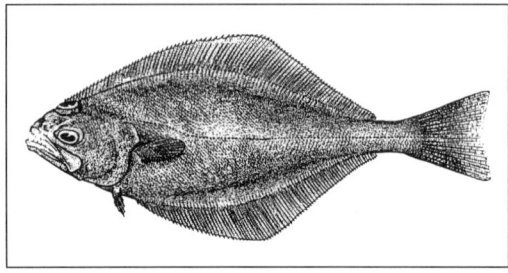

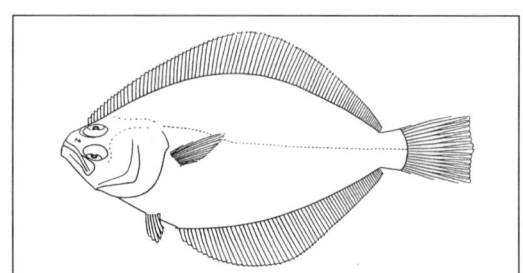

1126 PLEURONECTIDAE

SC	*Eopsetta grigorjewi* (Herzenstein, 1891)
ES	rodaballo de Japón
DA	japansk flynder
DE	Japanische Scholle
EL	καλκάνι της Ιαπωνίας
EN	North Pacific sole; roundnose flounder
FR	plie japonaise
IT	passera del Giappone
NL	Japanse schol
PT	solhão malhado do Japão
FI	kampela-laji
SV	rundnosflundra

1128 PLEURONECTIDAE WIT

SC	*Glyptocephalus cynoglossus* (Linnaeus, 1758); *Pleuronectes cynoglossus; Cynoglossus vulgaris*
ES	mendo
DA	skærising
DE	Rotzunge; Hundszunge; Zungenbutt; Echte Rotzunge
EL	καλκάνι
EN	witch; witch flounder
FR	plie grise; plie cynoglosse
IT	passera lingua di cane; passera
NL	witje
PT	solhão
FI	mustaveväkampela
SV	rödtunga

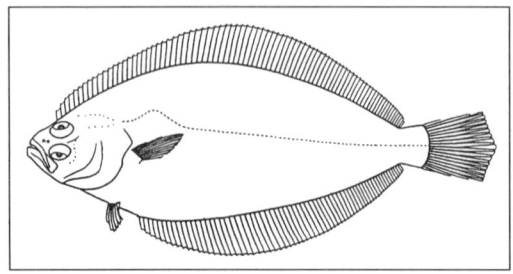

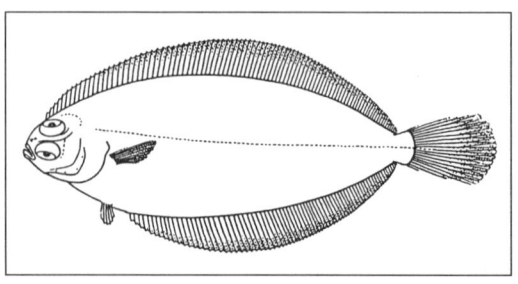

1129 PLEURONECTIDAE

SC *Glyptocephalus zachirus* (Lockington, 1879)
ES lenguado americano
DA amerikansk skærising
DE Amerikanische Scholle
EL καλκάνι της Αμερικής
EN rex sole
FR sole américaine
IT passera del Pacifico
NL Amerikaans witje
PT solhão americano
FI amerikankielikampela
SV stillahavsrödtunga

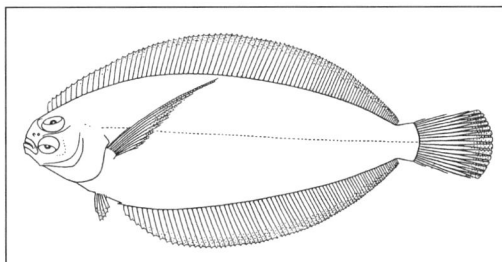

1131 PLEURONECTIDAE   FTS

SC *Hippoglossoides elassodon* (Jordan and Gilbert, 1880)
ES platija japonesa
DA nordpacifisk håising
DE Heilbuttscholle
EL γιαπωνέζικο καλκάνι
EN flathead sole
FR balai du Pacifique-Nord
IT passera giapponese
NL Japanse lange schar
PT solha japonesa
FI kampela-laji
SV japansk lerskädda

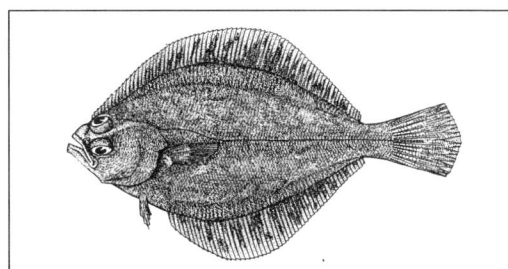

1130 PLEURONECTIDAE

SC *Hippoglossoides dubius* (Schmidt,1904)
ES platija japonesa
DA japansk håising
DE Japanischer Heilbutt
EL πλατυκέφαλο καλκάνι
EN Pacific false halibut; flathead flounder
FR balai japonais
IT passera giapponese
NL valse Japanse lange schar
PT solha-cabeça chata
FI kampela-laji
SV flathuvudsskädda

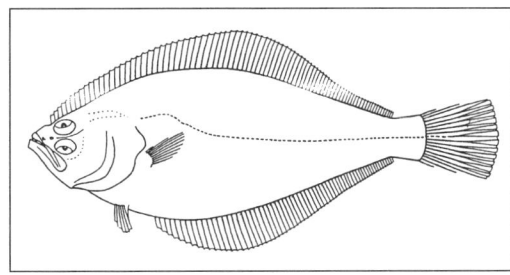

1132 PLEURONECTIDAE   PLA

SC *Hippoglossoides platessoides* (Fabricius, 1780);
   Drepanopsetta platessoides
ES platija americana; platija canadiense
DA almindelig håising
DE Doggerscharbe; Rauhe Scharbe; Rauhe Scholle;
   Scharbenzunge
EL καλκάνι του Καναδά
EN long rough dab; American plaice
FR balai; faux flétan; flétan nain; halibut faux;
   plie canadienne
IT passera canadese; passera
NL Amerikaanse schol
PT solha americana; solha-flanda
FI liejukampela
SV lerskädda

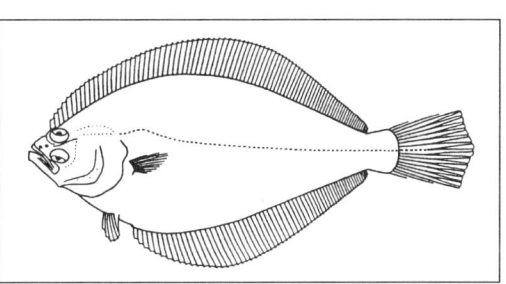

1133 PLEURONECTIDAE HAL

SC	*Hippoglossus hippoglossus* (Linnaeus, 1758); *Hippoglossus vulgaris*
ES	fletán; halibut; napoleón; hipogloso
DA	helleflynder
DE	Atlantischer Heilbutt; Weißer Heilbutt; Heilbutt
EL	χάλιμπατ· χάλιμπατ του Ατλαντικού
EN	Atlantic halibut; halibut
FR	flétan de l'Atlantique; flétan commun; halibut commun
IT	ippoglosso atlantico; halibut
NL	heilbot
PT	alabote do Atlântico
FI	ruijanpallas
SV	hälleflundra

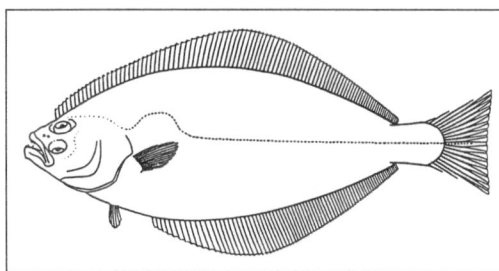

1135 PLEURONECTIDAE ROS

SC	*Lepidopsetta bilineata* (Ayres, 1855)
ES	lenguado del Pacífico
DA	pacifisk flynder
DE	Pazifische Scholle
EL	καλκάνι του Ειρηνικού
EN	rock sole
FR	sole du Pacifique; fausse limande du Pacifique
IT	passera del Pacifico
NL	Pacifische schol
PT	solha da rocha
FI	kampela-laji
SV	klippflundra

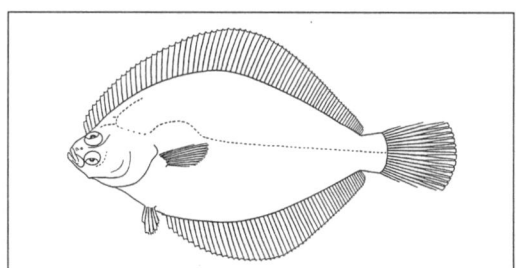

1134 PLEURONECTIDAE HAP

SC	*Hippoglossus stenolepis* (Schmidt, 1904)
ES	halibut; fletán del Pacífico
DA	Stillehavs-helleflynder
DE	Pazifischer Heilbutt
EL	χάλιμπατ του Ειρηνικού
EN	Pacific halibut
FR	flétan du Pacifique
IT	halibut del Pacifico
NL	Pacifische heilbot
PT	alabote do Pacífico
FI	tyynenmerenpallas
SV	stillahavshälleflundra

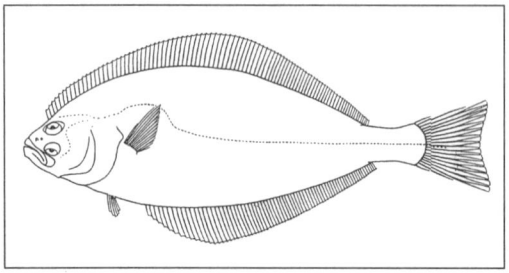

1136 PLEURONECTIDAE YES

SC	*Limanda aspera* (Pallas, 1811)
ES	limanda japonesa
DA	japansk ising
DE	Rauhe Kliesche
EL	γιαπωνέζικη χωματίδα
EN	yellowfin sole
FR	limande du Japon
IT	limanda giapponese
NL	Japanse schar
PT	solha áspera
FI	japaninhietakampela
SV	stillahavsskädda

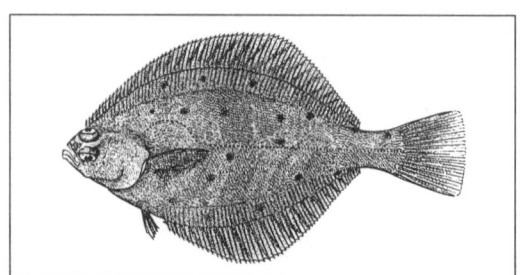

1137 PLEURONECTIDAE YEL

SC	*Limanda ferruginea* (Storer, 1839)
ES	limanda nórdica
DA	gulhalet ising
DE	Gelbschwanzflunder
EL	χωματίδα με κίτρινη ουρά
EN	yellowtail flounder; yellowfin sole
FR	limande à queue jaune
IT	limanda
NL	geelstaartschar
PT	solha dos mares do Norte
FI	ruostekampela
SV	gulstjärtsskädda

1139 PLEURONECTIDAE DAB

SC	*Limanda limanda* (Linnaeus, 1758)
ES	limanda
DA	ising; slette; almindelig ising
DE	Kliesche; Scharbe
EL	λιμάντα· χωματίδα
EN	common dab
FR	limande commune
IT	limanda
NL	schar
PT	solha escura do mar do Norte; limanda
FI	hietakampela
SV	sandskädda

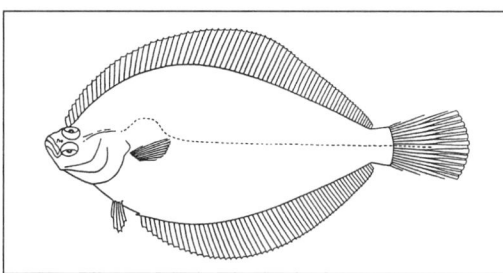

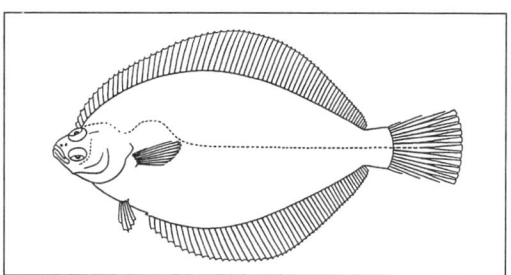

1138 PLEURONECTIDAE

SC	*Limanda herzensteini* (Jordan and Snydner, 1901); *Pseudopleuronectes herzensteini*
ES	limanda japonesa; acedía del Japón
DA	japansk flynder
DE	Japanische Flunder
EL	χωματίδα της Ιαπωνίας
EN	Japanese dab; yellow-striped flounder; magarei
FR	magare; limande du Japon; limande-pile du Japon
IT	limanda giapponese; limanda striata
NL	Japanse geelgestreepte schar
PT	solha castanha do Japão
FI	japaninhietakampela
SV	japansk sandskädda

1140 PLEURONECTIDAE

SC	*Liopsetta glacialis* (Pallas, 1776)
ES	solla ártica
DA	arktisk flynder
DE	Polarscholle
EL	καλκάνι της Αρκτικής
EN	Arctic flounder
FR	plie arctique
IT	passera artica
NL	ijsbot
PT	solha árctica
FI	napakampela
SV	arktisk flundra

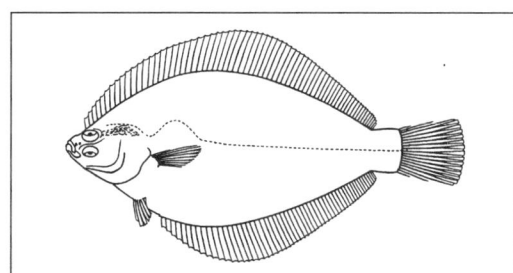

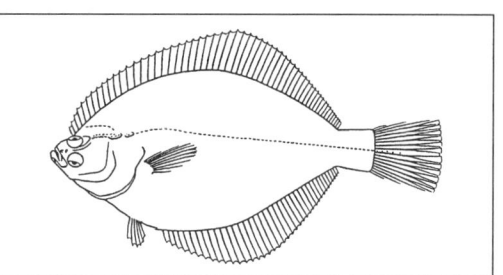

1141 PLEURONECTIDAE

SC	*Liopsetta putnami* (Gill, 1864)
ES	solla lisa
DA	glat flynder
DE	Glattflunder
EL	λείο καλκάνι
EN	smooth flounder
FR	plie lisse
IT	passera liscia
NL	gladde schol
PT	solha lisa
FI	napakampela-laji
SV	slätflundra

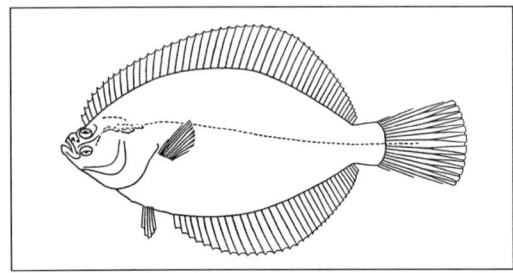

1143 PLEURONECTIDAE   FLE

SC	*Platichthys flesus* (Linnaeus, 1758)
ES	platija europea
DA	skrubbe
DE	Flunder; Struffbutt; Sandbutt
EL	καλκάνι
EN	flounder; European flounder; fluke
FR	flet; flet commun
IT	passera pianuzza; passera
NL	bot
PT	solha das pedras
FI	kampela
SV	flundra; skrubbskädda; skrubba

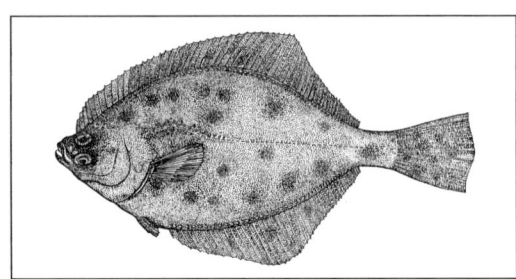

1142 PLEURONECTIDAE   LEM

SC	*Microstomus kitt* (Walbaum, 1792); *Pleuronectes microcephalus*
ES	falsa limanda
DA	rødtunge
DE	Limande; Echte Rotzunge
EL	λεμονόγλωσσα
EN	lemon sole
FR	limande sole
IT	limanda
NL	tongschar
PT	solha-limão
FI	pikkupääkampela
SV	bergtunga; bergskädda

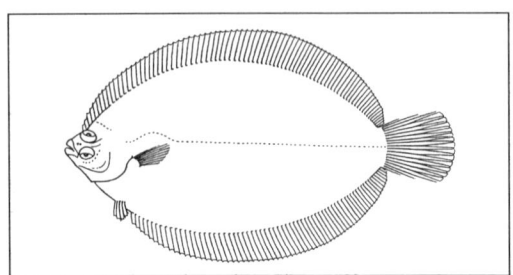

1144 PLEURONECTIDAE

SC	*Platichthys stellatus* (Pallas, 1787)
ES	platija del Pacífico
DA	stjerneflynder
DE	Sternflunder
EL	αστροκαλκάνι
EN	starry flounder
FR	plie du Pacifique
IT	passera stellata
NL	sterschol
PT	solha estrelada do Pacífico
FI	tähtikampela
SV	stjärnskrubba

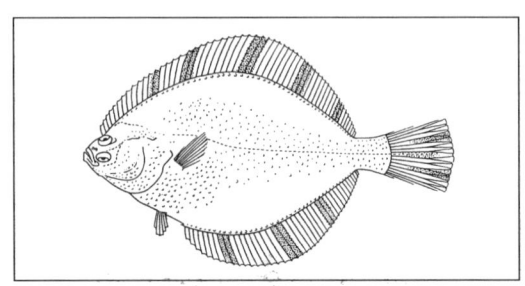

1145 PLEURONECTIDAE PLE

SC	*Pleuronectes platessa* (Linnaeus, 1758)
ES	solla; platija; platuxa; platura; platecha
DA	rødspætte
DE	Scholle; Goldbutt
EL	ευρωπαϊκή χωματίδα
EN	plaice; European plaice
FR	plie d'Europe; carrelet; plie franche
IT	passera di mare; platessa; passera
NL	schol
PT	solha; solha avessa
FI	punakampela
SV	rödspätta

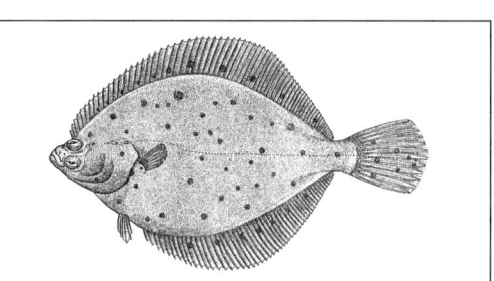

1146 PLEURONECTIDAE ALP

SC	*Pleuronectes quadrituberculatus* (Pallas, 1814); *Pleuronectes pallasii*
ES	solla de Alaska
DA	Alaska-rødspætte
DE	Alaska-Scholle
EL	χωματίδα της Αλάσκας
EN	Alaska plaice
FR	plie de l'Alaska
IT	platessa dell'Alasca
NL	Alaskaschol
PT	solha do Alasca
FI	alaskanpunakampela
SV	alaskarödspätta

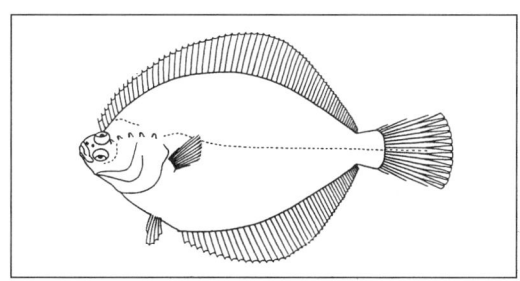

1147 PLEURONECTIDAE FLW

SC	*Pseudopleuronectes americanus* (Walbaum, 1792)
ES	solla roja
DA	vinter-flynder
DE	Amerikanische Winterflunder
EL	χωματίδα της Αμερικής
EN	winter flounder
FR	plie rouge; limande d'Amérique
IT	limanda americana
NL	Amerikaanse winterschol
PT	solha de Inverno
FI	mustaselkäkampela
SV	vinterflundra

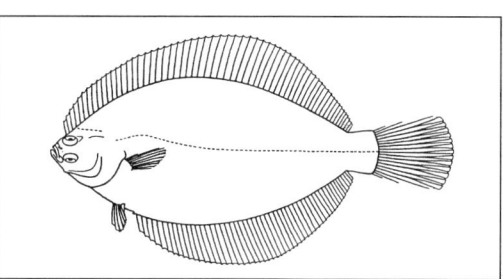

1148 PLEURONECTIDAE GHL

SC	*Reinhardtius hippoglossoides* (Walbaum, 1792); *Hippoglossus reinhardtius*
ES	halibut negro
DA	hellefisk
DE	Schwarzer Heilbutt; Grönland-Heilbutt
EL	χάλιμπατ της Γροιλανδίας· μαύρο χάλιμπατ
EN	Greenland halibut; black halibut
FR	flétan noir; halibut noir; flétan du Groenland
IT	ippoglosso nero; halibut; halibut di Groenlandia
NL	zwarte heilbot; Groenlandse heilbot
PT	alabote da Gronelândia
FI	grönlanninpallas
SV	liten hälleflundra; blåkveite

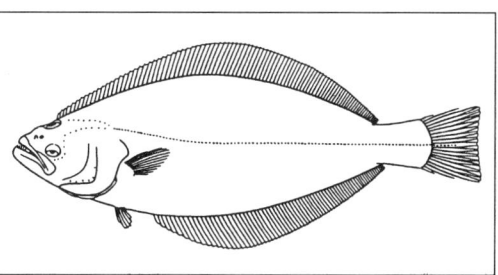

1149 PLEURONECTIDAE

SC	*Rhombosolea plebeia* (Richardson, 1843)
ES	solla de arena
DA	sandflynder
DE	Sandflunder
EL	ρομβόγλωσσα
EN	sand flounder
FR	sole de Nouvelle-Zélande; flet de Nouvelle-Zélande
IT	rombo australe
NL	Nieuw-Zeelandse bot
PT	solha da areia da Nova Zelândia
FI	kampela-laji
SV	sandflundra

1151 SOLEIDAE   SOX

SC	*Soleidae*
ES	lenguados
DA	tunge-familien
DE	Seezungen
EL	γλώσσες
EN	soles
FR	soles
IT	soleidi
NL	tongen
PT	soleídeos; linguados e afins
FI	kielikampelat; kielikampelat-heimo
SV	tungefiskar

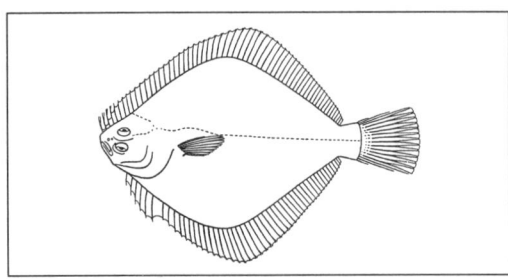

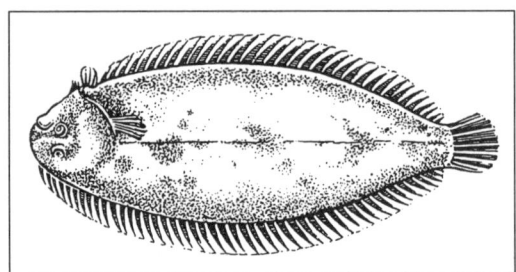

1150 PLEURONECTIDAE   FSA

SC	*Rhombosolea* spp.
ES	sollas de arena
DA	flynder-slægt
DE	Schollen
EL	ρομβόγλωσσες
EN	sand flounders
FR	rhombosoles; flets de Nouvelle-Zélande
IT	rombi australi
NL	Nieuw-Zeelandse botten
PT	solhas da areia
FI	kampelat
SV	sandflundror

1152 SOLEIDAE

SC	*Achirus lineatus* (Linnaeus, 1758)
ES	lenguado americano
DA	amerikansk tunge
DE	Pazifische Seezunge
EL	γραμμωτή γλώσσα
EN	lined sole
FR	sole américaine
IT	sogliola rotonda
NL	Amerikaanse tong
PT	linguado redondo riscado
FI	kielikampela-laji
SV	stillahavstunga

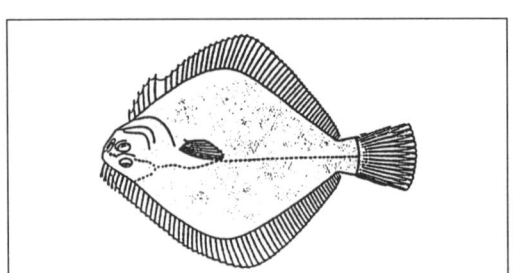

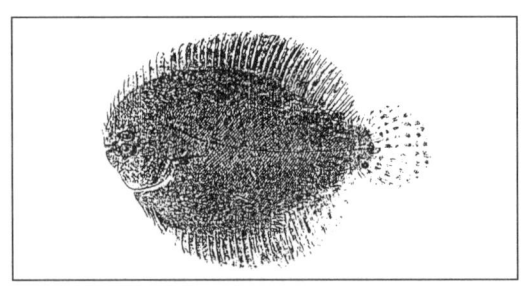

1153 SOLEIDAE SOW

SC	*Austroglossus microlepis* (Bleeker, 1863)
ES	lenguado austral oeste
DA	sydvestafrikansk tunge
DE	Westküsten-Seezunge
EL	γλώσσα της Δυτικής Αυστραλίας
EN	West Coast sole
FR	sole australe occidentale
IT	sogliola del Sudafrica
NL	Zuid-Afrikaanse Westkusttong
PT	linguado austral-oeste
FI	etelänkielikampela-laji
SV	sydafrikansk västkusttunga

1155 SOLEIDAE SOA

SC	*Austroglossus* spp.
ES	lenguados
DA	tunge-slægt
DE	Südost-Atlantik-Seezungen
EL	αυστραλογλώσσες
EN	Southeast Atlantic soles
FR	soles australes
IT	sogliole del Sudafrica
NL	Zuid-Afrikaanse tongen
PT	linguados austrais
FI	etelänkielikampela-suku
SV	tungor

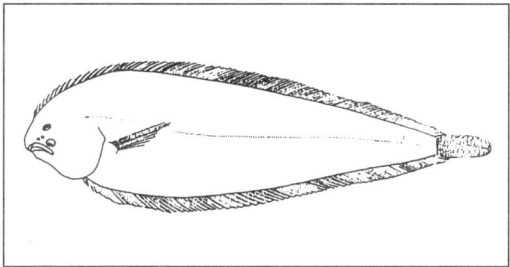

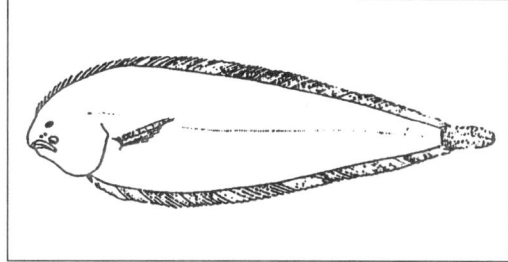

1154 SOLEIDAE SOE

SC	*Austroglossus pectoralis* (Kaup, 1858)
ES	lenguado austral este
DA	sydøstafrikansk tunge
DE	Ostküsten-Seezunge
EL	αυστραλογλώσσα
EN	East Coast sole; mud sole
FR	sole australe orientale
IT	sogliola del Sudafrica
NL	Zuid-Afrikaanse Oostkusttong
PT	linguado austral-este
FI	etelänkielikampela-laji
SV	sydafrikansk östkusttunga

1156 SOLEIDAE

SC	*Buglossidium luteum* (Risso, 1810); *Solea lutea*; *Pleuronectes luteus*
ES	tambor
DA	glastunge
DE	Zwergzunge; Zwergseezunge
EL	γλώσσα
EN	solenette
FR	petite sole jaune
IT	sogliola gialla
NL	dwergtong
PT	língua-de-gato
FI	pikkukielikampela
SV	småtunga

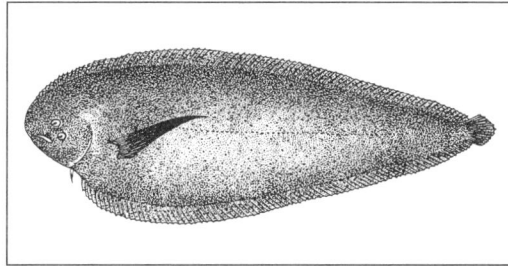

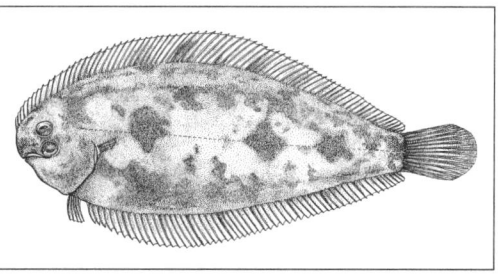

1157 SOLEIDAE

CET

SC	*Dicologlossa cuneata* (Moreau, 1881); *Dicologoglossa cuneata*; *Solea cuneata*
ES	acedia; lenguadillo
DA	Senegal-tunge
DE	Cuneata-Seezunge
EL	γλώσσα
EN	thickback sole; Senegal sole; wedge sole
FR	céteau; langue d'avocat
IT	sogliola cuneata
NL	wigtong
PT	língua
FI	kielikampela-laji
SV	tjocktunga

1159 SOLEIDAE

SC	*Microchirus theophila* (Risso, 1810); *Microchirus azevia*; *Zevaia theophila*
ES	golleta acevia
DA	azevia-tunge
DE	Azevia-Seezunge
EL	ψευδόγλωσσα
EN	Jewish sole
FR	sole-perdrix juive
IT	sogliola azevia
NL	joodse tong
PT	azevia
FI	kielikampela-laji
SV	azeviatunga

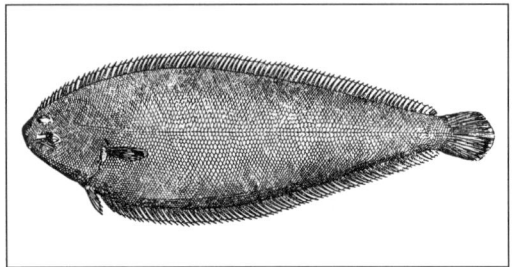 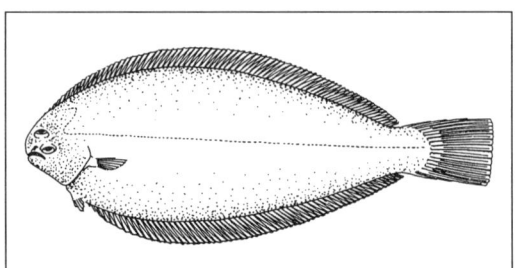

1158 SOLEIDAE

SC	*Microchirus ocellatus* (Linnaeus, 1758); *Solea ocellata*
ES	tambor real
DA	øjeplettet tunge
DE	Augenfleck-Seezunge; Augen-Seezunge
EL	γλώσσα
EN	four-eyed sole
FR	sole ocellée; sole-perdrix
IT	sogliola occhiuta
NL	ogentong
PT	azevia de malhas
FI	kielikampela-laji
SV	ögonfläckstunga

1160 SOLEIDAE

SC	*Microchirus variegatus* (Donovan, 1808); *Solea variegata*
ES	acedia
DA	stribet tunge
DE	Bastardzunge
EL	γλωσσάκι
EN	wedge sole
FR	sole panachée; sole-perdrix commune
IT	sogliola fasciata; sogliola variegata
NL	dikrugtong
PT	azevia raiada
FI	kielikampela-laji
SV	strimmad tunga

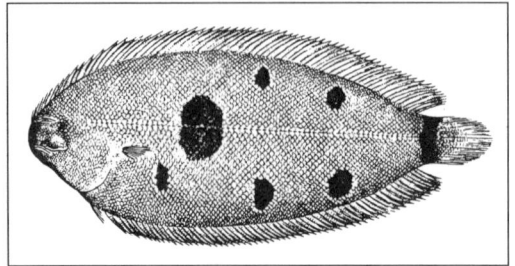 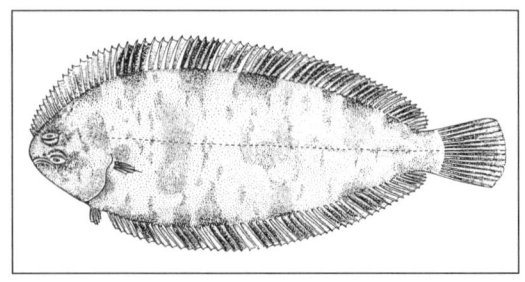

1161 SOLEIDAE

SC   *Microchirus* spp.
ES   golletas
DA   tunge-slægt
DE   Seezungen
EL   γλωσσάκια
EN   thickback soles
FR   soles-perdrix
IT   sogliole
NL   dikrugtongen
PT   azevias
FI   kielikampela-suku
SV   tungor

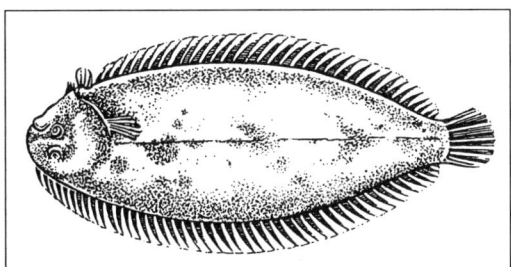

1163 SOLEIDAE

SC   *Solea kleinii* (Bonaparte, 1833); *Pegusa kleinii; Synapturichthys kleinii*
ES   lenguado turco
DA   Kleins tunge
DE   Schwarzrand-Seezunge
EL   γλώσσα
EN   Klein's sole
FR   sole; sole tachetée; sole de Klein
IT   sogliola turca
NL   Kleins tong
PT   linguado turco
FI   kielikampela-laji
SV   fläcktunga

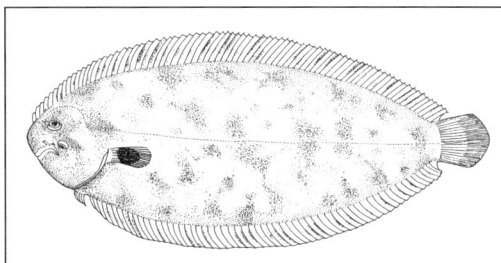

1162 SOLEIDAE

SC   *Monochirus hispidus* (Rafinesque, 1814)
ES   soldado
DA   håret tunge
DE   Pelz-Seezunge
EL   γλώσσα
EN   whiskered sole
FR   sole velue
IT   sogliola pelosa
NL   pelstong
PT   cascarra
FI   kielikampela-laji
SV   skäggtunga

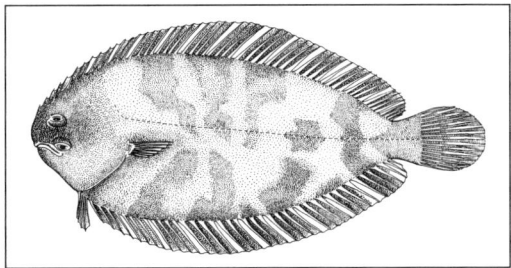

1164 SOLEIDAE

SC   *Solea lascaris* (Risso, 1810); *Pegusa lascaris*
ES   sortija
DA   sandtunge
DE   Sandzunge; Warzen-Seezunge
EL   γλώσσα
EN   sand sole
FR   sole de sable; sole-pole; sole-pelouse
IT   sogliola dal porro
NL   Franse tong
PT   linguado da areia
FI   hietakielikampela
SV   sandtunga

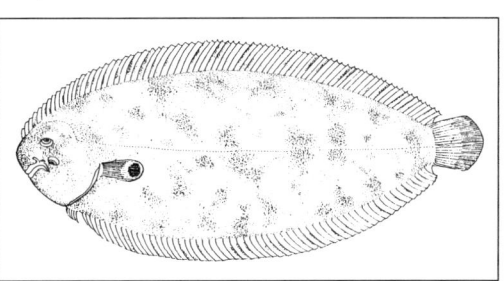

1165 SOLEIDAE

SOL

SC	*Solea vulgaris* (Quensel, 1806); *Solea solea*
ES	lenguado común
DA	tunge; søtunge; almindelig tunge
DE	Gemeine Seezunge
EL	γλώσσα
EN	sole; common sole
FR	sole commune; sole
IT	sogliola
NL	tong
PT	linguado legítimo; linguado
FI	kielikampela
SV	tunga

1167 SCOPHTHALMIDAE

SC	*Scophthalmidae*
ES	escoftálmidos
DA	pighvar-familien
DE	Butte
EL	καλκάνια· πησιά
EN	turbots
FR	turbots et cardines
IT	scoftalmidi; rombi
NL	tarbotten
PT	areeiros, pregados e rodovalhos
FI	silokampelat; silokampelat-heimo
SV	varar

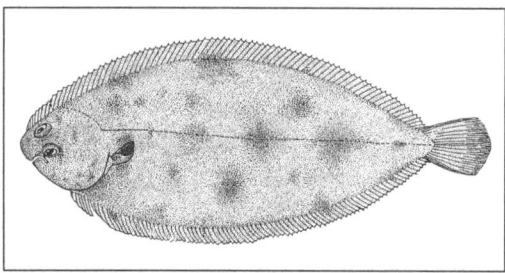

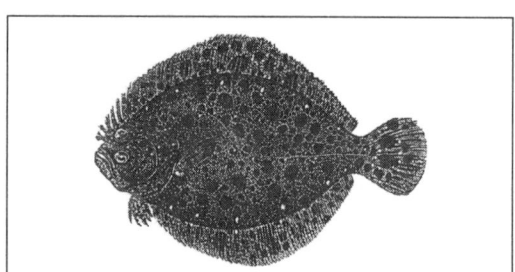

1166 CYNOGLOSSIDAE

TOX

SC	*Cynoglossidae*
ES	cinoglósidos; lenguas
DA	hundetunge-familien
DE	Hundszungen
EL	γλώσσες
EN	tongue-soles; tonguefishes
FR	cynoglosses; soles-langues
IT	cinoglossidi
NL	hondstongen
PT	línguas-de-cão
FI	etelänkielikampelat; etelänkielikampelat-heimo
SV	hundtungefiskar

1168 SCOPHTHALMIDAE

LDB

SC	*Lepidorhombus boscii* (Risso, 1810)
ES	gallo
DA	firplettet glashvarre
DE	Vierfleckbutt
EL	ζαγκέτα· γλώσσα
EN	four-spot megrim
FR	cardine à quatre taches
IT	rombo quattrocchi
NL	viervlekkkige scharretong
PT	areeiro de quatro manchas
FI	silokampela-laji
SV	fyrfläckig var

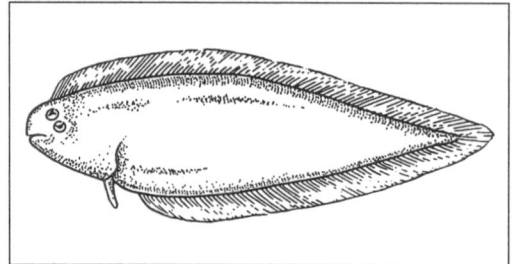

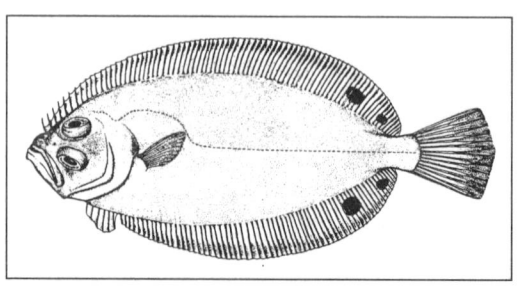

1169 SCOPHTHALMIDAE MEG

SC   *Lepidorhombus whiffiagonis* (Walbaum, 1792);
       Lepidorhombus megastoma
ES   gallo
DA   almindelig glashvarre
DE   Scheefsnut; Flügelbutt; Migram; Glasbutt;
       Blindling; Blendling
EL   ζαγκέτα· γλώσσα
EN   megrim; whiff; sail-fluke
FR   cardine franche; cardine; limande salope
IT   rombo giallo
NL   scharretong
PT   areeiro
FI   lasikampela
SV   glasvar

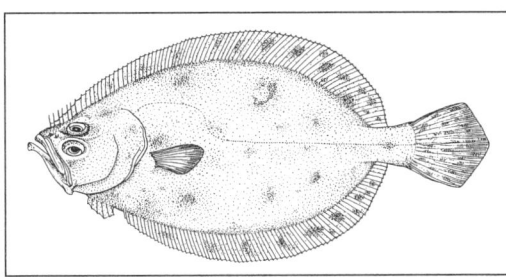

1171 SCOPHTHALMIDAE

SC   *Phrynorhombus norvegicus* (Günther, 1862)
ES   limanda noruega
DA   småhvarre
DE   Norwegischer Zwergbutt
EL   ζαγκέτα της Νορβηγίας
EN   Norwegian topknot
FR   phrynorhombe de Norvège; cardine de Norvège
IT   rombo peloso di Norvegia
NL   Noorse dwergbot
PT   bruxa norueguesa
FI   pikkukampela
SV   småvar

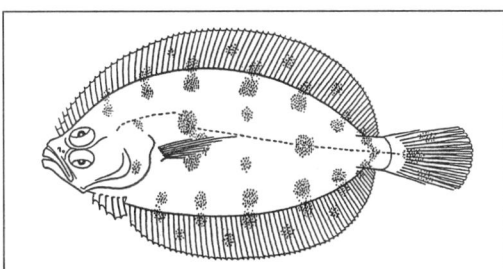

1170 SCOPHTHALMIDAE LEZ

SC   *Lepidorhombus* spp.
ES   gallos
DA   glashvarre-slægten
DE   Butte
EL   ζαγκέτες· γλώσσες
EN   megrims
FR   cardines
IT   lepidorombi
NL   scharretongen
PT   areeiros
FI   silokampela-suku
SV   glasvarar

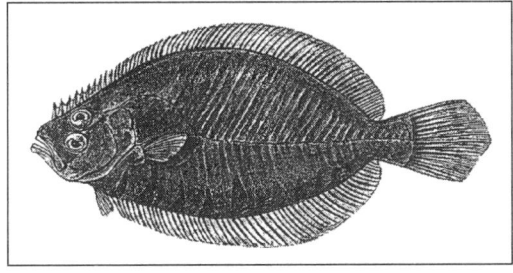

1172 SCOPHTHALMIDAE TUR

SC   *Psetta maxima* (Linnaeus, 1758); *Rhombus maximus*; *Scophthalmus maximus*
ES   rodaballo; parracho; rémol; corujo; sollo
DA   pighvar
DE   Steinbutt; Tarbutt; Turbutt; Turbot; Torbutt
EL   καλκάνι· σιάκι
EN   turbot
FR   turbot commun; turbot
IT   rombo chiodato; rombo
NL   tarbot
PT   pregado
FI   piikkikampela
SV   piggvar

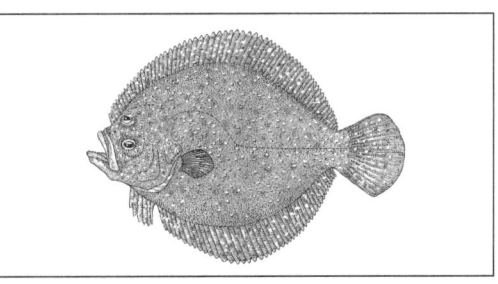

1173 SCOPHTHALMIDAE

SC   *Psetta maxima maeotica* (Pallas, 1811);
     Pleuronectes maeoticus; Scophthalmus maeoticus
ES   rodaballo del Mar Negro
DA   Sortehavs-pighvar
DE   Schwarzmeer-Steinbutt
EL   καλκάνι
EN   Black Sea turbot
FR   turbot de la mer Noire
IT   rombo chiodato del mar Nero
NL   Zwarte-Zeetarbot
PT   pregado do mar Negro
FI   mustanmerenpiikkikampela
SV   svartahavspiggvar

1175 SCOPHTHALMIDAE   BLL

SC   *Scophthalmus rhombus* (Linnaeus, 1758); *Rhombus*
     laevis; Rhombus vulgaris
ES   rémol; rodaballo; rapante; corujo; escamudo
DA   slethvar
DE   Glattbutt; Kleist; Tarbutt; Brill; Viereck
EL   καλκάνι· πησί
EN   brill
FR   barbue; turbot lisse
IT   rombo liscio; soaso
NL   griet
PT   rodovalho
FI   silokampela
SV   slätvar

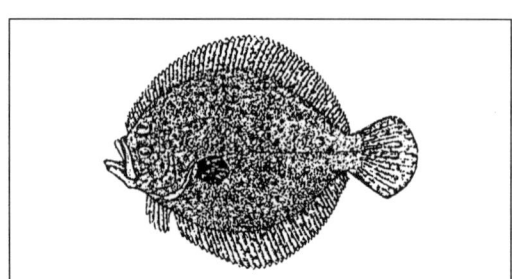

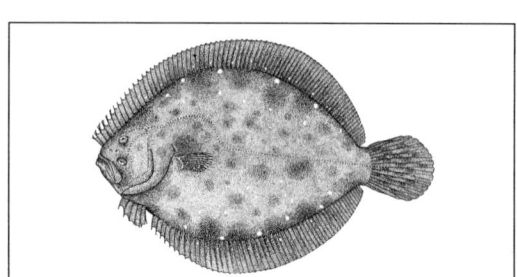

1174 SCOPHTHALMIDAE   FLD

SC   *Scophthalmus aquosus* (Mitchill, 1815)
ES   rodaballo americano
DA   amerikansk slethvar
DE   Sandbutt
EL   καλκάνι της Αμερικής· πησί της Αμερικής
EN   windowpane flounder
FR   barbue américaine
IT   rombo canadese
NL   Amerikaanse griet
PT   rodovalho americano
FI   amerikansilokampela
SV   fönsterglasvar

1176 SCOPHTHALMIDAE

SC   *Zeugopterus punctatus* (Bloch, 1787)
ES   rodaballo
DA   hårhvarre
DE   Zwergbutt; Müllers Zwergbutt; Haarbutt
EL   καλκάνι
EN   topknot
FR   targeur; sole de roche
IT   rombo camaso
NL   gevlekte griet
PT   rodovalho-bruxa
FI   kalliokampela
SV   bergvar

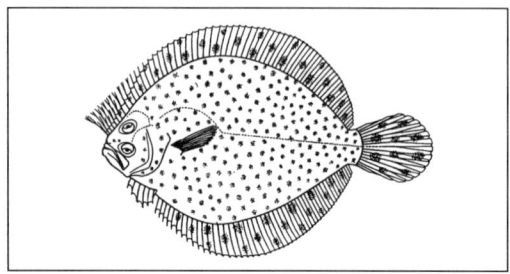

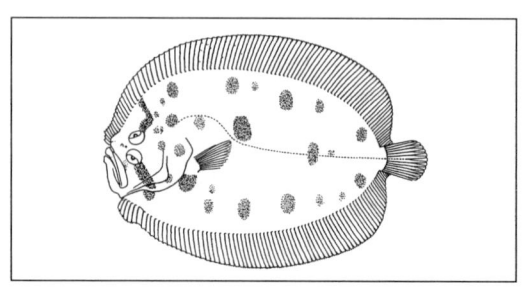

1177 CITHARIDAE

SC	*Citharus linguatula* (Linnaeus, 1758)
ES	solleta
DA	storskællet hvarre
DE	Großschuppige Scholle
EL	ζαγκέτα
EN	spotted flounder
FR	cithare feuille
IT	linguatula
NL	gevlekte bot
PT	carta de bico
FI	isosilmäkampela-laji
SV	fjällvar

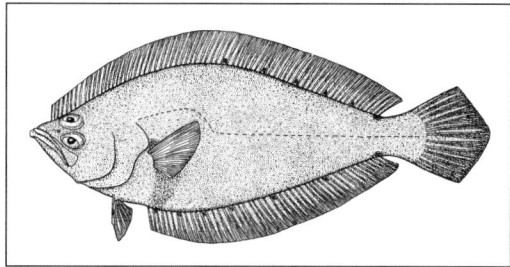

1179 BALISTIDAE   TRI

SC	*Balistidae*
ES	peces ballesta
DA	aftrækkerfisk-familien
DE	Drückerfische
EL	βαλιστές· μονόχειροι
EN	triggerfishes
FR	balistes; arbalétriers
IT	balistidi
NL	trekkervissen
PT	cangulos; balistas
FI	säppikalat; säppikalat-heimo
SV	tryckarfiskar

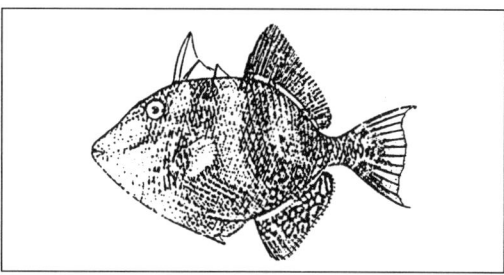

1178 ECHENEIDIDAE

SC	*Remora remora* (Linnaeus, 1758)
ES	rémora
DA	sugefisk
DE	Kleiner Schiffshalter; Küstensauger
EL	κολησόψαρο· κολαούζος
EN	shark sucker; suckerfish
FR	rémora; calfat; sucet
IT	remora nera
NL	remora
PT	pegador
FI	remora
SV	remora; sugfisk

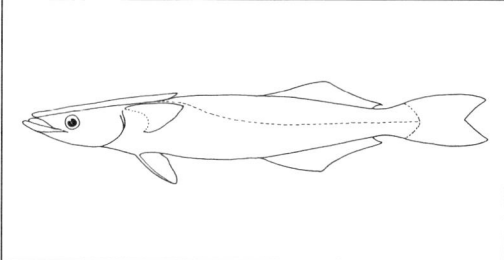

1180 BALISTIDAE   TRG

SC	*Balistes carolinensis* (Gmelin, 1789); *Balistes capriscus*
ES	pez ballesta; pejepuerco blanco
DA	aftrækkerfisk
DE	Drückerfisch; Schweinsdrückerfisch
EL	μονόχειρος· βαλιστής· γαϊδουρόψαρο
EN	grey triggerfish; clown triggerfish
FR	baliste; arbalétrier; baliste gris; baliste-cabri
IT	pesce balestra
NL	trekkervis
PT	cangulo cinzento
FI	porsassäppikala
SV	grå tryckarfisk

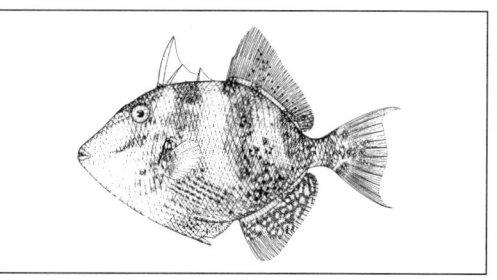

1181 MONACANTHIDAE

SC	*Aluterus schoepfi* (Walbaum, 1792)
ES	pez ballesta anaranjado
DA	orange filfisk
DE	Orangeroter Feilenfisch
EL	βαλιστής
EN	orange filefish
FR	bourse orange
IT	balistide arancio
NL	oranje vijlvis
PT	peixe-gatilho laranja
FI	oranssiviilakala
SV	orange filfisk

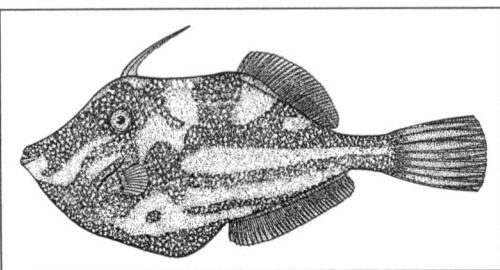

1183 MONACANTHIDAE — FIL

SC	*Stephanolepis cirrhifer* (Temminck and Schlegel, 1850)
ES	pez ballesta
DA	sejlfilfisk
DE	Segel-Feilenfisch
EL	βαλιστής
EN	thread-sail filefish
FR	bourse-fil
IT	pesce balestra
NL	draadzeilvijlvis
PT	peixe-gatilho de vela; cangulo de vela
FI	rihmaviilakala
SV	trådfilfisk

DRAWING NOT AVAILABLE

1182 MONACANTHIDAE — FLF

SC	*Cantherhines* spp.; *Navodon* spp.
ES	peces ballesta
DA	filfisk-slægt
DE	Feilenfische
EL	βαλιστής
EN	filefishes
FR	poissons-bourses
IT	pesci balestra
NL	vijlvissen
PT	peixes-gatilho; cangulos galhudos
FI	viilakala
SV	filfiskar

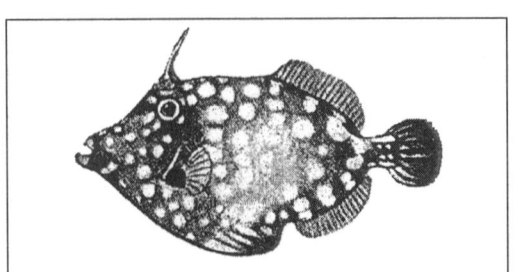

1184 TETRAODONTIDAE — PUX

SC	*Tetraodontidae*
ES	tamboriles
DA	kuglefisk-familien
DE	Kugelfische; Aufbläser
EL	τετράδοντοι
EN	puffers
FR	tétrodons; poissons-globes; poissons-ballons
IT	pesci palla; tetraodontidi
NL	kogelvissen
PT	tetraodontídeos
FI	pallokalat; pallokalat-heimo
SV	blåsfiskar; kulfiskar

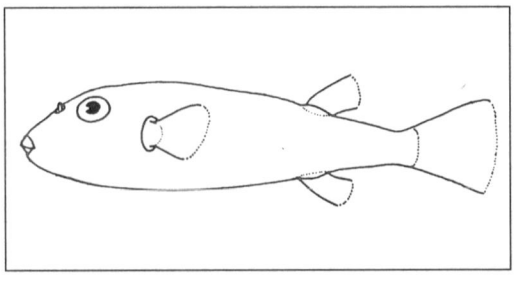

1185 TETRAODONTIDAE — PUF

SC   *Sphoeroides maculatus* (Bloch and Schneider, 1801)
ES   tamboril norteño; pez globo maculado
DA   nordlig kuglefisk
DE   Nördlicher Kugelfisch; Gefleckter Aufbläser
EL   γουρουνόψαρο
EN   Northern puffer; swellfish
FR   orbe-étoile; sphéroide du Nord; tétrodon bigarré
IT   pesce palla maculato; pesce palla
NL   noordelijke kogelvis
PT   peixe-bola do Norte
FI   pohjanpallokala
SV   fläckig blåsfisk

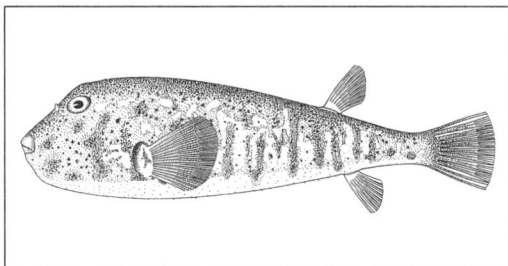

1187 MOLIDAE

SC   *Masturus lanceolatus* (Lienard, 1840)
ES   pez luna lanceolado
DA   spidshalet klumpfisk
DE   Spitzschwanz-Mondfisch
EL   φεγγαρόψαρο
EN   sharp-tail sunfish
FR   môle lancéolé; poisson-lune lancéolé
IT   pesce luna
NL   spitsstaartmaanvis
PT   peixe-lua rabudo
FI   möhkäkala-laji
SV   spetsstjärtsklumpfisk

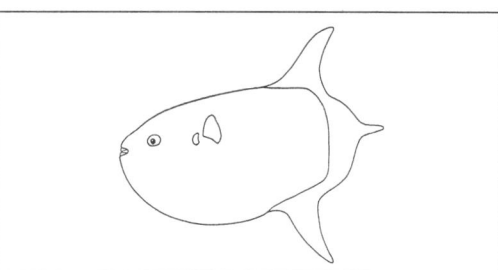

1186 TETRAODONTIDAE — PUA

SC   *Sphoeroides* spp.
ES   tamboriles
DA   kuglefisk-slægt
DE   Aufbläser
EL   γουρουνόψαρα
EN   puffers; Atlantic puffers
FR   tétrodons
IT   pesci palla
NL   Atlantische kogelvissen
PT   peixes-bola
FI   pallokala-suku
SV   blåsfiskar

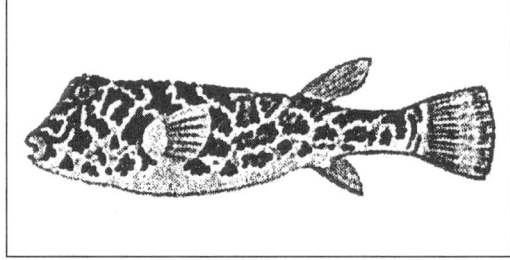

1188 MOLIDAE — MOX

SC   *Mola mola* (Linnaeus, 1758)
ES   pez luna; atalo
DA   klumpfisk; almindelig klumpfisk
DE   Mondfisch; Klumpfisch
EL   φεγγαρόψαρο
EN   ocean sunfish
FR   poisson-lune; môle commun
IT   pesce luna
NL   maanvis
PT   peixe-lua
FI   möhkäkala
SV   klumpfisk

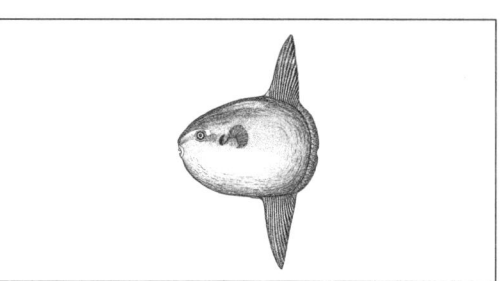

1189 MOLIDAE

SC	*Ranzania laevis* (Pennant, 1776); *Ranzania trunca-ta*
ES	pez luna alargado
DA	lang klumpfisk
DE	Stutz-Mondfisch; Schwimmender Kopf; Langer Mondfisch
EL	φεγγαρόψαρο
EN	slender sunfish
FR	môle voyageur; poisson-lune voyageur
IT	pesce luna troncato
NL	kleine maanvis
PT	peixe-lua comprido
FI	pikkumöhkäkala
SV	mindre klumpfisk

1191 BATRACHOIDIDAE

SC	*Batrachoididae*
ES	sapos
DA	paddefisk-familien
DE	Froschfische
EL	βατραχόψαρα
EN	toadfishes
FR	crapauds; poissons-crapauds
IT	batracoidi
NL	paddenvissen
PT	charrocos
FI	konnakalat; konnakalat-heimo
SV	paddfiskar

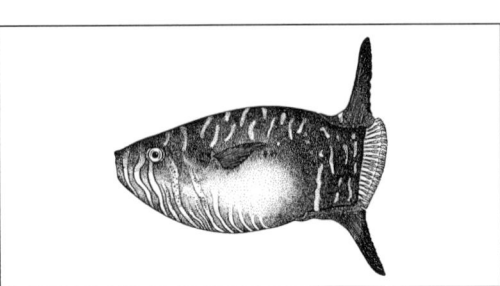

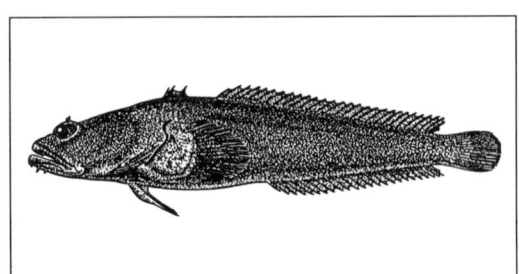

1190 GOBIESOCIDAE

SC	*Lepadogaster lepadogaster* (Bonnaterre, 1788); *Lepadogaster gouanii*
ES	pega roques; xucladit; chafarrocas
DA	dobbeltsuger
DE	Bandschild
EL	τζιτζίκι
EN	clingfish
FR	barbier; porte-écuelle; lépadogaster
IT	succiascoglio
NL	klipzuiger
PT	sugador
FI	imurikala
SV	dubbelsugare

1192 LOPHIIDAE   ANF

SC	*Lophiidae*
ES	rapes; sapos
DA	havtaske-familien
DE	Seeteufel; Anglerfische
EL	πεσκαντρίτσες· βατραχόψαρα· σκλεμπού
EN	anglerfishes
FR	baudroies; lottes
IT	rane pescatrici; lofidi
NL	zeeduivels
PT	tamboris
FI	merikrotit; merikrotit-heimo
SV	marulkfiskar

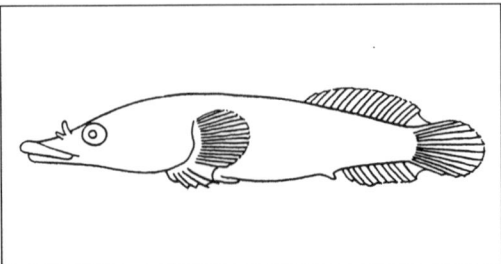

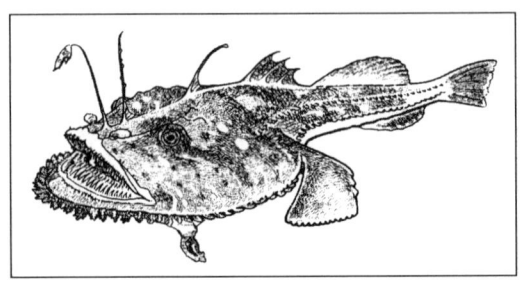

1193 LOPHIIDAE ANG

SC	*Lophius americanus* (Valenciennes, 1837)
ES	rape americano
DA	amerikansk havtaske
DE	Amerikanischer Seeteufel
EL	πεσκαντρίτσα της Αμερικής
EN	goosefish; American angler
FR	baudroie d'Amérique
IT	rana pescatrice americana
NL	Amerikaanse zeeduivel
PT	tamboril americano
FI	amerikanmerikrotti
SV	amerikansk marulk

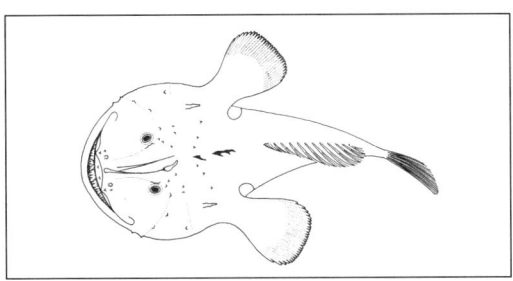

1195 LOPHIIDAE MON

SC	*Lophius piscatorius* (Linnaeus, 1758)
ES	rape; sapo; rana pescadora; rana marina; pez pescador; pez tamboril; pigotin; embarroco; pejesapo; peje armado; rape común; rape blanco; sapo blanco
DA	almindelig havtaske; bredflab
DE	Atlantischer Seeteufel; Seeteufel; Angler; Europäischer Seeteufel
EL	πεσκαντρίτσα· βατραχόψαρο· σκλεμπού
EN	anglerfish; frogfish; sea devil; monkfish
FR	baudroie commune; lotte; lotte commune
IT	rana pescatrice; rospo
NL	zeeduivel
PT	tamboril; tamboril branco
FI	merikrotti
SV	marulk

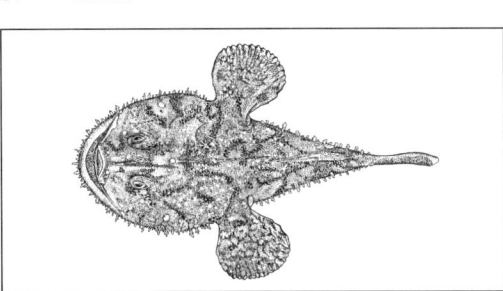

1194 LOPHIIDAE ANK

SC	*Lophius budegassa* (Spinola, 1807)
ES	rape negro
DA	sort havtaske
DE	Budegassa-Anglerfisch
EL	πεσκαντρίτσα· βατραχόψαρο· σκλεμπού
EN	black-bellied angler
FR	baudroie rousse
IT	rospo; rana pescatrice; budego
NL	zwarte zeeduivel
PT	tamboril sovaco-preto; tamboril preto
FI	merikrotti-laji
SV	mindre marulk

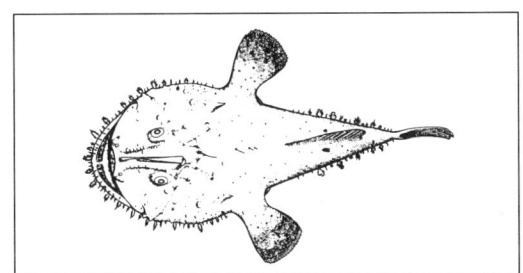

1196 LOPHIIDAE MOK

SC	*Lophius upsicephalus* (A. Smith, 1841)
ES	rape diablo
DA	vestafrikansk havtaske
DE	Westafrikanischer Anglerfisch
EL	ινδική πεσκαντρίτσα
EN	devil anglerfish; Cape monk
FR	baudroie-diable
IT	rana pescatrice indiana
NL	West-Afrikaanse zeeduivel
PT	tamboril da África do Sul
FI	merikrotti-laji
SV	västafrikansk marulk

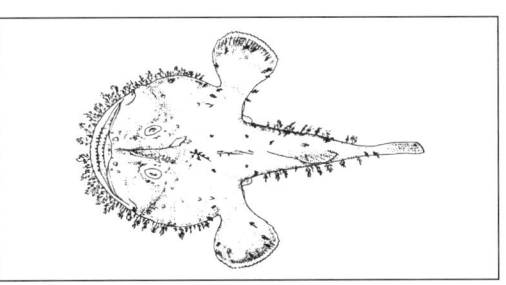

1197 SCALPELLIDAE PCB

SC	*Mitella pollicipes* (Gmelin, 1790); *Pollicipes cornucopia*
ES	percebe
DA	langhals
DE	Felsen-Entenmuschel
EL	
EN	goose barnacle
FR	pouce-pied; pied de biche
IT	lepade cornucopia
NL	eendenmossel
PT	perceve
FI	hanhenkaula-laji
SV	långhals

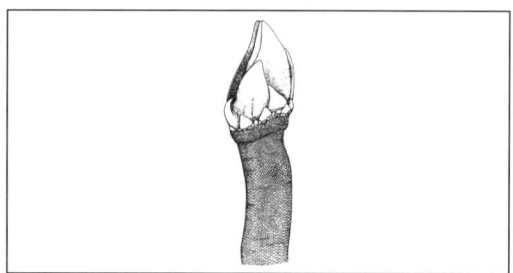

1199 BALANIDAE

SC	*Balanus* spp.
ES	balanos; clacas
DA	rur-slægt
DE	Seepocken
EL	βάλανοι
EN	barnacles
FR	balanes; bernaches; bernicles
IT	balani
NL	zeepokken
PT	cracas
FI	merirokko-suku
SV	havstulpaner

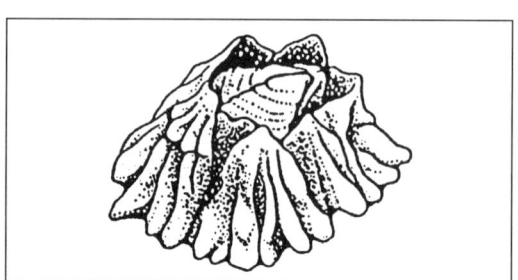

1198 LEPADIDAE GOO

SC	*Lepas* spp.
ES	falsos percebes
DA	langhalse-slægt
DE	Entenmuscheln
EL	λέπας
EN	goose barnacles
FR	anatifes
IT	lepadi
NL	eendenmossels
PT	perceves lisos; perceves bravos
FI	hanhenkaula-suku
SV	långhalsar

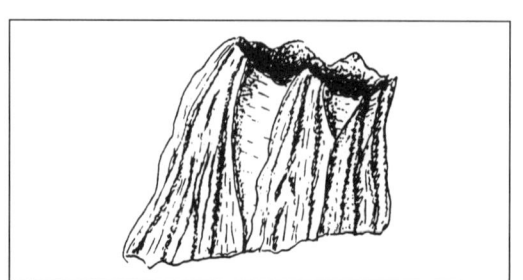

1200 BALANIDAE

SC	*Megabalanus psittacus* (Molina, 1782)
ES	balano gigante
DA	kæmperur
DE	Riesen-Seepocke
EL	μεγαβάλανος
EN	giant barnacle
FR	balane géante
IT	balano gigante
NL	reuzenzeepok
PT	craca bicuda
FI	jättimerirokko
SV	jättehavstulpan

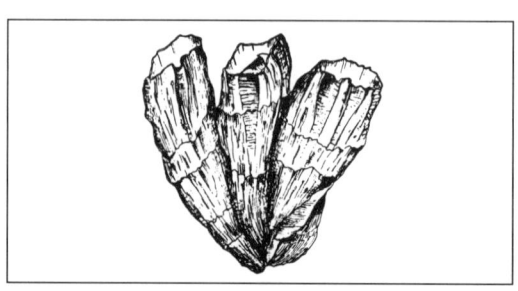

1201 SQUILLIDAE   SQY

SC	*Squillidae*
ES	galeras
DA	søknæler-familien
DE	Heuschreckenkrebse
EL	σκουλίδες
EN	squillids
FR	squilles
IT	pannocchie; squillidi
NL	bidsprinkhaankreeften
PT	zagaias
FI	sirkkaäyriäiset; sirkkaäyriäiset-heimo
SV	bönsyrseräka

1203 EUPHAUSIIDAE

SC	*Euphausiidae*
ES	eufasiáceos; krill
DA	lyskrebs-familien
DE	Leuchtkrebse und Krill
EL	ευφαυσεώδη
EN	euphausids; krill
FR	euphausiacés; krill; euphausiidés
IT	eufausiacei
NL	krill
PT	eufausídeos; krill
FI	krilliäyriäiset; krilliäyriäiset-heimo
SV	lysräkor; krill

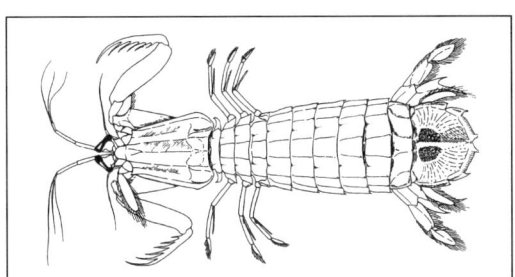

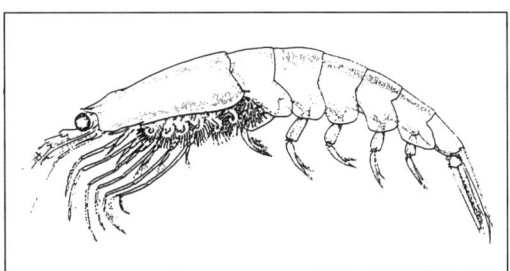

1202 SQUILLIDAE   MTS

SC	*Squilla mantis* (Linnaeus, 1758)
ES	galera
DA	søknæler
DE	Gemeiner Heuschreckenkrebs; Fangschreckenkrebs
EL	κατσαρίδα της θάλασσας
EN	mantis squillid
FR	squille; mante
IT	pannocchia
NL	bidsprinkhaankreeft
PT	zagaia-castanheta
FI	sirkkaäyriäinen
SV	mantis

1204 EUPHAUSIIDAE   KRI

SC	*Euphausia superba* (Dana, 1852)
ES	krill antártico
DA	antarktisk lyskrebs; krill
DE	Antarktischer Krill
EL	κριλ της Ανταρκτικής
EN	Antarctic krill
FR	krill antarctique
IT	krill antartico
NL	Antarctische krill
PT	krill do Antárctico
FI	etelänkrilli
SV	antarktisk krill

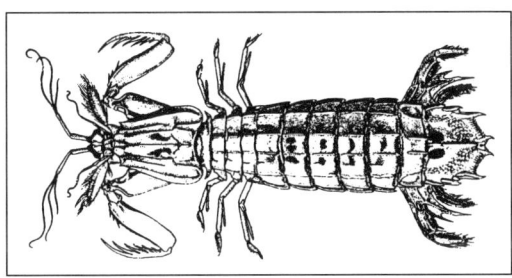

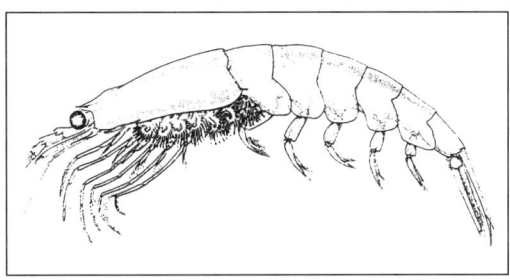

1205 EUPHAUSIIDAE   NKR

SC	*Meganyctiphanes norvegica* (Sars, 1857)
ES	krill de Noruega
DA	norsk lyskrebs; norsk krill
DE	Norwegischer Krill
EL	κριλ της Νορβηγίας
EN	Norwegian krill
FR	krill norvégien
IT	krill norvegese
NL	Noorse krill
PT	krill da Noruega
FI	norjankrilli
SV	norsk lysräka

1207 PENAEIDAE   ENS

SC	*Metapenaeus endeavouri* (Schmitt, 1926)
ES	camarón australiano
DA	brun reje
DE	Braune Geißelgarnele
EL	μαύρη γαρίδα
EN	endeavour shrimp
FR	crevette devo
IT	gamberone bruno
NL	Australische garnaal
PT	camarão da Austrália
FI	katkarapu-laji
SV	australisk räka

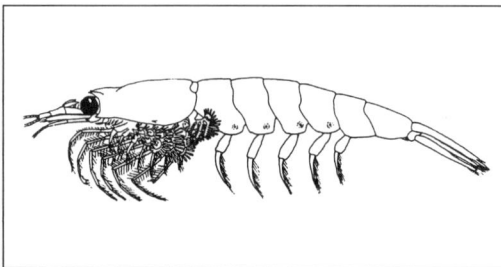

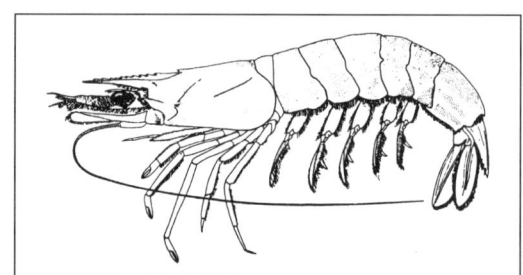

1206 PENAEIDAE

SC	*Artemesia longinaris* (Bate, 1888)
ES	camarón estilete argentino
DA	argentinsk stiletreje
DE	Argentinische Stilett-Garnele
EL	γαρίδα· στιλέτο της Αργεντινής
EN	Argentine stiletto shrimp
FR	crevette stylet d'Argentine
IT	gambero a lungo rostro
NL	Argentijnse stilettogarnaal
PT	camarão-estilete argentino
FI	katkarapu-laji
SV	argentinsk stiletträka

1208 PENAEIDAE   SHI

SC	*Metapenaeus joyneri* (Miers, 1880)
ES	camarón siba
DA	shiba-reje
DE	Shiba-Geißelgarnele
EL	γαρίδα σίμπα
EN	shiba shrimp
FR	crevette siba
IT	gamberone siba
NL	shiba-garnaal
PT	camarão-siba
FI	katkarapu-laji
SV	shibaräka

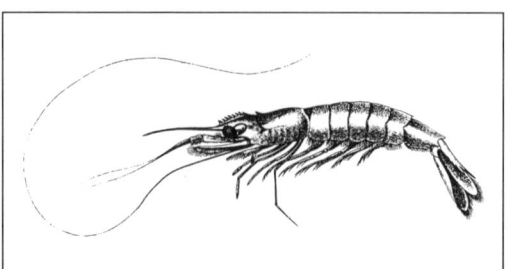

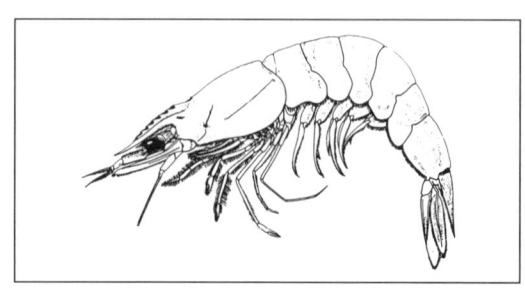

1209 PENAEIDAE — MET

SC	*Metapenaeus* spp.
ES	camarones metapenaeus
DA	reje-slægt
DE	Geißelgarnelen
EL	γαρίδες
EN	metapenaeus shrimps
FR	crevettes metapenaeus
IT	gamberoni
NL	metapeneide garnalen
PT	camarões «Metapenaeus»
FI	katkarapu-suku
SV	metapenaeusräkor

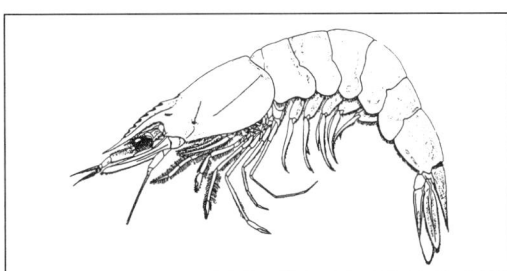

1211 PENAEIDAE — NPP

SC	*Parapenaeopsis* spp.
ES	camarones parapeneopsis
DA	reje-slægt
DE	Geißelgarnelen
EL	γαρίδες
EN	tropical shrimps; parapenacopsis shrimps
FR	crevettes tropicales
IT	gamberi
NL	diepzeegarnalen
PT	camarões «Parapenaeopsis»
FI	katkarapu-laji
SV	parapenaeopsisräkor

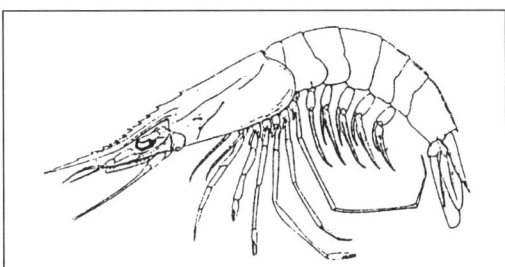

1210 PENAEIDAE — GUS

SC	*Parapenaeopsis atlantica* (Balss, 1914)
ES	camarón guineano
DA	Guinea-reje
DE	Guinea-Geißelgarnele
EL	γαρίδα της Γουινέας
EN	Guinea shrimp
FR	crevette guinéenne
IT	gambero di Guinea
NL	Atlantische diepzeegarnaal
PT	camarão guinéu
FI	katkarapu-laji
SV	guinearäka

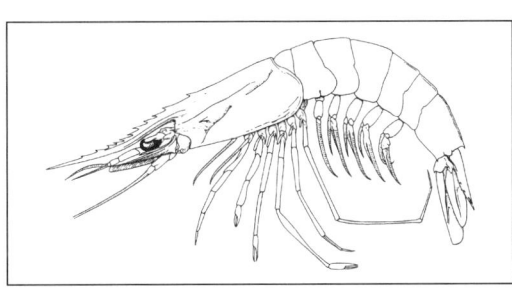

1212 PENAEIDAE — DPS

SC	*Parapenaeus longirostris* (Lucas, 1846)
ES	camarón de altura; gamba de altura
DA	dybvands-rosenreje
DE	Rosa Geißelgarnele
EL	κόκκινη γαρίδα
EN	deepwater rose shrimps
FR	crevette rose du large
IT	gambero rosa mediterraneo
NL	roze diepzeegarnaal
PT	gamba branca; gamba
FI	katkarapu-laji
SV	djuphavsräka

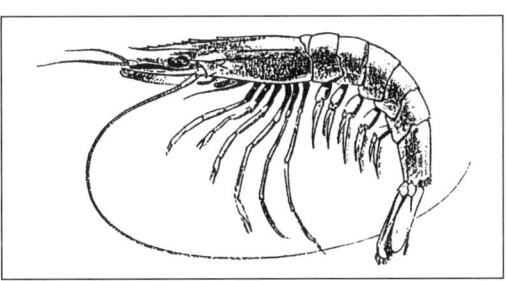

1213 PENAEIDAE   ABS

SC   *Penaeus aztecus* (Yves, 1891)
ES   camarón café norteño; langostino mejicano
DA   azteker-reje
DE   Azteken-Geißelgarnele
EL   γκρίζα γαρίδα
EN   Northern brown shrimp
FR   crevette royale grise
IT   mazzancolla caffè
NL   Azteken-garnaal
PT   camarão-café do Norte
FI   katkarapu-laji
SV   aztekräka

1215 PENAEIDAE   YPS

SC   *Penaeus californiensis* (Holmes, 1900)
ES   langostino patiamarillo
DA   californisk reje
DE   Kalifornia-Geißelgarnele
EL   γαρίδα της Καλιφόρνιας
EN   yellow-leg shrimp
FR   crevette à pattes jaunes
IT   mazzancolla californiana
NL   geelpootgarnaal
PT   camarão pata-amarela
FI   katkarapu-laji
SV   gulbensräka

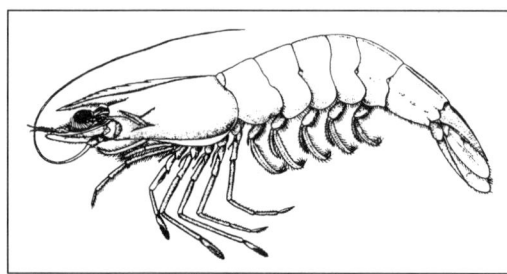

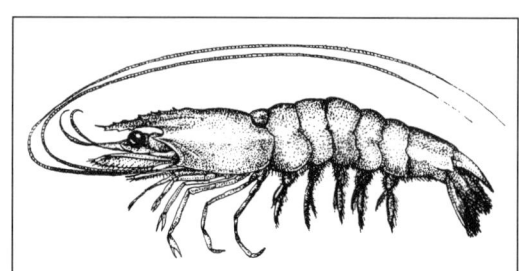

1214 PENAEIDAE   CSP

SC   *Penaeus brevirostris* (Kingsley, 1878)
ES   langostino cristal
DA   krystalreje
DE   Kristall-Geißelgarnele
EL   κρυσταλλογαρίδα
EN   crystal shrimp
FR   crevette cristal
IT   mazzancolla rossa
NL   kristalgarnaal
PT   camarão cristal
FI   katkarapu-laji
SV   kristallräka

1216 PENAEIDAE   FLP

SC   *Penaeus chinensis* (Osbeck, 1765)
ES   langostino carnoso
DA   kødet reje
DE   Hauptmannsgarnele
EL   σαρκογαρίδα
EN   fleshy prawn
FR   crevette charnue
IT   mazzancolla cinese
NL   vlezige garnaal
PT   camarão carnudo
FI   katkarapu-laji
SV   kötträka

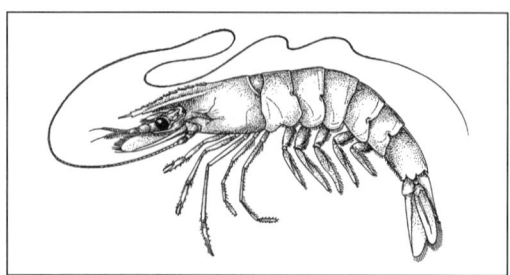

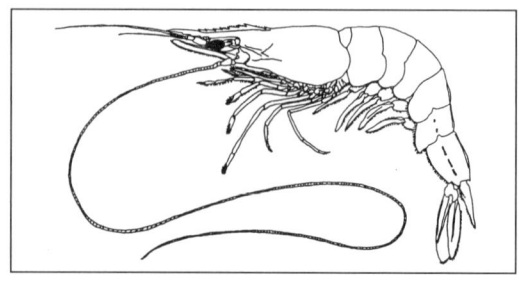

1217 PENAEIDAE APS

SC   *Penaeus duorarum* (Burkenroad, 1939)
ES   camarón norteño; langostino de Guinea
DA   nordlig rosenreje
DE   Nördliche Rosa Geißelgarnele
EL   ροζ γαρίδα της Βόρειας Θάλασσας
EN   Northern pink shrimp
FR   crevette rose du Nord; gamba du Nord
IT   mazzancolla; gamberone
NL   noordelijke roze garnaal
PT   camarão rosado do norte
FI   katkarapu-laji
SV   nordlig rosenräka

1219 PENAEIDAE KUP

SC   *Penaeus japonicus* (Bate, 1888)
ES   langostino kuruma
DA   kuruma-reje
DE   Radgarnele
EL   γαρίδα της Ιαπωνίας
EN   kuruma prawn
FR   crevette kuruma; gamba japonaise
IT   mazzancolla; gamberone; mazzancolla giapponese
NL   kurumagarnaal
PT   camarão japonês; camarão do Japão; camarão-
     -kuruma; camarão-tigre
FI   katkarapu-laji
SV   kurumaräka

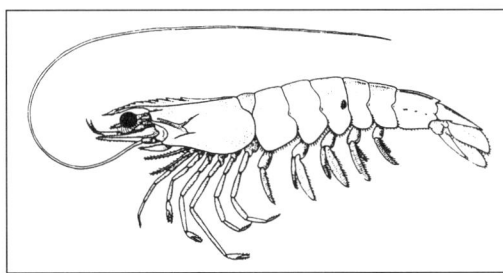

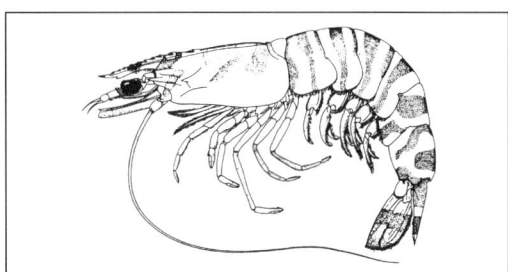

1218 PENAEIDAE PRB

SC   *Penaeus esculentus* (Haswell, 1879)
ES   langostino tigre marrón
DA   brun tigerreje
DE   Braune Tigergarnele
EL   γαρίδα-τίγρης
EN   brown tiger prawn
FR   crevette tigrée brune
IT   mazzancolla striata australiana
NL   bruine tijgergarnaal
PT   camarão-tigre-castanho
FI   katkarapu-laji
SV   brun tigerräka

1220 PENAEIDAE TGS

SC   *Penaeus kerathurus* (Forsskål, 1775); *Penaeus*
     caramote; *Penaeus trisulcatus*
ES   camarón; langostino español; langostino real
DA   rynket reje
DE   Furchengarnele
EL   γάμπαρι· γαρίδα
EN   caramote prawn
FR   crevette royale; crevette caramote
IT   mazzancolla; gambero imperiale; gamberone medi-
     terraneo
NL   caramotegarnaal
PT   gamba manchada; camarão da Quarteira
FI   katkarapu-laji
SV   gaffelräka

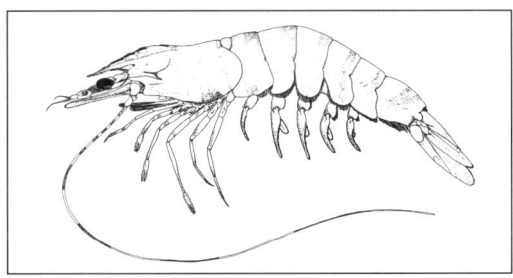

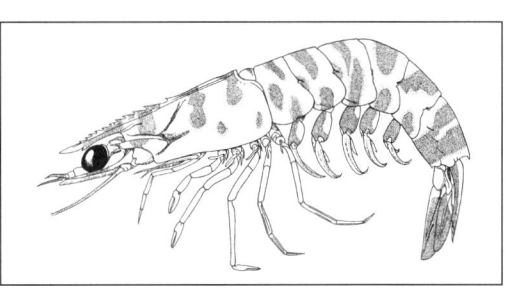

1221 PENAEIDAE WKP

SC	*Penaeus latisulcatus* (Kishinouye, 1896)
ES	langostino marfil
DA	kongereje
DE	Königsgeißelgarnele
EL	πρασινογαρίδα
EN	Western king prawn
FR	crevette royale occidentale
IT	mazzancolla verde
NL	westelijke koningsgarnaal
PT	camarão real
FI	katkarapu-laji
SV	västlig kungsräka

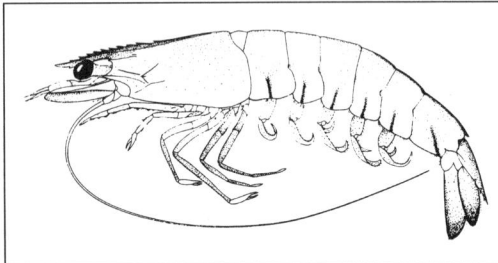

1223 PENAEIDAE GIT

SC	*Penaeus monodon* (Fabricius, 1798)
ES	langostino tigre gigante; langostino yumbo
DA	kæmpe tigerreje
DE	Bärengarnele; Schiffskielgarnele
EL	γάμπαρι-τίγρης
EN	giant tiger prawn
FR	crevette géante tigrée
IT	mazzancolla gigante
NL	grote tijgergarnaal
PT	camarão-tigre gigante
FI	katkarapu-laji
SV	tigerräka

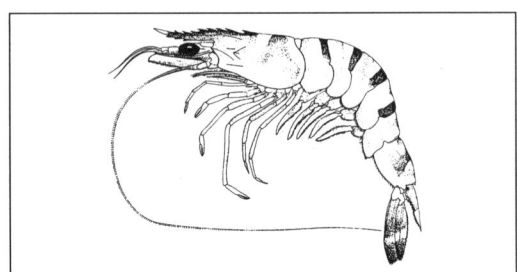

1222 PENAEIDAE PBA

SC	*Penaeus merguiensis* (de Man, 1888)
ES	langostino banana
DA	bananreje
DE	Bananen-Garnele
EL	μπανανογαρίδα
EN	banana prawn
FR	crevette-banane
IT	mazzancolla banana
NL	bananengarnaal
PT	camarão-banana
FI	katkarapu-laji
SV	bananräka

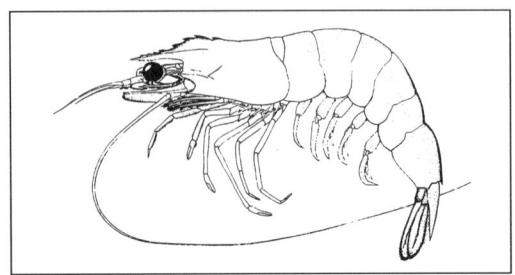

1224 PENAEIDAE SOP

SC	*Penaeus notialis* (Perez Farfante, 1967); *Penaeus duorarum notialis*
ES	langostino rosado sureño
DA	sydlig rosenreje
DE	Südliche Rosa Geißelgarnele
EL	ροζ γαρίδα
EN	Southern pink shrimp
FR	crevette rose du Sud; gamba
IT	mazzancolla rosa
NL	roze garnaal
PT	camarão rosado do Sul; gamba de Angola
FI	katkarapu-laji
SV	sydlig rosenräka

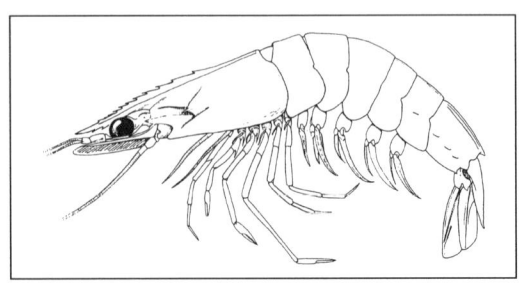

1225 PENAEIDAE

REP

SC	*Penaeus penicillatus* (Alcock, 1905)
ES	langostino de cola roja
DA	rødhalet reje
DE	Rotschwanzgarnele
EL	γαρίδα με κόκκινη ουρά
EN	red-tail prawn
FR	crevette à queue rouge
IT	mazzancolla coda rossa
NL	roodstaartgarnaal
PT	camarão-cauda vermelha
FI	katkarapu-laji
SV	rödstjärtad räka

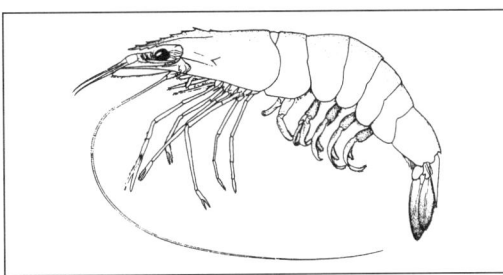

1227 PENAEIDAE

PST

SC	*Penaeus setiferus* (Linnaeus, 1767)
ES	langostino blanco norteño
DA	nordlig hvidreje
DE	Nördliche Weiße Geißelgarnele
EL	λευκογαρίδα
EN	Northern white shrimp
FR	crevette ligubam du Nord
IT	mazzancolla bianca atlantica
NL	noordelijke witte garnaal
PT	camarão branco nortenho
FI	katkarapu-laji
SV	nordlig viträka

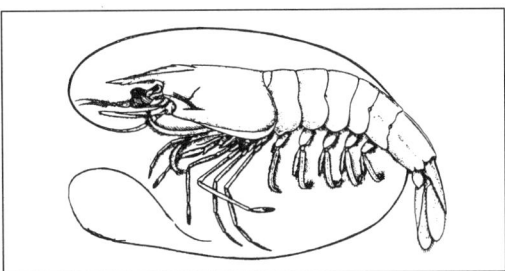

1226 PENAEIDAE

TIP

SC	*Penaeus semisulcatus* (de Haan, 1844)
ES	langostino tigre verde
DA	grøn tigerreje
DE	Grüne Tigergarnele
EL	γάμπαρι-τίγρης
EN	green tiger prawn
FR	crevette tigrée verte
IT	mazzancolla; gamberone
NL	groene tijgergarnaal
PT	camarão-tigre verde
FI	katkarapu-laji
SV	grön tigerräka

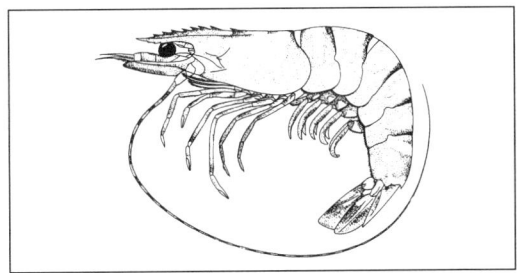

1228 PENAEIDAE

PNV

SC	*Penaeus vannamei* (Boone, 1931)
ES	langostino blanco; camarón patiblanco
DA	mellemamerikansk reje
DE	Zentralamerikanische Geißelgarnele
EL	γαρίδα της Κεντρικής Αμερικής
EN	Central American shrimp
FR	crevette d'Amérique centrale
IT	gamberone centramericano
NL	Midden-Amerikaanse garnaal
PT	camarão-pata branca
FI	katkarapu-laji
SV	viträka

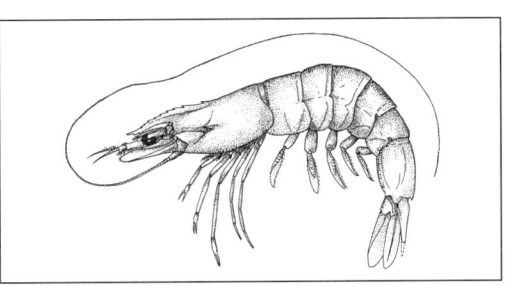

307

1229 PENAEIDAE

SC	*Penaeus* spp.
ES	langostinos; camarones
DA	reje-slægt
DE	Geißelgarnelen
EL	γαρίδες
EN	white shrimps
FR	crevettes; crevettes royales; gambas
IT	mazzancolle; gamberoni
NL	peneide garnalen
PT	camarões «Penaeus»
FI	katkaravut
SV	peneidaräkor

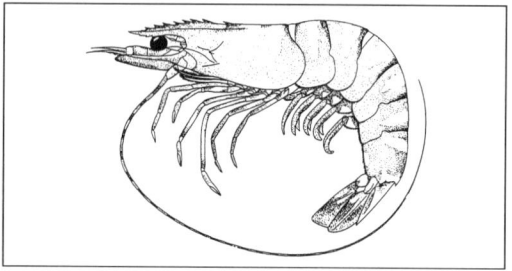

1231 PENAEIDAE

SC	*Xiphopenaeus kroyeri* (Heller, 1862)
ES	camarón siete barbas
DA	atlantisk reje
DE	Kroyers Geißelgarnele
EL	γαρίδα με μουστάκια
EN	Atlantic seabob
FR	crevette seabob de l'Atlantique
IT	gambero barbato
NL	Atlantische seabobgarnaal
PT	camarão barbudo
FI	katkarapu-laji
SV	skäggig räka

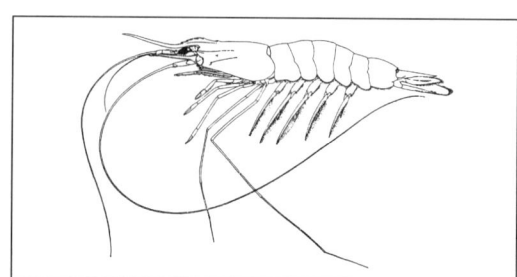

1230 PENAEIDAE

SC	*Trachypenaeus* spp.
ES	camaroncillos
DA	pacifiske reje-slægt
DE	Geißelgarnelen
EL	γαρίδες του Ειρηνικού
EN	Pacific seabobs
FR	crevettes gambri; crevettes seabob
IT	gamberi del Pacifico
NL	Pacifische seabobgarnalen
PT	camarões barbudos do Pacífico
FI	katkaravut
SV	gisselräkor

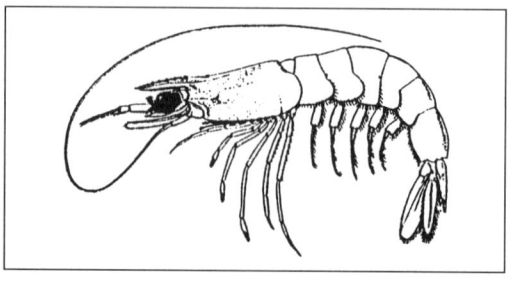

1232 ARISTEIDAE

SC	*Aristeidae*
ES	aristéidos
DA	dybhavsreje-familie
DE	Tiefseegarnelen
EL	αριστίδες· κόκκινες γαρίδες
EN	aristeid shrimps
FR	crevettes aristeus
IT	aristeidi; gamberi rossi
NL	aristeide garnalen; diepzeegarnalen
PT	camarões aristeídeos
FI	katkaravut; katkaravut-heimo
SV	aristeidräkor

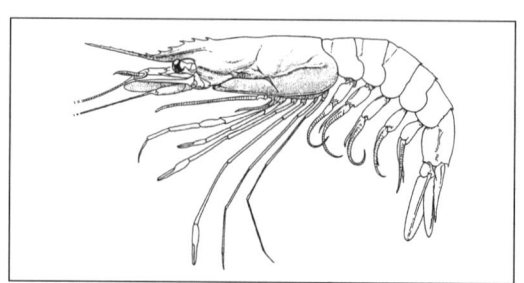

1233 ARISTEIDAE ARA

SC *Aristeus antennatus* (Risso, 1816)
ES gamba rosada
DA blårød reje
DE Afrikanische Tiefseegarnele
EL κόκκινη γαρίδα
EN blue-and-red shrimp
FR crevette rouge
IT gambero rosso mediterraneo
NL blauwrode diepzeegarnaal
PT camarão vermelho; carabineiro
FI katkarapu-laji
SV blåröd räka

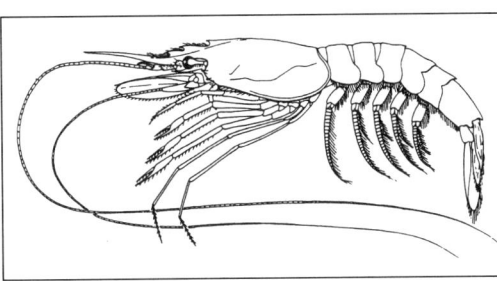

1234 ARISTEIDAE SSH

SC *Plesiopenaeus edwardsianus* (Johnson, 1867)
ES carabinero; gamba carabinero
DA skarlagen reje
DE Atlantische Rote Riesengarnele
EL κόκκινη γαρίδα
EN scarlet shrimp
FR gambon écarlate; crevette impériale;
    crevette rouge géante
IT gambero rosso
NL reuzendiepzeegarnaal
PT carabineiro cardeal
FI katkarapu-laji
SV rödräka

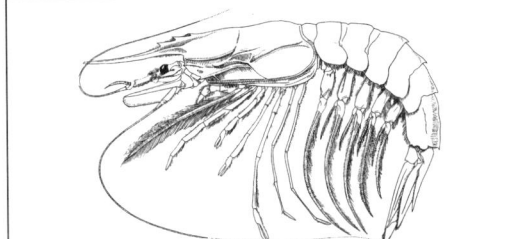

1235 PANDALIDAE CHS

SC *Heterocarpus reedi* (Bahamonde, 1955)
ES camarón nailon
DA chilensk nylonreje
DE Chile-Tiefseegarnele
EL τροπιδογαρίδα
EN Chilean nylon shrimp
FR crevette nylon chilienne
IT gambero carenato cileno
NL Chileense nylongarnaal
PT camarão-nailon chileno
FI katkarapu-laji
SV chilensk nylonräka

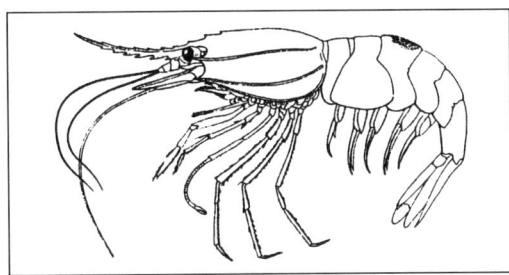

1236 PANDALIDAE

SC *Heterocarpus* spp.
ES camarones nailon
DA reje-slægt
DE Kanten-Tiefseegarnelen
EL τροπιδογαριδάκια
EN arrow shrimps
FR crevettes-flèches
IT gamberetti carenati
NL pijlgarnalen
PT camarões-nailon
FI katkaravut
SV nylonräkor

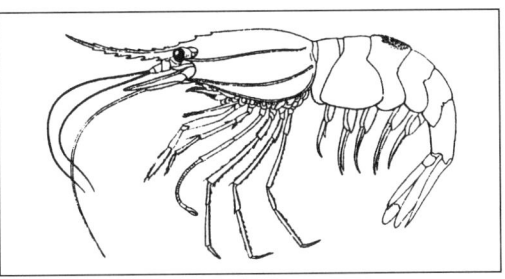

1237 PANDALIDAE   PRA

SC	*Pandalus borealis* (Kroyer, 1838)
ES	camarón boreal; camarón norteño
DA	dybvandsreje
DE	Grönlandgarnele; Tiefseegarnele; Nordische Garnele; Nordmeergarnele; Krabbe
EL	γαρίδα της Αρκτικής
EN	deepwater prawn; pink shrimp; Northern prawn
FR	crevette nordique; crevette rouge
IT	gamberello boreale; gambero
NL	Noorse garnaal
PT	camarão árctico
FI	pohjankatkarapu
SV	nordhavsräka

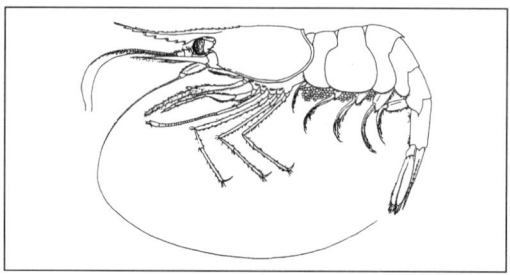

1238 PANDALIDAE   AES

SC	*Pandalus montagui* (Leach, 1814)
ES	camarón espico
DA	rejekonge
DE	Rosa Garnele
EL	γαρίδα
EN	Aesop shrimp
FR	crevette ésope
IT	gamberetto rosa
NL	ringsprietgarnaal
PT	camarão boreal
FI	katkarapu-laji
SV	karamellräka

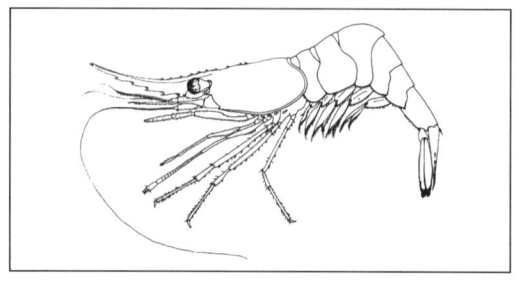

1239 PANDALIDAE   PSH

SC	*Pandalus* spp.
ES	camarones pandálidos
DA	reje-slægt
DE	Tiefseegarnelen
EL	κοκκινογαριδάκια
EN	pink shrimps; pandalid shrimps
FR	crevettes pandalides
IT	gamberetti rosa
NL	pandalide garnalen
PT	camarões pandalídeos
FI	katkarapu-suku
SV	räkor

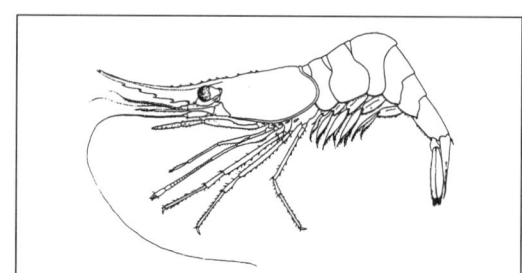

1240 PANDALIDAE

SC	*Plesionika* spp.
ES	camarones nórdicos
DA	reje-slægt
DE	Tiefseegarnelen im engeren Sinne
EL	γαρίδες
EN	pandalid shrimps; Northern prawns
FR	crevettes pandalides; crevettes nordiques
IT	gobetti; gamberi
NL	diepzee rode garnalen
PT	camarões marrecos
FI	katkaravut
SV	nordräkor

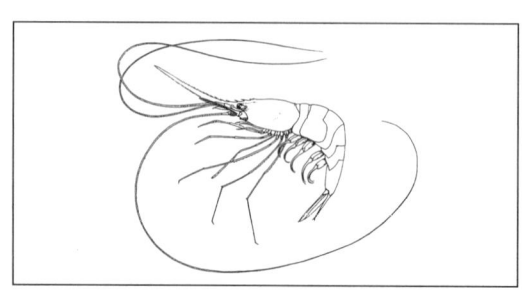

1241 SERGESTIDAE   SHS

SC	Sergestidae
ES	camarones sergestid
DA	højsøreje-familien
DE	Leuchtgarnelen
EL	γαρίδες· σεργκεστρίδες
EN	sergestid shrimps
FR	chevrettes sergestides
IT	sergestidi
NL	sergestide-garnalen
PT	camarões sergestídeos
FI	katkaravut
SV	sergestidräkor

1243 ATYIDAE

SC	Atyidae
ES	camarones atíidos
DA	ferskvandsreje-familien
DE	Atyiden-Garnelen
EL	
EN	atyid shrimps
FR	
IT	atidi
NL	Atyide-garnalen
PT	camarões «Atyidae»
FI	katkaravut-heimo
SV	atyidräkor

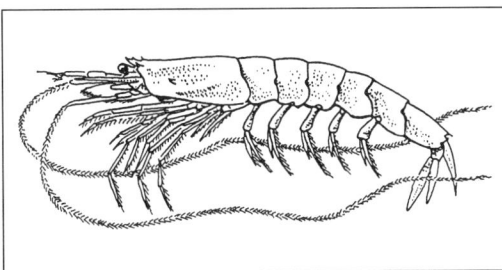

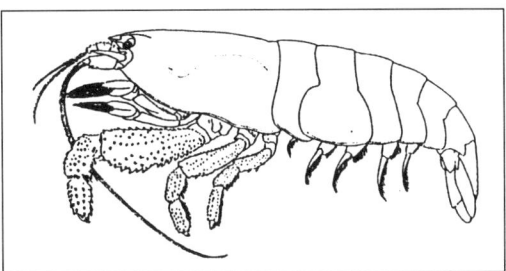

1242 SERGESTIDAE   AKS

SC	Acetes japonicus (Kishinouye, 1905)
ES	camaroncillo akiami
DA	akiami-reje
DE	Akiami-Garnele
EL	ακιάμι
EN	akiami paste shrimp
FR	crevette akiami
IT	akiami
NL	akiami-garnaal
PT	camarão-akiami
FI	katkarapu-laji
SV	akiamiräka

1244 PALAEMONIDAE   PPZ

SC	Palaemonidae
ES	gambas y camarones de agua dulce
DA	reje-familie
DE	Süßwassergarnelen
EL	γαρίδες ή καραβίδες των γλυκών νερών
EN	freshwater prawns and shrimps
FR	chevrettes; crevettes-bouquets
IT	palemonidi d'acqua dolce
NL	zoetwater steurgarnalen
PT	camarões palemonídeos de água doce
FI	katkaravut-heimo
SV	palaemonidräkor

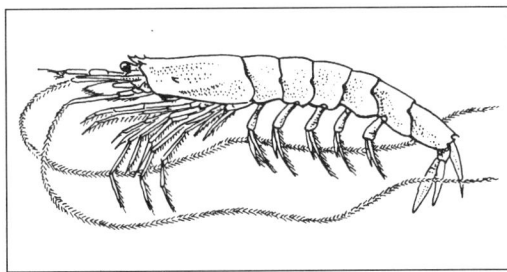

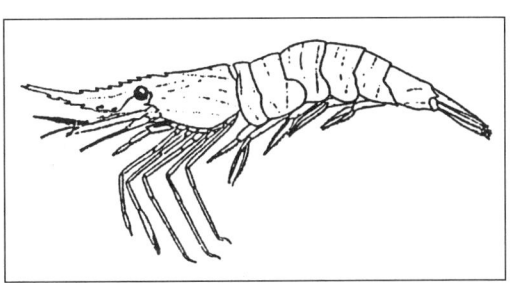

1245 PALAEMONIDAE PPZ

SC	*Palaemonidae*
ES	camarones palemónidos
DA	reje-familie
DE	Felsengarnelen
EL	γαρίδες
EN	palaemonid shrimps
FR	crevettes-bouquets
IT	palemonidi
NL	palaemonide garnalen; steurgarnalen
PT	camarões palemonídeos
FI	katkaravut
SV	palaemonidräkor

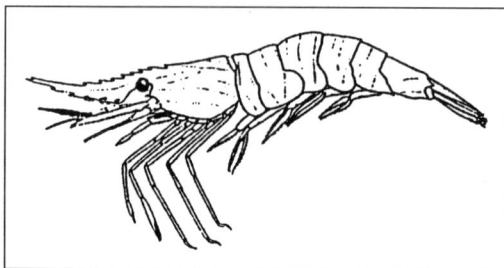

1247 PALAEMONIDAE PRF

SC	*Macrobrachium rosenbergii* (de Man, 1879)
ES	langostino de río
DA	kæmpe ferskvandsreje
DE	Rosenbergs Süßwassergarnele
EL	γιγαντογαρίδα των γλυκών νερών
EN	giant river prawn
FR	chevrette géante
IT	gambero blu
NL	Rosenberggarnaal; reuzenriviergarnaal
PT	camarão gigante do rio
FI	jättijokikatkarapu
SV	sötvattensräka

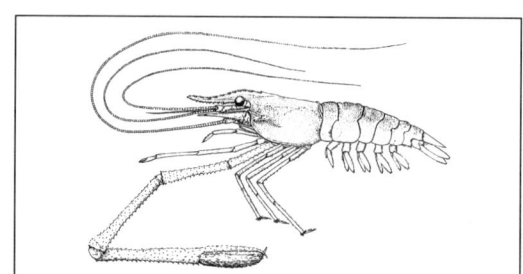

1246 PALAEMONIDAE

SC	*Macrobrachium carcinus* (Linnaeus, 1758)
ES	langostino de río americano
DA	sydamerikansk ferskvandsreje
DE	Südamerikanische Süßwassergarnele
EL	ποταμογαρίδα της Αμερικής
EN	freshwater prawn
FR	crevette d'eau douce; chevrette américaine
IT	gamberetto americano
NL	Zuid-Amerikaanse riviergarnaal
PT	camarão americano do rio
FI	jokikatkarapu-laji
SV	sydamerikansk sötvattensräka

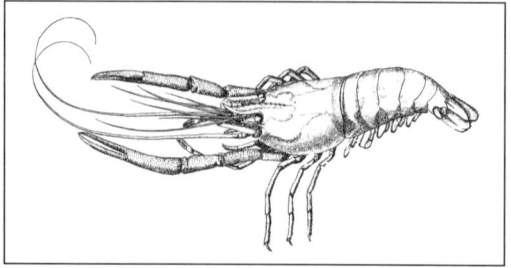

1248 PALAEMONIDAE PPF

SC	*Macrobrachium* spp.
ES	langostinos de agua dulce
DA	ferskvandsreje-slægt
DE	Süßwassergarnelen
EL	γαρίδες των γλυκών νερών
EN	river prawns
FR	crevettes
IT	gamberi a lunghe chele
NL	riviergarnalen
PT	camarões de água doce
FI	jokikatkarapu-suku
SV	sötvattensräkor

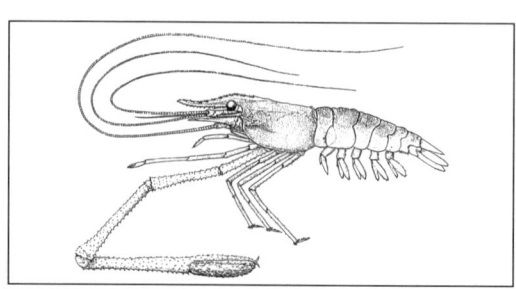

1249 PALAEMONIDAE CPR

SC	*Palaemon serratus* (Pennant, 1777); *Leander serratus*
ES	camarón común
DA	Roskildereje
DE	Sägegarnele; Steingarnele
EL	αραπογαρίδα· γαριδάκι
EN	common prawn
FR	bouquet commun; crevette rose
IT	gamberetto maggiore; gamberetto sega
NL	gewone steurgarnaal
PT	camarão branco legítimo
FI	katkarapu-laji
SV	tångräka

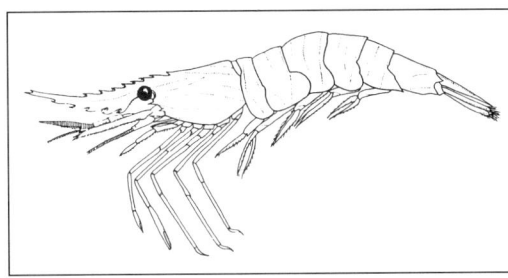

1251 CRANGONIDAE CSH

SC	*Crangon crangon* (Linnaeus, 1758)
ES	quisquilla de arena; camarón; quisquilla gris
DA	hestereje; sandhest; sandreje
DE	Granat; Sandgarnele
EL	σταχτογαρίδα
EN	common shrimp
FR	crevette grise
IT	gamberetto grigio; gambero grigio
NL	Noordzeegarnaal
PT	camarão negro; camarão do rio
FI	hietakatkarapu
SV	sandräka; hästräka

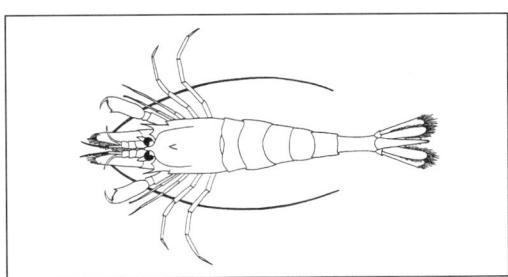

1250 CRANGONIDAE CRN

SC	*Crangonidae*
ES	camarones crangónidos
DA	reje-familie
DE	Sandgarnelen
EL	στατογαρίδες
EN	crangonid shrimps
FR	crevettes crangonides; crevettes grises
IT	crangonidi
NL	crangonide garnalen
PT	camarões crangonídeos
FI	hietakatkaravut; hietakatkaravut-heimo
SV	crangonidräkor

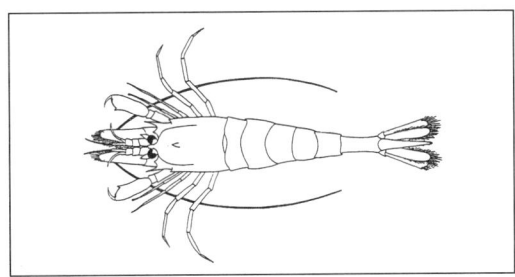

1252 SICYONIIDAE RSH

SC	*Sicyonia brevirostris* (Stimpson, 1874)
ES	camarón de piedra
DA	klippereje
DE	Furchengeißelgarnele
EL	πετρογαρίδα
EN	rock shrimp
FR	crevette ovetgernade; crevette de roche
IT	gambero duro
NL	rots-pantsergarnaal
PT	camarão da pedra
FI	katkarapu-laji
SV	stenräka

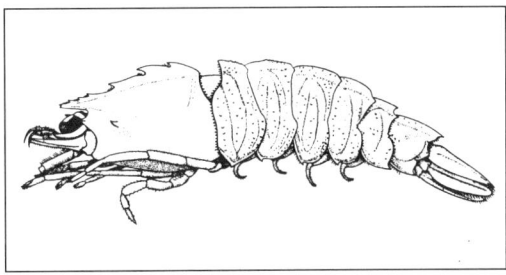

1253 SICYONIIDAE

SC	*Sicyonia* spp.
ES	camarones; siciones
DA	reje-slægt
DE	Furchengeißelgarnelen
EL	γαριδάκια
EN	rock shrimps
FR	boucots
IT	gamberi duri; sicione
NL	pantsergarnalen
PT	camarões da pedra
FI	katkaravut
SV	stenräkor

1255 SOLENOCERIDAE   KNI

SC	*Hymenopenaeus* spp.; *Haliporoides* spp.
ES	camarones navaja
DA	reje-slægter
DE	Messergarnelen
EL	μαχαιρογαρίδες
EN	knife shrimps
FR	salicoques; crevettes-couteaux
IT	gamberi coltello
NL	mesgarnalen
PT	camarões-navalha
FI	katkaravut
SV	knivräkor

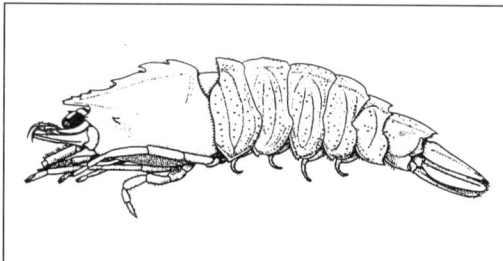

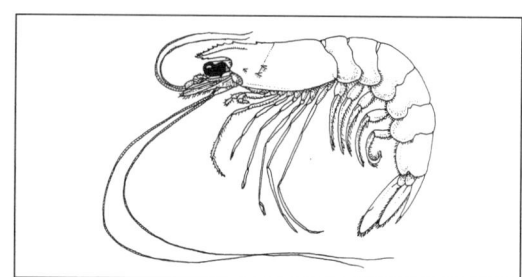

1254 SOLENOCERIDAE

SC	*Hymenopenaeus triarthrus* (Stebbing, 1914)
ES	camarón navaja
DA	knivreje
DE	Messergarnele
EL	μαχαιρογαρίδα
EN	knife shrimp
FR	salicoque navaja
IT	gambero coltello
NL	navaja-mesgarnaal
PT	camarão-navalha
FI	katkarapu-laji
SV	knivräka

1256 SOLENOCERIDAE   LAA

SC	*Pleoticus muelleri* (Bate, 1888)
ES	camarón argentino
DA	argentinsk rødreje
DE	Argentinische Rotgarnele
EL	κόκκινη γαρίδα της Αργεντινής
EN	Argentine red shrimp
FR	salicoque rouge d'Argentine
IT	gambero rosso argentino
NL	Argentijnse rode garnaal
PT	camarão vermelho argentino
FI	katkarapu-laji
SV	argentinsk rödräka

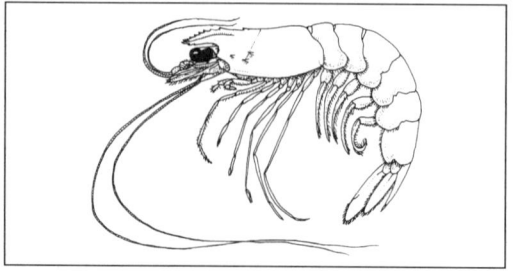

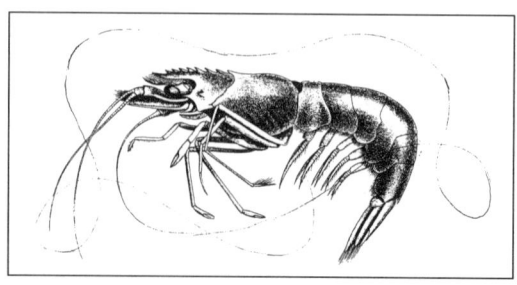

1257 SOLENOCERIDAE RRS

SC	*Pleoticus robustus* (Smith, 1885)
ES	camarón rojo real
DA	konge-rødreje
DE	Königs-Rotgarnele
EL	κόκκινη γαρίδα της Αμερικής
EN	royal red shrimp
FR	salicoque royale rouge
IT	gambero rosso americano
NL	koningsrode garnaal
PT	camarão vermelho real
FI	katkarapu-laji
SV	kungsröd räka

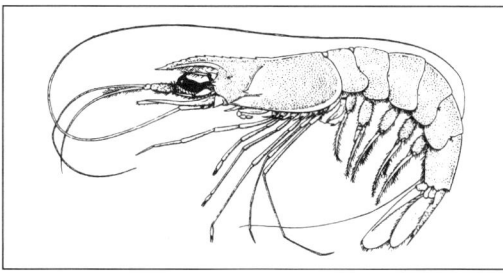

1258 SOLENOCERIDAE

SC	*Solenocera membranacea* (Risso, 1816)
ES	gamba de fango del Atlántico
DA	mudderreje
DE	Membran-Geißelgarnele
EL	λασπογαρίδα
EN	mud shrimp
FR	crevette de vase
IT	gambero
NL	Atlantische slibgarnaal
PT	camarão da vasa
FI	katkarapu-laji
SV	träskräka

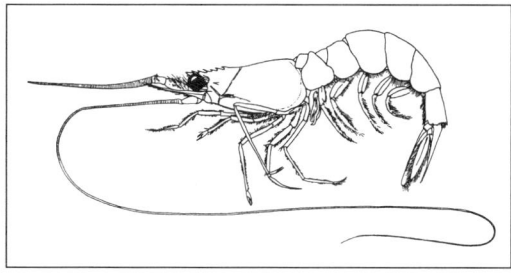

1259 PALINURIDAE VLO

SC	*Palinuridae*
ES	langostas
DA	languster-familien
DE	Langusten
EL	αστακοί
EN	spiny lobsters
FR	langoustes
IT	palinuridi; aragoste
NL	langoesten
PT	lagostas
FI	langustit; langustit-heimo
SV	languster

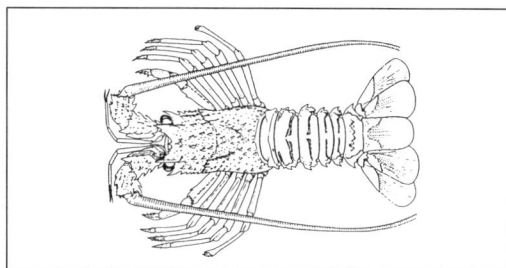

1260 PALINURIDAE LOR

SC	*Jasus edwardsii* (Hutton, 1875)
ES	langosta de Nueva Zelanda
DA	newzealandsk languster
DE	Edwards-Languste
EL	αστακός της Νέας Ζηλανδίας
EN	red rock lobster
FR	langouste de Nouvelle-Zélande
IT	aragosta
NL	Nieuw-Zeelandse langoest
PT	lagosta da Nova Zelândia
FI	langusti-laji
SV	nyazeeländsk langust

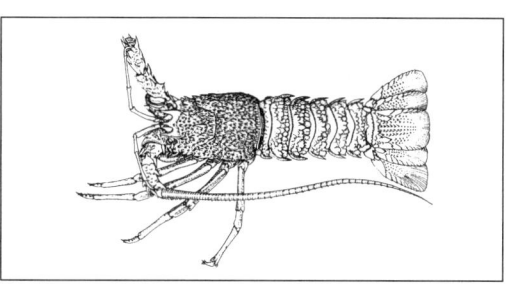

1261 PALINURIDAE LOF

SC	*Jasus frontalis* (Milne Edwards, 1837)
ES	langosta de Juan Fernández
DA	Juan Fernandez-languster
DE	Juan-Fernandez-Languste
EL	αστακός της Αυστραλίας
EN	Australian spiny lobster; Juan Fernandez rock lobster
FR	langouste de Juan Fernandez
IT	aragosta
NL	Juan-Fernandez-langoest
PT	lagosta Juan Fernández
FI	langusti-laji
SV	juanfernandezlangust

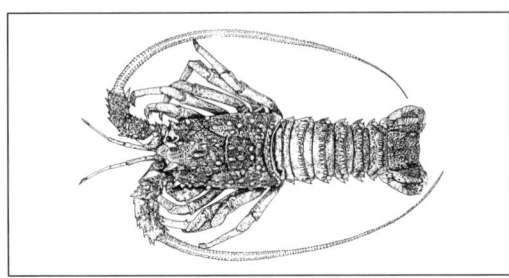

1262 PALINURIDAE LBC

SC	*Jasus lalandii* (Milne Edwards, 1837)
ES	langosta del Cabo
DA	Kap-languster
DE	Kap-Languste; Afrikanische Languste; Rote Languste
EL	αστακός του Ακρωτηρίου
EN	Cape rock lobster
FR	langouste du Cap
IT	aragosta; aragosta sudafricana
NL	Kaapse langoest
PT	lagosta do Cabo
FI	tyynenmerenlangusti
SV	kaplangust

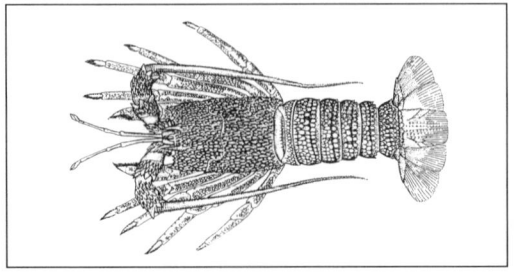

1263 PALINURIDAE LBT

SC	*Jasus tristani* (Holthuis, 1963)
ES	langosta de Tristán
DA	Tristan-languster
DE	Tristans-Languste
EL	αστακός
EN	Tristan rock lobster; Tristan da Cunha rock lobster
FR	langouste de Tristan
IT	aragosta
NL	Tristans langoest
PT	lagosta-tristão
FI	tristaninlangusti
SV	tristanlangust

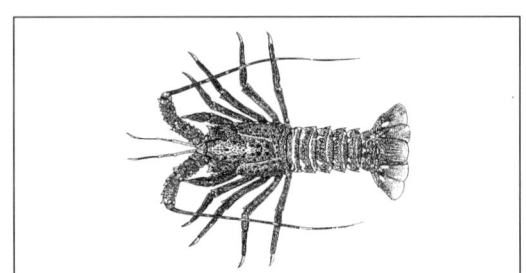

1264 PALINURIDAE LOG

SC	*Jasus verreauxi* (Milne Edwards, 1851)
ES	langosta de Oceanía
DA	oceanisk languster
DE	Ostaustralische Languste
EL	αστακός της Ωκεανίας
EN	green rock lobster
FR	langouste d'Océanie
IT	aragosta
NL	Oost-Australische langoest
PT	lagosta da Oceânia
FI	langusti-laji
SV	oceanisk langust

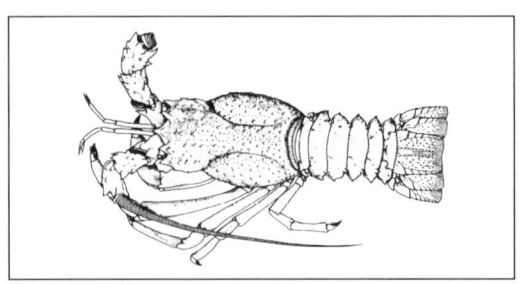

1265 PALINURIDAE SLN

SC	*Palinurus delagoae* (Barnard, 1926)
ES	langosta de Natal
DA	Natal-languster
DE	Natal-Languste
EL	αστακός του Νατάλ
EN	Natal spiny lobster
FR	langouste du Natal
IT	aragosta
NL	Natal-langoest
PT	lagosta do Natal
FI	natalinlangusti
SV	natallangust

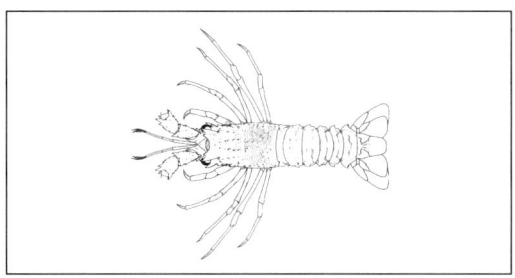

1267 PALINURIDAE SLS

SC	*Palinurus gilchristi* (Stebbing, 1900)
ES	langosta del sur
DA	sydlanguster
DE	Gilchrists-Languste
EL	αστακός
EN	South Coast spiny lobster
FR	langouste du Sud
IT	aragosta
NL	zuidkustlangoest
PT	lagosta de Moçambique
FI	langusti-laji
SV	sydlig langust

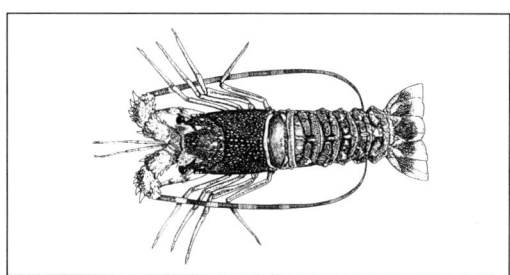

1266 PALINURIDAE SLO

SC	*Palinurus elephas* (Fabricius, 1787)
ES	langosta común
DA	europæisk languster
DE	Europäische Languste
EL	αστακός
EN	common spiny lobster
FR	langouste rouge
IT	aragosta mediterranea
NL	gewone langoest
PT	lagosta castanha; lagosta vulgar
FI	langusti
SV	europeisk langust

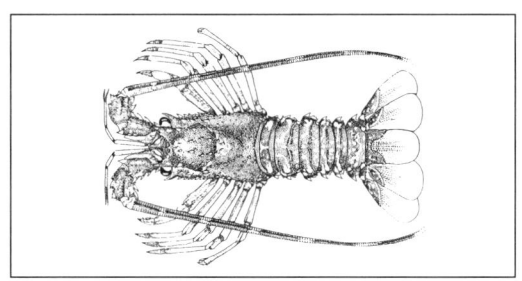

1268 PALINURIDAE PSL

SC	*Palinurus mauritanicus* (Gruvel, 1911)
ES	langosta rosada
DA	mauretansk languster
DE	Mauretanische Languste; Portugiesische Languste
EL	κοκκινοαστακός
EN	pink spiny lobster
FR	langouste rose
IT	aragosta di fondale; aragosta
NL	Portugese langoest
PT	lagosta rósea; lagosta vermelha africana
FI	langusti-laji
SV	mauretansk langust

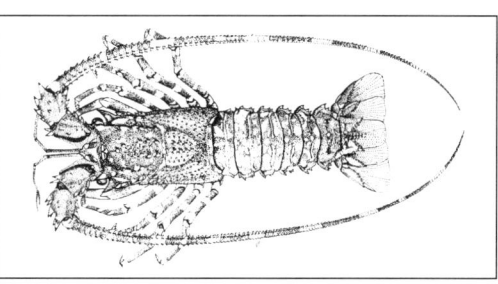

1269 PALINURIDAE　　　　　　SLC

SC	*Panulirus argus* (Latreille, 1804)
ES	langosta del Caribe
DA	caraibisk languster
DE	Amerikanische Languste
EL	αμερικάνικος ασταχός
EN	Caribbean spiny lobster
FR	langouste blanche; langouste de Cuba
IT	aragosta dei Caraibi
NL	Caraïbische langoest
PT	lagosta das Caraíbas
FI	karibianlangusti
SV	karibisk langust

1271 PALINURIDAE　　　　　　LOK

SC	*Panulirus homarus* (Linnaeus, 1758); *Panulirus burgeri*
ES	langosta de Transkei
DA	Transkei-languster
DE	Transkei-Languste
EL	ασταχός
EN	Transkei spiny lobster; scalloped spiny lobster
FR	langouste du Transkei
IT	aragosta
NL	Transkei-langoest
PT	lagosta de Transkei
FI	langusti-laji
SV	transkeilangust

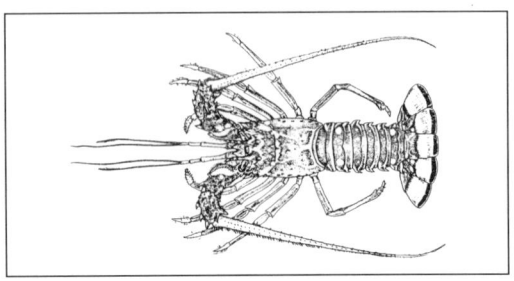

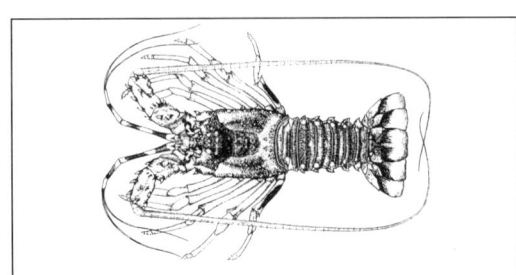

1270 PALINURIDAE

SC	*Panulirus cygnus* (George, 1962)
ES	langosta de Australia
DA	australsk languster
DE	Australische Languste
EL	ασταχός της Αυστραλίας
EN	Australian spiny lobster
FR	langouste d'Australie
IT	aragosta australiana
NL	Australische langoest
PT	lagosta da Austrália
FI	australianlangusti
SV	australisk langust

1272 PALINURIDAE

SC	*Panulirus japonicus* (Von Siebold, 1824)
ES	langosta japonesa
DA	japansk languster
DE	Japanische Languste
EL	γιαπωνέζικος ασταχός
EN	Japanese lobster
FR	langouste japonaise
IT	aragosta
NL	Japanse langoest
PT	lagosta japonesa
FI	japaninlangusti
SV	japansk langust

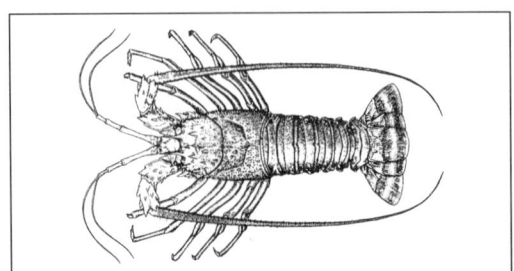

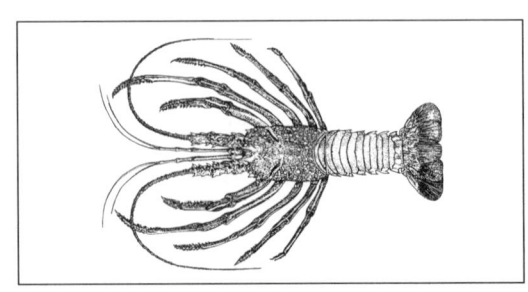

1273 PALINURIDAE

SC   *Panulirus regius* (de Brito Capello, 1864)
ES   langosta real
DA   grøn languster
DE   Königslanguste
EL   πρασινοαστακός
EN   Royal spiny lobster
FR   langouste royale; langouste verte
IT   aragosta verde
NL   koningslangoest
PT   lagosta verde; lagosta africana
FI   langusti-laji
SV   kungslangust

LOY

1275 PALINURIDAE

SC   *Puerulus* spp.
ES   langostas de fusta
DA   languster-slægt
DE   Pazifische Langusten
EL   αστακουδάκια
EN   whip lobsters
FR   langoustes-fouets
IT   aragostelle
NL   zweeplangoesten
PT   larvas de lagosta
FI   langustit
SV   pisklanguster

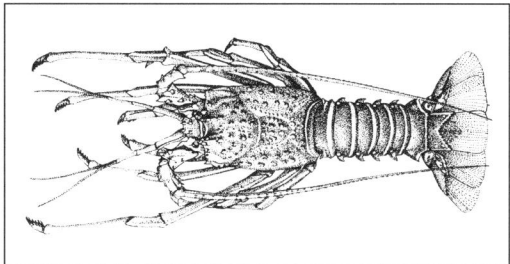

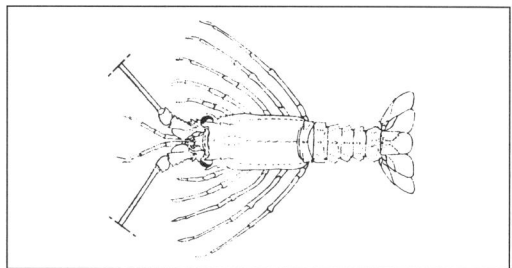

1274 PALINURIDAE

SC   *Panulirus* spp.; *Palinurus* spp.
ES   langostas
DA   languster-slægt
DE   Langusten
EL   αστακός
EN   crawfishes; spiny lobsters; tropical spiny lobster
FR   langoustes
IT   aragoste
NL   langoesten
PT   lagostas
FI   langusti
SV   languster

SLV

1276 ASTACIDAE

SC   *Astacus astacus* (Linnaeus, 1758)
ES   cangrejo de río de patas rojas
DA   krebs; almindelig krebs; flodkrebs
DE   Edelkrebs
EL   ποταμοκαραβίδα
EN   crayfish
FR   écrevisse commune
IT   gambero di fiume europeo
NL   Europese rivierkreeft
PT   lagostim do rio
FI   rapu; jokirapu
SV   flodkräfta

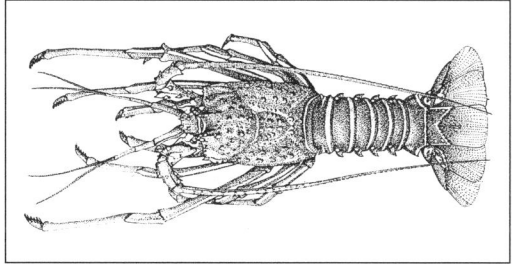

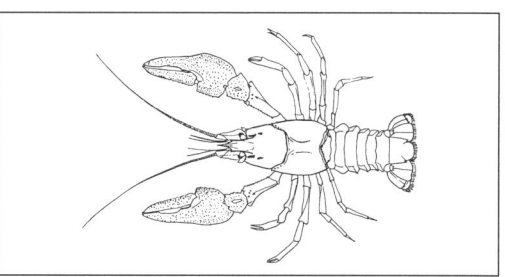

1277 ASTACIDAE   CRD

SC   *Astacus leptodactylus* (Eschscholtz, 1823)
ES   cangrejo de patas punteadas
DA   galizisk flodkrebs
DE   Sumpfkrebs; Galizier
EL   ποταμογαρίδα του Δούναβη
EN   Galician crayfish
FR   écrevisse du Danube; écrevisse turque
IT   gambero turco
NL   Galicische rivierkreeft; Turkse rivierkreeft
PT   lagostim do Danúbio
FI   kapeasaksirapu
SV   sumpkräfta; östlig flodkräfta; donauflodkräfta

1279 ASTACIDAE

SC   *Austropotamobius pallipes* (Lereboullet, 1858);
     Astacus pallipes
ES   cangrejo de río
DA   hvidfodet flodkrebs
DE   Dohlenkrebs
EL   ασπροπόδαρη καραβίδα· ποταμοκαραβίδα
EN   river crayfish
FR   écrevisse à pieds blancs
IT   gambero di fiume
NL   witvoetrivierkreeft
PT   lagostim do rio
FI   kolorapu
SV   vitfotad flodkräfta

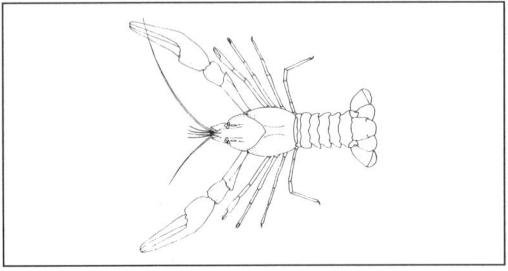

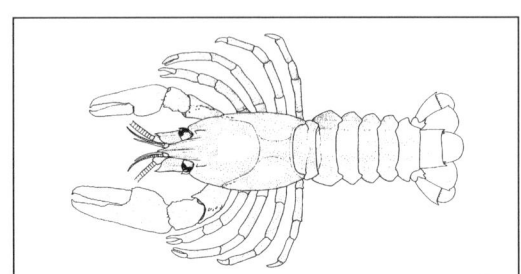

1278 ASTACIDAE

SC   *Astacus* spp.; *Austropotamobius* spp.
ES   cangrejos de río
DA   flodkrebs-slægter
DE   Flußkrebse
EL   αστακοί· ποταμοκαραβίδες
EN   crayfishes
FR   écrevisses
IT   astacidi
NL   rivierkreeftjes
PT   lagostins do rio
FI   jokiravut
SV   flodkräftor

1280 ASTACIDAE

SC   *Austropotamobius torrentium* (Schrank, 1803);
     Astacus torrentium
ES   cangrejo de torrente
DA   stenkrebs
DE   Steinkrebs
EL   ποταμοκαραβίδα
EN   torrent crayfish
FR   écrevisse des torrents
IT   gambero di torrente
NL   steenkreeft
PT   lagostim do monte
FI   kivirapu
SV   strömflodkräfta

DRAWING NOT AVAILABLE

1281 ASTACIDAE

SC	*Cambarus* spp.
ES	cangrejos de río
DA	amerikansk flodkrebs-slægt
DE	Flußkrebse; Edelkrebse
EL	ποταμοκαραβίδες
EN	crayfishes
FR	écrevisses
IT	gamberi di fiume americani
NL	rivierkreeften
PT	lagostins do rio
FI	rapu-suku
SV	flodkräftor

1283 ASTACIDAE

SC	*Orconectes* spp.
ES	langostinos de río americanos
DA	flodkrebs-slægt
DE	Amerikanische Flußkrebse
EL	ποταμοκάβουρες της Αμερικής
EN	American crayfishes
FR	écrevisses américaines
IT	gamberi di fiume americani
NL	rivierkreeften
PT	lagostins «Orconectes»
FI	kääpiörapu-suku
SV	amerikansk flodkräfta

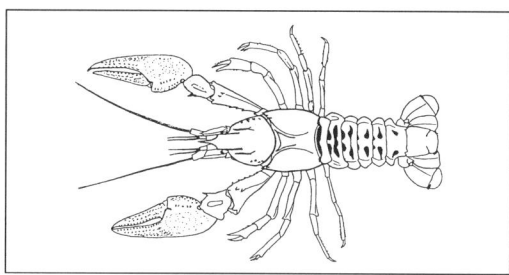 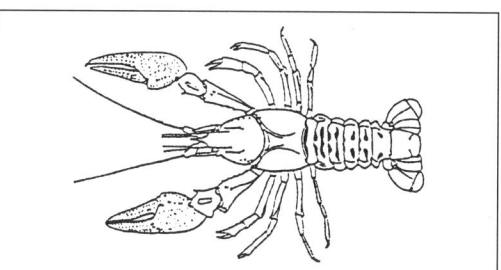

1282 ASTACIDAE

SC	*Orconectes limosus* (Rafinesque, 1817); *Cambarus affinis*
ES	langostino americano de agua dulce
DA	amerikansk krebs; amerikansk flodkrebs
DE	Amerikanischer Flußkrebs
EL	ποταμοκαραβίδα της Αμερικής
EN	American crayfish
FR	écrevisse américaine
IT	gambero d'acqua dolce americano
NL	Amerikaanse rivierkreeft
PT	lagostim americano de água doce
FI	kääpiörapu
SV	amerikansk flodkräfta

1284 ASTACIDAE

SC	*Procambarus clarckii* (Girard, 1852)
ES	cangrejo de río rojo
DA	Louisiana-flodkrebs
DE	Louisiana-Flußkrebs
EL	βαλτογαρίδες
EN	red swamp crawfish
FR	écrevisse rouge des marais
IT	gambero di palude
NL	Louisiana-rivierkreeft
PT	lagostim vermelho do rio; lagostim vermelho da Luisiana
FI	punarapu
SV	louisianaflodkräfta

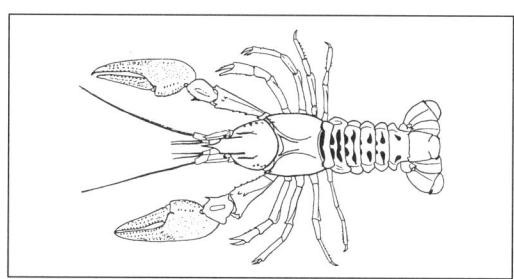 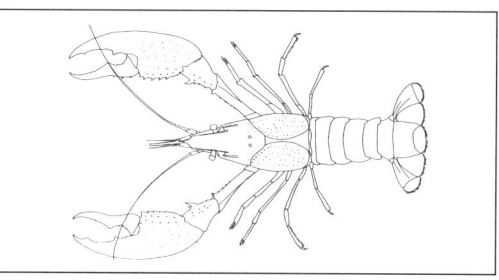

1285 REPTANTIA

CRA

SC *Reptantia*
ES crangrejos de mar
DA krybende tibenede krebsdyr
DE Panzerkrebse
EL καρκινοειδή
EN marine crabs
FR crabes de mer
IT crostacei reptanti
NL zeekrabbensoorten
PT caranguejos do mar
FI ravut
SV krypande tiofotade kräftdjur

```
DRAWING NOT AVAILABLE
```

1286 PAGURIDAE

SC *Paguridae*
ES cangrejos ermitaños
DA eremitkrebs-familie
DE Einsiedlerkrebse
EL Βερνάρδος ο ερημίτης
EN hermit crabs
FR pagures
IT paguri
NL heremietkreeften
PT casas alugadas; eremitas
FI erakkoravut-heimo
SV eremitkräftor

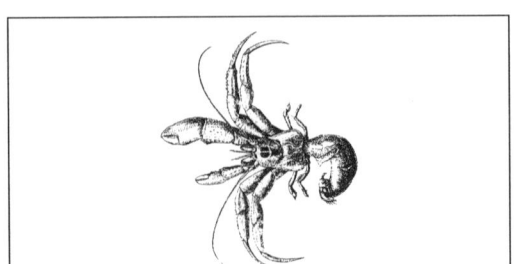

1287 CANCRIDAE

CAD

SC *Cancridae*
ES bueyes
DA taskekrabbe-familien
DE Taschenkrebse; Krabben; Kurzschwanzkrebse
EL κάβουρες
EN crabs
FR crabes dormeurs
IT granchi; granciporri
NL krabben
PT sapateiras
FI taskuravut-heimo
SV krabbtaskor

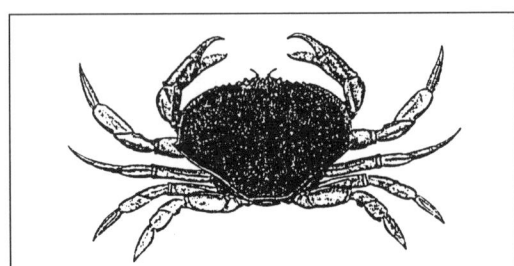

1288 CANCRIDAE

CRJ

SC *Cancer borealis* (Stimpson, 1859)
ES jaiba de roca Jonás
DA Jonaskrabbe
DE Jonahkrabbe
EL κάβουρας του Βορρά
EN Jonah crab
FR tourteau-jona
IT granciporro atlantico rosso
NL Jonaskrab
PT sapateira boreal
FI taskurapu-laji
SV nordlig krabbtaska

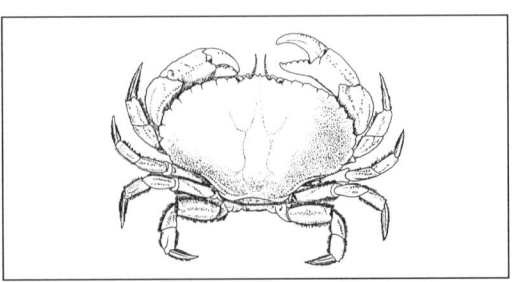

1289 CANCRIDAE CRK

SC	*Cancer irroratus* (Say, 1817)
ES	jaiba de roca amarilla
DA	østamerikansk taskekrabbe
DE	Felsenkrabbe
EL	κάβουρας του Ατλαντικού
EN	Atlantic rock crab
FR	tourteau poïnclos
IT	granciporro atlantico giallo
NL	Atlantische rotskrab
PT	sapateira de rocha do Atlântico
FI	kalliorapu
SV	klippkrabbtaska

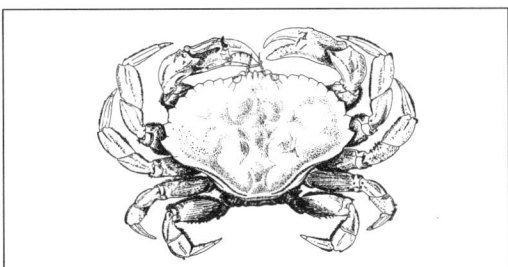

1290 CANCRIDAE DUN

SC	*Cancer magister* (Dana, 1852)
ES	buey del Pacífico
DA	Dungeness-taskekrabbe
DE	Pazifischer Taschenkrebs
EL	κάβουρας του Ειρηνικού
EN	Dungeness crab
FR	dormeur du Pacifique
IT	granciporro del Pacifico
NL	Dungenesskrab
PT	sapateira do Pacífico
FI	taskurapu-laji
SV	stillahavskrabbtaska

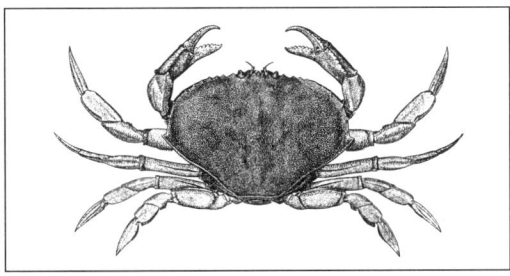

1291 CANCRIDAE CRE

SC	*Cancer pagurus* (Linnaeus, 1758)
ES	buey; buey de mar; paguro; sabago; pato; masera; jaiba de roca masera
DA	taskekrabbe; almindelig taskekrabbe
DE	Taschenkrebs
EL	κάβουρας
EN	edible crab
FR	tourteau; crabe dormeur
IT	granchio di mare; granciporro
NL	Noordzeekrab
PT	sapateira
FI	isotaskurapu
SV	krabbtaska

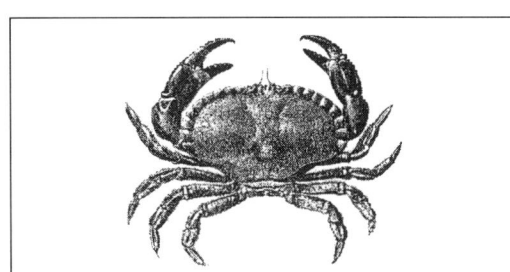

1292 CANCRIDAE ROC

SC	*Cancer productus* (Randall, 1839)
ES	jaiba del Pacífico
DA	pacifisk taskekrabbe
DE	Bogenkrabbe
EL	κάβουρας του Ειρηνικού
EN	Pacific rock crab
FR	tourteau du Pacifique
IT	granciporro del Pacifico
NL	Pacifische rotskrab
PT	sapateira de rocha do Pacífico
FI	taskurapu-laji
SV	stillahavskrabbtaska

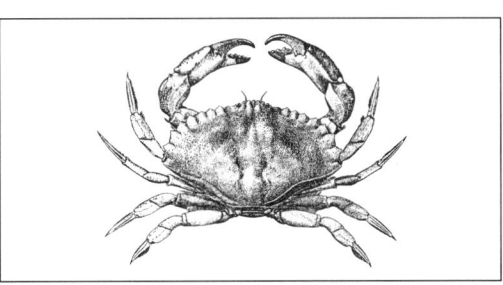

1293 XANTHIDAE

STC

SC	*Menippe mercenaria* (Say, 1818)
ES	cangrejo de piedra negro
DA	sort stenkrabbe
DE	Große Steinkrabbe
EL	κάβουρας των βράχων
EN	black stone crab
FR	crabes à pieds noirs
IT	granciporro della Florida
NL	zwarte steenkrab
PT	caranguejo negro da pedra
FI	taskurapu-laji
SV	svart stenkrabba

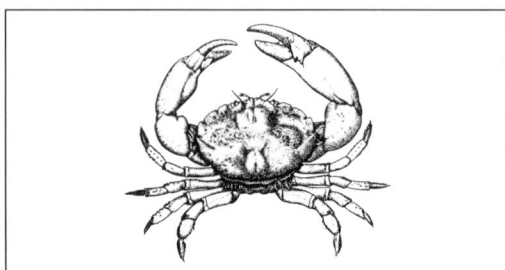

1295 PORTUNIDAE

CRB

SC	*Callinectes sapidus* (Rathbun, 1896)
ES	jaiba azul
DA	blå svømmekrabbe; blåkrabbe
DE	Blaue Schwimmkrabbe; Blaukrabbe
EL	κάβουρας· ιταλικός κάβουρας
EN	blue crab
FR	crabe bleu
IT	granchio nuotatore; granchio blu
NL	blauwe krab
PT	navalheira azul
FI	sinitaskurapu
SV	blå krabba

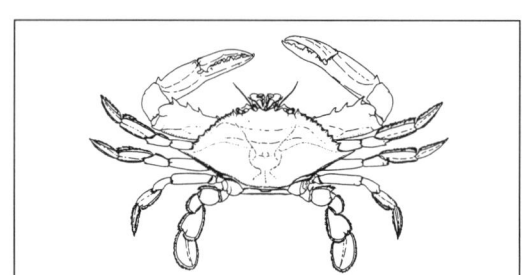

1294 PORTUNIDAE

CRZ

SC	*Callinectes danae* (Smith, 1869)
ES	jaiba siri
DA	Danas svømmekrabbe
DE	Dana-Blaukrabbe
EL	κάβουρας κολυμβητής
EN	Dana swimcrab
FR	crabe lénée
IT	granchio nuotatore purpureo
NL	Dana blauwe krab
PT	navalheira-dana
FI	taskurapu-laji
SV	danasimkrabba

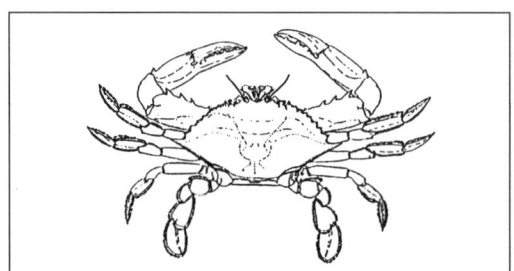

1296 PORTUNIDAE

CAL

SC	*Callinectes* spp.
ES	cangrejos nadadores
DA	svømmekrabbe-slægt
DE	Blaukrabben
EL	κάβουρες
EN	swimcrabs
FR	crabes nageurs
IT	granchi nuotatori
NL	blauwe krabben
PT	navalheiras
FI	sinitaskuravut
SV	simkrabbor

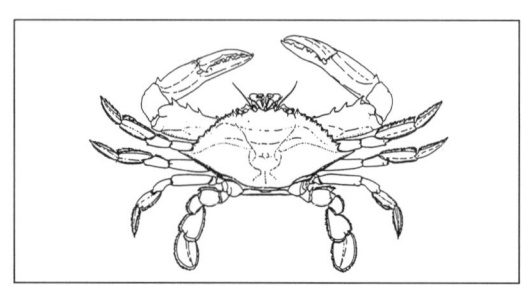

1297 PORTUNIDAE   CMR

SC	*Carcinus aestuarii* (Nardo, 1847); *Carcinus mediterraneus*
ES	cangrejo verde del Mediterráneo
DA	Middelhavsstrandkrabbe
DE	Mittelmeer-Strandkrabbe
EL	κάβουρας
EN	Mediterranean green crab
FR	crabe vert de Méditerranée
IT	granchio ripario; granchio comune
NL	Mediterrane strandkrab
PT	caranguejo verde do Mediterrâneo
FI	rantataskurapu-laji
SV	medelhavsstrandkrabba

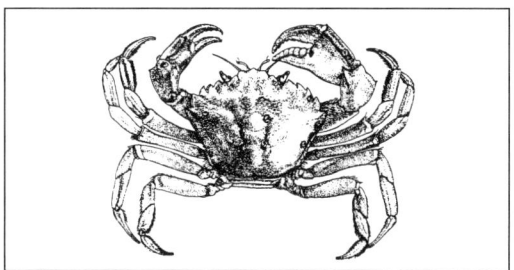

1299 PORTUNIDAE

SC	*Necora puber* (Linnaeus, 1767); *Liocarcinus puber*; *Macropipus puber*; *Portunus puber*
ES	nécora; andarica; jaiba nécora
DA	fløjlssvømmekrabbe
DE	Wollige Schwimmkrabbe
EL	βελουδοκάβουρας
EN	velvet swimcrab; velvet swimming-crab
FR	étrille
IT	necora
NL	fluwelen zwemkrab
PT	navalheira felpuda
FI	taskurapu-laji
SV	sammetssimkrabba

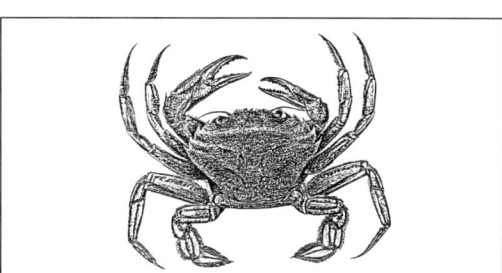

1298 PORTUNIDAE   CRA

SC	*Carcinus maenas* (Linnaeus, 1758)
ES	cangrejo atlántico; jaiba verde; cangrejo verde
DA	almindelig strandkrabbe
DE	Strandkrabbe
EL	πρασινοκάβουρας
EN	common shore crab; green shore crab; green crab
FR	crabe vert; crabe européen
IT	granchio comune; granchio ripario
NL	strandkrab
PT	caranguejo verde; caranguejo
FI	rantataskurapu
SV	strandkrabba

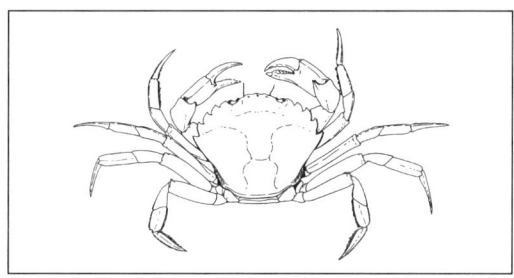

1300 PORTUNIDAE   GAZ

SC	*Portunus trituberculatus* (Miers, 1876)
ES	jaiba gazami
DA	gazami-svømmekrabbe
DE	Gazami-Schwimmkrabbe
EL	κάβουρας-γκαζάμι
EN	gazami crab
FR	étrille-gazami; étrille à trois pointes
IT	granchio nuotatore gazami
NL	gazamizwemkrab
PT	navalheira-gazami
FI	uimataskurapu-laji
SV	gazamikrabba

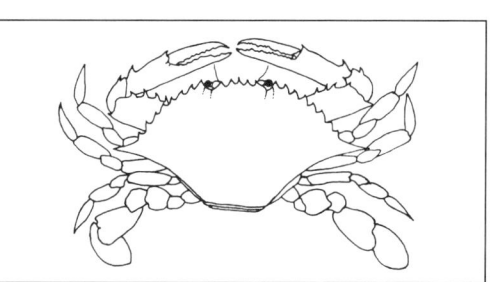

1301 PORTUNIDAE   CRS

SC	*Portunus* spp.
ES	jaibas
DA	svømmekrabbe-slægt
DE	Schwimmkrabben
EL	κολυμβητικά καβούρια
EN	swimcrabs
FR	crabes; étrilles
IT	granchi nuotatori
NL	zwemkrabben
PT	caranguejos nadadores
FI	uimataskuravut
SV	simkrabbor

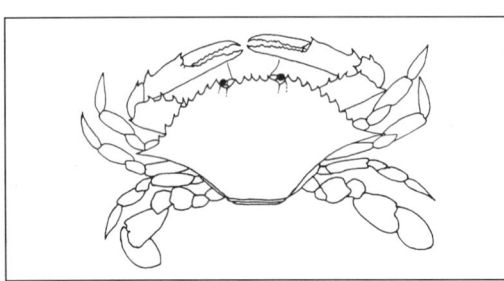

1303 GRAPSIDAE   ERS

SC	*Eriocheir sinensis* (Milne Edwards, 1853)
ES	cangrejo chino
DA	uldhåndskrabbe
DE	Chinesische Wollhandkrabbe
EL	κάβουρας της Κίνας
EN	mitten crab
FR	crabe chinois; crabe velu
IT	granchio cinese
NL	Chinese wolhandkrab
PT	caranguejo chinês
FI	villasaksirapu
SV	kinesisk ullhandskrabba

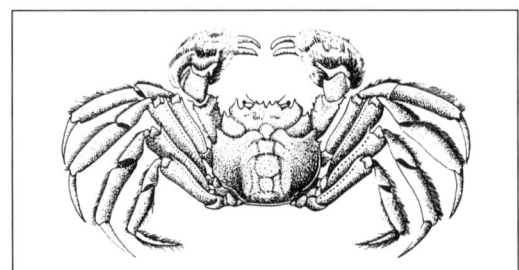

1302 PORTUNIDAE   MUD

SC	*Scylla serrata* (Forsskål, 1775)
ES	cangrejo de fango australiano
DA	mangrovekrabbe
DE	Mangrovenkrabbe
EL	λασποκάβουρας
EN	mud crab
FR	crabe de vase d'Australie
IT	granchio indiano
NL	mangrovekrab
PT	caranguejo da lama
FI	uimataskurapu-laji
SV	mangrovekrabba

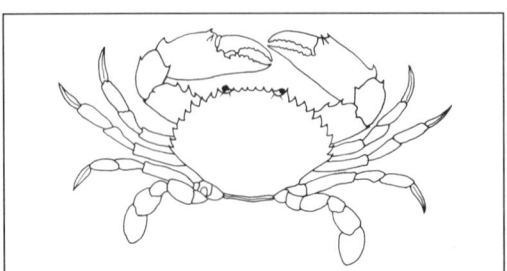

1304 SCYLLARIDAE   LOS

SC	*Scyllaridae*
ES	cigarros; santiaguiños
DA	bjørnekrebs-familien
DE	Bärenkrebse
EL	λασποκαβούρια
EN	slipper lobsters
FR	cigales de mer
IT	scillaridi; magnose; cicale
NL	beerkreeften
PT	cigarras e cavacos
FI	ravut
SV	toffelkräftor

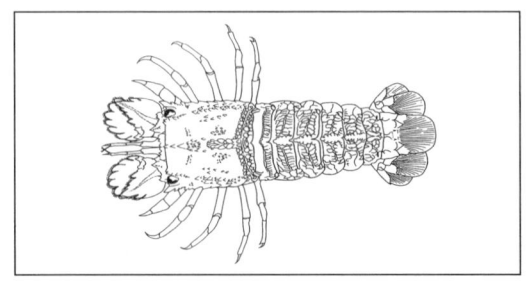

1305 SCYLLARIDAE SCY

SC	*Scyllarus arctus* (Linnaeus, 1758)
ES	santiaguiño
DA	lille bjørnekrebs
DE	Kleiner Bärenkrebs
EL	λύρα της θάλασσας
EN	lesser slipper lobster
FR	petite cigale
IT	magnosella; cicala di mare
NL	kleine beerkreeft
PT	cigarra do mar
FI	rapu-laji
SV	toffelhummer

1307 LITHODIDAE KCT

SC	*Lithodes maja* (Linnaeus, 1758)
ES	centolla de roca
DA	konge-troldkrabbe
DE	Nördliche Steinkrabbe
EL	βασιλικός κάβουρας
EN	stone king crab
FR	crabe royal de roche
IT	granchio reale
NL	Augustinuskrab
PT	caranguejo real da pedra
FI	piikikäs kivirapu
SV	stentrollkrabba

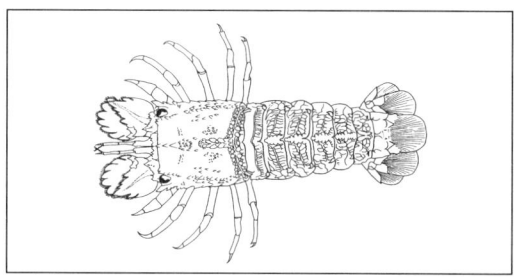

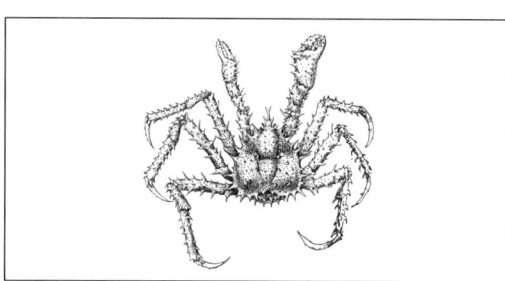

1306 GALATHEIDAE LOQ

SC	*Galatheidae*
ES	galateidos; langostinos
DA	troldhummer-familien
DE	Furchenkrebse
EL	γαλαθείδες
EN	squad lobsters; craylets
FR	galathées
IT	galateidi
NL	springkrabben
PT	galateídeos
FI	hummerit-heimo
SV	trollhumrar

1308 LITHODIDAE KCM

SC	*Lithodes murrayi* (Henderson, 1888)
ES	centolla real
DA	Murrays troldkrabbe
DE	Murray-Steinkrabbe; Subantarktische Steinkrabbe
EL	βασιλικός κάβουρας
EN	Murray king crab; Subantarctic stone crab
FR	crabe royal de Murray; crabe royal subantarctique
IT	granchio reale
NL	Murray koningskrab
PT	caranguejo real de Murray
FI	kivirapu-laji
SV	murraytrollkrabba

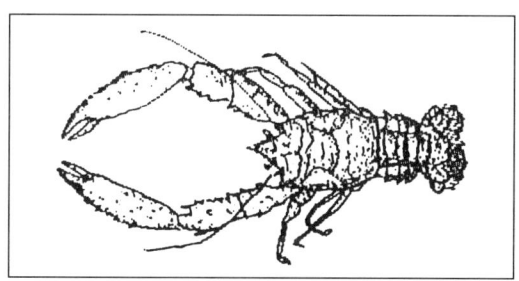

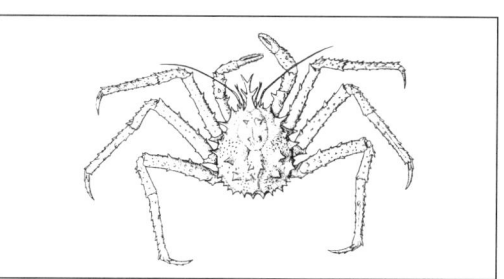

1309 LITHODIDAE   KCA

SC	*Lithodes santolla* (Molina, 1782); *Lithodes antarcticus*
ES	centolla patagónica
DA	antarktisk troldkrabbe
DE	Antarktische Königskrabbe
EL	βασιλικός κάβουρας της Ανταρκτικής
EN	Southern king crab
FR	crabe royal antarctique
IT	granchio reale australe
NL	zuidelijke koningskrab
PT	caranguejo real do sul
FI	kivirapu-laji
SV	antarktisk trollkrabba

DRAWING NOT AVAILABLE

1311 LITHODIDAE   KCS

SC	*Paralithodes* spp.
ES	cangrejos rusos
DA	troldkrabbe-slægt
DE	Königskrabben
EL	βασιλικά καβούρια
EN	king crabs
FR	crabes royaux
IT	granchi reali
NL	koningskrabben
PT	caranguejos reais
FI	kuningasravut
SV	trollkrabbor

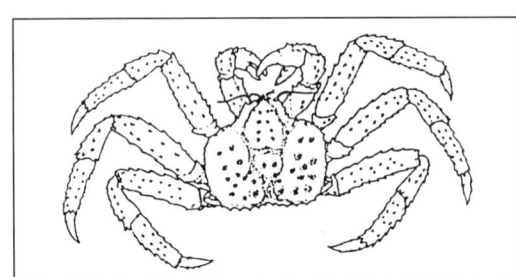

1310 LITHODIDAE

SC	*Paralithodes camtschatica* (Tilesius, 1815)
ES	cangrejo japonés
DA	Japan-krabbe; Kamtjatka-krabbe
DE	Kamschatka-Krabbe
EL	βασιλικός κάβουρας
EN	king crab; Japanese crab; Alaska deepsea crab
FR	crabe royal du Kamchatka
IT	granchio reale; grancevola del Camciatca
NL	Kamtsjatkakrab; koningskrab
PT	caranguejo real
FI	kuningasrapu
SV	japansk trollkrabba

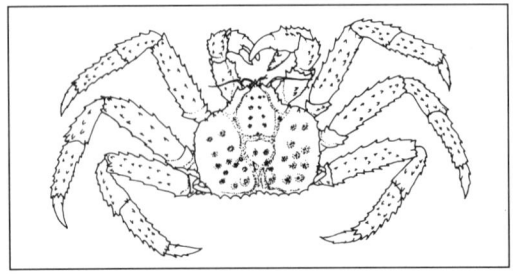

1312 MAJIDAE   CRQ

SC	*Chionoecetes opilio* (Fabricius, 1788)
ES	cangrejo de las nieves
DA	arktisk krabbe
DE	Arktische Seespinne
EL	κάβουρας της Αρκτικής
EN	queen crab
FR	crabes des neiges
IT	grancevola artica
NL	Arctische sneeuwkrab
PT	caranguejo das neves
FI	rapu-laji
SV	arktisk maskeringskrabba

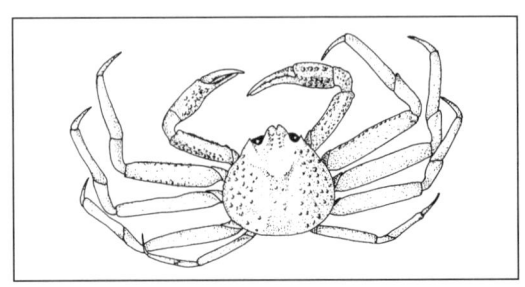

1313 MAJIDAE PCR

SC *Chionoecetes* spp.
ES cangrejos de las nieves
DA krabbe-slægt
DE Arktische Seespinnen
EL κάβουρες της Αρκτικής
EN Pacific snow crabs
FR crabes des neiges du Pacifique
IT grancevole artiche
NL Pacifische sneeuwkrabben
PT caranguejos das neves do Pacífico
FI ravut
SV maskeringskrabbor

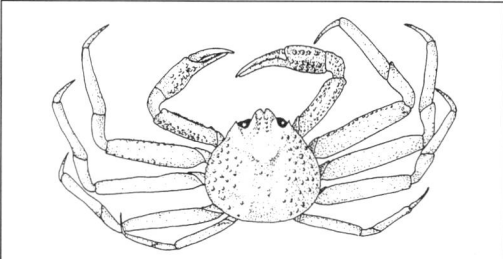

1315 MAJIDAE SCR

SC *Maja squinado* (Herbst, 1788)
ES centolla; centolla europea; centolla de roca; centollo
DA edderkopkrabbe
DE Große Seespinne; Seespinne
EL καβουρομάνα
EN sea spider; spinous spider crab; spider crab; spiny crab
FR araignée de mer; grande araignée de mer
IT grancevola; granceola; granseola
NL Europese spinkrab
PT santola europeia; santola
FI isohämähäkkirapu
SV spindelkrabba

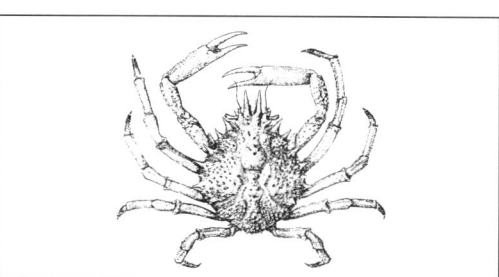

1314 MAJIDAE JAJ

SC *Jacquinotia edwardsii* (Jacquinot, 1853)
ES centolla del sur
DA syd-stankelbenskrabbe
DE Südliche Spinnenkrabbe
EL αραχνοκάβουρας της Νέας Ζηλανδίας
EN Southern spider crab
FR araignée du Sud
IT grancevola neozelandese
NL zuidelijke spinkrab
PT santola do sul
FI rapu-laji
SV sydlig spindelkrabba

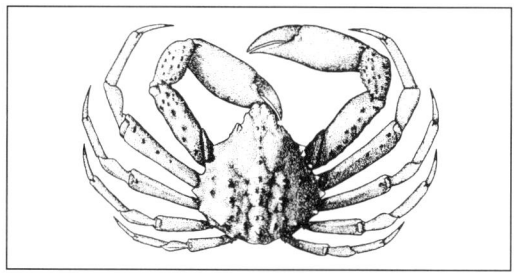

1316 PARASTACIDAE AYA

SC *Euastacus serratus* (Shaw, 1794)
ES cangrejo de río de Australia
DA australsk flodkrebs
DE Australischer Flußkrebs
EL γαρίδα των ποταμών της Αυστραλίας
EN Australian crayfish
FR écrevisse d'Australie
IT gambero di fiume australiano
NL Australische rivierkreeft
PT lagostim de rio da Austrália
FI rapu-laji
SV australisk flodkräfta

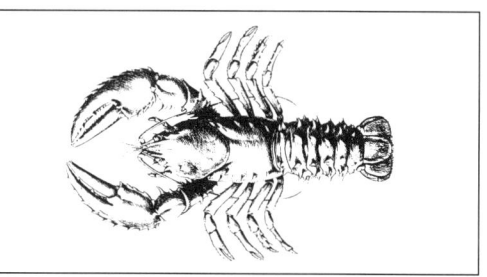

329

1317 NEPHROPIDAE   LBA

SC	*Homarus americanus* (Milne Edwards, 1837)
ES	bogavante americano
DA	amerikansk hummer
DE	Amerikanischer Hummer
EL	αμερικάνικος αστακός
EN	American lobster
FR	homard américain
IT	astice americano
NL	Amerikaanse kreeft
PT	lavagante americano
FI	amerikanhummeri
SV	amerikansk hummer

1319 NEPHROPIDAE   NEA

SC	*Metanephrops andamanicus* (Wood-Mason, 1891)
ES	langostino del Índico
DA	Andaman-hummer
DE	Andamanen-Schlankhummer
EL	καραβίδα του Ινδικού
EN	Southern langoustine; Andaman lobster
FR	langoustine andamane; langoustine de l'océan Indien
IT	scampo oceanico
NL	Andamanlangoestine
PT	lagostim do Índico
FI	hummeri-laji
SV	andamanhummer

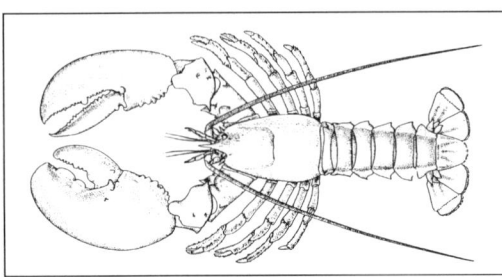

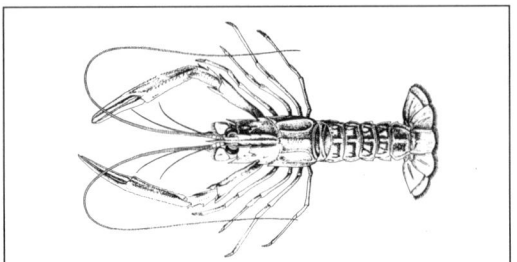

1318 NEPHROPIDAE   LBE

SC	*Homarus gammarus* (Linnaeus, 1758); *Homarus vulgaris*
ES	bogavante europeo
DA	europæisk hummer
DE	Europäischer Hummer; Hummer
EL	αστακογαρίδα· αστακός
EN	European lobster
FR	homard européen
IT	astice; lupicante
NL	kreeft
PT	lavagante
FI	hummeri
SV	hummer

1320 NEPHROPIDAE   NEP

SC	*Nephrops norvegicus* (Linnaeus, 1758)
ES	cigala; maganto; escamarlànc
DA	jomfruhummer; dybvandshummer
DE	Kaisergranat; Tiefseehummer; Norwegischer Schlankhummer; Scampi
EL	καραβίδα
EN	Norway lobster; scampi; langoustine; Dublin Bay prawn
FR	langoustine
IT	scampo
NL	Noorse kreeft; langoestine
PT	lagostim
FI	keisarihummeri
SV	havskräfta; kejsarhummer

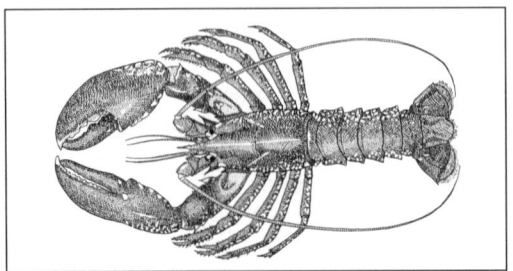

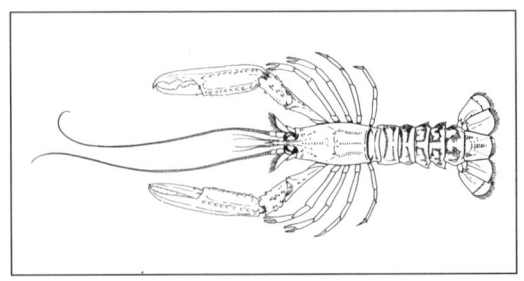

1321 GERYONIDAE　　　　　CRR

SC	*Chaceon maritae* (Manning and Holthius, 1981); *Geryon quinquedens*
ES	geriocangrejo rojo; cangrejo colorado
DA	femtandet dybvandskrabbe
DE	Rote Tiefseekrabbe
EL	κόκκινος κάβουρας
EN	red crab
FR	gérion ouest-africain
IT	granchio rosso di fondale
NL	rode diepzeekrab
PT	caranguejo vermelho da fundura
FI	isoabaloni
SV	djuphavsrödkrabba

1323 MOLLUSCA　　　　　MOL

SC	*Mollusca*
ES	moluscos marinos
DA	bløddyr
DE	Meeresweichtiere
EL	θαλάσσια μαλάκια
EN	marine molluscs
FR	mollusques marins
IT	molluschi marini
NL	weekdieren
PT	moluscos marinhos
FI	nilviäiset-lahko
SV	blötdjur

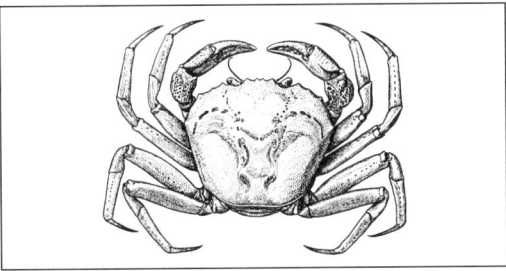

DRAWING NOT AVAILABLE

1322 GERYONIDAE　　　　　GER

SC	*Geryon* spp.
ES	cangrejos rojos
DA	dybvandskrabbe-slægt
DE	Tiefseekrabben
EL	κόκκινα καβούρια
EN	red crabs
FR	crabes rouges; gérions
IT	granchi di fondale rossi
NL	rode diepzeekrabben
PT	caranguejos da fundura
FI	ravut
SV	djuphavskrabbor

1324 BIVALVIA　　　　　CLX

SC	*Bivalvia*
ES	almejas
DA	muslinger
DE	Muscheln
EL	δίθυρα
EN	clams
FR	clams
IT	bivalvi
NL	tweekleppige schelpdieren
PT	bivalves
FI	simpukat
SV	musslor

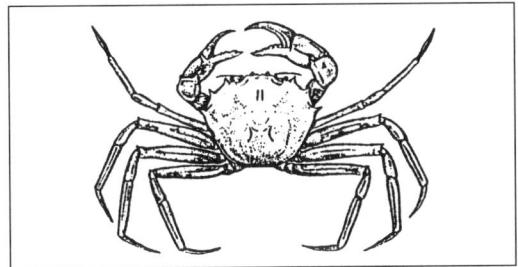

DRAWING NOT AVAILABLE

1325 HALIOTIDAE ABG

SC	*Haliotis gigantea* (Gmelin, 1791)
ES	oreja de mar gigante
DA	stor søøre
DE	Riesenmeerohr
EL	μεγαλαλιώτις· μεγάλο αυτί της θάλασσας
EN	giant abalone
FR	ormeau géant
IT	abalone
NL	reuzenzeeoor
PT	orelha gigante
FI	isoabaloni
SV	stort havsöra; abalone

1327 HALIOTIDAE ABR

SC	*Haliotis ruber* (Leach, 1814)
ES	oreja de mar de labios negros
DA	rød søøre
DE	Rotes Meerohr
EL	κόκκινο αυτί της θάλασσας
EN	blacklip abalone
FR	ormeau à lèvres noires
IT	abalone
NL	rode zeeoor
PT	orelha-lábio preto
FI	merikorva-laji
SV	rött havsöra

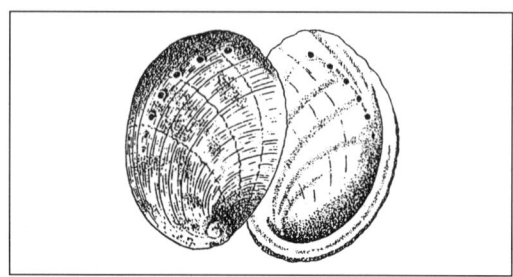

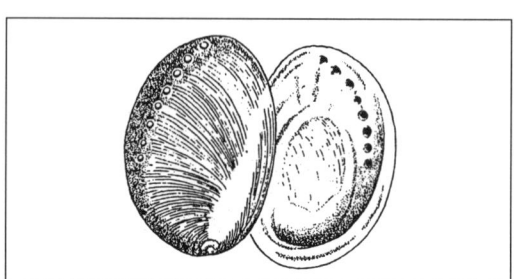

1326 HALIOTIDAE ABP

SC	*Haliotis midae* (Linnaeus, 1758)
ES	oreja de mar
DA	Midas søøre
DE	Südafrikanisches Meerohr
EL	αυτί της θάλασσας
EN	perlemoen abalone
FR	ormeau de Mida
IT	abalone
NL	parelmoerzeeoor
PT	orelha-pérola
FI	abaloni
SV	Midas havsöra

1328 HALIOTIDAE HLT

SC	*Haliotis tuberculata* (Linnaeus, 1758)
ES	oreja de mar
DA	europæisk søøre
DE	Gemeines Seeohr; Seeohr
EL	αυτί της θάλασσας· αχιβάδα· αλιώτις
EN	ormer; sea ear; ear shell; European abalone
FR	ormeau européen; oreille de mer
IT	orecchio di mare
NL	zeeoor
PT	orelha do mar
FI	merikorva; abaloni
SV	europeiskt havsöra

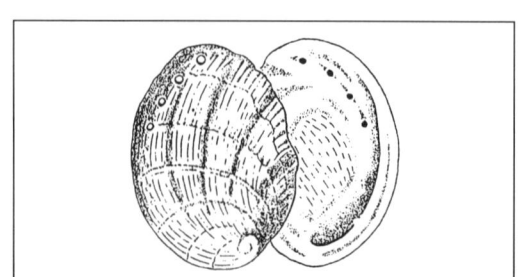

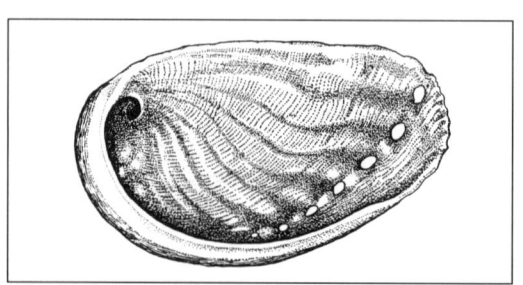

1329 HALIOTIDAE

ABX

SC	*Haliotis* spp.
ES	orejas de mar
DA	søøre-slægten
DE	Meerohren; Abalonen
EL	αυτί της θάλασσας
EN	abalones
FR	ormeaux; oreilles de mer
IT	orecchie di mare; abaloni
NL	zeeoren
PT	orelhas; orelhas-do-mar
FI	merikorvat
SV	havsöron

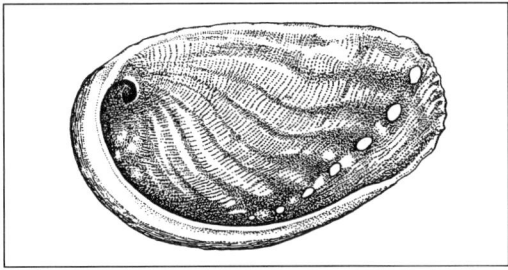

1331 TURBINIDAE

TOS

SC	*Turbo cornutus*
ES	peonza cornuda
DA	hornet turbansnegl
DE	Gehörnte Turbanschnecke
EL	
EN	horned turban
FR	turbo à cornes
IT	turbo
NL	hoornturbo
PT	turbo ornado
FI	kotilo-laji
SV	turbansnäcka

DRAWING NOT AVAILABLE

1330 TROCHIDAE

SC	*Trochus* spp.
ES	tróquidos
DA	knapsnegle-slægten
DE	Kreiselschnecken
EL	τροχοί
EN	trochus
FR	trocas; troques
IT	trocus
NL	tolhoorns
PT	trocos
FI	pyramidikotilo-suku
SV	trochus

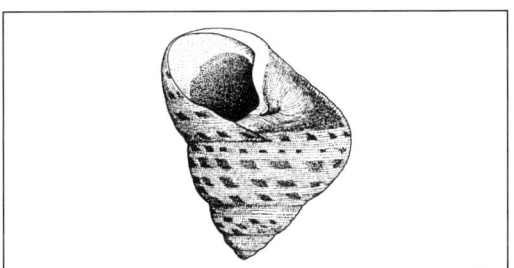

1332 TURBINIDAE

GIW

SC	*Turbo sarmaticus*
ES	peonza sudafricana
DA	sydafrikansk turbansnegl
DE	Südafrikanische Turbanschnecke
EL	
EN	giant pearlwinkle
FR	turbo d'Afrique du Sud
IT	turbo sudafricano
NL	Zuid-Afrikaanse turbo
PT	turbo da África do Sul
FI	kotilo-laji
SV	sydafrikansk turbansnäcka

DRAWING NOT AVAILABLE

1333 PATELLIDAE

SC	*Patella* spp.
ES	lapas
DA	albueskæl-slægt
DE	Napfschnecken
EL	πεταλίδες
EN	limpets
FR	patelles; berniques; arapèdes
IT	patelle
NL	schaalhoorns
PT	lapas
FI	maljakotilo-suku
SV	patellaskålsnäckor

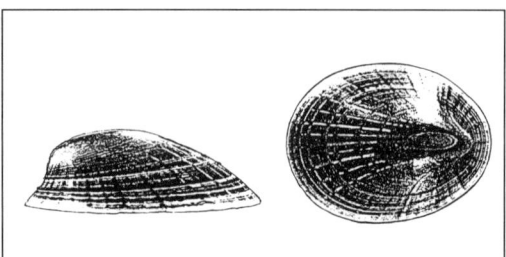

1335 BUCCINIDAE   WHE

SC	*Buccinum undatum* (Linnaeus, 1758)
ES	bocina
DA	konksnegl; konk, rødkonk
DE	Wellhornschnecke
EL	βούκινο
EN	whelk; buckie
FR	buccin; bulot
IT	buccina
NL	wulk
PT	buzo; búzio
FI	metsästystorvikotilo
SV	valthornssnäcka

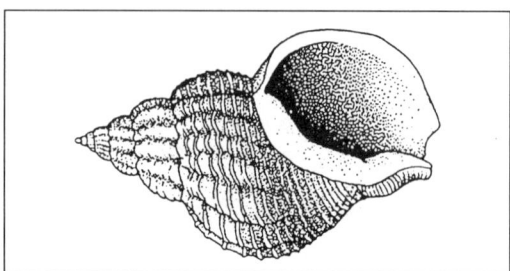

1334 MURICIDAE   MUE

SC	*Murex* spp.
ES	múrices, busanos
DA	pigsnegle-slægt
DE	Stachelschnecken
EL	μούρεξ
EN	murex
FR	rochers; murex
IT	murici
NL	brandhoornslakken
PT	búzios
FI	piikkikotilot
SV	taggsnäckor; purpursnäckor

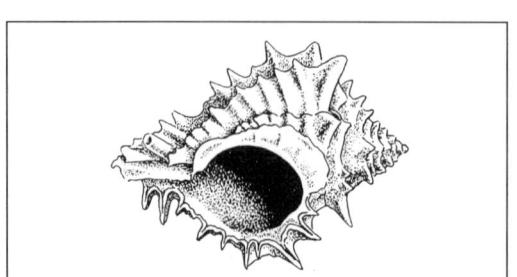

1336 MELONGENIDAE   WHX

SC	*Busycon* spp.
ES	busicones
DA	snegle-slægt
DE	Helmschnecken
EL	
EN	whelks
FR	busycons
IT	busici
NL	busyconwulken
PT	cornetinhas
FI	kotilo-suku
SV	hjälmsnäckor

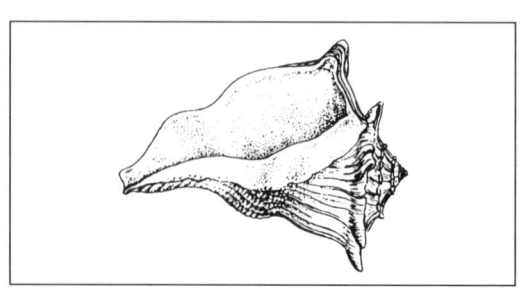

1337 LITTORINIDAE   PER

SC	*Littorina* spp.
ES	bígaros
DA	strandsnegle-slægt
DE	Strandschnecken
EL	
EN	periwinkles
FR	bigorneaux
IT	chiocciole di scogliera
NL	alikruiken
PT	borrelhos; litorinas
FI	rantakotilo-suku
SV	strandsnäckor; kubbongar

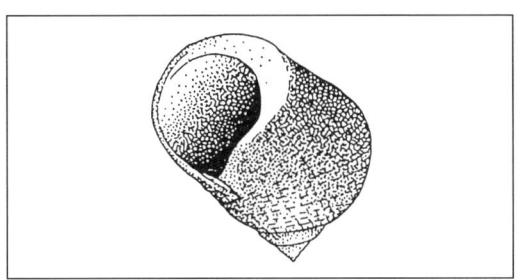

1339 ARCIDAE   BLC

SC	*Anadara granosa* (Linnaeus, 1758)
ES	arca del Pacífico occidental
DA	kornet arca
DE	Westpazifische Archenmuschel
EL	καλογνώμη του Ειρηνικού
EN	blood cockle
FR	arche du Pacifique
IT	sanguinaccio orientale
NL	West-Pacifische arkschelp
PT	arca do Pacífico ocidental
FI	arkkisimpukka-laji
SV	stillahavskistmussla

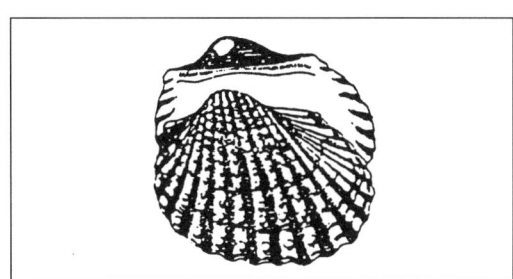

1338 STROMBIDAE   CON

SC	*Strombus* spp.
ES	cobos
DA	skorpionsnegle-slægt
DE	Flügelschnecken; Spinnenschnecken; Fingerschnecken
EL	στρόμβοι
EN	stromboid conchs
FR	strombes
IT	strombi
NL	strombushoorns
PT	estrombos
FI	siipikotilo-suku
SV	vingsnäckor; fäktarsnäckor

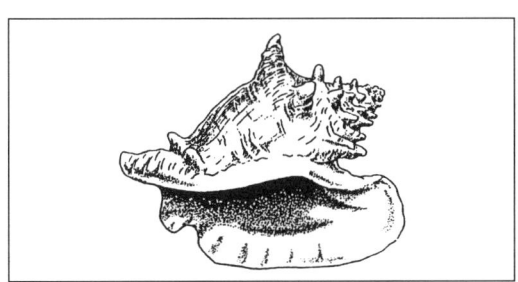

1340 ARCIDAE   MCL

SC	*Anadara subcrenata* (Lischke, 1869)
ES	arca japonesa
DA	japansk arca
DE	Japanische Archenmuschel
EL	γιαπωνέζικη καλογνώμη
EN	mogal clam
FR	arche du Japon; clam «mogal»
IT	sanguinaccio giapponese
NL	Japanse arkschelp
PT	arca japonesa
FI	japaninarkkisimpukka
SV	japansk kistmussla

1341 ARCIDAE BLS

SC	*Anadara* spp.
ES	arcas
DA	musling-slægt
DE	Rote Archenmuscheln
EL	καλογνώμες
EN	blood cockles
FR	arches
IT	sanguinacci
NL	arkschelpen
PT	arcas-vermelhas
FI	arkkisimpukka-suku
SV	kistmusslor

1343 UNIONIDAE

SC	*Anodonta cygnea* (Linnaeus, 1758)
ES	anodonta
DA	dammusling
DE	Teichmuschel
EL	
EN	goose barnacle
FR	moule d'étang; anodonte
IT	anodonta
NL	eendenmossel
PT	anodonta gigante
FI	isojärvisimpukka
SV	dammussla

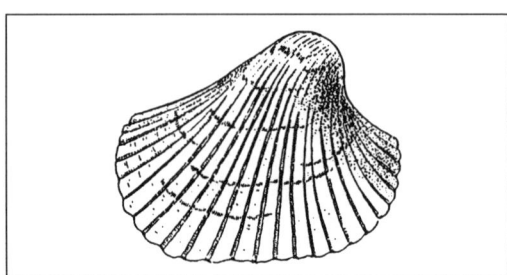

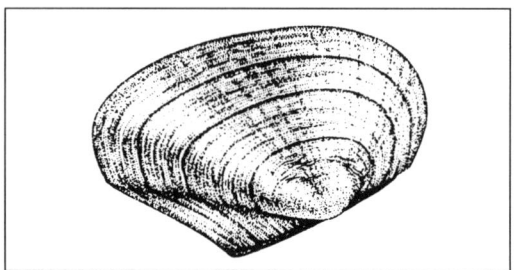

1342 ARCIDAE ARK

SC	*Arca* spp.
ES	arcas
DA	musling-slægt
DE	Archenmuscheln
EL	καλογνώμες
EN	blood cockles
FR	arches
IT	arche
NL	arkschelpen
PT	arcas
FI	nooanarkki-suku
SV	musslor

1344 UNIONIDAE

SC	*Unionidae*
ES	mejillones de agua dulce
DA	malermusling-familien
DE	Malermuscheln
EL	μύδια των γλυκών νερών
EN	freshwater mussels
FR	moules d'eau douce
IT	cozze d'acqua dolce
NL	schildermosselen
PT	mexilhões de água doce
FI	jokisimpukat; jokisimpukat-heimo
SV	sötvattensmusslor

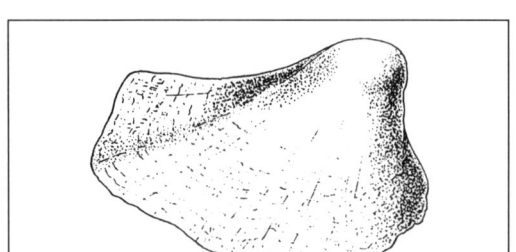

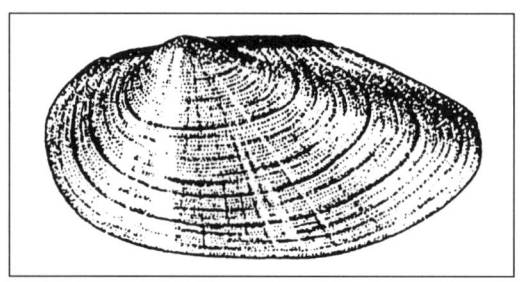

1345 UNIONIDAE

SC	*Unio pictorum* (Linnaeus, 1758); *Pollicepes pictorum*
ES	mejillón de río
DA	malermusling
DE	Malermuschel
EL	μύδι των γλυκών νερών
EN	river mussel; freshwater mussel
FR	moule de rivière; mulette
IT	cozza d'acqua dolce
NL	schildermossel
PT	mexilhão pintado do rio
FI	soukkojokisimpukka
SV	sötvattensmussla

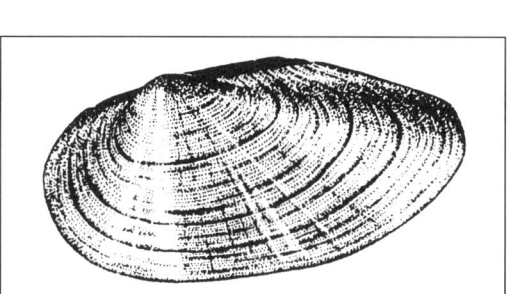

1346 PTERIIDAE

SC	*Pinctada maxima* (Jameson); *Pinctada martensii*
ES	ostra perlífera
DA	perlemusling
DE	Große Perlmuschel
EL	μαργαριτοφόρο στρείδι
EN	mother-of-pearl shell
FR	huître perlière
IT	ostrica perlifera
NL	parelmoerschelp
PT	ostra perlífera
FI	jalohelmisimpukka
SV	pärlostron

1347 OSTREIDAE   OYG

SC	*Crassostrea gigas* (Thunberg, 1793); *Crassostrea angulata*
ES	ostión del Pacífico; ostión; ostra portuguesa; ostra de pobre
DA	Stillehavsøsters; portugisisk østers
DE	Pazifische Felsenauster; Riesenauster; Portugiesische Auster; Felsenauster
EL	στρείδι της Ιαπωνίας· στρείδι
EN	Pacific cupped oyster; Portuguese cupped oyster; Portuguese oyster
FR	huître creuse japonaise; huître portugaise
IT	ostrica giapponese; ostrica concava; ostrica portoghese
NL	Japanse oester
PT	ostra portuguesa; ostra gigante
FI	tyynenmerenosteri
SV	japanskt jätteostron

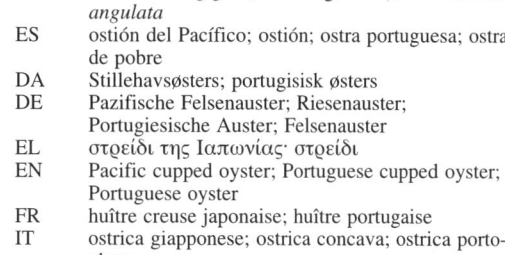

1348 OSTREIDAE   OYM

SC	*Crassostrea rhizophorae* (Guilding, 1828)
ES	ostión de manglar
DA	pacifisk mangroveøsters
DE	Pazifische Felsenauster
EL	στρείδι του Ειρηνικού
EN	Pacific cupped oyster; mangrove cupped oyster
FR	huître creuse de Palétuvier
IT	ostrica dei Caraibi
NL	mangrove-oester
PT	ostra das Caraíbas
FI	mangroveosteri
SV	mangroveostron

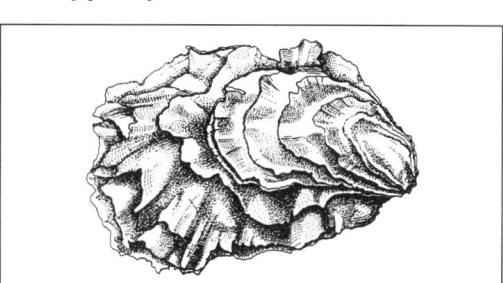

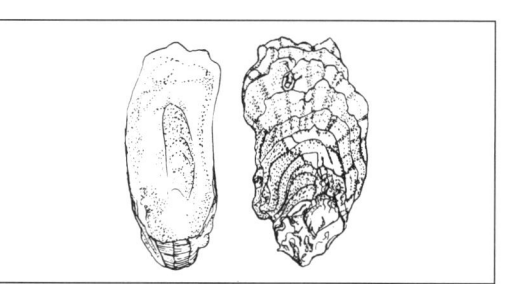

1349 OSTREIDAE　　　　　　　　OYA

SC   *Crassostrea virginica* (Gmelin, 1791)
ES   ostión americano
DA   amerikansk østers
DE   Amerikanische Auster
EL   αμερικάνικο στρείδι
EN   Blue Point oyster; American cupped oyster
FR   huître américaine; huître de Virginie
IT   ostrica della Virginia
NL   Noord-Amerikaanse oester
PT   ostra americana
FI   amerikanosteri
SV   amerikanskt ostron

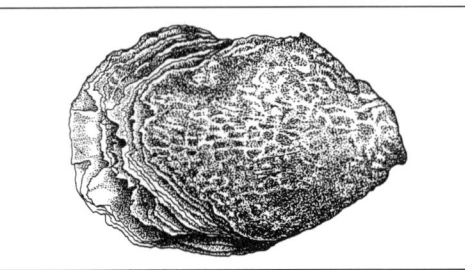

1351 OSTREIDAE　　　　　　　　OCH

SC   *Ostrea chilensis* (Philippi, 1847)
ES   ostra chilena
DA   chilensk østers
DE   Chilenische Plattauster
EL   στρείδι της Χιλής
EN   Chilean flat oyster
FR   huître plate du Chili
IT   ostrica cilena
NL   Chileense oester
PT   ostra plana chilena
FI   chilenosteri
SV   chilenskt ostron

DRAWING NOT AVAILABLE

1350 OSTREIDAE　　　　　　　　OYC

SC   *Crassostrea* spp.
ES   ostiones
DA   østers-slægt
DE   Felsenaustern
EL   στρείδια
EN   cupped oysters
FR   huîtres creuses
IT   ostriche
NL   holle oesters
PT   ostras
FI   osteri-suku
SV   ostron

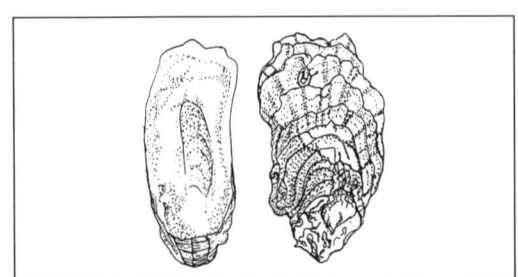

1352 OSTREIDAE　　　　　　　　ODE

SC   *Ostrea denticulata* (Born, 1778)
ES   ostra denticulada
DA   tandet østers
DE   Gezähnte Auster
EL   δαντελόστρειδο
EN   denticulate rock oyster
FR   huître plate de Guinée
IT   ostrica dentellata
NL   getande oester
PT   ostra da rocha da Guiné; ostra plana da Guiné
FI   hammasosteri
SV   tandat ostron

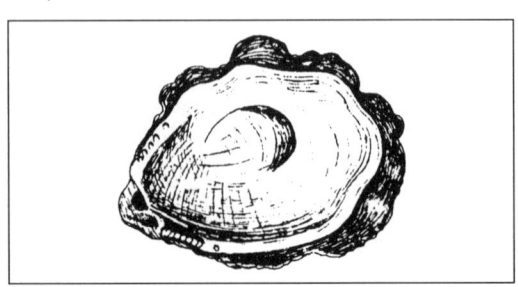

1353 OSTREIDAE OYF

SC	*Ostrea edulis* (Linnaeus, 1758)
ES	ostra plana; ostra europea; ostra común
DA	europæisk østers
DE	Europäische Auster; Auster
EL	στρείδι
EN	European flat oyster; common oyster
FR	huître plate; huître plate européenne; huître de Belon
IT	ostrica; ostrica europea piatta; ostrica piatta
NL	Europese platte oester; oester
PT	ostra plana europeia; ostra plana; ostra redonda
FI	osteri
SV	ostron

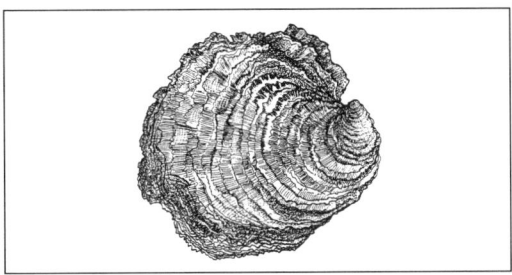

1354 OSTREIDAE OFO

SC	*Ostrea lurida* (Carpenter)
ES	ostra plana del Pacífico
DA	Stillehavsøsters
DE	Pazifische Plattauster
EL	στρείδι
EN	Olympia flat oyster
FR	huître plate du Pacifique
IT	ostrica piatta del Pacífico
NL	Olympia platte oester
PT	ostra plana do Pacífico
FI	olympiaosteri
SV	stillahavsplattostron

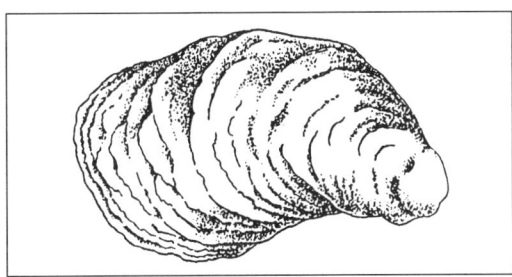

1355 OSTREIDAE DRY

SC	*Ostrea lutaria* (Hutton, 1873)
ES	ostra de Nueva Zelanda
DA	newzealandsk østers
DE	Neuseeland-Plattauster
EL	στρείδι της Νέας Ζηλανδίας
EN	New Zealand dredge oyster
FR	huître plate de Nouvelle-Zélande
IT	ostrica della Nuova Zelanda
NL	Nieuw-Zeelandse platte oester
PT	ostra plana da Nova Zelândia
FI	uudenseelanninosteri
SV	nyzeeländskt ostron

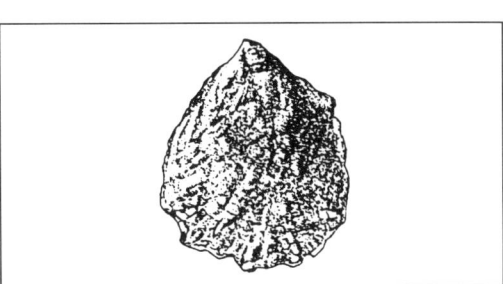

1356 OSTREIDAE OYX

SC	*Ostrea* spp.
ES	ostras
DA	østers-slægt
DE	Plattaustern
EL	στρείδια
EN	flat oysters
FR	huîtres plates
IT	ostriche piatte
NL	platte oesters
PT	ostras planas; ostras redondas
FI	osterit
SV	ostron

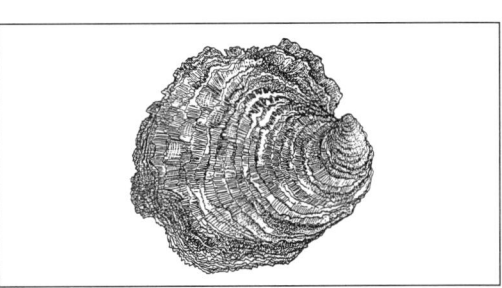

1357 OSTREIDAE

OYS

SC	*Saccostrea cuccullata* (Born, 1778); *Crassostrea commercialis*
ES	ostra australiana
DA	australsk østers
DE	Australische Auster; Sydney-Felsenauster
EL	στρείδι της Αυστραλίας
EN	Sydney cupped oyster
FR	huître creuse d'Australie
IT	ostrica concava australiana
NL	Sydney-oester
PT	ostra de Sydney
FI	osteri-laji
SV	australiskt ostron

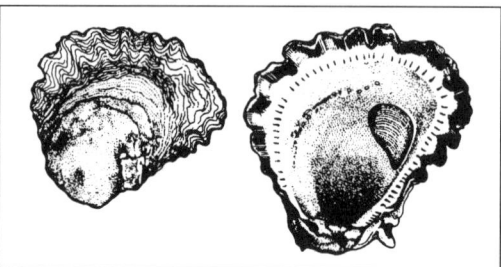

1359 PECTINIDAE

QSC

SC	*Aequipecten opercularis* (Linnaeus, 1758); *Chlamys opercularis*
ES	volandeira
DA	almindelig jomfruøsters
DE	Bunte Kammuschel
EL	χτένι· τηγανάκι
EN	queen scallop
FR	vanneau; pétoncle operculaire
IT	canestrello; pettine
NL	wijde mantel
PT	leque
FI	atlantinkampasimpukka
SV	drottningkammussla

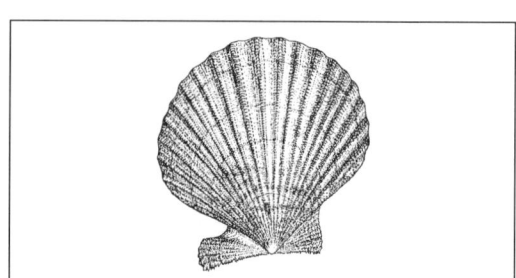

1358 PECTINIDAE

SCX

SC	*Pectinidae*
ES	vieiras; peines
DA	kammusling-familien
DE	Pilgermuscheln; Kammuscheln
EL	χτένια
EN	scallops
FR	coquilles Saint-Jacques; peignes et pétoncles
IT	pettinidi
NL	mantel en kamschelpen
PT	vieiras e leques
FI	kampasimpukat; kampasimpukat-heimo
SV	kammusslor

1360 PECTINIDAE

SCC

SC	*Argopecten gibbus* (Linnaeus, 1758)
ES	peine percal
DA	calico-kammusling
DE	Calico-Pilgermuschel
EL	χτένι
EN	calicot scallop
FR	peigne calicot
IT	canestrello calico
NL	calico-scallop
PT	vieira-percal
FI	kampasimpukka-laji
SV	kalikåkammussla

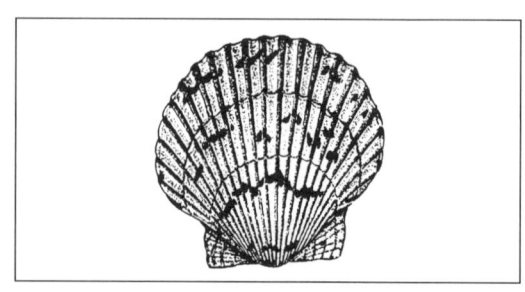

1361 PECTINIDAE   SCB

SC	*Argopecten irradians* (Lamarck, 1819)
ES	peine caletero
DA	amerikansk kammusling
DE	Karibik-Pilgermuschel
EL	αμερικάνικο χτένι
EN	bay scallop
FR	peigne baie de l'Atlantique
IT	canestrello americano
NL	kamschelp
PT	vieira de baía
FI	kampasimpukka-laji
SV	vikkammussla

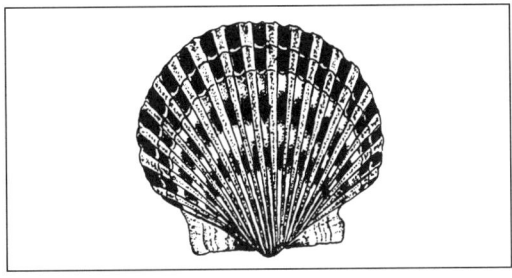

1363 PECTINIDAE   VSC

SC	*Chlamys varia* (Linnaeus, 1758); *Pecten varius*
ES	zamburiña
DA	nordlig jomfruøsters
DE	Bunte Kammuschel
EL	χτένι
EN	variegated scallop
FR	pétoncle
IT	canestrello; pettine
NL	bonte mantel
PT	leque variado
FI	kampasimpukka-laji
SV	brokig kammusla

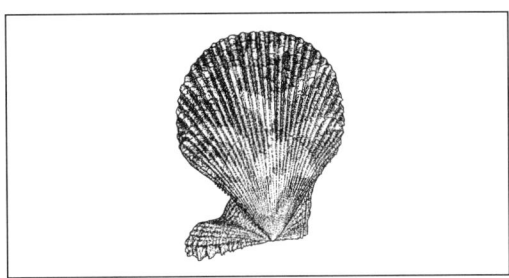

1362 PECTINIDAE   ISC

SC	*Chlamys islandica* (Müller, 1776)
ES	peine islándico
DA	islandsk kammusling
DE	Isländische Kammuschel
EL	χτένι της Ισλανδίας
EN	Icelandic scallop
FR	peigne islandais
IT	canestrello d'Islanda
NL	noordelijke kamschelp
PT	leque islandês
FI	islanninkampasimpukka
SV	isländsk kammussla

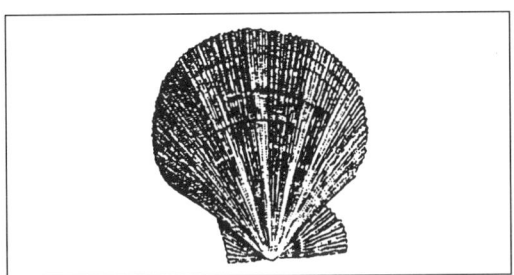

1364 PECTINIDAE   SJA

SC	*Pecten jacobaeus* (Linnaeus, 1758)
ES	concha de peregrino; venera
DA	Jacobsmusling; Jacobsskal
DE	Pilgermuschel; Jakobsmuschel
EL	μεγάλο χτένι
EN	great scallop
FR	coquille Saint-Jacques
IT	cappasanta; conchiglia di San Giacomo
NL	mediterrane Sint-Jacobsschelp
PT	vieira do Mediterrâneo
FI	kampasimpukka-laji
SV	jacobsmussla

1365 PECTINIDAE   SCE

SC	*Pecten maximus* (Linnaeus, 1758)
ES	vieira; aviñeira
DA	kæmpekammusling
DE	Atlantische Pilgermuschel; Große Jakobsmuschel; Große Pilgermuschel; Große Kammuschel
EL	χτένι του Ατλαντικού
EN	common scallop; coquille St-Jacques; scallop
FR	coquille Saint-Jacques
IT	ventaglio; pettine maggiore; conchiglia dei pellegrini; cappasanta atlantica
NL	Sint-Jacobsschelp
PT	vieira
FI	isokampasimpukka
SV	stor kammussla; egentlig pilgrimsmussla

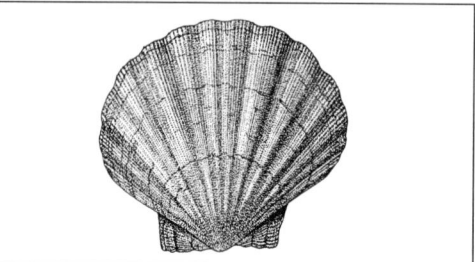

1367 PECTINIDAE   SCZ

SC	*Pecten novaezealandiae* (Reeve, 1852)
ES	vieira de Nueva Zelanda
DA	newzealandsk kammusling
DE	Neuseeländische Jakobsmuschel
EL	χτένι της Νέας Ζηλανδίας
EN	New Zealand scallop
FR	pecten de la Nouvelle-Zélande
IT	cappasanta della Nuova Zelanda
NL	Nieuw-Zeelandse grote mantel
PT	vieira da Nova Zelândia
FI	uudenseelanninkampasimpukka
SV	nyazeeländsk kammussla

DRAWING NOT AVAILABLE

1366 PECTINIDAE   ASC

SC	*Pecten meridionalis* (Tate, 1887)
ES	vieira de Australia
DA	australsk kammusling
DE	Australische Jakobsmuschel
EL	χτένι της Αυστραλίας
EN	Australian scallop
FR	pecten d'Australie
IT	cappasanta australiana
NL	Australische grote mantel
PT	vieira da Austrália
FI	australiankampasimpukka
SV	australisk kammussla

DRAWING NOT AVAILABLE

1368 PECTINIDAE   PSU

SC	*Pecten sulcicostatus* (Sowerby, 1842)
ES	vieira del Atlántico Sur
DA	ribbet kammusling
DE	Südatlantische Kammuschel
EL	χτένι
EN	South Atlantic scallop
FR	pecten de l'Atlantique Sud
IT	cappasanta sudatlantica
NL	Zuid-Atlantische mantel
PT	vieira do Atlântico sul
FI	kampasimpukka-laji
SV	sydatlantisk kammussla

DRAWING NOT AVAILABLE

1369 PECTINIDAE SCA

SC	*Placopecten magellanicus* (Gmelin, 1792)
ES	vieira americana
DA	atlantisk dybhavsmusling
DE	Atlantischer Tiefseescallop
EL	αμερικάνικο χτένι
EN	sea scallop
FR	pecten d'Amérique
IT	cappasanta americana
NL	Amerikaanse grote mantel
PT	vieira americana
FI	kampasimpukka-laji
SV	atlantisk (djupshavs-) mantelkammussla

1371 MYTILIDAE MSC

SC	*Aulacomya ater* (Molina, 1782)
ES	cholga; choro
DA	Magellan-musling
DE	Magellan-Miesmuschel
EL	κόλγκα
EN	Magellan mussel
FR	moule striée
IT	colga
NL	cholgamossel
PT	mexilhão-choro
FI	sinisimpukka-laji
SV	magellanmussla

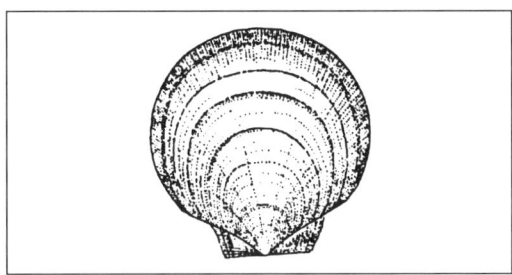

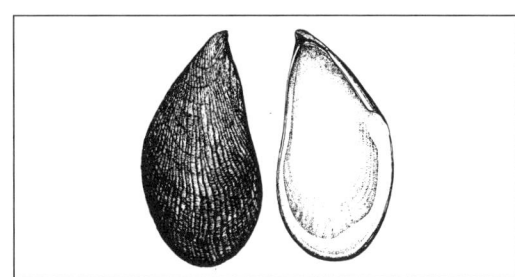

1370 MYTILIDAE MSX

SC	*Mytilidae*
ES	mejillones
DA	blåmusling-familien
DE	Miesmuscheln
EL	μύδια
EN	sea mussels
FR	moules
IT	mitilidi
NL	mosselen
PT	mexilhões
FI	sinisimpukat; sinisimpukat-heimo
SV	blåmusslor

1372 MYTILIDAE

SC	*Lithophaga lithophaga* (Linnaeus, 1758)
ES	dátil de mar
DA	daddelmusling
DE	Meerdattel; Seedattel; Steindattel
EL	δακτύλι· λιθοφάγος· σωλήνας
EN	date shell
FR	datte de mer
IT	dattero di mare
NL	dadelmossel
PT	mexilhão-tâmara
FI	kivitaateli
SV	stendadel

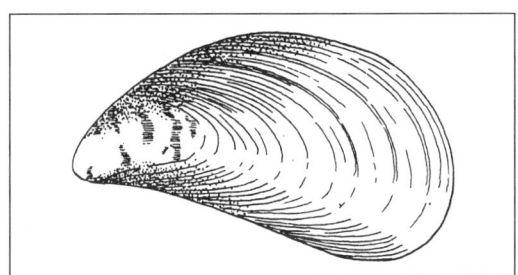

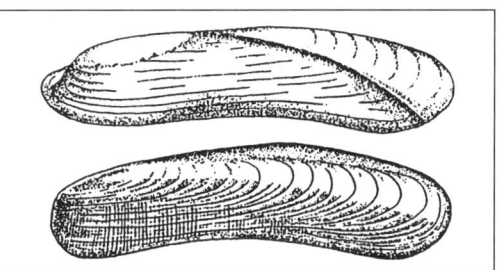

1373 MYTILIDAE

MOD

SC	Modiolus spp.
ES	modiolos
DA	hestemusling-slægt
DE	Bartmuscheln
EL	μυδάκια
EN	horse mussels
FR	modioles
IT	cozze pelose
NL	paardenmosselen
PT	mexilhões pata-de-cavalo
FI	hevossimpukka-suku
SV	hästmusslor; blåmusslor

1375 MYTILIDAE

MUK

SC	Mytilus crassitesta
ES	mejillón coreano
DA	koreansk blåmusling
DE	Koreanische Miesmuschel
EL	μύδι της Κορέας
EN	Korean mussel
FR	moule coréenne
IT	cozza coreana
NL	Koreaanse mossel
PT	mexilhão coreano
FI	koreansinisimpukka
SV	koreansk blåmussla

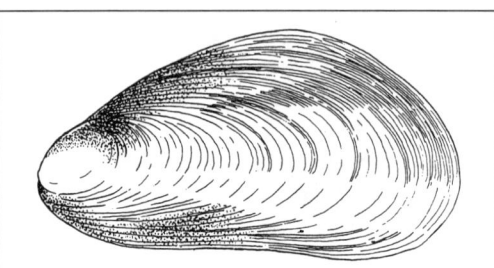

DRAWING NOT AVAILABLE

1374 MYTILIDAE

MYC

SC	Mytilus chilensis (Hupé, 1854)
ES	chorito
DA	chilensk blåmusling
DE	Chilenische Miesmuschel
EL	μύδι της Χιλής
EN	Chilean mussel
FR	moule du Chili; moule chilienne
IT	cozza cilena
NL	Chileense mossel
PT	mexilhão chileno
FI	chilensinisimpukka
SV	chilensk blåmussla

1376 MYTILIDAE

MUS

SC	Mytilus edulis (Linnaeus, 1758)
ES	mejillón; mejillón común
DA	europæisk blåmusling
DE	Miesmuschel; Pfahlmuschel
EL	μύδι
EN	common mussel; blue mussel
FR	moule commune
IT	mitilo; cozza
NL	mossel
PT	mexilhão vulgar
FI	sinisimpukka
SV	blåmussla

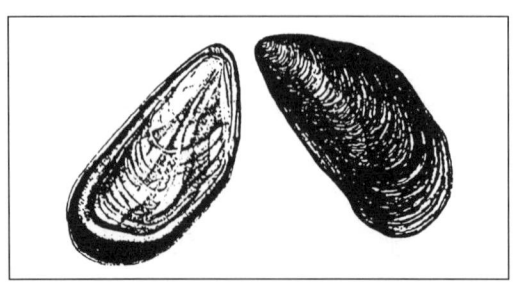

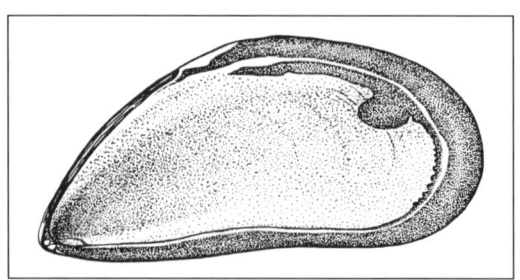

1377 **MYTILIDAE**   MSM

SC	*Mytilus galloprovincialis* (Lamarck, 1819)
ES	mejillón mediterráneo
DA	Middelhavsblåmusling
DE	Mittelmeer-Miesmuschel; Blaubartmuschel; Seemuschel
EL	μύδι
EN	Mediterranean mussel
FR	moule méditerranéenne
IT	cozza; mitilo
NL	Middellandse-Zeemossel
PT	mexilhão do Mediterrâneo
FI	välimerensinisimpukka
SV	medelhavsblåmussla

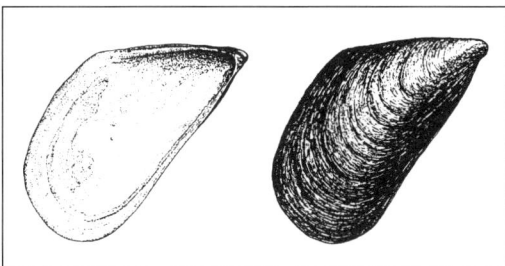

1378 **MYTILIDAE**   MYA

SC	*Mytilus planulatus*
ES	mejillón de Australia
DA	australsk blåmusling
DE	Australische Miesmuschel
EL	μύδι της Αυστραλίας
EN	Australian mussel
FR	moule d'Australie
IT	cozza australiana
NL	Australische mossel
PT	mexilhão da Austrália
FI	australiansinisimpukka
SV	australisk blåmussla

DRAWING NOT AVAILABLE

1379 **MYTILIDAE**   MSR

SC	*Mytilus platensis* (Orbigny, 1846)
ES	mejillón del Plata
DA	La Plata-blåmusling
DE	Rio-de-la-Plata-Miesmuschel
EL	μύδι της Αργεντινής
EN	River Plate mussel
FR	moule de la Plata
IT	cozza del Rio della Plata
NL	Platariviermossel
PT	mexilhão do rio da Prata
FI	platansinisimpukka
SV	riodelaplatablåmussla

DRAWING NOT AVAILABLE

1380 **MYTILIDAE**   MUG

SC	*Mytilus smaragdinus*
ES	mejillón verde
DA	grøn blåmusling
DE	Grüne Miesmuschel
EL	πρασινομύδι
EN	green mussel
FR	moule verte
IT	
NL	groene mossel
PT	mexilhão verde
FI	aasianvihersimpukka
SV	grön blåmussla

DRAWING NOT AVAILABLE

345

1381 MYTILIDAE MUZ

SC *Perna canaliculus* (Gmelin, 1791); *Mytilus canaliculus*
ES mejillón de Nueva Zelanda
DA newzealandsk blåmusling
DE Neuseeland-Miesmuschel
EL μύδι της Νέας Ζηλανδίας
EN New Zealand mussel
FR moule de la Nouvelle-Zélande
IT cozza della Nuova Zelanda
NL Nieuw-Zeelandse mossel
PT mexilhão verde da Nova Zelândia
FI vihersimpukka
SV nyazeeländsk blåmussla

1383 MARGARITANIDAE

SC *Margaritifera margaritifera* (Linnaeus, 1758)
ES mejillón de agua dulce
DA flodperlemusling
DE Flußperlmuschel
EL μαργαριτάρι
EN river pearl mussel
FR huître perlière; mulette
IT madreperea di fiume; margaritifera
NL rivierparelmossel
PT mexilhão perlífero do rio
FI jokihelmisimpukka
SV flodpärlemormussla

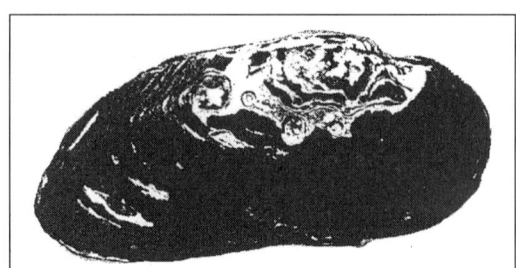

1382 MYTILIDAE MSL

SC *Perna perna* (Linnaeus, 1767)
ES mejillón de roca sudamericano
DA sydamerikansk blåmusling
DE Westatlantische Miesmuschel
EL μύδι των βράχων
EN South American rock mussel
FR moule de roche sud-américaine
IT cozza verde
NL Afrikaanse rotsmossel
PT mexilhão da rocha sul-americano
FI atlantinvihersimpukka
SV sydamerikansk klippmussla

1384 GLYCIMERIDAE

SC *Glycymeris glycymeris* (Linnaeus, 1767); *Pectunculus glycymeris*
ES rabioso
DA hundemusling
DE Gemeine Samtmuschel; Meermandel
EL γαϊδουροχτένι
EN scallop; little neck; dog cockle; comb shell
FR amande de mer; amande marbrée
IT piè d'asino
NL amande; kamschelp
PT castanhola-do-mar
FI samettisimpukka
SV kammussla

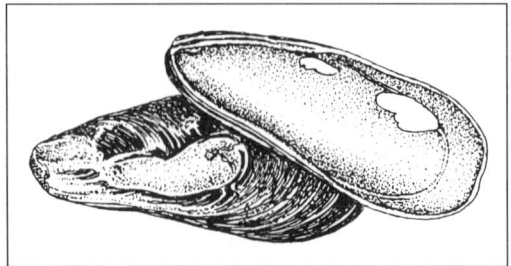

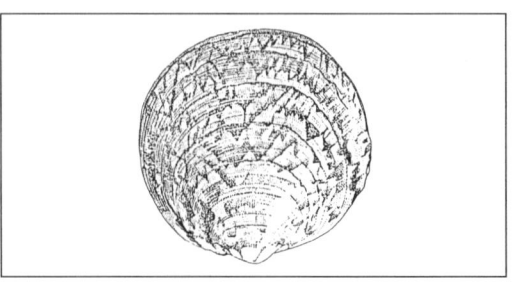

1385 PINNIDAE

SC	*Pinna nobilis* (Linnaeus, 1758)
ES	nácar
DA	skinkemusling
DE	Steckmuschel; Schinkenmuschel
EL	πίνα
EN	ham mussel
FR	jambonneau de mer; pinna; pinne géante
IT	pinna
NL	hammossel
PT	funil escamudo
FI	jalosilkkisimpukka
SV	skinkmussla

1387 CARDIIDAE   COZ

SC	*Cardiidae*
ES	berberechos; cárdidos
DA	hjertemusling-familien
DE	Herzmuscheln
EL	μεθύστρες
EN	clams; cockles
FR	coques
IT	cardidi; cuori
NL	kokkels
PT	berbigões
FI	sydänsimpukat; sydänsimpukat-heimo
SV	hjärtmusslor

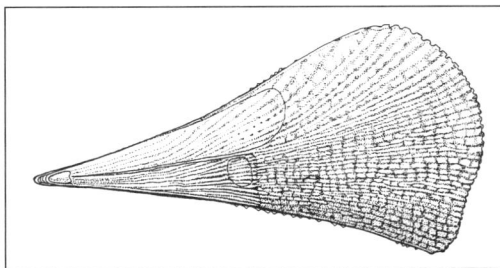

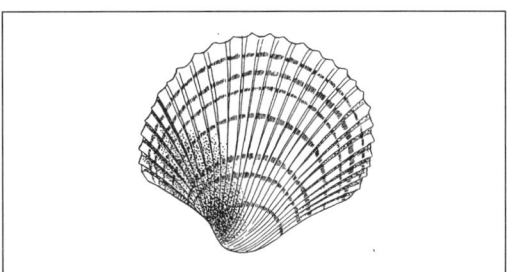

1386 ARCTICIDAE   CLQ

SC	*Arctica islandica* (Linnaeus, 1767); *Cyprina islandica*
ES	almeja de Islandia
DA	molboøsters
DE	Islandmuschel
EL	
EN	ocean quahaug; ocean quahog; cyprine; Iceland cyprine
FR	praire d'Islande
IT	cappa artica
NL	noordkromp
PT	clame islandesa
FI	islanninsimpukka
SV	islandsmussla

1388 CARDIIDAE   DIA

SC	*Acanthocardia aculeata* (Linnaeus, 1758); *Cardium aculeatum*
ES	marolo; berberecho espinoso
DA	tornet hjertemusling
DE	Stachelige Herzmuschel
EL	αγκαθομεθύστρα
EN	spiny cockle
FR	soudron-coque; bucarde épineuse
IT	cuore spinoso; cuore
NL	gedoornde hartschelp
PT	berbigão de bicos
FI	piikkisydänsimpukka
SV	taggig hjärtmussla

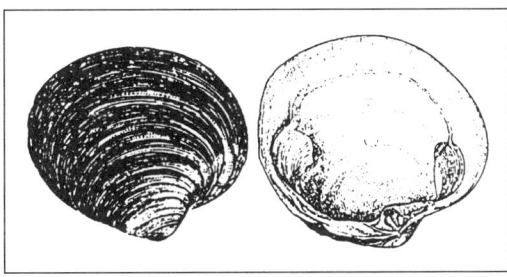

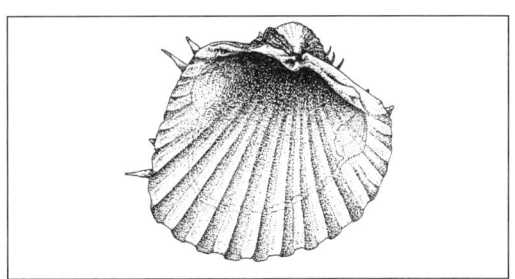

1389 CARDIIDAE COC

SC	*Cerastoderma edule* (Linnaeus, 1758); *Cardium edule*
ES	berberecho; verdigón; gurrimaño; berberecho común; gurrimaña; perdigón; chica
DA	hjertemusling; almindelig hjertemusling
DE	Herzmuschel
EL	μεθύστρα
EN	common cockle
FR	coque; coque commune; bucarde
IT	cuore edule; cuore
NL	kokkel
PT	berbigão vulgar
FI	sydänsimpukka
SV	ätlig hjärtmussla

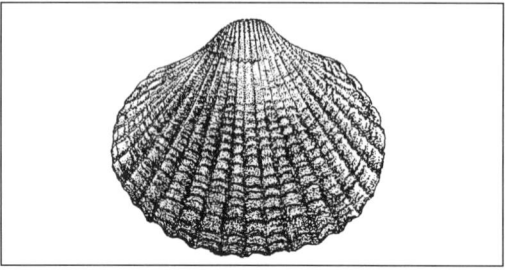

1391 MACTRIDAE HCL

SC	*Mactra sachalinensis*
ES	almeja lisa falsa
DA	Sakhalin-trugmusling
DE	Sachalin-Trogmuschel
EL	μικρή γυαλιστερή
EN	hen clam
FR	mactre; fausse palourde; fausse praire
IT	madia giapponese
NL	Sachalin strandschelp
PT	ameijola japonesa
FI	vinohammassimpukka-laji
SV	hönmussla

DRAWING NOT AVAILABLE

1390 MACTRIDAE MAG

SC	*Mactra glabrata* (Linnaeus, 1758)
ES	mactra lisa
DA	glat trugmusling
DE	Glatte Mactra
EL	γυαλιστερή
EN	smooth mactra
FR	mactre lisse
IT	madia glabra
NL	gladde strandschelp
PT	ameijola lisa
FI	vinohammassimpukka-laji
SV	glattmussla

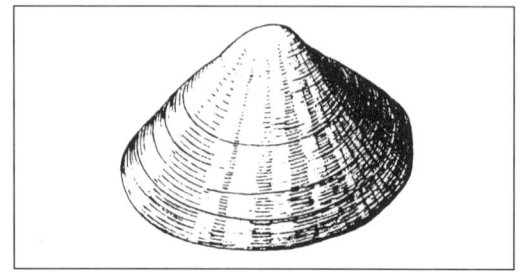

1392 MACTRIDAE CLB

SC	*Spisula solidissima* (Dillwyn, 1817)
ES	almeja blanca
DA	tykskallet trugmusling
DE	Riesentrogmuschel
EL	γυαλιστερή της Αμερικής
EN	surf clam
FR	mactre solide
IT	cappa americana
NL	stevige strandschelp
PT	amêijoa branca americana
FI	vinohammassimpukka-laji
SV	bränningsmussla

1393 VENERIDAE   CLV

SC	*Veneridae*
ES	almejas; venéridos
DA	venusmusling-familien
DE	Venusmuscheln
EL	κυδώνια
EN	venus clams
FR	praires; palourdes; vénus; clovisses
IT	veneridi
NL	venusschelpen
PT	venerídeos
FI	venussimpukat; venussimpukat-heimo
SV	venusmusslor

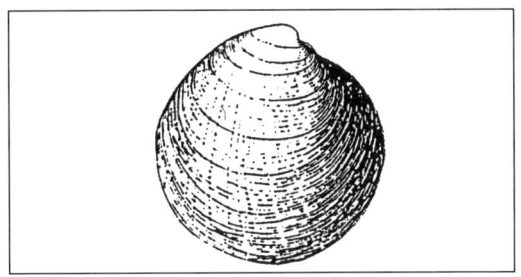

1395 VENERIDAE

SC	*Callista chione* (Linnaeus, 1758); *Meretrix chione*
ES	almejón de sangre; almejón brillante
DA	glat callista
DE	Glatte Venusmuschel; Braune Venusmuschel
EL	γυαλιστερή
EN	hard clam
FR	verni; coque rouge; palourde rouge; pelote
IT	fasolaro; cappa chione
NL	bruine venusschelp
PT	clame dura
FI	venussimpukka-laji
SV	brun venusmussla

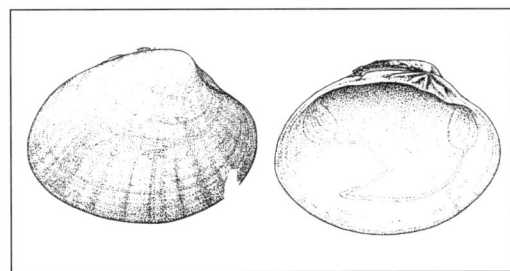

1394 VENERIDAE   TPS

SC	*Tapes* spp.
ES	almejas
DA	tæppemusling-slægt
DE	Teppichmuscheln
EL	κοχύλι
EN	carpet shells
FR	clovisses
IT	vongole
NL	tapijtschelpsoorten
PT	amêijoas
FI	venussimpukka-suku
SV	tapesmusslor

DRAWING NOT AVAILABLE

1396 VENERIDAE   SVE

SC	*Chamelea gallina* (Linnaeus, 1758); *Venus gallina*
ES	chirla
DA	stribet venusmusling
DE	Gestreifte Venusmuschel
EL	κυδώνι
EN	striped venus
FR	petite praire
IT	vongola
NL	venusschelp
PT	pé-de-burrinho
FI	raitavenussimpukka
SV	randig venusmussla

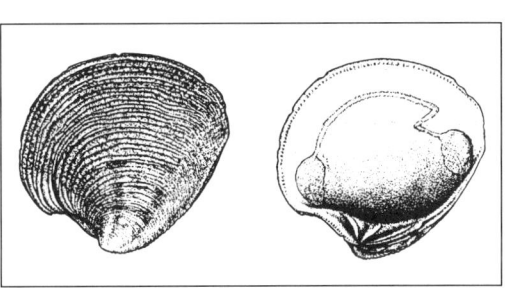

1397 VENERIDAE

DOR

SC	*Dosinia orbignyi* (Dunker, 1845)
ES	almeja de Orbignyi
DA	afrikansk venusmusling
DE	Orbignys Dosinia
EL	
EN	Orbigny mussel
FR	montre d'Orbigny
IT	lupino africano
NL	Orbigny-artemisschelp
PT	amêijoa redonda de Angola
FI	venussimpukka-laji
SV	d'Orbignys venusmussla

1399 VENERIDAE

HCJ

SC	*Meretrix lusoria* (Röding, 1798)
ES	mercenaria japonesa
DA	japansk musling
DE	Japanische Venusmuschel
EL	γιαπωνέζικη αχιβάδα
EN	Japanese hard clam
FR	vernis du Japon
IT	fasolaro orientale
NL	Japanse venusschelp
PT	clame dura japonesa
FI	japaninvenussimpukka
SV	japansk venusmussla

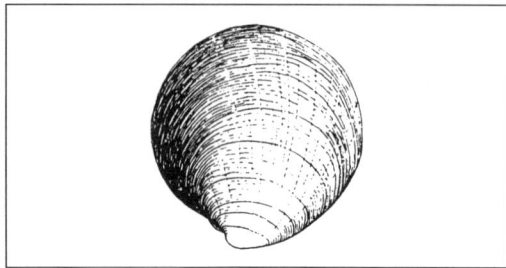

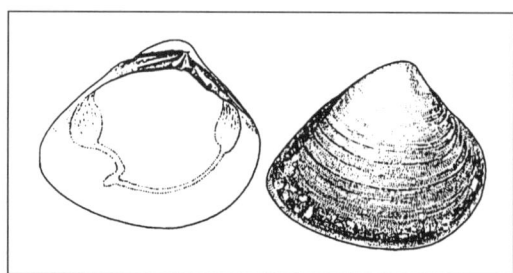

1398 VENERIDAE

CLH

SC	*Mercenaria mercenaria* (Linnaeus, 1758); *Venus mercenaria*
ES	mercenaria
DA	quahog-venusmusling
DE	Nördliche Venusmuschel; Venusmuschel; QuahogMuschel
EL	σκληρή γυαλιστερή
EN	hard clam; quahaug; quahog; round clam
FR	praire
IT	cappa dura
NL	Amerikaanse venusschelp; clam
PT	clame; clame redonda
FI	kampavenussimpukka
SV	hård venusmussla

1400 VENERIDAE

HCX

SC	*Meretrix* spp.
ES	mercenarias duras
DA	musling-slægt
DE	Venusmuscheln
EL	αχιβάδες
EN	hard clams
FR	vernis
IT	fasolari
NL	venusschelpen
PT	clames duras
FI	venussimpukka-suku
SV	venusmusslor

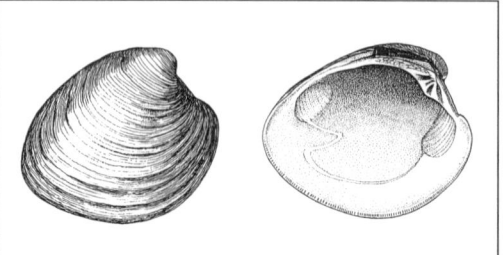

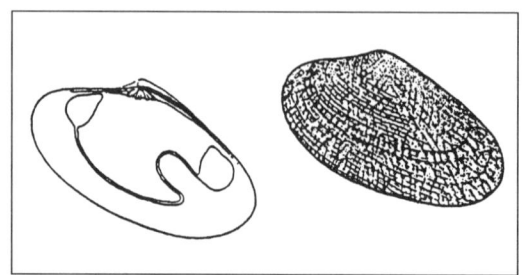

1401 VENERIDAE CTG

SC	*Ruditapes decussatus* (Linnaeus, 1758); *Venerupis decussata*; *Tapes decussa*
ES	almeja fina; amayuela; escupiña lisa; chirlía; petxina negra
DA	stor tæppemusling
DE	Große Teppichmuschel; Teppichmuschel; Kreuzmuster-Teppichmuschel
EL	κυδώνι· χάβαρο
EN	grooved carpetshell; carpetshell; clam; calico clam
FR	palourde
IT	vongola verace; vongola nera
NL	tapijtschelp
PT	amêijoa boa; amêijoa cristã
FI	mattosimpukka
SV	stor venusmussla

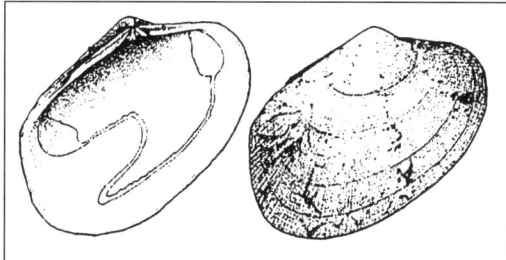

1403 VENERIDAE BCL

SC	*Saxidomus giganteus* (Deshayes, 1839)
ES	almeja amarilla
DA	Alaska-musling
DE	
EL	αχιβάδα της Αλάσκας
EN	butter clam
FR	coque jaune; clam de l'Alaska
IT	cappa dell'Alaska
NL	boterschelp
PT	clame amarela
FI	venussimpukka-laji
SV	smörvenusmussla

DRAWING NOT AVAILABLE

1402 VENERIDAE CLJ

SC	*Ruditapes philippinarum* (Adams and Reeve, 1850); *Tapes japonica*; *Tapes variegata*; *Tapes semidecussatus*
ES	almeja japonesa
DA	japansk tæppemusling
DE	Japanische Teppichmuschel
EL	κυδώνι της Ιαπωνίας
EN	short-necked clam; Japanese clam; Manilla clam
FR	palourde japonaise; clam japonais
IT	vongola verace
NL	Japanse tapijtschelp
PT	amêijoa japonesa
FI	japaninmattosimpukka
SV	japansk venusmussla

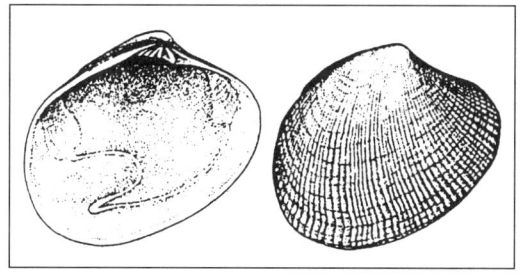

1404 VENERIDAE

SC	*Venerupis aurea* (Gmelin, 1791); *Tapes aureus*; *Paphia aurea*
ES	almeja dorada
DA	gylden tæppemusling
DE	Goldene Teppichmuschel
EL	κοχύλι
EN	golden carpetshell
FR	clovisse jaune; palourde dorée
IT	vongola; longona
NL	gouden tapijtschelp
PT	amêijoa bicuda
FI	kultamattosimpukka
SV	gyllene venusmussla

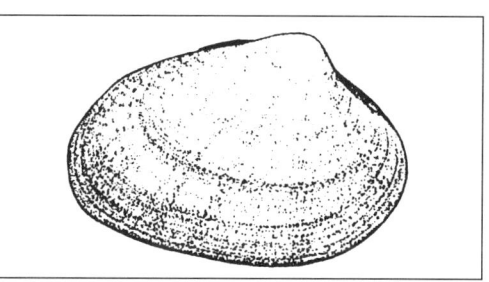

1405 VENERIDAE

CTS

SC	*Venerupis pullastra* (Montagu, 1803); *Tapes pullastra*
ES	almeja babosa; chocha; amayuela; chirlía; petxina negra
DA	almindelig tæppemusling
DE	Kleine Teppichmuschel; Teppichmuschel
EL	αχιβάδα
EN	carpetshell
FR	clovisse
IT	vongola; longona
NL	kleine tapijtschelp
PT	amêijoa macha; amêijoa judia
FI	pikkumattosimpukka
SV	liten venusmussla

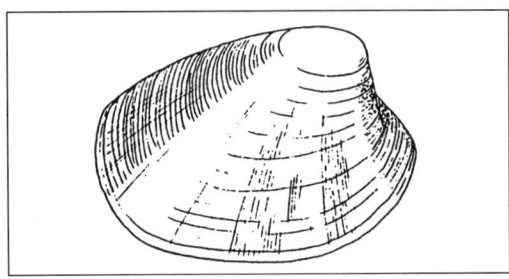

1407 SOLENIDAE

SOI

SC	*Solenidae*
ES	navajas
DA	knivmusling-familien
DE	Scheidenmuscheln; Meerscheiden
EL	σωλήνες
EN	razor clams; razor shells; knife clams
FR	couteaux; solens
IT	cannolicchi; manicai; cappelunghe
NL	messcheden
PT	longueirões; canivetes; facas; lingueirões; navalhas
FI	veitsisimpukat; veitsisimpukat-heimo
SV	knivmusslor

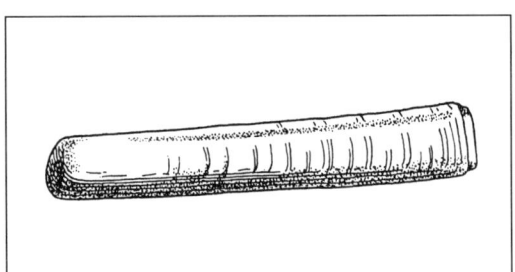

1406 DONACIDAE

DON

SC	*Donax* spp.
ES	coquinas
DA	kilemusling-slægt
DE	Sägezähnchen; Dreiecksmuscheln; Koffermuscheln
EL	κοχύλι
EN	donax clams
FR	olives de mer; cébettes
IT	telline
NL	zaagjes
PT	cadelinhas; conquilhas
FI	liemisimpukka-suku
SV	knivmusslor

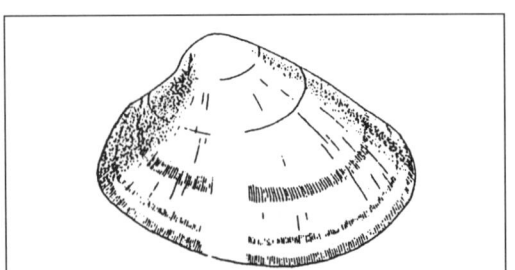

1408 SOLENIDAE

CLR

SC	*Ensis directus* (Conrad, 1843)
ES	navaja del Atlántico; solen del Atlántico
DA	amerikansk knivmusling
DE	Amerikanische Schwertmuschel
EL	σωλήνας του Ατλαντικού
EN	Atlantic razor clam
FR	couteau de l'Atlantique
IT	cannolicchio dell'Atlantico
NL	Amerikaanse zwaardschede
PT	longueirão da América do Norte
FI	veitsisimpukka-laji
SV	atlantisk knivmussla

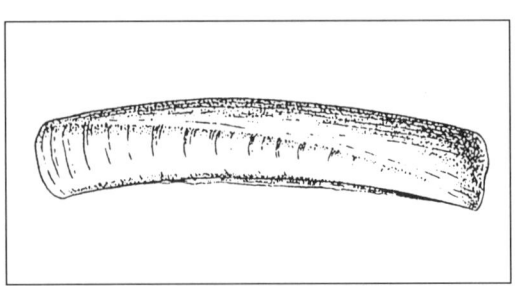

1409 SOLENIDAE

SC   *Ensis ensis* (Linnaeus, 1758); *Solen ensis*
ES   navaja
DA   lille knivmusling
DE   Schwertmuschel
EL   σωλήνας
EN   pod razor
FR   couteau courbé
IT   cappalunga
NL   kleine zwaardschede
PT   longueirão curvo
FI   veitsisimpukka
SV   böjd knivmussla

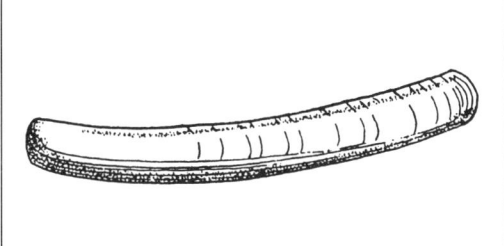

1411 SOLENIDAE   RAP

SC   *Siliqua patula* (Dixon, 1789)
ES   navaja del Pacífico
DA   pacifisk knivmusling
DE   Pazifische Messermuschel
EL   σωλήνας του Ειρηνικού
EN   Pacific razor clam
FR   couteau du Pacifique
IT   cannolicchio del Pacifico
NL   Pacifische messchede
PT   longueirão do Alasca
FI   veitsisimpukka-laji
SV   stillahavsknivmussla

DRAWING NOT AVAILABLE

1410 SOLENIDAE

SC   *Ensis siliqua* (Linnaeus, 1758)
ES   muergo
DA   barberknivmusling
DE   Messermuschel; Taschenmessermuschel
EL   στραβολοσωλήνας
EN   sword razor
FR   couteau droit
IT   cannolicchio
NL   tafelmesheft
PT   longueirão direito
FI   taskuveitsisimpukka
SV   rakknivsmussla

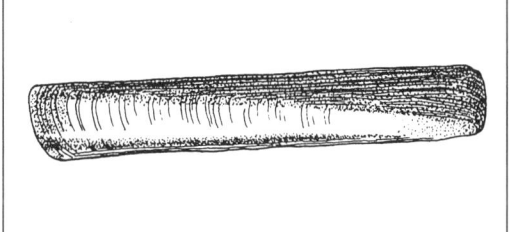

1412 SOLENIDAE   RAC

SC   *Solen capensis*
ES   navaja del Cabo
DA   Kap-knivmusling
DE   Kap-Messermuschel
EL   σωλήνας του Ακρωτηρίου
EN   Cape razor clam
FR   couteau du Cap
IT   cannolicchio sudafricano
NL   Kaapse messchede
PT   longueirão do Cabo
FI   kapinveitsisimpukka
SV   kapknivmussla

DRAWING NOT AVAILABLE

1413 SOLENIDAE RAE

SC	*Solen vagina* (Linnaeus, 1758)
ES	navaja europea
DA	europæisk knivmusling
DE	Scheidenmuschel; Meerscheide; Große Messermuschel
EL	σωλήνας της Ευρώπης
EN	European razor clam
FR	couteau d'Europe
IT	cannolicchio; cappalunga
NL	grote messchede
PT	longueirão direito europeu
FI	veitsisimpukka-laji
SV	europeisk knivmussla

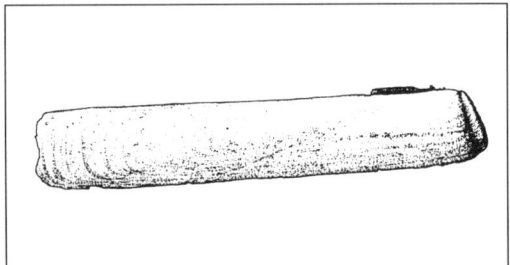

1414 SOLENIDAE RAZ

SC	*Solen* spp.
ES	navajas; longeirones; muergos
DA	knivmusling-slægt
DE	Meerscheiden; Scheidenmuscheln
EL	σωλήνες
EN	razor clams
FR	couteaux; manches de couteaux
IT	cannolicchi; cappelunghe
NL	messcheden; scheermessen
PT	longueirões
FI	veitsisimpukka-suku
SV	knivmusslor

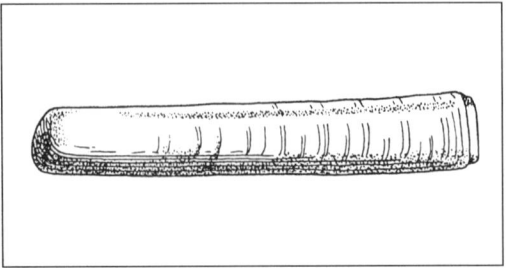

1415 MYIDAE CLS

SC	*Mya arenaria* (Linnaeus, 1758)
ES	almeja de río; almeja de can
DA	sandmusling
DE	Sandklaffmuschel; Sandmuschel; Schlickauster; Strandauster; Große Sandklaffmuschel
EL	αμμοκοχύλι
EN	clam; soft clam; soft-shell clam; long clam; gaper; longneck; mananose; old maid; sandgaper
FR	mye; clauque; bec de jars
IT	cappa molle
NL	grote strandgaper
PT	clame da areia
FI	hietasimpukka
SV	sandmussla

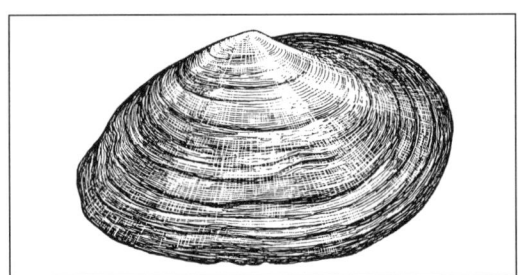

1416 PHOLADIDAE

SC	*Pholas dactylus* (Linnaeus, 1758)
ES	barrena
DA	almindelig boremusling
DE	Gemeine Bohrmuschel
EL	δάκτυλο
EN	common piddock
FR	pholade; datte de mer
IT	folade; dattero bianco
NL	boormossel
PT	taralhão
FI	nävertäjäsimpukka
SV	självlysande borrmussla

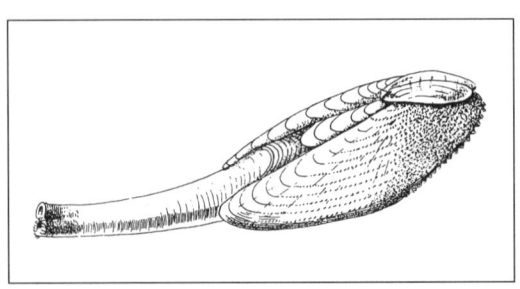

1417 PHOLADIDAE

SC	*Pholas* spp.
ES	barrena; mangones
DA	boremusling-slægt
DE	Bohrmuscheln; Dattelmuscheln
EL	δάκτυλα
EN	piddocks
FR	gites; pholades
IT	foladi
NL	boormosselen
PT	taralhões
FI	nävertäjäsimpukka-suku
SV	borrmusslor

1419 SEPIOLIDAE

SC	*Rossia macrosoma* (Delle Chiaje, 1829)
ES	sepiola; globito
DA	Ross' blæksprutte
DE	Große Rossie
EL	σουπιά
EN	Ross' cuttle
FR	sépiole melon
IT	seppiola grossa
NL	Ross' dwerginktvis
PT	chopo
FI	välimerensepia
SV	stor sepia

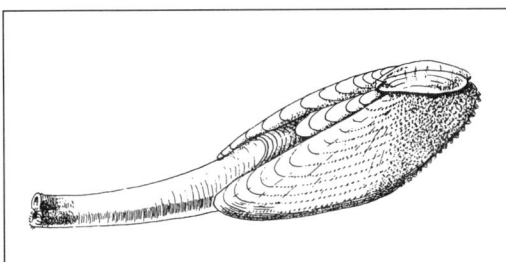

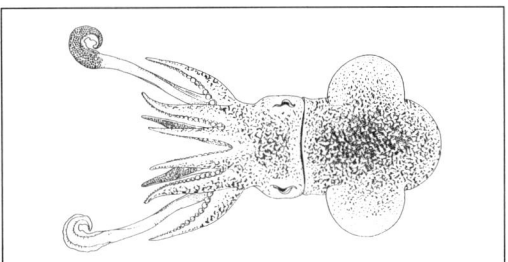

1418 SEPIIDAE CTC

SC	*Sepia officinalis* (Linnaeus, 1758)
ES	jibia; sepia común
DA	sepiablæksprutte
DE	Gemeiner Tintenfisch; Tintenfisch; Sepia; Sepie
EL	σουπιά
EN	common cuttlefish; cuttlefish
FR	seiche; seiche commune
IT	seppia; seppia mediterranea
NL	gewone zeekat
PT	choco vulgar; choco
FI	yleinen mustekala; sepia
SV	sepiabläckfisk

1420 SEPIOLIDAE

SC	*Sepiola rondeleti* (Steenstrup, 1856)
ES	globito
DA	sepiola-blæksprutte
DE	Mittelmeer-Sepiole; Zwergtintenfisch
EL	σουπιά· σουπίτσα
EN	lesser cuttlefish; little cuttlefish
FR	sépiole; sépion; souchet
IT	seppiola
NL	kleine zeekat
PT	chopo anão
FI	sepiola
SV	dvärgbläckfisk

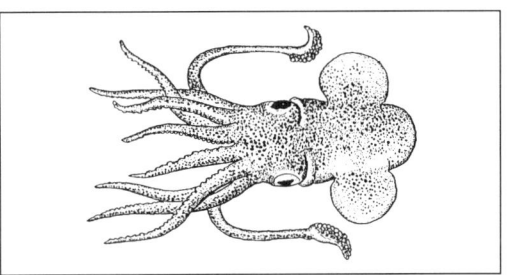

1421 LOLIGINIDAE, OMMASTREPHIDAE
SQU

SC	*Loliginidae, Ommastrephidae*
ES	calamares, jibias, potas
DA	blæksprutte-familier
DE	Kalmare, Tintenfische
EL	καλαμάρια
EN	squids
FR	calmars, encornets
IT	calamari, totani
NL	pijlinktvisachtigen
PT	lulas, potas
FI	kalmarit; kalmarit-heimo
SV	bläckfiskar

1423 LOLIGINIDAE

SC	*Alloteuthis media* (Linnaeus, 1758)
ES	calamarín; puntilla
DA	dværgblæksprutte
DE	Kleiner Kalmar; Mittelländischer Zwergkalmar; Großkeuliger Zwergkalmar
EL	καλαμαράκι· τέφτης
EN	little squid
FR	petit encornet; casseron-bambou
IT	totariello; calamaretto
NL	Middellandse-Zee kleine pijlinktvis
PT	lula bicuda menor
FI	pikkukalmari
SV	liten bläckfisk

DRAWING NOT AVAILABLE

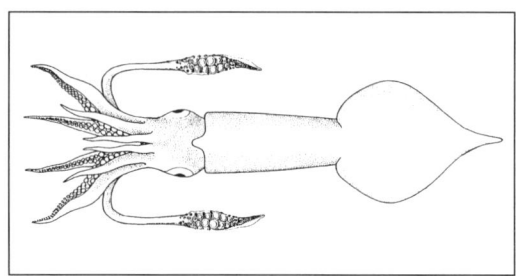

1422 LOLIGINIDAE   SQZ

SC	*Loliginidae*
ES	calamares
DA	tiarmet blæksprutte-familie
DE	Kalmare
EL	καλαμάρια
EN	inshore squids
FR	calmars
IT	loliginidi
NL	pijlinktvissen
PT	lulas
FI	kalmarit
SV	kalamarer; tioarmade bläckfiskar; strutar

1424 LOLIGINIDAE   SQL

SC	*Loligo pealei* (Lesueur, 1821)
ES	calamar de Boston
DA	langfinnet loligo
DE	Langflossen-Schelfkalmar
EL	καλαμάρι
EN	long-fin squid
FR	calmar-totam
IT	calamaro
NL	langvinpijlinktvis
PT	lula pálida
FI	kalmari-laji
SV	långfenad bläckfisk

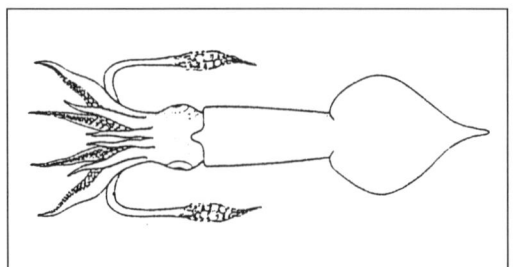

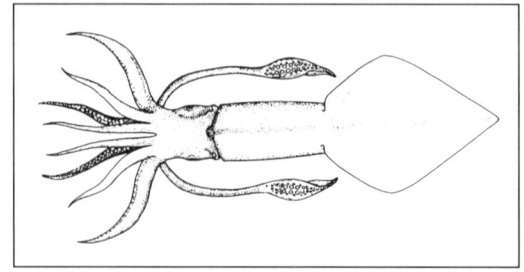

1425 LOLIGINIDAE CHO

SC	*Loligo reynaudi* (Orbigny, 1845)
ES	calamar del Cabo
DA	Kap-loligo
DE	Kap-Kalmar
EL	καλαμάρι του Ακρωτηρίου
EN	Cape Hope squid; chokker squid
FR	calmar du Cap
IT	calamaro del Sudafrica
NL	Kaapse pijlinktvis
PT	lula do Cabo
FI	kapinkalmari
SV	kapbläckfisk

1427 OMMASTREPHIDAE GIS

SC	*Dosidicus gigas* (Orbigny, 1835)
ES	jibia gigante
DA	ocean-blæksprutte
DE	Riesenkalmar
EL	ιπτάμενο καλαμάρι
EN	jumbo flying squid
FR	encornet géant
IT	totano gigante del Pacifico
NL	reuzenpijlinktvis
PT	pota gigante
FI	kalmari-laji
SV	jumbobläckfisk

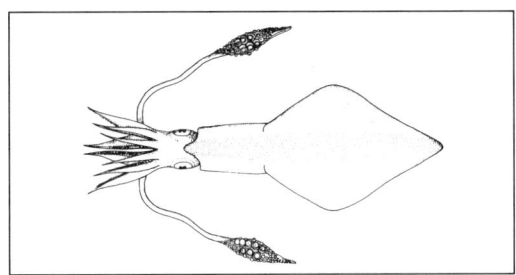

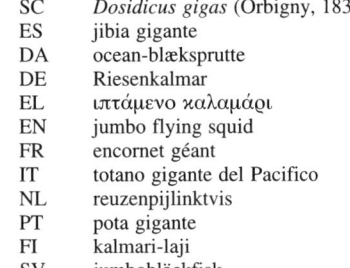

1426 LOLIGINIDAE

SC	*Loligo vulgaris* (Lamarck, 1798)
ES	calamar
DA	europæisk loligo
DE	Gewöhnlicher Kalmar; Kalmar; Tintenfisch
EL	καλαμάρι· τευθίς
EN	common squid; squid; inkfish; sea arrow
FR	calmar; calmar commun
IT	calamaro; calamaro mediterraneo
NL	gewone pijlinktvis
PT	lula vulgar; lula
FI	kalmari
SV	vanlig bläckfisk; havspil

1428 OMMASTREPHIDAE SQA

SC	*Illex argentinus* (Castellanos, 1960)
ES	pota argentina
DA	argentinsk blæksprutte
DE	Argentinischer Kurzflossenkalmar
EL	θράψαλο της Αργεντινής
EN	Argentine short-fin squid
FR	encornet rouge argentin
IT	totano
NL	Argentijnse rode pijlinktvis
PT	pota argentina
FI	argentiinankalmari
SV	argentinsk bläckfisk

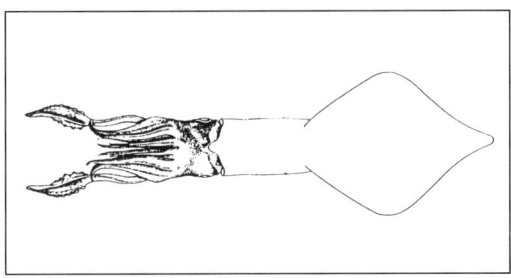

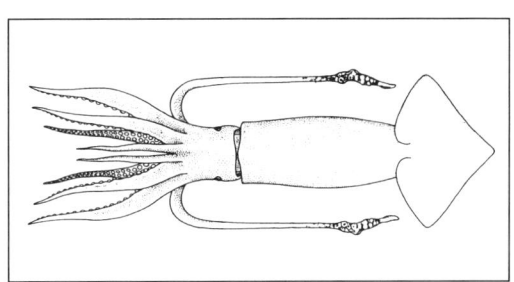

1429 OMMASTREPHIDAE SQM

SC	*Illex coindetii* (Verany, 1839)
ES	pota voladora
DA	rød blæksprutte
DE	Roter Kalmar
EL	θράψαλο
EN	broad-tail shortfin squid
FR	encornet rouge
IT	totano
NL	rode pijlinktvis
PT	pota voadora
FI	kalmari-laji
SV	bredstjärtfenad bläckfisk

1431 OMMASTREPHIDAE TSQ

SC	*Nototodarus sloani* (Gray, 1849)
ES	pota neozelandesa
DA	newzealandsk blæksprutte
DE	Wellington-Flugkalmar
EL	θράψαλο της Νέας Ζηλανδίας
EN	Wellington flying squid
FR	encornet minami
IT	totano neozelandese
NL	Nieuw-Zeelandse pijlinktvis
PT	pota da Nova Zêlandia
FI	kalmari-laji
SV	nyzeeländsk bläckfisk

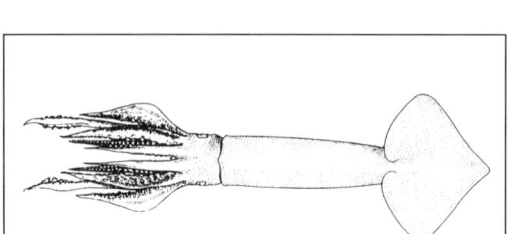

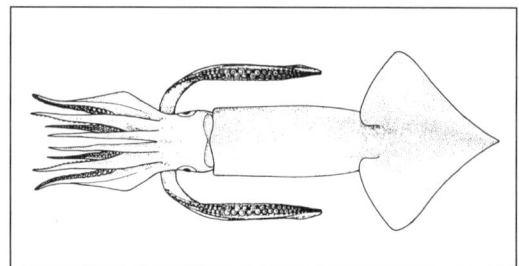

1430 OMMASTREPHIDAE SQI

SC	*Illex illecebrosus* (Lesueur, 1821)
ES	pota
DA	nordlig blæksprutte
DE	Nördlicher Kurzflossenkalmar
EL	θράψαλο του Βορρά
EN	shortfin squid; Northern squid
FR	encornet rouge nordique
IT	totano
NL	kortvinpijlinktvis
PT	pota do norte
FI	kalmari-laji
SV	nordlig stjärtfenad bläckfisk

1432 OMMASTREPHIDAE SQJ

SC	*Todarodes pacificus* (Steenstrup, 1880)
ES	pota japonesa
DA	japansk blæksprutte
DE	Japanischer Flugkalmar
EL	θράψαλο του Ειρηνικού
EN	Japanese flying squid
FR	toutenon japonais; calmar japonais
IT	totano giapponese
NL	Japanse vliegende pijlinktvis
PT	pota japonesa
FI	japaninliitokalmari
SV	japansk bläckfisk

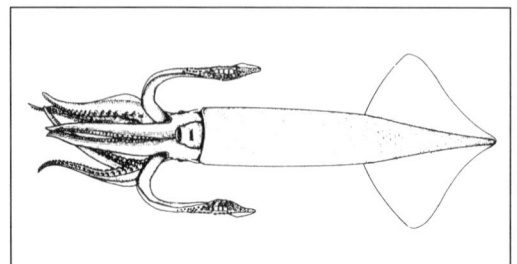

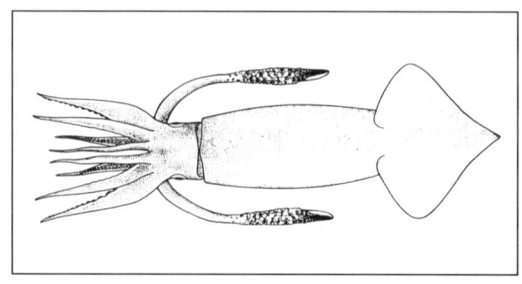

1433 OMMASTREPHIDAE   SQE

SC	*Todarodes sagittatus sagittatus* (Lamarck, 1798); *Ommastrephes sagittatus*
ES	pota europea
DA	flyveblæksprutte
DE	Pfeilkalmar
EL	θράψαλο· καλαμαράκι
EN	flying squid; sea squid; red squid; European flying squid
FR	calmar; encornet; toutenon commun
IT	totano; todaro
NL	grote pijlinktvis
PT	pota europeia; potra europeia
FI	euroopanliitokalmari
SV	flygbläckfisk

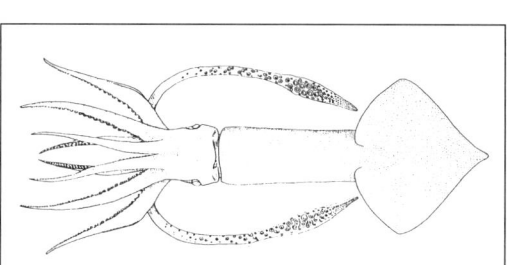

1435 OCTOPODIDAE

SC	*Eledone cirrosa* (Lamarck, 1798)
ES	cabezón; pulpo blanco
DA	eledoneblæksprutte
DE	Zirrenkrake; Kleiner Krake
EL	αλιδώνα· μοσχοχτάποδο· οχταπόδι
EN	curled octopus
FR	élédone commune
IT	moscardino; moscardino bianco
NL	kleine octopus
PT	polvo do alto
FI	pikkumyskitursas
SV	åttaarmad bläckfisk

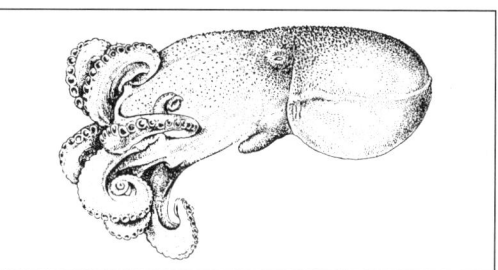

1434 OCTOPODIDAE   OCT

SC	*Octopodidae*
ES	pulpitos; pulpos
DA	ottearmet blæksprutte-familie
DE	Oktopusartige
EL	καλαμάρια της Νέας Ζηλανδίας
EN	octopuses
FR	pieuvres; poulpes
IT	ottopodi
NL	achtarmige inktvissen
PT	polvos
FI	myskitursaat; myskitursaat-heimo
SV	åttaarmade bläckfiskar

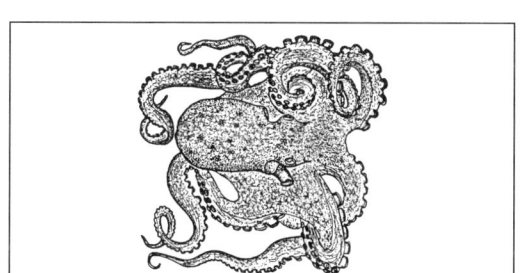

1436 OCTOPODIDAE

SC	*Eledone moschata* (Lamarck, 1798)
ES	pulpo amizclado
DA	moskusblæksprutte
DE	Moschuskrake; Moschuspolyp
EL	μοσχιός· μοσχοχτάποδο
EN	white octopus
FR	élédone musquée; élédone de la Méditerranée
IT	moscardino
NL	muskusoctopus
PT	polvo mosqueado
FI	myskitursas
SV	vit bläckfisk

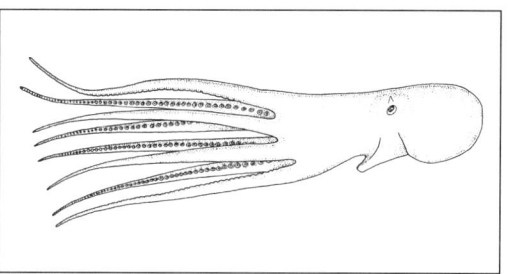

1437 OCTOPODIDAE   OCM

SC	*Eledone* spp.
ES	pulpos blancos
DA	ottearmet blæksprutte-slægt
DE	Zirrenkraken; Moschuskraken
EL	μοσχιοί· μοσχοχτάποδα
EN	horned and musky octopuses
FR	élédones
IT	moscardini
NL	muskusoctopussen
PT	polvos do alto
FI	myskitursas-suku
SV	åttaarmade bläckfiskar

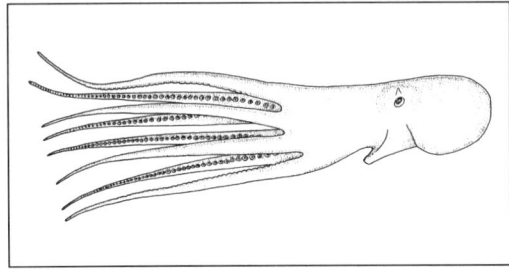

1439 OCTOPODIDAE   OCZ

SC	*Octopus* spp.
ES	pulpos
DA	ottearmet blæksprutte-slægt
DE	Kraken
EL	χταπόδι
EN	octopuses
FR	poulpes; pieuvres
IT	polpi
NL	octopussen
PT	polvos
FI	myskitursas-suku
SV	åttaarmade bläckfiskar

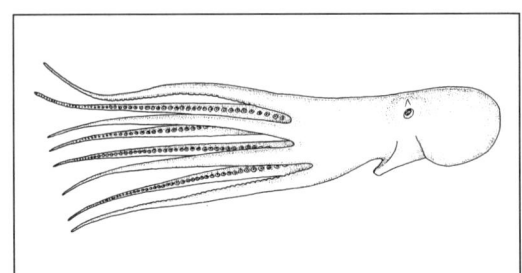

1438 OCTOPODIDAE   OCC

SC	*Octopus vulgaris* (Cuvier, 1797)
ES	pulpo común; pulpo
DA	almindelig ottearmet blæksprutte
DE	Gewöhnlicher Krake; Gemeiner Tintenfisch; Krake; Tintenfisch
EL	χταπόδι· οκτάπους
EN	common octopus; octopus
FR	poulpe; pieuvre commune
IT	polpo di scoglio; polpo
NL	octopus
PT	polvo vulgar; polvo
FI	meritursas
SV	vanlig åttaarmad bläckfisk

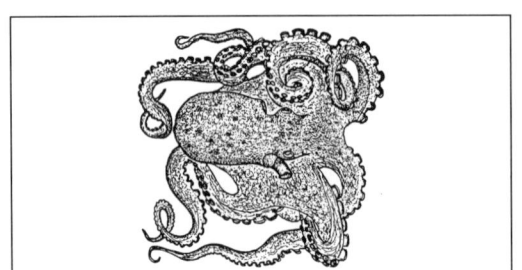

1440 OTARIIDAE   SXX

SC	*Otariidae*
ES	focas
DA	øresæler
DE	Ohrenrobben; Hundsrobben
EL	φώκιες
EN	sea lions; eared seals; fur seals
FR	otaries
IT	otarie
NL	oorrobben
PT	leões-marinhos
FI	turkishylkeet; turkishylkeet-heimo
SV	öronsälar

1441 OTARIIDAE

SEF

SC	*Arctocephalus australis* (Zimmermann, 1783)
ES	lobo marino de dos pelos
DA	sydamerikansk søbjørn
DE	Südamerikanischer Seebär
EL	φώκια της Νότιας Αμερικής
EN	South American fur seal
FR	otarie à fourrure des Falklands; otarie australe
IT	otaria da pelliccia sudamericana
NL	zuidelijke pelsrob
PT	leão-marinho de dois pêlos; otária de dois pêlos
FI	eteläamerikanmerikarhu
SV	sydamerikansk pälssäl

1443 OTARIIDAE

SC	*Eumetopias jubatus* (Schreber, 1776); *Otaria jubata*; *Emmenopias steller*
ES	león de mar
DA	Stellers søløve; nordlig søløve
DE	Stellerscher Seelöwe; Stellers Seelöwe
EL	λιοντάρι της θάλασσας
EN	Steller sea lion; Northern sea lion; sea king
FR	lion de mer à crinière; lion de mer de Steller; otarie à crinière
IT	otaria; leone marino di Steller
NL	zeeleeuw
PT	leão-marinho de juba
FI	stellerinmerileijona
SV	nordligt sjölejon; Stellers sjölejon

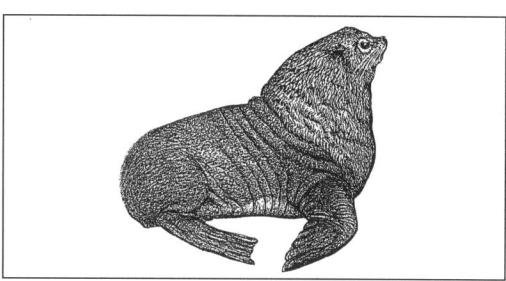

1442 OTARIIDAE

SEK

SC	*Arctocephalus pusillus* (Schreber, 1776)
ES	lobo marino del Cabo
DA	sydafrikansk søbjørn
DE	Südafrikanischer Seebär
EL	φώκια του Ακρωτηρίου
EN	Cape fur seal; South African fur seal
FR	otarie à fourrure du Cap; otarie d'Afrique du Sud
IT	otaria da pelliccia sudafricana
NL	dwergzeebeer; Zuid-Afrikaanse zeeleeuw
PT	leão-marinho do Cabo; otária do Cabo
FI	afrikanmerikarhu
SV	sydafrikansk pälssäl

1444 OTARIIDAE

SEL

SC	*Otaria flavescens* (Blainville, 1820)
ES	lobo marino de un pelo
DA	sydamerikansk søløve
DE	Südamerikanischer Seelöwe
EL	λιοντάρι της θάλασσας
EN	South American sea lion
FR	otarie à crinière; lion de mer d'Amérique du Sud
IT	leone marino sudamericano
NL	Zuid-Amerikaanse zeeleeuw
PT	leão-marinho de um pêlo; otária de um pêlo
FI	eteläamerikanmerileijona
SV	sydamerikanskt sjölejon; mansäl

1445 ODOBENIDAE

SC	*Odobenus rosmarus* (Linnaeus, 1758)
ES	morsa
DA	hvalros
DE	Walroß
EL	
EN	walrus
FR	morse
IT	tricheco
NL	walrus
PT	morsa
FI	mursu
SV	valross

1447 PHOCIDAE · SEZ

SC	*Cystophora cristata* (Erxleben, 1777)
ES	foca capuchina
DA	klapmyds
DE	Klappmütze
EL	κουκουλοφόρος φώκια
EN	hooded seal
FR	phoque à capuchon; phoque à crête
IT	foca dal cappuccio
NL	klapmuts
PT	foca de mitra
FI	kuplahylje
SV	klappmyts; blåssäl

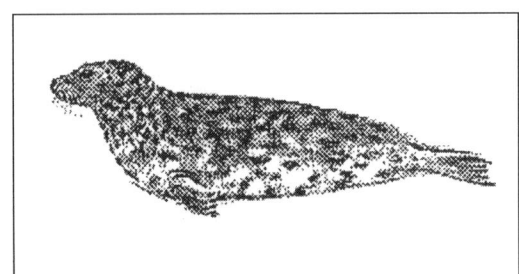

1446 PHOCIDAE · SXX

SC	*Phocidae*
ES	focas
DA	ægte sæler
DE	Hundsrobben
EL	φώκιες
EN	seals
FR	phoques
IT	foche
NL	zeehonden
PT	focas
FI	hylkeet; hylkeet-heimo
SV	egentliga sälar

1448 PHOCIDAE · SEB

SC	*Erignathus barbatus* (Erxleben, 1777)
ES	foca barbuda
DA	remmesæl
DE	Bartrobbe
EL	γενειοφόρος φώκια
EN	bearded seal
FR	phoque barbu
IT	foca barbuta
NL	baardrob
PT	foca barbuda
FI	partahylje
SV	storsäl

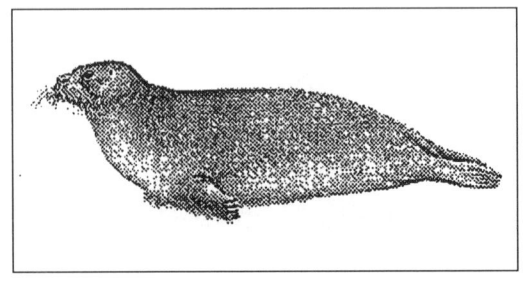

1449 PHOCIDAE   SEG

SC   *Halichoerus grypus* (Fabricius, 1791)
ES   foca gris
DA   gråsæl
DE   Kegelrobbe
EL   γκρίζα φώκια
EN   grey seal
FR   phoque gris
IT   foca grigia
NL   grijze zeehond
PT   foca cinzenta
FI   harmaahylje; halli
SV   gråsäl

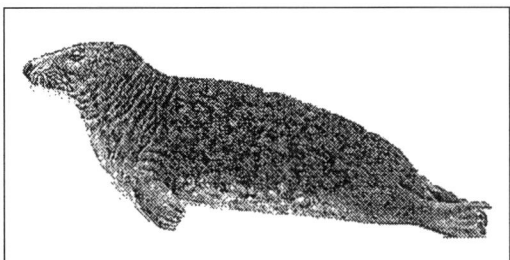

1450 PHOCIDAE

SC   *Monachus monachus* (Hermann, 1779); *Monachus albiventer*
ES   foca fraile
DA   munkesæl
DE   Mittelmeer-Mönchsrobbe; Seemönch; Seejungfrau
EL   φώκια
EN   Mediterranean monk seal; pied monk seal
FR   phoque-moine; phoque-moine à ventre blanc
IT   foca monaca
NL   monniksrob
PT   lobo-marinho; foca-monga
FI   munkkihylje
SV   havsmunk

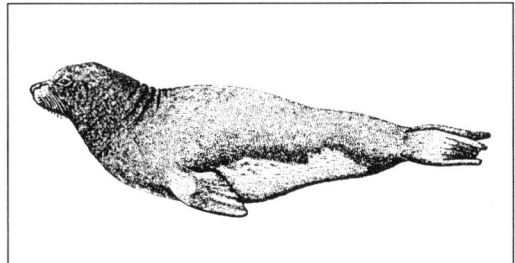

1451 PHOCIDAE   SEH

SC   *Pagophilus groenlandicus* (Erxleben, 1777)
ES   foca de Groenlandia
DA   grønlandssæl
DE   Sattelrobbe
EL   φώκια της Γροιλανδίας
EN   harp seal
FR   phoque du Groenland; phoque marin; phoque à selle
IT   foca di Groenlandia
NL   zadelrob
PT   foca da Gronelândia
FI   grönlanninhylje
SV   grönlandssäl

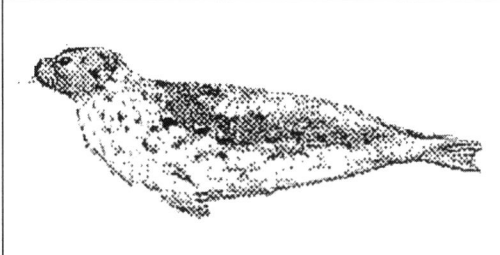

1452 PHOCIDAE   SER

SC   *Phoca hispida* (Schreber, 1775); *Pusa hispida*
ES   foca anillada
DA   ringsæl
DE   Eismeer-Ringelrobbe
EL   φώκια
EN   ringed seal
FR   phoque marbré; phoque annelé
IT   foca dagli anelli
NL   ringelrob
PT   foca marmoreada
FI   norppa
SV   vikaresäl

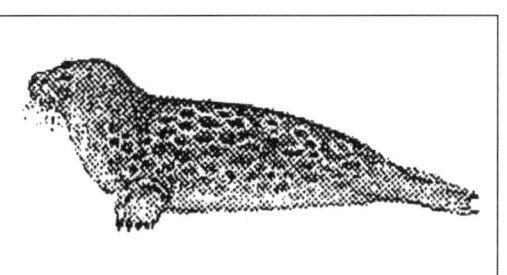

1453 PHOCIDAE SEC

SC	*Phoca vitulina* (Linnaeus, 1758)
ES	foca común
DA	spættet sæl
DE	Seehund
EL	φώκια
EN	common seal; harbour seal
FR	phoque commun; phoque-veau marin
IT	foca comune; foca
NL	zeehond
PT	foca vulgar
FI	kirjohylje
SV	knubbsäl

1455 TRICHECHIDAE SEW

SC	*Trichechus inunguis* (Natterer, 1883)
ES	vaca marina; manatí
DA	søko
DE	Fluß-Manati
EL	θαλάσσια αγελάδα
EN	sea-cow; manatee
FR	lamantin de l'Amazonie
IT	manato senza unghie; lamantino dell'Amazzonia
NL	Amazone-lamentijn
PT	manatim do Amazonas
FI	kynnetönmanaatti
SV	amazonmanat; amazon-sjöko

1454 DUGONGIDAE

SC	*Halicore dugong* (Erxleben, 1777); *Dugong dugong*
ES	dugongo
DA	dygong
DE	Dugong; Gabelschwanzseekuh
EL	ντουγκόνγκο
EN	dugong
FR	dugong
IT	dugongo
NL	dugong
PT	dugongue
FI	dugongi
SV	dugong

1456 TRICHECHIDAE

SC	*Trichechus manatus* (Linnaeus, 1758)
ES	manatí común
DA	lamantin
DE	Nagel-Manati
EL	θαλάσσια αγελάδα
EN	manatee
FR	lamantin d'Amérique
IT	manato comune; vacca marina; lamantino americano
NL	Amerikaanse lamentijn
PT	manatim das Caraíbas
FI	manaatti
SV	lamantin

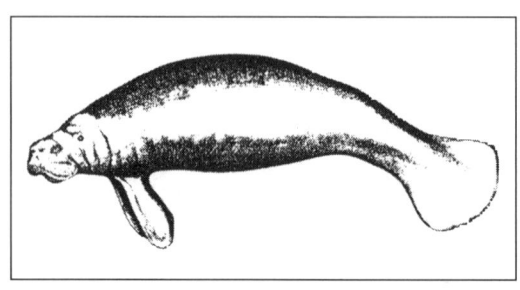

1457 TRICHECHIDAE

SC	*Trichechus senegalensis* (Link, 1795)
ES	manatí africano
DA	afrikansk manat
DE	Afrikanischer Manati
EL	θαλάσσια αγελάδα της Αφρικής
EN	African manatee
FR	lamantin d'Afrique; lamantin du Sénégal
IT	lamantino africano; manato africano
NL	Afrikaanse lamentijn
PT	manatim africano
FI	afrikanmanaatti
SV	senegalmanat

1459 ZIPHIIDAE BEW

SC	*Berardius bairdii* (Stejneger,1883)
ES	zífidos
DA	Bairds næbhval
DE	Baird-Wal
EL	ρυγχοφάλαινα
EN	beaked whale
FR	baleine de Baird
IT	berardo
NL	zwarte dolfijn
PT	baleia bicuda de Baird
FI	pohjoisennelihammasvalas
SV	Bairds näbbval

DRAWING NOT AVAILABLE

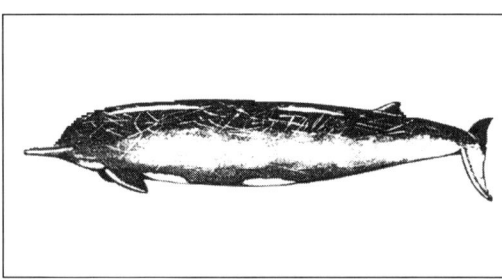

1458 ZIPHIIDAE

SC	*Ziphiidae*
ES	cifios; zífidos
DA	næbhvaler; næbhval-familien
DE	Spitzschnauzen-Delphine
EL	ζίφιοι
EN	beaked whales
FR	baleines à bec; baleine de Cuvier; baleine bécune de Cuvier
IT	zifidi
NL	snaveldolfijnen
PT	zifídeos
FI	nokkavalaat; nokkavalaat-heimo
SV	näbbvalar

1460 ZIPHIIDAE BOW

SC	*Hyperoodon ampullatus* (Ferster, 1770); *Hyperoodon rostratus*
ES	ballena hocico de botella
DA	døgling; nordlig døgling
DE	Entenwal; Nördlicher Entenwal
EL	φάλαινα
EN	bottlenosed whale
FR	hyperoodon boréal
IT	iperodonte
NL	butskop
PT	bico-de-garrafa
FI	pohjoisenpullokuonovalas
SV	vanlig näbbval; dögling

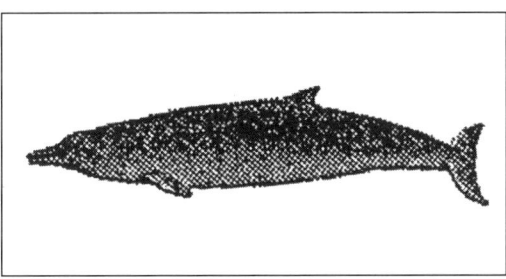

1461 ZIPHIIDAE

SC	*Ziphius cavirostris* (Cuvier, 1823)
ES	cifio vivíparo
DA	småhovedet hval
DE	Cuvier-Schnabelwal
EL	ζίφιος
EN	beaked whale
FR	baleine à bec d'oie; ziphius
IT	zifio
NL	spitsdolfijn van Cuvier
PT	bico-de-pato
FI	hanhennokkavalas
SV	småhuvudval

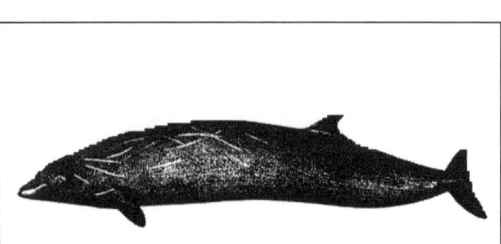

1462 PHYSETERIDAE   SPW

SC	*Physeter macrocephalus* (Linnaeus, 1758); *Physeter catodon*
ES	cachalote
DA	kaskelot
DE	Pottwal
EL	φυσητήρας
EN	sperm whale; cachalot; pot whale
FR	cachalot
IT	capodoglio
NL	potvis
PT	cachalote
FI	kaskelotti
SV	kaskelot; pottval

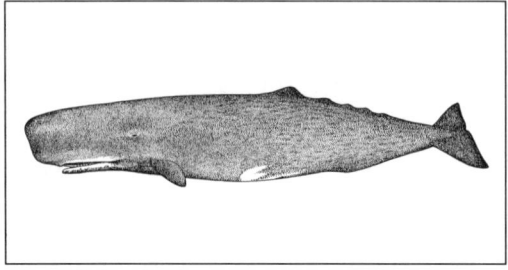

1463 DELPHINIDAE   DLP

SC	*Delphinidae*
ES	delfínidos
DA	delfin-familien
DE	Delphine
EL	δελφίνια
EN	dolphins
FR	dauphins
IT	delfinidi
NL	dolfijnen
PT	delfinídeos
FI	delfiinit; delfiinit-heimo
SV	delfiner

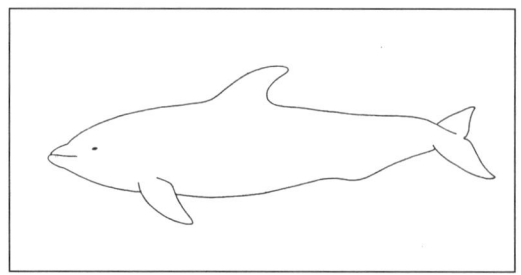

1464 DELPHINIDAE   DCO

SC	*Delphinus delphis* (Linnaeus, 1758)
ES	delfín común
DA	delfin; almindelig delfin
DE	Gemeiner Delphin
EL	δελφίνι
EN	common dolphin
FR	dauphin commun
IT	delfino comune; delfino
NL	gewone dolfijn
PT	golfinho
FI	delfiini
SV	delfin; springare

1465 DELPHINIDAE SHW

SC	*Globicephala macrorhyncha* (Gray, 1846)
ES	calderón tropical
DA	kortluffet grindehval
DE	Indischer Grindwal
EL	δελφίνι-πιλότος
EN	short-fin pilot whale
FR	globicéphale tropical; globicéphale à nageoire courte; globicéphale de Siebold
IT	delfino pilota indiano
NL	Indische griend
PT	caldeirão
FI	lyhyteväpallopää
SV	indisk grindval

1467 DELPHINIDAE DRR

SC	*Grampus griseus* (Cuvier, 1812)
ES	calderón gris
DA	halvgrindehval
DE	Rundkopfdelphin; Rissos-Delphin; Gramper
EL	στακτοδέλφινο
EN	Risso's dolphin
FR	dauphin de Risso; grampus
IT	grampo; delfino di Risso
NL	grijze dolfijn
PT	boto raiado
FI	harmaadelfiini
SV	Rissos delfin

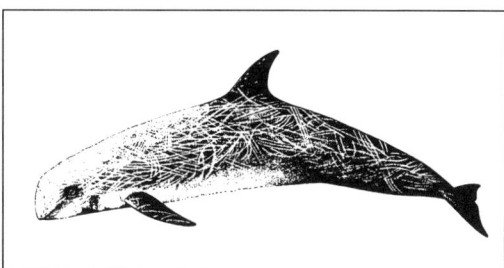

1466 DELPHINIDAE PIW

SC	*Globicephala melaena* (Traill, 1809)
ES	calderón
DA	grindehval; langluffet grindehval
DE	Grindwal; Gewöhnlicher Grindwal
EL	μαυροδέλφινο
EN	pilot whale; long-fin pilot whale
FR	globicéphale noir
IT	globicefalo; delfino pilota
NL	griend
PT	boca-de-panela
FI	pitkäeväpallopää; pallopäävalas
SV	grindval; svartval; pilotval

1468 DELPHINIDAE

SC	*Lagenorhynchus acutus* (Gray, 1828)
ES	delfín de lomo blanco
DA	hvideside; hvidskæving
DE	Weißseiten-Delphin
EL	ασπροπλευροδέλφινο
EN	white-sided dolphin
FR	lagénorhynque à flancs blancs
IT	delfino a fianchi bianchi
NL	witflankdolfijn
PT	golfinho branco do Atlântico
FI	valkokuvedelfiini
SV	vitsiding

1469 DELPHINIDAE

SC	*Lagenorhynchus albirostris* (Gray, 1846)
ES	delfin de hocico blanco
DA	hvidnæse
DE	Weißschnauziger Delphin; Weißschnauzen-Delphin; Weißschnauziger Springer
EL	ασπρορυγχοδέλφινο
EN	white-beaked dolphin
FR	lagénorhynque à bec blanc
IT	delfino muso bianco
NL	witsnuitdolfijn
PT	golfinho-focinho branco
FI	valkokuonodelfiini
SV	vitnos

1471 DELPHINIDAE   PHR

SC	*Phocoena phocoena* (Linnaeus, 1758)
ES	marsopa
DA	marsvin
DE	Kleiner Tümmler; Schweinswal
EL	φώκια· μαρσουάνος
EN	porpoise; harbour porpoise
FR	marsouin; cochon de mer
IT	focena; marsuino
NL	bruinvis
PT	boto
FI	pyöriäinen
SV	tumlare

1470 DELPHINIDAE   KIW

SC	*Orcinus orca* (Linnaeus, 1758); *Orca gladiator*
ES	orca
DA	spækhugger
DE	Schwertwal
EL	όρκα· φάλαινα δολοφόνος· σπαθοδέλφινο
EN	killer whale
FR	orque; épaulard
IT	orca
NL	zwaardwalvis; orka
PT	orca; orca gladiadora
FI	miekkavalas
SV	späckhuggare

1472 DELPHINIDAE

SC	*Phocoenoides dalli* (True, 1885)
ES	marsopa de Dall
DA	Dalls marsvin
DE	Dalls Hafenschweinswal
EL	φώκια
EN	Dall's porpoise
FR	marsouin de Dall
IT	marsuino di Dall
NL	Dalls dolfijn
PT	marsopa de Porto Dall
FI	suihkupyöriäinen
SV	stillahavstumlare

1473 DELPHINIDAE

SC	*Pseudorca crassidens* (Owen, 1846)
ES	orca negra; falsa orca
DA	halvspækhugger
DE	Kleiner Schwertwal; Unechter Schwertwal
EL	ψευτόρκα· μαύρη όρκα
EN	false killer whale
FR	faux orque; orque noire
IT	pseudorca
NL	zwarte zwaardwalvis
PT	falsa orca
FI	pikkumiekkavalas
SV	halvspäckhuggare

1475 DELPHINIDAE   DSI

SC	*Stenella longirostris* (Gray, 1828)
ES	delfín de pico largo
DA	langnæbbet delfin
DE	Langschnauzen-Delphin
EL	ακανθοδέλφινο
EN	spinner dolphin
FR	dauphin à long bec
IT	stenella rostrata
NL	langsnuitdolfijn
PT	golfinho fiandeiro
FI	pitkäkuonodelfiini
SV	långnäbbad delfin

1474 DELPHINIDAE   DST

SC	*Stenella coeruleoalba* (Meyen, 1833)
ES	delfín azul; delfín listado
DA	stribet delfin
DE	Blauweißer Delphin
EL	ζωνοδέλφινο
EN	striped dolphin
FR	dauphin bleu et blanc
IT	stenella striata
NL	gestreepte dolfijn
PT	golfinho riscado
FI	raitadelfiini
SV	strimmig delfin

1476 DELPHINIDAE   DSP

SC	*Stenella* spp.
ES	delfines moteados
DA	delfin-slægt
DE	Fleckendelphine
EL	δελφίνια
EN	spotted dolphins
FR	dauphins pélagiques; stenelles
IT	stenelle
NL	gevlekte dolfijnen
PT	golfinhos malhados
FI	delfiini-suku
SV	delfiner

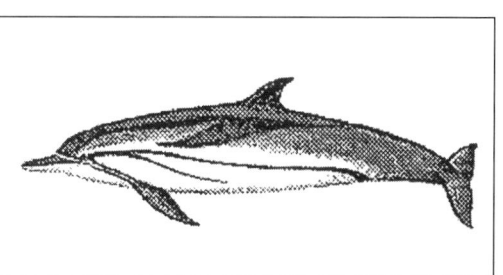

1477 DELPHINIDAE   DBO

SC	*Tursiops truncatus* (Montagu, 1821)
ES	mular
DA	øresvin
DE	Großer Tümmler; Tümmler
EL	ρινοδέλφινο
EN	bottle-nose dolphin
FR	souffleur; grand dauphin
IT	tursiope; tursione; delfino maggiore
NL	tuimelaar
PT	roaz corvineiro
FI	pullokuonodelfiini
SV	öresvin; flasknosdelfin

1479 MONODONTIDAE   NAR

SC	*Monodon monoceros* (Linnaeus, 1758)
ES	narval
DA	narhval
DE	Narwal
EL	μονόκερως· ναρβάλ
EN	narwhale; unicorn whale
FR	narval
IT	narvalo
NL	narwal
PT	narval
FI	sarvivalas
SV	narval

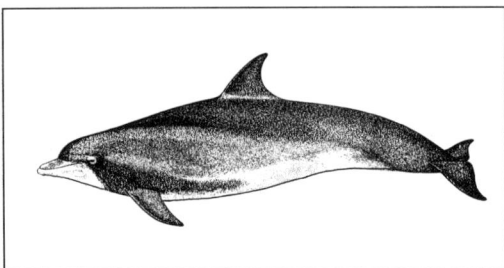

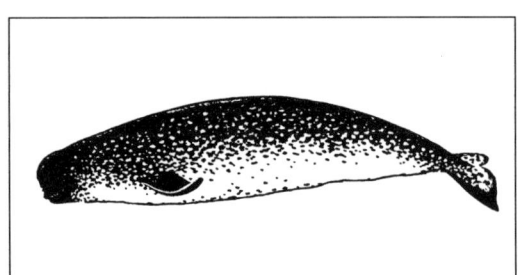

1478 MONODONTIDAE   BEL

SC	*Delphinapterus leucas* (Pallas, 1776)
ES	beluga; ballena blanca
DA	hvidhval
DE	Weißwal; Beluga
EL	ασπροδέλφινο
EN	white whale; beluga whale
FR	dauphin blanc; bélouga
IT	beluga
NL	beluga
PT	golfinho branco
FI	maitovalas; beluga
SV	vitval; beluga

1480 BALAENOPTERIDAE   MIW

SC	*Balaenoptera acutorostrata* (Lacepède, 1804)
ES	ballena enana; ballenato; rocual aliblanco
DA	vågehval
DE	Zwergwal
EL	ρυγχοφάλαινα
EN	minke whale; Davidson's whale; lesser rorqual; little piked whale; pikeheaded whale; sharp headed finner whale
FR	petit rorqual
IT	balenottera rostrata; balena minore
NL	dwergvinvis
PT	baleia anã
FI	lahtivalas
SV	vikval

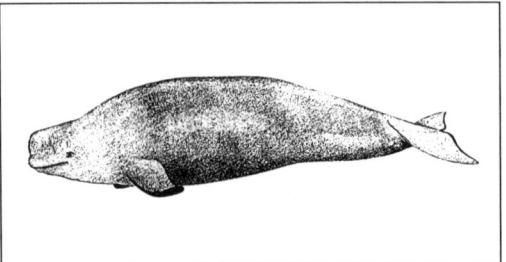

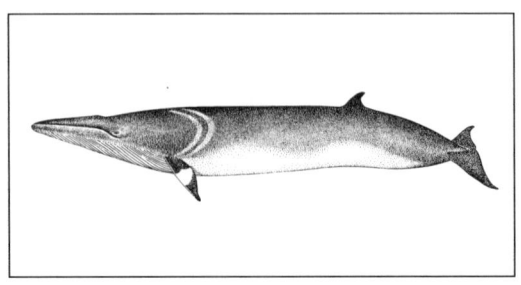

1481 BALAENOPTERIDAE   SIW

SC	*Balaenoptera borealis* (Lesson, 1828)
ES	rorcual negro; ballena boba; rorcual norteño
DA	sejhval
DE	Seiwal
EL	αρκτοφάλαινα
EN	sei whale; pollack whale; coalfish whale; Rudolph's rorqual
FR	rorqual boréal; rorqual de Rudolf; baleinoptère boréal; baleinoptère de Rudolf
IT	balenottera boreale
NL	noordse vinvis
PT	baleia boreal
FI	seitivalas
SV	sejval

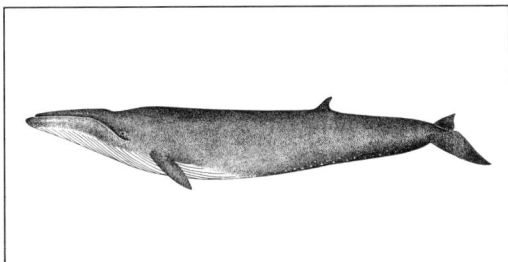

1483 BALAENOPTERIDAE   BLW

SC	*Balaenoptera musculus* (Linnaeus, 1758)
ES	ballena azul; rorcual azul
DA	blåhval
DE	Blauwal
EL	γαλάζια φάλαινα
EN	blue whale
FR	baleine bleue; rorqual bleu
IT	balenottera azzurra
NL	blauwe vinvis
PT	baleia azul; grande rorqual; rorqual azul
FI	sinivalas
SV	blåval

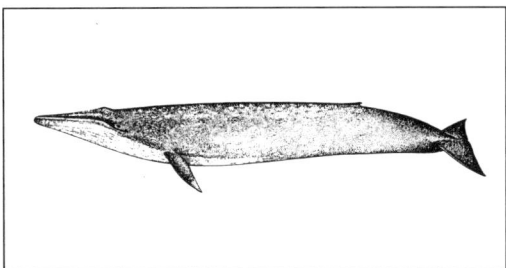

1482 BALAENOPTERIDAE   BRW

SC	*Balaenoptera edeni* (Anderson, 1878)
ES	rorcual de Bryde; ballena de Bryde
DA	Brydes hval
DE	Brydewal
EL	φάλαινα του Μπρυντ
EN	Bryde's whale
FR	baleine de Bryde; rorqual de Bryde
IT	balenottera di Bryde
NL	Bryde's vinvis
PT	baleia de Braide; rorqual de Braide
FI	tropiikinvalas; brydenvalas
SV	Brydes fenval

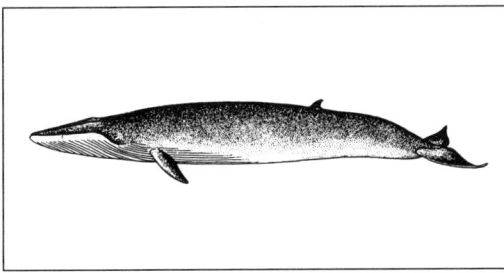

1484 BALAENOPTERIDAE   FIW

SC	*Balaenoptera physalus* (Linnaeus, 1758)
ES	rorcual común
DA	finhval
DE	Finnwal
EL	φάλαινα
EN	fin whale; common rorqual; finner; common finback; herring whale; razorback
FR	rorqual commun
IT	balenottera comune; rorqualo
NL	gewone vinvis
PT	baleia comum
FI	sillivalas
SV	sillval; fenval

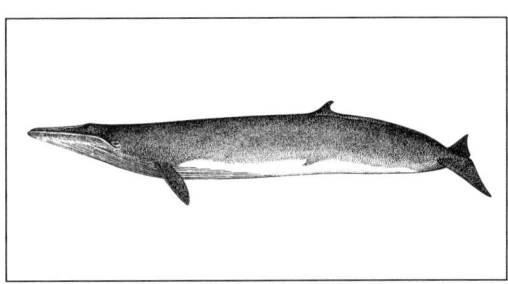

1485 BALAENOPTERIDAE   HUW

SC	*Megaptera novaeangliae* (Borowski, 1781); *Megaptera nodosa*
ES	ballena jorobada; yubarta
DA	pukkelhval
DE	Buckelwal
EL	μεγαπτεροφάλαινα
EN	humpback whale
FR	mégaptère; jubarte; baleine à bosse
IT	megattera; balenottera gobba
NL	bultrug
PT	baleia de bossas
FI	ryhävalas
SV	knölval

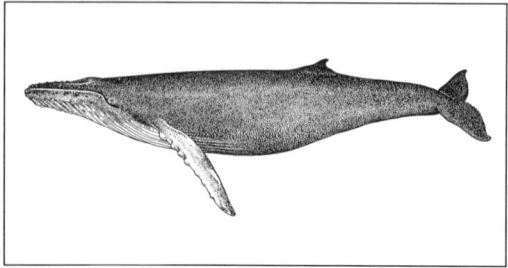

1487 BALAENIDAE

SC	*Balaena mysticetus* (Linnaeus, 1758); *Eubalaena mysticetus*
ES	ballena franca; ballena de Groenlandia
DA	grønlandshval; nordhval
DE	Grönlandwal
EL	φάλαινα της Γροιλανδίας
EN	Greenland right whale; Arctic right whale; great Polar whale
FR	baleine franche boréale; baleine franche du Groenland
IT	balena di Groenlandia
NL	Groenlandse walvis
PT	baleia franca boreal
FI	grönlanninvalas
SV	grönlandsval

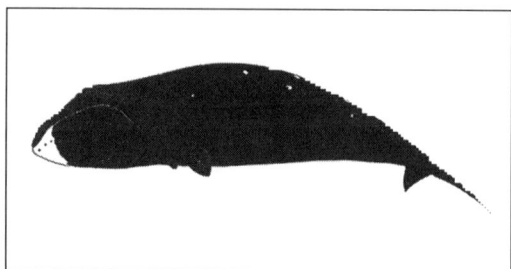

1486 BALAENIDAE

SC	*Balaenidae*
ES	ballenas; rorcuales
DA	rethval-familien
DE	Glattwale
EL	φάλαινες
EN	whales
FR	baleines franches
IT	balenidi; balene
NL	warewalvissen
PT	baleias francas
FI	silovalaat; silovalaat-heimo
SV	rätvalar

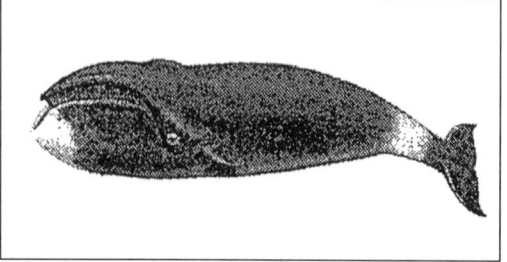

1488 BALAENIDAE

SC	*Caperea marginata* (Gray, 1846); *Neobalaena marginata*
ES	ballena enana; ballena pigmea
DA	dværgrethval
DE	Zwergglattwal
EL	νανοφάλαινα
EN	pigmy whale
FR	baleine naine
IT	balena marginata; balena pigmea
NL	dwerg-warewalvis
PT	baleia-pigmeu
FI	kääpiövalas
SV	dvärgrätval

DRAWING NOT AVAILABLE

1489 BALAENIDAE

SC   *Eubalaena australis*
ES   ballena antártica
DA   sydlig rethval
DE   Südlicher Glattwal
EL   μαύρη φάλαινα
EN   Southern right whale
FR   baleine franche australe
IT   balena antartica
NL   zuidkaper; Australische walvis
PT   baleia franca austral
FI   mustavalas
SV   sydkapare

1491 ESCHRICHTIIDAE

SC   *Eschrichtius gibbosus* (Erxleben, 1777); *Eschrichtius glaucus*; *Eschrichtius robustus*
ES   ballena gris de California
DA   gråhval
DE   Grauwal
EL   γκρίζα φάλαινα της Καλιφόρνιας
EN   Pacific grey whale
FR   baleine grise de Californie
IT   balenottera grigia della California
NL   grijze walvis
PT   baleia cinzenta do Pacífico
FI   harmaavalas
SV   gråval

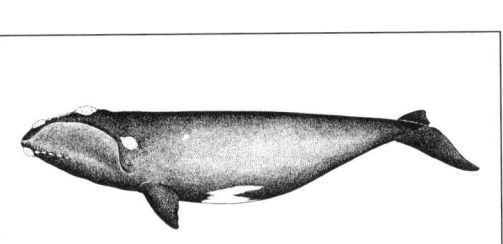

1490 BALAENIDAE

SC   *Eubalaena glacialis* (Müller, 1976); *Balaena biscayensis*; *Balaena glacialis*
ES   ballena vasca
DA   nordkaper
DE   Nordkaper
EL   μαύρη φάλαινα
EN   North Atlantic right whale; black right whale; Biscayan right whale; North Cape whale
FR   baleine franche noire; baleine de Biscaye; baleine des Basques
IT   balena nera; balena franca
NL   noordkaper
PT   baleia franca; baleia franca negra
FI   mustavalas
SV   nordkapare

1492 RANIDAE

SC   *Rana catesbyana* (Shaw, 1802)
ES   rana toro
DA   amerikansk oksefrø
DE   Amerikanischer Ochsenfrosch
EL   ταυροβάτραχος
EN   bull frog
FR   grenouille-taureau
IT   rana toro
NL   brulkikker
PT   rã japonesa; rã-toiro
FI   härkäsammakko
SV   oxgroda

DRAWING NOT AVAILABLE

1493 RANIDAE

FRG

SC	*Rana* spp.
ES	ranas
DA	springfrø-slægt
DE	Frösche
EL	βατράχια
EN	frogs
FR	grenouilles
IT	rane
NL	kikkers
PT	rãs
FI	sammakko-suku
SV	grodor

1495 DERMOCHELYIDAE

SC	*Dermochelys coriacea* (Vandelli, 1761)
ES	tortuga laúd
DA	læderskildpadde
DE	Lederschildkröte
EL	δερματοχελώνα
EN	leatherback turtle
FR	tortue-luth; tortue-cuir
IT	tartaruga liuto
NL	lederschildpad
PT	tartaruga gigante
FI	merinahkakilpikonna
SV	havsläderssköldpadda

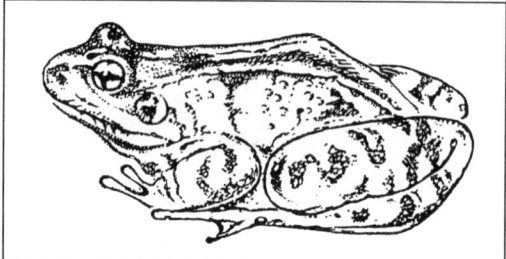

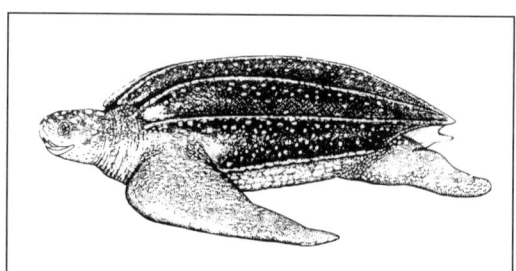

1494 TESTUDINATA

TTX

SC	*Testudinata*
ES	tortugas de mar
DA	skildpadder
DE	Meeresschildkröten
EL	θαλάσσιες χελώνες
EN	marine turtles
FR	tortues de mer
IT	testudinati, tartarughe
NL	zeeschildpadsoorten
PT	tartarugas do mar
FI	merikilpikonnat
SV	sköldpaddor

1496 CHELONIDAE

TTL

SC	*Caretta caretta* (Linnaeus, 1758)
ES	tortuga boba
DA	uægte karetteskildpadde
DE	Unechte Karettschildkröte
EL	καρέτα· θαλάσσια χελώνα
EN	loggerhead turtle
FR	caouanne
IT	tartaruga comune; tartaruga caretta
NL	valse karetschildpad
PT	tartaruga; tartaruga vulgar
FI	valekarettikilpikonna
SV	oäkta karettsköldpadda

DRAWING NOT AVAILABLE

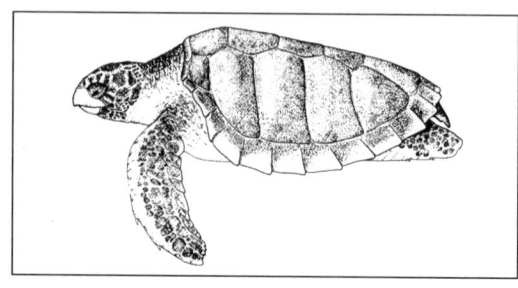

1497 CHELONIDAE

SC	*Chelonia mydas* (Linnaeus, 1758)
ES	tortuga verde
DA	suppeskildpadde
DE	Suppenschildkröte
EL	πρασινοχελώνα
EN	green turtle
FR	tortue verte
IT	tartaruga verde
NL	groene zeeschildpad
PT	tartaruga verde; tartaruga franca
FI	liemikilpikonna
SV	soppsköldpadda

1499 EMYDIDAE

SC	*Malaclemys* spp.
ES	tortugas de dorso diamantino
DA	spiselige sumpskildpadder
DE	Diamantschildkröten
EL	αμερικάνικη χελώνα
EN	terrapins; American terrapins
FR	tortues américaines; terrapènes
IT	terrapin
NL	moerasschildpadden
PT	tartarugas americanas
FI	kilpikonna-suku
SV	diamantsköldpaddor

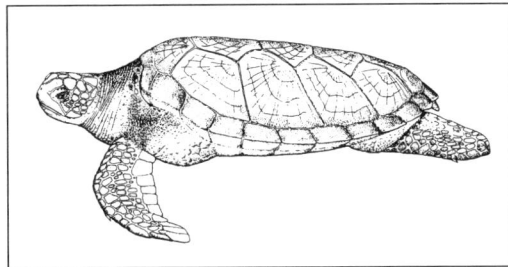

DRAWING NOT AVAILABLE

1498 CHELONIDAE

SC	*Lepidochelys olivacea* (Eschscholtz, 1829)
ES	tortuga bastarda
DA	olivengrøn ridley
DE	Bastardschildkröte
EL	θαλάσσια χελώνα
EN	Olive Ridley
FR	tortue olivâtre; tortue bâtarde
IT	tartaruga bastarda
NL	Kemps schildpad
PT	tartaruga das Guianas
FI	käärmeenkaulakilpikonna
SV	bastardsköldpadda

1500 CROCODYLIDAE

SC	*Crocodylidae*
ES	aligatores
DA	alligator-familien
DE	Alligatoren
EL	αλιγάτορες
EN	alligators
FR	alligators et caïmans
IT	alligatori
NL	alligators
PT	aligátores
FI	alligaattorit
SV	alligator

DRAWING NOT AVAILABLE

1501 CROCODYLIDAE CRO

SC	*Crocodylidae*
ES	cocodrílidos; cocodrilos
DA	krokodille-familien
DE	Krokodile
EL	κροκόδειλοι
EN	crocodiles
FR	crocodiles
IT	coccodrillidi; coccodrilli
NL	krokodillen
PT	crocodilos
FI	krokotiilit
SV	äkta krokodiler

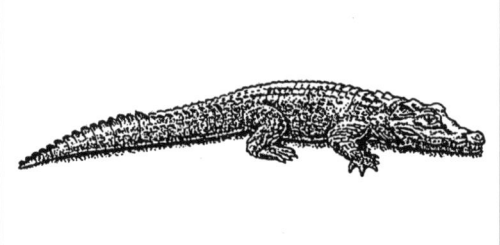

1503 CROCODYLIDAE CAI

SC	*Caiman crocodilus* (Linnaeus, 1758)
ES	caimán
DA	brillekaiman
DE	Krokodilkaiman
EL	καϊμάν· κροκόδειλος
EN	spectacled caiman
FR	caïman à lunettes
IT	caimano dagli occhiali
NL	brilkaaiman
PT	caimão
FI	kaimaani
SV	glasögonkajman

DRAWING NOT AVAILABLE

1502 CROCODYLIDAE AGM

SC	*Alligator mississippiensis* (Daudin, 1802)
ES	aligator del Misisipí
DA	nordamerikansk alligator
DE	Mississippi-Alligator
EL	αλιγάτορας του Μισσισσιππή
EN	alligator
FR	alligator du Mississippi; alligator américain
IT	alligatore del Mississipi
NL	Mississippi-alligator
PT	aligátor do Mississípi
FI	alligaattori
SV	mississipialligator

1504 CROCODYLIDAE CNG

SC	*Crocodylus novaeguineae* (Schmidt, 1928)
ES	cocodrilo de Nueva Guinea
DA	Ny Guinea-krokodille
DE	Neuguinea-Krokodil
EL	κροκόδειλος της Νέας Γουινέας
EN	New Guinea crocodile
FR	crocodile de Nouvelle-Guinée
IT	coccodrillo della Nuova Guinea
NL	Nieuw-Guinea-krokodil
PT	crocodilo da Nova Guiné
FI	uudenguineankrokotiili
SV	sydostasiatisk krokodil

DRAWING NOT AVAILABLE

1505 CROCODYLIDAE
CDP

SC   *Crocodylus porosus* (Schneider, 1801)
ES   cocodrilo marino
DA   deltakrokodille
DE   Leistenkrokodil
EL   θαλάσσιος κροκόδειλος
EN   estuarine crocodile
FR   crocodile marin
IT   coccodrillo marino
NL   zeekrokodil
PT   crocodilo marinho
FI   merikrokotiili
SV   saltvattenskrokodil

DRAWING NOT AVAILABLE

1507 ASCIDIACEA
SSX

SC   *Ascidiacea*
ES   ascidias
DA   søpunge
DE   Seescheiden; Aszidien
EL   ασκίδια
EN   sea squirts
FR   ascidies; violets
IT   ascidiacei
NL   manteldieren
PT   ascídias
FI   meritupet
SV   sjöpungar

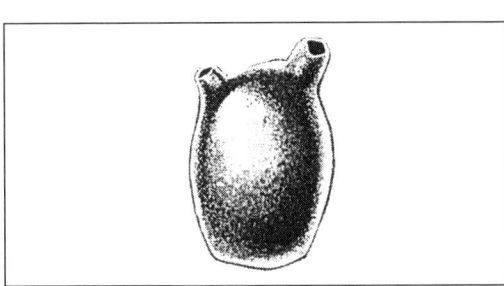

1506 CROCODYLIDAE
CDS

SC   *Crocodylus siamensis* (Schneider, 1801)
ES   cocodrilo del Siam
DA   siamesisk krokodille
DE   Siam-Krokodil
EL   κροκόδειλος του Σιάμ
EN   Siamese crocodile
FR   crocodile du Siam
IT   coccodrillo siamese
NL   Chinese krokodil
PT   crocodilo do Sião
FI   siaminkrokotiili
SV   siamesisk krokodil

DRAWING NOT AVAILABLE

1508 PYURIDAE
SSG

SC   *Microcosmus sabatieri* (Roule, 1885);
       Microcosmus sulcatus
ES   provecho; buñelo de mar
DA   violet søpung
DE   Seefeige
EL   φούσκα
EN   grooved sea squirt
FR   violet
IT   limone di mare; uovo di mare
NL   violet-zakpijp
PT   ascídia violeta
FI   merituppi-laji
SV   sjöpung

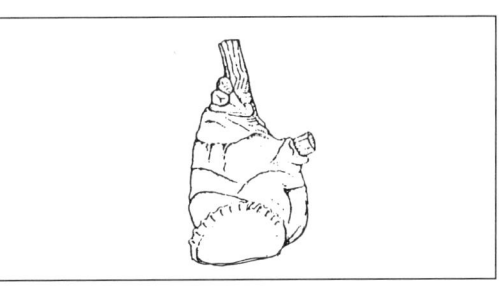

1509 PYURIDAE SSE

SC   *Pyura chilensis* (Molina, 1782)
ES   piura; ascidia de Chile
DA   chilensk søpung
DE   Chilenische Aszidie
EL   ασκίδια της Χιλής
EN   red sea squirt
FR   ascidie du Chili; violet du Chili
IT   ascidia cilena
NL   Chileense pyurazakpijp
PT   ascídia comestível; ascídia do Chile
FI   merituppi-laji
SV   chilensk sjöpung

DRAWING NOT AVAILABLE

1511 SPONGIDAE SPO

SC   *Spongidae*
ES   esponjas
DA   havsvampe-familie
DE   Schwämme
EL   σπόγγοι
EN   sponges
FR   éponges
IT   spugne
NL   sponzen
PT   esponjas
FI   sarveissienet; sarveissienet-heimo
SV   svampdjur

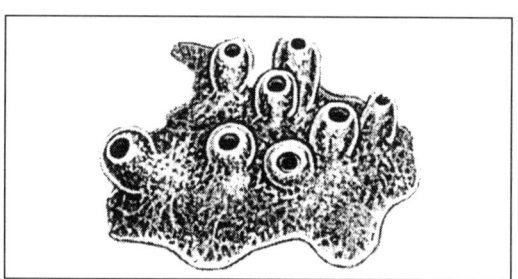

1510 PYURIDAE SSR

SC   *Pyura stolonifera* (Heller, 1878)
ES   piura; ascidia estolonífera
DA   en art søpung
DE   Mittelmeer-Aszidie
EL   ασκίδια
EN   red bait
FR   ascidie stolonifère; violet stolon
IT   ascidia stolonifera
NL   pyurazakpijp
PT   ascídia do Sudoeste Africano
FI   merituppi-laji
SV   röd sjöpung

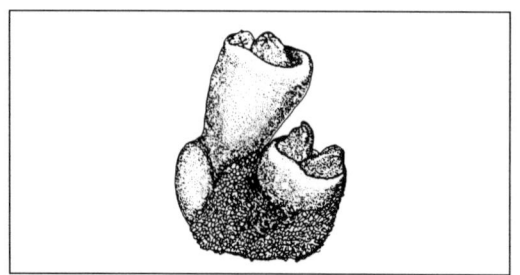

1512 RHIZOSTOMATIDAE JEL

SC   *Rhopilema* spp.
ES   medusas
DA   meduse-slægt
DE   Wurzelmund-Quallen
EL   μέδουσες
EN   jellyfishes
FR   méduses
IT   meduse
NL   kwallen
PT   medusas; alforrecas
FI   meduusa-suku
SV   lungmaneter

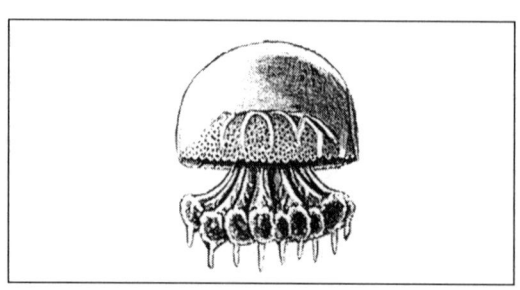

1513 CORALLIIDAE

SC	*Coralliidae*
ES	corales
DA	ædelkoral-familien
DE	Edelkorallen
EL	κοράλλια
EN	corals
FR	corail
IT	coralli preziosi
NL	koralen
PT	corais
FI	korallit; korallit-heimo
SV	koraller; äkta koraller

1515 LIMULIDAE   HSC

SC	*Limulus polyphemus* (Linnaeus, 1758)
ES	límulo; cangrejo cacerola
DA	dolkhale
DE	Atlantischer Schwertschwanz
EL	καβούρι
EN	horseshoe crab
FR	limule
IT	limulo; xifosuro
NL	degenkrab
PT	límulo
FI	molukkirapu
SV	dolksvans

DRAWING NOT AVAILABLE

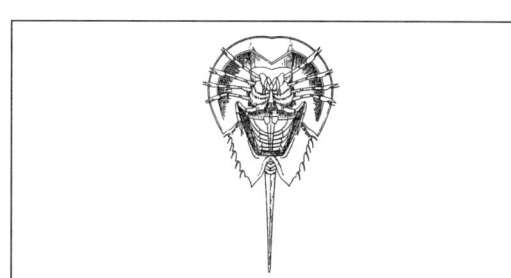

1514 CORALLIIDAE   COL

SC	*Corallium rubrum* (Linnaeus, 1758)
ES	coral rojo
DA	ædelkoral
DE	Rote Koralle
EL	κόκκινο κοράλλι
EN	red coral; Sardinia coral
FR	corail rouge
IT	corallo rosso
NL	rode koraal
PT	coral vermelho
FI	jalokoralli
SV	rödkorall; äkta korall

1516 ASTERIIDAE

SC	*Asteriidae*
ES	estrellas de mar
DA	søstjerne-familie
DE	Seesterne
EL	αστερίες
EN	starfishes
FR	étoiles de mer
IT	stelle di mare
NL	zeesterren
PT	estrelas-do-mar
FI	meritähdet; meritähdet-heimo
SV	sjöstjärnor

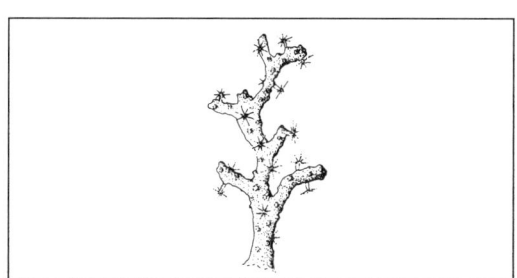

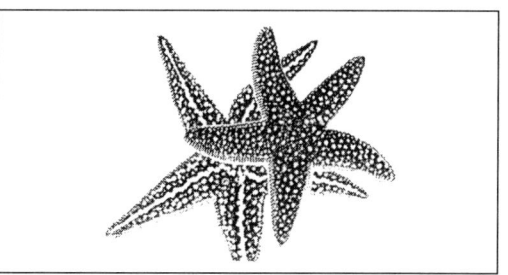

1517 ASTERIIDAE STH

SC   *Asterias rubens* (Linnaeus, 1758)
ES   estrella de mar
DA   almindelig søstjerne
DE   Gemeiner Seestern
EL   αστερίας
EN   starfish
FR   étoile de mer rouge
IT   stella marina
NL   zeester
PT   estrela-do-mar comum
FI   punameritähti
SV   vanlig sjöstjärna

1519 STRONGYLOCENTROTIDAE URC

SC   *Strongylocentrotus* spp.
ES   erizos
DA   søpindsvine-slægt
DE   Seeigel
EL   αχινός
EN   urchins; sea urchins
FR   oursins
IT   ricci di mare
NL   zee-egels
PT   ouriços-do-mar
FI   merisiili-suku
SV   sjöborrar

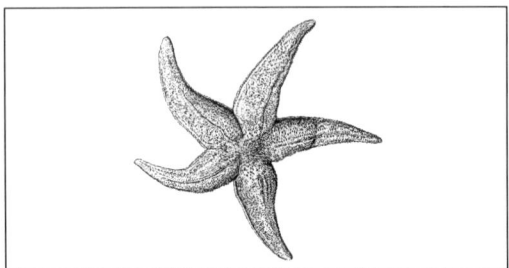

DRAWING NOT AVAILABLE

1518 STRONGYLOCENTROTIDAE URM

SC   *Paracentrotus lividus* (Lamarck, 1816);
       Strongylocentrotus lividus
ES   erizo común; erizo de roca; erizo de mar
DA   sten-søpindsvin
DE   Steinseeigel
EL   αχινός
EN   purple sea urchin; stony sea urchin
FR   oursin violet; oursin-pierre
IT   riccio di mare
NL   zeeappel
PT   ouriço-do-mar púrpura; ouriço-do-mar
FI   kalliomerisiili
SV   sjöborre

1520 ECHINIDAE

SC   *Echinidae*
ES   erizos de fondo
DA   søpindsvine-familie
DE   Seeigel
EL   αχινοί
EN   sea urchins
FR   oursins
IT   ricci di mare
NL   zee-egels
PT   ouriços-do-mar; oiriços-do-mar
FI   merisiilit; merisiilit-heimo
SV   sjöborrar

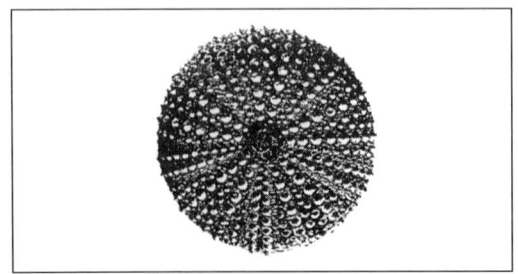

DRAWING NOT AVAILABLE

1521 ECHINIDAE URS

SC	*Echinus esculentus* (Linnaeus, 1758)
ES	erizo
DA	spiseligt søpindsvin
DE	Eßbarer Seeigel
EL	αχινός
EN	sea urchin
FR	oursin comestible
IT	riccio di mare
NL	eetbare zee-egel
PT	ouriço-do-mar
FI	merisiili
SV	ätlig sjöborre

1523 STICHOPODIDAE CUJ

SC	*Stichopus japonicus* (Selenka, 1867)
ES	cohombro de mar japonés
DA	japansk søpølse
DE	Japanische Seegurke
EL	γιαπωνέζικη ολοθούρια
EN	Japanese sea cucumber
FR	bêche-de-mer japonaise
IT	oloturia giapponese
NL	Japanse zeekomkommer
PT	holotúria japonesa
FI	japaninmerimakkara
SV	japansk sjögurka

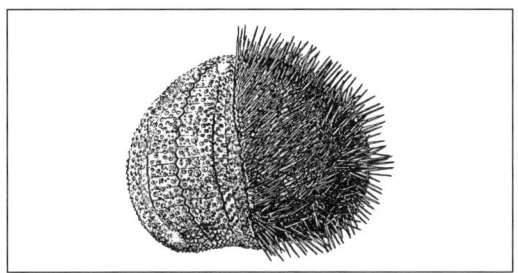

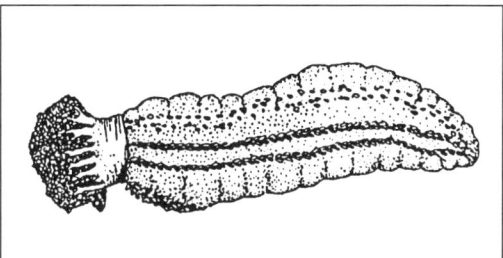

1522 ECHINIDAE UCH

SC	*Loxechinus albus* (Molina, 1782)
ES	erizo chileno; erizo blanco
DA	chilensk søpindsvin
DE	Chilenischer Seeigel
EL	αχινός της Χιλής
EN	Chilean sea urchin
FR	oursin chilien; oursin du Chili
IT	riccio di mare cileno
NL	Chileense zee-egel
PT	ouriço-do-mar chileno
FI	chilenmerisiili
SV	chilensk vit sjöborre

1524 ALGAE APL

SC	*Algae*
ES	algas
DA	alger
DE	Algen
EL	φύκια
EN	algae
FR	algues
IT	alghe
NL	algen
PT	algas
FI	levät
SV	alger

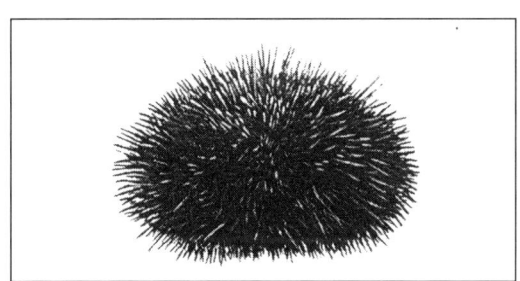

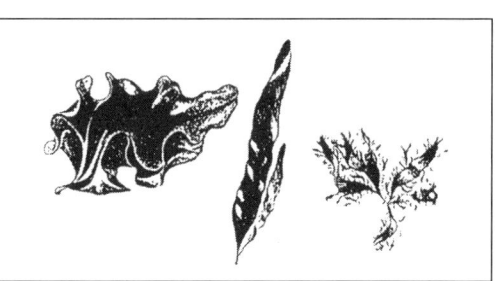

1525 PHAEOPHYCEAE

SC	*Phaeophyceae*
ES	algas pardas; algas oscuras
DA	brunalger
DE	Braunalgen
EL	φαιοφύκια· φαιοφύκη
EN	brown algae; brown seaweed
FR	algues brunes; phéophycées
IT	alghe brune; feoficee
NL	bruinwieren
PT	algas castanhas
FI	ruskolevät
SV	brunalger

DRAWING NOT AVAILABLE

1527 FUCALES

SC	*Fucales*
ES	fucales
DA	klørtang
DE	Tang; Seetang
EL	φυκίδες
EN	wracks
FR	fucales
IT	fucali
NL	blaasjeswieren
PT	fucales
FI	rakkolevät
SV	tång

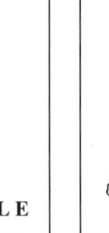

1526 LAMINARIACEAE

SC	*Laminariaceae*
ES	laminarias
DA	bladtang
DE	Tang; Kelp
EL	λαμινάριες
EN	kelps
FR	varech; laminaires; taly; anguiller
IT	laminarie
NL	laminariawieren
PT	laminárias
FI	laminariat
SV	bladtång

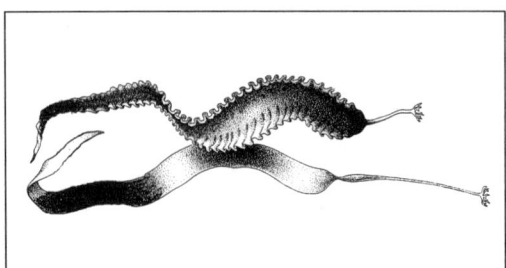

1528 RHODOPHYCEAE

SC	*Rhodophyceae*
ES	algas rojas
DA	rødalger
DE	Rotalgen
EL	ροδοφύκια· ερυθροφύκη
EN	red algae; red seaweed
FR	algues rouges
IT	alghe rosse; rodoficee
NL	roodwieren
PT	algas vermelhas
FI	punalevät
SV	rödalger

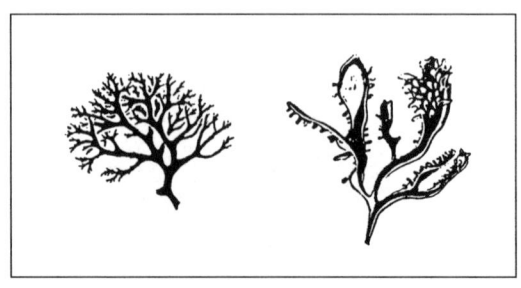

382

1529 GELIDIACEAE GEL

SC	*Gelidium* spp.
ES	gelidium
DA	rødalge-slægt
DE	Gelidium
EL	αγαρόφυκα
EN	gelidium
FR	gelidium
IT	gelidium
NL	gelidiumwieren
PT	gelídeos
FI	punalevä-suku
SV	gelidium

1531 GIGARTINACEAE GIG

SC	*Gigartina* spp.
ES	gigartinas
DA	vortetang
DE	Gigartina
EL	γιγαρτίνα
EN	gigartinas
FR	gigartines
IT	gigartine
NL	gigartinawieren
PT	gigartinas; musgos; botelhas
FI	punalevä-suku
SV	gigartiner

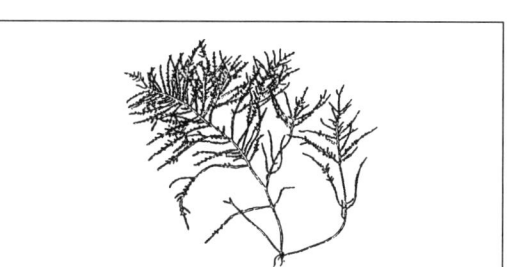

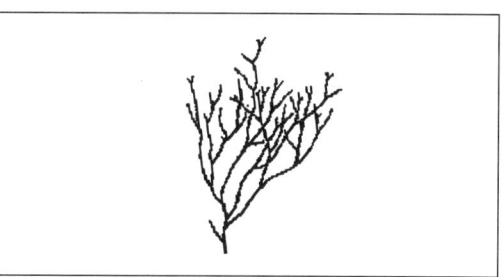

1530 CORALLINACEAE LIT

SC	*Lithothamnion* spp.
ES	algas calcáreas
DA	kalkalge-slægt
DE	Kalkalgen
EL	ασβεστοφύκη
EN	lithothamnion
FR	lithothamnies; algues calcaires; maërl
IT	alghe calcaree
NL	lithothamnion; kalkwieren
PT	algas calcárias
FI	punalevä-suku
SV	kalkformiga rödalger; stenhinna

1532 GIGARTINACEAE IMS

SC	*Chondrus crispus* (Linnaeus, 1767)
ES	condrus
DA	carrigeen tang
DE	Irischmoos; Irisches Moos
EL	καραγγινό
EN	carragheen; Irish moss
FR	chondrus; carraghéen; mousse d'Irlande
IT	muschio irlandese
NL	Iers mos
PT	musgo gordo; folha-de-alface
FI	punalevä-laji
SV	karragener

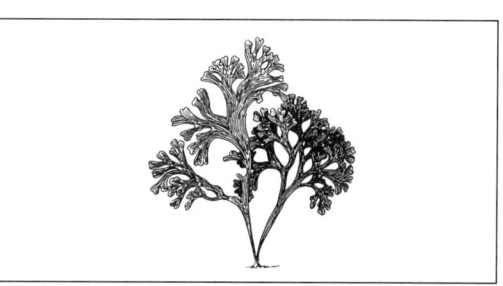

Indexes

Index of scientific names

A

Abramis spp.	307
A. ballerus	304
A. bjoerkna	319
A. brama	305
A. sapa	306
A. vimba	365
Acanthias vulgaris	64
Acanthistius brasilianus	568
Acanthocardia aculeata	1388
Acanthocybium solandri	962
Acanthopsis taenia	370
Acanthuridae	945
Acanthurus monroviae	946
Acerina cernua	614
A. schraetzer	615
Acetes japonicus	1242
Achirus lineatus	1152
Acipenser gueldenstaedti	131
A. medirostris	132
A. naccarii	133
A. nudiventris	134
A. ruthenus	135
A. stellatus	136
A. sturio	137
A. transmontanus	138
Acipenseridae	130
Aequipecten opercularis	1359
Agonus cataphractus	1103
Albula vulpes	152
Albulidae	151
Alburnoides bipunctatus	308
Alburnus albidus	309
A. alburnus	310
A. lucidus	310
Alectis alexandrinus	633
Alepes djedaba	634
Alepocephalus bairdii	223
Algae	1524
Alligator mississippiensis	1502
Alloteuthis media	1423
Allothunnus fallai	963
Alopias pelagicus	15
A. superciliosus	16
A. vulpinus	17
Alosa spp.	163
A. aestivalis	155
A. alosa	156
A. fallax	157
A. fallax fallax	157
A. fallax nilotica	158
A. finta	157
A. mediocris	159
A. pilchardus	189
A. pontica	160
A. pseudoharengus	161
A. sapidissima	162
A. vulgaris	156
Aluterus schoepfi	1181
Ambloplites rupestris	600
A. salmoides	603
Amblyraja radiata	97
Amia calva	142
Ammodytes spp.	937
A. americanus	932
A. cicerellus	938

A. dubius	933
A. hexapterus	934
A. lancea	936
A. lanceolatus	939
A. marinus	935
A. tobianus	936
Ammodytidae	931
Anabas testudineus	1039
Anadara spp.	1341
A. granosa	1339
A. subcrenata	1340
Anarhichadidae	913
Anarhichas spp.	917
A. denticulatus	914
A. latifrons	914
A. lupus	915
A. minor	916
Anchoa hepsetus	209
A. mitchilli	210
Anguilla spp.	401
A. anguilla	396
A. australis	397
A. diffenbachii	398
A. japonica	399
A. rostrata	400
A. vulgaris	396
Anguillidae	395
Anisotremus virginicus	702
Anodonta cygnea	1343
Anodontostoma chacunda	164
Anoplopoma fimbria	1094
Anthias anthias	569
Antimora rostrata	437
Aphanius fasciatus	513
A. iberus	514
Aphanopus carbo	956
Aphia minuta	1050
Aphya pellucida	1050
Aplodinotus grunniens	716
Apogon imberbis	610
Apogonidae	609
Apristurus brunneus	29
Arca spp.	1342
Archosargus probatocephalus	764
Arctica islandica	1386
Arctocephalus australis	1441
A. pusillus	1442
Arctoscopus japonicus	863
Argentina semifasciata	283
A. sialis	280
A. silus	281
A. sphyraena	282
Argentinidae	279
Argopecten gibbus	1360
A. irradians	1361
Argyrosoma regium	718
Argyrosomus hololepidotus	717
A. regius	718
Argyrozona argyrozona	765
Ariidae	374
Ariomma indica	1038
Aristeidae	1232
Aristeus antennatus	1233
Aristichys nobilis	336
Arius heudeloti	375
A. maculatus	376
Arnoglossus laterna	1115

A. moltoni	1116
A. thori	1116
Arripis georgiana	682
A. trutta	683
Artemesia longinaris	1206
Ascidiacea	1507
Aspitrigla cuculus	1079
Aspitriglia obscura	1080
Aspius aspius	311
Aspro asper	622
A. streber	623
A. zingel	624
Astacus spp.	1278
A. astacus	1276
A. leptodactylus	1277
A. pallipes	1279
A. torrentium	1280
Asterias rubens	1517
Asteriidae	1516
Astroconger myriaster	409
Atheresthes evermanni	1124
A. stomias	1125
Atherina boyeri	549
A. hepsetus	550
A. presbyter	551
Atherinidae	548
Atractoscion aequidens	719
A. atelodus	719
A. nobilis	720
Atrobucca nibe	721
Atyidae	1243
Aulacomya ater	1371
Aulopus filamentosus	389
Aulopyge huegelii	312
Austroglossus spp.	1155
A. microlepis	1153
A. pectoralis	1154
Austropotamobius spp.	1278
A. pallipes	1279
A. torrentium	1280
Auxis rochei	964
A. thazard	965

B

Balaena biscayensis	1490
B. glacialis	1490
B. mysticetus	1487
Balaenidae	1486
Balaenoptera acutorostrata	1480
B. borealis	1481
B. edeni	1482
B. musculus	1483
B. physalus	1484
Balanus spp.	1199
Balistes capriscus	1180
B. carolinensis	1180
Balistidae	1179
Barbus spp.	354
B. albanicus	313
B. barbus	314
B. comiza	315
B. javanicus	353
B. meridionalis petenyi	317
B. meridionalis	316
B. plebejus	318

Bathyraja spinicauda	99	C. djedaba	634	C. capensis	1081
Batrachoididae	1191	C. fusus	635	C. cuculus	1079
Belone acus	416	C. georgianus	652	C. kumu	1082
B. belone	416	C. hippos	636	Chelon labrosus	538
B. euxini	416	C. kalla	634	Chelonia mydas	1497
B. vulgaris	416	C. rhonchus	637	Chimaera monstrosa	123
Belonidae	415	C. ruber	638	Chimaeridae	122
Berardius bairdii	1459	Carapus acus	930	Chionodraco rastrospinosus	900
Berycidae	517	Carassius auratus	320	Chionoecetes spp.	1313
Beryx spp	519	C. carassius	321	C. opilio	1312
B. decadactylus	518	Carcharhinidae	34	Chirocentridae	224
Bivalvia	1324	Carcharhinus albimarginatus	35	Chirocentrus spp.	226
Blenniidae	904	C. brachyurus	36	C. dorab	225
Blennius spp.	906	C. glaucus	42	Chlamydoselachus anguineus	8
B. fluviatilis	911	C. longimanus	37	Chlamys islandica	1362
B. ocellaris	905	C. melanopterus	38	C. opercularis	1359
Blicca bjoerkna	319	C. obscurus	39	C. varia	1363
Boops boops	766	C. plumbeus	40	Chlorophthalmidae	298
Boreogadus saida	441	Carcharias ferox	14	Chlorophthalmus agassizi	299
Bothidae	1114	C. glaucus	42	C. productus	299
Bothus podas podas	1117	C. taurus	13	Chloroscombrus chrysurus	640
Box boops	766	Carcharodon carcharias	19	Chondrostoma genei	323
B. salpa	801	C. rondeletti	19	C. kneri	324
B. vulgaris	766	Carcinus aestuarii	1297	C. nasus	325
Brachydeuterus auritus	703	C. maenas	1298	C. phoxinus	326
Brachyochirus pellucidus	1050	C. mediterraneus	1297	C. polylepis	327
Brama brama	680	Cardiidae	1387	C. soetta	328
B. raii	680	Cardium aculeatum	1388	C. toxostoma	329
B. raji	680	C. edule	1389	Chondrus crispus	1532
Bramidae	679	Caretta caretta	1496	Chorinemus spp.	653
Bregmaceros mcclellandi	439	Caspialosa pontica	160	Christivomer namaycush	266
Brevoortia spp.	167	Catostomidae	301	Chromis chromis	836
B. patronus	165	Centracanthidae	810	Chrysoblephus spp.	770
B. tyrannus	166	Centrarchus hexacanthus	605	C. gibbiceps	769
Brosme brosme	442	Centrina salviani	70	Chrysophrys aurata	802
Brosmius brosme	442	Centrolophidae	1040	C. major	805
Brotula barbata	924	Centrolophus niger	1041	Cichlidae	831
Buccinum undatum	1335	Centronotus gunnellus	919	Ciliata mustela	443
Buglossidium luteum	1156	Centrophorus granulosus	58	Cirrhinus molitorella	330
Busycon spp.	1336	C. squamosus	59	Citharus linguatula	1177
		Centropomidae	560	Clarias spp.	388
		Centropomus spp.	562	Clinitrachus argentatus	918
C		C. undecimalis	561	Clupanodon thrissa	168
		Centropristis striata	571	Clupea alosa	156
Caesio spp.	686	Cepola macrophthalma	834	C. bentincki	204
Caiman crocodilus	1503	C. rubescens	834	C. finta	157
Calamus spp.	767	Cerastoderma edule	1389	C. harengus	169
Callanthias ruber	570	Cerna gigas	577	C. harengus pallasi	170
Callinectes spp.	1296	Cetengraulis edentulus	211	C. pallasii	170
C. danae	1294	C. mysticetus	212	C. pilchardus	189
C. sapidus	1295	Cetorhinus maximus	12	C. sprattus	202
Callionymus lyra	941	Chaceon maritae	1321	Clupeidae	154
C. maculatus	942	Chaenocephalus aceratus	896	Clupeonella cultriventris	171
Callista chione	1395	Chaenodraco wilsoni	897	C. delicatula	171
Callorhynchus spp.	126	Chaetodontidae	830	Cobitis aurata	366
C. capensis	125	Chalcalburnus chalcoides	322	C. barbatulus	372
Cambarus spp.	1281	Chamelea gallina	1396	C. conspersa	373
C. affinis	1282	Champsocephalus gunnari	898	C. elongata	367
Cancer borealis	1288	Chanda spp.	563	C. fossilis	371
C. irroratus	1289	Chanidae	227	C. larvata	368
C. magister	1290	Channa spp.	559	C. romanica	369
C. pagurus	1291	C. micropeltes	557	C. taenia	370
C. productus	1292	C. striatus	558	Coelorhynchus carminatus	493
Cancridae	1287	Channichthyidae	895	C. coelorhynchus	493
Cantharus cantharus	808	Channichthys rhinoceratus	899	Coelorinchus coelorinchus	493
C. griseus	808	Chanos chanos	228	Coilia mystus	213
C. lineatus	808	Characidae	300	Collichthys crocea	722
Cantherhines spp.	1182	Cheilodactylus bergi	857	Cololabis saira	421
Caperea marginata	1488	C. macropterus	859	Conger conger	410
Caproidae	526	C. variegatus	858	C. myriaster	409
Capros aper	527	Cheilopogon agoo	424	C. oceanicus	411
Carangidae	632	Cheimerius canariensis	772	C. orbignyanus	412
Caranx spp.	639	C. nufar	768	C. vulgaris	410
C. crysos	635	Chelidonichthys spp.	1083	Congridae	408

Conodon nobilis 704
Coracinidae 823
Coracinus capensis 824
Coralliidae 1513
Corallium rubrum 1514
Coregoninae 261
Coregonus spp. 241
    C. albula 229
    C. altior 230
    C. artedii 231
    C. clupeaformis 232
    C. elegans 230
    C. fera 236
    C. hiemalis 233
    C. lavaretus 234
    C. lavaretus oxyrinchus 237
    C. lavaretus pidschianoides 239
    C. macrophthalmus 235
    C. nasus 236
    C. oxyrinchus 237
    C. peled 238
    C. pidschianoides 239
    C. vandesius 229
    C. wartmanni 240
Coris julis 838
Corvina nigra 755
Coryphaena hippurus 681
Coryphaenoides rupestris 494
Cottus gobio 1097
Crangon crangon 1251
Crangonidae 1250
Crassostrea spp. 1350
    C. angulata 1347
    C. commercialis 1357
    C. gigas 1347
    C. rhizophorae 1348
    C. virginica 1349
Crenilabrus cinereus 845
    C. griseus 845
    C. mediterraneus 846
    C. melops 847
    C. ocellatus 848
    C. pavo 851
    C. quinquemaculatus 849
    C. roissali 849
    C. tinca 851
Crenimugil labrosus 538
Crocodylidae 1500, 1501
Crocodylus novaeguineae 1504
    C. porosus 1505
    C. siamensis 1506
Ctenolabrus rupestris 839
Ctenopharyngodon idella 331
Cybiosarda elegans 966
Cybium commersoni 989
Cyclopteridae 1104
Cyclopterus lumpus 1105
Cynoglossidae 1166
Cynoglossus vulgaris 1128
Cynoponticus ferox 405
Cynoscion spp. 729
    C. analis 723
    C. macdonaldi 724
    C. nebulosus 725
    C. nobilis 720
    C. regalis 726
    C. steindachneri 727
    C. striatus 728
Cyprina islandica 1386
Cyprinidae 303
Cyprinodon iberus 514
Cyprinodontidae 512
Cyprinus carpio 332
Cypselurus spp. 425

C. agoo 424
Cystophora cristata 1447

D

Dactylopterus volitans 1109
Dalatias licha 60
Dasyatidae 103
Dasyatis akajei 104
    D. centroura 105
    D. pastinaca 106
    D. violacea 107
Decapterus spp. 645
    D. macarellus 641
    D. maruadsi 642
    D. punctatus 643
    D. russeli 644
Delphinapterus leucas 1478
Delphinidae 1463
Delphinus delphis 1464
Dentex angolensis 771
    D. canariensis 772
    D. congoensis 773
    D. dentex 774
    D. gibbosus 775
    D. macrophthalmus 776
    D. maroccanus 777
    D. vulgaris 774
Dermochelys coriacea 1495
Dicentrarchus spp. 595
    D. labrax 593
    D. punctatus 594
Dicologlossa cuneata 1157
Dicologoglossa cuneata 1157
Diplectrum formosum 572
Diplodus spp. 784
    D. annularis 778
    D. cervinus 779
    D. holbrooki 780
    D. puntazzo 781
    D. sargus 782
    D. trifasciatus 779
    D. vulgaris 783
Dipturus batis 84
    D. oxyrinchus 96
Dissostichus spp. 879
    D. eleginoides 877
    D. mawsoni 878
Donax spp. 1406
Dorosoma cepedianum 172
Dosidicus gigas 1427
Dosinia orbignyi 1397
Drepane africana 827
    D. punctata 828
Drepanopsetta platessoides 1132
Dugong dugong 1454
Dussumieria acuta 173
    D. productissima 173

E

Echelus myrus 413
Echiichthys vipera 867
Echinidae 1520
Echinorhinus brucus 71
    E. spinosus 71
Echinus esculentus 1521
Elagatis bipinnulata 646
Eledone spp. 1437
    E. cirrosa 1435
    E. moschata 1436
Eleginops maclovinus 880
Eleginus gracilis 444

E. navaga 445
Eleotridae 1048
Eleutheronema tetradactylum 554
Elopidae 145
Elops spp. 147
    E. saurus 146
Emmelichthyidae 684
Emmelichthys nitidus 685
Emmenopias steller 1443
Engraulidae 208
Engraulis anchoita 214
    E. australis 215
    E. capensis 216
    E. encrasicolus 217
    E. hepsetus 209
    E. japonica 218
    E. japonicus 218
    E. mordax 219
    E. ringens 220
Enophris bubalis 1102
Ensis directus 1408
    E. ensis 1409
    E. siliqua 1410
Eopsetta grigorjewi 1126
    E. jordani 1127
Ephippidae 825
Epigonus telescopus 611
Epinephelus spp. 582
    E. aeneus 573
    E. alexandrinus 574
    E. analogus 575
    E. caninus 576
    E. gigas 577
    E. guaza 577
    E. itajara 578
    E. malabaricus 579
    E. morio 580
    E. striatus 581
Epinephlus marginatus 577
Erignathus barbatus 1448
Eriocheir sinensis 1303
Eschrichtius gibbosus 1491
    E. glaucus 1491
    E. robustus 1491
Esocidae 286
Esox americanus vermiculatus 287
    E. lucius 288
    E. masquinongy 289
    E. niger 290
    E. reicherti 291
    E. vermiculatus 287
Ethmalosa fimbriata 174
Ethmidium maculatum 175
Etmopterus spp. 62
    E. spinax 61
Etrumeus spp. 179
    E. acuminatus 176
    E. sadina 177
    E. teres 178
Euastacus serratus 1316
Eubalaena australis 1489
    E. glacialis 1490
    E. mysticetus 1487
Eucinostomus melanopterus 902
Eudontomyzon danfordi 2
    E. mariae 3
    E. vladykovi 4
Eugaleus galeus 50
Eugomphodus taurus 13
Eumetopias jubatus 1443
Euphausia superba 1204
Euphausiidae 1203
Eupomotis gibbosus 601
Euthynnus affinis 967

E. alletteratus	968
E. lineatus	969
E. pelamis	973
E. quadripunctatus	968
Eutrigla gurnardus	1084
Exocoetidae	423
Exocoetus volitans	426

F

Fiatola fasciata	1037
Fierasfer acus	930
Formio niger	677
Fucales	1527
Fundulus hispanicus	515

G

Gadiculus argenteus	449
G. thori	449
Gadidae	440
Gadiformes	435
Gadus aeglefinus	453
G. callarias	447
G. esmarki	471
G. luscus	472
G. macrocephalus	446
G. merlangus	454
G. minutus	473
G. morhua	447
G. morhua ogac	448
G. morrhua	447
G. navaga	445
G. ogac	448
G. pollachius	466
G. poutassou	459
G. virens	467
Gaidropsarus spp.	451
G. vulgaris	450
Galatheidae	1306
Galeichthys feliceps	377
Galeorhinus galeus	50
Galeus melastomus	30
G. vulgaris	50
Gambusia affinis	516
G. holbrooki	516
Gardonus rutilis	360
Gasterochisma melampus	970
Gasterosteidae	499
Gasterosteus aculeatus	500
Gelidium spp.	1529
Gempylidae	947
Genyonemus lineatus	730
Genypterus spp.	928
G. blacodes	925
G. capensis	926
G. chilensis	927
Germo alalunga	1007
Gerres spp.	903
Geryon quinquedens	1321
G. spp.	1322
Gigartina spp.	1531
Ginglymostoma cirratum	26
Globicephala macrorhyncha	1465
G. melaena	1466
Glossanodon semifasciatus	283
Glycymeris glycymeris	1384
Glyptocephalus cynoglossus	1128
G. zachirus	1129
Gobiidae	1049
Gobio fluviatilis	333
G. gobio	333
G. uranoscopus	334

Gobius spp.	1055
G. auratus	1051
G. capito	1052
G. cobitis	1052
G. jozo	1053
G. minutus	1059
G. niger jozo	1053
G. niger	1053
G. paganellus	1054
Grammatorcynus bicarinatus	971
G. bilineatus	972
Grampus griseus	1467
Gymnammodytes cicerellus	938
Gymnocephalus cernuus	614
G. schraetzer	615
Gymnocranius spp.	761
Gymnoecephalus cernua	614
Gymnothorax funebris	403
Gymnura spp.	116
G. altavela	115

H

Haemulidae	705
Halichoerus grypus	1449
Halicore dugong	1454
Haliotis spp.	1329
H. gigantea	1325
H. midae	1326
H. ruber	1327
H. tuberculata	1328
Haliporoides spp.	1255
Harengula spp.	180
H. zunasi	194
Harpadon nehereus	297
Harpadontidae	296
Hectoria oxygeneios	586
Helicolenus spp.	1064
H. dactylopterus	1063
Helostoma temminckii	1047
Hemiramphidae	427
Hemiramphus spp.	433
H. australensis	428
H. balao	429
H. brasiliensis	430
H. sajori	431
H. saltator	432
Hepsetia presbyter	551
Heptranchias perlo	10
Herklotsichthys punctatus	181
Heterocarpus spp.	1236
H. reedi	1235
Hexanchidae	9
Hexanchus griseus	11
Hilsa ilisha	205
H. kelee	182
H. reevesi	206
H. toli	207
Hippocampus spp.	505
Hippoglossoides dubius	1130
H. elassodon	1131
H. platessoides	1132
Hippoglossus hippoglossus	1133
H. reinhardtius	1148
H. stenolepis	1134
H. vulgaris	1133
Hirundichthys affinis	434
Holocentrus ruber	522
Homarus americanus	1317
H. gammarus	1318
H. vulgaris	1318
Hoplostethus atlanticus	520
H. mediterraneus	521

Hucho hucho	242
Huso huso	139
Hydrolagus colliei	124
Hymenocephalus italicus	495
Hymenopenaeus spp.	1255
H. triarthrus	1254
Hyperoglyphe antarctica	1043
Hyperoodon ampullatus	1460
H. rostratus	1460
Hyperoplus lanceolatus	939
Hypomesus spp.	277
H. olidus	271
H. pretiosus	272
Hypophthalmichthys molitrix	335
H. nobilis	336
Hypoptychus dybowskii	940
Hyporhamphus sajori	431

I

Ictaluridae	383
Ictalurus spp.	387
I. melas	384
I. nebulosus	385
I. punctatus	386
Ictiobus spp.	302
Idus idus	343
Ilisha africana	183
I. elongata	184
Illex argentinus	1428
I. coindetii	1429
I. illecebrosus	1430
Isacia conceptionis	706
Istiophoridae	1015
Istiophorus albicans	1016
I. americanus	1016
I. platypterus	1017
Isurus glaucus	20
I. oxyrhinchus	20
I. paucus	21

J

Jacquinotia edwardsii	1314
Jasus edwardsii	1260
J. frontalis	1261
J. lalandii	1262
J. tristani	1263
J. verreauxi	1264
Joturus pichardi	539

K

Katsuwonus pelamis	973
Knipowitschia panizzae	1056
Kryptopterus spp.	379

L

Labeo spp.	337
Labrax lupus	593
L. punctatus	594
Labridae	837
Labrus bergylta	840
L. bimaculatus	841
L. merula	842
L. mixtus	841

L. pavo	854	*L. putnami*	1141	*Maena vulgaris*	812
L. turdus	843	*Liparis atlanticus*	1107	*Maja squinado*	1315
L. viridis	843	*L. liparis*	1108	*Makaira* spp.	1020
Lachnolaimus maximus	844	*L. vulgaris*	1108	*M. albida*	1021
Lactarius lactarius	628	*Lipophrys pavo*	907	*M. audax*	1023
Lagenorhynchus acutus	1468	*L. pholis*	908	*M. indica*	1018
L. albirostris	1469	*Lithodes antarcticus*	1309	*M. marlina*	1021
Laminariaceae	1526	*L. maja*	1307	*M. nigricans*	1019
Lamna cornubica	23	*L. murrayi*	1308	*Malacanthidae*	626
L. ditropis	22	*L. santolla*	1309	*Malaclemys* spp.	1499
L. nasus	23	*Lithognathus* spp.	787	*Malapterurus electricus*	394
Lamnidae	18	*L. lithognatus*	785	*Mallotus villosus*	273
Lampanyctodes hectoris	295	*L. mormyrus*	786	*Manta birostris*	112
Lampetra fluviatilis	5	*Lithophaga lithophaga*	1372	*M. hamiltoni*	113
L. planeri	6	*Lithothamnion* spp.	1530	*Margaritifera margaritifera*	1383
Lampris guttatus	509	*Littorina* spp.	1337	*Masturus lanceolatus*	1187
L. luna	509	*Liza aurata*	540	*Megabalanus psittacus*	1200
L. regius	509	*L. haematochila*	541	*Megalaspis cordyla*	649
Lateolabrax japonicus	599	*L. ramada*	542	*Megalobrama amblycephala*	346
Lates spp.	566	*L. saliens*	543	*Megalopidae*	148
L. calcarifer	564	*Loliginidae*	1422	*Megalops atlanticus*	150
L. niloticus	565	*Loliginidae, Ommastrephidae*	1421	*M. cyprinoides*	149
Latridae	861	*Loligo pealei*	1424	*Meganyctiphanes norvegica*	1205
Latridopsis ciliaris	862	*L. reynaudi*	1425	*Megaptera nodosa*	1485
Latrunculus pellucidus	1050	*L. vulgaris*	1426	*M. novaeangliae*	1485
Leander serratus	1249	*Lophiidae*	1192	*Melanogrammus aeglefinus*	453
Leiognathidae	699	*Lophius americanus*	1193	*Mene maculata*	678
Leiognathus spp.	701	*L. budegassa*	1194	*Menidia menidia*	552
L. klunzingeri	700	*L. piscatorius*	1195	*M. notata*	552
Leiostomus xanthurus	731	*L. upsicephalus*	1196	*Menippe mercenaria*	1293
Lepadogaster gouanii	1190	*Lopholatilus chamaeleonticeps*	627	*Menticirrhus* spp.	736
L. lepadogaster	1190	*Lota lacustris*	452	*M. americanus*	733
Lepas spp.	1198	*L. lota*	452	*M. littoralis*	734
Lepidochelys olivacea	1498	*L. maculosa*	452	*M. saxatilis*	735
Lepidocybium spp.	948	*L. molva*	462	*Mercenaria mercenaria*	1398
Lepidopsetta bilineata	1135	*Loxechinus albus*	1522	*Meretrix* spp.	1400
Lepidopus caudatus	957	*Luciolates* spp.	566	*M. chione*	1395
L. xantusi	958	*Lucioperca lucioperca*	620	*M. lusoria*	1399
Lepidorhombus spp.	1170	*Lusciscus meidingeri*	357	*Merlangius merlangus*	454
L. boscii	1168	*Lutjanidae*	687	*M. merlangus euxinus*	455
L. megastoma	1169	*Lutjanus* spp.	693	*M. pollachius*	466
L. whiffiagonis	1169	*L. argentimaculatus*	688	*Merlangus virens*	467
Lepidotrigla asper	1085	*L. argentiventris*	689	*Merlucciidae*	478
L. cavillone	1085	*L. aya*	691	*Merluccius* spp.	492
L. dieuzeidei	1086	*L. campechanus*	690	*M. albidus*	481
Lepisosteus osseus	143	*L. purpureus*	691	*M. australis*	482
L. platystomus	144	*L. synagris*	692	*M. bilinearis*	483
Lepomis gibbosus	601	*Lycengraulis grossidens*	221	*M. cadenati*	489
Leptobarbus hoëveni	338	*L. olidus*	221	*M. capensis*	484
Lethrinidae	762	*Lycodes* spp.	921	*M. gayi*	485
Leucaspius delineatus	339	*Lycodontis funebris*	403	*M. hubbsi*	486
Leucichthys artedii	231			*M. merluccius*	487
Leuciscus borysthenicus	340			*M. paradoxus*	488
L. cephalus	341	**M**		*M. polli*	489
L. cephalus cabeda	342			*M. productus*	490
L. idus	343	*Macrobrachium* spp.	1248	*M. senegalensis*	491
L. leuciscus	344	*M. carcinus*	1246	*Metanephrops andamanicus*	1319
L. pigus	358	*M. rosenbergii*	1247	*Metapenaeus* spp.	1209
L. rostratus	344	*Macrodon ancylodon*	732	*M. endeavouri*	1207
L. rutilus	360	*Macropipus puber*	1299	*M. joyneri*	1208
L. souffia	345	*Macroramphosidae*	503	*Microchirus* spp.	1161
Leucoraja circularis	86	*Macroramphosus scolopax*	504	*M. azevia*	1159
L. fullonica	89	*Macrouridae*	496	*M. ocellatus*	1158
L. naevus	94	*Macrourus* spp.	498	*M. theophila*	1159
Lichia spp.	648	*M. atlanticus*	493	*M. variegatus*	1160
L. amia	647	*M. berglax*	497	*Microcosmus sabatieri*	1508
Limanda aspera	1136	*M. orhynchus*	493	*M. sulcatus*	1508
L. ferruginea	1137	*Macrozoarces americanus*	922	*Microgadus* spp.	457
L. herzensteini	1138	*Macruronus* spp.	477	*M. tomcod*	456
L. limanda	1139	*M. magellanicus*	479	*Micromesistius australis*	458
Limnothrissa spp.	203	*M. novaezealandiae*	480	*M. poutassou*	459
Limulus polyphemus	1515	*Macrurus rupestris*	494	*Micropogonias* spp.	739
Liocarcinus puber	1299	*Mactra glabrata*	1390	*M. furnieri*	737
Liopsetta glacialis	1140	*M. sachalinensis*	1391	*M. undulatus*	738

Micropterus dolomieui	602	M. chilensis	1374	Ophisurus serpens	414
M. salmoides	603	M. crassitesta	1375	Opisthonema spp.	187
Microstomus kitt	1142	M. edulis	1376	O. libertate	185
Misgurnus fossilis	371	M. galloprovincialis	1377	O. oglinum	186
Mitella pollicipes	1197	M. planulatus	1378	Orca gladiator	1470
Mobula hypostoma	114	M. platensis	1379	Orcinus orca	1470
M. mobular	114	M. smaragdinus	1380	Orconectes spp.	1283
M. olfersi	114			O. limosus	1282
Mobulidae	111	**N**		Orcynopsis unicolor	974
Modiolus spp.	1373			Orectolobus maculatus	25
Mola mola	1188	Naucrates ductor	650	Oreochromis mossambicus	832
Mollusca	1323	Navodon spp.	1182	O. niloticus	833
Molva byrkelange	460	Necora puber	1299	Oreosomatidae	528
M. dipterygia dipterygia	460	Negaprion brevirostris	41	Orthopristis chrysoptera	707
M. dipterygia macrophthalma	461	Nemacheilus barbatulus	372	Orthrias barbatulus	372
M. dypterygia	460	Nemachilus barbatula	372	Osmeridae	270
M. elongata	461	Nemadactylus spp.	860	Osmerus spp.	277
M. macrophthalma	461	N. macropterus	859	O. dentex	276
M. molva	462	Nemipteridae	695	O. eperlanus	274
Monachus albiventer	1450	Nemipterus spp.	697	O. mordax dentex	276
M. monachus	1450	N. virgatus	696	O. mordax	275
Monochirus hispidus	1162	Neobalaena marginata	1488	Osphronemus goramy	1045
Monodon monoceros	1479	Neoceratodus forsteri	127	Ostrea spp.	1356
Moridae	436	Neogobius kessleri	1057	O. chilensis	1351
Morone americana	596	Neothunnus albacora	1008	O. denticulata	1352
M. chrysops	597	N. macropterus	1008	O. edulis	1353
M. labrax	593	Nephrops norvegicus	1320	O. lurida	1354
M. punctata	594	Nibea albiflora	740	O. lutaria	1355
M. saxatilis	598	N. mitsukurii	741	Otaria flavescens	1444
Motella tricirrata	450	Notidanus griseus	11	O. jubata	1443
Mugil auratus	540	Notopterus spp.	292	Otariidae	1440
M. brasiliensis	546	Notothenia coriiceps neglecta	882	Otolithes ruber	742
M. capito	542	N. coriiceps	881	Oxynotidae	69
M. capurrii	544	N. gibberifrons	883	Oxynotus centrina	70
M. cephalus	545	N. guentheri	891	Oxyrhina spallanzani	20
M. chelo	538	N. kempi	884		
M. labeo	547	N. nudifrons	887	**P**	
M. labrosus labrosus	538	N. ramsayi	892		
M. liza	546	N. rossii	885	Pachychilon pictum	348
M. provensalis	538	N. squamifrons	886	Pachymetopon spp.	789
M. ramada	542	Nototheniidae	876	Pagellus spp.	795
M. saliens	543	Nototheniops nudifrons	887	P. acarne	790
Mugilidae	537	Nototodarus sloani	1431	P. bellotti natalensis	792
Mullidae	815			P. bellottii bellottii	791
Mullus spp.	819	**O**		P. bellottii	791
M. auratus	816			P. bogaraveo	793
M. barbatus	817	Oblada melanura	788	P. canariensis	794
M. surmuletus	818	Octopodidae	1434	P. centrodontus	793
Muraena helena	404	Octopus spp.	1439	P. erythrinus	794
Muraenesox spp.	407	O. vulgaris	1438	P. mormyrus	786
M. cinereus	406	Ocyurus chrysurus	694	P. natalensis	792
Muraenidae	402	Odobenus rosmarus	1445	Pagophilus groenlandicus	1451
Murex spp.	1334	Odontaspis ferox	14	Pagothenia hansoni	888
Mustelus spp.	55	O. taurus	13	Pagrus spp.	807
M. antarcticus	51	Oedalechilus labeo	547	P. auriga	803
M. asterias	52	Ommastrephes sagittatus	1433	P. caeruleostictus	804
M. laevis	53	Onchorynchus gilae	246	P. major	805
M. mustelus	53	Oncorhynchus aguabonita	243	P. pagrus	806
M. schmitti	54	O. apache	244	P. vulgaris	806
M. stellatus	52	O. chrysogaster	245	Paguridae	1286
M. vulgaris	52	O. gorbuscha	247	Palaemon serratus	1249
Mya arenaria	1415	O. keta	248	Palaemonidae	1244, 1245
Mycteroperca spp.	583	O. kisutch	249	Palinuridae	1259
Myctophidae	294	O. masou	250	Palinurus spp.	1274
Myliobatidae	108	O. masu	250	P. delagoae	1265
Myliobatis aquila	109	O. mykiss	251	P. elephas	1266
M. bovina	110	O. nerka	252	P. gilchristi	1267
Mylopharyngodon piceus	347	O. rhodurus	253	P. mauritanicus	1268
Myoxocephalus spp.	1101	O. tshawytscha	254	Pampus argenteus	1031
M. octodecemspinosus	1098	Onos tricirrata	450	P. chinensis	1032
M. quadricornis	1099	Ophicephalus spp.	558	Pandalus spp.	1239
M. scorpius	1100	Ophidion barbatum	929	P. borealis	1237
Mytilidae	1370	Ophiodon elongatus	1092	P. montagui	1238
Mytilus canaliculus	1381				

Panulirus spp.	1274	*P. simillimus*	1034	*Polysteganus macrophthalmus*	776
P. argus	1269	*P. triacanthus*	1035	*P. maroccanus*	777
P. burgeri	1271	*Perca flavescens*	616	*P. praeorbitalis*	797
P. cygnus	1270	*P. fluviatilis*	617	*P. undulosus*	798
P. homarus	1271	*Percidae*	613	*Pomacentridae*	835
P. japonicus	1272	*Percophidae*	872	*Pomadasys* spp.	714
P. regius	1273	*Percophis brasilianus*	873	*P. argenteus*	711
Paphia aurea	1404	*Peristedion cataphractum*	1106	*P. incisus*	712
Parablennius gattorugine	909	*Perna canaliculus*	1381	*P. jubelini*	713
P. sanguinolentus	910	*P. perna*	1382	*Pomatomidae*	629
Parabramis pekinensis	349	*Petromyzon fluviatilis*	5	*Pomatomus saltator*	630
Paracentrotus lividus	1518	*P. marinus*	7	*P. saltatrix*	630
Paralabrax humeralis	584	*Petromyzontidae*	1	*Pomatoschistus canestrinii*	1058
Paralichthys spp.	1121	*Petrus rupestris*	796	*P. minutus*	1059
P. dentatus	1118	*Phaeophyceae*	1525	*Pomolobus pseudoharengus*	161
P. oblongus	1119	*Phoca hispida*	1452	*Pomoxis annularis*	604
P. olivaceus	1120	*P. vitulina*	1453	*P. nigromaculatus*	605
Paralithodes spp.	1311	*Phocidae*	1446	*Pontinus kuhlii*	1065
P. camtschatica	1310	*Phocoena phocoena*	1471	*Portunus* spp.	1301
Paralonchurus peruanus	743	*Phocoenoides dalli*	1472	*P. puber*	1299
Paralosa fallax	157	*Pholas* spp.	1417	*P. trituberculatus*	1300
Paranotothenia magellanica	889	*P. dactylus*	1416	*Priacanthidae*	606
Parapenaeopsis spp.	1211	*Pholis gunnellus*	919	*Priacanthus* spp.	608
P. atlantica	1210	*Phoxinus percnurus*	351	*P. macracanthus*	607
Parapenaeus longirostris	1212	*P. phoxinus*	352	*Prionace glauca*	42
Parapercis colias	864	*Phrynorhombus norvegicus*	1171	*Prionotus* spp.	1087
Parapristipoma octolineatum	708	*Phycis* spp.	465	*Pristidae*	78
Parasilurus aristotelis	381	*P. blennoides*	463	*Pristis pectinata*	79
Parastromateus niger	677	*P. phycis*	464	*Pristiurus melanostomus*	30
Parathunnus obesus	1011	*Physeter catodon*	1462	*Procambarus clarckii*	1284
Parona signata	651	*P. macrocephalus*	1462	*Prognichthys agoo*	424
Patagonotothen brevicauda	890	*Physiculus bacchus*	438	*Promicrops itajara*	578
P. brevicauda guntheri	891	*Pinctada martensii*	1346	*Proterorhinus marmoratus*	1060
P. longipes ramsay	892	*P. maxima*	1346	*Protopterus* spp.	129
Patella spp.	1333	*Pinguipes* spp.	865	*Psenes indicus*	1038
Pecten jacobaeus	1364	*Pinna nobilis*	1385	*Psenopsis anomala*	1044
P. maximus	1365	*Placopecten magellanicus*	1369	*Psephurus gladius*	141
P. meridionalis	1366	*Platax* spp.	826	*Psetta maxima*	1172
P. novaezealandiae	1367	*Platichthys flesus*	1143	*P. maxima maeotica*	1173
P. sulcicostatus	1368	*P. stellatus*	1144	*Psettodes belcheri*	1111
P. varius	1363	*Platycephalidae*	1095	*P. bennetti*	1112
Pectinidae	1358	*Platycephalus indicus*	1096	*P. erumei*	1113
Pectunculus glycymeris	1384	*Plecoglossus altivelis*	269	*Pseudocaranx dentex*	652
Pegusa kleinii	1163	*Plectorhinchus macrolepis*	709	*Pseudochaenichthys georgianus*	901
P. lascaris	1164	*P. mediterraneus*	710	*Pseudophycis bacchus*	438
Pelamys bonapartei	974	*Pleoticus muelleri*	1256	*Pseudopleuronectes americanus*	1147
P. sarda	982	*P. robustus*	1257	*P. herzensteini*	1138
Pelecus cultratus	350	*Plesionika* spp.	1240	*Pseudorca crassidens*	1473
Pellona ditchela	188	*Plesiopenaeus edwardsianus*	1234	*Pseudorhombus cinnamoneus*	1122
Penaeus spp.	1229	*Pleuragramma antarcticum*	893	*Pseudosciaena crocea*	722
P. aztecus	1213	*Pleurogrammus azonus*	1093	*P. manchurica*	746
P. brevirostris	1214	*Pleuronectes cynoglossus*	1128	*P. polyactis*	747
P. californiensis	1215	*P. luteus*	1156	*Pseudotolithus* spp.	752
P. caramote	1220	*P. maeoticus*	1173	*P. brachygnatus*	748
P. chinensis	1216	*P. microcephalus*	1142	*P. elongatus*	749
P. duorarum	1217	*P. pallasii*	1146	*P. senegalensis*	750
P. duorarum notialis	1224	*P. platessa*	1145	*P. typus*	751
P. esculentus	1218	*P. quadrituberculatus*	1146	*Pseudupeneus prayensis*	820
P. japonicus	1219	*Pleuronectidae*	1123	*Pterogymnus laniarius*	799
P. kerathurus	1220	*Pleuronectiformes*	1110	*Pteromylaeus bovinus*	110
P. latisulcatus	1221	*Plotosus* spp.	378	*Pteroplatea altavela*	115
P. merguiensis	1222	*Pneumatophorus colias*	984	*Pteroscion peli*	753
P. monodon	1223	*Pogonias cromis*	745	*Pterosmaris melanura*	811
P. notialis	1224	*Pollachius pollachius*	466	*Pterothrissus belloci*	153
P. penicillatus	1225	*P. virens*	467	*Pterygotrigla polyommata*	1088
P. semisulcatus	1226	*Pollicepes pictorum*	1345	*Puerulus* spp.	1275
P. setiferus	1227	*Pollicipes cornucopia*	1197	*Pungitius pungitius*	501
P. trisulcatus	1220	*Polynemidae*	553	*Puntazzo puntazzo*	781
P. vannamei	1228	*Polynemus quadrifilis*	556	*Puntius* spp.	354
Pennahia argentata	744	*Polyodontidae*	140	*P. javanicus*	353
Pentanemus quinquarius	555	*Polyprion americanus*	585	*Pusa hispida*	1452
Peprilus spp.	1036	*P. cernium*	585	*Pygosteus pungitius*	501
P. alepidotus	1033	*P. oxygeneios*	586	*Pyura chilensis*	1509
P. paru	1033	*Polypterus bichir*	128	*P. stolonifera*	1510

R

Rachycentron canadum	631
Raja spp.	102
R. *alba*	82
R. *asterias*	83
R. *batis*	84
R. *brachyura*	85
R. *chagrine*	89
R. *circularis*	86
R. *clavata*	87
R. *erinacea*	88
R. *fullonica*	89
R. *lintea*	90
R. *macrorhynchus*	84
R. *marginata*	82
R. *microocellata*	91
R. *miraletus*	92
R. *montagui*	93
R. *naevus*	94
R. *ocellata*	95
R. *oxyrinchus*	96
R. *punctata*	83
R. *radiata*	97
R. *radula*	98
R. *stellulata*	100
R. *undulata*	101
Rajidae	81
Rajiformes	80
Rana spp.	1493
R. *catesbyana*	1492
Raniceps raninus	468
Ranzania laevis	1189
R. *truncata*	1189
Rastrelliger spp.	978
R. *brachysoma*	975
R. *faughni*	976
R. *kanagurta*	977
Rastrineobola argentea	355
Regalecus glesne	510
Regificola grandis	659
Reinhardtius hippoglossoides	1148
Remora remora	1178
Reptantia	1285
Rexea solandri	949
Rhabdosargus globiceps	800
Rhincodon typus	27
Rhiniodon typus	27
Rhinobatidae	72
Rhinobatos spp.	77
R. *albomaculatus*	73
R. *columnae*	76
R. *percellens*	74
R. *planiceps*	75
R. *rhinobatos*	76
Rhinonemus cimbrius	469
Rhizoprionodon longurio	43
R. *terraenovae*	44
Rhodeus amarus	356
R. *sericeus amarus*	356
Rhodophyceae	1528
Rhombosolea spp.	1150
R. *plebeia*	1149
Rhombus laevis	1175
R. *maximus*	1172
R. *vulgaris*	1175
Rhopilema spp.	1512
Roccus americanus	596
R. *chrysops*	597
R. *labrax*	593
R. *saxatilis*	598
Romanichtys valsanicola	618
Rossia macrosoma	1419
Rostroraja alba	82

Ruditapes decussatus	1401
R. *philippinarum*	1402
Rutilus spp.	361
R. *frisii*	357
R. *pigus*	358
R. *rubilio*	359
R. *rutilus*	360
Ruvettus pretiosus	950

S

Sabanejewia conspersa	373
Saccostrea cuccullata	1357
Salaria fluviatilis	911
Salmo spp.	260
S. *aguabonita*	243
S. *clarki*	255
S. *crysogaster*	245
S. *gairdneri*	251
S. *gilae*	246
S. *hucho*	242
S. *irideus*	251
S. *salar*	256
S. *salvelinus*	263
S. *trutta fario*	258
S. *trutta lacustris*	259
S. *trutta*	257
S. *trutta trutta*	257
Salmonidae	262
Salvelinus spp.	267
S. *alpinus*	263
S. *fontinalis*	264
S. *malma*	265
S. *namaycush*	266
Sander lucioperca	620
Sarda australis	979
S. *chiliensis*	980
S. *orientalis*	981
S. *sarda*	982
S. *unicolor*	974
Sardina pilchardus	189
Sardinella spp.	195
S. *aurita*	190
S. *eba*	193
S. *fimbriata*	191
S. *granigera*	193
S. *longiceps*	192
S. *maderensis*	193
S. *zunasi*	194
Sardinops caeruleus	196
S. *melanostictus*	197
S. *neopilchardus*	198
S. *ocellatus*	199
S. *sagax*	200
Sargocentron rubrum	522
Sarpa salpa	801
Saurida grandisquamis	392
S. *tumbil*	391
S. *undosquamis*	392
Saxidomus giganteus	1403
Scardinius erythrophthalmus	362
S. *graecus*	363
Scarus cretensis	856
Scatophagus spp.	829
Sciaena spp.	756
S. *antarctica*	754
S. *aquila*	718
S. *umbra*	755
Sciaenidae	715
Sciaenops ocellatus	757
Scolopsis spp.	698
Scomber spp.	986
S. *australasicus*	983

S. *colias*	984
S. *japonicus*	984
S. *scomber*	985
S. *scombrus*	985
Scomberesocidae	420
Scomberesox saurus	422
Scomberoides spp.	653
Scomberomorus spp.	1005
S. *brasiliensis*	987
S. *cavalla*	988
S. *commerson*	989
S. *concolor*	990
S. *guttatus*	991
S. *koreanus*	992
S. *lineolatus*	993
S. *maculatus*	994
S. *multiradiatus*	995
S. *munroi*	996
S. *niphonius*	997
S. *plurilineatus*	998
S. *queenslandicus*	999
S. *regalis*	1000
S. *semifasciatus*	1001
S. *sierra*	1002
S. *sinensis*	1003
S. *tritor*	1004
Scombridae	961
Scombroidei	960
Scophthalmidae	1167
Scophthalmus aquosus	1174
S. *maeoticus*	1173
S. *maximus*	1172
S. *rhombus*	1175
Scorpaena spp.	1069
S. *notata*	1066
S. *porcus*	1067
S. *scrofa*	1068
S. *ustulata*	1066
Scorpaenidae	1062
Scyliorhinidae	28
Scyliorhinus spp.	33
S. *canicula*	31
S. *stellaris*	32
Scylla serrata	1302
Scyllaridae	1304
Scyllarus arctus	1305
Scymnorhinus licha	60
Sebastes spp.	1077
S. *alutus*	1070
S. *capensis*	1071
S. *entomelas*	1072
S. *fasciatus*	1073
S. *flavidus*	1074
S. *marinus*	1075
S. *mentella*	1076
S. *norvegicus*	1075
Sebastodes alutus	1070
Selar crumenophthalmus	654
Selaroides leptolepis	655
Selene dorsalis	656
S. *setapinnis*	657
Sepia officinalis	1418
Sepiola rondeleti	1420
Sergestidae	1241
Seriola spp.	661
S. *dumerili*	658
S. *lalandi*	659
S. *quinqueradiata*	660
Seriolella spp.	1042
Serranidae	567
Serranus spp.	590
S. *cabrilla*	587
S. *gigas*	577
S. *hepatus*	588

S. scriba	589	*S. megalops*	66	*T. indicus*	1018
Seserinus microchirus	1037	*Squatina angelus*	68	*T. pfluegeri*	1026
Sicyonia spp.	1253	*S. squatina*	68	*Teuthis sigan*	943
S. brevirostris	1252	*Squatinidae*	67	*Thalassoma pavo*	854
Siganus spp.	944	*Squilla mantis*	1202	*Thaleichthys pacificus*	278
S. rivulatus	943	*Squillidae*	1201	*Theragra chalcogramma*	470
Siliqua patula	1411	*Stegostoma fasciatum*	24	*Theraponidae*	592
Sillaginidae	625	*S. tigrinum*	24	*Thunnidae*	1006
Siluridae	380	*Stenella* spp.	1476	*Thunnus* spp.	1014
Silurus aristotelis	381	*S. coeruleoalba*	1474	*T. alalunga*	1007
S. glanis	382	*S. longirostris*	1475	*T. albacares*	1008
Siphonostomus typhle	508	*Stenotomus chrysops*	809	*T. albacores*	1008
Smaris maurii	813	*Stephanolepis cirrhifer*	1183	*T. atlanticus*	1009
S. smaris	813	*Stereolepis gigas*	591	*T. maccoyii*	1010
S. vulgaris	813	*Stichopus japonicus*	1523	*T. obesus*	1011
Solea cuneata	1157	*Stizostedion canadense*	619	*T. thynnus*	1012
S. kleinii	1163	*S. lucioperca*	620	*T. tonggol*	1013
S. lascaris	1164	*S. vitreum glaucum*	620	*Thymallus thymallus*	268
S. lutea	1156	*S. vitreum vitreum*	621	*T. vexillifer*	268
S. ocellata	1158	*Stolephorus* spp.	222	*Thynnus pelamys*	973
S. solea	1165	*Stolothrissa* spp.	203	*T. thonina*	968
S. variegata	1160	*Strangomera bentincki*	204	*Thyrsites* spp.	952
S. vulgaris	1165	*Stromateidae*	1030	*T. atun*	951
Soleidae	1151	*Stromateus capensis*	1037	*Thyrsitoides marleyi*	953
Solen spp.	1414	*S. fiatola*	1037	*Thyrsitops lepidopoides*	954
S. capensis	1412	*Strombus* spp.	1338	*Tinca tinca*	364
S. ensis	1409	*Strongylocentrotus* spp.	1519	*Todarodes pacificus*	1432
S. vagina	1413	*S. lividus*	1518	*T. sagittatus sagittatus*	1433
Solenidae	1407	*Strongylura marina*	417	*Torpedinidae*	117
Solenocera membranacea	1258	*Symphodus cinereus*	845	*Torpedo* spp.	121
Somniosus microcephalus	63	*S. mediterraneus*	846	*T. marmorata*	118
Sparidae	763	*S. melops*	847	*T. nobiliana*	119
Sparisoma cretense	856	*S. ocellatus*	848	*T. torpedo*	120
Sparus alcedo	813	*S. pavo*	851	*Totoaba macdonaldi*	724
S. aurata	802	*S. quinquemaculatus*	849	*Trachinidae*	866
S. caeruleostictus	804	*S. roissali*	849	*Trachinotus* spp.	664
S. ehrenbergii	804	*S. rostratus*	850	*T. carolinus*	662
S. major	805	*S. tinca*	851	*T. ovatus*	663
S. pagrus	806	*Synagrops japonicus*	612	*Trachinus* spp.	871
Sparusa auriga	803	*Synapturichthys kleinii*	1163	*T. araneus*	868
Sphoeroides spp.	1186	*Syngnathus acus*	506	*T. draco*	869
S. maculatus	1185	*S. rostellatus*	507	*T. radiatus*	870
Sphyraena spp.	536	*S. typhle*	508	*T. vipera*	867
S. afra	530	*Synodontidae*	390	*Trachipterus taenia*	511
S. argentea	531	*Synodontis* spp.	293	*T. trachipterus*	511
S. barracuda	532	*Synodus saurus*	393	*Trachurus* spp.	675
S. chrysotaenia	533			*T. capensis*	673
S. jello	534			*T. declivis*	665
S. obtusata	533	**T**		*T. japonicus*	666
S. sphyraena	535			*T. lathami*	667
Sphyraenidae	529	*Tapes* spp.	1394	*T. mediterraneus*	668
Sphyrna spp.	48	*T. aureus*	1404	*T. murphyi*	669
S. lewini	46	*T. decussa*	1401	*T. picturatus*	670
S. zygaena	47	*T. japonica*	1402	*T. symmetricus*	671
Sphyrnidae	45	*T. pullastra*	1405	*T. trachurus*	672
Spicara spp.	814	*T. semidecussatus*	1402	*T. trachurus capensis*	673
S. maena	812	*T. variegata*	1402	*T. trecae*	674
Spinachia spinachia	502	*Tarpon atlanticus*	150	*Trachynotus glaucus*	663
Spinax acanthias	64	*Taurulus bubalis*	1102	*Trachypenaeus* spp.	1230
S. niger	61	*Tautoga onitis*	852	*Trematomus* spp.	894
Spirlinus bipunctatus	308	*Tautogolabrus adspersus*	853	*T. hansoni*	888
Spisula solidissima	1392	*Telestes agassizii*	345	*Triakidae*	49
Spondyliosoma cantharus	808	*Temnodon saltator*	630	*Triakis semifasciata*	56
Spongidae	1511	*Tenualosa ilisha*	205	*Trichechus inunguis*	1455
Spratelloides gracilis	201	*T. reevesi*	206	*T. manatus*	1456
Sprattus sprattus	202	*T. toli*	207	*T. senegalensis*	1457
Squalidae	57	*Testudinata*	1494	*Trichiuridae*	955
Squalius cephalus	341	*Tetraodontidae*	1184	*Trichiurus lepturus*	959
S. leuciscus	344	*Tetrapturus* spp.	1027	*Trichogaster pectoralis*	1046
S. rostratus	344	*T. albidus*	1021	*Trigla aspera*	1085
Squalus acanthias	64	*T. angustirostris*	1022	*T. corax*	1089
S. blainvillei	65	*T. audax*	1023	*T. cuculus*	1079
S. fernandinus	65	*T. belone*	1024	*T. gurnardus*	1084
S. glaucus	42	*T. georgei*	1025	*T. hirundo*	1089

T. lineata	1091	*Umbrina* spp.	760	**X**	
T. lucerna	1089	*U. canosai*	758		
T. lyra	1090	*U. cirrosa*	759	*Xiphias gladius*	1029
T. obscura	1080	*Unio pictorum*	1345	*Xiphiidae*	1028
T. pini	1079	*Unionidae*	1344	*Xiphopenaeus kroyeri*	1231
Triglidae	1078	*Upeneus* spp.	822	*Xyrichthys novacula*	855
Trigloporus lastoviza	1091	*U. asymmetricus*	821		
Tripterygion nasus	912	*Uranoscopus* spp.	875	**Z**	
T. tripteronotum	912	*U. scaber*	874		
Trisopterus esmarki	471	*Urophycis blennoides*	463	*Zeidae*	523
T. luscus	472	*U. brasiliensis*	474	*Zenopsis conchifera*	524
T. minutus	473	*U. chuss*	475	*Z. ocellata*	524
T. minutus capelanus	473	*U. tenuis*	476	*Zeugopterus punctatus*	1176
Trochus spp.	1330			*Zeus faber*	525
Trutta fluviatilis	258			*Z. japonicus*	525
T. marina	257			*Z. pungio*	525
Turbo cornutus	1331	**V**		*Zevaia theophila*	1159
T. sarmaticus	1332			*Zingel asper*	622
Tursiops truncatus	1477	*Valencia hispanica*	515	*Z. streber*	623
Tylosaurus acus melanotus	418	*Veneridae*	1393	*Z. zingel*	624
Tylosurus acus	418	*Venerupis aurea*	1404	*Ziphiidae*	1458
		V. decussata	1401	*Ziphius cavirostris*	1461
		V. pullastra	1405	*Zoarces viviparus*	923
U		*Venus gallina*	1396	*Zoarcidae*	920
		V. mercenaria	1398	*Zosterisessor ophiocephalus*	1061
Umbra krameri	284	*Vimba vimba*	365	*Zygaena malleus*	47
U. pygmaea	285	*Vomer* spp.	676		

Index of family names

Muraenidae	402-404
Muricidae	1334
Myctophidae	294-295
Myidae	1415
Myliobatidae	108-110
Mytilidae	1370-1382

N

Nemipteridae	695-698
Nephropidae	1317-1320
Nomeidae	1038
Notopteridae	292
Nototheniidae	876-894

O

Octopodidae	1434-1439
Odobenidae	1445
Odontaspididae	13-14
Ommastrephidae	1427-1433
Ophichthidae	413-414
Ophidiidae	924-929
Orectolobidae	25
Oreosomatidae	528
Osmeridae	270-278
Osphronemidae	1045
Ostreidae	1347-1357
Otariidae	1440-1444
Oxynotidae	69-70

P

Paguridae	1286
Palaemonidae	1244-1249
Palinuridae	1259-1275
Pandalidae	1235-1240
Parastacidae	1316
Patellidae	1333
Pectinidae	1358-1369
Penaeidae	1206-1231
Percichthyidae	599
Percidae	613-624
Percophidae	872-873
Peristediidae	1106
Petromyzontidae	1-7
Phocidae	1446-1453
Pholadidae	1416-1417
Pholidae	919
Physeteridae	1462
Pinnidae	1385
Platycephalidae	1095-1096

Plecoglossidae	269
Pleuronectidae	1123-1150
Plotosidae	378
Poeciliidae	516
Polynemidae	553-556
Polyodontidae	140-141
Polypteridae	128
Pomacentridae	835-836
Pomatomidae	629-630
Portunidae	1294-1302
Priacanthidae	606-608
Pristidae	78-79
Protopteridae	129
Psettodidae	1111-1113
Pteriidae	1346

R

Rachycentridae	631
Rajidae	81-102
Ranidae	1492-1493
Regalecidae	510
Rhincodontidae	27
Rhinobatidae	72-77
Rhizostomatidae	1512

S

Salmonidae	229-267
Scalpellidae	1197
Scaridae	856
Scatophagidae	829
Sciaenidae	715-760
Scomberesocidae	420-422
Scombridae	961-1014
Scophthalmidae	1167-1176
Scorpaenidae	1062-1077
Scyliorhinidae	28-33
Scyllaridae	1304-1305
Sepiidae	1418
Sepiolidae	1419-1420
Sergestidae	1241-1242
Serranidae	567-591
Sicyoniidae	1252-1253
Siganidae	943-944
Sillaginidae	625
Siluridae	379-382
Soleidae	1151-1165
Solenidae	1407-1414
Solenoceridae	1254-1258
Sparidae	763-809
Sphyraenidae	529-536
Sphyrnidae	45-48

Spongidae	1511
Squalidae	57-66
Squatinidae	67-68
Squillidae	1201-1202
Stegostomatidae	24
Strichopodidae	1523
Stromateidae	1030-1037
Strombidae	1338
Strongylocentrotidae	1518-1519
Syngnathidae	505-508
Synodontidae	390-393

T

Tetraodontidae	1184-1186
Theraponidae	592
Thymallidae	268
Torpedinidae	117-121
Trachichthyidae	520-521
Trachinidae	866-871
Trachipteridae	511
Triakidae	49-56
Trichechidae	1455-1457
Trichiuridae	955-959
Trichodontidae	863
Triglidae	1078-1091
Trochidae	1330
Turbinidae	1331-1332

U

Umbridae	284-285
Unionidae	1343-1345
Uranoscopidae	874-875

V

Veneridae	1393-1405

X

Xanthidae	1293
Xiphiidae	1028-1029
Yuridae	1508-1510

Z

Zeidae	523-525
Ziphiidae	1458-1461
Zoarcidae	920-923

English index

A

abalone, blacklip	1327
abalone, European	1328
abalone, giant	1325
abalone, perlemoen	1326
abalones	1329
Acadian redfish	1073
adalah	1113
Adriatic sturgeon	133
Aesop shrimp	1238
African lookdown	656
African lungfishes	129
African manatee	1457
African mullet, leaping	544
African sicklefish	827
African striped grunt	708
African threadfin, giant	556
African weakfish	719
agujon needlefish	418
air-breathing catfishes	388
akiami paste shrimp	1242
alache	190
Alaska deepsea crab	1310
Alaska plaice	1146
Alaska pollack	470
albacore	1007
albacore, false	968
albacore, long-finned	1007
albacore, Pacific	1007
Albanian barbel	313
Albanian roach	348
alewife	156, 161
Alexandria pompano	633
alfonsinos	517, 519
algae	1524
algae, brown	1525
algae, red	1528
allice shad	156
alligator	1502
alligators and crocodiles	1500
allis shad	156
allis shads	163
Allison's tuna	1008
amarillo snapper	689
amberjack, greater	658
amberjack, Japanese	660
amberjack, yellow-tail	659
amberjacks	661
American angler	1193
American conger	411
American crayfish	1282
American crayfishes	1283
American cupped oyster	1349
American eel	400
American gizzard shad	172
American John Dory	524
American lake char	266
American lake trout	266
American lobster	1317
American mud-minnow	285
American plaice	1132
American sandlance	932
American shad	162
American smelt	275
American terrapins	1499
American yellow perch	616

Amur pike	291
amur, white	331
anchoita	214
anchoita, river	221
anchoveta	220
anchoveta, Atlantic	211
anchoveta, Californian	219
anchoveta, Pacific	212
anchovies	208
anchovies, stolephorus	222
anchovy	217
anchovy, Argentine	214
anchovy, Australian	215
anchovy, bay	210
anchovy, broad-striped	209
anchovy, European	217
anchovy, Japanese	218
anchovy, North Pacific	219
anchovy, Northern	219
anchovy, Osbeck's grenadier	213
anchovy, Peruvian	220
anchovy, rat-tail	213
anchovy, Southern African	216
anchovy, stet	216
anchovy, striped	209
Andaman lobster	1319
angel shark	68
angel sharks	67
angelfish	68
angler, American	1193
angler, black-bellied	1194
anglerfish	1195
anglerfish, devil	1196
anglerfishes	1192
Angola dentex	771
angular rough shark	70
annular seabream	778
Antarctic butterfish	1043
Antarctic cods	876, 894
Antarctic icefish	898
Antarctic krill	1204
Antarctic rockcods	876, 894
Antarctic sidestripe	893
Antarctic silverfish	893
Antarctic toothfish	878
Antarctic toothfishes	879
antimora, blue	437
Araucanian herring	204
Arctic charr	263
Arctic cod	441, 445
Arctic flounder	1140
Arctic right whale	1487
Arctic smelt	276
Argentine anchovy	214
Argentine conger	412
Argentine croaker	758
Argentine hake	486
Argentine red shrimp	1256
Argentine seabass	568
Argentine squid short-fin	1428
Argentine stiletto shrimp	1206
argentine, Atlantic	281
argentine, Pacific	280, 283
argentines	279
aristeid shrimps	1232
Arizona trout	244
armed gurnard	1106

arrow-tooth flounder	1125
arrow shrimps	1236
arrow, sea	1426
Asian barbs	354
Asiatic smelt	276
asp	311
Atka mackerel	1093
Atlantic anchoveta	211
Atlantic argentine	281
Atlantic black skipjack	968
Atlantic blue marlin	1019
Atlantic bluefin tuna	1012
Atlantic bonito	982
Atlantic bumper	640
Atlantic butterfish	1035
Atlantic catfish	915
Atlantic cod	447
Atlantic croaker	738
Atlantic halibut	1133
Atlantic herring	169
Atlantic horse mackerel	672
Atlantic John Dory	525
Atlantic lizardfish	393
Atlantic mackerel	985
Atlantic manta	112
Atlantic menhaden	166
Atlantic moonfish	657
Atlantic navaga	445
Atlantic needlefish	417
Atlantic pomfret	680
Atlantic prickly skate	97
Atlantic puffers	1186
Atlantic razor clam	1408
Atlantic rock crab	1289
Atlantic round herring	177
Atlantic sailfish	1016
Atlantic salmon	256
Atlantic saury	422
Atlantic sea snail	1107
Atlantic seabob	1231
Atlantic searobins	1087
Atlantic sharp-nose shark	44
Atlantic silverside	552
Atlantic Spanish mackerel	994
Atlantic spotted grunt	713
Atlantic thread herring	186
Atlantic tomcod	456
Atlantic torpedo	119
Atlantic tuna	1012
Atlantic white marlin	1021
Atlantic wolffish	915
atyid shrimps	1243
Australian anchovy	215
Australian bonito	979
Australian crayfish	1316
Australian eel	397
Australian lungfish	127
Australian mussel	1378
Australian salmon	683
Australian scallop	1366
Australian spiny lobster	1261, 1270
Australian spotted mackerel	996
axillary seabream	790
axillary wrasse	846
ayu sweetfish	269
ayu	269
Azovtyulka	171

399

B

baggar, white 377
Baird's smooth-head 223
balao halfbeak 429
balao 429
Balkan loach 367
ballan wrasse 840
ballyhoo halfbeak 430
ballyhoo 430
banana prawn 1222
bandang 228
bandeng 228
bandfish, red 834
bar jack 638
barb, Java 353
barbel 314
barbel, Albanian 313
barbel, Danubian 317
barbel, Iberian 315
barbel, Italian 318
barbel, Mediterranean 316
barbel-gudgeon, Dalmatian 312
barberfish 570
barbs, Asian 354
barnacle, giant 1200
barnacle, goose 1197, 1343
barnacles 1199
barnacles, goose 1198
barracouta 951
barracoutas 952
barracuda 532, 535
barracuda, blunt-jaw 533
barracuda, California 531
barracuda, great 532
barracuda, Guinean 530
barracuda, Indian 534
barracuda, Pacific 531
barracuda, yellow-finned 533
barracudas 529, 536
barramundi 564
barred grunt 704
barrelfishes 1040
basking shark 12
bass 593
bass, giant sea 591
bass, large mouthed black 603
bass, large-mouth 603
bass, Peruvian rock 584
bass, rock 600
bass, sea spotted 594
bass, small-mouth 602
bass, stone 585
bass, striped 598
bass, white 597
bass, wreck 585
bastard grunt 712
bastard halibut 1120
bastard halibuts 1121
batfishes 826
bay anchovy 210
bay scallop 1361
beaked redfish 1076
beaked whale 1459, 1461
beaked whales 1458
bearded brotula 924
bearded seal 1448
belted bonito 982
beluga whale 1478
beluga 139
Benguela hake 489
Bergatino loach 368
bib 472
bibbaud loach 370

bichir 128
big powan 238
big skate 95
big-eye herring 179
big-eye scad 654
big-eye thresher 16
bigeye grunt 703
bigeye tuna 1011
bigeye, red 607
bigeyes 606, 608
bighead carp 336
bighead goby 1057
biglip grunt 709
billfish 416, 422
billfishes 1015
Biscayan right whale 1490
bitterling 356
Biwa 253
black bass, large mouthed 603
black bullhead 384
black cardinalfish 611
black carp 347
black centrina 61
black cod 467
black crappie 605
black croaker 721
black drum 745
black goby 1053
black hake 491
black halibut 1148
black marlin 1018
black pollack 467
black pomfret 677
black right whale 1490
black rockcod 881
black salmon 254
black scabbardfish 956
black scorpionfish 1067
black sea bream 808
Black Sea goby, mottled 1060
Black Sea roach 357
Black Sea shad 160
Black Sea sprat 171
Black Sea turbot 1173
Black Sea whiting 455
black seabass 571
black skipjack 969
black skipjack, Atlantic 968
black stone crab 1293
black-bellied angler 1194
black-belly rosefish 1063
blackfin icefish 896
blackfin tuna 1009
blackfish 1041
blacklip abalone 1327
black-mouth catshark 30
black-mouth croaker 721
black-mouthed dogfish 30
black-spot grenadier 493
blackspot picarel 811
blackspot seabream 793
black-striped goatfish 821
black-tip reef shark 38
Blainville's spurdog 65
bleak 310
bleak, Danube 322
bleak, Italian 309
blennie, Patagonian 880
blennies 906
blennies, combtooth 904
blenny, butterfly 905
blenny, freshwater 911
blenny, peacock 907
blenny, snake 929

blenny, tompot 909
blenny, triple-fin 912
blenny, viviparous 923
blond ray 85
blood cockle 1339
blood cockles 1341, 1342
blood-striped blennie 910
blotched picarel 812
blue antimora 437
blue butterfish 1037
blue cod, New Zealand 864
blue crab 1295
blue grenadier 480
blue grenadiers 477
blue jack mackerel 670
blue ling 460
blue mackerel 983
blue marlin 1019
blue marlin, Atlantic 1019
blue moki 862
blue mussel 1376
Blue Point oyster 1349
blue runner 635
blue shark 42
blue skate 84
blue whale 1483
blue whaler 42
blue whitefish 240
blue whiting 459
blue whiting, Southern 458
blue-and-red shrimp 1233
blue-back herring 155
blue-back shad 155
blueback 249, 252
blue-fin gurnard 1082
bluefin tuna 1012
bluefin tuna, Atlantic 1012
bluefin tuna, Northern 1012
bluefin tuna, Southern 1010
bluefish 630
bluefishes 629
blue-mouth 1063
bluenose warehou 1043
blunt-jaw barracuda 533
bluntnose six-gill shark 11
boarfish 527
boarfishes 526
bobo croaker 749
bobo mullet 539
bobyrez 340
boe drum 753
bogue 766, 801
Bombay duck 297
Bombay ducks 296
bonefish 152
bonefish, long-fin 153
bonefishes 151
bonga shad 174
bonito 973, 974
bonito, Atlantic 982
bonito, Australian 979
bonito, belted 982
bonito, Eastern Pacific 980
bonito, leaping 966
bonito, oceanic 973
bonito, Pacific 980
bonito, plain 964, 974
bonito, stripe-bellied 973
bonito, striped 981
bonnetfish, Cape 685
bonnetfishes 684
bonnetheads 45
bonnetmouth 685
bordered skate 82

boreal smelt	276	
bottle-nose dolphin	1477	
bottlenose skate	82	
bottlenosed whale	1460	
bottom pollan	239	
bowfin	142	
box-lip mullet	547	
Boyer's sandsmelt	549	
bramble shark	71	
Brazilian codling	474	
Brazilian flathead	873	
Brazilian groupers	583	
Brazilian mullet	546	
Brazilian sandperches	865	
bream	305	
bream, black sea-	808	
bream, common sea-	793	
bream, common	305	
bream, Couch's sea-	806	
bream, Danube	306	
bream, dog's tooth	774	
bream, freshwater	305	
bream, golden thread-fin	696	
bream, Ray's	680	
bream, red sea-	793	
bream, red	518, 793	
bream, saddled	788	
bream, sea-	774, 808	
bream, silver	319	
bream, Spanish sea-	794	
bream, striped sea-	786	
bream, striped sea	786	
bream, vimba	365	
bream, white amur	349	
bream, Wuchang	346	
breams	307, 679	
breams, copper	789	
breams, dwarf	695	
breams, freshwater	307	
breams, large-eye	761	
breams, monocle	695, 698	
breams, pargo	807	
breams, red	519	
breams, sargo	784	
breams, sea-	763	
breams, South Pacific	1042	
breams, thread-fin	695, 697	
brill	1175	
brisling	202	
broad-barred king mackerel	1001	
broadbill	1029	
broad-headed notothenia	881	
broad-striped anchovy	209	
broad-tail shortfin squid	1429	
brook charr	264	
brook lamprey	6	
brook trout	258, 264	
brotula, bearded	924	
brown algae	1525	
brown bullhead	385	
brown cat shark	29	
brown comber	588	
brown meagre	755	
brown ray	92	
brown seaweed	1525	
brown shrimps, Northern	1213	
brown tiger prawn	1218	
brown trout	257, 258	
brown wrasse	842	
brushtooth lizardfish	392	
Bryde's whale	1482	
buckie	1335	
buffalofishes	302	
bull frog	1492	

bull ray	110	
bull trout	265	
bullet mackerel	964	
bullet tuna	964	
bullhead catfishes	387	
bull-head perch, Romanian	618	
bullhead	1097	
bullhead, black	384	
bullhead, brown	385	
bumalo	297	
bummalow	297	
bumper, Atlantic	640	
bumphead notothenia	883	
bunker	165, 166	
burbot	452	
butter clam	1403	
butterfish	919, 1035, 1037	
butterfish, Antarctic	1043	
butterfish, Atlantic	1035	
butterfish, blue	1037	
butterfishes	1030	
butterfishes, Gulf	1036	
butterfly blenny	905	
butterfly kingfish	970	
butterfly ray, spiny	115	
butterfly rays	103, 116	
butterfly skate	94	
butterflyfishes	830	

C

cachalot	1462	
caiman, spectacled	1503	
calico salmon	248	
calico clam	1401	
calicot scallop	1360	
California barracuda	531	
California round herring	176	
Californian anchoveta	219	
Californian pilchard	196	
Californian sardine	196	
Canary dentex	772	
candlefish	278	
Canestrini's goby	1058	
cannadelle	850	
Cape bonnetfish	685	
Cape elephantfish	125	
Cape fur seal	1442	
Cape gurnard	1081	
Cape hake	484	
Cape hake, deepwater	488	
Cape Hope squid	1425	
Cape horse mackerel	673	
Cape monk	1196	
Cape razor clam	1412	
Cape redfish	1071	
Cape rock lobster	1262	
capelin	273	
caplin	273	
caramote prawn	1220	
carangids	632	
cardinalfish	610	
cardinalfish, black	611	
cardinalfishes	609	
Caribbean spiny lobster	1269	
carp	332	
carp, bighead	336	
carp, black	347	
carp, common	332	
carp, crucian	321	
carp, grass	331	
carp, Hoven's	338	
carp, koi	332	
carp, mirror	332	

carp, mud	330	
carp, sabre	350	
carp, silver	335	
Carpathian lamprey	2	
carpenter seabream	765	
carpenter	765	
carpet shells	1394	
carpetshell	1401, 1405	
carpetshell, golden	1404	
carpetshell, grooved	1401	
carps	303	
carragheen	1532	
Caspian sandsmelt	549	
cassava croaker	750	
castaneta	857	
cat shark, brown	29	
cat, spotted sea	916	
catfish eels	378	
catfish, Atlantic	915	
catfish, channel	386	
catfish, electric	394	
catfish, smooth-mouth sea-	375	
catfish, som	382	
catfish, spotted	376	
catfish, wels	380, 382	
catfishes	383, 387, 917	
catfishes, air-breathing	388	
catfishes, bullhead	387	
catfishes, glass	379	
catfishes, sea-	374	
catfishes, torpedo-shaped	388	
catfishes, upsidedown	293	
catshark, black-mouth	30	
catsharks	28, 33	
Central American shrimp	1228	
centrina shark	70	
centrina, black	61	
cero	1000	
chacunda gizzard shad	164	
chad	793	
chain pickerel	290	
channel catfish	386	
char	263, 266	
char, American lake	266	
characins	300	
charr, Arctic	263	
charr, brook	264	
charrs	267	
charrs, salmons, trouts and	262	
chars	267	
chat	453	
cherry salmon	250	
cherry salmon, Japanese	250	
Chilean flat oyster	1351	
Chilean hake	485	
Chilean herring	204	
Chilean jack mackerel	669	
Chilean mussel	1374	
Chilean nylon shrimp	1235	
Chilean pilchard	200	
Chilean sea urchin	1522	
chimaeras, plownose	126	
Chinese gizzard shad	168	
Chinese paddlefish	141	
Chinese pomfret	1032	
Chinese seerfish	1003	
Chinese silver pomfret	1032	
chinook salmon	254	
chinook	254	
chokker squid	1425	
chola guitarfish	74	
chub mackerel	984	
chub salmon	254	
chub	231, 341	

chub, Mediterranean	342	common dab	1139	crab, Murray king	1308
chum salmon	248	common dentex	774	crab, Pacific rock	1292
cichlids	831	common dolphin	1464	crab, queen	1312
cinnamon flounder	1122	common dolphinfish	681	crab, red	1321
cisco	231	common eel	396	crab, Southern king	1309
cisco, lake	231	common European skate	84	crab, Southern spider	1314
clam	1401, 1415	common finback	1484	crab, spider	1315
clam, Atlantic razor	1408	common grey mullet	545	crab, spinous spider	1315
clam, butter	1403	common guitarfish	76	crab, spiny	1315
clam, European razor	1413	common hammerhead	47	crab, stone king	1307
clam, hen	1391	common mackerel	985	crab, Subantarctic stone	1308
clam, Japanese hard	1399	common mussel	1376	crabs	1287
clam, Japanese	1402	common nase	325	crabs, hermit	1286
clam, long	1415	common octopus	1438	crabs, king	1311
clam, Manilla	1402	common oyster	1353	crabs, marine	1285
clam, mogal	1340	common pandora	794	crabs, red	1322
clam, Pacific razor	1411	common piddock	1416	crangonid shrimps	1250
clam, razor	1407	common pompano	662	crappie, black	605
clam, round	1398	common prawn	1249	crappie, white	604
clam, short-necked	1402	common rorqual	1484	crawfish, red swamp	1284
clam, soft	1415	common sawfish	79	crawfishes	1274
clam, surf	1392	common scallop	1365	crayfish	1276
clams	1324, 1387	common sea bream	793	crayfish, American	1282
clams, donax	1406	common seabream	806	crayfish, Australian	1316
clams, razor	1414	common seal	1453	crayfish, Galician	1277
clams, venus	1393	common shore crab	1298	crayfish, river	1279
cleaver wrasse	855	common shrimp	1251	crayfish, torrent	1280
climbing perch	1039	common snook	561	crayfishes	1278, 1281
clingfish	1190	common sole	1165	crayfishes, American	1283
clinid, Mediterranean	918	common spiny lobster	1266	craylets	1306
clown triggerfish	1180	common spiny	64	crevalle jack	636
clupeonella	171	common squid	1426	crevalles	639
coalfish whale	1481	common torpedo	120	croaker	718, 759
coalfish	467	common tunny	1012	croaker, Argentine	758
cobia	631	common two-banded seabream	783	croaker, Atlantic	738
cockle, blood	1339	common whitefish	232	croaker, black	721
cockle, common	1389	conchs, stromboid	1338	croaker, black-mouth	721
cockle, dog	1384	conger eel	410, 411	croaker, bobo	749
cockle, spiny	1388	conger eels	408	croaker, cassava	750
cockles	1387	conger	410	croaker, geelback	719
cockles, blood	1341, 1342	conger, American	411	croaker, honnibe	741
cod	447	conger, Argentine	412	croaker, large yellow	722
cod, Arctic	441, 445	conger, European	410	croaker, law	748
cod, Atlantic	447	conger, Guinean pike	405	croaker, lesser yellow	747
cod, black	467	conger, white-spotted	409	croaker, longneck	751
cod, gray	446	congers	408	croaker, long-tooth	742
cod, Greenland	448	Congo dentex	773	croaker, silver	744
cod, New Zealand blue	864	copper breams	789	croaker, spot	731
cod, New Zealand red	438	copper shark	36	croaker, tiger-tooth	742
cod, Northern	444	coquille St. Jacques	1365	croaker, white flower	740
cod, Pacific	446	coral, red	1514	croaker, white	730
cod, Polar	441	coral, Sardinia	1514	croaker, white-mouth	737
cod, poor	473	corals	1513	croaker, yellow	746
cod, saffron	444	corb	759	croakers	715, 739
cod, unicorn	439	corkwing wrasse	847	croakers, West African	752
cod, wachna	444, 445	Couch's sea bream	806	crocodile, estuarine	1505
codfish	447	cowsharks	9	crocodile, New Guinea	1504
codfishes	440	crab, Alaska deepsea	1310	crocodile, Siamese	1506
codling	447	crab, Atlantic rock	1289	crocodiles	1501
codling, Brazilian	474	crab, black stone	1293	crocodiles, alligators and	1500
cods, Antarctic	876, 894	crab, blue	1295	crucian carp	321
cods, morid	436	crab, common shore	1298	crystal shrimp	1214
coho salmon	249	crab, Dungeness	1290	cuckoo gurnard	1079
coho	249	crab, edible	1291	cuckoo ray	94
comb shell	1384	crab, gazami	1300	cuckoo wrasse	841
comber	587	crab, green shore	1298	cucumber, Japanese sea	1523
comber, brown	588	crab, green	1298	Cunene horse mackerel	674
comber, painted	589	crab, horseshoe	1515	cunner	853
combers	590	crab, Japanese	1310	cupped oyster, American	1349
combtooth blennies	904	crab, Jonah	1288	cupped oyster, mangrove	1348
common bream	305	crab, king	1310	cupped oyster, Pacific	1347, 1348
common carp	332	crab, Mediterranean green	1297	cupped oyster, Portuguese	1347
common cockle	1389	crab, mitten	1303	cupped oyster, Sydney	1357
common cuttlefish	1418	crab, mud	1302	cupped oysters	1350

curled octopus	1435
cusk	442
cusk-eel, red	927
cusk-eels	928
cutlassfish	959
cutlassfishes	955
cutthroat trout	255
cuttle, Ross'	1419
cuttlefish	1418
cuttlefish, common	1418
cuttlefish, lesser	1420
cuttlefish, little	1420
cyprine	1386
cyprine, Iceland	1386
cyprinid, silver	355

D

dab, common	1139
dab, Japanese	1138
dab, long rough	1132
dace	344
dagaas	203
dageraads	770
dagger-tooth pike-conger	406
Dall's porpoise	1472
Dalmatian barbel-gudgeon	312
Dalmatian nase	324
damselfish	836
damselfishes	835
Dana swimcrab	1294
Danube bleak	322
Danube bream	306
Danube gudgeon	334
Danube roach	358
Danube ruffe	615
Danube salmon	242
Danube sturgeon	131
Danubian barbel	317
Danubian wels	382
dark electric ray	119
dark torpedo	119
Darkie Charlie	60
date shell	1372
Davidson's whale	1480
dealfish, Mediterranean	511
deep-body thread herring	185
deepsea crab, Alaska	1310
deepsea smelt	283
deep-snouted pipefish	508
deepwater Cape hake	488
deepwater hake	488
deepwater prawn	1237
deepwater redfish	1076
deepwater rose shrimps	1212
dentex	774
dentex, Angola	771
dentex, Canary	772
dentex, common	774
dentex, Congo	773
dentex, large-eye	776
dentex, large-eyed	776
dentex, Morocco	777
dentex, pink	775
denticulate rock oyster	1352
derbio	663
devil anglerfish	1196
devil rays	111
devilfish	111, 114
digby	169
doctorfish, Monrovian	946
doctorfishes	945
dog cockle	1384
dog salmon	248

dogfish	31, 64
dogfish, black-mouthed	30
dogfish, piked	64
dogfish, prickly	70
dogfish, spiny	64
dogfish, spring	64
dogfishes	28
dog's tooth bream	774
dogtooth grouper	576
dog-tooth tuna	974
Dolly Varden trout	265
Dolly Varden	265
dolphin, bottle-nose	1477
dolphin, common	1464
dolphin, Risso's	1467
dolphin, spinner	1475
dolphin, striped	1474
dolphin, white-beaked	1469
dolphin, white-sided	1468
dolphinfish	681
dolphinfish, common	681
dolphins	1463
dolphins, spotted	1476
donax clams	1406
dorab wolf-herring	225
dorade	793
dories	523
dories, oreo	528
dory	525
double-lined mackerel	972
dragonet	941
dragonet, spotted	942
dredge oyster, New Zealand	1355
driftfish, Indian	1038
drum, black	745
drum, boe	753
drum, freshwater	716
drum, Peruvian	743
drum, red	757
drum, shi	759
drum, yellow	740, 741
drums	715, 756, 760
Dublin Bay prawn	1320
duck, Bombay	297
ducks, Bombay	296
dugong	1454
Dungeness crab	1290
dusky grouper	577
dusky sea perch	577
dusky shark	39
dwarf breams	695

E

eagle ray	109
eagle rays	108
ear shell	1328
eared seals	1440
East Coast sole	1154
Eastern mud-minnow	285
Eastern Pacific bonito	980
edible crab	1291
eel pout	452
eel	396
eel, American	400
eel, Australian	397
eel, common	396
eel, conger	410, 411
eel, European	396
eel, freshwater	396
eel, Japanese	399
eel, New Zealand	398
eel, river	396
eel, serpent	414

eel, sharp-toothed	406
eel, short-finned	397
eelpout	923
eelpout, viviparous	923
eelpouts	920, 921
eels	395, 401
eels, catfish	378
eels, conger	408
eels, moray	402
eels, river	395
electric catfish	394
electric ray, dark	119
electric ray, eyed	120
electric ray, marbled	118
electric rays	117
elephantfish, Cape	125
elephantfishes	126
elongate ilisha	184
emperors	762
endeavour shrimp	1207
escolars	947, 948
estuarine crocodile	1505
eulachon	278
euphausids	1203
European abalone	1328
European anchovy	217
European conger	410
European eel	396
European flat oyster	1353
European flounder	1143
European flying squid	1433
European hake	487
European lobster	1318
European moray	404
European mud-minnow	284
European nase, Southwest	329
European perch	617
European pilchard	189
European plaice	1145
European razor clam	1413
European sardine	189
European seabass	593
European skate, common	84
European smelt	274
European sprat	202
European whitefish	229
eyed electric ray	120
eyed skate	95

F

fall salmon	248
false albacore	968
false killer whale	1473
false trevally	628
Far Eastern navaga	444
fifteen-spined stickleback	502
filefish, orange	1181
filefish, thread-sail	1183
filefishes	1182
fin whale	1484
finback, common	1484
finner whale, sharp headed	1480
finner	1484
fish, Samson	636
five-bearded rockling	443
five-rayed threadfin	556
five-spot wrasse	849
flagfin mojarra	902
flagfin, Mediterranean	389
flake	50
flapper skate	84
flat oyster, Chilean	1351
flat oyster, European	1353

flat oyster, Olympia	1354
flat oysters	1356
flatfishes	1110
flathead flounder	1130
flathead gurnards	1095
flat-head grey mullet	545
flathead sole	1131
flathead, Brazilian	873
flathead, Indo-Pacific	1096
flatheads	872
fleshy prawn	1216
Florida pompano	662
flounder	1143
flounder, Arctic	1140
flounder, arrow-tooth	1125
flounder, cinnamon	1122
flounder, European	1143
flounder, flathead	1130
flounder, four-spot	1119
flounder, olive	1120
flounder, roundnose	1126
flounder, sand	1149
flounder, smooth	1141
flounder, spotted	1177
flounder, starry	1144
flounder, summer	1118
flounder, wide-eyed	1117
flounder, windowpane	1174
flounder, winter	1147
flounder, witch	1128
flounder, yellow-striped	1138
flounder, yellowtail	1137
flounders	1123
flounders, left-eye	1114
flounders, right-eye	1123
flounders, sand	1150
flower croaker, white	740
fluke	1143
flukes	1121
flying gurnard	1109
flying squid	1433
flying squid, European	1433
flying squid, Japanese	1432
flying squid, jumbo	1427
flying squid, Wellington	1431
flyingfish	426
flyingfish, four-wing	434
flyingfish, Japanese	424
flyingfishes	423, 425
forkbeard	464
forkbeard, greater	463
forkbeards	465
four-bearded rockling	469
four-eyed sole	1158
four-finger threadfin	554
four-horn sculpin	1099
four-spot flounder	1119
four-spot megrim	1168
four-wing flyingfish	434
freshwater blenny	911
freshwater bream	305
freshwater breams	307
freshwater drum	716
freshwater eel	396
freshwater gobies	1055
freshwater herring	230
freshwater lamprey	5
freshwater mussel	1345
freshwater mussels	1344
freshwater prawn	1246
freshwater prawns and shrimps	1244
frigate mackerel	964
frigate tuna	965
frilled shark	8

fringe-barbel sturgeon	134
fringe-scale sardinella	191
frog, bull	1492
frogfish	1195
frogs	1493
frostfish	957
Fuller's ray	89
fur seal, Cape	1442
fur seal, South African	1442
fur seal, South American	1441
fur seals	1440
fusiliers	686

G

gadiforms	435
Galician crayfish	1277
galjoen	824
galjoens	823
gaper	1415
gar	416
gar, long-nose	143
gar, sea	416
gar, short-nose	144
garfish	416
garfish, sea	428
garfishes	415
garpike	416
gazami crab	1300
geelback croaker	719
gelidium	1529
gemfish	949
giant abalone	1325
giant African threadfin	556
giant barnacle	1200
giant goby	1052
giant gourami	1045
giant pearlwinkle	1332
giant river prawn	1247
giant sea bass	591
giant sea perch	564
giant tiger prawn	1223
gigartina	1531
gila trout	246
gilt sardine	190
gilt-head seabream	802
gizzard shad	172
gizzard shad, American	172
gizzard shad, chacunda	164
gizzard shad, Chinese	168
glass catfishes	379
glassfishes	563
glory mullet	540
goatfish, black-striped	821
goatfish, golden	816
goatfish, golden-striped	821
goatfish, West African	820
goatfishes	815, 822
gobie, lagoon	1056
gobie, tubenose	1060
gobies	1049
gobies, freshwater	1055
gobies, sleeper	1048
goby, bighead	1057
goby, black	1053
goby, Canestrini's	1058
goby, giant	1052
goby, golden	1051
goby, grass	1061
goby, mottled Black Sea	1060
goby, Panizza's	1056
goby, rock	1054
goby, sand	1059
goby, transparent	1050

golden carpetshell	1404
golden goatfish	816
golden goby	1051
golden grey mullet	540
golden grouper	574
golden loach	366
golden mullet	540
golden redfish	1075
golden scad	634
golden thread-fin bream	696
golden trout	243
golden trout, Mexican	245
golden-striped goatfish	821
goldfish	320
goldline	801
goldsinny wrasse	839
goose barnacle	1197, 1343
goose barnacles	1198
goosefish	1193
gorbuscha	247
gourami, giant	1045
gourami, kissing	1047
gourami, snakeskin	1046
graining	341
grass carp	331
grass goby	1061
grass pickerel	287
gray cod	446
grayfish	446
grayfish	64
grayling	268
great barracuda	532
great blue shark	42
Great Lake trout	266
great pipefish	506
great Polar whale	1487
great scallop	1364
great silver smelt	281
great white shark	19
greater amberjack	658
greater forkbeard	463
greater lizardfish	391
greater sandeel	939
greater weever	869
Greek rudd	363
Greek sheatfish	381
green crab	1298
green crab, Mediterranean	1297
green moray	403
green mussel	1380
green pollack	466
green rock lobster	1264
green rockcod	888
green shore crab	1298
green sturgeon	132
green tiger prawn	1226
green turtle	1497
green wrasse	843
green-back horse mackerel	665
greenbone	416
green-eye	299
green-eye, short-nose	299
greeneyes	298
Greenland cod	448
Greenland halibut	1148
Greenland right whale	1487
Greenland shark	63
grenadier anchovy, Osbeck's	213
grenadier, black-spot	493
grenadier, blue	480
grenadier, Italian	495
grenadier, Patagonian	479
grenadier, rock	494
grenadier, rough-head	497

grenadier, round-nose	494
grenadiers	496, 498
grenadiers, blue	477
grey gurnard	1084
grey mullet, common	545
grey mullet, golden	540
grey mullet, leaping	543
grey mullet, lesser	547
grey mullet, long-finned	540
grey mullet, thick-lip	538
grey mullet, thick-lipped	538
grey mullet, thin-lip	542
grey mullet, thin-lipped	542
grey mullets	537
grey rockcod	886
grey seal	1449
grey triggerfish	1180
grey trout	266
grey weakfish	726
grey whale, Pacific	1491
grey wrasse	845
Grohman's scaldfish	1116
grooved carpetshell	1401
grooved sea squirt	1508
ground shark	63
groundling	370
grouper, dogtooth	576
grouper, dusky	577
grouper, golden	574
grouper, malabar	579
grouper, Nassau	581
grouper, red	580
grouper, spotted	575
grouper, white	573
groupers	567
groupers	582
groupers, Brazilian	583
grunt, African striped	708
grunt, barred	704
grunt, bastard	712
grunt, bigeye	703
grunt, biglip	709
grunt, lined silver	711
grunt, rubberlip	710
grunt, silver	711
grunt, sompat	713
grunt, Southeast Pacific	706
grunts	705, 714
gudgeon	333
gudgeon, Danube	334
Guinea shrimp	1210
Guinean barracuda	530
Guinean pike conger	405
guitarfish	76
guitarfish, chola	74
guitarfish, common	76
guitarfish, Peruvian	75
guitarfish, white-spotted	73
guitarfishes	72, 77
Gulf butterfishes	1036
Gulf kingcroaker	734
Gulf menhaden	165
gulper shark	58
gulper shark, leaf-scale	59
gummy shark	51
gunnard	1084
Gunther's notothenia	891
gurnard, armed	1106
gurnard, blue-fin	1082
gurnard, Cape	1081
gurnard, cuckoo	1079
gurnard, flying	1109
gurnard, grey	1084
gurnard, large-scaled	1085
gurnard, long-finned	1080
gurnard, piper	1090
gurnard, red	1079
gurnard, sharp-beak	1088
gurnard, spiny	1086
gurnard, streaked	1091
gurnard, tub	1089
gurnards	1078, 1083
gurnards, flathead	1095
Gwyniad	234

H

haddock	453
haddock, Norway	1075
hairtail, large-eyed	959
hairtail, large-head	959
hairtails	955
hake	487
hake, Argentine	486
hake, Benguela	489
hake, black	491
hake, Cape	484
hake, Chilean	485
hake, deepwater Cape	488
hake, deepwater	488
hake, European	487
hake, North Pacific	490
hake, offshore silver	481
hake, offshore	483
hake, Pacific	490
hake, Peruvian	485
hake, red	475
hake, Senegalese	491
hake, shallow-water	484
hake, silver	483
hake, South Pacific	485
hake, Southern	482
hake, Southwest Atlantic	486
hake, squirrel	475
hake, white	476
hakes	478, 492
halfbeak, balao	429
halfbeak, ballyhoo	430
halfbeak, Japanese	431
halfbeak, long-fin	432
halfbeaks	427
halfbeaks	433
halibut	1133
halibut, Atlantic	1133
halibut, bastard	1120
halibut, black	1148
halibut, Greenland	1148
halibut, Indian	1113
halibut, Kamchatka	1124
halibut, Pacific false	1130
halibut, Pacific	1134
halibuts, bastard	1121
ham mussel	1385
hammerhead shark	47, 45
hammerhead sharks	48
hammerhead	47
hammerhead, common	47
hammerhead, scalloped	46
hammerhead, smooth	47
hapuku wreckfish	586
harbour porpoise	1471
harbour seal	1453
hard clam	1395, 1398
hard clam, Japanese	1399
hard clams	1400
harp seal	1451
harvestfish	1033
harvestfish, North Atlantic	1033
harvestfishes	1030, 1036
hen clam	1391
henfish	1105
hermit crabs	1286
herring smelt	281
herring whale	1484
herring	169, 198
herring, Araucanian	204
herring, Atlantic round	177
herring, Atlantic thread	186
herring, Atlantic	169
herring, blue-back	155
herring, Chilean	204
herring, deep-body thread	185
herring, freshwater	230
herring, lake	231
herring, Pacific	170
herring, picton	198
herring, red-eye round	178
herring, river	161
herring, rock	156
herring, round	173, 178, 179
herring, sea	169
herring, shad	155
herring, silver-stripe round	201
herring, spotted	181
herrings	154
herrings, thread	187
hickory shad	159
hilsa shad	205
hogfish	844
hogfishes	837
hogsnapper	844
homelyn ray	93
honnibe croaker	741
hooded seal	1447
hook-nose	1103
horned and musky octopuses	1437
horned turban	1331
hornpike	416
horse mackerel, Atlantic	672
horse mackerel, Cape	673
horse mackerel, Cunene	674
horse mackerel, green-back	665
horse mackerel, Mediteranean	668
horse mackerel, yellow	637
horse mackerels, jack and	675
horse mackerels, scads and	675
horse mussels	1373
horseshoe crab	1515
houndsharks	49
houting	237
Hoven's carp	338
huchen	242
hum	248
humantin	70
humpback salmon	247
humpback whale	1485
humped rockcod	883

I

Iberian barbel	315
Iberian nase	327
icefish, Antarctic	898
icefish, blackfin	896
icefish, Kathleen's	900
icefish, long-snouted	899
icefish, mackerel	898
icefish, ocellated	900
icefish, Ramsay's	892
icefish, Scotia Sea	896
icefish, South Georgia	901

icefish, spiny 897
icefish, unicorn 899
icefish, Wilson 897
icefishes 895
Iceland cyprine 1386
Icelandic scallop 1362
ide 343
ilisha, elongate 184
ilisha, West African 183
ilkalupik 263
Indian barracuda 534
Indian driftfish 1038
Indian halibut 1113
Indian mackerel 977
Indian mackerels 978
Indian oil sardine 192
Indian pellona 188
Indian scad 644
Indonesian snakehead 557
Indo-Pacific flathead 1096
Indo-Pacific king mackerel 991
Indo-Pacific sailfish 1017
Indo-Pacific tarpon 149
inkfish 1426
inshore squids 1422
Irish moss 1532
island mackerel 976
Italian barbel 318
Italian bleak 309
Italian grenadier 495
Italian loach 373
Italian nase 328

J

jack and horse mackerels 675
jack mackerel, blue 670
jack mackerel, Chilean 669
jack mackerel, Japanese 666
jack mackerel, offshore 670
jack mackerel, Pacific 671
jack salmon 249
jack 288
jack, bar 638
jack, crevalle 636
jacks 632, 639
Jacopever 1063, 1071, 1523
Japanese amberjack 660
Japanese anchovy 218
Japanese cherry salmon 250
Japanese clam 1402
Japanese crab 1310
Japanese dab 1138
Japanese eel 399
Japanese flying squid 1432
Japanese flyingfish 424
Japanese halfbeak 431
Japanese hard clam 1399
Japanese jack mackerel 666
Japanese lobster 1272
Japanese pilchard 197
Japanese salmon 250
Japanese sandfish 863
Japanese sardine 197
Japanese sardinella 194
Japanese seabass 599
Japanese seabream 805
Japanese sea cucumber 1523
Japanese Spanish mackerel 997
Japanese splitfin 612
Java barb 353
jellyfishes 1512
jewfish 578

Jewish sole 1159
jobfishes 687
John Dories 523
John Dory 525
John Dory, American 524
Jonah crab 1288
Juan Fernandez rock lobster 1261
jumbo flying squid 1427
jumbo 453

K

Kamchatka halibut 1124
Kanadi kingfish 998
Kathleen's icefish 900
kawakawa 967
kelee shad 182
kelps 1526
keta salmon 248
kilka 171
killer whale 1470
killer whale, false 1473
killifishes 512
king crab 1310
king crab, Murray 1308
king crab, Southern 1309
king crab, stone 1307
king crabs 1311
king mackerel 988
king mackerel, broad-barred 1001
king mackerel, Indo-Pacific 991
king mackerel, narrow-barred 989
king prawn, Western 1221
king salmon 254
king weakfish 732
kingcroaker, Gulf 734
king-croakers 736
kingfish, butterfly 970
kingfish, Kanadi 998
kingfish, Northern 735
kingfish, sea 636
kingfish, Southern 733
kingfish, yellow-tail 659
kingklip 926
kissing gourami 1047
kitefin shark 60
Klein's sole 1163
knife shrimp 1254
knife shrimps 1255
knifefishes 292
kob 717
koi carp 332
Korean mussel 1375
Korean sandeel 940
Korean seerfish 992
krill 1203
krill, Antarctic 1204
krill, Norwegian 1205
Kuhl's scorpionfish 1065
kuruma prawn 1219

L

ladyfish 146, 152
ladyfishes 684
lagoon gobie 1056
lake char, American 266
lake cisco 231
Lake Geneva whitefish 233
lake herring 231
Lake Neuchâtel whitefish 235
lake trout 259, 266
lake trout, American 266

lake whitefish 232
lakefish 231
lampern 5
lamprey, brook 6
lamprey, Carpathian 2
lamprey, freshwater 5
lamprey, mud 5
lamprey, river 5
lamprey, sea 7
lamprey, Ukranian 3
lamprey, Vladikov's 4
lamprey-eel 5
lampreys 1
lane snapper 692
langoustine 1320
langoustine, Southern 1319
lantern shark 61
lantern sharks 62
lanternfish 295
lanternfishes 294
large mouthed black bass 603
large scale menhaden 165
large yellow croaker 722
large-bottom pollan 236
large-eye breams 761
large-eye dentex 776
large-eyed dentex 776
large-eyed hairtail 959
large-head hairtail 959
large-mouth bass 603
large-scale lizardfish 392
large-scaled gurnard 1085
latchet 1088
launce 939
law croaker 748
leaf-scale gulper shark 59
leaping African mullet 544
leaping bonito 966
leaping grey mullet 543
leatherback turtle 1495
leatherjack, parona 651
leerfish 647
leerfishes 648
left-eye flounders 1114
lemon shark 41
lemon sole 1142
leopard shark 56
lesser cuttlefish 1420
lesser grey mullet 547
lesser rorqual 1480
lesser silver smelt 282
lesser slipper lobster 1305
lesser weever 867
lesser yellow croaker 747
limpets 1333
lined silver grunt 711
lined sole 1152
ling 462
ling, blue 460
ling, Mediterranean 461
ling, Spanish 461
lingcod 1092
lion, Steller sea 1443
lithothamnion 1530
little cuttlefish 1420
little neck 1384
little piked whale 1480
little scorpionfish 1066
little skate 88
little squid 1423
little tuna 968
little tunny 968
lizardfish 393
lizardfish, Atlantic 393

lizardfish, brushtooth	392	mackerel guide	416	Manilla clam	1402
lizardfish, greater	391	mackerel icefish	898	manta, Atlantic	112
lizardfish, large-scale	392	mackerel scad	641	manta, Pacific	113
lizardfishes	390	mackerel shark	23	mantas	111
loach, Balkan	367	mackerel sharks	18	mantis squillid	1202
loach, Bergatino	368	mackerel tuna	968	marbled electric ray	118
loach, bibbaud	370	mackerel	985	marbled notothenia	885
loach, golden	366	mackerel, Atka	1093	marbled ray	101
loach, Italian	373	mackerel, Atlantic horse	672	marbled rockcod	885
loach, Romanian	369	mackerel, Atlantic Spanish	994	marbled spine-foot	943
loach, spined	370	mackerel, Atlantic	985	marine crabs	1285
loach, stone	372	mackerel, Australian spotted	996	marine molluscs	1323
lobster, American	1317	mackerel, blue jack	670	marine turtles	1494
lobster, Andaman	1319	mackerel, blue	983	Marley's snake mackerel	953
lobster, Australian spiny	1261, 1270	mackerel, broad-barred king	1001	marlin, Atlantic blue	1019
lobster, Cape rock	1262	mackerel, bullet	964	marlin, Atlantic white	1021
lobster, Caribbean spiny	1269	mackerel, Cape horse	673	marlin, black	1018
lobster, common spiny	1266	mackerel, Chilean jack	669	marlin, blue	1019
lobster, European	1318	mackerel, chub	984	marlin, striped	1023
lobster, green rock	1264	mackerel, common	985	marlin, white	1021
lobster, Japanese	1272	mackerel, Cunene horse	674	marling	454
lobster, lesser slipper	1305	mackerel, double-lined	972	marlins	1015
lobster, Natal spiny	1265	mackerel, frigate	964	marlins	1020
lobster, Norway	1320	mackerel, green-back horse	665	maskinonge	289
lobster, pink spiny	1268	mackerel, Indian	977	masu salmon	250
lobster, red rock	1260	mackerel, Indo-Pacific king	991	masu	250
lobster, Royal spiny	1273	mackerel, island	976	mattie	169
lobster, South Coast spiny	1267	mackerel, Japanese jack	666	meagre	718
lobster, Transkei spiny	1271	mackerel, Japanese Spanish	997	meagre, brown	755
lobster, Tristan da Cunha rock	1263	mackerel, king	988	meagre, Southern	717
lobster, Tristan rock	1263	mackerel, Marley's snake	953	meagres	715
lobster, tropical spiny	1274	mackerel, Mediterranean horse	668	Mediterranean barbel	316
lobsters, slipper	1304	mackerel, Monterey Spanish	990	Mediterranean chub	342
lobsters, spiny	1259, 1274	mackerel, narrow-barred king	989	Mediterranean clinid	918
lobsters, squad	1306	mackerel, narrow-barred Spanish	989	Mediterranean dealfish	511
lobsters, whip	1275	mackerel, offshore jack	670	Mediterranean flagfin	389
loggerhead turtle	1496	mackerel, Pacific jack	671	Mediterranean green crab	1297
long clam	1415	mackerel, Queensland school	999	Mediterranean horse mackerel	668
long rough dab	1132	mackerel, Serra Spanish	987	Mediterranean ling	461
long-bill spearfish	1026	mackerel, shark	971	Mediterranean monk seal	1450
long-fin bonefish	153	mackerel, short	975	Mediterranean moray	404
long-fin halfbeak	432	mackerel, Southern	984	Mediterranean mussel	1377
long-fin mako	21	mackerel, Spanish	984	Mediterranean sandeel	938
long-fin pilot whale	1466	mackerel, spotted chub	983	Mediterranean scad	668
long-fin squid	1424	mackerel, thimble-eyed	984	Mediterranean spearfish	1024
long-finned albacore	1007	mackerel, West African Spanish	1004	Mediterranean starry ray	83
long-finned grey mullet	540	mackerel, white snake	954	medium red salmon	249
long-finned gurnard	1080	mackerel, yellow horse	637	megrim	1169
long-finned tuna	1007	mackerel-pike	421	megrim, four-spot	1168
long-finned tunny	1007	mackerels	986	megrims	1170
long-horn sculpin	1098	mackerels, Indian	978	mendole	812
longneck croaker	751	mackerels, jack and horse	675	menhaden	166
longneck	1415	mackerels, scads and horse	675	menhaden, Atlantic	166
long-nose gar	143	mackerels, snake	947	menhaden, Gulf	165
long-nose skate	96	mackerels, tunas and	961	menhaden, large scale	165
long-nose spurdog	65	mactra, smooth	1390	menhaden, Pacific	175
long-nosed skate	96	madai	805	menhadens	167
long-snouted icefish	899	Madeiran sardinella	193	metapenaeus shrimps	1209
long-spine snipefish	504	magarei	1138	Mexican golden trout	245
longtail tuna	1013	Magellan mussel	1371	milkfish	228
long-tooth croaker	742	Magellanic notothenia	889	milkfishes	227
lookdown, African	656	Magellanic rock cod	889	miller's thumb	1097
lookdownfish	656	maigre	718	minke whale	1480
lumpfish	1105	mako shark	20	minnow	352
lumpfishes	1104	mako, long-fin	21	minnow, swamp	351
lumpsucker	1105	mako, short-finned	20	minnow-nase	326
lumpsuckers	1104	malabar grouper	579	minnows	303
lungfish, Australian	127	mananose	1415	mirror carp	332
lungfishes, African	129	manatee	1455, 1456	mitten crab	1303
lythe, pollack	466	manatee, African	1457	moderlieschen	339
		maneater shark	19	mogal clam	1340
M		maneater	19	mojarra, flagfin	902
		mangrove cupped oyster	1348	mojarras	903
Macdonald weakfish	724	mangrove red snapper	688	moki, blue	862

molluscs, marine	1323	mussel, ham	1385	Northern wolffish	914
monk seal, Mediterranean	1450	mussel, Korean	1375	Norway haddock	1075
monk seal, pied	1450	mussel, Magellan	1371	Norway lobster	1320
monk, Cape	1196	mussel, Mediterranean	1377	Norway pout	471
monkfish	68, 1195	mussel, New Zealand	1381	Norwegian krill	1205
monocle breams	695, 698	mussel, Orbigny	1397	Norwegian topknot	1171
Monrovian doctorfish	946	mussel, river pearl	1383	notie, yellow-fin	887
Monrovian surgeonfish	946	mussel, River Plate	1379	noties	876
Monterey Spanish mackerel	990	mussel, river	1345	notothenia, broad-headed	881
moonfish	678	mussel, South American rock	1382	notothenia, bumphead	883
moonfish, Atlantic	657	mussels, freshwater	1344, 1373	notothenia, Gunther's	891
moonfishes	676	mussels, sea	1370	notothenia, Magellanic	889
moranec	348			notothenia, marbled	885
moray eels	402	**N**		notothenia, naked-head	887
moray	404			notothenia, scaled	886
moray, European	404	naked-head notothenia	887	notothenia, smoothhead	882
moray, green	403	namaycush	266	notothenids	876
moray, Mediterranean	404	narrow-barred king mackerel	989	nurse shark	26
morays	402	narrow-barred Spanish mackerel	989	nursehound	32
morid cods	436	narrow-head mullet	544	nursehounds	33
Morocco dentex	777	narwhale	1479	nylon shrimp, Chilean	1235
morwongs	860	nase	325		
mosquitofish	516	nase, common	325	**O**	
moss, Irish	1532	nase, Dalmatian	324		
mossbunker	165, 166	nase, Iberian	327	oarfish	510
mother-of-pearl shell	1346	nase, Italian	328	ocean perch, Pacific	1070
mottled Black Sea goby	1060	nase, South European	323	ocean pout	922
mountain trout	263	nase, Southwest European	329	ocean quahaug	1386
Mozambique tilapia	832	Nassau grouper	581	ocean quahog	1386
mud carp	330	Natal pandora	792	ocean sunfish	1188
mud crab	1302	Natal spiny lobster	1265	oceanic bonito	973
mud lamprey	5	navaga	445	oceanic white-tip shark	37
mud shrimp	1258	navaga, Atlantic	445	ocellated icefish	900
mud sole	1154	navaga, Far Eastern	444	ocellated wrasse	848
mud-minnow, American	285	needlefish	416	octopus	1438, 1439
mud-minnow, Eastern	285	needlefish, agujon	418	octopus, common	1438
mud-minnow, European	284	needlefish, Atlantic	417	octopus, curled	1435
mullet, bobo	539	needlefishes	415, 419	octopus, white	1436
mullet, box-lip	547	needlenose	422	octopuses	1434
mullet, Brazilian	546	New Guinea crocodile	1504	octopuses, horned and musky	1437
mullet, common grey	545	New Zealand blue cod	864	offshore hake	483
mullet, glory	540	New Zealand dregde oyster	1355	offshore jack mackerel	670
mullet, golden grey	540	New Zealand eel	398	offshore silver hake	481
mullet, golden	540	New Zealand mussel	1381	oil sardine	192
mullet, leaping African	544	New Zealand red cod	438	oilfish	950
mullet, leaping grey	543	New Zealand scallop	1367	oilfishes	947
mullet, lesser grey	547	New Zealand trevally	652	old maid	1415
mullet, long-finned grey	540	Nile perch	565	old wife	808
mullet, narrow-head	544	Nile perches	566	olive flounder	1120
mullet, Patagonian	880	Nile tilapia	833	Olive Ridley	1498
mullet, red striped	818	Nile twaite shad	158	Olympia flat oyster	1354
mullet, red	817, 818	Nilsson's pipefish	507	opah	509
mullet, so-iny	541	nine-spined stickleback	501	orange filefish	1181
mullet, striped	545, 817	North Atlantic harvestfish	1033	orange roughy	520
mullet, thick-lip grey	538	North Atlantic right whale	1490	Orbigny mussel	1397
mullet, thick-lipped grey	538	North Cape whale	1490	oreo dories	528
mullet, thin-lip grey	542	North Pacific anchovy	219	orfe	343
mullet, thin-lipped grey	542	North Pacific hake	490	ormer	1328
mullets	537	North Pacific sole	1126	Osbeck's grenadier anchovy	213
mullets, grey	537	Northern anchovy	219	osetr	131
mullets, red	815, 819	Northern bluefin tuna	1012	ox-eye	766
mulloway	754	Northern brown shrimps	1213	oyster, American cupped	1349
murex	1334	Northern cod	444	oyster, Blue Point	1349
Murray king crab	1308	Northern kingfish	735	oyster, Chilean flat	1351
murruls	559	Northern pike	288	oyster, common	1353
murry	404	Northern pink shrimp	1217	oyster, denticulate rock	1352
muskellunge	289	Northern prawn	1237	oyster, European flat	1353
musky octopuses, horned and	1437	Northern prawns	1240	oyster, mangrove cupped	1348
mussel, Australian	1378	Northern puffer	1185	oyster, New Zealand dredge	1355
mussel, blue	1376	Northern red snapper	690	oyster, Olympia flat	1354
mussel, Chilean	1374	Northern sandlance	933	oyster, Pacific cupped	1347, 1348
mussel, common	1376	Northern sea lion	1443	oyster, Portuguese cupped	1347
mussel, freshwater	1345	Northern squid	1430	oyster, Portuguese	1347
mussel, green	1380	Northern white shrimp	1227	oyster, Sydney cupped	1357

oysters, cupped	1350	perch, giant sea	564	piper gurnard	1090
oysters, flat	1356	perch, Nile	565	piper	1090
		perch, Pacific ocean	1070	plaice	1145
		perch, Romanian bull-head	618	plaice, Alaska	1146

P

		perch, sand	572	plaice, American	1132
Pacific albacore	1007	perch, sea-	593	plaice, European	1145
Pacific anchoveta	212	perch, swallow-tail sea-	569	plain bonito	964, 974
Pacific argentine	280, 283	perch, white sea-	593	plownose chimaeras	126
Pacific barracuda	531	perch, white	596	pod razor	1409
Pacific bonito	980	perches	613	pogy	165, 166
Pacific cod	446	perches, Nile	566	pointer, white	19
Pacific cupped oyster	1347, 1348	perches, therapon	592	Polar cod	441
Pacific false halibut	1130	periwinkles	1337	Polar whale, great	1487
Pacific grey whale	1491	perlemoen abalone	1326	pollack lythe	466
Pacific hake	490	Peruvian anchovy	220	pollack whale	1481
Pacific halibut	1134	Peruvian drum	743	pollack	467
Pacific herring	170	Peruvian guitarfish	75	pollack, Alaska	470
Pacific jack mackerel	671	Peruvian hake	485	pollack, black	467
Pacific manta	113	Peruvian rock bass	584	pollack, green	466
Pacific menhaden	175	Peruvian sardine	200	pollack, walleye	470
Pacific ocean perch	1070	Peruvian weakfish	723	pollan	230, 234
Pacific pompano	1034	Peter fish	525	pollan, bottom	239
Pacific razor clam	1411	picarel	813	pollan, large-bottom	236
Pacific rock crab	1292	picarel, blackspot	811	pollock	466
Pacific rudderfish	1044	picarel, blotched	812	pomfret	1037
Pacific sailfish	1017	picarels	810, 814	pomfret, Atlantic	680
Pacific salmon	254	pickerel, chain	290	pomfret, black	677
Pacific sandlance	934	pickerel, grass	287	pomfret, Chinese silver	1032
Pacific saury	421	pickerels	286	pomfret, Chinese	1032
Pacific seabob	1230	picton herring	198	pomfret, silver	1031
Pacific sharp-nose shark	43	piddock, common	1416	pompano	663
Pacific sierra	1002	piddocks	1417	pompano, Alexandria	633
Pacific snow crabs	1313	pied monk seal	1450	pompano, common	662
Pacific thread herring	185	pigfish	707	pompano, Florida	662
paddlecock	1105	pigmy whale	1488	pompano, Pacific	1034
paddlefish, Chinese	141	pike conger, Guinean	405	pompanos	664
paddlefishes	140	pike	288	pond smelt	271
painted comber	589	pike, Amur	291	Pontic shad	160
painted ray	101, 91	pike, Northern	288	ponyfish	700
palaemonid shrimps	1245	pike, saury	422	ponyfishes	699, 701
pandalid shrimps	1239, 1240	pike-conger, dagger-tooth	406	poor cod	473
pandora	794	pike-congers	407	porbeagle shark	23
pandora, common	794	piked dogfish	64	porbeagle	23
pandora, Natal	792	piked whale, little	1480	porbeagles	18
pandora, red	791	pikeheaded whale	1480	porgies	763, 767
pandoras	795	pike-perch	620	porgy, red	806
panga	799	pikes	286	porkfish	702
Panizza's goby	1056	pikes, sea	529	porpoise	1471
Papuan seerfish	995	pilchard	189	porpoise, Dall's	1472
parapenacopsis shrimps	1211	pilchard, Californian	196	porpoise, harbour	1471
pargo breams	807	pilchard, Chilean	200	Portuguese cupped oyster	1347
parona leatherjack	651	pilchard, European	189	Portuguese oyster	1347
parrot fish	856	pilchard, Japanese	197	pot whale	1462
paste shrimp, akiami	1242	pilchard, South American	200	pout	472
Patagonian blennie	880	pilchard, Southern African	199	pout, eel	452
Patagonian grenadier	479	pilot whale	1466	pout, Norway	471
Patagonian mullet	880	pilot whale, long-fin	1466	pout, ocean	922
Patagonian rockcod	890	pilot whale, short-fin	1465	pout, silver	449
Patagonian smoothhound	54	pilotfish	650	poutassou	459
Patagonian toothfish	877	pinfish, spot-tail	780	pouting	472
peacock blenny	907	pink cusk-eel	925	powan	234
peacock wrasse	851	pink dentex	775	powan, big	238
pearl mussel, river	1383	pink salmon	247	prawn, banana	1222
pearlfish	930	pink shrimp	1237	prawn, brown tiger	1218
pearlwinkle, giant	1332	pink shrimp, Northern	1217	prawn, caramote	1220
pelagic sting ray	107	pink shrimp, Southern	1224	prawn, common	1249
pelagic thresher	15	pink shrimps	1239	prawn, deepwater	1237
pelamid	982	pink spiny lobster	1268	prawn, Dublin Bay	1320
pellona, Indian	188	pink	352	prawn, fleshy	1216
perch	617	pintadilla	858	prawn, freshwater	1246
perch, American yellow	616	pipefish, deep-snouted	508	prawn, giant river	1247
perch, climbing	1039	pipefish, great	506	prawn, giant tiger	1223
perch, dusky sea	577	pipefish, Nilsson's	507	prawn, green tiger	1226
perch, European	617			prawn, kuruma	1219

prawn, Northern 1237
prawn, red-tail 1225
prawn, Western king 1221
prawns and shrimps, freshwater 1244
prawns, Northern 1240
prawns, river 1248
prickly dogfish 70
prickly skate, Atlantic 97
puffer, Northern 1185
puffers 1184, 1186
puffers, Atlantic 1186
pumpkin-seed 601
purple sea urchin 1518

Q

quahaug 1398
quahaug, ocean 1386
quahog 1398
quahog, ocean 1386
qualla 248
queen crab 1312
queen scallop 1359
queenfishes 653
Queensland school mackerel 999
quinalt 252
quinnat salmon 254
quinnat 254

R

rabbitfish 123
rabbitfishes 122, 944
ragged-tooth shark 14
rainbow runner 646
rainbow sardine 173
rainbow sardine, slender 173
rainbow smelt 275, 276
rainbow trout 251
rainbow wrasse 838, 854
Raitt's sandeel 935
Ramsay's icefish 892
ratfish 123, 124
ratfish, spotted 124
ratfishes 122
rat-tail anchovy 213
rat-tail 493
rat-tails 496
ray, blond 85
ray, brown 92
ray, bull 110
ray, cuckoo 94
ray, dark electric 119
ray, eagle 109
ray, eyed electric 120
ray, Fuller's 89
ray, marbled electric 118
ray, marbled 101
ray, Mediterranean starry 83
ray, painted 91, 101
ray, pelagic sting 107
ray, rough 98
ray, sandy 86
ray, shagreen 89
ray, small-eyed 91
ray, spiny butterfly 115
ray, spotted 93
ray, sting 106
ray, thornback 87
ray, undulate 101
Ray's bream 680
rays, butterfly 103, 116
rays, devil 111
rays, eagle 108

rays, electric 117
rays, skates and 80, 81
rays, torpedo 117, 121
razor clam 1407
razor clam, Atlantic 1408
razor clam, Cape 1412
razor clam, European 1413
razor clam, Pacific 1411
razor clams 1414
razor shell 1407
razor, pod 1409
razor, sword 1410
razorback 1484
red algae 1528
red bait 1510
red bandfish 834
red bigeye 607
red bream 518, 793
red breams 519
red cod, New Zealand 438
red coral 1514
red crab 1321
red crabs 1322
red cusk-eel 927
red drum 757
red grouper 580
red gurnard 1079
red hake 475
red mullet 817, 818
red mullet, striped 818
red mullets 815, 819
red pandora 791
red porgy 806
red rock lobster 1260
red salmon 252
red salmon, medium 249
red scorpionfish 1068
red sea bream 793
red sea squirt 1509
red seabreams 795
red seaweed 1528
red shrimp, Argentine 1256
red shrimp, royal 1257
red snapper 690
red snapper, mangrove 688
red snapper, Northern 690
red snapper, Southern 691
red soldierfish 522
red squid 1433
red steenbras 796
red stumpnose 769
red stumpnoses 770
red swamp crawfish 1284
red tjor tjor 792
red trout 264
red-bait 685
red-banded seabream 803
red-eye round herring 178
red-eye 362, 600
redfish 1075
redfish, Acadian 1073
redfish, beaked 1076
redfish, Cape 1071
redfish, deepwater 1076
redfish, golden 1075
redfishes 1077
red-tail prawn 1225
reef shark, black-tip 38
Reeves' shad 206
requiem sharks 34
rex sole 1129
rhinofishes 337
Rhône streber 622
right whale, Arctic 1487

right whale, Biscayan 1490
right whale, black 1490
right whale, Greenland 1487
right whale, North Atlantic 1490
right whale, Southern 1489
right-eye flounders 1123
ringed seal 1452
Risso's dolphin 1467
river anchoita 221
river crayfish 1279
river eel 396
river eels 395
river herring 161
river lamprey 5
river mussel 1345
river pearl mussel 1383
River Plate mussel 1379
river prawn, giant 1247
river prawns 1248
river trout 258
roach 360
roach, Albanian 348
roach, Black Sea 357
roach, Danube 358
roach, South European 359
roaches 361
robalos 562
rock bass 600
rock bass, Peruvian 584
rock cod, Magellanic 889
rock crab, Atlantic 1289
rock crab, Pacific 1292
rock goby 1054
rock grenadier 494
rock herring 156
rock lobster, Cape 1262
rock lobster, green 1264
rock lobster, red 1260
rock lobster, Tristan da Cunha 1263
rock mussel, South American 1382
rock oyster, denticulate 1352
rock shrimp 1252, 1253
rock sole 1135
rockcod, black 881
rockcod, green 888
rockcod, grey 886
rockcod, humped 883
rockcod, marbled 885
rockcod, Patagonian 890
rockcod, striped 888
rockcod, striped-eyed 884
rockcod, yellow belly 882
rockcods, Antarctic 876, 894
rockfish 915
rockfish, widow 1072
rockfish, yellowtail 1074
rockfishes 1077
rockling, five-bearded 443
rockling, four-bearded 469
rockling, three-beard 450
rockling, three-bearded 450
rocklings 451
roker 87
Romanian bull-head perch 618
Romanian loach 369
rorqual, common 1484
rorqual, lesser 1480
rorqual, Rudolph's 1481
rose shrimps, deepwater 1212
rosefish, black-belly 1063
rosefishes 1064
Ross' cuttle 1419
rosy soldierfish 521
rough ray 98

rough scad	667	
rough shark	58	
rough shark, angular	70	
rough sharks	69	
rough-head grenadier	497	
rough-tail stingray	105	
roughy, orange	520	
round clam	1398	
round herring	173, 178, 179	
round herring, Atlantic	177	
round herring, California	176	
round herring, red-eye	178	
round herring, silver-stripe	201	
round sardinella	190	
round scad	642, 643	
roundnose flounder	1126	
round-nose grenadier	494	
round-scale spearfish	1025	
rovers	684	
royal red shrimp	1257	
Royal spiny lobster	1273	
royal threadfin	555	
rubberlip grunt	710	
rudd	362	
rudd, Greek	363	
rudderfish, Pacific	1044	
Rudolph's rorqual	1481	
ruff	682, 683	
ruffe	614	
ruffe, Danube	615	
ruffs	1040	
runner, blue	635	
runner, rainbow	646	

S

sablefish	1094	
sabre carp	350	
saddled bream	788	
saffron cod	444	
sail bearers	1015	
sailfin sandfish	863	
sailfish, Atlantic	1016	
sailfish, Indo-Pacific	1017	
sailfish, Pacific	1017	
sailfishes	1015	
sail-fluke	1169	
sailray	90	
saithe	467	
salema	801	
salmon	256	
salmon shark	22	
salmon trout	263, 265	
salmon, Atlantic	256	
salmon, Australian	683	
salmon, black	254	
salmon, calico	248	
salmon, cherry	250	
salmon, chinook	254	
salmon, chub	254	
salmon, chum	248	
salmon, coho	249	
salmon, Danube	242	
salmon, dog	248	
salmon, fall	248	
salmon, humpback	247	
salmon, jack	249	
salmon, Japanese cherry	250	
salmon, Japanese	250	
salmon, keta	248	
salmon, king	254	
salmon, masu	250	
salmon, medium red	249	
salmon, Pacific	254	

salmon, pink	247	
salmon, quinnat	254	
salmon, red	252	
salmon, silver	249	
salmon, sockeye	252	
salmon, spring	254	
salmons, trouts and charrs	262	
salmontrout	264	
Samson fish	636	
sand devils	67	
sand flounder	1149	
sand flounders	1150	
sand goby	1059	
sand perch	572	
sand shark	13	
sand sharks	34	
sand sole	1164	
sand steenbras	786	
sandbar shark	40	
sandeel	936	
sandeel, greater	939	
sandeel, Korean	940	
sandeel, Mediterranean	938	
sandeel, Raitt's	935	
sandeels	931, 937	
sandfish, Japanese	863	
sandfish, sailfin	863	
sandgaper	1415	
sandlance	939	
sandlance, American	932	
sandlance, Northern	933	
sandlance, Pacific	934	
sandlance, smooth	938	
sandlances	931, 937	
sandperches, Brazilian	865	
sand-smelt	550	
sandsmelt	551	
sandsmelt, Boyer's	549	
sandsmelt, Caspian	549	
sandsmelt, small	549	
sandsmelts	548	
sandy ray	86	
santer seabream	768	
sardine	189	
sardine, Californian	196	
sardine, European	189	
sardine, gilt	190	
sardine, Indian oil	192	
sardine, Japanese	197	
sardine, oil	192	
sardine, Peruvian	200	
sardine, rainbow	173	
sardine, short-bodied	193	
sardine, slender rainbow	173	
sardinella	190	
sardinella, fringe-scale	191	
sardinella, Japanese	194	
sardinella, Madeiran	193	
sardinella, round	190	
sardinella, short-body	193	
sardinellas	195	
sardines	154	
sardines, scaled	180	
Sardinia coral	1514	
sargo breams	784	
sauger	619	
saupe	801	
sauries	420	
saury pike	422	
saury, Atlantic	422	
saury, Pacific	421	
sawfish	79	
sawfish, common	79	
sawfish, smalltooth	79	

sawfishes	78	
scabbardfish	957, 958	
scabbardfish, black	956	
scabbardfish, silver	957	
scabbardfishes	955	
scad	672	
scad, golden	634	
scad, Indian	644	
scad, mackerel	641	
scad, Mediterranean	668	
scad, rough	667	
scad, round	642, 643	
scad, shrimp	634	
scad, torpedo	649	
scad, yellow-stripe	655	
scads and horse mackerels	675	
scads	645	
scaldfish	1115	
scaldfish, Grohman's	1116	
scaled notothenia	886	
scaled sardines	180	
scallop	1365, 1384	
scallop, Australian	1366	
scallop, bay	1361	
scallop, calicot	1360	
scallop, common	1365	
scallop, great	1364	
scallop, Icelandic	1362	
scallop, New Zealand	1367	
scallop, queen	1359	
scallop, sea	1369	
scallop, South Atlantic	1368	
scallop, variegated	1363	
scalloped hammerhead	46	
scalloped spiny lobster	1271	
scallops	1358	
scampi	1320	
scarlet shrimp	1234	
scats	829	
scavengers	762	
schneider	308	
school mackerel, Queensland	999	
scorpion, sea	1102	
scorpionfish, black	1067	
scorpionfish, Kuhl's	1065	
scorpionfish, little	1066	
scorpionfish, red	1068	
scorpionfishes	1062, 1069	
Scotia Sea icefish	896	
Scotsman seabream	797	
scourer	950	
sculpin, four-horn	1099	
sculpin, long-horn	1098	
sculpin, short-horn	1100	
sculpins	1101	
scup	809	
sea arrow	1426	
sea bass, giant	591	
sea bream	774, 808	
sea bream, black	808	
sea bream, common	793	
sea bream, Couch's	806	
sea bream, red	793	
sea bream, Spanish	794	
sea bream, striped	786	
sea breams	763	
sea cat, spotted	916	
sea catfish, smooth-mouth	375	
sea catfishes	374	
sea cucumber, Japanese	1523	
sea devil	1195	
sea ear	1328	
sea gar	416	
sea garfish	428	

411

sea hen	1105	
sea herring	169	
sea king	1443	
sea kingfish	636	
sea lamprey	7	
sea lion, Northern	1443	
sea lion, South American	1444	
sea lion, Steller	1443	
sea lions	1440	
sea mussels	1370	
sea needle	416	
sea perch	593	
sea perch, dusky	577	
sea perch, giant	564	
sea perch, swallow-tail	569	
sea perch, white	593	
sea pikes	529	
sea scallop	1369	
sea scorpion	1102	
sea snail	1108	
sea snail, Atlantic	1107	
sea spider	1315	
sea spotted bass	594	
sea squid	1433	
sea squirt, grooved	1508	
sea squirt, red	1509	
sea squirts	1507	
sea trout	257	
sea trout, spotted	725	
sea urchin	1521	
sea urchin, Chilean	1522	
sea urchin, purple	1518	
sea urchin, stony	1518	
sea urchins	1519, 1520	
seabass, Argentine	568	
seabass, black	571	
seabass, European	593	
seabass, Japanese	599	
seabass, white	720	
seabasses	567, 595	
seabob, Atlantic	1231	
seabob, Pacific	1230	
seabream	799	
seabream, annular	778	
seabream, axillary	790	
seabream, blackspot	793	
seabream, carpenter	765	
seabream, common two-banded	783	
seabream, common	806	
seabream, gilt-head	802	
seabream, Japanese	805	
seabream, red-banded	803	
seabream, santer	768	
seabream, Scotsman	797	
seabream, seventy-four	798	
seabream, sharp-snout	781	
seabream, spot-tail	780	
seabream, white	782	
seabream, zebra	779	
seabreams, red	795	
sea-cow	1455	
seahorses	505	
seal, bearded	1448	
seal, Cape fur	1442	
seal, common	1453	
seal, grey	1449	
seal, harbour	1453	
seal, hooded	1447	
seal, Mediterranean monk	1450	
seal, pied monk	1450	
seal, ringed	1452	
seal, South African fur	1442	
seal, South American fur	1441	
seals	1446	
seals, eared	1440	
seals, fur	1440	
seaperches	567	
searobins, Atlantic	1087	
seaweed, brown	1525	
seaweed, red	1528	
seerfish, Chinese	1003	
seerfish, Korean	992	
seerfish, Papuan	995	
seerfish, streaked	993	
seerfishes	1005	
sei whale	1481	
Senegal sole	1157	
Senegalese hake	491	
sergestid shrimps	1241	
serpent eel	414	
Serra Spanish mackerel	987	
seven-gill shark, sharpnose	10	
seven-gilled shark	10	
seventy-four seabream	798	
sevruga	136	
shad herring	155	
shad	165, 166	
shad, allice	156	
shad, allis	156	
shad, American gizzard	172	
shad, American	162	
shad, Black Sea	160	
shad, blue-back	155	
shad, bonga	174	
shad, chacunda gizzard	164	
shad, Chinese gizzard	168	
shad, gizzard	172	
shad, hickory	159	
shad, hilsa	205	
shad, kelee	182	
shad, Nile twaite	158	
shad, Pontic	160	
shad, Reeves'	206	
shad, toli	207	
shad, twaite	157	
shadefish	718	
shads	163	
shads, allis	163	
shads, twaite	163	
shagreen ray	89	
shallow-water hake	484	
shanny	908	
shark mackerel	971	
shark sucker	1178	
shark, angel	68	
shark, angular rough	70	
shark, Atlantic sharp-nose	44	
shark, basking	12	
shark, black-tip reef	38	
shark, blue	42	
shark, bluntnose six-gill	11	
shark, bramble	71	
shark, centrina	70	
shark, copper	36	
shark, dusky	39	
shark, frilled	8	
shark, great blue	42	
shark, great white	19	
shark, Greenland	63	
shark, ground	63	
shark, gulper	58	
shark, gummy	51	
shark, hammerhead	47	
shark, kitefin	60	
shark, lantern	61	
shark, leaf-scale gulper	59	
shark, lemon	41	
shark, leopard	56	
shark, mackerel	23	
shark, mako	20	
shark, maneater	19	
shark, nurse	26	
shark, oceanic white-tip	37	
shark, Pacific sharp-nose	43	
shark, porbeagle	23	
shark, ragged-tooth	14	
shark, rough	58	
shark, salmon	22	
shark, sand	13	
shark, sandbar	40	
shark, sharpnose seven-gill	10	
shark, silvertip	35	
shark, six-gill	11	
shark, six-gilled	11	
shark, sleeper	63	
shark, soupfin	50	
shark, spinous	71	
shark, spiny	71	
shark, thresher	17	
shark, tope	50	
shark, white	19	
shark, white-tip	37	
shark, zebra	24	
sharks, angel	67	
sharks, hammerhead	45, 48	
sharks, lantern	62	
sharks, mackerel	18	
sharks, requiem	34	
sharks, rough	69	
sharks, sand	34	
sharp headed finner whale	1480	
sharp-beak gurnard	1088	
sharpnose seven-gill shark	10	
sharp-nose shark, Atlantic	44	
sharp-nose shark, Pacific	43	
sharpnose skate	90	
sharp-snout seabream	781	
sharp-tail sunfish	1187	
sharp-toothed eel	406	
sheatfish	382	
sheatfish, Greek	381	
sheepshead	764	
shell, comb	1384	
shell, date	1372	
shell, mother-of-pearl	1346	
shells, carpet	1394	
shi drum	759	
shiba shrimp	1208	
ship	134	
shore crab, common	1298	
shore crab, green	1298	
short mackerel	975	
short-bill spearfish	1022	
short-bodied sardine	193	
short-body sardinella	193	
short-fin pilot whale	1465	
shortfin squid	1430	
short-fin squid, Argentine	1428	
shortfin squid, broad-tail	1429	
short-finned eel	397	
short-finned mako	20	
short-finned tunny	982	
short-horn sculpin	1100	
short-necked clam	1402	
short-nose gar	144	
short-nose green-eye	299	
short-nose spurdog	66	
shrimp scad	634	
shrimp, Aesop	1238	
shrimp, akiami paste	1242	
shrimp, Argentine red	1256	
shrimp, Argentine stiletto	1206	
shrimp, blue-and-red	1233	
shrimp, Central American	1228	
shrimp, Chilean nylon	1235	

412

shrimp, common	1251	skate, common European	84
shrimp, crystal	1214	skate, eyed	95
shrimp, endeavour	1207	skate, flapper	84
shrimp, Guinea	1210	skate, little	88
shrimp, knife	1254	skate, long-nose	96
shrimp, mud	1258	skate, long-nosed	96
shrimp, Northern pink	1217	skate, sharpnose	90
shrimp, Northern white	1227	skate, starry	100
shrimp, pink	1237	skate, tarry	97
shrimp, rock	1252, 1253	skate, thorny	97
shrimp, royal red	1257	skate, white	82
shrimp, scarlet	1234	skate, white-bellied	82
shrimp, shiba	1208	skate, winter	95
shrimp, Southern pink	1224	skates and rays	80, 81
shrimp, yellow-leg	1215	skates	102
shrimps, aristeid	1232	skelly	341
shrimps, arrow	1236	skipjack tuna	973
shrimps, atyid	1243	skipjack	973
shrimps, crangonid	1250	skipjack, Atlantic black	968
shrimps, deepwater rose	1212	skipjack, black	969
shrimps, freshwater prawns and	1244	skipper	421, 422
shrimps, knife	1255	skippers	420
shrimps, metapenaeus	1209	sleeper gobies	1048
shrimps, Northern brown	1213	sleeper shark	63
shrimps, palaemonid	1245	slender rainbow sardine	173
shrimps, pandalid	1239, 1240	slender snipefish	504
shrimps, parapenacopsis	1211	slender sunfish	1189
shrimps, pink	1239	slender tuna	963
shrimps, sergestid	1241	slipmouths	699, 701
shrimps, tropical	1211	slipper lobster, lesser	1305
shrimps, white	1229	slipper lobsters	1304
Siamese crocodile	1506	small sandsmelt	549
Siberian sterlet	135	small-eyed ray	91
sichel	350	small-mouth bass	602
sicklefish, African	827	smalltooth sawfish	79
sicklefish, spotted	828	small-tooth weakfish	727
sidestripe, Antarctic	893	smelt	274, 281
sierra	954	smelt, American	275
sierra, Pacific	1002	smelt, Arctic	276
sild	169	smelt, Asiatic	276
sillago-whitings	625	smelt, boreal	276
sillock	467	smelt, deepsea	283
silver bream	319	smelt, European	274
silver carp	335	smelt, great silver	281
silver croaker	744	smelt, herring	281
silver cyprinid	355	smelt, lesser silver	282
silver grunt	711	smelt, pond	271
silver grunt, lined	711	smelt, rainbow	275, 276
silver hake	483	smelt, surf	272
silver hake, offshore	481	smelts	270, 277
silver pomfret	1031	smooth flounder	1141
silver pomfret, Chinese	1032	smooth hammerhead	47
silver pout	449	smooth mactra	1390
silver salmon	249	smooth sandlance	938
silver scabbardfish	957	smoothhead notothenia	882
silver smelt, great	281	smooth-head, Baird's	223
silver smelt, lesser	282	smoothhound	53
silver-biddies	903	smoothhound, Patagonian	54
silverfish, Antarctic	893	smooth-hound, stellate	52
silverfishes	1030	smoothhounds	49, 55
silverside	249, 550	smooth-mouth sea catfish	375
silverside, Atlantic	552	snail, sea-	1108
silversides	548	snake blenny	929
silver-stripe round herring	201	snake mackerel, Marley's	953
silvertip shark	35	snake mackerel, white	954
six-gill shark	11	snake mackerels	947
six-gill shark, bluntnose	11	snake-eel	414
six-gilled shark	11	snakehead, Indonesian	557
skate	84	snakehead, striped	558
skate, Atlantic prickly	97	snakeheads	559
skate, big	95	snakeskin gourami	1046
skate, blue	84	snapper, amarillo	689
skate, bordered	82	snapper, lane	692
skate, bottlenose	82	snapper, mangrove red	688
skate, butterfly	94	snapper, Northern red	690

snapper, red	690		
snapper, Southern red	691		
snapper, yellow	689		
snapper, yellow-tail	694		
snappers	687, 693		
snipefish	504		
snipefish, long-spine	504		
snipefish, slender	504		
snipefishes	503		
snoek	951, 952		
snook	561		
snook, common	561		
snooks	560, 562		
snow crabs, Pacific	1313		
sockeye salmon	252		
soft clam	1415		
soft-shell clam	1415		
so-iny mullet	541		
soldier	768, 1079		
soldierfish, red	522		
soldierfish, rosy	521		
sole	1165		
sole, common	1165		
sole, East Coast	1154		
sole, flathead	1131		
sole, four-eyed	1158		
sole, Jewish	1159		
sole, Klein's	1163		
sole, lemon	1142		
sole, lined	1152		
sole, mud	1154		
sole, North Pacific	1126		
sole, petrale	1127		
sole, rock	1135		
sole, sand	1164		
sole, Senegal	1157		
sole, thickback	1157		
sole, wedge	1157, 1160		
sole, West Coast	1153		
sole, whiskered	1162		
sole, yellowfin	1136, 1137		
solenette	1156		
soles	1151		
soles, Southeast Atlantic	1155		
soles, thickback	1161		
som catfish	382		
sompat grunt	713		
souffie	323		
soufie	329		
soupfin shark	50		
South African fur seal	1442		
South American fur seal	1441		
South American pilchard	200		
South American rock mussel	1382		
South American sea lion	1444		
South Atlantic scallop	1368		
South Coast spiny lobster	1267		
South European nase	323		
South European roach	359		
South European toothcarp	513		
South Georgia icefish	901		
South Pacific breams	1042		
South Pacific hake	485		
Southeast Atlantic soles	1155		
Southeast Pacific grunt	706		
Southern African anchovy	216		
Southern African pilchard	199		
Southern blue whiting	458		
Southern bluefin tuna	1010		
Southern hake	482		
Southern king crab	1309		
Southern king-croaker	733		
Southern kingfish	733		
Southern langoustine	1319		
Southern mackerel	984		

Southern meagre	717	spotted grouper	575	stickleback, nine-spined	501
Southern pink shrimp	1224	spotted grunt, Atlantic	713	stickleback, three-spine	500
Southern red snapper	691	spotted herring	181	stickleback, three-spined	500
Southern right whale	1489	spotted mackerel, Australian	996	sticklebacks	499
Southern spider crab	1314	spotted ratfish	124	stiletto shrimp, Argentine	1206
Southwest Atlantic hake	486	spotted ray	93	sting ray	106
Southwest European nase	329	spotted sea cat	916	sting ray, pelagic	107
spadefishes	825	spotted sea trout	725	stingray, rough-tail	105
Spanish ling	461	spotted sicklefish	828	stingrays	103
Spanish mackerel	984	spotted weakfish	725	stolephorus anchovies	222
Spanish mackerel, Atlantic	994	spotted weever	868	stone bass	585
Spanish mackerel, Japanese	997	spotted wobbegong	25	stone crab, black	1293
Spanish mackerel, Monterey	990	spotted wrasse	848	stone crab, Subantarctic	1308
Spanish mackerel, narrow-barred	989	sprat	202	stone king crab	1307
Spanish mackerel, Serra	987	sprat, Black Sea	171	stone loach	372
Spanish mackerel, West African	1004	sprat, European	202	stony sea urchin	1518
Spanish sea bream	794	sprat, Tyulka	171	streaked gurnard	1091
Spanish toothcarp	514, 515	spring dogfish	64	streaked seerfish	993
sparling	274	spring salmon	254	streaked weever	870
spearfish, long-bill	1026	spurdog	64	streber	623
spearfish, Mediterranean	1024	spurdog, Blainville's	65	streber, Rhône	622
spearfish, round-scale	1025	spurdog, long-nose	65	strepie	801
spearfish, short-bill	1022	spurdog, short-nose	66	stripe-bellied bonito	973
spearfishes	1015, 1027	spurdogs	57	striped anchovy	209
speckled trout	264	squad lobsters	1306	striped bass	598
spectacled caiman	1503	squaretail	264	striped bonito	981
sperm whale	1462	squeteague	726	striped dolphin	1474
spider crab	1315	squid	1426	striped grunt, African	708
spider crab, Southern	1314	squid, Argentine short-fin	1428	striped marlin	1023
spider crab, spinous	1315	squid, broad-tail shortfin	1429	striped mullet	545, 817
spider, sea	1315	squid, Cape Hope	1425	striped red mullet	818
spiderfish	298	squid, chokker	1425	striped rockcod	888
spined loach	370	squid, common	1426	striped sea bream	786
spinefeet	944	squid, European flying	1433	striped snakehead	558
spine-foot, marbled	943	squid, flying	1433	striped tuna	973
spinner dolphin	1475	squid, Japanese flying	1432	striped venus	1396
spinous shark	71	squid, jumbo flying	1427	striped weakfish	728
spinous spider crab	1315	squid, little	1423	striped-eyed rockcod	884
spiny butterfly ray	115	squid, long-fin	1424	stroemling	345
spiny cockle	1388	squid, Northern	1430	stromboid conchs	1338
spiny crab	1315	squid, red	1433	stumpnose, red	769
spiny dogfish	64	squid, sea	1433	stumpnose, white	800
spiny gurnard	1086	squid, shortfin	1430	stumpnoses, red	770
spiny icefish	897	squid, Wellington flying	1431	sturgeon	137
spiny lobster, Australian	1261	squids	1421	sturgeon, Adriatic	133
spiny lobster, Caribbean	1269	squids, inshore	1422	sturgeon, Danube	131
spiny lobster, common	1266	squillid, mantis	1202	sturgeon, fringe-barbel	134
spiny lobster, Natal	1265	squillids	1201	sturgeon, green	132
spiny lobster, pink	1268	squirrel hake	475	sturgeon, starry	136
spiny lobster, Royal	1273	squirt, grooved sea	1508	sturgeon, white	138
spiny lobster, South Coast	1267	squirt, red sea	1509	sturgeons	130
spiny lobster, Transkei	1271	squirts, sea	1507	Subantarctic stone crab	1308
spiny lobster, tropical	1274	starfish	1517	sucker, shark	1178
spiny lobsters	1259, 1274	starfishes	1516	suckerfish	1178
spiny shark	71	stargazer	874	suckers	301
spiny turbot	1112	stargazers	875	summer flounder	1118
spiny turbot, spot-tail	1111	starry flounder	1144	sunfish, ocean	1188
spiny	64	starry ray, Mediterranean	83	sunfish, sharp-tail	1187
spiny, common	64	starry skate	100	sunfish, slender	1189
spiny-tail skate	99	starry smooth-hound	52	surf clam	1392
splitfin, Japanese	612	starry sturgeon	136	surf smelt	272
sponges	1511	steelhead trout	251	surgeonfish, Monrovian	946
spoonbills	140	steelhead	251	surgeonfishes	945
spot croaker	731	steenbras, red	796	surmullet	818, 819
spot	731	steenbras, sand	786	swallow-tail sea perch	569
spot-tail pinfish	780	steenbras, white	785	swamp crawfish, red	1284
spot-tail seabream	780	steenbrasses, white	787	swamp minnow	351
spot-tail spiny turbot	1111	stellate smooth-hound	52	sweetfish, ayu	269
spotted bass, sea	594	Steller sea lion	1443	sweetlips	705
spotted catfish	376	sterlet	135	swellfish	1185
spotted chub mackerel	983	sterlet, Siberian	135	swimcrab, Dana	1294
spotted dolphins	1476	stet anchovy	216	swimcrab, velvet	1299
spotted dragonet	942	stet	443	swimcrabs	1296, 1301
spotted flounder	1177	stickleback, fifteen-spined	502	swimming-crab, velvet	1299

sword razor 1410
swordfish 1029
swordfishes 1028
Sydney cupped oyster 1357

T

tadpole fish 468
tarakihi 859
tarpon 150
tarpon, Indo-Pacific 149
tarpons 145, 147, 148
tarry skate 97
tasselfishes 553
tautog 852
tench 364
ten-pounder 145
terrapins 1499
terrapins, American 1499
tetras 300
therapon perches 592
thickback sole 1157
thickback soles 1161
thick-lip grey mullet 538
thick-lipped grey mullet 538
thimble-eyed mackerel 984
thin-lip grey mullet 542
thin-lipped grey mullet 542
thornback ray 87
thorny skate 97
thread herring, Atlantic 186
thread herring, deep-body 185
thread herring, Pacific 185
thread herrings 187
thread-fin bream, golden 696
thread-fin breams 695, 697
threadfin 556
threadfin, five-rayed 556
threadfin, four-finger 554
threadfin, giant African 556
threadfin, royal 555
threadfins 553
thread-sail filefish 1183
three-beard rockling 450
three-bearded rockling 450
three-spine stickleback 500
three-spined stickleback 500
thresher shark 17
thresher, pelagic 15
tiger prawn, brown 1218
tiger prawn, giant 1223
tiger prawn, green 1226
tiger-tooth croaker 742
tilapia, Mozambique 832
tilapia, Nile 833
tilefish 627
tilefishes 626
tjor tjor, red 792
toadfishes 1191
togue 266
toli shad 207
tomcod 457
tomcod, Atlantic 456
tompot blenny 909
tonguefishes 1166
tongue-soles 1166
toothcarp, South European 513
toothcarp, Spanish 514, 515
toothfish, Antarctic 878
toothfish, Patagonian 877
toothfishes, Antarctic 879
tope shark 50
tope 50

topes 49
topknot 1176
topknot, Norwegian 1171
torpedo rays 117, 121
torpedo scad 649
torpedo, Atlantic 119
torpedo, common 120
torpedo, dark 119
torpedo-shaped catfishes 388
torrent crayfish 1280
torsk 442
touladi 266
Transkei spiny lobster 1271
transparent goby 1050
trevally, false 628
trevally, New Zealand 652
trevally, yellow-stripe 655
triggerfish, clown 1180
triggerfish, grey 1180
triggerfishes 1179
triple-fin blenny 912
Tristan da Cunha rock lobster 1263
Tristan rock lobster 1263
trochus 1330
tropical shrimps 1211
tropical spiny lobster 1274
trout, Arizona 244
trout, brook 258, 264
trout, brown 257, 258
trout, bull 265
trout, Dolly Varden 265
trout, gila 246
trout, golden 243
trout, Great Lake 266
trout, grey 266
trout, lake 259, 266
trout, Mexican golden 245
trout, mountain 263
trout, rainbow 251
trout, red 264
trout, river 258
trout, salmon 263, 265
trout, sea 257
trout, speckled 264
trout, spotted sea 725
trout, steelhead 251
trouts and charrs, salmons 262
trouts 260
trumpeters 861
tub gurnard 1089
tubenose gobie 1060
tullibee 231
tuna 1012
tuna, Allison's 1008
tuna, Atlantic bluefin 1012
tuna, Atlantic 1012
tuna, bigeye 1011
tuna, blackfin 1009
tuna, bluefin 1012
tuna, bullet 964
tuna, frigate 965
tuna, little 968
tuna, long-finned 1007
tuna, longtail 1013
tuna, mackerel 968
tuna, Northern bluefin 1012
tuna, skipjack 973
tuna, slender 963
tuna, Southern bluefin 1010
tuna, striped 973
tuna, white 1007
tuna-like fishes 960
tunas and mackerels 961
tunas 1006, 1014

tunny 1006, 1012
tunny, common 1012
tunny, little 968
tunny, long-finned 1007
tunny, short-finned 982
tunny, yellowfin 1008
turban, horned 1331
turbot 1172
turbot, Black Sea 1173
turbot, spiny 1112
turbot, spot-tail spiny 1111
turbots 1167
turtle, green 1497
turtle, leatherback 1495
turtle, loggerhead 1496
turtles, marine 1494
tusk 442
twaite shad 157
twaite shads 163
two-banded seabream, common 783
tyee 254
Tyulka sprat 171

U

Ukranian lamprey 3
undulate ray 101
unicorn cod 439
unicorn icefish 899
unicorn whale 1479
upsidedown catfishes 293
urchin, Chilean sea- 1522
urchin, purple sea- 1518
urchin, sea- 1521
urchin, stony sea- 1518
urchins 1519
urchins, sea- 1519, 1520

V

vairone 345
variegated scallop 1363
velvet belly 61
velvet swimcrab 1299
velvet swimming-crab 1299
vendace 229
venus clams 1393
venus, striped 1396
vimba bream 365
viviparous blenny 923
viviparous eelpout 923
Vladikov's lamprey 4

W

wachna cod 444, 445
wahoo 962
walleye pollack 470
walleye 621
walrus 1445
warehou, bluenose 1043
warman 1007
weakfish 726
weakfish, African 719
weakfish, grey 726
weakfish, king 732
weakfish, Macdonald 724
weakfish, Peruvian 723
weakfish, small-tooth 727
weakfish, spotted 725
weakfish, striped 728
weakfish, white 720
weakfishes 729

weatherfish 371
wedge sole 1157, 1160
weever fishes 866, 871
weever, greater 869
weever, lesser 867
weever, spotted 868
weever, streaked 870
weevers 866, 871
Wellington flying squid 1431
wels catfish 380, 382
wels 382
wels, Danubian 382
West African croakers 752
West African goatfish 820
West African ilisha 183
West African Spanish mackerel 1004
West Coast sole 1153
Western king prawn 1221
whale, Arctic right 1487
whale, beaked 1459, 1461
whale, beluga 1478
whale, Biscayan right 1490
whale, blue 1483
whale, bottlenosed 1460
whale, Bryde's 1482
whale, coalfish 1481
whale, Davidson's 1480
whale, false killer 1473
whale, fin 1484
whale, great Polar 1487
whale, Greenland right 1487
whale, herring 1484
whale, humpback 1485
whale, killer 1470
whale, little piked 1480
whale, long-fin pilot 1466
whale, minke 1480
whale, North Atlantic right 1490
whale, North Cape 1490
whale, Pacific grey 1491
whale, pigmy 1488
whale, pikeheaded 1480
whale, pilot 1466
whale, pollack 1481
whale, pot 1462
whale, sei 1481
whale, sharp headed finner 1480
whale, short-fin pilot 1465
whale, Southern right 1489
whale, sperm 1462
whale, unicorn 1479
whale, white 1478
whaler, blue 42
whales 1486
whales, beaked 1458
whale-shark 27
whelk 1335
whelks 1336
whiff 1169
whip lobsters 1275
whip ray 104
whip stingray 104
whiskered sole 1162

white amur bream 349
white amur 331
white baggar 377
white bass 597
white crappie 604
white croaker 730
white fish 628
white flower croaker 740
white grouper 573
white hake 476
white marlin 1021
white marlin, Atlantic 1021
white octopus 1436
white perch 596
white pointer 19
white sea perch 593
white seabass 720
white seabream 782
white shark 19
white shark, great 19
white shrimp, Northern 1227
white shrimps 1229
white skate 82
white snake mackerel 954
white steenbras 785
white steenbrasses 787
white stumpnose 800
white sturgeon 138
white tuna 1007
white weakfish 720
white whale 1478
white-beaked dolphin 1469
white-bellied skate 82
whitefish 229, 234
whitefish, blue 240
whitefish, common 232
whitefish, European 229
whitefish, Lake Geneva 233
whitefish, Lake Neuchâtel 235
whitefish, lake 232
whitefishes 241, 261
white-mouth croaker 737
white-sided dolphin 1468
white-spotted conger 409
white-spotted guitarfish 73
white-tip shark 37
white-tip shark, oceanic 37
whiting 454, 483
whiting, Black Sea 455
whiting, blue 459
whiting, Southern blue 458
whiting-pout 472
wide-eyed flounder 1117
widow rockfish 1072
Wilson icefish 897
windowpane flounder 1174
winter flounder 1147
winter skate 95
witch flounder 1128
witch 1128
wobbegong 25
wobbegong, spotted 25
wolffish 915

wolffish, Atlantic 915
wolffish, Northern 914
wolffishes 913, 917
wolf-herring 225
wolf-herring, dorab 225
wolf-herrings 224, 226
worm-eel 413
wracks 1527
wrasse, axillary 846
wrasse, ballan 840
wrasse, brown 842
wrasse, cleaver 855
wrasse, corkwing 847
wrasse, cuckoo 841
wrasse, five-spot 849
wrasse, goldsinny 839
wrasse, green 843
wrasse, grey 845
wrasse, ocellated 848
wrasse, peacock 851
wrasse, rainbow 838, 854
wrasse, spotted 848
wrasses 837
wreck bass 585
wreckfish 585
wreckfish, hapuku 586
Wuchang bream 346

Y

yawling 169
yellow belly rockcod 882
yellow croaker 746
yellow croaker, large 722
yellow croaker, lesser 747
yellow drum 740, 741
yellow horse mackerel 637
yellow perch, American 616
yellow snapper 689
yellow-fin notie 887
yellowfin sole 1136, 1137
yellowfin tunny 1008
yellow-finned barracuda 533
yellow-leg shrimp 1215
yellow-stripe scad 655
yellow-stripe trevally 655
yellow-striped flounder 1138
yellow-tail amberjack 659
yellowtail flounder 1137
yellow-tail kingfish 659
yellowtail rockfish 1074
yellow-tail snapper 694

Z

zander 620
zebra seabream 779
zebra shark 24
zingel 624
zope 304

Índice español

A

abadejo	466
abadejo de Alaska	470
abichón	550, 551
acedia	1157, 1160
acedía del Japón	1138
acerina	614
acerina del Danubio	615
agono	1103
aguacioso	933, 939
aguaciosos	931
aguijón	364
águila marina	109
águilas de mar	108
aguja	416, 432
aguja azul	1019
aguja blanca	1021
aguja de mar armada	507
aguja mula	508
aguja negra	1018
aguja paladar	957
aguja palar	1029
aguja picuda	1026
agujas	415
agujeta	416
agujeta austral	428
agujeta balaju	429
agujeta brasileña	430
agujeta del Japón	431
agujetas	427, 433
agujilla	422
agujón	422
agujón verde	417
agujones	415
alacha	190
alachas	195
albacora	968, 1007, 1012
albur	545
alburno	152, 310
alburno bimaculado	308
alburno del Danubio	322
alburno italiano	309
alburno rayado	339
alcaraz	142
alepocéfalo	223
alfaneca	464
alfonsinos	519
algas	1524
algas calcáreas	1530
algas oscuras	1525
algas pardas	1525
algas rojas	1528
aligator del Misisipí	1502
aligatores	1500
aligote	790
aligote bogaraveo	793
alitán	32
alitanes	33
almeja amarilla	1403
almeja babosa	1405
almeja blanca	1392
almeja de can	1415
almeja de Islandia	1386
almeja de Orbignyi	1397
almeja de río	1415
almeja dorada	1404
almeja fina	1401
almeja japonesa	1402
almeja lisa falsa	1391
almejas	1324, 1393, 1394
almejón brillante	1395
almejón de sangre	1395
alosa	155, 156, 157
alosa chata	168
alosa de Reeves	206
alosa mediocre	159
alosa mediterránea	158
amayuela	1401, 1405
amia	142
anchoa australiana	215
anchoa cola de rata	213
anchoa de caleta	210
anchoa del Cabo	216
anchoa del Pacífico	219
anchoa del Pacífico norte	219
anchoa europea	217
anchoa japonesa	218
anchoa listrada	209
anchoa tropical	209
anchoas	208
anchoas «stolephorus»	222
anchoita	214
anchoita de río	221
anchoveta	212
anchoveta del Perú	220
anchoveta rabo amarillo	211
andarica	1299
andoderro del Cabo	685
andoderros	684
angelote	68
angelotes	67
anguila	396
anguila americana	400
anguila australiana	397
anguila de Nueva Zelanda	398
anguila europea	396
anguila japonesa	399
anguilas	395, 401
anguilas peces-gato	378
anjora	630
anjova	630
anjovas	629
anodonta	1343
apogónidos	609
araña	868
araña blanca	869
arañas	866, 871
aranya	370
arbitán	460, 461
arca del Pacífico occidental	1339
arca japonesa	1340
arcas	1341, 1342
arencón	156
arencones	226
arenque	169
arenque de hebra	185
arenque de lago	231
arenque del Pacífico	170
arenque lobo de la India	225
arenque punteado	181
arenque redondo atlántico	177
arenque redondo de California	176
arenques lobo	224
arenques redondos	179
arete	1079
arete oscuro	1080
argentina	280, 281
argentina del Pacífico	283
argentinas	279
argós	829
aristéidos	1232
armado	1106
arroaz	1012
ascidia de Chile	1509
ascidia estolonífera	1510
ascidias	1507
aspio	311
aspro común	622
aspro mayor	624
aspro menor	623
atalo	1188
aterínidos	548
atuarro	1012
atún	1012
atún aleta amarilla	1008
atún aleta larga	1007
atún blanco	1007
atún chauchera	970
atún de aleta negra	1009
atún del sur	1010
atún lazón	963
atún rojo	1012
atún tongol	1013
atunes	1014
aviñeira	1365
ayu	269

B

babosa	909
babosa atlántica	908
babosa de río	911
babosa de tres aletas	912
babosa ocelada	905
babosa sanguinolenta	910
babosa vivípara	923
babosa vivípara americana	922
babosas	904, 906
bacalada	459
bacaladilla	459
bacalao	447
bacalao de Alaska	446
bacalao de Groenlandia	448
bacalao del Ártico	445
bacalao del Pacífico	446
bacalao largo	1092
bacalao negro	1094
bacalao oriental	444
bacalao polar	441
bacalaos antárticos	894
bacoreta	968
bacoreta oriental	967
bagaravel	793
bagre barba blanca	377
bagre bocalisa	375
bagre moteado	376
bagres	383, 387
bagres marinos	374
baila	594

balano gigante	1200
balanos	1199
balao saltador	432
ballena antártica	1489
ballena azul	1483
ballena blanca	1478
ballena boba	1481
ballena de Bryde	1482
ballena de Groenlandia	1487
ballena enana	1480, 1488
ballena franca	1487
ballena gris de California	1491
ballena hocico de botella	1460
ballena jorobada	1485
ballena pigmea	1488
ballena vasca	1490
ballenas	1486
ballenato	1480
banano	146
barbada	464
barbadas	451
barbito	314
barbo albanés	313
barbo comiza	315
barbo común	314
barbo de Hoven	338
barbo de Java	353
barbo de mar	817
barbo de montaña	316
barbo del Danubio	317
barbo italiano	318
barbos de Asia	354
barbuda de cuatro barbillas	469
barbudo del Indo-Pacífico	554
barbudo gigante africano	556
barbudo rayado	556
barbudo real	555
barbudos	553
barracuda	531, 535
barracuda india	534
barracuda oriental	533
barracudas	529
barrena	1416, 1417
barrilete negro	969
barrinaire	938
barruenda	462
bastangas	103
becut	314
bejel	1089
beluga	1478
berberecho	1389
berberecho común	1389
berberecho espinoso	1388
berberechos	1387
bermejuela	360
berruguete	759
besugo americano	518
besugo blanco	790
besugo chato	790
besugo de Laredo	793
besugo del Cantábrico	793
besugo del Norte	793
bichir	128
bígaros	1337
blanquillo camello	627
blanquillos	626
blenio de río	911
boba	766
bobo	539, 1079
boca blanca	53
boca negra	611
bocanegra	30
bocarte	217
bocina	1335

bodión verde	843
boga	766
boga de mar	766
boga de río	327
bogavante americano	1317
bogavante europeo	1318
bombay	296
bonito	982, 1007
bonito austral	979
bonito de altura	973
bonito del norte	1007
bonito del Pacífico oriental	980
bonito mono	981
bonito saltador	966
boquerón	217
boquerones	208
boquidulce	10
borracho	1084
borrico	123
borricos	122
brámidos	679
breca	791, 792, 794
brecas	795
bregmacero	439
brema blanca	319
brema común	305
brema de Pekín	349
brema del Danubio	306
bremas	307
brosmio	442
brótola brasileña	474
brótola de fango	463
brótola de roca	464
brótolas	465
brotolilla	438
brótula de barbas	924
buey	1291
buey de mar	1291
buey del Pacífico	1290
bueyes	1287
bumalo	297
buñelo de mar	1508
burro	710
burro labiogrueso	709
burro listado	708
burro ojón	703
burros	714
busicones	1336
buvego	950

C

caballa	985
caballa de la India	977
caballa del Atlántico	985
caballa isleña	976
caballa pintoja	983
caballa rechoncha	975
caballas	986
caballas del Indo-Pacífico	978
caballas reales	653
caballitos de mar	505
cabete	1085
cabete espinudo	1086
cabeza de serpiente cabrío	558
cabeza de serpiente rojo	557
cabezas de serpiente	559
cabezón	1435
cabracho	1068
cabrachos	1062
cabrilla	587
cabrilla loca	584
cabruza	909

cabuchino	1059
cachalote	1462
cacho	341, 342, 343
cachorreta	1012
cachucho	776
cachuelo	342, 343
caella	52
cailón marrajo	23
cailones	18
caimán	1503
calamar	1426
calamar de Boston	1424
calamar del Cabo	1425
calamares	1422
calamares, jibias, potas	1421
calamarín	1423
calandino	360
calderón	1466
calderón gris	1467
calderón tropical	1465
calé	776
callantias	570
caluga	547
camarón	1220, 1251
camarón argentino	1256
camarón australiano	1207
camarón boreal	1237
camarón café norteño	1213
camarón común	1249
camarón de altura	1212
camarón de piedra	1252
camarón espico	1238
camarón estilete argentino	1206
camarón guineano	1210
camarón nailon	1235
camarón navaja	1254
camarón norteño	1217, 1237
camarón patiblanco	1228
camarón rojo real	1257
camarón siba	1208
camarón siete barbas	1231
camaroncillo akiami	1242
camaroncillos	1230
camarones	1229, 1253
camarones atíidos	1243
camarones crangónidos	1250
camarones metapenaeus	1209
camarones nailon	1236
camarones navaja	1255
camarones nórdicos	1240
camarones palemónidos	1245
camarones pandálidos	1239
camarones parapeneopsis	1211
camarones sergestid	1241
cañabota	11
cañabota gris	11
cañabotas	9
cangrejo atlántico	1298
cangrejo cacerola	1515
cangrejo chino	1303
cangrejo colorado	1321
cangrejo de fango australiano	1302
cangrejo de las nieves	1312
cangrejo de patas punteadas	1277
cangrejo de piedra negro	1293
cangrejo de río	1279
cangrejo de río de Australia	1316
cangrejo de río de patas rojas	1276
cangrejo de río rojo	1284
cangrejo de torrente	1280
cangrejo japonés	1310
cangrejo verde	1298
cangrejo verde del Mediterráneo	1297
cangrejos de las nieves	1313

cangrejos de río	1278, 1281	cerdito marino	70	colas de rata	480
cangrejos ermitaños	1286	cerdo marino	70	colayo	30
cangrejos nadadores	1296	cerdos marinos	69	colín de Alaska	470
cangrejos rojos	1322	chacarona de Canarias	772	colmilleja	370
cangrejos rusos	1311	chafarrocas	1190	colmilleja de cejas	368
canutero	964	chancharros	1077	colmilleja dorada	366
capelán	273	chanos	228	colmilleja italiana	373
capelán del Mar Negro	455	chanquete	1050	colmilleja larga	367
capellán	473	chaparrudo	1053	colmilleja rumana	369
capitón	542, 774	chaparrudos	1055	colorada	791
carabinero	1234	charrasco	1100	coloradilla	360
carácidos	300	cherna	585	concertina	828
carajuelo	473	cherna criolla	581	concha de peregrino	1364
caramel	813	cherna de Juan Fernández	586	condrostoma común	325
carameles	814	cherna de ley	573	condrostoma dálmata	324
carángidos	632	cherna hapuku	586	condrostoma foxino	326
carbonero	467	chica	1389	condrostoma italiano	328
cardenal japonés	612	chicharra	1109	condrostoma sudeuropeo	323
cárdidos	1387	chicharro	670	condrus	1532
carite atlántico	994	chicharro de rayas amarillas	655	conejo	973
carite australiano	996	chicharro garetón	667	congrio	410
carite brasileño	987	chicharro ojón	654	congrio americano	411
carite chinigua	1000	chipa	352	congrio argentino	412
carite coreano	992	chirla	1396	congrio del Pacífico	409
carite de Monterrey	990	chirlía	1401, 1405	congrios	408
carite de Queensland	999	chiro	146	coquinas	1406
carite del Indo-Pacífico	991	chocha	1405	coral rojo	1514
carite estriado	989	cholga	1371	corales	1513
carite indochino	1003	chopa	808	corbina	718
carite kanadi	998	chorito	1374	corcón	538
carite lucio	988	choro	1371	corcones	537
carite oriental	997	chova	630	coregono	237
carite papuense	995	chucho pelágico	107	coregono azul	240
carite pintado	1005	chucla	812	coregono blanco	229
carite rayado	993	chuclas	810, 814	coregono de Artedi	231
carite sierra	1002	chucleto	550	coregono de lago	232
carite tigre	1001	chupadores	301	coregono de ojos grandes	235
carite-cazón	971	cíclidos	831	coregono de Pidchian	239
carite-cazón pintado	972	ciclóptero	1105	coregono del Leman	233
carites	1004	ciclópteros	1104	coregono elegante	230
carocho	60	cifio vivíparo	1461	coregono narigudo	236
carpa	332	cifios	1458	coregono peled	238
carpa cabezona	336	cigala	1320	coregonos	241, 261
carpa china	331	cigarros	1304	cornuda	47
carpa de fango	330	cimarrón	1012	cornuda común	46
carpa de Wuchang	346	cinoglósidos	1166	cornudilla	47
carpa gibel	321	cinta	834	corocoro burro	707
carpa negra	347	ciprínidos	303	corujo	1172, 1175
carpa plateada	335	ciprino plateado	355	corvallo	755
carpas	303	ciprinodóntidos	512	corvina	718
carpín	321	clacas	1199	corvina africana	717
casabe	640	clámide	8	corvina amarilla	740
casarte ojón	974	clarias	388	corvina bobo	749
castaña	680	clérigo	1084	corvina bosoro	751
castañeta	680, 857	clínido plateado	918	corvina casava	750
castañolas	679	cloroftálmidos	298	corvina honnibe	741
castañuela	836	cloroftalmo	299	corvina japonesa	722
catalufa	766	clulpeidos	154	corvina negra	721
catalufa del Pacífico	607	clupeonella	171	corvina plateada	744
catalufas	608	cobia	631	corvina reina	748
catemo africano	827	cobos	1338	corvina tigre	742
cavilat	1097	cochino	61	corvinas	715, 752, 756
cazón	50, 53	coco	743	corvinata ayanque	723
cazones	49	cocodrílidos	1501	corvinata blanca	720
centolla	1315	cocodrilo de Nueva Guinea	1504	corvinata pescada	727
centolla de roca	1307, 1315	cocodrilo del Siam	1506	corvinata pescadilla	728
centolla del sur	1314	cocodrilo marino	1505	corvinata pintada	725
centolla europea	1315	cocodrilos	1501	corvinata prieta	719
centolla patagónica	1309	cohombro de mar japonés	1523	corvinata real	726
centolla real	1308	cojinovas	1042	corvinata totoaba	724
centollo	1315	cojiúna carbonera	638	corvinatas	729
centrolófidos	1040	cojiúna negra	635	corvinato	718, 759
centrolofo	1041	cola de rata azul	479	corvinón brasileño	738
centropómidos	560	colaina	30	corvinón negro	745

corvinón ocelado	757	erizo común	1518	faneca	472
corvinón rayado	737	erizo de mar	1518	faneca noruega	471
coto común	1097	erizo de roca	1518	faneca plateada	449
coto escorpión	1100	erizos	1519	farfet común	514
coto negro	384	erizos de fondo	1520	farfet sudeuropeo	513
coto pardo	385	erla	785	ferrón	64
coto punteado	386	erlas	787	fletán	1133
cotos	383, 387	escacho	1079	fletán del Pacífico	1134
cova	630	escamarlànc	1320	foca anillada	1452
cangrejos de mar	1285	escamudo	1175	foca barbuda	1448
crapet	604	escardinio	362	foca capuchina	1447
crego	1084	escardinio griego	363	foca común	1453
cuchillo	1079	esciénidos	715, 739	foca de Groenlandia	1451
cuco	1079	escoftálmidos	1167	foca fraile	1450
cuco americano	1084	escolano	464	foca gris	1449
cuco de altura	1084	escolar	463, 950	focas	1440, 1446
cunas	583	escolar de Marley	953	fogonero	467
		escolar real	949	foxino común	352
D		escolar sierra	954	foxino de fangal	351
		escolares	947, 948	fraile	911
dabeta	540	escómbridos	961	fucales	1527
dagas	203	escorpión	867, 869	fura	463
damba	824	escorpión de cuatro cuernos	1099	fusileros	686
dambas	823	escorpión de dieciocho espinas	1098		
daplata	542	escorpión marino	1102		
dátil de mar	1372	escorpiones	866, 1101	**G**	
delfín azul	1474	escórpora	1066		
delfín común	1464	escrita	85	gádidos	440
delfin de hocico blanco	1469	escupiña lisa	1401	gadiformes	435
delfín de lomo blanco	1468	espada	957, 959	galateidos	1306
delfín de pico largo	1475	espadarte	1029	galeo	50
delfín listado	1474	espadilla del Caspio	171	galera	1202
delfines moteados	1476	espadín	202	galeras	1201
delfínidos	1463	espadín grácil	201	gallano	841
dentón	774	espadones	1015	gallego	341
dentón angolés	771	espáridos	763	gallerbo	907
dentón carpintero	765	espet	535	gallineta	1063, 1079
dentón común	774	espetón	535	gallineta cola amarilla	1074
dentón congolés	773	espetón de Guinea	530	gallineta del Cabo	1071
dentón de Canarias	772	espetones	529, 536	gallineta del Pacífico	1070
dentón listado	797	espinosillo	501	gallineta nórdica	1075, 1076
dentón nufar	768	espinoso	500, 502	gallineta rocote	1072
dentón ocelado	798	espinosos	499	gallinetas	1062, 1077
dentón rupestre	796	esponjas	1511	gallo	525, 1168, 1169
dentones de ojos grandes	761	esterlete	135	gallo cristo	525
diabelo antártico	893	estornino	984	gallos	1170
diablo de mar	114	estrella de mar	1517	galludo	64, 65
doncella	838	estrellas de mar	1516	galludo chato	66
doncella barbada	929	esturión	137	galludo de pintilla	64
dorada	802	esturión adriático	133	galludos	57
dorada americana	702	esturión barba de flecos	134	galúa	543
dorada gigante	805	esturión beluga	139	galúa africana	544
dragó	393	esturión blanco	138	galúa blanca	543
dugongo	1454	esturión de vientre desnudo	134	galúa negra	543
		esturión del Danubio	131	galupe	540
		esturión estrellado	136	gamba carabinero	1234
E		esturión verde	132	gamba de altura	1212
		esturiones	130	gamba de fango del Atlántico	1258
eglefino	453	eufasiáceos	1203	gamba rosada	1233
embarroco	1195	eulacón	278	gambas y camarones de agua dulce	1244
emelíchtidos	684			gambusia	516
emperador	1024, 1029			garapello	791, 794
enfú	938	**F**		garneo	1090
eperlán	274			garoupa	580
eperlano	274	falsa limanda	1142	gaspar chato	144
eperlano arco iris	275	falsa orca	1473	gasper picudo	143
eperlano ártico	276	falso halibut de cuatro ocelos	1119	gatet	370
eperlano de estanque	271	falso halibut del Canadá	1118	gato	50
eperlano del Pacífico	272	falso halibut del Japón	1120	gatuso	54
eperlanos	270, 277	falso salmonete de roca	821	gazapa	30
eperlón del Pacífico	278	falsos arenques	187	gelidium	1529
erizo	1521	falsos halibuts	1121	geriocangrejo rojo	1321
erizo blanco	1522	falsos percebes	1198	gigartinas	1531
erizo chileno	1522	falsos salmonetes	822	globito	1419, 1420

góbido italiano	1056	jorobado africano	656	langostino blanco	1228
góbidos	1049, 1055	jorobado lamparosa	657	langostino blanco norteño	1227
gobio	333	jorobados	676	langostino carnoso	1216
gobio dálmata	312	judío	964	langostino cristal	1214
gobio de arena	1059	juliana	462	langostino de cola roja	1225
gobio de Canestrini	1058	junco	1008	langostino de Guinea	1217
gobio de Kesler	1057	jurel	635, 670, 672	langostino de río	1247
gobio del Danubio	334	jurel amarillo	637	langostino de río americano	1246
gobio del Mar Negro	1060	jurel azul	635	langostino del Índico	1319
gobio dorado	1051	jurel chileno	669	langostino español	1220
gobio gigante	1052	jurel cunene	674	langostino kuruma	1219
gobio negro	1053	jurel de Alejandría	633	langostino marfil	1221
gobio ofiocéfalo	1061	jurel de Nueva Zelanda	652	langostino mejicano	1213
gobios durmientes	1048	jurel del Cabo	673	langostino patiamarillo	1215
golleta acevia	1159	jurel del Pacífico	671	langostino real	1220
golletas	1161	jurel japonés	666	langostino rosado sureño	1224
golondrina	426	jurel mediterráneo	668	langostino tigre gigante	1223
goraz	793	jurel verde	665	langostino tigre marrón	1218
granadero	494	jureles	639, 675	langostino tigre verde	1226
granadero de roca	497			langostino yumbo	1223
granaderos	496, 498	**K**		langostinos	1229, 1306
guardia civil	47			langostinos de agua dulce	1248
guitarra	75, 76, 462	keta	248	langostinos de río americanos	1283
guitarra chola	74	krill	1203	lansoya	939
guitarra pecosa	73	krill antártico	1204	lanzón	939
guitarras	72, 77	krill de Noruega	1205	lanzón americano	932
gurami besador	1047			lanzón de Corea	940
gurami gigante	1045	**L**		lanzón del norte	933, 935
gurami piel de serpiente	1046			lanzón del Pacífico	934
gurrimaña	1389	laban	540	lanzón mediterráneo	938
gurrimaño	1389	labeo	547	lanzones	931, 937
		labeos	337	lapas	1333
H		lábridos	837	lavareto	234
		lacha escamuda	165	lebranche	546
halibut	1133, 1134	lacha tirana	166	lenguadillo	1157
halibut del Pacífico	1125	lachas	167	lenguado americano	1129, 1152
halibut índico	1113	lagarto	391, 392, 393, 941	lenguado austral este	1154
halibut japonés	1124	lagarto escamoso	392	lenguado austral oeste	1153
halibut negro	1148	lagarto manchado	942	lenguado común	1165
herrera	786	lagarto real	389	lenguado del Pacífico	1135
herreras	787	lagartos	390	lenguado espinudo	1112
hipogloso	1133	lambe caletero	733	lenguado espinudo de altura	1111
hotentotes	789	lambe verrugato	734	lenguado turco	1163
huachinango	689	lambe zorro	735	lenguados	1151, 1155
hurta	803	lambes	736	lenguas	1166
		laminarias	1526	león de mar	1443
I		lamprea de mar	7	lepisósteo chato	144
		lamprea de río	5	lepisósteo óseo	143
igüelo	817	lamprea de Vladikov	4	letrínidos	762
ilisa	184	lamprea del Danubio	2	leucisco cabezudo	341
		lamprea marina	7	leucisco común	344
J		lamprea ucraniana	3	leucisco de Kessler	340
		lampreas	1	leucisco italiano	345
jaiba azul	1295	lampreilla	6	levirraya	92
jaiba de roca amarilla	1289	lampuga	681	liba	453, 454
jaiba de roca Jonás	1288	langosta común	1266	licodes	921
jaiba de roca masera	1291	langosta de Australia	1270	lija	50
jaiba del Pacífico	1292	langosta de Juan Fernández	1261	limanda	1139
jaiba gazami	1300	langosta de Natal	1265	limanda japonesa	1136, 1138
jaiba nécora	1299	langosta de Nueva Zelanda	1260	limanda nórdica	1137
jaiba siri	1294	langosta de Oceanía	1264	limanda noruega	1171
jaiba verde	1298	langosta de Transkei	1271	límulo	1515
jaibas	1301	langosta de Tristán	1263	liparis	1108
japuta	680	langosta del Cabo	1262	liparis atlántico	1107
jaqueta	53	langosta del Caribe	1269	lirio	459
jaquetón	19	langosta del sur	1267	lisa	538, 540
jaquetón blanco	19	langosta japonesa	1272	lisa dorada	540
jaquetones	18	langosta real	1273	lisa japonesa	541
jargueta	808	langosta rosada	1268	lisa negra	538
jibia	1418	langostas	1259, 1274	lisa pardete	545
jibia gigante	1427	langostas de fusta	1275	lisas	537
jonquillo	1050	langostino americano de agua		listado	973
		dulce	1282	lobo	372
		langostino banana	1222	lobo marino de dos pelos	1441

ES

421

lobo marino de un pelo 1444
lobo marino del Cabo 1442
locha 370, 463, 464, 475, 476
locha blanca 476
locha de roca 372
locha roja 475
lompo 1105
lompos 1104
longeirones 1414
lorcha 1053, 1092
lorcha de Atka 1093
loro gallo 844
loro viejo 856
lota 450
lota de río 452
lubina 593
lubina americana 598
lubina blanca 597
lubina blanca americana 596
lubina estriada 598
lubinas 595
lucio 288
lucio del Pacífico 291
lucio listado 287
lucio masquinongy 289
lucio negro 290
lucioperca 620
lucioperca americana 621
lucioperca canadiense 619
lucios 286
luna real 509
lutjánidos 687

M

macab badejo 153
macabijo 152
macarela caballa 641
macarela chuparaco 643
macarela de Russel 644
macarela japonesa 642
macarela salmón 646
macarelas 645
machete 175
machuelo 193
machuelo del Pacífico 185
machuelo hebra atlántico 186
machuelos 195
mactra lisa 1390
madrilla 329
magallón 593
maganto 1320
magnote 845
maira 459
makulu 677
malacho 146
malachos 147
manatí 1455
manatí africano 1457
manatí común 1456
mangones 1417
manta 112, 114
manta del Pacífico 113
mantas 111
mantenilla 115
maragota 840
marao ojón 418
maraos 419
marlín 1024
marlín peto 1025
marlín rayado 1023
marlín trompa corta 1022
marlines 1015, 1020, 1027
marolo 1388

marrajo 20
marrajo carite 21
marrajo gigante 12
marrajo sardinero 23
marsopa 1471
marsopa de Dall 1472
maruca 460, 462
maruca azul 460
masera 1291
medregal coronado 658
medusas 1512
mejillón 1376
mejillón común 1376
mejillón coreano 1375
mejillón de agua dulce 1383
mejillón de Australia 1378
mejillón de Nueva Zelanda 1381
mejillón de río 1345
mejillón de roca sudamericano 1382
mejillón del Plata 1379
mejillón mediterráneo 1377
mejillón verde 1380
mejillones 1370
mejillones de agua dulce 1344
melgacho 39
melva 964
melva tazard 965
mendo 1128
mercenaria 1398
mercenaria japonesa 1399
mercenarias duras 1400
merillo 588
merlán 454
merlo 842
merluza 487
merluza angolense 489
merluza argentina 486
merluza atlántica 483
merluza austral 482, 878
merluza blanca de altura 481
merluza chilena 485
merluza costera 484
merluza de altura 488
merluza de Benguela 489
merluza de Boston 483
merluza de cola 478
merluza del Cabo 484
merluza del Pacífico norte 490
merluza del Senegal 491
merluza europea 487
merluza negra 877
merluza norteamericana 483
merluza pacífica norteamericana 490
merluza senegalesa 491
merluza sudamericana 486
merluza sureña 482
merluzas 477, 492
merluzas australes 879
mero 577
mero americano 580
mero de Alejandría 574
mero gigante 578
mero gigante del Pacífico 591
mero malabárico 579
mero moteado 575
mero negro 576
mero sureño 568
meros 582
mielga 64
mielgas 57
mijarco 817
misgurno 371
modiolos 1373
mojarra, muxarra 783

mojarras 699, 701, 903
mojarrita de ley 902
molinero 341
mollareta 450
mollera 473
mollera azul 437
moluscos marinos 1323
moras 436
morena 404
morena verde 403
morenas 402
morenocio de Guinea 405
morenocio dentón 406
morenocios 407
morragute 542
morsa 1445
morwongos 860
muergo 1410
muergos 1414
mugil común 545
mugílidos 537
mugle 538
mujol 545
mula 506
mular 1477
mullosa 463
múrices, busanos 1334
musola 53
musola antártica 51
musola dentuda 52
musola pintada 53
musola vera 53
musolas 49, 55
mustela 443

N

nácar 1385
namorado 864
napoleon 1133
narval 1479
navaja 1409
navaja del Atlántico 1408
navaja del Cabo 1412
navaja del Pacífico 1411
navaja europea 1413
navajas 1407, 1414
navajón coniveto 946
navajones 945
nécora 1299
negra 60, 61
negrita 60, 61
negrito 61
negritos 62
nemiptéridos 695, 697
nemíptero 696
noriega 84
nototenia 887
nototenia amarilla 882
nototenia cabezota 883
nototenia de Magallanes 889
nototenia de ojos rayados 884
nototenia de Ramsay 892
nototenia gris 886
nototenia jaspeada 885
nototenia negra 881
nototénidos 876

O

oblada 788
ochavo 527
ochavos 528

ES

ojiverde	299	
olayo	30	
orca	1470	
orca negra	1473	
oreja de mar	1326, 1328	
oreja de mar de labios negros	1327	
oreja de mar gigante	1325	
orejas de mar	1329	
óreos	526	
ostión	1347	
ostión americano	1349	
ostión de manglar	1348	
ostión del Pacífico	1347	
ostiones	1350	
ostra australiana	1357	
ostra chilena	1351	
ostra común	1353	
ostra de Nueva Zelanda	1355	
ostra de pobre	1347	
ostra denticulada	1352	
ostra europea	1353	
ostra perlífera	1346	
ostra plana	1353	
ostra plana del Pacífico	1354	
ostra portuguesa	1347	
ostras	1356	

P

paganel	1054	
pagel	791, 792, 794	
pagualas	825	
paguro	1291	
pailona	60	
paire	959	
pajel	794	
palenca	472	
paletas	626	
palometa china	1032	
palometa del Pacífico	1034	
palometa negra	677, 680	
palometa pámpano	1033	
palometa plateada	1031	
palometa roja	518	
palometas	517, 664, 679	
palometón	647	
palometones	648	
palomida	973	
pámpano	1037	
pámpano amarillo	662	
pámpano del Pacífico	1044	
pámpano índico	1038	
pámpanos	639, 664, 1030	
pámpanos del Golfo	1036	
pancho	790, 793	
paneka	472	
panga	649, 799	
pañoso	808	
panzón	321	
papagallo	570	
paparda	422	
paparda del Atlántico	422	
paparda del Pacífico	421	
papardas	420	
paraperca azul	864	
pardete	545	
pardilla	360	
pardilleja	718	
pargo	806	
pargo biajaiba	692	
pargo colorado	691	
pargo de mangle	688	

pargo de pluma	844	
pargo del Cabo	769	
pargo del Golfo	690	
pargo dorado	802	
pargo testudo	774	
pargo zapata	804	
pargos	693, 770, 807	
parona	651	
parpada	680	
parracho	1172	
parrocha	189	
pastinaca	104, 106	
pastinaca espinosa	105	
pastinacas	103	
pato	1291	
patudo	1011	
peces ballesta	1179, 1182	
peces banano	151	
peces cuchillo	292	
peces de cristal	563	
peces de San Pedro	523	
peces espada	1028	
peces espátula	140	
peces gato de cristal	379	
peces hielo	895	
peces lagarto	145	
peces limón	661	
peces linterna	294	
peces luna	698	
peces mariposa	830	
peces martillo	45	
peces murciélago	826	
peces palo	872	
peces parecidos a los atunes	960	
peces perro	913	
peces planos	1110	
peces plata	279	
peces rata	122, 875	
peces sable	955	
peces sierra	78	
peces toro	302	
peces voladores	423, 425	
pega roques	1190	
pegatimón	7	
peine caletero	1361	
peine islándico	1362	
peine percal	1360	
peines	1358	
peje araña	869	
peje armado	1195	
pejepuerco blanco	1180	
pejerrey	280, 281, 549, 550, 551	
pejerrey del Atlántico	552	
pejerreyes	548	
pejesapo	1195	
peleco	350	
peluda	1115	
peludas	1114	
peludilla	1116	
peona	1079	
peonza cornuda	1331	
peonza sudafricana	1332	
pequeño lanzón	936	
perca	617	
perca americana	603	
perca canadiense	616	
perca de roca	600	
perca del Nilo	565	
perca gigante	564	
perca negra	602	
perca plateada	605	
perca rumana	618	
perca trepadora	1039	
percas	566, 613	

percas de arena brasileñas	865	
percebe	1197	
perdigón	1389	
peregrino	12	
perlita	459	
perlón	1084	
perritos del norte	917	
perro	1111	
perro azul	914	
perro del Norte	915	
perro pintado	916	
pescadilla	487	
pescadilla real	732	
peto	535, 962	
petxina negra	1401, 1405	
pez ángel	68	
pez ballesta	1180, 1183	
pez ballesta anaranjado	1181	
pez blanco	628	
pez cinto	957	
pez clavo	71	
pez de arena japonés	863	
pez de bacalao	462	
pez de cristal	930	
pez de Mahoma común	87	
pez de plata	280	
pez de plata grande	281	
pez de San Pedro	525	
pez de San Pedro americano	524	
pez dorado	320	
pez escorpión americano	1073	
pez espada	1029	
pez espátula chino	141	
pez globo maculado	1185	
pez hielo austral	896	
pez hielo común	898	
pez hielo de Georgia	901	
pez hielo de Wilson	897	
pez hielo narigudo	899	
pez hielo ocelado	900	
pez holoturia	930	
pez limón del Japón	660	
pez linterna	295	
pez lirio	459	
pez luna	1188	
pez luna alargado	1189	
pez luna de la India	678	
pez luna lanceolado	1187	
pez mantequilla	919	
pez mantequilla americano	1035	
pez martillo	47	
pez nariz azul antártico	1043	
pez obispo	110	
pez palo	873	
pez peine	50	
pez perro	844	
pez pescador	1195	
pez piloto	650	
pez plata	282	
pez plata pequeño	282	
pez poni	700	
pez rana	468	
pez rata	493	
pez remo	510	
pez rojo	320	
pez sable	957, 959	
pez sable del Pacífico	958	
pez sierra	79	
pez sol	601	
pez tachuela	71	
pez tamboril	1195	
pez vela del Atlántico	1016	
pez vela del Pacífico	1017	
pez verde	854	

picarel	813
picón	96
picos de pato	872
picuda	531
picuda barracuda	532
picudas	529, 536
picudo	535
pigotin	1195
pijota	487
pijotilla	487
pinchagua	161
pinchorro	64
pintadilla	858
pintadilla cola larga	859
pintarroja	31
pintarroja bocanegra	30
pintarrojas	28, 33
pique	53
piura	1509, 1510
platecha	1145
platicefálidos	1095
platicéfalo índico	1096
platija	1145
platija americana	1132
platija canadiense	1132
platija del Pacífico	1144
platija europea	1143
platija japonesa	1130, 1131
platura	1145
platuxa	1145
plegonegro	454
pleuronéctidos	1123
plumas	767
podas	1117
polaca austral	458
pomacéntridos	835
porredana	847
pota	1430
pota argentina	1428
pota europea	1433
pota japonesa	1432
pota neozelandesa	1431
pota voladora	1429
priacántidos	606
protópteros	129
provecho	1508
pulmonado australiano	127
pulpitos	1434
pulpo	1438
pulpo amizclado	1436
pulpo blanco	1435
pulpo común	1438
pulpos	1434, 1439
pulpos blancos	1437
puntilla	1423

Q

quelvacho	58
quelvacho negro	59
quelve	58
quimera	123
quimera americana	124
quimera del Cabo	125
quimeras	122, 126
quisquilla de arena	1251
quisquilla gris	1251

R

rabeta	753
rabil	1008
rabioso	1384

rabirrubia	694
rahó	855
rana marina	1195
rana pescadora	1195
rana toro	1492
ranas	1493
raó	855
raor	855
rapante	1175
rape	1195
rape americano	1193
rape blanco	1195
rape común	1195
rape diablo	1196
rape negro	1194
rapes	1192
rascacio	1067
rascacio de Kuhl	1065
rascacio rubio	1063
rascacios	1062, 1064, 1069
raspallón	778
rata	874
rata italiana	495
ratón	493
ratón de mar	1103
raya aguda	96
raya alba	82
raya áspera	98
raya basta	94
raya blanca	82
raya boca de rosa	85
raya bramante	82
raya cardadora	89
raya cimbreira	91
raya común	87
raya de Canadá	88
raya de clavos	87
raya de cola espinosa	99
raya de espejos	92
raya de San Pedro	94
raya del Pacífico	100
raya estrellada	83, 97
raya falsa-vela	86
raya fina	93
raya látigo del Pacífico	104
raya manchada americana	95
raya mosaica	101
raya picuda	96
raya pintada	93
raya radiada	83, 97
raya radiante	97
raya santiguesa	94
raya vela	90
rayas	80, 81, 102
rayas mariposa	116
rayón	96
rebeca	76
relazón	422
reloj	521
reloj anaranjado	520
rémol	1172, 1175
rémol del Índico	1122
_rémora	1178
reo	257
rey del besugo	518
reyezuelo	610
robaliza	593
róbalo	593
róbalo blanco	561
róbalo patagónico	880
róbalos	562
rocual aliblanco	1480
rocual negro	1481
rocual norteño	1481

rodaballo	1172, 1175, 1176
rodaballo americano	1174
rodaballo de California	1127
rodaballo de Japón	1126
rodaballo del Mar Negro	1173
ródeo	356
romerillo	1041
romero	473
roncador del Pacífico sudoriental	706
roncador plateado	711
roncadores	705, 714
ronco canario	704
ronco mestizo	712
ronco sompat	713
rorcual azul	1483
rorcual común	1484
rorcual de Bryde	1482
rorcuales	1486
rosada	925
rosada chilena	927
rosada del Cabo	926
rosadas	928
rubiel	776
rubio	1091
rubio del Cabo	1081
rubio kumu	1082
rubio picudo	1088
rubios	1078, 1083
rubios americanos	1087
rutilo	360
rutilo de Albania	348
rutilo del Danubio	358
rutilo del Mar Negro	357
rutilo sudeuropeo	359
rutilos	361

S

sabago	1291
sábalo	156
sábalo africano	174
sábalo americano	162
sábalo chacunda	164
sábalo común	156
sábalo del Canadá	155
sábalo del Mar Negro	160
sábalo hilsa	205
sábalo kelee	182
sábalo molleja	172
sábalo toli	207
sábalos comunes	163
sabalote	228
sabalotes	227
sable	957, 959
sable negro	956
sables	955
saboga	157
sabogas	163
salema	801
salmón	252, 256
salmón australiano	683
salmón chinook	254
salmón chum	248
salmón coho	249
salmón de Biwa	253
salmón de Georgia	682
salmón del Atlántico	256
salmón del Danubio	242
salmón gila	246
salmón japonés	250
salmón jorobado	247
salmón keta	248
salmón plateado	249
salmón real	254

salmón rojo	252	siciones	1253	tiburón toro	13
salmón rosado	247	sierra	951	tiburón zorro	17
salmones y truchas	262	sierras	952	tiburones	34
salmonete	817	siganos	944	tiburones martillo	45, 48
salmonete barbudo	820	siguro	943	tilapia de Mozambique	832
salmonete de buena casta	818	sillagínidos	625	tilapia del Nilo	833
salmonete de fango	817	siluro	382	tímalo	268
salmonete de malacasta	817	siluro eléctrico	394	tinca	364
salmonete de roca	818	siluro europeo	382	tintoleta	42
salmonete dorado	816	siluro griego	381	tintorera	42
salmonete rayado	818	siluros	380	tintorera-marrajo	42
salmonetes	815, 819	sinodontis	293	tolla	52
salpa	801	sobrajo	14	tolle	50
salpa xurel	500	soldado	1162	tollo	50, 53
salta murades	393	soldado rojo	522	tollo de Groenlandia	63
saltarín	422	solen del Atlántico	1408	tollos	49
saltón	416	solla	1145	tomcod	456, 457
salvelino	263, 264, 265	solla ártica	1140	toñina	968
salvelinos	267	solla de Alaska	1146	tordo	851
sama de pluma	775	solla de arena	1149	tordo de cinco manchas	849
sama dorada	774	solla lisa	1141	tordo de roca	848
sama marroquí	777	solla roja	1147	tordo picudo	850
samarugo	515	sollas de arena	1150	torillo	905
San Martín	525	solleta	1177	torpedo	118, 119
santiaguiño	1305	sollo	1172	torpedos	117, 121
santiaguiños	1304	sonso	938	tortuga bastarda	1498
sapo	1195	sortija	1164	tortuga boba	1496
sapo blanco	1195	sucla	811	tortuga laúd	1495
sapos	1191, 1192	sula	939	tortuga verde	1497
sarbito	372	sula de playa	531	tortugas de dorso diamantino	1499
sarbo	372	surela	643	tortugas de mar	1494
sarda	968, 982, 985			trama de Günther	891
sardina	189			trama patagónica	890
sardina arco iris	173	**T**		trama rayada	888
sardina australiana	198			trancho	202
sardina chilena	200	tabernero	839	traquínidos	866
sardina de California	196	tachuela	462	traquíptero	511
sardina de ley	190	tambor	1156	tremielga	118, 120
sardina del sur	204	tambor real	1158	tremielga negra	119
sardina europea	189	tamboril norteño	1185	tremielgas	117, 121
sardina japonesa	197	tamboriles	1184, 1186	tremolinas	117
sardina sudafricana	199	tarpón	150	tres colas	569
sardinela	191	tarpón atlántico	150	tríglidos	1078
sardinela de la India	192	tarpón Indo-Pacífico	149	trisa	156
sardineta	202	tarpones	145, 148	trompetero	504, 862
sardineta africana	183	tautoga americana	853	trompeteros	503, 861
sardineta canalera	178	tautoga negra	852	tróquidos	1330
sardineta del Japón	194	tembladera	118, 119	trucha	257
sardineta índica	188	tembladeras	117, 121	trucha alpina	263
sardinetas	180	temblón	118	trucha arco iris	251
sargo americano	780	tenca	364	trucha clarki	255
sargo austral	800	terapónidos	592	trucha común	258
sargo breado	779	tetenkel	649	trucha de Arizona	244
sargo chopa	764	tiburón alfombra	25	trucha de arroyo	258
sargo de América del Norte	809	tiburón azul	42	trucha de fontana	264
sargo marroquí	782	tiburón ballena	12, 27	trucha de mar	257
sargo mojarra	783	tiburón blanco	19	trucha de río	258
sargo picudo	781	tiburón boreal	63	trucha dorada mejicana	245
sargos, raspallones	784	tiburón cebra	24	trucha dorado	243
sarreta	454	tiburón cobrizo	36	trucha lacustre	259, 266
sensu	938	tiburón de aletas negras	38	trucha lacustre americana	266
sepia común	1418	tiburón de aletas plateadas	35	trucha marina	257
sepiola	1419	tiburón de peinetas	11	trucha marisca	257
seriola australiana	659	tiburón gato marrón	29	truchas	260
seriola caballo	636	tiburón gris	40	tuna	1011
seriola dorada	634	tiburón lagarto	8	túnidos	1006
serpiente de mar	413, 414	tiburón lamia	39		
serrano	589	tiburón leopardo	56		
serrano arenero	572	tiburón limón	41	**U**	
serrano estriado	571	tiburón martillo	47		
serrano japonés	599	tiburón narigudo atlántico	44	umbra común	284
serranos	567, 590	tiburón narigudo del Pacífico	43	umbra pigmea	285
serreta	466	tiburón nodriza	26	umbrina de agua dulce	716
serrucho	973	tiburón oceánico	37	umbrina del Pacífico	730
		tiburón salmón	22		

ES

V

vaca marina	114, 1455
vaqueta	846
venenoso	500
venera	1364
venéridos	1393
verdel	985
verdigón	1389
verrugato	759
verrugato austral	754
verrugato común	759
verrugato croca	731
verrugato de Japón	747
verrugato de Manchuria	746
verrugato de piedra	759
verrugato pargo	758
verrugatos	715, 760
víbora	870
vieira	1365
vieira americana	1369
vieira de Australia	1366
vieira de Nueva Zelanda	1367
vieira del Atlántico Sur	1368
vieiras	1358
vieja	856
villano	68
vimba	365
viola	76
volador golondrina	434
volador japonés	424
voladores	423
volandeira	1359

X

xasclet	550
xífidos	1028
xucladit	1190

Y

yonquillo	1050
yubarta	1485

Z

zamburiña	1363
zapatero	680
zarbo	372
zífidos	1458, 1459
zoárcidos	920
zope	304
zorro ojón	15, 16

ES

Dansk indeks

A

aborre	617
aborre, almindelig	617
aborre, argentinsk hav-	568
aborre, brasiliansk sand-	865
aborre, californisk hav-	591
aborre, grå hav-	576
aborre, gul	616
aborre, gylden hav-	574
aborre, hvid hav-	573
aborre, hvid sol-	604
aborre, japansk hav-	599
aborre, kæmpe-	577
aborre, Malabar-hav-	579
aborre, nil-	565
aborre, peruansk hav-	584
aborre, plettet hav-	575
aborre, rumænsk ulke-	618
aborre, rød hav-	580
aborre, sandhav-	572
aborre, småmundet ørred-	602
aborre, sol-	601
aborre, sort hav-	571
aborre, sort sol-	605
aborre, stormundet ørred-	603
aborre-familien	613
aborre-familien, hav-	567
aborre-slægt, hav-	582, 583, 590
adriatisk stør	133
afrikansk hestehoved	656
afrikansk lungefisk-slægt	129
afrikansk manat	1457
afrikansk multe	544
afrikansk spadefisk	827
afrikansk trommefisk	719
afrikansk venusmusling	1397
afrikansk ørnefisk	717
afrikansk ørnerokke	110
aftrækkerfisk	1180
aftrækkerfisk-familien	1179
akarnanisk blankesten	790
akiami-reje	1242
Alaska-musling	1403
Alaska-rødspætte	1146
Alaska-sej	470
albansk barbe	313
albansk skalle	348
albueskæl-slægt	1333
albulid-familien	151
Alexandria-trådfisk	633
alge-slægt, kalk-	1530
alge-slægt, rød-	1529
alger	1524
alger, brun-	1525
alger, rød-	1528
alligator, nordamerikansk	1502
alligator-familien	1500
almindelig aborre	617
almindelig bars	593
almindelig blankesten	806
almindelig boremusling	1416
almindelig delfin	1464
almindelig døbel	341
almindelig elektrisk rokke	120
almindelig gedde	288
almindelig glashvarre	1169

almindelig glathaj	53
almindelig grundling	333
almindelig guitarfisk	76
almindelig hammerhaj	47
almindelig havkat	915
almindelig havrude	808
almindelig havtaske	1195
almindelig havål	410
almindelig helt	234
almindelig hestemakrel	672
almindelig hjertemusling	1389
almindelig hork	614
almindelig hornfisk	416
almindelig håising	1132
almindelig ising	1139
almindelig jomfruøsters	1359
almindelig karpe	332
almindelig klumpfisk	1188
almindelig krebs	1276
almindelig løje	310
almindelig makrel	985
almindelig makrelgedde	422
almindelig ottearmet blæksprutte	1438
almindelig pighaj	64
almindelig pigsmerling	370
almindelig robalo	561
almindelig rudskalle	362
almindelig rævehaj	17
almindelig sandart	620
almindelig savgylte	847
almindelig sildehaj	23
almindelig skalle	360
almindelig skælbrosme	463
almindelig smerling	372
almindelig sortfisk	1041
almindelig strandkrabbe	1298
almindelig streber	623
almindelig stribefisk	551
almindelig strømskalle	344
almindelig stør	137
almindelig søstjerne	1517
almindelig tangkvabbe	908
almindelig tangnål	508
almindelig taskekrabbe	1291
almindelig trommefisk	759
almindelig tun	1012
almindelig tunge	1165
almindelig tungehvarre	1115
almindelig tæppemusling	1405
almindelig ulk	1100
almindelig ørnefisk	718
almindelig ørnerokke	109
almindelig ål	396
amerikansk bars	596
amerikansk flodkrebs	1282
amerikansk flodkrebs-slægt	1281
amerikansk havtaske	1193
amerikansk havål	411
amerikansk helt	231
amerikansk hummer	1317
amerikansk hundefisk	285
amerikansk kammusling	1361
amerikansk knivmusling	1408
amerikansk krebs	1282
amerikansk kråsesild	172
amerikansk læbefisk	853
amerikansk oksefrø	1492

amerikansk rødfisk	1073
amerikansk sandart	619
amerikansk sanktpetersfisk	524
amerikansk skærising	1129
amerikansk slethvar	1174
amerikansk smelt	275
amerikansk smørfisk	1033
amerikansk stamsild	162
amerikansk søørred	266
amerikansk tobis	932
amerikansk trommefisk	733
amerikansk tunge	1152
amerikansk ulk	1098
amerikansk østers	1349
amerikansk ål	400
amia	142
Amur-gedde	291
Andaman-hummer	1319
angolansk havrude	771
ansjos, argentinsk	214
ansjos, atlantisk	211
ansjos, australsk	215
ansjos, europæisk	217
ansjos, flod-	221
ansjos, japansk	218
ansjos, Kap-	216
ansjos, langhalet	213
ansjos, nordpacifisk	219
ansjos, nordvestatlantisk	210
ansjos, peruansk	220
ansjos, Stillehavs-	212
ansjos, stribet	209
ansjos-familien	208
ansjos-slægt	222
antarktisk glathaj	51
antarktisk isfisk	878
antarktisk lyskrebs	1204
antarktisk sortfisk	1043
antarktisk sølvfisk	893
antarktisk troldkrabbe	1309
antarktiske torsk	876
antimora, blå	437
arca, japansk	1340
arca, kornet	1339
argentinsk ansjos	214
argentinsk blæksprutte	1428
argentinsk havaborre	568
argentinsk havål	412
argentinsk rødreje	1256
argentinsk stiletreje	1206
argentinsk trommefisk	758
argusfisk-slægt	829
arktisk flynder	1140
arktisk krabbe	1312
arktisk smelt	276
asiatisk sardinel	191
asp	311
Atka-makrel	1093
atlantisk ansjos	211
atlantisk djævlerokke	112
atlantisk dybhavsmusling	1369
atlantisk hestehoved	657
atlantisk hornfisk	417
atlantisk kongemakrel	988
atlantisk laks	256
atlantisk menhaden	166
atlantisk navaga	445

DA

427

atlantisk pampano	662	bars, hav-	593	blåmusling, australsk	1378
atlantisk reje	1231	bars, hvid	597	blåmusling, chilensk	1374
atlantisk ringbug	1107	bars, Nassau-koral-	581	blåmusling, europæisk	1376
atlantisk rundsild	177	bars, newzealandsk flod-	864	blåmusling, grøn	1380
atlantisk sanktpetersfisk	525	bars, plettet	594	blåmusling, koreansk	1375
atlantisk sejlfisk	1016	bars, skrift-	589	blåmusling, La Plata-	1379
atlantisk sild	169	bars, sten-	600	blåmusling, Middelhavs-	1377
atlantisk spidssnudet haj	44	bars, stribet	598	blåmusling, newzealandsk	1381
atlantisk stribefisk	552	bars, svalehale-	569	blåmusling, sydamerikansk	1382
atlantisk tarpon	150	bars-familien, koral-	567	blåmusling-familien	1370
atlantisk tomcod	456	bars-slægt	595	blåplettet havrude	804
atlantisk torsk	447	bastardgryntefisk	712	blårød reje	1233
atlantisk trommefisk	738	Batavia-geddemuræne	406	blåsej	466
atlantisk trådsild	186	Benguela-kulmule	489	blåstak	841
atlantisk tun	1012	Bergatino pigsmerling	368	bobo-multe	539
atlantisk vragfisk	585	berggylt	840	bobo-trommefisk	749
atlantisk øglefisk	393	Bergs falkefisk	857	bobyrez	340
atun	951	berycid-familien	517	boe-trommefisk	753
australsk ansjos	215	berycid-slægt	519	Bombay-and	297
australsk blåmusling	1378	beryx, nordisk	518	Bombay-and-familien	296
australsk flodkrebs	1316	bikir	128	bonga-stamsild	174
australsk halvnæb	428	bitterling	356	bonit, bugstribet	973
australsk kammusling	1366	bjørnekrebs, lille	1305	bonit, chilensk	980
australsk kongemakrel	996	bjørnekrebs-familien	1304	bonit, plettet	966
australsk laks	683	bladtang	1526	bonit, Stillehavs-	981
australsk languster	1270	Blainvilles pighaj	65	boremusling, almindelig	1416
australsk lungefisk	127	blankesten	806	boremusling-slægt	1417
australsk pelamide	979	blankesten, akarnanisk	790	Boyers stribefisk	549
australsk ravfisk	659	blankesten, almindelig	806	brasen	305
australsk sardin	198	blankesten, Natal-	792	brasen, Donau-	306
australsk skæghaj	25	blankesten, rød	791, 794	brasen, guld-	802
australsk trommefisk	754	blankesten, spidstandet	793	brasen, hav-	680
australsk østers	1357	blankesten, stribet	786	brasen, hvid sten-	785
australsk ål	397	blankesten-familien	763	brasen, japansk guld-	805
auxide	965	blankesten-slægt	795	brasen, Peking-	349
ayu	269	blinkhindehaj-familien	34	brasen, rød sten-	796
azevia-tunge	1159	blodstribet slimfisk	910	brasen, tand-	774
azteker-reje	1213	blond rokke	85	brasen, Wuchang-	346
		blæksprutte, almindelig ottearmet	1438	brasen-familien, hav-	679
		blæksprutte, argentinsk	1428	brasen-slægten	307
B		blæksprutte, dværg-	1423	brasenflire	304
		blæksprutte, eledone-	1435	brasiliansk fladhoved	873
Bairds glathovedfisk	223	blæksprutte, flyve-	1433	brasiliansk halvnæb	430
Bairds næbhval	1459	blæksprutte, japansk	1432	brasiliansk kongemakrel	987
balao-halvnæb	429	blæksprutte, moskus-	1436	brasiliansk multe	546
balkansk pigsmerling	367	blæksprutte, newzealandsk	1431	brasiliansk sandaborre	865
bananreje	1222	blæksprutte, nordlig	1430	brasiliansk skægbrosme	474
bars, gede-	587	blæksprutte, ocean-	1427	bredbåndet kongemakrel	1001
barbe, albansk	313	blæksprutte, rød	1429	bredflab	1195
barbe, flod-	314	blæksprutte, Ross'	1419	bredpandet havkat	914
barbe, Hovens	338	blæksprutte, sepia-	1418	bredøjet hvarre	1117
barbe, iberisk	315	blæksprutte, sepiola-	1420	brillekaiman	1503
barbe, italiensk	318	blæksprutte-familie, ottearmet	1434	brisling	202
barbe, Java-	353	blæksprutte-familie, tiarmet	1422	brisling, europæisk	202
barbe, Middelhavs-	316	blæksprutte-familier	1421	broget rokke	101
barbe-grundling, dalmatisk	312	blæksprutte-slægt, ottearmet	1437, 1439	broget Sortehavs-kutling	1060
barbe-slægten, haj-	337	blødddyr	1323	brosme	442
barbe-slægter	354	blå antimora	437	brosme, almindelig skæl-	463
barberfisk, rød	570	blå havkat	914	brosme, brasiliansk skæg-	474
barberknivmusling	1410	blå hestemakrel	635, 670	brosme, hvid skæg-	476
barhovedet isfisk	887	blå marlin	1019	brosme, Middelhavs-skæl-	464
barracuda, europæisk	535	blå smørfisk	1037	brosme, rød skæg-	475
barracuda, Guinea-	530	blå svømmekrabbe	1295	brosme, skæl-	463
barracuda, gulfinnet	533	blå trompeterfisk	862	brosme-slægten, skæl-	465
barracuda, indisk	534	blåbars	630	brotula, skægget	924
barracuda, Stillehavs-	531	blåfinnet knurhane	1082	brugde	12
barracuda, stor	532	blåfisk	356	brun dværgmalle	385
barracuda-familien	529	blåhaj	42	brun kattehaj	29
barracuda-slægten	536	blåhelt	240	brun læbefisk	842
barramundi	564	blåhval	1483	brun reje	1207
bars, almindelig	593	blåhvilling	459	brun tigerreje	1218
bars, amerikansk	596	blåkæft	1063	brun ørnefisk	755
bars, blå-	630	blåkæft-slægten	1064	brunalger	1525
bars, dværg-	588	blåkrabbe	1295	brunhaj	40

Brydes hval	1482
bugstribet bonit	973
bundhelt	239
bundhelt, stor	236
byrkelange	460
bæklampret	6
bæknegenøje	6
bækørred	258
båndet isfisk	898
båndet knurhane	1091
båndfisk, rød	834

C

calico-kammusling	1360
California-flynder	1127
californisk havaborre	591
californisk reje	1215
californisk rundsild	176
californisk sardin	196
callista, glat	1395
Canada-rødding	266
canadisk stamsild	155
Canestrinis kutling	1058
caraibisk languster	1269
carrigeen tang	1532
catalufa, Stillehavs-	607
catalufa-slægt	608
centropomid-slægt	566
centropomider	560
chacunda-kråsesild	164
chilensk blåmusling	1374
chilensk bonit	980
chilensk hestemakrel	669
chilensk kingklip	927
chilensk kulmule	485
chilensk nylonreje	1235
chilensk sardin	200
chilensk sild	204
chilensk søpindsvin	1522
chilensk søpung	1509
chilensk østers	1351
chokolade-haj	60
cichlide, Mozambique-	832
cichlide, Nil-	833
cichlide-familien	831
citronhaj	41
crevalle	636
Cunene-hestemakrel	674
cutthroat-ørred	255

D

daddelmusling	1372
Dalls marsvin	1472
dalmatisk barbe-grundling	312
dalmatisk næsling	324
damefisk	152
damefisk, langfinnet	153
damkarusse	321
dammusling	1343
Danas svømmekrabbe	1294
delfin	1464
delfin, almindelig	1464
delfin, langnæbbet	1475
delfin, stribet	1474
delfin-familien	1463
delfin-slægt	1476
delta-krokodille	1505
djævlerokke, atlantisk	112
djævlerokke, lille	114
djævlerokke, Stillehavs-	113
djævlerokke-familien	111

dobbeltlinjet makrel	972
dobbeltsuger	1190
dolkhale	1515
dollarfisk	1035
Donau-brasen	306
Donau-grundling	334
Donau-hork	615
Donau-laks	242
Donau-lampret	2
Donau-løje	322
Donau-skalle	358
dragehovedfisk, Kuhls	1065
dragehovedfisk, lille	1066
dragehovedfisk, orange	1068
dragehovedfisk, sort	1067
dragehovedfisk-familien	1062
dragehovedfisk-slægt	1069
dronningetrommefisk	748
Dungeness-taskekrabbe	1290
dunkel knurhane	1080
dværg-blæksprutte	1423
dværgbars	588
dværgmakrel, indisk	977
dværgmakrel, kort	975
dværgmakrel-slægten	978
dværgmakrel, ø-	976
dværgmalle, brun	385
dværgmalle, plettet	386
dværgmalle, sort	384
dværgmalle-familien	383
dværgrethval	1488
dybhavsmusling, atlantisk	1369
dybhavsreje-familie	1232
dybhavsrødfisk	1076
dybhavstorske-familien	436
dybvands-rosenreje	1212
dybvandshummer	1320
dybvandskrabbe, femtandet	1321
dybvandskrabbe-slægt	1322
dybvandskulmule	488
dybvandsreje	1237
dygong	1454
dyndfisk	142
dyndsmerling	371
døbel	341
døbel, almindelig	341
døbel, Middelhavs-	342
døgling	1460
døgling, nordlig	1460

E

edderkopkrabbe	1315
egernfisk, rød	522
el-malle	394
el-rokke-familien	117
el-rokker	117
eledoneblæksprutte	1435
elefantfisk, Kap-	125
elefantfisk-slægten	126
elektrisk malle	394
elektrisk rokke, almindelig	120
elektrisk rokke, marmoreret	118
elektrisk rokke, sort	119
elektrisk rokke-slægt	121
elektriske rokker	117
elritse	352
elritse, sump-	351
elritsenæsling	326
emde	343
enke-rødfisk	1072
eremitkrebs-familie	1286
europæisk ansjos	217

europæisk barracuda	535
europæisk blåmusling	1376
europæisk brisling	202
europæisk ferskvandsål	396
europæisk hummer	1318
europæisk hundefisk	284
europæisk knivmusling	1413
europæisk kulmule	487
europæisk languster	1266
europæisk loligo	1426
europæisk malle	382
europæisk mulle	818
europæisk pilrokke	106
europæisk sardin	189
europæisk smelt	274
europæisk stjernekigger	874
europæisk søøre	1328
europæisk østers	1353

F

fakkelfisk	1094
falkefisk, Bergs	857
falkefisk, storfinnet	859
falkefisk, varieret	858
falkefisk-slægt	860
fanefisk	830
femplettet savgylte	849
femtandet dybvandskrabbe	1321
femtrådet havkvabbe	443
ferskvandsål, europæisk	396
ferskvandsåle-familien	395
ferskvandsåle-slægten	401
ferskvandskvabbe	452
ferskvandsreje, kæmpe-	1247
ferskvandsreje, sydamerikansk	1246
ferskvandsreje-familie	1243
ferskvandsreje-slægt	1248
ferskvandssmelt	271
ferskvandstangkvabbe	911
ferskvandsulk, hvidfinnet	1097
fierasfer	930
filfisk, orange	1181
filfisk-slægter	1182
finhval	1484
finnebræmmet ringbug	1108
firplettet glashvarre	1168
firplettet hvarre	1119
firstrålet trådfinnefisk	556
firtrådet havkvabbe	469
firvinget flyvefisk	434
fjeldørred	263
fjordtorsk	448
fjæsing, lille	867
fjæsing, plettet	868
fjæsing, stor	869
fjæsing-familien	866
fjæsing-slægten	871
fladfisk	1110
fladhoved, brasiliansk	873
fladhoved-familien	872
fladhoved-ulk, indisk	1096
fladhovedet ulke-familie	1095
flagermusfisk-slægten	826
flagfinnefisk	389
flire	319
flod-ansjos	221
flodbarbe	314
flodbars, newzealandsk	864
flodkrebs	1276
flodkrebs, amerikansk	1282
flodkrebs, australsk	1316
flodkrebs, galizisk	1277

flodkrebs, hvidfodet 1279
flodkrebs, Louisiana- 1284
flodkrebs-slægt 1283
flodkrebs-slægt, amerikansk 1281
flodkrebs-slægter 1278
flodlampret 5
flodnegenøje 5
flodniøje 5
flodperlemusling 1383
flodsild 161
flodtrommefisk 716
flodulk 1097
flynder, arktisk 1140
flynder, California- 1127
flynder, glat 1141
flynder, helle- 1133
flynder, indo-pacifisk pig- 1113
flynder, japansk 1126
flynder, japansk 1138
flynder, pacifisk 1135
flynder, plettet pig- 1111
flynder, sand- 1149
flynder, Stillehavs-helle- 1134
flynder, stjerne- 1144
flynder, vestafrikansk pig- 1112
flynder, vinter- 1147
flynder-slægt 1150
flynderfisk 1110
flyve-blæksprutte 1433
flyvefisk 426
flyvefisk, firvinget 434
flyvefisk, japansk 424
flyvefisk-familien 423
flyvefisk-slægt 425
flyveknurhane 1109
fløjfisk, plettet 942
fløjfisk, stribet 941
fløjls-svømmekrabbe 1299
fregat-makrel 964
fårehovedhavrude 764

G

gaffelmakrel 663
gaffelmakrel, stor 647
galeonsfisk, Kap- 824
galeonsfisk-familien 823
galizisk flodkrebs 1277
gambusia 516
gazami-svømmekrabbe 1300
gedde 288
gedde, almindelig 288
gedde, almindelig makrel- 422
gedde, Amur- 291
gedde, græs- 287
gedde, kortnæset panser- 144
gedde, langnæset panser- 143
gedde, makrel- 422
gedde, Stillehavs-makrel- 421
gedde, østamerikansk 290
gedde-familien 286
gedde-familien, makrel- 420
geddemuræne, Batavia- 406
geddemuræne, Guinea- 405
geddemuræne-slægt 407
gedebars 587
Georgia-isfisk 901
Georgia-laks 682
gerrid, sortfinnet 902
gerrid-slægt 903
gila-ørred 246
glansfisk 509
glasfisk-slægt 563

glashvarre, almindelig 1169
glashvarre, firplettet 1168
glashvarre-slægten 1170
glaskutling 1050
glasmalle-slægt 379
glastunge 1156
glat callista 1395
glat flynder 1141
glat trugmusling 1390
glatdick 134
glathaj, almindelig 53
glathaj, antarktisk 51
glathaj, patagonisk 54
glathaj, plettet 52
glathaj-familien 49
glathaj-slægt 55
glathovedfisk, Bairds 223
glatmundet havmalle 375
glyse 473
Golf-trommefisk 734
grindehval 1466
grindehval, halv- 1467
grindehval, kortluffet 1465
grindehval, langluffet 1466
grundling 333
grundling, almindelig 333
grundling, dalmatisk barbe- 312
grundling, Donau- 334
gryntefisk, bastard- 712
gryntefisk, gulfinnet 707
gryntefisk, Middelhavs- 710
gryntefisk, sølv- 711
gryntefisk, sompat- 713
gryntefisk, storøjet 703
gryntefisk, stribet 704, 708
gryntefisk, sydøstpacifisk 706
gryntefisk, tyklæbet 709
gryntefisk-familien 705
gryntefisk-slægt 714
græsgedde 287
græsk malle 381
græsk rudskalle 363
græskarpe 331
grøn blåmusling 1380
grøn læbefisk 843
grøn languster 1273
grøn muræne 403
grøn stør 132
grøn tigerreje 1226
Grønlandshaj 63
Grønlandshval 1487
Grønlandssæl 1451
grønøje 299
grønøje-familien 298
grønrygget hestemakrel 665
grå havaborre 576
grå isfisk 886
grå knurhane 1084
grå savgylte 845
gråhaj 50
gråhval 1491
gråsæl 1449
gråsej 467
Guinea-barracuda 530
Guinea-geddemuræne 405
Guinea-reje 1210
guitarfisk 72
guitarfisk, almindelig 76
guitarfisk, hvidplettet 73
guitarfisk, peruansk 75
guitarfisk, sydlig 74
guitarfisk-slægt 77
gul aborre 616
gul hestemakrel 637

gul snapper 689
gul trommefisk 722
gulbuget isfisk 882
guldbrasen 802
guldbrasen, japansk 805
guldfisk 320
guldkutling 1051
guldlaks 281
guldlaks, japansk 283
guldlaks, Stillehavs- 280
guldlaks-familien 279
guldmakrel 681
guldmulle 816
guldmulte 540
guldstribet havrude 801
guldstribet selar 655
guldørred 243
Gulf-menhaden 165
gulfinnet barracuda 533
gulfinnet gryntefisk 707
gulfinnet tun 1008
gulhalet hestemakrel 641
gulhalet ising 1137
gulhalet rødfisk 1074
gulhalet snapper 694
Günthers isfisk 891
gurami, kysse- 1047
gurami 1045
gurami, slangeskinds- 1046
gylden havaborre 574
gylden hestemakrel 634
gylden nemipterid 696
gylden pigsmerling 366
gylden tæppemusling 1404
gylte, almindelig sav- 847
gylte, femplettet sav- 849
gylte, grå sav- 845
gylte, junker- 838
gylte, langsnudet sav- 850
gylte, påfuglesav- 851
gylte, regnbue- 854
gylte, sav- 847
gylte, øjeplettet sav- 848
gøgerokke 89

H

haj, almindelig glat- 53
haj, brun- 40
haj, almindelig hammer- 47
haj, almindelig pig- 64
haj, almindelig ræve- 17
haj, almindelig silde- 23
haj, antarktisk glat- 51
haj, atlantisk spidssnudet 44
haj, australsk skæg- 25
haj, Blainvilles pig- 65
haj, blå- 42
haj, brun katte- 29
haj, chokolade- 60
haj, citron- 41
haj, Grønlands- 63
haj, grå- 50
haj, hval- 27
haj, hvidtippet 37
haj, indo-pacifisk hammer- 46
haj, is- 63
haj, kobber- 36
haj, krave- 8
haj, leopard- 56
haj, mako- 20
haj, makrel- 20, 21
haj, menneske- 19

DA

haj, mørk pig-	59	havgalt-familien	526	helt, bund-	239
haj, mørk-	39	havkal	63	helt, Léman-	233
haj, nurse-	26	havkarusse	839	helt, pollan-	230
haj, patagonisk glat-	54	havkat, almindelig	915	helt, sø-	232
haj, plettet glat-	52	havkat, blå	914	helt, storøjet	235
haj, ring-	30	havkat, bredpandet	914	helt-familien	261
haj, ru pig-	58	havkat, plettet	916	helt-slægt	241
haj, sandtiger-	13	havkat, stribet	915	heltling	229
haj, seksgællet	11	havkat-familien	913	hestehoved, afrikansk	656
haj, skjoldtand-	14	havkat-slægten	917	hestehoved, atlantisk	657
haj, småplettet rød-	31	havkvabbe, femtrådet	443	hestemakrel	672
haj, sort-	61	havkvabbe, firtrådet	469	hestemakrel, almindelig	672
haj, sorttippet	38	havkvabbe, tretrådet	450	hestemakrel, blå	635, 670
haj, Stillehavs-ræve-	15	havkvabbe-slægt	451	hestemakrel, chilensk	669
haj, Stillehavs-silde-	22	havlampret	7	hestemakrel, Cunene-	674
haj, Stillehavs-spidssnudet	43	havmalle, glatmundet	375	hestemakrel, en art	640
haj, Stillehavs-zebra-	24	havmalle, hvid	377	hestemakrel, grønrygget	665
haj, store hvide	19	havmalle, plettet	376	hestemakrel, gul	637
haj, storplettet rød-	32	havmalle-familie	374	hestemakrel, gulhalet	641
haj, storøjet pig-	66	havmus	123	hestemakrel, gylden	634
haj, storøjet ræve-	16	havmus-familien	122	hestemakrel, indisk	644
haj, syvgællet	10	havnegenøje	7	hestemakrel, japansk	666
haj, sølvtippet	35	havrotte	124	hestemakrel, langfinnet	642
haj, søm-	71	havrude, almindelig	808	hestemakrel, Middelhavs-	668
haj, trekant-	70	havrude, angolansk	771	hestemakrel, newzealandsk	652
haj-familien, blinkhinde-	34	havrude, blåplettet	804	hestemakrel, plettet	643
haj-familien, glat-	49	havrude, fårehoved-	764	hestemakrel, regnbue-	646
haj-familien, hammer-	45	havrude, guldstribet	801	hestemakrel, ru	667
haj-familien, pig-	57	havrude, Holbrooks	780	hestemakrel, sortrygget	638
haj-familien, rød-	28	havrude, kanarisk	772	hestemakrel, Stillehavs-	671
haj-familien, silde-	18	havrude, kongolesisk	773	hestemakrel, sydafrikansk	673
haj-familien, trekant-	69	havrude, marokkansk	777	hestemakrel, torpedo-	649
haj-slægt, glat-	55	havrude, Natal-	797	hestemakrel-familien	632
haj-slægt, pig-	62	havrude, nufar-	768	hestemakrel-slægt	639, 645, 648,
haj-slægt, rød-	33	havrude, panga-	799		664, 675, 676, 653
haj-slægten, hammer-	48	havrude, plettet	798	hestemusling-slægt	1373
hajbarbe-slægten	337	havrude, ring-	778	hestereje	1251
hajer, seksgællede	9	havrude, rød	775	hilsa	205
hajmakrel	971	havrude, rødbåndet	803	hjertemusling	1389
hajrokker	72	havrude, saddel-	788	hjertemusling, almindelig	1389
haleplettet picarel	811	havrude, sorthalet	778	hjertemusling, tornet	1388
halvgrindehval	1467	havrude, spidssnudet	781	hjertemusling-familien	1387
halvnæb, australsk	428	havrude, storøjet	776	Holbrooks havrude	780
halvnæb, balao-	429	havrude, stumpsnudet	769	hork	614
halvnæb, brasiliansk	430	havrude, sydlig	800	hork, almindelig	614
halvnæb, japansk	431	havrude, tømrer-	765	hork, Donau-	615
halvnæb, langfinnet	432	havrude, tvebåndet	783	hornet turbansnegl	1331
halvnæb-familien	427	havrude, zebra-	779	hornfisk	416
halvnæb-slægt	433	havrude-familien	763	hornfisk, almindelig	416
halvspækhugger	1473	havrude-slægt	767, 770, 784, 787,	hornfisk, atlantisk	417
hammerhaj, almindelig	47		789, 807	hornfisk, spids	418
hammerhaj, indo-pacifisk	46	havsvampe-familie	1511	hornfisk-familien	415
hammerhaj-familien	45	havtaske, almindelig	1195	hornfisk-slægt	419
hammerhaj-slægt	48	havtaske, amerikansk	1193	hornulk	1099
havaborre, argentinsk	568	havtaske, sort	1194	Hovens barbe	338
havaborre, californisk	591	havtaske, vestafrikansk	1196	huefisk, Kap-	685
havaborre, grå	576	havtaske-familien	1192	huefisk-familien	684
havaborre, gylden	574	havtobis	935	hummer, amerikansk	1317
havaborre, hvid	573	havørred	257	hummer, Andaman-	1319
havaborre, japansk	599	havål	410	hummer, dybvands-	1320
havaborre, Malabar-	579	havål, almindelig	410	hummer, europæisk	1318
havaborre, peruansk	584	havål, amerikansk	411	hummer, jomfru-	1320
havaborre, plettet	575	havål, argentinsk	412	hummer-familien, trold-	1306
havaborre, rød	580	havål, Stillehavs-	409	hundefisk, amerikansk	285
havaborre, sand-	572	havål-familien	408	hundefisk, europæisk	284
havaborre, sort	571	Hektors prikfisk	295	hundefisk, lille	285
havaborre-familien	567	hellefisk	1148	hundemusling	1384
havaborre-slægt	582, 583, 590	hellefisk, japansk	1124	hundestejle, nipigget	501
havbars	593	hellefisk, Stillehavs-	1125	hundestejle, stor	500
havbrasen	680	helleflynder	1133	hundestejle, trepigget	500
havbrasen-familien	679	helleflynder, Stillehavs-	1134	hundestejle-familien	499
havengel	68	helt, almindelig	234	hundetunge-familien	1166
havengel-familien	67	helt, amerikansk	231	hus	139
havgalt	527	helt, blå-	240	hval, blå-	1483

hval, Brydes 1482
hval, dværgret- 1488
hval, fin- 1484
hval, grinde- 1466
hval, Grønlands- 1487
hval, grå- 1491
hval, halvgrinde- 1467
hval, hvid- 1478
hval, kortluffet grinde- 1465
hval, langluffet grinde- 1466
hval, nar- 1479
hval, nord- 1487
hval, pukkel- 1485
hval, sej- 1481
hval, småhovedet 1461
hval, sydlig ret- 1489
hval, våge- 1480
hvalhaj 27
hvalros 1445
hvarre, almindelig glas- 1169
hvarre, almindelig tunge- 1115
hvarre, bredøjet 1117
hvarre, firplettet 1119
hvarre, firplettet glas- 1168
hvarre, hår- 1176
hvarre, kanel- 1122
hvarre, olivengrøn 1120
hvarre, små- 1171
hvarre, sommer- 1118
hvarre, storskællet 1177
hvarre, Thors tunge- 1116
hvarre-familien 1114
hvarre-slægt 1121
hvarre-slægten, glas- 1170
hvid bars 597
hvid havaborre 573
hvid havmalle 377
hvid marlin 1021
hvid skægbrosme 476
hvid slangemakrel 954
hvid solaborre 604
hvid stenbrasen 785
hvid stør 138
hvid trommefisk 720
hvid tun 1007
hvid-reje, nordlig 1227
hvidblomstret trommefisk 740
hvide haj, store 19
hvideside 1468
hvidfinnet ferskvandsulk 1097
hvidfisk 628
hvidfodet flodkrebs 1279
hvidhval 1478
hvidmundet trommefisk 737
hvidnæse 1469
hvidplettet guitarfisk 73
hvidrokke 90
hvidskæving 1468
hvidtippet haj 37
hvidøjet sandart 621
hvilling 454
hvilling, blå- 459
hvilling, Sortehavs- 455
højsøreje-familie 1241
håising, almindelig 1132
håising, japansk 1130
håising, nordpacifisk 1131
håret tunge 1162
hårhale 959
hårhale-familien 955
hårhvarre 1176

iberisk barbe 315
iberisk næsling 327
indisk barracuda 534
indisk dværgmakrel 977
indisk fladhoved-ulk 1096
indisk hestemakrel 644
indisk kongemakrel 989
indisk rovsild 225
indisk smørfisk 1038
indo-pacifisk hammerhaj 46
indo-pacifisk kongemakrel 991
indo-pacifisk makrel 983
indo-pacifisk pigflynder 1113
indo-pacifisk trådfinnefisk 554
indonesisk slangehovedfisk 557
isfisk, antarktisk 878
isfisk, barhovedet 887
isfisk, båndet 898
isfisk, Georgia- 901
isfisk, grå 886
isfisk, gulbuget 882
isfisk, Günthers 891
isfisk, korthalet 890
isfisk, langsnudet 899
isfisk, Magellan- 889
isfisk, marmor- 885
isfisk, patagonisk 880
isfisk, pukkel- 883
isfisk, Ramsays 892
isfisk, Scotia- 896
isfisk, sort 881
isfisk, sort patagonisk 877
isfisk, stribet 888
isfisk, Wilsons 897
isfisk, øjeplettet 900
isfisk, øjestribet 884
isfisk-familie 876, 895
isfisk-slægt 879, 894
ishaj 63
ising 1139
ising, almindelig 1139
ising, almindelig hå- 1132
ising, amerikansk skær- 1129
ising, gulhalet 1137
ising, japansk 1136
ising, japansk hå- 1130
ising, nordpacifisk hå- 1131
islandsk kammusling 1362
italiensk barbe 318
italiensk langhale 495
italiensk løje 309
italiensk næsling 328
italiensk pigsmerling 373

Jacobsmusling 1364
Jacobsskal 1364
Japan-krabbe 1310
Japan-laks 250
japansk ansjos 218
japansk arca 1340
japansk blæksprutte 1432
japansk flynder 1126
japansk flynder 1138
japansk flyvefisk 424
japansk guldbrasen 805
japansk guldlaks 283
japansk halvnæb 431
japansk havaborre 599
japansk hellefisk 1124

japansk hestemakrel 666
japansk håising 1130
japansk ising 1136
japansk kardinalfisk 612
japansk kongemakrel 997
japansk languster 1272
japansk multe 541
japansk musling 1399
japansk ravfisk 660
japansk sandfisk 863
japansk sardin 197
japansk sardinel 194
japansk sortfisk 1044
japansk søpølse 1523
japansk trommefisk 741
japansk trommefisk, lille 747
japansk tæppemusling 1402
japansk ål 399
Java-barbe 353
jomfrufisk 836
jomfrufisk-familien 835
jomfruhummer 1320
jomfruøsters, almindelig 1359
jomfruøsters, nordlig 1363
Jonaskrabbe 1288
Juan Fernandez-languster 1261
junkergylte 838
jødefisk 578

kaiman, brille- 1503
kalkalge-slægt 1530
kammusling, amerikansk 1361
kammusling, australsk 1366
kammusling, calico- 1360
kammusling, islandsk 1362
kammusling, kæmpe- 1365
kammusling, newzealandsk 1367
kammusling, ribbet 1368
kammusling-familien 1358
Kamtjatka-krabbe 1310
kanarisk havrude 772
kanelhvarre 1122
kaninfisk 943
kaninfisk-slægt 944
Kap-ansjos 216
Kap-elefantfisk 125
Kap-galeonsfisk 824
Kap-huefisk 685
Kap-knivmusling 1412
Kap-knurhane 1081
Kap-languster 1262
Kap-loligo 1425
kardinalfisk 610
kardinalfisk, japansk 612
kardinalfisk, teleskop- 611
kardinalfisk-familien 609
karetteskildpadde, uægte 1496
karpe 332
karpe, almindelig 332
karpe, græs- 331
karpe, marmor- 336
karpe, mudder- 330
karpe, sabel- 350
karpe, sort- 347
karpe, spansk tand- 514
karpe, sydeuropæisk tand- 513
karpe, sølv- 335
karpe, Valencia-tand- 515
karpe, Victoria- 355
karpe-familien 303
karpe-familien, suge- 301

karpefisk-slægt	361	kongelaks	254	krebs, amerikansk	1282
karpelaks-familien	300	kongemakrel, atlantisk	988	krebs, amerikansk flod-	1282
karper, æglæggende tand-	512	kongemakrel, australsk	996	krebs, antarktisk lys-	1204
karusse	321	kongemakrel, brasiliansk	987	krebs, australsk flod-	1316
karusse, dam-	321	kongemakrel, bredbåndet	1001	krebs, flod-	1276
karusse, hav-	839	kongemakrel, indisk	989	krebs, galizisk flod-	1277
karusse, sø-	321	kongemakrel, indo-pacifisk	991	krebs, hvidfodet flod-	1279
karusse, sølv-	320	kongemakrel, japansk	997	krebs, lille bjørne-	1305
kaskelot	1462	kongemakrel, kinesisk	1003	krebs, Louisiana-flod-	1284
kaspisk sild	171	kongemakrel, koreansk	992	krebs, norsk lys-	1205
kaspisk stamsild	160	kongemakrel, mangestribet	998	krebs, sten-	1280
kattehaj, brun	29	kongemakrel, Monterey-	990	krebs-familie, eremit-	1286
kelee-stamsild	182	kongemakrel, Papua-	995	krebs-familien, bjørne-	1304
Kesslers kutling	1057	kongemakrel, plettet	994	krebs-familien, lys-	1203
ketalaks	248	kongemakrel, prægtig	1000	krebs-slægt, amerikansk flod-	1281
kildeørred	264	kongemakrel, Queensland-	999	krebs-slægt, flod-	1283
kilemusling-slægt	1406	kongemakrel, Stillehavs-	1002	krebs-slægter, flod-	1278
kimærer	122	kongemakrel, stribet	993	krill	1204
kinesisk kongemakrel	1003	kongemakrel, vestafrikansk	1004	krill, norsk	1205
kinesisk kråsesild	168	kongemakrel-slægten	1005	krokodille, delta-	1505
kinesisk smørfisk	1032	kongereje	1221	krokodille, Ny Guinea-	1504
kinesisk spadestør	141	kongeslangemakrel	949	krokodille, siamesisk	1506
kingklip, chilensk	927	kongetrommefisk	726	krokodille-familien	1501
kingklip, rosa	925	kongolesisk havrude	773	krystalreje	1214
kingklip, sydafrikansk	926	konk	1335	kråsesild, amerikansk	172
kingklip-slægten	928	konksnegl	1335	kråsesild, chacunda-	164
kirurgfisk, Monrovia-	946	koral, ædel-	1514	kråsesild, kinesisk	168
kirurgfisk-familien	945	koral-familien, ædel-	1513	kuglefisk, nordlig	1185
klapmyds	1447	koralbars, Nassau-	581, 583	kuglefisk-familien	1184
klatrefisk	1039	koralbars-familien	567	kuglefisk-slægt	1186
Kleins tunge	1163	koralbars-slægt	583	Kuhls dragehovedfisk	1065
klippekutling	1054	koralmaller	378	kuller	453
klippereje	1252	koreansk blåmusling	1375	kulmule	487
klumpfisk	1188	koreansk kongemakrel	992	kulmule, Benguela-	489
klumpfisk, almindelig	1188	koreansk sandål	940	kulmule, chilensk	485
klumpfisk, lang	1189	kornet arca	1339	kulmule, dybvands-	488
klumpfisk, spidshalet	1187	kort dværgmakrel	975	kulmule, europæisk	487
Klunzingers ponyfisk	700	korthalet isfisk	890	kulmule, nordvestatlantisk	483
kløртang	1527	kortluffet grindehval	1465	kulmule, senegalsk	491
kløvefisk	855	kortnæbbet spydfisk	1022	kulmule, Stillehavs-	490
knapsnegle-slægt	1330	kortnæset pansergedde	144	kulmule, sydafrikansk	484
knivbladfisk-slægt	292	krabbe, almindelig strand-	1298	kulmule, sydlig	482
knivmusling, amerikansk	1408	krabbe, almindelig taske-	1291	kulmule, sydvestatlantisk	486
knivmusling, barber-	1410	krabbe, antarktisk trold-	1309	kulmule, sølv-	481
knivmusling, europæisk	1413	krabbe, arktisk	1312	kulmule-familien	478
knivmusling, Kap-	1412	krabbe, blå svømme-	1295	kulmule-slægten	492
knivmusling, lille	1409	krabbe, blå-	1295	kuruma-reje	1219
knivmusling, pacifisk	1411	krabbe, Danas svømme-	1294	kutling, broget Sortehavs-	1060
knivmusling-familien	1407	krabbe, Dungeness-taske-	1290	kutling, Canestrinis	1058
knivmusling-slægt	1414	krabbe, edderkop-	1315	kutling, glas-	1050
knivreje	1254	krabbe, femtandet dybvands-	1321	kutling, guld-	1051
knude	452	krabbe, fløjls-svømme-	1299	kutling, kæmpe-	1052
knurhane, blåfinnet	1082	krabbe, gazami-svømme-	1300	kutling, Kesslers	1057
knurhane, båndet	1091	krabbe, Japan-	1310	kutling, klippe-	1054
knurhane, dunkel	1080	krabbe, Jonas-	1288	kutling, Panizzas	1056
knurhane, flyve-	1109	krabbe, Kamtjatka-	1310	kutling, sand-	1059
knurhane, grå	1084	krabbe, konge-trold-	1307	kutling, sort	1053
knurhane, Kap-	1081	krabbe, mangrove-	1302	kutling-familien	1049
knurhane, langpigget	1090	krabbe, Middelhavs-strand-	1297	kutling-familien, søvn-	1048
knurhane, panser-	1106	krabbe, Murrays trold-	1308	kutling-slægt	1055
knurhane, pigget	1086	krabbe, østamerikansk taske-	1289	kvabbe, almindelig tang-	908
knurhane, ru	1085	krabbe, pacifisk taske-	1292	kvabbe, femtrådet hav-	443
knurhane, rød	1089	krabbe, sort sten-	1293	kvabbe, ferskvands-	452
knurhane, spidssnudet	1088	krabbe, syd-stankelbens-	1314	kvabbe, ferskvandstang-	911
knurhane, tværstribet	1079	krabbe, taske-	1291	kvabbe, firtrådet hav-	469
knurhane-familien	1078	krabbe, uldhånds-	1303	kvabbe, slange-	929
knurhane-slægt	1083, 1087	krabbe-familien, taske-	1287	kvabbe, tretrådet hav-	450
kobberhaj	36	krabbe-slægt, trold-	1311	kvabbe, vestatlantisk åle-	922
koho	249	krabbe-slægt	1313	kvabbe, øjeplettet tang-	905
kokanee-laks	252	krabbe-slægt, dybvands-	1322	kvabbe, åle-	923
konge, silde-	510	krabbe-slægt, svømme-	1296, 1301	kvabbe-familien, åle-	920
konge-rødreje	1257	kravehaj	8	kvabbe-slægt, åle-	921
konge-troldkrabbe	1307	krebs	1276	kvabbe-slægt, hav-	451
konge-trådfinnefisk	555	krebs, almindelig	1276	kvabber-slægt, tang-	906

kvabso	1105
kyssegurami	1047
kysttobis	936
kæmpe-ferskvandsreje	1247
kæmpe-kammusling	1365
kæmpe-tigerreje	1223
kæmpeaborre	577
kæmpekutling	1052
kæmperur	1200
kærtefisk	278
kødet reje	1216

L

La Plata-blåmusling	1379
laks, atlantisk	256
laks, australsk	683
laks, Donau-	242
laks, Georgia-	682
laks, guld-	281
laks, Japan-	250
laks, japansk guld-	283
laks, keta-	248
laks, kokanee-	252
laks, konge-	254
laks, masu	250
laks, pukkel-	247
laks, Stillehavs-guld-	280
laks, sølv-	249
laks-familien, guld-	279
laks-familien, karpe-	300
lakse-familien	262
lakse-slægt	260
lamantin	1456
lampret, bæk-	6
lampret, Donau-	2
lampret, flod-	5
lampret, hav-	7
lampret, ukrainsk	3
lampret, Vladykovs	4
lampret-familien	1
lang klumpfisk	1189
lange	462
lange, Middelhavs-	461
langfinnet damefisk	153
langfinnet halvnæb	432
langfinnet hestemakrel	642
langfinnet loligo	1424
langfinnet mako	21
langfinnet tun	1007
langhale, italiensk	495
langhale, newzealandsk	480
langhale, patagonisk	479
langhale, sortplettet	493
langhale-familien	496
langhale-slægt	477, 498
langhalet ansjos	213
langhals	1197
langhalse-slægt	1198
langhovedet trommefisk	751
langluffet grindehval	1466
langnæbbet delfin	1475
langnæbbet spydfisk	1026
langnæset pansergedde	143
langpigget knurhane	1090
langsnudet isfisk	899
langsnudet savgylte	850
langtornet ulk	1102
languster, australsk	1270
languster, caraibisk	1269
languster, europæisk	1266
languster, grøn	1273
languster, japansk	1272

languster, Juan Fernandez-	1261
languster, Kap-	1262
languster, mauretansk	1268
languster, Natal-	1265
languster, newzealandsk	1260
languster, oceanisk	1264
languster, syd-	1267
languster, Transkei-	1271
languster, Tristan-	1263
languster-familien	1259
languster-slægt	1274, 1275
Léman-helt	233
leopardhaj	56
lethrinid-familien	762
lethrinid-slægt	761
lille bjørnekrebs	1305
lille djævlerokke	114
lille dragehovedfisk	1066
lille fjæsing	867
lille hundefisk	285
lille japansk trommefisk	747
lille knivmusling	1409
lille stribefisk	549
lille tangnål	507
lille thunnin	967
lingcod	1092
lodde	273
lodsfisk	650
loligo, europæisk	1426
loligo, Kap-	1425
loligo, langfinnet	1424
Louisiana-flodkrebs	1284
lubbe	466
lungefisk, australsk	127
lungefisk-slægt, afrikansk	129
lyskrebs, antarktisk	1204
lyskrebs, norsk	1205
lyskrebs-familien	1203
lysspættet rokke	91
læbefisk, amerikansk	853
læbefisk, brun	842
læbefisk, grøn	843
læbefisk, Middelhavs-	846
læbefisk, tautog	852
læbefisk, vestatlantisk	844
læbefisk-familien	837
læbemulte	547
læderskildpadde	1495
løje	310
løje, almindelig	310
løje, Donau-	322
løje, italiensk	309
løje, regn-	339
løje, strøm-	308

M

Madeira-sardinel	193
Magellan-isfisk	889
Magellan-musling	1371
majsild	156
mako, langfinnet	21
mako-haj	20
makrel, koreansk konge-	992
makrel, almindelig	985
makrel, almindelig heste-	672
makrel, Atka-	1093
makrel, atlantisk konge-	988
makrel, australsk konge-	996
makrel, blå heste-	635, 670
makrel, brasiliansk konge-	987
makrel, bredbåndet konge-	1001
makrel, chilensk heste-	669

makrel, Cunene-heste-	674
makrel, dobbeltlinjet	972
makrel, en art heste-	640
makrel, fregat-	964
makrel, gaffel-	663
makrel, grønrygget heste-	665
makrel, gul heste-	637
makrel, guld-	681
makrel, gulhalet heste-	641
makrel, gylden heste-	634
makrel, haj-	971
makrel, heste-	672
makrel, hvid slange-	954
makrel, indisk dværg-	977
makrel, indisk heste-	644
makrel, indisk konge-	989
makrel, indo-pacifisk	983
makrel, indo-pacifisk konge-	991
makrel, japansk heste-	666
makrel, japansk konge-	997
makrel, kinesisk konge-	1003
makrel, kongeslange-	949
makrel, kort dværg-	975
makrel, langfinnet heste-	642
makrel, mangestribet konge-	998
makrel, Marleys slange-	953
makrel, Middelhavs-heste	668
makrel, Monterey-konge-	990
makrel, newzealandsk heste-	652
makrel, ø-dværg-	976
makrel, Papua-konge-	995
makrel, plettet heste-	643
makrel, plettet konge-	994
makrel, prægtig konge-	1000
makrel, Queensland-konge-	999
makrel, regnbue-heste-	646
makrel, ru heste-	667
makrel, sortrygget heste-	638
makrel, spansk	984
makrel, Stillehavs-heste-	671
makrel, Stillehavs-konge-	1002
makrel, stor gaffel-	647
makrel, stribet konge-	993
makrel, sydafrikansk heste-	673
makrel, torpedo-heste-	649
makrel, vestafrikansk konge-	1004
makrel- og tunfisk-familien	961
makrel-familien, heste-	632
makrel-familien, slange-	947
makrel-slægt, heste-	639, 645, 648, 664, 675, 676, 653
makrel-slægt, slange-	948, 952
makrel-slægten, dværg-	978
makrel-slægten, konge-	1005
makrelfisk-slægt	986
makrelgedde	422
makrelgedde, almindelig	422
makrelgedde, Stillehavs-	421
makrelgedde-familien	420
makrelhaj	20, 21
Malabar-havaborre	579
malermusling	1345
malermusling-familien	1344
malle, plettet dværg-	386
malle	382
malle, brun dværg-	385
malle, el-	394
malle, elektrisk	394
malle, europæisk	382
malle, glatmundet hav-	375
malle, græsk	381
malle, hvid hav-	377
malle, plettet dværg-	386
malle, plettet hav-	376

malle, sort dværg-	384	
malle-familie	380	
malle-familie, hav-	374	
malle-familien, dværg-	383	
malle-slægt	293	
malle-slægt, glas-	379	
malle-slægt, åle-	388	
malle-slægt, nordamerikansk	387	
maller, koral-	378	
Malma-ørred	265	
manat, afrikansk	1457	
manchurisk trommefisk	746	
mangestribet kongemakrel	998	
mangrove-østers, pacifisk	1348	
mangrovekrabbe	1302	
mangrovesnapper	688	
Marleys slangemakrel	953	
marlin, blå	1019	
marlin, hvid	1021	
marlin, sort	1018	
marlin, stribet	1023	
marlin-slægten	1020	
marliner	1015	
marmor-isfisk	885	
marmoreret elektrisk rokke	118	
marmorkarpe	336	
marokkansk havrude	777	
marsvin	1471	
marsvin, Dalls	1472	
masu laks	250	
mauretansk languster	1268	
meduse-slægt	1512	
mellemamerikansk reje	1228	
menhaden, atlantisk	166	
menhaden, Gulf-	165	
menhaden, Stillehavs-	175	
menhaden-slægten	167	
menneskehaj	19	
Midas søøre	1326	
Middelhavs-barbe	316	
Middelhavs-blåmusling	1377	
Middelhavs-døbel	342	
Middelhavs-gryntefisk	710	
Middelhavs-hestemakrel	668	
Middelhavs-læbefisk	846	
Middelhavs-lange	461	
Middelhavs-muræne	404	
Middelhavs-nøgentobis	938	
Middelhavs-savbug	521	
Middelhavs-skælbrosme	464	
Middelhavs-spydfisk	1024	
Middelhavs-strandkrabbe	1297	
Middelhavs-vågmær	511	
molboøsters	1386	
Monrovia-kirurgfisk	946	
Monterey-kongemakrel	990	
moskusblæksprutte	1436	
Mozambique-cichlide	832	
mudderkarpe	330	
mudderreje	1258	
mulle, europæisk	818	
mulle, guld-	816	
mulle, rød	817	
mulle, sortstribet	821	
mulle, stribet	818	
mulle, vestafrikansk	820	
mulle-familien	815	
mulle-slægt	819, 822	
multe, afrikansk	544	
multe, bobo-	539	
multe, brasiliansk	546	
multe, guld-	540	
multe, japansk	541	
multe, læbe-	547	

multe, spring-	543	
multe, stribet	545	
multe, tyklæbet	538	
multe, tyndlæbet	542	
multe-familien	537	
munkefisk	68	
munkesæl	1450	
Murrays troldkrabbe	1308	
muræne, Batavia-gedde-	406	
muræne, grøn	403	
muræne, Guinea-gedde-	405	
muræne, Middelhavs-	404	
muræne-familien	402	
muræne-slægt, gedde-	407	
muskellunge	289	
musketerfisk	686	
musling, afrikansk venus-	1397	
musling, Alaska-	1403	
musling, almindelig bore-	1416	
musling, almindelig hjerte-	1389	
musling, almindelig tæppe-	1405	
musling, amerikansk kam-	1361	
musling, amerikansk kniv-	1408	
musling, atlantisk dybhavs-	1369	
musling, australsk blå-	1378	
musling, australsk kam-	1366	
musling, barberkniv-	1410	
musling, calico-kam-	1360	
musling, chilensk blå-	1374	
musling, daddel-	1372	
musling, dam-	1343	
musling, europæisk blå-	1376	
musling, europæisk kniv-	1413	
musling, flodperle-	1383	
musling, glat trug-	1390	
musling, grøn blå-	1380	
musling, gylden tæppe-	1404	
musling, hjerte-	1389	
musling, hunde-	1384	
musling, islandsk kam-	1362	
musling, Jacobs-	1364	
musling, japansk	1399	
musling, japansk tæppe-	1402	
musling, kæmpe-kam-	1365	
musling, Kap-kniv-	1412	
musling, koreansk blå-	1375	
musling, La Plata-blå-	1379	
musling, lille kniv-	1409	
musling, Magellan-	1371	
musling, maler-	1345	
musling, Middelhavs-blå-	1377	
musling, newzealandsk blå-	1381	
musling, newzealandsk kam-	1367	
musling, pacifisk kniv-	1411	
musling, perle-	1346	
musling, quahog-venus-	1398	
musling, ribbet kam-	1368	
musling, Sakhalin-trug-	1391	
musling, sand-	1415	
musling, skinke-	1385	
musling, stor tæppe-	1401	
musling, stribet venus-	1396	
musling, sydamerikansk blå-	1382	
musling, tornet hjerte-	1388	
musling, tykskallet trug-	1392	
musling-familien, blå-	1370	
musling-familien, hjerte-	1387	
musling-familien, kam-	1358	
musling-familien, kniv-	1407	
musling-familien, maler-	1344	
musling-familien, venus-	1393	
musling-slægt	1341, 1342, 1400	
musling-slægt, bore-	1417	
musling-slægt, heste-	1373	

musling-slægt, kile-	1406	
musling-slægt, tæppe-	1394	
muslinger	1324	
mælkefisk	228	
mælkefisk-familien	227	
mørk pighaj	59	
mørkhaj	39	
månefisk, plettet	678	

N

narhval	1479	
Nassau-koralbars	581	
Natal-blankesten	792	
Natal-havrude	797	
Natal-languster	1265	
navaga, atlantisk	445	
navaga, østlig	444	
negenøje, bæk-	6	
negenøje, flod-	5	
negenøje, hav-	7	
negenøjer	1	
nemipterid, gylden	696	
nemipterid-familien	695	
nemipterid-slægt	697, 698	
newzealandsk blæksprutte	1431	
newzealandsk blåmusling	1381	
newzealandsk flodbars	864	
newzealandsk hestemakrel	652	
newzealandsk kammusling	1367	
newzealandsk langhale	480	
newzealandsk languster	1260	
newzealandsk rødtorsk	438	
newzealandsk vragfisk	586	
newzealandsk østers	1355	
newzealandsk ål	398	
nilaborre	565	
Nil-cichlide	833	
Nil-stavsild	158	
nipigget hundestejle	501	
niøje, flod-	5	
niøjer	1	
nordamerikansk alligator	1502	
nordamerikansk malle-slægt	387	
nordhval	1487	
nordisk beryx	518	
nordkaper	1490	
nordlig blæksprutte	1430	
nordlig døgling	1460	
nordlig hvid-reje	1227	
nordlig jomfruøsters	1363	
nordlig kuglefisk	1185	
nordlig rosenreje	1217	
nordlig skælfisk	809	
nordlig skolæst	497	
nordlig snapper	690	
nordlig søløve	1443	
nordlig tobis	933	
nordlig trommefisk	735	
nordpacifisk ansjos	219	
nordpacifisk håising	1131	
nordvestatlantisk ansjos	210	
nordvestatlantisk kulmule	483	
norsk krill	1205	
norsk lyskrebs	1205	
nufar-havrude	768	
nursehaj	26	
Ny Guinea-krokodille	1504	
nylonreje, chilensk	1235	
næbhval, Bairds	1459	
næbhvaler	1458	
næbhval-familien	1458	
næse	325	

næse, hvid-	1469
næsehornstorsk	439
næsling, dalmatisk	324
næsling, elritse-	326
næsling, iberisk	327
næsling, italiensk	328
næsling, sydeuropæisk	323
næsling, sydvesteuropæisk	329
nøgentobis, Middelhavs-	938

O

ocean-blæksprutte	1427
oceanisk languster	1264
oksefrø, amerikansk	1492
okseøjefisk	766
oliefisk	950
olivengrøn hvarre	1120
olivengrøn ridley	1498
orange dragehovedfisk	1068
orange filfisk	1181
orange savbug	520
ormeål	413
oreo-familien	528
ottearmet blæksprutte, almindelig	1438
ottearmet blæksprutte-familie	1434
ottearmet blæksprutte-slægt	1437, 1439

P

pacifisk flynder	1135
pacifisk knivmusling	1411
pacifisk mangrove-østers	1348
pacifisk reje-slægt	1230
pacifisk taskekrabbe	1292
paddefisk-familie	1191
pampano, atlantisk	662
panga-havrude	799
Panizzas kutling	1056
pansergedde, kortnæset	144
pansergedde, langnæset	143
panserknurhane	1106
panserulk	1103
papegøjefisk	856
Papua-kongemakrel	995
parona	651
patagonisk glathaj	54
patagonisk isfisk	880
patagonisk isfisk, sort	877
patagonisk langhale	479
patagonisk stentorsk	890
Peking-brasen	349
pelamide, australsk	979
pelamide, rygstribet	982
pelamide, ustribet	974
pellona-sild	188
perlemusling	1346
perlemusling, flod-	1383
peruansk ansjos	220
peruansk guitarfisk	75
peruansk havaborre	584
peruansk trommefisk	723, 743
picarel, haleplettet	811
picarel, slank	813
picarel, sortplettet	812
picarel-familien	810
picarel-slægt	814
pigflynder, indo-pacifisk	1113
pigflynder, plettet	1111
pigflynder, vestafrikansk	1112
pigget knurhane	1086
pighaj, almindelig	64
pighaj, Blainvilles	65

pighaj, mørk	59
pighaj, ru	58
pighaj, storøjet	66
pighaj-familien	57
pighaj-slægt	62
pighvar	1172
pighvar, Sortehavs-	1173
pighvar-familien	1167
pigrokke	106
pigrokke-familien	103
pigsmerling	370
pigsmerling, almindelig	370
pigsmerling, balkansk	367
pigsmerling, Bergatino	368
pigsmerling, gylden	366
pigsmerling, italiensk	373
pigsmerling, rumænsk	369
pigsnegle-slægt	1334
pilrokke	106
pilrokke, europæisk	106
pilrokke, ruhalet	105
pilrokke, Stillehavs-	104
pilrokke, violet	107
pilrokke-familien	103
pindsvinerokke	88
pletrokke	94
pletsnapper	692
plettet bars	594
plettet bonit	966
plettet dværgmalle	386
plettet fjæsing	868
plettet fløjfisk	942
plettet glathaj	52
plettet havaborre	575
plettet havkat	916
plettet havmalle	376
plettet havrude	798
plettet hestemakrel	643
plettet kongemakrel	994
plettet månefisk	678
plettet pigflynder	1111
plettet sild	181
plettet spadefisk	828
plettet trommefisk	725
plovjernsrokke	96
polartorsk	441
pollan-helt	230
pomatomider	629
pomfret, sort	677
ponyfisk, Klunzingers	700
ponyfisk-familien	699
ponyfisk-slægt	701
portugisisk østers	1347
prikfisk, Hektors	295
prikfisk-familien	294
prægtig kongemakrel	1000
pukkel-isfisk	883
pukkelhval	1485
pukkellaks	247
punkttrommefisk	731
påfuglesavgylte	851
påfugleslimfisk	907

Q

quahog-venusmusling	1398
Queensland-kongemakrel	999

R

Ramsays isfisk	892
ravfisk, australsk	659
ravfisk, japansk	660

ravfisk, stor	658
ravfisk-slægten	661
Reeves' stamsild	206
regnbuegylte	854
regnbue-hestemakrel	646
regnbueørred	251
regnbue-sardin	173
regnløje	339
reje, krystal-	1214
reje, mudder-	1258
reje, akiami-	1242
reje, argentinsk rød-	1256
reje, argentinsk stilet-	1206
reje, atlantisk	1231
reje, azteker-	1213
reje, banan-	1222
reje, blårød	1233
reje, brun	1207
reje, brun tiger-	1218
reje, californisk	1215
reje, chilensk nylon-	1235
reje, dybvands-	1237
reje, dybvands-rosen-	1212
reje, grøn tiger-	1226
reje, Guinea-	1210
reje, heste-	1251
reje, klippe-	1252
reje, kniv-	1254
reje, konge-	1221
reje, konge-rød-	1257
reje, kuruma-	1219
reje, kuruma-	1219
reje, kæmpe-ferskvands-	1247
reje, kæmpe-tiger-	1223
reje, kødet	1216
reje, mellemamerikansk	1228
reje, nordlig hvid-	1227
reje, nordlig rosen-	1217
reje, Roskilde-	1249
reje, rynket	1220
reje, rødhalet	1225
reje, sand-	1251
reje, shiba-	1208
reje, skarlagen	1234
reje, sydamerikansk ferskvands-	1246
reje, sydlig rosen-	1224
reje-familie	1244, 1245, 1250
reje-familie, dybhavs-	1232
reje-familie, ferskvands-	1243
reje-familie, højsø-	1241
rejekonge	1238
reje-slægt	1209, 1211, 1229, 1236, 1239, 1240, 1253
reje-slægt, ferskvands-	1248
reje-slægt, pacifisk	1230
reje-slægter	1255
remmesæl	1448
rethval, sydlig	1489
rethval-familien	1486
Rhône-streber	622
ribbet kammusling	1368
rimte	343
ringbug, atlantisk	1107
ringbug, finnebræmmet	1108
ringhaj	30
ringhavrude	778
ringsæl	1452
robalo, almindelig	561
robalo-slægten	562
rokke, afrikansk ørne-	110
rokke, almindelig elektrisk	120
rokke, almindelig ørne-	109
rokke, atlantisk djævle-	112
rokke, blond	85

rokke, broget	101	rynket reje	1220	sanktpetersfisk-familien	523
rokke, europæisk pil-	106	rævehaj, almindelig	17	sardin	189
rokke, gøge-	89	rævehaj, Stillehavs-	15	sardin, australsk	198
rokke, hvid-	90	rævehaj, storøjet	16	sardin, californisk	196
rokke, lille djævle-	114	rød båndfisk	834	sardin, chilensk	200
rokke, lysspættet	91	rød barberfisk	570	sardin, europæisk	189
rokke, marmoreret elektrisk	118	rød blæksprutte	1429	sardin, japansk	197
rokke, pig-	106	rød blankesten	791, 794	sardin, regnbue-	173
rokke, pil-	106	rød egernfisk	522	sardin, sydafrikansk	199
rokke, pindsvine-	88	rød havaborre	580	sardinel, asiatisk	191
rokke, plet-	94	rød havrude	775	sardinel, japansk	194
rokke, plovjerns-	96	rød knurhane	1089	sardinel, Madeira-	193
rokke, ru	98	rød mulle	817	sardinel, rund	190
rokke, ruhalet pil-	105	rød skægbrosme	475	sardinel, storhovedet	192
rokke, sand-	86	rød stenbrasen	796	sardineller	195
rokke, småtandet sav-	79	rød søøre	1327	savbug, Middelhavs-	521
rokke, småøjet	91	rød trommefisk	757	savbug, orange	520
rokke, sommerfugle-	115	rødalger	1528	savfisk-familien	78
rokke, sort elektrisk	119	rødalge-slægt	1529	savgylte	847
rokke, spejl-	92	rødbåndet havrude	803	savgylte, almindelig	847
rokke, spids-	82	rødding, Canada-	266	savgylte, femplettet	849
rokke, Stillehavs-	100	røddinger	267	savgylte, grå	845
rokke, Stillehavs-djævle-	113	rødfisk, amerikansk	1073	savgylte, langsnudet	850
rokke, Stillehavs-pil-	104	rødfisk, dybhavs-	1076	savgylte, øjeplettet	848
rokke, stjerne-	83	rødfisk, enke-	1072	savrokke, småtandet	79
rokke, storplettet	93	rødfisk, gulhalet	1074	savrokke-familien	78
rokke, søm-	87	rødfisk, Stillehavs-	1070	Scotia-isfisk	896
rokke, tornhalet	99	rødfisk, stor	1075	sej	467
rokke, vinter-	95	rødfisk, sydafrikansk	1071	sej, Alaska-	470
rokke, violet pil-	107	rødfisk-slægten	1077	sej, blå-	466
rokke, ørne-	109	rødhaj, småplettet	31	sej, grå-	467
rokke-familien	81	rødhaj, storplettet	32	sejhval	1481
rokke-familien, djævle-	111	rødhaj-familien	28	sejlfilfisk	1183
rokke-familien, el-	117	rødhaj-slægt	33	sejlfisk, atlantisk	1016
rokke-familien, pig-	103	rødhalet reje	1225	sejlfisk, Stillehavs-	1017
rokke-familien, pil-	103	rødkonk	1335	sejlfisk-familien	1015
rokke-familien, sav-	78	rødnæb (hun)	841	seksgællede hajer	9
rokke-familien, ørne-	108	rødreje, argentinsk	1256	seksgællet haj	11
rokker	80	rødreje, konge-	1257	selar, guldstribet	655
rokker, el-	117	rødskæg	817	selar, storøjet	654
rokker, elektriske	117	rødspætte	1145	semling	317
rokker, haj-	72	rødspætte, Alaska-	1146	senegalsk kulmule	491
rokke-slægt	102	rødspætte-familien	1123	Senegal-trommefisk	750
rokke-slægt, elektrisk	121	rødtorsk, newzealandsk	438	Senegal-tunge	1157
rokke-slægt, sommerfugle-	116	rødtunge	1142	sepiablæksprutte	1418
rosa kingklip	925	rødøjet rundsild	178	sepiola-blæksprutte	1420
rosenreje, dybvands-	1212			sergentfisk	631
rosenreje, nordlig	1217	**S**		shiba-reje	1208
rosenreje, sydlig	1224			siamesisk krokodille	1506
Roskildereje	1249	sabelfisk, sort	956	sild, kinesisk kråse-	168
Ross' blæksprutte	1419	sabelkarpe	350	sild	169
rovsild, indisk	225	saddel-havrude	788	sild, amerikansk kråse-	172
rovsild-familien	224	Sakhalin-trugmusling	1391	sild, amerikansk stam-	162
rovsild-slægten	226	sandaborre, brasiliansk	865	sild, atlantisk	169
ru fjæsing	870	sandart	620	sild, atlantisk rund-	177
ru hestemakrel	667	sandart, almindelig	620	sild, atlantisk tråd-	186
ru knurhane	1085	sandart, amerikansk	619	sild, bonga-stam-	174
ru pighaj	58	sandart, hvidøjet	621	sild, californisk rund-	176
ru rokke	98	sandfisk, japansk	863	sild, canadisk stam-	155
rudskalle	362	sandflynder	1149	sild, chacunda-kråse-	164
rudskalle, almindelig	362	sandhavaborre	572	sild, chilensk	204
rudskalle, græsk	363	sandhest	1251, 333	sild, flod-	161
ruhalet pilrokke	105	sandkutling	1059	sild, indisk rov-	225
rumænsk pigsmerling	369	sandmusling	1415	sild, kaspisk	171
rumænsk ulkeaborre	618	sandreje	1251	sild, kaspisk stam-	160
rund sardinel	190	sandrokke	86	sild, kelee-stam-	182
rundsild, atlantisk	177	sandtigerhaj	13	sild, maj-	156
rundsild, californisk	176	sandtunge	1164	sild, Nil-stav-	158
rundsild, rødøjet	178	sandål	936	sild, pellona-	188
rundsild, sølvstribet	201	sandål, koreansk	940	sild, plettet	181
rundsild-slægten	179	sanktpetersfisk	525	sild, Reeves' stam-	206
rundskællet spydfisk	1025	sanktpetersfisk, amerikansk	524	sild, rødøjet rund-	178
rur-slægt	1199	sanktpetersfisk, atlantisk	525	sild, slank stam-	184
rygstribet pelamide	982			sild, stav-	157

sild, Stillehavs-	170	slethvar	1175	sort sabelfisk	956
sild, strøm-	282	slethvar, amerikansk	1174	sort solaborre	605
sild, sølvstribet rund-	201	slette	1139	sort stenkrabbe	1293
sild, toli-stam-	207	slimfisk, blodstribet	910	sort thunnin	969
sild, vestafrikansk stam-	183	slimfisk, påfugle-	907	sort trommefisk	745
sild, vestatlantisk stam-	159	slimfisk, sølv-	918	Sortehavs-hvilling	455
silde-familien	154	slimfisk, stribet	909	Sortehavs-kutling, broget	1060
sildefisk-slægter	203	slimfisk, trehale-	912	Sortehavs-pighvarre	1173
sildehaj, almindelig	23	slimfisk-familie	904	Sortehavs-skalle	357
sildehaj, Stillehavs-	22	slimfisk-slægt	906	sortfinnet gerrid	902
sildehaj-familien	18	smelt	274	sortfinnet tun	1009
sildekonge	510	smelt, amerikansk	275	sortfisk	1041
silde-slægt	180	smelt, arktisk	276	sortfisk, almindelig	1041
sild-familien, rov-	224	smelt, europæisk	274	sortfisk, antarktisk	1043
sild-slægt, tråd-	187	smelt, ferskvands-	271	sortfisk, japansk	1044
sild-slægten, rov-	226	smelt, Stillehavs-	272	sortfisk-familien	1040
sild-slægten, rund-	179	smelt-familien	270	sorthaj	61
sild-slægten, stam-	163	smelt-slægter	277	sorthale	782
sillago-familien	625	smerling	372	sorthalet havrude	778
skade	84	smerling, almindelig	372	sortkarpe	347
skalle	360	smerling, almindelig pig-	370	sortmund	459
skalle, albansk	348	smerling, balkansk pig-	367	sortmund, sydlig	458
skalle, almindelig	360	smerling, Bergatino pig-	368	sortmundet trommefisk	721
skalle, almindelig rud-	362	smerling, dynd-	371	sortplettet langhale	493
skalle, almindelig strøm-	344	smerling, gylden pig-	366	sortplettet picarel	812
skalle, Donau-	358	smerling, italiensk pig-	373	sortrygget hestemakrel	638
skalle, græsk rud-	363	smerling, pig-	370	sortstribet mulle	821
skalle, rud-	362	smerling, rumænsk pig-	369	sorttippet haj	38
skalle, Sortehavs-	357	smørfisk, amerikansk	1033	sortvels	468
skalle, strøm-	344	smørfisk, blå	1037	spadefisk, afrikansk	827
skalle, sydeuropæisk	359	smørfisk, indisk	1038	spadefisk, plettet	828
skarlagen reje	1234	smørfisk, kinesisk	1032	spadefisk-familien	825
skildpadde, læder-	1495	smørfisk, Stillehavs-	1034	spadestør, kinesisk	141
skildpadde, suppe-	1497	smørfisk, sølv-	1031	spadestør-familien	140
skildpadde, uægte karette-	1496	smørfisk-familien	1030	spansk makrel	984
skildpadder	1494	smørfisk-slægt	1036, 1042	spansk tandkarpe	514
skildpadder, spiselige sump-	1499	småhovedet hval	1461	spejlrokke	92
skinkemusling	1385	småhvarre	1171	sperling	471
skjoldtandhaj	14	småmundet ørredaborre	602	spids hornfisk	418
skolæst	494	småøjet rokke	91	spidshalet klumpfisk	1187
skolæst, nordlig	497	småplettet rødhaj	31	spidsrokke	82
skorpionfisk	1062	småtandet savrokke	79	spidssnudet knurhane	1088
skorpionsnegle-slægt	1338	småtandet trommefisk	727	spidssnudet haj, atlantisk	44
skriftbars	589	snæbel	237	spidssnudet haj, Stillehavs-	43
skrubbe	1143	snapper, gul	689	spidssnudet havrude	781
skægbrosme, brasiliansk	474	snapper, gulhalet	694	spidstandet blankesten	793
skægbrosme, hvid	476	snapper, mangrove-	688	spiselige sumpskildpadder	1499
skægbrosme, rød	475	snapper, nordlig	690	spiseligt søpindsvin	1521
skægget brotula	924	snapper, plet-	692	springfrø-slægt	1493
skæghaj, australsk	25	snapper, sydlig	691	springmulte	543
skægtorsk	472	snapper-familien	687	spydfisk	1015
skælbrosme	463	snapper-slægt	693	spydfisk, kortnæbbet	1022
skælbrosme, almindelig	463	snarre, tang-	502	spydfisk, langnæbbet	1026
skælbrosme, Middelhavs-	464	snegle-slægt	1336	spydfisk, Middelhavs-	1024
skælbrosme-slægten	465	sneppefisk	504	spydfisk, rundskællet	1025
skælfinnefisk-familien	830	sneppefisk-familien	503	spydfisk-slægten	1027
skælfisk, nordlig	809	solaborre	601	spækhugger	1470
skærising	1128	solaborre, hvid	604	spættet sæl	1453
skærising, amerikansk	1129	solaborre, sort	605	stævhelt, stor	238
slangehovedfisk, indonesisk	557	sommerfuglerokke	115	stalling	268
slangehovedfisk, stribet	558	sommerfuglerokke-slægt	116	stamsild, amerikansk	162
slangehovedfisk-slægt	559	sommerfugletun	970	stamsild, bonga-	174
slangehoved-kutling	1061	sommerhvarre	1118	stamsild, canadisk	155
slangekvabbe	929	sompatgryntefisk	713	stamsild, kaspisk	160
slangemakrel, hvid	954	sort dragehovedfisk	1067	stamsild, kelee-	182
slangemakrel, konge-	949	sort dværgmalle	384	stamsild, Reeves'	206
slangemakrel, Marleys	953	sort elektrisk rokke	119	stamsild, slank	184
slangemakrel-familien	947	sort havaborre	571	stamsild, toli-	207
slangemakrel-slægt	948, 952	sort havtaske	1194	stamsild, vestafrikansk	183
slangeskinds-gurami	1046	sort isfisk	881	stamsild, vestatlantisk	159
slangeål	414	sort kutling	1053	stamsild-slægten	163
slank picarel	813	sort marlin	1018	stavsild	157
slank stamsild	184	sort patagonisk isfisk	877	stavsild, Nil-	158
slank tun	963	sort pomfret	677	Stellers søløve	1443

stenbars	600	storøjet rævehaj	16	sydamerikansk trommefisk	732
stenbider	1105	storøjet selar	654	sydeuropæisk næsling	323
stenbider-familien	1104	storøjet tun	1011	sydeuropæisk skalle	359
stenbrasen, hvid	785	strandkrabbe, almindelig	1298	sydeuropæisk tandkarpe	513
stenbrasen, rød	796	strandkrabbe, Middelhavs-	1297	sydlanguster	1267
stenkrabbe, sort	1293	strandsnegle-slægt	1337	sydlig guitarfisk	74
stenkrebs	1280	streber	623	sydlig havrude	800
sten-søpindsvin	1518	streber, almindelig	623	sydlig kulmule	482
stentorsk, patagonisk	890	streber, Rhône-	622	sydlig rethval	1489
sterlet	135	stribefisk	551	sydlig rosenreje	1224
stiletreje, argentinsk	1206	stribefisk, almindelig	551	sydlig snapper	691
Stillehavs-ansjos	212	stribefisk, atlantisk	552	sydlig sortmund	458
Stillehavs-barracuda	531	stribefisk, Boyers	549	sydlig tun	1010
Stillehavs-bonit	981	stribefisk, lille	549	sydøstafrikansk tunge	1154
Stillehavs-catalufa	607	stribefisk, stor	550	sydøstpacifisk gryntefisk	706
Stillehavs-djævlerokke	113	stribefisk-familien	548	syd-stankelbenskrabbe	1314
Stillehavs-guldlaks	280	stribet ansjos	209	sydvestafrikansk tunge	1153
Stillehavs-havål	409	stribet bars	598	sydvestatlantisk kulmule	486
Stillehavs-hellefisk	1125	stribet blankesten	786	sydvesteuropæisk næsling	329
Stillehavs-helleflynder	1134	stribet delfin	1474	syvgællet haj	10
Stillehavs-hestemakrel	671	stribet fløjfisk	941	sæl, grå-	1449
Stillehavs-kongemakrel	1002	stribet gryntefisk	704, 708	sæl, Grønlands-	1451
Stillehavs-kulmule	490	stribet havkat	915	sæl, munke-	1450
Stillehavs-makrelgedde	421	stribet isfisk	888	sæl, remme-	1448
Stillehavs-menhaden	175	stribet kongemakrel	993	sæl, ring-	1452
Stillehavs-pilrokke	104	stribet marlin	1023	sæl, spættet	1453
Stillehavs-rokke	100	stribet mulle	818	sæler, ægte	1446
Stillehavs-rævehaj	15	stribet multe	545	søbjørn, sydafrikansk	1442
Stillehavs-rødfisk	1070	stribet slangehovedfisk	558	søbjørn, sydamerikansk	1441
Stillehavs-sejlfisk	1017	stribet slimfisk	909	sø-helt	232
Stillehavs-sild	170	stribet trommefisk	728	søheste-slægt	505
Stillehavs-sildehaj	22	stribet tunge	1160	søkarusse	321
Stillehavs-smelt	272	stribet venusmusling	1396	søknæler	1202
Stillehavs-smørfisk	1034	strømling	345	søknæler-familien	1201
Stillehavs-spidssnudet haj	43	strømløje	308	søko	1455
Stillehavs-tarpon	149	strømpebåndsfisk	957, 958	søløve, nordlig	1443
Stillehavs-tobis	934	strømsild	282	søløve, Stellers	1443
Stillehavs-torsk	446	strømskalle	344	søløve, sydamerikansk	1444
Stillehavs-trommefisk	730	strømskalle, almindelig	344	sølvfisk, antarktisk	893
Stillehavs-zebrahaj	24	stumpsnudet havrude	769	sølv-gryntefisk	711
Stillehavs-østers	1347, 1354	stør, adriatisk	133	sølvkarpe	335
stjerneflynder	1144	stør, almindelig	137	sølvkarusse	320
stjernehus	136	stør, grøn	132	sølvkulmule	481
stjernekigger, europæisk	874	stør, hvid	138	sølvlaks	249
stjernekigger-slægt	875	stør, kinesisk spade-	141	sølv-slimfisk	918
stjernerokke	83	stør-familien	130	sølv-smørfisk	1031
stor barracuda	532	stør-familien, spade-	140	sølvstribet rundsild	201
stor bundhelt	236	suder	364	sølvtippet haj	35
stor fjæsing	869	sugefisk	1178	sølvtorsk	449
stor gaffelmakrel	647	sugekarpe-familien	301	sølvtrommefisk	744
stor hundestejle	500	sugekarpe-slægt	302	sømhaj	71
stor ravfisk	658	sump-elritse	351	sømrokke	87
stor rødfisk	1075	sumpskildpadder, spiselige	1499	søpindsvin, chilensk	1522
stor stævhelt	238	suppeskildpadde	1497	søpindsvin, spiseligt	1521
stor stribefisk	550	svalehale-bars	569	søpindsvin, sten-	1518
stor søøre	1325	svinefisk	702	søpindsvine-familie	1520
stor tæppemusling	1401	sværdfisk	1029	søpindsvine-slægt	1519
stor tangnål	506	sværdfisk-familien	1028	søpølse, japansk	1523
stor trådfinnefisk	556	svømmekrabbe, blå	1295	søpung, chilensk	1509
stor øglefisk	391	svømmekrabbe, Danas	1294	søpung, en art	1510
store hvide haj	19	svømmekrabbe, fløjls-	1299	søpung, violet	1508
storfinnet falkefisk	859	svømmekrabbe, gazami-	1300	søpunge	1507
storhovedet sardinel	192	svømmekrabbe-slægt	1296, 1301	søstjerne, almindelig	1517
stormundet ørredaborre	603	sydafrikansk hestemakrel	673	søstjerne-familie	1516
storplettet rokke	93	sydafrikansk kingklip	926	søtunge	1165
storplettet rødhaj	32	sydafrikansk kulmule	484	søvnkutling-familien	1048
storskællet hvarre	1177	sydafrikansk rødfisk	1071	søøre, europæisk	1328
storskællet øglefisk	392	sydafrikansk sardin	199	søøre, Midas	1326
stortandet trommefisk	742	sydafrikansk søbjørn	1442	søøre, rød	1327
storøjefisk	606	sydafrikansk turbansnegl	1332	søøre, stor	1325
storøjet gryntefisk	703	sydamerikansk blåmusling	1382	søøre-slægten	1329
storøjet havrude	776	sydamerikansk ferskvandsreje	1246	søørred	259
storøjet helt	235	sydamerikansk søbjørn	1441	søørred, amerikansk	266
storøjet pighaj	66	sydamerikansk søløve	1444		

DA

439

T

tandbrasen	774
tandet østers	1352
tandkarpe, spansk	514
tandkarpe, sydeuropæisk	513
tandkarpe, Valencia-	515
tandkarper, æglæggende	512
tang, blad-	1526
tang, carrigeen	1532
tang, klør-	1527
tangkvabbe, almindelig	908
tangkvabbe, øjeplettet	905
tangkvabbe-slægt	906
tangnål, almindelig	508
tangnål, lille	507
tangnål, stor	506
tangsnarre	502
tangspræl	919
tarpon, atlantisk	150
tarpon, Stillehavs-	149
tarpon-familien	148
taskekrabbe	1291
taskekrabbe, almindelig	1291
taskekrabbe, Dungeness-	1290
taskekrabbe, pacifisk	1292
taskekrabbe, østamerikansk	1289
taskekrabbe-familien	1287
tautog læbefisk	852
teglfisk	627
teglfisk-familien	626
teleskop-kardinalfisk	611
teraponid-familien	592
Thors tungehvarre	1116
thunnin	968
thunnin, lille	967
thunnin, sort	969
tiarmet blæksprutte-familie	1422
tigerhaj, sand-	13
tigerreje, brun	1218
tigerreje, grøn	1226
tigerreje, kæmpe-	1223
tipunder	146
tipunder-familien	145
tipunder-slægten	147
tobis, amerikansk	932
tobis, kyst-	936
tobis, Middelhavs-nøgen-	938
tobis, nordlig	933
tobis, Stillehavs-	934
tobis-familien	931
tobiskonge	939
tobis-slægt	937
toli-stamsild	207
tomcod	457
tomcod, atlantisk	456
tongol tun	1013
tornet hjertemusling	1388
tornhalet rokke	99
torpedo-hestemakrel	649
torsk	447
torsk, antarktiske	876
torsk, atlantisk	447
torsk, fjord-	448
torsk, næsehorns-	439
torsk, newzealandsk rød-	438
torsk, patagonisk sten-	890
torsk, polar-	441
torsk, skæg-	472
torsk, Stillehavs-	446
torsk, sølv-	449
torske-familien, dybhavs-	436
torskefisk	435
torskefisk-familien	440
totuava	724
Transkei-languster	1271
trehale-slimfisk	912
trekanthaj	70
trekanthaj-familien	69
trepigget hundestejle	500
tretrådet havkvabbe	450
Tristan-languster	1263
troldhummer-familien	1306
troldkrabbe, antarktisk	1309
troldkrabbe, konge-	1307
troldkrabbe, Murrays	1308
troldkrabbe-slægt	1311
trommefisk, afrikansk	719
trommefisk, almindelig	759
trommefisk, amerikansk	733
trommefisk, argentinsk	758
trommefisk, atlantisk	738
trommefisk, australsk	754
trommefisk, bobo-	749
trommefisk, boe-	753
trommefisk, dronninge-	748
trommefisk, flod-	716
trommefisk, Golf-	734
trommefisk, gul	722
trommefisk, hvid	720
trommefisk, hvidblomstret	740
trommefisk, hvidmundet	737
trommefisk, japansk	741
trommefisk, konge-	726
trommefisk, langhovedet	751
trommefisk, lille japansk	747
trommefisk, manchurisk	746
trommefisk, nordlig	735
trommefisk, peruansk	723, 743
trommefisk, plettet	725
trommefisk, punkt-	731
trommefisk, rød	757
trommefisk, Senegal-	750
trommefisk, småtandet	727
trommefisk, sort	745
trommefisk, sortmundet	721
trommefisk, Stillehavs-	730
trommefisk, stortandet	742
trommefisk, stribet	728
trommefisk, sydamerikansk	732
trommefisk, sølv-	744
trommefisk-familien	715
trommefisk-slægt	729, 736, 739, 752, 756, 760
trompeterfisk, blå	862
trompeterfisk-familien	861
trugmusling, glat	1390
trugmusling, Sakhalin-	1391
trugmusling, tykskallet	1392
trådfinnefisk, firstrålet	556
trådfinnefisk, indo-pacifisk	554
trådfinnefisk, konge-	555
trådfinnefisk, stor	556
trådfinnefisk-familien	553
trådfisk, Alexandria-	633
trådsild, atlantisk	186
trådsild, Stillehavs-	185
trådsild-slægt	187
tun, almindelig	1012
tun, atlantisk	1012
tun, gulfinnet	1008
tun, hvid	1007
tun, langfinnet	1007
tun, slank	963
tun, sommerfugle-	970
tun, sortfinnet	1009
tun, storøjet	1011
tun, sydlig	1010
tun, tongol	1013
tunfisk-familien	1006
tunge	1165
tunge, almindelig	1165
tunge, amerikansk	1152
tunge, azevia-	1159
tunge, glas-	1156
tunge, håret	1162
tunge, Kleins	1163
tunge, rød	1142
tunge, sand-	1164
tunge, Senegal-	1157
tunge, stribet	1160
tunge, sydøstafrikansk	1154
tunge, sydvestafrikansk	1153
tunge, sø-	1165
tunge, øjeplettet	1158
tunge-familie	1151
tunge-familien, hunde-	1166
tungehvarre, almindelig	1115
tungehvarre, Thors	1116
tunge-slægt	1155, 1161
tun-slægt	1014
turbansnegl, hornet	1331
turbansnegl, sydafrikansk	1332
tvebåndet havrude	783
tværstribet knurhane	1079
tyklæbet gryntefisk	709
tyklæbet multe	538
tykskallet trugmusling	1392
tyndlæbet multe	542
tæppemusling, almindelig	1405
tæppemusling, gylden	1404
tæppemusling, japansk	1402
tæppemusling, stor	1401
tæppemusling-slægt	1394
tærbe	97
tømrerhavrude	765

U

ukrainsk lampret	3
uldhåndskrabbe	1303
ulk, almindelig	1100
ulk, amerikansk	1098
ulk, flod-	1097
ulk, hvidfinnet ferskvands-	1097
ulk, indisk fladhoved-	1096
ulk, langtornet	1102
ulk, panser-	1103
ulkeaborre, rumænsk	618
ulke-familie, fladhovedet	1095
ulke-slægt	1101
ustribet pelamide	974
uvak	448
uægte karetteskildpadde	1496

V

Valencia-tandkarpe	515
varieret falkefisk	858
venusmusling, afrikansk	1397
venusmusling, quahog-	1398
venusmusling, stribet	1396
venusmusling-familien	1393
vestafrikansk havtaske	1196
vestafrikansk kongemakrel	1004
vestafrikansk mulle	820
vestafrikansk pigflynder	1112
vestafrikansk stamsild	183
vestatlantisk læbefisk	844
vestatlantisk stamsild	159
vestatlantisk ålekvabbe	922

Victoria-karpe	355
vimme	365
vinter-flynder	1147
vinterrokke	95
violet pilrokke	107
violet søpung	1508
Vladykovs lampret	4
vortetang	1531
vragfisk	585
vragfisk, atlantisk	585
vragfisk, newzealandsk	586
vågehval	1480
vågmær, Middelhavs-	511

W

wahoo	962
waxdick	131
Wilsons isfisk	897
wobbegong	25
Wuchang-brasen	346

Z

zebrahaj, Stillehavs-	24
zebrahavrude	779
zingel	624

Æ

ædelkoral	1514
ædelkoral-familien	1513
æglæggende tandkarper	512
ægte sæler	1446

Ø

ø-dværgmakrel	976
øglefisk, atlantisk	393
øglefisk, stor	391
øglefisk, storskællet	392
øglefisk-familien	390
øjeplettet isfisk	900
øjeplettet savgylte	848
øjeplettet tangkvabbe	905
øjeplettet tunge	1158
øjestribet isfisk	884
øresæler	1440
øresvin	1477
ørnefisk, afrikansk	717
ørnefisk, almindelig	718
ørnefisk, brun	755
ørnefisk-familien	715
ørnerokke	109
ørnerokke, afrikansk	110
ørnerokke, almindelig	109
ørnerokke-familien	108
ørred, amerikansk sø-	266
ørred, bæk-	258
ørred, cutthroat-	255
ørred, fjeld-	263
ørred, gila-	246
ørred, guld-	243
ørred, hav-	257
ørred, kilde-	264
ørred, Malma-	265
ørred, regnbue-	251
ørred, sø-	259
ørredaborre, småmundet	602
ørredaborre, stormundet	603
østamerikansk gedde	290
østamerikansk taskekrabbe	1289
østers, almindelig jomfru-	1359
østers, amerikansk	1349

østers, australsk	1357
østers, chilensk	1351
østers, europæisk	1353
østers, molbo-	1386
østers, newzealandsk	1355
østers, nordlig jomfru-	1363
østers, pacifisk mangrove-	1348
østers, portugisisk	1347
østers, Stillehavs-	1347, 1354
østers, tandet	1352
østers-slægt	1350, 1356
østlig navaga	444

Å

ål	396
ål, almindelig	396
ål, almindelig hav-	410
ål, amerikansk	400
ål, amerikansk hav-	411
ål, argentinsk hav-	412
ål, australsk	397
ål, europæisk ferskvands-	396
ål, hav-	410
ål, japansk	399
ål, newzealandsk	398
ål, slange-	414
ål, Stillehavs-hav-	409
åle-familien, ferskvands-	395
ålekvabbe	923
ålekvabbe, vestatlantisk	922
ålekvabbe-familien	920
ålekvabbe-slægt	921
ålemalle-slægt	388
åle-slægt, ferskvands-	401
ål-familien, hav-	408

DA

Deutscher Index

A

Aal 396
Aal, Amerikanischer 400
aal, Argentinischer Meer- 412
Aal, Australischer 397
aal, Conger- 410
Aal, Europäischer 396
aal, Europäischer Fluß- 396
aal, Fluß- 396
aal, Gemeiner Meer- 410
Aal, Japanischer 399
aal, Kurzschnauziger Schlangen- 413
aal, Meer- 410
aal, Mittelmeer-Schlangen- 414
Aal, Neuseeland- 398
aal, See- 410
aal, Weißflecken-Conger- 409
Aale 395
Aale, Echte 395, 401
aale, Fluß- 395, 401
aale, Meer- 408
aale, Messerzahn- 407
Aalmutter 923
Aalmutter, Nordamerikanische 922
Aalmuttern 920
Aalquappe 452
Aalrauke 452
Aalraupe 452
Aalruppe 452
Aalrutte 452
Abalonen 1329
Abt 341
Achselbrasse 790
Adlerfisch 718
Adlerfisch, Afrikanischer 717
Adlerlachs 718
Adlerrochen 108, 109
Adlerrochen, Afrikanischer 110
Adlerrochen, Gewöhnlicher 109
Adriastör 133
Afrikanische Languste 1262
Afrikanische Lungenfische 129
Afrikanische Meeräsche 544
Afrikanische Streifensüßlippe 708
Afrikanische Tiefseegarnele 1233
Afrikanischer Adlerfisch 717
Afrikanischer Adlerrochen 110
Afrikanischer Manati 1457
Afrikanischer Pferdekopf 656
Afrikanischer Sichelflosser 827
Afrikanischer Umberfisch 719
Ahlbock 240
Ährenfisch 551
Ährenfisch, Boyers 549
Ährenfisch, Gezeiten- 552
Ährenfisch, Großer 550
Ährenfisch, Kleiner 549
Ährenfische 548
Aitel 341
Akadei 805
Akadischer Rotbarsch 1073
Akiami-, Garnele 1242
Akiami-Garnele 1242
Aland 343
Alandblecke 308
Alaska-Pollack 470

Alaska-Scholle 1146
Albacore 1008
Albanische Barbe 313
Albanische Plötze 348
Albe 310
Albola 310
Alexandria-Fadenmakrele 633
Algen 1524
algen, Braun- 1525
algen, Kalk- 1530
algen, Rot- 1528
alligator, Mississipi- 1502
Alligatoren und Krokodile 1500
Alligatorfisch 143
Alligatorhai 71
Alose 156
Alpenforelle 258
Alse 156
Alse, Amerikanische 162
Alse, Chinesische 168
Alse, Fadenflossige 172
Alse, Gewöhnliche 156
Alse, Ilisha- 205
Alse, Kanadische 155
Alse, Kelee- 182
Alse, Reeves- 206
Alse, Toli- 207
Alse, Westatlantische 159
Alwe 310
Amaul 620
Amerikanische Alse 162
Amerikanische Auster 1349
Amerikanische Chimäre 124
Amerikanische Flußkrebse 1283
Amerikanische Forelle 251
Amerikanische Kleine Maräne 231
Amerikanische Languste 1269
Amerikanische Sardelle 219
Amerikanische Scholle 1129
Amerikanische Schwertmuschel 1408
Amerikanische Seeforelle 266
Amerikanische Spöke 124
Amerikanische Winterflunder 1147
Amerikanischer Aal 400
Amerikanischer Butterfisch 1033, 1035
Amerikanischer Flußbarsch 616
Amerikanischer Flußkrebs 1282
Amerikanischer Hummer 1317
Amerikanischer Hundsfisch 285
Amerikanischer Lippfisch 853
Amerikanischer Maifisch 162
Amerikanischer Meeraal 411
Amerikanischer Ochsenfrosch 1492
Amerikanischer Petersfisch 524
Amerikanischer Pfeilzahnheilbutt 1125
Amerikanischer Sandaal 932
Amerikanischer Schlammfisch 142
Amerikanischer Seesaibling 266
Amerikanischer Seeskorpion 1098
Amerikanischer Seeteufel 1193
Amerikanischer Streifenbarsch 596
Amerikanischer Umberfisch 733
Amerikanischer Wels 385
Amerikanischer Zander 621
Amerikanischer Zwergwels 385
Ammenhai 26
Ammenhai, Australischer 25

Amsellippfisch 843
Amur-Hecht 291
Anchoveta 220
Anchoveta, Atlantische 211
Anchoveta, Pazifische 212
Anchovis 208, 217, 222
anchovis, Grenadier- 213
Andamanen-Schlankhummer 1319
Angler 1195
Anglerfisch, Budegassa- 1194
Anglerfisch, Westafrikanischer 1196
Anglerfische 1192
Angola-Meerbrasse, Großaugen- 703
Angola-Zahnbrasse 771
Antarktische Königskrabbe 1309
Antarktischer Krill 1204
Antarktischer Schwarzfisch 1043
Antarktischer Silberfisch 893
Antarktischer Umberfisch 754
Antarktischer Zahnfisch 878
Antarktis-Eisfische 876
Archenmuschel, Japanische 1340
Archenmuschel, Westpazifische 1339
Archenmuscheln 1342
Archenmuscheln, Rote 1341
Argentina-Schattenfisch 758
Argentinische Rotgarnele 1256
Argentinische Sardelle 214
Argentinische Stilett-Garnele 1206
Argentinischer Kurzflossenkalmar 1428
Argentinischer Meeraal 412
Argentinischer Seehecht 486
Argusfische 829
Aristoteles-Wels 381
Arktische Seespinne 1312
Arktische Seespinnen 1313
Asche 268
äsche, Bobo-Meer- 539
äsche, Brasilianische Meer- 546
äsche, Dicklippige Meer- 538
äsche, Dünnlippige Meer- 542
Äsche, Europäische 268
äsche, Gemeine Meer- 542
äsche, Gestreifte Meer- 545
äsche, Gewöhnliche Meer- 545
äsche, Gold- 540
äsche, Goldmeer- 540
äsche, Grau- 538
äsche, Graue Meer- 547
äsche, Großkopf-Meer- 545
äsche, Kleine Meer- 543
äsche, Meer- 540
äsche, Springmeer- 543
äschen, Meer- 537
Asiatische Kleinsardine 191
Asiatischer Gelber Umberfisch 722
Asiatischer Pfeilzahnheilbutt 1124
Asiatischer Schwarzfisch 1044
Asiatischer Stint 276
Aszidie, Chilenische 1509
Aszidie, Mittelmeer- 1510
Aszidien 1507
Atlantische Anchoveta 211
Atlantische Pilgermuschel 1365
Atlantische Rote Riesengarnele 1234
Atlantische Säbelzahn-Sardelle 221

DE

Atlantischer Barrakuda 532
Atlantischer Braunhai 40
Atlantischer Eidechsenfisch 393
Atlantischer Fächerfisch 1016
Atlantischer Fadenhering 186
Atlantischer Heilbutt 1133
Atlantischer Hering 169
Atlantischer Hornhecht 417
Atlantischer Lachs 256
Atlantischer Makrelenhecht 422
Atlantischer Pfeilhecht 532
Atlantischer Pferdekopf 657
Atlantischer Rundhering 177
Atlantischer Sägebauch 520
Atlantischer Schleimfisch 908
Atlantischer Schwertschwanz 1515
Atlantischer Seeteufel 1195
Atlantischer Segelfisch 1016
Atlantischer Spitzmaulhai 44
Atlantischer Sternrochen 97
Atlantischer Tarpun 150
Atlantischer Tiefseescallop 1369
Atlantischer Tomcod 456
Atlantischer Umberfisch 738
Atlantischer Wrackbarsch 585
Atun 951
Atun, Weißer 954
Atyiden-Garnelen 1243
Auch 268
Aufbläser 1184, 1186
Aufbläser, Gefleckter 1185
Augenfleck-Eisfisch 900
Augenfleck-Lippfisch 848
Augenfleck-Seezunge 1158
Augenfleck-Umberfisch 757
Augenlippfisch 848
Augen-Seezunge 1158
Augenstreifen-Notothenia 884
Augenzitterrochen 120
Auster 1353
Auster, Amerikanische 1349
auster, Chilenische Platt- 1351
Auster, Europäische 1353
auster, Felsen- 1347
Auster, Gezähnte 1352
auster, Neuseeland-Platt- 1355
auster, Pazifische Felsen- 1347, 1348
auster, Pazifische Platt- 1354
Auster, Portugiesische 1347
auster, Riesen- 1347
auster, Schlick- 1415
auster, Strand- 1415
auster, Sydney-Felsen- 1357
austern, Felsen- 1350
austern, Platt- 1356
Austernfisch 852, 915
Australische Auster 1357
Australische Fleckenmakrele 996
Australische Gelbschwanzmakrele 659
Australische Jakobsmuschel 1366
Australische Languste 1270
Australische Miesmuschel 1378
Australische Sardelle 215
Australische Sardine 198
Australischer Aal 397
Australischer Ammenhai 25
Australischer Bonito 979
Australischer Flußkrebs 1316
Australischer Glatthai 51
Australischer Großaugenbarsch 607
Australischer Halbschnäbler 428
Australischer Lachs 683
Australischer Lungenfisch 127
Ayu 269

Azevia-Seezunge 1159
Azteken-Geißelgarnelen 1213

B

Bachforelle 258
Bachneunauge 6
Bachneunauge, Gewöhnliches 6
Bachneunauge, Ukrainisches 3
Bachneunauge, Vladikov- 4
Bachsaibling 264
Bachschmerle 372
Baird-Wal 1459
Balao-Halbschnäbler 429
Balkan-Steinbeißer 367
Baltischer Stör 137
Bananen-Garnele 1222
Bändereisfisch 898
Bänderrochen 101
Bandfisch 510, 834
Bandrochen 82
Bandschild 1190
Barbe 314
Barbe, Albanische 313
barbe, Gestreifte Meer- 818
barbe, Gewöhnliche Meer- 817
barbe, Gold-Meer- 821
barbe, Hunds- 316
Barbe, Iberische 315
Barbe, Italienische 318
Barbe, Java- 353
barbe, Rote Meer- 817
barbe, Steilstirnige Meer- 817
barbe, Streifen- 818
barbe, Westafrikanische Meer- 820
Barben 314
barben, Meer- 815, 819, 822
Barbengründling 312
barbenkönig, Meer- 610
Barbine 314
Bärengarnele 1223
Bärenkrebs, Kleiner 1305
Bärenkrebse 1304
Barrakuda, Atlantischer 532
Barrakuda, Guinea- 530
Barrakuda, Indischer 534
Barrakuda, Kalifornischer 531
Barrakuda, Mittelmeer- 535
Barrakudas 529, 536
Barramundi 564
Bars 617
Barsch 617
barsch, Amerikanischer Fluß- 616
barsch, Amerikanischer Streifen- 596
barsch, Argentinischer Zacken- 568
barsch, Atlantischer Wrack- 585
barsch, Australischer Großaugen- 607
barsch, Blau- 630
barsch, Blauer Ziegel- 627
barsch, Brauner Zacken- 577
barsch, Felsen- 598
barsch, Fluß- 617
barsch, Forellen- 603
barsch, Gefleckter Streifen- 594
barsch, Gelb- 616
barsch, Glasaugen- 621
barsch, Gold- 1075
barsch, Goldener Zacken- 574
barsch, Grauer Zacken- 576
barsch, Groppen- 618
barsch, Hapuku-Wrack- 586
barsch, Hecht- 620
Barsch, Japanischer 599

barsch, Kaiser- 518, 519
barsch, Kaul- 614
barsch, Klippen- 839
barsch, Königs- 631
barsch, Malabar-Zacken- 579
barsch, Marmor- 885
barsch, Meer- 593
barsch, Mittelmeer-Kaiser- 521
barsch, Moçambique-Bunt- 832
Barsch, Nassau- 581
barsch, Neuseeland-Fluß- 864
barsch, Nil- 565
barsch, Nil-Bunt- 833
barsch, Papageien-Fahnen- 570
barsch, Pazifischer Flecken- 575
barsch, Peruanischer Fels- 584
barsch, Riesen-Zacken- 577
barsch, Rot- 1075
barsch, Roter Fahnen- 569
barsch, Roter Zacken- 580
barsch, Rotz- 614
barsch, Sand- 572, 620
barsch, Schnabel- 1076
barsch, Schrift- 589
barsch, Schwarz- 602
barsch, Schwarzer Säge- 571
barsch, See- 593
barsch, Silber- 605
barsch, Sonnen- 601
barsch, Stein- 600
barsch, Streif- 617
barsch, Tiefen- 1076
barsch, Weißer Säge- 597
barsch, Weißer Zacken- 573
barsch, Wolfs- 593
barsch, Wrack- 585
barsch, Ziegen- 587
barsch, Zwerg- 588
Barsche 613
barsche, Blau- 629
barsche, Brasilianische Sand- 865
barsche, Bunt- 831
Barsche, Echte 613
barsche, Glas- 560, 562, 563
barsche, Gold- 1077
barsche, Großaugen- 606, 608
barsche, Nil- 566
barsche, Riff- 835
barsche, Rot- 1077
barsche, Säge- 567
barsche, Schaufelkopf- 562
barsche, See- 595
barsche, Tiger- 592
barsche, Zacken- 582, 583
barsche, Ziegel- 626
Bartgrundel 372
Bärtige Brotula 924
Bartmännchen 929
Bartmuscheln 1373
Bartrobbe 1448
Bastardmakrele 672
Bastardmakrele, Blaue 670
Bastardmakrele, Chilenische 669
Bastardmakrele, Cunene- 674
Bastardmakrele, Grünrücken- 665
Bastardmakrele, Japanische 666
Bastardmakrele, Kap- 673
Bastardmakrele, Mittelmeer- 668
Bastardmakrele, Pazifische 671
Bastardmakrele, Rauhe 667
Bastardmakrelen 632, 675
Bastardschildkröte 1498
Bastard-Süßlippe 712
Bastardzunge 1160

Batavia-Hechtmuräne	406
Batavia-Putjekanipa	406
Bauchstreifiger Bonito	973
Bauernkarpfen	321
Beluga	1478
Benguela Seehecht	489
Bergforelle	258
Bergs Morwong	857
Bernsteinfisch	658
Bersig	617
Biethai	50
Bindenbrasse	782
Binnenstint	274
Bitterling	356
Bitterling, Gewöhnlicher	356
Blainvilles Dornhai	65
Blaubarsch	630
Blaubarsche	629
Blaubartmuschel	1377
Blaue Bastardmakrele	670
Blaue Notothenia	889
Blaue Schwimmkrabbe	1295
Blaue Stachelmakrele	635
Bläuel	663
Blauer Katfisch	914
Blauer Marlin	1019
Blauer Seewolf	914
Blauer Stöcker	670
Blauer Wittling	459
Blauer Ziegelbarsch	627
Blaufelchen	234, 240
Blaufisch	467, 630
Blaufische	629
Blauflossen-Knurrhahn	1082
Blauflossenthun, Südlicher	1010
Blauforelle	259
Blaugefleckte Meerbrasse	804
Blauhai	42
Blauhai, Großer	42
Blauhaie	34
Blauhecht	437
Blaukrabbe	1295
Blaukrabbe, Dana-	1294
Blaukrabben	1296
Blauleng	460
Bläuling	240
Blaumaul	1063
Blaunase	365
Blaurücken	252
Blaurückenlachs	252
Blauwal	1483
Blauweißer Delphin	1474
Blauwittling	459
Blei	305
Bleier	360
Blendling	1169
Blicke	310, 319
Blindling	1169
Blonde	85
Blutstriemen-Schleimfisch	910
Bobo-Meeräsche	539
Bobo-Umberfisch	749
Bobyrez	340
Bodenrenke	234
Bodenrenke, Große	236
Bodenrenke, Kleine	239
Boe-Umberfisch	753
Bogenkrabbe	1292
Bohrmuschel, Gemeine	1416
Bohrmuscheln	1417
Bombay-Ente	297
Bombay-Enten	296
Bonga-Hering	174
Bonito	973, 982

Bonito, Australischer	979
Bonito, Bauchstreifiger	973
Bonito, Echter	973
Bonito, Gefleckter	966
Bonito, Unechter	965
Boretsch	321
Borstenzähner	830
Boyers Ährenfisch	549
Brachmen	305
Brachsen	305, 307
Brachsen, Chinesischer	346
brachsen, Donau-	306
brachsen, Halb-	306, 319
Brachsen, Peking-	349
Brachsenmakrele	680
Brachsenmakrelen	679
Brandbrasse	788, 808
Brasilianische Meeräsche	546
Brasilianische Sandbarsche	865
Brasilianischer Gabeldorsch	474
Brasilianischer Halbschnäbler	430
Brasilianischer Plattkopf	873
brasse, Achsel-	790
brasse, Angola-Zahn-	771
brasse, Binden-	782
brasse, Blaugefleckte Meer-	804
brasse, Geiß-	782
brasse, Gelbrote Meer-	796
brasse, Gewöhnliche Sack-	806
brasse, Gold-	802
brasse, Großaugen-Angola-Meer-	703
brasse, Großaugenzahn-	776
brasse, Große Geiß-	782
Brasse, Holbrooks-	780
brasse, Japanische Gold-	805
brasse, Kanarische Zahn-	772
brasse, Kongo-Zahn-	773
brasse, Marmor-	786
brasse, Marokko-Zahn-	777
brasse, Meer-	790
brasse, Natal-Meer-	797
Brasse, Nordamerikanische	809
brasse, Nordische Meer-	790
brasse, Nufar-See-	768
brasse, Ringel-	778
brasse, Rosa Zahn-	775
brasse, Rot-	790
brasse, Rote Stumpfnasen-	769
brasse, Rotgebänderte Meer-	803
Brasse, Schafskopf-	764
brasse, Spanische Meer-	790
brasse, Spitz-	781
brasse, Streifen-	808
brasse, Südafrikanische Marmor-	785
brasse, Tischler-See-	765
brasse, Zahn-	774
brasse, Zebra-	779
Brassen	305, 307
Brassen, Großaugen-	761
brassen, Meer-	763, 789, 795, 807
Brassen, Monokel-	698
brassen, Nordischer Meer-	793
brassen, Rot-	794
brassen, Roter Meer-	794
brassen, Schnauzen-	810, 814
brassen, See-	770, 784, 789, 795
brassen, Streifen-	787
brassen, Stumpfnasen-	770
Brassenplieten	305
Braunalgen	1525
Braune Geißelgarnele	1207
Braune Tigergarnele	1218
Braune Venusmuschel	1395
Brauner Katzenhai	29

Brauner Lippfisch	842
Brauner Zackenbarsch	577
Braunhai, Atlantischer	40
Breitling	202, 305, 321
Breitstreifensardelle	209
Bresen	305
Bressen	305
Brill	1175
Brisling	202
Brosme	442
Brotula, Bärtige	924
Brucko	105
Brutt	352
Brydewal	1482
Buckellachs	247
Buckelwal	1485
Budegassa-Anglerfisch	1194
Bunker	166
Buntbarsch, Moçambique-	832
Buntbarsch, Nil-	833
Buntbarsche	831
Bunte Kammuschel	1359
Bunte Kammuschel	1363
Bunter Lippfisch	841
Burretschel	321
Bürschling	617
butt, Flügel-	1169
butt, Glas-	1169
butt, Glatt-	1175
butt, Gold-	1145
butt, Haar-	1176
butt, Müllers Zwerg-	1176
butt, Norwegischer Zwerg-	1171
butt, Sand-	1143, 1174
butt, Struff-	1143
butt, Tar-	1172, 1175
butt, Tor-	1172
butt, Vierfleck-	1168
Butt, Weitäugiger	1117
butt, Zungen-	1128
butt, Zwerg-	1176
Butte	1114, 1167, 1170
Butterfisch	919
Butterfisch, Amerikanischer	1033, 1035
Butterfisch, Pazifischer	1034
Butterfische	1030, 1036
Buttermakrelen	952

C

Calico-Pilgermuschel	1360
Canestrini-Grundel	1058
Chacunda	164
Chagrinrochen	89
Chilenische Aszidie	1509
Chilenische Bastardmakrele	669
Chilenische Miesmuschel	1374
Chilenische Pelamide	980
Chilenische Plattauster	1351
Chilenischer Hering	204
Chilenischer Kingklip	927
Chilenischer Seehecht	485
Chilenischer Seeigel	1522
Chile-Tiefseegarnele	1235
Chimäre	124
Chimäre, Amerikanische	124
chimäre, Totenkopf-	125
Chimären	122
chimären, Pflugnasen-	126
Chinesische Alse	168
Chinesische Königsmakrele	1003
Chinesische Pampel	1032
Chinesische Wollhandkrabbe	1303

Chinesischer Brachsen	346	Dorsch, Fjord-	448	Eidechsenfisch, Großer	391
Chinesischer Schwertstör	141	dorsch, Franzosen-	472	Eidechsenfisch, Großschuppen-	392
Chirurgenfische	945	Dorsch, Französischer	472	Eidechsenfische	390
Cichliden	831	dorsch, Frosch-	468	Eigentliche Drachenfische	866
Clupeiden	154	dorsch, Gabel-	463	Einhorndorsch	439
Coco-Umberfisch	743	Dorsch, Grönland-	448	Einsiedlerkrebse	1286
Conger	410	dorsch, Hecht-	487	Eisfisch, Augenfleck-	900
Congeraal	410	dorsch, Klamotten-	466	eisfisch, Bänder-	898
Congeraal, Weißflecken-	409	dorsch, Mittelmeer-Gabel-	464	Eisfisch, Langschnauzen-	899
Coregone	234	dorsch, Nördlicher Silber-	449	Eisfisch, Neuseeland-	438
corvina, Kleine Gelb-	747	dorsch, Pazifischer Polar-	470	Eisfisch, Scotia-See-	896
Corvina, Königs-	726	dorsch, Polar-	441	Eisfisch, South-Georgia-	901
Corvina, Mandschurei-	746	dorsch, Roter Gabel-	475	Eisfisch, Wilsons	897
Crappie, Schwarzer	605	dorsch, Silber-	449	Eisfische	895
Crappie, Weißer	604	dorsch, Stint-	471	Eisfische, Antarktis-	876
Cuneata-Seezunge	1157	dorsch, Weißer Gabel-	476	Eishai	63
Cunene-Bastardmakrele	674	Dorschartige	435	Eismeer-Ringelrobbe	1452
Cutthroat-Forelle	255	dorsche, Ostatlantische Gabel-	465	Elbel	237
Cuvier-Schnabelwal	1461	dorsche, Tiefsee-	436	Elben	157
		Dorschfische	440	Elderitz	352
		Dosinia, Orbignys	1397	Elektrischer Wels	394
D		Drachenfische	866	Elf	157
		Drachenfische, Eigentliche	866	Elritze	352
Dagas	203	Drachenkopf, Gelbschwanz-	1074	Elritze, Sumpf-	351
Dall-Hafenschweinswal	1472	Drachenkopf, Großer	1068	Elritzennäsling	326
Dalmatinischer Näsling	324	Drachenkopf, Kap-	1071	Engelhai	68
Damenfisch	152	Drachenkopf, Kleiner	1067	Entenmuschel, Felsen-	1197
Damenfische	151	Drachenkopf, Kleiner Roter	1066	Entenmuscheln	1198
Dana-Blaukrabbe	1294	Drachenkopf, Kuhls	1065	Entenwal	1460
Dattelmuscheln	1417	Drachenkopf, Roter	1068	Entenwal, Nördlicher	1460
Deckfisch	1037	Drachenkopf, Witwen-	1072	Erling	352
Degenfisch	957, 958, 959	Drachenköpfe	1062, 1064, 1069	Erntefische	1030
Deibel	341	Dreibärtelige Seequappe	450	Escolar-Schlangenmakrele	948
Delfine	1463	Dreibärtelige Seequappen	451	Eßbarer Seeigel	1521
Delphin, Blauweißer	1474	Dreiecksmuscheln	1406	Europäische Äsche	268
Delphin, Gemeiner	1464	Dreiflossenschleimfisch	912	Europäische Auster	1353
Delphin, Langschnauzen-	1475	Dreistachliger Stichling	500	Europäische Languste	1266
Delphin, Rissos	1467	Drescher	17	Europäische Makrele	985
delphin, Rundkopf-	1467	Drescherhai	17	Europäische Meersau	1068
Delphin, Weißschnauzen-	1469	Drückerfisch	1180	Europäische Navaga	445
Delphin, Weißseiten-	1468	drückerfisch, Schweins-	1180	Europäische Sardelle	217
Delphine	1463	Drückerfische	1179	Europäischer Aal	396
delphine, Flecken-	1476	Dugong	1454	Europäischer Flußaal	396
Delphine, Spitzschnauzen-	1458	Dünnlippige Meeräsche	542	Europäischer Hausen	139
Diamantschildkröten	1499	Düsterer Dornhai	59	Europäischer Hornhecht	416
Dick	134			Europäischer Hummer	1318
dick, Glatt-	134			Europäischer Hundsfisch	284
Dickkopf	341	**E**		Europäischer Leierfisch	941
Dickleibiger Thun	1011			Europäischer Seehecht	487
Dicklippige Meeräsche	538	Ebarme, Gefleckter	1111	Europäischer Seeteufel	1195
Dicklippiger Grunzer	709	Ebarme, Indopazifischer	1113	Europäischer Steinbeißer	370
Döbel	341, 342	Ebarme, Westafrikanischer	1112		
Doggerscharbe	1132	Eberfisch	527		
Dohlenkrebs	1279	Eberfische	526, 844	**F**	
Doktorfisch, Monrovia-	946	Echte Aale	395, 401		
Doktorfische	945	Echte Barsche	613	Fächerfisch, Atlantischer	1016
Donaubrachsen	306	Echte Rochen	81	Fächerfisch, Pazifischer	1017
Donauhering	160	Echte Rotzunge	1128, 1142	Fächerfische	1015
Donaulachs	242	Echte Welse	380	fadenfisch, Königs-	555
Donau-Lamprete	2	Echte Zitterrochen	117, 121	fadenfisch, Riesen-	554
Donau-Neunauge	2	Echter Bonito	973	fadenfisch, Schaufel-	1046
Doppellinien-Makrele	972	Echter Lachs	256	Fadenfische	553
Dorngrundel	370	Edelfisch	240	Fadenflossige Alse	172
Dornhai	64	Edelkorallen	1513	Fadenhering, Atlantischer	186
Dornhai, Blainvilles	65	Edelkrebs	1276	Fadenhering, Pazifischer	185
Dornhai, Düsterer	59	Edelkrebse	1281	Fadenheringe	187
Dornhai, Großaugen-	66	Edelmaräne	234, 237	Fadenmakrele, Alexandria-	633
Dornhai, Rauher	58	Edler Tolstolob	336	Fadensegelfisch	389
Dornhai, Schwarzer	61	Edwards Languste	1260	Fähnchen-Messerfische	292
Dornhaie	57	Egli	617	Fahnenbarsch, Papageien-	570
Dornhaie, Schwarze	62	Eidechsenfisch	393	Fahnenbarsch, Roter	569
Dorsch	447	Eidechsenfisch, Atlantischer	393	Falsche Königsmakrele	1000
dorsch, Brasilianischer Gabel-	474	Eidechsenfisch, Gefleckter	392	Falscher, Bonito	968
dorsch, Einhorn-	439			Fangschreckenkrebs	1202

Feilenfisch, Orangeroter	1181	
Feilenfisch, Segel-	1183	
Feilenfische	1182	
Felchen	230, 231, 241, 261	
felchen, Blau-	234, 240	
felchen, Kropf-	239	
Felchen, Leman-	233	
Felchen, Nordamerikanisches	232	
felchen, Sand-	236	
felchen, Silber-	237	
felchen, Weiß-	236	
Felsbarsch, Peruanischer	584	
Felsenauster	1347	
Felsenauster, Pazifische	1347, 1348	
Felsenauster, Sydney-	1357	
Felsenaustern	1350	
Felsenbarsch	598	
Felsen-Entenmuschel	1197	
Felsengarnelen und		
Süßwassergarnelen	1244	
Felsengrundel	1054	
Felsenkrabbe	1289	
Felsgrundel	1054	
Ferchen	259	
Fernöstliche Navaga	444	
Fiederbartwelse	293	
Fingerfisch	556	
Fingerschnecken	1338	
Finnwal	1484	
Finte	157	
Finte, Mittelmeer-	158	
Fjord-Dorsch	448	
flachkopf, Sand-	1096	
Flachköpfe	1095	
Flecke	308	
Fleckenbarsch, Pazifischer	575	
Fleckendelphine	1476	
Fleckenmakrele, Australische	996	
Fleckenmondfisch	678	
Fleckhai	30	
Fleckrochen	93	
Fledermausfische	826	
Fliegende Fische	423, 425, 426	
Fliegender Fisch, Indopazifischer	434	
Fliegender Fisch, Japanischer	424	
Flixe	304	
Flösselhecht	128	
flösselhecht, Nil-	128	
Flügelbutt	1169	
Flügelschnecken	1338	
Flugfisch	426	
Flugfische	425	
Flughahn	1109	
Flugkalmar, Japanischer	1432	
Flugkalmar, Wellington-	1431	
Flunder	1143	
flunder, Amerikanische Winter-	1147	
flunder, Gelbschwanz-	1137	
flunder, Glatt-	1141	
Flunder, Japanische	1138	
Flunder, Olivgrüne	1120	
flunder, Sand-	1149	
flunder, Sommer-	1118	
flunder, Stern-	1144	
flunder, Vierfleck-	1119	
flunder, Zimt	1122	
Flußaal	396	
Flußaal, Europäischer	396	
Flußaale	395, 401	
Flußbarsch	617	
Flußbarsch, Amerikanischer	616	
Flußbarsch, Neuseeland-	864	
Flußforelle	258	
Flußhecht	288	

Flußhering, Nordamerikanischer	161	
Flußkarpfen	332	
Flußkrebs, Amerikanischer	1282	
Flußkrebs, Australischer	1316	
Flußkrebs, Louisiana-	1284	
Flußkrebse	1278, 1281	
Flußkrebse, Amerikanische	1283	
Fluß-Manati	1455	
Flußneunauge	5	
Flußperlmuschel	1383	
Flußpricke	5	
Fluß-Schleimfisch	911	
Flußwels	382	
forelle, Alpen-	258	
Forelle, Amerikanische	251	
forelle, Amerikanische See-	266	
forelle, Bach-	258	
forelle, Berg-	258	
forelle, Blau-	259	
Forelle, Cutthroat-	255	
forelle, Fluß-	258	
Forelle, Gila-	246	
forelle, Gold-	243	
forelle, Grund-	259	
forelle, Herbst-	259	
forelle, Lachs-	257, 259	
forelle, Mai-	259	
forelle, Meer-	257	
forelle, Purpur-	251	
forelle, Regenbogen-	251	
forelle, Rot-	263	
forelle, Schlei-	364	
forelle, Schwarz-	258	
forelle, Schweb-	259	
forelle, See-	259	
forelle, Silber-	259	
forelle, Stahlkopf-	251	
forelle, Stein-	258	
forelle, Teich-	251	
forelle, Weiß-	257, 258	
forelle, Wild-	258	
Forellen	260	
Forellen und Saiblinge, Lachse,	262	
Forellenbarsch	603	
Fransenlipper	337	
Franzosendorsch	472	
Französischer Dorsch	472	
Frauenfische	145	
Fregattmakrele	965	
Fregattmakrele, Melvera-	964	
Froschdorsch	468	
Frösche	1493	
Froschfische	1191	
Froschquabbe	468	
Froschquappe	468	
Frostfisch	456	
Fuchshai	17	
Fuchshai, Großäugiger	16	
Fuchshai, Indopazifischer	15	
Fünfbärtelige Seequappe	443	
Fünffleckiger Lippfisch	849	
Furchengarnele	1220	
Furchengeißelgarnele	1252	
Furchenkrebse	1306	
Füsilierfische	686	

G

Gabeldorsch	463	
Gabeldorsch, Brasilianischer	474	
Gabeldorsch, Mittelmeer-	464	
Gabeldorsch, Roter	475	
Gabeldorsch, Weißer	476	

Gabeldorsche, Ostatlantische	465	
Gabelmakrele	663	
Gabelmakrele, Große	647	
Gabelschwanzseekuh	1454	
Gabelwels, Getüpfelter	386	
Galizier	1277	
Galjoen, Südafrikanischer	824	
Galjoen-Fische	823	
Gambuse	516	
Gangfisch	235, 237, 345	
Gängling	343	
Gareisel	321	
Gareisle	321	
garnele, Argentinische Rot-	1256	
Garnele, Argentinische Stilett-	1206	
Garnele, Bananen-	1222	
garnele, Bären-	1223	
garnele, Braune Geißel-	1207	
garnele, Braune Tiger-	1218	
garnele, Chile-Tiefsee-	1235	
garnele, Furchen-	1220	
garnele, Furchengeißel-	1252	
garnele, Grönland-	1237	
garnele, Grüne Tiger-	1226	
garnele, Guinea-Geißel-	1210	
garnele, Hauptmanns-	1216	
garnele, Kalifornia-Geißel-	1215	
garnele, Königsgeißel-	1221	
garnele, Königs-Rot-	1257	
garnele, Kristall-Geißel-	1214	
garnele, Kroyers Geißel-	1231	
garnele, Membran-Geißel-	1258	
garnele, Messer-	1254	
Garnele, Nordische	1237	
garnele, Nördliche Rosa Geißel-	1217	
garnele, Nördliche Weiße Geißel-	1227	
garnele, Nordmeer-	1237	
garnele, Rad-	1219	
Garnele, Rosa	1238	
garnele, Rosa Geißel-	1212	
garnele, Rosenbergs Süßwasser-	1247	
garnele, Rotschwanz-	1225	
garnele, Säge-	1249	
garnele, Sand-	1251	
garnele, Schiffskiel-	1223	
garnele, Shiba-Geißel-	1208	
garnele, Stein-	1249	
garnele, Südamerikanische		
Süßwasser-	1246	
garnele, Südliche Rosa Geißel-	1224	
garnele, Tiefsee-	1237	
garnele, Zentralamerikanische		
Geißel-	1228	
Garnelen, Atyiden-	1243	
garnelen, Azteken-Geißel-	1213	
garnelen, Felsen- und Süßwasser-	1244	
garnelen, Geißel-	1209, 1211, 1229,	
	1230	
garnelen, Kanten-Tiefsee-	1236	
garnelen, Leucht-	1241	
garnelen, Messer-	1255	
garnelen, Sand-	1250	
garnelen, Süßwasser-	1248	
garnelen, Tiefsee-	1232, 1239	
Gazami-Schwimmkrabbe	1300	
Gebärfische	920	
Geese	343	
Geflecke Königsmakrele	994	
Geflecke Stachelmakrele	643	
Gefleckter Aufbläser	1185	
Gefleckter Bonito	966	
Gefleckter Ebarme	1111	
Gefleckter Eidechsenfisch	392	
Gefleckter Glatthai	52	

Gefleckter Katfisch	916	
Gefleckter Kreuzwels	376	
Gefleckter Leierfisch	942	
Gefleckter Lippfisch	840	
Gefleckter Rochen	93	
Gefleckter Seewolf	916	
Gefleckter Streifenbarsch	594	
Gefleckter Thunfisch	968	
Gefleckter Umberfisch	725	
Gefleckter Zitterrochen	120	
Gehörnte Turbanschnecke	1331	
Geibel	321	
Geigenrochen	72, 76, 77	
Geigenrochen, Gemeiner	76	
Geigenrochen, Peruanischer	75	
Geigenrochen, Südlicher	74	
Geigenrochen, Weißflecken-	73	
Geißbrasse	782	
Geißbrasse, Große	782	
Geißelgarnele, Braune	1207	
geißelgarnele, Furchen-	1252	
Geißelgarnele, Guinea-	1210	
geißelgarnele, Königs-	1221	
Geißelgarnele, Kristall-	1214	
Geißelgarnele, Kroyers	1231	
Geißelgarnele, Membran-	1258	
Geißelgarnele, Nördliche Rosa	1217	
Geißelgarnele, Nördliche Weiße	1227	
Geißelgarnele, Rosa	1212	
Geißelgarnele, Shiba-	1208	
Geißelgarnele, Südliche Rosa	1224	
Geißelgarnele, Zentralamerikanische	1228	
Geißelgarnelen	1209, 1211, 1229, 1230	
Geißelgarnelen, Azteken-	1213	
Gelbbarsch	616	
Gelbbauch-Notothenia	882	
Gelbcorvina, Kleine	747	
Gelbe Stachelmakrele	637	
Gelber Asiatischer Umberfisch	722	
Gelber Schnapper	689	
Gelbflossen-Notothenia	887	
Gelbflossen-Pfeilhecht	533	
Gelbflossen-Süßlippe	707	
Gelbflossenthun	1008	
Gelbrote Meerbrasse	796	
Gelbschwanz	658	
Gelbschwanz-Drachenkopf	1074	
Gelbschwanzflunder	1137	
Gelbschwanzmakrele	658	
Gelbschwanzmakrele, Australische	659	
Gelbschwanzschnapper	694	
Gelbschwanz-Stachelmakrele	641	
Gelbstriemen	766	
Gelidium	1529	
Gemeine Bohrmuschel	1416	
Gemeine Goldmakrele	681	
Gemeine Makrele	985	
Gemeine Meeräsche	542	
Gemeine Pampel	1037	
Gemeine Samtmuschel	1384	
Gemeine Seezunge	1165	
Gemeiner Delphin	1464	
Gemeiner Geigenrochen	76	
Gemeiner Heuschreckenkrebs	1202	
Gemeiner Himmelsgucker	874	
Gemeiner Leierfisch	941	
Gemeiner Meeraal	410	
Gemeiner Meerengel	68	
Gemeiner Pampano	662	
Gemeiner Rundhering	178	
Gemeiner Seestern	1517	
Gemeiner Stichling	500	
Gemeiner Tintenfisch	1418, 1438	
Gemeines Seeohr	1328	
Georgia-Lachs	682	
Geperlter Sichelflosser	828	
Germon	1007	
Gestreifte Makrele	993	
Gestreifte Meeräsche	545	
Gestreifte Meerbarbe	818	
Gestreifte Notothenia	888	
Gestreifte Süßlippe	704	
Gestreifte Venusmuschel	1396	
Gestreifter Katfisch	915	
Gestreifter Knurrhahn	1091	
Gestreifter Marlin	1023	
Gestreifter Schlangenkopffisch	558	
Gestreifter Schleimfisch	909	
Gestreifter Seehahn	1091	
Gestreifter Seewolf	915	
Gestreifter Thunfisch	973	
Gestreifter Umberfisch	728	
Getüpfelter Gabelwels	386	
Gewöhnliche Alse	156	
Gewöhnliche Makrele	985	
Gewöhnliche Meeräsche	545	
Gewöhnliche Meerbarbe	817	
Gewöhnliche Sackbrasse	806	
Gewöhnliche Schmerle	372	
Gewöhnlicher Adlerrochen	109	
Gewöhnlicher Bitterling	356	
Gewöhnlicher Grindwal	1466	
Gewöhnlicher Gründling	333	
Gewöhnlicher Kalmar	1426	
Gewöhnlicher Krake	1438	
Gewöhnlicher Maifisch	156	
Gewöhnlicher Stechrochen	106	
Gewöhnlicher Thunfisch	1012	
Gewöhnlicher Tolstolob	335	
Gewöhnliches Bachneunauge	6	
Gezähnte Auster	1352	
Gezeiten-Ährenfisch	552	
Ghiozzo	1056	
Giebel	321	
Giefen	333	
Giffer	333	
Gigartina	1531	
Gila-Forelle	246	
Gilchrists Languste	1267	
Glasauge	280, 281, 282	
Glasauge, Japanisches	283	
Glasauge, Pazifisches	280	
Glasaugen	279	
Glasaugenbarsch	621	
Glasbarsche	560, 562, 563	
Glasbutt	1169	
Glasgrundel	1050	
Glasküling	1050	
Glaswelse, Indische	379	
Glattbutt	1175	
Glattdick	134	
Glatte Mactra	1390	
Glatte Venusmuschel	1395	
Glatter Hammerhai	47	
Glattflunder	1141	
Glatthai, Australischer	51	
Glatthai, Gefleckter	52	
Glatthai, Mittelmeer-	53	
Glatthai, Nördlicher	52	
Glatthai, Patagonischer	54	
Glatthai, Südlicher	53	
Glatthaie	49, 55	
Glattkopf	223	
Glattmaul-Kreuzwels	375	
Glattrochen	84	
Glattwal, Südlicher	1489	
glattwal, Zwerg-	1488	
Glattwale	1486	
Goldäsche	540	
Goldband-Selar	655	
Goldbarsch	1075	
Goldbarsche	1077	
Goldbrasse	802	
Goldbrasse, Japanische	805	
Goldbutt	1145	
Goldene Stachelmakrele	634	
Goldene Teppichmuschel	1404	
Goldener Scheinschnapper	696	
Goldener Zackenbarsch	574	
Goldfisch	320	
Goldforelle	243	
Goldgrundel	1051	
Goldkarpfe	321	
Goldlachs	280, 281, 283	
Goldlachse	279	
Goldmaid	847	
Goldmakrele	681	
Goldmakrele, Gemeine	681	
Goldmakrelle	323	
Goldmeeräsche	540	
Gold-Meerbarbe	821	
Goldorfe	343	
Gold-Steinbeißer	366	
Goldstrieme	801	
Golf-Menhaden	165	
Golf-Umberfisch	734	
Gotteslachs	509	
Gramper	1467	
Granat	1251	
granat, Kaiser-	1320	
Graskarpfen	331	
Grasnadel	508	
Grätenfisch	152	
Grätenfisch, Großflossen-	153	
Grätenfische	151	
Grauäsche	538	
Graue Meeräsche	547	
Graue Notothenia	886	
Grauer Knurrhahn	1084	
Grauer Lippfisch	845	
Grauer Zackenbarsch	576	
Grauhai	11	
Grauhaie	9	
Grauwal	1491	
Grenadier, Neuseeländischer	480	
Grenadier, Nordatlantik-	497	
Grenadieranchovis	213	
Grenadiere	477	
Grenadierfisch	494	
Grenadierfisch, Kurzschnauzen-	495	
Grenadierfisch, Schwarzfleck-	493	
Grenadierfische	477, 496, 498	
Greßling	333	
greßling, Stein-	334	
Griechische Rotfeder	363	
Grieslauge	345	
Grindwal	1466	
Grindwal, Gewöhnlicher	1466	
Grindwal, Indischer	1465	
Grönland-Dorsch	448	
Grönlandgarnele	1237	
Grönlandhai	63	
Grönland-Heilbutt	1148	
Grönland-Kabeljau	448	
Grönlandrochen	99	
Grönlandwal	1487	
Groppe	1097	
Groppen	1102	
Groppenbarsch	618	
Großauge	281	
Großaugen-Angola-Meerbrasse	703	

Großaugenbarsch, Australischer 607
Großaugenbarsche 606, 608
Großaugen-Brassen 761
Großaugen-Dornhai 66
Großaugenthun 1011
Großaugenzahnbrasse 776
Großäugiger Fuchshai 16
Großäugiger Seehecht 481
Großäugiger Selar 654
Großäugiger Thun 1011
Große Bodenrenke 236
Große Gabelmakrele 647
Große Geißbrasse 782
Große Jakobsmuschel 1365
Große Kammuschel 1365
Große Maräne 234, 236
Große Meergrundel 1052
Große Messermuschel 1413
Große Perlmuschel 1346
Große Pilgermuschel 1365
Große Rossie 1419
Große Sandklaffmuschel 1415
Große Schwebrenke 234, 240
Große Seenadel 506
Große Seespinne 1315
Große Steinkrabbe 1293
Große Teppichmuschel 1401
Großer Ährenfisch 550
Großer Blauhai 42
Großer Drachenkopf 1068
Großer Eidechsenfisch 391
Großer Katzenhai 32
Großer Rotbarsch 1075
Großer Sandaal 939
Großer Sandspierling 939
Großer Scheibenbauch 1108
Großer Steinbeißer 367
Großer Teufelsrochen 112
Großer Thun 1012
Großer Tümmler 1477
Großer Wolfshering 225
Großes Petermännchen 869
Großflossen-Grätenfisch 153
Großflossen-Morwong 859
Großgefleckter Katzenhai 32
Großkeuliger Zwergkalmar 1423
Großkopf-Meeräsche 545
Großkopfsardine 192
Großschuppen-Eidechsenfisch 392
Großschuppige Scholle 1177
Grouper, Roter 580
Grümpel 352
Grünauge 299
Grünaugen 298
Grundel 333
grundel, Bart- 372
Grundel, Canestrini- 1058
grundel, Dorn- 370
grundel, Fels- 1054
grundel, Felsen- 1054
grundel, Glas- 1050
grundel, Gold- 1051
grundel, Große Meer- 1052
Grundel, Kessler- 1057
grundel, Marmor- 1060
grundel, Marmorierte Meer- 1060
Grundel, Panizza- 1056
grundel, Sand- 1059
Grundel, Schlangenkopf- 1061
grundel, Schwarz- 1053
Grundeln 1049, 1055
grundeln, Meer- 1049
grundeln, Schläfer- 1048
Grundforelle 259

Grundhai 63, 64
Gründling 333
gründling, Barben- 312
Gründling, Gewöhnlicher 333
gründling, Stein- 334
Grüne Miesmuschel 1380
Grüne Notothenia 883
Grüne Tigergarnele 1226
Grünel 658
Grüner Lippfisch 843
Grüner Stör 132
Grünknochen 416
Grünling, Langer 1092
Grünrücken-Bastardmakrele 665
Grünschleie 364
Grunzer 705
Grunzer, Dicklippiger 709
Guester 319
Guinea-Barrakuda 530
Guinea-Geißelgarnele 1210
Guinea-Messerzahnaal 405
Günthers-Notothenia 891
Gurami 1045
Gurami, Knurrender 1045
Gurami, Küssender 1047

H

hai, Großäugiger Fuchs- 16
Haarbutt 1176
Haarschwanz 959
Haarschwanz, Kurzflossen- 956
Haarschwänze 955
Hafenschweinswal, Dall- 1472
Hägling 240
hai, Alligator- 71
hai, Ammen- 26
hai, Atlantischer Braun- 40
hai, Atlantischer Spitzmaul- 44
hai, Australischer Ammen- 25
hai, Australischer Glatt- 51
hai, Biet- 50
hai, Blainvilles Dorn- 65
hai, Blau- 42
hai, Brauner Katzen- 29
hai, Dorn- 64
hai, Drescher- 17
hai, Düsterer Dorn- 59
hai, Eis- 63
hai, Engel- 68
hai, Fleck- 30
hai, Fuchs- 17
hai, Gefleckter Glatt- 52
hai, Glatter Hammer- 47
hai, Grau- 11
hai, Grönland- 63
hai, Großer Blau- 42
hai, Großer Katzen- 32
hai, Großgefleckter Katzen- 32
hai, Grund- 63, 64
hai, Hammer- 47
hai, Herings- 23
hai, Hunds- 50
hai, Indopazifischer Fuchs- 15
hai, Indopazifischer Hammer- 46
hai, Kalifornischer Leopard- 56
hai, Kleiner Katzen- 31
hai, Kleingefleckter Katzen- 31
hai, Kragen- 8
hai, Krausen- 8
hai, Kupfer- 36
Hai, Langflossen- 37
hai, Langflossen-Mako- 21

hai, Makrelen- 20
hai, Menschen- 19
hai, Mittelmeer-Glatt- 53
hai, Nagel- 71
hai, Nördlicher Glatt- 52
hai, Patagonischer Glatt- 54
hai, Pazifischer Herings- 22
hai, Pazifischer Spitzmaul- 43
hai, Rauher Dorn- 58
hai, Riesen- 12
hai, Sand- 13
hai, Sandbank- 39
hai, Sandtiger- 13
hai, Schildzahn- 14
hai, Schokoladen- 60
hai, Schwarzer Dorn- 61
hai, Schwarzspitzen-Riff- 38
hai, Schwein- 70
Hai, Silberspitzen- 35
hai, Stachel- 71
hai, Südlicher Glatt- 53
hai, Wal- 27
hai, Weiß- 19
hai, Weißspitzen-Menschen- 37
hai, Zebra- 24
hai, Zitronen- 41
haie, Blau- 34
haie, Dorn- 57
haie, Glatt- 49, 55
haie, Grau- 9
haie, Hammer- 45, 48
haie, Herings- 18
haie, Hunds- 49
haie, Katzen- 28, 33
haie, Marder- 49
haie, Schwarze Dorn- 62
Haimakrele 971
Halbbrachsen 306, 319
Halbfisch 305
Halbschnabelhechte 427
Halbschnäbler 433
Halbschnäbler, Australischer 428
Halbschnäbler, Balao- 429
Halbschnäbler, Brasilianischer 430
Halbschnäbler, Japanischer 431
Halbschnäbler, Langflossen- 432
Hammerhai 47
Hammerhai, Glatter 47
Hammerhai, Indopazifischer 46
Hammerhaie 45, 48
Hapuku-Wrackbarsch 586
Harr 268
Hartkopf 341
Hasel 344
hasel, Rot- 362
Häsling 344
Haubenschleimfisch 918
Hauptmannsgarnele 1216
Hausen 139
Hausen, Europäischer 139
Hecht 288
Hecht, Amur- 291
hecht, Argentinischer See- 486
hecht, Atlantischer Horn- 417
hecht, Atlantischer Makrelen- 422
hecht, Atlantischer Pfeil- 532
hecht, Benguela See- 489
hecht, Blau- 437
hecht, Chilenischer See- 485
hecht, Europäischer Horn- 416
hecht, Europäischer See- 487
hecht, Flössel- 128
hecht, Fluß- 288
hecht, Gelbflossen-Pfeil- 533

hecht, Großäugiger See- 481
hecht, Horn- 416
hecht, Indischer Pfeil- 534
hecht, Kahl- 142
hecht, Kap- 484
hecht, Ketten- 290
hecht, Knochen- 143
hecht, Kurznasen-Knochen- 144
hecht, Kurzschnabel-Makrelen- 421
hecht, Makrelen- 422
Hecht, Maß- 287
hecht, Mittelmeer-Pfeil- 535
hecht, Nadel-Horn- 418
hecht, Nilflössel- 128
hecht, Nordamerikanischer See- 483
hecht, Nordpazifischer See- 490
hecht, Patagonischer See- 486
hecht, Pazifischer See- 485
hecht, Schlanker Knochen- 143
hecht, Schwarzer See- 877
hecht, See- 487
hecht, Senegalesischer See- 491
hecht, Silber- 483
hecht, Südlicher See- 482
hecht, Tiefenwasser-Kapsee- 488
Hechtbarsch 620
Hechtdorsch 487
Hechte 286
hechte, Halbschnabel- 427
hechte, Horn- 415, 419
hechte, Makrelen- 420
hechte, Pfeil- 529, 536
hechte, See- 478, 492
Hechtmuräne 406
Hechtmuräne, Batavia- 406
Heilbutt 1133
heilbutt, Amerikanischer Pfeilzahn- 1125
heilbutt, Asiatischer Pfeilzahn- 1124
Heilbutt, Atlantischer 1133
Heilbutt, Grönland- 1148
Heilbutt, Japanischer 1130
Heilbutt, Pazifischer 1134
Heilbutt, Pfeilzahn- 1124
Heilbutt, Schwarzer 1148
Heilbutt, Weißer 1133
Heilbuttscholle 1131
Heilstint 274
Hektor-Laternenfisch 295
Heller Leng 462
Heller Seelachs 466
Hellfleckiger Rochen 91
Helmschnecken 1336
Herbstforelle 259
Hering 169
Hering, Atlantischer 169
hering, Atlantischer Faden- 186
hering, Atlantischer Rund- 177
Hering, Bonga- 174
Hering, Chilenischer 204
hering, Donau- 160
hering, Gemeiner Rund- 178
hering, Großer Wolfs- 225
Hering, Indischer 225
hering, Kalifornischer Rund- 176
hering, Nordamerikanischer Fluß- 161
Hering, Pazifischer 170
hering, Pazifischer Faden- 185
hering, Pazifischer Klein- 194
hering, Regenbogen-Rund- 173
Hering, Rotmeer- 181
hering, Silberstreifen-Rund- 201
hering, Staff- 157
Heringe 154
heringe, Faden- 187

heringe, Klein- 180
heringe, Rund- 179
heringe, Wolfs- 224, 226
Heringshai 23
Heringshai, Pazifischer 22
Heringshaie 18
Heringskönig 525
Herzmuschel 1389
Herzmuschel, Stachelige 1388
Herzmuscheln 1387
Heuerling 240
Heuschreckenkrebs, Gemeiner 1202
Heuschreckenkrebse 1201
Himmelsgucker 874, 875
Himmelsgucker, Gemeiner 874
Hochantarktische Nototheniiden 894
Holbrooks-Brasse 780
Hornfisch 416
Hornhecht 416
Hornhecht, Atlantischer 417
Hornhecht, Europäischer 416
Hornhecht, Nadel- 418
Hornhechte 415, 419
Hovens Karpfen 338
Huchen 242
Huchen, Sibirischer 242
Humback-Lachs 247
Hummer 1318
Hummer, Amerikanischer 1317
hummer, Andamanen-Schlank- 1319
Hummer, Europäischer 1318
hummer, Norwegischer Schlank- 1320
hummer, Tiefsee- 1320
Hundelachs 248
Hundfisch, Schwarzer 61
Hundsbarbe 316
Hundsfisch 284
Hundsfisch, Amerikanischer 285
Hundsfisch, Europäischer 284
hundsfisch, Zwerg- 285
Hundshai 50
Hundshaie 49
Hundslachs 248
Hundsrobben 1440, 1446
Hundszahn-Umberfisch 742
Hundszunge 1128
Hundszungen 1166

Iberische Barbe 315
Iberischer Nässling 327
Igelrochen 88
Ilisha, Schlank- 184
Ilisha, Westafrikanische 183
Ilisha-Alse 205
Illanke 259
Imperatorfische 761, 762
Indische Glaswelse 379
Indische Königsmakrele 989
Indische Makrelen 978
Indische Pellona 188
Indische Schwebmakrele 1038
Indische Stachelmakrele 644
Indische Zwergmakrele 977
Indischer Barrakuda 534
Indischer Grindwal 1465
Indischer Hering 225
Indischer Pfeilhecht 534
Indonesischer Schlangenkopffisch 557
Indopazifische Königsmakrele 991
Indopazifische Makrele 983
Indopazifische Zwergmakrelen 978

Indopazifischer Ebarme 1113
Indopazifischer Fliegender Fisch 434
Indopazifischer Fuchshai 15
Indopazifischer Hammerhai 46
Indopazifischer Segelfisch 1017
Insel-Zwergmakrele 976
Irisches Moos 1532
Irischmoos 1532
Isländische Kammuschel 1362
Islandmuschel 1386
Italienische Barbe 318
Italienischer Nässling 328
Italienischer Steinbeißer 368

J

Jacks 639
Jakobsmuschel 1364
Jakobsmuschel, Australische 1366
Jakobsmuschel, Große 1365
Jakobsmuschel, Neuseeländische 1367
Japanische Archenmuschel 1340
Japanische Bastardmakrele 666
Japanische Flunder 1138
Japanische Goldbrasse 805
Japanische Languste 1272
Japanische Makrele 997
Japanische Sardelle 218
Japanische Sardine 197
Japanische Scholle 1126
Japanische Seegurke 1523
Japanische Seriola 660
Japanische Teppichmuschel 1402
Japanische Venusmuschel 1399
Japanischer Aal 399
Japanischer Barsch 599
Japanischer Fliegender Fisch 424
Japanischer Flugkalmar 1432
Japanischer Halbschnäbler 431
Japanischer Heilbutt 1130
Japanischer Kardinalfisch 612
Japanischer Lachs 250
Japanischer Sandfisch 863
Japanisches Glasauge 283
Japan-Lachs 250
Java-Barbe 353
Jeese 343
Jessen 343
Jonahkrabbe 1288
Juan Fernandez-Languste 1261
Judenfisch 578
Judenfisch, Kalifornischer 591

K

Kabeljau 447, 448
Kabeljau, Grönland- 448
Kabeljau, Pazifischer 446
Kahlhecht 142
kaiman, Krokodil- 1503
Kaiserbarsch 518, 519
Kaiserbarsch, Mittelmeer- 521
Kaisergranat 1320
Kalifornia-Umberfisch, Weißer 730
Kalifornische Sardine 196
Kalifornische Scholle 1127
Kalifornischer Barrakuda 531
Kalifornischer Judenfisch 591
Kalifornischer Leopardhai 56
Kalifornischer Rundhering 176
Kalifornischer Seestint,
    Kleinmäuliger 272
Kalkalgen 1530

DE

Kalmar	1426	Katfisch, Gestreifter	915	Knurrhahn, Grauer	1084
kalmar, Argentinischer		Katfische	917	Knurrhahn, Kap-	1081
Kurzflossen-	1428	Katzenfisch	385	Knurrhahn, Kuckucks-	1079
Kalmar, Gewöhnlicher	1426	Katzenhai, Brauner	29	knurrhahn, Leier-	1090
kalmar, Großkeuliger Zwerg-	1423	Katzenhai, Großer	32	Knurrhahn, Rauher	1080
kalmar, Japanischer Flug-	1432	Katzenhai, Großgefleckter	32	Knurrhahn, Roter	1089
kalmar, Kap-	1425	Katzenhai, Kleiner	31	Knurrhahn, Spitzmaul-	1088
Kalmar, Kleiner	1423	Katzenhai, Kleingefleckter	31	Knurrhahn, Stachel-	1085
kalmar, Langflossen-Schelf-	1424	Katzenhaie	28, 33	Knurrhahn, Stachliger	1086
kalmar, Mittelländischer Zwerg-	1423	Katzen-Kreuzwels	377	Knurrhähne	1078, 1083
kalmar, Nördlicher Kurzflossen-	1430	Katzenwels	385	Knurrhähne, Nordamerikanische	1087
kalmar, Pfeil-	1433	Katzenwelse	387	Koboldkärpfling	516
kalmar, Riesen-	1427	Kaulbarsch	614	Koffermuscheln	1406
Kalmar, Roter	1429	Kaulkopf	1097	Kohlenfisch	1094
kalmar, Wellington-Flug-	1431	Kegelrobbe	1449	Köhler	467
Kalmare	1422	Kelee-Alse	182	köhler, Stein-	466
Kammuschel, Bunte	1359	Kelp	1526	Kongo-Zahnbrasse	773
Kammuschel, Große	1365	Kerzenfisch	278	Königinnen-Stachelmakrele	653
Kammuschel, Isländische	1362	Kessler-Grundel	1057	Königsbarsch	631
Kammuschel, Südatlantische	1368	Keta-Lachs	248	Königs-Corvina	726
Kammuscheln	1358	Kettenhecht	290	Königsfadenfisch	555
Kamschatka-Krabbe	1310	Keulenrochen	87	Königsgeißelgarnele	1221
Kanadi-Makrele	998	Kilch	239	Königskrabbe, Antarktische	1309
Kanadische Alse	155	Kilka	171	Königskrabben	1311
Kanadischer Zander	619	Killerwal	1470	Königslachs	254
Kanarische Zahnbrasse	772	Kingklip, Chilenischer	927	Königslanguste	1273
Kaninchenfisch	943	Kingklip, Rosa	925	Königsmakrele	988
Kaninchenfische	944	Kingklip, Südafrikanischer	926	Königsmakrele, Chinesische	1003
Kanov	306	Kingklips	928	Königsmakrele, Falsche	1000
Kanten-Tiefseegarnelen	1236	Kisutch-Lachs	249	Königsmakrele, Gefleckte	994
Kap-Bastardmakrele	673	klaffmuschel, Sand-	1415	Königsmakrele, Indische	989
Kap-Drachenkopf	1071	Klamottendorsch	466	Königsmakrele, Indopazifische	991
Kapelan	273	Klappmütze	1447	Königsmakrele, Ostatlantische	1004
Kaphecht	484	Kleinäugiger Rochen	91	Königsmakrele, Ostpazifische	1002
Kapitänsfisch	556	Kleine Bodenrenke	239	Königsmakrelen	1005
Kapkalmar	1425	Kleine Gelbcorvina	747	Königs-Rotgarnele	1257
Kap-Knurrhahn	1081	Kleine Maräne	229	Königs-Schlangenmakrele	949
Kaplanguste	1262	Kleine Meeräsche	543	Königs-Umberfisch	735
Kap-Messermuschel	1412	Kleine Schwebrenke	237	Königs-Umberfisch,	
Kapseehecht, Tiefenwasser-	488	Kleine Seenadel	507	Südamerikanischer	732
Karausche	321	Kleine Teppichmuschel	1405	Kopf, Schwimmender	1189
karausche, Silber-	320	Kleiner Ährenfisch	549	Koppe	1097
Kardinalfisch, Japanischer	612	Kleiner Bärenkrebs	1305	Koralle, Rote	1514
Kardinalfisch, Teleskop-	611	Kleiner Drachenkopf	1067	korallen, Edel-	1513
Kardinalfische	609	Kleiner Kalmar	1423	Korallenwelse	378
Karettschildkröte, Unechte	1496	Kleiner Katzenhai	31	Koreanische Makrele	992
Karibik-Pilgermuschel	1361	Kleiner Krake	1435	Koreanische Miesmuschel	1375
karpf, Messer-	350	Kleiner Roter Drachenkopf	1066	Koreanischer Sandaal	940
karpfe, Gold-	321	Kleiner Sandaal	936	Kotschebel	321
karpfe, Pot-	321	Kleiner Schiffshalter	1178	Krabbe	1237
karpfe, Schneider-	321	Kleiner Schwertwal	1473	krabbe, Antarktische Königs-	1309
karpfe, Stein-	321	Kleiner Teufelsrochen	114	krabbe, Blau-	1295
Karpfen	332	Kleiner Thun	968	krabbe, Blaue Schwimm-	1295
karpfen, Bauern-	321	Kleiner Tümmler	1471	krabbe, Bogen-	1292
karpfen, Fluß-	332	Kleines Petermännchen	867	krabbe, Chinesische Wollhand-	1303
karpfen, Gras-	331	Kleingefleckter Katzenhai	31	krabbe, Dana-Blau-	1294
Karpfen, Hovens	338	Kleinhering, Pazifischer	194	krabbe, Felsen-	1289
karpfen, Moor-	321	Kleinheringe	180	krabbe, Gazami-Schwimm-	1300
karpfen, Rohr-	341	Kleinmäuliger Kalifornischer		krabbe, Große Stein-	1293
karpfen, Rot-	360, 362	Seestint	272	krabbe, Jonah-	1288
Karpfen, Schlamm-	330	Kleinsardine, Asiatische	191	Krabbe, Kamschatka-	1310
Karpfen, Schwarzer	347	Kleist	1175	krabbe, Mangroven-	1302
karpfen, See-	808	Kletterfisch	1039	krabbe, Mittelmeer-Strand-	1297
Karpfen, Victoria-See-	355	Kliesche	1139	krabbe, Murray Stein-	1308
karpfen, Zahn-	512	Kliesche, Rauhe	1136	krabbe, Nördliche Stein-	1307
Karpfenfische	303	Klippenbarsch	839	krabbe, Rote Tiefsee-	1321
kärpfling, Kobold-	516	Klumpfisch	1188	krabbe, Strand-	1298
kärpfling, Spanien-	514	Klunzingers Ponyfisch	700	krabbe, Südliche Spinnen-	1314
kärpfling, Valencia-	515	Knochenhecht	143	krabbe, Wollige Schwimm-	1299
kärpfling, Zebra-	513	Knochenhecht, Kurznasen-	144	Krabben	1287
Karras	321	Knochenhecht, Schlanker	143	krabben, Blau-	1296
Katfisch	915	Knurrender Gurami	1045	krabben, Königs-	1311
Katfisch, Blauer	914	Knurrhahn, Blauflossen-	1082	krabben, Schwimm-	1301
Katfisch, Gefleckter	916	Knurrhahn, Gestreifter	1091	krabben, Tiefsee-	1322

Kragenhai 8
Krake 1438
Krake, Gewöhnlicher 1438
Krake, Kleiner 1435
krake, Moschus- 1436
krake, Zirren- 1435
Kraken 1439
kraken, Moschus- 1437
kraken, Zirren- 1437
Krätzer 617
Krausenhai 8
krebs, Amerikanischer Fluß- 1282
krebs, Australischer Fluß- 1316
krebs, Dohlen- 1279
krebs, Edel- 1276
krebs, Fangschrecken- 1202
krebs, Gemeiner Heuschrecken- 1202
krebs, Kleiner Bären- 1305
krebs, Louisiana-Fluß- 1284
krebs, Pazifischer Taschen- 1290
krebs, Stein- 1280
krebs, Sumpf- 1277
krebs, Taschen- 1291
krebse, Amerikanische Fluß- 1283
krebse, Bären- 1304
krebse, Edel- 1281
krebse, Einsiedler- 1286
krebse, Fluß- 1278, 1281
krebse, Furchen- 1306
krebse, Heuschrecken- 1201
krebse, Kurzschwanz- 1287
krebse, Panzer- 1285
krebse, Taschen- 1287
Kreiselschnecken 1330
Kreßling 333
Kretzer 617
Kreuzmuster-Teppichmuschel 1401
Kreuzwels, Gefleckter 376
Kreuzwels, Glattmaul- 375
Kreuzwels, Katzen- 377
Kreuzwelse 374
Krill, Antarktischer 1204
Krill, Leuchtkrebse und 1203
Krill, Norwegischer 1205
Kristall-Geißelgarnele 1214
krokodil, Leisten- 1505
Krokodil, Neuguinea- 1504
Krokodil, Siam- 1506
Krokodile 1501
Krokodile, Alligatoren und 1500
Krokodilfische 872
Krokodilkaiman 1503
Kropffelchen 239
Kröpfling 239
Kroyers Geißelgarnele 1231
Kuckucks-Knurrhahn 1079
Kuckucks-Lippfisch 841
Kuckucksrochen 94
Kugelfisch, Nördlicher 1185
Kugelfische 1184
Kühling 343
Kuhls Drachenkopf 1065
küling, Glas- 1050
küling, Sand- 1059
küling, Schwarz- 1053
küling, Swalt- 1053
Kupferhai 36
Kurze Zwergmakrele 975
Kurzflossen-Haarschwanz 956
Kurzflossenkalmar, Argentinischer 1428
Kurzflossenkalmar, Nördlicher 1430
Kurzkiefer-Umberfisch 748
Kurznasen-Knochenhecht 144
Kurzschnabel-Makrelenhecht 421
Kurzschnauzen-Grenadierfisch 495
Kurzschnauziger Schlangenaal 413
Kurzschnauziger Speerfisch 1022
Kurzschwanzkrebse 1287
Kurzschwanz-Notothenia 890
Küssender Gurami 1047
Küsser 1047
Küstensauger 1178
Kutt 614

L

Lachs 256
lachs, Adler- 718
Lachs, Atlantischer 256
Lachs, Australischer 683
lachs, Blaurücken- 252
lachs, Buckel- 247
lachs, Donau- 242
Lachs, Echter 256
Lachs, Georgia- 682
lachs, Gold- 280, 281, 283
lachs, Gottes- 509
lachs, Heller See- 466
Lachs, Humback- 247
lachs, Hunde- 248
lachs, Hunds- 248
Lachs, Japan- 250
Lachs, Japanischer 250
Lachs, Keta- 248
Lachs, Kisutch- 249
lachs, Königs- 254
Lachs, Masu- 250
Lachs, Rosa 247
lachs, Rot- 252
Lachs, Roter 247, 252
lachs, Schwarz- 257
lachs, See- 467
lachs, Silber- 249, 257
lachs, Strand- 257
Lachse, Forellen und Saiblinge 262
lachse, Gold- 279
Lachsforelle 257, 259
Lactarius 628
Lammzunge 1115
Lammzunge, Thors- 1116
Lamprete 7
Lamprete, Donau- 2
lamprete, See- 7
Lamprete, Ukrainische 3
Lamprete, Vladikov- 4
Lampreten 1
Langer Grünling 1092
Langer Mondfisch 1189
Langer Terpug 1092
Langflossen-Hai 37
Langflossen-Halbschnäbler 432
Langflossen-Makohai 21
Langflossen-Schelfkalmar 1424
Langflossen-Stachelmakrele 642
Langflossenthun 1007
Langkopf-Umberfisch 751
Langschnauzen-Delphin 1475
Langschnauzen-Eisfisch 899
Langschnauziger Speerfisch 1026
Langschwanz 494
Langschwänze, Tiefsee- 496
Langschwänziger Speerfisch 1024
Langschwanz-Thun 1013
Languste, Afrikanische 1262
Languste, Amerikanische 1269
Languste, Australische 1270
Languste, Edwards 1260
Languste, Europäische 1266
Languste, Gilchrists 1267
Languste, Japanische 1272
Languste, Juan Fernandez- 1261
languste, Kap- 1262
languste, Königs- 1273
Languste, Mauretanische 1268
Languste, Natal- 1265
Languste, Ostaustralische 1264
Languste, Pazifische 1275
Languste, Portugiesische 1268
Languste, Rote 1262
Languste, Transkei- 1271
Languste, Tristans 1263
Langusten 1259, 1274
Laternenfisch, Hektor- 295
Laternenfische 294
Lau 323
Laube 308, 309, 310
laube, Zwerg- 339
Lauel 310
Lauge 345
Laugele 310
Lavaret 234
Laxierfisch 812, 810
Lederschildkröte 1495
Leierfisch 941
Leierfisch, Europäischer 941
Leierfisch, Gefleckter 942
Leierfisch, Gemeiner 941
Leierhahn 1090
Leierknurrhahn 1090
Leistenkrokodil 1505
Leman-Felchen 233
Leng 462
leng, Blau- 460
Leng, Heller 462
Leng, Mittelmeer- 461
Lengfisch 462
Leopardhai, Kalifornischer 56
Leuchtgarnelen 1241
Leuchtkrebse und Krill 1203
Lichia 648
Limande 1142
ling, Bläu- 240
linge, Weiß- 625
Lippfisch, Amerikanischer 853
lippfisch, Amsel- 843
lippfisch, Augen- 848
Lippfisch, Augenfleck- 848
Lippfisch, Brauner 842
Lippfisch, Bunter 841
Lippfisch, Fünffleckiger 849
Lippfisch, Gefleckter 840
Lippfisch, Grauer 845
Lippfisch, Grüner 843
Lippfisch, Kuckucks- 841
lippfisch, Mittelmeer- 846
lippfisch, Pfauen- 851
lippfisch, Schnauzen- 850
Lippfisch, Westatlantischer 844
Lippfische 837
Lodde 273
Löffelstöre 140
Lotsenfisch 650
Louisiana-Flußkrebs 1284
Luftatmende Welse 388
Lumb 442
Lump 1105
Lumpfisch 1105
Lumpfische 1104
Lungenfisch, Australischer 127
Lungenfische, Afrikanische 129

Macdonalds Umberfische	724
Mactra, Glatte	1390
Madeira-Sardinelle	193
Madümäräne	234
Magellan-Miesmuschel	1371
Maiblecke	310
Maifisch	156, 157, 362
Maifisch, Amerikanischer	162
Maifisch, Gewöhnlicher	156
Maifische	163
Maiforelle	259
Mailing	268
Mairenke	322
Mako	20
Makohai, Langflossen-	21
Makrele	985
makrele, Australische Flecken-	996
makrele, Australische Gelbschwanz-	659
makrele, Bastard-	672
makrele, Blaue Bastard-	670
makrele, Blaue Stachel-	635
makrele, Brachsen-	680
makrele, Chilenische Bastard-	669
makrele, Chinesische Königs-	1003
makrele, Cunene-Bastard-	674
Makrele, Doppellinien-	972
makrele, Escolar-Schlangen-	948
Makrele, Europäische	985
makrele, Falsche Königs-	1000
makrele, Fregatt-	965
makrele, Gabel-	663
makrele, Gefleckte Königs-	994
makrele, Gefleckte Stachel-	643
makrele, Gelbe Stachel-	637
makrele, Gelbschwanz-	658
makrele, Gelbschwanz-Stachel-	641
Makrele, Gemeine	985
makrele, Gemeine Gold-	681
makrele, Gestreifte	993
Makrele, Gewöhnliche	985
makrele, Gold-	681
makrele, Goldene Stachel-	634
makrele, Große Gabel-	647
makrele, Grünrücken-Bastard-	665
makrele, Hai-	971
makrele, Indische Königs-	989
makrele, Indische Schweb-	1038
makrele, Indische Stachel-	644
makrele, Indische Zwerg-	977
Makrele, Indopazifische	983
makrele, Indopazifische Königs-	991
makrele, Insel-Zwerg-	976
Makrele, Japanische	997
makrele, Japanische Bastard-	666
Makrele, Kanadi-	998
makrele, Kap-Bastard-	673
makrele, Königinnen-Stachel-	653
makrele, Königs-	988
makrele, Königs-Schlangen-	949
Makrele, Koreanische	992
makrele, Kurze Zwerg-	975
makrele, Langflossen-Stachel-	642
makrele, Marleys Schlangen-	953
makrele, Melvera-Fregatt-	964
makrele, Mittelmeer-Bastard-	668
Makrele, Monterey-	990
makrele, Neuseeland-Stachel-	652
makrele, Ostatlantische Königs-	1004
makrele, Ostpazifische Königs-	1002
Makrele, Papua-	995
makrele, Pazifische Bastard-	671
makrele, Pferde-Stachel-	636
Makrele, Queensland-	999
makrele, Rauhe Bastard-	667
makrele, Regenbogen-Stachel-	646
makrele, Schlangen-	950
makrele, Schwanzfleck-Stachel-	640
makrele, Schwarzrücken-Stachel-	638
Makrele, Serra-	987
Makrele, Spanische	989
makrele, Tiger-	1001
makrele, Torpedo Stachel-	649
Makrelen	986
Makrelen und Thunfische	961
makrelen, Bastard-	632, 632, 675
makrelen, Brachsen-	679
makrelen, Butter-	952
Makrelen, Indische	978
makrelen, Indopazifische Zwerg-	978
makrelen, Königs-	1005
makrelen, Schlangen-	947
Makrelen, Spanische	1005
makrelen, Stachel-	632, 639, 645, 661
makrelen, Zwerg-	978
Makrelenhai	20
Makrelenhecht	422
Makrelenhecht, Atlantischer	422
Makrelenhecht, Kurzschnabel-	421
Makrelenhechte	420
makrelle, Gold-	323
Makrelthunfisch	965
Malabar-Zackenbarsch	579
Malermuschel	1345
Malermuscheln	1344
Malma-Saibling	265
Manati, Afrikanischer	1457
Manati, Fluß-	1455
Manati, Nagel-	1456
Mandschurei-Corvina	746
Mangrovenkrabbe	1302
Mangroven-Schnapper	688
Manta	112
Manta, Pazifischer	113
Mantas	111
Maräne	230, 231
Maräne, Amerikanische Kleine	231
maräne, Edel-	234, 237
Maräne, Große	234, 236
Maräne, Kleine	229
maräne, Madü-	234
maräne, Peipus-	237
maräne, Peled-	238
Maräne, Pollan-	230
maräne, Silber-	229
maräne, Strom-	268
maräne, Wander-	234
maräne, Zwerg-	229
Maränen	261
Marderhaie	49
Marenke	229
Marleys Schlangenmakrele	953
Marlin, Blauer	1019
Marlin, Gestreifter	1023
Marlin, Schwarzer	1018
Marlin, Weißer	1021
Marline	1020, 1027
Marmel-Zitterrochen	118
Marmorbarsch	885
Marmorbrasse	786
Marmorgrundel	1060
Marmorierte Meergrundel	1060
Marmorrochen	101
Marmor-Zitterrochen	118
Marokko-Zahnbrasse	777
Maß-Hecht	287
Masu-Lachs	250
Mauretanische Languste	1268
Meeraal	410
Meeraal, Amerikanischer	411
Meeraal, Argentinischer	412
Meeraal, Gemeiner	410
Meeraale	408
Meeräsche	540
Meeräsche, Afrikanische	544
Meeräsche, Bobo-	539
Meeräsche, Brasilianische	546
Meeräsche, Dicklippige	538
Meeräsche, Dünnlippige	542
Meeräsche, Gemeine	542
Meeräsche, Gestreifte	545
Meeräsche, Gewöhnliche	545
meeräsche, Gold-	540
Meeräsche, Graue	547
Meeräsche, Großkopf-	545
Meeräsche, Kleine	543
Meeräsche, So-iny-	541
meeräsche, Spring-	543
Meeräschen	537
Meerbarbe, Gestreifte	818
Meerbarbe, Gewöhnliche	817
Meerbarbe, Gold-	821
Meerbarbe, Rote	817
Meerbarbe, Steilstirnige	817
Meerbarbe, Westafrikanische	820
Meerbarben	815, 819, 822
Meerbarbenkönig	610
Meerbarsch	593
Meerbrasse	790
Meerbrasse, Blaugefleckte	804
Meerbrasse, Gelbrote	796
Meerbrasse, Großaugen-Angola-	703
Meerbrasse, Natal-	792
Meerbrasse, Nordische	790
Meerbrasse, Panga-	799
Meerbrasse, Rotgebänderte	803
Meerbrasse, Spanische	790
Meerbrassen	763, 789, 795, 807
Meerbrassen, Nordischer	793
Meerbrassen, Roter	794
Meerbraum	793
Meerdattel	1372
Meerdrossel	843
Meerengel	67
Meerengel, Gemeiner	68
Meeresweichtiere	1323
Meerforelle	257
Meergrundel, Große	1052
Meergrundel, Marmorierte	1060
Meerjunker	838
Meermandel	1384
Meernase	365
Meerneunauge	7
meerohr, Riesen-	1325
Meerohr, Rotes	1327
Meerohr, Südafrikanisches	1326
Meerohren	1329
Meerpfau	854
Meerpricke	7
Meerrabe	755
Meersau	1067, 70
Meersau, Europäische	1068
Meersäue	69
Meerscheide	1413
Meerscheiden	1407, 1414
Meerschlei	851
Meerschleie, Südliche	464
Meerstichling	502
Meertrüsche	463
Meerwelse	374

Melvera-Fregattmakrele	964	Monterey-Makrele	990	muschel, Scheiden-		1413
Membran-Geißelgarnele	1258	Moorkarpfen	321	muschel, Schinken-		1385
Menhaden	166, 167	moos, Irisch-	1532	muschel, Schwert-		1409
Menhaden, Golf-	165	Moos, Irisches	1532	muschel, See-		1377
Menhaden, Nordwestatlantischer	166	Moränen	241	muschel, Stachelige Herz-		1388
Menhaden, Pazifischer	175	Mordwal	1470	muschel, Steck-		1385
Menschenhai	19	Morwong, Bergs	857	muschel, Südatlantische Kamm-		1368
Menschenhai, Weißspitzen-	37	Morwong, Großflossen-	859	muschel, Taschenmesser-		1410
Merlan	454	Morwong, Prächtiger	858	muschel, Teich-		1343
Messerfisch	350, 919	Morwongs	860	muschel, Teppich-	1401, 1405
messerfisch, Scher-	855	Moschuskrake	1436	muschel, Venus-		1398
Messerfische, Fähnchen-	292	Moschuskraken	1437	muschel, Westatlantische Mies-		1382
Messergarnele	1254	Moschuspolyp	1436	muschel, Westpazifische Archen-		1339
Messergarnelen	1255	Müllers Zwergbutt	1176	Muscheln		1324
Messerkarpf	350	Mülpe	311	muscheln, Archen-		1342
Messermuschel	1410	muräne, Batavia-Hecht-	406	muscheln, Bart-		1373
Messermuschel, Große	1413	muräne, Hecht-	406	muscheln, Bohr-		1417
Messermuschel, Kap-	1412	Muräne, Mittelmeer-	404	muscheln, Dattel-		1417
Messermuschel, Pazifische	1411	Muräne, Westatlantische	403	muscheln, Dreiecks-		1406
Messerzahnaal, Guinea-	405	Muränen	402	muscheln, Enten-		1198
Messerzahnaale	407	Murray Steinkrabbe	1308	muscheln, Herz-		1387
Miesmuschel	1376	muschel, Atlantische Pilger-	1365	muscheln, Kamm-		1358
Miesmuschel, Australische	1378	muschel, Australische Jakobs-	1366	muscheln, Koffer-		1406
Miesmuschel, Chilenische	1374	muschel, Australische Mies-	1378	muscheln, Maler-		1344
Miesmuschel, Grüne	1380	muschel, Blaubart-	1377	muscheln, Mies-		1370
Miesmuschel, Koreanische	1375	muschel, Braune Venus-	1395	muscheln, Pilger-		1358
Miesmuschel, Magellan-	1371	muschel, Bunte Kamm-	1363	muscheln, Rote Archen-		1341
Miesmuschel, Mittelmeer-	1377	muschel, Calico-Pilger-	1360	muscheln, Scheiden-	1407, 1414
Miesmuschel, Neuseeland-	1381	muschel, Chilenische Mies-	1374	muscheln, Teppich-		1394
Miesmuschel, Rio-de-la-Plata-	1379	muschel, Felsen-Enten-	1197	muscheln, Venus-	1393, 1400
Miesmuschel, Westatlantische	1382	muschel, Flußperl-	1383	Muskellunge		289
Miesmuscheln	1370	muschel, Gemeine Bohr-	1416			
Migram	1169	muschel, Gemeine Samt-	1384			
Milchfisch	228	muschel, Gestreifte Venus-	1396	**N**		
Milchfische	227	muschel, Glatte Venus-	1395			
Mississippi-Alligator	1502	muschel, Goldene Teppich-	1404	Nacktsandaal		938
Mitsukuri-Umberfisch	741	muschel, Große Jakobs-	1365	Nadelfisch		930
Mittelländischer Zwergkalmar	1423	muschel, Große Kamm-	1365	Nadel-Hornhecht		418
Mittelmeer-Aszidie	1510	muschel, Große Messer-	1413	Nagelhai		71
Mittelmeer-Barrakuda	535	muschel, Große Perl-	1346	Nagel-Manati		1456
Mittelmeer-Bastardmakrele	668	muschel, Große Pilger-	1365	Nagelrochen		87
Mittelmeer-Finte	158	muschel, Große Sandklaff-	1415	Napfschnecken		1333
Mittelmeer-Gabeldorsch	464	muschel, Große Teppich-	1401	Narwal		1479
Mittelmeer-Glatthai	53	muschel, Grüne Mies-	1380	Nase	325, 365
Mittelmeer-Kaiserbarsch	521	muschel, Herz-	1389	nase, Meer-		365
Mittelmeer-Leng	461	muschel, Island-	1386	nase, Ruß-		365
Mittelmeerlippfisch	846	muschel, Isländische Kamm-	1362	nase, Weiße Stumpf-		800
Mittelmeer-Miesmuschel	1377	muschel, Jakobs-	1364	Näsling		365
Mittelmeer-Mönchsrobbe	1450	muschel, Japanische Archen-	1340	Näsling, Dalmatinischer		324
Mittelmeer-Muräne	404	muschel, Japanische Teppich-	1402	näsling, Elritzen-		326
Mittelmeer-Pfeilhecht	535	muschel, Japanische Venus-	1399	Näsling, Iberischer		327
Mittelmeersandaal	938	muschel, Kap-Messer-	1412	Näsling, Italienischer		328
Mittelmeer-Schlangenaal	414	muschel, Karibik-Pilger-	1361	Näsling, Südwesteuropäischer		329
Mittelmeer-Seenadel	506	muschel, Kleine Teppich-	1405	Nassau-Barsch		581
Mittelmeer-Sepiole	1420	muschel, Koreanische Mies-	1375	Natal-Languste		1265
Mittelmeer-Sternrochen	83	muschel, Kreuzmuster-Teppich-	1401	Natal-Meerbrasse		792
Mittelmeerstöcker	668	muschel, Magellan-Mies-	1371	Natal-Meerbrasse		792
Mittelmeerstör	133	muschel, Maler-	1345	Navaga		445
Mittelmeer-Strandkrabbe	1297	muschel, Messer-	1410	Navaga, Europäische		445
Mittelmeer-Trüsche	464	muschel, Mies-	1376	Navaga, Fernöstliche		444
Moçambique-Buntbarsch	832	muschel, Mittelmeer-Mies-	1377	Nerfling		343
Moderlieschen	339	muschel, Neuseeländische Jakobs-	1367	Nesling		344
Mojarra, Schwarzflossen-	902	muschel, Neuseeland-Mies-	1381	Neuguinea-Krokodil		1504
Mojarras	903	muschel, Nördliche Venus-	1398	neunauge, Bach-		6
Mölenke	321	muschel, Pazifische Messer-	1411	Neunauge, Donau-		2
Mönchfisch	836	muschel, Pfahl-	1376	neunauge, Fluß-		5
Mönchsrobbe, Mittelmeer-	1450	muschel, Pilger-	1364	neunauge, Gewöhnliches Bach-		6
Mondfisch	1188	Muschel, Quahog-	1398	neunauge, Meer-		7
mondfisch, Flecken-	678	muschel, Riesentrog-	1392	neunauge, Ukrainisches Bach-		3
Mondfisch, Langer	1189	muschel, Rio-de-la-Plata-Mies-	1379	neunauge, Vladikov-Bach-		4
Mondfisch, Spitzschwanz-	1187	muschel, Riesen-Trog-	1392	Neunaugen		1
Mondfisch, Stutz-	1189	muschel, Sachalin-Trog-	1391	Neunaugenkönig		7
Monokel-Brassen	698	muschel, Sand-	1415	Neunstachliger Stichling		501
Monrovia-Doktorfisch	946	muschel, Sandklaff-	1415	Neuseeland-Aal		398

Neuseeland-Eisfisch	438	Orfe	343	Pelz-Seezunge	1162
Neuseeland-Flußbarsch	864	orfe, Gold-	343	Pennahia, Silberne	744
Neuseeländische Jakobsmuschel	1367	Osmerus-Stinte	277	Perel	5
Neuseeländischer Grenadier	480	Ostatlantische Gabeldorsche	465	Pergel	157
Neuseeland-Miesmuschel	1381	Ostatlantische Königsmakrele	1004	perlmuschel, Fluß-	1383
Neuseeland-Plattauster	1355	Ostaustralische Languste	1264	Perlmuschel, Große	1346
Neuseeland-Stachelmakrele	652	Ostküsten-Seezunge	1154	Perlon	10
Nilbarsch	565	Ostpazifische Königsmakrele	1002	Perpel	268
Nilbarsche	566	Ostseeschnepel	234	Perschke	617
Nil-Buntbarsch	833			Peruanischer Felsbarsch	584
Nilflösselhecht	128			Peruanischer Geigenrochen	75
Nordamerikanische Aalmutter	922	**P**		Peruanischer Umberfisch	723
Nordamerikanische Brasse	809			Peru-Sardelle	220
Nordamerikanische Knurrhähne	1087	Pampano, Gemeiner	662	Peterfisch	525
Nordamerikanischer Flußhering	161	Pampanos	664	Petermännchen	868, 869, 871
Nordamerikanischer Seehecht	483	Pampel, Chinesische	1032	Petermännchen, Großes	869
Nordamerikanisches Felchen	232	Pampel, Gemeine	1037	Petermännchen, Kleines	867
Nordatlantik-Grenadier	497	Pampel, Silberne	1031	petermännchen, Strahlen-	870
Nordische Garnele	1237	Pampelfisch	1037	petermännchen, Zwerg-	867
Nordische Meerbrasse	790	Pandora, Rote	791	Petersfisch	525
Nordischer Meerbrassen	793	Panga-Meerbrasse	799	Petersfisch, Amerikanischer	524
Nordischer Sandaal	935	Panizza-Grundel	1056	Petersfische	523
Nordischer Schleimkopf	519, 519	Panzerfisch	1106	Petersmann	869
Nordkaper	1490	Panzerhahn	1106	Peto	962
Nördliche Rosa Geißelgarnele	1217	Panzerkrebse	1285	Pfaffenlaus	614
Nördliche Steinkrabbe	1307	Papageien-Fahnenbarsch	570	Pfahlmuschel	1376
Nördliche Venusmuschel	1398	Papua-Makrele	995	Pfauenlippfisch	851
Nördliche Weiße Geißelgarnele	1227	Parona	651	Pfauenschleimfisch	907
Nördlicher Entenwal	1460	Patagonischer Glatthai	54	Pfeifenfisch	508, 1090
Nördlicher Glatthai	52	Patagonischer Grenadier	479	Pfeiferl	623
Nördlicher Kugelfisch	1185	Patagonischer Seehecht	486	Pfeilhecht, Atlantischer	532
Nördlicher Kurzflossenkalmar	1430	Patagonischer Zahnfisch	880	Pfeilhecht, Gelbflossen-	533
Nördlicher Roter Thun	1012	Pazifische Anchoveta	212	Pfeilhecht, Indischer	534
Nördlicher Schnapper	690	Pazifische Bastardmakrele	671	Pfeilhecht, Mittelmeer-	535
Nördlicher Silberdorsch	449	Pazifische Felsenauster	1347, 1348	Pfeilhechte	529, 536
Nördlicher Ziegenfisch	816	Pazifische Languste	1275	Pfeilkalmar	1433
Nordmeergarnele	1237	Pazifische Messermuschel	1411	Pfeilzahn-Heilbutt	1124
Nordpazifischer Seehecht	490	Pazifische Plattauster	1354	Pfeilzahnheilbutt, Amerikanischer	1125
Nordseeschnäpel	237	Pazifische Sardine	196	Pfeilzahnheilbutt, Asiatischer	1124
Nordwestatlantische Sardelle	210	Pazifische Scholle	1135	Pferdekopf, Afrikanischer	656
Nordwestatlantischer Menhaden	166	Pazifische Seezunge	1152	Pferdekopf, Atlantischer	657
Norwegischer Krill	1205	Pazifische Thonine	967	Pferdeköpfe	676
Norwegischer Schlankhummer	1320	Pazifischer Butterfisch	1034	Pferde-Stachelmakrele	636
Norwegischer Zwergbutt	1171	Pazifischer Fächerfisch	1017	Pflugnasenchimären	126
Notothenia, Augenstreifen-	884	Pazifischer Fadenhering	185	Pfrille	352
Notothenia, Blaue	889	Pazifischer Fleckenbarsch	575	Pierling	352
Notothenia, Gelbbauch-	882	Pazifischer Heilbutt	1134	Pigo	358
Notothenia, Gelbflossen-	887	Pazifischer Hering	170	Pikarel	813
Notothenia, Gestreifte	888	Pazifischer Heringshai	22	Pikarel, Schwanzfleck-	811
Notothenia, Graue	886	Pazifischer Kabeljau	446	Pikarels	814
Notothenia, Grüne	883	Pazifischer Kleinhering	194	Pilchard	189
Notothenia, Günthers	891	Pazifischer Manta	113	Pilgermuschel	1364
Notothenia, Kurzschwanz-	890	Pazifischer Menhaden	175	Pilgermuschel, Atlantische	1365
Notothenia, Ramsay's	892	Pazifischer Peitschenrochen	104	Pilgermuschel, Calico-	1360
Notothenia, Schwarze	881	Pazifischer Polardorsch	470	Pilgermuschel, Große	1365
Nototheniiden, Hochantarktische	894	Pazifischer Pollack	470	Pilgermuschel, Karibik-	1361
Nufar-Seebrasse	768	Pazifischer Rochen	100	Pilgermuscheln	1358
		Pazifischer Rotbarsch	1070	Pilotfisch	650
		Pazifischer Sandaal	934	Plattauster, Chilenische	1351
O		Pazifischer Seehecht	485	Plattauster, Neuseeland-	1355
		Pazifischer Segelfisch	1017	Plattauster, Pazifische	1354
Ochsenauge	149, 766	Pazifischer Spitzmaulhai	43	Plattaustern	1356
Ochsenfrosch, Amerikanischer	1492	Pazifischer Stechrochen	104	Plattkopf, Brasilianischer	873
Ogac	448	Pazifischer Taschenkrebs	1290	Pleinzen	304, 306
Ogosch	620	Pazifisches Glasauge	280	Pliete	319
Ohrenrobben	1440	Peipusmaräne	237	Plötze	360
Ohrensardine	190	Peitschenrochen, Pazifischer	104	Plötze, Albanische	348
Oktopusartige	1434	Peking-Brachsen	349	plötze, Schwarzmeer-	357
Ölfisch	950	Pelamide	982	Plötze, Südeuropäische	359
Olivgrüne Flunder	1120	Pelamide, Chilenische	980	Polardorsch	441
Olivgrüner Snook	561	Pelamide, Ungestreifte	974	Polardorsch, Pazifischer	470
Orangeroter Feilenfisch	1181	Pelamide, Westpazifische	981	Polarscholle	1140
Orbignys Dosinia	1397	Peledmaräne	238	Pollack	466
Oreos	528	Pellona, Indische	188	Pollack, Alaska-	470

DE

455

Pollack, Pazifischer	470
Pollan-Maräne	230
Pomfret, Schwarzer	677
Ponyfisch, Klunzingers	700
Ponyfische	699, 701
Porgys	767
Portugiesische Auster	1347
Portugiesische Languste	1268
Potkarpfe	321
Pottwal	1462
Prächtiger Morwong	858
Prächtiger Umberfisch	720
Pricke	5
pricke, Fluß-	5
pricke, Meer-	7
Punkt-Umberfisch	731
Purpurforelle	251
Putjekanipa, Batavia-	406

Q

quabbe, Frosch-	468
Quahog-Muschel	1398
Quallen, Wurzelmund-	1512
Quappau	452
Quappe	452
quappe, Frosch-	468
quappe, Fünfbärtelige See-	443
quappe, See-	450
quappe, Vierbärtelige See-	469
Queensland-Makrele	999
Queise	869
queise, Viper-	867
queisen, Viper-	871
Quinnat	254

R

Raap	311
Radgarnele	1219
Ramsay's Notothenia	892
Rapfen	311
Rappe	311
Rattenschwänze	496
Raubwelse	388
Rauhe Bastardmakrele	667
Rauhe Kliesche	1136
Rauhe Scharbe	1132
Rauhe Scholle	1132
Rauher Knurrhahn	1080
Rauher Rochen	98
Reeves-Alse	206
Regenbogenforelle	251
Regenbogen-Rundhering	173
Regenbogen-Stachelmakrele	646
Regenbogen-Stint	275
Reinanke	240
Reitblecke	308
Renke	237
Renken	241, 261
Rheinank	237
Rheinanke	259
Rheinankel	365
Rhône-Streber	622
Ridder	360
Riemenfisch	510
Riemling	308
Riesenauster	1347
Riesenfadenfisch	554
Riesengarnele, Atlantische Rote	1234
Riesenhai	12
Riesenkalmar	1427
Riesenmeerohr	1325

Riesen-Seepocke	1200
Riesen-Totoaba	724
Riesentrogmuschel	1392
Riesen-Zackenbarsch	577
Rießling	345
Riffbarsche	835
Riffhai, Schwarzspitzen-	38
Ringelbrasse	778
Ringelrobbe, Eismeer-	1452
Rinkfische	955
Rio-de-la-Plata-Miesmuschel	1379
Rissos Delphin	1467
Ritter	263
Rizinusöl-Fisch	950
Robalos	562
robbe, Bart-	1448
robbe, Eismeer-Ringel-	1452
robbe, Kegel-	1449
robbe, Mittelmeer-Mönchs-	1450
robbe, Sattel-	1451
robben, Hunds-	1440, 1446
robben, Ohren-	1440
Rochen	80
Rochen im engeren Sinne	102
rochen, Adler-	108, 109
rochen, Afrikanischer Adler-	110
rochen, Atlantischer Stern-	97
rochen, Augenzitter-	120
rochen, Band-	82
rochen, Bänder-	101
rochen, Chagrin-	89
Rochen, Echte	81
rochen, Echte Zitter-	117, 121
rochen, Fleck-	93
Rochen, Geflecker	93
rochen, Geflecker Zitter-	120
rochen, Geigen-	72, 76, 77
rochen, Gemeiner Geigen-	76
rochen, Gewöhnlicher Adler-	109
rochen, Gewöhnlicher Stech-	106
rochen, Glatt-	84
rochen, Grönland-	99
rochen, Großer Teufels-	112
Rochen, Hellfleckiger	91
rochen, Igel-	88
rochen, Keulen-	87
Rochen, Kleinäugiger	91
rochen, Kleiner Teufels-	114
rochen, Kuckucks-	94
rochen, Marmel-Zitter-	118
rochen, Marmor-	101
rochen, Marmor-Zitter-	118
rochen, Mittelmeer-Stern-	83
rochen, Nagel-	87
Rochen, Pazifischer	100
rochen, Pazifischer Peitschen-	104
rochen, Pazifischer Stech-	104
rochen, Peruanischer Geigen-	75
Rochen, Rauher	98
rochen, Säge-	78, 79
rochen, Sand-	86
rochen, Schecken-	101
rochen, Schmetterlings-	115, 116
rochen, Schnabel-	89
rochen, Schwarzer Zitter-	119
rochen, Spiegel-	84
rochen, Spitz-	82, 96
rochen, Spitzschnauzen-	96
rochen, Stachel-	103
rochen, Stachelschwanz-Stech-	105
rochen, Stech-	103, 106
rochen, Stein-	87
rochen, Steken-	101
rochen, Stern-	83

rochen, Südlicher Geigen-	74
rochen, Teufels-	111
rochen, Vieräugiger Spiegel-	92
rochen, Violetter Stech-	107
rochen, Walker-	89
rochen, Weiß-	90
rochen, Weißflecken-Geigen-	73
rochen, Winter-	95
rochen, Zitter-	117
Rohrkarpfen	341
Rosa Garnele	1238
Rosa Geißelgarnele	1212
Rosa Geißelgarnele, Nördliche	1217
Rosa Geißelgarnele, Südliche	1224
Rosa Kingklip	925
Rosa Lachs	247
Rosa Zahnbrasse	775
Rosenbergs Süßwassergarnele	1247
Rossie, Große	1419
Rotalgen	1528
Rotauge	360
Rotaugen	361
Rotbarsch	1075
Rotbarsch, Akadischer	1073
Rotbarsch, Großer	1075
Rotbarsch, Pazifischer	1070
Rotbarsche	1077
Rotbart	818
Rotbrasse	790
Rotbrassen	794
Rote Archenmuscheln	1341
Rote Koralle	1514
Rote Languste	1262
Rote Meerbarbe	817
Rote Pandora	791
Rote Riesengarnele, Atlantische	1234
Rote Stumpfnasenbrasse	769
Rote Tiefseekrabbe	1321
Rötel	360, 362
Roter Drachenkopf	1068
Roter Drachenkopf, Kleiner	1066
Roter Fahnenbarsch	569
Roter Gabeldorsch	475
Roter Grouper	580
Roter Kalmar	1429
Roter Knurrhahn	1089
Roter Lachs	247, 252
Roter Meerbrassen	794
Roter Soldatenfisch	522
Roter Tai	805
Roter Thun	1012
Roter Thun, Nördlicher	1012
Roter Zackenbarsch	580
Rotes Meerohr	1327
Rotfeder	362
Rotfeder, Griechische	363
Rotfisch	242, 263
Rotflosser	362
Rotforelle	263
Rotgarnele, Argentinische	1256
Rotgarnele, Königs-	1257
Rotgebänderte Meerbrasse	803
Rothasel	362
Rotkarpfen	360, 362
Rotlachs	252
Rötling	569
Rotmeer-Hering	181
Rotschiedl	311
Rotschwanzgarnele	1225
Rotschwanzschnapper	692
Rottez	360
Rotzbarsch	614
Rotzunge	1128
Rotzunge, Echte	1128, 1142

Rufolken	452
Rufunkel	452
Rumänischer Steinbeißer	369
Rümpahen	352
Rundhering, Atlantischer	177
Rundhering, Gemeiner	178
Rundhering, Kalifornischer	176
Rundhering, Regenbogen-	173
Rundhering, Silberstreifen-	201
Rundheringe	179
Rundkopfdelphin	1467
Rundschuppen-Speerfisch	1025
Rüßling	344
Rußnase	365
Rutte	452

S

Säbelfisch	350
Säbelzahn-Sardelle, Atlantische	221
Sachalin-Trogmuschel	1391
Sackbrasse, Gewöhnliche	806
Sägebarsch	587, 590
Sägebarsch, Schwarzer	571
Sägebarsch, Weißer	597
Sägebarsche	567
Sägebauch, Atlantischer	520
Sägefisch	78, 79
Sägefisch, Westlicher	79
Sägegarnele	1249
Sägerochen	78, 79
Sägeschwanz	30
Sägezähnchen	1406
Saibling	263
saibling, Amerikanischer See-	266
saibling, Bach-	264
Saibling, Malma-	265
saibling, See-	263
saibling, Wander-	263
Saiblinge	267
Saiblinge, Lachse, Forellen und	262
Sako	250
Salm	256
Salmerin	263
Salmler i.e.S.	300
Samtmuschel, Gemeine	1384
Sandaal, Amerikanischer	932
Sandaal, Großer	939
Sandaal, Kleiner	936
Sandaal, Koreanischer	940
sandaal, Mittelmeer-	938
sandaal, Nackt-	938
Sandaal, Nordischer	935
Sandaal, Pazifischer	934
Sandaal, Westatlantischer	933
Sandaale	931, 937
Sandart	620
Sandbankhai	39
Sandbarsch	572, 620
Sandbarsche, Brasilianische	865
Sandbutt	1143, 1174
Sandfelchen	236
Sandfisch, Japanischer	863
Sandflachkopf	1096
Sandflunder	1149
Sandgarnele	1251
Sandgarnelen	1250
Sandgrundel	1059
Sandhai	13
Sandklaffmuschel	1415
Sandklaffmuschel, Große	1415
Sandküling	1059
Sandmuschel	1415

Sandrochen	86
Sandspierling	936, 939
Sandspierling, Großer	939
Sandspierlinge	931
Sandtigerhai	13
Sandzunge	1164
Sape	306
Sardelle	217
Sardelle, Amerikanische	219
Sardelle, Argentinische	214
Sardelle, Atlantische Säbelzahn-	221
Sardelle, Australische	215
sardelle, Breitstreifen-	209
Sardelle, Europäische	217
Sardelle, Japanische	218
Sardelle, Nordwestatlantische	210
Sardelle, Peru-	220
Sardelle, Südafrikanische	216
Sardelle, Tyulka-	171
Sardellen	208
Sardine	189
sardine, Asiatische Klein-	191
Sardine, Australische	198
sardine, Großkopf-	192
Sardine, Japanische	197
Sardine, Kalifornische	196
sardine, Ohren-	190
Sardine, Pazifische	196
Sardine, Südafrikanische	199
Sardine, Südamerikanische	200
Sardinelle	190
Sardinelle, Madeira-	193
Sardinellen	195
Sattelrobbe	1451
Sauger	301, 302
scallop, Atlantischer Tiefsee-	1369
Scampi	1320
Schafskopf-Brasse	764
Scharbe	1139
scharbe, Dogger-	1132
Scharbe, Rauhe	1132
Scharbenzunge	1132
Schattenfisch	718, 759
Schattenfisch, Argentina-	758
Schattenfische	760
Schaufelfadenfisch	1046
Schaufelkopfbarsche	562
Schaufelrüssler	140
Scheckenrochen	101
Scheefsnut	1169
Scheibenbauch, Großer	1108
Scheibenbauch, Westatlantischer	1107
Scheibpleinzen	305, 306
Scheidenmuschel	1413
Scheidenmuscheln	1407, 1414
Scheinschnapper	695, 697
Scheinschnapper, Goldener	696
Schelfkalmar, Langflossen-	1424
Schellfisch	453
Schemaja	322
Schermesserfisch	855
Schied	311
Schiffshalter, Kleiner	1178
Schiffskielgarnele	1223
schildkröte, Bastard-	1498
schildkröte, Leder-	1495
schildkröte, Suppen-	1497
schildkröte, Unechte Karett-	1496
schildkröten, Diamant-	1499
Schildzahnhai	14
Schill	620
Schillschrätzer	615
Schinkenmuschel	1385
Schläfergrundeln	1048

Schlaffke	305
Schlammbeißer	371
Schlammfisch, Amerikanischer	142
Schlammkarpfen	330
Schlammler	364
Schlammpeitzger	371
Schlangenaal, Kurzschnauziger	413
Schlangenaal, Mittelmeer-	414
Schlangenkopffisch, Gestreifter	558
Schlangenkopffisch, Indonesischer	557
Schlangenkopffische	559
Schlangenmakrele	950
Schlangenmakrele, Escolar-	948
Schlangenmakrele, Königs-	949
Schlangenmakrele, Marleys	953
Schlangenmakrelen	947
Schlanker Knochenhecht	143
Schlankhummer, Andamanen-	1319
Schlankhummer, Norwegischer	1320
Schlank-Ilisha	184
Schlankthun	963
Schlei	364
schlei, Meer-	851
Schleie	364
schleie, Grün-	364
schleie, Südliche Meer-	464
Schleiforelle	364
Schleihe	364
Schleimfisch, Atlantischer	908
Schleimfisch, Blutstriemen-	910
schleimfisch, Dreiflossen-	912
Schleimfisch, Fluß-	911
Schleimfisch, Gestreifter	909
schleimfisch, Hauben-	918
schleimfisch, Pfauen-	907
schleimfisch, Spitzkopf-	912
Schleimfisch, Süßwasser-	911
Schleimfische	904, 906
Schleimkopf, Nordischer	519, 518
Schleimköpfe	517
Schleischnäpel	234
Schlickauster	1415
Schlüpfling	364
Schmal	360
Schmalschnäuzige Seenadel	508
Schmerle	372
schmerle, Bach-	372
Schmerle, Gewöhnliche	372
Schmetterlingsfische	830
Schmetterlingsrochen	115, 116
Schmetterlings-Thunfisch	970
Schnabelbarsch	1076
Schnabelrochen	89
Schnabelwal, Cuvier-	1461
Schnäpel	237, 241, 261
schnäpel, Nordsee-	237
Schnapper	687, 693
schnapper, Gelbschwanz-	694
schnapper, Goldener Schein-	696
Schnapper, Mangroven-	688
Schnapper, Nördlicher	690
schnapper, Rotschwanz-	692
schnapper, Schein-	695, 697
Schnapper, Südlicher	691
Schnauzenbrassen	810, 814
Schnauzenlippfisch	850
schnecke, Gehörnte Turban-	1331
schnecke, Südafrikanische Turban-	1332
schnecke, Wellhorn-	1335
schnecken, Finger-	1338
schnecken, Flügel-	1338
schnecken, Helm-	1336
schnecken, Kreisel-	1330
schnecken, Napf-	1333

457

schnecken, Spinnen- 1338
schnecken, Stachel- 1334
schnecken, Strand- 1337
Schneider 308
Schneiderkarpfe 321
Schnepel 237
schnepel, Ostsee- 234
Schnepfenfisch 503, 504
Schnutt 344
Schokoladenhai 60
Scholle 1145
Scholle, Alaska- 1146
Scholle, Amerikanische 1129
Scholle, Großschuppige 1177
scholle, Heilbutt- 1131
Scholle, Japanische 1126
Scholle, Kalifornische 1127
Scholle, Pazifische 1135
scholle, Polar- 1140
Scholle, Rauhe 1132
Schollen 1123, 1150
Schraitzer 615
Schral 360
Schranzen 617
Schratz 615, 617
Schratzen 615
Schrätzer 615
schrätzer, Schill- 615
Schriftbarsch 589
Schroll 614
Schupper 310
Schusslaube 308
Schuster 364
Schwabe 304
schwalbenfisch, See- 1089
Schwall 360
Schwämme 1511
Schwanzfleck-Pikarel 811
Schwanzfleck-Stachelmakrele 640
Schwarzbarsch 602
Schwarze Dornhaie 62
Schwarze Notothenia 881
Schwarze Thonine 969
Schwarzer Crappie 605
Schwarzer Dornhai 61
Schwarzer Heilbutt 1148
Schwarzer Hundfisch 61
Schwarzer Karpfen 347
Schwarzer Marlin 1018
Schwarzer Pomfret 677
Schwarzer Sägebarsch 571
Schwarzer Seehecht 877
Schwarzer Zahnfisch 877
Schwarzer Zitterrochen 119
Schwarzer Zwergwels 384
Schwarzfisch 1040, 1041, 1042
Schwarzfisch, Antarktischer 1043
Schwarzfisch, Asiatischer 1044
Schwarzfleck-Grenadierfisch 493
Schwarzflossen-Mojarra 902
Schwarzflossen-Silberling 902
Schwarzflossenthun 1009
Schwarzforelle 258
Schwarzgrundel 1053
Schwarzküling 1053
Schwarzlachs 257
Schwarzmaul-Umberfisch 721
Schwarzmeerplötze 357
Schwarzmeer-Steinbutt 1173
Schwarzmeer-Wittling 455
Schwarzmerfling 343
Schwarzrand-Seezunge 1163
Schwarzreuter 263
Schwarzrücken-Stachelmakrele 638

Schwarzspitzen-Riffhai 38
Schwebforelle 259
Schwebmakrele, Indische 1038
Schwebrenke, Große 234, 240
Schwebrenke, Kleine 237
Schweinhai 70
Schweinsdrückerfisch 1180
Schweinsfisch 702
Schweinswal 1471
schweinswal, Dall-Hafen- 1472
Schwertfisch 1029
Schwertfische 1028
Schwertmuschel 1409
Schwertmuschel, Amerikanische 1408
Schwertschwanz, Atlantischer 1515
Schwertstör 141
Schwertstör, Chinesischer 141
Schwertwal 1470
Schwertwal, Kleiner 1473
Schwertwal, Unechter 1473
Schwimmender Kopf 1189
Schwimmkrabbe, Blaue 1295
Schwimmkrabbe, Gazami- 1300
Schwimmkrabbe, Wollige 1299
Schwimmkrabben 1301
Schwuppe 304
Scotia-See-Eisfisch 896
Sechskiemer 11
Seeaal 410
Seebader 945
Seebär, Südafrikanischer 1442
Seebär, Südamerikanischer 1441
Seebarsch 593
Seebarsche 595
Seebrasse, Nufar- 768
Seebrasse, Tischler- 765
Seebrassen 770, 784, 789, 795
Seebull 1102
Seedattel 1372
Seefeige 1508
Seeforelle 259
Seeforelle, Amerikanische 266
Seegurke, Japanische 1523
Seehahn, Gestreifter 1091
Seehase 1105
Seehasen 1104
Seehecht 487
Seehecht, Argentinischer 486
Seehecht, Benguela 489
Seehecht, Chilenischer 485
Seehecht, Europäischer 487
Seehecht, Großäugiger 481
Seehecht, Nordamerikanischer 483
Seehecht, Nordpazifischer 490
Seehecht, Pazifischer 485
Seehecht, Schwarzer 877
Seehecht, Senegalesischer 491
Seehecht, Südlicher 482
seehecht, Tiefenwasser-Kap- 488
Seehechte 478, 492
Seehund 1453
Seeigel 1519, 1520
Seeigel, Chilenischer 1522
Seeigel, Eßbarer 1521
seeigel, Stein- 1518
Seejungfrau 1450
Seekarpfen 808
Seekatze 123
Seekatzen 122
Seekuckuck 1079
seekuh, Gabelschwanz- 1454
Seelachs 467
Seelachs, Heller 466
Seelamprete 7

Seelen 240
Seelöwe, Stellers 1443
Seelöwe, Stellerscher 1443
Seelöwe, Südamerikanischer 1444
Seemönch 1450
Seemuschel 1377
Seenadel, Große 506
Seenadel, Kleine 507
Seenadel, Mittelmeer- 506
Seenadel, Schmalschnäuzige 508
Seeohr 1328
Seeohr, Gemeines 1328
Seepapagei 856
Seepferdchen 505
Seepocke, Riesen- 1200
Seepocken 1199
Seequappe 450
Seequappe, Dreibärtelige 450
Seequappe, Fünfbärtelige 443
Seequappe, Vierbärtelige 469
Seequappen, Dreibärtelige 451
Seerabe 755
Seeratte 123
Seeratten 122
Seerüßling 365
Seesaibling 263
Seesaibling, Amerikanischer 266
Seescheiden 1507
Seeschmetterling 905
Seeschwalbenfisch 1089
Seeskorpion 1100, 1102
Seeskorpion, Amerikanischer 1098
Seeskorpion, Vierhörniger 1099
Seeskorpione 1101
Seespinne 1315
Seespinne, Arktische 1312
Seespinne, Große 1315
Seespinnen, Arktische 1313
Seestern, Gemeiner 1517
Seesterne 1516
Seestichling 502
Seestint 274
Seestint, Kleinmäuliger
  Kalifornischer 272
Seetang 1527
Seeteufel 1192, 1195
Seeteufel, Amerikanischer 1193
Seeteufel, Atlantischer 1195
Seeteufel, Europäischer 1195
Seewolf 915
Seewolf, Blauer 914
Seewolf, Gefleckter 916
Seewolf, Gestreifter 915
Seewölfe 913, 917
Seezunge, Augen- 1158
Seezunge, Augenfleck- 1158
Seezunge, Azevia- 1159
Seezunge, Cuneata- 1157
Seezunge, Gemeine 1165
Seezunge, Ostküsten- 1154
Seezunge, Pazifische 1152
Seezunge, Pelz- 1162
Seezunge, Schwarzrand- 1163
Seezunge, Warzen- 1164
Seezunge, Westküsten- 1153
seezunge, Zwerg- 1156
Seezungen 1151, 1161
Seezungen, Südost-Atlantik- 1155
Segel-Feilenfisch 1183
Segelfisch, Atlantischer 1016
segelfisch, Faden- 389
Segelfisch, Indopazifischer 1017
Segelfisch, Pazifischer 1017
Segelfische 1015

DE

Seiwal	1481	Spitzmaulhai, Pazifischer	43	Stellers Seelöwe	1443
Selar, Goldband-	655	Spitzmaul-Knurrhahn	1088	Stellerscher Seelöwe	1443
Selar, Großäugiger	654	Spitzpleinzen	304	Sterlet	135
Semling	317	Spitzrochen	82, 96	Sternflunder	1144
Senegalesischer Seehecht	491	Spitzschnauzen-Delphine	1458	Sterngucker	874
Senegal-Umberfisch	750	Spitzschnauzenrochen	96	Sternhausen	136
Sepia	1418	Spitzschwanz-Mondfisch	1187	Sternrochen	83
Sepie	1418	Spöke	123	Sternrochen, Atlantischer	97
Sepiole, Mittelmeer-	1420	Spöke, Amerikanische	124	Sternrochen, Mittelmeer-	83
Seriola	661	Sporn	304	Stichling, Dreistachliger	500
Seriola, Japanische	660	Springer	268	Stichling, Gemeiner	500
Serra-Makrele	987	Springmeeräsche	543	stichling, Meer-	502
Shiba-Geißelgarnele	1208	Sprott	202	Stichling, Neunstachliger	501
Siam-Krokodil	1506	Sprotte	202	stichling, See-	502
Sibirischer Huchen	242	Stachelferzer	617	stichling, Zwerg-	501
Sichelfisch	350	Stachelhai	71	Stichlinge	499
Sichelflosser, Afrikanischer	827	Stachelige Herzmuschel	1388	Stilett-Garnele, Argentinische	1206
Sichelflosser, Geperlter	828	Stachel-Knurrhahn	1085	Stint	274
Sichling	350	Stachelmakrele, Blaue	635	Stint, Asiatischer	276
Silberbarsch	605	Stachelmakrele, Gefleckte	643	stint, Binnen-	274
Silberdorsch	449	Stachelmakrele, Gelbe	637	stint, Heil-	274
Silberdorsch, Nördlicher	449	Stachelmakrele, Gelbschwanz-	641	Stint, Regenbogen-	275
Silberfelchen	237	Stachelmakrele, Goldene	634	stint, See-	274
Silberfisch, Antarktischer	893	Stachelmakrele, Indische	644	Stint, Süßwasser-	271
Silberforelle	259	Stachelmakrele, Königinnen-	653	stint, Wander-	274
Silberhecht	483	Stachelmakrele, Langflossen-	642	Stintdorsch	471
Silberkarausche	320	Stachelmakrele, Neuseeland-	652	Stinte	270
Silberlachs	249, 257	Stachelmakrele, Pferde-	636	Stinte, Osmerus-	277
Silberling, Schwarzflossen-	902	Stachelmakrele, Regenbogen-	646	Stöcker	672
Silberlinge	903	Stachelmakrele, Schwanzfleck-	640	Stöcker, Blauer	670
Silbermaräne	229	Stachelmakrele, Schwarzrücken-	638	stöcker, Mittelmeer-	668
Silberne Pampel	1031	Stachelmakrele, Torpedo-	649	Stör	137
Silberne Pennahia	744	Stachelmakrelen	632, 639, 645, 661	Stör, Baltischer	137
Silberne Süßlippe	711	Stachelrochen	103	stör, Chinesischer Schwert-	141
Silberspitzen-Hai	35	Stachelschnecken	1334	Stör, Grüner	132
Silberstreifen-Rundhering	201	Stachelschwanz-Stechrochen	105	stör, Mittelmeer-	133
Skorpionsfische	1062	Stachliger Knurrhahn	1086	stör, Schwert-	141
Snoeks	952	Staffhering	157	Stör, Weißer	138
Snook, Olivgrüner	561	Stahlkopfforelle	251	Störe	130
Snooks	560, 562	Stechrochen	103, 106	störe, Löffel-	140
So-iny-Meeräsche	541	Stechrochen, Gewöhnlicher	106	Strahlenpetermännchen	870
Soldatenfisch, Roter	522	Stechrochen, Pazifischer	104	Strandauster	1415
Sommerflunder	1118	Stechrochen, Stachelschwanz-	105	Strandkrabbe	1298
Sompat-Süßlippe	713	Stechrochen, Violetter	107	Strandkrabbe, Mittelmeer-	1297
Sonnenbarsch	601	Steckmuschel	1385	Strandlachs	257
Sonnenfisch	525	Steilstirnige Meerbarbe	817	Strandschnecken	1337
Sonnenfischchen	339	Steinbarsch	600	Streber	623
South-Georgia-Eisfisch	901	Steinbeißer	370, 915	Streber, Rhône-	622
Spanfisch	511	Steinbeißer, Balkan-	367	Streifbarsch	617
Spanienkärpfling	514	Steinbeißer, Europäischer	370	Streifenbarbe	818
Spanische Makrele	989	Steinbeißer, Gold-	366	Streifenbarsch, Amerikanischer	596
Spanische Makrelen	1005	Steinbeißer, Großer	367	Streifenbarsch, Gefleckter	594
Spatenfische	825	Steinbeißer, Italienischer	368	Streifenbrasse	808
Speerfisch	1024	Steinbeißer, Rumänischer	369	Streifenbrassen	787
Speerfisch, Kurzschnauziger	1022	Steinbeißer, Venezianischer	373	Streifenfisch	551
Speerfisch, Langschnauziger	1026	Steinbutt	1172	Streifensüßlippe, Afrikanische	708
Speerfisch, Langschwänziger	1024	Steinbutt, Schwarzmeer-	1173	Streifling	308
Speerfisch, Rundschuppen-	1025	Steinbutte	1121	Streukatze	623
Speerfische	1027	Steindachner's Umberfisch	727	Strömer	329, 345
Spiegelrochen	84	Steindattel	1372	Strommaräne	268
Spiegelrochen, Vieräugiger	92	Steinforelle	258	Struffbutt	1143
Spierling	274	Steingarnele	1249	Strumpfbandfisch	957
spierling, Sand-	936, 939	Steingreßling	334	Stubben	240
spierlinge, Sand-	931	Steingründling	334	Stuhr	614
Spindelfisch	623	Steinkarpfe	321	Stumme Umberfische	736
Spinnenkrabbe, Südliche	1314	Steinköhler	466	Stumpfnase, Weiße	800
Spinnenqueise	868	Steinkrabbe, Große	1293	Stumpfnasenbrasse, Rote	769
Spinnenschnecken	1338	Steinkrabbe, Murray	1308	Stumpfnasenbrassen	770
Spitzbrasse	781	Steinkrabbe, Nördliche	1307	Stur	614
Spitzer	304	Steinkrebs	1280	Stutz-Mondfisch	1189
Spitzkopfschleimfisch	912	Steinpicker	1103	Subantarktische Steinkrabbe	1308
Spitzkopfsiebenkiemer	10	Steinrochen	87	Südafrikanische Marmorbrasse	785
Spitzlaube	310	Steinseeigel	1518	Südafrikanische Sardelle	216
Spitzmaulhai, Atlantischer	44	Stekenrochen	101	Südafrikanische Sardine	199

Südafrikanische Turbanschnecke	1332
Südafrikanischer Galjoen	824
Südafrikanischer Kingklip	926
Südafrikanischer Seebär	1442
Südafrikanisches Meerohr	1326
Südamerikanische Sardine	200
Südamerikanische Süßwassergarnele	1246
Südamerikanischer Königs-Umberfisch	732
Südamerikanischer Seebär	1441
Südamerikanischer Seelöwe	1444
Südatlantische Kammuschel	1368
Südeuropäische Plötze	359
Südliche Meerschleie	464
Südliche Rosa Geißelgarnele	1224
Südliche Spinnenkrabbe	1314
Südlicher Blauflossenthun	1010
Südlicher Geigenrochen	74
Südlicher Glatthai	53
Südlicher Glattwal	1489
Südlicher Schnapper	691
Südlicher Seehecht	482
Südlicher Wittling	458
Südost-Atlantik-Seezungen	1155
Südostpazifische Süßlippe	706
Südwesteuropäischer Näsling	329
Sumpf-Elritze	351
Sumpfkrebs	1277
Suppenschildkröte	1497
Süßlippe, Bastard-	712
Süßlippe, Gelbflossen-	707
Süßlippe, Gestreifte	704
Süßlippe, Silberne	711
Süßlippe, Sompat-	713
Süßlippe, Südostpazifische	706
Süßlippe, Westmediterrane	710
Süßlippen	705, 714
Süßwassergarnele, Rosenbergs	1247
Süßwassergarnele, Südamerikanische	1246
Süßwassergarnelen	1248
Süßwasser-Schleimfisch	911
Süßwasser-Stint	271
Süßwasser-Trommelfisch	716
Swaltküling	1053
Sydney-Felsenauster	1357

T

Tai, Roter	805
Tang	1526, 1527
tang, See-	1527
Tarbutt	1172, 1175
Tarpon	146
Tarpone	145, 147
Tarpun	150
Tarpun, Atlantischer	150
Tarpune	148
Taschenkrebs	1291
Taschenkrebs, Pazifischer	1290
Taschenkrebse	1287
Taschenmessermuschel	1410
Tautog	852
Teichforelle	251
Teichmuschel	1343
Teleskop-Kardinalfisch	611
Teppichmuschel	1401, 1405
Teppichmuschel, Goldene	1404
Teppichmuschel, Große	1401
Teppichmuschel, Japanische	1402
Teppichmuschel, Kleine	1405
Teppichmuschel, Kreuzmuster-	1401

Teppichmuscheln	1394
Terpug	1093
Terpug, Langer	1092
Teufelsrochen	111
Teufelsrochen, Großer	112
Teufelsrochen, Kleiner	114
Thonine	968
Thonine, Pazifische	967
Thonine, Schwarze	969
Thors-Lammzunge	1116
Thun	1006
Thun, Dickleibiger	1011
thun, Gelbflossen-	1008
thun, Großaugen-	1011
Thun, Großäugiger	1011
Thun, Großer	1012
Thun, Kleiner	968
thun, Langflossen-	1007
Thun, Langschwanz-	1013
Thun, Nördlicher Roter	1012
Thun, Roter	1012
thun, Schlank-	963
thun, Schwarzflossen-	1009
thun, Südlicher Blauflossen-	1010
Thun, Weißer	1007
Thunfisch	1012
Thunfisch, Gefleckter	968
Thunfisch, Gestreifter	973
Thunfisch, Gewöhnlicher	1012
thunfisch, Makrel-	965
Thunfisch, Schmetterlings-	970
Thunfisch, Weißer	1007
Thunfischartige	960
Thunfische	1014
Thunfische, Makrelen und	961
Tiefenbarsch	1075
Tiefenwasser-Kapseehecht	488
Tiefseedorsche	436
Tiefseegarnele	1237
Tiefseegarnele, Afrikanische	1233
Tiefseegarnele, Chile-	1235
Tiefseegarnelen	1232, 1239
Tiefseegarnelen im engeren Sinne	1240
Tiefseegarnelen, Kanten-	1236
Tiefseehummer	1320
Tiefseekrabbe, Rote	1321
Tiefseekrabben	1322
Tiefsee-Langschwänze	496
Tiefseescallop, Atlantischer	1369
Tigerbarsche	592
Tigergarnele, Braune	1218
Tigergarnele, Grüne	1226
tigerhai, Sand-	13
Tigermakrele	1001
Tintenfisch	1418, 1426, 1438
Tintenfisch, Gemeiner	1418, 1438
tintenfisch, Zwerg-	1420
Tischler-Seebrasse	765
Tobiasfisch	936
Tobieschen	936
Toli-Alse	207
Tolstolob, Edler	336
Tolstolob, Gewöhnlicher	335
Tomcod	457
Tomcod, Atlantischer	456
Torbutt	1172
Tork	614
Torpedo Stachelmakrele	649
Totenkopfchimäre	125
Totoaba, Riesen-	724
Transkei-Languste	1271
Treische	452
Tristans Languste	1263
Trogmuschel, Riesen-	1392

Trogmuschel, Sachalin-	1391
Trommelfisch	745
Trommelfisch, Süßwasser-	716
Trompeterfisch	862
Trompeterfische	861
Trüsche	452
Trüsche, Mittelmeer-	464
Tulzbull	1097
Tümmler	1477
Tümmler, Großer	1477
Tümmler, Kleiner	1471
Turbanschnecke, Gehörnte	1331
Turbanschnecke, Südafrikanische	1332
Turbot	1172
Turbutt	1172
Tyulka-Sardelle	171

U

Uckelin	310
Uhl	5
Uhlen	5
Ukelei	310
Ukelei, Weißer	309
Ukrainische Lamprete	3
Ukrainisches Bachneunauge	3
Umberfisch	759
Umberfisch, Afrikanischer	719
Umberfisch, Amerikanischer	733
Umberfisch, Antarktischer	754
Umberfisch, Asiatischer Gelber	722
Umberfisch, Atlantischer	738
Umberfisch, Augenfleck-	757
Umberfisch, Bobo-	749
Umberfisch, Boe-	753
Umberfisch, Coco-	743
Umberfisch, Gefleckter	725
Umberfisch, Gestreifter	728
Umberfisch, Golf-	734
Umberfisch, Hundszahn-	742
Umberfisch, Königs-	735
Umberfisch, Kurzkiefer-	748
Umberfisch, Langkopf-	751
Umberfisch, Mitsukuri-	741
Umberfisch, Peruanischer	723
Umberfisch, Prächtiger	720
Umberfisch, Punkt-	731
Umberfisch, Schwarzmaul-	721
Umberfisch, Senegal-	750
Umberfisch, Steindachner's	727
Umberfisch, Südamerikanischer Königs-	732
Umberfisch, Weißblumen-	740
Umberfisch, Weißer Kalifornia-	730
Umberfisch, Weißmaul-	737
Umberfische	715, 729, 739, 752, 756
Umberfische, Macdonalds	724
Umberfische, Stumme	736
Unechte Karettschildkröte	1496
Unechter Bonito	965
Unechter Schwertwal	1473
Ungestreifte Pelamide	974

V

Valencia Kärpfling	515
Venezianischer Steinbeißer	373
Venusmuschel	1398
Venusmuschel, Braune	1395
Venusmuschel, Gestreifte	1396
Venusmuschel, Glatte	1395
Venusmuschel, Japanische	1399
Venusmuschel, Nördliche	1398

Venusmuscheln	1393, 1400	Weißer Zackenbarsch	573	Wolfsheringe	224, 226
Victoria-See-Karpfen	355	Weißfelchen	236	Wollhandkrabbe, Chinesische	1303
Vielzähner	140	Weißfisch	360	Wollige Schwimmkrabbe	1299
Vieräugiger Spiegelrochen	92	Weißfische	303	Wrackbarsch	585
Vierbärtelige Seequappe	469	Weißflecken-Congeraal	409	Wrackbarsch, Atlantischer	585
Viereck	1175	Weißflecken-Geigenrochen	73	Wrackbarsch, Hapuku-	586
Vierfleckbutt	1168	Weißforelle	257, 258	Wurzelmund-Quallen	1512
Vierfleckflunder	1119	Weißhai	19		
Vierhörniger Seeskorpion	1099	Weißlinge	625	**Z**	
Violetter Stechrochen	107	Weißmaul-Umberfisch	737		
Viperqueise	867	Weißrochen	90	Zackenbarsch, Argentinischer	568
Viperqueisen	871	Weißschnauzen-Delphin	1469	Zackenbarsch, Brauner	577
Vladikov-Bachneunauge	4	Weißschnauziger Delphin	1469	Zackenbarsch, Goldener	574
Vladikov-Lamprete	4	Weißschnauziger Springer	1469	Zackenbarsch, Grauer	576
		Weißseiten-Delphin	1468	Zackenbarsch, Malabar-	579
		Weißspitzen-Menschenhai	37	Zackenbarsch, Riesen-	577
W		Weißwal	1478	Zackenbarsch, Roter	580
		Weitäugiger Butt	1117	Zackenbarsch, Weißer	573
Wachna	444, 445	Wellhornschnecke	1335	Zackenbarsche	582, 583
Wachnja	444, 445	Wellington-Flugkalmar	1431	zahnaal, Guinea-Messer-	405
Wahoo	962	Wels	382	zahnaale, Messer-	407
Wal, Baird-	1459	Wels, Amerikanischer	385	Zahnbrasse	774
wal, Blau-	1483	wels, Amerikanischer Zwerg-	385	Zahnbrasse, Angola-	771
wal, Bryde-	1482	Wels, Aristoteles-	381	zahnbrasse, Großaugen-	776
wal, Buckel-	1485	Wels, Elektrischer	394	Zahnbrasse, Kanarische	772
wal, Cuvier-Schnabel-	1461	wels, Fluß-	382	Zahnbrasse, Kongo-	773
wal, Dall-Hafenschweins-	1472	wels, Gefleckter Kreuz-	376	Zahnbrasse, Marokko-	777
wal, Enten-	1460	wels, Getüpfelter Gabel-	386	Zahnbrasse, Rosa	775
wal, Finn-	1484	wels, Glattmaul-Kreuz-	375	Zahnfisch, Antarktischer	878
wal, Gewöhnlicher Grind-	1466	wels, Katzen-	385	Zahnfisch, Schwarzer	877
wal, Grau-	1491	wels, Katzen-Kreuz-	377	Zahnfische	879
wal, Grind-	1466	wels, Schwarzer Zwerg-	384	Zahnkarpfen	512
wal, Grönland-	1487	wels, Zwerg-	385	Zährte	365
wal, Indischer Grind-	1465	Welse	380	Zalat	311
wal, Killer-	1470	Welse, Echte	380	Zander	620
wal, Kleiner Schwert-	1473	welse, Fiederbart-	293	Zander, Amerikanischer	621
wal, Mord-	1470	welse, Indische Glas-	379	Zander, Kanadischer	619
wal, Nar-	1479	welse, Katzen-	387	Zebrabrasse	779
wal, Nördlicher Enten-	1460	welse, Korallen-	378	Zebrahai	24
wal, Pott-	1462	welse, Kreuz-	374	Zebrakärpfling	513
wal, Schweins-	1471	Welse, Luftatmende	388	Zegel	623
wal, Schwert-	1470	welse, Meer-	374	Zeitasch	268
wal, Sei-	1481	welse, Raub-	388	Zentralamerikanische	
wal, Südlicher Glatt-	1489	welse, Zwerg-	383	Geißelgarnele	1228
wal, Unechter Schwert-	1473	Westafrikanische Ilisha	183	Zicke	350
wal, Weiß-	1478	Westafrikanische Meerbarbe	820	Ziege	350
wal, Zwerg-	1480	Westafrikanischer Anglerfisch	1196	Ziegelbarsch, Blauer	627
wal, Zwergglatt-	1488	Westafrikanischer Ebarme	1112	Ziegelbarsche	626
wale, Glatt-	1486	Westatlantische Alse	159	Ziegenbarsch	587
Walhai	27	Westatlantische Miesmuschel	1382	Ziegenfisch, Nördlicher	816
Walkerrochen	89	Westatlantische Muräne	403	Zierbarben	354
Waller	382	Westatlantischer Lippfisch	844	Zimtflunder	1122
Walroß	1445	Westatlantischer Sandaal	933	Zindel	624
Wandermaräne	234	Westatlantischer Scheibenbauch	1107	Zingel	624
Wandersaibling	263	Westküsten-Seezunge	1153	Zink	624
Wanderstint	274	Westlicher Sägefisch	79	Zinne	624
Warzen-Seezunge	1164	Westmediterrane Süßlippe	710	Zirrenkrake	1435
Wasserkatze	914	Westpazifische Archenmuschel	1339	Zirrenkraken	1437
Waxdick	131	Westpazifische Pelamide	981	Zitronenhai	41
weichtiere, Meeres-	1323	Wetterfisch	371	zitterrochen, Augen-	120
Weißblumen-Umberfisch	740	Wettling	352	Zitterrochen, Echte	117, 121
Weiße Geißelgarnele, Nördliche	1227	Wibling	352	Zitterrochen, Geflecker	120
Weiße Stumpfnase	800	Wildforelle	258	Zitterrochen, Marmel-	118
Weißer Atun	954	Wilsons Eisfisch	897	Zitterrochen, Marmor-	118
Weißer Crappie	604	Winterflunder, Amerikanische	1147	Zitterrochen, Schwarzer	119
Weißer Gabeldorsch	476	Winterrochen	95	Zobel	306
Weißer Heilbutt	1133	Wittling	454	Zollfisch	229
Weißer Kalifornia-Umberfisch	730	wittling, Blau-	459	Zope	304
Weißer Marlin	1021	Wittling, Blauer	459	Zumpel	309
Weißer Sägebarsch	597	Wittling, Schwarzmeer-	455	zunge, Bastard-	1160
Weißer Stör	138	Wittling, Südlicher	458	zunge, Hunds-	1128
Weißer Thun	1007	Witwen-Drachenkopf	1072	zunge, Lamm-	1115
Weißer Thunfisch	1007	Wolfsbarsch	593		
Weißer Ukelei	309	Wolfsfische	921		

DE

DE

Ελληνικό ευρετήριο

A

αγαρόφυκα	1529
άγγελοι	67
άγγελος	68
αγιού	269
αγκαθερά	499
αγκαθερό	500, 501
αγκαθερό της θάλασσας	502
αγκαθίτες	62
αγκαθίτης	58
αγκαθίτης του Ατλαντικού	59
αγκαθομεθύστρα	1388
αγριοκαρχαρίας	14
αγριόσαλπες	944
αγριοσαύριδο	667
αγριοτσιρώνι	362
αετοί	108
αετός	96, 109
αετόψαρα	108
αετόψαρο	109
αθερίνα	549, 550, 551
αθερίνες	548
ακανθόβατος	89
ακανθοδέλφινο	1475
ακερίνα	614
ακερίνα του Δούναβη	615
ασκάμι	1242
ακιπίσιος	137
ακτινόβατος	97
αλάια του Δούναβη	322
αλβουλίδες	151
αλεποκέφαλος	223
αλεπόσκυλος	17
αλεπόσκυλος μεγαλομάτης	16
αλεπόσκυλος του πελάγους	15
αλέτρι	9, 11
αλιγάτορας του Μισσισσιππή	1502
αλιγάτορες	1500
αλιδώνα	1435
αλιώτις	1328
αλουμινόψαρο	640
αμερικάνικα καπόνια	1087
αμερικάνικη χελώνα	1499
αμερικάνικο μυλοκόπι	733, 736
αμερικάνικο στρείδι	1349
αμερικάνικο χτένι	1361, 1369
αμερικάνικος αστακός	1269, 1317
αμερικάνικος σαργός	764
άμα	142
αμμοκοχύλι	1415
αμμόπερκα	572
αμμόπερκα της Βραζιλίας	865
αμμόχελα	931
αμμόχελο	935, 937, 938, 939
αμμόχελο της Αμερικής	932
αμμόχελο του Βορρά	933
αμμόχελο του Ειρηνικού	934
αμμόχελο της Κορέας	940
ανταρκτογαλέος	51
απογονίδες	609
αραπογαρίδα	1249
αραχνοκάβουρας της Νέας Ζηλανδίας	1314
αργυρόπερκα	605
αργυροσυναγρίδα	765
αριστίδες	1232

αρκτοσαλβελίνος	263
αρκτοσκόπος	863
αρκτοφάλαινα	1481
άσα	99
ασβεστοφύκη	1530
ασημένια λέστια	319
ασημένιος σολομός	249
ασημοβουτυρόψαρο	711
ασημοκυπρίνος	335, 355
ασημοσκιός	744
ασημόψαρα	955
ασημόψαρο	957, 958, 959
ασκίδια	1507, 1510
ασκίδια της Χιλής	1509
ασπιδοκαπόνι	1079
άσπρη αγριόσαλπα	943
ασπρίτσα	360
άσπρο μυλοκόπι	730
ασπρογόβαδο	311
ασπροδέλφινο	1478
ασπρολάβρακο	597
ασπρόπερκα	604
ασπροπλευροδέλφινο	1468
ασπροπόδαρη καραβίδα	1279
ασπρορυγχοδέλφινο	1469
ασπροσαύριδο	668
ασπροστούριονο	138
ασπρόψαρο	311, 628
αστακογαρίδα	1318
αστακοί	1278, 1259
αστακός	1263, 1266, 1267, 1271, 1274, 1318
αστακός της Αυστραλίας	1261, 1270
αστακός του Ακρωτηρίου	1262
αστακός του Νατάλ	1265
αστακός της Νέας Ζηλανδίας	1260
αστακός της Ωκεανίας	1264
αστακουδάκια	1275
αστερίας	1517
αστερίες	1516
αστρόβατος	83
αστρογαλέος	52
αστροκαλκάνι	1144
αστροστούριονο	136
αυλιάς	783
αυστραλογλώσσα	1154
αυστραλογλώσσες	1155
αυτί της θάλασσας	1326, 1328, 1329
αφρικανική μαρίδα	811
αφρικάνικο δρέπανο	827
αχιβάδα	1328, 1405
αχιβάδα της Αλάσκας	1403
αχιβάδες	1400
αχινοί	1520
αχινός	1518, 1519, 1521
αχινός της Χιλής	1522
αχινοσκυλόψαρο	71

B

βαβούκι	356
βάλανοι	1199
βαλιστές	1179
βαλιστής	1183, 1181, 1180, 1182
βαλτογαρίδες	1284
βαρελάκι	965, 964

βασιλάκης	527
βασιλιάς της ρέγγας	510
βασιλικά καβούρια	1311
βασιλικά σκουμπριά	1004
βασιλική ζαργάνα	1025, 1026
βασιλικό μυλοκόπι	726
βασιλικό μυστακόψαρο	555
βασιλικό σαυρίδι	653
βασιλικό σκουμπρί	988, 989, 998, 1000
βασιλικό στικτοσκουμπρί	991
βασιλικό φαγγρί	803
βασιλικός εσκολάρ	949
βασιλικός κάβουρας	1307, 1308, 1310
βασιλικός κάβουρας της Ανταρκτικής	1309
βασιλικός σκιός	732
βάτοι	102, 81
βάτος	82, 83, 84, 85, 86, 87, 88, 89, 90, 91, 92, 93, 94, 96, 97, 98, 99, 100, 101
βάτος του Ειρηνικού	100
βατράχια	1493
βατραχομπακαλιάρος	468
βατραχόψαρα	1191, 1192
βατραχόψαρο	1194, 1195
βελανίτσα	538
βελονίδα	422
βελονίδι	416
βελουδοκάβουρας	1299
Βερνάρδος ο ερημίτης	1286
βιολί	68
βλάχος	585
βλάχος της Νέας Ζηλανδίας	586
βλαχοσφυρίδα	576
βοϊδοκεφαλόψαρα	1101
βοϊδοκεφαλόψαρο	1097, 1100, 1102
βοϊδοκεφαλόψαρο της Αμερικής	1098
βοϊδοκεφαλόψαρο του Βορρά	1099
βοϊδόψαρο	302
βορεομπακαλιάρος	441
βούκινο	1335
βουτυρόψαρα	705, 714
βουτυρόψαρο	707, 710, 713
βουτυρόψαρο της Αφρικής	708
βουτυρόψαρο του Ειρηνικού	706
βραχοφιδόψαρο	372

Γ

γάδοι	435
γαϊδουροχτένι	1384
γαϊδουρόψαρα	451
γαϊδουρόψαρο	443, 1180
γαϊτανούρι	854
γαλάζια αντιμόρα	437
γαλάζια φάλαινα	1483
γαλάζιο μάρλιν	1019
γαλάζιος χορέγονος	234, 240
γαλάζιος μπακαλιάρος της Νέας Ζηλανδίας	864
γαλαθείδες	1306
γαλατόψαρο	228
γαλέοι	49, 55
γαλέος	50, 52, 53
γαλέος της Παταγωνίας	54
γάμπαρι	1220
γάμπαρι-τίγρης	1223, 1226

γαρίδα 1220, 1206, 1238
γαρίδα με κόκκινη ουρά 1225
γαρίδα με μουστάκια 1231
γαρίδα σίμπα 1208
γαρίδα της Αρκτικής 1237
γαρίδα της Γουινέας 1210
γαρίδα της Ιαπωνίας 1219
γαρίδα της Καλιφόρνιας 1215
γαρίδα της Κεντρικής Αμερικής 1228
γαρίδα-τίγρης 1218
γαρίδα των ποταμών
   της Αυστραλίας 1316
γαριδάκι 1249
γαριδάκια 1253
γαρίδες 1209, 1211, 1229, 1240,
   1241, 1245
γαρίδες ή καραβίδες
   των γλυκών νερών 1244
γαρίδες του Ειρηνικού 1230
γαρίδες των γλυκών νερών 1248
γάστρος 543
γάτοι 122
γάτος 30, 31, 32, 123
γατοστόματα 301
γατόψαρα 293, 380, 383, 387
γατοψαράκι 31, 32
γατόψαρο 384, 385, 386
γαύροι 208, 222
γαύρος 217
γαύρος της Αμερικής 210
γαύρος της Αργεντινής 214
γαύρος της Αυστραλίας 215
γαύρος της Ιαπωνίας 218
γαύρος του Ακρωτηρίου 216
γαύρος του Ειρηνικού 219
γαύρος του Περού 220
γελαδοσκιός 753
γενειοφόρος φώκια 1448
γερμανός 943
γιαπωνέζικη αχιβάδα 1399
γιαπωνέζικη καλογνώμη 1340
γιαπωνέζικη ολοθούρια 1523
γιαπωνέζικη χωματίδα 1136
γιαπωνέζικο καλκάνι 1131
γιαπωνέζικο κρεμμύδι 612
γιαπωνέζικο λαβράκι 599
γιαπωνέζικος αστακός 1272
γιαπωνέζικος ημίραμφος 431
γιαπωνέζικος σκιός 747
γιγαντογαρίδα των γλυκών
   νερών 1247
γιγαντοζαργάνα 418
γιγαντοζαργάνες 419
γιγαντόπερκα 564
γιγαντοροφός 578
γιγαντοροφός του Ειρηνικού 591
γιγαντοσκαρμός 391
γιγαρτίνα 1531
γιλάρι 929, 1092
γκίσου 153
γκουράμι 1045, 1046, 1047
γκρίζα γαρίδα 1213
γκρίζα φάλαινα
   της Καλιφόρνιας 1491
γκρίζα φώκια 1449
γκριζόβατος 84
γκριζογαλέος 53
γκριζοκαπόνι 1080
γλανίδι 381
γλανός 381
γλαντίκ 134
γλαυκοκαρχαρίας 42
γλήνι 364
γλινός 905, 907, 909, 910

γλώσσα 1115, 1156, 1157, 1158, 1162,
   1163, 1165, 1168, 1169
γλώσσα της Δυτικής Αυστραλίας 1153
γλωσσάκι 1160
γλωσσάκια 1161
γλώσσες 1123, 1151, 1166, 1170
γόπα 766
γοπόρεγγα 174
γουάχο 962
γουλιανός 382
γουργουρόψαρο 704
γουργουρόψαρο ματάς 703
γουρλομάτηδες 279, 298
γουρλομάτης 280, 282, 299, 389, 449
γουρλομάτης του Ατλαντικού 281
γουρλομάτης του Ειρηνικού 283
γουρουνομύτης 325
γουρουνόψαρα 69, 1186
γουρουνόψαρο 70, 844, 1185
γοφάρι 630
γοφάρια 629
γραμμοκαπόνι 1091
γραμμωτή γλώσσα 1152
γραμμωτή συναγρίδα 797, 798
γραμμωτό λαβράκι 598
γραμμωτό μάρλιν 1023
γραμμωτό μυλοκόπι 728
γραμμωτό σκουμπρί 993
γραμμωτός γαύρος 209
γραμμωτός οφιοκέφαλος 558
γραμμωτός σκιός 731
γρεναδιέροι 496, 498
γρεναδιέρος 497
γρεναδιέρος της Νέας Ζηλανδίας 479
γρεναδιέρος της Παταγωνίας 478
γρεναδιέρος των βράχων 494
γρέντζος 547
γριβάδι 332
γυαλιστερή 1395, 1390
γυαλιστερή της Αμερικής 1392
γύλος 838
γωβιοί 1049, 1055
γωβιός 1052, 1053, 1054, 1056,
   1057, 1059, 1061
γωβιός της Μαύρης Θάλασσας 1060
γωβιός του κανεστρίνι 1058
γωβιουδάκι άφια 1050

Δ
δάκτυλα 1417
δακτύλι 1372
δάκτυλο 1416
δαντελόστρειδο 1352
δεκάπτερο σαυρίδι 641
δελφίνι 1464
δελφίνι-πιλότος 1465
δελφίνια 1463, 1476
δερματοχελώνα 1495
διάβολος του Ειρηνικού 113
διαβολόψαρα 111
διαβολόψαρο 114
διαβολόψαρο του Ατλαντικού 112
δίγραμμο σκουμπρί 972
δίθυρα 1324
δίπνοος 127
δίπνοος της Αυστραλίας 127
δόλεμος 759
δράκαινα 867, 868, 869, 870
δράκαινες 866, 871
δρόγγος 410
δρομίτσα 359

δρομίτσες 361
δροσίνα 341
δροσίτης 50, 52, 53

Ε
ελεοτρίδοι 1048
εμμελιχθίδες 684
επερλάνοι 270, 277
επερλάνος 274, 278
επερλάνος της Αμερικής 275
επερλάνος της Αρκτικής 276
επερλάνος του Ειρηνικού 272
ερπετοκαρχαρίας καραγγινό 8
εσκολάρ 948
ευρωπαϊκή χωματίδα 1145
ευφαυσεώδη 1203
εφίπιδοι 825

Ζ
ζαγκέτα 1115, 1116, 1168, 1169, 1177
ζαγκέτα της Νορβηγίας 1171
ζαγκέτες 1170
ζαμπαρόλα 513
ζαμπαρόλα της Ισπανίας 514
ζαργάνα 416, 422
ζαργάνα του Ατλαντικού 417
ζαργάνες 415
ζεμπρακαρχαρίας 24
ζεμπρασαργός 779
ζεμπρασκυλόψαρο 24
ζίγγκελ 624
ξίφιοι 1458
ξίφιος 1461
ζόπα 304
ζουρνάς 515
ζύγαινα 47, 46
ζύγαινες 45, 48
ζωνοδέλφινο 1474

Η
ηλεκτρογατόψαρο 394
ηλιόπερκα 601
ημίραμφοι 427, 433
ημίραμφος 429
ημίραμφος της Αυστραλίας 428
ημίραμφος της Βραζιλίας 430

Θ
θαλάσσια αγελάδα 1455, 1456
θαλάσσια αγελάδα της Αφρικής 1457
θαλάσσια μαλάκια 1323
θαλάσσια χελώνα 1496, 1498
θαλάσσιες χελώνες 1494
θαλασσινά γατόψαρα 374
θαλασσινό γατόψαρο 375, 377
θαλάσσιο σαλιγκάρι 1108
θαλάσσιος κροκόδειλος 1505
θεραπονίδες 592
θράψαλο 1429, 1433
θράψαλο της Αργεντινής 1428
θράψαλο της Νέας Ζηλανδίας 1431
θράψαλο του Βορρά 1430
θράψαλο του Ειρηνικού 1432
θροφύκη 1528
θύμαλος 268

I

ινδική πεσκαντρίτσα	1196
ινδικό ταρπόνι	149
ιππόκαμποι	505
ιπτάμενο καλαμάρι	1427
ιριδίζουσα πέστροφα	251
ιριδίζων επερλάνος	275
ιστιοφόροι	1015
ιστιοφόρος του Ατλαντικού	1016
ιστιοφόρος του Ειρηνικού	1017
ιταλικός κάβουρας	1295

K

κάβουρας	1291, 1295, 1297
κάβουρας κολυμβητής	1294
κάβουρας της Αρκτικής	1312
κάβουρας της Κίνας	1303
κάβουρας του Ατλαντικού	1289
κάβουρας του Βορρά	1288
κάβουρας του Ειρηνικού	1290, 1292
κάβουρας των βράχων	1293
κάβουρας-γκαζάμι	1300
κάβουρες	1287, 1296
κάβουρες της Αρκτικής	1313
καβούρι	1515
καβουρομάνα	1315
καθρεπτόψαρο	521
καθρεπτόψαρο του Ατλαντικού	520
καϊμάν	1503
καλαμαράκι	1423, 1433
καλαμάρι	1424, 1426
καλαμάρι του Ακρωτηρίου	1425
καλαμάρια	1421, 1422
καλαμάρια της Νέας Ζηλανδίας	1434
καλαμίθρα	363
καλκάνι	1117, 1124, 1128, 1143, 1172, 1173, 1174, 1175, 1176
καλκάνι της Αμερικής	1129, 1174
καλκάνι της Αρκτικής	1140
καλκάνι της Ιαπωνίας	1126
καλκάνι της Καλιφόρνιας	1127
καλκάνι του Ειρηνικού	1125, 1135
καλκάνι του Καναδά	1132
καλκάνια	1110, 1123, 1167
καλκανόβατος	87
καλογνώμες	1341, 1342
καλογνώμη του Ειρηνικού	1339
καλογρίτσα	836
καπελάνος	273
καπίνες	837
καπόνι	1079, 1080, 1082, 1084, 1085, 1086, 1089, 1090, 1091
καπόνι του Ακρωτηρίου	1081
καπόνια	1078
καπόνια του Ινδικού	1083
καπροειδή	526
καραβίδα	1320
καραβίδα του Ινδικού	1319
καραγγινό	1532
καραγκιόζης	783
καρβούνι	968
καρέτα	1496
καρκινοειδή	1285
καρχαρίας	9, 10, 11, 12, 13, 14, 19, 20, 21, 22, 23, 26, 40, 42, 71
καρχαρίας ασπροπτερύγιος	35
καρχαρίας λεοπάρδαλη	56
καρχαρίες	18, 34, 57
καστανόψαρα	679
καστανόψαρο	680
κατέργαρος	855

κατουρλίδα	506, 507, 508
κατραβάνος	839
κατσαρίδα της θάλασσας	1202
κατσούλα	855
καφέ σκυλάκι	29
καψομούλα	588
κεντρακάνθινοι	810
κεντρόλοφοι	1040
κεντρόνι	58, 64, 65, 70
κεντροπομίδες	560
κερατάς	1103, 1106
κερατόψαρο	1106
κεφαλάς	793
κέφαλοι	537
κεφαλόπουλο	541, 545
κέφαλος	341, 342, 344, 545
κέφαλος μπόμπο	539
κέφαλος της Αφρικής	544
κέφαλος της Βραζιλίας	546
κέφαλος της Ιταλίας	345
κέφαλος της Μαύρης Θάλασσας	340
κηλιδογαϊδουρόψαρο	450
κηλιδοκαρχαρίας	25
κηλιδόμαυρος γρεναδιέρος	493
κηλιδόστικτη σφυρίδα	575
κηλιδόστικτο δρέπανο	828
κηλιδόστικτος σολομός	255
κιθάρα	73
κιθάρες	72
κιθαρόψαρο	74, 75, 76, 77
κιχλίδες	831
κίλκα	171
κινέζικος κυπρίνος	336
κιτρινονεμίπτερος	696
κιτρινόπερκα	616
κιτρινοπτέρυγο κοκκινόψαρο	1074
κίτρινος καρχαρίας	41
κίτρινος μπακαλιάρος	466
κίτρινος σκιός	746
κιτρινοσταύριδο	655
κλαρίας	388
κλένι	341, 342
κλουπεοειδή	154
κόγγρος	410
κοκκάλι	635, 636, 637, 639
κοκκάλι της Αλεξάνδρειας	633
κοκκάλι της Νέας Ζηλανδίας	652
κοκκάλια	632
κόκκινα καβούρια	1322
κόκκινες γαρίδες	1232
κόκκινη γαρίδα	1212, 1233, 1234
κόκκινη γαρίδα της Αμερικής	1257
κόκκινη γαρίδα της Αργεντινής	1256
κόκκινο αυτί της θάλασσας	1327
κόκκινο κοκκάλι	638
κόκκινο κοράλλι	1514
κοκκινοαστακός	1268
κοκκινογαριδάκια	1239
κοκκινόγαστρος	351, 352
κοκκινολαπίνα	846
κοκκινομάτα πέρκα	600
κοκκινομπαρμπούνια	822
κοκκινομπάρμπουνο	821
κόκκινος κάβουρας	1321
κόκκινος λουτιάνος	690, 691
κόκκινος μπακαλιάρος	475
κόκκινος μπακαλιάρος της Νέας Ζηλανδίας	438
κόκκινος μπέργξ	518
κόκκινος σολομός	252
κόκκινος ψευτοσαργός	857
κοκκινοφίδιο	925
κοκκινοφίδιο της Χιλής	927
κοκκινοφτέρα	362

κοκκινόχανος	569
κοκκινόψαρα	1077
κοκκινόψαρα του Ατλαντικού	1064
κοκκινόψαρο	522, 1070
κοκκινόψαρο της Αμερικής	1073
κοκκινόψαρο της Νορβηγίας	1075
κοκκινόψαρο του βυθού	1076
κοκκοαγκαθίτης	58
κολαούζος	650, 1178
κόλγκα	1371
κολησόψαρο	1178
κολιός	984
κολυμβητικά καβούρια	1301
κομήτης	646
κόμπια	631
κοντόρυγχη φρίσσα	168
κοντόρυγχο μάρλιν	1022
κοντόρυγχος αγκαθίτης	66
κοπανάκι	965
κοπάνι	964, 965, 974
κοράλλια	1513
κορδέλα	493, 494, 495, 497, 834
κορδέλες	496
κορέγονοι	241, 261
κορέγονος	230, 233, 235, 236, 238
κορέγονος του βυθού	239
κότα	527
κοτόψαρα	1104
κοτόψαρο	1105
κούκος	94, 1079, 1091
κουκουλοφόρος φώκια	1447
κούμου	1082
κουνουπόψαρο	516
κουτσομούρα	817
κόχο	249
κοχύλι	1394, 1404, 1406
κρεμμύδι	610
κρεμμύδια	609
κριλ της Ανταρκτικής	1204
κριλ της Νορβηγίας	1205
κροκοδείλι	941, 942
κροκόδειλοι	1501
κροκόδειλος	1503
κροκόδειλος της Νέας Γουινέας	1504
κροκόδειλος του Σιάμ	1506
κρυσταλλογαρίδα	1214
κρυσταλλόψαρο	930
κρυσταλλογατόψαρα	379
κυδώνι	1396, 1401
κυδώνι της Ιαπωνίας	1402
κυδώνια	1393
κυματόβατος	101
κυνηγός	658, 681
κυπρινόδοντοι	512
κυπρινοειδή	303
κυπρίνος	332
κυπρίνος του Χόβεν	338

Λ

λαβράκι	593, 594, 595
λαβράκι της Αμερικής	596
λακέρδα	973
λάμια	23
λάμια του Ειρηνικού	22
λαμινάριες	1526
λαμπούγγα	681
λάμπρενα	5, 7
λάμπρενα της Ουκρανίας	3
λάμπρενα του Βλαντύκωφ	4
λάμπρενα του Δούναβη	2
λάμπρενες	1
λάμπρινα	5, 7

λαπίνα   839, 841, 842, 843, 845, 846
      847, 848, 849, 850, 851
λαπίνα της Αμερικής   853
λασπογαρίδα   1258
λασποκάβουρας   1302
λασποκαβούρια   1304
λασποκυπρίνος   330
λασποκυπρίνος της Αμερικής   285
λασποκυπρίνος της Ευρώπης   284
λασποσαλούβαρδος   463
λεβέρα   905, 909, 910
λείο καλκάνι   1141
λειογναθίδες   699
λειψός   1063
λεκατίκι   968
λεμονόγλωσσα   1142
λέπας   1198
λεπιδοκαπόνι   1085, 1086
λεπιδοφόρα φρίσσα   165
λεπισόστεος   143, 144
λεπτότονος   963
λεπτουροσπαθόψαρο   959
λέστια   305
λέστια της Κίνας   346
λέστια του Δούναβη   306
λέστια του Πεκίνου   349
λέστιες   307
λευκό μάρλιν   1021
λευκό μυλοκόπι   750
λευκόβατος   82
λευκογαρίδα   1227
λευκοκορέγονος   229
λευκομυλοκόπι της Καλιφόρνιας   720
λευκός μπακαλιάρος   476
λευκόστομο μυλοκόπι   737
λιθοφάγος   1372
λιμάντα   1139
λιμνοκορέγονος   232
λιμνοπέστροφα   259, 264
λιμνοπέστροφα της Αμερικής   266
λιμνοπέστροφες   267
λιμνόρεγγα   231
λιοντάρι της θάλασσας   1443, 1444
λίτσα   647, 663
λίτσα με λέπια   649
λίτσα της Φλώριδας   662
λίτσες   648, 664
λουτιάνιδοι   687
λουτιάνοι   693
λουτιάνος   688, 689
λουτιάνος με κίτρινη ουρά   694
λουτσοζαργάνα   422
λουτσοζαργάνα του Ειρηνικού   421
λουτσοζαργάνες   420
λούτσοι   529
λούτσος   535
λοχίας   821
λοχίες   822
λυθρίνι   792, 794
λυθρίνι της Αφρικής   791
λυθρίνια   795
λυθρινίδες   762
λυκόδοντας   921
λυκόρεγγες   224, 225, 226
λυκόψαρα   913, 917
λυκόψαρο   914, 915
λυκόψαρο του Ατλαντικού   915
λύρα της θάλασσας   1305
λύτρι   10
λυχναρόψαρα   294
λυχναρόψαρο   295
λύχνοι   875
λύχνος   874

M

μαγιάτικο   658, 661, 717, 718
μαγιάτικο της Αυστραλίας   659
μαγιάτικο της Ιαπωνίας   660
μακροαγκαθής   61
μακρόπτερος ημίραμφος   432
μακροπτερυγοκαρχαρίας   37
μακρόσωμη σαρδέλα   184
μακρύουρος τόνος   1013
μακρυπτερυγο-ρυγχοκαρχαρίας   21
μαργαριτάρι   1383
μαργαριτοφόρο στρείδι   1346
μαρίδα   813
μαρίδες   814
μαρίτσα   313
μαρκόβα   911
μάρλιν   1020, 1027
μάρλιν της Μεσογείου   1024
μαρμαρομουδιάστρα   118
μαρσουάνος   1471
ματόβατος   92
μαυράκι   542
μαύρη γαρίδα   1207
μαύρη όρκα   1473
μαύρη τούρνα   290
μαύρη φάλαινα   1489, 1490
μαύρο αετόψαρο   109
μαύρο καρβούνι   969
μαύρο κρεμμύδι   611
μαύρο μάρλιν   1018
μαύρο μυλοκόπι   745
μαύρο σκυλόψαρο   60
μαύρο σπαθόψαρο   956
μαύρο χάλιμπατ   1148
μαυρογωβιός   1053
μαυροδέλφινο   1466
μαυροκυπρίνος   347
μαυρολαπίνα   848
μαυρομάτα   365
μαυρομουδιάστρα   119, 120
μαυρόπερκα   571
μαυροπτερυγοκαρχαρίας   38
μαύρος μπακαλιάρος   467, 877
μαύρος σολομός   254
μαυροσαύριδο της Αυστραλίας   670
μαυροσκυλόψαρο της Γροιλανδίας   63
μαυρόστομος σκιός   721
μαυροχειλού   842
μαυρόψαρα   1040
μαυρόψαρο   950, 1041
μαχαιρογαρίδα   1254
μαχαιρογαρίδες   1255
μαχαιρόψαρα   292
μεγαβάλανος   1200
μεγαλαλιώτις   1325
μεγάλο αυτί της θάλασσας   1325
μεγάλο χτένι   1364
μεγαλομάτης σούρος   654
μεγαλόστομη πέρκα   603
μεγαλόφθαλμοι   606, 608
μεγαλόφθαλμος σκιός   722
μεγαλόφθαλμος του Ειρηνικού   607
μεγαπτεροφάλαινα   1485
μέδουσες   1512
μεθύστρα   1389
μεθύστρες   1387
μελανόστομος   30
μελανούρι   788
μενάχεν   167
μένουλα   812
μερτζάνι   806
μικρή γυαλιστερή   1391
μικρό λυκόψαρο   916

μικροαμμόχελο   936
μικρόβατος   88
μικρολάμπρενα   6
μικροστιγματόβατος   91
μικρόστομη πέρκα   602
μικρόστομος επερλάνος   271
μοϊάρα   902
μοϊάρες   903
μονόκερως   1479
μονόχειροι   1179
μονόχειρος   1180
μορίδες   436
μοσχιοί   1437
μοσχιός   1436
μοσχοχτάποδα   1437
μοσχοχτάποδο   1435, 1436
μουγγρί   410
μουγγρί της Αμερικής   411
μουγγρί της Αργεντινής   412
μουγγρί της Γουινέας   405
μουγγρί της Ιαπωνίας   409
μουγγριά   408
μουδιάστρα   119, 121, 118, 120
μουδιάστρες   117
μουλίδες   815, 819
μούρεξ   1334
μουρμούρα   786
μουρμούρα της Αφρικής   785
μουρμούρες   787
μουρμουρίτσα   356
μουρούνα   132, 133, 134, 139, 460, 461
μουρούνα του Δούναβη   131
μουσμούλι   790
μουστακάτο   314, 318
μουστακογωβιός της Δαλματίας   312
μπακαλιαράκι   449
μπακαλιαράκι του Ατλαντικού   456
μπακαλιαράκια   457
μπακαλιάροι   440, 477
μπακαλιάροι
  της Ανταρκτικής   879, 894
μπακαλιάρος   447, 453, 481, 487, 492
μπακαλιάρος της Αγκόλας   489
μπακαλιάρος της Αλάσκας   470
μπακαλιάρος της Ανατολής   444
μπακαλιάρος της Ανταρκτικής   878
μπακαλιάρος της Αργεντινής   486
μπακαλιάρος της Αρκτικής   445
μπακαλιάρος της Αυστραλίας   482
μπακαλιάρος της Βραζιλίας   474
μπακαλιάρος της Γροιλανδίας   448
μπακαλιάρος της Νότιας Αφρικής   484
μπακαλιάρος της Σενεγάλης   491
μπακαλιάρος του Ακρωτηρίου   488
μπακαλιάρος του Ατλαντικού   483
μπακαλιάρος του Ειρηνικού   446, 490
μπακαλιάρος του Περού   485
μπαλάς   776
μπαμαλίδες   296
μπάμαλο   297
μπανάνα της θάλασσας   152
μπανανογαρίδα   1222
μπανανόψαρο   146
μπαρακούντα   529, 532
μπαρακούντα της Ανατολής   533
μπαρακούντα της Γουινέας   530
μπαρακούντα της Ινδίας   534
μπαρακούντα της Καλιφόρνιας   531
μπαρακούντες   536
μπαραμούντι   564
μπαρμπούνι   818
μπαρμπούνι της Ανατολής   821
μπαρμπούνι της Δυτικής Αφρικής   820
μπάφα   545

μπεκατσόψαρα 503
μπεκατσόψαρο 504
μπελοβίτσα 309, 310
μπελούγκα 139
μπερυτσίδες 517
μπέρυχες 519
μπίβα 253
μπιχίρ 128
μπλε γρεναδιέροι 480
μπλε φεγγαρόψαρο 678
μπλε τρομπέτα 862
μπλε τρυγόνα 107
μπρεγμακέρος 439
μπρένα 317
μπριάνα 314, 316, 318
μπριάνα της Ιάβας 353
μπριάνα της Ισπανίας 315
μπριάνες της Ασίας 354
μπρόσμιος 442
μπρότουλα με μουστάκια 924
μυδάκια 1373
μύδι 1376, 1377
μύδι της Αργεντινής 1379
μύδι της Αυστραλίας 1378
μύδι της Κορέας 1375
μύδι της Νέας Ζηλανδίας 1381
μύδι της Χιλής 1374
μύδι των βράχων 1382
μύδι των γλυκών νερών 1345
μύδια 1370
μύδια των γλυκών νερών 1344
μυλοκόπι 751, 760, 759
μυλοκόπι γκάπο 748
μυλοκόπι Μακ Ντόναλντ 724
μυλοκόπι μικρόδοντο 727
μυλοκόπι μπόμπο 749
μυλοκόπι της Αργεντινής 758
μυλοκόπι της Αφρικής 719
μυλοκόπι της Βραζιλίας 738
μυλοκόπι της Δ. Αφρικής 752
μυλοκόπι της Ιαπωνίας 740, 741
μυλοκόπι τίγρης 742
μυλοκόπι του Βορρά 735
μυλοκόπι του γλυκού νερού 716
μυλοκόπι του Ειρηνικού 739
μυλοκόπι του Κόλπου 734
μυλοκόπι του Περού 723, 743
μυλοκόπια 729
μυξινάρι 540
μυστακόψαρο γίγας 556
μυστακόψαρο του Ινδικού 554
μυτάκι 781

N

ναβάγκα 444, 445
νανοφάλαινα 1488
ναρβάλ 1479
νεμιπτερίδες 695
νεμίπτεροι 697
νηματόρεγγα του Ατλαντικού 186
νηματόρεγγα του Ειρηνικού 185
νονά 96
νοτοθένια 881, 882, 883, 884, 885, 886,
887, 888, 889, 890, 891, 892, 893
νοτοθενίδες 876
ντάμπα 823, 824
νταούκι 455
νταούκι του Ατλαντικού 454
ντόγκες 203
ντουγκόνγκο 1454

Ξ

ξανθόβατος 85
ξιφίας 1029
ξιφιοί 1028
ξιφιός 1029
ξυρήχι 133, 137

Ο

οκτάπους 1438
οξύρρυγχο καπόνι 1088
οξυρρυγχοκαρχαρίας
του Ατλαντικού 44
οξυρρυγχοκαρχαρίας
του Ειρηνικού 43
όρκα 1470
όρκυνος 1007
οσμερίδες 270
ούγαινα 781
ουρόστικτο τάρμποτ 1111
οφίδια 928
οφίδιο του Ακρωτηρίου 926
οφιοκέφαλοι 559
οφιοκέφαλος της Ινδονησίας 557
οχταπόδι 1435

Π

παγόψαρα 895
παγόψαρο 897, 901
παγόψαρο της Ανταρκτικής 898
παγόψαρο της Σκωτίας 896
παγόψαρο του Κάθλιν 900
παλαμίδα 973, 982
παλαμίδα με λέπια 970
παλαμίδα της Ανατολής 981
παλαμίδα της Αυστραλίας 979
παλαμίδα του Ειρηνικού 980
παλαμίδες 961
πάνγκα 799
παπαγάλος 855, 907
παπαγαλόψαρο 570
παπαλίνα 202
παπαλίνα της Μαύρης Θάλασσας 171
παπαλινάκι 201
παπάς 12
παπλωματάς 106
παρόνα 651
πατερίτσα 46, 47
πατερίτσες 48
παχύχελο 452
πελεκοκυπρίνος 350
πελόνα της Ινδίας 188
πέρδικα 589
πέρκα 588, 589
πέρκα της Ρουμανίας 618
πέρκα του Νείλου 565
πέρκα του Περού 584
πέρκες 590, 613
πέρκες του Νείλου 566
περκοφιδίδες 872
πεσκαντρίτσα 1194, 1195
πεσκαντρίτσα της Αμερικής 1193
πεσκαντρίτσες 1192
πέστροφα 251, 257, 258
πέστροφα της Αριζόνας 244
πέστροφα του Μεξικού 245
πέστροφες 260, 262
πεστροφοσσολομός 265
πεταλίδες 1333
πεταλούδα 115, 116, 321

Ρ

πετρογαρίδα 1252
πετροσαλούβαρδος 464
πησί 1114, 1117, 1175
πησί της Αμερικής 1174
πησιά 1167
πίγγες 583
πίνα 1385
πλακολεπιδόψαρα 626
πλακολεπιδόψαρο 627
πλατίτσες 361
πλατίτσα 308, 360
πλατίτσα της Αλβανίας 348
πλατίτσα της Μαύρης Θάλασσας 357
πλατίτσα του Δούναβη 358
πλατυκέφαλο καλκάνι 1130
πλατυκέφαλοι 1095
πλατυκέφαλος 1096
πλατυκέφαλος της Βραζιλίας 873
πλατυσέλαχο 115, 116
πλατύψαρα 1110
πολύδοντοι 140
πολύδοντος της Κίνας 141
πολυνημίδες 553
πολύπτερος 128
πομακεντρίδες 835
ποντίκι 462
ποντικόγαυρος 213
ποντικός 443, 463, 464
ποντικουρόψαρα 498, 496
ποντικουρόψαρο 493, 494, 495, 497
ποντικόψαρο 460, 461
ποταμογαρίδα της Αμερικής 1246
ποταμογαρίδα του Δούναβη 1277
ποταμόγαυρος 221
ποταμοκάβουρες της Αμερικής 1283
ποταμοκαραβίδα 1276, 1279, 1280
ποταμοκαραβίδα της Αμερικής 1282
ποταμοκαραβίδες 1278, 1281
ποταμολάβρακο 620
ποταμολάβρακο της Αμερικής 621
ποταμολάβρακο του Καναδά 619
ποταμολαύρακο 314
ποταμόπερκα 617
ποταμόρεγγα 161
ποταμοσαλάρα 911
πράσινη σμέρνα 403
πράσινη χωματίδα 1120
πρασινοαστακός 1273
πρασινοαθερίνα 552
πρασινογαρίδα 1221
πρασινοκάβουρας 1298
πρασινομάτηδες 298
πρασινομάτης 299
πρασινομύδι 1380
πρασινοχειλού 843
πρασινοχελώνα 1497
πράσινο σαυρίδι 665
πράσινο στουρνιόνι 132
πριακάνθιδοι 606
πριονόψαρα 78
πριονόψαρο 79
προβατόψαρο 922
προσκυνητής 12
προσφυγάκι 459
προσφυγάκι της Αυστραλίας 458
πρωτόπτεροι 129

Ρ

ραβδολουτιάνος 692
ράσα 82, 83, 84, 85, 86, 87, 88, 89, 90,
91, 92, 93, 94, 96, 97, 98, 100, 101
ράσες 81, 102

ρέγγα 169
ρέγγα της Χιλής 204
ρέγγα του Ειρηνικού 170
ρεγγάκια 187
ρετρογωβιός 1053
ρίκι 982
ρίνα 68
ρίνες 67
ρινόβατοι 72
ρινόβατος 73, 74, 75, 76, 77
ρινοδέλφινο 1477
ροδοφύκια 1528
ροζ γαρίδα 1224
ροζ γαρίδα της Βόρειας
   Θάλασσας 1217
ρός σολομός 247
ρομβόγλωσσα 1149
ρομβόγλωσσες 1150
ροφοί 582
ροφός 577, 579
ροφός του Νασσάου 581
ρυγχόβατος 90
ρυγχοκαρχαρίας 20
ρυγχοκορέγονος 237
ρυγχοπαγόψαρο 899
ρυγχοφάλαινα 1459, 1480

Σ

σαζάνι 332
σακοράφα 506, 507, 508
σαλβελίνοι 267
σαλβελίνος 264
σαλβελίνος του Ειρηνικού 265
σαλιάρα 908, 912, 919, 910, 905,
    907, 911, 909
σαλιάρα της Παταγωνίας 880
σαλιάρα των φυκών 918
σαλιάρες 904, 906
σαλιάρης 905, 906, 907, 908, 909, 910
σαλιγκάρι του Ατλαντικού 1107
σαλούβαρδοι 465
σαλούβαρδος 463, 464
σάλπα 801
σαμπανιός 668, 672, 675
σαπουνόψαρα 701
σαπουνόψαρο 700
σαραβάνος 464
σάργομος 416
σαργός 782
σαργός της Αμερικής 809
σαρδέλα 189
σαρδέλα της Ανατολής 173
σαρδέλα της Αυστραλίας 198
σαρδέλα της Ιαπωνίας 197
σαρδέλα της Καλιφόρνιας 196
σαρδέλα της Νότιας Αφρικής 199
σαρδέλα του Περού 200
σαρδέλες με λέπια 180
σαρδέλα της Δυτικής Αφρικής 183
σαρδελομάνα 156, 157, 158, 163
σαρκογαρίδα 1216
σαυρίδι 668, 672, 675
σαυρίδι κούνενε 674
σαυρίδι της Ιαπωνίας 642, 666
σαυρίδι της Χιλής 669
σαυρίδι του Ακρωτηρίου 673
σαυρίδι του Ειρηνικού 671
σαυρίδι του Ινδικού 644
σαυρίδια 645
σεβαστός 1063

σελάχι 73, 74, 75, 76, 77, 82, 83, 84,
   85, 86, 87, 88, 89, 90, 91, 92, 93,
   94, 96, 97, 98, 99, 100, 101, 104,
   105, 106, 107, 114, 115, 116
σελάχι της Γροιλανδίας 99
σελάχια 80, 81, 102, 103
σελαχοκεφαλόψαρο 114
σερανίδες 567
σεργκεστρίδες 1241
σέτσκα 333
σέτσκα του Δούναβη 334
σιανίδες 715
σιάκι 1172
συλλάγγοι 625
σίρκο 310
σκαθάρι 808
σκαρμοί 390
σκαρμός 393
σκαρμός της Ανατολής 392
σκάρος 856
σκατοφάγος 829
σκιός 755, 756
σκιός της Ανταρκτικής 754
σκλεμπού 1192, 1194, 1195
σκληρή γυαλιστερή 1398
σκομβροειδή 960
σκομπροκαρχαρίας 971
σκορπάκι του Ακρωτηρίου 1071
σκορπίνα 1066, 1068
σκορπίνες 1062
σκορπιοί 1062, 1069
σκορπιομάνα 1063
σκορπιός 1066, 1067, 1068
σκορπιός των κοραλλιών 1065
σκουλίδες 1201
σκουμπρί 985
σκουμπρί της Αυστραλίας 996
σκουμπρί της Βραζιλίας 987
σκουμπρί της Γουινέας 1005
σκουμπρί της Ιαπωνίας 997
σκουμπρί της Ισπανίας 994
σκουμπρί της Καλιφόρνιας 990
σκουμπρί της Κίνας 1003
σκουμπρί της Κορέας 992
σκουμπρί τίγρης 1001
σκουμπρί του Ειρηνικού 1002
σκουμπρί του Κουίνσλαντ 999
σκουμπρί των Ινδιών 975, 976, 977
σκουμπρί των Παπούα 995
σκουμπριά 961, 986
σκουμπριά των Ινδιών 978
σκούρο κοκκινόψαρο 1072
σκυλάκι 30, 31, 32
σκυλάκια 28, 33
σκυλογαλέος 50, 52, 53
σκύλος 50, 64, 65
σκυλόψαρα 33, 57
σκυλοψαράκι 31, 32
σκυλόψαρο 10, 12, 14, 17, 19, 21,
   40, 42, 60, 64, 65, 911
σκυμνοσκυλόψαρο 60
σμέρνα 404
σμέρνες 402
σμπρίλλιας 19
σμπρίλλιος 19
σολομοί 262
σολομός 246
σολομός κέτα 248
σολομός της Αυστραλίας 683
σολομός της Γεωργίας 682
σολομός της Ιαπωνίας 250
σολομός του Ατλαντικού 256
σολομός του Δούναβη 242
σολομός του Ειρηνικού 254

σουβλίτης 549, 550
σουβλομύτης 551
σουπιά 1418, 1419, 1420
σουπίτσα 1420
σπαθοδέλφινο 1470
σπαθόψαρα 955
σπαθόψαρο 957, 958, 959
σπαρίδες 763
σπαροειδή 784
σπάρος 778
σπάρος της Φλώριδας 780
σπόγγοι 1511
στακτογαρίδες 1250
στακτοδέλφινο 1467
σταχτογαρίδα 1251
σταχτοκαρχαρίας 39, 40
στείρα 574
στέρβα 366, 370
στέρλετ 135
στικτογατόψαρο 376
στικτοκεντρόνι 64, 65
στικτομυλοκόπι 725, 757
στικτοπαλαμίδα 966
στικτόρεγγα 181
στικτοροφός 580
στικτοσκουμπρί 983
στικτοχειλού 841
στιλέτο της Αργεντινής 1206
στουριόνι 130, 133, 134, 135,
    136, 137, 139
στουριόνι του Δούναβη 131
στραβολοσωλήνας 1410
στρείδι 1347, 1353, 1354
στρείδι της Αυστραλίας 1357
στρείδι της Ιαπωνίας 1347
στρείδι της Νέας Ζηλανδίας 1355
στρείδι της Χιλής 1351
στρείδι του Ειρηνικού 1348
στρείδια 1350, 1356
στρέμπερ 623
στρέμπερ του Ροδανού 622
στρογγυλόβατος 86
στρογγυλόρεγγα 178
στρογγυλόρεγγα της Καλιφόρνιας 176
στρογγυλόρεγγα του Ατλαντικού 177
στρογγυλόρεγγες 179
στρογγυλοσαυρίδο 643
στρόμβοι 1338
στρωσίδι 313
σύκο 473
σύκο της Νορβηγίας 471
σύκο του Ατλαντικού 472
συναγρίδα 774
συναγρίδα της Αγκόλας 771
συναγρίδα του Ακρωτηρίου 796
συναγρίδα του Κονγκό 773
συναγρίδα του Μαρόκου 777
συναγρίδα των Καναρίων 772
συναγρίδα-νούφαρ 768
συρτάρι 323, 325, 326, 329
συρτάρι της Δαλματίας 324
συρτάρι της Ισπανίας 327
συρτάρι της Ιταλίας 328
σύρτης 681
σφυρίδα 573, 576
σφυρίδες 582
σφύρνα 47
σωλήνας 1372, 1409
σωλήνας της Ευρώπης 1413
σωλήνας του Ακρωτηρίου 1412
σωλήνας του Ατλαντικού 1408
σωλήνας του Ειρηνικού 1411
σωλήνες 1407, 1414

T

ταινιοσαλάρα	909
τάουτογκ	852
τάρμποτ του Ατλαντικού	1112
τάρμποτ του Ινδικού	1113
ταρπόνι	146, 147, 150, 152
ταρπόνια	145, 148
τάσκα	968
ταυροβάτραχος	1492
ταυροκαρχαρίας	13
τετράδοντοι	1184
τετραμούστακο γαϊδουρόψαρο	469
τετραπτέρυγο χελιδονόψαρο	434
τετράστιχτη χωματίδα	1119
τευθίς	1426
τέφτης	1423
τζεμπιλίδες	947
τζιτζίκι	941, 942, 1190
τηγανάκι	1359
τιλάπια της Μοζαμβίκης	832
τιλάπια του Νείλου	833
τονάκι	1007
τονάκι της Ανατολής	967
τονίνα	968
τόνοι	961, 1006, 1014
τόνος	1010, 1012
τόνος κιτρινόπτερος	1008
τόνος μακρύπτερος	1007
τόνος μαυρόπτερος	1009
τόνος μεγαλόφθαλμος	1011
τούρνα	288, 289
τούρνα της Αμερικής	287
τούρνα του Ειρηνικού	291
τούρνες	286
τουφεκόψαρα	686
τραχιβάτος	98
τριχιοί	195
τριχιός	190, 191
τριχιός της Αμερικής	162
τριχιός της Αφρικής	193
τριχιός της Ιαπωνίας	194
τριχιός της Ινδίας	192
τρομπέτα	504
τροπιδογαρίδα	1235
τροπιδογαριδάκια	1236
τροχοί	1330
τρυγόνα	104, 105, 106
τρυγόνες	103
τσαούσης	775
τσίμα	339
τσιμπλάκι	459
τσιμπούκι	610
τσιπούρα	802
τσιπούρα της Αμερικής	702
τσιρωνάκι	339
τσιρώνι	309

Y

υαλόψαρα	563

Φ

φαγγρί	806
φαγγρί κεφαλάς	800
φαγγρί κορωνάτο	775
φαγγρί με μπλε στίγματα	804
φαγγρί της Ιαπωνίας	805
φαγγρί του Ναπολέοντα	769

φαγγριά	767, 807
φαγγριά του Ναπολέοντα	770
φαιοφύκη	1525
φαιοφύκια	1525
φάλαινα	1460, 1484
φάλαινα δολοφόνος	1470
φάλαινα της Γροιλανδίας	1487
φάλαινα του Μπρυντ	1482
φάλαινες	1486
φαλαινοκαρχαρίας	27
φεγγαρόψαρα	676
φεγγαρόψαρο	509, 1187, 1188, 1189
φεγγαρόψαρο της Αφρικής	656
φεγγαρόψαρο του Ατλαντικού	657
φιάμουλα	854
φίδι της θάλασσας	406, 413, 414
φίδια της θάλασσας	407
φιδόχελο	414
φιδόψαρο	366, 370, 371
φιδόψαρο της Ιταλίας	373
φιδόψαρο της Ρουμανίας	369
φιδόψαρο του Μπεργαντίνου	368
φιδόψαρο των Βαλκανίων	367
φούσκα	1508
φουσκοχειλόχελα	920
φουσκοχειλόχελο	923
φρίσσα	156, 157, 158, 159, 163, 190, 191, 195
φρίσσα-κέλε	182
φρίσσα της Αμερικής	172
φρίσσα της Χιλής	175
φρίσσα του Ατλαντικού	166
φρίσσα του καλοκαιριού	155
φρίσσα του Πόντου	160
φρίσσα των Ινδιών	205, 206, 207
φύκια	1524
φυκίδες	1527
φυσητήρας	1462
φώκια	1450, 1452, 1453, 1471, 1472
φώκια της Γροιλανδίας	1451
φώκια της Νότιας Αμερικής	1441
φώκια του Ακρωτηρίου	1442
φώκιες	1440, 1446

X

χάβαρο	1401
χακούντα	164
χάλιμπατ	1124, 1125, 1133
χάλιμπατ της Γροιλανδίας	1148
χάλιμπατ του Ατλαντικού	1133
χάλιμπατ του Ειρηνικού	1134
χαλκοτσιπούρα	789
χαλκοχροοκαρχαρίας	36
χαμουζαίος	316
χανάκι	588
χανίδες	227
χάνος	587
χάνος της Αργεντινής	568
χαρακίδες	300
χαρτόψαρο	511
χειλού	841, 842, 843, 845, 847, 848, 849, 850, 851
χειλού παπαγάλος	840
χειλούδες	837
χειλόψαρα	337
χειμωνιάτικη ράσα	95
χειμωνιάτικο σελάχι	95
χειμωνιάτικος βάτος	95
χέλι	396

χέλι της Αμερικής	400
χέλι της Αυστραλίας	397
χέλι της Ιαπωνίας	399
χέλι της Νέας Ζηλανδίας	398
χέλια	395, 401
χελιδονάς	1089
χελιδόνι	110
χελιδονόψαρα	423, 425
χελιδονόψαρο	426, 1109
χελιδονόψαρο της Ιαπωνίας	424
χελογατόψαρα	378
χίμαιρα	123, 124, 126
χίμαιρα του Ακρωτηρίου	125
χίμαιρες	122
χιάμι	1120
χονδρόχειλο βουτυρόψαρο	709
χορτοκυπρίνος	331
χριστόψαρα	523
χριστόψαρο	525
χριστόψαρο της Αμερικής	524
χρύσκος	333
χρυσό κοκκάλι	634
χρυσογωβιός	1051
χρυσοκέφαλος	343
χρυσομπάρμπουνο	816
χρυσός σολομός	243
χρυσόψαρο	320
χταπόδι	1438, 1439
χτένι	1359, 1360, 1363, 1368
χτένι της Αυστραλίας	1366
χτένι της Ιολανδίας	1362
χτένι της Νέας Ζηλανδίας	1367
χτένι του Ατλαντικού	1365
χτένια	1358
χωματίδα	1139
χωματίδα με κίτρινη ουρά	1137
χωματίδα της Αλάσκας	1146
χωματίδα της Αμερικής	1147
χωματίδα της Αυστραλίας	1122
χωματίδα της Ιαπωνίας	1138
χωματίδα του καλοκαιριού	1118
χωματίδες	1121, 1123

Ψ

ψάρι ορειβάτης	1039
ψάρι χειρουργός	946
ψάρια αυτοκράτορες	761
ψάρια νυχτερίδες	826
ψάρια πεταλούδες	830
ψάρια τρομπέτες	503, 861
ψάρια χειρουργοί	945
ψευδόγλωσσα	1159
ψευδολαβράκι	562
ψευδολαβράκι της Αμερικής	561
ψευδολίτσα	1031, 1037
ψευδολίτσα αμερικάνικη	1033
ψευδολίτσα της Καλιφόρνιας	1034
ψευδολίτσα της Κίνας	1032
ψευδολίτσα της Φλόριντας	1036
ψευδολίτσα του Ατλαντικού	1035
ψευδολίτσες	1030
ψευτο-βουτυρόψαρο	712
ψευτόγαυρος του Ατλαντικού	211
ψευτόγαυρος του Ειρηνικού	212
ψευτόκοτες	528
ψευτόρκα	1473
ψευτοσαργοί	860
ψευτοσαργός	858, 859

Index français

A

abadèche du Cap	926
abadèche du Chili	927
abadèche rose	925
abadèche rouge	927
abadèches	928
able de Heckel	339
ablette	310
ablette blanche	309
ablette de rivière	308
ablette du Danube	322
ablette grise	308
ablette italienne	309
acanthuridés	945
achigan	603
acoupa à petites dents	727
acoupa blanc	720
acoupa chasseur	732
acoupa de Macdonald	724
acoupa de Steindachner	727
acoupa du Pérou	723
acoupa pintade	725
acoupa rayé	728
acoupa royal	726
acoupas	729
aigle	718
aigle de mer	109
aigle vachette	110
aiglefin	453
aigles de mer	108
aiguillat	64
aiguillat à nez court	66
aiguillat commun	64
aiguillat galludo	65
aiguillat noir	61
aiguillat-coq	65
aiguillats	57
aiguillat-sagre	61
aiguille de mer	422, 416, 508
aiguille de mer commune	506
aiguille de mer, grande	506
aiguille de mer, petite	507
aiguille voyeuse	418
aiguilles	415, 419
aiguillette	416
aiguillette verte	417
aiguillettes	415
aileron argenté	1031
aileron blanc du large	37
aileron chinois	1032
albacore	1008
albule commun	152
albules	151
alépocéphale	223
algues	1524
algues brunes	1525
algues calcaires	1530
algues rouges	1528
allache	190
allache, grande	193
allaches	195
alligator américain	1502
alligator du Mississippi	1502
alligators et caïmans	1500
alose	156

alose à gésier	172
alose à museau court	168
alose américaine	172
alose canadienne	162
alose chacunda	164
alose de la mer Noire	160
alose de Reeves	206
alose d'été	155
alose d'été du Canada	155
alose feinte	157
alose feinte du Nil	158
alose feinte nilotique	158
alose gracile	184
alose hilsa	205
alose kelee	182
alose médiocre	159
alose noyer	172
alose rasoir	183
alose savoureuse	162
alose toli	207
alose vraie	156
alose vulgaire	156
alose, grande	156
alose-caille indienne	188
aloses feintes	163
aloses vraies	163
altacke	156
amande de mer	1384
amande marbrée	1384
amia	647
amie chauve	142
ammodyte	936
anabas	1039
anatifes	1198
anchois	208
anchois	217
anchois à queue jaune	211
anchois américain	210
anchois baie	210
anchois commun	217
anchois d'Argentine	214
anchois d'Australie	215
anchois de Guinée	209
anchois de l'Atlantique Sud	209
anchois d'eau douce	221
anchois du Cap	216
anchois du Japon	218
anchois du Nord	219
anchois du Pacifique	219
anchois du Pacifique Nord	219
anchois du Pérou	212
anchois du Pérou	220
anchois européen	217
anchois grenadier	213
anchois indopacifique	222
anchois japonais	218
anchois péruvien	220
anchois tropical	209
anchovetta	220
anchovette	212
andorrève du Cap	685
andouille	396
ange de mer	68
ange de mer commun	68
angel	68
angelot	68

anges de mer	67
anguille	396
anguille américaine	400
anguille argentée	396
anguille australienne	397
anguille commune	396
anguille d'Australie	397
anguille de Nouvelle-Zélande	398
anguille de rivière	396
anguille demoiselle	413
anguille d'Europe	396
anguille du Japon	399
anguille japonaise	399
anguille serpent	414
anguiller	1526
anguilles	395, 401
anodonte	1343
anoli de mer	391, 392
anolis	390
anolis de mer	390
ânon	453
anthias	569
antimore bleue	437
apocalle	63
apogon	610
apogon noir	611
apogons	609
apron	624
apron du Rhône	622
araignée de mer	1315
araignée de mer, grande	1315
araignée du Sud	1314
arapèdes	1333
arbalétrier	1180
arbalétriers	1179
arbusseau	550
arc-en-ciel	646
arche du Japon	1340
arche du Pacifique	1339
arches	1341, 1342
argentin	511
argentine	282
argentine du Japon	283
argentine du Pacifique	280
argentine, grande	281
argentine, petite	282
argentines	279
argus	829
ariome indienne	1038
arnoglosse	1115
arnoglosse de Thor	1116
arselet	500
ascidie du Chili	1509
ascidie stolonifère	1510
ascidies	1507
aspe	311
aspidophore armé	1103
assiette	678
assiettes-mussolini	676
athérine	551
athérine, petite	549
athérines	548
aubour	345
aurin	930
auxide	964, 965
ayu	269

B

bachelier noir 605
badèches 583
balai 1132
balai du Pacifique Nord 1131
balai japonais 1130
balane géante 1200
balanes 1199
balaou 422
balaou atlantique 422
balaou de l'Atlantique 422
balaou du Japon 421
balaou japonais 421
balaous 420
baleine à bec d'oie 1461
baleine à bosse 1485
baleine bécune de Cuvier 1458
baleine bleue 1483
baleine de Baird 1459
baleine de Biscaye 1490
baleine de Bryde 1482
baleine de Cuvier 1458
baleine des Basques 1490
baleine franche australe 1489
baleine franche boréale 1487
baleine franche du Groenland 1487
baleine franche noire 1490
baleine grise de Californie 1491
baleine naine 1488
baleines à bec 1458
baleines franches 1486
baleinoptère boréal 1481
baleinoptère de Rudolf 1481
baliste 1180
baliste cabri 1180
baliste gris 1180
balistes 1179
banane à longues nageoires 153
banane de mer 152
banane gisu 153
bananes 147
bananes de mer 151
bar 593
bar américain 596
bar blanc 597
bar blanc d'Amérique 596
bar commun 593
bar d'Amérique 598
bar du Japon 599
bar du Pérou 584
bar tacheté 594
barbeau 314
barbeau commun 314
barbeau d'Albanie 313
barbeau de Java 353
barbeau du Danube 317
barbeau ibérique 315
barbeau italien 318
barbeau méridional 316
barbeaux d'Asie 354
barbette 372
barbier 1190
barbier rose 569
barbier rouge 570
barbillon blanc 377
barbue 1175
barbue américaine 1174
barbur de l'Indo-pacifique 554
barburs 553
barbus d'Hoven 338
barp 316
barracuda 532
barracuda atlantique 532

barracuda de Guinée 530
barracuda du Pacifique 531
barracuda européen 535
barracudas 529, 536
barramunda 127
barreau 593
bars 595
bars, serrans et mérous 567
baudroie commune 1195
baudroie d'Amérique 1193
baudroie diable 1196
baudroie rousse 1194
baudroies 1192
bavard 1097
baveuse 907, 909, 911
baveuse papillon 905
baveuses 904, 906
beauclaire du Pacifique 607
beauclaires 606, 608
bec 288
bec de jars 1415
bécasse de mer 504
bécasses 503
bêche-de-mer japonaise 1523
becquet 288
becs de canard 872
bécune 532, 535
bécune de Guinée 530
bécune indienne 534
bécune orientale 533
bécunes 529, 536
béliers 767
belouga 1478
béluga 139
bernaches 1199
bernicles 1199
berniques 1333
bérycidés 517
béryx 517, 518, 519
béryx commun 518
béryx rouge 518
bezole 234
bichir 128
bigorneaux 1337
black bass à grande bouche 603
blageon 329, 345
blanche drapeau 902
blanches 699, 700, 701, 903
blennie de Patagonie 880
blennie fluviatile 911
blennie gattorugine 909
blennie ocellée 905
blennie palmicorne 908
blennie papillon 905
blennie sanguine 910
blennie vivipare 923
blennie-paon 907
blennies 904, 906
bobyrez 340
bocasette dégarnie 887
bocasse aux yeux rayés 884
bocasse bossue 883
bocasse de Magellan 889
bocasse de Patagonie 891
bocasse de Patagonie
   à queue courte 890
bocasse grise 886
bocasse jaune 882
bocasse marbrée 885
bocasse noire 881
bocasses 876
bocassette de Ramsay 892
bocassettes 876
bocasson rayé 888

bocassons 894
bogaravelle 790, 793
bogue 766
bombay 296
bombay duck 297
bondelle 235, 237
bonite à dos rayé 982
bonite à dos tacheté 966
bonite à ventre rayé 973
bonite commune 982
bonite d'Australie 979
bonite du Pacifique 980
bonite élégante 966
bonite grosyeux 974
bonite orientale 974, 981
bonite vraie 973
bonites et thazards 961
bonitou 964
bonnat 957
boucots 1253
boulereau 1053
bouquet commun 1249
bourgette 1059
bourrugue de crique 733
bourrugue du golfe 734
bourrugue renard 735
bourrugues 736
bourse fil 1183
bourse orange 1181
bouvière 356
branchostégidés malacanthidés 626
bregmaceros de l'océan Indien 439
brème bordelière 319
brème commune 305
brème de mer 680, 793
brème de Pékin 349
brème d'eau douce 305
brème du Danube 306
brèmes 307
brèmes d'eau douce 307
brisures 529
brochet à points noirs 291
brochet commun 288
brochet de mer 535
brochet du Nord 288
brochet du Pacifique 291
brochet européen 288
brochet lance à long nez 143
brochet lance à nez plat 144
brochet maillé 290
brochet noir américain 290
brochet vermiculé 287
brochet, grand 288
brochets 286
brochets de mer 529
brochets de mer 536
brochets de mer 560
brosmes 442
brotule barbe 924
bucarde 1389
bucarde épineuse 1388
buccin 1335
buché 288
bulot 1335
bumalo 297
bummalow 297
busycons 1336

C

cabasson 550
cabassons 548
cabillaud 447
cabot 1097

cabotin	344	carreau	321	chinchard à queue jaune	668
cachalot	1462	carrelet	1145	chinchard bleu	670
cafard	1097	casseron bambou	1423	chinchard commun	672
cagna rayé	704	castagnole rouge	610	chinchard de la Méditerranée	668
cagnette	911	castagnole, grande	680	chinchard d'Europe	672
caïman à lunettes	1503	castagnole, petite	836	chinchard du Cap	673
calandre antarctique	893	castagnoles, grandes	679	chinchard du Chili	669
calfat	1178	castanette «Tarakihi»	859	chinchard du Japon	666
callionyme lyre	941	castanette pontude	857	chinchard du large	670
callionyme tacheté	942	castanette variable	858	chinchard du Pacifique	671
calmar	1426, 1433	castanoline noire	677	chinchard frappeur	667
calmar commun	1426	catalufas	608	chinchard noir	674
calmar du Cap	1425	cavillone commun	1085	chinchards	675
calmar japonais	1432	cébettes	1406	chinchards ronds	645
calmar totam	1424	ceintures d'argent	955	chinchards royaux	653
calmars	1422	centrine	70	chinook	254
calmars, encornets	1421	centrine commune	70	chiqueur	941
canin	316	centrines	69	chirocentre dorab	225
cannadelle	850	centrolophe noir	1041	chirocentres	226
canthare gris	808	centrolophes	1040	chirurgien chas-chas	946
caouanne	1496	centrophore	58	chondrus	1532
capelan	273, 473	centrophore granuleux	58	christivomer	266
capelan atlantique	273	cépole rougeâtre	834	chuscle	812
capelan de France	473	cernier atlantique	585	cicerelle	938
capelan de Méditerranée	473	cernier brun	585	ciclidés	831
capitaine	844	cernier commun	585	cigale, petite	1305
capitaine à quatre rayons	556	cernier de Nouvelle-Zélande	586	cigales de mer	1304
capitaine de l'Indo-Pacifique	554	céteau	1157	cisco de lac	231
capitaine royal	555	chabot	341	cisco de l'Est	231
capitaine, gros	556	chabot à dix-huit épines	1098	cithare feuille	1177
capitaines	553	chabot à quatre cornes	1099	clam «mogal»	1340
capitaines de mer	752	chabot de mer	1102	clam de l'Alaska	1403
capucette	552	chabot de rivière	1097	clam japonais	1402
carangidés	632	chabot scorpion	1100	clams	1324
carangue cheval	636	chabots	1101	clarias	388
carangue commune	635	chanidé	228	clauque	1415
carangue coubali	635	chanidés	227	cline argenté	918
carangue de Nouvelle-Zélande	652	chanos	228	clovisse	1405
carangue jaune	637	chapon	1068	clovisse jaune	1404
carangue rouge	638	characidés piranhas et hydrocyons	300	clovisses	1393, 1394
carangue, grande	636	charax bec fin	781	clupéidés	154
carangues	639	chardin	186	clupéonelle	171
carangues du Pacifique Sud	1042	chardin du Pacifique	185	clupes	154
carangues et chinchards	632	charin	362	cochon de mer	1471
carassin	321	chat de mer, grand	32	cohana	695, 697
carassin commun	321	chat de mer, petit	31	cohana doré	696
carassin doré	320	chatouille	6, 372	colin	487
carassin rouge	320	chenille	71	colin austral	886
carassin vulgaire	320	chevaine commun	341	colin jaune	466
carcharodonte	19	chevaine de la mer Noire	340	colin noir	467
cardeau à quatre ocelles	1119	chevaine méridional	342	colinot	487
cardeau d'été	1118	chevaux de mer	505	colins	477
cardeau olivâtre	1120	chevaux marins	505	comète	643
cardeaux	1114, 1121	chevesne commun	341	comète de Russel	644
cardinal à bouche noire	612	chevesne méridional	342	comète japonaise	642
cardinaux	609	chèvres de mer	825	comète maquereau	641
cardine	1169	chevrette américaine	1246	comète saumon	646
cardine à quatre taches	1168	chevrette géante	1247	comtes	645
cardine de Norvège	1171	chevrettes	1244	congre	410
cardine franche	1169	chevrettes sergestides	1241	congre à points blancs	409
cardines	1167, 1170	chien	64	congre argentin	412
carida	545	chien à dard	64	congre brochet de Guinée	405
carouche	321	chien de mer	50, 64	congre commun	410
carpe	332	chien de mer antarctique	51	congre d'Amérique	411
carpe à grosse tête	336	chien espagnol	30	congre du Pacifique nord-ouest	409
carpe argentée	335	chien piquet	64	congres	408
carpe blanche	713	chiens de mer	55	coque	1389
carpe commune	332	chimère commune	123	coque commune	1389
carpe de vase	330	chimère d'Amérique	124	coque jaune	1403
carpe de Wuchang	346	chimère du Cap	125	coque rouge	1395
carpe herbivore	331	chimères	122	coques	1387
carpe miroir	332	chimères éléphants	126	coquette	841
carpe noire	347	chinchard	672	coquille Saint-Jacques	1364, 1365
carraghéen	1532	chinchard à dos vert	665	coquilles Saint-Jacques	1358

FR

corail	1513
corail rouge	1514
corb	755
corb commun	755
corbeau	755
corbs	715
cordonnier bossu	633
corégone	237
corégone blanc	229
corégone bleue	240
corégone de lac	232
corégone du Léman	233
corégone élégante	230
corégone lavaret	234
corégone oxyringue	237
corégones	241, 261
coryphène	681
coryphène, grande	681
cou coupé	255
coupa tident	727
coupe fil	629, 630
courbine	718
courbine australe	754
courbine de Mandchourie	746
courbine dorée	717
courbines	715, 756
courlazo	847
couteau courbé	1409
couteau de l'Atlantique	1408
couteau d'Europe	1413
couteau droit	1410
couteau du Cap	1412
couteau du Pacifique	1411
couteaux	1407, 1414
coutelas	957
crabe bleu	1295
crabe chinois	1303
crabe de vase d'Australie	1302
crabe dormeur	1291
crabe européen	1298
crabe lénée	1294
crabe royal antarctique	1309
crabe royal de Murray	1308
crabe royal de roche	1307
crabe royal du Kamchatka	1310
crabe royal subantarctique	1308
crabe velu	1303
crabe vert	1298
crabe vert de Méditerranée	1297
crabes	1301
crabes à pieds noirs	1293
crabes de mer	1285
crabes des neiges	1312
crabes des neiges du Pacifique	1313
crabes dormeurs	1287
crabes nageurs	1296
crabes rouges	1322
crabes royaux	1311
crapauds	1191
crapet calicot	604
crénilabre à cinq taches	849
crénilabre cendré	845
crénilabre commun	847
crénilabre de Roissal	849
crénilabre méditerranéen	846
crénilabre melops	847
crénilabre ocellé	848
crénilabre paon	851
crénilabre tanche	851
crénilabre tigré	845
crevette à pattes jaunes	1215
crevette à queue rouge	1225
crevette akiami	1242
crevette banane	1222

crevette caramote	1220
crevette charnue	1216
crevette cristal	1214
crevette d'Amérique centrale	1228
crevette de roche	1252
crevette de vase	1258
crevette d'eau douce	1246
crevette devo	1207
crevette ésope	1238
crevette géante tigrée	1223
crevette grise	1251
crevette guinéenne	1210
crevette impériale	1234
crevette kuruma	1219
crevette ligubam du Nord	1227
crevette nordique	1237
crevette nylon chilienne	1235
crevette ovetgernade	1252
crevette rose	1249
crevette rose du large	1212
crevette rose du Nord	1217
crevette rose du Sud	1224
crevette rouge	1233, 1237
crevette rouge géante	1234
crevette royale	1220
crevette royale grise	1213
crevette royale occidentale	1221
crevette seabob de l'Atlantique	1231
crevette siba	1208
crevette stylet d'Argentine	1206
crevette tigrée brune	1218
crevette tigrée verte	1226
crevettes	1229, 1248
crevettes aristeus	1232
crevettes bouquets	1244, 1245
crevettes couteaux	1255
crevettes crangonides	1250
crevettes flèches	1236
crevettes gambri	1230
crevettes grises	1250
crevettes metapenaeus	1209
crevettes nordiques	1240
crevettes pandalides	1239, 1240
crevettes royales	1229
crevettes seabob	1230
crevettes tropicales	1211
croco	712
crocodile de Nouvelle-Guinée	1504
crocodile du Siam	1506
crocodile marin	1505
crocodiles	1501
crossies	562
cycloptères	1104
cycloptéridés	1104
cynoglosses	1166
cyprin	321
cyprin argenté	355
cyprin doré	320
cyprinidés	303
cyprinodonte de Valence	515
cyprinodonte d'Espagne	514
cyprinodonte d'Europe du Sud	513
cyprinodontes	512
cyprins sucets	301, 302

D

dagaas	203
dard	344
datte de mer	1372, 1416
daubenets	767
dauphin à long bec	1475
dauphin blanc	1478
dauphin bleu et blanc	1474

dauphin commun	1464
dauphin de Risso	1467
dauphin, grand	1477
dauphins	1463
dauphins pélagiques	1476
daurade américaine	702
daurade géante	805
daurade japonaise	805
daurade vraie	802
daurades	763
demi-bec à longues nageoires	432
demi-bec d'Australie	428
demi-bec balaou	429
demi-bec brésilien	430
demi-bec du Japon	431
demi-becs	427, 433
demoiselle	834
demoiselles	835
denté	774
denté à tache rouge	772
denté angolais	771
denté aux gros yeux	776
denté bossu	775
denté canarien	772
denté charpentier	765
denté commun	774
denté congolais	773
denté du Cap	796
denté du Maroc	777
denté du Natal	797
denté maculé	798
denté nufar	768
diable de mer	114, 1098, 1102
diables de mer	111, 1101
diagramme à grosses lèvres	709
diagramme gris	710
diagrammes	705
disque africain	827
disque ponctué	828
dolly varden	265
donzelle	929
dorade	793
dorade commune	793
dorade dorée	802
dorade grise	808
dorade rose	793
dorade royale	802
dorade vraie	802
dorades	763
doré commun	621
doré jaune	621
dormeur	26
dormeur du Pacifique	1290
doucet	941
dragonet	941
dragonet lyre	941
dragonet tacheté	942
drépane ailé	827
dugong	1454
durgan	316

E

écaille, grande	150
écailles, grandes	145
écrevisse à pieds blancs	1279
écrevisse américaine	1282
écrevisse commune	1276
écrevisse d'Australie	1316
écrevisse des torrents	1280
écrevisse du Danube	1277
écrevisse rouge des marais	1284
écrevisse turque	1277
écrevisses	1278, 1281

FR

écrevisses américaines	1283
églefin	453
élédone commune	1435
élédone de la Méditerranée	1436
élédone musquée	1436
élédones	1437
émissole	53
émissole commune	53
émissole de Patagonie	54
émissole gatuso	54
émissole lisse	53
émissole tachetée	52
émissoles	49, 55
empereurs	762
empereurs à gros yeux	761
encornet	1433
encornet géant	1427
encornet minami	1431
encornet rouge	1429
encornet rouge argentin	1428
encornet rouge nordique	1430
encornet, petit	1423
encornets	1421
engraulidés	208
épaulard	1470
éperlan	274, 276
éperlan à petite bouche	271
éperlan arc-en-ciel	275
éperlan arctique	276
éperlan d'Amérique	275
éperlan de l'Arctique	276
éperlan d'Europe	274
éperlan du large	299
éperlan du Pacifique	272
éperlans	270, 277
épineux de fond	61
épinoche	502
épinoche à trois épines	500
épinoche aiguillonée	500
épinoche de mer	502
épinoches	499
épinochette	501
éponges	1511
équille	936, 938
équille du Nord	935
équille élancée	939
équilles	931, 937
escolar	951
escolier	951
escolier blanc	954
escolier espagnol	950
escolier gracile	953
escolier royal	949
escoliers	947, 948
espadon	1029
espadons	1028
esprot	202
esturgeon à barbillons frangés	134
esturgeon à ventre nu	134
esturgeon blanc	138
esturgeon commun	137
esturgeon de l'Adriatique	133
esturgeon de Naccare	133
esturgeon d'Europe occidentale	137
esturgeon du Danube	135
esturgeon étoilé	136
esturgeon russe	131
esturgeon vert	132
esturgeon, grand	139
esturgeons	130
ethmalose	191
ethmalose d'Afrique	174
étoile de mer rouge	1517
étoiles de mer	1516
étrille	1299
étrille à trois pointes	1300
étrille gazami	1300
étrilles	1301
eulachon	278
eulakane	278
euphausiacés	1203
euphausiidés	1203
exocet hirondelle	434
exocet volant	426
exocets	423, 425

F

fanfre	585
fanfre noir	571
fausse badèche	574
fausse bonite	968
fausse courbine	746
fausse courbine du Japon	747
fausse limande	1115, 1116
fausse limande du Pacifique	1135
fausse palourde	1391
fausse praire	1391
fausse-vieille	839
faux flétan	1132
faux flétan du Japon	1124
faux flétan du Pacifique	1125
faux hareng de l'Atlantique	186
faux hareng du Pacifique	185
faux harengs	187
faux orque	1473
faux rouget de roche	821
faux rougets	822
faux Saint-Pierre	524
faux turbot	1117
faux turbots	1114
féra	234
féra, grande	236
féra, petite	233
fiatole	1037
fiatoles	1030
fierasfer	930
flaneur	12
flet	1143
flet commun	1143
flet de Nouvelle-Zélande	1149
flétan commun	1133
flétan d'Afrique	1112
flétan de l'Atlantique	1133
flétan du Groenland	1148
flétan du Japon, faux	1124
flétan du Pacifique	1124
flétan du Pacifique	1134
flétan du Pacifique, faux	1125
flétan nain	1132
flétan noir	1148
flétan tacheté d'Afrique	1111
flétan tropical indopacifique	1113
flétan, faux	1132
flets de Nouvelle-Zélande	1150
fondules	512
forgeron ailé	827
friture	753
friture à écaille	703
fucales	1527
fundule d'Espagne	515

G

gades	440
gadicule	449
gadiformes	435
galathées	1306
galinette	1089, 1106
galjoin franc	824
galjoins	823
gambas	1224, 1229
gambas du Nord	1217
gambas japonaise	1219
gambon écarlate	1234
gambusie	516
gandoise	345
garde écueil	1066
gardon	360
gardon commun	360
gardon d'Albanie	348
gardon de la mer Noire	357
gardon de l'Europe du Sud	359
gardon galant	358
gardon rouge	362
gardons	361
gaspareau	161
gasparot	161
gatte	60
gavaron	813
gelidium	1529
gempylidés	947
gérion ouest-africain	1321
gérions	1322
germanos	943
germon	1007
germon atlantique	1007
gigartines	1531
girelle commune	838
girelle paon	854
girelles et vieilles	837
gites	1417
gladiateur	1029
glatdick	134
globicéphale à nageoire courte	1465
globicéphale de Siebold	1465
globicéphale noir	1466
globicéphale tropical	1465
gobie buhotte	1059
gobie céphalote	1052
gobie de Canestrini	1058
gobie de Kessler	1057
gobie de la mer Noire	1060
gobie de lagune	1056
gobie de Panizza	1056
gobie de sable	1059
gobie doré	1051
gobie lotte	1061
gobie noir	1053
gobie paganel	1054
gobies	1049, 1055
gobies à tête plate	872
gobies dormeurs	1048
gode	472
goiffon	333
gonelle	919
gonzui	378
goret mule	707
goujon commun	333
goujon du Danube	334
goujon-barbeau de Dalmatie	312
gourami	1045
gourami géant	1045
gourami peau de serpent	1046
grampus	1467
grand brochet	288
grand chat de mer	32
grand dauphin	1477
grand esturgeon	139
grand lançon	939
grand pollan	236

grand powan	238
grand requin blanc	19
grand tambour	745
grande aiguille de mer	506
grande allache	193
grande alose	156
grande araignée de mer	1315
grande argentine	281
grande carangue	636
grande castagnole	680
grande coryphène	681
grande écaille	150
grande féra	236
grande gueule	899
grande gueule antarctique	896
grande gueule épineuse	897
grande gueule ocellée	900
grande lamproie marine	7
grande lingue	462
grande roussette	32
grande sébaste	1075
grande verrue tigrée	742
grande vieille	840
grande vive	869
grandes castagnoles	679
grandes écailles	145
gravanche	236
grémille	614
grémille du Danube	615
grenadier	493
grenadier à tête rude	497
grenadier bleu	
de Nouvelle-Zélande	479
grenadier de Patagonie	478
grenadier de roche	494
grenadier d'Italie	495
grenadiers	498
grenadiers bleus	480
grenadiers et rats	496
grenouille de mer	468
grenouille taureau	1492
grenouilles	1493
griset	808
grondeur argenté	711
grondeur du Pacifique Sud-Est	706
grondeur métis	712
grondeur noir	745
grondeur rayé	708
grondeurs	705, 714
grondin à museau pointu	1088
grondin aile bleue	1082
grondin de Dieuzeide	1086
grondin du Cap	1081
grondin gris	1084
grondin lyre	1090
grondin morrude	1080
grondin obscur	1080
grondin perlon	1089
grondin pin	1079
grondin rose	1079
grondin rouge	1079, 1089
grondin sombre	1080
grondin strié	1091
grondin volant	1109
grondins	1078, 1083
grondins à tête plate	1095
grondins d'Amérique	1087
gros capitaine	556
gros seigneur	1105
gueule antarctique, grande	896
gueule épineuse, grande	897
gueule ocellée, grande	900
gueule, grande	899
guidon blanc	360

guignol	1066
guinée machète	146
guinées	145, 147
guitare	76
guitare de mer commune	76
guitares de mer	72
gurnard	1084

H

hâ	50
halibut commun	1133
halibut faux	1132
halibut noir	1148
happeurs nains	686
hareng commun	169
hareng de lac	231
hareng de l'Atlantique	169
hareng de l'Atlantique, faux	186
hareng du Chili	204
hareng du Pacifique	170
hareng du Pacifique, faux	185
hareng tacheté	181
hareng-loup des Indes	225
harengs loup	224, 226
harengs, faux	187
harengule du Japon	194
harengules	180
haut-bar	718
hauting	237
hexanchidés	9
hippocampes	505
hirame	1120
hirondelle	680
hirondelles de mer	679
holbiche brune	29
homard américain	1317
homard européen	1318
hoplostète de Méditerranée	521
hoplostète orange	520
hotu	325
huchon	242
huchon du Danube	242
huiron	139
huître américaine	1349
huître creuse d'Australie	1357
huître creuse de Palétuvier	1348
huître creuse japonaise	1347
huître de Belon	1353
huître de Virginie	1349
huître perlière	1346, 1383
huître plate	1353
huître plate de Guinée	1352
huître plate de Nouvelle-Zélande	1355
huître plate du Chili	1351
huître plate du Pacifique	1354
huître plate européenne	1353
huître portugaise	1347
huîtres creuses	1350
huîtres plates	1356
hurlin	617
hyperoodon boréal	1460

I

ide mélanote	343

J

jambonneau de mer	1385
jarret	813
jarretière	834, 957, 958
Jean Doré	525

joël	549
joëls	548
jubarte	1485
juifs	606, 608

K

kilch	239
kilka	171
killi	512
kissing gourami vert	1047
krill	1203
krill antarctique	1204
krill norvégien	1205

L

labéos	337
labre mêlé	841
labre rupestre	839
labre vert	843
labres	837
lagénorhynque à bec blanc	1469
lagénorhynque à flancs blancs	1468
laimargue	63
laimargue du Groenland	63
lamantin d'Afrique	1457
lamantin d'Amérique	1456
lamantin de l'Amazonie	1455
lamantin du Sénégal	1457
lamie à nez pointu	20
laminaires	1526
lamnie	23
lampris	509
lamproie de Planer	6
lamproie de rivière	5
lamproie de Vladykov	4
lamproie du Danube	2
lamproie d'Ukraine	3
lamproie marine	7
lamproie marine, grande	7
lamproie, petite	6
lamproies	1
lançon	938
lançon d'Amérique	932
lançon de Corée	940
lançon du Nord	933
lançon du Pacifique	934
lançon nordique	935
lançon, grand	939
lançon, petit	936
lançons	931, 937
langouste blanche	1269
langouste d'Australie	1270
langouste de Cuba	1269
langouste de Juan Fernandez	1261
langouste de Nouvelle-Zélande	1260
langouste de Tristan	1263
langouste d'Océanie	1264
langouste du Cap	1262
langouste du Natal	1265
langouste du Sud	1267
langouste du Transkei	1271
langouste japonaise	1272
langouste rose	1268
langouste rouge	1266
langouste royale	1273
langouste verte	1273
langoustes	1259, 1274
langoustes-fouet	1275
langoustine	1320
langoustine andamane	1319
langoustine de l'océan Indien	1319

FR

langue d'avocat	1157
lavaret	234
lavaret du Bourget	234
lavette	308
légine antarctique	878
légine australe	877
légines	879
légines colandres	876
lépadogaster	1190
lépidope argentin	957
lépisostée à nez plat	144
lépisostée osseux	143
lethrinidés	762
lézard	393
lézard à dents en brosse	392
lézard géant	391
liche amie	647
liche glauque	663
liches	648
lieu de l'Alaska	470
lieu jaune	466
lieu noir	467
lièvre de mer	1105
limace barrée	1108
limace de mer	1108
limace de mer atlantique	1107
limande à queue jaune	1137
limande commune	1139
limande d'Amérique	1147
limande du Japon	1136
limande du Japon	1138
limande salope	1169
limande sole	1142
limande-pile du Japon	1138
limandes, plies et flétans	1123
limbert	389
limule	1515
lingue	462
lingue bleue	460
lingue commune	462
lingue espagnole	461
lingue méditerranéenne	461
lingue, grande	462
lion de mer à crinière	1443
lion de mer d'Amérique du Sud	1444
lion de mer de Steller	1443
lithothamnies	1530
loche de Bergatino	368
loche de rivière	370
loche des Balkans	367
loche d'étang	371
loche dorée	366
loche épineuse	370
loche franche	372
loche italienne	373
loche roumaine	369
lompe	1105
lompes	1104
loque	372
loquette	923
loquette d'Amérique	922
loquettes	920, 921
lotte	1195
lotte commune	1195
lotte de rivière	452
lottes	1192
loubine	561, 593
loup	593
loup anarhique à peau mince	915
loup atlantique	915
loup de mer	915
loup de mer à peau mince	915
loup de mer bleu	914
loup de mer tacheté	916

loup de mer, petit	916
loup des roches	755
loup gélatineux	914
loup marin	915
loups de mer	917
lure	678
lutjanidés	687
lycodes	920

M

machoiron banderille	375
machoiron tacheté	376
machoirons	374
mactre	1391
mactre lisse	1390
mactre solide	1392
maërl	1530
mafou	631
magare	1138
maigre	718
maigre argenté	744
maigre commun	718
maigre doré	717
maigre noir	721
maigres	715, 756
maillet	47
makaire à rostre court	1022
makaire bécune	1026
makaire blanc	1021
makaire bleu	1019
makaire épée	1025
makaire noir	1018
makaire strié	1023
makaires	1015, 1020
malachigan d'eau douce	716
malarmat	1106
mallarmat	1106
mamilas	698
manches de couteaux	1414
mante	1202
mante atlantique	112
mante du Pacifique	113
mante méditerranéenne	114
mantes	111
maquereau	985
maquereau bonite	1005
maquereau commun	985
maquereau de Atka	1093
maquereau des îles	976
maquereau des Indes	977
maquereau espagnol	984
maquereau royal	988
maquereau tacheté	983
maquereau trapu	975
maquereau unicolore	974
maquereaux	961, 986
maquereaux bonites	1004
maquereaux du Pacifique	978
maraîche	23
marbré	785, 786
marbrés	787
marène, petite	229
marlin	1024
marlin de Méditerranée	1024
marlins	1015, 1020, 1027
marmotte	1108
marsouin	1471
marsouin de Dall	1472
marteau	47
maskinongé	289
matowacca	159
méduses	1512

mégaptère	1485
melva	964
mendole commune	812
mendoles	810, 814
mendoles et picarels	684
menhaden	166, 167
menhaden du Pacifique Sud-Est	175
menhaden écailleux	165
menhaden tyran	166
merlan	454
merlan argenté	449
merlan bleu	459
merlan bleu austral	458
merlan commun	454
merlan de la mer Noire	455
merle	842
merlu	487
merlu argenté	483
merlu argenté d'Amérique du Nord	483
merlu argenté du large	481
merlu austral	482
merlu commun	487
merlu côtier	484
merlu d'Afrique tropicale	489
merlu d'Argentine	486
merlu du Cap	484
merlu du Chili	485
merlu du Pacifique	490
merlu du Pacifique Nord	490
merlu du Sénégal	491
merlu européen	487
merlu noir	491
merlu profond	488
merlu sud-américain	486
merluche	487
merluche blanche	476
merluche brésilienne	474
merluche écureuil	475
merluchon	487
merlus	477, 492
mérou blanc	573
mérou cabrilla	575
mérou commun	577
mérou d'Alexandrie	574
mérou de malabar	579
mérou de Méditerranée	577
mérou des Basques	585
mérou des provençaux	577
mérou géant du Pacifique	591
mérou noir	576
mérou rayé	581
mérou rouge	580
mérou sombre	577
mérous	582
meunier	341, 342
meunier argenté	344
milandre	50
mirandelle	309
modioles	1373
moki bleu	862
mokis	861
môle commun	1188
môle lancéolé	1187
môle voyageur	1189
mollet	1105
mollusques marins	1323
montre d'Orbigny	1397
mordocet	908
mord-pierre	370
mores	436
morse	1445
morue arctique	445
morue bleue de Nouvelle-Zélande	864
morue charbonnière	1094

morue de roche	448	
morue du Groenland	448	
morue du Pacifique	446	
morue du Pacifique occidental	470	
morue fraîche	447	
morue grise	446	
morue longue	462	
morue longue commune	462	
morue noire	453	
morue ogac	448	
morue polaire	441	
morue rouge de Nouvelle-Zélande	438	
morue Saint-Pierre	453	
morue-lingue	1092	
morues antarctiques	876, 894	
mostelle brésilienne	474	
mostelle de fond	463	
mostelle de roche	464	
mostelle de vase	463	
mostelles	465	
motelle à cinq barbillons	443	
motelle à quatre barbillons	469	
motelle à trois barbillons	450	
motelle commune	450	
motelles	451	
moule chilienne	1374	
moule commune	1376	
moule coréenne	1375	
moule d'Australie	1378	
moule de la Nouvelle-Zélande	1381	
moule de la Plata	1379	
moule de rivière	1345	
moule de roche sud-américaine	1382	
moule d'étang	1343	
moule du Chili	1374	
moule méditerranéenne	1377	
moule striée	1371	
moule verte	1380	
moules	1370	
moules d'eau douce	1344	
mourine	109	
mourine bâtarde	115	
mourine vachette	110	
mourme	785	
mousse d'Irlande	1532	
moutelle commune	53	
muge à grosse lèvre	538	
muge à grosses lèvres	538	
muge cabot	545	
muge capiton	542	
muge céphale	545	
muge commun	545	
muge daurin	540	
muge doré	540	
muge labéon	547	
muge ramada	542	
muge sauteur	543	
muges	537	
mugilidés	537	
mulet	541	
mulet à grosse tête	545	
mulet bobo	539	
mulet cabot	545	
mulet capiton	542	
mulet céphale	545	
mulet commun	545	
mulet daurin	540	
mulet doré	540	
mulet du Brésil	546	
mulet labéon	547	
mulet lippu	538	
mulet porc	542	
mulet ramada	542	
mulet sauteur	543	

mulet sauteur d'Afrique	544	
mulets	537	
mulette	1345, 1383	
murène	404	
murène brochet cendrée	406	
murène de Méditerranée	404	
murène verte	403	
murènes	402	
murènes brochets	407	
murènèsoce du Japon	406	
murènèsoce-dague	406	
murex	1334	
musso	678	
mussolini	657	
mussolini africain	656	
mustelle à cinq barbillons	443	
mustelle à trois barbillons	450	
mustellus gommé	51	
muzeraille	23	
mye	1415	

N

narval	1479	
nase commun	325	
nase dalmate	324	
nase de l'Europe du Sud	323	
nase de l'Europe du Sud-Ouest	329	
nase d'Italie	328	
nase ibérique	327	
nase-vairon	326	
navaga jaune	444	
nonnat	1050	
notothenia noir	881	
notothéniidés	876	
nounat	1050	

O

oblade	788	
odontaspide taureau	13	
ogac	448	
olives de mer	1406	
omble	263, 266	
omble chevalier	263	
omble de Canada	266	
omble de fontaine	264	
omble du Pacifique	265	
omble moucheté	264	
ombles	262, 267	
ombre commun	268	
ombrine	759	
ombrine commune	759	
ombrine côtière	759	
ombrine d'Argentine	758	
ombrine royale	735	
ombrines	715, 756, 760	
orbe étoile	1185	
oreille de mer	1328	
oreilles de mer	1329	
oréos	528	
ormeau à lèvres noires	1327	
ormeau de Mida	1326	
ormeau européen	1328	
ormeau géant	1325	
ormeaux	1329	
orphie	416	
orphies	415	
orque	1470	
orque noire	1473	
orque, faux	1473	
osetr	131	

otarie à crinière	1443, 1444	
otarie à fourrure des Falklands	1441	
otarie à fourrure du Cap	1442	
otarie australe	1441	
otarie d'Afrique du Sud	1442	
otaries	1440	
otolithe bossu	749	
otolithe commun	751	
otolithe du Sénégal	750	
otolithe gabo	748	
otolithe nanka	751	
otolithe tigre	742	
otolithes	752	
oumbre	268	
oursin chilien	1522	
oursin comestible	1521	
oursin du Chili	1522	
oursin pierre	1518	
oursin violet	1518	
oursins	1519, 1520	
ovelle	309	

P

pageau commun	794	
pageot acarné	790	
pageot bâtard	790	
pageot blanc	790	
pageot du Natal	792	
pageot rouge	794	
pageot rouge africain	791	
pageots	795	
pagre	804	
pagre à points bleus	804	
pagre commun	806	
pagre rayé	803	
pagre royal	803	
pagres	807	
pagures	1286	
palomète	974	
palomine	663	
palourde	1401	
palourde dorée	1404	
palourde japonaise	1402	
palourde rouge	1395	
palourdes	1393	
panga	799	
papillon de mer	919	
pastenague ailée	115	
pastenague commune	106	
pastenague du Pacifique	104	
pastenague épineuse	105	
pastenague violette	107	
pastenagues	103	
pastenagues ailées	116	
patelles	1333	
patudo	1011	
peau bleue	42	
pêche-madames	625	
pecten d'Amérique	1369	
pecten d'Australie	1366	
pecten de la Nouvelle-Zélande	1367	
pecten de l'Atlantique Sud	1368	
peigne baie de l'Atlantique	1361	
peigne calicot	1360	
peigne islandais	1362	
peignes et pétoncles	1358	
pélamide sarde	982	
pèlerin	12	
pelon	703	
pelote	1395	
perce-pierre	370	
perchat	617	

perchaude	617	picarels	810, 814
perche argentée	605	pichette	617
perche barramundi	564	picot	500
perche blanche	596	pied de biche	1197
perche brochet	620	pieuvre commune	1438
perche canadienne	616	pieuvres	1434, 1439
perche commune	617	pilchard d'Afrique du Sud	199
perche de mer	593	pilchard d'Australie	198
perche des roches	600	pilchard de Californie	196
perche du Nil	565	pilchard du Chili	200
perche du Rhin	620	pilchard du Japon	197
perche goujonnière	614	pinna	1385
perche noire	602	pinne géante	1385
perche noire d'Amérique	603	piquitinga	209
perche soleil	601	pironneau	793
perche-chabot roumaine	618	platelle	362
perchelle	617	platête du Brésil	873
perches	613	platophrys	1117
perches africaines	566	platycéphale indien	1096
perches cristal	560	platycéphales	1095
perches de sable	865	pleuronectidés	1123
perches de verre	560	plie arctique	1140
perdrix de mer	785	plie canadienne	1132
perdrix de rivière	617	plie cynoglosse	1128
perlon cendre	10	plie d'Europe	1145
perroquet de Méditerranée	856	plie de Californie	1127
perroquet vieillard	856	plie de l'Alaska	1146
péteuse	356	plie du Pacifique	1144
petit chat de mer	31	plie franche	1145
petit encornet	1423	plie grise	1128
petit lançon	936	plie japonaise	1126
petit loup de mer	916	plie lisse	1141
petit poisson-chien d'Amérique	285	plie rouge	1147
petit requin taupe	21	plouse	472
petit rorqual	1480	pocheteau blanc	84
petit syngnathe	507	pocheteau gris	84
petite aiguille de mer	507	pocheteau noir	96
petite argentine	282	pocheteaux	102
petite athérine	549	poisson blanc	628
petite castagnole	836	poisson ceinture argenté	959
petite cigale	1305	poisson chameau	827
petite féra	233	poisson-chat d'Amérique	384
petite lamproie	6	poisson-cochon	702
petite marène	229	poisson d'argent	551
petite praire	1396	poisson de mai	156, 157
petite roussette	31	poisson de sable japonais	863
petite sole jaune	1156	poisson des glaces à long nez	899
petite vieille	841	poisson des glaces	
petite vive	867	à nageoires noires	896
pétoncle	1363	poisson des glaces antarctique	898
pétoncle operculaire	1359	poisson des glaces épineux	897
phéophycées	1525	poisson des glaces ocellé	900
pholade	1416	poisson-glace de Georgie	901
pholades	1417	poisson juif	591
phoque à capuchon	1447	poisson-lanterne	295
phoque à crête	1447	poisson-lapin	943
phoque à selle	1451	poisson-lune	1188
phoque annelé	1452	poisson-montre	520, 521
phoque barbu	1448	poisson pilote	650
phoque commun	1453	poisson pulmoné	127
phoque du Groenland	1451	poisson rouge	320
phoque gris	1449	poisson-ruban	511
phoque marbré	1452	poisson-scie	79
phoque marin	1451	poisson-scie trident	79
phoque moine	1450	poisson-soldat	522
phoque moine à ventre blanc	1450	poisson-tétard	1102
phoque veau marin	1453	poisson-tonnerre	394
phoques	1446	poisson visqueux	700
phrynorhombe de Norvège	1171	poisson volant	426
phycis de fond	463	poisson volant du Japon	424
phycis de roche	464	poisson-chat à barbillon blanc	377
picarel	813	poisson-chat électrique	394
picarel à points noirs	811	poisson-chat noir	385
picarel de l'Atlantique Sud-Est	811	poisson-chat tacheté	386

poisson-chien d'Amérique, petit	285
poisson-chien d'Europe	284
poisson-épée	1029
poisson-lézard à grandes écailles	392
poisson-lézard de l'Atlantique	393
poisson-loup	915
poisson-lune lancéolé	1187
poisson-lune voyageur	1189
poisson-pique	1024
poissons-battoires	826
poissons-bourses	1182
poissons-chats nettoyeurs	293
poissons-chats anguilles	378
poissons-chauves-souris	826
poissons-chirurgiens	945
poissons-couteaux	292
poissons-crapauds	1191
poissons de verre	563
poissons des glaces	895
poissons-disques	825
poissons-éléphants	126
poissons-lanternes	294
poissons-lapins	944
poissons-papillons	830
poissons plats	1110
poissons-scorpions	1062
poissons-taureaux	302
poissons-têtes de serpent	559
poissons-tigres	592
poissons type thon	960
poissons volants	423, 425
poisson-sabre commun	959
poissons-ballons	1184
poissons-chat de mer	374
poissons-chats	383, 388
poissons-chats américains	387
poissons-chats de verre	379
poissons-globes	1184
poissons-loups	913
poissons-sabres	955
poissons-scies	78
pollan	239
pollan, grand	236
polyodon de Chine	141
pomacentridés	835
pomadasydés	705
pomatomidés	629
pomfret noir	677
pompano de Floride	662
pompano sole	662
pompanos	664
porte-écuelle	1190
pouce-pied	1197
poulamon	457
poulamon atlantique	456
poule de mer	1105
poule de mer dorée	525
poulpe	1438
poulpes	1434, 1439
poutassou	459
powan, grand	238
praire	1398
praire d'Islande	1386
praire, petite	1396
praires	1393
prêtre	549, 550, 551
prêtre capucette	552
prêtres	548
pristipome à quatre bandes	708
protoptères	129

FR

Q

quinnat	254

R

raie à queue épineuse	99
raie bâtarde	91
raie blanche	82
raie bordée	82
raie bouclée	87
raie brunette	101
raie capucin	96
raie cendrée	84
raie chardon	89
raie circulaire	86
raie cornue	114
raie de montague	93
raie douce	93
raie du Pacifique	100
raie électrique marbrée	118
raie électrique noire	119
raie épineuse	97
raie étoilée	83
raie fleurie	94
raie grise	84
raie guitare tachetée	73
raie hérisson	88
raie lisse	85
raie manta	112
raie mêlée	91
raie miroir	92
raie mosaïque	101
raie ocellée	95
raie ondulée	101
raie radiée	97
raie rape	98
raie ronde	86
raie voile	90
raie-guitare chola	74
raie-guitare du Pérou	75
raies	80, 81, 102
raies électriques	121
raies guitares	77
rameur	37
rascasse blanche	874
rascasse brune	1067
rascasse charbon	1094
rascasse de Kuhl	1065
rascasse de Norvège	1075
rascasse du Cap	1071
rascasse du Nord	1075
rascasse noire	1067, 1094
rascasse pustuleuse	1066
rascasse rouge	1068
rascasse verte	1092
rascasse verte de Atka	1093
rascasse verte de l'Antarctique	893
rascasses	1062, 1069
rascasses blanches	875
rasoir	350
rason	855
rat	493, 495
rat de mer	123
ravil	968
régalec	510
rémora	1178
renard à gros yeux	16
renard de mer	15
requin	19
requin à longues nageoires	37
requin à nez pointu de l'Atlantique	44
requin à nez pointu du Pacifique	43
requin à pointes blanches des récifs	35
requin à pointes noires	38
requin-aiguille	44
requin-baleine	27
requin-bironche	43
requin blanc	19
requin blanc, grand	19
requin bleu	42
requin bouclé	71
requin-chagrin	58
requin-chat brun	29
requin-citron	41
requin cuivré	36
requin de sable	39
requin du Groenland	63
requin épineux	64
requin féroce	14
requin gris	11, 40
requin griset	11
requin-hâ	50
requin-léopard	56
requin-lézard	8
requin-mangeur d'homme	19
requin-nourrice	26
requin obscur	39
requin océanique	37
requin-pèlerin	12
requin-perlon	10
requin-renard	17
requin-renard pélagique	15
requin-scie	79
requin-tapis	25
requin-tapis tacheté	25
requin-taupe bleu	20
requin-taupe commun	23
requin-taupe saumon	22
requin-taupe, petit	21
requin zébré	24
requin-marteau	47
requin-marteau commun	47
requin-marteau halicorne	46
requin-marteau lisse	47
requin-raie	68
requins	34
requins épineux	57
requins-marteaux	48
requins-taupes	18
requins-tigres	34
requins-marteaux	45
requin-taureau	13
rhombosoles	1150
robalos	562
rochers	1334
roi des harengs	510
roi des rougets	610
rombou d'Australie	1122
rondeau mouton	764
rorqual bleu	1483
rorqual boréal	1481
rorqual commun	1484
rorqual de Bryde	1482
rorqual de Rudolf	1481
rorqual, petit	1480
rotengle	362
rotengle grec	363
rouffe à nez bleu	1043
rouffes	1040
rouget-barbet	817
rouget-barbet de roche	818
rouget de roche	818
rouget de roche, faux	821
rouget de vase surmulet	817
rouget doré	816
rouget du Sénégal	820
rouget-grondin	1079
rougets	819
rougets-grondins	1078
rougets, faux	822
rougets-barbets	815
roupier	847
rousse	360
rousseau	793
roussette, grande	32
roussette, petite	31
roussettes	28, 33
rouvet	950

S

sabornié	547
sabre argenté	957
sabre du Pacifique	958
sabre noir	956
sabres	955
sagre commun	61
sagres	62
saïda	441
Saint-Christophe	525
Saint-Pierre	525
Saint-Pierre, faux	524
Saints-Pierres	523
salicoque navaja	1254
salicoque rouge d'Argentine	1256
salicoque royale rouge	1257
salicoques	1255
salut	382
samma	421
sandart	620
sandre	620
sandre américain	621
sandre canadien	619
sanglier	527
sangliers	526
sapater	640
sapsap	699, 700, 701
sar à grosses lèvres	779
sar à museau pointu	781
sar à tête noire	783
sar austral	800
sar commun	782
sar cotonnier	780
sar tambour	779
sardine	189
sardine arc-en-ciel	173
sardine australienne	198
sardine chilienne	200
sardine commune	189
sardine d'Afrique du Sud	199
sardine du Pacifique	196
sardine européenne	189
sardinelle des Indes	192
sardinelle jaune	174, 191
sardinelle plate	193
sardinelle ronde	190
sardinelles	195
sargue commun	782
sars, sparaillons	784
saumon à bosse	247
saumon argenté	249
saumon atlantique	256
saumon australien	683
saumon australien de Georgie	682
saumon bossu	246
saumon chien	248
saumon chum	248
saumon coho	249
saumon commun	256

FR

saumon de fontaine	264	silure d'Amérique	384	stromatés	1030
saumon de l'Atlantique	256	silure électrique	394	stromatés du Golfe	1036
saumon du Danube	242	silure glane	382	strombes	1338
saumon du Japon	250	silure grec	381	sublet	850
saumon du Pacifique	254	silure noir	385	sucet	1108, 1178
saumon japonais	250	silure tacheté	386	suiffe	329, 345
saumon keta	248	silures	380	surmulet	818
saumon masou	250	silures de verre	379	syngnathe aiguille	506
saumon rose	247	siluridés	380	syngnathe commun	506
saumon rouge	252	siouclet	550	syngnathe de Dumeril	507
saumon royal	254	siphonostome	508	syngnathe, petit	507
saumon sockeye	252	soffie	329		
saumons, truites et ombles	262	soldat rouge	522		
saupe	801	sole	1163, 1165	**T**	
saurel	645, 668, 670, 672	sole américaine	1129, 1152		
saurel de Russel	644	sole australe occidentale	1153	tacaud	472
saurel japonais	642	sole australe orientale	1154	tacaud commun	472
saurel maquereau	641	sole commune	1165	tacaud norvégien	471
saurel tacheté	643	sole de Klein	1163	taly	1526
saurel torpille	649	sole de Nouvelle-Zélande	1149	tambour	588
sautereau	545	sole de roche	1176	tambour à gros yeux	722
sauteur parone	651	sole de sable	1164	tambour croca	731
savetier	500	sole du Pacifique	1135	tambour du Brésil	738
scare de Grèce	856	sole jaune, petite	1156	tambour du Japon	741
scatophages	829	sole ocellée	1158	tambour du Pacifique	730
sciaenidés	715	sole panachée	1160	tambour du Pérou	743
sciène du Pacifique	730	sole-perdrix commune	1160	tambour honnibe	741
sciènes	729	sole-perdrix juive	1159	tambour jaune	740
scopélidé	297	sole tachetée	1163	tambour rayé	737
scorpènes	1062	sole velue	1162	tambour rouge	757
scorpion de mer	1098, 1102	solens	1407	tambour, grand	745
scorpions de mer	1101	sole-pelouse	1164	tambours	705, 714, 739
scyris d'Alexandrie	633	sole-perdrix	1158	tanche	364
sébaste	1075, 1076	sole-pole	1164	tanche commune	364
sébaste à queue jaune	1074	soles	1151	tanche-tautogue	853
sébaste atlantique	1075	soles australes	1155	targeur	1176
sébaste chèvre	1063	soles-langues	1166	tarpon argenté	150
sébaste du Cap	1071	soles-perdrix	1161	tarpon banane	146
sébaste du large	1076	sompat	713	tarpon de l'Atlantique	150
sébaste du Pacifique	1070	souchet	1420	tarpon du Pacifique	149
sébaste rose	1073	soudron-coque	1388	tarpons	148
sébaste veuf	1072	souffleur	1477	tassergal	630
sébaste, grande	1075	souris de mer	1103	tassergals	629
sébastes	1064, 1077	sparaillon	778	taupe	23
seiche	1418	sparaillons	784	taupe commun	23
seiche commune	1418	spare doré	809	tautogue noir	852
seigneur, gros	1105	spare gibbeux	769	tenca	364
sélar à rayures jaunes	655	spare panga	799	téraglin	719
sélar coulisson	654	spares	770	terpug	1093
sélar subari	634	spatules	140	terrapènes	1499
sépiole	1420	spet	535	terre	106
sépiole melon	1419	sphéroide du Nord	1185	testard	1097
sépion	1420	spinax	64	tétarde	578
sériole	658	spirlin	308	tétards	1101
sériole australienne	659	sprat	202	tête de serpent d'Indonésie	557
sériole cheval	636	sprat gracile	201	tête de serpent strié	558
sériole couronne	658	squale bouclé	71	tétra	300
sériole du Japon	660	squale liche	60	tétrodon bigarré	1185
sérioles	661	squale-chagrin commun	58	tétrodons	1184, 1186
serpent de mer	414	squale-chagrin de l'Atlantique	59	thazard	965, 988
serran chevrette	587	squales	57	thazard atlantique	994
serran d'Argentine	568	squales méditérranéens	64	thazard australien	996
serran de sable	572	squille	1202	thazard bâtard	962
serran écriture	589	squilles	1201	thazard cirrus	993
serran hépate	588	stenelles	1476	thazard coréen	992
serranidés	567	sterlet	135	thazard d'Afrique de l'Ouest	1005
serrans	590	streber	623	thazard de Commerson	989
seuffe	329	stromaté à fossettes	1035	thazard de Monterey	990
sevruga	136	stromaté argenté	1031	thazard du Brésil	987
shadine à yeux rouges	178	stromaté chinois	1032	thazard du Japon	997
shadine de Californie	176	stromaté du Japon	1044	thazard du Pacifique	1002
shadine de l'Atlantique	177	stromaté du Pacifique	1034	thazard du Queensland	999
shadines	179	stromaté lune	1033	thazard franc	1000
sigans	944	stromatéidés	1030		

FR

thazard kanadi	998	
thazard kusara	972	
thazard nébuleux	1003	
thazard oriental	997	
thazard papou	995	
thazard ponctué de l'Indo-Pacifique	991	
thazard requin	971	
thazard sierra	1002	
thazard tigre	1001	
thazards	1004	
thon à gros oeil	1011	
thon à nageoires jaunes	1008	
thon à nageoires noires	1009	
thon albacore	1008	
thon blanc	1007	
thon élégant	963	
thon mignon	1013	
thon obèse	1011	
thon papillon	970	
thon rouge	1012	
thon rouge commun	1012	
thon rouge du Nord	1012	
thon rouge du Sud	1010	
thonine	968	
thonine noire	969	
thonine orientale	967	
thons	961, 1006, 1014	
thyrsite	951	
thyrsites	952	
tiche	364	
tilapia du Mozambique	832	
tilapia du Nil	833	
tile	627	
tile chameau	627	
tiles	626	
tiof	573	
togue	266	
torpille marbrée	118	
torpille noire	119	
torpille ocellée	120	
torpille tachetée	120	
torpilles	117, 121	
tortue bâtarde	1498	
tortue cuir	1495	
tortue luth	1495	
tortue olivâtre	1498	
tortue verte	1497	
tortues américaines	1499	
tortues de mer	1494	
touille	23	
touladi	266	
tourdero	843	
touret	333	
tourteau	1291	
tourteau du Pacifique	1292	
tourteau jona	1288	
tourteau poïnclos	1289	
toutenon commun	1433	
toutenon japonais	1432	
trachynotes	664	
trachyptère	511	

tregan	333	
tremble	118	
trichiure argenté	959	
trichiures	955	
trident	468	
trigle gris	1084	
trigle hirondelle	1089	
trigles	1078	
triptérygion à bec	912	
trocas	1330	
troques	1330	
truite américaine	251	
truite apache	244	
truite arc-en-ciel	251	
truite biwamasou	253	
truite brune	257	
truite commune	258	
truite de Dieppe	257	
truite de lac	259, 264, 266	
truite de lac d'Amérique	266	
truite de mer	257	
truite de rivière	258	
truite de ruisseau	264	
truite dorée	243	
truite dorée mexicaine	245	
truite fardée	255	
truite fario	258	
truite grise	266	
truite mouchetée	264	
truite rouge	264	
truites	260, 262	
turbo à cornes	1331	
turbo d'Afrique du Sud	1332	
turbot	1172	
turbot commun	1172	
turbot de la mer Noire	1173	
turbot lisse	1175	
turbot tropical	1112	
turbot tropical tacheté	1111	
turbot, faux	1117	
turbots et cardines	1167	
turbots, faux	1114	
tusk	442	

U

umbra	268	
uranoscope	874	
uranoscopes	875	

V

vairon	352	
vairon des marais	351	
valet	454	
vandoise	344	
vanneau	1359	
varech	1526	
vénus	1393	
verni	1395	

vernis	1400	
vernis du Japon	1399	
véron	352	
verrue tigrée, grande	742	
vieille américaine	844	
vieille commune	840	
vieille de l'Ouest-Atlantique	844	
vieille verte	843	
vieille, grande	840	
vieille, petite	841	
vieilles	837	
violet	1508	
violet du Chili	1509	
violet stolon	1510	
violets	1507	
violon de mer	76	
vipère de mer	506, 508	
vivaneau à queue jaune	694	
vivaneau à queue rouge	692	
vivaneau à ventre d'argent	689	
vivaneau campèche	690	
vivaneau des mangroves	688	
vivaneau rouge	691	
vivaneaux	687, 693	
vivaneaux nains	686	
vivaneaux royaux d'Australie	860	
vivanette	686	
vive à tête rayonnée	870	
vive araignée	868	
vive commune	869	
vive, grande	869	
vive, petite	867	
vives	866, 871	
voilier de l'Atlantique	1016	
voilier du Pacifique	1017	
voiliers	1015	

W

wahoo	962	

X

xiphiidés	1028	

Y

yeux vert à nez court	299	
yeux verts	298	

Z

zaerthe	365	
zée	525	
zéé bouclée d'Amérique	524	
zéidés	523	
ziphius	1461	
zope	304	

Indice italiano

A

abadechi	928
abadeco	925
abadeco del Sudafrica	926
abadeco rosso	927
abalone	1325, 1326, 1327
abaloni	1329
abramide	305
abramide cinese	349
abramide del Danubio	306
abramide russo	365
abramidi	307
acciuga	214, 216, 217, 218
acciuga d'Argentina	214
acciuga d'Australia	215
acciuga del Cile	220
acciuga del Giappone	218
acciuga del Nord Pacifico	219
acciuga del Sudafrica	216
acciuga dentata	221
acciughe	208
acerina	614
acerina del Danubio	615
acipenseridi, storioni	130
acoupa di Macdonald	724
acquadella	549
acquadelle	548
agone	157
agora	416
agostinella	817
aguglia	416
aguglia americana	417
aguglia imperiale	1021, 1024, 1025, 1026
aguglia imperiale mediterranea	1024
aguglia maggiore	418
aguglie	415
aguglie imperiali	1027
aguglie maggiori	419
akiami	1242
alacce	195
alacce centramericane	180
alacce vessillifere	187
alaccia	156, 190, 191, 192, 193
alaccia americana	162
alaccia americana	166
alaccia cilena	175
alaccia larga	174
alaccia vessillifera	185, 186
alaccia zunasi	194
alalunga	1007
alborella	310
alborella del Vulture	309
alburno danubiano	322
alburno di fiume	308
alburno di Heckel	339
alepocefalo	223
alghe	1524
alghe brune	1525
alghe calcaree	1530
alghe rosse	1528
alice	217
alici	208
alletterato	968
alligatore del Mississipi	1502
alligatori	1500

alosa	156, 157, 159
alosa americana	172
alosa canadese	155
alosa cinese	168
alosa del Mar Nero	160
alosa indiana	205, 206, 207
alosa kelee	182
alose	163
altavela	115
altavele	116
amia	142
ammodite lanceolato	939
ammodite tobiano	936
ammoditidi	931
ammoniosgombro	971, 972
amur	331
anaricadidi	913
anguilla	396
anguilla americana	400
anguilla argentina	396
anguilla australiana	397
anguilla gialla	396
anguilla giapponese	399
anguilla neozelandese	398
anguille	395, 401
anguillidi	395
anodonta	1343
antimora blu	437
apogonidi	609
aquila di mare	109
aquile di mare	108
aragosta	1260, 1261, 1262, 1263, 1264, 1265, 1267, 1268, 1271, 1272
aragosta australiana	1270
aragosta dei Caraibi	1269
aragosta di fondale	1268
aragosta mediterranea	1266
aragosta sudafricana	1262
aragosta verde	1273
aragoste	1259, 1274
aragostelle	1275
arche	1342
argentina	280, 281, 282, 283
argentine	279
argentinidi	279
argentone	355
aringa	169, 170
aringa cilena	204
aringa del Pacifico	170
aringa tonda	178
aringa tonda dell'Atlantico	177
aringa tonda di California	176
aringhe tonde	179
aristeidi	1232
ascidia cilena	1509
ascidia stolonifera	1510
ascidiacei	1507
aspidoforo corazzato	1103
aspio	311
aspro	624
asprone	622
astacidi	1278
astice	1318
astice americano	1317
aterinidi	548
atidi	1243
austromerluzzi	879

austromerluzzo	877
austromerluzzo	878
ayu	269

B

baciatore	1047
balani	1199
balano gigante	1200
balena antartica	1489
balena di Groenlandia	1487
balena franca	1490
balena marginata	1488
balena minore	1480
balena nera	1490
balena pigmea	1488
balene	1486
balenidi	1486
balenottera azzurra	1483
balenottera boreale	1481
balenottera comune	1484
balenottera di Bryde	1482
balenottera gobba	1485
balenottera grigia della California	1491
balenottera rostrata	1480
balistide arancio	1181
balistidi	1179
barbatello	372
barbi asiatici	354
barbo	314, 318
barbo albanese	313
barbo canino	316
barbo dalmata	312
barbo del Danubio	317
barbo di Giava	353
barbo di Hoven	338
barbo iberico	315
barracuda	529, 535, 536
barracuda della California	531
barracuda di Guinea	530
barracuda indiano	534
barracuda maggiore	532
barracuda orientale	533
barramundi	564
batracoidi	1191
bavosa	907, 908, 909, 919
bavosa lupa	914, 915, 916
bavosa occhiuta	905
bavosa pavone	907
bavosa peparuolo	912
bavosa ruggine	909
bavosa sanguigna	910
bavose	904, 906
bavose lupe	917
bavosella d'alga	918
belonidi	415
beluga	1478
berardo	1459
berice rosso	518
berici	517, 519
bichir	128
biscia di mare	414
biso	964, 965
bivalvi	1324
blennidi	904
blennio viviparo	923

IT

blennio viviparo americano 922
blicca 319
boba 766
bobirez 340
bocca d'oro 717, 718
boccanegra 30
boccanera 30
boga 766
bondella 235
bonito 980
bosega 538
botidi 1114
botolo 542
bottatrice 452
branzino 593
bregmacero 439
brosme 442
brosmio 442
brotola 924
buccina 1335
budego 1194
bumali 296
bumalo 297
burri 705
burro della Virginia 702
busbana 473
busbana francese 472
busbana neozelandese 438
busbana norvegese 471
busici 1336

C

cabinza 706
cagnaccio 14
cagnetto 911
caimano dagli occhiali 1503
calamaretto 1423
calamari, totani 1421
calamaro 1424, 1426
calamaro del Sudafrica 1425
calamaro mediterraneo 1426
calamita 542
canario rotondo 570
canesca 50
canestrello 1359, 1363
canestrello americano 1361
canestrello calico 1360
canestrello d'Islanda 1362
cannittidi 895
cannolicchi 1407, 1414
cannolicchio 1410, 1413
cannolicchio del Pacifico 1411
cannolicchio dell'Atlantico 1408
cannolicchio sudafricano 1412
cantaro 808
capelin 273
capellano 473
capitani 553
capitano orientale 554
capitano reale 555
capitone 396
capodoglio 1462
capone 1081, 1082
capone cavota 1080
capone coccio 1079
capone gallinella 1089
capone gorno 1084
capone imperiale 1079
capone pizzuto 1088
capone ubriaco 1091
caponi 1078, 1083
caponi americani 1087

cappa americana 1392
cappa artica 1386
cappa chione 1395
cappa dell'Alaska 1403
cappa dura 1398
cappa molle 1415
cappalunga 1409, 1413
cappasanta 1364
cappasanta americana 1369
cappasanta atlantica 1365
cappasanta australiana 1366
cappasanta della Nuova Zelanda 1367
cappasanta sudatlantica 1368
cappelunghe 1407, 1414
cappone 1079, 1080, 1084, 1089, 1090, 1091
cappone lira 1090
caproidi 526
caracidi 300
carangi 639
carangidi 632
carango bandanera 638
carango calla 634
carango cavallo 636
carango d'Alessandria 633
carango di Nuova Zelanda 652
carango dorato 635
carango mediterraneo 635
carango piatto 656, 657
carango ronco 637
carassio 321
carassio comune 321
carassio dorato 320
carcarinidi 34
cardidi 1387
carpa 332
carpa a specchi 332
carpa argentata 335
carpa di fango 330
carpa erbivora 331
carpa nera 347
carpa testa grossa 336
carpa Wuchang 346
castagnola 836
castagnola rossa 569
catalufa 607
catalufe 606, 608
catostomidi 301
caustel 542
cavallucci marini 505
cavedano 341, 342
cavedano russo 340
caviglione 1085, 1086
cefali 537
cefalo 538, 540, 542, 543, 545
cefalo boccasotto 539
cefalo bosega 538
cefalo calamita 542
cefalo dorato 540
cefalo labbrone 547
cefalo lebranche 546
cefalo saltatore 544
cefalo so-iny 541
cefalo verzelata 543
cefalone 228
cefaloni 227
centracantidi 810
centrolofidi 1040
centropomidi 560
cepola 834
cerina 538
cernia 568, 573, 574, 575, 576, 577, 579, 580, 585
cernia bianca 573

cernia di fondale 585
cernia di Nassau 581
cernia dorata 574
cernia gigante 578
cernia gigante del Pacifico 591
cernia mediterranea 577
cernia nera 576
cernie 582, 583
cheppia 158
chimera 123
chimera elefante 124
chimera tapiro 125
chimere 122
chimere tapiro 126
chiocciole di scogliera 1337
ciacunda 164
cicala di mare 1305
cicale 1304
cicerelli 931, 937
cicerello 938, 939
cicerello americano 932
cicerello boreale 933
cicerello coreano 940
cicerello del largo 935
cicerello del Pacifico 934
cicerello minore 936
ciclidi 831
ciclotteri 1104
ciclottero 1105
cinoglossidi 1166
cinopontico 405
ciprinidi 303
ciprino dorato 320
ciprinodonti 512
civetta di mare 1109
clupeidi 154
cobia 631
cobice 133
cobite 373
cobite barbatello 372
cobite comune 370
cobite dei Balcani 367
cobite di stagno 371
cobite dorato 366
cobite mascherato 368
cobite rumeno 369
coccodrilli 1501
coccodrillidi 1501
coccodrillo della Nuova Guinea 1504
coccodrillo marino 1505
coccodrillo siamese 1506
colga 1371
cometa 646
conchiglia dei pellegrini 1365
conchiglia di San Giacomo 1364
congridi 408
coracinidi 823
coralli preziosi 1513
corallo rosso 1514
coregone 230, 233, 236, 237, 238, 239
coregone americano 231
coregone azzurro 240
coregone bianco 229
coregone dei Grandi Laghi 232
coregone lavarello 234
coregone musino 237
coregoni 241, 261
corifena 681
corvina 755
corvina argentea 744
corvina australe 754
corvina boccanegra 721
corvina giapponese 722, 746, 747
corvina lupo 732

corvina striata	731
corvina tonda	753
corvine	715, 756
corvo	755
costardella	422
costardella balau	421
costardella saira	421
costardelle	420
cozza	1376, 1377
cozza australiana	1378
cozza cilena	1374
cozza coreana	1375
cozza d'acqua dolce	1345
cozza del Rio della Plata	1379
cozza della Nuova Zelanda	1381
cozza verde	1382
cozze d'acqua dolce	1344
cozze pelose	1373
crangonidi	1250
crostacei reptanti	1285
cuore	1388, 1389
cuore edule	1389
cuore spinoso	1388
cuori	1387

D

daga	203
damba	823, 824
dattero bianco	1416
dattero di mare	1372
delfinidi	1463
delfino	1464
delfino a fianchi bianchi	1468
delfino comune	1464
delfino di Risso	1467
delfino maggiore	1477
delfino muso bianco	1469
delfino pilota	1466
delfino pilota indiano	1465
dentale	774
dentice	771, 772, 773, 774, 775, 776
dentice argentato	765
dentice corazziere	775
dentice lupo	796
dentice marocchino	777
dentice mediterraneo	774
dentice nufar	768
dentice occhione	776
dentice striato	797, 798
diavoli di mare	111
diavolo di mare	112, 114
dipnoo australiano	127
donzella	838
donzella pavonina	854
dorab	224, 225, 226
dotto	585
dotto neozelandese	586
dragoncello	941
dragoncello macchiato	942
drepana	827, 828
dugongo	1454
dussumeria	173

E

efipidi	825
eglefino	453
eleotridi	1048
emelittidi	684
emiranfidi	427
emulidi	705
engraulidi	208

eperlano	274, 275, 276
eufausiacei	1203

F

falsa aringa atlantica	161
fanfano	650
fasolari	1400
fasolaro	1395
fasolaro orientale	1399
feoficee	1525
fieti	1030
fieti americani	1036
fieto	1037
fieto americano	1033, 1035
fieto della California	1034
fieto indiano	1038
fieto nero	677
fimbria	1094
foca	1453
foca barbuta	1448
foca comune	1453
foca dagli anelli	1452
foca dal cappuccio	1447
foca di Groenlandia	1451
foca grigia	1449
foca monaca	1450
focena	1471
foche	1446
folade	1416
foladi	1417
fragolino	794
fucali	1527

G

gadidi	440
gadiformi	435
galateidi	1306
galiotto	930
galletto	929
gallinella	1079, 1080, 1081, 1082, 1084, 1089, 1090, 1091
gamberello boreale	1237
gamberetti carenati	1236
gamberetti rosa	1239
gamberetto americano	1246
gamberetto grigio	1251
gamberetto maggiore	1249
gamberetto rosa	1238
gamberetto sega	1249
gamberi	1211, 1240
gamberi a lunghe chele	1248
gamberi coltello	1255
gamberi del Pacifico	1230
gamberi di fiume americani	1281, 1283
gamberi duri	1253
gamberi rossi	1232
gambero	1237, 1258
gambero a lungo rostro	1206
gambero barbato	1231
gambero blu	1247
gambero carenato cileno	1235
gambero coltello	1254
gambero d'acqua dolce americano	1282
gambero di fiume	1279
gambero di fiume australiano	1316
gambero di fiume europeo	1276
gambero di Guinea	1210
gambero di palude	1284
gambero di torrente	1280
gambero duro	1252
gambero grigio	1251

gambero imperiale	1220
gambero rosa mediterraneo	1212
gambero rosso	1234
gambero rosso americano	1257
gambero rosso argentino	1256
gambero rosso mediterraneo	1233
gambero turco	1277
gamberone	1217, 1219, 1226
gamberone bruno	1207
gamberone centramericano	1228
gamberone mediterraneo	1220
gamberone siba	1208
gamberoni	1209, 1229
gambusia	516
gasterosteidi	499
gattomare	913, 915
gattopardo	32
gattucci	28, 33
gattuccio	31
gattuccio bruno	29
gelidium	1529
ghiozzetto cenerino	1058
ghiozzetto di laguna	1056
ghiozzetto minuto	1059
ghiozzi	1049, 1055
ghiozzo del Mar Nero	1060
ghiozzo di Kessler	1057
ghiozzo dorato	1051
ghiozzo gò	1061
ghiozzo nero	1053
ghiozzo paganello	1054
ghiozzo testone	1052
gigartine	1531
gisu africano	153
glatdick	134
globicefalo	1466
gobetti	1240
gobidi	1049
gobione	333
gobione del Danubio	334
gourami	1045
grampo	1467
granatiere	494, 497
granatieri	498
grancapitano	556
granceola	1315
grancevola	1315
grancevola artica	1312
grancevola del Camciatca	1310
grancevola neozelandese	1314
grancevole artiche	1313
granchi	1287
granchi di fondale rossi	1322
granchi nuotatori	1296, 1301
granchi reali	1311
granchio blu	1295
granchio cinese	1303
granchio comune	1297, 1298
granchio di mare	1291
granchio indiano	1302
granchio nuotatore	1295
granchio nuotatore gazami	1300
granchio nuotatore purpureo	1294
granchio reale	1307, 1308, 1310
granchio reale australe	1309
granchio riparo	1297, 1298
granchio rosso di fondale	1321
granciporri	1287
granciporro	1291
granciporro atlantico giallo	1289
granciporro atlantico rosso	1288
granciporro del Pacifico	1290
granciporro del Pacifico	1292
granciporro della Florida	1293

granseola	1315	lemargo	63
gronghi	408	leone marino di Steller	1443
grongo	410	leone marino sudamericano	1444
grongo americano	411	lepade cornucopia	1197
grongo canino	405	lepadi	1198
grongo d'Argentina	412	lepidorombi	1170
grongo giapponese	409	letrinidi	762
grugnoli	714	leucisco	344
grugnolo	711, 712, 713	licodi	921
grugnolo spinoso	704	limanda	1137, 1139, 1142
gurami	1045	limanda americana	1147
gurami scaglioso	1046	limanda giapponese	1136, 1138
		limanda striata	1138
		limone di mare	1508

H

halibut	1133, 1148	limulo	1515
halibut del Pacifico	1134	linguatula	1177
halibut di Groenlandia	1148	lipara atlantica	1107
hirame	1120, 1121	lipara striata	1108
		lira	1090
		lofidi	1192

I

ido	343	loliginidi	1422
ilissa africana	183	lompo	1105
ilissa maggiore	184	longona	1404, 1405
iperodonte	1460	losanga atlantica	1111, 1112
ippoglosso atlantico	1133	losanga indiana	1113
ippoglosso nero	1148	lotregano	540
istioforidi	1015	lucci	286
		lucci marini	529

K

krill antartico	1204	luccio	288
krill norvegese	1205	luccio alligatore	143, 144
		luccio del Pacifico	291
		luccio di mare	535

L

labei	337	luccio marino	535
labridi	837	luccio nordamericano	287, 289, 290
lacerto	389, 984	lucciomurene	407
laggion	851	luccioperca	620
lamantino africano	1457	lucerna mediterranea	874
lamantino americano	1456	luna blu	678
lamantino dell'Amazzonia	1455	lupi di mare	913
laminarie	1526	lupicante	1318
lampreda del Danubio	2	lupino africano	1397
lampreda dell'Ucraina	3	lupo di mare	915
lampreda di fiume	5	lutiani	687, 693
lampreda di mare	7	lutianidi	687
lampreda di ruscello	6	lutiano	689
lampreda di Vladikov	4	lutiano coda gialla	694
lampreda marina	7	lutiano delle mangrovie	688
lamprede	1	lutiano rosso	690, 691
lampuga	681	lutiano striato	692
lanzardo	984	luvaro	794
lasca	323, 326, 329		
lasca dalmatina	324		
lasca iberica	327		

M

latridi	861	maccarelli reali	1004
lattario	628	maccarello	985
latterini	548	maccarello reale	988, 989
latterino	549, 550, 551	maccarello reale atlantico	1000
latterino menidia	552	maccarello reale australiano	1001
latterino sardaro	550	maccarello reale cinese	1003
lecce	648, 664	maccarello reale coreano	992
lecce salterine	653	maccarello reale del Queensland	999
leccia	647	maccarello reale di California	990
leccia dei Caraibi	662	maccarello reale di Guinea	1005
leccia scagliosa	649	maccarello reale giapponese	997
leccia stella	663	maccarello reale Kanadi	998
leiognatidi	699	maccarello reale maculato	1002
		maccarello reale maculato	987, 991,
			994, 996
		maccarello reale papuano	995
		maccarello reale striato	993
		maccarello striato	962
		macruridi	496
		madia giapponese	1391

madia glabra	1390
madreperia di fiume	1383
magnose	1304
magnosella	1305
manato africano	1457
manato comune	1456
manato senza unghie	1455
manicai	1407
manta	112, 113
mante	111
margaritifera	1383
marlin	1015, 1020, 1027
marlin azzurro	1019
marlin bianco	1021
marlin inerme	1022
marlin nero	1018
marlin striato	1023
marmora	786
marsione	1059
marsuino	1471
marsuino di Dall	1472
mazzancolla	1217, 1219, 1220, 1226
mazzancolla banana	1222
mazzancolla bianca atlantica	1227
mazzancolla caffè	1213
mazzancolla californiana	1215
mazzancolla cinese	1216
mazzancolla coda rossa	1225
mazzancolla giapponese	1219
mazzancolla gigante	1223
mazzancolla rosa	1224
mazzancolla rossa	1214
mazzancolla striata australiana	1218
mazzancolla verde	1221
mazzancolle	1229
meduse	1512
megattera	1485
melù	459
melù australe	458
menhaden	166, 167
menhaden messicana	165
mennola	812
mennole	810, 814
menola	812, 813
menola africana	811
merlano	454, 455
merluccidi	477
merluzzetto	473
merluzzetto bruno	472
merluzzi	477, 492
merluzzi granatieri	480
merluzzo	447, 482, 483, 484, 485, 486,
	488, 489, 490, 491
merluzzo argentato	487
merluzzo artico	441
merluzzo bianco	446, 447, 448
merluzzo capellano	473
merluzzo carbonaro	467
merluzzo del Pacifico	446
merluzzo dell'Alasca	1094, 470
merluzzo gallo	466
merluzzo granatiere	478, 479
merluzzo senegalese	491
mezzobecchi	427, 433
mezzobecco	428, 429, 430, 432
mezzobecco brasiliano	430
mezzobecco giapponese	431
mictofidi	294
miro	413
mitilidi	1370
mitilo	1376, 1377
mojarra	902
mojarre	903
moki	861, 862

molluschi marini 1323
molo 454, 455
molva 461, 462
molva azzurra 460
molva occhiona 461
morati 865
moridi 436
mormora 786
mormora africana 785
mormore 787
morwong 860
moscardini 1437
moscardino 1435, 1436
moscardino bianco 1435
mostella 464, 468
motella a cinque baffi 443
motella a quattro baffi 469
motella maculata 450
motelle 451
muggine 545
muggine dorato 540
muggini 537
mugilidi 537
mullidi 815
murena 404
murena del Giappone 406
murena verde 403
murene 402
murici 1334
muschio irlandese 1532
musdea 464, 468
musdea americana 476
musdea atlantica 475
musdea bianca 463, 468
musdea brasiliana 474
musdee 465
musino 543
mussolini 676
mustella 463

N

narvalo 1479
naselli 477, 492
nasello 481, 482, 483, 484, 485, 486, 487, 488, 489, 490, 491
nasello atlantico 483
nasello del Capo 484
nasello del Cile 485
nasello del Pacifico 490
naso 325
navaga 444, 445
neceratodo di Forster 127
necora 1299
nemipteri 697
nemipteridi 695
nemiptero 696
nono 513
nono andaluso 515
nono iberico 514
nototenia 881, 882, 883, 884, 885, 886, 887, 888, 889, 890, 891, 892, 893
nototenidi 876
nototenie 894

O

occhialone 793
occhiata 788
occhione 299
occhioni 298
occhiverdi 298, 299
ofiodonte 1092

oloturia giapponese 1523
ombrina 728, 730, 737, 738, 748, 749, 750, 751, 759
ombrina americana 733, 734, 735
ombrina bianca 719, 720, 750
ombrina bocca d'oro 718
ombrina d'acqua dolce 716
ombrina d'Argentina 758
ombrina dentata 723, 725, 726, 727
ombrina giapponese 740, 741
ombrina nera 745
ombrina ocellata 757
ombrina peruviana 743
ombrina tigre 742
ombrine 715, 739, 752, 760
ombrine americane 736
ombrine dentate 729
orata 802
orata del Giappone 805
orca 1470
orecchie di mare 1329
orecchio di mare 1328
orei 528
ostrica 1353
ostrica cilena 1351
ostrica concava 1347
ostrica concava australiana 1357
ostrica dei Caraibi 1348
ostrica della Nuova Zelanda 1355
ostrica della Virginia 1349
ostrica dentellata 1352
ostrica europea piatta 1353
ostrica giapponese 1347
ostrica perlifera 1346
ostrica piatta 1353
ostrica piatta del Pacífico 1354
ostrica portoghese 1347
ostriche 1350
ostriche piatte 1356
otaria 1443
otaria da pelliccia sudafricana 1442
otaria da pelliccia sudamericana 1441
otarie 1440
otoperca 703
ottopodi 1434

P

pagelli 795
pagello 790, 793, 794
pagello bastardo 790
pagello fragolino 794
pagello mafrone 790
pagello rosso 791, 792
pagri 807
pagri dei Caraibi 767
pagri napoleoni 770
pagro 803, 804, 806
pagro azzurro 804
pagro mediterraneo 806
pagro napoleone 769
pagro reale 803
pagro testatonda 800
paguri 1286
palamita 982
palamita bianca 974
palamita cilena 980
palamita maculata 966
palamita orientale 979, 981
palamita squamosa 970
palemonidi 1245
palemonidi d'acqua dolce 1244
palinuridi 1259

palombi 49, 55
palombo 52, 53
palombo antartico 51
palombo atlantico 54
palombo maculato 56
palombo stellato 52
pampo argenteo 1031
pampo cinese 1032
pampo giapponese 1044
panga 799
pannocchia 1202
pannocchie 1201
papalina 202
papalina del Caspio 171
paraperca blu 864
parona 651
passera 1128, 1132, 1143, 1145
passera artica 1140
passera canadese 1132
passera canina 1124, 1125
passera del Giappone 1126
passera del Pacifico 1129, 1135
passera della California 1127
passera di mare 1145
passera giapponese 1130, 1131
passera lingua di cane 1128
passera liscia 1141
passera pianuzza 1143
passera stellata 1144
pastinaca 104, 106
pataracce 1114
patelle 1333
pellona indiana 188
peperoncino 912
perca 617
perca rumena 618
perchia 587
perchia americana 572
perchia del Perù 584
perchia nera 571
perchia striata 571
percidi 613
persici tropicali 566
persico argentato 605
persico bianco 604
persico del Nilo 565
persico di Dolomieu 602
persico dorato 616
persico occhi rossi 600
persico sole 601
persicospigola americana 596
persicospigola bianca 597
persicospigola striata 598
persicotrota 603
pesce ago 506
pesce ago cavallino 508
pesce ago minore 507
pesce angelo 68
pesce balestra 1180, 1183
pesce bandiera 17
pesce burro 709, 710
pesce burro maculato 707
pesce burro striato 708
pesce castagna 680
pesce chirurgo 946
pesce chitarra 73, 74, 75
pesce civetta 1109
pesce coltello 959
pesce del ghiaccio 896, 897, 898, 899, 901
pesce del ghiaccio ocellato 900
pesce diavolo 61
pesce fico 449
pesce forca 1106

pesce gatto 384, 385
pesce gatto marino 377
pesce gatto puntado 386
pesce lancia striato 1023
pesce lanterna 295
pesce lucertola 393
pesce luna 1187, 1188
pesce luna troncato 1189
pesce martello 46, 47
pesce nastro 511
pesce palla 1185
pesce palla maculato 1185
pesce palo 873
pesce pappagallo 856
pesce persico 617
pesce pettine 855
pesce pilota 650
pesce pony 700
pesce porco 70
pesce prete 874
pesce ramarro 393
pesce ramarro indiano 391
pesce ramarro orientale 392
pesce rampicante 1039
pesce rasoio 350
pesce ré 509
pesce rosso 320
pesce San Pietro 525
pesce sciabola 957
pesce sciabola del Pacifico 958
pesce sciabola nero 956
pesce sega 79
pesce serra 630
pesce sorcio 493
pesce spada 1029
pesce spatola cinese 141
pesce specchio 521
pesce specchio atlantico 520
pesce tamburo 527
pesce testa di serpente 557, 558
pesce topino 495
pesce topo pelagico 213
pesce trombetta 504
pesce vela atlantico 1016
pesce vela del Pacifico 1017
pesce violino 76
pesce volante 426
pesce volpe 17
pescecane 19
pesce-gatto marino 375
pescegatto marino 376
pesci angelo 830
pesci argo 829
pesci ascia 676
pesci azzurri 154
pesci balestra 1182
pesci bisonte 302
pesci burri 705
pesci castagna 679
pesci chirurgo 945
pesci chitarra 72, 77
pesci coltello 292
pesci cristallo 379, 563
pesci del ghiaccio 895
pesci diavolo 62
pesci fucilieri 686
pesci gatto 383, 387
pesci imperatore 761, 762
pesci lancia 1015, 1020
pesci lanterna 294
pesci lucertola 390
pesci martello 45, 48
pesci palla 1184, 1186
pesci palo 872

pesci pipistrello 826
pesci pony 701
pesci porco 69
pesci prete 875
pesci pulitori 293
pesci ragno 866, 871
pesci ramarro 390
pesci ratti 496
pesci rondine 423
pesci San Pietro 523
pesci sciabola 955
pesci sega 78
pesci serra 629
pesci spada 1028
pesci spatola 140
pesci tamburo 526
pesci testa di serpente 559
pesci trombetta 503
pesci vela 1015
pesci violino 72
pesci volanti 423, 425
pescigatto di mare 374
pettine 1359, 1363
pettine maggiore 1365
pettinidi 1358
piè d'asino 1384
pigo 358
pinna 1385
platax 826
platessa 1145
platessa dell'Alasca 1146
platicefalidi 1095
pleuronettidi 1123
pleuronettiformi, pesci piatti 1110
plutosi 378
polinemidi 553
pollack 466
pollack d'Alasca 470
polpi 1439
polpo 1438
polpo di scoglio 1438
pomacentridi 835
pomatomidi 629
potassolo 459
priacantidi 606
prototteri 129
pseudacciuga atlantica 211
pseudacciuga del Pacifico 212
pseudorca 1473
pseudosaraghi 860
pseudosarago 858, 859
pseudosarago rosa 857
pseudospigola americana 561
pseudospigole 562

R

ragano 396
ragno 869, 870
raiformi 80
rajidi 81
rana pescatrice 1194, 1195
rana pescatrice americana 1193
rana pescatrice indiana 1196
rana toro 1492
rane 1493
rane pescatrici 1192
razza 83, 87, 88, 90, 91, 92, 95, 99, 100
razza a coda corta 85
razza bavosa 84
razza bianca 82
razza bianca atlantica 90

razza chiodata 87
razza cuculo 94
razza fiorita 94
razza maculata 93
razza monaca 96
razza occhiata 95
razza ondulata 101
razza quattrocchi 92
razza rotonda 86
razza scuffina 98
razza spinosa 89, 98
razza stellata 83, 97
razze 81, 102
re di aringhe 510
re di triglie 609, 610
re di triglie nero 611
re di triglie orientale 612
remora nera 1178
ricci di mare 1519, 1520
riccio di mare 1518, 1521
riccio di mare cileno 1522
ricciola 658
ricciola australe 659
ricciola del Sudafrica 659
ricciola di fondale 1041
ricciola di fondale australe 1043
ricciola giapponese 660
ricciole 661
robalo patagonico 880
rodeo amaro 356
rodoficee 1528
rombi 1167
rombi australi 1150
rombo 1172
rombo australe 1149
rombo camaso 1176
rombo canadese 1174
rombo chiodato 1172
rombo chiodato del Mar Nero 1173
rombo dentato 1118
rombo di rena 1117
rombo giallo 1169
rombo liscio 1175
rombo peloso di Norvegia 1171
rombo quadripuntato 1119
rombo quattrocchi 1168
ronco 71
ronco spinoso 71
rondine di mare 434
rondone di mare giapponese 424
rorqualo 1484
rospo 1194
rospo 1195
rossetto 1050
rovella 359
rovello 793
ruvetti 947, 948
ruvetto 950

S

sacchetto 588
sagrì 58, 59
sagrì nero 61
salmerini 267
salmerino 265
salmerino alpino 263
salmerino di fontana 264
salmerino di lago 266
salmone 256
salmone argentato 249
salmone atlantico 256
salmone australiano 682, 683

IT

salmone del Danubio	242
salmone del Pacifico	247, 248, 249, 250, 252, 254
salmone giapponese	250
salmone keta	248
salmone reale	254
salmone rosa	247
salmone rosso	252
salmoni, trote e salmerini	262
salmonidi	262
salmotrota	257
salpa	801
San Pietro d'America	524
sandra	620
sandra americana	621
sandra canadese	619
sanguinacci	1341
sanguinaccio giapponese	1340
sanguinaccio orientale	1339
sanguinerola	326, 352
sanguinerola di palude	351
saraghi	784
sarago	779, 782, 783
sarago americano	764, 809
sarago faraone	779
sarago fasciato	783
sarago maggiore	782
sarago pizzuto	781
sarago sparaglione	778
sarago testa nera	783
sardella	189
sardella d'Africa	190
sardelletta macchiata	181
sardina	189
sardina australiana	198
sardina del Cile	200
sardina del Sudafrica	199
sardina di California	196
sardina giapponese	197
sardoncino americano	209, 210
sardone	217
sargo	779, 782
sargo pizzuto	781
savetta	328
sbaro	778
scampo	1320
scampo oceanico	1319
scardola	362
scardola greca	363
scaro	856
scarpene	1062
scazzone	1097
scazzone americano	1098
scazzone boreale	1099
scazzone marino	1100, 1102
scazzoni	1101
sciarrani	590
sciarrano	587, 588, 589
scienidi	715
scillaridi	1304
scoftalmidi	1167
scoloperi	698
scomberesocidi	420
scomberoidi	960
scombridi, sgombri, tonni, palamite e maccarelli	961
scorfani	1062, 1069, 1077
scorfani di fondale	1064
scorfano atlantico	1076
scorfano corallino	1065
scorfano di fondale	1063
scorfano di Norvegia	1075
scorfano mediterraneo	1068
scorfano nero	1067

scorfano rosso	1068
scorfanotto	1066
scorfanotto sudafricano	1071
scorpenidi	1062
sebaste	1070, 1073, 1076
sebaste a pinne gialle	1074
sebaste bruno	1072
sebaste di Norvegia	1075
sebasti	1077
seppia	1418
seppia mediterranea	1418
seppiola	1420
seppiola grossa	1419
sergente rosso	522
sergestidi	1241
seriolelle	1042
serpente di mare	414
serranidi	567
sfirenidi	529
sgombri	986
sgombri indiani	978
sgombro	984, 985
sgombro indiano	975, 976, 977
sgombro maculato	983
sicione	1253
sigani	944
sigano	943
sillaginidi	625
siluri	380
siluro	382
siluro greco	381
smeriglio	23
smeriglio del Pacifico	22
smeriglio mako	20, 21
soaso	1175
sogliola	1165
sogliola azevia	1159
sogliola cuneata	1157
sogliola dal porro	1164
sogliola del Sudafrica	1153, 1154
sogliola fasciata	1160
sogliola gialla	1156
sogliola occhiuta	1158
sogliola pelosa	1162
sogliola rotonda	1152
sogliola turca	1163
sogliola variegata	1160
sogliole	1161
sogliole del Sudafrica	1155
soleidi	1151
sparaglione	778
sparaglione americano	780
sparidi	763
sperlani	270, 277
sperlano	271, 274, 275, 276
sperlano del Pacifico	272
sperlano striato	278
spigola	593
spigola giapponese	599
spigola macchiata	594
spigola puntata	594
spigole	595
spinarelli	499
spinarello	500
spinarello marino	502
spinarello minore	501
spinaroli	57
spinarolo	64, 65
spinarolo bruno	65
spinarolo muso corto	66
spratello	201
spratto	202
spugne	1511
squadri	67

squadro	68
squali	18, 9
squali martello	45
squalidi	57
squalo alalunga	37
squalo balena	27
squalo bronzeo	36
squalo capopiatto	11
squalo di Groenlandia	63
squalo elefante	12
squalo grigio	39, 40
squalo limone	41
squalo mako	20
squalo manzo	10
squalo musoguzzo	43, 44
squalo nutrice	25, 26
squalo pinne bianche	35
squalo pinne nere	38
squalo serpente	8
squalo striato	24
squalo toro	13
squalo volpe	17
squalo volpe occhione	16
squalo volpe pelagico	15
squatinidi	67
squillidi	1201
stella marina	1517
stelle di mare	1516
stenella rostrata	1475
stenella striata	1474
stenelle	1476
sterleto	135
stolefori	222
storione	137
storione bianco	138
storione comune	137
storione danubiano	131
storione del Pacifico	132
storione ladano	139
storione stellato	136
storione verde	132
streber	623
strombi	1338
stronazza	316
suacia	1115
suacia mora	1116
suacia orientale	1122
suacie	1114
succiascoglio	1190
sugarelli	675
sugarello	641, 643, 666, 668, 671, 672
sugarello australe	665
sugarello cileno	669
sugarello giapponese	666
sugarello pittato	670
sugarotti	645
sugarotto	641, 642, 643, 644
suri	675
suro	641, 643, 666, 668, 670, 671, 672
suro americano	667
suro bandagialla	655
suro cunene	674
suro del Pacifico	671
suro del Sudafrica	673
suro occhione	654

T

talpa	23
tanuta	808
tanute sudafricane	789
tarpone	146, 150, 152
tarpone indiano	149

IT

489

tarponi	145, 147, 148, 151
tartaruga bastarda	1498
tartaruga caretta	1496
tartaruga comune	1496
tartaruga liuto	1495
tartaruga verde	1497
tautoga	852
telline	1406
temolo	268
teraponidi	592
terpugo	1093
terrapin	1499
testapiatta	1096
testudinati, tartarughe	1494
tetraodontidi	1184
tilapia del Mozambico	832
tilapia del Nilo	833
tile gibboso	627
tili	626
tinca	364
tiraghlin	719
tirsite	949, 951, 954
tirsiti	952
todaro	1433
tombarello	964, 965
tomcod	456, 457
tonnetto	967, 968, 969, 981, 982
tonnetto cileno	980
tonnetto orientale	979
tonnetto striato	973
tonni	1006, 1014
tonnina	963
tonno	1007, 1008, 1010, 1011, 1012, 1013, 973
tonno albacora	1008
tonno bianco	1007
tonno obeso	1011
tonno pinna nera	1009
tonno pinnagialla	1008
tonno rosso	1012
tordo	843
tordo americano	853
tordo dorato	839
tordo elettrico	394
tordo fischietto	841
tordo grigio	845
tordo marvizzo	840
tordo musolungo	850
tordo nero	842
tordo occhionero	847
tordo ocellato	848
tordo pavone	851
tordo piumato	844
tordo rosso	846

tordo verde	849
torpedine marezzata	118
torpedine nera	119
torpedine occhiuta	120
torpedini	117, 121
totano	1428, 1429, 1430, 1433
totano giapponese	1432
totano gigante del Pacifico	1427
totano neozelandese	1431
totariello	1423
totoaba	724
trachinidi	866
tracina	869, 870
tracina drago	869
tracina raggiata	870
tracina ragno	867, 868
tracina vipera	867
tracine	866, 871
tremoli	117
triachidi	49
tricheco	1445
trichiuridi	955
triglia	817
triglia bianca	817
triglia canina	820
triglia dentata	820
triglia di fango	817
triglia di scoglio	818
triglia dorata	816
triglia rossa	821
triglidi	1078
triglie	815, 819
triglie rosse	822
trigone	104
trigone del Pacifico	104
trigone spinoso	105
trigone viola	107
trigoni	103
triotti	361
triotto albanese	348
triotto rosso	360
triotto russo	357
trocus	1330
trota	243, 246, 255, 258
trota americana	251
trota apache	244
trota arcobaleno	251
trota di fiume	258
trota di lago	259
trota di mare	257
trota dorata messicana	245
trota fario	258
trota giapponese	253
trota iridea	251

trote	260
turbo	1331
turbo sudafricano	1332
tursione	1477
tursiope	1477

U

umbra	284
umbra americana	285
uovo di mare	1508

V

vacca marina	1456
vaccarella	110
vairone	345
veneridi	1393
ventaglio	1365
verdesca	42
verzelata	543
vimba	365
volpina	545
vongola	1396, 1404, 1405
vongola nera	1401
vongola verace	1401, 1402
vongole	1394
vopa	766

W

waho	962

X

xifosuro	1515

Z

zanchetta	1115
zanchette	1114
zeidi	523
zerri	810, 814
zerro	813
zifidi	1458
zifio	1461
zigrino	60
zinghel	624
zoarcidi	920
zope	304

IT

Nederlandse index

A

aal	396, 401
aal, Amerikaanse conger-	411
aal, Amerikaanse puit-	922
aal, Argentijnse conger-	412
aal, conger-	410
aal, Guinea-snoek-	405
aal, kwab-	452
aal, puit-	923
aal, worm	413
achtarmige inktvissen	1434
Adriatische blankvoorn	359
Adriatische steur	133
afkijkvis	656
afkijkvis, Atlantische	657
Afrikaanse gestreepte knorvis	708
Afrikaanse harder	544
Afrikaanse lamentijn	1457
Afrikaanse longvissen	129
Afrikaanse meervallen	388
Afrikaanse rotsmossel	1382
Afrikaanse sikkelvis	827
agujon-geep	418
akiami-garnaal	1242
Alaska-pollak	470
Alaska-schol	1146
Albanese barbeel	313
Albanese blankvoorn	348
alen	395
Alexandrië-pompano	633
algen	1524
alikruiken	1337
alligator, Mississippi-	1502
alligators	1500
alver	310
alver, Donau-	322
alver, gestippelde	308
alver, witte	309
amande	1384
Amazone-lamentijn	1455
Amerikaans witje	1129
Amerikaanse ansjovis	210
Amerikaanse baars	616
Amerikaanse congeraal	411
Amerikaanse draakvis	124
Amerikaanse elft	162
Amerikaanse grassnoek	287
Amerikaanse griet	1174
Amerikaanse grootbek	1035
Amerikaanse grote mantel	1369
Amerikaanse hondsvis	285
Amerikaanse knorvis	702
Amerikaanse koningsombervis	733
Amerikaanse kreeft	1317
Amerikaanse lamentijn	1456
Amerikaanse lipvis	853
Amerikaanse marene	232
Amerikaanse meerforel	266
Amerikaanse paling	400
Amerikaanse ponen	1087
Amerikaanse puitaal	922
Amerikaanse rivierkreeft	1282
Amerikaanse roodbaars	1073
Amerikaanse schol	1132
Amerikaanse snoekbaars	621
Amerikaanse spiering	275

Amerikaanse tong	1152
Amerikaanse venusschelp	1398
Amerikaanse winterschol	1147
Amerikaanse zandspiering	932
Amerikaanse zeebaars	596
Amerikaanse zeedonderpad	1098
Amerikaanse zeeduivel	1193
Amerikaanse zonnevis	524
Amerikaanse zwaardschede	1408
Amoer-snoek	291
Andaman-langoestine	1319
Angola-tandbrasem	771
ansjovis	217
ansjovis, Amerikaanse	210
ansjovis, Argentijnse	214
ansjovis, Atlantische	211
ansjovis, Australische	215
ansjovis, grenadier-	213
ansjovis, Japanse	218
ansjovis, Kaapse	216
ansjovis, Kaapse	218
ansjovis, Noord-Pacifische	219
ansjovis, Pacifische	212
ansjovis, Peruaanse	220
ansjovis, rivier-	221
ansjovis, Zuid-Atlantische	209
ansjovissen	208
ansjovissen, stoleforus-	222
Antarctisch krill	1204
Antarctische gladde haai	51
Antarctische ijsheek	878
Antarctische ijsheken	879
Antarctische kabeljauw	894
Antarctische ombervis	754
Antarctische zilvervis	893
Antarctische zwarte vis	1043
Arctische houting	239
Arctische sneeuwkrab	1312
Arctische spiering	276
arendrog	109
Argentijnse ansjovis	214
Argentijnse congeraal	412
Argentijnse heek	486
Argentijnse ombervis	758
Argentijnse rode garnaal	1256
Argentijnse rode pijlinktvis	1428
Argentijnse stilettogarnaal	1206
Argentijnse zeebaars	568
argusvissen	829
aristeide garnalen	1232
Aristoteles-meerval	381
arkschelp, Japanse	1340
arkschelp, West-Pacifische	1339
arkschelpen	1341, 1342
artemisschelp, Orbigny-	1397
asper	622
asprete	618
Atka	1093
Atlantische afkijkvis	657
Atlantische ansjovis	211
Atlantische diepzeegarnaal	1210
Atlantische draadvinnige haring	186
Atlantische gaffelkabeljauw	475
Atlantische geep	417
Atlantische hagedisvis	393
Atlantische kogelvissen	1186
Atlantische koornaarvis	552

Atlantische melkhaai	44
Atlantische ombervis	738
Atlantische ronde haring	177
Atlantische rotskrab	1289
Atlantische seabobgarnaal	1231
Atlantische sidderrog	119
Atlantische slibgarnaal	1258
Atlantische slijmkop	520
Atlantische tomcod	456
Atlantische wrakbaars	585
Atlantische zeilvis	1016
Atyide-garnalen	1243
Augustinuskrab	1307
Australische ansjovis	215
Australische bonito	979
Australische bot	1122
Australische garnaal	1207
Australische grote mantel	1366
Australische halfsnavelbek	428
Australische heek	482
Australische koningsmakreel	996
Australische langoest	1270
Australische longvis	127
Australische mossel	1378
Australische paling	397
Australische pelser	198
Australische rivierkreeft	1316
Australische sardine	198
Australische walvis	1489
Australische zalm	683
ayu	269
Aziatische barbeel	354
Aziatische sardinella	191
Azteken-garnaal	1213

B

baardmannetje	929
baardrob	1448
baars	617
baars, Amerikaanse	616
baars, Amerikaanse rood-	1073
baars, Amerikaanse snoek-	621
baars, Amerikaanse zee-	596
baars, Argentijnse zee-	568
baars, Atlantische wrak-	585
baars, barramundi-	564
baars, blauw-	630
baars, Braziliaanse zand-	865
baars, Californische grote zaag-	591
baars, Canadese snoek-	619
baars, diepzeerood-	1076
baars, driedoornige stekel-	500
baars, dwergzaag-	588
baars, forel-	603
baars, geelstaart-rood-	1074
baars, geiten-	587
baars, gestreepte zee-	598
baars, gevlekte zee-	594
baars, gouden zaag-	574
baars, grote zaag-	577
baars, hapuku-wrak-	586
baars, Japanse zee-	599
baars, Kaapse rood-	1071
baars, Malabar-zaag-	579

baars, Nassau-zaag-	581	
baars, nijl-	565	
baars, Pacifische gevlekte zaag-	575	
baars, Pacifische grootoog-	607	
baars, Pacifische rood-	1070	
baars, Patagonische ijs-	880	
baars, Peruaanse zee-	584	
baars, reuzenzaag-	578	
baars, rode vlag-	569	
baars, rode zaag-	580	
baars, rood-	1075	
baars, schrift-	589	
baars, snoek-	620	
baars, steen-	600	
baars, tiendoornige stekel-	501	
baars, weduwerood-	1072	
baars, witte steen-	785	
baars, witte zaag-	573	
baars, witte zee-	597	
baars, zadel-	788	
baars, zand-	572	
baars, zandsteen-	786	
baars, zee-	593, 595	
baars, zeestekel-	502	
baars, zilver-	605	
baars, zonne-	601	
baars, zwarte	602	
baars, zwarte zaag-	576	
baars, zwarte zee-	571	
baarzen, blauw-	629	
baarzen, glas-	563	
baarzen, grootoog-	608	
baarzen, koraal-	583	
baarzen, nijl-	566	
baarzen, rood-	1064, 1077	
baarzen, steen-	787	
baarzen, stekel-	499	
baarzen, zaag-	567, 582, 590	
baarzen, zoetwater-	613	
balao-halfsnavelbek	429	
Balkan-modderkruiper	367	
bananengarnaal	1222	
bandvis	511	
bandvis, rode	834	
barbeel	314	
barbeel, Albanese	313	
barbeel, Aziatische	354	
barbeel, Donau-	317	
barbeel, gestreepte zee-	817	
barbeel, Iberische	315	
barbeel, Italiaanse	318	
barbeel, Javaanse	353	
barbeel, mediterrane	316	
barbeelgrondel, Dalmatische	312	
barbelen, franjelip-	337	
barbelen, zee-	815, 819	
barbiervis, rode	570	
barracuda	532	
barracuda, Californische	531	
barracuda, Guinese	530	
barracuda, Indische	534	
barracuda, Middellandse-Zee-	535	
barracuda, oostelijke	533	
barracuda's	529, 536	
barramundibaars	564	
bastaardheilbot, olijfgroene	1120	
bastaardheilbot, viervlekkige	1119	
bastaardheilbotten	1121	
bastaardknorvis	712	
Batavia-snoekaal	406	
beekforel	258	
beekprik	6	
beensnoek, kortneus-	144	
beensnoek, langneus-	143	

beerkreeft, kleine	1305	
beerkreeften	1304	
beluga	139, 1478	
Benguela-heek	489	
Bergatino-modderkruiper	368	
Bergs vingervin	857	
bermpje	372	
beryciden	517	
beryx	519, 518	
bichir	128	
bidsprinkhaankreeft	1202	
bidsprinkhaankreeften	1201	
bijlhorsmakreel	640	
bittervoorn	356	
blaasjeswieren	1527	
Blainville's doornhaai	65	
blankvoorn	360	
blankvoorn, Adriatische	359	
blankvoorn, Albanese	348	
blankvoorn, parel-	357	
blankvoorns	361	
blauwbaars	630	
blauwbaarzen	629	
blauwe diepzeekabeljauw	437	
blauwe grenadier	480	
blauwe haai	42	
blauwe haaien	34	
blauwe horsmakreel	635, 670	
blauwe kabeljauw, Nieuw-Zeelandse	864	
blauwe krab	1295	
blauwe krab, Dana	1294	
blauwe krabben	1296	
blauwe leng	460	
blauwe marlijn	1019	
blauwe Nieuw-Zeelandse trompetvis	862	
blauwe tegelvis	627	
blauwe vinvis	1483	
blauwe wijting	459	
blauwe wijting, zuidelijke	458	
blauwe zeewolf	914	
blauwgevlekte zeebrasem	804	
blauwkeeltje	1063	
blauwneus	365	
blauwrode diepzeegarnaal	1233	
blauwvinpoon	1082	
blauwvintonijn, zuidelijke	1010	
blei, brasem-	304	
blei, kol-	319	
blei, roof-	311	
bliek, sabel	350	
blonde rog	85	
boboharder	539	
bobo-ombervis	749	
Bodenmeer-houting	240	
boe-ombervis	753	
bokvis	766	
bokvis, gestreepte	801	
bolk, dwerg-	473	
bolk, steen-	472	
Bombay-eend	297	
Bombay-eenden	296	
bonga-elft	174	
bonito	982	
bonito, Australische	979	
bonito, gestreepte	981	
bonito, gevlekte	966	
bonito, ongestreepte	974	
bonito, Pacifische	980	
bonte mantel	1363	
boormossel	1416	
boormosselen	1417	
bot	1143	
bot, Australische	1122	
bot, gevlekte	1177	

bot, grootoog-	1117	
bot, ijs-	1140	
bot, Nieuw-Zeelandse	1149	
bot, Noorse dwerg-	1171	
bot, zomer-	1118	
botachtigen	1114	
boterschelp	1403	
botervis	919	
botten, Nieuw-Zeelandse	1150	
braam	680	
braamhaai	71	
bramen	679	
brandhoornslakken	1334	
brasem	305	
brasem, Angola-tand-	771	
brasem, blauwgevlekte zee-	804	
brasem, Donau-	306	
brasem, gevlekte zee-	798	
brasem, gewone zee-	806	
brasem, goud-	802	
brasem, gouden vlinder-	696	
brasem, grootoog-tand-	776	
brasem, Holbrooks ring-	780	
brasem, Japanse zee-	805	
brasem, Kanarische tand-	772	
brasem, Kongo-tand-	773	
brasem, Marokkaanse tand-	777	
brasem, Natal-zee-	792	
brasem, nufar-zee-	768	
brasem, Peking-	349	
brasem, ring-	778	
brasem, rode steen-	796	
brasem, rode zee-	794	
brasem, roodgestreepte zee-	803	
brasem, roze tand-	775	
brasem, schaapskop-zee-	764	
brasem, Schot-zee-	797	
brasem, Spaanse zee-	790	
brasem, spitse ring-	781	
brasem, tand-	774	
brasem, witte ring-	782	
brasem, Wuchang-	346	
brasem, zebra-ring-	779	
brasem, zee-	793	
brasem, zwartkop-ring-	783	
brasems, vlinder	695, 697	
brasemachtigen, ring-	784	
brasemblei	304	
brasems	307	
brasems, hottentot-zee-	789	
brasems, zee-	763, 795, 807	
Braziliaanse eendenbek	873	
Braziliaanse gaffelkabeljauw	474	
Braziliaanse halfsnavelbek	430	
Braziliaanse harder	546	
Braziliaanse zandbaars	865	
brilkaaiman	1503	
bronforel	264	
brotula	924	
bruine dwergmeerval	385	
bruine hondshaai	29	
bruine lipvis	842	
bruine schorpioenvis	1067	
bruine tijgergarnaal	1218	
bruine venusschelp	1395	
bruinvis	1471	
bruinwieren	1525	
brulkikkker	1492	
Bryde's vinvis	1482	
bulrog	110	
bultkopvis	844	
bultrug	1485	
busyconwulken	1336	
butskop	1460	

C

calico-scallop	1360
Californische barracuda	531
Californische grote zaagbaars	591
Californische ronde haring	176
Californische schol	1127
Californische spiering	272
Californische witte ombervis	730
Canadese elft	155
Canadese rog	88
Canadese snoekbaars	619
Canestrini-grondel	1058
Caraïbische langoest	1269
caramotegarnaal	1220
chacunda-elft	164
channel-dwergmeerval	386
Chileense haring	204
Chileense heek	485
Chileense horsmakreel	669
Chileense koningsklip	927
Chileense mossel	1374
Chileense nylongarnaal	1235
Chileense oester	1351
Chileense pyurazakpijp	1509
Chileense sardine	200
Chileense zee-egel	1522
Chinese elft	168
Chinese koningsmakreel	1003
Chinese krokodil	1506
Chinese lepelsteur	141
Chinese pomfret	1032
Chinese wolhandkrab	1303
chinook-zalm	254
chola-gitaarrog	74
cholgamossel	1371
chumzalm	248
cichliden	831
ciscomarene	231
citroenhaai	41
clam	1398
cobia	631
cohozalm	249
conger, Pacifische	409
congeraal	410
congeraal, Amerikaanse	411
congeraal, Argentijnse	412
congeralen	408
crangonide garnalen	1250
crappie, witte	604
cristalgarnaal	1214
Cunene-horsmakreel	674
cutthroatforel	255

D

dadelmossel	1372
dagaas	203
dageraads	770
Dalls dolfijn	1472
Dalmatische barbeelgrondel	312
Dalmatische sneep	324
Dana blauwe krab	1294
degenkrab	1515
degenspeervis	1025
degenvis	959
diepwaterheek	488
diepzee rode garnalen	1240
diepzeegarnaal, Atlantische	1210
diepzeegarnaal, blauwrode	1233
diepzeegarnaal, roze	1212
diepzeegarnalen	1211, 1232
diepzeekabeljauw, blauwe	437
diepzeekabeljauw, Nieuw-Zeelandse	438

diepzeekabeljauwen	436
diepzeekrab, rode	1321
diepzeekrabben, rode	1322
diepzeeroodbaars	1076
diepzeezilverheek	481
dikkopje	1059
diklipharder	538
diklipknorvis	709
dikrugtong	1160
dikrugtongen	1161
doktersvis, Monrovische	946
doktersvissen	945
dolfijn van Cuvier, spits-	1461
dolfijn, Dalls	1472
dolfijn, gestreepte	1474
dolfijn, gewone	1464
dolfijn, grijze	1467
dolfijn, langsnuit-	1475
dolfijn, witflank-	1468
dolfijn, witsnuit-	1469
dolfijn, zwarte	1459
dolfijnen	1463
dolfijnen, gevlekte	1476
dolfijnen, snavel-	1458
dolfijnvis	681
Donaualver	322
Donaubarbeel	317
Donaubrasem	306
Donaugrondel	334
Donauprik	2
Donausteur	131
Donauvoorn	358
Donauzalm	242
donderpad, Amerikaanse zee-	1098
donderpad, gehoornde rivier-	1099
donderpad, groene zee-	1102
donderpad, rivier-	1097
donderpad, zee-	1100
donderpadden, zee-	1101
donkere doornhaai	59
donkere haai	39
donkere poon	1080
doornhaai	64
doornhaai, Blainville's	65
doornhaai, donkere	59
doornhaai, kortsnuit-	66
doornhaai, ruwe	58
doornhaai, zwarte	61
doornhaaien, haaien en	57
doornhaaien, zwarte	62
doornige heilbot	1112
doornige heilbot, gevlekte	1111
doornvissen	592
draadvinnige elft	172
draadvinnige haring, Atlantische	186
draadvinnige haring, Pacifische	185
draadvinnige haringen	187
draadzeilvijlvis	1183
draakvis	123
draakvis, Amerikaanse	124
draakvissen	122
driedoornige stekelbaars	500
driedradige meun	450
drievinnige slijmvis	912
drijfvis, Indische	1038
dubbellijnmakreel	972
dugong	1454
duivelsrog, kleine	114
Dungeness-krab	1290
dunlipharder	542
dwergbolk	473
dwergbot, Noorse	1171
dwerginktvis, Ross'	1419
dwergmakrelen	978

dwergmeerval, bruine	385
dwergmeerval, channel-	386
dwergmeerval, zwarte	384
dwergmeervallen	383, 387
dwergtong	1156
dwergvinvis	1480
dwerg-warewalvis	1488
dwergzaagbaars	588
dwergzeebeer	1442

E

eendenbek	872
eendenbek, Braziliaanse	873
eendenmossel	1197, 1343
eendenmossels	1198
eenhoornkabeljauw	439
eetbare zee-egel	1521
eierleggende tandkarpers	512
eilandmakreel	976
elft	156
elft, Amerikaanse	162
elft, bonga-	174
elft, Canadese	155
elft, chacunda-	164
elft, Chinese	168
elft, draadvinnige	172
elft, hilsa-	205
elft, Reeves'	206
elft, toli-	207
elft, Zwarte-Zee-	160
elften	163
elrits	352
elrits, moeras-	351
elritssneep	326
Engelse poon	1079
Europese hondsvis	284
Europese platte oester	1353
Europese rivierkreeft	1276
Europese spinkrab	1315
evervis	527
evervis, valse	528
evervissen	526

F

fint	157
fint, Middellandse-Zee-	158
fint, West-Atlantische	159
finten	163
fluwelen zwemkrab	1299
forel, Amerikaanse meer-	266
forel, beek-	258
forel, bron-	264
forel, cutthroat-	255
forel, gila-	246
forel, goud-	243
forel, Malma-	265
forel, meer-	259
forel, regenboog-	251
forel, zee-	257
forelbaars	603
forellen, zalmen en	262
franjehaai	8
franjelipbarbelen	337
Franse sneep	329
Franse tong	1164
fregattonijn	965
fuseliers	686

G

gaffelkabeljauw	463

gaffelkabeljauw, Atlantische	475	geelstaarten	661	gewone dolfijn	1464
gaffelkabeljauw, Braziliaanse	474	geelstaart-roodbaars	1074	gewone langoest	1266
gaffelkabeljauw, Middellandse-Zee-	464	geelstaartschar	1137	gewone pijlinktvis	1426
gaffelkabeljauwen	465	geelstaartsnapper	694	gewone steurgarnaal	1249
gaffelmakreel	663	geelstaart-stekelmakreel	641	gewone vinvis	1484
gaffelmakreel, grote	647	geelvintonijn	1008	gewone zeebrasem	806
Galicische rivierkreeft	1277	geep	416	gewone zeekat	1418
galjoen	824	geep, agujon-	418	giebel	320
galjoenvissen	823	geep, Atlantische	417	gigartinawieren	1531
gambusia	516	geep, Japanse makreel-	421	gilaforel	246
garnaal, akiami-	1242	geep, makreel-	422	gitaarrog	76
garnaal, Argentijnse rode	1256	gehoornde riverdonderpad	1099	gitaarrog, chola-	74
garnaal, Argentijnse stiletto-	1206	gehoornde slijmvis	909	gitaarrog, gevlekte	73
garnaal, Atlantische diepzee-	1210	geitenbaars	587	gitaarrog, Peruaanse	75
garnaal, Atlantische seabob-	1231	gele horsmakreel	637	gitaarroggen	72, 77
garnaal, Atlantische slib-	1258	gele ombervis	740	gladbek-zeemeerval	375
garnaal, Australische	1207	gele pompano	662	gladde haai	53
garnaal, Azteken-	1213	gele snapper	689	gladde haai, Antarctische	51
garnaal, bananen-	1222	gele trommelvis	722	gladde haai, gevlekte	52
garnaal, blauwrode diepzee-	1233	gele zuidpoolkabeljauw	882	gladde haai, Patagonische	54
garnaal, bruine tijger-	1218	gelidiumwieren	1529	gladde haaien	55
garnaal, caramote-	1220	gemarmerde sidderrog	118	gladde haaien, ruwe en	49
garnaal, Chileense nylon-	1235	gemarmerde zuidpoolkabeljauw	885	gladde schol	1141
garnaal, geelpoot-	1215	Georgia-ijsvis	901	gladde strandschelp	1390
garnaal, gewone steur-	1249	Georgia-zalm	682	glasbaarzen	563
garnaal, groene tijger-	1226	gepantserde poon	1106	glasgrondel	1050
garnaal, grote tijger-	1223	gepen	415, 419	glasmeerval	379
garnaal, koningsrode	1257	gepen, makreel-	420	glatdicksteur	134
garnaal, kristal-	1214	geschulpte hamerhaai	46	glijkop	223
garnaal, kuruma-	1219	gespikkelde knorvis	713	glipvis, Klunzingers	700
garnaal, Midden-Amerikaanse	1228	gespikkelde scheurtandhaai	13	glipvissen	699, 701
garnaal, navaja-mes-	1254	gestippelde alver	308	goerami	1045
garnaal, noordelijke roze	1217	gestreepte bokvis	801	goerami, slangenhuid-	1046
garnaal, noordelijke witte	1227	gestreepte bonito	981	goerami, zoen-	1047
garnaal, Noordzee-	1251	gestreepte dolfijn	1474	Golf-koningsombervis	734
garnaal, Noorse	1237	gestreepte knorvis	704	golfrog	101
garnaal, pijl-	1236	gestreepte knorvis, Afrikaanse	708	goudbandmul	821
garnaal, reuzendiepzee-	1234	gestreepte koningsmakreel	993	goudbandmullen	822
garnaal, reuzenrivier-	1247	gestreepte marlijn	1023	goudbrasem	802
garnaal, ringspriet-	1238	gestreepte ombervis	728	gouden horsmakeel	634
garnaal, roodstaart-	1225	gestreepte pieterman	870	gouden modderkruiper	366
garnaal, Rosenberg-	1247	gestreepte poon	1091	gouden sardinella	190
garnaal, rots-pantser-	1252	gestreepte pos	615	gouden vlinderbrasem	696
garnaal, roze	1224	gestreepte rotskabeljauw	888	gouden zaagbaars	574
garnaal, roze diepzee-	1212	gestreepte selar	655	gouden tapijtschelp	1404
garnaal, shiba-	1208	gestreepte slangenkopvis	558	goudforel	243
garnaal, vlezige	1216	gestreepte zeebaars	598	goudgrondel	1051
garnaal, westelijke konings-	1221	gestreepte zeebarbeel	817	goudharder	540
garnaal, Zuid-Amerikaanse rivier-	1246	getande oester	1352	goudmakreel	681
garnalen, aristeide	1232	gevlekte bonito	966	goudmul	816
garnalen, Atyide-	1243	gevlekte bot	1177	goudvis	320
garnalen, crangonide	1250	gevlekte dolfijnen	1476	graskarper	331
garnalen, diepzee-	1211, 1232	gevlekte doornige heilbot	1111	grassnoek, Amerikaanse	287
garnalen, diepzee rode	1240	gevlekte gitaarrog	73	gratenvis	152
garnalen, mes-	1255	gevlekte gladde haai	52	gratenvis, grootvin-	153
garnalen, metapeneide	1209	gevlekte grenadiervis	493	gratenvissen	151
garnalen, Pacifische seabob-	1230	gevlekte griet	1176	grauwe lipvis	845
garnalen, palaemonide	1245	gevlekte haring	181	grauwe poon	1084
garnalen, pandalide	1239	gevlekte koningsmakreel	994	grenadier, blauwe	480
garnalen, pantser-	1253	gevlekte lipvis	840	grenadier, Patagonische	479
garnalen, peneide	1229	gevlekte maanvis	678	grenadieransjovis	213
garnalen, rivier-	1248	gevlekte makreel	983	grenadiers	477
garnalen, sergestide-	1241	gevlekte ombervis	725	grenadiervis	494
garnalen, steur-	1245	gevlekte pieterman	868	grenadiervis, gevlekte	493
garnalen, zoetwater-steur-	1244	gevlekte pitvis	942	grenadiervis, Italiaanse	495
gazami-zwemkrab	1300	gevlekte rog	93	grenadiervis, noordelijke	497
gedoornde hartschelp	1388	gevlekte sidderrog	120	grenadiervissen	496, 498
gedoornde poon	1086	gevlekte sikkelvis	828	Griekse papegaaivis	856
geelbek	719	gevlekte stekelmakreel	643	Griekse rietvoorn	363
geelgestreepte schar, Japanse	1138	gevlekte zaagbaars, Pacifische	575	griend	1466
geelpootgarnaal	1215	gevlekte zeebaars	594	griend, Indische	1465
geelstaart, grote	658	gevlekte zeebrasem	798	griet	1175
geelstaart, Japanse	660	gevlekte zeemeerval	376	griet, Amerikaanse	1174
geelstaart, reuzen-	659	gevlekte zeewolf	916	griet, gevlekte	1176

grijze dolfijn	1467	grote tijgergarnaal	1223	haaimakreel	971
grijze haai	40	grote zaagbaars	577	haarstaart, Pacifische	958
grijze harder	547	grote zaagbaars, Californische	591	haarstaart, zilveren	957
grijze walvis	1491	grote zeenaald	506	haarstaart, zwarte	956
grijze zeehond	1449	grote zilversmelt	281	haarstaarten	955
grijze zuidpoolkabeljauw	886	Guinea-snoekaal	405	hagedisvis	391
groene horsmakreel	665	Guinese barracuda	530	hagedisvis, Atlantische	393
groene lipvis	843	Gunthers Patagonische		hagedisvis, grote schubben	392
groene mossel	1380	rotskabeljauw	891	hagedisvissen	390
groene murene	403			halfsnavelbek, Australische	428
groene steur	132			halfsnavelbek, balao-	429
groene tijgergarnaal	1226	**H**		halfsnavelbek, Braziliaanse	430
groene zeedonderpad	1102			halfsnavelbek, Japanse	431
groene zeeschildpad	1497	haai, Antarctische gladde	51	halfsnavelbek, langvin-	432
groene zuidpoolkabeljauw	883	haai, Atlantische melk-	44	halfsnavelbekken	427, 433
Groenlandse haai	63	haai, Blainville's doorn-	65	hamerhaai	47
Groenlandse heilbot	1148	haai, blauwe	42	hamerhaai, geschulpte	46
Groenlandse kabeljauw	448	haai, braam-	71	hamerhaaien	45, 48
Groenlandse rog	99	haai, bruine honds-	29	hammossel	1385
Groenlandse walvis	1487	haai, citroen-	41	hapuku-wrakbaars	586
groenogen	298	haai, donkere	39	harder, Afrikaanse	544
groenoog	299	haai, donkere doorn-	59	harder, bobo-	539
grondel, Canestrini-	1058	haai, doorn-	64	harder, Braziliaanse	546
grondel, Dalmatische barbeel-	312	haai, franje-	8	harder, diklip-	538
grondel, Donau-	334	haai, geschulpte hamer-	46	harder, dunlip-	542
grondel, glas-	1050	haai, gespikkelde scheurtand-	13	harder, goud-	540
grondel, goud-	1051	haai, gevlekte gladde	52	harder, grijze	547
grondel, grote	1052	haai, gladde	53	harder, grootkop-	545
grondel, Kesslers	1057	haai, grijze	40	harder, so-iny	541
grondel, Panizza's	1056	haai, Groenlandse	63	harder, spring-	543
grondel, rivier-	333	haai, grootoog-vos-	16	harders	537
grondel, rots-	1054	haai, hamer-	47	haring	169
grondel, slangenkop-	1061	haai, haring-	23	haring, Atlantische draadvinnige	186
grondel, zwarte	1053	haai, honds-	31	haring, Atlantische ronde	177
grondel, Zwarte-Zee-	1060	haai, kat-	32	haring, Californische ronde	176
grondels	1049, 1055	haai, kleintand-vos-	15	haring, Chileense	204
grondels, slaap-	1048	haai, knopstaart-scheurtand-	14	haring, gevlekte	181
grootbek	1033	haai, koper-	36	haring, Indische wolf-	225
grootbek, Amerikaanse	1035	haai, kortsnuit-doorn-	66	haring, kelee-	182
grootbek, Pacifische	1034	haai, langvin-makreel-	21	haring, Pacifische	170
grootbekken	1030, 1036	haai, luipaard-	56	haring, Pacifische draadvinnige	185
grootkopharder	545	haai, makreel-	20	haring, regenboog-ronde-	173
grootkopkarper	336	haai, nurse-	26	haring, rivier-	161
grootoog marene	235	haai, Pacifische haring-	22	haring, ronde	178
grootoogbaars, Pacifische	607	haai, Pacifische melk-	43	haring, zilverstreep-ronde-	201
grootoogbaarzen	608	haai, Patagonische gladde	54	haringen	154
grootoogbot	1117	haai, reuzen-	12	haringen, draadvinnige	187
grootoog-ijsvis	900	haai, ruwe	50	haringen, klein-	180
grootoog-knorvis	703	haai, ruwe doorn-	58	haringen, ronde	179
grootoogmarene	235	haai, scherpneus-koe-	10	haringen, wolf-	224, 226
grootoogrog	94	haai, tapijt-	25	haringhaai	23
grootoogselar	654	haai, vos-	17	haringhaai, Pacifische	22
grootoog-tandbrasem	776	haai, walvis-	27	haringhaaien	18
grootoogtonijn	1011	haai, witpunt-	37	haringkoning	510
grootoogvissen	606	haai, witte	19	harnasmannetje	1103
grootoog-voshaai	16	haai, zebra-	24	hartschelp, gedoornde	1388
grootvin-gratenvis	153	haai, zeevarken-	70	heek	487
grootvinnige vingervin	859	haai, zeskieuwige koe-	11	heek, Antarctische ijs-	878
grote gaffelmakreel	647	haai, zilvertip-	35	heek, Argentijnse	486
grote geelstaart	658	haai, zwarte	60	heek, Australische	482
grote grondel	1052	haai, zwarte doorn-	61	heek, Benguela-	489
grote houting	236	haai, zwartmond-honds-	30	heek, Chileense	485
grote kapiteinvis	556	haai, zwartvin-rif-	38	heek, diepwater-	488
grote koornaarvis	550	haaien en doornhaaien	57	heek, diepzeezilver-	481
grote mantel, Amerikaanse	1369	haaien, blauwe	34	heek, ondiepwater-	484
grote mantel, Australische	1366	haaien, gladde	55	heek, Pacifische	490
grote mantel, Nieuw-Zeelandse	1367	haaien, hamer-	45, 48	heek, Senegalese	491
grote marene	234	haaien, haring-	18	heek, witte	476
grote messchede	1413	haaien, honds-	28, 33	heek, zilver-	483
grote modderkruiper	371	haaien, makreel-	18	heilbot	1133
grote pieterman	869	haaien, ruwe en gladde	49	heilbot, doornige	1112
grote pijlinktvis	1433	haaien, zeevarken-	69	heilbot, gevlekte doornige	1111
grote schubben hagedisvis	392	haaien, zeskieuwige	9	heilbot, Groenlandse	1148
grote strandgaper	1415	haaien, zwarte doorn-	62	heilbot, Indo-Pacifische	1113

heilbot, Japanse 1124
heilbot, olijfgroene bastaard- 1120
heilbot, Pacifische 1125, 1134
heilbot, viervlekkige bastaard- 1119
heilbot, zwarte 1148
heilbotten, bastaard- 1121
heken 478, 492
heken, Antarctische ijs- 879
heremietkreeften 1286
hilsa-elft 205
Holbrooks ringbrasem 780
holle oesters 1350
hondshaai 31
hondshaai, bruine 29
hondshaai, zwartmond- 30
hondshaaien 28, 33
hondstongen 1166
hondsvis, Amerikaanse 285
hondsvis, Europese 284
hoornturbo 1331
horsmakeel, gouden 634
horsmakreel 672
horsmakreel, bijl- 640
horsmakreel, blauwe 635, 670
horsmakreel, Chileense 669
horsmakreel, Cunene- 674
horsmakreel, gele 637
horsmakreel, groene 665
horsmakreel, Japanse 666
horsmakreel, Kaapse 673
horsmakreel, koninginne- 653
horsmakreel, Middellandse-Zee- 668
horsmakreel, Nieuw-Zeelandse 652
horsmakreel, paarden- 636
horsmakreel, Pacifische 671
horsmakreel, ruwe 667
horsmakreel, zwarte 638
horsmakrelen 632, 639, 675
hottentot-zeebrasems 789
houting 237
houting, Arctische 239
houting, Bodenmeer- 240
houting, grote 236
houtingen 241
houtingen en marenen 261
Hovens karper 338
huso 139

I

Iberische barbeel 315
Iberische sneep 327
Iers mos 1532
ijsbaars, Patagonische 880
ijsbot 1140
ijsheek, Antarctische 878
ijsheek, zwarte Patagonische 877
ijsheken, Antarctische 879
IJslandse zandspiering 933
ijsvis 898
ijsvis, Georgia- 901
ijsvis, grootoog- 900
ijsvis, langsnuit- 899
ijsvis, Scotiazee- 896
ijsvis, Wilsons 897
ijsvissen 895
ilisha, slanke 184
ilisha, West-Afrikaanse 183
Indische barracuda 534
Indische drijfvis 1038
Indische griend 1465
Indische koningsmakreel 989
Indische makreel 977

Indische pellona 188
Indische platkop 1096
Indische sardinella 192
Indische wolfharing 225
Indonesische slangenkopvis 557
Indo-Pacifische heilbot 1113
Indo-Pacifische koningsmakreel 991
Indo-Pacifische ponen 1083
inktvis, Argentijnse rode pijl- 1428
inktvis, gewone pijl- 1426
inktvis, grote pijl- 1433
inktvis, Japanse vliegende pijl- 1432
inktvis, Kaapse pijl- 1425
inktvis, kortvin-pijl- 1430
inktvis, langvin-pijl- 1424
inktvis, Middellandse-Zee-kleine
pijl- 1423
inktvis, Nieuw-Zeelandse pijl- 1431
inktvis, reuzenpijl- 1427
inktvis, rode pijl- 1429
inktvis, Ross' dwerg- 1419
inktvisachtigen, pijl- 1421
inktvissen, achtarmige 1434
inktvissen, pijl- 1422
Italiaanse barbeel 318
Italiaanse grenadiervis 495
Italiaanse sneep 328

J

Japanse ansjovis 218
Japanse arkschelp 1340
Japanse geelgestreepte schar 1138
Japanse geelstaart 660
Japanse halfnavelbek 431
Japanse heilbot 1124
Japanse horsmakreel 666
Japanse kardinaalvis 612
Japanse koningsmakreel 997
Japanse lange schar 1131
Japanse lange schar, valse 1130
Japanse langoest 1272
Japanse makreelgeep 421
Japanse oester 1347
Japanse ombervis 741
Japanse paling 399
Japanse sardine 197
Japanse schar 1136
Japanse schol 1126
Japanse stekelmakreel 642
Japanse tapijtschelp 1402
Japanse trommelvis, kleine 747
Japanse venusschelp 1399
Japanse vliegende pijlinktvis 1432
Japanse vliegende vis 424
Japanse zalm 250
Japanse zandvis 863
Japanse zeebaars 599
Japanse zeebrasem 805
Japanse zeekomkommer 1523
Japanse zilversmelt 283
Japanse zwarte vis 1044
Javaanse barbeel 353
jonaskrab 1288
joodse tong 1159
josef 125
Juan-Fernandez-langoest 1261

K

kaaiman, bril- 1503
kaalneuzen 761
Kaapse ansjovis 216, 218
Kaapse horsmakreel 673

Kaapse koningsklip 926
Kaapse langoest 1262
Kaapse messchede 1412
Kaapse pijlinktvis 1425
Kaapse poon 1081
Kaapse roodbaars 1071
kaardrog 89
kaarsvis 278
kabeljauw 447
kabeljauw, Antarctische 894
kabeljauw, Atlantische gaffel- 475
kabeljauw, blauwe diepzee- 437
kabeljauw, Braziliaanse gaffel- 474
kabeljauw, eenhoorn- 439
kabeljauw, gaffel- 463
kabeljauw, gele zuidpool- 882
kabeljauw, gemarmerde zuidpool- 885
kabeljauw, gestreepte rots- 888
kabeljauw, grijze zuidpool- 886
kabeljauw, groene zuidpool- 883
kabeljauw, Groenlandse 448
kabeljauw, Gunthers
    Patagonische rots- 891
kabeljauw, kale zuidpool- 887
kabeljauw, kortestaart
    Patagonische rots- 890
kabeljauw, Magellaan-rots- 889
kabeljauw, Middellandse-Zee-gaffel- 464
kabeljauw, navaga- 445
kabeljauw, Nieuw-Zeelandse blauwe 864
kabeljauw, Nieuw-Zeelandse
diepzee- 438
kabeljauw, Pacifische 446
kabeljauw, pool- 441
kabeljauw, Ramsay's rots- 892
kabeljauw, streepoog-zuidpool- 884
kabeljauw, zwarte 1094
kabeljauw, zwarte zuidpool- 881
kabeljauwachtigen 435
kabeljauwen 440
kabeljauwen, diepzee- 436
kabeljauwen, gaffel- 465
kabeljauwen, zuidpool- 876
kale zuidpoolkabeljauw 887
kalkwieren 1530
kamschelp 1361, 1384
kamschelp, noordelijke 1362
kamschelpen, mantel en 1358
Kamsjatka-krab 1310
Kanadi-koningsmakreel 998
Kanarische tandbrasem 772
kapenaar 765
kapiteinvis, grote 556
kapiteinvis, konings- 555
kapiteinvis, reuzen- 554
kapiteinvissen 553
kardinaalvis 610
kardinaalvis, Japanse 612
kardinaalvis, zwarte 611
kardinaalvissen 609
karetschildpad, valse 1496
karper 332
karper, gras- 331
karper, grootkop- 336
karper, Hovens 338
karper, kroes- 321
karper, modder- 330
karper, Spaanse tand- 514, 515
karper, spiegel- 332
karper, Victoriameer- 355
karper, zee- 808
karper, zilver- 335
karper, Zuid-Europese tand- 513
karper, zuig- 302

NL

karper, zwarte	347
karpers	303
karpers, eierleggende tand-	512
karpers, zuig-	301
karperzalmen	300
kathaai	32
keizers	762
kelee-haring	182
Kemps schildpad	1498
Kesslers grondel	1057
ketazalm	248
kever	471
kikkers	1493
kilka	171
klapmuts	1447
kleine beerkreeft	1305
kleine duivelsrog	114
kleine Japanse trommelvis	747
kleine koornaarvis	549
kleine maanvis	1189
kleine marene	229
kleine modderkruiper	370
kleine octopus	1435
kleine pieterman	867
kleine pijlinktvis, Middellandse-Zee	1423
kleine schorpioenvis	1066
kleine tapijtschelp	1405
kleine tonijn	968
kleine zeekat	1420
kleine zeenaald	507
kleine zilversmelt	282
kleine zwaardschede	1409
kleinharingen	180
kleinoogrog	91
Kleins tong	1163
kleintand-voshaai	15
kliplipvis	839
klipzuiger	1190
Klunzingers glipvis	700
knopstaart-scheurtandhaai	14
knorvis, Afrikaanse gestreepte	708
knorvis, Amerikaanse	702
knorvis, bastaard-	712
knorvis, diklip-	709
knorvis, gespikkelde	713
knorvis, gestreepte	704
knorvis, grootoog-	703
knorvis, Middellandse-Zee-	710
knorvis, zilver-	711
knorvis, Zuidoost-Pacifische	706
knorvissen	705, 714
kob	717
kob, langvin-	721
koehaai, zeskieuwige	11
koekoeklipvis	841
kogeltonijn	964
kogelvis, noordelijke	1185
kogelvissen	1184
kogelvissen, Atlantische	1186
kokkel	1389
kokkels	1387
kolblei	319
Kongo-tandbrasem	773
konijnvis	943
konijnvissen	944
koninginne-horsmakreel	653
koningsgarnaal, westelijke	1221
koningskapiteinvis	555
koningsklip, Chileense	927
koningsklip, Kaapse	926
koningsklip, roze	925
koningsklippen	928
koningskrab	1310
koningskrab, Murray	1308
koningskrab, zuidelijke	1309
koningskrabben	1311
koningslangoest	1273
koningsmakreel	988
koningsmakreel, Australische	996
koningsmakreel, Chinese	1003
koningsmakreel, gestreepte	993
koningsmakreel, gevlekte	994
koningsmakreel, Indische	989
koningsmakreel, Indo-Pacifische	991
koningsmakreel, Japanse	997
koningsmakreel, Kanadi-	998
koningsmakreel, Koreaanse	992
koningsmakreel, Monterey-	990
koningsmakreel, Oost-Atlantische	1004
koningsmakreel, Papua-	995
koningsmakreel, Queensland-	999
koningsmakreel, serra-	987
koningsmakreel, sierra-	1002
koningsmakreel, tijger-	1001
koningsmakreel, valse	1000
koningsmakrelen	1005
koningsombervis	726, 735
koningsombervis, Amerikaanse	733
koningsombervis, Golf-	734
koningsombervis, Zuid-Amerikaanse	732
koningsombervissen	736
koningsrode garnaal	1257
koningsslangmakreel	949
koningsvis	509
koolvis	467
koornaarvis	551
koornaarvis, Atlantische	552
koornaarvis, grote	550
koornaarvis, kleine	549
koornaarvissen	548
koperhaai	36
kopvis, bult-	844
kopvoorn	341
kopvoorn, Middellandse-Zee-	342
kopvoorn, Zwarte-Zee-	340
koraal, rode	1514
koraalbaarzen	583
koraalmeervallen	378
koralen	1513
Koreaanse koningsmakreel	992
Koreaanse mossel	1375
Koreaanse zandspiering	940
kortbek-speervis	1022
korte makreel	975
kortestaart Patagonische rotskabeljauw	890
kortkaak-ombervis	748
kortneus-beensnoek	144
kortsnuit-doornhaai	66
kortvin-pijlinktvis	1430
krab, Arctische sneeuw-	1312
krab, Atlantische rots-	1289
krab, Augustinus-	1307
krab, blauwe	1295
krab, Chinese wolhand-	1303
krab, Dana blauwe	1294
krab, degen-	1515
krab, Dungeness-	1290
krab, Europese spin-	1315
krab, fluwelen zwem-	1299
krab, gazami-zwem-	1300
krab, jonas-	1288
krab, Kamsjatka-	1310
krab, konings-	1310
krab, mangrove-	1302
krab, mediterrane strand-	1297
krab, Murray konings-	1308
krab, Noordzee-	1291
krab, Pacifische rots-	1292
krab, rode diepzee-	1321
krab, strand-	1298
krab, zuidelijke konings-	1309
krab, zuidelijke spin-	1314
krab, zwarte steen-	1293
krabben	1287
krabben, blauwe	1296
krabben, konings-	1311
krabben, Pacifische sneeuw-	1313
krabben, rode diepzee-	1322
krabben, spring-	1306
krabben, zwem-	1301
krabbensoorten, zee-	1285
kreeft	1318
kreeft, Amerikaanse	1317
kreeft, Amerikaanse rivier-	1282
kreeft, Australische rivier-	1316
kreeft, bidsprinkhaan-	1202
kreeft, Europese rivier-	1276
kreeft, Galicische rivier-	1277
kreeft, kleine beer-	1305
kreeft, Louisiana-rivier-	1284
kreeft, Noorse	1320
kreeft, rivier-	1281
kreeft, steen-	1280
kreeft, Turkse rivier-	1277
kreeft, witvoet-rivier-	1279
kreeften, beer-	1304
kreeften, bidsprinkhaan-	1201
kreeften, heremiet-	1286
kreeften, rivier-	1283
kreeftjes, rivier-	1278
krill	1203
krill, Antarctisch	1204
krill, Noorse	1205
kristalgarnaal	1214
kroeskarper	321
krokodil, Chinese	1506
krokodil, Nieuw-Guinea-	1504
krokodil, zee-	1505
krokodillen	1501
kruiper, Balkan-modder-	367
kruiper, Bergatino-modder-	368
kruiper, gouden modder-	366
kruiper, grote modder-	371
kruiper, kleine modder-	370
kruiper, Roemeense modder-	369
kruiper, Venetiaanse modder-	373
Kuhls schorpioenvis	1065
kurumagarnaal	1219
kwabaal	452
kwallen	1512

L

labyrintvissen	1039
lactariusvis	628
lamentijn, Afrikaanse	1457
lamentijn, Amazone-	1455
lamentijn, Amerikaanse	1456
laminariawieren	1526
langbek-speervis	1026
lange schar, Japanse	1131
lange schar, valse Japanse	1130
langnek-ombervis	751
langneus-beensnoek	143
langoest, Australische	1270
langoest, Caraïbische	1269
langoest, gewone	1266
langoest, Japanse	1272
langoest, Juan-Fernandez-	1261
langoest, Kaapse	1262

NL

langoest, konings-	1273	makreel, eiland-	976	mantel, Nieuw-Zeelandse grote	1367
langoest, Natal-	1265	makreel, gaffel-	663	mantel, wijde	1359
langoest, Nieuw-Zeelandse	1260	makreel, geelstaart-stekel-	641	mantel, Zuid-Atlantische	1368
langoest, Oost-Australische	1264	makreel, gele hors-	637	manteldieren	1507
langoest, Portugese	1268	makreel, gestreepte konings-	993	marene, Amerikaanse	232
langoest, Transkei-	1271	makreel, gevlekte	983	marene, cisco-	231
langoest, Tristans	1263	makreel, gevlekte konings-	994	marene, grootoog	235
langoest, zuidkust-	1267	makreel, gevlekte stekel-	643	marene, grote	234
langoesten	1259, 1274	makreel, goud-	681	marene, kleine	229
langoesten, zweep-	1275	makreel, groene hors-	665	marene, pollan-	230
langoestine	1320	makreel, grote gaffel-	647	marene, Zwitserse	233
langoestine, Andaman-	1319	makreel, haai-	971	marenen, houtingen en	261
langsnuitdolfijn	1475	makreel, hors-	672	marlijn, blauwe	1019
langsnuit-ijsvis	899	makreel, Indische	977	marlijn, gestreepte	1023
langvin-halfsnavelbek	432	makreel, Indische konings-	989	marlijn, witte	1021
langvinkob	721	makreel, Indo-Pacifische konings-	991	marlijn, zwarte	1018
langvin-makreelhaai	21	makreel, Japanse hors-	666	marlijnen	1020, 1027
langvin-pijlinktvis	1424	makreel, Japanse konings-	997	Marokkaanse tandbrasem	777
lantaarnvis	295	makreel, Japanse stekel-	642	maskulonge	289
lantaarnvissen	294	makreel, Kaapse hors-	673	masouzalm	250
Laskasneep	323	makreel, Kanadi-konings-	998	mediterrane barbeel	316
lederschildpad	1495	makreel, koninginne-hors-	653	mediterrane sint-jakobsschelp	1364
leng	462	makreel, konings-	988	mediterrane strandkrab	1297
leng, blauwe	460	makreel, koningsslang-	949	meerforel	259
leng, Middellandse-Zee-	461	makreel, Koreaanse konings-	992	meerforel, Amerikaanse	266
lepelsteur, Chinese	141	makreel, korte	975	meerval	382
lepelsteuren	140	makreel, Middellandse-Zee-hors-	668	meerval, Aristoteles-	381
lichia	648	makreel, Monterey-konings-	990	meerval, bruine dwerg-	385
lierpoon	1090	makreel, Nieuw-Zeelandse hors-	652	meerval, channel-dwerg-	386
lingcod	1092	makreel, Oost-Atlantische konings-	1004	meerval, gevlekte zee-	376
lipvis, Amerikaanse	853	makreel, paardenhors-	636	meerval, gladbek-zee-	375
lipvis, bruine	842	makreel, Pacifische hors-	671	meerval, glas-	379
lipvis, gevlekte	840	makreel, Papua-konings-	995	meerval, rugzwemmende	293
lipvis, grauwe	845	makreel, Queensland-konings-	999	meerval, sidder-	394
lipvis, groene	843	makreel, regenboog-stekel-	646	meerval, witte zee-	377
lipvis, klip	839	makreel, Russels stekel-	644	meerval, zwarte dwerg-	384
lipvis, koekoek	841	makreel, ruwe hors-	667	meervallen	380
lipvis, Middellandse-Zee-	846	makreel, serra-konings-	987	meervallen, Afrikaanse	388
lipvis, oog-	848	makreel, sierra-konings-	1002	meervallen, dwerg-	383, 387
lipvis, pauw-	851, 854	makreel, snoek-	951	meervallen, koraal-	378
lipvis, regenboog-	838	makreel, Spaanse	984	meervallen, zee-	374
lipvis, snavel-	850	makreel, tijger-konings-	1001	melkhaai, Atlantische	44
lipvis, tautog-	852	makreel, torpedo-stekel-	649	melkhaai, Pacifische	43
lipvis, vijfvlekkige	849	makreel, valse konings-	1000	melkvis	228
lipvis, zwartoog-	847	makreel, witte snoek-	954	melkvissen	227
lipvissen	837	makreel, zwarte hors-	638	mendole-pikarel	812
lithothamnion	1530	makreel, zwarte snoek-	953	menhaden	166, 167
lodde	273	makreelgeep	422	menhaden, Mexicaanse	165
longvis, Australische	127	makreelgeep, Japanse	421	menhaden, Pacifische	175
longvissen, Afrikaanse	129	makreelgepen	420	mesgarnaal, navaja-	1254
loodsmannetje	650	makreelhaai	20	mesgarnalen	1255
Louisiana-rivierkreeft	1284	makreelhaai, langvin-	21	messchede, grote	1413
luipaardhaai	56	makreelhaaien	18	messchede, Kaapse	1412
		makrelen	986	messchede, Pacifische	1411
		makrelen en tonijnen	961	messcheden	1407, 1414
M		makrelen, dwerg-	978	messen, scheer-	1414
		makrelen, hors-	632, 639, 675	mesvissen	292
maanvis	1188	makrelen, konings-	1005	metapeneide garnalen	1209
maanvis, gevlekte	678	makrelen, slang-	947	meun, driedradige	450
maanvis, kleine	1189	makrelen, snoek-	952	meun, vierdradige	469
maanvis, spitsstaart-	1187	makrelen, stekel-	645	meun, vijfdradige	443
maasbanker	673	Malabar-zaagbaars	579	meunen	451
Macdonalds ombervis	724	Malma-forel	265	Mexicaanse menhaden	165
Madeira-sardinella	193	Mandschoeria-trommelvis	746	Middellandse-Zee-barracuda	535
Magellaan-rotskabeljauw	889	mangrovekrab	1302	Middellandse-Zee-fint	158
makeel, gouden hors-	634	mangrovensnapper	688	Middellandse-Zee-gaffelkabeljauw	464
makreel	985	mangroveoester	1348	Middellandse-Zee-horsmakreel	668
makreel, Australische konings-	996	manta	112	Middellandse-Zee-kleine pijlinktvis	1423
makreel, bijlhors-	640	manta, Pacifische	113	Middellandse-Zee-knorvis	710
makreel, blauwe hors-	635, 670	manta's	111	Middellandse-Zee-kopvoorn	342
makreel, Chileense hors-	669	mantel en kamschelpen	1358	Middellandse-Zee-leng	461
makreel, Chinese konings-	1003	mantel, Amerikaanse grote	1369	Middellandse-Zee-lipvis	846
makreel, Cunene-hors-	674	mantel, Australische grote	1366	Middellandse-Zee-mossel	1377
makreel, dubbellijn-	972	mantel, bonte	1363	Middellandse-Zee-parelvis	930

NL

Middellandse-Zee-slijmkop 521
Middellandse-Zee-sterrog 83
Midden-Amerikaanse garnaal 1228
Mississippi-alligator 1502
modderkarper 330
modderkruiper, Balkan- 367
modderkruiper, Bergatino- 368
modderkruiper, gouden 366
modderkruiper, grote 371
modderkruiper, kleine 370
modderkruiper, Roemeense 369
modderkruiper, Venetiaanse 373
moddersnoek 142
moeras-elrits 351
moerasschildpadden 1499
monniksrob 1450
monnikvis 836
monocle-stekelwang 698
Monrovische doktersvis 946
Monterey-koningsmakreel 990
mossel 1376
mossel, Afrikaanse rots- 1382
mossel, Australische 1378
mossel, boor- 1416
mossel, Chileense 1374
mossel, cholga- 1371
mossel, dadel- 1372
mossel, eenden- 1197, 1343
mossel, groene 1380
mossel, ham- 1385
mossel, Koreaanse 1375
mossel, Middellandse-Zee 1377
mossel, Nieuw-Zeelandse 1381
mossel, Platarivier- 1379
mossel, rivierparel- 1383
mossel, schilder- 1345
mosselen 1370
mosselen, boor- 1417
mosselen, paarden- 1373
mosselen, schilder- 1344
mossels, eenden- 1198
Mozambique-tilapia 832
mul 818
mul, goud- 816
mul, goudband- 821
mul, West-Afrikaanse 820
mullen, goudband- 822
murene 404
murene, groene 403
murenen 402
Murray-koningskrab 1308
muskusoctopus 1436
muskusoctopussen 1437

N

naakte slijmvissen 904
naakte zandspiering 938
narwal 1479
Nassau-zaagbaars 581
Natal-langoest 1265
Natal-zeebrasem 792
navaga, oostelijke 444
navaga-kabeljauw 445
navaja-mesgarnaal 1254
Nieuw-Guinea-krokodil 1504
Nieuw-Zeelandse blauwe kabeljauw 864
Nieuw-Zeelandse bot 1149
Nieuw-Zeelandse botten 1150
Nieuw-Zeelandse diepzeekabeljauw 438
Nieuw-Zeelandse grote mantel 1367
Nieuw-Zeelandse horsmakreel 652
Nieuw-Zeelandse langoest 1260

Nieuw-Zeelandse mossel 1381
Nieuw-Zeelandse paling 398
Nieuw-Zeelandse pijlinktvis 1431
Nieuw-Zeelandse platte oester 1355
Nieuw-Zeelandse trompetvis, blauwe 862
nijlbaars 565
nijlbaarzen 566
nijltilapia 833
Noord-Amerikaanse oester 1349
noordelijke grenadiervis 497
noordelijke kamschelp 1362
noordelijke kogelvis 1185
noordelijke rode snapper 690
noordelijke roze garnaal 1217
noordelijke witte garnaal 1227
noordkaper 1490
noordkromp 1386
Noord-Pacifische ansjovis 219
noordse vinvis 1481
Noordzeegarnaal 1251
Noordzeekrab 1291
Noorse dwergbot 1171
Noorse garnaal 1237
Noorse kreeft 1320
Noorse krill 1205
Noorse zandspiering 935
nufar-zeebrasem 768
nursehaai 26
nylongarnaal, Chileense 1235

O

octopus 1438
octopus, kleine 1435
octopus, muskus- 1436
octopussen 1439
octopussen, muskus- 1437
Oekraïense prik 3
oester 1353
oester, Chileense 1351
oester, Europese platte 1353
oester, getande 1352
oester, Japanse 1347
oester, mangrove- 1348
oester, Nieuw-Zeelandse platte 1355
oester, Noord-Amerikaanse 1349
oester, Olympia platte 1354
oester, Sydney- 1357
oesters, holle 1350
oesters, platte 1356
ogentong 1158
olievis 950
olifantzeeratten 126
olijfgroene bastaardheilbot 1120
olijfgroene zeesnoek 561
Olympia platte oester 1354
ombervis 718
ombervis, Amerikaanse konings- 733
ombervis, Antarctische 754
ombervis, Argentijnse 758
ombervis, Atlantische 738
ombervis, bobo- 749
ombervis, boe- 753
ombervis, Californische witte 730
ombervis, gele 740
ombervis, gestreepte 728
ombervis, gevlekte 725
ombervis, Golf-konings- 734
ombervis, Japanse 741
ombervis, konings- 726, 735
ombervis, kortkaak- 748
ombervis, langnek- 751
ombervis, Macdonalds 724

ombervis, Peruaanse 723
ombervis, punt- 731
ombervis, rode 757
ombervis, Senegalese 750
ombervis, Shi- 759
ombervis, Steindachners 727
ombervis, tijger- 742
ombervis, witmond- 737
ombervis, witte 720
ombervis, Zuid-Amerikaanse
    konings- 732
ombervissen 715, 729, 739,
    752, 756, 760
ombervissen, konings- 736
ondiepwaterheek 484
ongestreepte bonito 974
ooglipvis 848
oorrobben 1440
Oost-Atlantische koningsmakreel 1004
Oost-Australische langoest 1264
oostelijke barracuda 533
oostelijke navaga 444
Oostkusttong, Zuid-Afrikaanse 1154
oranje vijlvis 1181
Orbigny-artemisschelp 1397
orka 1470
ossenoog-tarpoen 149

P

paardenhorsmakreel 636
paardenkop 676
paardenmosselen 1373
Pacifische ansjovis 212
Pacifische bonito 980
Pacifische conger 409
Pacifische draadvinnige haring 185
Pacifische gevlekte zaagbaars 575
Pacifische grootbek 1034
Pacifische grootoogbaars 607
Pacifische haarstaart 958
Pacifische haring 170
Pacifische haringhaai 22
Pacifische heek 490
Pacifische heilbot 1125, 1134
Pacifische horsmakreel 671
Pacifische kabeljauw 446
Pacifische manta 113
Pacifische melkhaai 43
Pacifische menhaden 175
Pacifische messchede 1411
Pacifische pijlstaartrog 104
Pacifische rog 100
Pacifische roodbaars 1070
Pacifische rotskrab 1292
Pacifische sardine 196
Pacifische sardinella 194
Pacifische schol 1135
Pacifische seabobgarnalen 1230
Pacifische sneeuwkrabben 1313
Pacifische tonijn 967
Pacifische trompetvissen 861
Pacifische zandspiering 934
Pacifische zeilvis 1017
Pacifische zilversmelt 280
paddenvissen 1191
palaemonide garnalen 1245
paling 396, 401
paling, Amerikaanse 400
paling, Australische 397
paling, Japanse 399
paling, Nieuw-Zeelandse 398
palingen 395

pandalide garnalen	1239
pandora, rode	791
panga	799
Panizza's grondel	1056
pantsergarnaal, rots-	1252
pantsergarnalen	1253
papegaaivis, Griekse	856
Papua-koningsmakreel	995
parel-blankvoorn	357
parelmoerschelp	1346
parelmoerzeeoor	1326
parelscheermesvis	855
parelvis, Middellandse-Zee-	930
parona	651
Patagonische gladde haai	54
Patagonische grenadier	479
Patagonische ijsbaars	880
Patagonische ijsheek, zwarte	877
Patagonische rotskabeljauw,	
Gunthers	891
Patagonische rotskabeljauw,	
kortestaart	890
pauwlipvis	851, 854
pauwslijmvis	907
Pekingbrasem	349
pelagische pijlstaartrog	107
peled	238
pellona, Indische	188
pelser	189
pelser, Australische	198
pelser, Zuid-Afrikaanse	199
pelsrob, zuidelijke	1441
pelstong	1162
peneide garnalen	1229
Peruaanse ansjovis	220
Peruaanse gitaarrog	75
Peruaanse ombervis	723
Peruaanse trommelvis	743
Peruaanse zeebaars	584
pieterman, gestreepte	870
pieterman, gevlekte	868
pieterman, grote	869
pieterman, kleine	867
pietermannen	866, 871
pijlgarnaal	1236
pijlinktvis, Argentijnse rode	1428
pijlinktvis, gewone	1426
pijlinktvis, grote	1433
pijlinktvis, Japanse vliegende	1432
pijlinktvis, Kaapse	1425
pijlinktvis, kortvin-	1430
pijlinktvis, langvin-	1424
pijlinktvis, Middellandse-Zee-kleine	1423
pijlinktvis, Nieuw-Zeelandse	1431
pijlinktvis, reuzen-	1427
pijlinktvis, rode	
1429	
pijlinktvisachtigen	1421
pijlinktvissen	1422
pijlstaartrog	106
pijlstaartrog, Pacifische	104
pijlstaartrog, pelagische	107
pijlstaartrog, stekel-	105
pijlstaartroggen	103
pikarel	813
pikarel, mendole-	812
pikarel, zwartgevlekte	811
pikarellen	810, 814
pink zalm	247
pitvis	941
pitvis, gevlekte	942
Plata-riviermossel	1379
platkop, Indische	1096
platkoppen	1095

platte oester, Europese	1353
platte oester, Nieuw-Zeelandse	1355
platte oester, Olympia	1354
platte oesters	1356
platvissen	1110
plooibekken	903
pollak	466
pollak, Alaska-	470
pollanmarene	230
pomacentriden	835
pomfret	1037
pomfret, Chinese	1032
pomfret, zilver-	1031
pomfret, zwarte	677
pompano, Alexandrië-	633
pompano, gele	662
pompano's	664
ponen	1078
ponen, Amerikaanse	1087
ponen, Indo-Pacifische	1083
poolkabeljauw	441
poon, blauwvin-	1082
poon, donkere	1080
poon, Engelse	1079
poon, gedoornde	1086
poon, gepantserde	1106
poon, gestreepte	1091
poon, grauwe	1084
poon, Kaapse	1081
poon, lier-	1090
poon, rode	1089
poon, schub-	1085
poon, spitssnuit-	1088
poon, vliegende	1109
porgy's	767
Portugese langoest	1268
pos	614
pos, gestreepte	615
potvis	1462
prik, beek-	6
prik, Donau-	2
prik, Oekraïense	3
prik, rivier-	5
prik, Vladikovs	4
prik, zee-	7
prikken	1
puitaal	923
puitaal, Amerikaanse	922
puitalen	920, 921
puntombervis	731
pyurazakpijp	1510
pyurazakpijp, Chileense	1509

Q

Queensland-koningsmakreel	999

R

Ramsay's rotskabeljauw	892
Reeves' elft	206
regenboogforel	251
regenboog-lipvis	838
regenboog-rondeharing	173
regenboog-stekelmakreel	646
remora	1178
reuzendiepzeegarnaal	1234
reuzengeelstaart	659
reuzenhaai	12
reuzenkapiteinvis	554
reuzenpijlinktvis	1427
reuzenriviergarnaal	1247
reuzenzaagbaars	578

reuzenzeeoor	1325
reuzenzeepok	1200
riddervis	263
riddervissen	267
rietvoorn	362
rietvoorn, Griekse	363
rifhaai, zwartvin-	38
ringbrasem	778
ringbrasem, Holbrooks	780
ringbrasem, spitse	781
ringbrasem, witte	782
ringbrasem, zebra-	779
ringbrasem, zwartkop-	783
ringbrasemachtigen	784
ringelrob	1452
ringsprietgarnaal	1238
rivieransjovis	221
rivierdonderpad	1097
rivierdonderpad, gehoornde	1099
riviergarnaal, reuzen-	1247
riviergarnaal, Zuid-Amerikaanse	1246
riviergarnalen	1248
riviergrondel	333
rivierharing	161
rivierkreeft, Amerikaanse	1282
rivierkreeft, Australische	1316
rivierkreeft, Europese	1276
rivierkreeft, Galicische	1277
rivierkreeft, Louisiana-	1284
rivierkreeft, Turkse	1277
rivierkreeft, witvoet-	1279
rivierkreeften	1281, 1283
rivierkreeftjes	1278
rivierparelmossel	1383
rivierprik	5
rivierslijmvis	911
rob, baard-	1448
rob, monniks-	1450
rob, ringel-	1452
rob, zadel-	1451
rob, zuidelijke pels-	1441
robben, oor-	1440
rode bandvis	834
rode barbiervis	570
rode diepzeekrab	1321
rode diepzeekrabben	1322
rode garnaal, Argentijnse	1256
rode garnalen, diepzee	1240
rode koraal	1514
rode ombervis	757
rode pandora	791
rode pijlinktvis	1429
rode pijlinktvis, Argentijnse	1428
rode poon	1089
rode schorpioenvis	1068
rode snapper, noordelijke	690
rode snapper, zuidelijke	691
rode soldaatvis	522
rode steenbrasem	796
rode stompneus	769
rode vlagbaars	569
rode zaagbaars	580
rode zalm	252
rode zeebrasem	794
rode zeeoor	1327
roeffen	1040
Roemeense modderkruiper	369
rog, arend-	109
rog, Atlantische sidder-	119
rog, blonde	85
rog, bul	110
rog, Canadese	88
rog, chola-gitaar-	74
rog, gemarmerde sidder-	118

rog, gevlekte 93
rog, gevlekte gitaar- 73
rog, gevlekte sidder- 120
rog, gitaar- 76
rog, golf- 101
rog, Groenlandse 99
rog, grootoog- 94
rog, kaard- 89
rog, kleine duivels- 114
rog, kleinoog- 91
rog, Middellandse-Zee-ster- 83
rog, Pacifische 100
rog, Pacifische pijlstaart- 104
rog, pelagische pijlstaart- 107
rog, Peruaanse gitaar- 75
rog, pijlstaart- 106
rog, ruwe 98
rog, scherpsnuit- 96
rog, spiegel- 92
rog, spitsneus- 82
rog, stekel- 87
rog, stekelpijlstaart- 105
rog, ster- 97
rog, vlinder- 115
rog, winter- 95
rog, witte 90
rog, zand- 86
rogachtigen 80
roggen 102, 81
roggen, gitaar- 72, 77
roggen, pijlstaart- 103
roggen, sidder- 117, 121
roggen, vleermuis- 108
roggen, vlinder- 116
ronde haring 178
ronde haring, Atlantische 177
ronde haring, Californische 176
ronde haringen 179
rondeharing, regenboog- 173
rondeharing, zilverstreep- 201
rondloper, zuidelijke 685
rondlopers 684
roodbaars 1075
roodbaars, Amerikaanse 1073
roodbaars, diepzee- 1076
roodbaars, geelstaart- 1074
roodbaars, Kaapse 1071
roodbaars, Pacifische 1070
roodbaars, weduwe- 1072
roodbaarzen 1064, 1077
roodgestreepte slijmvis 910
roodgestreepte zeebrasem 803
roodstaartgarnaal 1225
roodstaartsnapper 692
roodwieren 1528
roofblei 311
Rosenberggarnaal 1247
Ross' dwerginktvis 1419
rotsgrondel 1054
rotskabeljauw, gestreepte 888
rotskabeljauw, Gunthers
   Patagonische 891
rotskabeljauw, kortestaart
   Patagonische 890
rotskabeljauw, Magellaan- 889
rotskabeljauw, Ramsay's 892
rotskrab, Atlantische 1289
rotskrab, Pacifische 1292
rotsmossel, Afrikaanse 1382
rots-pantsergarnaal 1252
roze diepzeegarnaal 1212
roze garnaal 1224
roze garnaal, noordelijke 1217
roze koningsklip 925

roze tandbrasem 775
roze zalm 247
rugzwemmende meerval 293
ruisvoorn 362
Russels stekelmakreel 644
ruwe doornhaai 58
ruwe en gladde haaien 49
ruwe haai 50
ruwe horsmakreel 667
ruwe rog 98

S

sabelbliek 350
Sachalin strandschelp 1391
sardien 189
sardine, Australische 198
sardine, Chileense 200
sardine, Japanse 197
sardine, Pacifische 196
sardine, Zuid-Afrikaanse 199
sardinella, Aziatische 191
sardinella, gouden 190
sardinella, Indische 192
sardinella, Madeira- 193
sardinella, Pacifische 194
sardinella's 195
scallop, calico- 1360
schaalhoorn 1333
schaapskop-zeebrasem 764
schar 1139
schar, geelstaart- 1137
schar, Japanse 1136
schar, Japanse geelgestreepte 1138
schar, Japanse lange 1131
schar, tong- 1142
schar, valse Japanse lange 1130
scharretong 1169
scharretong, viervlekkige 1168
scharretongen 1170
schede, Amerikaanse zwaard- 1408
schede, kleine zwaard- 1409
schede, Pacifische mes- 1411
scheden, mes- 1407, 1414
scheermessen 1414
schelp, Amerikaanse venus- 1398
schelp, boter- 1403
schelp, bruine venus- 1395
schelp, gedoornde hart- 1388
schelp, gladde strand- 1390
schelp, gouden tapijt- 1404
schelp, Japanse ark- 1340
schelp, Japanse tapijt- 1402
schelp, Japanse venus- 1399
schelp, kam- 1361, 1384
schelp, kleine tapijt- 1405
schelp, mediterrane sint-jakobs- 1364
schelp, noordelijke kam- 1362
schelp, Orbigny-artemis- 1397
schelp, parelmoer- 1346
schelp, Sachalin strand- 1391
schelp, sint-jakobs- 1365
schelp, stevige strand- 1392
schelp, tapijt- 1401
schelp, venus- 1396
schelp, West-Pacifische ark- 1339
schelpdieren, tweekleppige 1324
schelpen, ark- 1341, 1342
schelpen, venus- 1393, 1400
schelpsoorten, tapijt- 1394
schelvis 453
scherpneus-koehaai 10
scherpsnuitrog 96
scheurtandhaai, gespikkelde 13

scheurtandhaai, knopstaart- 14
schildermossel 1345
schildermosselen 1344
schildpad, groene zee- 1497
schildpad, Kemps 1498
schildpad, leder- 1495
schildpad, valse karet- 1496
schildpadden, moeras- 1499
schildpadsoorten, zee- 1494
schol 1145
schol, Alaska- 1146
schol, Amerikaanse 1132
schol, Amerikaanse winter- 1147
schol, Californische 1127
schol, gladde 1141
schol, Japanse 1126
schol, Pacifische 1135
schol, ster- 1144
scholieren 948
schollen 1123
schorpioenvis, bruine 1067
schorpioenvis, kleine 1066
schorpioenvis, Kuhls 1065
schorpioenvis, rode 1068
schorpioenvissen 1062, 1069
Schot-zeebrasem 797
schriftbaars 589
schubben hagedisvis, grote 392
schubpoon 1085
schurftvis 1115
schurftvis, Thors- 1116
Scotiazee-ijsvis 896
scup 809
seabobgarnaal, Atlantische 1231
seabobgarnalen, Pacifische 1230
selar, gestreepte 655
selar, grootoog- 654
Senegalese heek 491
Senegalese ombervis 750
sergestide-garnalen 1241
serpeling 344
serra-koningsmakreel 987
shiba-garnaal 1208
Shi-ombervis 759
siddermeerval 394
sidderrog, Atlantische 119
sidderrog, gemarmerde 118
sidderrog, gevlekte 120
sidderroggen 117, 121
sierra-koningsmakreel 1002
sikkelvis, Afrikaanse 827
sikkelvis, gevlekte 828
sillaginiden 625
sint-jakobsschelp 1365
sint-jakobsschelp, mediterrane 1364
skipjack 973
skipjack, zwarte 969
slaapgrondels 1048
slakdolf 1108
slakdolf, West-Atlantische 1107
slakdolven 1104
slakken, brandhoorn 1334
slangenhuidgoerami 1046
slangenkopgrondel 1061
slangenkopvis, gestreepte 558
slangenkopvis, Indonesische 557
slangenkopvissen 559
slangmakreel, konings- 949
slangmakrelen 947
slanke ilisha 184
slanke tonijn 963
slibgarnaal, Atlantische 1258
slijmkop, Atlantische 520
slijmkop, Middellandse-Zee- 521

slijmvis	906, 908	
slijmvis, drievinnige	912	
slijmvis, gehoornde	909	
slijmvis, pauw-	907	
slijmvis, rivier-	911	
slijmvis, roodgestreepte	910	
slijmvis, vlinder-	905	
slijmvis, zilver-	918	
slijmvissen, naakte	904	
smelt	939	
smelt, grote zilver-	281	
smelt, Japanse zilver-	283	
smelt, kleine zilver-	282	
smelt, Pacifische zilver-	280	
smelten, zilver-	279	
snapper, geelstaart-	694	
snapper, gele	689	
snapper, mangroven-	688	
snapper, noordelijke rode	690	
snapper, roodstaart-	692	
snapper, zuidelijke rode	691	
snappers	687, 693	
snaveldolfijnen	1458	
snavellipvis	850	
sneep	325	
sneep, Dalmatische	324	
sneep, elrits-	326	
sneep, Franse	329	
sneep, Iberische	327	
sneep, Italiaanse	328	
sneep, Laska-	323	
sneeuwkrab, Arctische	1312	
sneeuwkrabben, Pacifische	1313	
snipvis	504	
snipvissen	503	
snoek	288	
snoek, Amoer-	291	
snoek, kortneus-been-	144	
snoek, langneus-been-	143	
snoek, modder-	142	
snoek, olijfgroene zee-	561	
snoek, zwarte	290	
snoekaal, Batavia-	406	
snoekaal, Guinea-	405	
snoekalen	407	
snoekbaars	620	
snoekbaars, Amerikaanse	621	
snoekbaars, Canadese	619	
snoeken	286	
snoeken, zee-	560, 562	
snoekmakreel	951	
snoekmakreel, witte	954	
snoekmakreel, zwarte	953	
snoekmakrelen	952	
snotolf	1105	
snotolven	1104	
sockeye zalm	252	
so-iny harder	541	
soldaatvis, rode	522	
Spaanse makreel	984	
Spaanse tandkarper	514, 515	
Spaanse zeebrasem	790	
spadevissen	825	
speervis	1024	
speervis, degen-	1025	
speervis, kortbek-	1022	
speervis, langbek-	1026	
speervissen	1027	
spiegelkarper	332	
spiegelrog	92	
spiering	274	
spiering, Amerikaanse	275	
spiering, Amerikaanse zand-	932	
spiering, Arctische	276	
spiering, Californische	272	
spiering, IJslandse zand-	933	
spiering, Koreaanse zand-	940	
spiering, naakte zand-	938	
spiering, Noorse zand-	935	
spiering, Pacifische zand-	934	
spiering, vijver-	271	
spiering, zand-	936	
spieringen	270, 277	
spieringen, zand-	931, 937	
spinkrab, Europese	1315	
spinkrab, zuidelijke	1314	
spitsdolfijn van Cuvier	1461	
spitse ringbrasem	781	
spitsneusrog	82	
spitssnuitpoon	1088	
spitsstaartmaanvis	1187	
sponzen	1511	
springharder	543	
springkrabben	1306	
sprot	202	
steenbaars	600	
steenbaars, witte	785	
steenbaars, zand-	786	
steenbaarzen	787	
steenbolk	472	
steenbrasem, rode	796	
steenkrab, zwarte	1293	
steenkreeft	1280	
Steindachners ombervis	727	
stekelbaars, driedoornige	500	
stekelbaars, tiendoornige	501	
stekelbaars, zee-	502	
stekelbaarzen	499	
stekelmakreel, geelstaart-	641	
stekelmakreel, gevlekte	643	
stekelmakreel, Japanse	642	
stekelmakreel, regenboog-	646	
stekelmakreel, Russels	644	
stekelmakreel, torpedo-	649	
stekelmakrelen	645	
stekelpijlstaartrog	105	
stekelrog	87	
stekelwang, monocle-	698	
sterlet	135	
sterrenkijker	874	
sterrenkijkers	875	
sterrog	97	
sterrog, Middellandse-Zee-	83	
sterschol	1144	
stersteur	136	
steur	137	
steur, Adriatische	133	
steur, Chinese lepel-	141	
steur, Donau-	131	
steur, glatdick-	134	
steur, groene	132	
steur, ster-	136	
steur, witte	138	
steuren	130	
steuren, lepel-	140	
steurgarnaal, gewone	1249	
steurgarnalen	1245	
steurgarnalen, zoetwater	1244	
stevige strandschelp	1392	
stilettogarnaal, Argentijnse	1206	
stoleforus-ansjovissen	222	
stompneus, rode	769	
stompneus, witte	800	
stompneuzen	770	
strandgaper, grote	1415	
strandkrab	1298	
strandkrab, mediterrane	1297	
strandschelp, gladde	1390	
strandschelp, Sachalin	1391	
strandschelp, stevige	1392	
streber	623	
streepoog-zuidpoolkabeljauw	884	
strombushoorn	1338	
sufia-voorn	345	
Sydney-oester	1357	

T

tafelmesheft	1410	
tandbrasem	774	
tandbrasem, Angola-	771	
tandbrasem, grootoog-	776	
tandbrasem, Kanarische	772	
tandbrasem, Kongo-	773	
tandbrasem, Marokkaanse	777	
tandbrasem, roze	775	
tandkarper, Spaanse	514, 515	
tandkarper, Zuid-Europese	513	
tandkarpers, eierleggende	512	
tapijthaai	25	
tapijtschelp	1401	
tapijtschelp, gouden-	1404	
tapijtschelp, Japanse	1402	
tapijtschelp, kleine	1405	
tapijtschelpsoorten	1394	
tarbot	1172	
tarbot, Zwarte-Zee-	1173	
tarbotten	1167	
tarpoen	150	
tarpoen, ossenoog-	149	
tarpoenen	148	
tautog-lipvis	852	
tegelvis, blauwe	627	
tegelvissen	626	
Thors-schurftvis	1116	
tiendoornige stekelbaars	501	
tienponder	146, 147	
tienponders	145	
tijgergarnaal, bruine	1218	
tijgergarnaal, groene	1226	
tijgergarnaal, grote	1223	
tijger-koningsmakreel	1001	
tijgerombervis	742	
tilapia, Mozambique-	832	
tilapia, Nijl-	833	
tolhoorn	1330	
toli-elft	207	
tomcod	457	
tomcod, Atlantische	456	
tong	1165	
tong, Amerikaanse	1152	
tong, dikrug-	1160	
tong, dwerg-	1156	
tong, Franse	1164	
tong, joodse	1159	
tong, Kleins	1163	
tong, ogen-	1158	
tong, pels-	1162	
tong, scharre-	1169	
tong, viervlekkige scharre-	1168	
tong, wig-	1157	
tong, Zuid-Afrikaanse Oostkust-	1154	
tong, Zuid-Afrikaanse Westkust-	1153	
tongen	1151	
tongen, dikrug-	1161	
tongen, honds-	1166	
tongen, scharre-	1170	
tongen, Zuid-Afrikaanse	1155	
tongoltonijn	1013	
tongschar	1142	
tonijn	1012	
tonijn, fregat-	965	

NL

tonijn, geelvin-	1008
tonijn, grootoog-	1011
tonijn, kleine	968
tonijn, kogel-	964
tonijn, Pacifische	967
tonijn, slanke	963
tonijn, tongol-	1013
tonijn, vlinder-	970
tonijn, witte	1007
tonijn, zuidelijke blauwvin-	1010
tonijn, zwartvin-	1009
tonijnachtigen	960
tonijnen	1006, 1014
tonijnen, makrelen en	961
torpedo-stekelmakreel	649
torsk	442
Transkei-langoest	1271
trekkervis	1180
trekkervissen	1179
Tristans langoest	1263
trommelvis, gele	722
trommelvis, kleine Japanse	747
trommelvis, Mandschoeria-	746
trommelvis, Peruaanse	743
trommelvis, zilver-	744
trommelvis, zoetwater-	716
trommelvis, zwarte	745
trompetterzeenaald	508
trompetvis, blauwe Nieuw-Zeelandse	862
trompetvissen, Pacifische	861
tuimelaar	1477
turbo, hoorn-	1331
turbo, Zuid-Afrikaanse	1332
Turkse rivierkreeft	1277
tweekleppige schelpdieren	1324

V

valse evervis	528
valse Japanse lange schar	1130
valse karetschildpad	1496
valse koningsmakreel	1000
varkenvis	707
veelkleurige vingervin	858
Venetiaanse modderkruiper	373
venusschelp	1396
venusschelp, Amerikaanse	1398
venusschelp, bruine	1395
venusschelp, Japanse	1399
venusschelpen	1393, 1400
vetje	339
Victoriameer-karper	355
vierdradige meun	469
viervlekkkige scharretong	1168
viervleugelige vliegende vis	434
vijfdradige meun	443
vijfvlekkige lipvis	849
vijlvis, oranje	1181
vijlvissen	1182
vijverspiering	271
vingervin, Bergs	857
vingervin, grootvinnige	859
vingervin, veelkleurige	858
vingervinnen	860
vinvis, blauwe	1483
vinvis, Bryde's	1482
vinvis, gewone	1484
vinvis, noordse	1481
violet-zakpijp	1508
Vladikovs prik	4
vlagbaars, rode	569
vlagvinvis	389
vlagzalm	268
vleermuisroggen	108

vleermuisvissen	826
vleet	84
vlezige garnaal	1216
vliegende pijlinktvis, Japanse	1432
vliegende poon	1109
vliegende vis	426
vliegende vis, Japanse	424
vliegende vis, viervleugelige	434
vliegende vissen	423, 425
vlinderbrasem, gouden	696
vlinderbrasems	695, 697
vlinderrog	115
vlinderroggen	116
vlinderslijmvis	905
vlindertonijn	970
vlindervissen	830
voorn, Adriatische blank-	359
voorn, Albanese blank-	348
voorn, bitter-	356
voorn, blank-	360
voorn, Donau-	358
voorn, Griekse riet-	363
voorn, Middellandse-Zee-kop-	342
voorn, parel-blank-	357
voorn, riet-	362
voorn, ruis-	362
voorn, sufia-	345
voorns, blank-	361
vorskwab	468
voshaai	17
voshaai, grootoog-	16
voshaai, kleintand-	15

W

wahoo	962
walrus	1445
walvis, Australische	1489
walvis, dwerg-ware-	1488
walvis, grijze	1491
walvis, Groenlandse	1487
walvis, zwaard-	1470
walvis, zwarte zwaard-	1473
walvishaai	27
walvissen, ware-	1486
warewalvis, dwerg-	1488
warewalvissen	1486
weduweroodbaars	1072
weekdieren	1323
West-Afrikaanse ilisha	183
West-Afrikaanse mul	820
West-Afrikaanse zeeduivel	1196
West-Atlantische fint	159
West-Atlantische slakdolf	1107
West-Pacifische arkschelp	1339
westelijke koningsgarnaal	1221
Westkusttong, Zuid-Afrikaanse	1153
wieren, blaasjes-	1527
wieren, bruin-	1525
wieren, gelidium-	1529
wieren, gigartina-	1531
wieren, kalk-	1530
wieren, laminaria-	1526
wieren, rood-	1528
wigtong	1157
wijde mantel	1359
wijting	454
wijting, blauwe	459
wijting, zilver-	449
wijting, zuidelijke blauwe	458
wijting, Zwarte-Zee-	455
Wilsons ijsvis	897
winde	343
winterrog	95

winterschol, Amerikaanse	1147
witflankdolfijn	1468
witje	1128
witje, Amerikaans	1129
witmondombervis	737
witpunthaai	37
witsnuitdolfijn	1469
witte alver	309
witte crappie	604
witte garnaal, noordelijke	1227
witte haai	19
witte heek	476
witte marlijn	1021
witte ombervis	720
witte ombervis, Californische	730
witte ringbrasem	782
witte rog	90
witte snoekmakreel	954
witte steenbaars	785
witte steur	138
witte stompneus	800
witte tonijn	1007
witte zaagbaars	573
witte zeebaars	597
witte zeemeerval	377
witvoet-rivierkreeft	1279
wolfharing, Indische	225
wolfharingen	224, 226
wolhandkrab, Chinese	1303
wormaal	413
wrakbaars, Atlantische	585
wrakbaars, hapuku-	586
Wuchang-brasem	346
wulk	1335
wulken, busycon-	1336

Z

zaagbaars, Californische grote	591
zaagbaars, dwerg-	588
zaagbaars, gouden	574
zaagbaars, grote	577
zaagbaars, Malabar	579
zaagbaars, Nassau	581
zaagbaars, Pacifische gevlekte	575
zaagbaars, reuzen-	578
zaagbaars, rode	580
zaagbaars, witte	573
zaagbaars, zwarte	576
zaagbaarzen	567, 582, 590
zaagjes	1406
zaagvis	79
zaagvissen	78
zadelbaars	788
zadelrob	1451
zakpijp, violet-	1508
zalm	256
zalm, Australische	683
zalm, chinook	254
zalm, chum-	248
zalm, coho-	249
zalm, Donau-	242
zalm, Georgia-	682
zalm, Japanse	250
zalm, keta-	248
zalm, masou-	250
zalm, pink	247
zalm, rode	252
zalm, roze	247
zalm, sockeye	252
zalm, vlag-	268
zalmachtigen	260
zalmen en forellen	262
zalmen, karper-	300

zandbaars	572	
zandbaars, Braziliaanse	865	
zandrog	86	
zandspiering	936	
zandspiering, Amerikaanse	932	
zandspiering, IJslandse	933	
zandspiering, Koreaanse	940	
zandspiering, naakte	938	
zandspiering, Noorse	935	
zandspiering, Pacifische	934	
zandspieringen	931, 937	
zandsteenbaars	786	
zandvis, Japanse	863	
zebrahaai	24	
zebra-ringbrasem	779	
zeeappel	1518	
zeebaars	593, 595	
zeebaars, Amerikaanse	596	
zeebaars, Argentijnse	568	
zeebaars, gestreepte	598	
zeebaars, gevlekte	594	
zeebaars, Japanse	599	
zeebaars, Peruaanse	584	
zeebaars, witte	597	
zeebaars, zwarte	571	
zeebarbeel, gestreepte	817	
zeebarbelen	815, 819	
zeebeer, dwerg-	1442	
zeebrasem	793	
zeebrasem, blauwgevlekte	804	
zeebrasem, gevlekte	798	
zeebrasem, gewone	806	
zeebrasem, Japanse	805	
zeebrasem, Natal-	792	
zeebrasem, nufar-	768	
zeebrasem, rode	794	
zeebrasem, roodgestreepte	803	
zeebrasem, schaapskop-	764	
zeebrasem, Schot-	797	
zeebrasem, Spaanse	790	
zeebrasems	763, 795, 807	
zeebrasems, hottentot-	789	
zeedonderpad	1100	
zeedonderpad, Amerikaanse	1098	
zeedonderpad, groene	1102	
zeedonderpadden	1101	
zeeduivel	1195	
zeeduivel, Amerikaanse	1193	
zeeduivel, West-Afrikaanse	1196	
zeeduivels	1192	
zee-egel, Chileense	1522	
zee-egel, eetbare	1521	
zee-egels	1519, 1520	
zee-engel	68	
zee-engelen	67	
zeeforel	257	
zeehond	1453	
zeehond, grijze	1449	
zeehonden	1446	
zeekarper	808	
zeekat, gewone	1418	
zeekat, kleine	1420	
zeekomkommer, Japanse	1523	
zeekrabbensoorten	1285	
zeekrokodil	1505	
zeeleeuw	1443	
zeeleeuw, Zuid-Afrikaanse	1442	
zeeleeuw, Zuid-Amerikaanse	1444	
zeelt	364	
zeemeerval, gevlekte	376	
zeemeerval, gladbek-	375	
zeemeerval, witte	377	
zeemeervallen	374	
zeenaald, grote	506	
zeenaald, kleine	507	
zeenaald, trompetter-	508	
zeeoor	1328	
zeeoor, parelmoer-	1326	
zeeoor, reuzen-	1325	
zeeoor, rode	1327	
zeeoren	1329	
zeepaardjes	505	
zeepok, reuzen-	1200	
zeepokken	1199	
zeeprik	7	
zeeraaf	755	
zeeratten, olifant-	126	
zeeschildpad, groene	1497	
zeeschildpadsoorten	1494	
zeeslang	414	
zeesnoek, olijfgroene	561	
zeesnoeken	560, 562	
zeestekelbaars	502	
zeester	1517	
zeesterren	1516	
zeevarkenhaai	70	
zeevarkenhaaien	69	
zeewolf	915	
zeewolf, blauwe	914	
zeewolf, gevlekte	916	
zeewolven	913, 917	
zeilvinplooibek	902	
zeilvis, Atlantische	1016	
zeilvis, Pacifische	1017	
zeilvissen	1015	
zeskieuwige haaien	9	
zeskieuwige koehaai	11	
zilverbaars	605	
zilveren haarstaart	957	
zilverheek	483	
zilverheek, diepzee-	481	
zilverkarper	335	
zilverknorvis	711	
zilverpomfret	1031	
zilverslijmvis	918	
zilversmelt, grote	281	
zilversmelt, Japanse	283	
zilversmelt, kleine	282	
zilversmelt, Pacifische	280	
zilversmelten	279	
zilverstreep-rondeharing	201	
zilvertiphaai	35	
zilvertrommelvis	744	
zilvervis, Antarctische	893	
zilverwijting	449	
zingel	624	
zoengoerami	1047	
zoetwater-steurgarnalen	1244	
zoetwaterbaarzen	613	
zoetwatertrommelvis	716	
zomerbot	1118	
zonnebaars	601	
zonnevis	525	
zonnevis, Amerikaanse	524	
zonnevissen	523	
Zuid-Afrikaanse Oostkusttong	1154	
Zuid-Afrikaanse pelser	199	
Zuid-Afrikaanse sardine	199	
Zuid-Afrikaanse tongen	1155	
Zuid-Afrikaanse turbo	1332	
Zuid-Afrikaanse Westkusttong	1153	
Zuid-Afrikaanse zeeleeuw	1442	
Zuid-Amerikaanse koningsombervis	732	
Zuid-Amerikaanse riviergarnaal	1246	
Zuid-Amerikaanse zeeleeuw	1444	
Zuid-Atlantische ansjovis	209	
Zuid-Atlantische mantel	1368	
Zuid-Europese tandkarper	513	
Zuid-Pacifische zwarte vis	1042	
zuidelijke blauwe wijting	458	
zuidelijke blauwvintonijn	1010	
zuidelijke koningskrab	1309	
zuidelijke pelsrob	1441	
zuidelijke rode snapper	691	
zuidelijke rondloper	685	
zuidelijke spinkrab	1314	
zuidkaper	1489	
zuidkustlangoest	1267	
Zuidoost-Pacifische knorvis	706	
zuidpoolkabeljauw, gele	882	
zuidpoolkabeljauw, gemarmerde	885	
zuidpoolkabeljauw, grijze	886	
zuidpoolkabeljauw, groene	883	
zuidpoolkabeljauw, kale	887	
zuidpoolkabeljauw, streepoog-	884	
zuidpoolkabeljauw, zwarte	881	
zuidpoolkabeljauwen	876	
zuiger, klip-	1190	
zuigkarper	302	
zuigkarpers	301	
zwaardschede, Amerikaanse	1408	
zwaardschede, kleine	1409	
zwaardvis	1029	
zwaardvissen	1028	
zwaardwalvis	1470	
zwaardwalvis, zwarte	1473	
zwarte baars	602	
zwarte dolfijn	1459	
zwarte doornhaai	61	
zwarte doornhaaien	62	
zwarte dwergmeerval	384	
zwarte grondel	1053	
zwarte haai	60	
zwarte haarstaart	956	
zwarte heilbot	1148	
zwarte horsmakreel	638	
zwarte kabeljauw	1094	
zwarte kardinaalvis	611	
zwarte karper	347	
zwarte marlijn	1018	
zwarte Patagonische ijsheek	877	
zwarte pomfret	677	
zwarte skipjack	969	
zwarte snoek	290	
zwarte snoekmakreel	953	
zwarte steenkrab	1293	
zwarte trommelvis	745	
zwarte vis	1041	
zwarte vis, Antarctische	1043	
zwarte vis, Japanse	1044	
zwarte vis, Zuid-Pacifische	1042	
zwarte zaagbaars	576	
zwarte zeebaars	571	
zwarte zeeduivel	1194	
zwarte zuidpoolkabeljauw	881	
zwarte zwaardwalvis	1473	
Zwarte-Zee-elft	160	
Zwarte-Zee-grondel	1060	
Zwarte-Zee-kopvoorn	340	
Zwarte-Zee-tarbot	1173	
Zwarte-Zee-wijting	455	
zwartgevlekte pikarel	811	
zwartkop-ringbrasem	783	
zwartmond-hondshaai	30	
zwartoog-lipvis	847	
zwartvin-rifhaai	38	
zwartvintonijn	1009	
zweeplangoesten	1275	
zwemkrab, fluwelen	1299	
zwemkrab, gazami-	1300	
zwemkrabben	1301	
Zwitserse marene	233	

NL

Índice português

A

abadejo do Cabo	926
abadejo rosado	925
abadejo vermelho	927
abadejos	928
abrótea branca	476
abrótea brasileira	474
abrótea da costa	464
abrótea da Nova Zelândia	438
abrótea do alto	463
abrótea sul-africana	926
abrótea vermelha	475
abróteas	465
achigã	603
achigã-boca pequena	602
agulha	416
agulhão	422
agulhão do Japão	421
agulhas e agulhetas	415
agulheta imperial	418
agulheta verde	417
agulhetas	419
agulhões	420
alabote da Gronelândia	1148
alabote do Atlântico	1133
alabote do Pacífico	1134
alabote japonês	1124
alabote-dente curvo	1125
albacora	1008
albafar	11
alburno	310
alburno do Danúbio	322
alburno italiano	309
alburno pintado	308
alburno raiado	339
alcarraz	610
alfonsim	518
alfonsins	519
alforrecas	1512
algas	1524
algas calcárias	1530
algas castanhas	1525
algas vermelhas	1528
aligátor do Mississípi	1502
aligátores	1500
alosa azul	155
alosa cinzenta	161
amêijoa bicuda	1404
amêijoa boa	1401
amêijoa branca americana	1392
amêijoa cristã	1401
amêijoa japonesa	1402
amêijoa judia	1405
amêijoa macha	1405
amêijoa redonda de Angola	1397
amêijoas	1394
amêijola japonesa	1391
amêijola lisa	1390
anchova	217, 630
anchovas	629
anchoveta argentina	214
anchoveta do Peru	220
anequim	20
anjo	68
anjos	67
anodonta gigante	1343

apara-lápis	504
apogonídeos	609
aranha grande	869
arca do Pacífico Ocidental	1339
arca japonesa	1340
arcas	1342
arcas vermelhas	1341
areeiro	1169
areeiro de quatro manchas	1168
areeiros	1170
areeiros, pregados e rodovalhos	1167
arenque	169
arenque do Pacífico	170
arenque pontuado	181
arenque redondo	178
arenque redondo da Califórnia	176
arenque redondo·do Atlântico	177
arenque redondo prateado	201
arenque-de-lago	231
arenques redondos	179
argentina branca	282
argentina do Japão	283
argentina do Pacífico	280
argentina dourada	281
argentinídeos	279
arinca	453
ascídia comestível	1509
ascídia do Chile	1509
ascídia do Sudoeste Africano	1510
ascídia violeta	1508
ascídias	1507
áspio	311
atum albacora	1008
atum do Sul	1010
atum patudo	1011
atum rabilho	1012
atum tongol	1013
atum voador	1007
atum	1012
atum-barbatana negra	1009
atum-foguete	963
atuns	1014
atuns e afins	960
azevia	1159
azevia de malhas	1158
azevia raiada	1160
azevias	1161

B

babosa da Patagónia	880
bacalhau	447
bacalhau árctico	445
bacalhau da Gronelândia	448
bacalhau do Alasca	446
bacalhau do Árctico	445
bacalhau do Atlântico	447
bacalhau do Extremo Oriente	444
bacalhau do Pacífico	446
bacalhau do Pacífico Norte	446
bacalhau polar	441
badejinho do norte	449
badejinho	449
badejo	454
badejo do mar Negro	455
bagre malhado	376

bagre-barba branca	377
bagre-boca lisa	375
bagres marinhos	374
bagres	374
baleia anã	1480
baleia azul	1483
baleia bicuda de Baird	1459
baleia boreal	1481
baleia cinzenta do Pacífico	1491
baleia comum	1484
baleia de bossas	1485
baleia de Braide	1482
baleia franca austral	1489
baleia franca boreal	1487
baleia franca negra	1490
baleia franca	1490
baleia-pigmeu	1488
baleias francas	1486
balistas	1179
barbinhos	553
barbo	314
barbo da Albânia	313
barbo da Dalmácia	312
barbo de Hoven	338
barbo de Java	353
barbo do Danúbio	317
barbo italiano	318
barbo meridional	316
barbo-cumba	315
barbos da Ásia	354
barbudo do Indo-Pacífico	554
barbudo gigante	556
barbudo real	555
barbudos	553
barracuda	535
barracuda amarela	533
barracuda da Califórnia	531
barracuda da Guiné	530
barracuda de barras	534
barracuda gigante	532
barracudas	529, 536
barroso	58
beicinho-prata	902
beicinhos	903
beiçudos	337
berbigão de bicos	1388
berbigão vulgar	1389
berbigões	1387
besugo	790
besugo do Cabo	792
besugos, bicas e gorazes	795
bica	794
bica do Cabo	792
bica-buço	791
bico de garrafa	1460
bico de pato	1461
bicuda amarela	533
bicuda da Califórnia	531
bicuda da Guiné	530
bicuda de barras	534
bicuda gigante	532
bicuda	535
bicudas	529, 536
bicudo de focinho comprido	143
bicudo de focinho curto	144
biqueirão argentino	214
biqueirão branco	282

PT

biqueirão da Austrália	215	búzio	1335	camarão real	1221
biqueirão de baía	210	búzios	1334	camarão rosado do norte	1217
biqueirão do Cabo	216	buzo	1335	camarão rosado do sul	1224
biqueirão do Indo-Pacífico	216			camarão vermelho argentino	1256
biqueirão do Pacífico Central	212			camarão vermelho real	1257
biqueirão do Pacífico norte	219	**C**		camarão vermelho	1233
biqueirão do Peru	220			camarão-akiami	1242
biqueirão do rio	221	cabaço	1089	camarão-banana	1222
biqueirão japonês	218	cabeça chata do Brasil	873	camarão-café do norte	1213
biqueirão listrado	209	cabeça de cobra da Indonésia	557	camarão-cauda vermelha	1225
biqueirão	217	cabeça de cobra listado	558	camarão-estilete argentino	1206
biqueirão-cauda de rato	213	cabeças de cobra	559	camarão-kuruma	1219
biqueirão-rabo amarelo	211	cabeçudo	874	camarão-nailon chileno	1235
biqueirões «Stolephorus»	222	cabeçudos	875	camarão-navalha	1254
biqueirões	208	caboz cabeçudo	1052	camarão-pata amarela	1215
bivalves	1324	caboz da areia	1059	camarão-pata branca	1228
blénios	904	caboz da rocha	1054	camarão-siba	1208
boca de panela	1466	caboz de Kessler	1057	camarão-tigre	1219
boca doce	10	caboz do Adriático	1058	camarão-tigre castanho	1218
bodião cinzento	845	caboz do mar Negro	1060	camarão-tigre gigante	1223
bodião da ostra	852	caboz dourado	1051	camarão-tigre verde	1226
bodião das ervas	850	caboz italiano	1056	camarões «Atyidae»	1243
bodião de pinta	848	caboz negro	1053	camarões «Metapenaeus»	1209
bodião de pluma	844	caboz transparente	1050	camarões «Parapenaeopsis»	1211
bodião do Mediterrâneo	846	caboz-cabeça de cobra	1061	camarões «Penaeus»	1229
bodião do norte	853	cabozes	1049, 1055	camarões aristeídeos	1232
bodião fusco	842	cabra bicuda	1088	camarões barbudos do Pacífico	1230
bodião manchado	849	cabra de bandeira	1080	camarões crangonídeos	1250
bodião reticulado	840	cabra de casca	1106	camarões da pedra	1253
bodião rupestre	839	cabra do Cabo	1081	camarões de água doce	1248
bodião verde	854	cabra morena	1084	camarões marrecos	1240
bodião vulgar	847	cabra riscada	1091	camarões palemonídeos	1245
bodião-canário	841	cabra vermelha	1079	camarões palemonídeos de água	
bodião-pavão	851	cabra-cabaço	1089	doce	1244
bodião-tordo	843	cabra-kumu	1082	camarões pandalídeos	1239
bodiões	837	cabra-lira	1090	camarões sergestídeos	1241
boga	766	cabras	1083	camarões-nailon	1236
boga da Dalmácia	324	cabras e ruivos	1078	camarões-navalha	1255
boga da Jugoslávia	326	cabrinha de leque	1109	canarinho do mar	570
boga de boca curva	329	cabrito	912	canário do mar	569
boga de boca direita	327	cação antárctico	51	caneja	53
boga do Danúbio	325	cação da Patagónia	54	canejas	55
boga do mar	766	cação galhudo	64	cangueira do Golfo	734
boga italiana	328	cação liso	53	cangueira-cachorro	733
boga sul-europeia	323	cação pintado	52	cangueiras	736
bogardo do Danúbio	358	cachalote	1462	cangueira-zorra	735
bogardo do mar Negro	357	cachucho	776	cangulo cinzento	1180
bolota	442	cachucho-dentão	777	cangulo de vela	1183
bonito austral	979	cações e pernas de moça	49	cangulos galhudos	1182
bonito de barriga listada	973	cações	55	cangulos	1179
bonito de dorso listado	982	cadelinhas	1406	canivetes	1407
bonito do Indo-Pacífico	981	caimão	1503	cantarilho americano	1073
bonito do Pacífico	980	calafate da Argentina	758	cantarilho do Cabo	1071
bonito saltador	966	calafate de riscas	759	cantarilho do Pacífico	1070
bonito	982	calafates	760	cantarilho dos mares do Norte	1075
bonito-dente de cão	974	caldeirão	1465	cantarilho legítimo	1063
borboletas	830	camarão americano do rio	1246	cantarilho viúvo	1072
bordalinho da Albânia	348	camarão árctico	1237	cantarilho-rabo amarelo	1074
borrelhos	1337	camarão barbudo	1231	cantarilho-requeime	1065
botelhas	1531	camarão boreal	1238	cantarilhos	1064
boto	1471	camarão branco legítimo	1249	cantarilhos do Norte	1077
boto raiado	1467	camarão branco nortenho	1227	cantarilhos e rascassos	1062
brema	305	camarão carnudo	1216	capasseca	183
brema azul	304	camarão cristal	1214	capatão de bandeira	775
brema branca	349	camarão da Austrália	1207	capatão legítimo	774
brema chinesa	346	camarão da pedra	1252	capelim	273
brema do Danúbio	306	camarão da Quarteira	1220	carabineiro	1233
brema do rio Weser	365	camarão da vasa	1258	carabineiro cardeal	1234
brema prateada	319	camarão do Japão	1219	caracinídeos	300
bremas	307	camarão do rio	1251	carangídeos	632
bruxa norueguesa	1171	camarão gigante do rio	1247	caranguejo	1298
buço	791	camarão guinéu	1210	caranguejo chinês	1303
bumblim	297	camarão japonês	1219	caranguejo da lama	1302
bumblins	296	camarão negro	1251	caranguejo das neves	1312

PT

caranguejo negro da pedra	1293	charro preto	654	corvinão de pintas	757
caranguejo real	1310	charrocos	1191	corvinão negro	745
caranguejo real da pedra	1307	charro-olho largo	641	corvinas e afins	715
caranguejo real de Murray	1308	charros	645	corvinas-rainha	752
caranguejo real do sul	1309	charros ingleses	684	corvinata branca	720
caranguejo verde do Mediterrâneo	1297	charuteiro do Japão	660	corvinata do Peru	723
caranguejo verde	1298	charuteiro-catarino	658	corvinata gigante	724
caranguejo vermelho da fundura	1321	charuteiro-limão	659	corvinata pintada	725
caranguejos da fundura	1322	charuteiros	661	corvinata real	726
caranguejos do mar	1285	cherne	585	corvinata riscada	728
caranguejos das neves do Pacífico	1313	cherne da Nova Zelândia	586	corvinata-pescada	727
caranguejos nadadores	1301	chicharro	672	corvinatas	729
caranguejos reais	1311	choco	1418	corvinata-totuava	724
carapau chileno	669	choco vulgar	1418	corvina-teraglim	719
carapau do Cabo	673	chocolate	950	craca bicuda	1200
carapau do Cunene	674	chopo	1419	cracas	1199
carapau do Japão	666	chopo anão	1420	crocodilo da Nova Guiné	1504
carapau do Mediterrâneo	668	choupa	808	crocodilo do Sião	1506
carapau do Pacífico	671	choupaque	184	crocodilo marinho	1505
carapau negrão	670	chupadores	301	crocodilos	1501
carapau rugoso	667	ciclídeos	831		
carapau verde	665	cigarra do mar	1305	**D**	
carapau	672	cigarras e cavacos	1304		
carapaus	675	ciprinídeos	303	dagas	203
carcarinídeos	34	ciprinodontídeos	512	delfinídeos	1463
carpa cabeçuda	336	clame amarela	1403	dentão	774
carpa comum	332	clame da areia	1415	dentão das Canárias	772
carpa da lama	330	clame dura	1395	dentão de Angola	771
carpa do limo	331	clame dura japonesa	1399	dentão de bandeira	775
carpa espelho	332	clame islandesa	1386	dentão de Marrocos	777
carpa negra	347	clame redonda	1398	dentão do Cabo	796
carpa prateada	335	clame	1398	dentão do Congo	773
carpa	332	clames duras	1400	dentão do Natal	797
carpinteiro	765	clárias	388	dentão-carpinteiro	765
carta de bico	1177	clupeídeos	154	dentão-quissanga	772
carta de olhos grandes	1117	cobra de orelhas	413	dentinho do Cabo	612
carta de quatro olhos	1119	cobra do mar	414	diabo-do-mar	114
carta de verão	1118	colo-colo	703	diabos-do-mar	111
carta do Mediterrâneo	1115	congro americano	411	dobradiça	788
carta japonesa	1122	congro argentino	412	doirada	802
carta pontuada	1116	congro bicudo do Japão	406	doirado	681
carta-alabote de quatro olhos	1119	congro branco	405	donzela azul	460
carta-alabote de verão	1118	congro branco africano	405	donzela do Mediterrâneo	461
carta-alabote japonesa	1120	congro malhado do Pacífico	409	donzela	462
cartas	1114	congro	410	dormideira	118
cartas-alabote	1121	congros	408	dormidores	1048
casas alugadas	1286	congros bicudos	407	dourada do Japão	805
cascarra	1162	conquilhas	1406	dourada	802
castanheta	836	corais	1513	dugongue	1454
castanhetas e afins	835	coral vermelho	1514		
castanhola do mar	1384	corcovado africano	656	**E**	
castanholas	693	corcovado do Golfo	657		
cavala	984	corcovados	676	efipídeos	825
cavala curta	975	coregono azul	240	eirós	396
cavala das ilhas	976	coregono bicudo	237	elopídeos	145
cavala do Índico	977	coregono branco	229	enchova argentina	214
cavala pintada	983	coregono da Sibéria	238	enchova da Austrália	215
cavalas do Índico	978	coregono da vasa	236	enchova do Pacífico Central	212
cavalas e sardas	986	coregono das ilhas	230	enchova do rio	221
cavalas gigantes	1004	coregono de artedi	231	enchova	217
cavalas reais	653	coregono de lago	232	enchova-cauda de rato	213
cavalos marinhos	505	coregono de olhos grandes	235	enchova-rabo amarelo	211
celindra	223	coregono do fundo	239	enchovas	208
centracantídeos	810	coregono do lago Leman	233	enguia americana	400
centrolofídeos	1040	coregono lavareda	234	enguia australiana	397
centropomídeos	560	coregonos	241, 261	enguia da Nova Zelândia	398
chaputa	680	cornetinhas	1336	enguia europeia	396
chaputas	679	cornudas	45	enguia japonesa	399
charro amarelo	637	cornudas	48	enguia	396
charro de riscas amarelas	655	corvina	718	enguias	395, 401
charro de Russel	644	corvina africana	717	enxada africana	827
charro francês	637	corvina de água doce	716	enxada do Indo-Pacífico	828
charro japonês	642	corvina de boca amarela	719	enxadas e afins	825
charro moiro	643	corvina legítima	718		

eperlano arco-íris	275	esturjão-ventre nu	134	galo negro	525
eperlano árctico	276	esturjões	130	galos	523
eperlano de tanque	271	eufausídeos	1203	galucha	174
eperlano do Pacífico	272	eulacom	278	gamba	1212
eperlano europeu	274			gamba branca	1212
eperlanos	270, 277			gamba de Angola	1224
eremitas	1286	**F**		gamba manchada	1220
escalo	341			gambúsia	516
escalo do mar Negro	340	facas	1407	garoupa legítima	573
escalo do norte	342	falsa abrótea	924	garoupa malabar	579
escalo do sul	344	falsa orca	1473	garoupa mosqueada	575
escalo italiano	345	falso alabote de Canadá	1118	garoupas e meros	582
escalo prateado	343	falso alabote japonês	1120	garoupas, meros	567
escamudo do Alasca	470	falso badejo	153	garoupas-badejo	583
escamudo	467	falso besugo doirado	696	gata	60
escardínio grego	363	falso pampo	677	gata-lixa	60
escardínio-olho vermelho	362	falso robalo branco	561	gatos cabeçudos	293
escienídeos	715	falsos alabotes	1121	gatos da América do Norte	387
escocês de pinta	798	falsos besugos	697	gatos de cabeça chata	388
escocês	797	falsos pimpins	528	gatos marinhos	378
escolar real	949	falsos robalos	562	gatos transparentes	379
escolar	950	faneca	472	gelídeos	1529
escolares e senucas	947	faneca-noruega	471	gigartinas	1531
escolares pretos	948	fanecão	473	góbio	333
escolar-serra	954	fateixa	146	góbio do Danúbio	334
escombrídeos	961	fateixas	147	golfinho	1464
escorpião	1100	fateixa-torpedo	146	golfinho branco	1478
escorpião americano	1098	ferreira	786	golfinho branco do Atlântico	1468
escorpião de água doce	1097	ferreira branca	785	golfinho fiandeiro	1475
escorpião de quatro cornos	1099	ferreiras	787	golfinho riscado	1474
escorpião roco	1102	flecha	152	golfinho-focinho branco	1469
escorpiões	1101	flechas	151	golfinhos malhados	1476
esgana-gata marinha	502	foca barbuda	1448	gonela da rocha	919
esgana-gata	500	foca cinzenta	1449	goraz	793
esgana-gatas	499	foca da Gronelândia	1451	granadeiro azul	479
espada preto	956	foca de mitra	1447	granadeiro da Patagónia	478
espada-lírio	959	foca marmoreada	1452	granadeiros	480
espadarte	1029	foca vulgar	1453	grande rorqual	1483
espadartes	1028	foca-monga	1450	guelha	42
espadarte-serra	79	focas	1446	guerreiro de barras	768
espadartes-serra	78	foguete	951	guitarras	77
espadela	225	fogueteiro arco-íris	646	gurami beijador	1047
espadelas	224, 226	fogueteiro galego	631	gurami gigante	1045
espadilha chilena	204	folha de alface	1532	gurami-pele de cobra	1046
espadilha do mar Negro	171	frachão	939		
espadilha	202	frachões	931	**H**	
espadim azul do Atlântico	1019	fucales	1527		
espadim bicudo	1026	funil escamudo	1385	holotúria japonesa	1523
espadim branco do Atlântico	1021	fura-vasos	606, 608	hotentotes	789
espadim de bico curto	1022	fura-vasos do Pacífico	607		
espadim do Mediterrâneo	1024	fuzileiros	686	**I**	
espadim negro	1018				
espadim raiado	1023	**G**		ilhalvo	545
espadim-peto	1025			imperador	518
espadins	1027	gadídeos	440	imperadores	517
espadins e veleiros	1015	gadiformes	435	imperadores	519
espadins	1020	gaiado	973	irós	396
esparídeos	763	galateídeos	1306		
espinelha	500	galeão	824	**J**	
espinho	501	galeões	823		
esponjas	1511	galeota americana	932	jamanta	114
esqualídeos	57	galeota da areia	938	jamantas	111
esquilo vermelho	522	galeota da Coreia	940	judeu	964
estrela do mar comum	1517	galeota da Islândia	933	judeu liso	965
estrelas do mar	1516	galeota do norte	935	judia	838
estrombos	1338	galeota do Pacífico	934	juliana	466
esturjão	137	galeota maior	939		
esturjão adriático	133	galeota menor	936	**K**	
esturjão branco	138	galeotas	937		
esturjão do Danúbio	131	galeotas	931	krill	1203
esturjão do Volga	135	galhudo	65	krill da Noruega	1205
esturjão estrelado	136	galhudo de focinho curto	66	krill do Antárctico	1204
esturjão verde	132	galhudo malhado	64		
esturjão-beluga	139	galo branco	524		

PT

L

ladrão armado 1103
lagartixa da rocha 494
lagartixa do mar 493
lagartixa prateada 495
lagartixa-cabeça áspera 497
lagartixas 498
lagartixas e granadeiros 496
lagarto da costa 393
lagarto do mar 389
lagarto escamudo 392
lagarto verde 391
lagartos 390
lagartos-do-mar 390
lagosta africana 1273
lagosta castanha 1266
lagosta da Austrália 1270
lagosta da Nova Zelândia 1260
lagosta da Oceânia 1264
lagosta das Caraíbas 1269
lagosta de Moçambique 1267
lagosta de Transkei 1271
lagosta do Cabo 1262
lagosta do Natal 1265
lagosta japonesa 1272
lagosta Juan Fernandez 1261
lagosta rósea 1268
lagosta verde 1273
lagosta vermelha africana 1268
lagosta vulgar 1266
lagostas 1259, 1274
lagosta-tristão 1263
lagostim 1320
lagostim americano de água doce 1282
lagostim de rio da Austrália 1316
lagostim do Danúbio 1277
lagostim do Índico 1319
lagostim do monte 1280
lagostim do rio 1276, 1279
lagostim vermelho da Luisiana 1284
lagostim vermelho do rio 1284
lagostins «Orconectes» 1283
lagostins do rio 1278, 1281
laibeque de cinco barbilhos 443
laibeque de três barbilhos 450
laibeques 451
laminárias 1526
lampreia da Ucrânia 3
lampreia de esteiro 6
lampreia de Vladykov 4
lampreia do Danúbio 2
lampreia do mar 7
lampreia do rio 5
lampreias 1
lapas 1333
larotes 451
larvas de lagosta 1275
lavadilha 202
lavagante 1318
lavagante americano 1317
leão marinho de dois pêlos 1441
leão marinho de juba 1443
leão marinho de um pêlo 1444
leão-marinho do Cabo 1442
leitão 30
leões marinhos 1440
leque 1359
leque islandês 1362
leque variado 1363
limanda 1139
límulo 1515
língua 1157
língua de gato 1156

linguado 1165
linguado austral-este 1154
linguado austral-oeste 1153
linguado da areia 1164
linguado legítimo 1165
linguado redondo riscado 1152
linguado turco 1163
linguados austrais 1155
linguados e afins 1151
línguas de cão 1166
lingue 462
lingue azul 460
lingue da pedra 461
lingue de Atka 1093
lingueirão 939
lingueirões 1407
linguiça 475
lírio 959
liro antárctico 1043
liro preto 1041
litorinas 1337
lixa de escama 59
lixa de lei 58
lixa de pau 60
lixa 59
lixinha 61
lixinha da fundura 61
lixinhas da fundura 62
lobo 63
lobo marinho 1450
longueirão curvo 1409
longueirão da América do Norte 1408
longueirão direito 1410
longueirão direito europeu 1413
longueirão do Alasca 1411
longueirão do Cabo 1412
longueirões 1414
longueirões 1407
lorcha 1092
lorcha de Atka 1093
lota do rio 452
luciano do Golfo 690
luciano do mangal 688
luciano riscado 692
luciano vermelho 691
luciano-cauda amarela 694
lucianos 693
luciano-ventre prateado 689
lúcio 288
lúcio do Pacífico 291
lúcio listado 287
lúcio malhado 290
lúcio mosqueado 289
lucioperca 620
lúcios 286
lula 1426
lula bicuda menor 1423
lula do Cabo 1425
lula pálida 1424
lula vulgar 1426
lulas 1422
lulas, potas 1421
lunetas 698
lutjanídeos 687

M

machete do Atlântico 186
machete do Pacífico 185
machetes 187
macua 943
macuas 944
manatim africano 1457

manatim das Caraíbas 1456
manatim do Amazonas 1455
manta 112
manta do Pacífico 113
marachomba babosa 909
marachomba de água doce 911
marachomba salpicado 910
marachomba-borboleta 905
marachomba-frade 908
marachomba-pavão 907
marachombas 906
marachombas 904
marinha comum 506
marinha-cabeça chata 507
marinha-focinho grosso 508
marlim bicudo 1026
marlim de bico curto 1022
marlim negro 1018
marlim-peto 1025
marlins 1020
marlonga do Antárctico 878
marlonga negra 877
marlongas 879
marmota 487
marreco do Cabo 769
marrecos 770
marsopa de Porto Dall 1472
martelo 47
maruca azul 460
maruca da África do Sul 926
maruca da Argentina 925
maruca da pedra 461
maruca do Chile 927
maruca 462
marucas 928
masca-tabaco 875
medusas 1512
meia-agulha austral 428
meia-agulha brasileira 430
meia-agulha do Japão 431
meia-agulha preta 429
meia-agulha saltadora 432
meias-agulhas 427
meias-agulhas 433
meios-bicos 433
melga 64
menhadem 166
menhadem do Pacífico Sudeste 175
menhadem escamudo 165
menhadens 167
merma 968
merma negra 969
merma oriental 967
mero 577
mero amarelo 574
mero americano 580
mero crioulo 581
mero gigante 576
mero gigante do Pacífico 591
mero legítimo 577
mero-tigre 578
mexilhão chileno 1374
mexilhão coreano 1375
mexilhão da Austrália 1378
mexilhão da rocha sul-americano 1382
mexilhão do Mediterrâneo 1377
mexilhão do rio da Prata 1379
mexilhão perlífero do rio 1383
mexilhão pintado do rio 1345
mexilhão verde 1380
mexilhão verde da Nova Zelândia 1381
mexilhão vulgar 1376
mexilhão-choro 1371
mexilhão-tâmara 1372

mexilhões	1370	
mexilhões de água doce	1344	
mexilhões-pata de cavalo	1373	
mini-saias	526	
moluscos marinhos	1323	
mora azul	437	
moras	436	
mordedor	855	
moreia	404	
moreia verde	403	
moreias	402	
morsa	1445	
mugem	545	
muginha	545	
musgo gordo	1532	
musgos	1531	

N

narval	1479
navalhas	1407
navalheira azul	1295
navalheira felpuda	1299
navalheira-dana	1294
navalheira-gazami	1300
navalheiras	1296
nedopa da Nova Zelândia	864
nedopas do Brasil	865
nemipterídeos	695
nototénia da Patagónia	889
nototénia de Günther	891
nototénia de Ramsay	892
nototénia do Antárctico	882
nototénia escamuda	886
nototénia marmoreada	885
nototénia negra	881
nototénia-cabeça chata	883
nototénia-cabeça lisa	887
nototénia-olhos raiados	884
nototénia-rabo curto	890
nototénias e afins	876
nototenídeos	876

O

oió	173
oiriços-do-mar	1520
olho de vidro	521
olho de vidro laranja	520
olho verde	299
olhos verdes	298
olhudo	611
orca gladiadora	1470
orca	1470
orelha do mar	1328
orelha gigante	1325
orelha-lábio preto	1327
orelha-pérola	1326
orelhas do mar	1329
orelhas	1329
ostra americana	1349
ostra da rocha da Guiné	1352
ostra das Caraíbas	1348
ostra de Sydney	1357
ostra gigante	1347
ostra perlífera	1346
ostra plana chilena	1351
ostra plana da Guiné	1352
ostra plana da Nova Zelândia	1355
ostra plana do Pacífico	1354
ostra plana europeia	1353
ostra plana	1353
ostra portuguesa	1347

ostra redonda	1353
ostras	1350
ostras planas	1356
ostras redondas	1356
otária de dois pêlos	1441
otária de um pêlo	1444
otária do Cabo	1442
ouriço do mar	1521
ouriço do mar chileno	1522
ouriço-do-mar	1518
ouriço-do-mar púrpura	1518
ouriços-do-mar	1519
ouriços-do-mar	1520

P

pala	205
pala chata	182
pala de estação	206
pala-toli	207
palma	1112
palma do Índico	1113
palma espinhosa	1111
palmeta	974
paloco	467
palombeta	647
palombetas	648
pâmpano do Pacífico	1034
pâmpano-lua	1033
pâmpanos do Golfo	1036
pampo chinês	1032
pampo prateado	1031
pampo-godinho	1037
pampos e pâmpanos	1030
panga	799
papagaio velho	856
pardelha dos Alpes	360
pardelhas, bogardos e ruivacas	361
pargo de bandeira	775
pargo de pintas azuis	804
pargo de riscas	803
pargo legítimo	806
pargo marreco do Cabo	769
pargo mulato	710
pargo ruço	804
pargo verdadeiro	806
pargo	806
pargos	807
pargos azuis	767
pargos marrecos	770
pargo-sêmola	803
pargo-tereso	803
parona	651
passarinhos e ronquinhas	762
pata-roxa	31
pata-roxa castanha	29
pata-roxa-gata	32
pata-roxas	33
pata-roxas e leitões	28
patudo	1011
pé de burrinho	1396
pegador	1178
peixe amargo	356
peixe doce	269
peixe dourado	320
peixe encarnado	320
peixe-espinho	500
peixe vermelho da fundura	1076
peixe vermelho	1075
peixe-agulha	416
peixe-algodão	628
peixe-aranha maior	869
peixe-aranha menor	867

peixe-aranha pontuado	868
peixe-aranha raiado	870
peixe-areia japonês	863
peixe-banana	152
peixe-bobo «taraki»	859
peixe-bobo bicudo	857
peixe-bobo pintado	858
peixe-bola do norte	1185
peixe-calhandra antárctico	893
peixe-caracol comum	1108
peixe-caracol do Atlântico	1107
peixe-carneiro americano	922
peixe-carneiro europeu	923
peixe-carvão do Pacífico	1094
peixe-castor	142
peixe-casulo australiano	127
peixe-choque	394
peixe-chumbo	469
peixe-cobrelo	929
peixe-cravo	509
peixe-deriva do Índico	1038
peixe-elefante	125
peixe-espada	957
peixe-espada do Pacífico	958
peixe-espada preto	956
peixe-espátula chinês	141
peixe-frade	12
peixe-fuso	622
peixe-fuso maior	624
peixe-fuso menor	623
peixe-galo	525
peixe-galo branco	524
peixe-gatilho de vela	1183
peixe-gatilho laranja	1181
peixe-gato castanho	385
peixe-gato negro	384
peixe-gato pontuado	386
peixe-gelo austral	896
peixe-gelo bicudo	899
peixe-gelo catarino	900
peixe-gelo da Geórgia do Sul	901
peixe-gelo de Wilson	897
peixe-gelo do Antárctico	898
peixe-gonela	919
peixe-grosa australiano	683
peixe-grosa da Geórgia	682
peixe-lanterna	295
peixe-lapa	1105
peixe-leite	228
peixe-leme do Pacífico	1044
peixe-lobo azul	914
peixe-lobo malhado	916
peixe-lobo riscado	915
peixe-lua	1188
peixe-lua comprido	1189
peixe-lua rabudo	1187
peixe-macaco	918
peixe-manel	128
peixe-manteiga americano	1035
peixe-paleta camelo	627
peixe-pau malhado	942
peixe-pau-lira	941
peixe-pérola	930
peixe-piloto	650
peixe-pónei	700
peixe-porco	70
peixe-prata do Atlântico	150
peixe-prego	71
peixe-rei	551
peixe-rei do alto	550
peixe-rei do Mediterrâneo	549
peixe-rei verde	552
peixe-roda	678
peixe-rubi do Índico	685

peixes cabeçudos	875	periquitos	526	raia da Gronelândia	99
peixes chatos	1110	perna de moça	50	raia de dois olhos	94
peixes vermelhos do Atlântico	1077	pescada argentina	486	raia de quatro olhos	92
peixe-sabre	350	pescada branca	487	raia de São Pedro	86
peixes-aranha	866, 871	pescada da África do Sul	484	raia de verão	88
peixes-barba	553	pescada da Califórnia	490	raia do Pacífico	100
peixes-bobo	860	pescada da Nova Zelândia	482	raia estrelada	83
peixes-bola	1186	pescada de Angola	489	raia inverneira	95
peixes-borboleta	830	pescada do Alasca	490	raia lenga	87
peixes-búfalo	302	pescada do Chile	485	raia manchada	93
peixes-canivete	945	pescada do Pacífico Norte	490	raia nevoeira	90
peixes-carneiro	920	pescada do Sudoeste Africano	488	raia oirega	84
peixes-carneiro do Árctico	921	pescada negra	491	raia pintada	83
peixes-casulo africanos	129	pescada prateada	483	raia pontuada	85
peixes-cirurgião	945	pescada prateada do alto	481	raia pregada	89
peixes-elefante	126	pescada	487	raia repregada	97
peixes-enxada e afins	825	pescadas	477, 492	raia taigora	82
peixe-serra	79	pescadinha	487	raia zimbreira	91
peixes-espada e lírios	955	piarda	550	raia-pinta	87
peixes-espátula	140	picão canadiano	619	raias	102, 81
peixes-faca	292	picão verde	621	raias e afins	80
peixes-galo	523	pimpão comum	321	rainha branca	751
peixes-gatilho	1182	pimpim	527	rainha de lei	748
peixes-gato	383	pimpins	526	rainha dentuda	742
peixes-gato americanos	387	pleuronectídeos	1123	rainha-bobo	749
peixes-gelo	895	plumas	767	rainhas	752
peixes-lanterna	294	plumuda do Atlântico	186	rainha-senegal	750
peixes-lapa	1104	plumuda do Pacífico	185	rainúnculo negro	468
peixes-leite	227	plumudas	187	rajiformes	80
peixes-lobo	913, 917	polvo	1438	raposo olhudo	16
peixes-luneta	698	polvo do alto	1435	rãs	1493
peixes-martelo	48	polvo mosqueado	1436	rascasso de pintas	1067
peixes-morcego	826	polvo vulgar	1438	rascasso vermelho	1068
peixe-sombra	268	polvos	1434, 1439	rascasso-escorpião	1066
peixes-paleta	626	polvos do alto	1437	rascassos	1062, 1069
peixes-pónei	699, 701	pomacentrídeos	835	ratão-águia	109
peixes-porco	69	pombo	710	ratão-bispo	110
peixes-rato	122	ponteiros	433	ratazana	123
peixes-rei	548	porco	70	ratazana americana	124
peixes-rubi	684	porco marinho	70	ratazanas	122
peixes-serra	78	pota argentina	1428	ratões	108
peixes-vidro	563	pota da Nova Zêlandia	1431	rã-toiro	1492
peixes-voadores	423	pota do norte	1430	rebecas	77
peixes-voadores	425	pota europeia	1433	relangueiro	510
peixe-tábua	511	pota gigante	1427	remexidos	829
peixe-voador	426	pota japonesa	1432	requeime	1065
peixe-voador do Japão	424	pota voadora	1429	riscado	708
peixe-voador-andorinha	434	potra europeia	1433	roaz corvineiro	1477
peixinho da lama americano	285	prato de alumínio	640	robalo	593
peixinho da lama europeu	284	pregado	1172	robalo branco	597
peixinho de engodo	352	pregado do mar Negro	1173	robalo do norte	596
peixinho do pântano	351			robalo do Peru	584
peixinho ligeiro andaluz	515			robalo japonês	599
peixinho ligeiro espanhol	514	**Q**		robalo legítimo	593
peixinho ligeiro sul-europeu	513			robalo riscado	598
penacheiro índico	439	quilucha	174	robalo-baila	594
perca americana	616	quimera americana	124	robalo-muge	598
perca anelada	604			robalos	595
perca da rocha	600			rodovalho	1175
perca do Danúbio	615	**R**		rodovalho americano	1174
perca do Nilo	565			rodovalho-bruxa	1176
perca europeia	617	rã japonesa	1492	roncadeira amarela	740
perca gigante	564	rabeta africana	753	roncadeira amarela maior	722
perca prateada	605	rabeta branca	730	roncadeira amarela menor	747
perca romena	618	rabeta brasileira	738	roncadeira austral	754
perca trepadora	1039	rabeta caçadora	732	roncadeira da Manchúria	746
perca-caboz	614	rabeta marisqueira	737	roncadeira de pinta	731
percas africanas	566	rabetas	739	roncadeira do Peru	743
percas e afins	613	rabilho	1012	roncadeira japonesa	741
perca-sol	601	rabilo	1012	roncadeira negra	721
perceve	1197	raboto do Antárctico	888	roncadeira prateada	744
perceves bravos	1198	rabotos	894	roncadeira preta	755
perceves lisos	1198	raia áspera	98	roncadeiras	756
percofidídeos	872	raia bicuda	96	roncador colo-colo	703
		raia curva	101		

PT

roncador de olhos grandes	703
roncador de pintas	713
roncador do Pacífico Sudeste	706
roncador listado americano	702
roncador mexicano	707
roncador prateado	711
roncador riscado	708
roncador-batata	709
roncador-bravura	712
roncador-canário	704
roncadores	705, 714
ronquinhas	761
rorqual azul	1483
rorqual de Braide	1482
ruivaca	360
ruivaca do sul	359
ruivo	1085
ruivo espinhoso	1086
ruivos americanos	1087

S

saboga	157
safio	410
salema	801
salmão de Biwa	253
salmão do Atlântico	256
salmão do Danúbio	242
salmão japonês	250
salmão prateado	249
salmão real	254
salmão rosa	247
salmão vermelho	252
salmão-cão	248
salmonete barbudo	820
salmonete branco	820
salmonete da rocha	821
salmonete da vasa	817
salmonete dourado	816
salmonete legítimo	818
salmonete vermelho	818
salmonetes	815, 819
salmonetes da rocha	822
salmonídeos	262
salvelino árctico	263
salvelino do Pacífico	265
salvelino lacustre	266
salvelinos	267
sandilho da areia	938
sandilho do norte	935
sandilhos	931
santola	1315
santola do sul	1314
santola europeia	1315
sapateira	1291
sapateira boreal	1288
sapateira de rocha do Atlântico	1289
sapateira de rocha do Pacífico	1292
sapateira do Pacífico	1290
sapateiras	1287
sapateiro do Indo-Pacífico	1096
sapateiros	1095
sapatorra	681
sarda	985
sardinata Índica	188
sardinela da Índia	192
sardinela da Madeira	193
sardinela franjada	191
sardinela lombuda	190
sardinelas	195
sardineta do Japão	194
sardinetas	180
sardinha da África do Sul	199

sardinha do Cabo	199
sardinha do Chile	200
sardinha europeia	189
sardinha	189
sardinopa chilena	200
sardinopa da África do Sul	199
sardinopa da Austrália	198
sardinopa da Califórnia	196
sardinopa japonesa	197
sardinopa do Japão	197
sardo	23
sargo austral	800
sargo bicudo	781
sargo da América do Norte	809
sargo da Florida	780
sargo legítimo	782
sargo-alcorraz	778
sargo-choupa	764
sargos	784
sargo-safia	783
sargo-veado	779
sarrajão	982
sáveis e savelhas	163
sável	156
sável americano	162
sável chacunda	164
sável chato	168
sável de papo	172
sável de salto	159
sável do mar Negro	160
savelha do Mediterrâneo	158
savelha	157
senuca preta	953
senuca	951
senucas	952
sereia da Florida	662
sereia-camochilo	663
sereias	664
seríolas	661
seriolelas	1042
serra australiana	996
serra branca	1005
serra brasileira	987
serra chinesa	1003
serra coreana	992
serra da Índia	962
serra de faixas	1001
serra de Monterrey	990
serra do Indo-Pacífico	999
serra do Pacífico	1002
serra espanhola	994
serra malhada	1000
serra oriental	997
serra papuense	995
serra raiada	993
serra real	988
serra riscada	998
serra-borboleta	970
serra-cação	971
serra-cação pintada	972
serrajão	982
serra-leopardo	991
serranídeos	567
serrano argentino	568
serrano da areia	572
serrano estriado	571
serrano riscado	589
serrano-alecrim	587
serrano-ferreiro	588
serranos	590
serras	1004
serra-tigre	989
serrões	590
silaginídeos	625

siluro europeu	382
siluro grego	381
siluros	380
sipa do Lago Victória	355
soleídeos	1151
solha americana	1132
solha árctica	1140
solha áspera	1136
solha avessa	1145
solha castanha do Japão	1138
solha da areia da Nova Zelândia	1149
solha da Califórnia	1127
solha da rocha	1135
solha das pedras	1143
solha de inverno	1147
solha do Alasca	1146
solha dos mares do Norte	1137
solha escura do mar do Norte	1139
solha estrelada do Pacífico	1144
solha japonesa	1131
solha lisa	1141
solha	1145
solha-cabeça chata	1130
solha-flanda	1132
solha-limão	1142
solhão	1128
solhão americano	1129
solhão malhado do Japão	1126
solhas da areia	1150
solhas e alabotes	1123
sugador	1190
suspensório	834

T

tainha africana	544
tainha brasileira	546
tainha de salto	543
tainha-amarela	540
tainha-bobo	539
tainha-fataça	542
tainha-garrento	540
tainha-lábio vermelho	541
tainha-liça	538
tainha-negrão	538
tainha-olhalvo	545
tainhas	537
tainha-sabão	547
tamboril americano	1193
tamboril branco	1195
tamboril da África do Sul	1196
tamboril preto	1194
tamboril	1195
tamboril-sovaco preto	1194
tamboris	1192
taralhão	1416
taralhões	1417
tarpão do Atlântico	150
tarpão do Indo-Pacífico	149
tarpão	150
tarpões	148
tartaruga das Guianas	1498
tartaruga franca	1497
tartaruga gigante	1495
tartaruga verde	1497
tartaruga vulgar	1496
tartaruga	1496
tartarugas americanas	1499
tartarugas do mar	1494
tenca	364
teraponídeos	592
tetraodontídeos	1184
tico-tico	791

tilápia de Moçambique 832
tilápia do Nilo 833
tilápias e afins 831
tintureira 42
tomecode 456, 457
torpedo 649
tremedeira 119
tremelga de olhos 120
tremelga marmoreada 118
tremelga negra 119
tremelgas 117, 121
triglídeos 1078
triste-linda 223
trocos 1330
trombeiro-boga 813
trombeiro-choupa 812
trombeiro-malha redonda 811
trombeiros 814
trombeteiro 504
trombeteiros 503
trompete da Nova Zelândia 862
trompetes 861
trotas 260
truta arco-íris 251
truta comum 258
truta das fontes 264
truta do Arizona 244
truta do lago 259, 266
truta dourada 243
truta dourada mexicana 245
truta feroz 255
truta marinha 257
truta marisca 257
truta sapeira 257
truta-gila 246
tubarão bicudo 44
tubarão bicudo do Pacífico 43
tubarão corre-costa 40
tubarão da Gronelândia 63
tubarão de pontas brancas 37
tubarão de pontas prateadas 35
tubarão de São Tomé 19
tubarão dormedor 26
tubarão negro 38
tubarão sardo 23
tubarão-albafar 11
tubarão-anequim de gadanha 21
tubarão-anequim 20

tubarão-areia 14
tubarão-baleia 27
tubarão-cobra 8
tubarão-cobre 36
tubarão-faqueta 39
tubarão-frade 12
tubarão-leopardo 56
tubarão-limão 41
tubarão-martelo recortado 46
tubarão-martelo 47
tubarão-perna de moça 50
tubarão-prego 71
tubarão-raposo 17
tubarão-raposo do Índico 15
tubarão-raposo olhudo 16
tubarão-sardo do Japão 22
tubarão-tapete 25
tubarão-toiro 13
tubarão-zebra 24
tubarões-albafar 9
tubarões-marracho 34
tubarões-martelo 45, 48
tubarões-sardo 18
tunídeos 1006
turbo da África do Sul 1332
turbo ornado 1331

U

uge 106
uge de cardas 105
uge do Pacífico 104
uge violeta 107
uges 103
uje-manta 115
ujes-manta 116
unha 946
unhas 945

V

veleiro do Atlântico 1016
veleiro do Indo-Pacífico 1017
veleiro do Pacífico 1017
venerídeos 1393
verdemã 370
verdemã da pedra 372

verdemã de Bergatino 368
verdemã de tanque 371
verdemã dos Balcãs 367
verdemã dourada 366
verdemã italiana 373
verdemã romana 369
verdinho 459
verdinho austral 458
vermelhão 796
vieira 1365
vieira americana 1369
vieira da Austrália 1366
vieira da Nova Zelândia 1367
vieira de baía 1361
vieira do Atlântico Sul 1368
vieira do Mediterrâneo 1364
vieira-percal 1360
vieiras e leques 1358
viola 76
viola do Golfo 74
viola do Peru 75
viola malhada 73
violas 72
violas 77
voador 1007
voadores 425

X

xaputa 680
xaputas 679
xara branca 59
xaréu azul 635
xaréu bicudo 652
xaréu carvoeiro 638
xaréu-enxada 633
xaréu-macoa 636
xaréu-rei 634
xaréus e charros 639

Z

zagaia-castanheta 1202
zagaias 1201
zifídeos 1458
zorro de olhos grandes 16

PT

513

Suomenkielinen hakemisto

A

aaltorausku	101
aaltorumpukala	738
aasianvihersimpukka	1380
abaloni	1326, 1328
abaloni, iso-	1321, 1325
adriansampi	133
afrikanhilsa	183
afrikankeltti	544
afrikankeuhkokala-suku	129
afrikanlevykala	827
afrikanmanaatti	1457
afrikanmerikarhu	1442
afrikanmurjottajakala	656
ahven	617
ahven, amerikanhammas-	809
ahven, angolanhammas-	771
ahven, aurinko-	601
ahven, hammas-	774
ahven, hammasjää-	877
ahven, harlekiini-	589
ahven, hietameri-	572
ahven, hopea-	604
ahven, hylky-	585
ahven, isosilmähammas-	776
ahven, japaninkardinaali-	612
ahven, kalliomeri-	571
ahven, kanarianhammas-	772
ahven, kapinpuna-	1071
ahven, kelta-	616
ahven, kivi-	600
ahven, kongonhammas-	773
ahven, malabarinmeri-	579
ahven, marokonhammas-	777
ahven, mustapilkku-	605
ahven, nassaunmeri-	581
ahven, niilin-	565, 566
ahven, pastelli-	587
ahven, perunmeri-	584
ahven, puna-	1075
ahven, punahammas-	805
ahven, punameri-	580
ahven, raitameri-	578
ahven, ruskomeri-	588
ahven, saha-	569
ahven, simppu-	618
ahven, tummameri-	577, 591
ahven, tyynenmerenpuna-	1070
ahven, valkomeri-	573
ahven-laji, hammas-	764, 765, 768, 769, 775, 780, 781, 796, 797, 798, 799, 800, 803, 804
ahven-laji, hylky-	586
ahven-laji, jää-	878, 880, 881, 882, 883, 884, 885, 886, 887, 888, 889, 890, 891, 892, 893
ahven-laji, kardinaali-	610, 611
ahven-laji, lima-	700
ahven-laji, meri-	568, 570, 574, 575, 576
ahven-laji, puna-	1072, 1074, 1076
ahven-laji, valko-	628
ahven-laji, viiri-	696
ahven-suku, hammas-	767, 770, 787, 789, 807
ahven-suku, jää-	894
ahven-suku, lima-	701

ahven-suku, meri-	582, 583, 590
ahven-suku, puna-	1077
ahven-suku, viiri-	697, 698
ahvenet	612, 613
ahvenet, antarktiset	876
ahvenet, hammas-	763
ahvenet, kardinaali-	609
ahvenet, kirjo-	831
ahvenet, koralli-	835
ahvenet, kuore-	625
ahvenet, lima-	699
ahvenet, meri-	567, 595
ahvenet, tiikeri-	592
ahvenet, viiri-	695
ahvenet-heimo	613
ahvenet-heimo, antarktiset	876
ahvenet-heimo, hammas-	763
ahvenet-heimo, kardinaali-	609
ahvenet-heimo, kirjo-	831
ahvenet-heimo, koralli-	835
ahvenet-heimo, kuore-	625
ahvenet-heimo, lima-	699
ahvenet-heimo, meri-	567
ahvenet-heimo, tiikeri-	592
ahvenet-heimo, viiri-	695
ahvenet-suku, antarktiset	879
airokala	510
aju	269
alaskanpunakampela	1146
albanianbarbi	313
albaniansärki	348
alexandrianrihmapompano	633
alkukampela	1113
alligaattori	1502
allikkosalakka	339
amerikanankerias	400
amerikanbassi	596
amerikanhammasahven	809
amerikanhärkäsimppu	1098
amerikanhauki	287
amerikanhummeri	1317
amerikankantasilli	162
amerikankielikampela	1129
amerikankivinilkka	922
amerikankuore	275
amerikanmeriankerias	411
amerikanmerikrotti	1193
amerikanmuikku	231
amerikanosteri	1349
amerikanpietarikala	524
amerikansillikuningas	124
amerikansilokampela	1174
amerikantuulenkala	932
amerikanvoikala	1033, 1035
amia	142
amiapompano	647
angolanhammasahven	771
ankeriaat	395
ankeriaat, meri-	408
ankeriaat-heimo	395
ankeriaat-heimo, meri-	408
ankerias	396
ankerias, amerikan-	400
ankerias, amerikanmeri-	411
ankerias, argentiinanmeri-	412
ankerias, australian-	397
ankerias, hopeahauki-	406

ankerias, japanin-	399
ankerias, meri-	410
ankerias, uudenseelannin-	398
ankerias-laji, hauki-	405
ankerias-laji, käärme-	413, 414
ankerias-laji, meri-	409
ankerias-suku	401
ankerias-suku, hauki-	407
ankeriasmonni-suku	388
antarktiset ahvenet	876
antarktiset ahvenet-heimo	876
antarktiset ahvenet-suku	879
argentiinankalmari	1428
argentiinankummeliturska	486
argentiinanmeriankerias	412
argentiinanrumpukala	758
argentiinansardelli	214
arguskala-suku	829
arkkisimpukka, japanin-	1340
arkkisimpukka-laji	1339
arkkisimpukka-suku	1341
arripi	683
arripi-laji	682
atlantinimukala	1107
atlantinkampasimpukka	1359
atlantinnokkakala	417
atlantinpartahai	26
atlantinpurjekala	1016
atlantinpyörösilli	177
atlantinvihersimpukka	1382
auksidi	965
auksidi-laji	964
auloppi-laji	389
aurinkoahven	601
australianankerias	397
australiankampasimpukka	1366
australiankeuhkokala	127
australianlangusti	1270
australianmakrilli	996
australiannokkakala	428
australianpiikkimakrilli	659
australiansardelli	215
australiansardiini	198
australiansinisimpukka	1378

B

balao	429
balkaninnuoliainen	367
barbi	354
barbi, albanian-	313
barbi, espanjan-	315
barbi, italian-	318
barbi, jaavan-	353
barbi, joki-	314
barbi, siamin-	338
barbi, tonavan-	317
barbi, välimeren-	316
barbi-suku, huuli-	337
barrakuda	532
barrakuda, guinean-	530
barrakuda, hopea-	531
barrakuda, keihäs-	534
barrakuda, keltaevä-	533
barrakuda, pikku-	535
barrakuda-suku	536
barrakudat	529

barrakudat-heimo	529	eteläntonnikala	1010	hammasahven, puna-	805
barramundi	564	euroopanliitokalmari	1433	hammasahven-laji	764, 765, 768, 769,
bassi, amerikan-	596				775, 780, 781, 796, 797, 798,
bassi, iso-	603	**G**			799, 800, 803, 804
bassi, japanin-	599			hammasahven-suku	767, 770, 787,
bassi, juova-	598	Geneve-järven siika	233		789, 807
bassi, meri-	593	gilataimen	246	hammasahvenet	763
bassi, pikku-	602	grönlanninhylje	1451	hammasahvenet-heimo	763
bassi, täplä-	594	grönlanninpallas	1148	hammasjääahven	877
bassi, valko-	597	grönlanninturska	448	hammaskarpit	512
beluga	1478	grönlanninvalas	1487	hammaskarpit-heimo	512
bergatinonnuoliainen	368	guineanbarrakuda	530	hammaskarppi, espanjan-	515
bobyretsi	340	gurami	1045	hammaskarppi, seepra-	513
boga	766			hammaskarppi, turkoosi-	514
boga, juova-	801	**H**		hammasosteri	1352
boniitti	973			hammassimpukka-laji, vino-	1390,
boniitti, musta-	969	hai, atlantinparta-	26		1391, 1392
boniitti-laji	966, 967	hai, harmaa-	50	hanhenkaula-laji	1197
boyerinhopeakylki	549	hai, hieta-	13	hanhenkaula-suku	1198
brasiliankeltti	546	hai, hopeatippa-	35	hanhennokkavalas	1461
brasilianlattapääkala	873	hai, jättiläis-	12	harjus	268
brasiliannokkakala	430	hai, kampavasara-	46	harlekiiniahven	589
brasiliansuomuturska	474	hai, kapeapää-	10	harmaadelfiini	1467
brydenvalas	1482	hai, karibiankettu-	16	harmaahai	50
bummalo	297	hai, kärppä-	53	harmaahuulikala	845
bummalot	296	hai, kaulus-	8	harmaahylje	1449
bummalot-heimo	296	hai, kettu-	17	harmaalimakala	910
		hai, kidus-	11	harmaanieriä	266
C		hai, kupari-	36	harmaasilli	161
		hai, leopardi-	56	harmaaturska	471
chilenmerisiili	1522	hai, makrilli-	20	harmaavalas	1491
chilenosteri	1351	hai, mustaevä-	38	hauet	286
chilenpiikkimakrilli	669	hai, nokka-	44	hauet-heimo	286
chilenrihmanilkka	927	hai, oka-	71	hauki	288
chilensarda	980	hai, patagoniansilo-	54	hauki, amerikan-	287
chilensilli	204	hai, piikki-	64	hauki, jätti-	289
chilensinisimpukka	1374	hai, pikku-	61	hauki, luu-	143
		hai, pistepuna-	31	hauki, musta-	290
D		hai, pitkäevämakrilli-	21	hauki, pikkuluu-	144
		hai, rengas-	30	hauki, täplä-	291
dalmatiannokkasärki	324	hai, rengasparta-	25	haukiankerias-laji	405
dalmatiantörö	312	hai, seepraparta-	24	haukiankerias-suku	407
delfiini	1464	hai, silli-	23	heimo, kidushait-	9
delfiini, harmaa-	1467	hai, sini-	42	helmisärki	357
delfiini, pitkäkuono-	1475	hai, sitruuna-	41	hevospiikkimakrilli	636
delfiini, pullokuono-	1477	hai, sumu-	39	hevossimpukka-suku	1373
delfiini, raita-	1474	hai, täpläpuna-	32	hietahai	13
delfiini, valkokuono-	1469	hai, tyynenmerennokka-	43	hietahai-laji	14
delfiini, valkokuve-	1468	hai, tyynenmerensilli-	22	hietakala-laji	864
delfiini-suku	1476	hai, ulappakettu-	15	hietakala-suku	865
delfiinit	1463	hai, valas-	27	hietakampela	1139
delfiinit-heimo	1463	hai, valko-	19	hietakampela, japanin-	1136, 1138
dolfiini	681	hai, valkopilkka-	37	hietakatkarapu	1251
dugongi	1454	hai, vasara-	47	hietakatkaravut	1250
		hai-heimo	69	hietakatkaravut-heimo	1250
E		hai-laji	70	hietakielikampela	1164
		hai-laji, ihmis-	40	hietakuha	619
erakkoravut-heimo	1286	hai-laji, piikki-	58, 59, 60, 65, 66	hietameriahven	572
espanjanbarbi	315	hai-laji, silo-	51, 52	hietarausku	86
espanjanhammaskarppi	515	hai-suku, piikki	62	hietasimpukka	1415
espanjannokkasärki	327	hai-suku, puna-	33	hietatokko	1059
eteläamerikanmerikarhu	1441	hai-suku, silo-	55	hikkorisilli	159
eteläamerikanmerileijona	1444	hai-suku, vasara-	48	hilsa	182, 205
etelänkielikampela-laji	1153, 1154	haimakrilli	971	hilsa, afrikan-	183
etelänkielikampela-suku	1155	haimakrilli-laji	972	hilsa-laji	184, 206, 207
etelänkielikampelat	1166	halli	1449, 1449	hoki	480
etelänkielikampelat-heimo	1166	hammasahven	774	hoki, patagonian-	479
etelänkrilli	1204	hammasahven, amerikan-	809	hoki-suku	477
etelänkummelliturska	482	hammasahven, angolan-	771	holkeri	63
etelänmeduusakala	1043	hammasahven, isosilmä-	776	hopea-ahven	604
etelänmustakitaturska	458	hammasahven, kanarian-	772	hopeabarrakuda	531
etelänsarda	979	hammasahven, kongon-	773	hopeahaukiankerias	406
etelänsärki	359	hammasahven, marokon-	777	hopeahuotrakala	957
				hopeakala	146

FI

hopeakala-suku	147
hopeakalat	145
hopeakalat-heimo	145
hopeakummeliturska	483
hopeakuore	282
hopeakuoreet	279
hopeakuoreet-heimo	279
hopeakyljet	548
hopeakyljet-heimo	548
hopeakylki	551
hopeakylki-laji	552
hopealohi	249
hopeamurisija	711
hopeapaksuotsa	335
hopearumpukala	744
hopeatippahai	35
hopeaturska	449
hopeavoikala	1031
hummeri	1318
hummeri, amerikan-	1317
hummeri, keisari-	1320
hummeri-laji	1319
hummerit-heimo	1306
huotrakala	959
huotrakala, hopea-	957
huotrakala, musta-	956
huotrakala-laji	958
huotrakalat	955
huotrakalat-heimo	955
huulibarbi-suku	337
huulikala, harmaa-	845
huulikala, kivi-	839
huulikala, koru-	838
huulikala, kuono-	850
huulikala, muste-	851
huulikala, ranta-	847
huulikala, riikinkukko-	854
huulikala, rusko-	842
huulikala, silmä-	848
huulikala, sini-	841
huulikala, täplä-	849
huulikala, välimeren-	846
huulikala, viher-	840, 843
huulikala-laji	852, 853, 855
huulikalat	837
huulikalat-heimo	837
huulisärki	323
hylje, grönlannin-	1451
hylje, harmaa-	1449
hylje, kirjo-	1453
hylje, kupla-	1447
hylje, munkki-	1450
hylje, parta-	1448
hylkeet	1446
hylkeet-heimo	1446
hylkyahven	585
hylkyahven-laji	586
hyppijätsingeli	622
härkänieriä	265
härkäsammakko	1492
härkäsimppu	1099
härkäsimppu, amerikan-	1098

I

idännavaga	444
ihmishai-laji	40
ihmishait	34
ihmishait-heimo	34
imukala	1108
imukala, atlantin-	1107
imukalat	1104
imukalat-heimo	1104
imukarpit	301

imukarpit-heimo	301
imukarppi-suku	302
imurikala	1190
intiaanilohi	252
intianmakrilli	977
intianpolyyppikala	1038
islanninkampasimpukka	1362
islanninsimpukka	1386
isoabaloni	1321
isoabaloni	1325
isobassi	603
isohämähäkkirapu	1315
isohopeakylki	550
isojärvisimpukka	1343
isokampasimpukka	1365
isokurnusimppu	1089
isolestikala	497
isolimakala	909
isomerikokki	941
isoneula	506
isoperhosrauku	115
isopiikkimakrilli	658
isopikarelli	812
isosargi	782
isosilmähammasahven	776
isosilmäkampela-laji	1177
isosilmäpiikkimakrilli	654
isosilmätonnikala	1011
isosimppu	1100
isotaskurapu	1291
isotokko	1052
isotuulenkala	939
isoveltto	724
italianbarbi	318
italianlestikala	495
italiannokkasärki	328
italiannuoliainen	373
italiansalakka	309

J

jaavanbarbi	353
jalohelmisimpukka	1346
jalokoralli	1514
jalosilkkisimpukka	1385
japaninankerias	399
japaninarkkisimpukka	1340
japaninbassi	599
japaninhietakampela	1136
japaninhietakampela	1138
japaninhietatähystäjä	863
japaninkardinaaliahven	612
japaninlangusti	1272
japaninliitokala	424
japaninliitokalmari	1432
japaninmakrilli	984, 997
japaninmattosimpukka	1402
japaninmeduusakala	1044
japaninmerimakkara	1523
japaninnokkakala	431
japaninpiikkimakrilli	666
japaninsardelli	218
japaninsardiini	197
japaninvenussimpukka	1399
jokibarbi	314
jokihelmisimpukka	1383
jokikatkarapu-laji	1246
jokikatkarapu-suku	1248
jokilimakala	911
jokilohi	242
jokinahkiainen	5
jokirapu	1276
jokiravut	1278
jokirumpukala	716

jokisimpukat	1344
jokisimpukat-heimo	1344
juovabassi	598
juovaboga	801
juovakäärmeenpääkala	557
juovakeltti	545
juovakotkarausku	110
juovamakrilli	989
juovamarliini	1023
juovanapsija	692
juovaoravakala	522
juovapiikkimakrilli	638
juovasarda	981
juovaton sarda	974
juovaveltto	728
järvisiika	237
järvitaimen	259
jättihauki	289
jättijokikatkarapu	1247
jättiläishai	12
jättimerirokko	1200
jättirihmaevä	554
jääahven, hammas-	877
jääahven-laji	878, 880, 881, 882, 883,
	884, 885, 886, 887, 888, 889,
	890, 891, 892, 893
jääahven-suku	894
jääkala	896
jääkala, makrilli-	898
jääkala, tumma-	901
jääkala-laji	899, 900
jääkalat	895
jääkalat-heimo	895
jäämerenseiti	441
jääturska	456
jääturska-suku	457

K

kaimaani	1503
käkirausku	89
kaliforniankummeliturska	490
kalifornianpyrösilli	176
kaliforniansardelli	219
kaliforniansardiini	196
kaljuunakala	824
kaljuunakalat	823
kaljuunakalat-heimo	823
kalliokampela	1176
kalliomeriahven	571
kalliomerisiili	1518
kalliorapu	1289
kalmari	1426
kalmari, argentiinan-	1428
kalmari, euroopanliito-	1433
kalmari, japaninliito-	1432
kalmari, kapin-	1425
kalmari, pikku-	1423
kalmari-laji	1424, 1427, 1429,
	1430, 1431
kalmarit	1421, 1422
kalmarit-heimo	1421
kampasimpukat	1358
kampasimpukat-heimo	1358
kampasimpukka, atlantin-	1359
kampasimpukka, australian-	1366
kampasimpukka, islannin-	1362
kampasimpukka, iso-	1365
kampasimpukka, uudenseelannin-	1367
kampasimpukka-laji	1360, 1361, 1363,
	1364, 1368, 1369
kampavasarahai	46
kampavenussimpukka	1398

517

kampela	1143	karibianlangusti	1269	kielikampela-laji	1152, 1157, 1158,
kampela, alaskanpuna-	1146	karjukala	527		1159, 1160, 1162, 1163
kampela, alku-	1113	karjukalat	526	kielikampela-laji, etelän-	1153, 1154
kampela, amerikankieli-	1129	karjukalat-heimo	526	kielikampela-suku	1161
kampela, amerikansilo-	1174	karpit, hammas-	512	kielikampela-suku, etelän-	1155
kampela, hieta-	1139	karpit-heimo, hammas-	512	kielikampelat	1151
kampela, hietakieli-	1164	karppi	332	kielikampelat, etelän-	1166
kampela, japaninhieta-	1136, 1138	karppi, espanjanhammas-	515	kielikampelat-heimo	1151
kampela, kallio-	1176	karppi, musta-	347	kielikampelat-heimo, etelän-	1166
kampela, kesä-	1118	karppi, ruoho-	331	kiiltolahna	509
kampela, kieli-	1165	karppi, seeprahammas-	513	kiinankuningasmakrilli	1003
kampela, kukka-	1117	karppi, seeprahammas-	513	kiinankupusilli	168
kampela, lasi-	1169	karppi, turkoosihammas-	514	kiinanvoikala	1032
kampela, lieju-	1132	kärsäkuningas-laji	125	kiipijäkala	1039
kampela, mustaevä-	1128	kärsäkuningas-suku	126	kiiski	614
kampela, mustanmerenpiikki-	1173	kaskelotti	1462	kiiski, tonavan-	615
kampela, mustaselkä-	1147	katkarapu	1255	kilohaili	202
kampela, napa-	1140	katkarapu, hieta-	1251	kilpikonna, käärmeenkaula-	1498
kampela, pikku-	1171, 1172	katkarapu, jättijoki-	1247	kilpikonna, liemi-	1497
kampela, pikkukieli-	1156	katkarapu, pohjan-	1237	kilpikonna, merinahka-	1495
kampela, pikkupää-	1142	katkarapu-heimo	1241, 1243	kilpikonna, valekaretti-	1496
kampela, puna-	1145	katkarapu-laji	1206, 1207, 1208, 1210,	kilpikonna-suku	1499
kampela, ruoste-	1137		1211, 1212, 1213, 1214, 1215, 1216,	kilpikonnat, meri-	1494
kampela, silo-	1175		1217, 1218, 1219, 1220, 1221, 1222,	kirjoahvenet	831
kampela, suomu-	1115		1223, 1224, 1225, 1226, 1227, 1228,	kirjoahvenet-heimo	831
kampela, tähti-	1144		1230, 1231, 1233, 1234, 1235, 1236,	kirjohylje	1453
kampela-laji	1124, 1125, 1126, 1127,		1238, 1242, 1249, 1252, 1254, 1256,	kirjolohi	251
	1130, 1131, 1135, 1149		1257, 1258	kirjomerikissa	916
kampela-laji, etelänkieli-	1153, 1154	katkarapu-laji, joki-	1246	kirjosardiini	173
kampela-laji, isosilmä-	1177	katkarapu-suku	1209, 1229, 1239,	kirveskala	678
kampela-laji, kieli-	1152, 1157,		1240, 1253	kitararausku	76
	1158, 1159, 160, 1162, 1163	katkarapu-suku, joki-	1248	kitararausku, perun-	75
kampela-laji, napa-	1141	katkaravut	1232, 1245	kitararausku, valkopilkku-	73
kampela-laji, piikki-	1111, 1112,	katkaravut, hieta-	1250	kitararausku-laji	74
	1119, 1122	katkaravut-heimo	1232, 1244	kitararausku-suku	77
kampela-laji, silo-	1168	katkaravut-heimo, hieta-	1250	kitararauskut	72
kampela-laji, töyhtö-	1116, 1120	katkerokala	356	kitararauskut-heimo	72
kampela-suku	1150	kaulushai	8	kitasampi	139
kampela-suku kieli-	1161	kaulussargi	783	kivennuoliainen	372
kampela-suku, etelänkieli-	1155	keihäsbarrakuda	534	kiviahven	600
kampela-suku, silo-	1170	keihäsrausku	106	kivihuulikala	839
kampela-suku, vasensilmä-	1121	keihäsrauskut	103	kivinilkat	920
kampelakalat-lahko	1110	keihäsrauskut-heimo	103	kivinilkat-heimo	920
kampelat, etelänkieli-	1166	keila	442	kivinilkka	923
kampelat, kieli-	1151	keisarihummeri	1320	kivinilkka, amerikan-	922
kampelat, oikeasilmä-	1123	kelta-ahven	616	kivinilkka-suku	921
kampelat, silo-	1167	keltaeväbarrakuda	533	kivirapu	1280
kampelat, vasensilmä-	1114	keltaevätonnikala	1008	kivirapu, piikikäs	1307
kampelat-heimo, etelänkieli-	1166	keltajuovamullo	818	kivirapu-laji	1308, 1309
kampelat-heimo, kieli-	1151	keltapiikkimakrilli	637	kivisimppu	1097
kampelat-heimo, silo-	1167	keltapyrstönapsija	694	kivitaateli	1372
kampelat-heimo, vasensilmä-	1114	keltaroussi	520	kivitörö	334
kanarianhammasahven	772	keltit	537	koirakala	284
kaniinikala-laji	943	keltit-heimo	537	koirakala, pikku-	285
kaniinikala-suku	944	keltti, afrikan-	544	koiralohi	248
kantasilli, amerikan-	162	keltti, brasilian-	546	kolja	453
kantasilli-laji	174	keltti, juova-	545	kolmipiikki	500
kantasilli-suku	163	keltti, kulta-	540	kolmiviiksimade	450
kapeapäähai	10	keltti, ohuthuuli-	542	kolorapu	1279
kapeasaksirapu	1277	keltti, paksuhuuli-	538	kongonhammasahven	773
kapinkalmari	1425	keltti-laji	539, 541, 543, 547	konnakalat	1191
kapinkummeliturska	484	keromakrilli	1000	konnakalat-heimo	1191
kapinkurnusimppu	1081	kesäkampela	1118	koralli, jalo-	1514
kapinpiikkimakrilli	673	kettuhai	17	koralliahvenet	835
kapinpuna-ahven	1071, 1071	kettuhai, karibian-	16	koralliahvenet-heimo	835
kapinrihmanilkka	926	kettuhai, ulappa-	15	korallimonni-suku	378
kapinrumpukala	719	keuhkokala, australian-	127	korallit	1513
kapinsardelli	216	keuhkokala-suku, afrikan-	129	korallit-heimo	1513
kapinveitsisimpukka	1412	kidushai	11	koreanmakrilli	992
kardinaaliahven, japanin-	612	kidushait	9	koreansinisimpukka	1375
kardinaaliahven-laji	610, 611	kidushait-heimo	9	koreantuulenkala	940
kardinaaliahvenet	609	kielikampela	1165	korokoro	707
kardinaaliahvenet-heimo	609	kielikampela, hieta-	1164	koruhuulikala	838
karibiankettuhai	16	kielikampela, pikku-	1156	kotilo, metsästystorvi-	1335

kotilo-laji 1331, 1332
kotilo-suku 1336
kotilo-suku, malja- 1333
kotilo-suku, piikki- 1334
kotilo-suku, pyramidi- 1330
kotilo-suku, ranta- 1337
kotilo-suku, siipi- 1338
kotkakala 718
kotkarausku 109
kotkarausku, juova- 110
kotkarauskut 108, 109
kotkarauskut-heimo 108
kreikanmonni 381
kreikansorva 363
krilliäyriäiset 1203
krilliäyriäiset-heimo 1203
krokotiili, meri- 1505
krokotiili, siamin- 1506
krokotiili, uudenguinean- 1504
krokotiilikala 897
krokotiilit 1500, 1501
krokotiilit-heimo 1500
kuha 620
kuha, hieta- 619
kuha, valkosilmä- 621
kukkakampela 1117
kultakala 320
kultakeltti 540
kultakuore 281
kultamattosimpukka 1404
kultamullo 816
kultanuoliainen 366
kultaotsa-ahven 802
kultataimen 243
kultatokko 1051
kummeliturska 487
kummeliturska, argentiinan- 486
kummeliturska, etelän- 482
kummeliturska, hopea- 483
kummeliturska, kalifornian- 490
kummeliturska, kapin- 484
kummeliturska, perun- 485
kummeliturska, senegalin- 491
kummeliturska-laji 481, 488, 489
kummeliturska-suku 492
kummeliturskat 478
kummeliturskat-heimo 478
kuningaslohi 254
kuningasmakrilli 988
kuningasmakrilli, kiinan- 1003
kuningasmakrilli, papuan- 995
kuningasmakrilli, tyynenmeren- 1002
kuningasmakrilli-laji 987, 990, 991,
993, 998, 1004
kuningasmakrilli-suku 1005
kuningasrapu 1310
kuningasrapu-suku 1311
kuonohuulikala 850
kuonosiika 236
kuore 274, 277
kuore, amerikan- 275
kuore, hopea- 282
kuore, kulta- 281
kuore, kynttilä- 278
kuore, lampi- 271
kuore, tyrsky- 272
kuore, villa- 273
kuore-laji 276, 280, 283
kuoreahvenet 625
kuoreahvenet-heimo 625
kuoreet 270
kuoreet, hopea- 279
kuoreet-heimo 270
kuoreet-heimo, hopea- 279

kuparihai 36
kuplahylje 1447
kupusilli 172
kupusilli, kiinan- 168
kupusilli-laji 164
kupusuu-laji 685
kupusuut 684
kupusuut-heimo 684
kurnusimppu, iso- 1089
kurnusimppu, kapin- 1081
kurnusimppu, kyhmy- 1084
kurnusimppu, panssari- 1106
kurnusimppu, piikki- 1086, 1090
kurnusimppu, puna- 1079
kurnusimppu, sinievä- 1082
kurnusimppu, töyhtö- 1080
kurnusimppu-laji 1085, 1088, 1091
kurnusimppu-suku 1083, 1087
kurnusimput 1078
kurnusimput-heimo 1078
kuta 951
kuultotokko 1050
kyhmykurnusimppu 1084
kyljet, hopea- 548
kyljet-heimo, hopea- 548
kylki, boyerinhopea- 549
kylki, hopea- 551
kylki, isohopea- 550
kylki-laji, hopea- 552
kymmenpiikki 501
kynnetönmanaatti 1455
kynsirausku 97
kynttiläkuore 278
kyttyrälohi 247
kärppähai 53
kääpiömakrilli-laji 976
kääpiömakrilli-suku 978
kääpiörapu 1282
kääpiörapu-suku 1283
kääpiövalas 1488
käärmeankerias-laji 413, 414
käärmeenkaulakilpikonna 1498
käärmeenpääkala 559
käärmekala-laji 953
käärmekala-suku 952
käärmemakrilli-laji 949
käärmemakrilli-suku 948
käärmemakrillit-heimo 947

L

lahna 305
lahna, meri- 680
lahna, pekingin- 349
lahna-suku 307
lahnat, meri- 679
lahnat-heimo, meri- 679
lahtimenhaden 165
lahtivalas 1480
laminariat 1526
lampikuore 271
langusti 1266, 1274
langusti, australian- 1270
langusti, japanin- 1272
langusti, karibian- 1269
langusti, natalin- 1265
langusti, tristanin- 1263
langusti, tyynenmeren- 1262
langusti-laji 1260, 1261, 1264,
1267, 1268, 1271, 1273
langusti-suku 1275
langustit 1259
langustit-heimo 1259
lapasammet 140

lapasammet-heimo 140
lasiahven-suku 562, 563
lasiahvenet 560
lasiahvenet-heimo 560
lasikampela 1169
lasimonni-suku 379
lattapääkala, brasilian- 873
lattapääkalat 872
lattapääkalat-heimo 872
leopardihai 56
lestikala 494
lestikala, iso- 497
lestikala, italian- 495
lestikala-laji 493
lestikala-suku 498
lestikalat 496
lestikalat-heimo 496
levät 1524
levykala, afrikan- 827
levykala, pilkku- 828
levykala-suku 826
levykalat 825
levykalat-heimo 825
levätokko 1061
liejukampela 1132
liekkikala 834
liemikilpikonna 1497
liemisimpukka-suku 1406
liitokala 426
liitokala, japanin- 424
liitokala-laji 434
liitokala-suku 425
liitokalat 423
liitokalat-heimo 423
liitokalmari, euroopan- 1433
liitokalmari, japanin- 1432
lima-ahven-laji 700
lima-ahven-suku 701
lima-ahvenet 699
lima-ahvenet-heimo 699
limakala 908
limakala, harmaa- 910
limakala, iso- 909
limakala, joki- 911
limakala, purje- 905
limakala, riikinkukko- 907
limakala-laji 912
limakala-suku 906
limakalat 904
limakalat-heimo 904
limapää, pohjan- 518
limapää-suku 519
limapäät 517
limapäät-heimo 517
lohet 262
lohet-heimo 262
lohi 256
lohi, hopea- 249
lohi, intiaani- 252
lohi, joki- 242
lohi, kirjo- 251
lohi, koira- 248
lohi, kuningas- 254
lohi, kyttyrä- 247
lohi, masu- 250
lohi, puna- 252
lohi, punakurkku- 255
lohi-suku 260
louhikala 869
louhikala, pikku- 867
louhikala, säde- 870
louhikala, täplä- 868
louhikala-suku 871
louhikalat 866

louhikalat-heimo 866
luikero-laji 918
luikeromolva 461
luikeroturska 464
luikeroturska-suku 465
luototokko 1054
luotsikala 650
luuhauki 143
lyhyteväpallopää 1465
lyhytkääpiömakrilli 975
lyhytnokkamarliini 1022
lyyraturska 466
länsiafrikanmullo 820
länsiatlantinsardelli 210
lättäsimppu-laji 1096
lättäsimput 1095
lättäsimput-heimo 1095

M

made 452
made, kolmiviiksi- 450
made, neliviiksi- 469
made-suku, meri- 451
maitokala 228
maitokalat 227
maitokalat-heimo 227
maitovalas 1478
makrilli 985
makrilli, australian- 996
makrilli, australianpiikki- 659
makrilli, chilenpiikki- 669
makrilli, hai- 971
makrilli, hevospiikki- 636
makrilli, intian- 977
makrilli, isopiikki- 658
makrilli, isosilmäpiikki- 654
makrilli, japanin- 984, 997
makrilli, japaninpiikki- 666
makrilli, juova- 989
makrilli, juovapiikki- 638
makrilli, kapinpiikki- 673
makrilli, keltapiikki- 637
makrilli, kero- 1000
makrilli, kiinankuningas- 1003
makrilli, korean- 992
makrilli, kuningas- 988
makrilli, lyhytkääpiö- 975
makrilli, oka- 677
makrilli, papuankuningas- 995
makrilli, piikki- 653, 672
makrilli, pikku- 994
makrilli, pilkkupiikki- 643
makrilli, punapyrstöpiikki- 641
makrilli, queenslandin- 999
makrilli, raita- 962
makrilli, tiikeri- 1001
makrilli, tyynenmerenkuningas- 1002
makrilli, tyynenmerenpiikki- 671
makrilli, uudenseelanninpiikki- 652
makrilli, välimerenpiikki- 668
makrilli-laji 983
makrilli-laji, hai- 972
makrilli-laji, kääpiö- 976
makrilli-laji, käärme- 949
makrilli-laji, kuningas- 987, 990, 991, 993, 998, 1004
makrilli-laji, piikki- 642, 644, 649, 651, 655, 660, 665, 667, 670, 674
makrilli-suku 986
makrilli-suku, kääpiö- 978
makrilli-suku, käärme- 948
makrilli-suku, kuningas- 1005

makrilli-suku, piikki- 639, 645, 661, 675, 676
makrillihai 20
makrillihai, pitkäevä- 21
makrillihauet 420
makrillihauet-heimo 420
makrillihauki 422
makrillijääkala 898
makrillit 960, 961
makrillit, käärme- 947
makrillit, piikki- 632
makrillit-heimo 961
makrillit-heimo, käärme- 947
makrillit-heimo, piikki- 632
malabarinmeriahven 579
maljakotilo-suku 1333
manaatti 1456
manaatti, afrikan- 1457
manaatti, kynnetön- 1455
mangrovenapsija 688
mangroveosteri 1348
mantsurianrumpukala 746
marliini 1024
marliini, juova- 1023
marliini, lyhytnokka- 1022
marliini, musta- 1018
marliini, pitkäkuono- 1026
marliini, purje- 1019
marliini, pyörösuomu- 1025
marliini, valko- 1021
marliini-suku 1020
marliini-suku 1027
marmoripagelli 786
marmoripaksuotsa 336
marmorirausku 94
marmorisähkörausku 118
marokonhammasahven 777
masulohi 250
mattosimpukka 1401
mattosimpukka, japanin- 1402
mattosimpukka, kulta- 1404
mattosimpukka, pikku- 1405
meduusa-suku 1512
meduusakala, etelän- 1043
meduusakala, japanin- 1044
meduusakala, musta- 1041
meduusakala-suku 1042
meduusakalat 1040
meduusakalat-heimo 1040
menhaden 166
menhaden, lahti- 165
menhaden, tyynenmeren- 175
menhaden-suku 167
meriahven, hieta- 572
meriahven, kallio- 571
meriahven, malabarin- 579
meriahven, nassaun- 581
meriahven, perun- 584
meriahven, puna- 580
meriahven, raita- 578
meriahven, rusko- 588
meriahven, tumma- 577, 591
meriahven, valko- 573
meriahven-laji 568, 570, 574, 575, 576
meriahven-suku 582, 583, 590
meriahvenet 567, 595
meriahvenet-heimo 567
meriankeriaat 408
meriankeriaat-heimo 408
meriankerias 410
meriankerias-laji 409
meribassi 593
merienkeli 68
merienkelit 67

merienkelit-heimo 67
merihevonen-suku 505
merikarhu, afrikan- 1442
merikarhu, eteläamerikan- 1441
merikilpikonnat 1494
merikissa 915
merikissa, kirjo- 916
merikissa, sini- 914
merikissa-suku 917
merikissat 913
merikissat-heimo 913
merikokki, iso- 941
merikokki, pikku- 942
merikorva 1328
merikorva-laji 1327
merikorva-suku 1329
merikrokotiili 1505
merikrotit 1192
merikrotit-heimo 1192
merikrotti 1195
merikrotti, amerikan- 1193
merikrotti-laji 1194, 1196
merilahna 680
merilahnat 679
merilahnat-heimo 679
merileijona, eteläamerikan- 1444
merileijona, stellerin- 1443
merimade-suku 451
merimakkara, japanin- 1523
merimonni-laji 375, 376, 377
merimonnit 374
merimonnit-heimo 374
merinahkakilpikonna 1495
merinahkiainen 7
merirokko, jätti- 1200
merirokko-suku 1199
meriruutana 808
merisiili 1521
merisiili, chilen- 1522
merisiili, kallio- 1518
merisiili-suku 1519
merisiilit 1520
merisiilit-heimo 1520
meritähdet 1516
meritähdet-heimo 1516
meritähti, puna- 1517
meritaimen 257
meritupet 1507
merituppi-laji 1508, 1509, 1510
meritursas 1438
merituulenkala 935
metsästystorvikotilo 1335
miekkakala 1029
miekkakalat 1028
miekkakalat-heimo 1028
miekkasampi 141
miekkasärki 350
miekkavalas 1470
miekkavalas, pikku- 1473
möhkäkala 1188
möhkäkala, pikku- 1189
möhkäkala-laji 1187
mojara-laji 902
mojara-suku 903
molukkirapu 1515
molva 462
molva, luikero- 461
molva, tylppäpyrstö- 460
monni 382
monni, kreikan- 381
monni, piikki- 384, 385
monni, pilkkupiikki- 386
monni-laji, meri- 375, 376, 377
monni-suku, koralli- 378

monni-suku, ankerias- 388
monni-suku, lasi- 379
monni-suku, piikki- 387
monnit 380
monnit, meri- 374
monnit, piikki- 383
monnit-heimo 380
monnit-heimo, meri- 374
monnit-heimo, piikki- 383
monrovianvälskärikala 946
mosambikintilapia 832
moskiittokala 516
muikku 229
mullo, keltajuova- 818
mullo, kulta- 816
mullo, länsiafrikan- 820
mullo, pikku- 817
mullo-laji 821
mullo-suku 819
mullo-suku 822
mullot 815
mullot-heimo 815
munkkihylje 1450
mureena 404
mureena, viher- 403
mureenat 402
mureenat-heimo 402
murisija, hopea- 711
murisija, paksuhuuli- 709
murisija, välimeren- 710
murisija-laji 706, 708, 712, 713
murisija-suku 714
murisijat 705
murisijat-heimo 705
murjottajakala, afrikan- 656
murjottakala-laji 657
murokas 240
mursu 1445
mustaboniitti 969
mustaevähai 38
mustaeväkampela 1128
mustaevätonnikala 1009
mustahauki 290
mustahuotrakala 956
mustakarppi 347
mustakitaturska 459
mustamarliini 1018
mustameduusakala 1041
mustanmerenpiikkikampela 1173
mustanmerensilli 160
mustanmerenvalkoturska 455
mustapilkkuahven 605
mustarumpukala 745
mustaselkäkampela 1147
mustatäpläpikarelli 811
mustatokko 1053
mustaturska 468
mustavalas 1489, 1490
mustehuulikala 851
mutakala 371
mutu 352
mutu, suo- 351
myskitursaat 1434
myskitursaat-heimo 1434
myskitursas 1436
myskitursas, pikku- 1435
myskitursas-suku 1437, 1439

N

nahkiainen 5
nahkiainen, joki- 5
nahkiainen, meri- 7
nahkiainen, pikku- 6

nahkiainen, tonavan- 2
nahkiainen, ukrainan- 3
nahkiainen, vladikovin- 4
naiskala 152
naiskala-laji 153
naiskalat 151
naiskalat-heimo 151
napakampela 1140
napakampela-laji 1141
napsija, juova- 692
napsija, keltapyrstö- 694
napsija, mangrove- 688
napsija, purppura- 691
napsija-laji 689, 690
napsija-suku 693
napsijat 687
napsijat-heimo 687
nassaunmeriahven 581
natalinlangusti 1265
navaga 445
neliviiksimade . 469
neula, iso- 506
neula, pikku- 507
neula, särmä- 508
neulakala 930
nieriä 263
nieriä, härkä- 265
nieriä, harmaa- 266
nieriä, puro- 264
nieriä-suku 267
niilinahven 565, 566
niilinhauki 128
niilintäpläsilli 158
niilintilapia 833
nilviäiset-lahko 1323
nokkahai 44
nokkahai, tyynenmeren- 43
nokkakala 416
nokkakala, atlantin- 417
nokkakala, australian- 428
nokkakala, brasilian- 430
nokkakala, japanin- 431
nokkakala, sini- 418
nokkakala-laji 432
nokkakala-suku 419, 433
nokkakalat 415, 427
nokkakalat-heimo 415, 427
nokkasärki 325
nokkasärki, dalmatian- 324
nokkasärki, espanjan- 327
nokkasärki, italian- 328
nokkasärki, pikku- 329
nokkasärki, pikkusuomu- 326
nokkavalaat 1458
nokkavalaat-heimo 1458
nooanarkki-suku 1342
norjankrilli 1205
norppa 1452
nuijapompano 663
nuoliainen, balkanin- 367
nuoliainen, bergatinon- 368
nuoliainen, italian- 373
nuoliainen, kiven- 372
nuoliainen, kulta- 366
nuoliainen, romanian- 369
nävertäjäsimpukka 1416
nävertäjäsimpukka-suku 1417

O

ohuthuulikeltti 542
oikeasilmäkampelat 1123
oikeasilmäkampelat-heimo 1123

okahai 71
okakala 631
okamakrilli 677
okarausku 87
okaskorpionikala 1066
olympiaosteri 1354
oranssiviilakala 1181
oreot 528
oreot-heimo 528
osteri 1353
osteri, amerikan- 1349
osteri, chilen- 1351
osteri, hammas- 1352
osteri, mangrove- 1348
osteri, olympia- 1354
osteri, tyynenmeren- 1347
osteri, uudenseelannin- 1355
osteri-laji 1357
osteri-suku 1350, 1356

P

pagelli, marmori- 786
pagelli, pilkku- 793
pagelli, puna- 794
pagelli-laji 785, 790, 791, 792
pagelli-suku 795
paholaisrausku 112
paholaisrauskut 111
paholaisrauskut-heimo 111
paksuhuulikeltti 538
paksuhuulimurisija 709
paksuotsa, hopea- 335
paksuotsa, marmori- 336
palettirausku 91
pallas, grönlannin- 1148
pallas, ruijan- 1133
pallas, tyynenmeren- 1134
pallokala, pohjan- 1185
pallokala-suku 1186
pallokalat 1184
pallokalat-heimo 1184
pallopäävalas 1466
panssarikurnusimppu 1106
papuankuningasmakrilli 995
papukaijakala 856
pargo 806
partahai, atlantin- 26
partahai, rengas- 25
partahai, seepra- 24
partahylje 1448
partanilkka 924, 929
partanilkka-suku 928
partarumpukala 759
partasimppu 1103
partaturska 472
pastelliahven 587
pasuri 319
patagonianhoki 479
patagoniansilohai 54
peilirausku 92
pekinginlahna 349
peledsiika 238
perhokalat 830
perhokalat-heimo 830
perhosrausku-suku 116
perhossimppu 1109
perunkitararausku 75
perunkummeliturska 485
perunmeriahven 584
perunrumpukala 743
perunsardelli 220
perunsardiini 200

FI

perunveltto	723	pilkkapyrstösardiini	193	purppuranapsija	691
pietarikala, amerikan-	524	pilkkulevykala	828	puskurikala	640
pietarinkala	525	pilkkumakrilli	994	pusukala	1047
pietarinkalat	523	pilkkupagelli	793	putsari-suku	761
pietarinkalat-heimo	523	pilkkupiikkimakrilli	643	putsarit	762
piikikäs kivirapu	1307	pilkkupiikkimonni	386	putsarit-heimo	762
piikkihai	64	pilkkurausku	85	pyöriäinen	1471
piikkihai-laji	58, 59, 60, 65, 66	pilkkusilli	156	pyramidikotilo-suku	1330
piikkihai-suku	62	pilkkuveltto	725	pyöriäinen, suihku-	1472
piikkihait	57	pistepunahai	31	pyörösardiini	190
piikkihait-heimo	57	pisterausku	93	pyörösilli, atlantin-	177
piikkikalat	499	pitkäevämakrillihai	21	pyörösilli, kalifornian-	176
piikkikalat-heimo	499	pitkäeväpallopää	1466	pyörösilli-suku	179
piikkikampela, mustanmeren-	1173	pitkäkuonodelfiini	1475	pyörösuomumarliini	1025
piikkikampela-laji	1111, 1112,	pitkäkuonomarliini	1026	pääkampela, pikku-	1142
	1119, 1122	platansinisimpukka	1379		
piikkikotilo-suku	1334	pohjankatkarapu	1237	**Q**	
piikkikurnusimppu	1086, 1090	pohjankeihäsrausku	105		
piikkimakrilli	653, 672	pohjanlimapää	518	queenslandinmakrilli	999
piikkimakrilli, australian-	659	pohjanpallokala	1185		
piikkimakrilli, chilen-	669	pohjanrumpukala	735	**R**	
piikkimakrilli, hevos-	636	pohjasiika	239		
piikkimakrilli, iso-	658	pohjoisennelihammasvalas	1459	raitadelfiini	1474
piikkimakrilli, isosilmä-	654	pohjoisenpullokuonovalas	1460	raitakäärmeenpääkala	558
piikkimakrilli, japanin-	666	pollansiika	230	raitamakrilli	962
piikkimakrilli, juova-	638	pompano	662	raitameriahven	578
piikkimakrilli, kapin-	673	pompano, alexandrianrihma-	633	raitasardelli	209
piikkimakrilli, kelta-	637	pompano, amia-	647	raitavenussimpukka	1396
piikkimakrilli, pilkku-	643	pompano, nuija-	663	rakkolevät	1527
piikkimakrilli, punapyrstö-	641	pompano-laji, rihma-	634	rantahuulikala	847
piikkimakrilli, tyynenmeren-	671	pompano-suku	648, 664	rantakotilo-suku	1337
piikkimakrilli, uudenseelannin-	652	porsaskala	844	rantaneula	370
piikkimakrilli, välimeren-	668	porsassäppikala	1180	rantataskurapu	1298
piikkimakrilli-laji	642, 644, 649, 651,	pullokuonodelfiini	1477	rantataskurapu-laji	1297
	655, 660, 665, 667, 670, 674	pullonnokkarausku	82	rapu	1276, 1322
piikkimakrilli-suku	639, 645, 661,	puna-ahven	1075	rapu, isohämähäkki-	1315
	675, 676	puna-ahven, kapin-	1071	rapu, isotasku-	1291
piikkimakrillit	632	puna-ahven-laji	1072, 1074, 1076	rapu, joki-	1276, 1278
piikkimakrillit-heimo	632	puna-ahven-suku	1077	rapu, kääpiö-	1282
piikkimonni	384, 385	punahai, piste-	31	rapu, kallio-	1289
piikkimonni-suku	387	punahai, täplä-	32	rapu, kivi-	1280
piikkimonnit	383	punahai-suku	33	rapu, kolo-	1279
piikkimonnit-heimo	383	punahait	28	rapu, kuningas-	1310
piikkisimppu	1102	punahait-heimo	28	rapu, puna-	1284
piikkisydänsimpukka	1388	punahammasahven	805	rapu, rantatasku-	1298
pikarelli, iso-	812	punakampela	1145	rapu, sinitasku-	1295
pikarelli, mustatäplä-	811	punakampela, alaskan-	1146	rapu, villasaksi-	1303
pikarelli-laji	813	punakeihäsrausku	104	rapu-heimo	1304
pikarelli-suku	814	punakurkkulohi	255	rapu-laji	1305, 1312, 1314, 1316
pikarellit	810	punakurnusimppu	1079	rapu-laji, rantatasku-	1297
pikarellit-heimo	810	punalevä-laji	1532	rapu-laji, tasku-	1288, 1290, 1292,
pikkubarrakuda	535	punalevä-suku	1529, 1530, 1531		1293, 1294, 1299
pikkubassi	602	punalevät	1528	rapu-laji, uimatasku-	1300, 1302
pikkuhai	61	punalohi	252	rapu-suku	1281, 1313
pikkukalmari	1423	punameriahven	580	rapu-suku, kääpiö-	1283
pikkukampela	1171, 1172	punameritähti	1517	rapu-suku, kuningas-	1311
pikkukielikampela	1156	punapagelli	794	rapu-suku, sinitasku-	1296
pikkukoirakala	285	punapyrstöpiikkimakrilli	641	rapu-suku, uimatasku-	1301
pikkulouhikala	867	punarapu	1284	rasvakala	1105
pikkuluuhauki	144	punarihmanilkka	925	rausku, aalto-	101
pikkumattosimpukka	1405	punarumpukala	757	rausku, hieta-	86
pikkumerikokki	942	punasalakka	308	rausku, isoperhos-	115
pikkumiekkavalas	1473	punasilmäsilli	178	rausku, juovakotka-	110
pikkumöhkäkala	1189	punasimppu	1073, 1075	rausku, käki-	89
pikkumullo	817	punasimppu-suku	1077	rausku, keihäs-	106
pikkumyskitursas	1435	punasuomuturska	475	rausku, kitara-	76
pikkunahkiainen	6	purjekala	1017	rausku, kotka-	109
pikkuneula	507	purjekala, atlantin-	1016	rausku, kynsi-	97
pikkunokkasärki	329	purjekalat	1015	rausku, marmori-	94
pikkupääkampela	1142	purjekalat-heimo	1015	rausku, marmorisähkö-	118
pikkusuomunokkasärki	326	purjelimakala	905	rausku, oka-	87
pikkutarponi	149	purjemarliini	1019	rausku, paholais-	112
pikkuturska	473	puronieriä	264	rausku, paletti-	91
pikkutuulenkala	936	purotaimen	258		

FI

rausku, peili- 92
rausku, perunkitara- 75
rausku, pilkku- 85
rausku, piste- 93
rausku, pohjankeihäs- 105
rausku, pullonnokka- 82
rausku, punakeihäs- 104
rausku, saha- 79
rausku, sarvi- 114
rausku, silmäsähkö- 120
rausku, silo- 84
rausku, sinikeihäs- 107
rausku, sysisähkö- 119
rausku, tähti- 83
rausku, täplä- 95
rausku, tyynenmerenpaholais- 113
rausku, valko- 90
rausku, valkopilkkukitara- 73
rausku, vannas- 96
rausku-laji 88, 98, 99, 100
rausku-laji, kitara- 74
rausku-suku 102
rausku-suku, kitara- 77
rausku-suku, perhos- 116
rausku-suku, sähkö- 121
rauskut 80, 81
rauskut, keihäs- 103
rauskut, kitara- 72
rauskut, kotka- 108, 109
rauskut, paholais- 111
rauskut, saha- 78
rauskut, sähkö- 117
rauskut-heimo 81
rauskut-heimo, keihäs- 103
rauskut-heimo, kitara- 72
rauskut-heimo, kotka- 108
rauskut-heimo, paholais- 111
rauskut-heimo, saha- 78
rauskut-heimo, sähkö- 117
ravut 1285
ravut-heimo, erakko- 1286
ravut-heimo, tasku- 1287
reeska 240
remora 1178
rengashai 30
rengaspartahai 25
rengassargi 778
rihmaevä-laji 555, 556
rihmaevät 553
rihmapompano, alexandrian- 633
rihmapompano-laji 634
rihmaviilakala 1183
riika 240
riikinkukkohuulikala 854
riikinkukkolimakala 907
ripsimonni-suku 293
robalo 561
romaniannuoliainen 369
roussi, kelta- 520
roussi, välimeren- 521
ruijanpallas 1133
rumpukala 755
rumpukala, aalto- 738
rumpukala, argentiinan- 758
rumpukala, hopea- 744
rumpukala, joki- 716
rumpukala, kapin- 719
rumpukala, mantsurian- 746
rumpukala, musta- 745
rumpukala, parta- 759
rumpukala, perun- 743
rumpukala, pohjan- 735
rumpukala, puna- 757
rumpukala, senegalin- 750

rumpukala-laji 717, 720, 721, 722,
730, 731, 732, 733, 734, 737,
740, 741, 742, 747, 748, 749,
751, 753, 754
rumpukala-suku 736, 739, 752, 756, 760
rumpukalat 715
rumpukalat-heimo 715
ruohokarppi 331
ruostekampela 1137
ruskohuulikala 842
ruskolevät 1525
ruskomeriahven 588
ruskoneitokala 836
ruutana 321
ryhävalas 1485
röhkijäkala 702
röhkijäkala-laji 703, 704

S

saha-ahven 569
saharausku 79
saharauskut 78
saharauskut-heimo 78
saira 421
salakka 310
salakka, allikko- 339
salakka, italian- 309
salakka, puna- 308
salakka, tonavan- 322
samettisimpukka 1384
sammakko, härkä- 1492
sammakko-suku 1493
sammet-heimo 130
sampi 137
sampi, adrian- 133
sampi, kita- 139
sampi, miekka- 141
sampi, tähti- 136
sampi, valko- 138
sampi, venäjän- 131
sampi, viher- 132
sampi-laji 134
sapa 306
sarda 982
sarda, chilen- 980
sarda, etelän- 979
sarda, juova- 981
sardelli 217
sardelli, argentiinan- 214
sardelli, australian- 215
sardelli, japanin- 218
sardelli, kalifornian- 219
sardelli, kapin- 216
sardelli, länsiatlantin- 210
sardelli, perun- 220
sardelli, raita- 209
sardelli-laji 211, 212, 213, 221
sardelli-suku 222
sardellit 208
sardellit-heimo 208
sardiini 189
sardiini, australian- 198
sardiini, japanin- 197
sardiini, kalifornian- 196
sardiini, kirjo- 173
sardiini, perun- 200
sardiini, pilkkapyrstö- 193
sardiini, pyörö- 190
sardiini, silmä- 199
sardiini-laji 191, 192, 194
sardiini-suku 195
sardiini-suku, vale- 180

sargi, iso- 782
sargi, kaulus- 783
sargi, rengas- 778
sargi, satula- 788
sargi, seepra- 779
sargi-suku 784
sarveissienet 1511
sarveissienet-heimo 1511
sarvirausku 114
sarvivalas 1479
sateenkaariharppoja 646
satulasargi 788
seeprahammaskarppi 513
seeprapartahai 24
seeprarihmakala 1046
seeprasargi 779
seipi 344
seipi, sini- 345
seiti 467
seitivalas 1481
senegalinkummeliturska 491
senegalinrumpukala 750
sepia 1418
sepia, välimeren- 1419
sepiola 1420
siaminbarbi 338
siaminkrokotiili 1506
siiat 261
siika 234
siika, Geneve-järven 233
siika, järvi- 237
siika, kuono- 236
siika, peled- 238
siika, pohja- 239
siika, silli- 232
siika, tuppi- 240
siika-laji 235
siika-suku 241
siimasilli 186
siimasilli, tyynenmeren- 185
siimasilli-suku 187
siimaturska, sini- 437
siimaturska-laji 438
siimaturskat 436
siimaturskat-heimo 436
siipikotilo-suku 1338
silli 169
silli 203
silli, amerikankanta- 162
silli, atlantinpyörö- 177
silli, chilen- 204
silli, harmaa- 161
silli, hikkori- 159
silli, kalifornianpyörö- 176
silli, kiinankupu- 168
silli, kupu- 172
silli, mustanmeren- 160
silli, niilintäplä- 158
silli, pilkku- 156
silli, punasilmä- 178
silli, siima- 186
silli, sini- 155
silli, täplä- 157
silli, tyynenmeren- 170
silli, tyynenmerensiima- 185
silli-laji 171, 181, 188, 201
silli-laji, kanta- 174
silli-laji, kupu- 164
silli-suku, kanta- 163
silli-suku, pyörö- 179
silli-suku, siima- 187
sillihai 23
sillihai, tyynenmeren- 22
sillihait 18

sillihait-heimo	18
sillikuningas	123
sillikuningas, amerikan-	124
sillikuninkaat	122
sillikuninkaat-heimo	122
sillisiika	232
sillit	154
sillit-heimo	154
sillivalas	1484
silmä-laji, suuri-	607
silmä-suku, suuri-	608
silmähuulikala	848
silmäsähkörausku	120
silmäsardiini	199
silmät, suuri-	606
silmät-heimo, suuri-	606
silohai, patagonian-	54
silohai-laji	51, 52
silohai-suku	55
silohait	49
silohait-heimo	49
silokampela	1175
silokampela, amerikan-	1174
silokampela-laji	1168
silokampela-suku	1170
silokampelat	1167
silokampelat-heimo	1167
silorausku	84
silosimppu	1094
silovalaat	1486
silovalaat-heimo	1486
simppu, amerikanhärkä-	1098
simppu, härkä-	1099
simppu, iso-	1100
simppu, isokurnu-	1089
simppu, kapinkurnu-	1081
simppu, kivi-	1097
simppu, kyhmykurnu-	1084
simppu, panssarikurnu-	1106
simppu, parta-	1103
simppu, perhos-	1109
simppu, piikki-	1102
simppu, piikkikurnu-	1086
simppu, piikkikurnu-	1090
simppu, puna-	1073, 1075
simppu, punakurnu-	1079
simppu, silo-	1094
simppu, sinieväkurnu-	1082
simppu, sinisuu-	1063
simppu, töyhtökurnu-	1080
simppu-laji, kurnu-	1085, 1088, 1091
simppu-laji, lättä-	1096
simppu-laji, viher-	1092, 1093
simppu-suku	1101
simppu-suku, kurnu-	1083, 1087
simppu-suku, puna-	1077
simppuahven	618
simpukat	1324
simpukat, joki-	1344
simpukat, kampa-	1358
simpukat, sydän-	1387
simpukat, veitsi-	1407
simpukat, venus-	1393
simpukat-heimo, joki-	1344
simpukat-heimo, kampa-	1358
simpukat-heimo, sydän-	1387
simpukat-heimo, veitsi-	1407
simpukat-heimo, venus-	1393
simpukka, aasianviher-	1380
simpukka, atlantinkampa-	1359
simpukka, atlantinviher-	1382
simpukka, australiankampa-	1366
simpukka, australiansini-	1378
simpukka, chilensini-	1374
simpukka, hieta-	1415
simpukka, islannin-	1386
simpukka, islanninkampa-	1362
simpukka, isojärvi-	1343
simpukka, isokampa-	1365
simpukka, jalohelmi-	1346
simpukka, jalosilkki-	1385
simpukka, japaninarkki-	1340
simpukka, japaninmatto-	1402
simpukka, japaninvenus-	1399
simpukka, jokihelmi-	1383
simpukka, kampavenus-	1398
simpukka, kapinveitsi-	1412
simpukka, koreansini-	1375
simpukka, kultamatto-	1404
simpukka, matto-	1401
simpukka, nävertäjä-	1416
simpukka, piikkisydän-	1388
simpukka, pikkumatto-	1405
simpukka, platansini-	1379
simpukka, raitavenus-	1396
simpukka, sametti-	1384
simpukka, sini-	1376
simpukka, soukkojoki-	1345
simpukka, sydän-	1389
simpukka, taskuveitsi-	1410
simpukka, uudenseelanninkampa-	1367
simpukka, välimerensini-	1377
simpukka, veitsi-	1409
simpukka, viher-	1381
simpukka-laji, arkki-	1339
simpukka-laji, kampa-	1360, 1361, 1363, 1364, 1368, 1369
simpukka-laji, veitsi-	1408, 1411, 1413
simpukka-laji, venus-	1395, 1397, 1403
simpukka-laji, vinohammas-	1390, 1391, 1392
simpukka-suku, arkki-	1341
simpukka-suku, hevos-	1373
simpukka-suku, liemi-	1406
simpukka-suku, nävertäjä-	1417
simpukka-suku, veitsi-	1414
simpukka-suku, venus-	1394, 1400
simput, kurnu-	1078
simput, lättä-	1095
simput-heimo, kurnu-	1078
simput-heimo, lättä-	1095
sinieväkurnusimppu	1082
sinihai	42
sinihuulikala	841
sinikala	630
sinikalat	629
sinikalat-heimo	629
sinikeihäsrausku	107
sinimerikissa	914
sininokkakala	418
sinipiikkimakrilli	635
siniseipi	345
sinisiimaturska	437
sinisilli	155
sinisimpukat	1370
sinisimpukat-heimo	1370
sinisimpukka	1376
sinisimpukka, australian-	1378
sinisimpukka, platan-	1379
sinisimpukka, välimeren-	1377
sinisimpukka-laji	1371
sinisuusimppu	1063
sinitaskurapu	1295
sinitaskurapu-suku	1296
sinivalas	1483
sirkkaäyriäinen	1202
sirkkaäyriäiset	1201
sirkkaäyriäiset-heimo	1201
sisiliskokala	392, 393
sisiliskokala-laji	391
sisiliskokalat	390
sitruunahai	41
skorpionikala	1068
skorpionikala, oka-	1066
skorpionikala, tumma-	1067
skorpionikala-laji	1065
skorpionikala-suku	1064, 1069
skorpionikalat	1062
skorpionikalat-heimo	1062
sorva	362
sorva, kreikan-	363
soukkojokisimpukka	1345
stellerinmerileijona	1443
sterletti	135
suihkupyöriäinen	1472
sulkava	304
sumuhai	39
suomukampela	1115
suomutu	351
suomuturska	463
suomuturska, brasilian-	474
suomuturska, puna-	475
suomuturska, valko-	476
susisilli	225
susisilli-suku	226
susisillit	224
susisillit-heimo	224
suurisilmä-laji	607
suurisilmä-suku	608
suurisilmät	606
suurisilmät-heimo	606
suutari	364
sydänsimpukat	1387
sydänsimpukat-heimo	1387
sydänsimpukka	1389
sydänsimpukka, piikki-	1388
sysisähkörausku	119
sädelouhikala	870
sähkömonni	394
sähkörausku-suku	121
sähkörauskut	117
sähkörauskut-heimo	117
säppikala, porsas-	1180
säppikalat	1179
säppikalat-heimo	1179
särki	360
särki, albanian-	348
särki, dalmatiannokka-	324
särki, espanjannokka-	327
särki, etelän-	359
särki, helmi-	357
särki, huuli-	323
särki, italiannokka-	328
särki, nokka-	325
särki, pikkunokka-	329
särki, pikkusuomunokka-	326
särki, tonavan-	358
särki-suku	361
särkikala-laji	330, 346, 355
särkikalat	303
särkikalat-heimo	303
särmäneula	508
säyne	343

T

taimen	257
taimen, gila-	246
taimen, järvi-	259
taimen, kulta-	243
taimen, meri-	257

taimen, puro-	258	
taivaantähystäjä	874	
taivaantähystäjä-suku	875	
tarponi	150	
tarponi, pikku-	149	
tarponit	148	
tarponit-heimo	148	
taskurapu, iso-	1291	
taskurapu, ranta-	1298	
taskurapu, sini-	1295	
taskurapu-laji	1288, 1290, 1292,	
1293, 1294, 1299		
taskurapu-laji, ranta-	1297	
taskurapu-laji, uima-	1300, 1302	
taskurapu-suku, sini-	1296	
taskurapu-suku, uima-	1301	
taskuravut-heimo	1287	
taskuveitsisimpukka	1410	
teisti	919	
teräkala-suku	292	
tetrat	300	
tetrat-heimo	300	
tiikeriahvenet	592	
tiikeriahvenet-heimo	592	
tiikerimakrilli	1001	
tiilikala-laji	627	
tiilikalat	626	
tiilikalat-heimo	626	
tilapia, mosambikin-	832	
tilapia, niilin-	833	
tokko, hieta-	1059	
tokko, iso-	1052	
tokko, kulta-	1051	
tokko, kuulto-	1050	
tokko, levä-	1061	
tokko, luoto-	1054	
tokko, musta-	1053	
tokko-laji	1056, 1057, 1058, 1060	
tokko-suku	1055	
tokot	1049	
tokot-heimo	1049	
tonavanbarbi	317	
tonavankiiski	615	
tonavannahkiainen	2	
tonavansalakka	322	
tonavansärki	358	
tonavantsingeli	623	
tonnikala	1012	
tonnikala, etelän-	1010	
tonnikala, isosilmä-	1011	
tonnikala, keltaevä-	1008	
tonnikala, mustaevä-	1009	
tonnikala, valko-	1007	
tonnikala-laji	1013, 963, 970	
tonnikala-suku	1014	
tonnikalat	1	
tonnikalat; tonnikalat-heimo	1006	
torkkujatokot	1048	
torkkujatokot-heimo	1048	
torvikala	504	
torvikalat	503	
torvikalat-heimo	503	
toutain	311	
tristaninlangusti	1263	
tropiikinvalas	1482	
tsingeli	624	
tsingeli, hyppijä-	622	
tsingeli, tonavan-	623	
tummajääkala	901	
tummameriahven	577, 591	
tummaskorpionikala	1067	
tunniina	968	
tuppisiika	240	
turkishylkeet	1440	
turkishylkeet-heimo	1440	
turkoosihammaskarppi	514	
turpa	341, 342	
turska	447	
turska, argentiinankummeli-	486	
turska, brasiliansuomu-	474	
turska, etelänkummeli-	482	
turska, etelänmustakita-	458	
turska, grönlannin-	448	
turska, harmaa-	471	
turska, hopea-	449	
turska, hopeakummeli-	483	
turska, jää-	456	
turska, kaliforniankummeli-	490	
turska, kapinkummeli-	484	
turska, kummeli-	487	
turska, luikero-	464	
turska, lyyra-	466	
turska, musta-	468	
turska, mustakita-	459	
turska, mustanmerenvalko-	455	
turska, parta-	472	
turska, perunkummeli-	485	
turska, pikku-	473	
turska, punasuomu-	475	
turska, senegalinkummeli-	491	
turska, sinisiima-	437	
turska, suomu-	463	
turska, täpläantenni-	439	
turska, tyynenmeren-	446	
turska, valko-	454	
turska, valkosuomu-	476	
turska-laji, kummeli-	481	
turska-laji, siima-	438	
turska-suku, jää-	457	
turska-suku, kummeli-	492	
turska-suku, luikero-	465	
turskakalat-lahko	435	
turskat	440	
turskat, kummeli-	478	
turskat, siima-	436	
turskat-heimo	440	
turskat-heimo, kummeli-	478	
turskat-heimo, siima-	436	
tuulenkala, amerikan-	932	
tuulenkala, iso-	939	
tuulenkala, korean-	940	
tuulenkala, meri-	935	
tuulenkala, pikku-	936	
tuulenkala, tyynenmeren-	934	
tuulenkala, välimeren-	938	
tuulenkala-laji	933	
tuulenkala-suku	937	
tuulenkalat	931	
tuulenkalat-heimo	931	
tylppäpyrstömolva	460	
tyrskykuore	272	
tyynenmerenkuningasmakrilli	1002	
tyynenmerenlangusti	1262	
tyynenmerenmenhaden	175	
tyynenmerennokkahai	43	
tyynenmerenosteri	1347	
tyynenmerenpaholaisrausku	113	
tyynenmerenpallas	1134	
tyynenmerenpiikkimakrilli	671	
tyynenmerenpuna-ahven	1070	
tyynenmerenseiti	470	
tyynenmerensiimasilli	185	
tyynenmerensilli	170	
tyynenmerensillihai	22	
tyynenmerenturska	446	
tyynenmerentuulenkala	934	
tyynenmerenvoikala	1034	
tähtikampela	1144	
tähtirausku	83	
tähtisampi	136	
täpläantenniturska	439	
täpläbassi	594	
täplähauki	291	
täplähuulikala	849	
täplälouhikala	868	
täpläpunahai	32	
täplärausku	95	
täpläsilli	157	
törö	333	
törö, dalmatian-	312	
törö, kivi-	334	
töyhtökampela-laji	1116, 1120	
töyhtökurnusimppu	1080	

U

uimataskurapu-laji	1300, 1302	
uimataskurapu-suku	1301	
ukrainannahkiainen	3	
ulappakettuhai	15	
uudenguineankrokotiili	1504	
uudenseelanninankerias	398	
uudenseelanninkampasimpukka	1367	
uudenseelanninosteri	1355	
uudenseelanninpiikkimakrilli	652	

V

valas, bryden-	1482	
valas, grönlannin-	1487	
valas, hanhennokka-	1461	
valas, harmaa-	1491	
valas, kääpiö-	1488	
valas, lahti-	1480	
valas, maito-	1478	
valas, miekka-	1470	
valas, musta-	1489, 1490	
valas, pallopää-	1466	
valas, pikkumiekka-	1473	
valas, pohjoisennelihammas-	1459	
valas, pohjoisenpullokuono-	1460	
valas, ryhä-	1485	
valas, sarvi-	1479	
valas, seiti-	1481	
valas, silli-	1484	
valas, sini-	1483	
valas, tropiikin-	1482	
valashai	27	
valekarettikilpikonna	1496	
valesardiini-suku	180	
valkoahven-laji	628	
valkobassi	597	
valkohai	19	
valkokuonodelfiini	1469	
valkokuta	954	
valkokuvedelfiini	1468	
valkomarliini	1021	
valkomeriahven	573	
valkopilkkahai	37	
valkopilkkukitararausku	73	
valkorausku	90	
valkosampi	138	
valkosilmäkuha	621	
valkosuomuturska	476	
valkotonnikala	1007	
valkoturska	454	
valokala-laji	295	
valokalat	294	
valokalat-heimo	294	
vannasrausku	96	
vasarahai	47	

FI

525

vasarahai, kampa-	46	
vasarahai-suku	48	
vasarahait	45	
vasarahait-heimo	45	
vasensilmäkampela-suku	1121	
vasensilmäkampelat	1114	
vasensilmäkampelat-heimo	1114	
vaskikala	502	
veitsisimpukat	1407	
veitsisimpukat-heimo	1407	
veitsisimpukka	1409	
veitsisimpukka, kapin-	1412	
veitsisimpukka, tasku-	1410	
veitsisimpukka-laji	1408, 1411, 1413	
veitsisimpukka-suku	1414	
veltto	726	
veltto, iso-	724	
veltto, juova-	728	
veltto, perun-	723	
veltto, pilkku-	725	
veltto-laji	727	
veltto-suku	729	
venäjänsampi	131	
venussimpukat	1393	
venussimpukat-heimo	1393	
venussimpukka, japanin-	1399	
venussimpukka, kampa-	1398	
venussimpukka, raita-	1396	

venussimpukka-laji	1395, 1397, 1403	
venussimpukka-suku	1394, 1400	
viherhuulikala	840, 843	
vihermureena	403	
vihersampi	132	
vihersilmä	299	
vihersilmät	298	
vihersilmät-heimo	298	
vihersimppu-laji	1092, 1093	
vihersimpukka	1381	
vihersimpukka, atlantin-	1382	
viikatekala-laji	511	
viilakala	1182	
viilakala, oranssi-	1181	
viilakala, rihma-	1183	
viiriahven-laji	696	
viiriahven-suku	697, 698	
viiriahvenet	695	
viiriahvenet-heimo	695	
viisiviiksimade	443	
villakuore	273	
villasaksirapu	1303	
vimpa	365	
vinohammassimpukka-laji	1390, 1391, 1392	
vladikovinnahkiainen	4	
voikala, amerikan-	1033, 1035	
voikala, hopea-	1031	

voikala, kiinan-	1032	
voikala, tyynenmeren-	1034	
voikala-laji	1037	
voikala-suku	1036	
voikalat	1030	
voikalat-heimo	1030	
välimerenbarbi	316	
välimerenhuulikala	846	
välimerenmurisija	710	
välimerenpiikkimakrilli	668	
välimerenroussi	521	
välimerensepia	1419	
välimerensinisimpukka	1377	
välimerentuulenkala	938	
välskärikala, monrovian-	946	
välskärikalat	945	
välskärikalat-heimo	945	

Y

yleinen mustekala	1418

Ö

öljykala	950

FI

Svenskt register

A

abalone	1325
abborrar, djuphavs-	609
abborrar, glas-	563
abborrar, havs-	590
abborrar, havs-	595
abborrar, nil-	566
abborrar, sand-	865
abborre	617
abborre, blå sand-	864
abborre, blå tegel-	627
abborre, brun havs-	577
abborre, brun medelhavs-	588
abborre, fläckig havs-	594
abborre, fors-	618
abborre, georgia-lax-	682
abborre, gul-	616
abborre, havs-	593
abborre, japansk djuphavs-	612
abborre, japansk havs-	599
abborre, kalifornisk havs-	591
abborre, klipp-	600
abborre, lax-	683
abborre, nil-	565
abborre, öring-	603
abborre, papegoj-	570
abborre, sandhavs-	572
abborre, silver-	628
abborre, skrift-	589
abborre, sol-	601
abborre, större medelhavs-	587
abborre, strimmig havs-	598
abborre, sval-	569
abborre, svart-	602
abborre, svart havs-	571
abborre, svart sol-	605
abborre, teleskop-	611
abborre, vit-	596
abborre, vit havs-	597
abborre, vit sol-	604
abborrfiskar	613
abborrfiskar, havs-	567
abborrfiskar, nil-	560
abborrfiskar, tegel-	626
aborre, argentinsk havs-	568
aborre, grå havs-	576
adriamört	359
adriastubb	1058
adriatisk stör	133
adriatisk stubb	1058
afrikansk barrakuda	530
afrikansk buksill	183
afrikansk hästhuvudmakrill	656
afrikansk havsgös	717
afrikansk kungsmakrill	1005
afrikansk multe	544
afrikansk sardinell	193
afrikansk spadfisk	827
afrikansk vekfisk	719
afrikanska lungfiskar	129
akadisk kungsfisk	1073
akiami-räka	1242
alaskapollack	470
alaskarödspätta	1146
albacor	1007
albacora	1008

albansk barb	313
alexandria-trådmakrill	633
alger	1524
alger, alkformiga röd-	1530
alger, brun-	1525
alger, röd-	1528
alkformiga rödalger	1530
alligator	1500
alligator, mississippi-	1502
amazon-manat	1455
amazon-sjöko	1455
amerikansk ål	400
amerikansk ansjovis	210
amerikansk flodkräfta	1282, 1283
amerikansk gös	621
amerikansk havsål	411
amerikansk hummer	1317
amerikansk marulk	1193
amerikansk siklöja	231
amerikansk tobis	932
amerikanskt ostron	1349
amhaj	26
amurgädda	291
anchoita	214
anchoveta	220
andaman-hummer	1319
angolansk tandbraxen	771
anka, bombay-	297
anknäbbfisk, brasiliansk	873
anknäbbfiskar	872
ankor, bombay-	296
ansjoveller	222
ansjovis	217
ansjovis, amerikansk	210
ansjovis, atlantisk grov-	211
ansjovis, australisk	215
ansjovis, brasiliansk rov-	221
ansjovis, bredrandig	209
ansjovis, japansk	218
ansjovis, kap-	216
ansjovis, stillahavs-	219
ansjovis, stillahavsgrov-	212
ansjovisfiskar	208
antarktisk krill	1204
antarktisk svartfisk	1043
antarktisk tandnoting	878
antarktisk trollkrabba	1309
antimora, blå	437
apachelax	244
apacheöring	244
argentinsk bläckfisk	1428
argentinsk havsaborre	568
argentinsk havsål	412
argentinsk hundhaj	54
argentinsk kummel	486
argentinsk rödräka	1256
argentinsk skuggfisk	758
argentinsk stiletträka	1206
argusfiskar	829
aristeid-räkor	1232
aristotelesmal	381
arktisk flundra	1140
arktisk maskeringskrabba	1312
arktisk regnbågsnors	276
asp	311
asper	622
atka-grönlånga	1093

atlantisk (djupshavs-) mantelkam-	
mussla	1369
atlantisk frostfisk	456
atlantisk grovansjovis	211
atlantisk hästhuvudmakrill	657
atlantisk knivmussla	1408
atlantisk menhaden	166
atlantisk nålgädda	417
atlantisk pompano	662
atlantisk ringbuk	1107
atlantisk rundsill	177
atlantisk segelfisk	1016
atlantisk soldatfisk	520
atlantisk trådsill	186
atyid-räkor	1243
australisk ål	397
australisk ansjovis	215
australisk blåmussla	1378
australisk flodkräfta	1316
australisk halvnäbb	428
australisk kammussla	1366
australisk kungsmakrill	996
australisk langust	1270
australisk lungfisk	127
australisk räka	1207
australisk sardin	198
australiskt ostron	1357
auxid	964
ayu	269
azeviatunga	1159
aztek-räka	1213
azursik	240

B

bahyttfiskar	684
Bairds näbbval	1459
Bairds släthuvudfisk	223
balao-halvnäbb	429
balkannissöga	367
bananräka	1222
bandfisk	834
bandgrymta	704
bandmulle	821
barb, albansk	313
barb, donau-	317
barb, flod-	314
barb, italiensk	318
barb, java-	353
barb, medelhavs-	316
barb, naken-	312
barb, slank-	338
barber	354
barrakuda	535
barrakuda, afrikansk	530
barrakuda, gulfenad	533
barrakuda, indisk	534
barrakuda, jätte-	532
barrakuda, stillahavs-	531
barrakudafiskar	529
barrakudor	536
barramundi	564
bastardsköldpadda	1498
beluga	1478
belugastör	139
bengalisk sardinell	191

bengädda, kortnosad	144	blåröd räka	1233	brosme, klipp-	464
bengädda, långnosad	143	blåsfisk, fläckig	1185	brosme, skägg-	475
bening	152	blåsfiskar	1184, 1186	brosme, vit-	476
beningar	151	blåssäl	1447	brotula, skägg-	924
bergantino-nissöga	368, 373	blåtobis	936	brugd	12
berggylta	840	blåval	1483	brun dvärgmal	385
bergskädda	1142	blåvitling	459	brun havsabborre	577
bergtunga	1142	blåvitling, sydlig	458	brun kattfisk	385
bergvar	1176	bläckfisk, argentinsk	1428	brun medelhavsabborre	588
beryx, nordisk	518	bläckfisk, åttaarmad	1435	brun pigghaj	59
beryxar	519	bläckfisk, bredstjärtfenad	1429	brun tigerräka	1218
beryxfiskar	517	bläckfisk, dvärg-	1420	brun venusmussla	1395
bichir	128	bläckfisk, japansk	1432	brunalger	1525
birkelånga	460	bläckfisk, långfenad	1424	brungylta	842
bitterling	356	bläckfisk, liten	1423	brunveka	755
Biwa-lax	253	bläckfisk, nordlig stjärtfenad	1430	Brydes fenval	1482
björkna	319	bläckfisk, nyzeeländsk	1431	buffelfiskar	302
bladtång	1526	bläckfisk, sepia-	1418	bukfisk, indisk	678
blandgrymta	712	bläckfisk, vanlig	1426	buksill, afrikansk	183
blank pelamid	974	bläckfisk, vanlig åttaarmad	1438	buksill, indisk	188
blecka, fläck-	798	bläckfisk, vit	1436	buksill, kinesisk	184
blecka, natal-	797	bläckfiskar	1421	bult, gräs-	1061
blecka, nufar-	768	bläckfiskar, åttaarmade 1434, 1437, 1439		bult, guld-	1051
blecka, panga-	799	bläckfiskar, tioarmade	1422	bågfena	142
blecka, randig sand-	786	blötdjur	1323	bäcknäsling	326
blecka, röd knöl-	769	bobo-kväkare	749	bäcknejonöga	6
blecka, sebra-	779	bobyrez	340	bäckröding	264
blecka, snickar-	765	bodensik	236, 239	böjd knivmussla	1409
blecka, spetsnos-	781	boga	327	bönsyrseräka	1201
blecka, sten-	796	boganäsling	327		
blecka, stjärtbands-	778	bombay-anka	297		
blecka, syd-	800	bombay-ankor	296	**C**	
blecka, tvåbands-	783	bondelle-sik	235		
blecka, vit-	782	bongasill	174	chacunda	164
blecka, vit sand-	785	bonit	973	chilegrymta	706
bleckor, stjärtfläcks-	784	bonit, chilensk	980	chilensk blåmussla	1374
bleka	466	bonit, fläckig	966	chilensk bonit	980
blodslemfisk	910, 911	bonit, strimmig	981	chilensk hoki	478
blomnibe	740	borrmussla, självlysande	1416	chilensk kummel	485
blomrocka	94	borrmusslor	1417	chilensk nylonräka	1235
blå antimora	437	bränningskväkare	734	chilensk sill	204
blå gaffelmakrill	663	bränningsmussla	1392	chilensk sjöpung	1509
blå havsgädda	951	brasiliansk anknäbbfisk	873	chilensk taggmakrill	669
blå havskatt	914	brasiliansk brosme	474	chilensk vit sjöborre	1522
blå krabba	1295	brasiliansk halvnäbb	430	chilenskt ostron	1351
blå marlin	1019	brasiliansk kungsmakrill	987	chokladhaj	60
blå moki	862	brasiliansk multe	546	cichlider	831
blå sandabborre	864	brasiliansk rovansjovis	221	ciklider	831
blå staksill	155	braxar	307	citronhaj	41
blå taggmakrill	670	braxen	305	cobia	631
blå tegelabborre	627	braxen, angolansk tand-	771	coho	249
blåfenad knorrhane	1082	braxen, blåfläckig röd-	804	coholax	249
blåfenad knot	1082	braxen, donau-	306	crangonid-räkor	1250
blåfisk	630	braxen, havs-	680	cunene-taggmakrill	674
blåfiskar	629	braxen, japansk röd-	805		
blåfläckig rödbraxen	804	braxen, kanarie-tand-	772		
blågylta	841	braxen, peking-	349	**D**	
blåhaj	42	braxen, röd-	806		
blåhuvad noting	888	braxen, röd tand-	777	daga-sillar	203
blåkäft	1063	braxen, rödbandad röd-	803	dalmatisk näsling	324
blåkäftar	1064	braxen, rosa tand-	775	dammussla	1343
blåkäxa	61	braxen, storögd tand-	776	dana-simkrabba	1294
blåkäxor	62	braxen, tand-	774	darrmal	394
blåkveite	1148	braxen, wuchang-	346	darrocka	119
blålöpare	635	braxnar, havs-	679	darrocka, marmorerad	118
blåmussla	1376	braxnar, röd-	807	darrockor	117, 121
blåmussla, australisk	1378	bredbandad kungsmakrill	1001	delfin	1464
blåmussla, chilensk	1374	bredrandig ansjovis	209	delfin, långnäbbad	1475
blåmussla, grön	1380	bredstjärtfenad bläckfisk	1429	delfin, Rissos	1467
blåmussla, koreansk	1375	brokig kammusla	1363	delfin, strimmig	1474
blåmussla, medelhavs-	1377	brokrocka	101	delfiner	1463, 1476
blåmussla, nyzeeländsk	1381	brosmar, fjäll-	465	diamantsköldpaddor	1499
blåmussla, riodelaplata-	1379	brosme, brasiliansk	474	dick, glatt-	134
blåmusslor	1370, 1373	brosme, fjäll-	463	djuphavsabborrar	609
				djuphavsabborre, japansk	612

SV

djuphavskrabbor	1322	
djuphavskungsfisk	1076	
djuphavsräka	1212	
djuphavsrödkrabba	1321	
djupkapkummel	488	
djuplånga, indisk	439	
djävulsrocka, mindre	114	
djävulsrockor	111	
dolkfisk	956	
dolksvans	1515	
dolly varden	265	
donaubarb	317	
donaubraxen	306	
donau-flodkräfta	1277	
donaugärs	615	
donaulöja	322	
donaunejonöga	2, 4	
donausandkrypare	334	
drakhuvudfiskar	1062	
drivfisk, indisk	1038	
drottningkammussla	1359	
drottningkväkare	748	
dubbellinjemakrill	972	
dubbelsugare	1190	
dugong	1454	
dummulte	539	
dvärgbläckfisk	1420	
dvärghundfisk	285	
dvärgmakrill, indisk	977	
dvärgmakrillar	978	
dvärgmal, brun	385	
dvärgmal, prickig	386	
dvärgmal, svart	384	
dvärgmalar	383, 387	
dvärgrätval	1488	
dykarp	330	
dögling	1460	

E

egentlig pilgrimsmussla	1365	
egentliga malar	380	
egentliga sälar	1446	
elritsa	352	
eremitkräftor	1286	
europeisk knivmussla	1413	
europeisk langust	1266	
europeiskt havsöra	1328	

F

faren	304	
femtömmad skärlånga	443	
fengädda, nil-	128	
fenknot	1089	
fenor, finger-	860	
fenval	1484	
fenval, Brydes	1482	
filfisk, orange	1181	
filfiskar	1182	
filippinermakrill	976	
fingerfena, vit	857	
fingerfenor	860	
fiskar, makrillika	960	
fiskar, torskartade	435	
fladdermusfiskar	826	
flaggödlefisk	389	
flasknosdelfin	1477	
flatfiskar	1110	
flathuvudsskädda	1130	
flerhornig hammarhaj	46	
flodbarb	314	
flodkräfta	1276	

flodkräfta, amerikansk	1282, 1283	
flodkräfta, australisk	1316	
flodkräfta, donau-	1277	
flodkräfta, louisiana-	1284	
flodkräfta, ström-	1280	
flodkräftor	1278, 1281	
flodnejonöga	5	
flodpärlemormussla	1383	
flundra	1143	
flundra, arktisk	1140	
flundra, hälle-	1133	
flundra, japansk	1124	
flundra, kalifornisk	1127	
flundra, klipp-	1135	
flundra, liten hälle-	1148	
flundra, piltands-	1125	
flundra, rundnos-	1126	
flundra, sand-	1149	
flundra, slät-	1141	
flundra, stillahavshälle-	1134	
flundra, vinter-	1147	
flundrefiskar	1123	
flundror, sand-	1150	
flygbläckfisk	1433	
flygfisk	426	
flygfisk, fyrvingad	434	
flygfisk, japansk	424	
flygfiskar	423, 425	
flygknot	1109	
flygsimpa	1109	
fjädertömsmalar	293	
fjällbrosmar	465	
fjällbrosme	463	
fjällknot	1085	
fjällröding	263	
fjällvar	1177	
fjäril, havs-	905	
fjärilsfiskar	830	
fjärilsrocka, marmorerad	115	
fjärilsrockor	116	
fjärilstonfisk	970	
fjärsing	869	
fjärsing, fläckig	868	
fjärsing, mindre	867	
fjärsing, randig	870	
fjärsingar	871	
fjärsingfiskar	866	
fläckblecka	798	
fläckig blåsfisk	1185	
fläckig bonit	966	
fläckig fjärsing	868	
fläckig grouper	575	
fläckig havsabborre	594	
fläckig havskatt	916	
fläckig havsmal	376	
fläckig isfisk	897	
fläckig kungsmakrill	994	
fläckig pelamid	966	
fläckig sjökock	942	
fläckig smörfisk	1035	
fläckig taggmakrill	643	
fläckig tandvar	1111	
fläckig wobbegong	25	
fläckiga smörfiskar	1036	
fläckpagell	793	
fläckrocka	93	
fläcksnultra	849	
fläcktunga	1163	
fläskläpp	709	
forsabborre	618	
fregattauxid	965	
frostfisk, atlantisk	456	
frostfiskar	457	
frökenfisk	836	

frökenfiskar	835	
fyrfläckig var	1119, 1168	
fyrtömmad skärlånga	469	
fyrvingad flygfisk	434	
fysiljärfiskar	686	
fårhuvudfisk	764	
fäktarsnäckor	1338	
färna	341, 342	
fönsterglasvar	1174	
fönsterglasvar	1174	

G

gaffelmakrill, blå	663	
gaffelmakrillar	664	
gaffelräka	1220	
galjonsfisk, kap-	824	
galjonsfiskar	823	
galludohaj	65	
garda-stubb	1056	
gazami-krabba	1300	
gelidium	1529	
genevesik	233	
georgia-laxabborre	682	
gigartiner	1531	
gila-öring	246	
gisselräkor	1230	
glansfisk	509	
glasabborrar	563	
glashuvudfisk	495	
glasmalar	379	
glasvar	1169	
glasvarar	1170	
glasögonkajman	1503	
glattdick	134	
glatthajar	49	
glattmussla	1390	
glogrymta	703	
glyskolja	473	
golden kingklip	925	
grekisk sarv	363	
grindval	1466	
grindval, indisk	1465	
grisfisk	702	
grisgylta	844	
grodor	1493	
groplöja	339	
grouper, fläckig	575	
grouper, guld-	574	
grouper, malabar-	579	
grouper, nassau-	581	
grouper, peruansk	584	
grouper, röd	580	
grouper, vit	573	
groupers	582, 583	
grovansjovis, atlantisk	211	
grovansjovis, stillahavs-	212	
gruntfiskar	687	
grymta, åsne-	707	
grymta, band-	704, 712	
grymta, glo-	703	
grymta, prick-	713	
grymta, silver-	711	
grymta, strim-	708	
grymtor	705	
grå havsabborre	576	
grå multe	545	
grå tryckarfisk	1180	
gråhaj	50	
gråhajar	34	
grånoting	886	
grårocka	82	
gråsäl	1449	

SV

gråsej	467	gädda, kortnosad ben-	144	hajar, glatt-	49
gråsnultra	845	gädda, kungshavs-	949	hajar, grå-	34
gråval	1491	gädda, långnosad ben-	143	hajar, hammar-	45
gräsbult	1061	gädda, makrill-	422	hajar, hammar-	48
gräskarp	331	gädda, näbb-	416	hajar, hund-	55
gräspickerell	287	gädda, nål-	418	hajar, jätte-	18
grön blåmussla	1380	gädda, nilfen-	128	hajar, kamtand-	9
grön muräna	403	gädda, stillahavsmakrill-	421	hajar, pigg-	57
grön noting	888	gädda, svart havs-	953	hajar, röd-	28, 33
grön stör	132	gädda, vit havs-	954	hajar, trekants-	69
grön taggmakrill	665	gäddfiskar	286	hajmakrill	971
grön tigerräka	1226	gäddfiskar, havs-	947	hajrocka	73, 76
grönfisk	1092	gäddfiskar, makrill-	420	hajrocka, peruansk	75
gröngylta	843	gäddmuräna	406	hajrocka, sydlig	74
grönlandssäl	1451	gäddmuränor	407	hajrockor	72, 77
grönlandsval	1487	gäddor, havs-	948	halvnäbb, australisk	428
grönlånga, atka-	1093	gäddor, havs-	952	halvnäbb, balao-	429
grönling	372	gäddor, nål-	419	halvnäbb, brasiliansk	430
grönögonfisk	299	gäddsfiskar, näbb-	415	halvnäbb, japansk	431
grönögonfiskar	298	gärs	614	halvnäbb, långfenad	432
guinea-räka	1210	gärs, donau-	615	halvnäbbar	433
gul snapper	689	gökrocka	94	halvnäbbfiskar	427
gul taggmakrill	637	gös	620	halvspäckhuggare	1473
gul vimpelsnapper	696	gös, afrikansk havs-	717	hammarhaj	47
gulabborre	616	gös, amerikansk	621	hammarhajar	45, 48
gulbandad hästmakrill	652	gös, havs-	718	harr	268
gulbensräka	1215	gös, kanada-	619	havsabborrar	590, 595
gulbuksnoting	882			havsabborre	593
guldbandsselar	655			havsabborre, brun	577
guldbult	1051	**H**		havsabborre, fläckig	594
guldfisk	320			havsabborre, japansk	599
guldgrouper	574	haj, am-	26	havsabborre, kalifornisk	591
guldkroppar	528	haj, argentinsk hund-	54	havsabborre, sand-	572
guldlax	281	haj, blå-	42	havsabborre, strimmig	598
guldlax, japansk	283	haj, brun pigg-	59	havsabborre, svart	571
guldlax, stillahavs-	280	haj, choklad-	60	havsabborre, vit	597
guldlaxfiskar	279	haj, citron-	41	havsabborrfiskar	567
guldmakrill	681	haj, flerhornig hammar-	46	havsaborre, argentinsk	568
guldmulle	816	haj, galludo-	65	havsaborre, grå	576
guldmulte	540	haj, grå-	50	havsbraxen	680
guldnissöga	366	haj, gummi-	51	havsbraxnar	679
guldöring	243	haj, hammar-	47	havsfjäril	905
guldöring, mexikansk	245	haj, högfenad	40	havsgädda, blå	951
guldsparid	802	haj, koppar-	36	havsgädda, svart	953
gulfenad barrakuda	533	haj, kortnosad pigg-	66	havsgädda, vit	954
gulfenad kungsfisk	1074	haj, krås-	8	havsgäddfiskar	947
gulfenad noting	887	haj, lax-	22	havsgäddor	948, 952
gulfenad tonfisk	1008	haj, leopard-	56	havsgös	718
gulfmenhaden	165	haj, makrill-	20	havsgös, afrikansk	717
gullöpare	636	haj, mörk-	39	havsgösfiskar	715
gulssjärtssnapper	694	haj, ox-	13	havskatt	915
gulstjärtsskädda	1137	haj, pärl-	10	havskatt, blå	914
gulveka	722	haj, pigg-	64	havskatt, fläckig	916
gummihaj	51	haj, räv-	17	havskatter	917
gummiläpp	710	haj, rov-	14	havskattfiskar	913
gumsill	161	haj, sebra-	24	havskräfta	1320
Gunnars isfisk	898	haj, sexbågig kamtand-	11	havslädersköldpadda	1495
gurami, japansk	1044	haj, sill-	23	havsmal, fläckig	376
gurami, jätte-	1045	haj, silverspets-	35	havsmal, västafrikansk	375
gurami, kyss-	1047	haj, småfläckig röd-	31	havsmalar	374
gurami, ormskinns-	1046	haj, sorg-	58	havsmunk	1450
gyllene venusmussla	1404	haj, stillahavslax-	22	havsmus	123
gylta, berg-	840	haj, stillahavsräv-	15	havsmus, vitfläckig	124
gylta, blå-	841	haj, stillahavsspetsnos-	43	havsmusfiskar	122
gylta, brun-	842	haj, storfläckig röd-	32	havsmusfiskar, plognosade	126
gylta, gris-	844	haj, storögd räv-	16	havsnejonöga	7
gylta, grön-	843	haj, svartspets-	38	havsöra, europeiskt	1328
gylta, regnbågs-	854	haj, sydlig hund-	51, 53	havsöra, midas	1326
Günthers noting	891	haj, tagg-	71	havsöra, rött	1327
gädda	288	haj, trekants-	70	havsöra, stort	1325
gädda, amur-	291	haj, val-	27	havsöron	1329
gädda, atlantisk nål-	417	haj, vit-	19	havspil	1426
gädda, blå havs-	951	haj, vitprickig spetsnos-	44	havsruda	808
gädda, horn-	416	haj, årfens-	37	havsrudefiskar	763

havssugga 1068
havstobis 935
havstulpan, jätte- 1200
havstulpaner 1199
havsål 410
havsål, amerikansk 411
havsål, argentinsk 412
havsålsfiskar 408
havsängel 68
havsänglar 67
hector-prickfisk 295
hickorysill 159
hjälmsnäckor 1336
hjärtmussla, ätlig 1389
hjärtmussla, taggig 1388
hjärtmusslor 1387
högfenad haj 40
hoki 479, 480
hoki, chilensk 478
hoppmulte 543
horngädda 416
hornsimpa 1099
huchen 242
hummer 1318
hummer, amerikansk 1317
hummer, andaman- 1319
hummer, kejsar- 1320
hummer, toffel- 1305
hundfisk 284
hundhaj, argentinsk 54
hundhaj, nordlig 52
hundhaj, sydlig 51, 53
hundhajar 55
hundlax 248
hundfisk, dvärg- 285
hundtungefiskar 1166
hus 139
håbrand 23
hågäl 29, 30
håkäring 63
hård venusmussla 1398
hårstjärt 959
hårstjärtfiskar 955
hälleflundra 1133
hälleflundra, liten 1148
hälleflundra, stillahavs- 1134
hästhuvudmakrill, afrikansk 656
hästhuvudmakrill, atlantisk 657
hästmakrill, gulbandad 652
hästmusslor 1373
hästräka 1251
hönmussla 1391

I

iberisk barb 315
id 343
igelkottsrocka 88
indianlax 252
indisk barrakuda 534
indisk bukfisk 678
indisk buksill 188
indisk djuplånga 439
indisk drivfisk 1038
indisk dvärgmakrill 977
indisk grindval 1465
indisk platthuvudfisk 1096
indisk sardinell 192
indisk staksill 205
indisk taggmakrill 644
isfisk, fläckig 897
isfisk, Gunnars 898

isfiskar 895
isländsk kammussla 1362
islandsmussla 1386
italiensk barb 318

J

jacobsmussla 1364
japansk ansjovis 218
japansk bläckfisk 1432
japansk djuphavsabborre 612
japansk flundra 1124
japansk flygfisk 424
japansk guldlax 283
japansk gurami 1044
japansk halvnäbb 431
japansk havsabborre 599
japansk kistmussla 1340
japansk kungsmakrill 997
japansk langust 1272
japansk lerskädda 1131
japansk multe 541
japansk rödbraxen 805
japansk sandfisk 863
japansk sandskädda 1138
japansk sardin 197
japansk sardinell 194
japansk seriola 660
japansk sjögurka 1523
japansk taggmakrill 642, 666
japansk trollkrabba 1310
japansk var 1120
japansk venusmussla 1399
japansk venusmussla 1402
japansk ål 399
japanskt jätteostron 1347
java-barb 353
juanfernandez-langust 1261
judefisk 578
jumbobläckfisk 1427
junkergirella 838
jättebarrakuda 532
jättegurami 1045
jättehajar 18
jättehavstulpan 1200
jätteostron, japanskt 1347
jättetrådfisk 554
jätteveka 724

K

Kahawai 683
kalamarer 1422
kalifornisk flundra 1127
kalifornisk havsabborre 591
kalifornisk rundsill 176
kalifornisk sardin 196
kalifornisk vekfisk 724
kalikå-kammussla 1360
kammusla, brokig 1363
kammussla 1384
kammussla, australisk 1366
kammussla, drottning- 1359
kammussla, isländsk 1362
kammussla, kalikå- 1360
kammussla, nyzeeländsk 1367
kammussla, stor 1365
kammussla, sydatlantisk 1368
kammussla, vik- 1361
kammusslor 1358
kamtandhaj, sexbågig 11
kamtandhajar 9
kanadagös 619

kanadaröding 266
kanadasik 232
kanadi-kungsmakrill 998
kanarie-tandbraxen 772
kanelvar 1122
kaninfisk 943
kaninfiskar 944
kantnål, mindre 507
kantnål, större 506
kapansjovis 216
kapare, nord- 1490
kapare, syd- 1489
kap-bahytt 685
kap-bläckfisk 1425
kapenta 203
kap-galjonsfisk 824
kap-kingklip 926
kap-knivmussla 1412
kapknorrhane 1081
kapknot 1081
kapkummel 484
kap-kungsfisk 1071
kap-langust 1262
kapmal 377
kap-plognos 125
kaptensfiskar 556
karamellräka 1238
kardinalfisk 610
karettsköldpadda, oäkta 1496
karibisk kungsmakrill 1000
karibisk langust 1269
karibisk tonfisk 1009
karp 332
karp, dy- 330
karp, gräs- 331
karp, marmor- 336
karp, silver- 335
karp, svart- 347
karp, valencia-tand- 515
karpar, lax- 300
karpar, sug- 301
karpar, tand- 512
karpatiskt nejonöga 2
karpfiskar 303
karragener 1532
kaskelot 1462
kaspisk skarpsill 171
kassava-kväkare 750
katt, havs- 915
katter, havs- 917
kattfisk, brun 385
kattfisk, svart 384
kattfiskar, havs- 913
kawakawa 967
kedjepickerell 290
kejsarhummer 1320
kelee-staksill 182
ketalax 248
killi, medelhavs- 513
killi, spansk 514
kinesisk buksill 184
kinesisk kungsmakrill 1003
kinesisk smörfisk 1032
kinesisk staksill 206
kinesisk ullhandskrabba 1303
kingklip 928
kingklip, golden 925
kingklip, kap- 926
kingklip, röd 927
kirurgfisk, monrovia- 946
kirurgfiskar 945
kistmussla, japansk 1340
kistmussla, stillahavs- 1339
kistmusslor 1341

klappmyts	1447
klarbult	1050
klippabborre	600
klippbrosme	464
klippflundra	1135
klippkrabbtaska	1289
klippmussla, sydamerikansk	1382
klippsmörbult	1054
klorocka	97
klumpfisk	1188
klumpfisk, mindre	1189
klumpfisk, spetsstjärts-	1187
klätterfisk	1039
knaggrocka	87
knivfiskar	292
knivmussla, atlantisk	1408
knivmussla, böjd	1409
knivmussla, europeisk	1413
knivmussla, kap-	1412
knivmussla, stillahavs-	1411
knivmusslor	1406, 1407, 1414
knivräka	1254
knivräkor	1255
knorrhanar	1083, 1087
knorrhane	1084
knorrhane, blåfenad	1082
knorrhane, kap-	1081
knorrhanefiskar	1078
knot	1084
knot, blåfenad	1082
knot, fen-	1089
knot, fjäll-	1085
knot, flyg-	1109
knot, kap-	1081
knot, lyr-	1090
knot, röd-	1079
knot, silverbandad	1080
knot, spets-	1088
knot, tagg-	1086
knot, tvärbandad	1091
knotar	1083
knotfiskar	1078
knubbig trumfisk	753
knubbsäl	1453
knölblecka, röd	769
knölbleckor	770
knölval	1485
kolfisk	1094
kolgapskväkare	721
kolja	453
kolmule	459
kometmakrill	641
kongo-tandbraxen	773
kopparbleckor	789
kopparhaj	36
korall, äkta	1514
korall, röd-	1514
koraller	1513
koraller, äkta	1513
korallmalar	378
koreansk blåmussla	1375
koreansk kungsmakrill	992
koreaspigg	940
korocka	110
kortmakrill	975
kortnosad bengädda	144
kortnosad pigghaj	66
kortnosad spjutfisk	1022
kortstjärtsnoting	890
kötträka	1216
krabba, antarktisk troll-	1309
krabba, arktisk maskerings-	1312
krabba, blå	1295
krabba, dana-sim-	1294

krabba, gazami-	1300
krabba, japansk troll-	1310
krabba, kinesisk ullhands-	1303
krabba, mangrove-	1302
krabba, spindel-	1315
krabba, stentroll-	1307
krabba, strand-	1298
krabba, svart sten-	1293
krabba, sydlig spindel-	1314
krabbor, djuphavs-	1322
krabbor, maskerings-	1313
krabbor, sim-	1296, 1301
krabbor, troll-	1311
krabbtaska	1291
krabbtaska, klipp-	1289
krabbtaska, nordlig	1288
krabbtaska, stillahavs-	1290, 1292
krabbtaskor	1287
krill	1203
krill, antarktisk	1204
kristallräka	1214
krokodil, saltvattens-	1505
krokodil, siamesisk	1506
krokodil, sydostasiatisk	1504
krokodiler, äkta	1501
krokodilisfisk	901
kroppar, guld-	528
krypande tiofotade kräftdjur	1285
krypare, sand-	333
kråshaj	8
kräfta, amerikansk flod-	1282, 1283
kräfta, australisk flod-	1316
kräfta, donau-flod-	1277
kräfta, flod-	1276
kräfta, havs-	1320
kräfta, louisiana-flod-	1284
kräfta, östlig flod-	1277
kräfta, sump-	1277
kräfta, vitfotad flod-	1279
kräftdjur, krypande tiofotade	1285
kräftor, eremit-	1286
kräftor, flod-	1278, 1281
kräftor, toffel-	1304
kubamulte	539
kubbongar	1337
kulfiskar	1184
kummel	487
kummel, argentinsk	486
kummel, chilensk	485
kummel, djupkap-	488
kummel, kap-	484
kummel, senegal-	491
kummel, silver-	483
kummel, stillahavs-	490
kummel, svart	489
kummel, syd-	482
kummel, vit-	481
kummelfiskar	477
kumlar	492
kungsfisk, akadisk	1073
kungsfisk, djuphavs-	1076
kungsfisk, gulfenad	1074
kungsfisk, kap-	1071
kungsfisk, stillahavs-	1070
kungsfisk, större	1075
kungsfiskar	1077
kungshavsgädda	949
kungskäggsfisk	555
kungskväkare	736
kungskväkare, randig	735
kungskväkare, slät	733
kungslangust	1273
kungslax	254
kungsmakrill	988

kungsmakrill, afrikansk	1005
kungsmakrill, australisk	996
kungsmakrill, brasiliansk	987
kungsmakrill, bredbandad	1001
kungsmakrill, fläckig	994
kungsmakrill, japansk	997
kungsmakrill, kanadi-	998
kungsmakrill, karibisk	1000
kungsmakrill, kinesisk	1003
kungsmakrill, koreansk	992
kungsmakrill, papua-	995
kungsmakrill, prickig	991
kungsmakrill, streckad	993
kungsmakrill, strimmig	989
kungsmakrillar	1004
kungsräka, västlig	1221
kungsröd räka	1257
kungstaggmakrillar	653
kungsveka	726, 732
kuruma-räka	1219
kusttobis	936
kveite, blå-	1148
kvidd	352
kväkare	752
kväkare, bobo-	749
kväkare, brännings-	734
kväkare, drottning-	748
kväkare, kassava-	750
kväkare, kolgaps-	721
kväkare, kungs-	736
kväkare, långhals-	751
kväkare, manchurisk	746
kväkare, orange	747
kväkare, randig kungs-	735
kväkare, silver-	744
kväkare, slät	731
kväkare, stubb-	738
kväkare, vit-	730
kväkare, vitmunnad	737
kväkarfiskar	739
kyssgurami	1047
käft, blå-	1063
käftar, blå-	1064
käxa, blå-	61
käxa, blå-	61
käxor, blå-	62

L

labeo	337
ladyfisk	146
lake	452
lake, tång-	923
lamantin	1456
langust, australisk	1270
langust, europeisk	1266
langust, japansk	1272
langust, juanfernandez-	1261
langust, kap-	1262
langust, karibisk	1269
langust, kungs-	1273
langust, mauretansk	1268
langust, natal-	1265
langust, nyzeeländsk	1260
langust, oceanisk	1264
langust, sydlig	1267
langust, transkei-	1271
langust, tristan-	1263
languster	1259, 1274
languster, pisk-	1275
lasca-näsling	323
lavaretsik	234

lax	256			malar, dvärg-	383, 387
lax, apache-	244			malar, glas-	379
lax, Biwa-	253	magellan-mussla	1371	malar, havs-	374
lax, coho-	249	magellan-noting	889	malar, korall-	378
lax, guld-	281	majfisk	156	manat, amazon-	1455
lax, hund-	248	majsill	156	manat, senegal-	1457
lax, indian-	252	mako	20	manchurisk kväkare	746
lax, japansk guld-	283	mako, långfenad	21	maneter, lung-	1512
lax, keta-	248	makrill	985	mangrove-krabba	1302
lax, kungs-	254	makrill, afrikansk hästhuvud-	656	mangrove-ostron	1348
lax, pink-	247	makrill, afrikansk kungs-	1005	mangrovesnapper	688
lax, puckel-	247	makrill, alexandria-tråd-	633	mansäl	1444
lax, silver-	249	makrill, atlantisk hästhuvud-	657	manta	112
lax, sockeye-	252	makrill, australisk kungs-	996	manta, stillahavs-	113
lax, stillahavsguld-	280	makrill, blå gaffel-	663	mantis	1202
laxabborre	683	makrill, blå tagg-	670	marlin, blå	1019
laxabborre, georgia-	682	makrill, brasiliansk kungs-	987	marlin, medelhavs-	1024
laxar och öringar	260	makrill, bredbandad kungs-	1001	marlin, randig	1023
laxfiskar	262	makrill, chilensk tagg-	669	marlin, svart	1018
laxfiskar, guld-	279	makrill, cunene-tagg-	674	marlin, vit	1021
laxhaj	22	makrill, dubbellinje-	972	marlins	1020
laxhaj, stillahavs-	22	makrill, filippiner-	976	marmorerad darrocka	118
laxkarpar	300	makrill, fläckig kungs-	994	marmorerad fjärilsrocka	115
leopardhaj	56	makrill, fläckig tagg-	643	marmorerad noting	885
lerskädda	1132	makrill, grön tagg-	665	marmorkarp	336
lerskädda, japansk	1131	makrill, gul tagg-	637	marmormört	348
liten bläckfisk	1423	makrill, gulbandad häst-	652	marmorsmörbult	1060
liten hälleflundra	1148	makrill, guld-	681	marulk	1195
liten silversida	549	makrill, haj-	971	marulk, amerikansk	1193
liten venusmussla	1405	makrill, indisk dvärg-	977	marulk, mindre	1194
ljusfisk	278	makrill, indisk tagg-	644	marulk, västafrikansk	1196
ljusrocka	85	makrill, japansk kungs-	997	marulkfiskar	1192
lodda	273	makrill, japansk tagg-	642, 666	maskalung	289
lotsfisk	650	makrill, kanadikungs-	998	maskeringskrabba, arktisk	1312
louisiana-flodkräfta	1284	makrill, karibisk kungs-	1000	maskeringskrabbor	1313
lubb	442	makrill, kinesisk kungs-	1003	Masulax	250
lungfisk, australisk	127	makrill, komet-	641	mauretansk langust	1268
lungfiskar, afrikanska	129	makrill, koreansk kungs-	992	medelhavsabborre, brun	588
lungmaneter	1512	makrill, kort-	975	medelhavsabborre, större	587
luring	853	makrill, kungs-	988	medelhavsbarb	316
lyrknot	1090	makrill, monterrey-	990	medelhavsblåmussla	1377
lyrtorsk	466	makrill, papua-kungs-	995	medelhavskilli	513
lysräka, norsk	1205	makrill, parona-	651	medelhavslånga	461
lysräkor	1203	makrill, prickig kungs-	991	medelhavsmarlin	1024
långa	462	makrill, queensland-	999	medelhavspigghaj	65
långa, femtömmad skär-	443	makrill, räk-	634	medelhavssoldatfisk	521
långa, fyrtömmad skär-	469	makrill, spansk	984	medelhavsspjutfisk	1024
långa, indisk djup-	439	makrill, stillahavs-	983	medelhavsstaksill	158
långa, medelhavs-	461	makrill, stillahavstagg-	671	medelhavsstrandkrabba	1297
långa, tretömmad skär-	450	makrill, sträv tagg-	667	medelhavstaggmakrill	668
långfenad bläckfisk	1424	makrill, streckad kungs-	993	medelhavstobis	938
långfenad halvnäbb	432	makrill, strimmig kungs-	989	menhaden	167
långfenad mako	21	makrill, svansfläckstagg-	640	menhaden, atlantisk	166
långfenad pompano	663	makrill, svartryggad tagg-	638	menhaden, gulf-	165
långfenad tonfisk	1007	makrill, sydafrikansk tagg-	673	menhaden, stillahavs-	175
långfensbening	153	makrill, tagg-	672	menidia	552
långhals	1197	makrillar, dvärg-	978	menolapikarell	812
långhalsar	1198	makrillar, gaffel-	664	metapenaeus-räkor	1209
långhalskväkare	751	makrillar, kungs-	1004	mexikansk guldöring	245
långnäbbad delfin	1475	makrillar, kungstagg-	653	Midas havsöra	1326
långnosad bengädda	143	makrillar, mån-	676	mindre djävulsrocka	114
långnosad spjutfisk	1026	makrillar, tagg-	639, 645, 675	mindre fjärsing	867
långsnultra	850	makrillfiskar	961, 986	mindre kantnål	507
långstjärt	497	makrillfiskar, tagg-	632	mindre klumpfisk	1189
långstjärtad tonfisk	1013	makrillgädda	422	mindre marulk	1194
långstjärtar	498	makrillgädda, stillahavs-	421	mississippialligator	1502
läppfiskar	837	makrillgäddfiskar	420	mjölkfisk	228
läppmulte	547	makrillhaj	20	mjölkfiskar	227
löja	310	makrillika fiskar	960	moçambique-tilapia	832
löja, donau-	322	mal	382	moki, blå	862
löpare, blå-	635	mal, kap-	377	mokifiskar	861
löpare, regnbågs-	646	mal, svart dvärg-	384	monokelsnappers	698
		mal, västafrikansk havs-	375	monrovia-kirurgfisk	946
		malabargrouper	579	monterrey-makrill	990

M

SV

533

moratorsk, röd	438
moratorskfiskar	436
moskitfisk	516
mukene	355
mullar	819
mullar, stillahavs-	822
mulle	818
mulle, band-	821
mulle, guld-	816
mulle, röd	817
mulle, västafrikansk	820
mullusfiskar	815
multe, afrikansk	544
multe, brasiliansk	546
multe, grå	545
multe, guld-	540
multe, japansk	541
multe, kuba-	539
multe, läpp-	547
multe, tjockläppad	538
multe, tunnläppad	542
multefiskar	537
multenoting, patagonisk	880
munk, havs-	1450
muräna	404
muräna, gädd-	406
muräna, grön	403
muräna, sorgkants-	405
muränfiskar	402
muränor, gädd-	407
murray-trollkrabba	1308
mus, havs-	123
musfiskar, havs-	122
musfiskar, plognosade havs-	126
mussla, atlantisk	
(djupshavs-) mantelkam-	1369
mussla, atlantisk kniv-	1408
mussla, australisk blå-	1378
mussla, australisk kam-	1366
mussla, blå-	1376
mussla, brokig kam-	1363
mussla, brun venus-	1395
mussla, brännings-	1392
mussla, böjd kniv-	1409
mussla, chilensk blå-	1374
mussla, dam-	1343
mussla, drottningkam-	1359
mussla, egentlig pilgrims-	1365
mussla, europeisk kniv-	1413
mussla, flodpärlemor-	1383
mussla, glatt-	1390
mussla, grön blå-	1380
mussla, gyllene venus-	1404
mussla, hön-	1391
mussla, islands-	1386
mussla, isländsk kam-	1362
mussla, jacobs-	1364
mussla, japansk kist-	1340
mussla, japansk venus-	1399, 1402
mussla, kalikå-kam-	1360
mussla, kam-	1384
mussla, kap-kniv-	1412
mussla, koreansk blå-	1375
mussla, liten venus-	1405
mussla, magellan-	1371
mussla, medelhavsblå-	1377
mussla, nyzeeländsk blå-	1381
mussla, nyzeeländsk kam-	1367
mussla, rakknivs-	1410
mussla, randig venus-	1396
mussla, riodelaplata-blå-	1379
mussla, sand-	1415
mussla, självlysande borr-	1416
mussla, skink-	1385

mussla, smörvenus-	1403
mussla, sötvattens-	1345
mussla, stillahavskist-	1339
mussla, stillahavskniv-	1411
mussla, stor kam-	1365
mussla, stor venus-	1401
mussla, sydamerikansk klipp-	1382
mussla, sydatlantisk kam-	1368
mussla, taggig hjärt-	1388
mussla, vikkam-	1361
mussla, ätlig hjärt-	1389
musslor	1324
musslor, blå-	1370, 1373
musslor, borr-	1417
musslor, häst-	1373
musslor, hjärt-	1387
musslor, kam-	1358
musslor, kist-	1341
musslor, kniv-	1406, 1407, 1414
musslor, sötvattens-	1344
musslor, tapes-	1394
musslor, venus-	1393, 1400
månmakrillar	676
mörkhaj	39
mört	360
mört, adria-	359
mört, pigo-	358
mört, svartahavs-	357
mörtar	361

N

nakenbarb	312
narval	1479
nassaugrouper	581
natal-blecka	797
natal-langust	1265
natal-pagell	792
navagatorsk	445
nejonöga, bäck-	6
nejonöga, donau-	2
nejonöga, donau-	4
nejonöga, flod-	5
nejonöga, havs-	7
nejonöga, karpatiskt	2
nejonöga, ukrainskt	3
nejonögon	1
nibe	741
nibe, blom-	740
nilabborrar	566
nilabborre	565
nilabborrfiskar	560
nilfengädda	128
nil-tilapia	833
nissöga	370
nissöga, balkan-	367
nissöga, bergantino-	368, 373
nissöga, guld-	366
nissöga, rumänskt	369
nordhavsräka	1237
nordisk beryx	518
nordkapare	1490
nordlig hundhaj	52
nordlig krabbtaska	1288
nordlig rosenräka	1217
nordlig stjärtfenad bläckfisk	1430
nordlig tobis	933
nordlig viträka	1227
nordligt sjölejon	1443
nordräkor	1240
nordsjösik	237
nors	274
nors, arktisk regnbågs-	276

norsar	277
norsfiskar	270
norsk lysräka	1205
noshörningsfisk	899
noting, antarktisk tand-	878
noting, blåhuvad	888
noting, grå-	886
noting, grön	888
noting, gulbuks-	882
noting, gulfenad	887
noting, Günthers	891
noting, kortstjärts-	890
noting, magellan-	889
noting, marmorerad	885
noting, patagonisk multe-	880
noting, Ramsays	892
noting, silver-	893
noting, slank-	883
noting, streckögd	884
noting, svart-	881
noting, tand-	877
notingar	876, 894
notingar, tand-	879
nufar-blecka	768
nultra, grås-	845
nultra, ögons-	848
nyzeeländsk bläckfisk	1431
nyzeeländsk blåmussla	1381
nyzeeländsk kammussla	1367
nyzeeländsk langust	1260
nyzeeländsk vrakfisk	586
nyzeeländsk ål	398
nyzeeländskt ostron	1355
nylonräka, chilensk	1235
nylonräkor	1236
nålfisk	930
nålgädda	418
nålgädda, atlantisk	417
nålgäddor	419
näbbar, halv-	433
näbbgädda	416
näbbgäddsfiskar	415
näbbrocka	89
näbbfiskar, halv-	427
näbbval, bairds	1459
näbbval, vanlig	1460
näbbvalar	1458
näsling	325
näsling, bäck-	326
näsling, boga-	327
näsling, dalmatisk	324
näsling, lasca-	323
näsling, trubb-	329

O

oblada	788
oceanisk langust	1264
oljefisk	950
orange filfisk	1181
orange kväkare	747
d'Orbignys venusmussla	1397
ormfisk	929
ormhuvudfisk, randig	559
ormhuvudfisk, röd	557
ormhuvudfiskar	558
ormskinnsgurami	1046
ormål, spetsnosig	414
ormål, trubbnosig	413
ostron	1350, 1353, 1356
ostron, amerikanskt	1349
ostron, australiskt	1357
ostron, chilenskt	1351

ostron, japanskt jätte-	1347
ostron, mangrove-	1348
ostron, nyzeeländskt	1355
ostron, tandat	1352
oxgroda	1492
oxhaj	13
oxsimpa	1102
oxögonfisk	766
oäkta karettsköldpadda	1496

P

paddfiskar	1191
paddtorsk	468
pagell	790
pagell, fläck-	793
pagell, natal-	792
pagell, prick-	791
pagell, röd-	794
pageller	795
palaemonid-räkor	1244, 1245
pangablecka	799
pansarhane	1106
papegojabborre	570
papegojfisk	856
papua-kungsmakrill	995
parapenaeopsis-räkor	1211
parfläckig rocka	89
paronamakrill	651
patagonisk multenoting	880
patella-skålsnäcker	1333
pekingbraxen	349
pelamid, fläckig	966
pelamid, ryggstrimmig	982
pelamid, sydlig	979
peledsik	238
peneida-räkor	1229
peruansk grouper	584
peruansk hajrocka	75
peruansk trumfisk	743
peru-veka	723
pickerell, gräs-	287
pickerell, kedje-	290
pigg, stors-	500
pigg, tångs-	502
pigghaj	64
pigghaj, brun	59
pigghaj, kortnosad	66
pigghaj, medelhavs-	65
pigghajar	57
piggvar	1172
piggvar, svartahavs-	1173
pigomört	358
pikarell	813
pikarell, menola-	812
pikarell, svansfläcks-	811
pikareller	810, 814
pil, havs-	1426
pilotval	1466
piltandsflundra	1125
pinklax	247
pintadilla	858
pisklanguster	1275
plattfiskar	1110
platthuvudfisk, indisk	1096
platthuvudfiskar	1095
plattostron, stillahavs-	1354
plogjärnsrocka	96
plognos, kap-	125
plognosade havsmusfiskar	126
polartorsk	441
pollack, alaska-	470
pollan	230
pomfret, svart	677

pompano, atlantisk	662
pompano, långfenad	663
ponnyfiskar	699, 701
porgier	767
pottval	1462
prickfiskar	294
prickgrymta	713
prickig dvärgmal	386
prickig kungsmakrill	991
prickig spadfisk	828
prickpagell	791
prickveka	725
prästfisk	551
puckellax	247
purpursnäckor	1334
påfågelslemfisk	907
påfågelssnultra	851
pälssäl, sydafrikansk	1442
pälssäl, sydamerikansk	1441
pärlemormussla, flod-	1383
pärlhaj	10
pärlostron	1346

Q

queensland-makrill	999

R

Ramsays noting	892
randig fjärsing	870
randig kungskväkare	735
randig marlin	1023
randig ormhuvudfisk	559
randig sandblecka	786
randig sjökock	941
randig snapper	692
randig venusmussla	1396
regnbåge	251
regnbågsgylta	854
regnbågslöpare	646
regnbågsnors	275
regnbågsnors, arktisk	276
regnbågssardin	173
remora	1178
revbleckefiskar	762
revbleckor	761
ringbuk	1108
ringbuk, atlantisk	1107
riodelaplata-blåmussla	1379
Rissos delfin	1467
rocka, blom-	94
rocka, brok-	101
rocka, darr-	119
rocka, fläck-	93
rocka, gök-	94
rocka, grå-	82
rocka, haj-	73, 76
rocka, igelkotts-	88
rocka, klo-	97
rocka, knagg-	87
rocka, ljus-	85
rocka, marmorerad darr-	118
rocka, marmorerad fjärils-	115
rocka, mindre djävuls-	114
rocka, näbb-	89
rocka, parfläckig	89
rocka, peruansk haj-	75
rocka, plogjärns-	96
rocka, sand-	86
rocka, slät-	84
rocka, smååögd	91
rocka, spegel-	92

rocka, spjut-	106
rocka, stillahavs-	100
rocka, stillahavsspjut-	104
rocka, stjärn-	83
rocka, sträv-	98
rocka, sydlig haj-	74
rocka, taggspjuts-	105
rocka, taggsvans-	99
rocka, vinter-	95
rocka, violett spjut-	107
rocka, vit-	90
rocka, ögondarr-	120
rocka, örn-	109
rockor	80, 81, 102
rockor, darr-	117, 121
rockor, djävuls-	111
rockor, fjärils-	116
rockor, haj-	72, 77
rockor, örn-	108
rockor, spjut-	103
rosa tandbraxen	775
rosenräka, nordlig	1217
rosenräka, sydlig	1224
rovansjovis, brasiliansk	221
rovhaj	14
ruda	321
ruda, havs-	808
rudefiskar, havs-	763
rumänskt nissöga	369
rund sardinell	190
rundfjällig spjutfisk	1025
rundnosflundra	1126
rundsill	178
rundsill, atlantisk	177
rundsill, kalifornisk	176
rundsill, silverrandig	201
rundsillar	179
ryggstrimmig pelamid	982
rysk stör	131
råttsvans	213
räka, akiami-	1242
räka, argentinsk röd-	1256
räka, argentinsk stilett-	1206
räka, australisk	1207
räka, aztek-	1213
räka, banan-	1222
räka, brun tiger-	1218
räka, chilensk nylon-	1235
räka, djuphavs-	1212
räka, gaffel-	1220
räka, grön tiger-	1226
räka, guinea-	1210
räka, gulbens-	1215
räka, häst-	1251
räka, karamell-	1238
räka, kniv-	1254
räka, kött-	1216
räka, kristall-	1214
räka, kungsröd,	1257
räka, kuruma-	1219
räka, nordhavs-	1237
räka, nordlig vit-	1227
räka, norsk lys-	1205
räka, röd-	1234
räka, rödstjärtad	1225
räka, sand-	1251
räka, shiba-	1208
räka, sötvattens-	1247
räka, sten-	1252
räka, sydamerikansk sötvattens-	1246
räka, sydlig rosen-	1224
räka, tång-	1249
räka, tiger-	1223
räka, västlig kungs-	1221

SV

räka, vit-	1228	
rakfisk	855	
rakknivsmussla	1410	
räkmakrill	634	
räkor	1239	
räkor, aristeid-	1232	
räkor, atyid-	1243	
räkor, crangonid-	1250	
räkor, gissel-	1230	
räkor, kniv-	1255	
räkor, lys-	1203	
räkor, metapenaeus-	1209	
räkor, nord-	1240	
räkor, nylon-	1236	
räkor, palaemonid-	1244, 1245	
räkor, parapenaeopsis-	1211	
räkor, peneida-	1229	
räkor, sergestid-	1241	
räkor, sötvattens-	1248	
räkor, sten-	1253	
rätval, dvärg-	1488	
rätvalar	1486	
rävhaj	17	
rävhaj, stillahavs-	15	
rävhaj, storögd	16	
röd grouper	580	
röd kingklip	927	
röd knölblecka	769	
röd moratorsk	438	
röd mulle	817	
röd ormhuvudfisk	557	
röd sjöpung	1510	
röd snapper	690	
röd soldatfisk	522	
röd storögonfisk	607	
röd tandbraxen	777	
röd trumfisk	757	
rödahavssill	181	
rödalger	1528	
rödalger, alkformiga	1530	
rödbandad rödbraxen	803	
rödbraxen	806	
rödbraxen, blåfläckig	804	
rödbraxen, japansk	805	
rödbraxen, rödbandad	803	
rödbraxnar	807	
rödfisk	1075	
rödhaj, småfläckig	31	
rödhaj, storfläckig	32	
rödhajar	28, 33	
röding, bäck-	264	
röding, fjäll-	263	
röding, kanada-	266	
rödingar	267	
rödknot	1079	
rödkorall	1514	
rödkrabba, djuphavs-	1321	
rödpagell	794	
rödräka	1234	
rödräka, argentinsk	1256	
rödsnapper, sydlig	691	
rödsnultra	846	
rödspätta	1145	
rödspätta, alaska-	1146	
rödstjärtad räka	1225	
rödtunga	1128	
rödtunga, stillahavs-	1129	
rötsimpa	1100	
rötsimpa, tagg-	1098	
rötsimpor	1101	
rött havsöra	1327	

S

saffranstorsk	444	
salpa	801	
saltvattenskrokodil	1505	
sammetssimkrabba	1299	
sandabborrar	865	
sandabborre, blå	864	
sandblecka, randig	786	
sandblecka, vit	785	
sandbleckor	787	
sandfisk, japansk	863	
sandflundra	1149	
sandflundror	1150	
sandhavsabborre	572	
sandkrypare	333	
sandkrypare, donau-	334	
sandmussla	1415	
sandräka	1251	
sandrocka	86	
sandskädda	1139	
sandskädda, japansk	1138	
sandstubb	1059	
sandtunga	1164	
sanktpersfisk	525	
sanktpersfisk, vit	524	
sanktpersfiskar	523	
sardell	217	
sardin	189	
sardin, australisk	198	
sardin, japansk	197	
sardin, kalifornisk	196	
sardin, regnbågs-	173	
sardin, sydafrikansk	199	
sardin, sydamerikansk	200	
sardinell, afrikansk	193	
sardinell, bengalisk	191	
sardinell, indisk	192	
sardinell, japansk	194	
sardinell, rund	190	
sardineller	195	
sarv	362	
sarv, grekisk	363	
savetta	328	
savetta-näsling	328	
scup	809	
sebrablecka	779	
sebrahaj	24	
segelfisk	1017	
segelfisk, atlantisk	1016	
segelfiskar	1015	
sej, grå-	467	
sejval	1481	
selar, guldbands-	655	
senegalkummel	491	
senegal-manat	1457	
sepia, stor	1419	
sepia-bläckfisk	1418	
sergestid-räkor	1241	
seriola	658	
seriola, japansk	660	
seriola, syd-	659	
serioler	661	
sexbågig kamtandhaj	11	
shiba-räka	1208	
siamesisk krokodil	1506	
siding, vit-	1468	
sierra	1002	
sik, bondelle-	235	
sik, geneve-	233	
sik, kanada-	232	
sik, lavaret-	234	
sik, peled-	238	
sikar	241	

sikfiskar	261	
siklöja	229	
siklöja, amerikansk	231	
sill	169	
sill, atlantisk rund-	177	
sill, atlantisk tråd-	186	
sill, blå stak-	155	
sill, bonga-	174	
sill, chilensk	204	
sill, gum-	161	
sill, hickory-	159	
sill, indisk stak-	205	
sill, kalifornisk rund-	176	
sill, kaspisk skarp-	171	
sill, kelee-stak-	182	
sill, kinesisk buk-	184	
sill, maj-	156	
sill, rödahavs-	181	
sill, silverrandig rund-	201	
sill, skarp-	202	
sill, stak-	157	
sill, stillahavs-	170	
sill, stillahavstråd-	185	
sill, ström-	281	
sill, svartahavsstak-	160	
sill, toli-stak-	207	
sill, varg-	225	
sill, vimpel-	168, 172	
sill, vit stak-	162	
sillaginider	625	
sillar, daga-	203	
sillar, rund-	179	
sillar, stak-	163	
sillar, varg-	224, 226	
sillfiskar	154	
sillhaj	23	
sillingar	180	
sillkung	510	
sillval	1484	
silveller	903	
silverabborre	628	
silverbandad knot	1080	
silverfisk	282	
silvergrymta	711	
silverkarp	335	
silverkummel	483	
silverkväkare	744	
silverlax	249	
silvernoting	893	
silverrandig rundsill	201	
silversida, liten	549	
silversida, stor	550	
silversidefiskar	548	
silverslemfisk	918	
silversmörfisk	1031	
silverspetshaj	35	
silvertorsk	449	
simkrabba, dana-	1294	
simkrabba, sammets-	1299	
simkrabbor	1296, 1301	
simpa, horn-	1099	
simpa, ox-	1102	
simpa, röt-	1100	
simpa, skägg-	1103	
simpa, sten-	1097	
simpa, taggröt-	1098	
simpor, röt-	1101	
sjurygg	1105	
sjuryggsfiskar	1104	
självlysande borrmussla	1416	
sjöborrar	1519, 1520	
sjöborre	1518	
sjöborre, ätlig	1521	
sjögräsblecka	780	

SV

sjögurka, japansk	1523	småfläckig rödhaj	31	spindelkrabba	1315
sjöhästar	505	småhuvudval	1461	spindelkrabba, sydlig	1314
sjöko, amazon-	1455	småögd rocka	91	spiritist	493
sjökock, fläckig	942	småspigg	501	spjutfisk, kortnosad	1022
sjökock, randig	941	småtandad sågfisk	79	spjutfisk, långnosad	1026
sjölejon, nordligt	1443	småtandad veka	727	spjutfisk, medelhavs-	1024
sjölejon, Stellers	1443	småtunga	1156	spjutfisk, rundfjällig	1025
sjölejon, sydamerikanskt	1444	småvar	1171	spjutfiskar	1015
sjöpung	1508	smörbult, klipp-	1054	spjutfiskar	1027
sjöpung, chilensk	1509	smörbult, större	1052	spjutrocka	106
sjöpung, röd	1510	smörbult, svart-	1053	spjutrocka, stillahavs-	104
sjöpungar	1507	smörbultar	1049, 1055	spjutrocka, violett	107
sjösik, nord-	237	smörfisk	1037	spjutrockor	103
sjöstjärna, vanlig	1517	smörfisk, fläckig	1035	spjutsrocka, tagg-	105
sjöstjärnor	1516	smörfisk, kinesisk	1032	springare	1464
skarpsill	202	smörfisk, silver-	1031	späckhuggare	1470
skarpsill, kaspisk	171	smörfisk, stillahavs-	1034	spätta, röd-	1145
skärsnultra	847	smörfiskar	1030	staksill	157
skedstör	140	smörfiskar, fläckiga	1036	staksill, blå	155
skinkmussla	1385	smörvenusmussla	1403	staksill, indisk	205
skoläst	494	snapper, gul	689	staksill, kelee-	182
skolästfiskar	496	snapper, gul vimpel-	696	staksill, kinesisk	206
skorpionfisk, svart	1067	snapper, gulssjärts-	694	staksill, medelhavs-	158
skorpionfiskar	1069	snapper, randig	692	staksill, svartahavs-	160
skriftabborre	589	snapper, röd	690	staksill, toli-	207
skrubba	1143	snapper, sydlig röd-	691	staksill, vit	162
skrubba, stjärn-	1144	snapperfiskar	687	staksillar	163
skrubbskädda	1143	snappers	693	Stellers sjölejon	1443
skuggfisk	759	snappers, monokel-	698	stenbit	1105
skuggfisk, argentinsk	758	snappers, vimpel-	697	stenblecka	796
skuggfiskar	760	snäppfisk	504	stendadel	1372
skyggfisk	908	snäppfiskar	503	stenhinna	1530
skäckponnyfisk	700	snegelfisk	647	stenkrabba, svart	1293
skädda, berg-	1142	snegelfiskar	648	stenräka	1252
skädda, flathuvuds-	1130	snickarblecka	765	stenräkor	1253
skädda, gulstjärts-	1137	snook	561	stensimpa	1097
skädda, japansk ler-	1131	snookar	562	stensnultra	839
skädda, japansk sand-	1138	snors, regnbåg-	275	stentrollkrabba	1307
skädda, sand-	1139	snultra, fläck-	849	sterlett	135
skädda, stillahavs-	1136	snultra, påfågels-	851	stiletträka, argentinsk	1206
skäggbrosme	475	snultra, röd-	846	stillahavsansjovis	219
skäggbrotula	924	snultra, skär-	847	stillahavsbarrakuda	531
skäggig räka	1231	snultra, sten-	839	stillahavsgrovansjovis	212
skäggsimpa	1103	snäcka, sydafrikansk turban-	1332	stillahavsguldlax	280
skäggtorsk	472	snäckor, tagg-	1334	stillahavshälleflundra	1134
skäggtunga	1162	snäckor, ving-	1338	stillahavskistmussla	1339
skärkniv	350	sockeyelax	252	stillahavsknivmussla	1411
skärlånga, femtömmad	443	solabborre	601	stillahavskrabbtaska	1290, 1292
skärlånga, fyrtömmad	469	solabborre, svart	605	stillahavskummel	490
skärlånga, tretömmad	450	solabborre, vit	604	stillahavskungsfisk	1070
skärlångor	451	soldatfisk, atlantisk	520	stillahavslaxhaj	22
sköldpadda, bastard-	1498	soldatfisk, medelhavs-	521	stillahavsmakrill	983
sköldpadda, havsläder-	1495	soldatfisk, röd	522	stillahavsmakrillgädda	421
sköldpadda, oäkta karett-	1496	sommarvar	1118	stillahavsmanta	113
sköldpadda, sopp-	1497	soppsköldpadda	1497	stillahavsmenhaden	175
sköldpaddor	1494	sorghaj	58	stillahavsmullar	822
sköldpaddor, diamant-	1499	sorgkantsmuräna	405	stillahavsplattostron	1354
skördefisk	1033	sovfiskar	1048	stillahavsrävhaj	15
slampiskare	371	spadfisk, afrikansk	827	stillahavsrocka	100
slank tonfisk	963	spadfisk, prickig	828	stillahavsrödtunga	1129
slankbarb	338	spadfiskar	825	stillahavssill	170
slanknoting	883	spansk killi	514	stillahavsskädda	1136
slemfisk	909	spansk makrill	984	stillahavssmörfisk	1034
slemfisk, blod-	910	sparid, guld-	802	stillahavsspetsnoshaj	43
slemfisk, blod-	911	spegelrocka	92	stillahavspjutrocka	104
slemfisk, påfågels-	907	spetshaj, svart	38	stillahavsstrumpebandsfisk	958
slemfisk, silver-	918	spetsknot	1088	stillahavsstaggmakrill	671
slemfiskar	904, 906	spetsnosblecka	781	stillahavstarpon	149
slät kungskväkare	733	spetsnoshaj, stillahavs-	43	stillahavstobis	934
slätflundra	1141	spetsnoshaj, vitprickig	44	stillahavstorsk	446
släthuvudfisk, bairds	223	spetsnosig ormål	414	stillahavsstrådsill	185
slätväkare	731	spetsstjärtsklumpfisk	1187	stillahavstumlare	1472
slätrocka	84	spigg, korea-	940	stillahavstunga	1152
slätvar	1175	spiggfiskar	499	stjärnkikare	874, 875

537

stjärnrocka	83	surfnors	272	sömnfiskar	1048
stjärnskrubba	1144	sutare	364	sötläppar	714
stjärnstör	136	svalabborre	569	sötnors	271
stjärtbandsblecka	778	svampdjur	1511	sötvattensmussla	1345
stjärtfenad bläckfisk, nordlig	1430	svansfläckspikarell	811	sötvattensmusslor	1344
stjärtfläcksbleckor	784	svansfläckstaggmakrill	640	sötvattensräka	1247
stor kammussla	1365	svansrocka, tagg-	99	sötvattensräka, sydamerikansk	1246
stor sepia	1419	svart dvärgmal	384	sötvattensräkor	1248
stor silversida	550	svart havsabborre	571	sötveka	716
stor venusmussla	1401	svart havsgädda	953		
stor ödlefisk	391	svart kattfisk	384		
storfjällad ödlefisk	392	svart kummel	489	**T**	
storfläckig rödhaj	32	svart marlin	1018		
storhuvudsbult	1057	svart pomfret	677	tagghaj	71
storspigg	500	svart skorpionfisk	1067	taggig hjärtmussla	1388
storsäl	1448	svart solabborre	605	taggknot	1086
stort havsöra	1325	svart stenkrabba	1293	taggmakrill	672
storögd rävhaj	16	svart tonfisk	1009	taggmakrill, blå	670
storögd tandbraxen	776	svart trumfisk	745	taggmakrill, chilensk	669
storögd tonfisk	1011	svart tunnina	969	taggmakrill, cunene-	674
storögd var	1117	svartabborre	602	taggmakrill, fläckig	643
storögonfisk, röd	607	svartahavsmört	357	taggmakrill, grön	665
storögonfiskar	606, 608	svartahavspiggvar	1173	taggmakrill, gul	637
storögonselar	654	svartahavsstaksill	160	taggmakrill, indisk	644
strandkrabba	1298	svartfenad isfisk	896	taggmakrill, japansk	642, 666
strandkrabba, medelhavs-	1297	svartfisk	1041	taggmakrill, medelhavs-	668
strandsnäckor	1337	svartfisk, antarktisk	1043	taggmakrill, stillahavs-	671
streber	623	svartfiskar	1040, 1042	taggmakrill, sträv	667
streckad kungsmakrill	993	svartkarp	347	taggmakrill, svansfläcks-	640
streckögd noting	884	svartnoting	881	taggmakrill, svartryggad	638
strimgrymta	708	svartryggad taggmakrill	638	taggmakrill, sydafrikansk	673
strimmad tunga	1160	svartsmörbult	1053	taggmakrillar	639, 645, 675
strimmig bonit	981	svartspetshaj	38	taggmakrillfiskar	632
strimmig delfin	1474	svartval	1466	taggrötsimpa	1098
strimmig havsabborre	598	svärdfisk	1029	taggsnäckor	1334
strimmig kungsmakrill	989	svärdfiskar	1028	taggspjutsrocka	105
strimveka	728	svärdstör	141	taggsvansrocka	99
strumpebandsfisk	957	sydafrikansk östkusttunga	1154	taggvar	1112
strumpebandsfisk, stillahavs-	958	sydafrikansk pälssäl	1442	tandat ostron	1352
strupsnittsöring	255	sydafrikansk sardin	199	tandbraxen	774
strutar	1422	sydafrikansk taggmakrill	673	tandbraxen, kanarie-	772
sträv taggmakrill	667	sydafrikansk turbansnäcka	1332	tandbraxen, röd	777
strävrocka	98	sydafrikansk västkusttunga	1153	tandbraxen, rosa	775
strömflodkräfta	1280	sydamerikansk klippmussla	1382	tandbraxen, storögd	776
strömling	345	sydamerikansk pälssäl	1441	tandhajar, kam-	9
strömlöja	308	sydamerikansk sardin	200	tandkarp, valencia-	515
strömming	169	sydamerikansk sötvattensräka	1246	tandkarpar	512
strömsill	281	sydamerikanskt sjölejon	1444	tandnoting	877
stubb, adria-	1058	sydatlantisk kammussla	1368	tandnoting, antarktisk	878
stubb, adriatisk	1058	sydblecka	800	tandnotingar	879
stubb, garda-	1056	sydkapare	1489	tandvar	1113
stubb, sand-	1059	sydkummel	482	tandvar, fläckig	1111
stubbkväkare	738	sydlig blåvitling	458	tapesmusslor	1394
stäm	344	sydlig hajrocka	74	tarakihi	859
stör	137	sydlig hundhaj	51, 53	tarpon	150
stör, adriatisk	133	sydlig langust	1267	tarpon, stillahavs-	149
stör, beluga-	139	sydlig pelamid	979	tarponfiskar	148
stör, grön	132	sydlig rödsnapper	691	tautog	852
stör, rysk	131	sydlig rosenräka	1224	tegelabborre, blå	627
stör, sked-	140	sydlig spindelkrabba	1314	tegelabborrfiskar	626
stör, svärd-	141	sydlig tonfisk	1010	tejstefisk	919
stör, vit	138	sydostasiatisk krokodil	1504	teleskopabborre	611
störfiskar	130	sydseriola	659	teraponfiskar	592
större kantnål	506	sydveka	754	tetror	300
större kungsfisk	1075	sågfisk, småtandad	79	Thorsvar	1116
större medelhavsabborre	587	sågfiskar	78	tigerhandveka	742
större smörbult	1052	säl, grå-	1449	tigerräka	1223
sugare, dubbel-	1190	säl, grönlands-	1451	tigerräka, brun	1218
sugfisk	1178	säl, knubb-	1453	tigerräka, grön	1226
sugga, havs-	1068	säl, man-	1444	tilapia, moçambique-	832
sugkarpar	301	säl, stor-	1448	tioarmade bläckfiskar	1422
sumpelritsa	351	säl, sydafrikansk päls-	1442	tiofotade kräftdjur, krypande	1285
sumpkräfta	1277	säl, sydamerikansk päls-	1441	tiopundare	145
sunda-barb	353	säl, vikare-	1452		

tiopundare	147
tjockläppad multe	538
tjocktunga	1157
tobis, amerikansk	932
tobis, blå-	936
tobis, havs-	935
tobis, kust-	936
tobis, medelhavs-	938
tobis, nordlig	933
tobis, stillahavs-	934
tobisar	937
tobisfiskar	931
tobiskung	939
toffelhummer	1305
toffelkräftor	1304
toli-staksill	207
tonfisk	1012
tonfisk, fjärils-	970
tonfisk, gulfenad	1008
tonfisk, karibisk	1009
tonfisk, långfenad	1007
tonfisk, långstjärtad	1013
tonfisk, slank	963
tonfisk, storögd	1011
tonfisk, svart	1009
tonfisk, sydlig	1010
tonfiskar	1006, 1014
torpedlöpare	649
torsk	447
torsk, lyr-	466
torsk, navaga-	445
torsk, padd-	468
torsk, polar-	441
torsk, röd mora-	438
torsk, saffrans-	444
torsk, silver-	449
torsk, skägg-	472
torsk, stillahavs-	446
torskartade fiskar	435
torskfiskar	440
transkei-langust	1271
trekantshaj	70
trekantshajar	69
tretömmad skärslånga	450
trippelfena	912
tristan-langust	1263
trochus	1330
trollhumrar	1306
trollkrabba, antarktisk	1309
trollkrabba, japansk	1310
trollkrabba, murray-	1308
trollkrabbor	1311
trubbnäsling	329
trubbnosig ormål	413
trumfisk, knubbig	753
trumfisk, peruansk	743
trumfisk, röd	757
trumfisk, svart	745
trumfiskar	756
trumpetarfiskar	861
tryckarfisk, grå	1180
tryckarfiskar	1179
trynfisk	527
trynfiskar	526
trådfensfiskar	553
trådfilfisk	1183
trådmakrill, alexandria-	633
trådsill, atlantisk	186
trådsill, stillahavs-	185
trådsillar	187
träskräka	1258
tulpaner, havs-	1199
tumlare	1471
tumlare, stillahavs-	1472

tunga	1165
tunga, berg-	1142
tunga, fläck-	1163
tunga, röd-	1128
tunga, sand-	1164
tunga, skägg-	1162
tunga, små-	1156
tunga, stillahavs-	1152
tunga, stillahavsröd-	1129
tunga, strimmad	1160
tunga, sydafrikansk östkust-	1154
tunga, sydafrikansk västkust-	1153
tunga, tjock-	1157
tunga, ögonfläcks-	1158
tungefiskar	1151
tungefiskar, hund-	1166
tungevar	1115
tungevarar	1114
tungor	1155, 1161
tunnina	968
tunnina, svart	969
tunnläppad multe	542
turbansnäcka	1331
turbansnäcka, sydafrikansk	1332
tvåbandsblecka	783
tvärbandad knot	1091
tång	1527
tång, blad-	1526
tånglake	923
tånglakefiskar	920
tångräka	1249
tångsnälla	508
tångspigg	502

U

uer	1075
ukrainskt nejonöga	3
ullhandskrabba, kinesisk	1303
uvak	448

V

val, Bairds näbb-	1459
val, blå-	1483
val, Brydes fen-	1482
val, dvärgrät-	1488
val, fen-	1484
val, grind-	1466
val, grönlands-	1487
val, grå-	1491
val, indisk grind-	1465
val, knöl-	1485
val, nar-	1479
val, pilot-	1466
val, pott-	1462
val, sej-	1481
val, sill-	1484
val, småhuvud-	1461
val, svart-	1466
val, vanlig näbb-	1460
val, vik-	1480
val, vit-	1478
valar, näbb-	1458
valar, rät-	1486
valencia-tandkarp	515
valhaj	27
valross	1445
valthornssnäcka	1335
vanlig bläckfisk	1426
vanlig näbbval	1460

vanlig sjöstjärna	1517
vanlig åttaarmad bläckfisk	1438
var, fjäll-	1177
var, fyrfläckig	1119, 1168
var, glas-	1169
var, japansk	1120
var, kanel-	1122
var, pigg-	1172
var, slät-	1175
var, små-	1171
var, sommar-	1118
var, storögd	1117
var, tagg-	1112
var, tand-	1113
varar	1121, 1167
varar, glas-	1170
vargsill	225
vargsillar	224, 226
vassbuk	202
veka, brun-	755
veka, peru-	723
veka, prick-	725
veka, småtandad	727
veka, syd-	754
veka, tigerhand-	742
veka, vit-	720
vekfisk	726
vekfisk, afrikansk	719
vekfisk, kalifornisk	724
vekfiskar	715, 729
vekor	715
venusmussla, brun	1395
venusmussla, gyllene	1404
venusmussla, hård	1398
venusmussla, japansk	1399, 1402
venusmussla, liten	1405
venusmussla, d'Orbignys	1397
venusmussla, randig	1396
venusmussla, smör-	1403
venusmussla, stor	1401
venusmusslor	1393, 1400
vikaresäl	1452
vikkammussla	1361
vikval	1480
vimma	365
vimpeldrakhuvudfisk	1065
vimpelsill	168, 172
vimpelsilvell	902
vimpelsnapper, gul	696
vimpelsnapperfiskar	695
vimpelsnappers	697
vingsnäcker	1338
vinterflundra	1147
vinterrocka	95
violett spjutrocka	107
vit bläckfisk	1436
vit fingerfena	857
vit grouper	573
vit havsabborre	597
vit havsgädda	954
vit marlin	1021
vit sandblecka	785
vit sanktpersfisk	524
vit solabborre	604
vit staksill	162
vit stör	138
vitabborre	596
vitblecka	782
vitbrosme	476
vitfläckig havsmus	124
vitfläckshavsål	409
vitfotad flodkräfta	1279
vithaj	19
vitkummel	481

SV

539

vitkväkare 730
vitling 454, 455
vitling, blå- 459
vitlinglyra 471
vitlöja 309
vitmunnad kväkare 737
vitnos 1469
vitprickig spetsnoshaj 44
vitrocka 90
viträka 1228
viträka, nordlig 1227
vitsiding 1468
vitval 1478
vitveka 720
vrakfisk 585
vrakfisk, nyzeeländsk 586
vågmär 511
vårthavssugga 1066
västafrikansk havsmal 375
västafrikansk marulk 1196
västafrikansk mulle 820
västatlantisk ålbrosme 922
västkusttunga, sydafrikansk 1153
västlig kungsräka 1221

W

wahoo 962
wobbegong, fläckig 25
wuchangbraxen 346

Z

zingel 624

Å

ål 396
ål, amerikansk 400
ål, amerikansk havs- 411
ål, argentinsk havs- 412
ål, australisk 397
ål, havs- 410
äl, håg- 29, 30
ål, japansk 399
ål, nyzeeländsk 398
ål, vitfläckshavs- 409
ålar 401
ålbrosmar 921
ålbrosme, västatlantisk 922
ålfiskar 395
ålmalar 388
årfenshaj 37
åsnegrymta 707
åttaarmad bläckfisk 1435
åttaarmad bläckfisk, vanlig 1438
åttaarmade bläckfiskar 1434, 1437, 1439

Ä

äkta korall 1514
äkta koraller 1513
äkta krokodiler 1501

ängel, havs- 68
änglar, havs- 67
änkeuer 1072
ätlig hjärtmussla 1389
ätlig sjöborre 1521

Ö

ödlefisk 393
ödlefisk, stor 391
ödlefisk, storfjällad 392
ödlefiskar 390
ögondarrocka 120
ögonfläckstunga 1158
ögonisfisk 900
ögonsnultra 848
öra, europeiskt havs- 1328
öra, rött havs- 1327
öra, stort havs- 1325
öresvin 1477
öring 257, 258, 259
öring, apache- 244
öring, gila- 246
öring, guld- 243
öring, mexikansk guld- 245
öringabborre 603
öringar, laxar och 260
örnrocka 109
örnrockor 108
öron, havs- 1329
öronsälar 1440
östkusttunga, sydafrikansk 1154
östlig flodkräfta 1277

SV

540

Index of inter-agency 3-alpha identifiers

A

ABG	1326
ABP	1327
ABR	1328
ABS	1213
ABX	1329
ACH	263
AES	1239
AGM	1500
AGN	68
AKS	1242
ALA	632
ALB	1007
ALC	223
ALE	161
ALF	519
ALK	470
ALP	1145
ALS	35
ALU	151
AMB	658
AMJ	660
AMU	291
AMX	661
ANA	214
ANB	210
ANC	216
AND	418
ANE	217
ANF	1192
ANG	1193
ANI	897
ANK	1194
ANN	778
ANR	221
ANS	893
ANT	437
APG	131
APL	1524
APO	610
APR	135
APS	1217
ARA	1233
ARF	1124
ARI	1232
ARK	1343
ASA	683
ASC	1366
ASD	156
ASK	67
ASM	995
ATK	1093
AVA	211
AWE	718
AYA	1317

B

BAG	529
BAH	1119
BAJ	599
BAL	430
BAP	583
BAR	535
BAS	590
BAT	826
BAU	978
BAX	1120
BAZ	536
BBH	155
BBM	1000
BBS	1066
BCL	1403
BDM	744
BEG	1087
BEL	1479
BEN	415
BEP	979
BET	1011
BEW	1459
BFT	1013
BGR	712
BHA	429
BIB	472
BIC	336
BIG	608
BIL	1015
BIP	980
BIR	607
BIS	654
BKC	347
BKJ	968
BLB	1037
BLC	1340
BLE	904
BLF	1009
BLI	460
BLL	1175
BLM	1018
BLP	879
BLR	38
BLS	1341
BLT	963
BLU	630
BLW	1484
BMO	862
BNF	153
BOA	174
BOB	1231
BOF	152
BOG	765
BON	981
BOP	973
BOR	526
BOW	1460
BPI	812
BRB	808
BRD	924
BRF	1062
BRG	704
BRO	36
BRS	986
BRW	1483
BRX	517
BRZ	680
BSB	570
BSC	804
BSE	593
BSF	955
BSH	42
BSK	12
BSP	1042
BSS	594

BSX	586
BSZ	567
BTG	1034
BTH	16
BUA	640
BUC	297
BUF	302
BUK	969
BUM	1019
BUP	1044
BUR	713
BUS	830
BUT	1035
BUX	1036
BWA	1043

C

CAA	915
CAD	1293
CAE	378
CAF	387
CAG	379
CAI	1501
CAL	1297
CAO	376
CAP	273
CAS	916
CAT	917
CAX	374
CBA	631
CBM	755
CBR	587
CDP	1505
CDS	1506
CDX	756
CEN	1040
CER	999
CET	1156
CEX	928
CEZ	810
CGO	320
CGX	634
CHA	300
CHE	248
CHG	164
CHI	252
CHM	125
CHO	1426
CHP	200
CHR	267
CHS	1236
CHU	246
CHY	1002
CIF	1122
CIS	231
CIX	831
CJM	669
CKA	738
CKI	204
CKL	747
CKM	736
CKW	750
CKY	758
CLA	171
CLB	1393
CLH	1398

CLI	1092
CLJ	1402
CLP	154
CLQ	1387
CLR	1408
CLS	1416
CLV	1404
CLX	1325
CMR	1298
CNG	1504
COA	411
COB	759
COC	1390
COD	447
COE	410
COH	247
COL	1514
CON	1339
COP	455
COS	412
COT	823
COW	445
COX	408
COZ	1389
CPI	196
CPO	1031
CPP	788
CPR	1248
CRA	1286
CRA	1299
CRB	1296
CRD	1278
CRE	1291
CRJ	1288
CRK	1289
CRL	720
CRN	1252
CRO	1503
CRQ	1313
CRR	1322
CRS	1301
CRV	743
CRX	737
CRY	745
CRZ	1295
CSH	1251
CSP	1214
CSY	293
CTA	857
CTC	1419
CTG	1401
CTO	388
CTS	1406
CUC	927
CUJ	1523
CUN	853
CUS	925
CUT	958
CVJ	636
CXR	638

D

DAB	1138
DAG	203
DAS	168
DBM	971

DBO	1478	FGS	1046	GPS	574	**I**	
DCO	1465	FGX	1055	GPW	572		
DEA	770	FID	343	GPX	580	ICX	898
DEC	773	FIL	1183	GRA	708	ILI	183
DEL	775	FIS	557	GRB	703	IMS	1532
DEM	776	FIW	1485	GRC	448	IOS	192
DEN	771	FJB	353	GRE	298	ISC	1362
DEP	774	FKN	292	GRL	711	ITE	385
DES	283	FLA	873	GRM	479	ITM	384
DGS	64	FLD	1174	GRN	480	ITP	386
DGX	57	FLE	1142	GRP	706		
DIA	114	FLF	1182	GRS	477	**J**	
DIA	1388	FLH	1095	GRV	498		
DIA	208	FLI	1096	GRX	705	JAA	670
DIH	780	FLP	1216	GSK	63	JAJ	1315
DLP	1464	FLS	1118	GTF	72	JAN	218
DNC	772	FLU	129	GUB	73	JAP	197
DOB	225	FLW	1147	GUC	1080	JAS	863
DOL	681	FLX	1110	GUD	74	JAX	671
DON	1407	FLY	423	GUF	75	JEL	1512
DOP	66	FOR	464	GUG	1083	JFL	424
DOR	1397	FOT	553	GUI	1082	JJM	666
DOS	226	FOX	465	GUN	1089	JOD	525
DPC	406	FPC	1039	GUP	58	JOS	524
DPS	1212	FPE	616	GUQ	59	JPG	805
DRI	1038	FPI	288	GUR	1078	JSS	194
DRR	1468	FPP	620	GUS	1210		
DRS	752	FPY	615	GUT	990	**K**	
DRU	754	FRG	1493	GUU	1088		
DRY	1356	FRI	964	GUX	1090	KAF	1123
DSI	1476	FRO	360	GUZ	77	KAK	997
DSP	1477	FRS	191			KAW	966
DST	1475	FRX	361	**H**		KCM	1309
DUN	1290	FSA	1150			KCP	926
DUS	39	FSC	350	HAD	453	KCS	1312
		FSN	559	HAI	1113	KCT	1308
E		FSS	558	HAJ	431	KGB	732
		FTE	364	HAL	1132	KGF	734
EAG	108	FTS	1130	HAP	1133	KGG	733
EIL	184	FUS	686	HAS	649	KGM	987
ELA	400	FVE	229	HAW	858	KGX	1003
ELE	396			HAX	433	KIC	729
ELF	126	**G**		HCJ	1399	KIF	900
ELJ	399			HCL	1392	KIW	1471
ELP	922	GAD	435	HCX	1400	KIX	735
ELS	409	GAG	50	HEP	170	KLA	940
ELU	397	GAJ	824	HER	169	KNI	1255
ELX	401	GAR	416	HIL	205	KOB	716
ELZ	920	GAT	377	HIX	182	KOS	991
EMM	685	GAZ	1302	HKB	489	KRI	1203
EMP	762	GBA	531	HKC	484	KUG	1081
EMT	684	GBL	709	HKE	487	KUP	1219
EMX	179	GBN	1053	HKM	491		
ENA	355	GBR	710	HKN	482	**L**	
ENP	209	GEL	1529	HKO	488		
ENS	1207	GEP	947	HKP	486	LAA	1257
EPI	611	GER	1323	HKR	475	LAD	146
EUL	278	GFB	463	HKS	483	LAN	295
		GHL	1148	HKU	474	LAR	5
F		GIG	1531	HKW	476	LAS	1
		GIP	564	HKX	492	LAT	266
FAB	354	GIS	1428	HKZ	478	LBA	1318
FBM	305	GIT	1223	HMC	674	LBC	1262
FBR	307	GIW	1333	HMG	665	LBE	1319
FBU	452	GLA	563	HMM	668	LBR	761
FCC	321	GOA	820	HMZ	675	LBT	1263
FCG	331	GOO	1198	HOC	740	LDB	1167
FCH	338	GOX	822	HOF	481	LEB	965
FCP	332	GPA	1050	HOM	673	LEE	647
FCY	303	GPB	582	HOU	237	LEF	1116
FFV	434	GPC	405	HSC	1515	LEM	1141
FGB	1048	GPD	576	HUS	689	LEP	1085
FGG	1045	GPN	581	HUW	1486	LES	56
FGO	1047	GPR	579	HVF	1032		

LEX	648	MUA	538	PAN	1235	**Q**	
LEZ	1168	MUB	545	PAO	651		
LHT	959	MUC	330	PAP	994	QSC	1359
LIB	392	MUD	1303	PAR	790	QUM	998
LIC	899	MUE	1335	PAS	933		
LIG	391	MUF	544	PAX	794	**R**	
LIN	462	MUG	1381	PBA	1222		
LIT	1530	MUI	402	PCA	269	RAA	213
LIX	390	MUK	1376	PCB	1197	RAB	974
LKR	741	MUL	546	PCO	446	RAC	1412
LMA	21	MUM	815	PCR	1314	RAE	1414
LOF	1261	MUO	543	PCX	407	RAF	975
LOG	1264	MUR	819	PDR	742	RAG	976
LOK	1272	MUS	1377	PEN	1228	RAJ	81
LOQ	1307	MUT	817	PEO	188	RAP	1411
LOR	1260	MUX	818	PER	1338	RAS	173
LOS	1305	MUY	753	PES	571	RAT	124
LOT	1014	MUZ	1382	PET	554	RAX	977
LOY	1274	MYA	1379	PEW	596	RAZ	1413
LSY	746	MYC	1375	PEX	566	RBY	116
LTA	967			PGA	798	RDM	757
LUK	656	**N**		PHA	485	REA	803
LUM	1105			PHR	1472	REB	1076
LXX	294	NAR	1480	PIA	199	REC	1071
LYC	721	NEA	1320	PIC	814	RED	1077
		NEB	864	PIG	707	REE	206
M		NEC	438	PIL	189	REG	1075
		NED	419	PIN	245	REN	1073
MAA	982	NEP	1321	PIW	1467	REP	1225
MAC	984	NFA	417	PJM	672	RER	795
MAG	1391	NHA	490	PLA	1131	RES	688
MAN	111	NIP	565	PLE	1144	RHG	497
MAR	578	NKR	1205	PLN	234	RHI	337
MAS	983	NOC	880	PLZ	1146	RJB	84
MAW	1005	NOD	887	PNV	1229	RJC	90
MAZ	985	NOG	882	POA	679	RJF	87
MCL	1342	NOK	883	POB	677	RJM	93
MDO	946	NON	881	POC	441	RJN	88
MEG	1169	NOP	471	POD	473	RNG	494
MES	175	NOR	884	POI	1064	ROB	561
MET	1209	NOS	885	POK	467	ROC	1292
MFF	902	NOT	890	POL	466	ROK	1063
MGA	539	NOX	886	POM	662	ROL	451
MGC	541	NPA	219	PON	699	ROS	1134
MGR	717	NPH	996	POP	663	RPG	806
MHA	166	NPP	1211	POR	23	RRH	178
MHG	165			POS	458	RRS	1258
MIL	228	**O**		POT	629	RRU	646
MIW	1481			POX	664	RSA	642
MLS	1023	OCC	1440	POY	701	RSC	667
MOA	657	OCH	1352	PPF	1247	RSE	1067
MOB	698	OCM	1437	PPO	1033	RSH	1253
MOD	1373	OCS	37	PPZ	1250	RSK	34
MOJ	903	OCT	1438	PRA	1238	RSN	768
MOK	1196	OCZ	1439	PRB	1218	RSP	1025
MOL	1324	ODE	1353	PRF	1246	RSX	769
MON	1195	OFO	1355	PRG	766	RUB	635
MOO	678	OIL	950	PRI	606	RUF	682
MOR	436	OPT	921	PRR	856	RUS	644
MOS	989	ORD	528	PSB	1112		
MOW	860	ORY	520	PSE	748	**S**	
MOX	1188	OXY	70	PSH	1240		
MPS	603	OYA	1351	PSL	1269	SAA	190
MSC	1371	OYC	1350	PSM	271	SAB	1094
MSD	641	OYF	1354	PSS	749	SAE	193
MSF	1114	OYG	1348	PST	1227	SAF	444
MSK	18	OYM	1349	PSU	1368	SAI	1016
MSL	1383	OYS	1358	PTB	314	SAL	257
MSM	1378	OYX	1357	PTH	15	SAN	935
MSP	1024			PTY	751	SAP	421
MSR	1380	**P**		PUA	1185	SAS	180
MSX	1374			PUF	1184	SAU	422
MTS	1202	PAC	792	PUX	1186	SAW	78
MTX	1004						

Code	No.	Code	No.	Code	No.	Code	No.
SAX	420	SLV	1275	STT	103	**U**	
SBA	789	SMA	20	STU	130		
SBF	1010	SMC	375	STV	621	UBS	760
SBG	802	SMD	53	STW	786	UCH	1522
SBL	11	SME	274	SUR	945	UNC	439
SBR	791	SMR	275	SUS	272	URA	875
SBS	787	SMX	277	SVC	335	URC	1519
SBX	801	SNA	692	SVE	1396	URM	1518
SBZ	779	SNC	691	SVF	264	URS	1521
SCA	1370	SNI	503	SWA	782	USK	442
SCB	1361	SNK	951	SWF	724		
SCC	1360	SNL	693	SWO	1028	**V**	
SCE	1365	SNO	562	SXX	1445		
SCK	60	SNR	690	SXX	1447	VAR	265
SCL	33	SNS	504	SYN	612	VEP	212
SCM	796	SNW	784	SYT	32	VET	220
SCO	1069	SNX	687	SYX	28	VLO	1265
SCP	809	SNY	694			VSC	1363
SCR	1316	SOA	1154				
SCS	1068	SOC	250			**W**	
SCT	829	SOE	1153	**T**			
SCU	1101	SOI	1415			WAB	349
SCX	1369	SOL	1164	TAI	149	WAH	961
SCY	1306	SOM	382	TAK	859	WEC	643
SCZ	1367	SOP	1224	TAR	150	WEG	869
SDP	54	SOS	1163	TAU	852	WEM	723
SDS	52	SOT	1111	TGS	1220	WEP	722
SDV	55	SOW	1152	THA	186	WEX	871
SDX	645	SOX	1165	THB	696	WHA	585
SEB	1449	SOY	540	THD	695	WHB	459
SEC	1454	SPA	825	THF	555	WHE	1336
SEF	1441	SPB	865	THG	697	WHF	241
SEG	1450	SPF	1026	THP	185	WHG	454
SEH	1452	SPH	763	THS	1158	WHL	232
SEK	1442	SPI	944	THX	187	WHM	1021
SEL	1444	SPL	46	TIL	626	WHS	625
SER	1453	SPN	48	TIP	1226	WHX	1337
SEV	797	SPO	1511	TJO	793	WIC	896
SEW	1456	SPR	202	TLM	832	WIT	1127
SEZ	1448	SPS	828	TLN	833	WKB	727
SFA	1017	SPT	730	TLV	268	WKK	731
SFS	956	SPU	595	TOA	877	WKP	1221
SGI	901	SPW	1463	TOD	117	WKS	728
SHA	162	SPY	45	TOE	121	WKX	726
SHB	71	SPZ	47	TOL	207	WRA	839
SHC	160	SQA	1429	TOM	456	WRF	584
SHD	163	SQE	1434	TOP	876	WRO	1072
SHG	172	SQI	1431	TOS	1332	WSH	19
SHH	159	SQJ	1433	TOT	878	WSM	954
SHI	1208	SQL	1425	TOX	1166	WSN	799
SHL	62	SQM	1430	TPS	1394	WST	104
SHM	970	SQU	1422	TRA	867	WUB	346
SHO	30	SQY	1201	TRE	639		
SHR	781	SQZ	1424	TRF	628	**X**	
SHS	1243	SRA	1086	TRG	1179		
SHW	1466	SRG	777	TRH	888	XIP	1029
SIC	827	SRH	201	TRI	1180		
SIE	1001	SRX	80	TRK	49	**Y**	
SIL	551	SSA	552	TRO	253		
SIP	1030	SSB	785	TRR	249	YED	739
SIW	1482	SSE	1509	TRS	258	YEL	1136
SIX	195	SSG	1508	TRT	894	YES	1135
SJA	1364	SSH	1234	TRU	861	YFT	1008
SKA	102	SSI	895	TRY	655	YPS	1215
SKJ	972	SSM	993	TRZ	652	YRO	1074
SLC	1270	SSP	1022	TSD	157	YTC	659
SLD	767	SSR	1510	TSQ	1432		
SLF	764	SSX	1507	TTG	1499	**Z**	
SLI	461	STB	598	TTL	1496		
SLM	800	STC	1294	TTX	1494	ZEX	523
SLN	1266	STG	725	TUG	1497		
SLO	1267	STH	1516	TUR	1171		
SLS	1268	STO	222	TUS	1012		
SLT	962	STS	992	TUX	960		

Picture credits

The European Commission wishes to record its appreciation to the publishers of the following publications for permission to reproduce the illustrations used in this dictionary. This applies particularly to the Food and Agriculture Organisation of the United Nations for permission to reproduce many of the illustrations which have appeared in the series of FAO Fish Identification Sheets.

Other valuable sources of material have been:

ARMITAGE, R. O., PAYNE, D. A., LOCKLEY, G. J. et al.
(1981)
Guidebook to New Zealand commercial fish species, New Zealand Fishing Industry Board, 216 pp.

BANARESCU, P.
(1964)
Pisces Osteichthyes. Fauna Republicii Populare Romine, 13, Academia Republicii Romania, Bucarest, 962 pp.

BLANC, M., BANARESCU, P., GAUDET, J. L. and HUREAU, J. C.
(1971)
European Inland Water Fish. A multilingual catalogue, Fishing News Books, Oxford.

BOONE, L.
(1931)
A collection of anomuran and macruran Crustacea from the Bay of Panama and the fresh waters of the Canal zone, Bull. American Museum of Natural History, New York, 63, pp.137-189.

BOUDAREL, N.
(1948)
Les richesses de la mer – Encyclopédie biologique, éditions Lechevalier, 29, 549.

CASTAGNOLO, L., FRANCHINI, D. and GIUSTI, F.
(1980)
Bivalvi. Guide per il riconoscimento delle specie animali delle acque interne italiane, CNR AQ/1/49, 10, 64 pp.

CORBET, G. and OVENDEN, D.
(1980)
The mammals of Britain and Europe, Collins, London, 253 pp.

DORE, I. and FRIMONDT, C.
(1987)
An illustrated guide to shrimps of the world, Scandinavian Fishing Year Book, Hedehusene, 229 pp.

ESCHMEYER, W. N. and HERALD, E. S.
(1983)
A field guide to Pacific Coast fishes of North America, Houghton Mifflin, Boston, 336 pp.

FIGUEIREDO, J. L.
(1977)
Manual de Peixes Marinhos do Sudeste do Brasil. I. Introdução. Cações, raias e quimeras, Museu de Zoologia, Univ. São Paulo, Brasil, 104 pp.

FIGUEIREDO, J. L. ard MENEZES, N. A.
(1978)
Manual de Peixes Marinhos do Sudeste do Brasil. II. Teleostei (1), Museu de Zoologia, Univ. São Paulo, Brasil, 110 pp.

FROGLIA, C.
(1978)
Decapoda. Guide per il riconoscimento delle specie animali delle acque interne italiane, CNR AQ/1/9, 4, 41 pp.

GOSNER, K. L.
(1978)
A field guide to the Atlantic seashore, Houghton Mifflin, Boston, 329 pp.

GRIFFIN, D. J. G.
(1966)
The marine fauna of New Zealand: spider crabs, family Majidae (Crustacea, Brachyura), New Zealand Department of Scientific and Industrial Research Bulletin, 172, 111 pp.

HART, J. L.
(1973)
Pacific fishes of Canada, Bull. Fish. Res. Board of Canada 180, Department of Fisheries and Oceans, Ottawa, 740 pp.

HAYWARD, P. J. and RYLAND, J. S.
(1990)
Marine fauna of the British Isles and Northeast Europe, Oxford Science Publ., Vols 1-2, Oxford Science Publications, 996 pp.

KIRKEGAARD, I. and WALKER, R. H.
(1969)
Synopsis of biological data on the tiger prawn Penaeus esculentus, *Haswell, 1879,* CSIRO Fisheries Synopsis (Australia), 3.

LEIM and SCOTT
(1966)
Fishes of the Atlantic coast of Canada, Bull. Fish. Res. Board Canada, 155, Department of Fisheries and Oceans, Ottawa, 485 pp.

LINDBERG, G. U. and KRASYUKOVA, Z. V.
(1971)
Fishes of the Sea of Japan and the adjacent areas of the Sea of Okhotsk and the Yellow Sea. Part 3, Israel Program Scientific Translations, Zoological Institute, Academy of Sciences, Leningrad, 498 pp.

LINDBERG, G. U. and LEGEZA, M. I.
(1969)
Fishes of the Sea of Japan and the adjacent areas of the Sea of Okhotsk and the Yellow Sea. Part 2. Israel Program Scientific Translations Zoological Institute, Academy of Sciences, Leningrad. 389 pp.

LLORIS, D. and RUCABADO, J.
(1991)
Ictiofauna del Canal Beagle (Tierra de Fuego), aspectos ecológicos y análisis biogeográfico, Publ. Espec. Inst. Esp. Oceanogr., Madrid, 8, 182 pp.

MAJOR, A.
(1974)
Collecting world sea shells, John Bartholomew & Son, Edinburgh, 187 pp.

MAUCHLINE, J.
(1971)
Euphausiaceae *adults,* ICES Fiches d'identification du zooplankton, 134, ICES, Copenhagen.

MENNI, R. C., RINGUELET, R. A. and ARAMBURU, R. H.
(1984)
Peces marinhos de la Argentina y Uruguay, Editorial Hemisferio Sur, Buenos Aires.

MIQUEL, J. C.
(1982)
'Le genre *Metapenaeus (Crustacea, Penaeidae)*: taxonomie, biologie et pêches mondiales',
Zoologische Verhandelingen, Leiden, 195, pp. 1-137.

NICKELS, M.
(1950)
Mollusques testacés marins de la côte occidentale d'Afrique, Manuels ouest-africains, Vol. 2,
Editions Lechevalier, 269 pp.

NORMAN, J. R.
(1934)
A systematic monograph of the flatfishes (Heterosomata). *Vol. I.* Psettodidae, Bothidae, Pleuro-
nectidae. British Museum (N.H.), London, 459 pp.

OJEDA, F. P.
(1982)
*Iconografia de los Principales Recursos Pesqueros de Chile. Moluscos, Crustaceos, Equinoder-
mos y Tunicados,* Subsecretaria de Pesca (Chile), 3, 87 pp.

PAGE, L. M. and BURR, B. M.
(1991)
A field guide to freshwater fishes of North America north of Mexico, Houghton Mifflin, Boston,
432 pp.

PIVNICKA, K. and CERNY, K.
(1990)
Poissons, Gründ, Paris, 304 pp.

ROBINS, C. R.
(1992)
Saltwater fish. Canadian Nature Guides, Smithmark Publishers Inc., New York, 192 pp.

ROBINS, C. R. and RAY, G. C.
(1986)
A field guide to Atlantic Coast fishes of North America, Houghton Mifflin Co., Boston, 354 pp.

ROBISON, H. W.
(1992)
Freshwater fish, Canadian Nature Guides, Smithmark Publishers Inc., New York, 192 pp.

RUYIU, LIU (J. Y. LIU)
(1988)
Penaeoid shrimps of the South China Sea, Agricultural Publishing House, Chinese Academy of
Sciences, p. 130.

SANCHES, J. G.
(1989)
Nomenclatura Portuguesa de Organismos Aquáticos (Proposta para Normalização Estatística),
Publicações Avul das INIP, 14, 322 pp.

SCOTT, W. B. and CROSSMAN, E. J.
(1973)
Freshwater fishes of Canada, Bull. Fish. Res. Board of Canada, 184.

SMALDON, G.
(1979)
British coastal shrimps and prawns, Synopses of the British Fauna, 15, Academic Press Ltd., London.

STERBA, G.
(1983)
The aquarist's encyclopaedia, Blandford Press, Poole, 605 pp.

TREWAVAS, E.
(1983)
Tilapiine fishes of the genera Sarotherodon, Oreochromis *and* Danakilia, Publications British Museum (N.H.), London, 878, 583 pp.

WHEELER, A.
(1975)
Fishes of the World: an illustrated dictionary, Ferndale Editions, London, 366 pp.

WHITEHEAD, P. J. P., BAUCHOT, M.-L., HUREAU, J. C. et al.
(1984-86)
Fishes of the Northeastern Atlantic and the Mediterranean, Unesco, Paris, 1473 pp.

European Commission

Multilingual illustrated dictionary of aquatic animals and plants

Second edition

Luxembourg: Office for Official Publications of the European Communities

1998 – LI + 548 pp., num. illustrations – 17.6 × 25 cm

ISBN 92-828-1886-1

Price (excluding VAT) in Luxembourg: ECU 90